Lecture Notes in Computer Science   11476

Commenced Publication in 1973
Founding and Former Series Editors:
Gerhard Goos, Juris Hartmanis, and Jan van Leeuwen

More information about this series at http://www.springer.com/series/7410

Yuval Ishai · Vincent Rijmen (Eds.)

Advances in Cryptology – EUROCRYPT 2019

38th Annual International Conference on the Theory
and Applications of Cryptographic Techniques
Darmstadt, Germany, May 19–23, 2019
Proceedings, Part I

 Springer

Editors
Yuval Ishai
Technion
Haifa, Israel

Vincent Rijmen
COSIC Group
KU Leuven
Heverlee, Belgium

ISSN 0302-9743   ISSN 1611-3349   (electronic)
Lecture Notes in Computer Science
ISBN 978-3-030-17652-5   ISBN 978-3-030-17653-2   (eBook)
https://doi.org/10.1007/978-3-030-17653-2

LNCS Sublibrary: SL4 – Security and Cryptology

This Springer imprint is published by the registered company Springer Nature Switzerland AG
The registered company address is: Gewerbestrasse 11, 6330 Cham, Switzerland

Preface

Eurocrypt 2019, the 38th Annual International Conference on the Theory and Applications of Cryptographic Techniques, was held in Darmstadt, Germany, during May 19–23, 2019. The conference was sponsored by the International Association for Cryptologic Research (IACR). Marc Fischlin (Technische Universität Darmstadt, Germany) was responsible for the local organization. He was supported by a local organizing team consisting of Andrea Püchner, Felix Günther, Christian Janson, and the Cryptoplexity Group. We are deeply indebted to them for their support and smooth collaboration.

The conference program followed the now established parallel track system where the works of the authors were presented in two concurrently running tracks. The invited talks and the talks presenting the best paper/best young researcher spanned over both tracks.

We received a total of 327 submissions. Each submission was anonymized for the reviewing process and was assigned to at least three of the 58 Program Committee members. Committee members were allowed to submit at most one paper, or two if both were co-authored. Submissions by committee members were held to a higher standard than normal submissions. The reviewing process included a rebuttal round for all submissions. After extensive deliberations the Program Committee accepted 76 papers. The revised versions of these papers are included in these three volume proceedings, organized topically within their respective track.

The committee decided to give the Best Paper Award to the paper "Quantum Lightning Never Strikes the Same State Twice" by Mark Zhandry. The runner-up was the paper "Compact Adaptively Secure ABE for NC^1 from k Lin" by Lucas Kowalczyk and Hoeteck Wee. The Best Young Researcher Award went to the paper "Efficient Verifiable Delay Functions" by Benjamin Wesolowski. All three papers received invitations for the *Journal of Cryptology*.

The program also included an IACR Distinguished Lecture by Cynthia Dwork, titled "Differential Privacy and the People's Data," and invited talks by Daniele Micciancio, titled "Fully Homomorphic Encryption from the Ground Up," and François-Xavier Standaert, titled "Toward an Open Approach to Secure Cryptographic Implementations."

We would like to thank all the authors who submitted papers. We know that the Program Committee's decisions can be very disappointing, especially rejections of very good papers that did not find a slot in the sparse number of accepted papers. We sincerely hope that these works eventually get the attention they deserve.

We are also indebted to the members of the Program Committee and all external reviewers for their voluntary work. The committee's work is quite a workload. It has been an honor to work with everyone. The committee's work was tremendously simplified by Shai Halevi's submission software and his support, including running the service on IACR servers.

Finally, we thank everyone else—speakers, session chairs, and rump-session chairs—for their contribution to the program of Eurocrypt 2019. We would also like to thank the many sponsors for their generous support, including the Cryptography Research Fund that supported student speakers.

May 2019   Yuval Ishai
                                                             Vincent Rijmen

Eurocrypt 2019

**The 38th Annual International Conference
on the Theory and Applications of Cryptographic Techniques**

Sponsored by *the International Association for Cryptologic Research*

May 19–23, 2019
Darmstadt, Germany

General Chair

Marc Fischlin   Technische Universität Darmstadt, Germany

Program Co-chairs

Yuval Ishai   Technion, Israel
Vincent Rijmen   KU Leuven, Belgium and University of Bergen,
                                 Norway

Program Committee

Michel Abdalla	CNRS and ENS Paris, France
Adi Akavia	University of Haifa, Israel
Martin Albrecht	Royal Holloway, UK
Elena Andreeva	KU Leuven, Belgium
Paulo S. L. M. Barreto	University of Washington Tacoma, USA
Amos Beimel	Ben-Gurion University, Israel
Alex Biryukov	University of Luxembourg, Luxembourg
Nir Bitansky	Tel Aviv University, Israel
Andrej Bogdanov	Chinese University of Hong Kong, SAR China
Christina Boura	University of Versailles and Inria, France
Xavier Boyen	QUT, Australia
David Cash	University of Chicago, USA
Melissa Chase	MSR Redmond, USA
Kai-Min Chung	Academia Sinica, Taiwan
Dana Dachman-Soled	University of Maryland, USA
Ivan Damgård	Aarhus University, Denmark
Itai Dinur	Ben-Gurion University, Israel
Stefan Dziembowski	University of Warsaw, Poland
Serge Fehr	Centrum Wiskunde & Informatica (CWI) and Leiden University, The Netherlands
Juan A. Garay	Texas A&M University, USA
Sanjam Garg	UC Berkeley, USA

Christina Garman	Purdue University, USA
Siyao Guo	New York University Shanghai, China
Iftach Haitner	Tel Aviv University, Israel
Shai Halevi	IBM Research, USA
Brett Hemenway	University of Pennsylvania, USA
Justin Holmgren	Princeton University, USA
Stanislaw Jarecki	UC Irvine, USA
Dakshita Khurana	Microsoft Research New England, USA
Ilan Komargodski	Cornell Tech, USA
Gregor Leander	Ruhr-Universität Bochum, Germany
Huijia Lin	UCSB, USA
Atul Luykx	Visa Research, USA
Mohammad Mahmoody	University of Virginia, USA
Bart Mennink	Radboud University, The Netherlands
Tal Moran	IDC Herzliya, Israel
Svetla Nikova	KU Leuven, Belgium
Claudio Orlandi	Aarhus University, Denmark
Rafail Ostrovsky	UCLA, USA
Rafael Pass	Cornell University and Cornell Tech, USA
Krzysztof Pietrzak	IST Austria, Austria
Bart Preneel	KU Leuven, Belgium
Christian Rechberger	TU Graz, Austria
Leonid Reyzin	Boston University, USA
Guy N. Rothblum	Weizmann Institute, Israel
Amit Sahai	UCLA, USA
Christian Schaffner	QuSoft and University of Amsterdam, The Netherlands
Gil Segev	Hebrew University, Israel
abhi shelat	Northeastern University, USA
Martijn Stam	Simula UiB, Norway
Marc Stevens	CWI Amsterdam, The Netherlands
Stefano Tessaro	UCSB, USA
Mehdi Tibouchi	NTT, Japan
Frederik Vercauteren	KU Leuven, Belgium
Brent Waters	UT Austin, USA
Mor Weiss	Northeastern University, USA
David J. Wu	University of Virginia, USA
Vassilis Zikas	University of Edinburgh, UK

Additional Reviewers

Divesh Aggarwal	Prabhanjan Ananth	Christian Badertscher
Shashank Agrawal	Gilad Asharov	Saikrishna
Gorjan Alagic	Tomer Ashur	Badrinarayanan
Abdelrahaman Aly	Arash Atashpendar	Shi Bai
Andris Ambainis	Benedikt Auerbach	Josep Balasch

Marshall Ball
James Bartusek
Balthazar Bauer
Carsten Baum
Christof Beierle
Fabrice Benhamouda
Iddo Bentov
Mario Berta
Ward Beullens
Ritam Bhaumik
Jean-François Biasse
Koen de Boer
Dan Boneh
Xavier Bonnetain
Charlotte Bonte
Carl Bootland
Jonathan Bootle
Joppe Bos
Adam Bouland
Florian Bourse
Benedikt Bünz
Wouter Castryck
Siu On Chan
Nishanth Chandran
Eshan Chattopadhyay
Yi-Hsiu Chen
Yilei Chen
Yu Long Chen
Jung-Hee Cheon
Mahdi Cheraghchi
Celine Chevalier
Nai-Hui Chia
Ilaria Chillotti
Chongwon Cho
Wutichai Chongchitmate
Michele Ciampi
Ran Cohen
Sandro Coretti
Ana Costache
Jan Czajkowski
Yuanxi Dai
Deepesh Data
Bernardo David
Alex Davidson
Thomas Debris-Alazard
Thomas De Cnudde

Thomas Decru
Luca De Feo
Akshay Degwekar
Cyprien Delpech de Saint Guilhem
Ioannis Demertzis
Ronald de Wolf
Giovanni Di Crescenzo
Christoph Dobraunig
Jack Doerner
Javad Doliskani
Leo Ducas
Yfke Dulek
Nico Döttling
Aner Ben Efraim
Maria Eichlseder
Naomi Ephraim
Daniel Escudero
Saba Eskandarian
Thomas Espitau
Pooya Farshim
Prastudy Fauzi
Rex Fernando
Houda Ferradi
Dario Fiore
Ben Fisch
Mathias Fitzi
Cody Freitag
Georg Fuchsbauer
Benjamin Fuller
Tommaso Gagliardoni
Steven Galbraith
Nicolas Gama
Chaya Ganesh
Sumegha Garg
Romain Gay
Peter Gazi
Craig Gentry
Marios Georgiou
Benedikt Gierlichs
Huijing Gong
Rishab Goyal
Lorenzo Grassi
Hannes Gross
Jens Groth
Paul Grubbs

Divya Gupta
Felix Günther
Helene Haagh
Björn Haase
Mohammad Hajiabadi
Carmit Hazay
Pavel Hubáček
Andreas Huelsing
Ilia Iliashenko
Muhammad Ishaq
Joseph Jaeger
Eli Jaffe
Aayush Jain
Abhishek Jain
Stacey Jeffery
Zhengfeng Ji
Yael Kalai
Daniel Kales
Chethan Kamath
Nathan Keller
Eike Kiltz
Miran Kim
Sam Kim
Taechan Kim
Karen Klein
Yash Kondi
Venkata Koppula
Mukul Kulkarni
Ashutosh Kumar
Ranjit Kumaresan
Rio LaVigne
Virginie Lallemand
Esteban Landerreche
Brandon Langenberg
Douglass Lee
Eysa Lee
François Le Gall
Chaoyun Li
Wei-Kai Lin
Qipeng Liu
Tianren Liu
Alex Lombardi
Julian Loss
Yun Lu
Vadim Lyubashevsky
Fermi Ma

Saeed Mahloujifar
Christian Majenz
Rusydi Makarim
Nikolaos Makriyannis
Nathan Manohar
Antonio Marcedone
Daniel Masny
Alexander May
Noam Mazor
Willi Meier
Rebekah Mercer
David Mestel
Peihan Miao
Brice Minaud
Matthias Minihold
Konstantinos Mitropoulos
Tarik Moataz
Hart Montgomery
Andrew Morgan
Pratyay Mukherjee
Luka Music
Michael Naehrig
Gregory Neven
Phong Nguyen
Jesper Buus Nielsen
Ryo Nishimaki
Daniel Noble
Adam O'Neill
Maciej Obremski
Sabine Oechsner
Michele Orrù
Emmanuela Orsini
Daniel Ospina
Giorgos Panagiotakos
Omer Paneth
Lorenz Panny
Anat Paskin-Cherniavsky
Alain Passelègue
Kenny Paterson
Chris Peikert
Geovandro Pereira
Léo Perrin
Edoardo Persichetti
Naty Peter

Rachel Player
Oxana Poburinnaya
Yuriy Polyakov
Antigoni Polychroniadou
Eamonn Postlethwaite
Willy Quach
Ahmadreza Rahimi
Sebastian Ramacher
Adrián Ranea
Peter Rasmussen
Shahram Rasoolzadeh
Ling Ren
Joao Ribeiro
Silas Richelson
Thomas Ricosset
Tom Ristenpart
Mike Rosulek
Dragos Rotaru
Yann Rotella
Lior Rotem
Yannis Rouselakis
Arnab Roy
Louis Salvail
Simona Samardziska
Or Sattath
Guillaume Scerri
John Schanck
Peter Scholl
André Schrottenloher
Sruthi Sekar
Srinath Setty
Brian Shaft
Ido Shahaf
Victor Shoup
Jad Silbak
Mark Simkin
Shashank Singh
Maciej Skórski
Caleb Smith
Fang Song
Pratik Soni
Katerina Sotiraki
Florian Speelman
Akshayaram Srinivasan

Uri Stemmer
Noah
    Stephens-Davidowitz
Alan Szepieniec
Gelo Noel Tabia
Aishwarya
    Thiruvengadam
Sergei Tikhomirov
Rotem Tsabary
Daniel Tschudy
Yiannis Tselekounis
Aleksei Udovenko
Dominique Unruh
Cédric Van Rompay
Prashant Vasudevan
Muthu
    Venkitasubramaniam
Daniele Venturi
Benoît Viguier
Fernando Virdia
Ivan Visconti
Giuseppe Vitto
Petros Wallden
Alexandre Wallet
Qingju Wang
Bogdan Warinschi
Gaven Watson
Hoeteck Wee
Friedrich Wiemer
Tim Wood
Keita Xagawa
Sophia Yakoubov
Takashi Yamakawa
Arkady Yerukhimovich
Eylon Yogev
Nengkun Yu
Yu Yu
Aaram Yun
Thomas Zacharias
Greg Zaverucha
Liu Zeyu
Mark Zhandry
Chen-Da Liu Zhang

Abstracts of Invited Talks

Differential Privacy and the People's Data

IACR DISTINGUISHED LECTURE

Cynthia Dwork[1]

Harvard University
dwork@seas.harvard.edu

Abstract. Differential Privacy will be the confidentiality protection method of the 2020 US Decennial Census. We explore the technical and social challenges to be faced as the technology moves from the realm of information specialists to the large community of consumers of census data.

Differential Privacy is a definition of privacy tailored to the statistical analysis of large datasets. Roughly speaking, differential privacy ensures that anything learnable about an individual could be learned independent of whether the individual opts in or opts out of the data set under analysis. The term has come to denote a field of study, inspired by cryptography and guided by theoretical lower bounds and impossibility results, comprising algorithms, complexity results, sample complexity, definitional relaxations, and uses of differential privacy when privacy is not itself a concern.

From its inception, a motivating scenario for differential privacy has been the US Census: data of the people, analyzed for the benefit of the people, to allocate the people's resources (hundreds of billions of dollars), with a legal mandate for privacy. Over the past 4–5 years, differential privacy has been adopted in a number of industrial settings by Google, Microsoft, Uber, and, with the most fanfare, by Apple. In 2020 it will be the confidentiality protection method for the US Decennial Census.

Census data are used throughout government and in thousands of research studies every year. This mainstreaming of differential privacy, the transition from the realm of technically sophisticated information specialists and analysts into much broader use, presents enormous technical and social challenges. The Fundamental Theorem of Information Reconstruction tells us that overly accurate estimates of too many statistics completely destroys privacy. Differential privacy provides a measure of privacy loss that permits the tracking and control of cumulative privacy loss as data are analyzed and re-analyzed. But provably no method can permit the data to be explored without bound. How will the privacy loss "budget" be allocated? Who will enforce limits?

More pressing for the scientific community are questions of how the multitudes of census data consumers will interact with the data moving forward. The Decennial Census is simple, and the tabulations can be handled well with existing technology. In contrast, the annual American Community Survey, which covers only a few million households yearly, is rich in personal details on subjects from internet access in the home to employment to ethnicity, relationships among persons in the home, and fertility. We are not (yet?) able to

---

[1] Supported in part by NSF Grant 1763665 and the Sloan Foundation.

offer differentially private algorithms for every kind of analysis carried out on these data. Historically, confidentiality has been handled by a combination of data summaries, restricted use access to the raw data, and the release of public-use microdata, a form of noisy individual records. Summary statistics are the bread and butter of differential privacy, but giving even trusted and trustworthy researchers access to raw data is problematic, as their published findings are a vector for privacy loss: think of the researcher as an arbitrary non-differentially private algorithm that produces outputs in the form of published findings. The very *choice* of statistic to be published is inherently not privacy-preserving! At the same time, past microdata noising techniques can no longer be considered to provide adequate privacy, but generating synthetic public-use microdata while ensuring differential privacy is a computationally hard problem. Nonetheless, combinations of exciting new techniques give reason for optimism.

Towards an Open Approach to Secure Cryptographic Implementations

François-Xavier Standaert[1]

UCL Crypto Group, Université Catholique de Louvain, Belgium

Abstract. In this talk, I will discuss how recent advances in side-channel analysis and leakage-resilience could lead to both stronger security properties and improved confidence in cryptographic implementations. For this purpose, I will start by describing how side-channel attacks exploit physical leakages such as an implementation's power consumption or electromagnetic radiation. I will then discuss the definitional challenges that these attacks raise, and argue why heuristic hardware-level countermeasures are unlikely to solve the problem convincingly. Based on these premises, and focusing on the symmetric setting, securing cryptographic implementations can be viewed as a tradeoff between the design of modes of operation, underlying primitives and countermeasures.

Regarding modes of operation, I will describe a general design strategy for leakage-resilient authenticated encryption, propose models and assumptions on which security proofs can be based, and show how this design strategy encourages so-called leveled implementations, where only a part of the computation needs strong (hence expensive) protections against side-channel attacks.

Regarding underlying primitives and countermeasures, I will first emphasize the formal and practically-relevant guarantees that can be obtained thanks to masking (i.e., secret sharing at the circuit level), and how considering the implementation of such countermeasures as an algorithmic design goal (e.g., for block ciphers) can lead to improved performances. I will then describe how limiting the leakage of the less protected parts in a leveled implementations can be combined with excellent performances, for instance with respect to the energy cost.

I will conclude by putting forward the importance of sound evaluation practices in order to empirically validate (by lack of falsification) the assumptions needed both for leakage-resilient modes of operation and countermeasures like masking, and motivate the need of an open approach for this purpose. That is, by allowing adversaries and evaluators to know implementation details, we can expect to enable a better understanding of the fundamentals of physical security, therefore leading to improved security and efficiency in the long term.

---

[1] The author is a Senior Research Associate of the Belgian Fund for Scientific Research (FNRS-F.R.S.). This work has been funded in part by the ERC Project 724725.

Fully Homomorphic Encryption
from the Ground Up

Daniele Micciancio (ID)

University of California, Mail Code 0404, La Jolla,
San Diego, CA, 92093, USA
daniele@cs.ucsd.edu
http://cseweb.ucsd.edu/~daniele/

Abstract. The development of fully homomorphic encryption (FHE), i.e., encryption schemes that allow to perform arbitrary computations on encrypted data, has been one of the main achievements of theoretical cryptography of the past 20 years, and probably the single application that brought most attention to lattice cryptography. While lattice cryptography, and fully homomorphic encryption in particular, are often regarded as a highly technical topic, essentially all constructions of FHE proposed so far are based on a small number of rather simple ideas. In this talk, I will try highlight the basic principles that make FHE possible, using lattices to build a simple private key encryption scheme that enjoys a small number of elementary, but very useful properties: a simple decryption algorithm (requiring, essentially, just the computation of a linear function), a basic form of circular security (i.e., the ability to securely encrypt its own key), and a very weak form of linear homomorphism (supporting only a bounded number of addition operations.)

All these properties are easily established using simple linear algebra and the hardness of the Learning With Errors (LWE) problem or standard worst-case complexity assumptions on lattices. Then, I will use this scheme (and its abstract properties) to build in a modular way a tower of increasingly more powerful encryption schemes supporting a wider range of operations: multiplication by arbitrary constants, multiplication between ciphertexts, and finally the evaluation of arithmetic circuits of arbitrary, but a-priory bounded depth. The final result is a *leveled*[1] FHE scheme based on standard lattice problems, i.e., a scheme supporting the evaluation of arbitrary circuits on encrypted data, as long as the depth of the circuit is provided at key generation time. Remarkably, lattices are used only in the construction (and security analysis) of the basic scheme: all the remaining steps in the construction do not make any direct use of lattices, and can be expressed in a simple, abstract way, and analyzed using solely the weakly homomorphic properties of the basic scheme.

Keywords: Lattice-based cryptography · Fully homomorphic encryption · Circular security · FHE bootstrapping

---

[1] The "leveled" restriction in the final FHE scheme can be lifted using "circular security" assumptions that have become relatively standard in the FHE literature, but that are still not well understood. Achieving (non-leveled) FHE from standard lattice assumptions is the main theoretical problem still open in the area.

Contents – Part I

Block Ciphers

Differential Privacy

Bounds for Symmetric Cryptography

Non-malleability

Blockchain and Consensus

Contents – Part II

Secure Computation

Quantum I

Secure Computation and NIZK

Lattice-Based Cryptography

Contents – Part III

Foundations II

Signatures II

ABE and CCA Security

Compact Adaptively Secure ABE for NC¹ from k-Lin

Lucas Kowalczyk[1](✉) and Hoeteck Wee[2]

[1] Columbia University, New York, USA
luke@cs.columbia.edu
[2] CNRS, ENS, PSL, Paris, France
wee@di.ens.fr

Abstract. We present compact attribute-based encryption (ABE) schemes for NC¹ that are adaptively secure under the k-Lin assumption with polynomial security loss. Our KP-ABE scheme achieves ciphertext size that is linear in the attribute length and independent of the policy size even in the many-use setting, and we achieve an analogous efficiency guarantee for CP-ABE. This resolves the central open problem posed by Lewko and Waters (CRYPTO 2011). Previous adaptively secure constructions either impose an attribute "one-use restriction" (or the ciphertext size grows with the policy size), or require q-type assumptions.

1 Introduction

Attribute-based encryption (ABE) [17,31] is a generalization of public-key encryption to support fine-grained access control for encrypted data. Here, ciphertexts and keys are associated with descriptive values which determine whether decryption is possible. In a key-policy ABE (KP-ABE) scheme for instance, ciphertexts are associated with attributes like '(author:Waters), (inst:UT), (topic:PK)' and keys with access policies like ((topic:MPC) OR (topic:SK)) AND (NOT(inst:UCL)), and decryption is possible only when the attributes satisfy the access policy. A ciphertext-policy (CP-ABE) scheme is the dual of KP-ABE with ciphertexts associated with policies and keys with attributes.

Over past decade, substantial progress has been made in the design and analysis of ABE schemes, leading to a large families of schemes that achieve various

L. Kowalczyk—Supported in part by an NSF Graduate Research Fellowship DGE-16-44869; The Leona M. & Harry B. Helmsley Charitable Trust; ERC Project aSCEND (H2020 639554); the Defense Advanced Research Project Agency (DARPA) and Army Research Office (ARO) under Contract W911NF-15-C-0236; and NSF grants CNS-1445424, CNS-1552932 and CCF-1423306. Any opinions, findings and conclusions or recommendations expressed are those of the authors and do not necessarily reflect the views of the Defense Advanced Research Projects Agency, Army Research Office, the National Science Foundation, or the U.S. Government.
H. Wee—Supported in part by ERC Project aSCEND (H2020 639554).

Y. Ishai and V. Rijmen (Eds.): EUROCRYPT 2019, LNCS 11476, pp. 3–33, 2019.
https://doi.org/10.1007/978-3-030-17653-2_1

trade-offs between efficiency, security and underlying assumptions. Meanwhile, ABE has found use as a tool for providing and enhancing privacy in a variety of settings from electronic medical records to messaging systems and online social networks. Moreover, we expect further deployment of ABE, thanks to the recent standardization efforts of the European Telecommunications Standards Institute (ETSI).

In this work, we consider KP-ABE schemes for access policies in NC^1 that simultaneously:

(1) enjoy compact ciphertexts whose size grows only with the length of the attribute and is independent of the policy size, even for complex policies that refer to each attribute many times;
(2) achieve adaptive security (with polynomial security loss);
(3) rely on simple hardness assumptions in the standard model;
(4) can be built with asymmetric prime-order bilinear groups.

We also consider the analogous question for CP-ABE schemes with compact keys. In both KP and CP-ABE, all four properties are highly desirable from both a practical and theoretical stand-point and moreover, properties (1), (2) and (4) are crucial for many real-world applications of ABE. In addition, properties (2), (3) and (4) are by now standard cryptographic requirements pertaining to speed and efficiency, strong security guarantees under realistic and natural attack models, and minimal hardness assumptions. There is now a vast body of works on ABE (e.g. [17,23,26,27], see Fig. 1) showing how different combinations of (1)–(4), culminating in several unifying frameworks that provide a solid understanding of the design and analysis of these schemes [1–3,6,34]. Nonetheless, prior to this work, it was not known how to even simultaneously realize (1)–(3) for NC^1 access policies[1]; indeed, this is widely regarded one of the main open problems in pairing-based ABE.

Our Results. We present the first KP-ABE and CP-ABE schemes for NC^1 that simultaneously realize properties (1)–(4). Our KP-ABE scheme achieves ciphertext size that is linear in the attribute length and independent of the policy size even in the many-use setting; the same holds for the key size in our CP-ABE. Both schemes achieve adaptive security under the k-Lin assumption in asymmetric prime-order bilinear groups with polynomial security loss. We also present an "unbounded" variant of our compact KP-ABE scheme with constant-size public parameters.

As an immediate corollary, we obtain delegation schemes for NC^1 with public verifiability and adaptive soundness under the k-Lin assumption [9,26,30].

---

[1] Note that there exist constructions of ABE for more general access policies like monotone span programs/Boolean formulas with threshold gates [17], and even polynomial-sized Boolean circuits [14,16], but all such constructions sacrifice at least one of the properties (1)–(3).

Our construction leverages a refinement of the recent "partial selectivization" framework for adaptive security [20] (which in turn builds upon [12,13,18,21]) along with the classic dual system encryption methodology [26,33].

reference	adaptive	compact	assumption	unbounded
GPSW [17]		✓	static	✓
LOSTW [23,27]	✓		static	✓
LW [26]	✓	✓	q-type	
OT [28]	✓		2-Lin	✓
Att [3]	✓	✓	q-type	✓
CGKW [7]	✓		k-Lin	✓
ours, Section 6	✓	✓	static	✓
ours, [22]	✓	✓	static	✓

Fig. 1. Summary of KP-ABE schemes for NC1

1.1 Technical Overview

Our starting point is the Lewko–Waters framework for constructing compact adaptively secure ABE [26] based on the dual system encryption methodology[2] [23,25,33]. Throughout, we focus on monotone NC^1 circuit access policies, and note that the constructions extend readily to the non-monotone setting[3]. Let (G_1, G_2, G_T) be an asymmetric bilinear group of prime order p, where g, h are generators of G_1, G_2 respectively.

Warm-Up. We begin with the prior compact KP-ABE for monotone NC^1 [17,23,26]; this is an adaptively secure scheme that comes with the downside of relying on q-type assumptions (q-type assumptions are assumptions of size that grows with some parameter q. It is known that many q-type assumptions become stronger as q grows [10], and in general such complex and dynamic assumptions are not well-understood). The construction uses composite-order groups, but here we'll suppress the distinction between composite-order and prime-order groups for simplicity. We associate ciphertexts $\mathsf{ct_x}$ with attribute vectors[4] $\mathbf{x} \in \{0,1\}^n$ and keys sk_f with Boolean formulas f:

---

[2] Essentially, the dual system proof method provides guidance for transforming suitably-designed functional encryption schemes which are secure for one adversarial secret key request to the multi-key setting where multiple keys may be requested by the adversary. Our main technical contribution involves the analysis of the initial single-key-secure component, which we refer to later as our "Core 1-ABE" component.

[3] Most directly by pushing all NOT gates to the input nodes of each circuit and using new attributes to represent the negation of each original attribute. It is likely that the efficiency hit introduced by this transformation can be removed through more advanced techniques à la [24,29], but we leave this for future work.

[4] Some works associate ciphertexts with a set $S \subseteq [n]$ where $[n]$ is referred to as the attribute universe, in which case $\mathbf{x} \in \{0,1\}^n$ corresponds to the characteristic vector of S.

$$\mathsf{msk} := (\mu, w_1, \ldots, w_n) \tag{1}$$
$$\mathsf{mpk} := (g, g^{w_1}, \ldots, g^{w_n}, e(g,h)^\mu),$$
$$\mathsf{ct_x} := (g^s, \{g^{sw_i}\}_{x_i=1}, e(g,h)^{\mu s} \cdot M)$$
$$\mathsf{sk}_f := (\{h^{\mu_j + r_j w_{\rho(j)}}, h^{r_j}\}_{j \in [m]}), \rho : [m] \to [n]$$

where $\mu_1, \ldots, \mu_m$ are shares of $\mu \in \mathbb{Z}_p$ w.r.t. the formula f; the shares satisfy the requirement that for any $\mathbf{x} \in \{0,1\}^n$, the shares $\{\mu_j\}_{x_{\rho(j)}=1}$ determine μ if $\mathbf{x}$ satisfies f (i.e., $f(\mathbf{x}) = 1$), and reveal nothing about μ otherwise; and ρ is a mapping from the indices of the shares (in $[m]$) to the indices of the attributes (in $[n]$) to which they are associated. For decryption, observe that we can compute $\{e(g,h)^{\mu_j s}\}_{x_i=1}$, from which we can compute the blinding factor $e(g,h)^{\mu s}$ via linear reconstruction "in the exponent".

Here, m is polynomial in the formula size, and we should think of $m = \mathrm{poly}(n) \gg n$. Note that the ciphertext consists only of $O(n)$ group elements and therefore satisfies our compactness requirement.

Proving Adaptive Security. The crux of the proof of adaptive security lies in proving that μ remains computationally hidden given just a single ciphertext and a single key and no mpk (the more general setting with mpk and multiple keys follows via what is by now a textbook application of the dual system encryption methodology). In fact, it suffices to show that μ is hidden given just

$$\mathsf{ct_x'} := (\{w_i\}_{x_i=1}) \quad // \text{ ``stripped down'' } \mathsf{ct_x}$$
$$\mathsf{sk}_f := (\{h^{\mu_j + r_j w_{\rho(j)}}, h^{r_j}\}_{j \in [m]})$$

where $\mathbf{x}, f$ are adaptively chosen subject to the constraint $f(\mathbf{x}) = 0$. Henceforth, we refer to $(\mathsf{ct_x'}, \mathsf{sk}_f)$ as our "core 1-ABE component". Looking ahead to our formalization of adaptive security for this core 1-ABE, we actually require that μ is hidden even if the adversary sees $h^{w_1}, \ldots, h^{w_n}$; this turns out to be useful for the proof of our KP-ABE (for improved concrete efficiency).

Core Technical Contribution. The technical novelty of this work lies in proving adaptive security of the core 1-ABE component under the DDH assumption. Previous analysis either relies on a q-type assumption [1,2,4,26], or imposes the one-use restriction (that is, ρ is injective and $m = n$, in which case security can be achieved unconditionally) [23,34]. Our analysis relies on a piecewise guessing framework which refines and simplifies a recent framework of Jafargholi et al. for proving adaptive security via pebbling games [20] (which in turn builds upon [12,13,18,21]).

Let G_0 denote the view of the adversary $(\mathsf{ct_x'}, \mathsf{sk}_f)$ in the real game, and G_1 denote the same thing except we replace $\{\mu_j\}$ in sk_f with shares of a random value independent of μ. Our goal is to show that $\mathsf{G}_0 \approx_c \mathsf{G}_1$. First, let us define an additional family of games $\{\mathsf{H}^U\}$ parameterized by $U \subseteq [m]$: H^U is the same as G_0 except we replace $\{\mu_j : j \in U\}$ in sk_f with uniformly random values. In particular, $\mathsf{H}^\emptyset = \mathsf{G}_0$.

We begin with the "selective" setting, where the adversary specifies $\mathbf{x}$ at the start of the game. Suppose we can show that $G_0 \approx_c G_1$ in this simpler setting via a series of $L+1$ hybrids of the form:

$$G_0 = H^{h_0(\mathbf{x})} \approx_c H^{h_1(\mathbf{x})} \approx_c \cdots \approx_c H^{h_L(\mathbf{x})} = G_1$$

where $h_0, \ldots, h_L : \{0,1\}^n \to \{U \subseteq [m] : |U| \leq R'\}$ are functions of the adversary's choices $\mathbf{x}$. Then, the piecewise guessing framework basically tells us that $G_0 \approx_c G_1$ in the adaptive setting with a security loss roughly $m^{R'} \cdot L$, where the factor L comes from the hybrid argument and the factor $m^{R'}$ comes from guessing $h_i(\mathbf{x})$ (a subset of $[m]$ of size at most R'). Ideally, we would want $m^{R'} \ll 2^n$, where 2^n is what we achieve from guessing $\mathbf{x}$ itself.

First, we describe a straight-forward approach which achieves $L = 2$ and $R' = m$ implicit in [26] (but incurs a huge security loss $2^m \gg 2^n$) where

$$h_1(\mathbf{x}) = \{j : x_{\rho(j)} = 0\}.$$

That is, $H^{h_1(\mathbf{x})}$ is G_0 with μ_j in sk_f replaced by fresh $\mu_j' \leftarrow \mathbb{Z}_p$ for all j satisfying $x_{\rho(j)} = 0$. Here, we have

- $G_0 \approx_c H^{h_1(\mathbf{x})}$ via DDH, since $h^{\mu_j + w_{\rho(j)} r_j}, h^{r_j}$ computationally hides μ_j whenever $x_{\rho(j)} = 0$ and $w_{\rho(j)}$ is not leaked in $\mathsf{ct}_{\mathbf{x}}$;
- $H^{h_1(\mathbf{x})} \approx_s G_1$ via security of the secret-sharing scheme since the shares $\{\mu_j : x_{\rho(j)} = 1\}$ leak no information about μ whenever $f(\mathbf{x}) = 0$.

This approach is completely generic and works for any secret-sharing scheme.

In our construction, we use a variant of the secret-sharing scheme for NC^1 in [20] (which is in turn a variant of Yao's secret-sharing scheme [19,32]), for which the authors also gave a hybrid argument achieving $L = 8^d$ and $R' = O(d \log m)$ where d is the depth of the formula; this achieves a security loss $2^{O(d \log m)}$. Recall that the circuit complexity class NC^1 is captured by Boolean formulas of logarithmic depth and fan-in two, so the security loss here is quasi-polynomial in n. We provide a more detailed analysis of the functions $h_0, h_1, \ldots, h_L$ used in their scheme, and show that the subsets of size $O(d)$ output by these functions can be described only $O(d)$ bits instead of $O(d \log m)$ bits. Roughly speaking, we show that the subsets are essentially determined by a path of length d from the output gate to an input gate, which can be described using $O(d)$ bits since the gates have fan-in two. Putting everything together, this allows us to achieve adaptive security for the core 1-ABE component with a security loss $2^{O(d)} = \mathrm{poly}(n)$.

Our ABE Scheme. To complete the overview, we sketch our final ABE scheme which is secure under the k-Linear Assumption in prime-order bilinear groups.

To obtain prime-order analogues of the composite-order examples, we rely on the previous framework of Chen et al. [5,6,15] for simulating composite-order groups in prime-order ones. Let (G_1, G_2, G_T) be a bilinear group of prime order p. We start with the KP-ABE scheme in (1) and carry out the following substitutions:

$$g^s \mapsto [\mathbf{s}^\top \mathbf{A}]_1, \ h^{r_j} \mapsto [\mathbf{r}_j]_2, \ w_i \mapsto \mathbf{W}_i \leftarrow \mathbb{Z}_p^{(k+1)\times k}, \ \mu \mapsto \mathbf{v} \leftarrow \mathbb{Z}_p^{k+1} \qquad (2)$$

where $\mathbf{A} \leftarrow \mathbb{Z}_p^{k\times(k+1)}, \mathbf{s}, \mathbf{r}_j \leftarrow \mathbb{Z}_p^k$, k corresponds to the k-Lin Assumption desired for security[5], and $[\cdot]_1, [\cdot]_2$ correspond respectively to exponentiations in the prime-order groups G_1, G_2. We note that the naive transformation following [6] would have required $\mathbf{W}_i$ of dimensions at least $(k+1) \times (k+1)$; here, we incorporated optimizations from [5,15]. This yields the following prime-order KP-ABE scheme for NC^1:

$$\mathsf{msk} := (\mathbf{v}, \mathbf{W}_1, \ldots, \mathbf{W}_n)$$
$$\mathsf{mpk} := ([\mathbf{A}]_1, [\mathbf{A}\mathbf{W}_1]_1, \ldots, [\mathbf{A}\mathbf{W}_n]_1, \ e([\mathbf{A}]_1, [\mathbf{v}]_2)),$$

$$\mathsf{ct}_{\mathbf{x}} := \left([\mathbf{s}^\top \mathbf{A}]_1, \{[\mathbf{s}^\top \mathbf{A}\mathbf{W}_i]_1\}_{x_i=1}, \ e([\mathbf{s}^\top \mathbf{A}]_1, [\mathbf{v}]_2) \cdot M \right)$$

$$\mathsf{sk}_f := (\{[\mathbf{v}_j + \mathbf{W}_{\rho(j)}\mathbf{r}_j]_2, [\mathbf{r}_j]_2\}_{j\in[m]})$$

where $\mathbf{v}_j$ is the j'th share of $\mathbf{v}$. Decryption proceeds as before by first computing

$$\{e([\mathbf{s}^\top \mathbf{A}]_1, [\mathbf{v}_j]_2)\}_{\rho(j)=0 \vee x_{\rho(j)}=1}$$

and relies on the associativity relations $\mathbf{A}\mathbf{W}_i \cdot \mathbf{r}_j = \mathbf{A} \cdot \mathbf{W}_i \mathbf{r}_j$ for all i, j [8].

In the proof, in place of the DDH assumption which allows us to argue that $(h^{w_i r_j}, h^{r_j})$ is pseudorandom, we will rely on the fact that by the k-Lin assumption, we have

$$(\mathbf{A}, \mathbf{A}\mathbf{W}_i, [\mathbf{W}_i \mathbf{r}_j]_2, [\mathbf{r}_j]_2) \approx_c (\mathbf{A}, \mathbf{A}\mathbf{W}_i, [\mathbf{W}_i \mathbf{r}_j + \delta_{ij}\mathbf{a}^\perp]_2, [\mathbf{r}_j]_2)$$

where $\mathbf{A} \leftarrow \mathbb{Z}_p^{k\times(k+1)}, \mathbf{W}_i \leftarrow \mathbb{Z}_p^{(k+1)\times 2k}, \mathbf{r}_j \leftarrow \mathbb{Z}_p^{2k}$ and $\mathbf{a}^\perp \in \mathbb{Z}_p^{k+1}$ satisfies $\mathbf{A} \cdot \mathbf{a}^\perp = \mathbf{0}$.

Organization. We describe the piecewise guessing framework for adaptive security in Sect. 3 and a pebbling strategy (used to define $h_0, \ldots, h_L$) in Sect. 4. We describe a secret-sharing scheme and prove adaptive security of the core 1-ABE component in Sect. 5. We present our full KP-ABE scheme in Sect. 6, and present a CP-ABE and unbounded KP-ABE scheme in the full version of this paper [22].

2 Preliminaries

Notation. We denote by $s \leftarrow S$ the fact that s is picked uniformly at random from a finite set S. By PPT, we denote a probabilistic polynomial-time algorithm. Throughout this paper, we use 1^λ as the security parameter. We use lower case boldface to denote (column) vectors and upper case boldcase to denote

---

[5] E.g.: $k = 1$ corresponds to security under the Symmetric External Diffie-Hellman Assumption (SXDH), and $k = 2$ corresponds to security under the Decisional Linear Assumption (DLIN).

matrices. We use $\equiv$ to denote two distributions being identically distributed, and $\approx_c$ to denote two distributions being computationally indistinguishable. For any two finite sets (also including spaces and groups) S_1 and S_2, the notation "$S_1 \approx_c S_2$" means the uniform distributions over them are computationally indistinguishable.

2.1 Monotone Boolean Formulas and NC^1

Monotone Boolean Formula. A monotone Boolean formula $f : \{0,1\}^n \rightarrow \{0,1\}$ is specified by a directed acyclic graph (DAG) with three kinds of nodes: input gate nodes, gate nodes, and a single output node. Input nodes have in-degree 0 and out-degree 1, AND/OR nodes have in-degree (fan-in) 2 and out-degree (fan-out) 1, and the output node has in-degree 1 and out-degree 0. We number the edges (wires) $1, 2, \ldots, m$, and each gate node is defined by a tuple (g, a_g, b_g, c_g) where $g : \{0,1\}^2 \rightarrow \{0,1\}$ is either AND or OR, a_g, b_g are the incoming wires, c_g is the outgoing wire and $a_g, b_g < c_g$. The size of a formula m is the number of edges in the underlying DAG and the depth of a formula d is the length of the longest path from the output node.

NC^1 *and Log-Depth Formula.* A standard fact from complexity theory tells us that the circuit complexity class monotone NC^1 is captured by monotone Boolean formulas of log-depth and fan-in two. This follows from the fact that we can transform any depth d circuit with fan-in two and unbounded fan-out into an equivalent circuit with fan-in two and fan-out one (for all gate nodes) of the same depth, and a 2^d blow-up in the size. To see this, note that one can start with the root gate of an NC^1 circuit and work downward by each level of depth. For each gate g considered at depth i, if either of its two input wires are coming from the output wire of a gate (at depth $i-1$) with more than one output wire, then create a new copy of the gate at depth $i-1$ with a single output wire going to g (note that this copy may increase the output wire multiplicity of gates at depth strictly lower than $i-1$). This procedure preserves the functionality of the original circuit, and has the result that at its end, each gate in the circuit has input wires which come from gates with output multiplicity 1. The procedure does not increase the depth of the circuit (any duplicated gates are added at a level that already exists), so the new circuit is a formula (all gates have fan-out 1) of depth d with fan-in 2, so its size is at most 2^d. d is logarithmic in the size of the input for NC^1 circuits, so the blowup from this procedure is polynomial in n. Hence we will consider the class NC^1 as a set of Boolean formulas (where gates have fan-in 2 and fan-out 1) of depth $O(\log n)$ and refer to $f \in$ NC^1 formulas.

2.2 Secret Sharing

A secret sharing scheme is a pair of algorithms (share, reconstruct) where share on input $f : \{0,1\}^n \rightarrow \{0,1\}$ and $\mu \in \mathbb{Z}_p$ outputs $\mu_1, \ldots, \mu_m \in \mathbb{Z}_p$ together with $\rho : [m] \rightarrow \{0, 1, \ldots, n\}$.

- Correctness stipulates that for every $x \in \{0,1\}^n$ such that $f(x) = 1$, we have

$$\mathsf{reconstruct}(f, x, \{\mu_j\}_{\rho(j)=0 \vee x_{\rho(j)}=1}) = \mu.$$

- Security stipulates that for every $x \in \{0,1\}^n$ such that $f(x) = 0$, the shares $\{\mu_j\}_{\rho(j)=0 \vee x_{\rho(j)}=1}$ perfectly hide μ.

Note the inclusion of $\rho(j) = 0$ in both correctness and security. All the secret sharing schemes in this work will in fact be linear (in the standard sense): share computes a linear function of the secret μ and randomness over $\mathbb{Z}_p$, and reconstruct computes a linear function of the shares over $\mathbb{Z}_p$, that is, $\mu = \sum_{\rho(j)=0 \vee x_{\rho(j)}=1} \omega_j \mu_j$.

2.3 Attribute-Based Encryption

An attribute-based encryption (ABE) scheme for a predicate $\mathsf{pred}(\cdot, \cdot)$ consists of four algorithms (Setup, Enc, KeyGen, Dec):

Setup($1^\lambda, \mathcal{X}, \mathcal{Y}, \mathcal{M}$) $\rightarrow$ (mpk, msk). The setup algorithm gets as input the security parameter λ, the attribute universe $\mathcal{X}$, the predicate universe $\mathcal{Y}$, the message space $\mathcal{M}$ and outputs the public parameter mpk, and the master key msk.

Enc(mpk, x, m) $\rightarrow$ ct$_x$. The encryption algorithm gets as input mpk, an attribute $x \in \mathcal{X}$ and a message $m \in \mathcal{M}$. It outputs a ciphertext ct$_x$. Note that x is public given ct$_x$.

KeyGen(mpk, msk, y) $\rightarrow$ sk$_y$. The key generation algorithm gets as input msk and a value $y \in \mathcal{Y}$. It outputs a secret key sk$_y$. Note that y is public given sk$_y$.

Dec(mpk, sk$_y$, ct$_x$) $\rightarrow$ m. The decryption algorithm gets as input sk$_y$ and ct$_x$ such that $\mathsf{pred}(x, y) = 1$. It outputs a message m.

Correctness. We require that for all $(x, y) \in \mathcal{X} \times \mathcal{Y}$ such that $\mathsf{pred}(x, y) = 1$ and all $m \in \mathcal{M}$,

$$\Pr[\mathsf{Dec}(\mathsf{mpk}, \mathsf{sk}_y, \mathsf{Enc}(\mathsf{mpk}, x, m)) = m] = 1,$$

where the probability is taken over (mpk, msk) $\leftarrow$ Setup($1^\lambda, \mathcal{X}, \mathcal{Y}, \mathcal{M}$), sk$_y$ $\leftarrow$ KeyGen(mpk, msk, y), and the coins of Enc.

Security Definition. For a stateful adversary $\mathcal{A}$, we define the advantage function

$$\mathsf{Adv}_{\mathcal{A}}^{\mathrm{ABE}}(\lambda) := \Pr\left[b = b' : \begin{array}{l} (\mathsf{mpk}, \mathsf{msk}) \leftarrow \mathsf{Setup}(1^\lambda, \mathcal{X}, \mathcal{Y}, \mathcal{M}); \\ (x^*, m_0, m_1) \leftarrow \mathcal{A}^{\mathsf{KeyGen}(\mathsf{msk}, \cdot)}(\mathsf{mpk}); \\ b \leftarrow_{\mathrm{R}} \{0, 1\}; \mathsf{ct}_{x^*} \leftarrow \mathsf{Enc}(\mathsf{mpk}, x^*, m_b); \\ b' \leftarrow \mathcal{A}^{\mathsf{KeyGen}(\mathsf{msk}, \cdot)}(\mathsf{ct}_{x^*}) \end{array} \right] - \frac{1}{2}$$

with the restriction that all queries y that $\mathcal{A}$ makes to KeyGen(msk, $\cdot$) satisfy $\mathsf{pred}(x^*, y) = 0$ (that is, sk$_y$ does not decrypt ct$_{x^*}$). An ABE scheme is *adaptively secure* if for all PPT adversaries $\mathcal{A}$, the advantage $\mathsf{Adv}_{\mathcal{A}}^{\mathrm{ABE}}(\lambda)$ is a negligible function in λ.

2.4 Prime-Order Bilinear Groups and the Matrix Diffie-Hellman Assumption

A generator $\mathcal{G}$ takes as input a security parameter λ and outputs a group description $\mathbb{G} := (p, G_1, G_2, G_T, e)$, where p is a prime of $\Theta(\lambda)$ bits, G_1, G_2 and G_T are cyclic groups of order p, and $e : G_1 \times G_2 \to G_T$ is a non-degenerate bilinear map. We require that the group operations in G_1, G_2 and G_T as well the bilinear map e are computable in deterministic polynomial time with respect to λ. Let $g_1 \in G_1$, $g_2 \in G_2$ and $g_T = e(g_1, g_2) \in G_T$ be the respective generators. We employ the *implicit representation* of group elements: for a matrix $\mathbf{M}$ over $\mathbb{Z}_p$, we define $[\mathbf{M}]_1 := g_1^{\mathbf{M}}, [\mathbf{M}]_2 := g_2^{\mathbf{M}}, [\mathbf{M}]_T := g_T^{\mathbf{M}}$, where exponentiation is carried out component-wise. Also, given $[\mathbf{A}]_1, [\mathbf{B}]_2$, we let $e([\mathbf{A}]_1, [\mathbf{B}]_2) = [\mathbf{AB}]_T$.

We define the matrix Diffie-Hellman (MDDH) assumption on G_1 [11]:

Definition 1 (MDDH$_{k,\ell}^m$ Assumption). *Let $\ell > k \geq 1$ and $m \geq 1$. We say that the MDDH$_{k,\ell}^m$ assumption holds if for all PPT adversaries $\mathcal{A}$, the following advantage function is negligible in λ.*

$$\mathsf{Adv}_{\mathcal{A}}^{\mathrm{MDDH}_{k,\ell}^m}(\lambda) := \left| \Pr[\mathcal{A}(\mathbb{G}, [\mathbf{M}]_1, [\mathbf{MS}]_1) = 1] - \Pr[\mathcal{A}(\mathbb{G}, [\mathbf{M}]_1, [\mathbf{U}]_1) = 1] \right|$$

where $\mathbf{M} \leftarrow_R \mathbb{Z}_p^{\ell \times k}$, $\mathbf{S} \leftarrow_R \mathbb{Z}_p^{k \times m}$ and $\mathbf{U} \leftarrow_R \mathbb{Z}_p^{\ell \times m}$.

The MDDH assumption on G_2 can be defined in an analogous way. Escala *et al.* [11] showed that

$$k\text{-Lin} \Rightarrow \mathrm{MDDH}_{k,k+1}^1 \Rightarrow \mathrm{MDDH}_{k,\ell}^m \ \forall \ell > k, m \geq 1$$

with a tight security reduction (that is, $\mathsf{Adv}_{\mathcal{A}}^{\mathrm{MDDH}_{k,\ell}^m}(\lambda) = \mathsf{Adv}_{\mathcal{A}'}^{k\text{-LIN}}(\lambda)$). In fact, the MDDH assumption is a generalization of the k-Lin Assumption, such that the k-Lin Assumption is equivalent to the MDDH$_{k,k+1}^1$ Assumption as defined above.

Definition 2 (k-Lin Assumption). *Let $k \geq 1$. We say that the k-Lin Assumption holds if for all PPT adversaries $\mathcal{A}$, the following advantage function is negligible in λ.*

$$\mathsf{Adv}_{\mathcal{A}}^{k\text{-LIN}}(\lambda) := \mathsf{Adv}_{\mathcal{A}}^{\mathrm{MDDH}_{k,k+1}^1}(\lambda)$$

Henceforth, we will use MDDH$_k$ to denote MDDH$_{k,k+1}^1$. Lastly, we note that the k-Lin Assumption itself is a generalization, where setting $k = 1$ yields the Symmetric External Diffie-Hellman Assumption (SXDH), and setting $k = 2$ yields the standard Decisional Linear Assumption (DLIN).

3 Piecewise Guessing Framework for Adaptive Security

We now refine the adaptive security framework of [20], making some simplifications along the way to yield the piecewise guessing framework that will support our security proof. We use $\langle A, G \rangle$ to denote the output of an adversary A in an interactive game G, and an adversary wins if the output is 1, so that the winning probability is denoted by $\Pr[\langle A, G \rangle = 1]$.

Suppose we have two adaptive games G_0 and G_1 which we would like to show to be indistinguishable. In both games, an adversary A makes some adaptive choices that define $z \in \{0,1\}^R$. Informally, the piecewise guessing framework tells us that if we can show that G_0, G_1 are ϵ-indistinguishable in the selective setting where all choices defining z are committed to in advance via a series of $L + 1$ hybrids, where each hybrid depends only on at most $R' \ll R$ bits of information about z, then G_0, G_1 are $2^{2R'} \cdot L \cdot \epsilon$-indistinguishable in the adaptive setting.

Overview. We begin with the selective setting where the adversary commits to $z = z^*$ in advance. Suppose we can show that $G_0 \approx_c G_1$ in this simpler setting via a series of $L + 1$ hybrids of the form:

$$G_0 = H^{h_0(z^*)} \approx_c H^{h_1(z^*)} \approx_c \cdots \approx_c H^{h_L(z^*)} = G_1$$

where $h_0, \ldots, h_L : \{0,1\}^R \to \{0,1\}^{R'}$ and $\{H^u\}_{u \in \{0,1\}^{R'}}$ is a family of games where the messages sent to the adversary in H^u depend on u.[6] In particular, the ℓ'th hybrid only depends on $h_\ell(z^*)$ where $|h_\ell(z^*)| \ll |z^*|$.

Next, we describe how to slightly strengthen this hybrid sequence so that we can deduce that $G_0 \approx_c G_1$ even for an adaptive choice of z. Note that $\{H^u\}_{u \in \{0,1\}^{R'}}$ is now a family of adaptive games where z is adaptively defined as the game progresses. We have two requirements:

The first, *end-point equivalence*, just says the two equivalences

$$G_0 = H^{h_0(z^*)}, \quad G_1 = H^{h_L(z^*)}$$

hold even in the adaptive setting, that is, even if the adversary's behavior defines an z different from z^*. In our instantiation, h_0 and h_L are constant functions, so this equivalence will be immediate.

The second, *neighbor indistinguishability*, basically says that for any $\ell \in [L]$, we have

$$H^{u_0} \approx_c H^{u_1}, \; \forall u_0, u_1 \in \{0,1\}^{R'}$$

as long as the adversary chooses z such that $h_{\ell-1}(z) = u_0 \wedge h_\ell(z) = u_1$ It is easy to see that this is a generalization of $H^{h_{\ell-1}(z^*)} \approx_c H^{h_\ell(z^*)}$ if we require $z = z^*$. To formalize this statement, we need to formalize the restriction on the adversary's choice of z by having the game output 0 whenever the restriction is violated. That is, we define a pair of "selective" games $\widehat{H}_{\ell,0}(u_0, u_1), \widehat{H}_{\ell,1}(u_0, u_1)$ for any $u_0, u_1 \in \{0,1\}^{R'}$, where

---

[6] Informally, $\{H^u\}$ describes the simulated games used in the security reduction, where the reduction guesses R' bits of information described by u about some choices z made by the adversary; these R' bits of information are described by $h_\ell(z)$ in the ℓ'th hybrid. In the ℓ'th hybrid, the reduction guesses a $u \in \{0,1\}^{R'}$ and simulates the game according to H^u and hopes that the adversary will pick an z such that $h_\ell(z) = u$; note that the adversary is not required to pick such an z. One way to think of H^u is that the reduction is committed to u, but the adversary can do whatever it wants.

$\widehat{\mathsf{H}}_{\ell,b}(u_0, u_1)$ is the same as H^{u_b}, except we replace the output with 0 whenever $(h_{\ell-1}(z), h_\ell(z)) \neq (u_0, u_1)$.

That is, in both games, the adversary "commits" in advance to u_0, u_1. Proving indistinguishability here is easier because the reduction knows u_0, u_1 and only needs to handle adaptive choices of z such that $(h_{\ell-1}(z), h_\ell(z)) = (u_0, u_1)$.

Adaptive Security Lemma. The next lemma tells us that the two requirements above implies that $\mathsf{G}_0 \approx_c \mathsf{G}_1$ with a security loss $2^{2R'} \cdot L$ (stated in the contra-positive). In our applications, $2^{R'}$ and L will be polynomial in the security parameter.

Lemma 1 (adaptive security lemma). *Fix* $\mathsf{G}_0, \mathsf{G}_1$ *along with* $h_0, h_1, \ldots, h_L :$ $\{0,1\}^R \to \{0,1\}^{R'}$ *and* $\{\mathsf{H}^u\}_{u \in \{0,1\}^{R'}}$ *such that*

$$\forall \, z^* \in \{0,1\}^R : \mathsf{H}^{h_0(z^*)} = \mathsf{G}_0, \; \mathsf{H}^{h_L(z^*)} = \mathsf{G}_1$$

Suppose there exists an adversary A *such that* $\Pr[\langle \mathsf{A}, \mathsf{G}_0 \rangle = 1] - \Pr[\langle \mathsf{A}, \mathsf{G}_1 \rangle = 1] \geq \epsilon$ *then there exists* $\ell \in [L]$ *and* $u_0, u_1 \in \{0,1\}^{R'}$ *such that*

$$\Pr[\langle \mathsf{A}, \widehat{\mathsf{H}}_{\ell,0}(u_0, u_1) \rangle = 1] - \Pr[\langle \mathsf{A}, \widehat{\mathsf{H}}_{\ell,1}(u_0, u_1) \rangle = 1] \geq \frac{\epsilon}{2^{2R'} L}$$

This lemma is essentially a restatement of the main theorem of [20, Theorem 2]; we defer a comparison to the end of this section.

Proof. For the proof, we need to define the game $\mathsf{H}_\ell(z^*)$ for all $\ell = 0, 1, \ldots, L$ and all $z^* \in \{0,1\}^R$

$\mathsf{H}_\ell(z^*)$ is the same as $\mathsf{H}^{h_\ell(z^*)}$, except we replace the output with 0 whenever $z \neq z^*$.

Roughly speaking, in $\mathsf{H}_\ell(z^*)$, the adversary "commits" to making choices $z = z^*$ in advance.

– Step 1. We begin the proof by using "random guessing" to deduce that

$$\Pr_{z^* \leftarrow \{0,1\}^R}[\langle \mathsf{A}, \mathsf{H}_0(z^*) \rangle = 1] - \Pr_{z^* \leftarrow \{0,1\}^R}[\langle \mathsf{A}, \mathsf{H}_L(z^*) \rangle = 1] \geq \frac{\epsilon}{2^R}$$

This follows from the fact that $\mathsf{H}^{h_0(z)} = \mathsf{G}_0, \mathsf{H}^{h_L(z)} = \mathsf{G}_1$ which implies

$$\Pr_{z^* \leftarrow \{0,1\}^R}[\langle \mathsf{A}, \mathsf{H}_0(z^*) \rangle = 1] = \frac{1}{2^R} \Pr[\langle \mathsf{A}, \mathsf{G}_0 \rangle = 1]$$

$$\Pr_{z^* \leftarrow \{0,1\}^R}[\langle \mathsf{A}, \mathsf{H}_L(z^*) \rangle = 1] = \frac{1}{2^R} \Pr[\langle \mathsf{A}, \mathsf{G}_1 \rangle = 1].$$

– Step 2. Via a standard hybrid argument, we have that there exists ℓ such that

$$\Pr_{z^* \leftarrow \{0,1\}^R}[\langle A, H_{\ell-1}(z^*)\rangle = 1] - \Pr_{z^* \leftarrow \{0,1\}^R}[\langle A, H_\ell(z^*)\rangle = 1] \geq \frac{\epsilon}{2^R L}$$

which implies that:

$$\sum_{z' \in \{0,1\}^R} [\langle A, H_{\ell-1}(z')\rangle = 1] - \sum_{z' \in \{0,1\}^R} [\langle A, H_\ell(z')\rangle = 1] \geq \frac{\epsilon}{L}$$

– Step 3. Next, we relate $\widehat{H}_{\ell,0}, \widehat{H}_{\ell,1}$ and $H_{\ell-1}, H_\ell$. First, we define the set

$$\mathcal{U}_\ell := \{(h_{\ell-1}(z'), h_\ell(z')) : z' \in \{0,1\}^R\} \subseteq \{0,1\}^{R'} \times \{0,1\}^{R'}, \ell \in [L]$$

Observe that for all $(u_0, u_1) \in \mathcal{U}_\ell$, we have

$$\Pr[\langle A, \widehat{H}_{\ell,1}(u_0, u_1)\rangle = 1] = \sum_{z':(h_{\ell-1}(z'),h_\ell(z'))=(u_0,u_1)} \Pr[\langle A, H_\ell(z')\rangle = 1]$$

Then, we have

$$\sum_{z' \in \{0,1\}^R} \Pr[\langle A, H_\ell(z')\rangle = 1]$$

$$= \sum_{(u_0,u_1)\in\mathcal{U}_\ell} \left(\sum_{z':(h_{\ell-1}(z'),h_\ell(z'))=(u_0,u_1)} \Pr[\langle A, H_\ell(z')\rangle = 1] \right)$$

$$= \sum_{(u_0,u_1)\in\mathcal{U}_\ell} \Pr[\langle A, \widehat{H}_{\ell,1}(u_0, u_1)\rangle = 1]$$

By the same reasoning, we also have

$$\sum_{z' \in \{0,1\}^R} \Pr[\langle A, H_{\ell-1}(z')\rangle = 1] = \sum_{(u_0,u_1)\in\mathcal{U}_\ell} \Pr[\langle A, \widehat{H}_{\ell,0}(u_0, u_1)\rangle = 1]$$

This means that

$$\sum_{(u_0,u_1)\in\mathcal{U}_\ell} \left(\Pr[\langle A, \widehat{H}_{\ell,0}(u_0, u_1)\rangle = 1] - \Pr[\langle A, \widehat{H}_{\ell,1}(u_0, u_1)\rangle = 1] \right)$$

$$= \sum_{z' \in \{0,1\}^R} \Pr[\langle A, H_{\ell-1}(z')\rangle = 1] - \sum_{z' \in \{0,1\}^R} \Pr[\langle A, H_\ell(z')\rangle = 1] \geq \frac{\epsilon}{L}$$

where the last inequality follows from Step 2.

– Step 4. By an averaging argument, and using the fact that $|\mathcal{U}_\ell| \leq 2^{2R'}$, there exists $(u_0, u_1) \in \mathcal{U}_\ell$ such that

$$\Pr[\langle A, \widehat{H}_{\ell,0}(u_0, u_1)\rangle = 1] - \Pr[\langle A, \widehat{H}_{\ell,1}(u_0, u_1)\rangle = 1] \geq \frac{\epsilon}{2^{2R'} L}$$

This completes the proof. Note that $2^{2R'}$ can be replaced by $\max_\ell |\mathcal{U}_\ell|$. $\qquad\square$

Comparison with [20]. Our piecewise guessing framework makes explicit the game H^u which are described implicitly in the applications of the framework in [20]. Starting from H^u and $h_0, \ldots, h_L$, we can generically specify the intermediate games $\widehat{\mathsf{H}}_{\ell,0}, \widehat{\mathsf{H}}_{\ell,1}$ as well as the games $\mathsf{H}_0, \ldots, \mathsf{H}_L$ used in the proof of security. The framework of [20] does the opposite: it starts with the games $\mathsf{H}_0, \ldots, \mathsf{H}_L$, and the theorem statement assumes the existence of $h_0, \ldots, h_L$ and $\widehat{\mathsf{H}}_{\ell,0}, \widehat{\mathsf{H}}_{\ell,1}$ that are "consistent" with $\mathsf{H}_0, \ldots, \mathsf{H}_L$ (as defined via a "selectivization" operation). We believe that starting from H^u and $h_0, \ldots, h_L$ yields a simpler and clearer framework which enjoys the advantage of not having to additionally construct and analyze $\widehat{\mathsf{H}}_{\ell,0}, \widehat{\mathsf{H}}_{\ell,1}$ and H_ℓ in the applications.

Finally, we point out that the sets $\mathcal{U}$ and $\mathcal{W}$ in [20, Theorem 2] corresponds to $\mathcal{U}_\ell$ and $\{0,1\}^R$ over here (that is, we do obtain the same bounds), and the i'th function h_i corresponds to the ℓ'th function $h_{\ell-1} \circ h_\ell$ over here.

4   Pebbling Strategy for NC^1

We now define a pebbling strategy for NC^1 which will be used to define the functions $h_0, \ldots, h_L$ we'll use in the piecewise guessing framework. Fix a formula $f : \{0,1\}^n \to \{0,1\}$ of size m and an input $x \in \{0,1\}^n$ for which $f(x) = 0$. A pebbling strategy specifies a sequence of L subsets of $[m]$, corresponding to subsets of input nodes and gates in f that are pebbled. We refer to each subset in the sequence as a pebbling configuration and the i'th term in this sequence is the output of $h_i(f, x)$ (where the combination of f, x correspond to the adaptive choices z made in our security game that will be later analyzed in the piecewise guessing framework).

Our pebbling strategy is essentially the same as that in [20, Section 4]; the main difference is that we provide a better bound on the size of the description of each pebbling configuration in Theorem 1.

4.1   Pebbling Rules

Fix a formula $f : \{0,1\}^n \to \{0,1\}$ and an input $x \in \{0,1\}^n$ for which $f(x) = 0$. We are allowed to place or remove pebbles on input nodes and gates in f, subject to some rules. The goal of a pebbling strategy is to find a sequence of pebbling instructions that follow the rules and starting with the initial configuration (in which there are no pebbles at all), will end up in a configuration where only the root gate has a pebble. Intuitively, the rules say that we can place a pebble a node or a gate if we know that the out-going wire will be 0. More formally,

Definition 3. (Pebbling Rules)

1. *Can place or remove a pebble on any AND gate for which (at least) one input wire comes out of a node with a pebble on it.*
2. *Can place or remove a pebble on any OR gate for which all of the incoming wires come out of nodes which have pebbles on them.*
3. *Can place or remove a pebble on any input node for which $x_i = 0$.*

Given (f, x), a pebbling strategy returns a sequence of pebbling instructions of the form PEBBLE g or unPEBBLE g for some gate g, with the property that each successively applied instruction follows the pebbling rules in Definition 3.

4.2 Pebbling Strategy

Given an NC^1 formula f (recall Sect. 2.1) and an input x on which the formula evaluates to 0, consider the pebbling instruction sequence returned by the following recursive procedure, which maintains the invariant that the output wire evaluates to 0 for each gate that the procedure is called upon. The strategy is described in Fig. 2 and begins by calling Pebble(f, x, g^*) on the root gate g^*. We give an example in Fig. 3.

Pebble(f, x, g):

Input: A node g of an NC^1 formula f with children g_L and g_R along with input x defining values along the wires of f.

1. (Base Case) If g is an input node, Return "PEBBLE g".
2. (Recursive Case) If $g = OR$, first call Pebble(f, x, g_L) to get a list of operations Λ_L, then call Pebble(f, x, g_R) to get a second list of operations λ_R.
   Return $\Lambda_L \circ \Lambda_R \circ$ "PEBBLE g" $\circ$ Reverse(Λ_R) $\circ$ Reverse(Λ_L)
3. (Recursive Case) If $g = AND$, call Pebble$(f, x, \cdot)$ on the first child gate whose output wire evaluates to 0 on input x to get a list of operations Λ.
   Return $\Lambda \circ$ "PEBBLE g" $\circ$ Reverse(Λ)

Reverse(Λ):

Input: A list of instructions of the form "PEBBLE g" or "unPEBBLE g" for a gate g.

1. Return the list Λ in the reverse order, additionally changing each original "PEBBLE " instruction to "unPEBBLE " and each original "unPEBBLE " instruction to "PEBBLE ".

Fig. 2. NC^1 formula pebbling strategy.

Note that if this procedure is called on the root gate of a formula f with an input x such that $f(x) = 0$, then every AND gate on which the Pebble() procedure is called will have *at least one* child node with an output wire which evaluates to 0, and every OR gate on which the Pebble() procedure is called will have child nodes with output wires which *both* evaluate to 0. Furthermore, by inspection, Pebble(f, x, g^*) returns a sequence of pebbling instructions for the circuit that follows the rules in Definition 3.

4.3 Analysis

To be useful in the piecewise guessing framework, we would like for the sequence of pebbling instructions to have the property that each configuration formed by

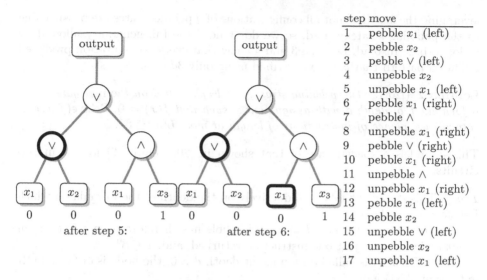

step	move
1	pebble x_1 (left)
2	pebble x_2
3	pebble $\vee$ (left)
4	unpebble x_2
5	unpebble x_1 (left)
6	pebble x_1 (right)
7	pebble $\wedge$
8	unpebble x_1 (right)
9	pebble $\vee$ (right)
10	pebble x_1 (right)
11	unpebble $\wedge$
12	unpebble x_1 (right)
13	pebble x_1 (left)
14	pebble x_2
15	unpebble $\vee$ (left)
16	unpebble x_2
17	unpebble x_1 (left)

Fig. 3. Intermediate pebbling configurations on input $x = 001$. The thick black outline around a node corresponds to having a pebble on the node. Note that steps 10–17 correspond to "undoing" steps 1–8 so that at the end of step 17, there is exactly one pebble on the $\vee$ node leading to the output node.

successive applications of the instructions in the sequence is as short to describe as possible (i.e., minimize the maximum representation size R'). One way to achieve this is to have, at any configuration along the way, as few pebbles as possible. An even more succinct representation can be obtained if we allow many pebbles but have a way to succinctly represent their location. Additionally, we would like to minimize the worst-case length, L, of any sequence produced. We achieve these two goals in the following theorem.

Theorem 1 (pebbling NC^1). *For every input $x \in \{0,1\}^n$ and any monotone formula f of depth d and fan-in two for which $f(x) = 0$, there exists a sequence of $L(d) = 8^d$ pebbling instructions such that every intermediate pebbling configuration can be described using $R'(d) = 3d$ bits.*

Proof. Follows from the joint statements of Lemmas 2 and 4 applied to the pebbling strategy in Fig. 2.

Comparison with [20]. Note that the strategy reproduced in Fig. 2 is essentially the same as one analyzed by [20], which argued that every configuration induced by the pebbling instruction sequence it produces can be described using $d(\log m + 2)$ bits, where m is the number of wires in the formula. This follows from the fact that each such pebbling configuration has at most d gates with pebbled children, and we can specify each such gate using $\log m$ bits and the pebble-status of its two children using an additional two bits. Our Lemma 4 analyzes the same pebbling strategy but achieves a more succinct representation by leveraging the fact that not all configurations of d pebbled gates are possible due to the pebbling strategy used, so we don't need the full generality allowed by $d \cdot \log m$ bits. Instead, Lemmas 3 and 4 show that every configuration produced follows a pattern that can be described using only $3d$ bits.

Lemma 2 ([20]). *The pebbling strategy in Fig. 2 called on the root gate g^* for a formula f of depth d with assignment x such that $f(x) = 0$, Pebble(f, x, g^*), returns a sequence of instructions of length at most $L(d) \leq 8^d$.*

This bound is a special case of that shown in [20, Lemma 2] for fan-in two circuits.

Proof. This statement follows inductively on the depth of the formula on which Pebble() is called.

For the base case, when $d = 0$ (and Pebble has therefore been called on an input node) there is just one instruction returned, and: $1 \leq 8^0$

When Pebble() is called on a node at depth $d > 0$, the node is either an OR gate or an AND gate.

When Pebble() is called on an OR gate, using our inductive hypothesis for the instructions returned for the subformula of depth $d - 1$, notice that the number of instructions returned is:

$$L(d-1)+L(d-1)+1+L(d-1)+L(d-1) = 8^{d-1}+8^{d-1}+1+8^{d-1}+8^{d-1} = 4\cdot 8^{d-1}+1 \leq 8^d$$

When Pebble() is called on an AND gate, using our inductive hypothesis for the instructions returned for the subformula of depth $d - 1$, notice that the number of instructions returned is:

$$L(d - 1) + 1 + L(d - 1) = 8^{d-1} + 1 + 8^{d-1} = 2 \cdot 8^{(d-1)} + 1 \leq 8^d \qquad \square$$

We note that the following lemma is new to this work and will be used to bound the representation size $R(d)$ of any configuration produced by application of the instructions output by the pebbling strategy.

Lemma 3 (structure of pebbling configuration). *Every configuration induced by application of the instructions produced by the pebbling strategy in Fig. 2 called on the root gate g^* of a formula f of depth d with assignment x such that $f(x) = 0$, Pebble(f, x, g^*), has the following property for all gates g in f with children g_L, g_R:*

If any node in the sub-tree rooted at g_R is pebbled, then there exists at most one pebble on the sub-tree rooted at g_L, namely a pebble on g_L itself

Proof. Call a node "good" if it satisfies the property above. First, we make the following observation about the behavior of Reverse(): Applying Reverse() to a list of instructions inducing a list of configurations for which all nodes are "good" produces a new list for which this is true. This holds since Reverse() does not change the configurations induced by a list of instructions, just the ordering (which is reversed). This follows from a simple proof by induction on the length of the input instruction list and the fact that for an input list of instructions parsed as $L_1 \circ L_2$ for two smaller-length lists, we can implement Reverse($L_1 \circ L_2$) as Reverse(L_2) $\circ$ Reverse(L_1).

We proceed with our original proof via a proof by induction on the depth of the formula upon which Pebble() is called.

Inductive Hypothesis: For formulas f of depth $d-1$ with root gate g^* and assignment x such that $f(x) = 0$, Pebble(f, x, g^*) returns a sequence of instructions that induces a sequence of configurations that (1) end with a configuration where g^* is the only pebbled node, and satisfies: (2) in every configuration all nodes are "good."

Base Case: when Pebble(f, x, g^*) is called on a formula of depth 0, the formula consists of just an input node g^*. The (single) returned instruction PEBBLE g^* then satisfies that in both the initial and final configuration, the single node g^* is good. Also, the sequence ends in the configuration where g^* is the only pebbled node.

Inductive Step: when Pebble(f, x, g^*) is called on formula of depth $d > 0$. Let g_L^*, g_R^* denote the children of the root gate g^* (either an AND or OR gate). Note that the sub-formulas $f_{g_L^*}$ and $f_{g_R^*}$ rooted at g_L^* and g_R^* have depth $d-1$. We proceed via a case analysis:

If g^* is an AND gate, then suppose the sequence of instructions returned is

$$\text{Pebble}(f_{g_R^*}, x, g_R^*) \circ \text{PEBBLE } g^* \circ \text{Reverse}(\text{Pebble}(f_{g_R^*}, x, g_R))$$

(The case with g_L^* instead of g_R^* is handled analogously, even simpler). Suppose Pebble($f_{g_R^*}, x, g_R^*$) (and thus Reverse(Pebble($f_{g_R^*}, x, g_R^*$))) produces L_0 instructions. We proceed via a case analysis:

- Take any of the first L_0 configurations (starting from 0'th). Here, all pebbles are in the subformula rooted at g_R^*. We can then apply part (2) of the inductive hypothesis to the subformula $f_{g_R^*}$ rooted at g_R^* (of depth $d-1$) to deduce that property "good" holds for all nodes in $f_{g_R^*}$. All nodes in $f_{g_L^*}$ are unpebbled in all configurations, so they are automatically good. Lastly, the root gate g^* has no pebbled nodes in the subformula rooted at g_L, so it is also good.
- For the ($L_0 + 1$)'th configuration reached after PEBBLE g^*, there are only two pebbles, one on g^* (from the PEBBLE g^* instruction) and another on g_R^* (from part (1) of our inductive hypothesis applied to the (depth $d-1$) subformula $f_{g_R^*}$). It is clear that all nodes in this configuration are good.
- For the last L_0 configurations, there is one pebble on g^* and all remaining pebbles are in the subformula rooted at g_R^*. Clearly, g^* is good. All nodes in $f_{g_L^*}$ are unpebbled in all configurations, so they are also good. Moreover, we can apply the inductive hypothesis to $f_{g_R^*}$ combined with our observation that Reverse preserves property (2) of this hypothesis to deduce that all nodes in the subformula are also good for all configurations.

Lastly, notice that since the last L_0 instructions undo the first L_0 instructions, the final configuration features a single pebble on g^*.

If g^* is an OR gate, then the sequence of instructions returned is

$$\mathsf{Pebble}(f_{g_L^*}, x, g_L^*) \circ \mathsf{Pebble}(f_{g_R^*}, x, g_R^*) \circ \mathrm{PEBBLE}\ g^* \circ \mathsf{Reverse}(\mathsf{Pebble}(f_{g_R^*}, x, g_R^*))$$
$$\circ \mathsf{Reverse}(\mathsf{Pebble}(f_{g_L^*}, x, g_L^*))$$

Suppose $\mathsf{Pebble}(f_{g_R^*}, x, g_R^*), \mathsf{Pebble}(f_{g_L^*}, x, g_L^*)$, and thus $\mathsf{Reverse}(\mathsf{Pebble}(f_{g_R^*}, x, g_R^*))$, $\mathsf{Reverse}(\mathsf{Pebble}(f_{g_L^*}, x, g_L^*))$, produces L_0, L_1 instructions. We proceed via a case analysis:

- Take any of the first L_0 configurations (starting from 0'th). Here, all pebbles are in the subformula $f_{g_L^*}$ rooted at g_L^*. We can then apply part (2) of the inductive hypothesis to (depth $d-1$) $f_{g_L^*}$ to deduce that property "good" holds for all nodes in $f_{g_L^*}$. All nodes in the subformula rooted at g_R^*, $f_{g_R^*}$, are unpebbled in all configurations, so they are automatically good. Lastly, the root gate g^* has no pebbled nodes in the subformula rooted at g_R^*, so it is also good. Finally, by part (1) of this application of the inductive hypothesis, we know that L_0th configuration features a single pebble on g_L^*.
- Take any of the next L_1 configurations (starting from the L_0'th). Here, all pebbles are in the subformula rooted at g_R^* except for the single pebble on g_L^*. We can then apply part (2) of the inductive hypothesis to (depth $d-1$) $f_{g_R^*}$ (of depth $d-1$) to deduce that property "good" holds for all nodes in $f_{g_R^*}$. All nodes in the subformula rooted at g_L^* have no pebbles in their own subformulas, so they are automatically good. Lastly, the root gate g^* may have pebbled nodes in the subformula rooted at g_R^* but the only pebbled node in the subformula rooted at g_L^* is g_L^* itself, so it is also good. Finally, we know that the $L_0 + L_1$th configuration features two pebbles: a pebble on g_L^* (from the first L_0 instructions), and a pebble on g_R^* (by part (1) of this application of the inductive hypothesis).
- For the $(L_0 + L_1 + 1)$'th configuration reached after PEBBLE g^*, there are only three pebbles, one on g^* (from the PEBBLE g^* instruction), one on g_L^* (from the first L_0 instructions), and another on g_R^* (from the next L_1 instructions). It is clear that all nodes in this configuration are good.
- For the next L_1 configurations (reversing the instructions of the set of size L_1), there is one pebble on g^*, one pebble on g_L^*, and all remaining pebbles are in the subformula rooted at g_R^*, $f_{g_R^*}$. g^* is good, since it only has one pebble in the subformula rooted at g_L^*, on g_L^* itself. All nodes in the subformula rooted at g_L^* have no pebbles in their own subformulas, so they are also good. Moreover, we can apply the inductive hypothesis to (depth $d-1$) $f_{g_R^*}$ combined with our observation that $\mathsf{Reverse}$ preserves property (2) of this hypothesis to deduce that all nodes in $f_{g_R^*}$ are also good for all configurations. Note the final configuration in this sequence then contains two pebbles, one of g^* and one on g_L^*.
- For the final L_0 configurations (reversing the instructions of the set of size L_0), there is one pebble on g^*, and all remaining pebbles are in the subformula

rooted at g_L^*. g^* is good, since it has no pebbles in the subformula rooted at g_R^*. Similarly, all nodes in the subformula rooted at g_R^* are also good. Moreover, we can apply the inductive hypothesis to (depth $d-1$) $f_{g_L^*}$ combined with our observation that Reverse preserves property (2) of this hypothesis to deduce that all nodes in $f_{g_L^*}$ are also good for all configurations.

Lastly, notice that since the last $L_0 + L_1$ instructions undo the first $L_0 + L_1$ instructions, the final configuration features a single pebble on g^*. □

Lemma 4 ($R'(d) = 3d$). *Every configuration induced by application of the instructions produced by the pebbling strategy in Fig. 2 for a formula f of depth d with assignment x such that $f(x) = 0$ can be described using $R'(d) = 3d$ bits.*

Proof. We can interpret $3d$ bits in the following way to specify a pebbling: the first d bits specify a path down the formula starting at the root gate (moving left or right based on the setting of each bit), the next $2(d-1)$ bits specify, for each of the $(d-1)$ non-input nodes along the path, which of its children are pebbled. Finally one of the last 2 bits is used to denote if the root node is pebbled.

From Lemma 3, we know that for all gates g with children g_L, g_R, if any node in the sub-tree rooted at g_R is pebbled, then there exists at most one pebble on the sub-tree rooted at g_L, namely a pebble on g_L itself. So, given a pebbling configuration, we can start at the root node and describe the path defined by taking the child with more pebbles on its subtree using d bits. All pebbles in the configuration are either on the root node or on children of nodes on this path and therefore describable in the remaining $2d$ bits. □

5 Core Adaptive Security Component

In this section, we will describe the secret-sharing scheme (share, reconstruct) used in our ABE construction. In addition, we describe a core component of our final ABE, and prove adaptive security using the pebbling strategy defined and analyzed in Sect. 4 to define hybrids in the piecewise guessing framework of Sect. 3.

Overview. As described in the overview in Sect. 1.1, we will consider the following "core 1-ABE component":

$$\mathsf{ct}'_\mathbf{x} := (\{w_i\}_{x_i=1}) \quad // \text{ "stripped down" } \mathsf{ct}_\mathbf{x}$$
$$\mathsf{sk}_f := (\{h^{\mu_j}\}_{\rho(j)=0} \cup \{h^{\mu_j + r_j w_{\rho(j)}}, h^{r_j}\}_{\rho(j) \neq 0})$$

where $(\{\mu_j\}, \rho) \leftarrow \mathsf{share}(f, \mu)$. We want to show that under the DDH assumption, μ is hidden given just $(\mathsf{ct}'_\mathbf{x}, \mathsf{sk}_f)$ where $\mathbf{x}, f$ are adaptively chosen subject to the constraint $f(\mathbf{x}) = 0$. We formalize this via a pair of games $\mathsf{G}_0^{\text{1-ABE}}, \mathsf{G}_1^{\text{1-ABE}}$ and the requirement $\mathsf{G}_0^{\text{1-ABE}} \approx_c \mathsf{G}_1^{\text{1-ABE}}$. In fact, we will study a more abstract construction based on any CPA-secure encryption with:

$$\mathsf{ct}'_\mathbf{x} := (\{w_i\}_{x_i=1}) \quad // \text{ "stripped down" } \mathsf{ct}_\mathbf{x}$$
$$\mathsf{sk}'_f := \{\mu_j\}_{\rho(j)=0} \cup \{\mathsf{CPA.Enc}(w_{\rho(j)}, \mu_j)\}_{\rho(j) \neq 0} \text{ where } (\{\mu_j\}, \rho) \leftarrow \mathsf{share}(f, \mu)$$

5.1   Linear Secret Sharing for NC^1

We first describe a linear secret-sharing scheme for NC^1; this is essentially the information-theoretic version of Yao's secret-sharing for NC^1 in [19,20,32]. It suffices to work with Boolean formulas where gates have fan-in 2 and fan-out 1, thanks to the transformation in Sect. 2.1. We describe the scheme in Fig. 4, and give an example in Fig. 5. Note that our non-standard definition of secret-sharing in Sect. 2.2 allows the setting of $\rho(j) = 0$ for shares that are available for reconstruction for all x. We remark that the output of share satisfies $|\{\mu_j\}| \leq 2m$ since each of the m nodes adds a single μ_j to the output set, except for OR gates which add two: μ_{j_a} and μ_{j_b}.

---

share(f, μ):

Input: A formula $f : \{0,1\}^n \to \{0,1\}$ of size m and a secret $\mu \in \mathbb{Z}_p$.

1. For each non-output wire $j = 1, ..., m - 1$, pick a uniformly random $\hat{\mu}_j \leftarrow \mathbb{Z}_p$. For the output wire, set $\hat{\mu}_m = \mu$
2. For each outgoing wire j from input node i, add $\mu_j = \hat{\mu}_j$ to the output set of shares and set $\rho(j) = i$.
3. For each AND gate g with input wires a, b and output wire c, add $\mu_c = \hat{\mu}_c + \hat{\mu}_a + \hat{\mu}_b \in \mathbb{Z}_p$ to the output set of shares and set $\rho(c) = 0$.
4. For each OR gate g with input wires a, b and output wire c, add $\mu_{c_a} = \hat{\mu}_c + \hat{\mu}_a \in \mathbb{Z}_p$ and $\mu_{c_b} = \hat{\mu}_c + \hat{\mu}_b \in \mathbb{Z}_p$ to the output set of shares and set $\rho(c_a) = 0$ and $\rho(c_b) = 0$.
5. Output $\{\mu_j\}, \rho$.

Fig. 4. Information-theoretic linear secret sharing scheme share for NC^1

The reconstruction procedure reconstruct of the scheme is essentially applying the appropriate linear operations to get the output wire value $\hat{\mu}_c$ at each node starting from the leaves of the formula to get to the root $\hat{\mu}_m = \mu$.

- Given $\hat{\mu}_a, \hat{\mu}_b$ associated with the input wires of an AND gate, we recover the gate's output wire value $\hat{\mu}_c$ by subtracting their values from μ_c (which is available since $\rho(c) = 0$).
- Given one of $\hat{\mu}_a, \hat{\mu}_b$ associated with the input wires of an OR gate, we recover the gate's output wire value $\hat{\mu}_c$ by subtracting it from the appropriate choice of μ_{c_a} or μ_{c_b} (which are both available since $\rho(c_a) = \rho(c_b) = 0$).

Note that reconstruct$(f, x, \{\mu_j\}_{\rho(j)=0 \vee x_{\rho(j)}=1})$ computes a linear operation with respect to the shares μ_j. This follows from the fact that the operation at each gate in reconstruction is a linear operation, and the composition of linear operations is itself a linear operation. Therefore, reconstruct$(f, x, \{\mu_j\}_{\rho(j)=0 \vee x_{\rho(j)}=1})$ is equivalent to identifying the coefficients ω_j of this linear function, where $\mu = \sum_{\rho(j)=0 \vee x_{\rho(j)}=1} \omega_j \mu_j$.

As with any linear secret-sharing scheme, share and reconstruct can be extended in the natural way to accommodate vectors of secrets. Specifically, for a vector of secrets $\mathbf{v} \in \mathbb{Z}_p^k$, define:

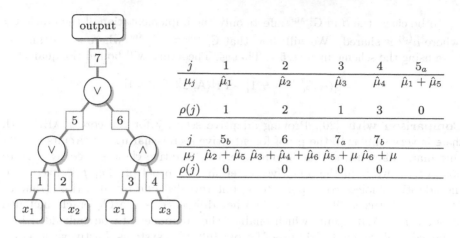

j	1	2	3	4	5_a
μ_j	$\hat{\mu}_1$	$\hat{\mu}_2$	$\hat{\mu}_3$	$\hat{\mu}_4$	$\hat{\mu}_1 + \hat{\mu}_5$
$\rho(j)$	1	2	1	3	0

j	5_b	6	7_a	7_b
μ_j	$\hat{\mu}_2 + \hat{\mu}_5$	$\hat{\mu}_3 + \hat{\mu}_4 + \hat{\mu}_6$	$\hat{\mu}_5 + \mu$	$\hat{\mu}_6 + \mu$
$\rho(j)$	0	0	0	0

Fig. 5. Left: Formula $(x_1 \vee x_2) \vee (x_1 \wedge x_3)$, where the wires are numbered $1, 2, \ldots, 7$. Right: Shares $(\mu_1, \ldots, \mu_{7_b})$ and mapping ρ for the formula corresponding to secret $\mu \in \mathbb{Z}_p$

$$\mathsf{share}(f, \mathbf{v}) := (\{\mathbf{v}_j := (v_{1,j}, ..., v_{k,j}))\}, \rho) \text{ where } (\{v_{i,j}\}, \rho) \leftarrow \mathsf{share}(f, v_i)$$

(note that ρ is identical for all i). reconstruct can also be defined component-wise:

$$\mathsf{reconstruct}(f, x, \{\mathbf{v}_j\}_{\rho(j)=0 \vee x_{\rho(j)}=1}) := \sum_{\rho(j)=0 \vee x_{\rho(j)}=1} \omega_j \mathbf{v}_j \text{ where } \omega_j \text{ are computed as above}$$

Our final ABE construction will use this extension.

5.2 Core 1-ABE Security Game

Definition 4 (core 1-ABE security $\mathsf{G}_0^{1\text{-abe}}, \mathsf{G}_1^{1\text{-abe}}$). *For a stateful adversary $\mathcal{A}$, we define the following games $\mathsf{G}_\beta^{1\text{-ABE}}$ for $\beta \in \{0, 1\}$.*

$$\langle \mathcal{A}, \mathsf{G}_\beta^{1\text{-ABE}} \rangle := \mathbb{I} \left\{ \begin{array}{l} \mu^{(0)}, \mu^{(1)} \leftarrow \mathbb{Z}_p; w_i \leftarrow \mathsf{CPA.Setup}(\lambda) \\ b' \leftarrow \mathcal{A}^{\mathcal{O}_\mathsf{F}(\cdot), \mathcal{O}_\mathsf{X}(\cdot), \mathcal{O}_\mathsf{E}(\cdot,\cdot)}(\mu^{(0)}) \end{array} \right\}$$

where the adversary $\mathcal{A}$ adaptively interacts with three oracles:

$$\mathcal{O}_\mathsf{F}(f) := \{\mathsf{sk}'_f = \{\mu_j\}_{\rho(j)=0} \cup \{\mathsf{CPA.Enc}(w_{\rho(j)}, \mu_j)\}_{\rho(j)\neq 0} \text{ where } (\{\mu_j\}, \rho) \leftarrow \mathsf{share}(f, \mu^{(\beta)})$$
$$\mathcal{O}_\mathsf{X}(x) := (\mathsf{ct}'_x = \{w_i\}_{x_i=1})$$
$$\mathcal{O}_\mathsf{E}(i, m) := \mathsf{CPA.Enc}_{w_i}(m)$$

with the restrictions that (i) only one query is made to each of $\mathcal{O}_\mathsf{F}(\cdot)$ and $\mathcal{O}_\mathsf{X}(\cdot)$, and (ii) the queries f and x to $\mathcal{O}_\mathsf{F}(\cdot), \mathcal{O}_\mathsf{X}(\cdot)$ respectively, satisfy $f(x) = 0$.

To be clear, the β in $\mathsf{G}_\beta^{\text{1-ABE}}$ affects only the implementation of the oracle $\mathcal{O}_\mathsf{F}$ (where $\mu^{(\beta)}$ is shared). We will show that $\mathsf{G}_0^{\text{1-ABE}} \approx_c \mathsf{G}_1^{\text{1-ABE}}$ where we instantiate share using the scheme in Sect. 5.1. That is, Theorem 2 will bound the quantity:

$$\Pr[\langle \mathsf{A}, \mathsf{G}_0^{\text{1-ABE}} \rangle = 1] - \Pr[\langle \mathsf{A}, \mathsf{G}_1^{\text{1-ABE}} \rangle = 1]$$

Comparison with [20]. Proving adaptive security for the core 1-ABE with share is very similar to the proof for adaptive secret-sharing for circuits in [20]. One main difference is that in our case, the adaptive choices z correspond to both (f, x), while in the adaptive secret-sharing proof of [20], f is fixed, and the adaptive choices correspond to x, but revealed one bit at a time (that is, $\mathcal{O}_\mathsf{X}(i, x_i)$ returns w_i if $x_i = 1$). Another difference is the $\mathcal{O}_\mathsf{E}$ oracle included in our core 1-ABE game, which enables the component to be embedded in a standard dual-system hybrid proof for our full ABE systems. Lastly, we leverage our improved analysis in Lemmas 3 and 4 to achieve polynomial security loss, rather than the quasi-polynomial loss we would get from following their proof more directly.

5.3   Adaptive Security for Core 1-ABE Component

We will show that $\mathsf{G}_0^{\text{1-ABE}} \approx_c \mathsf{G}_1^{\text{1-ABE}}$ as defined in Definition 4 using the piecewise guessing framework. To do this, we need to first define a family of games $\{\mathsf{H}^u\}$ along with functions $h_0, \ldots, h_L$, using the pebbling strategy in Sect. 4. First, we will describe share^u, which will be used to define H^u.

Defining share^u. Recall that Lemma 4 describes how to parse a $u \in \{0, 1\}^{3d}$ as a pebbling configuration: a subset of the nodes of f. Further, note that each node contains one output wire, so we can equivalently view u as a subset of $[m]$ denoting the output wires of pebbled gates. Given a pebbling configuration u of an NC^1 formula, the shares are generated as in the secret-sharing scheme in Fig. 4, except for each pebbled node with output wire c, we replace μ_c with an independent random $\mu_c \leftarrow \mathbb{Z}_p$ (in the case of a pebbled OR gate, we replace both associated μ_{c_a} and μ_{c_b} with independent random $\mu_{c_a}, \mu_{c_b} \leftarrow \mathbb{Z}_p$, i.e: both μ_{c_a}, μ_{c_b} are associated with wire c.). In particular, we get the procedure share$^u(f, \mu)$ defined in Fig. 6.

Hybrid Distribution H^u. We now define our hybrid games, and remark that Sect. 3 used $z \in \{0, 1\}^R$ to denote the adaptive choices made by an adversary, and the functions h_ℓ that define our hybrid games will depend on the adaptive choices of both the $f \in \mathsf{NC}^1$ and $x \in \{0, 1\}^n$ chosen during the game, so in our application of the piecewise guessing framework of Sect. 3, z will be (f, x). Note that the conclusion of the framework is independent of the size of the adaptive input ($R = |f| + n$), and the framework allows its x to be defined in parts over time, though in our application, x will be defined in one shot.

---

share^u(f, μ):

Input: A formula $f : \{0,1\}^n \to \{0,1\}$, a secret $\mu \in \mathbb{Z}_p$, and a pebbling configuration u of the nodes of f.

1. Compute $(\{\mu'_j\}, \rho) \leftarrow$ share(f, μ) as defined in Figure 4
2. For each μ'_j, if $j \in u$ (i.e: if j is the output wire of a pebbled node), then sample $\mu_j \leftarrow \mathbb{Z}_p$. Otherwise, set $\mu_j := \mu'_j$.
3. Output $\{\mu_j\}, \rho$.

Fig. 6. Pebbling-modified secret sharing scheme share^u

Definition 5 (H^u and h_ℓ). _Let H^u be $\mathsf{G}_0^{\text{1-ABE}}$ with_ $\boxed{\text{share}^u}$ _($f, \mu^{(0)}$) used in the implementation of oracle $\mathcal{O}_\mathsf{F}(f)$ (replacing_ $\boxed{\text{share}}$ _($f, \mu^{(0)}$)). Let $h_\ell : \mathsf{NC}^1 \times \{0,1\}^n \to \{0,1\}^{R'}$ denote the function that on formula f with root gate g^* and input $x \in \{0,1\}^n$ where $f(x) = 0$, outputs the pebbling configuration created from following the first ℓ instructions from $\mathsf{Pebble}(f, x, g^*)$ of Fig. 2._

Note that the first 0 instructions specify a configuration with no pebbles, so h_0 is a constant function for all f, x. Also, from the inductive proof in Lemma 3, we know that all sequences of instructions from $\mathsf{Pebble}(f, x, g^*)$ when $f(x) = 0$ result in a configuration with a single pebble on the root gate g^*, so h_L is a constant function for all f, x where $f(x) = 0$. Furthermore, note that for all such f, x:

- $\mathsf{H}^{h_0(f,x)}$ is equivalent to $\mathsf{G}_0^{\text{1-ABE}}$ (since share$^{h_0(f,x)}$($f, \mu^{(0)}$) = share($f, \mu^{(0)}$));
- $\mathsf{H}^{h_L(f,x)}$ is equivalent to $\mathsf{G}_1^{\text{1-ABE}}$ (since share$^{h_L(f,x)}$($f, \mu^{(0)}$) = share($f, \mu^{(1)}$) for an independently random $\mu^{(1)}$ which is implicitly defined by the independently random value associated with the output wire of the pebbled root gate: μ_m).

We now have a series of hybrids $\mathsf{G}_0^{\text{1-ABE}} \equiv \mathsf{H}^{h_0(f,x)}, \mathsf{H}^{h_1(f,x)}, ..., \mathsf{H}^{h_L(f,x)} \equiv \mathsf{G}_1^{\text{1-ABE}}$ which satisfy end-point equivalence and, according to the piecewise guessing framework described in Sect. 3, define games $\widehat{\mathsf{H}}_{\ell,0}(u_0, u_1), \widehat{\mathsf{H}}_{\ell,1}(u_0, u_1)$ for $\ell \in [0, L]$.

Lemma 5 (neighboring indistinguishability). _For all $\ell \in [L]$ and $u_0, u_1 \in \{0,1\}^{R'}$, $\Pr[\langle \mathsf{A}, \widehat{\mathsf{H}}_{\ell,0}(u_0, u_1) \rangle = 1] - \Pr[\langle \mathsf{A}, \widehat{\mathsf{H}}_{\ell,1}(u_0, u_1) \rangle = 1] \leq n \cdot \mathsf{Adv}_\mathsf{B}^{\mathrm{CPA}}(\lambda)$._

Proof. First, observe that the difference between $\widehat{\mathsf{H}}_{\ell,0}(u_0, u_1)$ and $\widehat{\mathsf{H}}_{\ell,1}(u_0, u_1)$ lies in $\mathcal{O}_\mathsf{F}(\cdot)$: the former uses $\boxed{\text{share}^{u_0}}$ and the latter uses $\boxed{\text{share}^{u_1}}$. Now, fix the adaptive query f to $\mathcal{O}_\mathsf{F}$. We consider two cases.

First, suppose there does not exist $x' \in \{0,1\}^n$ such that $h_{\ell-1}(f, x') = u_0$ and $h_\ell(f, x') = u_1$. Then, both $\langle \mathsf{A}, \widehat{\mathsf{H}}_{\ell,0}(u_0, u_1) \rangle$ and $\langle \mathsf{A}, \widehat{\mathsf{H}}_{\ell,1}(u_0, u_1) \rangle$ output 0 (i.e., abort) with probability 1 and then we are done.

In the rest of the proof, we deal with the second case, namely there exists $x' \in \{0,1\}^n$ such that $h_{\ell-1}(f, x') = u_0$ and $h_\ell(f, x') = u_1$. This means that u_0

and u_1 are neighboring pebbling configurations in $\mathsf{Pebble}(f, x', g^*)$, so they differ by a pebbling instruction that follows one of the rules in Definition 3. We proceed via a case analysis depending on what the instruction taking configuration u_0 to u_1 is (the instruction is uniquely determined given u_0, u_1, f):

- pebble/unpebble input node with out-going wire $\boxed{j}$: Here, the only difference from $\mathsf{share}^{u_0}(f, \mu^{(0)})$ to $\mathsf{share}^{u_1}(f, \mu^{(0)})$ is that we change $\boxed{\mu_j}$ to a random element of $\mathbb{Z}_p$ (or vice-versa). The pebbling rule for an input node requires that the input x to $\mathcal{O}_\mathsf{X}(\cdot)$ in both $\widehat{\mathsf{H}}_{\ell,0}(u_0, u_1)$ and $\widehat{\mathsf{H}}_{\ell,1}(u_0, u_1)$ satisfies $x_{\rho(j)} = 0$. Indistinguishability then follows from the CPA security of $(\mathsf{CPA.Setup}, \mathsf{CPA.Enc}, \mathsf{CPA.Dec})$ under key $w_{\rho(j)}$; this is because $x_{\rho(j)} = 0$ and therefore $w_{\rho(j)}$ will not need to be supplied in the answer to the query to $\mathcal{O}_\mathsf{X}(x)$. In fact, the two hybrids are computationally indistinguishable even if the adversary sees all $\{w_i : i \neq \rho(j)\}$ (as may be provided by $\mathcal{O}_\mathsf{X}(x)$).
- pebble/unpebble AND gate with out-going wire $\boxed{c}$ and input wires a, b corresponding to nodes g_a, g_b. Here, the only difference from $\mathsf{share}^{u_0}(f, \mu^{(0)})$ to $\mathsf{share}^{u_1}(f, \mu^{(0)})$ is that we change $\boxed{\mu_c}$ from an actual share $\hat{\mu}_a + \hat{\mu}_b + \hat{\mu}_c$ to a random element of $\mathbb{Z}_p$ (or vice-versa). The pebbling rules for an AND gate require that there is a pebble on either g_a or g_b, say g_a. Therefore, μ_a is independent and uniformly random in both distributions $\mathsf{share}^{u_0}(f, \mu^{(0)})$ and $\mathsf{share}^{u_1}(f, \mu^{(0)})$, and thus $\hat{\mu}_a$ is fresh and independently random in both distributions (this uses the fact that g_a has fan-out 1) and makes the distribution of $\mu_c = \hat{\mu}_a + \hat{\mu}_b + \hat{\mu}_c$ in hybrid $\ell - 1$ independently random. We may then deduce that $\mathsf{share}^{u_0}(f, \mu^{(0)})$ and $\mathsf{share}^{u_1}(f, \mu^{(0)})$ are identically distributed, and therefore so is the output $\mathcal{O}_\mathsf{F}(f)$. (This holds even if the adversary receives all of $\{w_i : i \in [n]\}$ from its query to $\mathcal{O}_\mathsf{X}(x)$).
- pebble/unpebble OR gate with out-going wire $\boxed{c}$ and input wires a, b corresponding to nodes g_a, g_b. Here, the only difference from $\mathsf{share}^{u_0}(f, \mu^{(0)})$ to $\mathsf{share}^{u_1}(f, \mu^{(0)})$ is that we change $\boxed{\mu_{c_a}, \mu_{c_b}}$ from actual shares $(\hat{\mu}_a + \hat{\mu}_c, \hat{\mu}_b + \hat{\mu}_c)$ to random elements of $\mathbb{Z}_p$ (or vice-versa). The pebbling rules for an OR gate require that there are pebbles on both g_a and g_b. Therefore, μ_a and μ_b are independent and uniformly random in both distributions $\mathsf{share}^{u_0}(f, \mu^{(0)})$ and $\mathsf{share}^{u_1}(f, \mu^{(0)})$, and thus $\hat{\mu}_a, \hat{\mu}_b$ are fresh and independently random in both distributions (using the fact that g_a, g_b have fan-out 1), and make the distributions of $\mu_{c_a} = \hat{\mu}_a + \hat{\mu}_c$, $\mu_{c_b} = \hat{\mu}_a + \hat{\mu}_b$ in hybrid $\ell - 1$ both independently random. We may then deduce that $\mathsf{share}^{u_0}(f, \mu^{(0)})$ and $\mathsf{share}^{u_1}(f, \mu^{(0)})$ are identically distributed, and therefore so is the output $\mathcal{O}_\mathsf{F}(f)$. (This holds even if the adversary receives all of $\{w_i : i \in [n]\}$ in its query to $\mathcal{O}_\mathsf{X}(x)$).

In all cases, the simulator can return an appropriately distributed answer to $\mathcal{O}_\mathsf{X}(x) = \{w_i\}_{x_i=1}$ since it has all w_i except in the first case, where it is missing only a w_i such that $x_i = 0$. Additionally, we note that in all cases, a simulator can return appropriately distributed answers to queries to the encryption oracle $\mathcal{O}_\mathsf{E}(i, m) = \mathsf{Enc}_{w_i}(m)$, since only in the first case (an input node being pebbled or unpebbled) is there a w_i not directly available to be used to simulate the oracle, and in that case, the simulator has oracle access to an $\mathsf{Enc}_{w_i}(\cdot)$ function in the CPA symmetric-key security game, and it can uniformly guess which of the n variables is associated with the input node being pebbled and answer $\mathcal{O}_E$ requests to that variable with the CPA $\mathsf{Enc}_{w_i}(\cdot)$ oracle (the factor of n due to guessing is introduced here since the simulator may not know which variable is associated with the input node at the time of the oracle request, e.g.: for requests to $\mathcal{O}_E$ made before $\mathcal{O}_X$, so the simulator must guess uniformly and take a security loss of n).

In all but the input node case, the two distributions $\langle A, \widehat{H}_{\ell,0}(u_0, u_1)\rangle$ and $\langle A, \widehat{H}_{\ell,1}(u_0, u_1)\rangle$ are identical, and in the input node case, we've bounded the difference by the distinguishing probability of the symmetric key encryption scheme, the advantage function $\mathsf{Adv}_{\mathcal{B}}^{\mathrm{CPA}}(\lambda)$, conditioned on a correct guess of which of the n input variables corresponds to the pebbled/unpebbled input node. Therefore, $\Pr[\langle A, \widehat{H}_{\ell,0}(u_0, u_1)\rangle = 1] - \Pr[\langle A, \widehat{H}_{\ell,1}(u_0, u_1)\rangle = 1] \leq n \cdot \mathsf{Adv}_{\mathcal{B}}^{\mathrm{CPA}}(\lambda)$ $\square$

5.4 CPA-Secure Symmetric Encryption

We will instantiate (CPA.Setup, CPA.Enc, CPA.Dec) in our Core 1-ABE of Definition 4 with a variant of the standard CPA-secure symmetric encryption scheme based on k-Lin from [11] that supports messages $[M]_2 \in G_2$ of an asymmetric prime-order bilinear group $\mathbb{G}$:

CPA.Setup(1^λ): Run $\mathbb{G} \leftarrow \mathcal{G}(1^\lambda)$. Sample $\mathbf{M}_0 \leftarrow \mathbb{Z}_p^{k \times k}$, $\mathbf{m}_1 \leftarrow \mathbb{Z}_p^k$,
  output $\mathsf{sk} = (\mathsf{sk}_0, \mathsf{sk}_1) := (\mathbf{M}_0, \mathbf{m}_1^\top)$
CPA.Enc($\mathsf{sk}, [M]_2$): Sample $\mathbf{r} \leftarrow \mathbb{Z}_p^k$, output $(\mathsf{ct}_0, \mathsf{ct}_1) := ([M + \mathbf{m}_1^\top \mathbf{r}]_2, [\mathbf{M}_0 \mathbf{r}]_2)$
CPA.Dec($(\mathsf{sk}_0, \mathsf{sk}_1), (\mathsf{ct}_0, \mathsf{ct}_1)$): Output $\mathsf{ct}_0 \cdot \mathsf{sk}_1 \cdot \mathsf{sk}_0^{-1} \cdot \mathsf{ct}_1$.

Correctness. Note that: $\mathsf{ct}_0 \cdot \mathsf{sk}_1 \cdot \mathsf{sk}_0^{-1} \cdot \mathsf{ct}_1 = [M + \mathbf{m}_1^\top \mathbf{r} - \mathbf{m}_1^\top \mathbf{r}]_2 = [M]_2$.

Lemma 6. $\mathsf{Adv}_{\mathcal{B}}^{\mathrm{CPA}}(\lambda) \leq \mathsf{Adv}_{\mathcal{B}'}^{\mathrm{K\text{-}LIN}}(\lambda)$.

Proof. Proof is contained in the full version of this paper [22] and omitted here for brevity.

Theorem 2. *The Core 1-ABE component of Definition 4 implemented with* (share, reconstruct) *from Sect. 5.1 and the CPA-secure symmetric encryption scheme* (CPA.Setup, CPA.Enc, CPA.Dec) *from Sect. 5.4 satisfies:*

$$\Pr[\langle A, G_0^{1\text{-}ABE}\rangle = 1] - \Pr[\langle A, G_1^{1\text{-}ABE}\rangle = 1] \leq 2^{6d} \cdot 8^d \cdot n \cdot \mathsf{Adv}_{\mathcal{B}^*}^{k\text{-}LIN}(\lambda)$$

Proof. Recall the hybrids $G_0^{1\text{-}ABE} \equiv H^{h_0(f,x)}, H^{h_1(f,x)}, ..., H^{h_L(f,x)} \equiv G_1^{1\text{-}ABE}$ defined in Sect. 5.3. Lemma 5 tells us that: for all $\ell \in [L]$ and $u_0, u_1 \in \{0, 1\}^{R'}$,

$$\Pr[\langle A, \widehat{H}_{\ell,0}(u_0, u_1)\rangle = 1] - \Pr[\langle A, \widehat{H}_{\ell,1}(u_0, u_1)\rangle = 1] \leq n \cdot \mathsf{Adv}_{\mathcal{B}}^{\mathrm{CPA}}(\lambda)$$

These hybrids satisfy the end-point equivalence requirement, so Lemma 1 then tells us that:

$$\Pr[\langle A, G_0^{1\text{-}ABE}\rangle = 1] - \Pr[\langle A, G_1^{1\text{-}ABE}\rangle = 1] \leq 2^{2R'} \cdot L \cdot n \cdot \mathsf{Adv}_{\mathcal{B}}^{\mathrm{CPA}}(\lambda)$$

Lemma 4 tells us that $R' \leq 3d$, and Lemma 2 tells us that $L \leq 8^d$, where d is the depth of the formula. Finally, Lemma 6 tells us that $\mathsf{Adv}_B^{\mathrm{CPA}}(\lambda) \leq \mathsf{Adv}_{B^*}^{\text{K-LIN}}(\lambda)$. So: $\Pr[\langle \mathsf{A}, \mathsf{G}_0^{\text{1-ABE}} \rangle = 1] - \Pr[\langle \mathsf{A}, \mathsf{G}_1^{\text{1-ABE}} \rangle = 1] \leq 2^{6d} \cdot 8^d \cdot n \cdot \mathsf{Adv}_{B^*}^{k\text{-LIN}}(\lambda)$ □

6   Our KP-ABE Scheme

In this section, we present our compact KP-ABE for NC^1 that is adaptively secure under the MDDH_k assumption in asymmetric prime-order bilinear groups. For attributes of length n, our ciphertext comprises $O(n)$ group elements, independent of the formula size, while simultaneously allowing attribute reuse in the formula. As mentioned in the overview in Sect. 1.1, we incorporated optimizations from [5,15] to shrink $\mathbf{W}_i$ and thus the secret key, and hence the need for the $\mathcal{O}_E$ oracle in the core 1-ABE security game.

6.1   The Scheme

Our KP-ABE scheme is as follows:

$\mathsf{Setup}(1^\lambda, 1^n)$: Run $\mathbb{G} = (p, G_1, G_2, G_T, e) \leftarrow \mathcal{G}(1^\lambda)$. Sample

$$\mathbf{A} \leftarrow \mathbb{Z}_p^{k \times (k+1)}, \mathbf{W}_i \leftarrow \mathbb{Z}_p^{(k+1) \times k} \ \forall i \in [n], \mathbf{v} \leftarrow \mathbb{Z}_p^{k+1}$$

and output:

$$\mathsf{msk} := (\mathbf{v}, \mathbf{W}_1, \ldots, \mathbf{W}_n)$$
$$\mathsf{mpk} := ([\mathbf{A}]_1, [\mathbf{A}\mathbf{W}_1]_1, \ldots, [\mathbf{A}\mathbf{W}_n]_1, \ e([\mathbf{A}]_1, [\mathbf{v}]_2))$$

$\mathsf{Enc}(\mathsf{mpk}, x, M)$: Sample $\mathbf{s} \leftarrow \mathbb{Z}_p^k$. Output:

$$\mathsf{ct_x} = (\mathsf{ct}_1, \{\mathsf{ct}_{2,i}\}_{x_i=1}, \mathsf{ct}_3)$$
$$:= \left([\mathbf{s}^\top \mathbf{A}]_1, \{[\mathbf{s}^\top \mathbf{A}\mathbf{W}_i]_1\}_{x_i=1}, \quad e([\mathbf{s}^\top \mathbf{A}]_1, [\mathbf{v}]_2) \cdot M \right)$$

$\mathsf{KeyGen}(\mathsf{mpk}, \mathsf{msk}, f)$: Sample $(\{\mathbf{v}_j\}, \rho) \leftarrow \mathsf{share}(f, \mathbf{v})$, $\mathbf{r}_j \leftarrow \mathbb{Z}_p^k$. Output:

$$\mathsf{sk}_f = (\{\mathsf{sk}_{1,j}, \mathsf{sk}_{2,j}\})$$
$$:= (\{[\mathbf{v}_j + \mathbf{W}_{\rho(j)}\mathbf{r}_j]_2, [\mathbf{r}_j]_2\})$$

where $\mathbf{W}_0 = \mathbf{0}$.

$\mathsf{Dec}(\mathsf{mpk}, \mathsf{sk}_f, \mathsf{ct_x})$: Compute ω_j such that $\mathbf{v} = \displaystyle\sum_{\rho(j)=0 \vee x_{\rho(j)}=1} \omega_j \mathbf{v}_j$ as described in

Sect. 5.1. Output:

$$\mathsf{ct}_3 \cdot \prod_{\rho(j)=0 \vee x_{\rho(j)}=1} \left(\frac{e(\mathsf{ct}_{2,\rho(j)}, \mathsf{sk}_{2,j})}{e(\mathsf{ct}_1, \mathsf{sk}_{1,j})} \right)^{\omega_j}$$

6.2   Correctness

Correctness relies on the fact that for all j, we have

$$\frac{e(\mathsf{ct}_1, \mathsf{sk}_{1,j})}{e(\mathsf{ct}_{2,\rho(j)}, \mathsf{sk}_{2,j})} = [\mathbf{s}^\top \mathbf{A} \mathbf{v}_j]_T$$

which follows from the fact that

$$\mathbf{s}^\top \mathbf{A} \mathbf{v}_j = \underbrace{\mathbf{s}^\top \mathbf{A}}_{\mathsf{ct}_1} \cdot \underbrace{(\mathbf{v}_j + \mathbf{W}_{\rho(j)} \mathbf{r}_j)}_{\mathsf{sk}_{1,j}} - \underbrace{\mathbf{s}^\top \mathbf{A} \mathbf{W}_{\rho(j)}}_{\mathsf{ct}_{2,\rho(j)}} \cdot \underbrace{\mathbf{r}_j}_{\mathsf{sk}_{2,j}}$$

Therefore, for all f, x such that $f(x) = 1$, we have:

$$\mathsf{ct}_3 \cdot \prod_{\rho(j)=0 \vee x_{\rho(j)}=1} \left(\frac{e(\mathsf{ct}_{2,\rho(j)}, \mathsf{sk}_{2,j})}{e(\mathsf{ct}_1, \mathsf{sk}_{1,j})} \right)^{\omega_j} = M \cdot [\mathbf{s}^\top \mathbf{A} \mathbf{v}]_T \cdot \prod_{\rho(j)=0 \vee x_{\rho(j)}=1} [\mathbf{s}^\top \mathbf{A} \mathbf{v}_j]_T^{-\omega_j}$$

$$= M \cdot [\mathbf{s}^\top \mathbf{A} \mathbf{v}]_T \cdot [-\mathbf{s}^\top \mathbf{A} \sum_{\rho(j)=0 \vee x_{\rho(j)}=1} \omega_j \mathbf{v}_j]_T$$

$$= M \cdot [\mathbf{s}^\top \mathbf{A} \mathbf{v}]_T \cdot [-\mathbf{s}^\top \mathbf{A} \mathbf{v}]_T$$

$$= M$$

6.3   Adaptive Security

Description of Hybrids. To describe the hybrid distributions, it would be helpful to first give names to the various forms of ciphertext and keys that will be used. A ciphertext can be in one of the following forms:

– Normal: generated as in the scheme.
– SF: same as a Normal ciphertext, except $\mathbf{s}^\top \mathbf{A}$ replaced with $\mathbf{c}^\top \leftarrow \mathbb{Z}_p^{k+1}$. That
  is, $\mathsf{ct_x} := \left([\boxed{\mathbf{c}^\top}]_1, \{[\boxed{\mathbf{c}^\top} \mathbf{W}_i]_1\}_{x_i=1}, \quad e([\boxed{\mathbf{c}^\top}]_1, [\mathbf{v}]_2) \cdot M \right)$

A secret key can be in one of the following forms:

– Normal: generated as in the scheme.
– SF: same as a Normal key, except $\mathbf{v}$ replaced with $\mathbf{v} + \delta \mathbf{a}^\perp$, where a fresh
  $\delta \leftarrow \mathbb{Z}_p$ is chosen per SF key and $\mathbf{a}^\perp$ is any fixed $\mathbf{a}^\perp \in \mathbb{Z}_p^{k+1} \setminus \{\mathbf{0}\}$ such that
  $\mathbf{A} \mathbf{a}^\perp = \mathbf{0}$. That is, $\mathsf{sk}_f := (\{[\mathbf{v}_j + \mathbf{W}_{\rho(j)} \mathbf{r}_j]_2, [\mathbf{r}_j]_2\})$
  where $(\{\mathbf{v}_j\}, \rho) \leftarrow \mathsf{share}(f, \boxed{\mathbf{v} + \delta \mathbf{a}^\perp})$, $\mathbf{r}_j \leftarrow \mathbb{Z}_p^k$.

SF stands for semi-functional following the terminology in previous works [25,33].

Hybrid Sequence. Suppose the adversary A makes at most Q secret key queries. The hybrid sequence is as follows:

- H_0: real game
- H_1: same as H_0, except we use a SF ciphertext.
- $H_{2,\ell}, \ell = 0, \ldots, Q$: same as H_1, except the first ℓ keys are SF and the remaining $Q - \ell$ keys are Normal.
- H_3: replace M with random $\widetilde{M}$.

Proof Overview

- We have $H_0 \approx_c H_1 \equiv H_{2,0}$ via k-Lin, which tells us $([\mathbf{A}]_1, [\mathbf{s}^\top\mathbf{A}]_1) \approx_c ([\mathbf{A}]_1, [\mathbf{c}^\top]_1)$. Here, the security reduction will pick $\mathbf{W}_1, \ldots, \mathbf{W}_n$ and $\mathbf{v}$ so that it can simulate the mpk, the ciphertext and the secret keys.
- We have $H_{2,\ell-1} \approx_c H_{2,\ell}$, for all $\ell \in [Q]$. The difference between the two is that we switch the ℓ'th sk_f from Normal to SF using the adaptive security of our core 1-ABE component in $\mathsf{G}^{\text{1-ABE}}$ from Sect. 5. The idea is to sample

$$\mathbf{v} = \tilde{\mathbf{v}} + \mu\mathbf{a}^\perp, \mathbf{W}_i = \widetilde{\mathbf{W}}_i + \mathbf{a}^\perp\mathbf{w}_i^\top$$

so that mpk can be computed using $\tilde{\mathbf{v}}, \widetilde{\mathbf{W}}_i$ and perfectly hide $\mu, \mathbf{w}_1, \ldots, \mathbf{w}_n$. Roughly speaking: the reduction
  - uses $\mathcal{O}_X(x)$ in $\mathsf{G}^{\text{1-ABE}}$ to simulate the challenge ciphertext
  - uses $\mathcal{O}_F(f)$ in $\mathsf{G}^{\text{1-ABE}}$ to simulate ℓ'th secret key
  - uses $\mu^{(0)}$ from $\mathsf{G}^{\text{1-ABE}}$ together with $\mathcal{O}_E(i, \cdot) = \mathsf{Enc}(w_i, \cdot)$ to simulate the remaining $Q - \ell$ secret keys
- We have $H_{2,Q} \equiv H_3$. In $H_{2,Q}$, the secret keys only leak $\mathbf{v} + \delta_1\mathbf{a}^\perp, \ldots, \mathbf{v} + \delta_Q\mathbf{a}^\perp$. This means that $\mathbf{c}^\top\mathbf{v}$ is statistically random (as long as $\mathbf{c}^\top\mathbf{a}^\perp \neq 0$).

Theorem 3 (adaptive KP-ABE). *The KP-ABE construction in Sect. 6.1 is adaptively secure under the $MDDH_k$ assumption.*

Proof. The detailed proof is contained in the full version of this paper [22] and omitted here for brevity.

Acknowledgments. We thank Allison Bishop, Sanjam Garg, Rocco Servedio, and Daniel Wichs for helpful discussions.

References

1. Agrawal, S., Chase, M.: Simplifying design and analysis of complex predicate encryption schemes. In: Coron, J.-S., Nielsen, J.B. (eds.) EUROCRYPT 2017, Part I. LNCS, vol. 10210, pp. 627–656. Springer, Cham (2017). https://doi.org/10.1007/978-3-319-56620-7_22
2. Attrapadung, N.: Dual system encryption via doubly selective security: framework, fully secure functional encryption for regular languages, and more. In: Nguyen, P.Q., Oswald, E. (eds.) EUROCRYPT 2014. LNCS, vol. 8441, pp. 557–577. Springer, Heidelberg (2014). https://doi.org/10.1007/978-3-642-55220-5_31

3. Attrapadung, N.: Dual system encryption framework in prime-order groups via computational pair encodings. In: Cheon, J.H., Takagi, T. (eds.) ASIACRYPT 2016, Part II. LNCS, vol. 10032, pp. 591–623. Springer, Heidelberg (2016). https://doi.org/10.1007/978-3-662-53890-6_20

4. Bethencourt, J., Sahai, A., Waters, B.: Ciphertext-policy attribute-based encryption. In: IEEE Symposium on Security and Privacy, pp. 321–334. IEEE Computer Society Press, May 2007

5. Blazy, O., Kiltz, E., Pan, J.: (Hierarchical) identity-based encryption from affine message authentication. In: Garay, J.A., Gennaro, R. (eds.) CRYPTO 2014, Part I. LNCS, vol. 8616, pp. 408–425. Springer, Heidelberg (2014). https://doi.org/10.1007/978-3-662-44371-2_23

6. Chen, J., Gay, R., Wee, H.: Improved dual system ABE in prime-order groups via predicate encodings. In: Oswald, E., Fischlin, M. (eds.) EUROCRYPT 2015, Part II. LNCS, vol. 9057, pp. 595–624. Springer, Heidelberg (2015). https://doi.org/10.1007/978-3-662-46803-6_20

7. Chen, J., Gong, J., Kowalczyk, L., Wee, H.: Unbounded ABE via bilinear entropy expansion, revisited. In: Nielsen, J.B., Rijmen, V. (eds.) EUROCRYPT 2018, Part I. LNCS, vol. 10820, pp. 503–534. Springer, Cham (2018). https://doi.org/10.1007/978-3-319-78381-9_19

8. Chen, J., Wee, H.: Fully, (almost) tightly secure IBE and dual system groups. In: Canetti, R., Garay, J.A. (eds.) CRYPTO 2013, Part II. LNCS, vol. 8043, pp. 435–460. Springer, Heidelberg (2013). https://doi.org/10.1007/978-3-642-40084-1_25

9. Chen, J., Wee, H.: Semi-adaptive attribute-based encryption and improved delegation for Boolean formula. In: Abdalla, M., De Prisco, R. (eds.) SCN 2014. LNCS, vol. 8642, pp. 277–297. Springer, Cham (2014). https://doi.org/10.1007/978-3-319-10879-7_16

10. Cheon, J.H.: Security analysis of the strong Diffie-Hellman problem. In: Vaudenay, S. (ed.) EUROCRYPT 2006. LNCS, vol. 4004, pp. 1–11. Springer, Heidelberg (2006). https://doi.org/10.1007/11761679_1

11. Escala, A., Herold, G., Kiltz, E., Ràfols, C., Villar, J.: An algebraic framework for Diffie-Hellman assumptions. In: Canetti, R., Garay, J.A. (eds.) CRYPTO 2013, Part II. LNCS, vol. 8043, pp. 129–147. Springer, Heidelberg (2013). https://doi.org/10.1007/978-3-642-40084-1_8

12. Fuchsbauer, G., Jafargholi, Z., Pietrzak, K.: A Quasipolynomial Reduction for Generalized Selective Decryption on Trees. In: Gennaro, R., Robshaw, M. (eds.) CRYPTO 2015. LNCS, vol. 9215, pp. 601–620. Springer, Heidelberg (2015). https://doi.org/10.1007/978-3-662-47989-6_29

13. Fuchsbauer, G., Konstantinov, M., Pietrzak, K., Rao, V.: Adaptive security of constrained PRFs. In: Sarkar, P., Iwata, T. (eds.) ASIACRYPT 2014, Part II. LNCS, vol. 8874, pp. 82–101. Springer, Heidelberg (2014). https://doi.org/10.1007/978-3-662-45608-8_5

14. Garg, S., Gentry, C., Halevi, S., Sahai, A., Waters, B.: Attribute-based encryption for circuits from multilinear maps. In: Canetti, R., Garay, J.A. (eds.) CRYPTO 2013, Part II. LNCS, vol. 8043, pp. 479–499. Springer, Heidelberg (2013). https://doi.org/10.1007/978-3-642-40084-1_27

15. Gong, J., Dong, X., Chen, J., Cao, Z.: Efficient IBE with tight reduction to standard assumption in the multi-challenge setting. In: Cheon, J.H., Takagi, T. (eds.) ASIACRYPT 2016, Part II. LNCS, vol. 10032, pp. 624–654. Springer, Heidelberg (2016). https://doi.org/10.1007/978-3-662-53890-6_21

16. Gorbunov, S., Vaikuntanathan, V., Wee, H.: Attribute-based encryption for circuits. In: Boneh, D., Roughgarden, T., Feigenbaum, J. (eds.) 45th ACM STOC, pp. 545–554. ACM Press, New York (2013)

17. Goyal, V., Pandey, O., Sahai, A., Waters, B.: Attribute-based encryption for fine-grained access control of encrypted data. In: Juels, A., Wright, R.N., Vimercati, S. (eds.) ACM CCS 2006, pp. 89–98. ACM Press, New York (2006). Available as Cryptology ePrint Archive Report 2006/309

18. Hemenway, B., Jafargholi, Z., Ostrovsky, R., Scafuro, A., Wichs, D.: Adaptively secure garbled circuits from one-way functions. In: Robshaw, M., Katz, J. (eds.) CRYPTO 2016, Part III. LNCS, vol. 9816, pp. 149–178. Springer, Heidelberg (2016). https://doi.org/10.1007/978-3-662-53015-3_6

19. Ishai, Y., Kushilevitz, E.: Perfect constant-round secure computation via perfect randomizing polynomials. In: Widmayer, P., Eidenbenz, S., Triguero, F., Morales, R., Conejo, R., Hennessy, M. (eds.) ICALP 2002. LNCS, vol. 2380, pp. 244–256. Springer, Heidelberg (2002). https://doi.org/10.1007/3-540-45465-9_22

20. Jafargholi, Z., Kamath, C., Klein, K., Komargodski, I., Pietrzak, K., Wichs, D.: Be Adaptive, Avoid Overcommitting. In: Katz, J., Shacham, H. (eds.) CRYPTO 2017, Part I. LNCS, vol. 10401, pp. 133–163. Springer, Cham (2017). https://doi.org/10.1007/978-3-319-63688-7_5

21. Jafargholi, Z., Wichs, D.: Adaptive security of Yao's garbled circuits. In: Hirt, M., Smith, A. (eds.) TCC 2016, Part I. LNCS, vol. 9985, pp. 433–458. Springer, Heidelberg (2016). https://doi.org/10.1007/978-3-662-53641-4_17

22. Kowalczyk, L., Wee, H.: Compact adaptively secure ABE for NC1 from k-Lin. IACR Cryptology ePrint Archive, 2019:224 (2019)

23. Lewko, A., Okamoto, T., Sahai, A., Takashima, K., Waters, B.: Fully secure functional encryption: attribute-based encryption and (hierarchical) inner product encryption. In: Gilbert, H. (ed.) EUROCRYPT 2010. LNCS, vol. 6110, pp. 62–91. Springer, Heidelberg (2010). https://doi.org/10.1007/978-3-642-13190-5_4

24. Lewko, A.B., Sahai, A., Waters, B.: Revocation systems with very small private keys. In: IEEE Symposium on Security and Privacy, pp. 273–285. IEEE Computer Society Press, May 2010

25. Lewko, A., Waters, B.: New techniques for dual system encryption and fully secure HIBE with short ciphertexts. In: Micciancio, D. (ed.) TCC 2010. LNCS, vol. 5978, pp. 455–479. Springer, Heidelberg (2010). https://doi.org/10.1007/978-3-642-11799-2_27

26. Lewko, A., Waters, B.: New proof methods for attribute-based encryption: achieving full security through selective techniques. In: Safavi-Naini, R., Canetti, R. (eds.) CRYPTO 2012. LNCS, vol. 7417, pp. 180–198. Springer, Heidelberg (2012). https://doi.org/10.1007/978-3-642-32009-5_12

27. Okamoto, T., Takashima, K.: Fully secure functional encryption with general relations from the decisional linear assumption. In: Rabin, T. (ed.) CRYPTO 2010. LNCS, vol. 6223, pp. 191–208. Springer, Heidelberg (2010). https://doi.org/10.1007/978-3-642-14623-7_11

28. Okamoto, T., Takashima, K.: Fully secure unbounded inner-product and attribute-based encryption. In: Wang, X., Sako, K. (eds.) ASIACRYPT 2012. LNCS, vol. 7658, pp. 349–366. Springer, Heidelberg (2012). https://doi.org/10.1007/978-3-642-34961-4_22

29. Ostrovsky, R., Sahai, A., Waters, B.: Attribute-based encryption with non-monotonic access structures. In: Ning, P., di Vimercati, S.D.C., Syverson, P.F. (eds.) ACM CCS 2007, pp. 195–203. ACM Press, New York (2007)

30. Parno, B., Raykova, M., Vaikuntanathan, V.: How to delegate and verify in public: verifiable computation from attribute-based encryption. In: Cramer, R. (ed.) TCC 2012. LNCS, vol. 7194, pp. 422–439. Springer, Heidelberg (2012). https://doi.org/10.1007/978-3-642-28914-9_24

31. Sahai, A., Waters, B.: Fuzzy identity-based encryption. In: Cramer, R. (ed.) EURO-CRYPT 2005. LNCS, vol. 3494, pp. 457–473. Springer, Heidelberg (2005). https://doi.org/10.1007/11426639_27

32. Vinod, V., Narayanan, A., Srinathan, K., Rangan, C.P., Kim, K.: On the power of computational secret sharing. In: Johansson, T., Maitra, S. (eds.) INDOCRYPT 2003. LNCS, vol. 2904, pp. 162–176. Springer, Heidelberg (2003). https://doi.org/10.1007/978-3-540-24582-7_12

33. Waters, B.: Dual system encryption: realizing fully secure IBE and HIBE under simple assumptions. In: Halevi, S. (ed.) CRYPTO 2009. LNCS, vol. 5677, pp. 619–636. Springer, Heidelberg (2009). https://doi.org/10.1007/978-3-642-03356-8_36

34. Wee, H.: Dual system encryption via predicate encodings. In: Lindell, Y. (ed.) TCC 2014. LNCS, vol. 8349, pp. 616–637. Springer, Heidelberg (2014). https://doi.org/10.1007/978-3-642-54242-8_26

Unbounded Dynamic Predicate Compositions in Attribute-Based Encryption

Nuttapong Attrapadung[✉]

National Institute of Advanced Industrial Science and Technology (AIST),
Tokyo, Japan
n.attrapadung@aist.go.jp

Abstract. We present several transformations that combine a set of attribute-based encryption (ABE) schemes for simpler predicates into a new ABE scheme for more expressive composed predicates. Previous proposals for predicate compositions of this kind, the most recent one being that of Ambrona *et al.* at Crypto'17, can be considered *static* (or partially dynamic), meaning that the policy (or its structure) that specifies a composition must be fixed at the setup. Contrastingly, our transformations are *dynamic* and *unbounded*: they allow a user to specify an arbitrary and unbounded-size composition policy right into his/her own key or ciphertext. We propose transformations for three classes of composition policies, namely, the classes of any monotone span programs, any branching programs, and any deterministic finite automata. These generalized policies are defined over arbitrary predicates, hence admitting *modular* compositions. One application from modularity is a new kind of ABE for which policies can be "nested" over ciphertext and key policies. As another application, we achieve the first fully secure completely unbounded key-policy ABE for non-monotone span programs, in a modular and clean manner, under the q-ratio assumption. Our transformations work inside a generic framework for ABE called symbolic pair encoding, proposed by Agrawal and Chase at Eurocrypt'17. At the core of our transformations, we observe and exploit an unbounded nature of the symbolic property so as to achieve unbounded-size policy compositions.

1 Introduction

Attribute-based encryption (ABE), introduced by Sahai and Waters [32], is a paradigm that generalizes traditional public key encryption. Instead of encrypting to a target recipient, a sender can specify in a more general way about who should be able to view the message. In ABE for predicate $P : \mathcal{X} \times \mathcal{Y} \to \{0, 1\}$, a ciphertext encrypting message M is associated with a ciphertext attribute, say, $y \in \mathcal{Y}$, while a secret key, issued by an authority, is associated with a key attribute, say, $x \in \mathcal{X}$, and the decryption will succeed if and only if $P(x, y) = 1$. From an application point of view, we can consider one kind of attributes as *policies*, and the other kind as inputs to policies. In this sense, we have two basic

© International Association for Cryptologic Research 2019
Y. Ishai and V. Rijmen (Eds.): EUROCRYPT 2019, LNCS 11476, pp. 34–67, 2019.
https://doi.org/10.1007/978-3-030-17653-2_2

forms of ABE called key-policy (KP) and ciphertext-policy (CP), depending on which side has a policy associated to.

Predicate Compositions. A central theme to ABE has been to expand the expressiveness by constructing new ABE for more powerful predicates (*e.g.*, [12,20,21,28–30]). In this work, we continue this theme by focusing on how to construct ABE for *compositions* of predicates. We are interested in devising *transformations* that combine ABE schemes for based predicates to a new ABE scheme for their composed predicate. To motivate that this can be powerful in the first place, we introduce an example primitive called Nested-policy ABE.

Example: Nested-policy ABE. As the name suggests, it allows a key policy and a ciphertext policy to be nested to each other. This might be best described by an example. Suppose there are three categories for attributes: PERSON, PLACE, CONTENT. Attached to a key, we could have attribute sets/policies categorized to three categories, PERSON:{TRAINEE, DOCTOR}, PLACE:{PARIS, ZIP:75001}, CONTENT:'(KIDNEY AND DISEASE) OR EMERGENCY', with a "composition policy" such as 'PERSON OR (PLACE AND CONTENT)', which plays the role of concluding the whole policy. A ciphertext could be associated to PERSON:'SENIOR AND DOCTOR', PLACE:'PARIS OR LONDON', CONTENT:{KIDNEY, DISEASE, CANCER}. Now we argue that the above key can be used to decrypt the ciphertext since the attribute set for PLACE satisfies the corresponding policy in the ciphertext, while the policy for CONTENT is satisfied by the corresponding attribute sets in the ciphertext, and the concluding policy (attached to the key) states that if both PLACE and CONTENT categories are satisfied, then it can decrypt.

We can consider this as a *composition* of two CP-ABE sub-schemes for the first two categories and KP-ABE for the last category, while on the top of that, a KP-ABE scheme over the three categories is then applied. To the best of our knowledge, no ABE with nested-policy functionality has been proposed so far, and it is not clear in the first place how to construct even for specific policies.

Our Design Goal. We aim at constructing *unbounded*, *dynamic*, and *generic* transformations for predicate compositions. *Dynamicity* refers to the property that one can choose *any* composition policy (defined in some sufficiently large classes) when composing predicates. In the above example, this translates to the property that the concluding policy is not fixed-once-and-for-all, where, for instance, one might want to define it instead as '(PERSON AND CONTENT) OR PLACE', when a key is issued. Moreover, we aim at *modular* compositions where we can recursively define policies over policies, over and over again. Furthermore, for highest flexibility, we focus on *unbounded* compositions, meaning that the sizes of composition policies and attribute sets are not a-priori bounded at the setup. *Generality* refers to that we can transform *any* ABE for *any* based predicates. This level of generality might be too ambitious, since this would imply an attempt to construct ABE from ID-based Encryption (IBE), of which no transformation is known. We thus confine our goal to within some well-defined ABE framework and/or a class of predicates. Towards this, we first confine our attention to ABE based on *bilinear groups*, which are now considerably efficient and have always been the main tool for constructing ABE since the original papers [21,32].

Previous Work on Predicate Compositions. We categorize as follows.

- **Static & Specific.** Dual-policy ABE (DP-ABE), introduced in [4], is the AND composition of KP-ABE and CP-ABE (both fixed for the Boolean formulae predicate). The fixed AND means that it is static. The underlying ABE schemes are also specific schemes, namely, those of [21,33].

- **Static & Small-class & Generic.** Attrapadung and Yamada [10] proposed a more general conversion that can combine ABE for any predicates that can be interpreted in the so-called *pair encoding* framework [1,2,5,6], but again, fixed for only the AND connector. A generic DUAL conversion, which swaps key and ciphertext attribute, was also proposed in [5,10]. All in all, only a small class of compositions were possible at this point.

- **Static/Partially-dynamic & Large-class & Generic.** Most recently, at Crypto'17, Ambrona, Barthe, and Schmidt [3] proposed general transformations for DUAL, AND, OR, and NOT connectors, hence complete any Boolean formulation, and thus enable a large class of combinations. Their scheme is generic and can combine ABE for any predicates in the so-called *predicate encoding* framework [16,36]. However, their compositions are static ones, where such a composition policy has to be fixed at the setup. A more flexible combination (§2 of [3]), which we call *partially dynamic*, is also presented, where the *structure* of the boolean combination must be fixed.

Our Contributions: Dynamic & Large-Class & Generic. We propose *unbounded, dynamic,* and *generic* transformations for predicate compositions that contain a large class of policies. They are generic in the sense that applicable ABE schemes can be any schemes within the generic framework of pair encoding, see below. These transformation convert ABE schemes for a set of "atomic" predicates $\mathcal{P} = \{P_1, \ldots, P_k\}$ to an ABE scheme for what we call *policy-augmented predicate over* $\mathcal{P}$. Both key-policy and ciphertext-policy augmentations are possible. In the key-policy case, the dynamicity allows a key issuer to specify a policy over atomic predicates, like the concluding policy over three sub-schemes in the above nested example. In the ciphertext-policy case, it allows an encryptor to specify such a policy. Below, we focus on the key-policy variant for illustrating purpose.

We propose the following four composition transformations.

1. **Span Programs over Predicates.** In this class, we let a composition policy be dynamically defined as any *monotone span program* (MSP) [22] where each of their Boolean inputs comes from each evaluation of atomic predicate. This is illustrated in Fig. 1. A key attribute is a tuple $M = (\mathbf{A}, (i_1, x_1), \ldots, (i_m, x_m))$ depicted on the left, where $\mathbf{A}$ is a span program (or, think of it as a boolean formula). A ciphertext attribute is a set

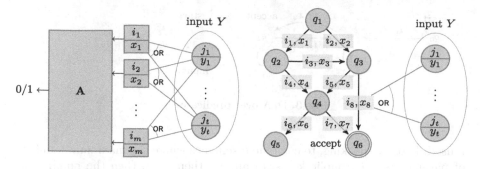

Fig. 1. Span program over predicates

Fig. 2. Branching program over predicates

$Y = \{(j_1, y_1), \ldots, (j_t, y_t)\}$. The indexes i_d and j_h specify the index of predicates in $\mathcal{P}$, that is, $i_d, j_h \in [1, k]$. To evaluate M on Y, we proceed as follows. First, we evaluate a "link" between node (i_d, x_d) and node (j_h, y_h) to on if $i_d = j_h =: i$ and $P_i(x_d, y_h) = 1$. Then, if one of the edges adjacent to the d-th node is on, then we input 1 as the d-th input to $\mathbf{A}$, and evaluate $\mathbf{A}$. Our transformation is unbounded, meaning that m and t can be arbitrary. Note that since span programs imply boolean formulae, we can think of it as boolean formula over atomic predicates.

2. **Branching Programs over Predicates.** In this class, we let a composition policy be dynamically defined as any *branching program* (BP) where each edge is evaluated in a similar manner as in each link in the case of span program composition above. This is depicted in Fig. 2. A branching program is described by a direct acyclic graph (DAG) with labels. It accepts Y if the on edges include a directed path from the start node to an accept node. A direct application for this is a predicate that comprises if-then clauses. We achieve this by a general implication from the first transformation, similarly to the implication from ABE for span programs to ABE for BP in [6].

3. **DFA over Predicates.** In this class, a composition policy can be defined as any *deterministic finite automata (DFA)* where each transition in DFA is defined based on atomic predicates. Such a DFA has an input as a vector $\mathbf{y} = ((j_1, y_1), \ldots, (j_t, y_t))$ which it reads in sequence. It allows any direct graph, even contains directed cycles and loops (as opposed to DAG for branching programs), and can read arbitrarily long vectors $\mathbf{y}$. This transformation fully generalizes ABE for regular languages [5,35], which can deal only with the equality predicate at each transition, to any predicates.

4. **Bundling ABE with Parameter Reuse.** We propose a generic way to bundle ABE schemes (without a policy over them, and where each scheme works separately) so that almost all of their parameters can be set to the same set of values among those ABE schemes. This is quite surprising in the first place since usually parameters for different schemes would play different roles (in both syntax and security proof). Nevertheless, we show that they can be

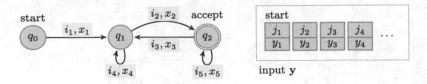

Fig. 3. DFA over predicates

reused. Loosely speaking, to combine k schemes where the maximum number of parameters (*i.e.*, public key size) among them is n, then the number of parameters for the combined scheme is $n + 2k$. Trivially combining them would yield $O(nk)$ size. We call this as the *direct sum with parameter reuse*.

We denote the above first three key-policy-augmented predicates over $\mathcal{P}$ as KP[$\mathcal{P}$], KB[$\mathcal{P}$], KA[$\mathcal{P}$], respectively. For ciphertext-policy case, we use C instead of K. Also, we call the generalized machines in the above classes as *predicative* machines.

Scope of Our Transformations. Our conversions apply to ABE that can be interpreted in the *pair encoding* framework, which is a generic framework for achieving fully secure ABE from a primitive called Pair Encoding Scheme (PES), proposed by Attrapadung [5]. PESs for many predicates have been proposed [2,5,6,10], notably, including regular language functionality [5,35]. Agrawal and Chase [2], at Eurocrypt'17, recently extended such a framework by introducing a notion called *symbolic security* for PES, which greatly simplifies both designing and security analysis of PES and ABE. A symbolically secure PES for predicate P can be used to construct fully secure ABE for the same predicate under the k-linear and the q-ratio assumption [2] in (prime-order) bilinear groups. Our conversions indeed work by converting PESs for a set $\mathcal{P}$ of predicates to a PES for KP[$\mathcal{P}$], KB[$\mathcal{P}$], and KA[$\mathcal{P}$], that preserves symbolic security.

Applications. Among many applications, we obtain:

- ABE with multi-layer/multi-base functionalities and nested-policy. The generality of our transformations make it possible to augment ABE schemes in a *modular* and *recursive* manner. This enables multi-layer functionalities in one scheme, *e.g.*, ABE for predicate KP[KB[KA[$\mathcal{P}$]]], which can deal with first checking regular expression (over predicates) via DFA, then inputting to an if-clause in branching program, and finally checking the whole policy. By skewing key and ciphertext policy, we can obtain a nested-policy ABE, *e.g.*, predicate KP[CP[$\mathcal{P}$]]. Moreover, the fact that we combine a *set* of predicates into a composed one enables multiple based functionalities, *e.g.*, revocation [3,37], range/subset membership [8], regular string matching [35], etc. This level of "plug-and-play" was not possible before this work.

- The first fully secure *completely-unbounded* KP-ABE for *non-monotone* span programs (NSP) over large universe.[1] Previous ABE for NSP is either only selectively secure [9,28,38] or has some bounded attribute reuse [29,30]. See Table 1 in Sect. 9.2 for a summary. Our approach is simple as we can obtain this modularly. As a downside, we have to rely on the q-type assumption inherited from the Agrawal-Chase framework [2]. Nevertheless, all the current *completely unbounded* KP-ABE for even *monotone* span programs still need q-type assumptions [2,5,31], even selectively secure one [31].
- Mixed-policy ABE. In nested-policy ABE, the nesting structure is fixed. Mixed-policy ABE generalizes it so as to be able to deal with arbitrary nesting structure in one scheme. The scheme crucially uses the direct sum with parameter reuse, so that its parameter size will not blow up exponentially.

Comparing to ABS17 [3]. Here, we compare our transformations to those of Ambrona *et al.* [3]. The most distinguished features of our transformations are finite automata based, and branching program based compositions. Moreover, all of our transformations are unbounded. For monotone Boolean formulae over predicates, our framework allows dynamic compositions, as opposed to static or partially-dynamic (thus, bounded-size) ones in ABS. As for applicability to based predicates, ours cover a larger class due to the different based frameworks (ours use symbolic pair encoding of [2], while ABS use predicate encoding of [16]). Notable differences are that pair encodings cover unbounded ABE for MSP, ABE for MSP with constant-size keys or ciphertexts, ABE for regular languages, while these are not known for predicate encodings. One drawback of using symbolic pair encoding is that we have to rely on q-type assumptions. A result in ABS also implies (static) *non-monotone* Boolean formulae composition (via their negation conversion). Although we do not consider negation conversion, we can use known pair encoding for negation of some common predicates such as IBE and negated of IBE (as we will do in Sect. 9). In this sense, non-monotone formulae composition can be done in our framework albeit in a semi-generic (but dynamic) manner.

We provide more related works and some future directions in the full version.

2   Intuition and Informal Overview

This section provides some intuition on our approaches in an informal manner.

Pair Encoding. We first informally describe PES [5] as refined in [2]. It consists of two encoding algorithms as the main components. The ciphertext encoding EncCt encodes $y \in \mathcal{Y}$ to a vector $\mathbf{c} = \mathbf{c}(\mathbf{s}, \hat{\mathbf{s}}, \mathbf{b}) = (c_1, \ldots, c_{w_3})$ of polynomials in variables $\mathbf{s} = (s_0, \ldots, s_{w_1})$, $\hat{\mathbf{s}} = (\hat{s}_1, \ldots, \hat{s}_{w_2})$, and $\mathbf{b} = (b_1, \ldots, b_n)$. The key encoding EncKey encodes $x \in \mathcal{X}$ to a vector $\mathbf{k} = \mathbf{k}(\mathbf{r}, \hat{\mathbf{r}}, \mathbf{b}) = (k_1, \ldots, k_{m_3})$ of

---

[1] For large universe ABE, there is no known conversion from ABE for monotone span programs. Intuitively, one would have to include negative attributes for all of the complement of a considering attribute set, which is of exponential size.

polynomials in variables $\mathbf{r} = (r_1, \ldots, r_{m_1})$, $\hat{\mathbf{r}} = (\alpha, \hat{r}_1, \ldots, \hat{r}_{m_2})$, and $\mathbf{b}$. The correctness requires that if $P(x, y) = 1$, then we can "pair" $\mathbf{c}$ and $\mathbf{k}$ to to obtain αs_0, which refers to the property that there exists a linear combination of terms $c_i r_u$ and $k_j s_t$ that is αs_0. Loosely speaking, to construct ABE from PES, we use a bilinear group $\mathbb{G} = (\mathbb{G}_1, \mathbb{G}_2)$ that conforms to dual system groups [1,2,17]. Let g_1, g_2 be their generators. The public key is $(g_2^{\mathbf{b}}, e(g_1, g_2)^{\alpha})$, a ciphertext for y encrypting a message M consists of $g_2^{\mathbf{c}}, g_2^{\mathbf{s}}$, and $e(g_1, g_2)^{\alpha s_0} \cdot M$, and a key for x consists of $g_1^{\mathbf{k}}, g_1^{\mathbf{r}}$. (In particular, the hatted variables are only internal to each encoding.) Decryption is done by pairing $\mathbf{c}$ and $\mathbf{k}$ to obtain αs_0 in the exponent.

Symbolic Security. In a nutshell, the symbolic security [2] of PES involves "substitution" of scalar variables in PES to vectors/matrices so that all the substituted polynomials in the two encodings $\mathbf{c}$ and $\mathbf{k}$ will evaluate to zero for any pair x, y such that $P(x, y) = 0$. The intuition for zero evaluation is that, behind the scene, there are some cancellations going on over values which cannot be computed from the underlying assumptions. To rule out the trivial all-zero substitutions, there is one more rule that the inner product of the substituted vectors for special variables that define correctness, namely, α and s_0, cannot be zero. In some sense, this can be considered as a generalization of the already well-known Boneh-Boyen cancellation technique for IBE [13].

Note that one has to prove two flavors of symbolic security: *selective* and *co-selective*. The former allows the substitutions of variables in $\mathbf{b}, \mathbf{c}$ to depend only on y, while those in $\mathbf{k}$ to depend on both x, y. In the latter, those in $\mathbf{b}, \mathbf{k}$ can depend only on x, while those in $\mathbf{c}$ can depend on both x, y. Intuitively, the framework of [2] uses each flavor in the two different phases—pre and post challenge—in the dual system proof methodology [2,5,23,26,34,36].

Our Modular Approach. In constructing a PES for KP[$\mathcal{P}$], we first look into the predicate definition itself and decompose to simpler ones as follows. Instead of dealing with predicates in the set $\mathcal{P}$ all at once, we consider its "direct sum", which allows us to view $\mathcal{P}$ as a single predicate, say P. Intuitively, this reduces KP[$\mathcal{P}$] of Fig. 1 to KP[P] of Fig. 4a. We then observe that KP[P] of Fig. 4a is,

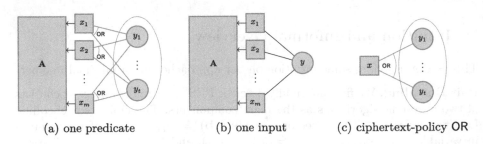

(a) one predicate   (b) one input   (c) ciphertext-policy OR

Fig. 4. Simpler variants of span program over predicates, for modular approach

in fact, already a *nested* predicate. It contains ciphertext-policy with the OR policy in the lower layer, followed by key-policy augmentation in the upper layer, as decomposed and shown in Fig. 4c and Fig. 4b, respectively. Hence, we can consider a much simpler variant that deal with only one input at a time.

Our Starting Point: Agrawal-Chase Unbounded ABE. To illustrate the above decomposition, we consider a concrete predicate, namely, unbounded KP-ABE for monotone span program (MSP), along with a concrete PES, namely, an instantiation by Agrawal and Chase [2], which is, in fact, our starting point towards generalization. First we recall this PES (Appendix B.2 of [2])[2]:

$$
\begin{aligned}
\mathbf{c}_Y &= \left(b_1 s_0 + (y_j b_2 + b_3) s_1^{(j)}\right)_{j\in[q]} \\
\mathbf{k}_{(\mathbf{A},\pi)} &= \left(\mathbf{A}_i \hat{\mathbf{r}}^\top + r_1^{(i)} b_1, \; r_1^{(i)}(\pi(i) b_2 + b_3)\right)_{i\in[m]}
\end{aligned}
\tag{1}
$$

where $(\mathbf{A}, \pi)$ is an MSP with $\mathbf{A} \in \mathbb{Z}_N^{m\times\ell}$, $\mathbf{A}_i$ is its i-th row, $\hat{\mathbf{r}} = (\alpha, \hat{r}_1, \ldots, \hat{r}_{\ell-1})$, and $Y = \{y_1, \ldots, y_q\}$. (The exact definition for MSP is not important for now.) We now attempt to view this as being achieved by two consecutive transformations. We view the starting PES as the following PES for IBE ($P^{\mathsf{IBE}}(x, y) = 1$ iff $x = y$):

$$
\begin{aligned}
\mathbf{c}_y &= b_1 s_0 + (y b_2 + b_3) s_1 \\
\mathbf{k}_x &= \left(\alpha + r_1 b_1, \; r_1(x b_2 + b_3)\right)
\end{aligned}
\tag{2}
$$

denoted as Γ_{IBE}, which is first transformed to the following PES for IBBE (ID-based broadcast encryption, $P^{\mathsf{IBBE}}(x, Y) = 1$ iff $x \in Y$), denoted as Γ_{IBBE}:

$$
\begin{aligned}
\mathbf{c}_Y &= \left(b_1 s_0 + (y_j b_2 + b_3) s_1^{(j)}\right)_{j\in[q]} = (c_j)_{j\in[q]} \\
\mathbf{k}_x &= \left(\alpha + r_1 b_1, \; r_1(x b_2 + b_3)\right)
\end{aligned}
\tag{3}
$$

which is then finally transformed to the above PES for KP-ABE. We aim to generalize this process to any PES for arbitrary predicate.

The two transformations already comprise a nested policy augmentation process: the first (IBE to IBBE) is a ciphertext-policy one with the policy being simply the OR policy, while the second (IBBE to KP-ABE for MSP) is a key-policy one with policy $(\mathbf{A}, \pi)$. To see an intuition on a policy augmentation, we choose to focus on the first one here which is simpler since it is the OR policy. To see the relation of both PESs, we look into their matrix/vector substitutions in showing symbolic property. We focus on selective symbolic property here. It can

---

[2] This encoding or closed variants are utilized in many works, *e.g.,* [5,18,25,31]. Rouselakis and Waters [31] were the first to (implicitly) use this exact encoding. Attrapadung [5] formalized it as PES. Agrawal and Chase [2] gave its symbolic proof.

be argued by showing matrix/vector substitutions that cause zero evaluations in all encodings, when $x \neq y$. For the base PES Γ_{IBE}, this is:[3]

$$\mathbf{c}_y : \overbrace{\boxed{\begin{smallmatrix}1\\0\end{smallmatrix}}}^{B_1} \overset{(s_0)^\top}{\underset{\uparrow}{1}} + \left(y \overbrace{\boxed{\begin{smallmatrix}0\\-1\end{smallmatrix}}}^{B_2} + \overbrace{\boxed{\begin{smallmatrix}-1\\y\end{smallmatrix}}}^{B_3} \right) \overset{(s_1)^\top}{\underset{\uparrow}{1}} = \boxed{\begin{smallmatrix}0\\0\end{smallmatrix}} \tag{4}$$

$$\mathbf{k}_x : \left(1 + \boxed{-1, -\tfrac{1}{y-x}} \boxed{\begin{smallmatrix}1\\0\end{smallmatrix}} = 0, \quad \boxed{-1, -\tfrac{1}{y-x}} \left(x \boxed{\begin{smallmatrix}0\\-1\end{smallmatrix}} + \boxed{\begin{smallmatrix}-1\\y\end{smallmatrix}} \right) = 0 \right)$$

where each rectangle box represents a matrix of size 1×2 or 2×1. On the other hand, the selective symbolic property for the PES Γ_{IBBE} can be shown below, where we let $\mathbf{1}_j$ be the length-q row vector with 1 at the j-th entry and $\mathbf{1}_{1,1}$ be the $(q+1) \times q$ matrix with 1 at the entry $(1,1)$ (and all the other entries are 0).

$$\mathbf{c}_Y : \overset{B_1'}{\underset{\uparrow}{\mathbf{1}_{1,1}}} \overset{(s_0')^\top}{\underset{\uparrow}{(\mathbf{1}_1)^\top}} + \left(y_j \overbrace{\begin{pmatrix} 0 & \cdots & 0 \\ -1 & & \\ & \ddots & \\ & & -1 \end{pmatrix}}^{B_2'} + \overbrace{\begin{pmatrix} -1 & \cdots & -1 \\ y_1 & & \\ & \ddots & \\ & & y_q \end{pmatrix}}^{B_3'} \right) \overset{(s_1'^{(j)})^\top}{\underset{\uparrow}{(\mathbf{1}_j)^\top}} = 0 \tag{5}$$

$$\mathbf{k}_x : \mathbf{1}_1 + \left(-1, -\tfrac{1}{y_1-x}, \ldots, -\tfrac{1}{y_q-x} \right) \mathbf{1}_{1,1} = 0,$$

$$\left(-1, -\tfrac{1}{y_1-x}, \ldots, -\tfrac{1}{y_q-x} \right) \left(x \begin{pmatrix} 0 & \cdots & 0 \\ -1 & & \\ & \ddots & \\ & & -1 \end{pmatrix} + \begin{pmatrix} -1 & \cdots & -1 \\ y_1 & & \\ & \ddots & \\ & & y_q \end{pmatrix} \right) = 0.$$

Our Observation on Unboundedness. We now examine the relation of substituted matrices/vectors between the two PESs: we observe that those for Γ_{IBBE} contains those for Γ_{IBE} as sub-matrices/vectors. For example, $\mathbf{B}_3$ for the substituted $\mathbf{c}_y$ in Eq. (4) is "embedded" in $\mathbf{B}_3'$ for the substituted $\mathbf{c}_Y$ in Eq. (5), for $y \in Y$. We denote such a sub-matrix as $\mathbf{B}_3^{(j)} = \begin{pmatrix} -1 \\ y_j \end{pmatrix}$.

We crucially observe that the unbounded property (of IBBE) stems from such an ability of embedding all the matrices from the base PES—$(\mathbf{B}_3^{(j)})_{j \in [q]}$— *regardless of size* q, into the corresponding matrix in the converted PES—$\mathbf{B}_3'$ in this case. Our aim is unbounded-size policy augmentation for *any* PES. We thus attempt to generalize this embedding process to work for any sub-matrices.

Difficulty in Generalizing to Any PES. Towards generalization, we could hope that such an embedding of sub-matrices/vectors has some patterns to follow. However, after a quick thought, we realize that the embedding here is quite

---

[3] As a convention throughout the paper, the substitution matrices/vectors are written in the exact order of appearance in their corresponding encodings (here is Eq. (3)).

specialized in many ways. The most obvious specialized form is the way that sub-matrices $\mathbf{B}_3^{(j)}$ are placed in $\mathbf{B}_3'$: the first row of $\mathbf{B}_3^{(j)}$ are placed in the same row in $\mathbf{B}_3'$, while the other row are placed in all different rows in $\mathbf{B}_3'$. Now the question is that such a special placement of sub-matrices into the composed matrices also applies to *any* generic PES. An answer for now is that this seems unlikely, if we do not restrict any structure of PES at all (which is what we aim).

We remark that, on the other hand, such a special embedding seems essential in our example here since, in each c_j, in order to cancel out the substitution of $b_1 s_0$, which is the same for all j, we must have the substitution for $(y_j b_2 + b_3) s_1^{(j)}$ to be the same for all $j \in [q]$. Therefore, we somehow must have a "projection" mechanism; this is enabled exactly by the placement in the first row of $\mathbf{B}_2', \mathbf{B}_3'$.

Our First Approach: Layering. Our first approach is to modify the transformed PES so that sub-matrices can be placed in a "generic" manner into the composed matrices. (It will become clear shortly what we mean by "generic".) In the context of IBBE, we consider the following modified PES, denoted as $\bar{\Gamma}_{\mathsf{IBBE}}$:

$$
\begin{aligned}
\mathbf{c}_Y &= \left(f_2 s_{\mathrm{new}} + f_1 s_0^{(j)}, \ b_1 s_0^{(j)} + (y_j b_2 + b_3) s_1^{(j)} \right)_{j \in [q]} \\
\mathbf{k}_x &= \left(\alpha_{\mathrm{new}} + r_{\mathrm{new}} f_2, \ r_{\mathrm{new}} f_1 + r_1 b_1, \ r_1 (x b_2 + b_3) \right)
\end{aligned}
\tag{6}
$$

This is modified from the PES in Eq. (3) by introducing one more layer involving the first element in each encoding, where f_1, f_2 are two new parameters. The main purpose is to modify the element $b_1 s_0$ to $b_1 s_0^{(j)}$ so that it varies with j, which, in turn, eliminating the need for "projection" as previously. This becomes clear in the following assessment for its selective symbolic property:

$\mathbf{c}_Y : \hat{\mathbf{1}}_{1,1}(\mathbf{1}_1)^\top + \mathbf{F}_1(\mathbf{1}_j)^\top = 0,$

$$
\left(\begin{array}{ccc} \boxed{\begin{smallmatrix}1\\0\end{smallmatrix}} & & \\ & \ddots & \\ & & \boxed{\begin{smallmatrix}1\\0\end{smallmatrix}} \end{array} \right)(\mathbf{1}_j)^\top + \left(y_j \left(\begin{array}{ccc} \boxed{\begin{smallmatrix}0\\-1\end{smallmatrix}} & & \\ & \ddots & \\ & & \boxed{\begin{smallmatrix}0\\-1\end{smallmatrix}} \end{array} \right) + \left(\begin{array}{ccc} \boxed{\begin{smallmatrix}-1\\y_1\end{smallmatrix}} & & \\ & \ddots & \\ & & \boxed{\begin{smallmatrix}-1\\y_q\end{smallmatrix}} \end{array} \right) \right)(\mathbf{1}_j)^\top = 0
$$

$\mathbf{k}_x : \mathbf{1}_1 + (-\hat{\mathbf{1}}_1)\hat{\mathbf{1}}_{1,1} = 0,$

$$
(-\hat{\mathbf{1}}_1)\mathbf{F}_1 + \left(\boxed{-1, -\tfrac{1}{y_1 - x}}, \ \dots, \ \boxed{-1, -\tfrac{1}{y_q - x}} \right) \left(\begin{array}{ccc} \boxed{\begin{smallmatrix}1\\0\end{smallmatrix}} & & \\ & \ddots & \\ & & \boxed{\begin{smallmatrix}1\\0\end{smallmatrix}} \end{array} \right) = 0,
$$

$$
\left(\boxed{-1, -\tfrac{1}{y_1 - x}}, \ \dots, \ \boxed{-1, -\tfrac{1}{y_q - x}} \right) \left(x \left(\begin{array}{ccc} \boxed{\begin{smallmatrix}0\\-1\end{smallmatrix}} & & \\ & \ddots & \\ & & \boxed{\begin{smallmatrix}0\\-1\end{smallmatrix}} \end{array} \right) + \left(\begin{array}{ccc} \boxed{\begin{smallmatrix}-1\\y_1\end{smallmatrix}} & & \\ & \ddots & \\ & & \boxed{\begin{smallmatrix}-1\\y_q\end{smallmatrix}} \end{array} \right) \right) = 0.
\tag{7}
$$

where we let $\hat{\mathbf{1}}_{1,1}$ be of size $(2q) \times q$ and $\hat{\mathbf{1}}_1$ be of length $2q$ (defined similarly to $\mathbf{1}_{1,1}, \mathbf{1}_1$, resp.), and let $\mathbf{F}_1$ be the $(2q) \times q$ matrix with all entries in the first row being -1 (and all the other entries are 0). Here, we observe that all the composed matrices regarding the parameters (b_1, b_2, b_3) of the PES Γ_{IBBE} are formed exactly by including the substituted matrices of the base PES in the "diagonal blocks", namely, we can now "generically" define, for $i \in [n]$,

$$\mathbf{B}'_i = \begin{pmatrix} \mathbf{B}^{(1)}_i & & \\ & \ddots & \\ & & \mathbf{B}^{(q)}_i \end{pmatrix}.$$

Moreover, arranging the vector substitutions in their corresponding slots will result in exactly the zero evaluation of each substituted equation of the base PES. This approach is naturally generalized to any base PES. Put in other words, intuitively, we can obtain the proof of symbolic property of the composed PES from that of the base PES generically, via this conversion. Such a conversion, transforming any PES $(\mathbf{c}_y, \mathbf{k}_x)$ for predicate P to its ciphertext-policy augmentation (with OR policy), can be described by

$$\mathbf{c}'_Y = \left(f_2 s_{\mathrm{new}} + f_1 s_0^{(i)}, \ \mathbf{c}_{y_j} \right)_{j \in [q]}, \quad \mathbf{k}'_x = \left(\alpha_{\mathrm{new}} + r_{\mathrm{new}} f_2, \ (\mathbf{k}_x)|_{\alpha \mapsto r_{\mathrm{new}} f_1} \right) \quad (8)$$

where the variables s_u in $\mathbf{c}_{y_j}$ are superscripted as $s_u^{(j)}$, and "$\mapsto$" denotes the variable replacement. This PES is for the predicate of "ciphertext-OR-policy" over P—returning true iff $\exists j \, P(x, y_j) = 1$. In fact, one can observe that Eq. (8) is a generalization of Eq. (6).

Our Second Approach: Admissible PES. One disadvantage with our first approach is the inefficiency due to the additional terms. Comparing PES $\bar{\Gamma}_{\mathsf{IBBE}}$ to Γ_{IBBE}, the former requires $2q$ more elements than the latter (note that we include also $(s_0^{(j)})_{j \in [q]}$ when counting overall ciphertext elements). However, we already knew that the additional terms are not necessary for some specific PESs and predicates, notably our Γ_{IBE} for IBE.

We thus turn to the second approach which takes the following two steps. First, we find a class of "admissible" PESs where there exists a conversion for ciphertext-policy augmentation without additional terms. Second, we provide a conversion from any PES to a PES that is admissible.

As a result of our finding, the admissible class of PESs turns out to have a simple structure: $\mathbf{k}$ consists of $k_1 = \alpha + r_1 b_1$, and α, b_1 do not appear elsewhere in $\mathbf{k}$, while in $\mathbf{c}$, we allow b_1, s_0 only if they are multiplied—$b_1 s_0$. Intuitively, this "isolation" of b_1, α, s_0 somewhat provides a sufficient structure[4] where the "projection" can be enabled, but without mitigating to additional elements as done in the above first approach. The ciphertext-OR-policy augmentation can then be done by simply setting

$$\mathbf{c}'_Y = \left((\mathbf{c}_{y_j})|_{s_0^{(j)} \mapsto s_{\mathrm{new}}} \right)_{j \in [q]}, \quad \mathbf{k}'_x = \mathbf{k}_x. \quad (9)$$

---

[4] Note that we indeed require a few more simple requirements in order for the proof to go through: see Definition 4.

One can observe that this is a generalization of Eq. (3), and that there is no additional terms as in Eq. (8). Our conversion from any PES to an admissible one (for the same predicate) is also simple: we set

$$\mathbf{c}'_y = \Big(f_2 s_{\text{new}} + f_1 s_0, \ \mathbf{c}_y \Big), \quad \mathbf{k}'_x = \Big(\alpha_{\text{new}} + r_{\text{new}} f_2, \ (\mathbf{k}_x)|_{\alpha \mapsto r_{\text{new}} f_1} \Big) \quad (10)$$

where s_0 is the variable in $\mathbf{y}$, while s_{new} is the new special variable (that defines correctness). It is easy to see also that combining both conversions, that is, Eq. (10) followed by Eq. (9), we obtain the conversion of the first approach (Eq. (8)). But now, for any PES that is already admissible such as Γ_{IBE}, we do not have to apply the conversion of Eq. (10), which requires additional terms.

Towards General Policies. Up to now, we only consider the OR policy. It ensures that $P'(x, Y) = 0$ implies $P(x, y_j) = 0$ for all j. However, for general policies, this is not the case, that is, if we let $\bar{P}$ be such a ciphertext-policy augmented predicate over P (this will be formally given in Definition 5), $\bar{P}(x, (\mathbf{A}, \pi)) = 0$ may hold even if $P(x, \pi(j)) = 1$ for some j. Consequently, we have no available substituted matrices/vectors for the key encoding for such problematic j. Another important issue is how to embed the policy $(\mathbf{A}, \pi)$ without knowledge of x (cf. the selective property), but be able to deal with any x such that $\bar{P}(x, (\mathbf{A}, \pi)) = 0$.

We solve both simultaneously by a novel way of embedding $(\mathbf{A}, \pi)$ so that, intuitively, only the "non-problematic" blocks will turn "on", whatever x will be, together with a novel way of defining substituted vectors for $\mathbf{k}$ so that all the "problematic" blocks will turn "off". To be able to deal with any x, the former has to be done in the "projection" part, while the latter is done in the "non-projection" part of matrices. By combining both, we will have only the non-problematic blocks turned on, and thus can use the base symbolic property.

Towards Other Predicative Machines: Automata. At the core of the above mechanism is the existence of "mask" vectors which render problematic blocks to 0. We crucially observe that such "mask" vectors depend on *and only on* $(x, (\mathbf{A}, \pi))$ and the sole fact that $\bar{P}(x, (\mathbf{A}, \pi)) = 0$, *i.e.*, the non-acceptance condition of MSP. Notably, it does not depend on the actual PES construction. This feature provides an insight to extend our approach to other types of predicative machines—finite automata in particular—by finding appropriate combinatorial vectors that encode non-acceptance conditions. (See more discussions in the full version.)

Wrapping Up. Up to now, we mainly consider the selective symbolic property. The co-selective property (for the ciphertext-policy case) is simpler to achieve, since each substitution matrix of the converted PES is now required to embed only one matrix from the base PES, as our modular approach allows to consider one input at a time (for key attribute). The situation becomes reversed for the key-policy case: the co-selective property is harder. Nonetheless, we can always use the DUAL conversion to convert from ciphertext-policy to key-policy type.

Comparing to Unboundedness Approach in CGKW [18]. Chen *et al.* [18] recently proposed unbounded ABE for MSP. Their approach conceptually converts a specific bounded scheme [27] to an unbounded one for the *same* specific predicate—MSP. This is already semantically different to our conversion, which takes any pair encoding for a predicate P and outputs another for a *different* predicate—namely, the (unbounded) policy-augmented predicate over P.

3   Preliminaries

Notations. $\mathbb{N}$ denotes the set of positive integers. For $a, b \in \mathbb{N}$ such that $a \leq b$, let $[a, b] = \{a, \ldots, b\}$. For $m \in \mathbb{N}$, let $[m] = \{1, \ldots, m\}$ and $[m]^+ = \{0, 1, \ldots, m\}$. For a set S, we denote by 2^S the set of all subsets of S. Denote by S^* the set of all (unbounded-length) sequences where each element is in S. For $N \in \mathbb{N}$, we denote by $\mathbb{Z}_N^{m \times \ell}$ the set of all matrices of dimension $m \times \ell$ with elements in $\mathbb{Z}_N$. For a matrix $\mathbf{M} \in \mathbb{Z}_N^{m \times \ell}$, its i-th row vector is denoted by $\mathbf{M}_{i:}$ (in $\mathbb{Z}_N^{1 \times \ell}$). Its (i, j)-element is $\mathbf{M}_{i,j}$. Its transpose is denoted as $\mathbf{M}^\top$. For vectors $\mathbf{a} \in \mathbb{Z}_N^{1 \times c}$, $\mathbf{b} \in \mathbb{Z}_N^{1 \times d}$, we denote $(\mathbf{a}, \mathbf{b}) \in \mathbb{Z}_N^{1 \times (c+d)}$ as the concatenation. The i-th entry of $\mathbf{a}$ is denoted as $\mathbf{a}[i]$. For $i < j$, denote $\mathbf{a}[i, j] := (\mathbf{a}[i], \mathbf{a}[i+1], \ldots, \mathbf{a}[j])$. Let $\mathbb{M}(\mathbb{Z}_N)$ be the set of all matrices (of any sizes) in $\mathbb{Z}_N$, and $\mathbb{M}_m(\mathbb{Z}_N)$ be the set of those with m rows. For a set S of vectors of the same length (say, in $\mathbb{Z}_N^\ell$), we denote span(S) as the set of all linear combinations of vectors in S. For polynomials $p = p(x_1, \ldots, x_n)$ and $g = g(y_1, \ldots, y_n)$, we denote a new polynomial $p|_{x_1 \mapsto g} := p(g(y_1, \ldots, y_n), x_2, \ldots, x_n)$. Matrices and vectors with all 0's are simply denoted by 0, of which the dimension will be clear from the context. We define some useful fixed vectors and matrices.

- $\mathbf{1}_i^\ell$ is the (row) vector of length ℓ with 1 at position i where all others are 0.
- $\mathbf{1}_{i,j}^{m \times \ell}$ is the matrix of size $m \times \ell$ with 1 at position (i, j) and all others are 0.

3.1   Definitions for General ABE

Predicate Family. Let $P = \{P_\kappa : \mathcal{X}_\kappa \times \mathcal{Y}_\kappa \to \{0, 1\} \mid \kappa \in \mathcal{K}\}$ be a predicate family where $\mathcal{X}_\kappa$ and $\mathcal{Y}_\kappa$ denote "key attribute" and "ciphertext attribute" spaces. The index κ or "parameter" denotes a list of some parameters such as the universes of attributes, and/or bounds on some quantities, hence its domain $\mathcal{K}$ will depend on that predicate. We will often omit κ when the context is clear.

General ABE Syntax. Let $\mathcal{M}$ be a message space. An ABE scheme[5] for predicate family P is defined by the following algorithms:

- Setup($1^\lambda, \kappa$) $\to$ (PK, MSK): takes as input a security parameter 1^λ and a parameter κ of predicate family P, and outputs a master public key PK and a master secret key MSK.

---

[5] It is also called public-index predicate encryption, classified in the definition of Functional Encryption [15]. It is simply called predicate encryption in [2].

- Encrypt$(y, M, \mathsf{PK}) \to \mathsf{CT}$: takes as input a ciphertext attribute $y \in \mathcal{Y}_\kappa$, a message $M \in \mathcal{M}$, and public key PK. It outputs a ciphertext CT. We assume that Y is implicit in CT.
- KeyGen$(x, \mathsf{MSK}, \mathsf{PK}) \to \mathsf{SK}$: takes as input a key attribute $x \in \mathcal{X}_\kappa$ and the master key MSK. It outputs a secret key SK.
- Decrypt$(\mathsf{CT}, \mathsf{SK}) \to M$: given a ciphertext CT with its attribute y and the decryption key SK with its attribute x, it outputs a message M or $\bot$.

Correctness. Consider all parameters κ, all $M \in \mathcal{M}$, $x \in \mathcal{X}_\kappa$, $y \in \mathcal{Y}_\kappa$ such that $P_\kappa(x, y) = 1$. If Encrypt$(y, M, \mathsf{PK}) \to \mathsf{CT}$ and KeyGen$(x, \mathsf{MSK}, \mathsf{PK}) \to \mathsf{SK}$ where $(\mathsf{PK}, \mathsf{MSK})$ is generated from Setup$(1^\lambda, \kappa)$, then Decrypt$(\mathsf{CT}, \mathsf{SK}) \to M$.

Security. We use the standard notion for ABE, called full security. We omit it here and refer to *e.g.*, [5] (or the full version of this paper), as we do not work directly on it but will rather infer the implication from pair encoding scheme (*cf.* Sect. 3.3).

Duality of ABE. For a predicate $P : \mathcal{X} \times \mathcal{Y} \to \{0, 1\}$, we define its dual as $\bar{P} : \mathcal{Y} \times \mathcal{X} \to \{0, 1\}$ by setting $\bar{P}(Y, X) = P(X, Y)$. In particular, if P is considered as key-policy type, then its dual, $\bar{P}$, is the corresponding ciphertext-policy type.

3.2 Pair Encoding Scheme Definition

Definition 1. Let $P = \{ P_\kappa \}_\kappa$ where $P_\kappa : \mathcal{X}_\kappa \times \mathcal{Y}_\kappa \to \{ 0, 1 \}$, be a predicate family, indexed by $\kappa = (N, \mathsf{par})$, where par specifies some parameters. A *Pair Encoding Scheme* (PES) for a predicate family P is given by four deterministic polynomial-time algorithms as described below.

- Param$(\mathsf{par}) \to n$. When given par as input, Param outputs $n \in \mathbb{N}$ that specifies the number of *common* variables, which we denote by $\mathbf{b} := (b_1, \ldots, b_n)$.
- EncCt$(y, N) \to (w_1, w_2, \mathbf{c}(\mathbf{s}, \hat{\mathbf{s}}, \mathbf{b}))$. On input $N \in \mathbb{N}$ and $y \in \mathcal{Y}_{(N, \mathsf{par})}$, EncCt outputs a vector of polynomial $\mathbf{c} = (c_1, \ldots, c_{w_3})$ in *non-lone* variables $\mathbf{s} = (s_0, s_1, \ldots, s_{w_1})$ and *lone* variables $\hat{\mathbf{s}} = (\hat{s}_1, \ldots, \hat{s}_{w_2})$. For $p \in [w_3]$, the p-th polynomial is given as follows, where $\eta_{p,z}, \eta_{p,t,j} \in \mathbb{Z}_N$:

$$\sum_{z \in [w_2]} \eta_{p,z} \hat{s}_z + \sum_{t \in [w_1]^+, j \in [n]} \eta_{p,t,j} b_j s_t.$$

- EncKey$(x, N) \to (m_1, m_2, \mathbf{k}(\mathbf{r}, \hat{\mathbf{r}}, \mathbf{b}))$. On input $N \in \mathbb{N}$ and $x \in \mathcal{X}_{(N, \mathsf{par})}$, EncKey outputs a vector of polynomial $\mathbf{k} = (k_1, \ldots, k_{m_3})$ in *non-lone* variables $\mathbf{r} = (r_1, \ldots, r_{m_1})$ and *lone* variables $\hat{\mathbf{r}} = (\alpha, \hat{r}_1, \ldots, \hat{r}_{m_2})$. For $p \in [m_3]$, the p-th polynomial is given as follows, where $\phi_p, \phi_{p,u}, \phi_{p,v,j} \in \mathbb{Z}_N$:

$$\phi_p \alpha + \sum_{u \in [m_2]} \phi_{p,u} \hat{r}_u + \sum_{v \in [m_1], j \in [n]} \phi_{p,v,j} r_v b_j.$$

- Pair$(x, y, N) \to (\mathbf{E}, \overline{\mathbf{E}})$. On input N, and both x, and y, Pair outputs two matrices $\mathbf{E}, \overline{\mathbf{E}}$ of sizes $(w_1 + 1) \times m_3$ and $w_3 \times m_1$, respectively. $\Diamond$

Correctness. A PES is said to be correct if for every $\kappa = (N, \mathsf{par})$, $x \in \mathcal{X}_\kappa$ and $y \in \mathcal{Y}_\kappa$ such that $P_\kappa(x, y) = 1$, the following holds symbolically:

$$\mathbf{sEk}^\top + \mathbf{c\overline{E}r}^\top = \alpha s_0. \tag{11}$$

The left-hand side is indeed a linear combination of $s_t k_p$ and $c_q r_v$, for $t \in [w_1]^+, p \in [m_3], q \in [w_3], v \in [m_1]$. Hence, an equivalent (and somewhat simpler) way to describe Pair and correctness together at once is to show such a linear combination that evaluates to αs_0. We will use this approach throughout the paper. (The matrices $\mathbf{E}, \overline{\mathbf{E}}$ will be implicitly defined in such a linear combination).

Terminology. In the above, following [2], a variable is called *lone* as it is not multiplied with any b_j (otherwise called *non-lone*). Furthermore, since α, s_0 are treated distinguishably in defining correctness, we also often call them the *special* lone and non-lone variable, respectively. In what follows, we use ct-enc and key-enc as a shorthand for polynomials and variables output by EncCt (ciphertext-encoding) and EncKey (key-encoding), respectively. We often omit writing w_1, w_2 and m_1, m_2 in the output of EncCt and EncKey.

3.3 Symbolic Property of PES

We now describe the symbolic property of PES, introduced in [2]. As in [2], we use $a : b$ to denote that a variable a is substituted by a matrix/vector b.

Definition 2. A PES $\Gamma = (\mathsf{Param}, \mathsf{EncCt}, \mathsf{EncKey}, \mathsf{Pair})$ for predicate family P satisfies (d_1, d_2)-*selective symbolic property* for some $d_1, d_2 \in \mathbb{N}$ if there exists three deterministic polynomial-time algorithms $\mathsf{EncB}, \mathsf{EncS}, \mathsf{EncR}$ such that for all $\kappa = (N, \mathsf{par})$, $x \in \mathcal{X}_\kappa$, $y \in \mathcal{Y}_\kappa$ with $P_\kappa(x, y) = 0$,

- $\mathsf{EncB}(y) \to \mathbf{B}_1, \ldots, \mathbf{B}_n \in \mathbb{Z}_N^{d_1 \times d_2}$;
- $\mathsf{EncS}(y) \to \mathbf{s}_0, \ldots, \mathbf{s}_{w_1} \in \mathbb{Z}_N^{1 \times d_2}$, $\hat{\mathbf{s}}_1, \ldots, \hat{\mathbf{s}}_{w_2} \in \mathbb{Z}_N^{1 \times d_1}$;
- $\mathsf{EncR}(x, y) \to \mathbf{r}_1, \ldots, \mathbf{r}_{m_1} \in \mathbb{Z}_N^{1 \times d_1}$, $\mathbf{a}, \hat{\mathbf{r}}_1, \ldots, \hat{\mathbf{r}}_{m_2} \in \mathbb{Z}_N^{1 \times d_2}$;

we have that:

(P1) $\mathbf{as}_0^\top \neq 0$.
(P2) if we substitute, for all $j \in [n]$, $t \in [w_1]^+$, $z \in [w_2]$, $v \in [m_1]$, $u \in [m_2]$,

$$\hat{s}_z : \hat{\mathbf{s}}_z^\top, \quad b_j s_t : \mathbf{B}_j \mathbf{s}_t^\top, \quad \alpha : \mathbf{a}, \quad \hat{r}_u : \hat{\mathbf{r}}_u, \quad r_v b_j : \mathbf{r}_v \mathbf{B}_j,$$

into all the polynomials output by $\mathsf{EncCt}(y)$ and $\mathsf{EncKey}(x)$, then they evaluate to 0.
(P3) $\mathbf{a} = \mathbf{1}_1^{d_2}$.

Similarly, a PES satisfies (d_1, d_2)-*co-selective symbolic property* if there exists $\mathsf{EncB}, \mathsf{EncS}, \mathsf{EncR}$ satisfying the above properties but where EncB and EncR depends only on x, and EncS depends on both x and y.

Finally, a PES satisfies (d_1, d_2)-*symbolic property* if it satisfies both (d_1', d_2')-selective and (d_1'', d_2'')-co-selective properties for some $d_1', d_1'' \leq d_1$, $d_2', d_2'' \leq d_2$. $\Diamond$

Terminology. The original definition in [2] consists of only (P1) and (P2); we refer to this as Sym-Prop, as in [2]. We newly include (P3) here, and refer to the full definition with all (P1)–(P3) as Sym-Prop$^+$. This is w.l.o.g. since one can convert any PES with Sym-Prop to another with Sym-Prop$^+$, with minimal cost. Such a conversion, which we denote as Plus-Trans, also appears in [2]; we recap it in the full version.

For convenience, for the case of selective property, we use EncBS(y) to simply refer to the concatenation of EncB(y) and EncS(y). Similarly, we use EncBR(x) for referring EncB(x) and EncR(x) for the case of co-selective property.

Implication to Fully Secure ABE. Agrawal and Chase [2] show that a PES satisfying (d_1, d_2)-Sym-Prop implies fully secure ABE. They use an underlying assumption called (D_1, D_2)-q-ratio, which can be defined in the dual system groups [17] and can consequently be instantiated in the prime-order bilinear groups. Note that parameter (D_1, D_2) are related to (d_1, d_2). Since their theorem is not used explicitly in this paper, we recap it in the full version.

3.4 Definitions for Some Previous Predicates

ABE for Monotone Span Program. We recap the predicate definition for KP-ABE for monotone span program (MSP) [21]. We will mostly focus on *completely unbounded* variant [2,5], where the family index is simply $\kappa = N \in \mathbb{N}$, that is, any additional parameter par is not required.[6] Below, we also state a useful lemma which is implicit in *e.g.*, [21,27].

Definition 3. The predicate family of *completely unbounded KP-ABE for monotone span programs*, $P^{\mathsf{KP\text{-}MSP}} = \{\, P_\kappa : \mathcal{X}_\kappa \times \mathcal{Y}_\kappa \to \{0,1\} \,\}_\kappa$, is indexed by $\kappa = (N)$ and is defined as follows. Recall that $\mathbf{A}_{i:}$ denotes the i-th row of $\mathbf{A}$.

- $\mathcal{X}_\kappa = \{\, (\mathbf{A}, \pi) \mid \mathbf{A} \in \mathbb{M}(\mathbb{Z}_N),\ \pi : [m] \to \mathbb{Z}_N \,\}$.
- $\mathcal{Y}_\kappa = 2^{(\mathbb{Z}_N)}$.
- $P_\kappa((\mathbf{A}, \pi), Y) = 1 \iff \mathbf{1}_1^\ell \in \mathrm{span}(\mathbf{A}|_Y)$, where $\mathbf{A}|_Y := \{\, \mathbf{A}_{i:} \mid \pi(i) \in Y \,\}$.

where $m \times \ell$ is the size of the matrix $\mathbf{A}$. $\Diamond$

Proposition 1. *Consider a matrix $\mathbf{A} \in \mathbb{Z}_N^{m \times \ell}$. Let $Q \subseteq [m]$ be a set of row indexes. If $\mathbf{1}_1^\ell \notin \mathrm{span}\{\, \mathbf{A}_{i:} \mid i \in Q \,\}$, then there exists $\boldsymbol{\omega} = (w_1, \ldots, w_\ell) \in \mathbb{Z}_N^\ell$ such that $w_1 = 1$ and $\mathbf{A}_{i:}\boldsymbol{\omega}^\top = 0$ for all $i \in Q$.*

Specific Policies. It is well known that ABE for MSP implies ABE for monotone Boolean formulae [11,21]. The procedure of embedding a boolean formula as a span program can be found in *e.g.*, §C of [24]. We will be interested in the OR and the AND policy, for using as building blocks later on. For the OR

---

[6] Bounded schemes would use par for specifying some bounds, *e.g.*, on policy or attribute set sizes, or the number of attribute multi-use in one policy. The term "Unbounded ABE" used in the literature [18,25,30] still allows to have a bound for the number of attribute multi-use in one policy (or even a one-use restriction).

policy, the access matrix is of the form $\mathbf{A}_{\mathsf{OR},m} = (1, \ldots, 1)^\top \in \mathbb{Z}_N^{m \times 1}$. For the AND policy, it is $\mathbf{A}_{\mathsf{AND},m} = \sum_{i=1} \mathbf{1}_{i,i}^{m \times m} - \sum_{j=2} \mathbf{1}_{1,j}^{m \times m}$. For further use, we let $\mathbb{M}_{\mathsf{OR}}(\mathbb{Z}_N) = \{\, \mathbf{A}_{\mathsf{OR},m} \mid m \in \mathbb{N} \,\}$ and $\mathbb{M}_{\mathsf{AND}}(\mathbb{Z}_N) = \{\, \mathbf{A}_{\mathsf{AND},m} \mid m \in \mathbb{N} \,\}$.

Embedding Lemma. To argue that a PES for predicate P can be used to construct a PES for predicate P', intuitively, it suffices to find mappings that map attributes in P' to those in P, and argue that the predicate evaluation for P' is preserved to that for P on the mapped attributes. In such a case, we say that P' *can be embedded into* P. This is known as the embedding lemma, used for general ABE in [7,14]. We prove the implication for the case of PES in the full version.

4   Admissible Pair Encodings

We first propose the notion of *admissible PES*. It is a class of PESs where a conversion to a new PES for its policy-augmented predicate exists without additional terms, as motivated in the second approach in Sect. 2. We then provide a conversion from *any* PES to an admissible PES of the same predicate (this, however, poses additional terms).[7] Together, these thus allow us to convert any PES to a new PES for its policy-augmented predicate.

Definition 4. A PES is (d_1, d_2)-admissible if it satisfies (d_1, d_2)-Sym-Prop$^+$ with the following additional constraints.

(P4) In the key encoding $\mathbf{k}$, the first polynomial has the form $k_1 = \alpha + r_1 b_1$ and α, b_1 do not appear elsewhere in $\mathbf{k}$.

(P5) In the ciphertext encoding $\mathbf{c}$, the variables b_1 and s_0 can only appear in the term $b_1 s_0$.[8]

(P6) In the symbolic property (both selective and co-selective), we have that $\mathbf{B}_1 = \mathbf{1}_{1,1}^{d_1 \times d_2}$, $\mathbf{s}_0 = \mathbf{1}_1^{d_2}$, and $\mathbf{r}_v[1] \neq 0$ for all $v \in [m_1]$.   ◇

We will use the following for the correctness of our conversion in Sect. 5.

Corollary 1. *For any admissible PES, let $\mathbf{c}, \mathbf{k}, \mathbf{s}, \mathbf{r}, \mathbf{E}, \overline{\mathbf{E}}$ be defined as in Definition 1 with $P_\kappa(x, y) = 1$. Let $\tilde{\mathbf{s}} = (s_1, \ldots, s_{w_1})$. There exists a PPT algorithm that takes $\mathbf{E}$ and outputs a matrix $\tilde{\mathbf{E}}$ of size $w_1 \times m_3$ such that $\tilde{\mathbf{s}}\tilde{\mathbf{E}}\mathbf{k}^\top + \mathbf{c}\overline{\mathbf{E}}\mathbf{r}^\top = -r_1 b_1 s_0$.*

Proof. We re-write Eq. (11) as $s_0 k_1 + T + \mathbf{c}\overline{\mathbf{E}}\mathbf{r}^\top = \alpha s_0$ (where T is a sum of $s_t k_j$ with coefficients from $\mathbf{E}$). Note that $s_0 k_1$ has coefficient 1 since α appears only in k_1 and we match the monomial αs_0 to the right hand side. Substituting $k_1 = \alpha + r_1 b_1$, we have $T + \mathbf{c}\overline{\mathbf{E}}\mathbf{r}^\top = -r_1 b_1 s_0$. We claim that s_0 is not in T, which would prove the corollary. To prove the claim, we first see that k_1 is not in T, since α is not in the right hand side. Thus b_1 is also not in T (as b_1 only appears in k_1). Hence, s_0 is not in T, since otherwise $b_j s_0$ where $j \geq 2$ appears in T, but in such a case, it cannot be cancelled out since such term is not allowed in $\mathbf{c}$. □

---

[7] Interestingly, this conversion already appears in [2] but for different purposes.

[8] That is, $b_j s_0$ and $b_1 s_t$ for $j \in [2, n], t \in [1, n]$ are not allowed in $\mathbf{c}$.

Construction 1. Let Γ be a PES construction for P. We construct another PES Γ' for also the same P as follows. We denote this Γ' by Layer-Trans(Γ).

- Param$'$(par). If Param(par) returns n, then output $n + 2$. Denote $\mathbf{b} = (b_1, \ldots, b_n)$ and $\mathbf{b}' = (f_1, f_2, \mathbf{b})$.
- EncCt$'(y, N)$. Run EncCt$(y, N) \to \mathbf{c}$. Let s_0 be the special variable in $\mathbf{c}$. Let s_{new} be the new special variable. Output $\mathbf{c}' = (f_1 s_{\text{new}} + f_2 s_0, \mathbf{c})$.
- EncKey$'(x, N)$. Run EncKey$(x, N) \to \mathbf{k}$. Let r_{new} be a new non-lone variable and α_{new} be the new special lone variable. Let $\tilde{\mathbf{k}}$ be exactly $\mathbf{k}$ but with α being replaced by $r_{\text{new}} f_2$. Output $(\alpha_{\text{new}} + r_{\text{new}} f_1, \tilde{\mathbf{k}})$.

Pair/Correctness. Suppose $P(x, y) = 1$. From the correctness of Γ we have a linear combination that results in $\alpha s_0 = r_{\text{new}} f_2 s_0$. From then, we have $(\alpha_{\text{new}} + r_{\text{new}} f_1) s_{\text{new}} - r_{\text{new}}(f_1 s_{\text{new}} + f_2 s_0) + r_{\text{new}} f_2 s_0 = \alpha_{\text{new}} s_{\text{new}}$, as required.

Lemma 1. *Suppose that Γ for P satisfies (d_1, d_2)-Sym-Prop$^+$. Then, the PES Layer-Trans(Γ) for P is $(d_1 + 1, d_2)$-admissible.* (The proof is deferred to the full version.)

5   Ciphertext-Policy Augmentation

We now describe the notion of ciphertext-policy-span-program-augmented predicate over a *single* predicate family. We then construct a conversion that preserves admissibility. The case for a *set* of predicate families will be described in Sect. 7. The key-policy case will be in the next section Sect. 6.

Definition 5. Let $P = \{ P_\kappa \}_\kappa$ where $P_\kappa : \mathcal{X}_\kappa \times \mathcal{Y}_\kappa \to \{ 0, 1 \}$, be a predicate family. We define the *ciphertext-policy-span-program-augmented predicate* over P as $\mathsf{CP1}[P] = \{ \bar{P}_\kappa \}_\kappa$ where $\bar{P}_\kappa : \bar{\mathcal{X}}_\kappa \times \bar{\mathcal{Y}}_\kappa \to \{ 0, 1 \}$ by letting

- $\bar{\mathcal{X}}_\kappa = \mathcal{X}_\kappa$.
- $\bar{\mathcal{Y}}_\kappa = \{ (\mathbf{A}, \pi) \mid \mathbf{A} \in \mathbb{M}(\mathbb{Z}_N), \pi : [m] \to \mathcal{Y}_\kappa \}$.
- $\bar{P}_\kappa(x, (\mathbf{A}, \pi)) = 1 \iff \mathbf{1}_1^\ell \in \text{span}(\mathbf{A}|_x)$, where $\mathbf{A}|_x := \{ \mathbf{A}_{i:} \mid P_\kappa(x, \pi(i)) = 1 \}$.

where $m \times \ell$ is the size of the matrix $\mathbf{A}$. $\Diamond$

Construction 2. Let Γ be a PES construction for P satisfying admissibility. We construct a PES Γ' for $\mathsf{CP1}[P]$ as follows. Denote this Γ' by $\mathsf{CP1}$-Trans(Γ).

- Param$'$(par) = Param(par) = n. Denote $\mathbf{b} = (b_1, \ldots, b_n)$.
- EncKey$'(x, N)$ = EncKey(x, N).
- EncCt$'((\mathbf{A}, \pi), N)$. Parse $\mathbf{A} \in \mathbb{Z}_N^{m \times \ell}$.
  - For $i \in [m]$, run EncCt$(\pi(i), N)$ to obtain a vector $\mathbf{c}^{(i)} = \mathbf{c}^{(i)}(\mathbf{s}^{(i)}, \hat{\mathbf{s}}^{(i)}, \mathbf{b})$ of polynomials in variables $\mathbf{s}^{(i)} = (s_0^{(i)}, s_1^{(i)}, \ldots, s_{w_{1,i}}^{(i)})$, $\hat{\mathbf{s}}^{(i)} = (\hat{s}_1^{(i)}, \ldots, \hat{s}_{w_{2,i}}^{(i)})$, and $\mathbf{b}$. Denote $\bar{\mathbf{s}}^{(i)} = (s_1^{(i)}, \ldots, s_{w_{1,i}}^{(i)})$.

- Let s_{new} be the new special non-lone variable. Let $v_2, \ldots, v_\ell$ be new lone variables. Denote $\mathbf{v} = (b_1 s_{\text{new}}, v_2, \ldots, v_\ell)$.
- For $i \in [m]$, define a modified vector by variable replacement as

$$\mathbf{c}'^{(i)} := \mathbf{c}^{(i)}\big|_{b_1 s_0^{(i)} \mapsto \mathbf{A}_{i:}\mathbf{v}^\top}. \tag{12}$$

Finally, output $\mathbf{c}' = \mathbf{c}'(\mathbf{s}', \hat{\mathbf{s}}', \mathbf{b}')$ as $\mathbf{c}' = \big(\mathbf{c}'^{(i)}\big)_{i \in [m]}$. It contains variables $\mathbf{s}' = \big(s_{\text{new}}, (\tilde{\mathbf{s}}^{(i)})_{i \in [m]}\big)$, $\hat{\mathbf{s}}' = \big(v_2, \ldots, v_\ell, (\hat{\mathbf{s}}^{(i)})_{i \in [m]}\big)$, and $\mathbf{b}'$.

Pair/Correctness. For proving correctness, we suppose $\bar{P}_\kappa(x, (\mathbf{A}, \pi)) = 1$. Let $S := \{ i \in [m] \mid P_\kappa(x, \pi(i)) = 1 \}$. For $i \in S$, we can run $\mathsf{Pair}(x, \pi(i), N) \to (\mathbf{E}, \overline{\mathbf{E}})$. From the correctness of Γ, we derive $\tilde{\mathbf{E}}$ from $\mathbf{E}$ via Corollary 1, and obtain a linear combination $\tilde{\mathbf{s}}^{(i)}\tilde{\mathbf{E}}\mathbf{k}^\top + \mathbf{c}^{(i)}\overline{\mathbf{E}}\mathbf{r}^\top = -r_1 b_1 s_0^{(i)}$. With the variable replacement in Eq. (12), this becomes $\tilde{\mathbf{s}}^{(i)}\tilde{\mathbf{E}}\mathbf{k}^\top + \mathbf{c}'^{(i)}\overline{\mathbf{E}}\mathbf{r}^\top = -r_1 \mathbf{A}_{i:}\mathbf{v}^\top$. Now since $\mathbf{1}_1^\ell \in \text{span}(\mathbf{A}|_x)$, we have linear combination coefficients $\{ t_i \}_{i \in S}$ such that $\sum_{i \in S} t_i \mathbf{A}_{i:} = \mathbf{1}_1^\ell$. Hence we have the following linear combination, as required:[9] $k_1 s_{\text{new}} + \sum_{i \in S} t_i \big(- r_1 \mathbf{A}_{i:}\mathbf{v}^\top\big) = (\alpha + r_1 b_1)s_{\text{new}} - r_1 b_1 s_{\text{new}} = \alpha_{\text{new}} s_{\text{new}}$.

Theorem 1. *Suppose a PES Γ for P is (d_1, d_2)-admissible. Then, $\mathsf{CP1\text{-}Trans}(\Gamma)$ for $\mathsf{CP1}[P]$ is $(\ell + m(d_1 - 1), md_2)$-admissible, where $m \times \ell$ is the size of policy.*

Proof. We prove symbolic property of Γ' from that of Γ as follows.

Selective Symbolic Property. We define the following algorithms. $\boxed{\mathsf{EncBS}'(\mathbf{A}, \pi)}$: For each $i \in [m]$, run

$$\mathsf{EncBS}(\pi(i)) \to \Big(\mathbf{B}_1^{(i)}, \ldots, \mathbf{B}_n^{(i)}; \ s_0^{(i)}, \ldots, s_{w_{1,i}}^{(i)}; \ \hat{s}_1^{(i)}, \ldots, \hat{s}_{w_{2,i}}^{(i)}\Big),$$

where $\mathbf{B}_j^{(i)} \in \mathbb{Z}_N^{d_1 \times d_2}$, $s_t^{(i)} \in \mathbb{Z}_N^{1 \times d_2}$, $\hat{s}_z^{(i)} \in \mathbb{Z}_N^{1 \times d_1}$. For $j \in [2, n]$, we parse $\mathbf{B}_j^{(i)} =: \begin{pmatrix} \mathbf{e}_j^{(i)} \\ \tilde{\mathbf{B}}_j^{(i)} \end{pmatrix}$ where $\mathbf{e}_j^{(i)} \in \mathbb{Z}_N^{1 \times d_2}$ and $\tilde{\mathbf{B}}_j^{(i)} \in \mathbb{Z}_N^{(d_1 - 1) \times d_2}$ (*i.e.*, decomposing into the first row and the rest). Let $d_1' = \ell + m(d_1 - 1)$ and $d_2' = md_2$. Any vector of length d_2' can be naturally divided into m blocks, each with length d_2. Any d_1'-length vectors consists of the first ℓ positions which are then followed by m blocks of length $d_1 - 1$.[10] Let $\mathbf{B}_1' = \mathbf{1}_{1,1}^{d_1' \times d_2'}$, $s_{\text{new}} = \mathbf{1}_1^{d_2'}$, $\mathbf{v}_\iota' = \mathbf{1}_\iota^{d_1'}$ for $\iota \in [2, \ell]$, and

---

[9] Note that, since $\mathbf{s}'$ does not contain $s_0^{(i)}$, it is crucial that we use Corollary 1 where the linear combination relies only on $\tilde{\mathbf{s}}^{(i)} = (s_1^{(i)}, \ldots, s_{w_{1,i}}^{(i)})$.

[10] That is, the i-th block of a vector $\mathbf{h} \in \mathbb{Z}_N^{1 \times d_1'}$ is $\mathbf{h}[\ell + (d_1 - 1)(i - 1) + 1, \ell + (d_1 - 1)i]$.

$$\mathbf{B}_j' = \left(\begin{array}{ccc} \mathbf{e}_j^{(1)}\mathbf{A}_{1,1} & \cdots & \mathbf{e}_j^{(m)}\mathbf{A}_{m,1} \\ \vdots & & \vdots \\ \mathbf{e}_j^{(1)}\mathbf{A}_{1,\ell} & \cdots & \mathbf{e}_j^{(m)}\mathbf{A}_{m,\ell} \\ \hline \tilde{\mathbf{B}}_j^{(1)} & & \\ & \tilde{\mathbf{B}}_j^{(2)} & \\ & & \ddots \\ & & \tilde{\mathbf{B}}_j^{(m)} \end{array} \right) \in \mathbb{Z}_N^{d_1' \times d_2'}, \tag{13}$$

$$\mathbf{s}_t'^{(i)} = (0,\ldots,0,\ \overset{\text{block }i}{\underset{\downarrow}{\mathbf{s}_t^{(i)}}}\ ,0,\ldots,0) \qquad\qquad \in \mathbb{Z}_N^{1\times d_2'},$$

$$\hat{\mathbf{s}}_z'^{(i)} = \big(\ \overset{\text{block }i}{\underset{\downarrow}{\hat{\mathbf{s}}_z^{(i)}[1]\mathbf{A}_{i:}}},\ 0,\ldots,0,\ \hat{\mathbf{s}}_z^{(i)}[2,d_1],\ 0,\ldots,0\ \big) \in \mathbb{Z}_N^{1\times d_1'}, \tag{14}$$

for $j \in [2,n]$, $i \in [m]$, $t \in [w_{1,i}]$, $z \in [w_{2,i}]$. Output

$$\Big((\mathbf{B}_j')_{j\in[n]};\ \mathbf{s}_{\text{new}},\big(\mathbf{s}_1'^{(i)},\ldots,\mathbf{s}_{w_{1,i}}'^{(i)}\big)_{i\in[m]};\ \mathbf{v}_2',\ldots,\mathbf{v}_\ell',\big(\hat{\mathbf{s}}_1'^{(i)},\ldots,\hat{\mathbf{s}}_{w_{2,i}}'^{(i)}\big)_{i\in[m]}\Big).$$

$\boxed{\mathsf{EncR}'(x,(\mathbf{A},\pi))}$: First note that we have the condition $\bar{P}_\kappa(x,(\mathbf{A},\pi)) = 0$. Let $S = \{\, i \in [m] \mid P_\kappa(x,\pi(i)) = 1 \,\}$.

1. From $\bar{P}_\kappa(x,(\mathbf{A},\pi)) = 0$ and from Proposition 1, we can obtain a vector $\boldsymbol{\omega} = (\omega_1,\ldots,\omega_\ell) \in \mathbb{Z}_N^{1\times\ell}$ such that $\omega_1 = 1$ and $\mathbf{A}_{i:}\boldsymbol{\omega}^\top = 0$ for all $i \in S$.
2. For each $i \notin S$, we can run $\mathsf{EncR}(x,\pi(i)) \to \big(\mathbf{r}_1^{(i)},\ldots,\mathbf{r}_{m_1}^{(i)};\ \mathbf{a},\hat{\mathbf{r}}_1^{(i)},\ldots,\hat{\mathbf{r}}_{m_2}^{(i)}\big)$, where $\mathbf{r}_v^{(i)} \in \mathbb{Z}_N^{1\times d_1}$, $\hat{\mathbf{r}}_u^{(i)} \in \mathbb{Z}_N^{1\times d_2}$, and $\mathbf{a} = \mathbf{1}_1^{d_2} \in \mathbb{Z}_N^{1\times d_2}$.
3. For $i \in [m]$, let $g_i = \mathbf{A}_{i:}\boldsymbol{\omega}^\top/\mathbf{r}_v^{(i)}[1]$. Note that $\mathbf{r}_v^{(i)}[1] \neq 0$ due to admissibility.
4. Let $\mathbf{a}_{\text{new}} = \mathbf{1}_1^{d_2'}$, and for $v \in [m_1]$, $u \in [m_2]$ let

$$\mathbf{r}_v' = -\Big(\boldsymbol{\omega},\ g_1\mathbf{r}_v^{(1)}[2,d_1],\ldots,g_m\mathbf{r}_v^{(m)}[2,d_1]\Big) \in \mathbb{Z}_N^{1\times d_1'}, \tag{15}$$

$$\hat{\mathbf{r}}_u' = -(g_1\hat{\mathbf{r}}_u^{(1)},\ldots,g_m\hat{\mathbf{r}}_u^{(m)}) \qquad\qquad \in \mathbb{Z}_N^{1\times d_2'}. \tag{16}$$

5. Output $(\mathbf{r}_1',\ldots,\mathbf{r}_{m_1}';\ \mathbf{a}_{\text{new}},\hat{\mathbf{r}}_1',\ldots,\hat{\mathbf{r}}_{m_2}')$.

Verifying Properties (sketch). Properties (P1), (P3)–(P6) are straightforward. Due to limited space, we provide a sketch in verifying (P2)—zero evaluation of substituted polynomials—here, and defer the full details to the full version.

In ct-enc $\mathbf{c}'$, the p-th polynomial in $\mathbf{c}'^{(i)}$ is

$$c_p'^{(i)} = \sum_{z \in [w_{2,i}]} \eta_{p,z}^{(i)} \hat{s}_z'^{(i)} + \eta_{p,0,1}^{(i)} (\mathbf{A}_{i,1} b_1 s_{\text{new}} + \sum_{\iota=2}^{\ell} \mathbf{A}_{i,\iota} v_\iota) + \sum_{\substack{t \in [w_{1,i}] \\ j \in [2,n]}} \eta_{p,t,j}^{(i)} b_j s_t'^{(i)}.$$

(17)

Substituting $\hat{s}_z'^{(i)} : (\hat{\mathbf{s}}_z'^{(i)})^\top$, $b_1 s_{\text{new}} : \mathbf{B}_1'(\mathbf{s}_{\text{new}})^\top$, $v_\iota : (\mathbf{v}_\iota')^\top$, $b_j s_t'^{(i)} : \mathbf{B}_j'(\mathbf{s}_t'^{(i)})^\top$, into $c_p'^{(i)}$ will result in a column vector of length $d_1' = \ell + m(d_1 - 1)$. We denote it as $\mathbf{w}^\top$. We claim that $\mathbf{w}^\top = 0$. We use the symbolic property of the base PES, Γ, which ensures that the substitution of $c_p^{(i)}$ via $\mathsf{EncBS}(\pi(i))$, denoted $\mathbf{u}^\top$, evaluates to 0. In fact, via elementary linear algebra, one can verify that for $j \in [\ell]$, $\mathbf{w}[j]$ is $\mathbf{u}[1]$ scaled by $\mathbf{A}_{i,j}$, and that the i-th block of $\mathbf{w}$ is exactly $\mathbf{u}[2, d_1]$, while the rest of $\mathbf{w}$ is already 0 by construction. Hence the claim holds.

In key-enc $\mathbf{k}$, the substitution for k_1 is straightforward. For $p \in [2, m_3]$, we have $k_p = \sum_{u \in [m_2]} \phi_{p,u} \hat{r}_u + \sum_{v \in [m_1], j \in [2,n]} \phi_{p,v,j} r_v b_j$. Substituting $\hat{r}_u : \hat{\mathbf{r}}_u'$, $r_v b_j : \mathbf{r}_v' \mathbf{B}_j'$ into k_p will result in a row vector of length $d_2' = m d_2$. We denote it as $\mathbf{w}$. We claim that $\mathbf{w} = 0$. Let $\mathbf{u}_i$ be the substitution result for k_p via $\mathsf{EncR}(x, \pi(i))$. One can eventually verify that the i-th block of $\mathbf{w}$ is $g_i \mathbf{u}_i$, which evaluates to 0 since, if $i \in S$ we have $g_i = 0$, while if $i \notin S$ we have $\mathbf{u}_i = 0$ due to the symbolic property of the base PES. Hence the claim holds.

Co-selective Symbolic Property. Let $\mathsf{EncBR}'(x) = \mathsf{EncBR}(x)$.

$\boxed{\mathsf{EncS}'(x, (\mathbf{A}, \pi))}$: First note that we have the condition $\bar{P}_\kappa(x, (\mathbf{A}, \pi)) = 0$. Let $S = \{ i \in [m] \mid P_\kappa(x, \pi(i)) = 1 \}$.

1. For each $i \notin S$, we have $P_\kappa(x, \pi(i)) = 0$. Thus, we can run $\mathsf{EncS}(x, \pi(i)) \to \left(\mathbf{s}_0^{(i)}, \ldots, \mathbf{s}_{w_{1,i}}^{(i)}; \hat{\mathbf{s}}_1^{(i)}, \ldots, \hat{\mathbf{s}}_{w_{2,i}}^{(i)} \right)$, where $\mathbf{s}_t^{(i)} \in \mathbb{Z}_N^{1 \times d_2}$, and $\hat{\mathbf{s}}_z^{(i)} \in \mathbb{Z}_N^{1 \times d_1}$.

2. From $\bar{P}_\kappa(x, (\mathbf{A}, \pi)) = 0$ and Proposition 1, we can obtain a vector $\boldsymbol{\omega} = (\omega_1, \ldots, \omega_\ell)$ such that $\omega_1 = 1$ and $\mathbf{A}_{i:} \boldsymbol{\omega}^\top = 0$ for all $i \in S$. Let $q_i = \mathbf{A}_{i:} \boldsymbol{\omega}^\top$.

3. Let $\mathbf{s}_{\text{new}} = \mathbf{1}_1^{d_2}$, $\mathbf{s}_t'^{(i)} = q_i \mathbf{s}_t^{(i)}$, $\hat{\mathbf{s}}_z'^{(i)} = q_i \hat{\mathbf{s}}_z^{(i)}$, and $\mathbf{v}_\iota' = \omega_\iota \mathbf{1}_1^{d_1}$, for $i \in [m]$, $t \in [w_{1,i}]$, $\iota \in [2, \ell]$, $z \in [w_{2,i}]$.

4. Output $\left(\mathbf{s}_{\text{new}}, \left(\mathbf{s}_1'^{(i)}, \ldots, \mathbf{s}_{w_{1,i}}'^{(i)} \right)_{i \in [m]} ; \mathbf{v}_2', \ldots, \mathbf{v}_\ell', \left(\hat{\mathbf{s}}_1'^{(i)}, \ldots, \hat{\mathbf{s}}_{w_{2,i}}'^{(i)} \right)_{i \in [m]} \right)$.

Verifying Properties. First we can verify that $\mathbf{a}_{\text{new}} \mathbf{s}_{\text{new}}^\top = \mathbf{1}_1^{d_2} (\mathbf{1}_1^{d_2})^\top = 1 \neq 0$, as required. Next, since we define $\mathsf{EncBR}'(x) = \mathsf{EncBR}(x)$, the substitution for key-enc is trivially evaluated to 0, due to the co-selective symbolic property of Γ. It remains to consider the substitution for ct-enc $\mathbf{c}'$. For $i \in [m]$, $p \in [w_{3,i}]$, the polynomial $c_p^{(i)}$ is depicted in Eq. (17). We have that the middle sum term $\mathbf{A}_{i:} \mathbf{v}^\top$ is substituted and evaluated to $q_i (\mathbf{1}_1^{d_2})^\top$. Let $\mathbf{u}_i^\top \in \mathbb{Z}_N^{d_1 \times 1}$ denote the substitution result for $c_p^{(i)}$ (as a part of $\mathbf{c}^{(i)}$) via $\mathsf{EncS}(x, \pi(i))$ (and $\mathsf{EncBR}(x)$). By our constructions of $\mathbf{s}_t'^{(i)}$ and $\hat{\mathbf{s}}_z'^{(i)}$, it is straightforward to see that the substitution for $c_p'^{(i)}$ (as a part of $\mathbf{c}'^{(i)}$) via $\mathsf{EncS}'(x, (\mathbf{A}, \pi))$ (and $\mathsf{EncBR}'(x)$)

is indeed $q_i\mathbf{u}_i^\top$. Note that $\mathbf{u}_i^\top$ contains $\mathbf{B}_1\mathbf{s}_0^\top = \mathbf{1}_1^{d_2}$: this corresponds to the substitution of $\mathbf{A}_{i:}\mathbf{v}^\top$. Finally, we can see that $q_i\mathbf{u}_i^\top = 0$ since if $i \in S$ then $q_i = 0$, while if $i \notin S$, we have $\mathbf{u}_i^\top = 0$ due to the co-selective property of Γ. $\square$

Intuition. Due to an abstract manner of our scheme, it might be useful to relate the above *selective* proof to the idea described in Sect. 2. Intuitively, the upper part of $\mathbf{B}_j'$ of Eq. (13) acts as a "projection", generalizing $\mathbf{B}_j'$ of Eq. (5) in Sect. 2, but now we also embed the policy $\mathbf{A}$ in a novel way. Consider the multiplication $\mathbf{r}_v'\mathbf{B}_j'$. Here, only "non-problematic" blocks (the i-th block where $i \notin S$) are turned "on" by $\boldsymbol{\omega}$ from $\mathbf{r}_v'$. All "problematic" blocks ($i \in S$) are turned "off" by the "mask" vector $(\mathbf{A}_{1:}\boldsymbol{\omega}^\top, \ldots, \mathbf{A}_{m:}\boldsymbol{\omega}^\top)$. We also note that this "mask" vector encodes the non-acceptance condition as per Proposition 1. All in all, this gives us the relation: $\mathbf{r}_v'\mathbf{B}_j' = -\big(g_1\mathbf{r}_v^{(1)}\mathbf{B}_j^{(1)}, \ldots, g_m\mathbf{r}_v^{(m)}\mathbf{B}_j^{(m)}\big)$, where we recover the substitution vectors of the base PES, namely, $\mathbf{r}_v^{(i)}\mathbf{B}_j^{(i)}$, and thus can use the base symbolic property. We succeed in doing so despite having the "projection" part, which seems to hinder the independency among blocks in the first place.

6 Key-Policy Augmentation

For a predicate family P, we define its key-policy-span-program-augmented predicate—denoted as $\mathsf{KP1}[P]$—as the dual of $\mathsf{CP1}[P']$ where P' is the dual of P. Therefore, we can use the dual conversion [2,10]—applying two times–sandwiching CP1-Trans, to obtain a PES conversion for $\mathsf{KP1}[P]$. However, this would incur additional elements for encodings (from dual conversions). Below, we provide a direct conversion without additional elements.

Construction 3. Let Γ be a PES construction for a P satisfying admissibility. We construct a PES Γ' for $\mathsf{KP1}[P]$ as follows. Denote this Γ' by KP1-Trans(Γ).

- Param$'$(par) = Param(par) = n. Denote $\mathbf{b} = (b_1, \ldots, b_n)$.
- EncCt$'$(y, N) = EncCt(y, N) = $\mathbf{c}(\mathbf{s}, \hat{\mathbf{s}}, \mathbf{b})$.
- EncKey$'$(($\mathbf{A}, \pi$), N). Parse $\mathbf{A} \in \mathbb{Z}_N^{m \times \ell}$. Let $\mathbf{v} := (\alpha_{\mathrm{new}}, v_2, \ldots, v_\ell)$ be new lone variables. For all $i \in [m]$, do as follows.
  - Run EncKey($\pi(i), N$) to obtain a vector $\mathbf{k}^{(i)} = \mathbf{k}^{(i)}(\mathbf{r}^{(i)}, \hat{\mathbf{r}}^{(i)}, \mathbf{b})$ of polynomials in variables $\mathbf{r}^{(i)} = (r_1^{(i)}, \ldots, r_{m_{1,i}}^{(i)})$, $\hat{\mathbf{r}}^{(i)} = (\alpha^{(i)}, \hat{r}_1^{(i)}, \ldots, \hat{r}_{m_{2,i}}^{(i)})$, $\mathbf{b}$.
  - Define a modified vector by variable replacement as

$$\mathbf{k}'^{(i)} := \mathbf{k}^{(i)}\big|_{\alpha^{(i)} \mapsto \mathbf{A}_{i:}\mathbf{v}^\top}.$$

  In fact, this only modifies $k_1^{(i)} = \alpha^{(i)} + r_1^{(i)}b_1$ to $k_1'^{(i)} = \mathbf{A}_{i:}\mathbf{v}^\top + r_1^{(i)}b_1$. Finally, output $\mathbf{k}' = \mathbf{k}'(\mathbf{r}', \hat{\mathbf{r}}', \mathbf{b})$ as $\mathbf{k}' := \big(\mathbf{k}'^{(i)}\big)_{i \in [m]}$. It contains variables $\mathbf{r}' := (\mathbf{r}^{(i)})_{i \in [m]}$, $\hat{\mathbf{r}}' := (\alpha_{\mathrm{new}}, v_2, \ldots, v_\ell, (\hat{\mathbf{r}}^{(i)})_{i \in [m]})$, and $\mathbf{b}$.

Pair/Correctness. For proving correctness, we suppose $\bar{P}_\kappa((\mathbf{A}, \pi), y) = 1$. Let $S := \{\, i \in [m] \mid P_\kappa(\pi(i), y) = 1 \,\}$. For $i \in S$, we can run Pair($\pi(i), y, N$) $\to (\mathbf{E}, \overline{\mathbf{E}})$

and obtain a linear combination $\mathbf{s}\mathbf{E}(\mathbf{k}'^{(i)})^\top + \mathbf{c}\overline{\mathbf{E}}(\mathbf{r}^{(i)})^\top = \alpha^{(i)}s_0 = \mathbf{A}_{i:}\mathbf{v}^\top s_0$.
Now since $1_1^\ell \in \mathrm{span}(\mathbf{A}|_y)$, we have linear combination coefficients $\{ t_i \}_{i \in S}$ such
that $\sum_{i \in S} t_i \mathbf{A}_{i:} = 1_1^\ell$. Therefore, the above terms can be linearly combined to
$\sum_{i \in S} t_i (\mathbf{A}_{i:}\mathbf{v}^\top)s_0 = \alpha_{\mathrm{new}}s_0$, as required.

Theorem 2. *Suppose a PES Γ for P is (d_1, d_2)-admissible. Then, the the PES*
KP1-Trans(Γ) *for* KP1$[P]$ *satisfies* $(md_1, m'd_2)$-Sym-Prop$^+$, *where $m \times \ell$ is the*
size of policy and $m' = \max\{m, \ell\}$.

The proof is analogous to CP1-Trans, and is deferred to the full version. Note
that, unlike CP1-Trans, KP1-Trans does not preserve admissibility, by construction.

7   Direct Sum and Augmentation over Predicate Set

In this section, we explore policy augmentations over a *set* of predicate families.
We will also introduce the *direct sum* predicate as an intermediate notion, which
is of an independent interest in its own right.

Notation. Throughout this section, let $\mathcal{P} = \{P^{(1)}, \ldots, P^{(k)}\}$ be a set of pred-
icate families. Each family $P^{(j)} = \{P_{\kappa_j}^{(j)}\}_{\kappa_j}$ is indexed by $\kappa_j = (N, \mathrm{par}_j)$.
The domain for each predicate is specified by $P_{\kappa_j}^{(j)} : \mathcal{X}_{\kappa_j}^{(j)} \times \mathcal{Y}_{\kappa_j}^{(j)} \to \{0, 1\}$.
Unless specified otherwise, we define the combined index as $\kappa = (N, \mathrm{par}) = (N, (\mathrm{par}_1, \ldots, \mathrm{par}_k))$. Let $\mathbb{X}_\kappa := \bigcup_{i \in [k]}(\{i\} \times \mathcal{X}_{\kappa_i}^{(i)})$ and $\mathbb{Y}_\kappa := \bigcup_{i \in [k]}(\{i\} \times \mathcal{Y}_{\kappa_i}^{(i)})$.

Definition 6. We define the *key-policy-span-program-augmented predicate over*
set $\mathcal{P}$ as KP$[\mathcal{P}] = \{ \bar{P}_\kappa \}_\kappa$ where $\bar{P}_\kappa : \bar{\mathbb{X}}_\kappa \times \bar{\mathbb{Y}}_\kappa \to \{0, 1\}$ by letting

- $\bar{\mathbb{X}}_\kappa = \{ (\mathbf{A}, \pi) \mid \mathbf{A} \in \mathbb{M}(\mathbb{Z}_N), \ \pi : [m] \to \mathbb{X}_\kappa \}$.
- $\bar{\mathbb{Y}}_\kappa = 2^{\mathbb{Y}_\kappa}$.
- $\bar{P}_\kappa((\mathbf{A}, \pi), Y) = 1 \iff 1_1^\ell \in \mathrm{span}(\mathbf{A}|_Y)$, where[11]

$$\mathbf{A}|_Y := \left\{ \mathbf{A}_{i:} \ \middle| \ \exists (\pi_1(i), y) \in Y \text{ s.t. } P^{(\pi_1(i))}(\pi_2(i), y) = 1 \right\}.$$

where $\pi(i) = (\pi_1(i), \pi_2(i)) \in \mathbb{X}_\kappa$, and $m \times \ell$ is the size of the matrix $\mathbf{A}$.   ◇

Remark 1. When $\mathcal{P}$ has one element, say $\mathcal{P} = \{P\}$, we abuse the notation and
write KP$[P] := $ KP$[\{P\}]$. Note that KP$[P]$ is still more powerful than KP1$[P]$,
defined in Sect. 6, as it allows a ciphertext attribute to be a set.

Unbounded/Dynamic/Static/OR/AND. We consider (confined) variants
of the predicate KP$[\mathcal{P}]$ as follows. We will confine the domain of $(\mathbf{A}, \pi_1)$, which
specifies a policy over predicates. Their full domain, inferred from Definition 6, is
$D := \bigcup_{m \in \mathbb{N}} \mathbb{M}_m(\mathbb{Z}_N) \times F_{m,k}$, where $F_{m,k}$ denotes the set of all functions that map
$[m]$ to $[k]$. For a class $C \subseteq D$, the predicate KP$[\mathcal{P}]$ with the domain of $(\mathbf{A}, \pi_1)$

---

[11] In the bracket, we write $P^{(\pi_1(i))}$ instead of $P_{\kappa_{\pi_1(i)}}^{(\pi_1(i))}$ for simplicity.

being confined to C is denoted by $\mathsf{KP}_C[\mathcal{P}]$ and is also called *dynamic span-program composition with class C*. It is called *unbounded* if $C = D$. It is called *static* if $|C| = 1$. We denote $\mathsf{KP}_{\mathsf{OR}}[\mathcal{P}]$ as the shorthand for $\mathsf{KP}_C[\mathcal{P}]$ where $C = \bigcup_{m \in \mathbb{N}} \{\mathbf{A}_{\mathsf{OR},m}\} \times F_{m,k}$, and call it the *key-OR-policy-augmented* predicate over $\mathcal{P}$. (Recall that $\mathbf{A}_{\mathsf{OR},m}$ is the matrix for the OR policy, see Sect. 3.4.) Analogous notations go for the cases of $\mathsf{KP1}_{\mathsf{OR}}$, $\mathsf{KP}_{\mathsf{AND}}$, $\mathsf{CP}_{\mathsf{OR}}$, and so on.

Definition 7. We define the predicate called the *direct sum of $\mathcal{P}$* as $\mathsf{DS}[\mathcal{P}] = \{ \bar{P}_\kappa \}_\kappa$ where we let the predicate be $\bar{P}_\kappa : \mathbb{X}_\kappa \times \mathbb{Y}_\kappa \to \{0,1\}$ with

$$\bar{P}_\kappa\big((i,x),\,(j,y)\big) = 1 \quad \Longleftrightarrow \quad (i = j) \,\wedge\, \big(P^{(j)}_{\kappa_j}(x,y) = 1\big).$$

For notational convenience, we also denote it as $P^{(1)} \odot \cdots \odot P^{(k)} = \mathsf{DS}[\mathcal{P}]$. $\Diamond$

We are now ready to state a lemma for constructing $\mathsf{KP}[\mathcal{P}]$. The implication is quite straightforward from definitions. We defer the proof to the full version.

Lemma 2. $\mathsf{KP}[\mathcal{P}]$ *can be embedded into* $\mathsf{KP1}[\mathsf{CP1}_{\mathsf{OR}}[\mathsf{DS}[\mathcal{P}]]]$.

Constructing PES for $\mathsf{KP}[\mathcal{P}]$. Now, since $\mathsf{DS}[\mathcal{P}]$ is a *single* predicate family (rather than a *set* of them), we can apply the CP1-Trans and KP1-Trans to a PES for $\mathsf{DS}[\mathcal{P}]$ to obtain a PES for $\mathsf{KP}[\mathcal{P}]$. Note that we apply Layer-Trans for admissibility if necessary.

Constructing PES for Direct Sum. In the next two subsections, we provide two constructions of PESs for direct sum of a set $\mathcal{P}$ of predicate families. The first is a simpler one that simply "concatenates" all the base PESs for each predicate family in $\mathcal{P}$. The second is superior as the same parameter variables $\mathbf{b}$ can be "reused" for all predicate families in $\mathcal{P}$.

7.1 Simple Direct Sum by Parameter Concatenation

Construction 4. Let $\Gamma^{(j)}$ be a PES for $P^{(j)}$. Also let $\Gamma = (\Gamma^{(1)}, \dots, \Gamma^{(k)})$. We construct a PES Γ' for $\mathsf{DS}[\mathcal{P}]$, where $\mathcal{P} = \{P^{(1)}, \dots, P^{(k)}\}$, as follows. For further use, we denote this Γ' by $\mathsf{Concat\text{-}Trans}(\Gamma)$.

- Param$'$(par). For $j \in [k]$, run Param$^{(j)}$(par$_j$) to obtain n_j. Denote $\mathbf{b}^{(j)} = (b_1^{(j)}, \dots, b_{n_j}^{(j)})$. Output $n = n_1 + \dots + n_k$. Denote $\mathbf{b}' = (\mathbf{b}^{(1)}, \dots, \mathbf{b}^{(k)})$.
- EncCt$'((j,y), N)$. Run EncCt$^{(j)}(y, N) \to \mathbf{c} = \mathbf{c}(\mathbf{s}, \hat{\mathbf{s}}, \mathbf{b}^{(j)})$ and output $\mathbf{c}$.
- EncKey$'((i,x), N)$. Run EncKey$^{(i)}(x, N) \to \mathbf{k} = \mathbf{k}(\mathbf{r}, \hat{\mathbf{r}}, \mathbf{b}^{(i)})$ and output $\mathbf{k}$.

Pair/Correctness. This is straightforward from the base schemes. More precisely, for proving correctness, we suppose $\bar{P}_\kappa\big((i,x),(j,y)\big) = 1$. That is, $i = j$ and $P^{(j)}_{\kappa_j}(x,y) = 1$. Hence, we can run Pair$^{(j)}(x,y,N) \to (\mathbf{E}, \bar{\mathbf{E}})$ and obtain a linear combination $\mathbf{s}\mathbf{E}\mathbf{k}^\top + \mathbf{c}\bar{\mathbf{E}}\mathbf{r}^\top = \alpha s_0$, as required.

To prove symbolic security of Concat-Trans(Γ), we use one more intermediate constraint for the underlying PESs, called Sym-Prop^{++}, which, in turn, can be converted from PES with normal Sym-Prop via Plus-Trans. We defer these proofs to the full version. Below, we let $\bot$ be a special symbol which is not in $\mathcal{Y}_\kappa$, $\mathcal{X}_\kappa$, and abuse notation by letting any predicate evaluate to 0 if at least one input is the symbol $\bot$.

Definition 8. A PES Γ for predicate family P satisfies (d_1, d_2)-Sym-Prop^{++} if it satisfies (d_1, d_2)-Sym-Prop$^+$ with the following further requirement.

(P7) In the selective symbolic property definition, the zero evaluation property of key-enc (P2) also holds for EncB($\bot$), EncR($x, \bot$) for all $x \in \mathcal{X}_\kappa$.   ◇

Lemma 3. *Suppose that, for all $j \in [k]$, the PES $\Gamma^{(j)}$ for predicate family $P^{(j)}$ satisfies (d_1, d_2)-Sym-Prop^{++}. Then, the PES Concat-Trans(Γ) for predicate family DS[$\mathcal{P}$], where $\mathcal{P} = \{P^{(1)}, \ldots, P^{(k)}\}$, satisfies (d_1, d_2)-Sym-Prop$^+$.*

7.2 Efficient Direct Sum with Parameter Reuse

Construction 5. Let $\Gamma^{(j)}$ be a PES for $P^{(j)}$. Also let $\Gamma = (\Gamma^{(1)}, \ldots, \Gamma^{(k)})$. We construct a PES Γ' for DS[$\mathcal{P}$], where $\mathcal{P} = \{P^{(1)}, \ldots, P^{(k)}\}$, as follows. We denote this scheme by Reuse-Trans(Γ). The intuition is to use two new parameters g_j, h_j specific to $\Gamma^{(j)}$, where in the proof, their substituted matrices serve as the "switches" that turn on only the j-th scheme, and that is why we can reuse the same based parameters $\mathbf{b}$ (since the others are rendered zero by the switches).

- Param'(par). For $j \in [k]$, run Param$^{(j)}$(par$_j$) to obtain n_j. Let $n = \max_{j \in [k]} n_j$. Output $n' = n + 2k$. Denote $\mathbf{b} = (b_1, \ldots, b_n, g_1, \ldots, g_k, h_1, \ldots, h_k)$. Also denote $\mathbf{b}_j = (b_1, \ldots, b_{n_j})$.
- EncCt'($(j, y), N$). Run EncCt$^{(j)}(y, N) \to \mathbf{c} = \mathbf{c}(\mathbf{s}, \hat{\mathbf{s}}, \mathbf{b}_j)$. Let s_{new} be the new special non-lone variable. Output $\mathbf{c}' = \left(\mathbf{c}, \ g_j s_0 + h_j s_{\text{new}} \right)$.
- EncKey'($(i, x), N$). Run EncKey$^{(i)}(x, N) \to \mathbf{k} = \mathbf{k}(\mathbf{r}, \hat{\mathbf{r}}, \mathbf{b}_i)$. Let r_{new} be a new non-lone variable and α_{new} be the new special lone variable. Let $\tilde{\mathbf{k}}$ be exactly $\mathbf{k}$ but with α being replaced by $r_{\text{new}} g_i$. Output $\mathbf{k}' = \left(\tilde{\mathbf{k}}, \ \alpha_{\text{new}} + r_{\text{new}} h_i \right)$.

Pair/Correctness. Suppose $\bar{P}_\kappa\big((i, x), (j, y)\big) = 1$. Thus, $i = j$ and $P_{\kappa_j}^{(j)}(x, y) = 1$. Hence, we can run Pair$^{(j)}(x, y, N) \to (\mathbf{E}, \overline{\mathbf{E}})$ and obtain a linear combination $\mathbf{sEk}^\top + \mathbf{c}\overline{\mathbf{E}}\mathbf{r}^\top = \alpha s_0 = (r_{\text{new}} g_j) s_0$. Hence, we have the following, as required: $\left(\alpha_{\text{new}} + r_{\text{new}} h_j \right) s_{\text{new}} - r_{\text{new}} \big(g_j s_0 + h_j s_{\text{new}} \big) + (r_{\text{new}} g_j) s_0 = \alpha_{\text{new}} s_{\text{new}}$.

Lemma 4. *Suppose that PES $\Gamma^{(j)}$ for $P^{(j)}$ satisfies (d_1, d_2)-Sym-Prop$^+$, for all $j \in [k]$. Then, the PES Reuse-Trans(Γ) for predicate family DS[$\mathcal{P}$], where $\mathcal{P} = \{P^{(1)}, \ldots, P^{(k)}\}$, satisfies (d_1, d_2)-Sym-Prop$^+$. (The proof is deferred to the full version.)*

8   Predicative Automata

This section presents an augmentation via DFA over predicates. Due to direct sum transformations, it is again sufficient to consider a single predicate variant.

Let $P = \{ P_\kappa \}_\kappa$ where $P_\kappa : \mathfrak{X}_\kappa \times \mathcal{Y}_\kappa \to \{ 0, 1 \}$, be a predicate family. A *Predicative Automata* (PA) over P_κ is a 4-tuple $(Q, \mathcal{T}, q_0, F)$ where Q is the set of states, $\mathcal{T} \subseteq Q \times Q \times \mathfrak{X}_\kappa$ is the transition table, $q_0 \in Q$ is the start state, and $F \subseteq Q$ is the set of accept states. For simplicity and w.l.o.g., we can assume that there is only one accept state, and it has no outgoing transition. An input to such an automata is a sequence $Y = (y_1, \ldots, y_\ell) \in (\mathcal{Y}_\kappa)^*$, where ℓ is unbounded. A predicative automata $M = (Q = \{q_0, \ldots, q_{-1}\}, \mathcal{T}, q_0, q_{-1})$ accepts Y if there exists a sequence of states $(q^{(1)}, \ldots, q^{(\ell)}) \in Q^\ell$ such that for all $i \in [1, \ell]$, it holds that there exists $(q^{(i-1)}, q^{(i)}, x^{(i)}) \in \mathcal{T}$ such that $P_\kappa(x^{(i)}, y_i) = 1$, and that $q^{(0)} = q_0$ and $q^{(\ell)} = q_{-1}$. Following the predicate for deterministic finite automata (DFA) [2,5,35], we will assume *determinism* of such a predicative automata. (So we may call it predicative DFA.) In our context, this is the restriction that for any different transitions with the same outgoing state, namely (q, q', x') and (q, q'', x'') with $q' \neq q''$, we require that for all $y \in \mathcal{Y}_\kappa$, it must be that $P_\kappa(x', y) \neq P_\kappa(x'', y)$. We can observe that if P is the equality predicate (IBE), then the resulting predicative DFA over P is exactly the definition of DFA.

Example. We provide an example of languages. Suppose we have a list of words which are considered BAD. There exists a simple predicative DFA, depicted in Fig. 5, that accepts exactly any sentences that start with a BAD word and contain an even number of the total BAD words. This seems not possible with span programs, since a sentence can be arbitrarily long.

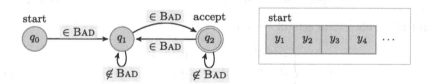

Fig. 5. Predicative DFA for language of sentences that start with a bad word and have an even number of the total bad words. Based predicates for testing membership/non-membership can use IBBE, IBR, defined in Sect. 9.2, respectively.

Definition 9. Let $P = \{ P_\kappa \}_\kappa$ where $P_\kappa : \mathfrak{X}_\kappa \times \mathcal{Y}_\kappa \to \{ 0, 1 \}$, be a predicate family, indexed by $\kappa = (N, \mathsf{par})$. We define the *Key-policy-Automata-augmented predicate* over P as $\mathsf{KA1}[P] = \{ \bar{P}_\kappa \}_\kappa$ where $\bar{P}_\kappa : \bar{\mathfrak{X}}_\kappa \times \bar{\mathcal{Y}}_\kappa \to \{ 0, 1 \}$ by letting

- $\bar{\mathfrak{X}}_\kappa = \{ M \mid M$ is a predicative automata over $P_\kappa \}$.
- $\bar{\mathcal{Y}}_\kappa = (\mathcal{Y}_\kappa)^*$.
- $\bar{P}_\kappa(M, Y) = 1 \iff M$ accepts Y. $\hfill \Diamond$

Intuition. The intuition for constructing PESs for DFA over predicates is similar to that of span program over predicates in that we follow the blueprint of generalizing PESs for X over IBE to X over any predicates, where X is either DFA or span program. Note that this blueprint was explained in Sect. 2 for the case of span programs. Here, for the DFA case, the starting PES is the ABE for regular languages (which can be considered as DFA over IBE) of [5], of which a symbolic proof was given in §B.5 of [2]. In our construction below, one may notice that the structure of PES contains "two copies" of the underlying PES. This feature is inherited from the PES for ABE for regular languages of [5], which already utilizes two copies of IBE encodings.

We note some differences from the case of span programs. For the constructions, while our conversions for span programs use the second approach in Sect. 2 (based on admissible PES), we will base our conversion for DFA instead on the first approach (using the layering technique). This is done for simplicity. For the proofs, we note that span programs and DFAs have completely different combinatorial properties and thus different kinds of substituted matrices. See more discussions below.

Construction 6. Let Γ be a PES construction for P. We construct a PES Γ' for $\mathsf{KA1}[P]$ as follows. For further use, we denote this Γ' by $\mathsf{KA1\text{-}Trans}(\Gamma)$.

– Param$'$(par). If Param(par) returns n, then output $2n + 5$. Denote $\mathbf{b}_1 = (b_{1,1}, \ldots, b_{1,n})$, $\mathbf{b}_2 = (b_{2,1}, \ldots, b_{2,n})$, and $\mathbf{b}' = (\mathbf{b}_1, \mathbf{b}_2, h_0, g_1, h_1, g_2, h_2)$.
– EncCt$'(Y, N)$. Parse $Y = (y_1, \ldots, y_\ell)$. For $i \in [\ell]$, run EncCt(y_i, N) to obtain a vector $\mathbf{c}^{(i)}$ of polynomials. We will use two copies of it, with two different sets of variables, written as:

$$\mathbf{c}^{(1,i)} := \mathbf{c}^{(i)}(\mathbf{s}^{(1,i)}, \hat{\mathbf{s}}^{(1,i)}, \mathbf{b}_1), \qquad \mathbf{c}^{(2,i)} := \mathbf{c}^{(i)}(\mathbf{s}^{(2,i)}, \hat{\mathbf{s}}^{(2,i)}, \mathbf{b}_2),$$

and relate these two sets of variables via:

$$s_0'^{(i)} := \begin{cases} s_0^{(1,i+1)} & \text{if } i = 0 \\ s_0^{(1,i+1)} = s_0^{(2,i)} & \text{if } i = 1, \ldots, \ell - 1 \\ s_0^{(2,i)} & \text{if } i = \ell \end{cases} \tag{18}$$

We then define $c_0' := h_0 s_{\mathrm{new}}^{(0)}$ and, for $i \in [\ell]$,

$$c_i' := h_1 s_{\mathrm{new}}^{(i-1)} + g_1 s_0'^{(i-1)} + h_2 s_{\mathrm{new}}^{(i)} + g_2 s_0'^{(i)},$$

where $s_{\mathrm{new}}^{(0)}, \ldots, s_{\mathrm{new}}^{(\ell)}$ are new non-lone variables with $s_{\mathrm{new}}^{(\ell)}$ being special. Finally, it outputs $\mathbf{c}' := \left(c_0', c_1', \ldots, c_\ell', \left(\mathbf{c}^{(1,i)}, \mathbf{c}^{(2,i)} \right)_{i \in [\ell]} \right)$.
– EncKey$'(M, N)$. Parse $M = (Q, \mathfrak{T}, q_0, q_{-1})$ and parse $\mathfrak{T} = \{ (q_{v_t}, q_{w_t}, x_t) \}_{t \in [m]}$ where each $v_t, \omega_t \in [0, -1]$. [12] Let $u_0, u_1, \ldots, u_{-1}$ be new lone variables with u_{-1} being special. For all $t \in [m]$, run EncKey(x_t, N) to obtain a vector $\mathbf{k}^{(t)}$

---

[12] v_t, ω_t indicate the "from" and the "to" state of the t-th transition in $\mathfrak{T}$, respectively.

of polynomials. We use two copies of it, with two different sets of variables. We then modify them via variable replacement as follows.

$$\mathbf{k}^{(1,t)} := \mathbf{k}^{(t)}(\mathbf{r}^{(1,t)}, \hat{\mathbf{r}}^{(1,t)}, \mathbf{b}_1), \qquad \mathbf{k}^{(2,t)} := \mathbf{k}^{(t)}(\mathbf{r}^{(2,t)}, \hat{\mathbf{r}}^{(2,t)}, \mathbf{b}_2),$$
$$\mathbf{k}'^{(1,t)} := \mathbf{k}^{(1,t)}\big|_{\alpha^{(1,t)} \mapsto r_{\text{new}}^{(t)} g_1}, \qquad \mathbf{k}'^{(2,t)} := \mathbf{k}^{(2,t)}\big|_{\alpha^{(2,t)} \mapsto r_{\text{new}}^{(t)} g_2},$$

where $r_{\text{new}}^{(t)}$ is a new non-lone variable (the same one for both). We then define

$$\tilde{k}_0 := -u_0 + r_{\text{new}}^{(0)} h_0, \qquad \tilde{k}_{1,t} := u_{v_t} + r_{\text{new}}^{(t)} h_1, \qquad \tilde{k}_{2,t} := -u_{\omega_t} + r_{\text{new}}^{(t)} h_2.$$

for $t \in [m]$. Finally, it outputs $\mathbf{k}' := \left(\tilde{k}_0, \left(\tilde{k}_{1,t}, \tilde{k}_{2,t}, \mathbf{k}'^{(1,t)}, \mathbf{k}'^{(2,t)}, \right)_{t \in [m]} \right).$

Pair/Correctness. Suppose $\bar{P}_\kappa(M, Y) = 1$. That is, there exists a sequence $(q^{(1)}, \ldots, q^{(\ell)}) \in Q^\ell$ such that for all $i \in [1, \ell]$, it holds that $P_\kappa(x^{(i)}, y_i) = 1$ and $(q^{(i-1)}, q^{(i)}, x^{(i)}) \in \mathcal{T}$, and that $q^{(0)} = q_0$, while $q^{(\ell)} = q_{-1}$. For $i \in [\ell]$, we proceed as follows. Denote $t_i \in [m]$ as the transition index that corresponds to the i-th move; that is, let $(q_{v_{t_i}}, q_{\omega_{t_i}}, x_{t_i}) = (q^{(i-1)}, q^{(i)}, x^{(i)})$. From this, we have $q_{v_{t_i}} = q_{\omega_{t_{i-1}}}$ for all $i \in [\ell]$. Now since $P_\kappa(x_{t_i}, y_i) = 1$, we can run $\mathsf{Pair}(x_{t_i}, y_i, N)$ to obtain linear combinations that are equal to

$$D_{1,i} := \alpha^{(1,t_i)} s_0^{(1,i)} = \left(r_{\text{new}}^{(t_i)} g_1 \right) s_0'^{(i-1)},$$
$$D_{2,i} := \alpha^{(2,t_i)} s_0^{(2,i)} = \left(r_{\text{new}}^{(t_i)} g_2 \right) s_0'^{(i)}.$$

We have $Q_i := D_{1,i} + D_{2,i} + s_{\text{new}}^{(i-1)} \tilde{k}_{1,t_i} + s_{\text{new}}^{(i)} \tilde{k}_{2,t_i} - c_i' r_{\text{new}}^{(t_i)} = s_{\text{new}}^{(i-1)} u_{\omega_{t_{i-1}}} - s_{\text{new}}^{(i)} u_{\omega_{t_i}}$. Let $Q_0 := s_{\text{new}}^{(0)} \tilde{k}_0 - r_{\text{new}}^{(0)} c_0' = -s_{\text{new}}^{(0)} u_0$. Combining them, we obtain $-\sum_{i=0}^{\ell} Q_i = s_{\text{new}}^{(\ell)} u_{-1}$, as required.

Theorem 3. *Suppose a PES Γ for P satisfies (d_1, d_2)-Sym-Prop^{++}. Then, the the PES $\mathsf{KA1}\text{-}\mathsf{Trans}(\Gamma)$ for $\mathsf{KA1}[P]$ satisfies $(\psi_1 d_1, \psi_2 d_2)$-Sym-Prop$^+$, where $\psi_1 = \max\{\ell + 1, m\}$, $\psi_2 = \max\{\ell + 1, 2m\}$, where ℓ is the size of ciphertext attribute Y, and m is the size of transition table $\mathcal{T}$ for predicative automata M.*

We defer the proof to the full version. At the core, we point out combinatorial vectors that encode the non-acceptance condition of predicative DFA and use them as the "mask" vectors in the proof. Since the combinatorial properties here is richer than the KP1 case, the proof is somewhat more complex.

9 Applications

We provide applications from our framework. Due to limited space, we offer more discussions in the full version, where we also motivate for real-world applications.

9.1   ABE for New Predicates

Predicative Branching Program. This is similar to and might be less power-ful than predicative DFA but may serve an independent interest, since its defini-tion and construction are simpler. A *Predicative Branching Program* (PBP) over a predicate $P_\kappa : \mathcal{X}_\kappa \times \mathcal{Y}_\kappa \to \{0,1\}$ is a 4-tuple $(\Gamma, q_1, q_\sigma, L)$ where $\Gamma = (V, E)$ is a directed acyclic graph (DAG) with a set of nodes $V = \{q_1, \ldots, q_\sigma\}$ and a set of directed edges $E \subseteq V^2$, q_1 is a distinguished terminal node (a node with no outgoing edge) called the accept node, q_σ is the unique start node (the node with no incoming edge), and $L : E \to \mathcal{X}_\kappa$ is an edge labelling function. An input to a PBP $M = (\Gamma, q_1, q_\sigma, L)$ is $y \in \mathcal{Y}_\kappa$. Let Γ_y be an induced subgraph of Γ that con-tains exactly all the edges e such that $P_\kappa(L(e), y) = 1$. Such a PBP M accepts y if Γ_y contains a directed path from the start node, q_σ, to the accept node, q_1. Following the deterministic characteristic of boolean branching programs, we will assume *determinism* of PBP: for any node v, for any two outgoing edges e_1, e_2 from the same node v, we require that $P_\kappa(L(e_1), y) \neq P_\kappa(L(e_2), y)$ for any $y \in \mathcal{Y}_\kappa$. We denote the key-policy-augmented predicate using PBP over $\mathcal{P}$ as KB1[$\mathcal{P}$]. We show that it can be embedded into KP1[$\mathcal{P}$] by using almost the same proof as in the case for the implication ABE for span programs to ABE for BP in [6]. We provide this in the full version.

Nested-Policy/Mixed-Policy ABE. We can define new type of ABE that nests policies. Nested-policy ABE is ABE for predicate CP[KP[$\mathcal{P}$]] or KP[CP[$\mathcal{P}$]], or any arbitrarily hierarchically nested ones. In these schemes, however, the *structure of nesting* is fixed. We define what we call *Mixed-policy ABE* to free up this restriction altogether. It is defined in a recursive manner to make sure that at level ℓ, it includes all the possible nesting structures that have at most ℓ layers. To construct a transformation for this, we observe that a trivial scheme using *parameter concatenation* would be inefficient as when going from level $\ell - 1$ to ℓ, the number of parameters will become at least d times of level $\ell - 1$, where d is the number of transformations plus one (*e.g.*, if we want only KP[·] and CP[·], then $d = 3$). Hence, the overall size at level ℓ would be $O(d^\ell)$. Fortunately, thanks to our construction for direct sum with *parameter reuse*, Reuse-Trans, the parameter size (which will correspond to the public key size for ABE) can be kept small. For ℓ-level scheme, the parameter size is $O(n + k + d\ell)$, where n is the maximum parameter size among k based predicates in $\mathcal{P}$. We explore this in more details in the full version.

9.2   Revisiting Known Predicates

Known Predicates and Modular Constructions. We describe some known predicates and how they are related to more basic predicates via the policy aug-mented predicate notions (*e.g.*, KP1[·], KP[·]). These relations directly suggest what transformations (*e.g.*, KP1-Trans) can be used so as to achieve PES for more expressive predicates from only PESs for basic predicates, namely, IBE and its negation (NIBE), in a modular way. We note that the ciphertext-policy variants can be considered analogously, and can be obtained simply by applying the dual conversion [2,5]. Let $\mathcal{U} = \mathbb{Z}_N$ be the attribute universe.

We consider the following predicates.

- $P^{\mathsf{IBE}} : \mathcal{U} \times \mathcal{U} \to \{0,1\}$ is defined as $P^{\mathsf{IBE}}(x,y) = 1 \Leftrightarrow x = y$.
- $P^{\mathsf{NIBE}} : \mathcal{U} \times \mathcal{U} \to \{0,1\}$ is defined as $P^{\mathsf{NIBE}}(x,y) = 1 \Leftrightarrow x \neq y$.
- $P^{\mathsf{IBBE}} : \mathcal{U} \times 2^{\mathcal{U}} \to \{0,1\}$ is defined as $P^{\mathsf{IBBE}}(x,Y) = 1 \Leftrightarrow x \in Y$.[13]
  - It is clear that P^{IBBE} can be embedded into $\mathsf{CP1_{OR}}[P^{\mathsf{IBE}}]$.
- $P^{\mathsf{IBR}} : \mathcal{U} \times 2^{\mathcal{U}} \to \{0,1\}$ is defined as $P^{\mathsf{IBR}}(x,Y) = 1 \Leftrightarrow x \notin Y$.
  - It is clear that P^{IBR} can be embedded into $\mathsf{CP1_{AND}}[P^{\mathsf{NIBE}}]$.
- $P^{\mathsf{TIBBE}} : (\{1,2\} \times \mathcal{U}) \times 2^{\mathcal{U}} \to \{0,1\}$ is defined as $P^{\mathsf{TIBBE}}((i,x),Y) = 1 \Leftrightarrow (i = 1 \wedge x \in Y) \vee (i = 2 \wedge x \notin Y)$.[14]
  - It is clear that P^{TIBBE} can be embedded into $\mathsf{CP1_{OR}}[P^{\mathsf{IBBE}} \odot P^{\mathsf{IBR}}]$.
- The predicate for completely-unbounded KP-ABE for monotone span program $P^{\mathsf{KP\text{-}MSP}}$ (as defined in [5] and recapped in Sect. 3.4) is the same as $\mathsf{KP1}[P^{\mathsf{IBBE}}]$, or equivalently, $\mathsf{KP}[P^{\mathsf{IBE}}]$.
- The predicate for completely-unbounded KP-ABE for non-monotone span program $P^{\mathsf{KP\text{-}NSP}}$ corresponds to exactly the definition of $\mathsf{KP1}[P^{\mathsf{TIBBE}}]$.

For self-containment, we provide PES constructions for P^{IBE} and P^{NIBE} in the full version.

On ABE for Non-monotone Span Programs. To the best of our knowledge, fully secure completely-unbounded large-universe KP-ABE for non-monotone span program (NSP) had not been achieved before this work. We achieve a scheme in prime-order groups, in a modular and clean manner from simple PESs for P^{IBE} and P^{NIBE}. An explicit description of our PES for it is given in the full version. We have to rely on the q-ratio assumption, inherited from the framework of [2][15]; nevertheless, all the current *completely unbounded* ABE for even *monotone* span programs still also need q-type assumptions [2,5,31], even *selectively* secure one [31]. We provide a comparison to known KP-ABE schemes for NSP in prime-order groups in Table 1. We further discuss why large-universe ABE for NSP is generally a more difficult task to achieve than ABE for MSP in the full version.

For the CP-ABE case, a fully secure completely-unbounded scheme for NSP was recently and independently reported in [39]. Their scheme is constructed in composite-order groups. Our instantiated CP-ABE for NSP is in prime-order groups, and unlike [39] of which proof is complex and specific, ours can be obtained in a modular manner. We defer a comparison table for CP-ABE for NSP to the full version.

On Constant-Size Schemes. One huge further advantage in using the symbolic PES framework of [2] is that any symbolically secure PES can be transformed to constant-size schemes (in ciphertext or key sizes) by bounding

---

[13] IBBE is for ID-based broadcast encryption [19]; IBR is for ID-based revocation [9].

[14] This is a unified notion for IBBE and IBR, and is called two-mode IBBE in [38].

[15] In defense, we also provide a positive remark towards the q-ratio assumption in the full version.

corresponding terms and trading-off with the parameter size (n from Param). In particular, any of our transformed PESs in this paper, *e.g.*, KP[$\mathcal{P}$], can be made constant-size. We include such ABE for NSP in Table 1. More discussions on their detail complexities are in the full version.

Table 1. Summary for KP-ABE for non-monotone span programs with large universe.

Schemes		|PK|	|SK|	|CT|	Unbounded			Security Assumption
					|policy|	/multi- use/	|attrib. set|	
OSW07 [28]	I	$O(T)$	$O(m)$	$O(T)$	✓	✓		selective DBDH
	II	$O(T)$	$O(m\log(T))$	$O(t)$	✓	✓		selective DBDH
OT10 [29]		$O(TR)$	$O(m)$	$O(tR)$	✓			full DLIN
OT12 [30]		$O(1)$	$O(m)$	$O(tR)$	✓		✓	full DLIN
ALP11 [9]		$O(T)$	$O(Tm)$	$O(1)$	✓	✓		selective T-DBDHE†
YAHK14 [38]	I	$O(T)$	$O(Tm)$	$O(1)$	✓	✓		selective T-DBDHE†
	II	$O(T)$	$O(m)$	$O(T)$	✓	✓		selective DBDH
	III	$O(T)$	$O(m\log(T))$	$O(t)$	✓	✓		selective DBDH
	IV	$O(1)$	$O(m)$	$O(t)$	✓	✓	✓	selective t-A†
Our KP-NSP	I	$O(1)$	$O(m)$	$O(t)$	✓	✓	✓	full qratio†
	II	$O(T^2)$	$O(T^3m)$	$O(1)$	✓	✓		full qratio†
	III	$O(M^2 + ML)$	$O(1)$	$O(t(M^3 + M^2L))$	✓	✓		full qratio†

Note: $t = $ |attribute set|, $m \times \ell$ is the span program size, R is the attribute multi-use bound, T, M, L are the maximum bound for t, m, ℓ, respectively (if required). Assumptions with † are q-type assumptions.

Revisiting the Okamoto-Takashima Definition. The Okamoto-Takashima type ABE [29,30] for non-monotone span program was defined differently. We recast it here in our terminology, and explain how to achieve a PES for it in a modular manner in the full version.

Acknowledgement. This work was partially supported by JST CREST Grant No. JPMJCR1688.

References

1. Agrawal, S., Chase, M.: A study of pair encodings: predicate encryption in prime order groups. In: Kushilevitz, E., Malkin, T. (eds.) TCC 2016-A. LNCS, vol. 9563, pp. 259–288. Springer, Heidelberg (2016). https://doi.org/10.1007/978-3-662-49099-0_10

2. Agrawal, S., Chase, M.: Simplifying Design and Analysis of Complex Predicate Encryption Schemes. In: Coron, J.-S., Nielsen, J.B. (eds.) EUROCRYPT 2017. LNCS, vol. 10210, pp. 627–656. Springer, Cham (2017). https://doi.org/10.1007/978-3-319-56620-7_22

3. Ambrona, M., Barthe, G., Schmidt, B.: Generic transformations of predicate encodings: constructions and applications. In: Katz, J., Shacham, H. (eds.) CRYPTO 2017, Part I. LNCS, vol. 10401, pp. 36–66. Springer, Cham (2017). https://doi.org/10.1007/978-3-319-63688-7_2

4. Attrapadung, N., Imai, H.: Dual-policy attribute based encryption. In: Abdalla, M., Pointcheval, D., Fouque, P.-A., Vergnaud, D. (eds.) ACNS 2009. LNCS, vol. 5536, pp. 168–185. Springer, Heidelberg (2009). https://doi.org/10.1007/978-3-642-01957-9_11

5. Attrapadung, N.: Dual system encryption via doubly selective security: framework, fully secure functional encryption for regular languages, and more. In: Nguyen, P.Q., Oswald, E. (eds.) EUROCRYPT 2014. LNCS, vol. 8441, pp. 557–577. Springer, Heidelberg (2014). https://doi.org/10.1007/978-3-642-55220-5_31

6. Attrapadung, N.: Dual system encryption framework in prime-order groups via computational pair encodings. In: Cheon, J.H., Takagi, T. (eds.) ASIACRYPT 2016. LNCS, vol. 10032, pp. 591–623. Springer, Heidelberg (2016). https://doi.org/10.1007/978-3-662-53890-6_20

7. Attrapadung, N., Hanaoka, G., Yamada, S.: Conversions among several classes of predicate encryption and applications to ABE with various compactness tradeoffs. In: Iwata, T., Cheon, J.H. (eds.) ASIACRYPT 2015. LNCS, vol. 9452, pp. 575–601. Springer, Heidelberg (2015). https://doi.org/10.1007/978-3-662-48797-6_24

8. Attrapadung, N., Hanaoka, G., Ogawa, K., Ohtake, G., Watanabe, H., Yamada, S.: Attribute-based encryption for range attributes. In: Zikas, V., De Prisco, R. (eds.) SCN 2016. LNCS, vol. 9841, pp. 42–61. Springer, Cham (2016). https://doi.org/10.1007/978-3-319-44618-9_3

9. Attrapadung, N., Libert, B., de Panafieu, E.: Expressive key-policy attribute-based encryption with constant-size ciphertexts. In: Catalano, D., Fazio, N., Gennaro, R., Nicolosi, A. (eds.) PKC 2011. LNCS, vol. 6571, pp. 90–108. Springer, Heidelberg (2011). https://doi.org/10.1007/978-3-642-19379-8_6

10. Attrapadung, N., Yamada, S.: Duality in ABE: converting attribute based encryption for dual predicate and dual policy via computational encodings. In: Nyberg, K. (ed.) CT-RSA 2015. LNCS, vol. 9048, pp. 87–105. Springer, Cham (2015). https://doi.org/10.1007/978-3-319-16715-2_5

11. Beimel, A.: Secure schemes for secret sharing and key distribution. Ph.D. thesis, Israel Institute of Technology, Technion, Haifa, Israel (1996)

12. Bethencourt, J., Sahai, A., Waters, B.: Ciphertext-policy attribute-based encryption. In: IEEE S&P 2007, pp. 321–334 (2007)

13. Boneh, D., Boyen, X.: Efficient selective-ID secure identity-based encryption without random Oracles. J. Cryptol. 24(4), 659–693 (2011). Extended abstract in Eurocrypt 2004. LNCS, pp. 223–238 (2004)

14. Boneh, D., Hamburg, M.: Generalized identity based and broadcast encryption schemes. In: Pieprzyk, J. (ed.) ASIACRYPT 2008. LNCS, vol. 5350, pp. 455–470. Springer, Heidelberg (2008). https://doi.org/10.1007/978-3-540-89255-7_28

15. Boneh, D., Sahai, A., Waters, B.: Functional encryption: definitions and challenges. In: Ishai, Y. (ed.) TCC 2011. LNCS, vol. 6597, pp. 253–273. Springer, Heidelberg (2011). https://doi.org/10.1007/978-3-642-19571-6_16

16. Chen, J., Gay, R., Wee, H.: Improved dual system ABE in prime-order groups via predicate encodings. In: Oswald, E., Fischlin, M. (eds.) EUROCRYPT 2015. LNCS, vol. 9057, pp. 595–624. Springer, Heidelberg (2015). https://doi.org/10.1007/978-3-662-46803-6_20

17. Chen, J., Wee, H.: Fully, (almost) tightly secure IBE and dual system groups. In: Canetti, R., Garay, J.A. (eds.) CRYPTO 2013. LNCS, vol. 8043, pp. 435–460. Springer, Heidelberg (2013). https://doi.org/10.1007/978-3-642-40084-1_25

18. Chen, J., Gong, J., Kowalczyk, L., Wee, H.: Unbounded ABE via bilinear entropy expansion, revisited. In: Nielsen, J.B., Rijmen, V. (eds.) EUROCRYPT 2018. LNCS, vol. 10820, pp. 503–534. Springer, Cham (2018). https://doi.org/10.1007/978-3-319-78381-9_19

19. Delerablée, C.: Identity-based broadcast encryption with constant size ciphertexts and private keys. In: Kurosawa, K. (ed.) ASIACRYPT 2007. LNCS, vol. 4833, pp. 200–215. Springer, Heidelberg (2007). https://doi.org/10.1007/978-3-540-76900-2_12

20. Gorbunov, S., Vaikuntanathan, V., Wee, H.: Attribute-based encryption for circuits. In: STOC 2013, pp. 545–554 (2013)

21. Goyal, V., Pandey, O., Sahai, A., Waters, B.: Attribute-based encryption for fine-grained access control of encrypted data. In: ACM CCS 2006, pp. 89–98 (2006)

22. Karchmer, M., Wigderson, A.: On span programs. In: Proceedings of the Eighth Annual Structure in Complexity Theory Conference, pp. 102–111. IEEE (1993)

23. Lewko, A., Waters, B.: New techniques for dual system encryption and fully secure HIBE with short ciphertexts. In: Micciancio, D. (ed.) TCC 2010. LNCS, vol. 5978, pp. 455–479. Springer, Heidelberg (2010). https://doi.org/10.1007/978-3-642-11799-2_27

24. Lewko, A., Waters, B.: Decentralizing attribute-based encryption. In: Paterson, K.G. (ed.) EUROCRYPT 2011. LNCS, vol. 6632, pp. 568–588. Springer, Heidelberg (2011). https://doi.org/10.1007/978-3-642-20465-4_31

25. Lewko, A., Waters, B.: Unbounded HIBE and Attribute-Based Encryption. In: Paterson, K.G. (ed.) EUROCRYPT 2011. LNCS, vol. 6632, pp. 547–567. Springer, Heidelberg (2011). https://doi.org/10.1007/978-3-642-20465-4_30

26. Lewko, A., Waters, B.: New proof methods for attribute-based encryption: achieving full security through selective techniques. In: Safavi-Naini, R., Canetti, R. (eds.) CRYPTO 2012. LNCS, vol. 7417, pp. 180–198. Springer, Heidelberg (2012). https://doi.org/10.1007/978-3-642-32009-5_12

27. Lewko, A., Okamoto, T., Sahai, A., Takashima, K., Waters, B.: Fully secure functional encryption: attribute-based encryption and (hierarchical) inner product encryption. In: Gilbert, H. (ed.) EUROCRYPT 2010. LNCS, vol. 6110, pp. 62–91. Springer, Heidelberg (2010). https://doi.org/10.1007/978-3-642-13190-5_4

28. Ostrovsky, R., Sahai, A., Waters, B.: Attribute-based encryption with non-monotonic access structures. In: ACM CCS 2007, pp. 195–203 (2007)

29. Okamoto, T., Takashima, K.: Fully secure functional encryption with general relations from the decisional linear assumption. In: Rabin, T. (ed.) CRYPTO 2010. LNCS, vol. 6223, pp. 191–208. Springer, Heidelberg (2010). https://doi.org/10.1007/978-3-642-14623-7_11

30. Okamoto, T., Takashima, K.: Fully secure unbounded inner-product and attribute-based encryption. In: Wang, X., Sako, K. (eds.) ASIACRYPT 2012. LNCS, vol. 7658, pp. 349–366. Springer, Heidelberg (2012). https://doi.org/10.1007/978-3-642-34961-4_22

31. Rouselakis, Y., Waters, B.: Practical constructions and new proof methods for large universe attribute-based encryption. In: ACM CCS 2013, pp. 463–474 (2013)

32. Sahai, A., Waters, B.: Fuzzy identity-based encryption. In: Cramer, R. (ed.) EUROCRYPT 2005. LNCS, vol. 3494, pp. 457–473. Springer, Heidelberg (2005). https://doi.org/10.1007/11426639_27

33. Waters, B.: Ciphertext-Policy Attribute-Based Encryption: An Expressive, Efficient, and Provably Secure Realization. In: Catalano, D., Fazio, N., Gennaro, R., Nicolosi, A. (eds.) PKC 2011. LNCS, vol. 6571, pp. 53–70. Springer, Heidelberg (2011). https://doi.org/10.1007/978-3-642-19379-8_4

34. Waters, B.: Dual system encryption: realizing fully secure IBE and HIBE under simple assumptions. In: Halevi, S. (ed.) CRYPTO 2009. LNCS, vol. 5677, pp. 619–636. Springer, Heidelberg (2009). https://doi.org/10.1007/978-3-642-03356-8_36
35. Waters, B.: Functional encryption for regular languages. In: Safavi-Naini, R., Canetti, R. (eds.) CRYPTO 2012. LNCS, vol. 7417, pp. 218–235. Springer, Heidelberg (2012). https://doi.org/10.1007/978-3-642-32009-5_14
36. Wee, H.: Dual system encryption via predicate encodings. In: Lindell, Y. (ed.) TCC 2014. LNCS, vol. 8349, pp. 616–637. Springer, Heidelberg (2014). https://doi.org/10.1007/978-3-642-54242-8_26
37. Yamada, K., Attrapadung, N., Emura, K., Hanaoka, G., Tanaka, K.: Generic constructions for fully secure revocable attribute-based encryption. In: Foley, S.N., Gollmann, D., Snekkenes, E. (eds.) ESORICS 2017, Part II. LNCS, vol. 10493, pp. 532–551. Springer, Cham (2017). https://doi.org/10.1007/978-3-319-66399-9_29
38. Yamada, S., Attrapadung, N., Hanaoka, G., Kunihiro, N.: A framework and compact constructions for non-monotonic attribute-based encryption. In: Krawczyk, H. (ed.) PKC 2014. LNCS, vol. 8383, pp. 275–292. Springer, Heidelberg (2014). https://doi.org/10.1007/978-3-642-54631-0_16
39. Yang, D., Wang, B., Ban, X.: Fully secure non-monotonic access structure CP-ABE scheme. In: KSII Transactions on Internet and Information Systems, pp. 1315–1329 (2018)

(R)CCA Secure Updatable Encryption with Integrity Protection

Michael Klooß[1]([✉]), Anja Lehmann[2], and Andy Rupp[1]

[1] Karlsruhe Institute for Technology, Karlsruhe, Germany
{michael.klooss,andy.rupp}@kit.edu
[2] IBM Research – Zurich, Rüschlikon, Switzerland
anj@zurich.ibm.com

Abstract. An updatable encryption scheme allows a data host to update ciphertexts of a client from an old to a new key, given so-called update tokens from the client. Rotation of the encryption key is a common requirement in practice in order to mitigate the impact of key compromises over time. There are two incarnations of updatable encryption: One is ciphertext-*dependent*, i.e. the data owner has to (partially) download all of his data and derive a dedicated token per ciphertext. Everspaugh et al. (CRYPTO'17) proposed CCA and CTXT secure schemes in this setting. The other, more convenient variant is ciphertext-*independent*, i.e., it allows a single token to update *all* ciphertexts. However, so far, the broader functionality of tokens in this setting comes at the price of considerably weaker security: the existing schemes by Boneh et al. (CRYPTO'13) and Lehmann and Tackmann (EUROCRYPT'18) only achieve CPA security and provide no integrity protection. Arguably, when targeting the scenario of outsourcing data to an untrusted host, plaintext integrity should be a minimal security requirement. Otherwise, the data host may alter or inject ciphertexts arbitrarily. Indeed, the schemes from BLMR13 and LT18 suffer from this weakness, and even EPRS17 only provides integrity against adversaries which cannot arbitrarily inject ciphertexts. In this work, we provide the first ciphertext-*independent* updatable encryption schemes with security beyond CPA, in particular providing strong integrity protection. Our constructions and security proofs of updatable encryption schemes are surprisingly modular. We give a generic transformation that allows key-rotation and confidentiality/integrity of the scheme to be treated almost separately, i.e., security of the updatable scheme is derived from simple properties of its static building blocks. An interesting side effect of our generic approach is that it immediately implies the unlinkability of ciphertext updates that was introduced as an essential additional property of updatable encryption by EPRS17 and LT18.

1 Introduction

Updatable encryption was introduced by Boneh et al. [1] as a convenient solution to enable key rotation for symmetric encryption. Rotating secret keys is considered good practice to realize proactive security: Periodically changing the

© International Association for Cryptologic Research 2019
Y. Ishai and V. Rijmen (Eds.): EUROCRYPT 2019, LNCS 11476, pp. 68–99, 2019.
https://doi.org/10.1007/978-3-030-17653-2_3

cryptographic key that is used to protect the data reduces the risk and impact of keys being compromised over time. For instance, key rotation is mandated when storing encrypted credit card data by the PCI DSS standard [21], and several cloud storage providers, such as Google and Amazon, offer data-at-rest encryption with rotatable keys [8].

The challenge with key rotation is how to efficiently update the existing ciphertexts when the underlying secret key is refreshed. The straightforward solution is to decrypt all old ciphertexts and re-encrypt them from scratch using the new key. Clearly, this approach is not practical in the typical cloud storage scenario where data is outsourced to a (potentially untrusted) host, as it would require the full download and upload of all encrypted data.

An *updatable encryption scheme* is a better solution to this problem: it extends a classic symmetric encryption scheme with integrated key rotation and update capabilities. More precisely, these schemes allow to derive a short update token from an old and new key, and provide an additional algorithm that re-encrypts ciphertexts using such a token. A crucial property for updatable encryption is that learning an update token does not impact the confidentiality and also the integrity of the ciphertexts. Thus, the procedure for re-encrypting all existing ciphertexts can be securely outsourced to the data host.

State of the Art. There are two different variants of updatable encryption, depending on whether the update tokens are generated for a specific ciphertext or are ciphertext-independent. The former type – called *ciphertext-dependent* updatable encryption – has been introduced by Boneh et al. [2] and requires the data owner to (partially) download all outsourced ciphertexts, derive a dedicated token for each ciphertext, and return all tokens to the host. Everspaugh et al. [8] provide a systematic treatment for such schemes i.e., defining the desirable security properties and presenting provably secure solutions. Their focus is on *authenticated* encryption schemes, and thus CCA security and ciphertext integrity (CTXT) are required and achieved by their construction.

While ciphertext-dependent schemes allow for fine-grained control of which ciphertexts should be re-encrypted towards the new key, they are clearly far less efficient and convenient for the data owner than *ciphertext-independent* ones. In ciphertext-independent schemes, the update token only depends on the new and old key and allows to re-encrypt *all* ciphertexts. The idea of ciphertext-independent schemes was informally introduced by Boneh et al. [1] and recently Lehmann and Tackmann [17] provided a rigorous treatment of their formal security guarantees. The broader applicability of update tokens in ciphertext-independent schemes is an inherent challenge for achieving strong security properties though: as a single token can be used to update *all* ciphertexts, the corruption of such a token gives the adversary significantly more power than the corruption of a ciphertext-dependent token. As a consequence, the ciphertext-independent schemes proposed so far only achieve CPA security instead of CCA, and did not guarantee any integrity protection [17].

	Encrypt-and-MAC (E&M, Sec. 3)	Naor-Yung (NYUAE, Sec. 4)
Confidentiality	CCA	RCCA
Integrity	ciphertext integrity	plaintext integrity
ReEnc algorithm	deterministic	probabilistic
ReEnc oracle	honestly derived ciphertexts only	arbitrary ciphertexts

Fig. 1. Overview of the core differences of our two main schemes and considered settings.

Updatable Encryption Needs (Stronger) Integrity Protection. Given that updatable encryption targets a cloud-based deployment setting where encrypted data is outsourced to an (untrusted) host, neglecting the integrity protection of the outsourced data is a dangerous shortcoming. For instance, the host might hold encrypted financial or medical data of the data owner. Clearly, a temporary security breach into the host should not allow the adversary to create new and valid ciphertexts that will temper with the owners' records. For the targeted setting of ciphertext-independent schemes no notion of (ciphertext) integrity was proposed so far, and the encryption scheme presented in [17] is extremely vulnerable to such attacks: their symmetric updatable encryption scheme (termed RISE) is built from (public-key) ElGamal encryption, which only uses the public key in the update token. However, a single corruption of the update token will allow the data host to create valid ciphertexts of arbitrary messages of his choice.

For the ciphertext-dependent setting, the scheme by Everspaugh et al. [8] does provide ciphertext-integrity, but only against a weak form of attacks: the security definition for their CTXT notion does not allow the adversary to obtain re-encryptions of *maliciously* formed ciphertexts. That is, the model restricts queries to the re-encryption oracle to honestly generated ciphertexts that the adversary has received from previous (re)encryption oracle queries. Thus, integrity protection is only guaranteed against passive adversaries. Again, given the cloud deployment setting in which updatable encryption is used in, assuming that an adversary that breaks into the host will behave honestly and does not temper with any ciphertexts is a critical assumption.

Our Contributions. In this work we address the aforementioned shortcomings for ciphertext-*independent* updatable encryption and present schemes that provide significantly stronger security than existing solutions. First, we formally define the desirable security properties of (R)CCA security, ciphertext (CTXT) and plaintext integrity (PTXT) for key-evolving encryption schemes. Our definitions allow the adversary to *adaptively* corrupt the secret keys or update tokens of the current and past epochs, as long as it does not empower him to trivially win the respective security game. We then propose two constructions: the first achieves CCA and CTXT security (against passive re-encryption attacks), and the second scheme realizes RCCA and PTXT security against active attacks. Both schemes make use of a generic (proof) strategy that derives the security of the updatable scheme from simple properties of the underlying *static* primitives,

Scheme	Assumption	Ciph. indep.	arbitr. ReEnc	IND	INT	UN- LINK	$\|c\|$	(Re)Enc	Dec
BLMR [2]	DDH (+ ROM)	✗	✗	(?)	✗	(?)	$2\mathbb{G}^*$	$2\mathbb{G}$	$2\mathbb{G}$
EPRS [8]	DDH + ROM	✗	(✗)	CPA	CTXT	✓	$2\mathbb{G}^*$	$2\mathbb{G}$	$2\mathbb{G}$
RISE [17]	DDH	✓	✗	CPA	✗	✓	$2\mathbb{G}$	$2\mathbb{G}$	$2\mathbb{G}$
E&M Sec. 3	DDH + ROM	✓	✗	CCA	CTXT	✓	$3\mathbb{G}$	$3\mathbb{G}$	$3\mathbb{G}$
NYUE Sec. 4	SXDH	✓	✓	RCCA	✗	✓	$(34, 34)$	$(60, 70)$	$22e$
NYUAE Sec. 4	SXDH	✓	✓	RCCA	PTXT	✓	$(58, 44)$	$(110, 90)$	$29e$

Fig. 2. Comparison of ciphertext-independent and -dependent updatable encryption schemes. The second set of columns states the achieved security notions, and whether security against arbitrary (opposed to honest) re-encryption attacks is achieved. For EPRS, security against arbitrary re-encryption attacks is only considered for confidentiality, not for integrity. For BLMR, it was shown that a security proof for confidentiality is unlikely to exist [8], and the formal notion of unlinkability of re-encryptions was only introduced later. The final set of columns states the efficiency in terms of ciphertext size and costs for (re-)encryption and decryption in the number of exponentiations and pairings. Tuples (x, y) specify x (resp. y) elements/exponentiations in $\mathbb{G}_1$ (resp. $\mathbb{G}_2$) in case of underlying pairing groups, and a pairing is denoted by e. (Re)encryption and decryption costs for NYUE and NYUAE are approximate. The ciphertext size is given for messages represented as a single group element (in $\mathbb{G}$ or $\mathbb{G}_1$). BLMR and EPRS support encryption of arbitrary size message with the ciphertext size growing linearly with the message blocks.

which greatly simplifies the design for such updatable encryption schemes. In more detail, our contributions are as follows:

CCA and CTXT Secure Ciphertext-Independent Updatable Encryption. Our first updatable encryption applies the Encrypt-and-MAC (E&M) transformation to primitives which are key-rotatable and achieves CCA and CTXT security. Using Encrypt-*and*-MAC is crucial for the updatability as we need direct access to both the ciphertext and the MAC. In order to use E&M, which is *not* a secure transformation for authenticated encryption in general, we require a one-to-one mapping between message-randomness pairs and ciphertexts as well as the decryption function to be randomness-recoverable. By applying a PRF on both, the message and the encryption randomness, we obtain the desired ciphertext integrity. Interestingly, we only need the underlying encryption and PRF to be secure w.r.t. their standard, static security notions and derive security for the updatable version of E&M from additional properties we introduce for the update token generation.

An essential property of this first scheme is that its re-encryptions are *deterministic*. This enables us to define and realize a meaningful CCA security notion, as the determinism allows the challenger to keep track of re-encryptions of the challenge ciphertext and prevent decryption of such updates. Similar to the CCA-secure (ciphertext-dependent) scheme of [8], we only achieve security against passive re-encryption attacks, i.e., where the re-encryption oracle in the security game can only be queried on honestly generated ciphertexts.

RCCA and PTXT Security Against Malicious Re-encryption Attacks. Our second scheme then provides strong security against active re-encryption attacks. On a high-level, we use the Naor-Yung approach [20] that lifts (public-key) CPA to CCA security by encrypting each message under two public keys and appending a NIZK that both ciphertexts encrypt the same message. The crucial benefit of this approach is that it allows for *public verifiability* of ciphertexts, and thus for any re-encryption it can first be checked that the provided ciphertext is valid — which then limits the power of malicious re-encryption attacks. To lift the approach to an updatable encryption scheme, we rely on the key-rotatable CPA-secure encryption RISE [17] and GS proofs [5,12] that exhibit the malleability necessary for rotating the associated NIZK proof.

A consequence of this approach is that re-encryptions are now probabilistic (as in RISE) and ciphertexts are re-randomizable in general. Therefore, CCA and CTXT are no longer achievable, and we revert to Replayable CCA (RCCA) and plaintext integrity. Informally, RCCA is a relaxed variant of CCA security that ensures confidentiality for all ciphertexts that are not re-randomization of the challenge ciphertext [3]. Plaintext integrity is a weaker notion than ciphertext integrity, as forging ciphertexts is now trivial, but still guarantees that an adversary can not come up with valid ciphertexts for *fresh* messages.

In Fig. 1 we provide an overview of both solutions and their settings, and Fig. 2 gives a compact comparison between our new schemes and the existing ones.

Generic (Proof) Transformation & Unlinkability of Re-encryption. The security models for updatable encryption are quite involved, which in turn makes proving security in these models rather cumbersome [8,17]. A core contribution of our work is a generic transformation that yields a surprisingly simple blueprint for building updatable encryption: We show that it is sufficient to consider the underlying encryption and the key-rotation capabilities (almost) separately. That is, we only require the underlying scheme – provided by the Enc, Dec algorithms in isolation – to satisfy standard security. In addition we need re-encryption to produce ciphertexts that are indistinguishable from fresh encryption and token generation to be *simulatable*. The latter allows us to produce "fake" tokens when we are dealing with a static CCA/RCCA game, and the former is used to answer re-encryption oracle calls in the security game with decrypt-then-encrypt calls. Further, we leverage the fact that all ciphertext-independent schemes so far are *bi-directional*, i.e., ciphertexts can also be downgraded. This property comes in very handy in the security proof as it essentially allows to embed a static-CCA/RCCA challenger in one epoch, and handle queries in all other epochs by rotating ciphertexts back-and-forth to this "challenge" epoch.

The notion of indistinguishability of re-encryptions and fresh encryptions (termed *perfect re-encryption*) that we define also has another very nice side-effect: it implies the property of re-encryption unlinkability as introduced in [8,17]. Both works propose a security notion that guarantees that a re-encrypted ciphertext can no longer be linked to its old version, which captures that the full ciphertext must get refreshed during an update. We adapt this unlinkability notion to the CCA and RCCA setting of our work and show that perfect re-encryption (in combination with CCA resp. CPA security) implies such unlinkability. Both of our schemes satisfy this strong security notion.

Other Related Work. Recently, Jarecki et al. [14] proposed an updatable and CCA secure encryption scheme in the context of an Oblivious Key Management Systems (KMS). The KMS is an external service that hosts the secret key, whereas the data owner stores all ciphertexts and adaptively decrypts them with the help of the KMS. Thus, their setting is considerably different to our notion of updatable encryption where the ciphertexts are outsourced, and the secret is managed by the data owner.

Another primitive that is highly related to updatable encryption is proxy re-encryption (PRE). In a recent work, Fuchsbauer et al. [10] show how to lift selectively secure PRE to adaptive security without suffering from an exponential loss when using straightforward approaches. Their overall idea is similar to our generic transformation, as it also relies on additional properties of the re-encryption procedure that facilitate the embedding of the static challenger. The different overall setting makes their work rather incomparable to ours: we exploit bi-directional behaviour of updates, whereas [10] focuses on uni-directional schemes, and we consider a symmetric key setting whereas the PRE's are public-key primitives. In fact, our security proofs are much tighter (partially due to these differences). We conjecture that our techniques can be applied to obtain adaptive security with polynomial security loss for a class of PRE schemes, cf. [16]. This would improve upon the superpolynomial loss in [10].

Organisation. We start our paper by recalling the necessary standard building blocks and the generic syntax of updatable encryption in Sect. 2. In Sect. 3, we then present our formal definitions for CCA and CTXT secure updatable encryption, tailored to our setting of schemes with deterministic re-encryption and covering passive re-encryption attacks. This section also contains our generic transformation for achieving these notions from the static security of the underlying building blocks, and our Encrypt-and-MAC construction that utilizes this generic approach. In Sect. 4 we then introduce RCCA and PTXT security against active re-encryption attacks and present our Naor-Yung inspired scheme. Since our generic transformation immediately implies the unlinkability property UP-REENC introduced in [8,17] we refer the formal treatment of this notion to [16].

2   Preliminaries

In this section we introduce our notational conventions and all necessary (standard) building blocks along with their security definitions.

2.1   Notation

We denote the security parameter by κ. All schemes and building blocks in this paper make use of some implicit PPT algorithm $pp \leftarrow \mathsf{GenPP}(1^\kappa)$ which on input of the security parameter 1^κ outputs some public parameters pp.

The public parameters, e.g., include a description of the cyclic groups and generators we use. We assume for our security definitions that pp also contains the security parameter. For the sake of simplicity, we omit GenPP in all definitions including security experiments. When composing building blocks as in our Encrypt-*and*-MAC construction, for example, the same GenPP algorithm is assumed for all those building blocks and the output pp is shared between them.

By $\mathbb{G}$ we denote a commutative group and by $(e, \mathbb{G}_1, \mathbb{G}_2, \mathbb{G}_T)$ a pairing group. All groups are of prime order p. The integers modulo p are denoted $\mathbb{F}_p$. We use *additive* notation for groups, in particular the well-established *implicit* representation introduced in [6]. That is, we write [1] for the generator $g \in \mathbb{G}$ and $[x] = xg$ (in multiplicative notation, g^x). For pairing groups, we write $[1]_1$, $[1]_2$ and $[1]_T$ and we require that $e([1]_1, [1]_2) = [1]_T$. We define $\mathbb{G}^\times := \mathbb{G} \setminus \{[0]\}$. By $\mathrm{supp}(X)$ we denote the support of a random variable X, i.e. the set of outcomes with positive probability.

2.2 Symmetric and Tidy Encryption

We use the following definition of a symmetric encryption scheme, where the existence of a system parameter generation algorithm GenSP reflects the fact, that we partially rely on primitives with public parameters (like a Groth-Sahai CRS) for our constructions.

Definition 1. *A symmetric encryption scheme* SKE $=$ (GenSP, GenKey, Enc, Dec) *is defined by the following PPT algorithms*

SKE.GenSP(pp) *returns system parameters sp. We treat sp as implicit inputs for the following algorithms.*
SKE.GenKey(sp) *returns a key k.*
SKE.Enc($k, m; r$) *returns a ciphertext c for message m, key k and randomness r.*
SKE.Dec(k, c) *returns the decryption m of c. ($m = \bot$ indicates failure.)*

We assume that the system parameters fix not only the key space $\mathcal{K}_{sp}$, but also the ciphertext space $\mathcal{C}_{sp}$, message space $\mathcal{M}_{sp}$ and randomness space $\mathcal{R}_{sp}$. Also, we assume that membership in $\mathcal{C}_{sp}$ and $\mathcal{M}_{sp}$ can be efficiently tested.

Tidy Encryption. Our construction of an updatable encryption scheme with deterministic reencryption resorts to tidy encryption. For this purpose, we use the following definition which is a slightly adapted version of the definition in [18].

Definition 2. *A symmetric encryption scheme* SKE *is called* **randomness-recoverable** *if there is an associated efficient deterministic algorithm* RDec(k, c) *such that*

$$\forall k, m, r: \quad \mathrm{RDec}(k, \mathrm{Enc}(k, m; r)) = (m, r). \tag{1}$$

We call a randomness-recoverable SKE *tidy if*

$$\forall k, c: \ \mathsf{RDec}(k, c) = (m, r) \implies \mathsf{Enc}(k, m; r) = c. \qquad (2)$$

In other words, SKE *is tidy if* Enc *and* RDec *are bijections (for a fixed key) between message-randomness pairs and valid ciphertexts (i.e. ciphertexts which do not decrypt to $\bot$).*[1]

Indistinguishability Notions. For our constructions, we consider a number of slight variations of the standard security notions IND-CPA and IND-CCA security.

One such variation is IND-RCCA security [3] which relaxes IND-CCA in the sense that it is not considered an attack if a ciphertext can be transformed into a new ciphertext of the same message. Hence, the RCCA decryption oracle refuses to decrypt any ciphertext containing one of the challenge messages.

Furthermore, we consider CPA, CCA, and RCCA security under key-leakage. Here the adversary is additionally given $\mathsf{leak}(k)$ as input, where leak is some function on the key space. This leakage reflects the fact that in one of our constructions (Sect. 4.2), that actually relies on public-key primitives, the corresponding public keys need to be leaked to the adversary. So we would have $k = (sk, pk)$ and $\mathsf{leak}(k) = pk$ in this case. For the deterministic construction in Sect. 3.2 we do not consider key-leakage, i.e., $\mathsf{leak}(k) = \bot$.

Finally, we can define (stronger) real or random variants (IND\$-CPA/CCA) of the former notions. Here, the adversary provides a single challenge message and the challenger responds with either an encryption of this message or a randomly chosen ciphertext.

Definition 3 compactly formalizes the security notions sketched above.

Definition 3. *Let* SKE *be a secret key encryption scheme. Let* $\mathsf{leak} \colon \mathcal{K} \to \mathcal{L}$ *be a leakage function. We call* SKE *IND-X secure, where* $X \in \{CPA, CCA, RCCA\}$, *under key-leakage* leak, *if for every efficient PPT adversary $\mathcal{A}$, the advantage*

$$\mathsf{Adv}^{ind\text{-}X}_{\mathsf{SKE}, \mathcal{A}}(\kappa) := \left| \Pr[\mathsf{Exp}^{ind\text{-}X}_{\mathsf{SKE}, \mathcal{A}}(\kappa, 0) = 1] - \Pr[\mathsf{Exp}^{ind\text{-}X}_{\mathsf{SKE}, \mathcal{A}}(\kappa, 1) = 1] \right|$$

in the experiment described in Fig. 3 is negligible. Analogous to IND-X, we define IND\$-X security for $X \in \{CPA, CCA\}$, with the experiments also described in Fig. 3. IND\$-X is the strictly stronger notion, i.e., it implies $IND\text{-}X$.

Integrity Notions. We consider both plaintext (PTXT) and ciphertext (CTXT) integrity. In the PTXT experiment, the adversary wins if it is able to output a valid ciphertext for a fresh plaintext, i.e., a ciphertext that decrypts to a plaintext for which it has not queried the encryption oracle before. In the CTXT experiment, in order to win, the adversary just needs to output a valid

---

[1] Since encryption of $\bot$ also yields $\bot$, Eq. 2 trivially holds for invalid ciphertexts.

Experiment $\mathrm{Exp}_{\mathsf{SKE},\mathcal{A}}^{\mathsf{ind\text{-}X}}(\kappa, b)$	Experiment $\mathrm{Exp}_{\mathsf{SKE},\mathcal{A}}^{\mathsf{ind\$\text{-}X}}(\kappa, b)$
$sp \leftarrow \mathsf{GenSP}(pp); \; k \leftarrow \mathsf{GenKey}(sp);$	$sp \leftarrow \mathsf{GenSP}(pp); \; k \leftarrow \mathsf{GenKey}(sp);$
$(m_0^*, m_1^*, state) \leftarrow \mathcal{A}^{\mathsf{Enc},\mathsf{Dec}}(pp, sp, \mathsf{leak}(k));$	$(m^*, state) \leftarrow \mathcal{A}^{\mathsf{Enc},\mathsf{Dec}}(pp, sp, \mathsf{leak}(k));$
abort if $\lvert m_0^* \rvert \neq \lvert m_1^* \rvert$ or $m_0^*, m_1^* \notin \mathcal{M}_{sp}$	abort if $m^* \notin \mathcal{M}_{sp}$
$c^* \leftarrow \mathsf{Enc}(k, m_b^*);$	$c_0^* \leftarrow \mathsf{Enc}(k, m^*); \; c_1^* \leftarrow_{\mathrm{R}} \mathcal{C}; \; c^* := c_b^*$
$\textbf{return } \mathcal{A}^{\mathsf{Enc},\mathsf{Dec}}(state, c^*) \overset{?}{=} b$	$\textbf{return } \mathcal{A}^{\mathsf{Enc},\mathsf{Dec}}(state, c^*) \overset{?}{=} b$

Fig. 3. The encryption oracle $\mathsf{Enc}(m)$ returns $c \leftarrow_{\mathrm{R}} \mathsf{Enc}(k, m)$. The decryption oracle $\mathsf{Dec}(m)$ computes $m \leftarrow_{\mathrm{R}} \mathsf{Dec}(k, c)$ but then behaves differently depending on the notion. For CPA, $\mathsf{Dec}(c)$ always returns $\bot$. For CCA, $\mathsf{Dec}(c)$ returns m except if $c = c^*$, in which case it returns $\bot$. For RCCA, $\mathsf{Dec}(c)$ returns m except if $m \in \{m_0^*, m_1^*\}$ in which case it returns $\mathtt{invalid}$. Note that $\mathtt{invalid} \neq \bot$, i.e. $\mathcal{A}$ learns that (one of) the challenge messages is encrypted in c. Everything else is unchanged.

and fresh ciphertext, i.e., one not resulting from a previous call to the encryption oracle. In both experiments, the adversary is equipped with a decryption oracle instead of an oracle that just tests the validity of ciphertexts. For CTXT, this actually makes no difference. For PTXT, however, there are (pathological) insecure schemes which are only secure w.r.t. validity oracles. Again, we consider variants of these integrity notions under key-leakage. Definition 4 formalizes these notions.

Definition 4. *Let* SKE *be a symmetric encryption scheme. Let* $\mathsf{leak}: \mathcal{K} \to \mathcal{L}$ *be a leakage function. The INT-CTXT as well as the INT-PTXT experiments are defined in Fig. 4. We call* SKE *INT-CTXT secure under (key-)leakage* leak *if the advantage* $\mathsf{Adv}_{\mathsf{SKE},\mathcal{A}}^{int\text{-}ctxt}(\kappa) := \Pr[\mathrm{Exp}_{\mathsf{SKE},\mathcal{A}}^{int\text{-}ctxt}(\kappa) = 1]$ *is negligible. Similarly, we call* SKE *INT-PTXT secure under (key-)leakage* leak *if the advantage* $\mathsf{Adv}_{\mathsf{SKE},\mathcal{A}}^{int\text{-}ptxt}(\kappa) := \Pr[\mathrm{Exp}_{\mathsf{SKE},\mathcal{A}}^{int\text{-}ptxt}(\kappa) = 1]$ *is negligible.*

Experiment $\mathrm{Exp}_{\mathsf{SKE},\mathcal{A}}^{\mathsf{int\text{-}ptxt}}(\kappa)$	Experiment $\mathrm{Exp}_{\mathsf{SKE},\mathcal{A}}^{\mathsf{int\text{-}ctxt}}(\kappa)$
$\mathbf{M} = \emptyset; \; \mathbf{Q} = \emptyset;$	$\mathbf{M} = \emptyset; \; \mathbf{Q} = \emptyset;$
$sp \leftarrow \mathsf{GenSP}(pp); \; k \leftarrow \mathsf{GenKey}(sp);$	$sp \leftarrow \mathsf{GenSP}(pp); \; k \leftarrow \mathsf{GenKey}(sp);$
$c \leftarrow \mathcal{A}^{\mathsf{Enc},\mathsf{Dec}}(pp, sp, \mathsf{leak}(k));$	$c \leftarrow \mathcal{A}^{\mathsf{Enc},\mathsf{Dec}}(pp, sp, \mathsf{leak}(k));$
$m \leftarrow \mathsf{Dec}(k, c);$	$m \leftarrow \mathsf{Dec}(k, c);$
$\textbf{return } 0 \text{ iff } m \in \mathbf{M} \text{ or } m = \bot$	$\textbf{return } 0 \text{ iff } c \in \mathbf{Q} \text{ or } m = \bot$

Fig. 4. The INT-PTXT (left) and INT-CTXT (right) games. The encryption oracle $\mathsf{Enc}(m)$ returns $c \leftarrow \mathsf{Enc}(k, m)$ and adds m to the list of queried messages $\mathbf{M}$ and adds c to the list of queried ciphertexts $\mathbf{Q}$. The oracle $\mathsf{Dec}(c)$ returns $\mathsf{Dec}(k, c)$.

2.3 Updatable Encryption

Roughly, an updatable encryption scheme is a symmetric encryption scheme which offers an additional re-encryption functionality that moves ciphertexts from an old to a new key.

The encryption key evolves with *epochs*, and the data is encrypted with respect to a specific epoch e, starting with $e = 0$. When moving from epoch e to epoch $e + 1$, one first creates a new key k_{e+1} via the UE.GenKey algorithm and then invokes the token generation algorithm UE.GenTok on the old and new key, k_e and k_{e+1}, to obtain the update token Δ_{e+1}. The update token Δ_{e+1} allows to update all previously received ciphertexts from epoch e to $e + 1$ using the re-encryption algorithm UE.ReEnc.

Definition 5 (Updatable Encryption). *An **updatable encryption scheme** UE is a tuple* (GenSP, GenKey, GenTok, Enc, Dec, ReEnc) *of PPT algorithms defined as:*

UE.GenSP(pp) *is given the public parameters and returns some system parameters sp. We treat the system parameters as implicit input to all other algorithms.*

UE.GenKey(sp) *is the key generation algorithm which on input of the system parameters outputs a key $k \in \mathcal{K}_{sp}$.*

UE.GenTok(k_e, k_{e+1}) *is given two keys k_e and k_{e+1} and outputs some update token Δ_{e+1}.*

UE.Enc(k_e, m) *is given a key k_e and a message $m \in \mathcal{M}_{sp}$ and outputs some ciphertext $c_e \in \mathcal{C}_{sp}$.*

UE.Dec(k_e, c_e) *is given a key k_e and a ciphertext c_e and outputs some message $m \in \mathcal{M}_{sp}$ or $\perp$.*

UE.ReEnc(Δ_e, c_{e-1}) *is given an update token Δ_e and a ciphertext c_{e-1} and returns an updated ciphertext c_e.*

Given UE, *we call* SKE $=$ (GenSP, GenKey, Enc, Dec) *the **underlying (standard) encryption scheme**.*

*UE is called **correct** if* SKE *is correct and* $\forall sp \leftarrow$ GenSP(pp), $\forall k^{old}, k^{new} \leftarrow$ GenKey(sp), $\forall \Delta \leftarrow$ GenTok(k^{old}, k^{new}), $\forall c \in \mathcal{C}$: Dec($k^{new}$, ReEnc($\Delta, c$)) $=$ Dec(k^{old}, c).

We will use both notations, i.e., k_e, k_{e+1} and k^{old}, k^{new} synonymous throughout the paper, where the latter omits the explicit epochs e whenever they are not strictly necessary and we simply want to refer to keys for two consecutive epochs.

In our first construction, the re-encryption algorithm UE.ReEnc will be a deterministic algorithm, whereas for our second scheme the ciphertexts are updated in a probabilistic manner. We define the desired security properties (UP-IND-CCA, UP-INT-CTXT) for updatable encryption schemes with deterministic re-encryption and (UP-IND-RCCA, UP-INT-PTXT) for schemes with a probabilistic UE.ReEnc algorithm in the following sections.

RISE. In [17], Lehmann and Tackmann proposed an updatable encryption scheme called RISE which is essentially (symmetric) ElGamal encryption with added update functionality. We use RISE as a building block in our RCCA and PTXT secure scheme. Please refer to [16] for a description of RISE in our setting.

3   CCA and CTXT Secure Updatable Encryption

In this section, we first introduce the considered confidentiality and integrity definitions for updatable encryption with deterministic re-encryption (Sect. 3.1). This is followed by a generic transformation that allows to realize these notions from simple, static security properties (Sect. 3.2). Finally, we describe a Encrypt-and-MAC construction that can be used in this transformation and give instantiations of its building blocks (Sect. 3.3).

	CCA	CTXT	RCCA	PTXT
Next()	moves to the next epoch $e + 1$ by generating new key and update token			
Enc(m)	returns encryption c of message m under current epoch key k_e			
	stores ciphertext (e, c) in $\mathbf{Q}$		stores (e, m, c) in $\mathbf{Q}$	stores (e, m) in $\mathbf{Q}$
Dec(c)	returns decryption m of ciphertext c under current epoch key k_e			
	ignores c if it is the challenge c^* or a re-encryption of c^*	—	ignores c if it decrypts to m_0 or m_1	—
ReEnc(i, c)	returns re-encryption c_e of ciphertext c from epoch i into current epoch e			
	allows only ciphertexts in $\mathbf{Q}$ and derivations of c^*	allows only ciphertexts in $\mathbf{Q}$	allows arbitrary ciphertexts	
	if c is c^* or a re-encryption of c^* it adds epoch e to $\mathbf{C}^*$	—	if c decrypts to m_0 or m_1 and $(i, *, c) \notin \mathbf{Q}$ it adds epoch e to $\mathbf{C}^*$	—
Corrupt(x, i)	returns either k_i (if $\mathsf{x} =$ key) or Δ_i (if $\mathsf{x} =$ token) for $0 \leq i \leq e$			

Fig. 5. Overview of oracles and their restrictions in our different security games. $\mathbf{C}^*$ is the set of challenge-equal epochs used in the CCA and RCCA games, c^* denotes the challenge ciphertext in the CCA game, and m_0, m_1 are the two challenge plaintexts chosen by $\mathcal{A}$ in the RCCA game. $\mathbf{Q}$ is the set of queried (re)encryptions.

3.1   Security Model

We follow the previous work on updatable encryption and require confidentiality of ciphertexts in the presence of temporary key and token corruption, covering both forward and post-compromise security. This is formalized through the indistinguishability-based security notion UP-IND-CCA which can be seen as the extension of the standard CCA game to the context of updatable encryption. In addition to confidentiality, we also require *integrity* of ciphertexts, which we formulate via our UP-INT-CTXT definition.

Both security notions are defined through experiments run between a challenger and an adversary $\mathcal{A}$. Depending on the notion, the adversary may issue queries to different oracles. At a high level, $\mathcal{A}$ is allowed to adaptively corrupt arbitrary choices of secret keys and update tokens, as long as they do not allow him to trivially win the respective security game.

Oracles and CCA Security. Our UP-IND-CCA notion is essentially the regular IND-CCA definition where the adversary is given additional oracles that capture the functionality inherent to an updatable encryption scheme.

These oracles are defined below and are roughly the same in all our security definitions. We describe the oracles in the context of our UP-IND-CCA security game, which needs some extra restrictions and care in order to prevent a decryption of the challenge ciphertext. When introducing our other security notions, we explain the differences w.r.t. the oracles presented here. An overview of all oracles and their differences in our security games is given in Fig. 5.

The oracles may access the global state $(sp, k_e, \Delta_e, \mathbf{Q}, \mathbf{K}, \mathbf{T}, \mathbf{C}^*)$ which is initialized via $\mathsf{Init}(pp)$ as follows:

$\mathsf{Init}(pp)$: This initializes the state of the challenger as $(sp, k_0, \Delta_0, \mathbf{Q}, \mathbf{K}, \mathbf{T}, \mathbf{C}^*)$ where $e \leftarrow 0$, $sp \leftarrow_R \mathsf{UE.GenSP}(pp)$ $k_0 \leftarrow_R \mathsf{UE.GenKey}(sp)$, $\Delta_0 \leftarrow \perp$, $\mathbf{Q} \leftarrow \emptyset$, $\mathbf{K} \leftarrow \emptyset$, $\mathbf{T} \leftarrow \emptyset$ and $\mathbf{C}^* \leftarrow \emptyset$.

The current epoch is denoted as e, and the list $\mathbf{Q}$ contains "honest" ciphertexts which the adversary has obtained entirely through the Enc or ReEnc oracles. The challenger also keeps sets $\mathbf{K}, \mathbf{T}$ and $\mathbf{C}^*$ (all initially set to $\emptyset$) that are used to keep track of the epochs in which $\mathcal{A}$ corrupted a secret key ($\mathbf{K}$), token ($\mathbf{T}$), or obtained a re-encryption of the challenge-ciphertext ($\mathbf{C}^*$). These will later be used to check whether the adversary has made a combination of queries that trivially allow him to decrypt the challenge ciphertext. For our integrity notions UP-INT-CTXT and UP-INT-PTXT we will omit the set $\mathbf{C}^*$ that is related to the challenge ciphertext. Moreover, the predicate isChallenge, which identifies challenge-related ciphertexts, unnecessary for integrity notions. We implicitly assume that the oracles only proceed when the input is valid, e.g., for the epoch i it must hold that $0 \le i < e$ for re-encryption queries, and $0 \le i \le e$ for corruption queries. The decryption or re-encryption oracle will only proceed when the input ciphertext is "valid" (which will become clear in the oracle definitions given below). For incorrect inputs, the oracles return `invalid`.

$\mathsf{Next}()$: Runs $k_{e+1} \leftarrow_R \mathsf{UE.GenKey}(sp)$, $\Delta_{e+1} \leftarrow_R \mathsf{UE.GenTok}(k_e, k_{e+1})$, adds (k_{e+1}, Δ_{e+1}) to the global state and updates the current epoch to $e \leftarrow e + 1$.

$\mathsf{Enc}(m)$: Returns $c \leftarrow_R \mathsf{UE.Enc}(k_e, m)$ and sets $\mathbf{Q} \leftarrow \mathbf{Q} \cup \{(e, c)\}$.

$\mathsf{Dec}(c)$: If $\mathsf{isChallenge}(k_e, c) = \mathtt{false}$, it returns $m \leftarrow \mathsf{UE.Dec}(k_e, c)$.

$\mathsf{ReEnc}(i, c)$: The oracle returns the re-encryption of c from the i-th into the current epoch e. That is, it returns c_e that is computed iteratively through $c_\ell \leftarrow \mathsf{UE.ReEnc}(\Delta_\ell, c_{\ell-1})$ for $\ell = i + 1, \ldots, e$ and $c_i \leftarrow c$. The oracle accepts only ciphertexts c that are honestly generated, i.e., either $(i, c) \in \mathbf{Q}$ or $\mathsf{isChallenge}(k_i, c) = \mathtt{true}$. It also updates the global state depending on whether the query is a challenge ciphertext or not:
  - If $(i, c) \in \mathbf{Q}$, set $\mathbf{Q} \leftarrow \mathbf{Q} \cup \{(e, c_e)\}$.
  - If $\mathsf{isChallenge}(k_i, c) = \mathtt{true}$, set $\mathbf{C}^* \leftarrow \mathbf{C}^* \cup \{e\}$.

$\mathsf{Corrupt}(\{\mathsf{key}, \mathsf{token}\}, i)$: This oracle models adaptive and retroactive corruption of keys and tokens, respectively:
  - Upon input (key, i), the oracle sets $\mathbf{K} \leftarrow \mathbf{K} \cup \{i\}$ and returns k_i.
  - Upon input (token, i), the oracle sets $\mathbf{T} \leftarrow \mathbf{T} \cup \{i\}$ and returns Δ_i.

Finally, we define UP-IND-CCA security as follows, requesting the adversary after engaging with the oracles defined above, to detect whether the challenge ciphertext $c^* \leftarrow_R$ UE.Enc(k_e, m_b) is an encryption of m_0 or m_1. The adversary wins if he correctly guesses the challenge bit b and has not corrupted the secret key in any challenge-equal epoch. In the following we explain how we define the set of challenge-equal epochs $\widehat{\mathbf{C}}^*$ and prevent trivial wins.

Definition 6. *An updatable encryption scheme* UE *(with deterministic re-encryption) is called UP-IND-CCA secure if for any PPT adversary $\mathcal{A}$ the advantage*

$$\mathsf{Adv}_{\mathsf{UE},\mathcal{A}}^{up\text{-}ind\text{-}cca}(pp) := \left| \Pr[\mathsf{Exp}_{\mathsf{UE},\mathcal{A}}^{up\text{-}ind\text{-}cca}(pp, 0) = 1] - \Pr[\mathsf{Exp}_{\mathsf{UE},\mathcal{A}}^{up\text{-}ind\text{-}cca}(pp, 1) = 1] \right|$$

is negligible in κ.

Experiment $\mathsf{Exp}_{\mathsf{UE},\mathcal{A}}^{up\text{-}ind\text{-}cca}(pp, b)$
   $(sp, k_0, \Delta_0, \mathbf{Q}, \mathbf{K}, \mathbf{T}, \mathbf{C}^*) \leftarrow \mathsf{Init}(pp)$
   $(m_0, m_1, state) \leftarrow_R \mathcal{A}^{\mathsf{Enc,Dec,Next,ReEnc,Corrupt}}(sp)$
   proceed only if $|m_0| = |m_1|$ and $m_0, m_1 \in \mathcal{M}_{sp}$
   $c^* \leftarrow_R$ UE.Enc(k_e, m_b), $\mathbf{C}^* \leftarrow \{e\}$, $e^* \leftarrow e$
   $b' \leftarrow_R \mathcal{A}^{\mathsf{Enc,Dec,Next,ReEnc,Corrupt}}(c^*, state)$
   return b' if $\mathbf{K} \cap \widehat{\mathbf{C}}^* = \emptyset$, i.e. $\mathcal{A}$ did not trivially win. (Else abort.)

Preventing Decryption of an Updated Challenge Ciphertext. We use a predicate isChallenge(k_i, c) to detect attempts of decrypting the challenge ciphertext c^* or a re-encryption thereof. Whether a given ciphertext is a re-encryption of the challenge c^* can be tested efficiently by exploiting the deterministic behaviour of the re-encryption algorithm, and the fact that all secret keys and token are known to the challenger. This approach has also been used to define CCA-security for ciphertext-*dependent* schemes by Everspaugh et al. [8].

For the following definition, recall that c^* is the challenge ciphertext obtained in epoch e^*, or $c^* = \bot$ if the adversary has not made the challenge query yet.

isChallenge(k_i, c) :
   – If $i = e^*$ and $c^* = c$, return **true**.
   – If $i > e^*$ and $c^* \neq \bot$, return **true** if $c_i^* = c$ where c_i^* for epoch i is computed iteratively as $c_\ell^* \leftarrow$ UE.ReEnc$(\Delta_{\ell+1}, c_\ell^*)$ for $\ell = e^*, \ldots, i$.
   – Else return **false**.

Defining Trivial Wins. A crucial part of the definition is to capture the information the adversary has learned through his oracle queries. In particular, any corruption of the token Δ_{e+1} in an epoch after where the adversary has learned the challenge ciphertext c_e^* (directly or via a re-encryption) will enable to adversary to further update the challenge ciphertext into the next epoch $e + 1$. The goal of capturing this inferable information, is to exclude adversaries following a trivial winning strategy such as, e.g., corrupting a key under which a given challenge ciphertext has been (re-)encrypted.

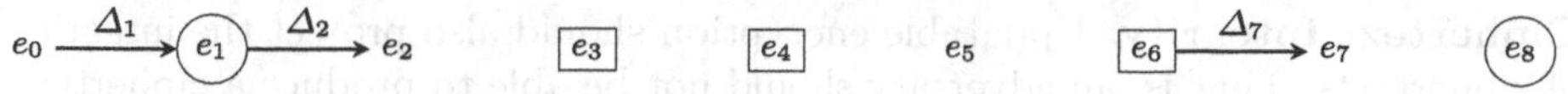

$$e_0 \xrightarrow{\Delta_1} \boxed{e_1} \xrightarrow{\Delta_2} e_2 \quad \boxed{e_3} \quad \boxed{e_4} \quad e_5 \quad \boxed{e_6} \xrightarrow{\Delta_7} e_7 \quad \boxed{e_8}$$

(Note: in the figure e_1, e_8 are circled; e_3, e_4, e_6 are boxed.)

Fig. 6. Example of corrupted tokens, keys (boxed) and challenge-equal epochs (circled) in a UP–IND–CCA game. Corrupting Δ_3 and Δ_8 is forbidden, as they would allow to re-encrypt the challenge ciphertext into an epoch where $\mathcal{A}$ knows the secret key.

We use the notation from [17] to define the information the adversary may trivially derive. We focus on schemes that are bi-directional, i.e., we assume up and downgrades of ciphertexts. That is, we assume that a token Δ_e may enable *downgrades* of ciphertexts from epoch e into epoch $e - 1$. While bi-directional security and schemes are not preferable from a security point of view, all currently known (efficient) solutions exhibit this additional property.[2] Thus, for the sake of simplicity we state all our definitions for this setting. As a consequence, it is sufficient to consider only the inferable information w.r.t. ciphertexts: [17] also formulate inference of keys, which in the case of bi-directional schemes has no effect on the security notions though (Fig. 6).

For the (R)CCA game, we need to capture all the epochs in which the adversary knows a version of the challenge ciphertext, which we define through the set $\widehat{\mathbf{C}}^*$ containing all *challenge-equal* epochs. Recall that $\mathbf{C}^*$ denotes the set of epochs in which the adversary has obtained an updated version of the ciphertext via the challenge query or by updating the challenge ciphertext via the ReEnc oracle. The set $\mathbf{T}$ contains all epochs in which the adversary has corrupted the update tokens, and e_{end} denotes the last epoch of the game. The set $\widehat{\mathbf{C}}^*$ of all challenge-equal ciphertexts is defined via the recursive predicate challenge-equal:

$$\widehat{\mathbf{C}}^* \leftarrow \{e \in \{0, \dots, e_{\mathsf{end}}\} \mid \mathsf{challenge\text{-}equal}(e) = \mathtt{true}\}$$
$$\text{and } \mathtt{true} \leftarrow \mathsf{challenge\text{-}equal}(e) \text{ iff: } (e \in \mathbf{C}^*) \vee$$
$$(\mathsf{challenge\text{-}equal}(e - 1) \wedge e \in \mathbf{T}) \vee (\mathsf{challenge\text{-}equal}(e + 1) \wedge e + 1 \in \mathbf{T})$$

Note that $\widehat{\mathbf{C}}^*$ is efficiently computable (e.g. via fixpoint iteration).

Re-encryptions of the Challenge Ciphertext. Note that we allow ReEnc to *skip* keys, as we let $\mathcal{A}$ give the starting epoch i as an additional parameter and return the re-encryption from any old key k_i to the current one. This is crucial for obtaining a meaningful security model: any ReEnc query where the input ciphertext is a derivation of the challenge ciphertext (that the adversary will receive in the CCA game), marks the current target epoch e as challenge-equal by adding e to $\mathbf{C}^*$. In our UP-IND-CCA security game defined below we disallow the adversary from corrupting the key of any challenge-equal epoch to prevent trivial wins. Calling the ReEnc oracle for a re-encryption of the challenge ciphertext from some epoch i to e will still allow $\mathcal{A}$ to corrupt keys between i and e.

---

[2] Note that bi-directionality is a property of the *security model*, not the scheme per se. That is, uni-directional schemes are evidently also bi-directional secure, even though they do *not* allow ciphertext downgrades.

Ciphertext Integrity. Updatable encryption should also protect the integrity of ciphertexts. That is, an adversary should not be able to produce a ciphertext himself that correctly decrypts to a message $m \neq \bot$. Our definition adapts he classic INT-CTXT notion to the setting of updatable encryption. We use the same oracles as in the UP-IND-CCA game defined above, but where isChallenge always returns false (as there is no challenge ciphertext). Again, the tricky part of the definition is to capture the set of trivial wins – in this case trivial forgeries – that the adversary can make given the secret keys and update tokens he corrupts. For simplicity, we only consider forgeries that the adversary makes in the current and final epoch e_{end}, but not in the past. This matches the idea of updatable encryption where the secret keys and update tokens of old epochs will (ideally) be deleted, and thus a forgery for an old key is meaningless anyway.

Clearly, when the adversary corrupted the secret key at some previous epoch and since then learned *all* update tokens until the final epoch e_{end}, then all ciphertexts in this last epoch can easily be forged. This is captured by the first case in the definition of UP-INT-CTXT security.

Definition 7. *An updatable encryption scheme* UE *is called UP-INT-CTXT secure if for any PPT adversary $\mathcal{A}$ the following advantage is negligible in κ:*
$$\mathsf{Adv}_{\mathsf{UE},\mathcal{A}}^{up\text{-}int\text{-}ctxt}(pp) := \Pr[\mathsf{Exp}_{\mathsf{UE},\mathcal{A}}^{up\text{-}int\text{-}ctxt}(pp) = 1].$$

Experiment $\mathsf{Exp}_{\mathsf{UE},\mathcal{A}}^{up\text{-}int\text{-}ctxt}(pp)$
  $(sp, k_0, \Delta_0, \mathbf{Q}, \mathbf{K}, \mathbf{T}) \leftarrow \mathsf{Init}(pp)$
  $c^* \leftarrow_{\mathrm{R}} \mathcal{A}^{\mathsf{Next},\mathsf{Enc},\mathsf{Dec},\mathsf{ReEnc},\mathsf{Corrupt}}(sp)$
  return 1 if $\mathsf{UE}.\mathsf{Dec}(k_{e_{end}}, c^*) \neq \bot$ and $(e_{end}, c^*) \notin \mathbf{Q}^*$ and
    $\nexists e \in \mathbf{K}$ where $i \in \mathbf{T}$ for $i = e$ to e_{end}; i.e. $\mathcal{A}$ did not trivially win.

Defining Trivial Ciphertext Updates. When defining the set of trivial ciphertexts $\mathbf{Q}^*$ for the UP–INT–CTXT game defined above, we now move from general epochs to concrete ciphertexts, i.e., we capture all ciphertexts that the adversary could know, either through queries to the Enc or ReEnc oracle or through updating such ciphertexts himself. We exploit that ReEnc is deterministic to define the set of trivial forgeries $\mathbf{Q}^*$ as narrow as possible. More precisely, $\mathbf{Q}^*$ is defined by going through the ciphertexts $(e, c) \in \mathbf{Q}$ the adversary has received through oracle queries and iteratively update them into the next epoch $e + 1$ whenever the adversary has corrupted Δ_{e+1}. The latter information is captured in the set $\mathbf{T}$ that contains all epochs in which the adversary learned the update token. We start with $\mathbf{Q}^* \leftarrow \emptyset$ and amend the set as follows:

  for each $(e, c) \in \mathbf{Q}$:
    set $\mathbf{Q}^* \leftarrow \mathbf{Q}^* \cup (e, c)$, and $i \leftarrow e + 1$, $c_{i-1} \leftarrow c$
    while $i \in \mathbf{T}$:
      set $\mathbf{Q}^* \leftarrow \mathbf{Q}^* \cup (i, c_i)$ where $c_i \leftarrow \mathsf{UE}.\mathsf{ReEnc}(\Delta_i, c_{i-1})$, and $i \leftarrow i + 1$

On the Necessity of the "Queried Restriction". Restricting ReEnc queries to honestly generated ciphertexts seems somewhat unavoidable, as the ability of ciphertext-independent key-rotation seems to require homomorphic properties on the encryption. In our construction, an adversary could exploit this homomorphism to "blind" the challenge ciphertext before sending it to the ReEnc oracle, and later "unblind" the re-encrypted ciphertext. This blinding would prevent us from recognizing that the challenge ciphertext was re-encrypted, and thus the target epoch would no longer be marked as challenge-equal, allowing the adversary to corrupt the secret key in the new epoch and trivially win by decrypting the re-encrypted challenge. A similar restriction is used in the CTXT definition for ciphertext-dependent schemes in [8] as well[3]. In Sect. 4 we overcome this challenge by making ciphertexts publicly verifiable. The above "blinding" trick then no longer works as it would invalidate the proof of ciphertext correctness.

3.2 Generic Transformation for Secure Updatable Encryption

In the following we prove UP-IND-CCA and UP-INT-CTXT security for a class of updatable encryption schemes satisfying some mild requirements. The goal is that given an updatable encryption scheme $UE = (Gen, GenKey, GenTok, Enc, Dec, ReEnc)$, we can prove the security of UE based only on classical security of the underlying encryption scheme $SKE = (Gen, GenKey, Enc, Dec)$ and simple properties satisfied by $GenTok$ and $ReEnc$.

Properties of the (Re-)encryption and Token Generation. Now we define the additional properties that are needed to lift static IND-CCA and INT-CTXT security to their updatable version with adaptive key and token corruptions as just defined.

Tidy Encryption & Strong CCA/CTXT. When re-encryptions are deterministic, we need the underlying standard encryption scheme SKE of an updatable scheme to be tidy (cf. Definition 2), so there is a one-to-one correspondence between ciphertexts and message-randomness pairs. Further, we need slightly stronger variants of the standard security definitions IND-CCA and INT-CTXT in the deterministic setting where the encryption oracle additionally reveals the used encryption randomness. We denote these stronger experiments by S-IND-CCA and S-INT-CTXT, or simply by saying **strong** IND-CCA/INT-CTXT.

Definition 8. *Strong IND-X, IND\$-X and INT-Y notions are defined as sketched above. (See also [16].)*

Simulatable & Reverse Tokens. We need further properties (Definitions 9 and 10) that are concerned with the token generation of an updatable encryption scheme. It should be possible to simulate perfectly indistinguishable tokens as well as reverse tokens, inverting the effect of the former ones, without knowing any key.

---

[3] The CTXT definition in the proceedings version of their paper did not have such a restriction, however the revised ePrint version [7] later showed that the original notion is not achievable and a weaker CTXT definition is introduced instead.

Definition 9. *We call a token Δ' a* **reverse token** *of a token Δ if for every pair of keys $k^{\text{old}}, k^{\text{new}} \in \mathcal{K}$ with $\Delta \in \text{supp}(\text{UE.GenTok}(k^{\text{old}}, k^{\text{new}}))$ we have $\Delta' \in \text{supp}(\text{UE.GenTok}(k^{\text{new}}, k^{\text{old}}))$.*

Definition 10. *Let* UE *be an updatable encryption scheme. We say that* UE *has* **simulatable** *token generation if it has the following property: There is a PPT algorithm* SimTok(sp) *which samples a pair (Δ, Δ') of token and reverse token. Furthermore, for arbitrary (fixed) $k^{\text{old}} \leftarrow \text{UE.GenKey}(sp)$ following distributions of Δ are identical: The distribution of Δ*

- *induced by $(\Delta, \_) \leftarrow_R \text{SimTok}(sp)$.*
- *induced by $\Delta \leftarrow_R \text{UE.GenTok}(k^{\text{old}}, k^{\text{new}})$ where $k^{\text{new}} \leftarrow_R \text{UE.GenKey}(sp)$.*

In other words, honest token generation and token simulation are perfectly indistinguishable.

Re-encryption = decrypt-then-encrypt. The final requirement states that the re-encryption of a ciphertext $c = \text{UE.Enc}(k^{\text{old}}, m; r)$ looks like a fresh encryption of m under k^{new} where UE.Enc uses the same randomness r. To formalize this, we make use of UE.RDec, the randomness-recoverable decryption algorithm of the underlying encryption scheme (Definition 2), where we have $(m, r) \leftarrow \text{UE.RDec}(k, c)$ for $c \leftarrow \text{UE.Enc}(k, m; r)$.

Definition 11. *Let* UE *be an updatable encryption scheme with deterministic re-encryption. We say that re-encryption (for* UE*) is* **randomness-preserving** *if the following holds: First, as usually assumed,* UE *encrypts with uniformly chosen randomness (i.e.,* UE.Enc(k, m) *and* UE.Enc($k, m; r$) *for uniformly chosen r are identically distributed). Second, for all $sp \leftarrow_R \text{UE.GenSP}(pp)$, all keys $k^{\text{old}}, k^{\text{new}} \leftarrow_R \text{UE.GenKey}(sp)$, tokens $\Delta \leftarrow_R \text{UE.GenTok}(k^{\text{old}}, k^{\text{new}})$, and all valid ciphertexts c under k^{old}, we have*

$$\text{UE.Enc}(k^{\text{new}}, \text{UE.RDec}(k^{\text{old}}, c)) = \text{UE.ReEnc}(\Delta, c).$$

More precisely, UE.Enc(k^{new}, UE.RDec(k^{old}, c)) *is defined as* UE.Enc($k^{\text{new}}, m; r$) *where $(m, r) \leftarrow \text{UE.RDec}(k^{\text{old}}, c)$.*

In [16], we argue that this randomness-preserving property additionally guarantees unlinkability of re-encrypted ciphertexts (UP-REENC security) as considered by prior work [8,17].

UP-IND-CCA and UP-INT-CTXT Security. We are now ready to state our generic transformation for achieving security of the updatable encryption scheme. The proofs for both properties are very similar, and below we describe the core ideas of our proof strategy. The detailed proofs are given in [16].

Theorem 1. *Let* UE $=$ (Gen, GenKey, GenTok, Enc, Dec, ReEnc) *be an updatable encryption scheme with deterministic re-encryption. Suppose that* UE *has randomness-preserving re-encryption and simulatable token generation and the underlying encryption scheme* SKE $=$ (Gen, GenKey, Enc, Dec) *is tidy.*

- If SKE *is S-IND-CCA-secure, then* UE *is UP-IND-CCA-secure.*
- If SKE *is S-INT-CTXT-secure, then* UE *is UP-INT-CTXT-secure.*

Proof (sketch). In the following, we illustrate the main challenges occurring in our security proofs as well as how we can cope with these using the properties we just introduced. Let us consider the problems that arise when we embed a static challenge, say an IND-CCA challenge, into an UP-IND-CCA game. Let us assume the UP-IND-CCA adversary $\mathcal{A}$ asks for its challenge under key k_{e^*} and we want to embed our IND-CCA challenge there. Then k_{e^*} is unknown to us but we can answer $\mathcal{A}$'s encryption and decryption queries under k_{e^*} using our own IND-CCA oracles.

However, the token Δ_{e^*+1} might be corrupted by $\mathcal{A}$. Note that in this case, k_{e^*+1} cannot be corrupted, since $\mathcal{A}$ could trivially win. Now, the question is how Δ_{e^*+1} can be generated without knowing k_{e^*}. For this purpose, we make use of the simulatable token generation property (Definition 10) that ensures that well-distributed tokens can be generated even without knowing keys. So we can hand over a simulated Δ_{e^*+1} to $\mathcal{A}$ if it asks for it. But when simulating tokens in this way, we do not know the corresponding keys. This is a potential problem as we need to be able to answer encryption and decryption queries under the unknown key k_{e^*+1}. To cope with this problem, we use the corresponding IND-CCA oracle for k_{e^*} and update or downgrade the ciphertexts from/to epoch e^*. That means, if we are asked to encrypt under k_{e^*+1}, we actually encrypt under k_{e^*} and update the resulting ciphertext to epoch $e^* + 1$ using Δ_{e^*+1}. Now, we need to ensure that ciphertexts created in this way look like freshly encrypted ciphertexts under key k_{e^*+1}. This is what Definition 11 requires. Similarly, if we are asked to decrypt under k_{e^*+1}, we downgrade the ciphertext using the reverse token Δ'_{e^*+1} (Definition 9) that was generated along with Δ_{e^*+1} (Definition 10). Note that in this case, we do not need the downgraded ciphertext to look like a fresh one as $\mathcal{A}$ never sees it. Assuming the next token Δ_{e^*+2} gets also corrupted we can do the same to handle encryption and decryption queries for epoch $e^* + 2$.

Now let us assume that not the token Δ_{e^*+1} but the key k_{e^*+1} gets corrupted. In this case we can neither generate Δ_{e^*+1} regularly as we do not know k_{e^*} nor simulate it as k_{e^*+1} is known to the adversary. As we know k_{e^*+1}, we have no problems in handling encryption and decryption queries for epoch $e^* + 1$. But it is not clear how we can re-encrypt a (non-challenge) ciphertext c freshly generated in epoch e^* to $e^* + 1$ without knowing Δ_{e^*+1}. As we called our IND-CCA encryption oracle to generate c, we certainly know the contained message m. So we could just encrypt m under key k_{e^*+1} yielding ciphertext c'. However, now the freshly encrypted ciphertext c' and a ciphertext c'' resulting from regularly updating c' to epoch $e^* + 1$ may look different as they involve different randomness. To circumvent this problem, we require the IND-CCA encryption oracle to additionally output the randomness r which has been used to generate c. Computing c' using randomness r then yields perfect indistinguishability assuming Definition 11. Hence, we need SKE to be S-IND-CCA (and S-INT-CTXT) and not only IND-CCA (and INT-CTXT) secure.

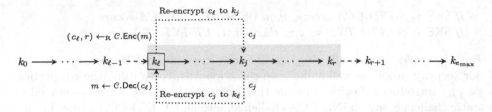

Fig. 7. Encryption and decryption in the insulated region. The keys in the grey area (k_ℓ to k_r) are not known in the reduction. Encryption and decryption for other keys is unchanged. The S-INT-CTXT resp. S-IND-CCA challenger $\mathcal{C}$ is embedded in epoch ℓ.

Finally, let us consider how to handle queries to the left of the challenge epoch. For this, let us assume that k_{e^*-1} gets corrupted and Δ_{e^*} is uncorrupted but unknown to us. Then again we can easily handle encrypt/decrypt queries for epoch $e^* - 1$ but cannot re-encrypt a ciphertext c from epoch $e^* - 1$ to e^* in a straightforward manner. Now, as c needs to result from a previous query the corresponding message-randomness pair (m, r) (due to tidyness there is only one such pair) is known. So, as before, we would like to replace the re-encryption by a fresh encryption under key k_{e^*}. Unfortunately, the S-IND-CCA encryption oracle we would use for this purpose only accepts the message but not the randomness as input. We cope with this as follows: when we are asked to encrypt a message m under key k_{e^*-1} (or prior keys), we will always first call the S-IND-CCA oracle to encrypt m yielding a ciphertext c' and randomness r. Then we would encrypt (m, r) under key k_{e^*-1} yielding c. The ciphertext c' can then be stored until a re-encryption of c is needed. Again Definition 11 ensures perfect indistinguishability from a real re-encryption. (Here, we use that encryption randomness is chosen uniformly, independent of the key.)

Note that the case that Δ_{e^*} is corrupted could actually be handled analogous to the case that Δ_{e^*+1} is corrupted by additionally demanding randomness-preserving re-encryption for reverse tokens but we can get around this.

Overall, this solves the main challenges when embedding an S-IND-CCA challenge into an UP-IND-CCA game.

Key Insulation. Our key insulation technique aims at coping with the problems when embedding challenges and follows the ideas just described. However, instead of guessing the challenge epoch *and* the region to the left and to the right in which the adversary corrupted all of the tokens (and none of keys) and embed our S-IND-CCA/S-INT-CTXT challenge there, we rather do the following: we only guess the boundaries of this region $\{\ell, \ldots, r\}$ (containing the challenge epoch) and embed the S-IND-CCA/S-INT-CTXT challenge at epoch ℓ. Note that the tokens Δ_ℓ and Δ_r entering and leaving the boundaries of this "insulated" region are not corrupted.

Now we change the inner workings in this region and the way it can be entered from the left using the ideas described before. Namely, only key k_ℓ in the region is generated. Recall, in the reduction we have S-IND-CCA/S-INT-CTXT

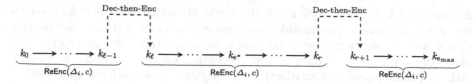

Fig. 8. Entering and leaving the insulated region. Re-encryption in the underbraced regions is done using the known tokens. The two missing tokens are "emulated" by decrypt-then-encrypt.

oracles at our disposal to replace this key. The tokens $\Delta_{\ell+1}, \ldots, \Delta_{r+1}$ along with corresponding reverse tokens are generated using SimTok (cf. Definition 10). For encryption in the region, we encrypt under k_ℓ and update the ciphertext to the desired epoch. For decryption, we the use reverse tokens to downgrade the ciphertext to k_ℓ and decrypt with this key. This is illustrated in Fig. 7. Leaving and entering the region which was originally done by re-encryption, is now essentially done by retrieving the plaintext and randomness of the ciphertext that should be reencrypted (so we sort of decrypt the queried ciphertext) and use it to generate a fresh ciphertext inside or outside the region by encryption. This is depicted in Fig. 8.

3.3   An Encrypt-and-MAC Construction

We construct an UE scheme with *deterministic* re-encryption that achieves UP-IND-CCA, UP-REENC, and UP-INT-CTXT security. For this, we use generic building blocks which can be securely instantiated from the DDH assumption.

High-Level Idea. Our idea is to do a Encrypt-and-MAC (E&M) construction with primitives which are key-rotatable. Using Encrypt-*and*-MAC instead of the more standard Encrypt-*then*-MAC approach is crucial for the updatability as we need "direct access" to both the ciphertext and the MAC.

It is well-known that, in general, E&M is *not* a secure transformation for authenticated encryption, as the MAC could leak information about the plaintext and does not authenticate the ciphertext. However, when using a *tidy* encryption scheme SKE (cf. Definition 2) and a pseudorandom function PRF as MAC, then E&M *does* provide (static) CCA and CTXT security. Recall that tidy encryption means that decryption is randomness-recoverable, i.e., it also outputs the randomness r used in the encryption. This allows to apply the PRF on both, the message and the randomness r, which then guarantees the integrity of ciphertexts.

We start with such tidy E&M for static primitives but also require that SKE and PRF support key-rotation and updates of ciphertexts and PRF values. Then, for yielding the updatable version of the E&M transform, one simply relies on the key-rotation capabilities of SKE and PRF and updates the individual parts of the authenticated ciphertext. Security of the UE scheme obtained in this way follows since the properties from Sect. 3.2 are satisfied.

Encrypt-and-MAC. First we recall the E&M transformation and its security for tidy (randomness recoverable) encryption. To make it clear that decryption recovers the encryption randomness we write RDec for decryption and make the randomness chosen in the encryption explicit as $\mathsf{Enc}(k, m; r)$. Let $\mathsf{SKE} = (\mathsf{GenSP}, \mathsf{GenKey}, \mathsf{Enc}, \mathsf{RDec})$ be a *tidy* encryption scheme and $\mathsf{PRF} = (\mathsf{GenSP}, \mathsf{GenKey}, \mathsf{Eval})$ be a pseudorandom function, then the E&M transform of SKE and PRF is defined as follows:

- $\mathsf{AE.GenSP}(pp)$ returns $sp = (sp_{\mathsf{SKE}}, sp_{\mathsf{PRF}})$ where $sp_{\mathsf{SKE}} \leftarrow_{\mathrm{R}} \mathsf{SKE.GenSP}(pp)$ and $sp_{\mathsf{PRF}} \leftarrow_{\mathrm{R}} \mathsf{PRF.GenSP}(pp)$.
- $\mathsf{AE.GenKey}(sp)$ returns $k = (k_{\mathsf{SKE}}, k_{\mathsf{PRF}})$, where $k_{\mathsf{SKE}} \leftarrow_{\mathrm{R}} \mathsf{SKE.GenKey}(sp_{\mathsf{SKE}})$ and $k_{\mathsf{PRF}} \leftarrow_{\mathrm{R}} \mathsf{PRF.GenKey}(sp_{\mathsf{PRF}})$.
- $\mathsf{AE.Enc}(k, m; r)$ returns (c, τ) where $c \leftarrow \mathsf{SKE.Enc}(k_{\mathsf{SKE}}, m; r)$ and $\tau \leftarrow \mathsf{PRF.Eval}(k_{\mathsf{PRF}}, (m, r))$.
- $\mathsf{AE.RDec}(k, (c, \tau))$ computes $(m, r) \leftarrow \mathsf{SKE.RDec}(k_{\mathsf{SKE}}, c)$. It returns (m, r) if $\mathsf{PRF.Eval}(k_{\mathsf{PRF}}, (m, r)) = \tau$, and $\perp$ otherwise.

Lemma 1 essentially follows from [18] where, however, a slightly different definition of tidy was used. But the adaption to our setting is straightforward.

Lemma 1. *If* SKE *is a tidy encryption scheme satisfying S-IND-CPA security, and* PRF *is a secure pseudorandom function (with domain $\mathcal{M} \times \mathcal{R}$), then* AE *as defined above is a S-IND-CCA and S-INT-CTXT secure tidy encryption scheme. The same holds for IND\$ instead of IND.*

Updatable Encrypt-and-MAC. To make this E&M construction a secure updatable encryption scheme, we need that both underlying primitives support key-rotation satisfying certain properties. That means, for SKE we assume that additional algorithms $\mathsf{GenTok}(k^{\mathrm{old}}, k^{\mathrm{new}})$ and $\mathsf{ReEnc}(\Delta, c)$ as in Definition 5 are given satisfying simulatable token generation [16] and randomness-preserving re-encryption (Definition 11). Likewise, we need similar algorithms $\mathsf{GenTok}(k^{\mathrm{old}}, k^{\mathrm{new}})$ and $\mathsf{Upd}(\Delta, \tau)$ for the PRF satisfying similar properties, i.e., a straightforward adaption of simulatable token generation (see [16]) and correctness in the sense that $\mathsf{Upd}(\Delta, \mathsf{Eval}(k^{\mathrm{old}}, (m, r))) = \mathsf{Eval}(k^{\mathrm{new}}, (m, r))$.

We now obtain our secure UE scheme by extending the AE scheme defined above with the following GenTok and ReEnc algorithms:

- $\mathsf{AE.GenTok}(k^{\mathrm{old}}, k^{\mathrm{new}})$ computes $\Delta_{\mathsf{SKE}} \leftarrow_{\mathrm{R}} \mathsf{SKE.GenTok}(k^{\mathrm{old}}_{\mathsf{SKE}}, k^{\mathrm{new}}_{\mathsf{SKE}})$ and $\Delta_{\mathsf{PRF}} \leftarrow_{\mathrm{R}} \mathsf{PRF.GenTok}(k^{\mathrm{old}}_{\mathsf{PRF}}, k^{\mathrm{new}}_{\mathsf{PRF}})$ and returns $\Delta := (\Delta_{\mathsf{SKE}}, \Delta_{\mathsf{PRF}})$.
- $\mathsf{AE.ReEnc}(\Delta, (c, \tau))$ computes $c' \leftarrow \mathsf{SKE.ReEnc}(\Delta_{\mathsf{SKE}}, c)$ and $\tau' \leftarrow \mathsf{PRF.Upd}(\Delta_{\mathsf{PRF}}, \tau)$ and returns (c', τ').

UP-IND-CCA and UP-INT-CTXT security directly follows from Theorem 1 and UP-REENC-CCA follows from [16] (where we also state the definition for UP-REENC security adapted to the CCA setting).

Corollary 1. *Suppose* AE *is the E&M construction as in Lemma 1, in particular S-IND-CCA and S-INT-CTXT secure. Suppose* AE *supports randomness-preserving reencryption and simulatable token generation as described above, i.e.*

AE *constitutes an updatable encryption scheme. Then* AE *is UP-IND-CCA and UP-INT-CTXT secure. Moreover, if* AE *is S-IND\$-CCA secure, then it is also UP-REENC-CCA secure.*

Instantiating the Key-Rotatable Building Blocks. We now show how the key-rotatable building blocks SKE and PRF can be securely instantiated. First we construct the encryption scheme which is S-IND\$-CPA secure under the DDH assumption and also tidy. Then we present the key-rotatable PRF that is secure under the DDH assumption in the random oracle model.

SKE_{DDH}. Since we need a tidy, and hence randomness recoverable encryption scheme, we must pick the encryption randomness $[r] \leftarrow_R \mathbb{G}$ from $\mathbb{G}$ if discrete logarithms are hard. A straightforward choice is to use $[r]sk$ instead of $r[pk]$ in RISE/ElGamal. However, our result which gives UP-REENC security (i.e., the unlinkability of re-encryptions) for free, c.f. [16], requires *strong* IND\$-CCA security. Thus, we instead use following variation of the mentioned approach:

$SKE_{DDH}.GenSP(pp)$ does nothing. That is, it returns $sp = pp$.

$SKE_{DDH}.GenKey(sp)$ returns $k = (k_1, k_2) \leftarrow_R \mathbb{F}_p^* \times \mathbb{F}_p = \mathcal{K}$.

$SKE_{DDH}.GenTok(k^{old}, k^{new})$ returns $\Delta = (\Delta_1, \Delta_2) = (\frac{k_1^{new}}{k_1^{old}}, \frac{k_2^{new} - k_2^{old}}{k_1^{old}}) \in \mathcal{D} = \mathcal{K}$.

$SKE_{DDH}.Enc(k, [m]; [r])$ returns $[c]$, encryption of a message $[m] \in \mathbb{G}$ with randomness $[r] \leftarrow_R \mathbb{G}$ as $[c] = (k_1[r], k_2[r] + [m]) \in \mathbb{G}^2 = \mathcal{C}$.

$SKE_{DDH}.RDec(k, [c])$ returns $([r], [m])^\top$ via $[r] = \frac{1}{k_1}[c_1]$, $[m] = [c_2] - k_2[r]$.

$SKE_{DDH}.ReEnc(\Delta, [c^{old}])$ returns $[c^{new}] = [\Delta_1 c_1^{old}, \Delta_2 c_1^{old} + c_2^{old}]$.

It is easy to see that the scheme is correct with deterministic re-encryption.

Lemma 2. *The scheme* SKE_{DDH} *is tidy, has simulatable token generation, and randomness-preserving deterministic re-encryption. The underlying encryption of* SKE_{DDH} *is strong IND\$-CPA secure under the DDH assumption over* $\mathbb{G}$.

It is evident, that the scheme is tidy. Randomness-preserving re-encryption follows from straightforward calculations. For simulatable token generation, note that any two of k^{old}, Δ, k^{new}, determine the third uniquely (and it is efficiently computable). Moreover, if we define $invert((\Delta_1, \Delta_2)) = (\frac{1}{\Delta_1}, -\frac{\Delta_2}{\Delta_1})$ then $invert(\Delta)$ is a token which downgrades ciphertexts from k^{new} to k^{old} With this, token simulation is easy to see. S-IND\$-CPA security follows from a straightforward adaptation of the standard ElGamal security proof. Note that we do not allow key-leakage, i.e. $leak(k) = \bot$.

PRF_{DDH}. Using a hash function $H: \{0, 1\}^* \to \mathbb{G}$, we instantiate the key-rotatable PRF as $PRF_{DDH}: \mathbb{F}_p^\times \times \{0, 1\}^* \to \mathbb{G}$. The core part of the PRF is the classical DDH-based construction from [2,19]. We show that it can also be extended to allow for key-rotation for which it enjoys token simulation.

$PRF_{DDH}.GenSP(pp)$ does nothing, i.e. returns $sp = pp$.

$PRF_{DDH}.GenKey(sp)$ returns $k \leftarrow_R \mathbb{F}_p = \mathcal{K}$.

$\mathsf{PRF_{DDH}.GenTok}(k^{\mathrm{old}}, k^{\mathrm{new}})$ returns $\Delta = \frac{k^{\mathrm{new}}}{k^{\mathrm{old}}}$.
$\mathsf{PRF_{DDH}.Eval}(k, x)$ returns $[\tau] = k\,\mathsf{H}(x) \in \mathbb{G}$.
$\mathsf{PRF_{DDH}.Upd}(\Delta, [\tau])$ returns $\Delta[\tau]$.

Lemma 3. *The* $\mathsf{PRF_{DDH}} = (\mathsf{GenSP}, \mathsf{GenKey}, \mathsf{Eval})$ *scheme defined above (without* GenTok *and* Upd*) is secure under the DDH assumption on* $\mathbb{G}$ *if* H *is a (programmable) random oracle.* $\mathsf{PRF_{DDH}}$ *has simulatable token generation.*

The security of $\mathsf{PRF_{DDH}}$ was shown in [19], and the simulatable properties of the token generation follow from the same observations as for $\mathsf{SKE_{DDH}}$.

4   RCCA and PTXT Secure Updatable Encryption

In this section, we first define RCCA and PTXT security for updatable encryption under active re-encryption attacks (Sect. 4.1). In Sect. 4.2 we then present our Naor-Yung inspired scheme that satisfies these strong security notions.

4.1   Security Model

We now present our definitions for updatable encryption with Replayable CCA (RCCA) security and plaintext integrity (PTXT). The oracles used in these definitions are mostly equivalent to the ones introduced for CCA security in Sect. 3.1, and thus we focus on the parts that have changed.

The most important difference is that the ReEnc oracle can be invoked on *arbitrary* ciphertexts in both definitions, whereas our CCA and CTXT definitions only allowed re-encryptions of ciphertexts that had been obtained through oracle queries themselves. This strengthening to arbitrary inputs is much closer to the reality of updatable encryption, where ciphertexts and the update procedure are outsourced to potentially untrusted data hosts. All previous definitions cover only passive corruptions of such a host, whereas our notions in this section even guarantee security against active adversaries.

RCCA Security. Standard RCCA is a relaxed variant of CCA security which is identical to CCA with the exception that the decryption oracle will not respond with `invalid` whenever a ciphertext decrypts to either of the challenge messages m_0 or m_1. This includes ciphertexts that are different from the challenge ciphertext c^* the adversary has obtained. RCCA is a suitable definition in particular for schemes where ciphertexts can be re-randomized, and thus cannot achieve the standard CCA notion. Our setting allows similar public re-randomization as ciphertext updates are now *probabilistic* instead of deterministic. Thus, as soon as the adversary has corrupted an update token we can no longer trace re-encryptions of the challenge ciphertexts (as we did in the UP-IND-CCA definition for deterministic schemes) in order to prevent the adversary from decrypting the challenge ciphertext.

Thus, instead of tracing the challenge ciphertext we now follow the RCCA approach. Our definition of UP-IND-RCCA security is essentially the standard

RCCA definition adapted for updatable encryption by giving the adversary access to a re-encryption oracle and allowing him to adaptively corrupt secret keys and tokens in the current or any past epoch.

In Enc and ReEnc described below we still keep track of honestly generated ciphertexts (and their plaintexts) which allows us to be less restrictive when a ciphertext-query can be traced down to a non-challenge ciphertext. We explain this modelling choice in more detail below.

Next(), Corrupt($\{$key, token$\}, i$): as in CCA game
Enc(m): Returns $c \leftarrow_R$ UE.Enc(k_e, m) and sets $\mathbf{Q} \leftarrow \mathbf{Q} \cup \{(e, m, c)\}$.
Dec(c): If isChallenge(k_e, c) = false, the oracle returns $m \leftarrow$ UE.Dec(k_e, c).
ReEnc(c, i): The oracle returns c_e which it iteratively computes as $c_\ell \leftarrow_R$
    UE.ReEnc($\Delta_\ell, c_{\ell-1}$) for $\ell = i + 1, \ldots, e$ and $c_i \leftarrow c$. It also updates the
    global state depending on whether the queried ciphertext is the challenge
    ciphertext or not:
    – If $(i, m, c) \in \mathbf{Q}$ (for some m), then set $\mathbf{Q} \leftarrow \mathbf{Q} \cup \{(e, m, c_e)\}$.
    – Else, if isChallenge(k_i, c) = true, then set $\mathbf{C}^* \leftarrow \mathbf{C}^* \cup \{e\}$.

As for UP-IND-CCA security, the challenge is to prevent trivial wins where an adversary tries to exploit the update capabilities of such schemes. We again achieve this by capturing the indirect knowledge of the adversary through the recursive predicate that defines all challenge-equal epochs $\widehat{\mathbf{C}}^*$. This set (which is as defined in Sect. 3.1) contains all epochs in which the adversary trivially knows a version of the challenge ciphertext, either through oracle queries or by up/downgrading the challenge ciphertext himself. The adversary wins UP-IND-RCCA if he can determine the challenge bit b used to compute $c^* \leftarrow_R$ UE.Enc(k_e, m_b) and does not corrupt the secret key in any challenge-equal epoch.

Definition 12. *An updatable encryption scheme* UE *is called UP-IND-RCCA secure if for any PPT adversary $\mathcal{A}$ the following advantage is negligible in κ :*
$$\text{Adv}_{\text{UE},\mathcal{A}}^{up\text{-}ind\text{-}rcca}(pp) := \left| \Pr[\text{Exp}_{\text{UE},\mathcal{A}}^{up\text{-}ind\text{-}rcca}(pp, 0) = 1] - \Pr[\text{Exp}_{\text{UE},\mathcal{A}}^{up\text{-}ind\text{-}rcca}(pp, 1) = 1] \right|.$$

Experiment $\text{Exp}_{\text{UE},\mathcal{A}}^{up\text{-}ind\text{-}rcca}(pp, b)$
  $(sp, k_0, \Delta_0, \mathbf{Q}, \mathbf{K}, \mathbf{T}, \mathbf{C}^*) \leftarrow$ Init(pp)
  $(m_0, m_1, state) \leftarrow_R \mathcal{A}^{\text{Enc,Dec,Next,ReEnc,Corrupt}}(sp)$
  proceed only if $|m_0| = |m_1|$ and $m_0, m_1 \in \mathcal{M}_{sp}$
  $c^* \leftarrow_R$ UE.Enc(k_e, m_b), $\mathbf{M}^* \leftarrow (m_0, m_1)$, $\mathbf{C}^* \leftarrow \{e\}$, $e^* \leftarrow e$
  $b' \leftarrow_R \mathcal{A}^{\text{Enc,Dec,Next,ReEnc,Corrupt}}(c^*, state)$
  return b' if $\mathbf{K} \cap \widehat{\mathbf{C}}^* = \emptyset$, i.e. $\mathcal{A}$ did not trivially win. (Else abort.)

Handling Queries of (Potential) Challenge Ciphertexts. As in the standard RCCA definition, we do not allow any decryption of ciphertexts that decrypts to either of the two challenge plaintexts m_0, or m_1. This is expressed via the isChallenge predicate that is checked for every Dec and ReEnc query and is defined as follows:

isChallenge(k_i, c):
  – If UE.Dec(k_i, c) = m_b where $m_b \in \mathbf{M}^*$, return true. Else, return false.

Whereas the decryption oracle will ignore any query where isChallenge $(k_e, c) = \mathtt{true}$, the re-encryption oracle is more generous: When ReEnc is invoked on (i, c) where isChallenge$(k_i, c) = \mathtt{true}$, it will still update the ciphertext into the current epoch e. The oracle might mark the epoch e as challenge-equal though, preventing the adversary from corrupting the secret key of epoch e. However, this is only done when c is *not* a previous oracle response from an encryption query (or re-encryption of such a response). That is, the re-encryption oracle will treat ciphertexts normally when they can be traced down to a honest encryption query, even when they encrypt one of the challenge messages. This added "generosity" is crucial for re-encryptions, as otherwise an adversary would not be able to see *any* re-encryption from a ciphertext that encrypts the same message as the challenge *and* corrupt the secret key in such an epoch.

Plaintext Integrity. Another impact of having a probabilistic instead of a deterministic ReEnc algorithm is that ciphertext integrity can no longer be guaranteed: When the adversary has corrupted an update token it can create various valid ciphertexts by updating an old ciphertext into the new epoch. Thus, instead we aim for the notion of plaintext integrity and request the adversary to produce a ciphertext that decrypts to a message for which he does not trivially know an encryption of.

The oracles used in this game are as in the UP-IND-RCCA definition above, except that we no longer need the isChallenge predicate and the set of honest queries $\mathbf{Q}$ only records the plaintexts but not the ciphertexts.

Next(), Corrupt($\{$key, token$\}, i$): as in CCA game
Enc(m): Returns $c \leftarrow_R$ UE.Enc(k_e, m) and sets $\mathbf{Q} \leftarrow \mathbf{Q} \cup \{(e, m)\}$.
Dec(c): Returns $m \leftarrow$ UE.Dec(k_e, c) and sets $\mathbf{Q} \leftarrow \mathbf{Q} \cup \{(e, m)\}$.
ReEnc(c, i): Returns c_e, the re-encryption of c from epoch i to the current epoch e. It also sets $\mathbf{Q} \leftarrow \mathbf{Q} \cup \{(e, m)\}$ where $m \leftarrow$ UE.Dec(k_e, c_e).

As in our definition of UP-INT-CTXT, we have to capture all plaintexts for which the adversary can easily create ciphertexts, based on the information he learned through the oracles and by exploiting his knowledge of some of the secret keys and update tokens. Again, the first case in our definition of UP-INT-PTXT security excludes adversaries that have corrupted a secret key and all tokens from then on, as this allows to create valid ciphertexts for *all* plaintexts

Definition 13. *An updatable encryption scheme* UE *is called UP-INT-PTXT secure if for any PPT adversary $\mathcal{A}$ the following advantage is negligible in* κ:$\mathsf{Adv}_{\mathsf{UE},\mathcal{A}}^{up\text{-}int\text{-}ptxt}(pp) := \Pr[\mathsf{Exp}_{\mathsf{UE},\mathcal{A}}^{up\text{-}int\text{-}ptxt}(pp) = 1].$

Experiment $\mathsf{Exp}_{\mathsf{UE},\mathcal{A}}^{up\text{-}int\text{-}ptxt}(pp)$
   $(sp, k_0, \Delta_0, \mathbf{Q}, \mathbf{K}, \mathbf{T}) \leftarrow$ Init(pp)
   $c^* \leftarrow_R \mathcal{A}^{\mathsf{Enc},\mathsf{Dec},\mathsf{Next},\mathsf{ReEnc},\mathsf{Corrupt}}(sp)$
   return 1 if UE.Dec($k_{e_{\mathsf{end}}}, c^*$) $= m^* \neq \bot$ and $(e_{\mathsf{end}}, m^*) \notin \mathbf{Q}^*$,
      and $\nexists e \in \mathbf{K}$ where $i \in \mathbf{T}$ for $i = e$ to e_{end}; i.e. if $\mathcal{A}$ does not trivially win.

Our definition of trivial plaintext forgeries $\mathbf{Q}^*$ is to the one for CTXT security. That is, when the adversary has received a ciphertext for a message m in epoch e (which is recorded in $\mathbf{Q}$) and the update token Δ_{e+1} (which is recorded in $\mathbf{T}$) for the following epoch, then we (iteratively) declare m to be a trivial forgery for epoch $e + 1$ as well. We start with $\mathbf{Q}^* \leftarrow \emptyset$ and amend the set as follows:

for each $(e, m) \in \mathbf{Q}$:
  set $\mathbf{Q}^* \leftarrow \mathbf{Q}^* \cup (e, m)$, and $i \leftarrow e + 1$
  while $i \in \mathbf{T}$: set $\mathbf{Q}^* \leftarrow \mathbf{Q}^* \cup (i, m)$ and $i \leftarrow i + 1$

4.2   RCCA and PTXT Secure Construction

We now present our construction with probabilistic re-encryption that achieves our definition of RCCA and PTXT security (under leakage). The main idea is to use the Naor-Yung (NY) CCA-transform [20] (for public-key schemes). That is, a message is encrypted under two (public) keys of a CPA-secure encryption scheme and accompanied with a NIZK proof that both ciphertexts indeed encrypt the same message. By relying on building blocks that support key-rotation, we then lift this approach into the setting of updatable encryption. For the key-rotatable CPA-secure encryption we use the RISE scheme as presented in [17], and NIZKs are realized with Groth–Sahai (GS) proofs which provide the malleability capabilities that are necessary for key rotation. As in the case of our deterministic scheme presented in Sect. 3, we prove the full security of the updatable scheme based on static properties of the underlying building blocks and simulation-based properties of their token generation and update procedures.

A downside of this NY approach is that it yields a *public key* encryption scheme in disguise. That is, we expose the resulting public key scheme in a symmetric key style and only use the "public key" for key rotation. However, the corruption of an update token then allows the adversary to create valid ciphertexts for messages of his choice. Thus, this scheme would not achieve the desirable PTXT security yet. We therefore extend the NY approach and let each encryption also contain a proof that one knows a valid signature on the underlying plaintext. This combined scheme then satisfies both RCCA and PTXT security.

The crucial feature of this overall approach is that it allows for *public verifiability* of well-formedness of ciphertexts, and thus provides security under arbitrary (as opposed to queried) re-encryption attacks.

Structure of the Rest of This Section. We start with an overview of GS proofs systems and their essential properties. We continue with *perfect* re-encryption, a stronger definition than randomness-preserving. Then, we give the intuition and definition of the basic NY-based RCCA-secure updatable encryption scheme. Finally we describe how to add plaintext integrity.

Linearly Malleable Proofs. As our proof system, we use Groth–Sahai proofs which is a so-called *commit-and-prove system* [5,12]. That is, one (first) commits to a witness w (with randomness r) and then proves statement(s) $stmt$ about the committed witness by running $\pi \leftarrow \mathsf{Prove}(crs, stmt, w, r)$. The statement(s) $stmt$ are "quadratic" equations, e.g. pairing product equations. See [16] for details.

Groth–Sahai proofs are a so-called *dual-mode* proof system, which has two setups: $\mathsf{GS.SetupH}(pp)$ (resp. $\mathsf{GS.SetupB}(pp)$) generates a hiding (resp. binding) crs for which commitments are perfectly hiding (resp. perfectly binding) and the proof π is perfectly zero-knowledge (resp. perfectly sound). Moreover, binding commitments to groups are *extractable*.

Groth–Sahai proofs offer extra-functionality. They are perfectly *rerandomis-able*, i.e. the commitments and proofs can be re-randomised. Also, they are *linearly malleable*. Roughly, given a set of "quadratic" equations, one can apply (certain) linear transformations to the witness and statement (i.e. the constants in the equation), which map satisfying assignments to satisfying assignments, and compute adapted commitments and proofs. In particular, the commitments are homomorphic. See [16] or [4,9] for more.

Perfect Re-encryption. Perfect re-encryption is a strengthening of randomness-preserving re-encryption. It assures that decrypt-then-encrypt has the same distribution as re-encryption, without any exceptions. In particular, it does neither require the encryption randomness, nor is it restricted to valid ciphertexts.

Definition 14. *Let* UE *be an updatable encryption scheme where* $\mathsf{UE.ReEnc}$ *is probabilistic. We say that re-encryption (of* UE*) is* **perfect***, if for all* $sp \leftarrow_R \mathsf{UE.GenSP}(pp)$*, all keys* $k^{old}, k^{new} \leftarrow_R \mathsf{UE.GenKey}(sp)$*, token* $\Delta \leftarrow_R \mathsf{UE.GenTok}(k^{old}, k^{new})$*, and all ciphertexts* c*, we have*

$$\mathsf{UE.Enc}(k^{new}, \mathsf{UE.Dec}(k^{old}, c)) \stackrel{dist}{\equiv} \mathsf{UE.ReEnc}(\Delta, c).$$

Note that $\mathsf{Enc}(k, \bot) = \bot$ *by definition.*

The General Idea: RCCA Security via NY Transform. Our first goal is to build a UP-IND-RCCA-secure updatable encryption scheme. which we achieve via the double-encryption technique of Naor-Yung [20] using key-rotatable building blocks: we use a linear encryption with a *linearly malleable* NIZK, namely RISE (i.e. ElGamal-based updatable encryption) with Groth–Sahai proofs [12]. The malleability and re-randomizability of GS proofs allow for key rotation and ciphertext re-randomisation (as part of the re-encryption procedure).

A double-encryption with a *simulation sound* consistency proof (as formalized in [11,22]) is too rigid and yields CCA security. We must allow certain transformations of the ciphertext, namely re-randomisation and re-encryption. Thus, we weaken our security to RCCA and rely on a relaxation of simulation soundness, which still ensures that the adversary cannot maul the message, but allows re-randomisation and re-encryption.

We achieve this property by the following variation of a standard technique, which was previously used in conjunction with Groth–Sahai proofs, e.g. in [13].

Our NIZK proves that either the NY statement holds, i.e., two ciphertexts $c_1 = \text{Enc}(pk_1, m_1)$ and $c_2 = \text{Enc}(pk_2, m_2)$ encrypt the same message $m_1 = m_2$, or m_1, m_2 (possibly being different) are signed under a signature verification key which is part of the system parameters. In the security proof the simulator will be privy of the signing key and thus can produce valid NIZK proofs for inconsistent ciphertexts. Further, the signature scheme is *structure-preserving*, which allows to hide the signature σ and its verification $\text{Verify}(vk, m_1, m_2, \sigma)$ in the NIZK proof. Note that the signature scheme does *not* have to be key-rotatable as the key is fixed throughout all epochs.

Definition 15 (NYUE). *Our Naor–Yung-like transformation* NYUE *of the key-rotatable encryption* RISE, *using GS proofs and a structure-preserving signature* SIG, *is defined as:*

NYUE.GenSP(pp): *Run* $\text{crs}_{GS} \leftarrow_R \text{GS.SetupH}(pp)$, $sp_{Enc} \leftarrow_R \text{RISE.GenSP}(pp)$, $sp_{SIG} \leftarrow_R \text{SIG.GenSP}(pp)$ *and* $(\_, vk_{SIG}) \leftarrow_R \text{SIG.GenKey}(sp_{SIG})$. *Return* $sp = (\text{crs}_{GS}, sp_{Enc}, (sp_{SIG}, vk_{SIG}))$.

NYUE.GenKey(sp): *Run* $k_i \leftarrow_R \text{RISE.GenKey}(sp_{Enc})$ *for* $i = 1, 2$ *and parse* $k_i = (sk_i, pk_i)$. *Let* $sk = (sk_1, sk_2)$ *and* $pk = (pk_1, pk_2)$. *Return* $k = (sk, pk)$.

NYUE.Enc($k, m; r_1, r_2$): *Parse* $k = (sk, pk)$. *Compute* $c_i = \text{RISE.Enc}(pk_i, m; r_i)$ *for* $i = 1, 2$ *and the following proof* $\pi \leftarrow_R \text{NIZK}(\text{OR}(S_{NY}, S_{SIG}))$ *with common input* sp, pk_1, pk_2, c_1, c_2 *where*[4]
  - S_{NY}: $\exists \widehat{m}, \widehat{r_1}, \widehat{r_2}$: $\text{RISE.Enc}(pk_1, \widehat{m}; \widehat{r_1}) = c_1, \wedge \text{RISE.Enc}(pk_2, \widehat{m}; \widehat{r_2}) = c_2$
  - S_{SIG}: $\exists \widehat{m_1}, \widehat{m_2}, \widehat{r_1}, \widehat{r_2}, \widehat{\sigma}$: $\text{RISE.Enc}(pk_1, \widehat{m_1}; \widehat{r_2}) = c_1 \wedge \text{RISE.Enc}(pk_2, \widehat{m_2}; \widehat{r_2}) = c_2 \wedge \text{SIG.Verify}(vk_{SIG}, (\widehat{m_1}, \widehat{m_2}), \widehat{\sigma}) = 1$

Return (c_1, c_2, π).

NYUE.Dec($k, (c_1, c_2, \pi)$): *Parse* $k = (sk, pk)$ *and verify the proof* π *w.r.t.* $pk = (pk_1, pk_2)$. *If* π *is valid, return* $\text{RISE.Dec}(sk_1, c_1)$, *and* $\perp$ *otherwise.*

NYUE.GenTok($k^{\text{old}}, k^{\text{new}}$): *Compute* $\Delta_i \leftarrow_R \text{RISE.GenTok}(k_i^{\text{old}}, k_i^{\text{new}})$ *for* $i = 1, 2$ *where* k^{old} *and* k^{new} *is parsed as in* NYUE.GenKey. *Return* $\Delta = (\Delta_1, \Delta_2)$.

NYUE.ReEnc(Δ, c): *is sketched below.*

We use a hiding crs_{GS} in the above construction to attain *perfect* re-encryption. just like RISE, c.f. [16].

For the ease of exposition, we use RISE for both encryptions in the NY transform. If one follows the classical NY approach that immediately deletes sk_2 (in epoch 0), it would be sufficient to require key-rotatable encryption only for c_1, whereas encryption for c_2 merely needs to be re-randomizable (as we also aim for UP-REENC security).

Re-encryption for NYUE. The high-level idea of the re-encryption is using the linear malleability, and re-randomisability of RISE and GS proofs. For NYUE.ReEnc(Δ, c) with $c = (c_1, c_2, \pi)$ we proceed in four steps. Steps 2 and 3 constitute a computation of RISE.ReEnc, separated into key-rotation and re-randomisation, c.f. [16].

---

[4] Here we exploit the public key nature of the construction, i.e., we only need pk_i (not sk_i) for verifying consistency proofs.

(1) **Verify ciphertext.** Note that the re-encryption tokens of RISE (and therefore NYUE) contain essentially the old and new public keys. We use this to let NYUE.ReEnc first verify the consistency proof of a ciphertext before starting the update procedure. Thus, re-encryption only works for well-formed, *decryptable* ciphertexts.

(2) **Key rotation.** We use the key rotation of RISE on the ciphertexts parts c_1 and c_2 of $c = (c_1, c_2, \pi)$, but *without* the implicit *re-randomisation*. Additionally, we use malleability of GS proofs to adapt the proof π.

(3) **Re-randomise** c_1, c_2. We re-randomise the RISE (i.e. ElGamal) ciphertexts c_1, c_2, thus completing the computation of $\mathsf{RISE.ReEnc}(\Delta_i, c_i)$ for $i = 1, 2$. Additionally, we use malleability of GS proofs to adapt the proof π.

(4) **Re-randomise** π. We re-randomise the proof π using re-randomisability of GS proofs.

Thus, we first switch to the new key, and then ensure that the ciphertext is distributed identically to a fresh encryption by re-randomising the RISE ciphertexts and the GS proofs (both of which are perfectly re-randomisable).

UP-IND-RCCA Security of NYUE. We now argue how NYUE achieves our notion of UP–IND–RCCA security that captures arbitrary re-encryption attacks. First, we observe that NYUE has *perfect* re-encryption, i.e., a re-encrypted ciphertext (c_1', c_2', π') has the same distribution as a fresh encryption (Definition 14). This follows because RISE has perfect re-encryption and GS proofs with hiding CRS have perfect re-randomisation. Furthermore, NYUE satisfies simulatable token generation *under (key-)leakage*, see [16].

Lemma 4. *The updatable encryption scheme* NYUE *has perfect re-encryption and simulatable token generation under leakage* $\mathsf{leak}(k) = pk$, *c.f. [16].*

Lemma 4 follows easily from token simulation for RISE, see [16]. The UP-IND-RCCA security of NYUE is shown analogous to UP-IND-CCA security in Theorem 1. That is, we bootstrap the UP-IND-RCCA security from the (static) IND-RCCA security of NYUE, perfect re-encryption and token simulation. By a standard reduction, the underlying encryption of NYUE is IND-RCCA secure (under leakage $\mathsf{leak}(k) = pk$), see [16]. There are three major differences compared to UP-IND-CCA:

First, NYUE.ReEnc uses the public verifiability of ciphertexts to reject invalid inputs, i.e., it updates only ciphertexts for which NYUE.Dec will not return $\bot$. Hence, the decrypt-then-encrypt strategy (used in the proof of Theorem 1) is not impacted by allowing arbitrary requests in the ReEnc oracle. Consequently, the queried restriction is not giving the adversary any additional advantage.

Second, re-encryption is *perfect*, which is stronger than randomness-preserving re-encryption. This simplifies the proof strategy slightly. Third, $\mathsf{leak}(k) = pk$ is non-trivial, unlike for the deterministic schemes. All in all, we obtain:

Proposition 1 ([16]). *Suppose the SXDH assumption holds in* $(e, \mathbb{G}_1, \mathbb{G}_2, \mathbb{G}_T)$, *and* SIG *is (one-time) EUF-CMA secure. Then the updatable encryption scheme* NYUE *from Definition 15 is UP-IND-RCCA secure.*

The SXDH assumption guarantees the security of RISE and GS proofs.

NYUAE Construction: Adding PTXT Security. As discussed, NYUE is a public key encryption scheme in disguise (with the public key "hidden" in the update token). Thus, a (corrupt) data host can trivially create new ciphertext to chosen messages, and thus we do not achieve the desired PTXT security yet.

To obtain such plaintext integrity, using a structure-preserving *key-rotatable* MAC [15] on the plaintext seems a straightforward solution. However, for proving security against arbitrary re-encryption attacks, we need that ciphertext validity is *publicly verifiable*. Thus, we use the signature from [15] instead (which is constructed from the MAC). Furthermore, we hide the signature (and its verification) behind a GS proof, to ensure confidentiality.

Updatable Signatures. Opposed to the signature SIG used in NYUE for the simulatability of the main GS proof, we need the signature scheme which ensures integrity of the plaintext to be key-rotatable and updatable as well. The definition of an updatable signature scheme USIG is straightforward and given in [16]. We stress that we will not require USIG to be secure in the updatable setting, but only need standard static (one-time) EUF-CMA security in combination with generic properties of the token generation (c.f. Definition 10).

We now incorporate plaintext integrity into the NYUE construction using such a key-rotatable signature USIG. For encryption, we additionally sign the plaintext with USIG and include this signature in the main NY statement of the GS proof π. That is, S_{NY+I} now asserts that c_1 and c_2 encrypt the same USIG-*signed* message. As before, we use concrete instantiations of all key-rotatable building blocks to avoid a cumbersome abstraction of malleability properties. We use the one-time signature OTS from [15, Fig. 2] for USIG with simulatable token generation and malleable signature verification. In [16] we recall their scheme, define its key-rotation capabilities, and show that it satisfies all required properties (OTS is one-time EUF-CMA secure under the SXDH assumption).

In the following we describe our final construction NYUAE. For the sake of brevity, we refer to the NYUE scheme whenever we use it in an unchanged way.

Definition 16 (NYUAE). *The Naor-Yung transformation with plaintext integrity from key-rotatable encryption* RISE, *GS proofs and structure-preserving signature* SIG *(with* RISE *and* SIG *being abstracted away in the* NYUE *scheme), and a key-rotatable structure-preserving signature* OTS *(c.f. [16]) is defined as follows:*

NYUAE.GenSP(pp): *Run* $sp_{NYUE} \leftarrow_R$ NYUE.GenSP(pp), *and* $sp_{OTS} \leftarrow_R$ OTS.GenSP(pp). *Return* $sp = (sp_{NYUE}, sp_{OTS})$.

NYUAE.GenKey(sp): *Run* $k_{NYUE} \leftarrow_R$ NYUE.GenKey(sp_{NYUE}), *and* $(sk_{OTS}, vk) \leftarrow_R$ OTS.GenKey(sp_{OTS}). *Let* $sk = (sk_{NYUE}, sk_{OTS})$, $pk = (pk_{NYUE}, vk)$. *Return* $k = (sk, pk)$.

NYUAE.Enc($k, m; r_1, r_2$): *Parse* $k = ((sk_{NYUE}, sk_{OTS}), (pk_{NYUE}, vk))$ *compute* c_1, c_2 *as in* NYUE, $\sigma \leftarrow$ OTS.Sign(sk_{OTS}, m) *and a proof* $\pi \leftarrow_R$ NIZK(OR(S_{NY+I}, S_{SIG})) *where*

- S_{NY+1}: $\exists\, \widehat{m}, \widehat{r_1}, \widehat{r_2}, \widehat{\sigma}$: OTS.Verify$(pk_{OTS}, \widehat{m}, \widehat{\sigma}) = 1 \wedge S_{NY}$

and with S_{NY}, S_{SIG} defined as in NYUE (Definition 15). Return (c_1, c_2, π).

NYUAE.Dec$(k, (c_1, c_2, \pi))$: If π is valid, return RISE.Dec(k_1, c_1), and $\perp$ else.

NYUAE.GenTok(k^{old}, k^{new}): Run $\Delta_{NYUE} \leftarrow_R$ NYUE.GenTok$(k^{old}_{NYUE}, k^{new}_{NYUE})$ and $\Delta_{OTS} \leftarrow_R$ OTS.GenTok$(k^{old}_{OTS}, k^{new}_{OTS})$. Return $\Delta = (\Delta_{NYUE}, \Delta_{OTS})$.

NYUAE.ReEnc(Δ, c): is as NYUE.ReEnc (Definition 15), but also adapts the proof of knowledge of an OTS-signature.

The details for generating, verifying and updating the proof π are given in [16]. The proof of security as an updatable encryption scheme follows the usual blueprint. As for NYUE, UP-REENC security follows from [16].

Theorem 2. *Suppose* SIG *is unbounded EUF-CMA secure, and SXDH holds in* $(\mathbb{G}_1, \mathbb{G}_2, \mathbb{G}_T, e)$. *Then* NYUAE *is UP-IND-RCCA and UP-INT-PTXT secure.*

Acknowledgments. We thank Kenny Paterson for fruitful discussions at early stages of this work. We also thank the reviewers for helpful suggestions. The first author is supported by the German Federal Ministry of Education and Research within the framework of the project "Sicherheit kritischer Infrastrukturen (SKI)" in the Competence Center for Applied Security Technology (KASTEL). The second author was supported by the European Union's Horizon 2020 research and innovation program under Grant Agreement No. 786725 (OLYMPUS). The third author is supported by DFG grant RU 1664/3-1 and KASTEL.

References

1. Boneh, D., Lewi, K., Montgomery, H., Raghunathan, A.: Key homomorphic PRFs and their applications. In: Canetti, R., Garay, J.A. (eds.) CRYPTO 2013, Part I. LNCS, vol. 8042, pp. 410–428. Springer, Heidelberg (2013). https://doi.org/10.1007/978-3-642-40041-4_23
2. Boneh, D., Lewi, K., Montgomery, H., Raghunathan, A.: Key homomorphic PRFs and their applications. Cryptology ePrint Archive, Report 2015/220 (2015). http://eprint.iacr.org/2015/220
3. Canetti, R., Krawczyk, H., Nielsen, J.B.: Relaxing chosen-ciphertext security. In: Boneh, D. (ed.) CRYPTO 2003. LNCS, vol. 2729, pp. 565–582. Springer, Heidelberg (2003). https://doi.org/10.1007/978-3-540-45146-4_33
4. Chase, M., Kohlweiss, M., Lysyanskaya, A., Meiklejohn, S.: Malleable proof systems and applications. Cryptology ePrint Archive, Report 2012/012 (2012). http://eprint.iacr.org/2012/012
5. Escala, A., Groth, J.: Fine-tuning Groth-Sahai proofs. In: Krawczyk, H. (ed.) PKC 2014. LNCS, vol. 8383, pp. 630–649. Springer, Heidelberg (2014). https://doi.org/10.1007/978-3-642-54631-0_36
6. Escala, A., Herold, G., Kiltz, E., Ràfols, C., Villar, J.: An algebraic framework for Diffie-Hellman assumptions. In: Canetti, R., Garay, J.A. (eds.) CRYPTO 2013, Part II. LNCS, vol. 8043, pp. 129–147. Springer, Heidelberg (2013). https://doi.org/10.1007/978-3-642-40084-1_8
7. Everspaugh, A., Paterson, K., Ristenpart, T., Scott, S.: Key rotation for authenticated encryption. Cryptology ePrint Archive, Report 2017/527 (2017). http://eprint.iacr.org/2017/527

8. Everspaugh, A., Paterson, K., Ristenpart, T., Scott, S.: Key rotation for authenticated encryption. In: Katz, J., Shacham, H. (eds.) CRYPTO 2017, Part III. LNCS, vol. 10403, pp. 98–129. Springer, Cham (2017). https://doi.org/10.1007/978-3-319-63697-9_4

9. Fuchsbauer, G.: Commuting signatures and verifiable encryption. In: Paterson, K.G. (ed.) EUROCRYPT 2011. LNCS, vol. 6632, pp. 224–245. Springer, Heidelberg (2011). https://doi.org/10.1007/978-3-642-20465-4_14

10. Fuchsbauer, G., Kamath, C., Klein, K., Pietrzak, K.: Adaptively secure proxy re-encryption. Cryptology ePrint Archive, Report 2018/426 (2018). https://eprint.iacr.org/2018/426

11. Groth, J.: Simulation-sound NIZK proofs for a practical language and constant size group signatures. In: Lai, X., Chen, K. (eds.) ASIACRYPT 2006. LNCS, vol. 4284, pp. 444–459. Springer, Heidelberg (2006). https://doi.org/10.1007/11935230_29

12. Groth, J., Sahai, A.: Efficient noninteractive proof systems for bilinear groups. SIAM J. Comput. **41**(5), 1193–1232 (2012)

13. Hofheinz, D., Jager, T.: Tightly secure signatures and public-key encryption. In: Safavi-Naini, R., Canetti, R. (eds.) CRYPTO 2012. LNCS, vol. 7417, pp. 590–607. Springer, Heidelberg (2012). https://doi.org/10.1007/978-3-642-32009-5_35

14. Jarecki, S., Krawczyk, H., Resch, J.: Threshold partially-oblivious PRFs with applications to key management. Cryptology ePrint Archive, Report 2018/733 (2018). https://eprint.iacr.org/2018/733

15. Kiltz, E., Pan, J., Wee, H.: Structure-preserving signatures from standard assumptions, revisited. In: Gennaro, R., Robshaw, M. (eds.) CRYPTO 2015, Part II. LNCS, vol. 9216, pp. 275–295. Springer, Heidelberg (2015). https://doi.org/10.1007/978-3-662-48000-7_14

16. Klooß, M., Lehmann, A., Rupp, A.: (R)CCA secure updatable encryption with integrity protection. IACR ePrint 2019/222. http://eprint.iacr.org/2019/222

17. Lehmann, A., Tackmann, B.: Updatable Encryption with Post-Compromise Security. In: Nielsen, J.B., Rijmen, V. (eds.) EUROCRYPT 2018, Part III. LNCS, vol. 10822, pp. 685–716. Springer, Cham (2018). https://doi.org/10.1007/978-3-319-78372-7_22

18. Namprempre, C., Rogaway, P., Shrimpton, T.: Reconsidering generic composition. In: Nguyen, P.Q., Oswald, E. (eds.) EUROCRYPT 2014. LNCS, vol. 8441, pp. 257–274. Springer, Heidelberg (2014). https://doi.org/10.1007/978-3-642-55220-5_15

19. Naor, M., Pinkas, B., Reingold, O.: Distributed pseudo-random functions and KDCs. In: Stern, J. (ed.) EUROCRYPT 1999. LNCS, vol. 1592, pp. 327–346. Springer, Heidelberg (1999). https://doi.org/10.1007/3-540-48910-X_23

20. Naor, M., Yung, M.: Public-key Cryptosystems Provably Secure against Chosen Ciphertext Attacks. In: 22nd ACM STOC, May 1990

21. PCI Security Standards Council: Requirements and security assessment procedures. PCI DSS v3.2 (2016)

22. Sahai, A.: Non-malleable non-interactive zero knowledge and adaptive chosen-ciphertext security. In: 40th FOCS, October 1999

Succinct Arguments and Secure Messaging

Aurora: Transparent Succinct Arguments for R1CS

Eli Ben-Sasson[1], Alessandro Chiesa[2(✉)], Michael Riabzev[1], Nicholas Spooner[2], Madars Virza[3], and Nicholas P. Ward[2]

[1] Technion/STARKWare, Haifa, Israel
{eli,mriabzev}@cs.technion.ac.il
[2] UC Berkeley, Berkeley, USA
{alexch,nick.spooner,npward}@berkeley.edu
[3] MIT Media Lab, Cambridge, USA
madars@mit.edu

Abstract. We design, implement, and evaluate a zero knowledge succinct non-interactive argument (SNARG) for Rank-1 Constraint Satisfaction (R1CS), a widely-deployed NP language undergoing standardization. Our SNARG has a transparent setup, is plausibly post-quantum secure, and uses lightweight cryptography. A proof attesting to the satisfiability of n constraints has size $O(\log^2 n)$; it can be produced with $O(n \log n)$ field operations and verified with $O(n)$. At 128 bits of security, proofs are less than 250 kB even for several million constraints, more than 10× shorter than prior SNARGs with similar features.

A key ingredient of our construction is a new Interactive Oracle Proof (IOP) for solving a *univariate* analogue of the classical sumcheck problem [LFKN92], originally studied for *multivariate* polynomials. Our protocol verifies the sum of entries of a Reed–Solomon codeword over any subgroup of a field.

We also provide libiop, a library for writing IOP-based arguments, in which a toolchain of transformations enables programmers to write new arguments by writing simple IOP sub-components. We have used this library to specify our construction and prior ones, and plan to open-source it.

Keywords: Zero knowledge · Interactive Oracle Proofs · Succinct arguments · Sumcheck protocol

1 Introduction

A zero knowledge proof is a protocol that enables one party (the *prover*) to convince another (the *verifier*) that a statement is true without revealing any information beyond the fact that the statement is true. Since their introduction [49], zero knowledge proofs have become fundamental tools not only in the theory of cryptography but also, more recently, in the design of real-world systems with strong privacy properties.

© International Association for Cryptologic Research 2019
Y. Ishai and V. Rijmen (Eds.): EUROCRYPT 2019, LNCS 11476, pp. 103–128, 2019.
https://doi.org/10.1007/978-3-030-17653-2_4

For example, zero knowledge proofs are the core technology in Zcash [1,18], a popular cryptocurrency that preserves a user's payment privacy. While in Bitcoin [65] users broadcast their private payment details in the clear on the public blockchain (so other participants can check the validity of the payment), users in Zcash broadcast *encrypted* transaction details and *prove*, in zero knowledge, the validity of the payments without disclosing what the payments are.

Many applications, including the aforementioned, require that proofs are *succinct*, namely, that proofs scale *sublinearly* in the size of the witness for the statement, or perhaps even in the size of the computation performed to check the statement. This strong efficiency requirement cannot be achieved with statistical soundness (under standard complexity assumptions) [47], and thus one must consider proof systems that are merely computationally sound, known as *argument systems* [34]. Many applications further require that a proof consists of a single non-interactive message that can be verified by anyone; such proofs are cheap to communicate and can be stored for later use (e.g., on a public ledger). Constructions that satisfy these properties are known as (publicly verifiable) *succinct non-interactive arguments* (SNARGs) [46].

In this work we present Aurora, a zero knowledge SNARG for (an extension of) arithmetic circuit satisfiability whose argument size is polylogarithmic in the circuit size. Aurora also has attractive features: it uses a transparent setup, is plausibly post-quantum secure, and only makes black-box use of fast symmetric cryptography (any cryptographic hash function modeled as a random oracle).

Our work makes an exponential asymptotic improvement in argument size over Ligero [4], a recent zero knowledge SNARG with similar features but where proofs scale as the *square root* of the circuit size. For example, Aurora's proofs are 20× smaller than Ligero's for circuits with a million gates (which already suffices for representative applications such as Zcash).

Our work also complements and improves on Stark [13], a recent zero knowledge SNARG that targets computations expressed as bounded halting problems on random access machines. While Stark was designed for a different computation model, we can still study its efficiency when applied to arithmetic circuits. In this case Aurora's prover is faster by a logarithmic factor (in the circuit size) and Aurora's proofs are concretely much shorter, e.g., 15× smaller for circuits with a million gates.

The efficiency features of Aurora stem from a new Interactive Oracle Proof (IOP) that solves a *univariate* analogue of the celebrated sumcheck problem [61], in which query complexity is *logarithmic* in the degree of the polynomial being summed. This is an *exponential* improvement over the original multi-variate protocol, where communication complexity is (at least) *linear* in the degree of the polynomial. We believe this protocol and its analysis are of independent interest.

1.1 The Need for a Transparent Setup

The first succinct argument is due to Kilian [57], who showed how to use collision-resistant hashing to compile any Probabilistically Checkable Proof (PCP) [5,6,9,43] into a corresponding interactive argument. Micali then showed how a similar construction, in the random oracle model, yields succinct *non*-interactive arguments (SNARGs) [63]. Subsequent work [55] noted that if the underlying PCP is zero knowledge then so is the SNARG. Unfortunately, PCPs remain very expensive, and this approach has not led to SNARGs with good concrete efficiency.

In light of this, a different approach was initially used to achieve SNARG implementations with good concrete efficiency [19,67]. This approach, pioneered in [29,45,50,60], relied on combining certain linearly homomorphic encodings with lightweight information-theoretic tools known as linear PCPs [29,54,71]; this approach was refined and optimized in several works [22,23,30,40,51,52]. These constructions underlie widely-used open-source libraries [70] and deployed systems [1], and their main feature is that proofs are very short (a few hundred bytes) and very cheap to verify (a few milliseconds).

Unfortunately, the foregoing approach suffers from a severe limitation, namely, the need for a central party to generate system parameters for the argument system. Essentially, this party must run a probabilistic algorithm, publish its output, and "forget" the secret randomness used to generate it. This party must be trustworthy because knowing these secrets allows forging proofs for false assertions. While this may sound like an inconvenience, it is a *colossal* challenge to real-world deployments. When using cryptographic proofs in distributed systems, relying on a central party negates the benefits of distributed trust and, even though it is invoked only once in a system's life, a party trusted by all users typically *does not exist*!

The responsibility for generating parameters can in principle be shared across multiple parties via techniques that leverage secure multi-party computation [20,32,33]. This was the approach taken for the launch of Zcash [2], but it also demonstrated how unwieldy such an approach is, involving a costly and logistically difficult real-world multi-party "ceremony". Successfully running such a multi-party protocol was a singular feat, and systems without such expensive setup are decidedly preferable.

Some setup is unavoidable because if SNARGs without *any* setup existed then so would sub-exponential algorithms for SAT [81]. Nevertheless, one could still aim for a "transparent setup", namely one that consists of *public randomness*, because in practice it is cheaper to realize. Recent efforts have thus focused on designing SNARGs with transparent setup (see discussion in Sect. 1.4).

1.2 Our Goal

The goal of this paper is to obtain *transparent SNARGs* that satisfy the following desiderata.

- *Post-quantum security.* Practitioners, and even standards bodies [66], have a strong interest in cryptographic primitives that are plausibly secure against efficient quantum adversaries. This is motivated by the desire to ensure long-term security of deployed systems and protocols.

- *Concrete efficiency.* We seek argument systems that not only exhibit good asymptotics (in argument size and prover/verifier time) but also demonstrably offer good efficiency via a prototype.

The second bullet warrants additional context. Most argument systems support an NP-complete problem, so they are in principle equivalent under polynomial-time reductions. Yet, whether such protocols can be efficiently used in practice actually depends on: (a) the particular NP-complete problem "supported" by the protocol; (b) the concrete efficiency of the protocol relative to this problem. This creates a complex tradeoff.

Simple NP-complete problems, like boolean circuit satisfaction, facilitate simple argument systems; but reducing the statements we wish to prove to boolean circuits is often expensive. On the other hand, one can design argument systems for rich problems (e.g., an abstract computer) for which it is cheap to express the desired statements; but such argument systems may use expensive tools to support these rich problems.

Our goal is concretely-efficient argument systems for *rank-1 constraint satisfaction* (R1CS), which is the following natural NP-complete problem: given a vector $v \in \mathbb{F}^k$ and three matrices $A, B, C \in \mathbb{F}^{m \times n}$, can one augment v to $z \in \mathbb{F}^n$ such that $Az \circ Bz = Cz$? (We use "$\circ$" to denote the entry-wise product.)

We choose R1CS because it strikes an attractive balance: it generalizes circuits by allowing "native" field arithmetic and having no fan-in/fan-out restrictions, but it is simple enough that one can design efficient argument systems for it. Moreover, R1CS has demonstrated *strong empirical value*: it underlies real-world systems [1] and there are compilers that reduce program executions to it (see [80] and references therein). This has led to efforts to standardize R1CS formats across academia and industry [3].

1.3 Our Contributions

In this work we study *Interactive Oracle Proofs* (IOPs) [21,69], a notion of "multi-round PCPs" that has recently received much attention [12–15,17,25]. These types of interactive proofs can be compiled into non-interactive arguments in the random oracle model [21], and in particular can be used to construct transparent SNARGs. Building on this approach, we present several contributions: (1) an IOP protocol for R1CS with attractive efficiency features; (2) the design, implementation, and evaluation of a transparent SNARG for R1CS, based on our IOP protocol; (3) a generic library for writing IOP-based non-interactive arguments. We now describe each contribution.

(1) IOP for R1CS. We construct a zero knowledge IOP protocol for rank-1 constraint satisfaction (R1CS) with *linear* proof length and *logarithmic* query complexity.

Given an R1CS instance $\mathcal{C} = (A, B, C)$ with $A, B, C \in \mathbb{F}^{m \times n}$, we denote by $N = \Omega(m+n)$ the total number of non-zero entries in the three matrices and by $|\mathcal{C}|$ the number of bits required to represent these; note that $|\mathcal{C}| = \Theta(N \log |\mathbb{F}|)$. One can view N as the number of "arithmetic gates" in the R1CS instance.

Theorem 1 (informal). *There is an $O(\log N)$-round IOP protocol for R1CS with proof length $O(N)$ over alphabet $\mathbb{F}$ and query complexity $O(\log N)$. The prover uses $O(N \log N)$ field operations, while the verifier uses $O(N)$ field operations. The IOP protocol is public coin and is a zero knowledge proof.*

The core of our result is a solution to a *univariate* analogue of the classical sumcheck problem [61]. Our protocol (including zero knowledge and soundness error reduction) is relatively simple: it is specified in a single page (see Fig. 12 in Sect. 6), given a low-degree test as a subroutine. The low degree test that we use is a recent highly-efficient IOP for testing proximity to the Reed–Solomon code [14].

(2) SNARG for R1CS. We design, implement, and evaluate **Aurora**, a zero knowledge SNARG of knowledge (zkSNARK) for R1CS with several notable features: (a) it only makes black-box use of fast symmetric cryptography (any cryptographic hash function modeled as a random oracle); (b) it has a transparent setup (users merely need to "agree" on which cryptographic hash function to use); (c) it is plausibly post-quantum secure (there are no known efficient quantum attacks against this construction). These features follow from the fact that Aurora is obtained by applying the transformation of [21] to our IOP for R1CS. This transformation preserves both zero knowledge and proof of knowledge of the underlying IOP.

In terms of asymptotics, given an R1CS instance $\mathcal{C}$ over $\mathbb{F}$ with N gates (and here taking for simplicity $\mathbb{F}$ to have size $2^{O(\lambda)}$ where λ is the security parameter), Aurora provides proofs of length $O_\lambda(\log^2 N)$; these can be produced in time $O_\lambda(N \log N)$ and checked in time $O_\lambda(N)$.

For example, setting our implementation to a security level of 128 bits over a 192-bit finite field, proofs range from 50 kB to 250 kB for instances of up to millions of gates; producing proofs takes on the order of several minutes and checking proofs on the order of several seconds. (See Sect. 4 for details.)

Overall, as indicated in Fig. 2, we achieve the smallest argument size among (plausibly) post-quantum non-interactive arguments for circuits, *by more than an order of magnitude*. Other approaches achieve smaller argument sizes by relying on (public-key) cryptography that is insecure against quantum adversaries.

(3) libiop :a library for non-interactive arguments. We provide libiop, a codebase that enables the design and implementation of non-interactive arguments based on IOPs. The codebase uses the C++ language and has three main components: (1) a library for writing IOP protocols; (2) a realization of [21]'s transformation, mapping any IOP written with our library to a corresponding non-interactive argument; (3) a portfolio of IOP protocols, including Ligero [4], Stark [13], and ours.

We plan to open-source libiop under a permissive software license for the community, so that others may benefit from its portfolio of IOP-based arguments, and may even write new IOPs tailored to new applications. We believe that our library will serve as a powerful tool in meeting the increasing demand by practitioners for transparent non-interactive arguments.

1.4 Prior Implementations of Transparent SNARGs

We summarize prior work that has designed *and implemented* transparent SNARGs; see Fig. 2.[1]

Based on Asymmetric Cryptography. *Bulletproofs* [31,35] proves the satisfaction of an N-gate arithmetic circuit via a recursive use of a low-communication protocol for inner products, achieving a proof with $O(\log N)$ group elements. *Hyrax* [79] proves the satisfaction of a *layered* arithmetic circuit of depth D and width W via proofs of $O(D \log W)$ group elements; the construction applies the Cramer–Damgård transformation [41] to doubly-efficient Interactive Proofs [39,48]. Both approaches use Pedersen commitments, and so are *vulnerable to quantum attacks*. Also, in both approaches the verifier performs *many expensive cryptographic operations*: in the former, the verifier uses $O(N)$ group exponentiations; in the latter, the verifier's group exponentiations are linear in the circuit's witness size. (Hyrax allows fewer group exponentiations but with longer proofs; see [79].)

Based on Symmetric Cryptography. The "original" SNARG construction of Micali [55,63] has advantages beyond transparency. First, it is unconditionally secure given a random oracle, which can be instantiated with extremely fast symmetric cryptography.[2] Second, it is plausibly post-quantum secure, in that there are no known efficient quantum attacks. But the construction relies on PCPs, which remain expensive.

IOPs are "multi-round PCPs" that can also be compiled into non-interactive arguments in the random oracle model [21]. This compilation retains the foregoing advantages (transparency, lightweight cryptography, and plausible post-quantum security) and, *in addition*, facilitates greater efficiency, as IOPs have superior efficiency compared to PCPs [12–15,17].

In this work we follow the above approach, by constructing a SNARG based on a new IOP protocol. Two recent works have also taken the same approach, but with different underlying IOP protocols, which have led to different features. We provide both of these works as part of our library (Sect. 5), and experimentally compare them with our protocol (Sect. 4). The discussion below is a qualitative comparison.

- **Ligero** [4] is a non-interactive argument that proves the satisfiability of an N-gate circuit via proofs of size $O(\sqrt{N})$ that can be verified in $O(N)$ cryptographic operations. As summarized in Fig. 1, the IOP underlying Ligero

---

[1] We omit a discussion of prior works without implementations, or that study non-transparent SNARGs; we refer the reader to the survey of Walfish and Blumberg [80] for an overview of sublinear argument systems. We also note that recent work [11] has used lattice cryptography to achieve sublinear zero knowledge arguments that are plausibly post-quantum secure, which raises the exciting question of whether these recent protocols can lead to efficient implementations.

[2] Some cryptographic hash functions, such as BLAKE2, can process almost $1\,\mathrm{GB/s}$ [8].

achieves the same oracle proof length, prover time, and verifier time as our IOP. However, we reduce query complexity from $O(\sqrt{N})$ to $O(\log N)$, which is an exponential improvement, at the expense of increasing round complexity from 2 to $O(\log N)$. The arguments that we obtain are still non-interactive, but our smaller query complexity translates into shorter proofs (see Fig. 2).

– **Stark** [13] is a non-interactive argument for bounded halting problems on a random access machine. Given a program P and a time bound T, it proves that P accepts within T steps on a certain abstract computer (when given suitable nondeterministic advice) via succinct proofs of size polylog(T). Moreover, verification is *also* succinct: checking a proof takes time only $|P| + \text{polylog}(T)$, which is polynomial in the size of the statement and much better than "naive verification" which takes time $\Omega(|P| + T)$. The main difference between Stark and Aurora is the computational models that they support. While Stark supports *uniform* computations specified by a program and a time bound, Aurora supports *non-uniform* computations specified by an explicit circuit (or constraint system). Despite this difference, we can compare the cost of Stark and Aurora with respect to the explicit circuit model, since one can reduce a given N-gate circuit (or N-constraint system) to a corresponding bounded halting problem with $|P|, T = \Theta(N)$.

In this case, Stark's verification time is the same as Aurora's, $O(N)$; this is best possible because just *reading* an N-gate circuit takes time $\Omega(N)$. But Stark's prover is a logarithmic factor more expensive because it uses a switching network to verify a program's accesses to memory. Stark's prover uses an IOP with oracles of size $O(N \log N)$, leading to an arithmetic complexity of $O(N \log^2 N)$. (See Figs. 1 and 2.) Both Stark and Aurora have argument size $O(\log^2 N)$, but additional costs in Stark (e.g., due to switching networks) result in Stark proofs being *one order of magnitude larger* than Aurora proofs. That said, we view Stark and Aurora as complementing each other: Stark offers savings in verification time for succinctly represented programs, while Aurora offers savings in argument size for explicitly represented circuits.

2 Techniques

Our main technical contribution is a linear-length logarithmic-query IOP for R1CS (Theorem 1), which we use to design, implement, and evaluate a transparent SNARG for R1CS. Below we summarize the main ideas behind our protocol, and postpone to Sects. 4 and 5 discussions of our system. In Sect. 2.1, we describe our approach to obtain the IOP for R1CS; this approach leads us to solve the univariate sumcheck problem, as discussed in Sect. 2.2; finally, in Sect. 2.3, we explain how we achieve zero knowledge. In Sect. 2.4 we conclude with a wider perspective on the techniques used in this paper.

2.1 Our Interactive Oracle Proof for R1CS

The R1CS relation consists of instance-witness pairs $((A, B, C, v), w)$, where A, B, C are matrices and v, w are vectors over a finite field $\mathbb{F}$, such that $(Az) \circ (Bz) = Cz$ for $z := (1, v, w)$ and "$\circ$" denotes the entry-wise product.[3] For example, R1CS captures arithmetic circuit satisfaction: A, B, C represent the circuit's gates, v the circuit's public input, and w the circuit's private input and wire values.[4]

We describe the high-level structure of our IOP protocol for R1CS, which has linear proof length and logarithmic query complexity. The protocol tests satisfaction by relying on two building blocks, one for testing the entry-wise vector product and the other for testing the linear transformations induced by the matrices A, B, C. Informally, we thus consider protocols for the following two problems.

- **Rowcheck:** given vectors $x, y, z \in \mathbb{F}^m$, test whether $x \circ y = z$, where "$\circ$" denotes entry-wise product.
- **Lincheck:** given vectors $x \in \mathbb{F}^m, y \in \mathbb{F}^n$ and a matrix $M \in \mathbb{F}^{m \times n}$, test whether $x = My$.

One can immediately obtain an IOP for R1CS when given IOPs for the rowcheck and lincheck problems. The prover first sends four oracles to the verifier: the satisfying assignment z and its linear transformations $y_A := Az, y_B := Bz, y_C := Cz$. Then the prover and verifier engage in four IOPs in parallel:

- An IOP for the lincheck problem to check that "$y_A = Az$". Likewise for y_B and y_C.
- An IOP for the rowcheck problem to check that "$y_A \circ y_B = y_C$".

	protocol type	round complexity	proof length (field elts)	query complexity	prover time (field ops)	verifier time (field ops)
Ligero	IPCP †	2	$O(N)$	$O(\sqrt{N})$	$O(N \log N)$	$O(N)$
Stark	IOP	$O(\log N)$	$O(N \log N)$	$O(\log N)$	$O(N \log^2 N)$	$O(N)$
Aurora	IOP	$O(\log N)$	$O(N)$	$O(\log N)$	$O(N \log N)$	$O(N)$

Fig. 1. Asymptotic comparison of the information-theoretic proof systems underlying Ligero, Stark, and Aurora, when applied to an N-gate arithmetic circuit. † An IPCP [56] is a PCP oracle that is checked via an Interactive Proof; it is a special case of an IOP.

---

[3] Throughout, we assume that $\mathbb{F}$ is "friendly" to FFT algorithms, i.e., $\mathbb{F}$ is a binary field or its multiplicative group is smooth.

[4] The reader may be familiar with a standard arithmetization of circuit satisfaction (used, e.g., in the inner PCP of [5]). Given an arithmetic circuit with m gates and n wires, each addition gate $x_i \leftarrow x_j + x_k$ is mapped to the linear constraint $x_i = x_j + x_k$ and each product gate $x_i \leftarrow x_j \cdot x_k$ is mapped to the quadratic constraint $x_i = x_j \cdot x_k$. The resulting system of equations can be written as $A \cdot ((1, x) \otimes (1, x)) = b$ for suitable $A \in \mathbb{F}^{m \times (n+1)^2}$ and $b \in \mathbb{F}^m$. However, this reduction results in a quadratic blowup in the instance size. There is an alternative reduction due to [45,62] that avoids this.

Finally, the verifier checks that z is consistent with the public input v. Clearly, there exist z, y_A, y_B, y_C that yield valid rowcheck and lincheck instances if and only if (A, B, C, v) is a satisfiable R1CS instance.

The foregoing reduces the goal to designing IOPs for the rowcheck and lincheck problems.

As stated, however, the rowcheck and lincheck problems only admit "trivial" protocols in which the verifier queries all entries of the vectors in order to check the required properties. In order to allow for sublinear query complexity, we need the vectors x, y, z to be *encoded* via some error-correcting code. We use the Reed–Solomon (RS) code because it ensures constant distance with constant rate while at the same time it enjoys efficient IOPs of Proximity [14].

Given an evaluation domain $L \subseteq \mathbb{F}$ and rate parameter $\rho \in [0, 1]$, RS$[L, \rho]$ is the set of all codewords $f \colon L \to \mathbb{F}$ that are evaluations of polynomials of degree less than $\rho |L|$. Then, the encoding of a vector $v \in \mathbb{F}^S$ with $S \subseteq \mathbb{F}$ and $|S| < \rho |L|$ is $\hat{v}|_L \in \mathbb{F}^L$ where $\hat{v}$ is the unique polynomial of degree $|S| - 1$ such that $\hat{v}|_S = v$. Given this encoding, we consider "encoded" variants of the rowcheck and lincheck problems.

- **Univariate rowcheck**: given a subset $H \subseteq \mathbb{F}$ and codewords $f, g, h \in$ RS$[L, \rho]$, check that $\hat{f}(a) \cdot \hat{g}(a) - \hat{h}(a) = 0$ for all $a \in H$. (This is a special case of the definition that we use later.)
- **Univariate lincheck**: given subsets $H_1, H_2 \subseteq \mathbb{F}$, codewords $f, g \in$ RS$[L, \rho]$, and a matrix $M \in \mathbb{F}^{H_1 \times H_2}$, check that $\hat{f}(a) = \sum_{b \in H_2} M_{a,b} \cdot \hat{g}(b)$ for all $a \in H_1$.

Given IOPs for the above problems, we can now get an IOP protocol for R1CS roughly as before. Rather than sending z, Az, Bz, Cz, the prover sends

	name	setup	post quantum?	argument size asymptotic	$N = 10^6$	verifier time	non-interactivity technology
[50][45] [60][29]...	various	private	no	$O_\lambda(1)$	128 B	$O_\lambda(k)$ †	linear PCP + linear encoding
[83]	ZK-vSQL	private	no	$O_\lambda(d \log N)$	N/A	$O_\lambda(N)$	apply [41]-transform to doubly-efficient IP [48, 39]
[79]	Hyrax	public	no	$O_\lambda(d \log N)$ ‡	50 kB	$O_\lambda(N)$	as above (but using a different polynomial commitment)
[31] [35]	Bulletproofs	public	no	$O_\lambda(\log N)$	1.5 kB	$O_\lambda(N)$	recursive inner product argument
[4]	Ligero	public	yes	$O_\lambda(\sqrt{N})$	4.0 MB	$O_\lambda(N)$	apply [21]-transform to IPCP
[13]	Stark	public	yes	$O_\lambda(\log^2 N)$	3.2 MB	$O_\lambda(N)$	apply [21]-transform to IOP
this work	Aurora	public	yes	$O_\lambda(\log^2 N)$	220 kB	$O_\lambda(N)$	apply [21]-transform to IOP

Fig. 2. Comparison of some non-interactive zero knowledge arguments for proving statements of the form "there exists a secret w such that $\mathcal{C}(x, w) = 1$" for a given arithmetic circuit $\mathcal{C}$ of N gates (and depth d) and public input x of size k. The table is grouped by "technology", and for simplicity assumes that the circuit's underlying field has size $2^{O(\lambda)}$ where λ is the security parameter. Approximate argument sizes are given for $N = 10^6$ gates over a cryptographically-large field, and a security level of 128 bits; some argument sizes may differ from those reported in the cited works because size had to be re-computed for the security level and N used here; also, [83] reports no implementation. † Given a per-circuit preprocessing step. ‡ A tradeoff between argument size and verifier time is possible; see [79].

their encodings $f_z, f_{Az}, f_{Bz}, f_{Cz}$. The prover and verifier then engage in rowcheck and lincheck protocols as before, but with respect to these encodings.

For these encoded variants, we achieve IOP protocols with linear proof length and logarithmic query complexity, as required. We obtain a protocol for rowcheck via standard techniques from the probabilistic checking literature [27]. As for lincheck, we do not use any routing and instead use a technique (dating back at least to [9]) to reduce the given testing problem to a *sumcheck instance*. However, since we are not working with multivariate polynomials, we cannot rely on the usual (multivariate) sumcheck protocol. Instead, we present a novel protocol that realizes a univariate analogue of the classical sumcheck protocol, and use it as the testing "core" of our IOP protocol for R1CS. We discuss univariate sumcheck next.

Remark 1. The verifier receives as input an explicit (non-uniform) description of the set of constraints, namely, the matrices A, B, C. In particular, the verifier runs in time that is at least linear in the number of non-zero entries in these matrices (if we consider a sparse-matrix representation for example).

2.2   A Sumcheck Protocol for Univariate Polynomials

A key ingredient in our IOP protocol is a *univariate* analogue of the classical (multivariate) sumcheck protocol [61]. Recall that the classical sumcheck protocol is an IP for claims of the form "$\sum_{a \in H^m} f(a) = 0$", where f is a given polynomial in $\mathbb{F}[X_1, \dots, X_m]$ of individual degree d and H is a subset of $\mathbb{F}$. In this protocol, the verifier runs in time $\text{poly}(m, d, \log |\mathbb{F}|)$ and accesses f at a single (random) location. The sumcheck protocol plays a fundamental role in computational complexity (it underlies celebrated results such as IP = PSPACE [72] and MIP = NEXP [10]) and in efficient proof protocols [39,48,73–77,79,82,83].

We work with univariate polynomials instead, and need a univariate analogue of the sumcheck protocol (see previous subsection): *how can a prover convince the verifier that "$\sum_{a \in H} f(a) = 0$" for a given polynomial $f \in \mathbb{F}[X]$ of degree d and subset $H \subseteq \mathbb{F}$?* Designing a "univariate sumcheck" is not straightforward because univariate polynomials (the Reed–Solomon code) do not have the tensor structure used by the sumcheck protocol for multivariate polynomials (the Reed–Muller code). In particular, the sumcheck protocol has m rounds, each of which reduces a sumcheck problem to a simpler sumcheck problem with one variable fewer. When there is only one variable, however, it is not clear to what simpler problems one can reduce.

Using different ideas, we design a natural protocol for univariate sumcheck in the cases where H is an additive or multiplicative coset in $\mathbb{F}$ (i.e., a coset of an additive or multiplicative subgroup of $\mathbb{F}$).

Theorem 2 (informal). *The univariate sumcheck protocol over additive or multiplicative cosets has a $O(\log d)$-round IOP with proof complexity $O(d)$ over alphabet $\mathbb{F}$ and query complexity $O(\log d)$. The IOP prover uses $O(d \log |H|)$ field operations and the IOP verifier uses $O(\log d + \log^2 |H|)$ field operations.*

We now provide the main ideas behind the protocol, when H is an *additive coset* in $\mathbb{F}$.

Suppose for a moment that the degree d of f is less than $|H|$ (we remove this restriction later). A theorem of Byott and Chapman [36] states that the sum of f over (an additive coset) H is zero if and only if the coefficient of $X^{|H|-1}$ in f is zero. In particular, $\sum_{a \in H} f(a)$ is zero if and only if f has degree less than $|H| - 1$. Thus, the univariate sumcheck problem over H when $d < |H|$ is equivalent to low-degree testing.

The foregoing suggests a natural approach: test that f has degree less than $|H| - 1$. Without any help from the prover, the verifier would need at least $|H|$ queries to f to conduct such a test, which is as expensive as querying all of H. However, the prover can help by engaging with the verifier in an IOP of Proximity for the Reed–Solomon code. For this we rely on the recent construction of Ben-Sasson et al. [14], which has proof length $O(d)$ and query complexity $O(\log d)$.

In our setting, however, we need to also handle the case where the degree d of f is larger than $|H|$. For this case, we observe that we can split any polynomial f into two polynomials g and h such that $f(x) \equiv g(x) + \prod_{\alpha \in H}(x - \alpha) \cdot h(x)$ with $\deg(g) < |H|$ and $\deg(h) < d - |H|$; in particular, f and g agree on H, and thus so do their sums on H. This observation suggests the following extension to the prior approach: the prover sends g (as an oracle) to the verifier, and then the verifier performs the prior protocol with g in place of f. Of course, a cheating prover may send a polynomial g that has nothing to do with f, and so the verifier must also ensure that g is consistent with f. To facilitate this, we actually have the prover send h rather than g; the verifier can then "query" $g(x)$ as $f(x) - \prod_{\alpha \in H}(x - \alpha) \cdot h(x)$; the prover then shows that f, g, h are all of the correct degrees.

A similar reasoning works when H is a multiplicative coset in $\mathbb{F}$. It remains an interesting open problem to establish whether the foregoing can be extended to any subset H in $\mathbb{F}$.

2.3   Efficient Zero Knowledge from Algebraic Techniques

The ideas discussed thus far yield an IOP protocol for R1CS with linear proof length and logarithmic query complexity. However these by themselves do not provide zero knowledge.

We achieve zero knowledge by leveraging recent algebraic techniques [17]. Informally, we adapt these techniques to achieve efficient zero knowledge variants of key sub-protocols, including the univariate sumcheck protocol and low-degree testing, and combine these to achieve a zero knowledge IOP protocol for R1CS.

We summarize the basic intuition for how we achieve zero knowledge in our protocols.

First, we use *bounded independence*. Informally, rather than encoding a vector $z \in \mathbb{F}^H$ by the unique polynomial of degree $|H| - 1$ that matches z on H, we instead sample uniformly at random a polynomial of degree, say, $|H| + 9$ conditioned on matching z on H. Any set of 10 evaluations of such a polynomial are independently and uniformly distributed in $\mathbb{F}$ (and thus reveal no information about z), *provided these evaluations are outside of H*. To ensure this latter condition, we choose the evaluation domain L of Reed–Solomon codewords to be disjoint from H. Thus, for example, if H is a linear space (an additive subgroup of $\mathbb{F}$) then we choose L to be an affine subspace (a coset of some additive subgroup), since the underlying machinery for low-degree testing (e.g., [14]) requires codewords to be evaluated over algebraically-structured domains. All of our protocols are robust to these variations.

Bounded independence alone does not suffice, though. For example, in the sumcheck protocol, consider the case where the input vector $z \in \mathbb{F}^H$ is all zeroes. The prover samples a random polynomial $\hat{f}$ of degree $|H|+9$, such that $\hat{f}(a) = 0$ for all $a \in H$, and sends its evaluation f over L disjoint from H to the verifier. As discussed, any ten queries to f result in ten independent and uniformly random elements in $\mathbb{F}$. Observe, however, that when we run the sumcheck protocol on f, the polynomial g (the remainder of $\hat{f}$ when divided by $\prod_{\alpha \in H}(x - \alpha)$) is the zero polynomial: all randomness is removed by the division.

To remedy this, we use *self-reducibility* to reduce a sumcheck claim about the polynomial f to a sumcheck claim about a random polynomial. The prover first sends a random Reed–Solomon codeword r, along with the value $\beta := \sum_{a \in H} r(a)$. The verifier sends a random challenge $\rho \in \mathbb{F}$. Then the prover and verifier engage in the univariate sumcheck protocol with respect to the new claim "$\sum_{a \in H} \rho f(a) + r(a) = \beta$". Since r is uniformly random, $\rho f + r$ is uniformly random for any ρ, and thus the sumcheck protocol is performed on a random polynomial, which ensures zero knowledge. Soundness is ensured by the fact that if f does not sum to 0 on H then the new claim is true with probability $1/|\mathbb{F}|$ over the choice of ρ.

2.4 Perspective on Our Techniques

A linear-length logarithmic-query IOP for a "circuit-like" NP-complete relation like R1CS (Theorem 1) may come as a surprise. We wish to shed some light on our IOP construction by connecting the ideas behind it to prior ideas in the probabilistic checking literature, and use these connections to motivate our construction.

A significant cost in all known PCP constructions with good proof length is using *routing networks* to reduce combinatorial objects (circuits, machines, and so on) to structured algebraic ones;[5] routing also plays a major role in many IOPs [12,13,15,17]. While it is plausible that one could adapt routing techniques to

---

[5] Polishchuk and Spielman [68] reduce boolean circuit satisfaction to a trivariate algebraic coloring problem with "low-degree" neighbor relations, by routing the circuit's wires over an arithmetized routing network. Ben-Sasson and Sudan [27] reduce non-deterministic machine computations to a univariate algebraic satisfaction problem by routing the machine's memory accesses over another arithmetized routing network. Routing is again a crucial component in the linear-size sublinear-query PCPs of [24].

route the constraints of an R1CS instance (similarly to [68]), such an approach would likely incur logarithmic-factor overheads, precluding *linear*-size IOPs.

A recent work [16] achieves linear-length constant-query IOPs for boolean circuit satisfaction *without routing the input circuit*. Unfortunately, [16] relies on other expensive tools, such as algebraic-geometry (AG) codes and quasilinear-size PCPs of proximity [27]; moreover, it is not zero knowledge. Informally, [16] tests arbitrary (unstructured) constraints by invoking a sumcheck protocol [61] on a $O(1)$-wise tensor product of AG codes; this latter is then locally tested via tools in [26, 27].

One may conjecture that, to achieve an IOP for R1CS like ours, it would suffice to merely replace the AG codes in [16] with the Reed–Solomon code, since both codes have constant rate. But taking a tensor product exponentially deteriorates rate, and testing proximity to that tensor product would be expensive.

An alternative approach is to solve a sumcheck problem *directly* on the Reed–Solomon code. Existing protocols are not of much use here: the multivariate sumcheck protocol relies on a tensor structure that is *not* available in the Reed–Solomon code, and recent IOP implementations either use routing [12, 13] or achieve only sublinear query complexity [4].

Instead, we design a completely new IOP for a sumcheck problem on the Reed–Solomon code. We then combine this solution with ideas from [16] (to avoid routing) and from [17] (to achieve zero knowledge) to obtain our linear-length logarithmic-query IOP for R1CS. Along the way, we rely on recent efficient proximity tests for the Reed–Solomon code [14].

3 Roadmap

In Sect. 4 we evaluate Aurora, and compare it to other IOP-based SNARGs. In Sect. 5 we describe the implementation. In Sect. 6 we present the underlying IOP for R1CS. Figure 3 summarizes the structure of this protocol. For details of this construction, including proofs of theorems, we refer the reader to the full version.

Throughout, we focus on the case where all relevant domains are *additive* cosets (affine subspaces) in $\mathbb{F}$. The case where domains are *multiplicative* cosets is similar, with only minor modifications. Moreover, while for convenience we limit

IOP for R1CS (Section 6)

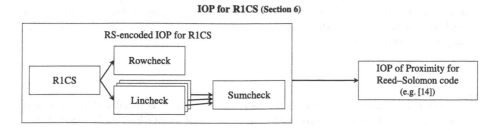

Fig. 3. Structure of our IOP for R1CS in terms of key sub-protocols.

our discussions to establishing soundness, all protocols described in this paper are easily seen to satisfy the stronger notion of proof of knowledge. Informally, this is because we prove soundness by showing that oracles sent by convincing provers can be decoded to valid witnesses.

4 Evaluation

In Sect. 4.1 we evaluate the performance of Aurora. Then, in Sect. 4.2 we compare Aurora with Ligero [4] and Stark [13], two other IOP-based SNARGs. Our experiments not only demonstrate that Aurora's performance matches the theoretical predictions implied by the protocol but also that Aurora achieves the smallest argument size of any IOP-based SNARG, *by more than an order of magnitude.*

That said, there is still a sizable gap between the argument sizes of IOP-based SNARGs and other SNARGs that use public-key cryptographic assumptions vulnerable to quantum adversaries; see Fig. 2 for how argument sizes vary across these. It remains an exciting open problem to close this gap.

Experiments ran on a machine with an Intel Xeon W-2155 3.30 GHz 10-core processor and 64 GB of RAM.

4.1 Performance of Aurora

We consider Aurora at the standard security level of 128 bits, over the binary field $\mathbb{F}_{2^{192}}$. We report data on key efficiency measures of a SNARG: the time to generate a proof (running time of the prover), the length of a proof, and the time to check a proof (running time of the verifier). We also indicate how much of each cost is due to the IOP protocol, and how much is due to the BCS transformation [21].

In Aurora, all of these quantities depend on the number of constraints m in an R1CS instance.[6] Our experiments report how these quantities change as we vary m over the range $\{2^{10}, 2^{11}, \ldots, 2^{20}\}$.

Prover Running Time. In Fig. 4 we plot the running time of the prover, as absolute cost (top graph) and as relative cost when compared to native execution (bottom graph). For R1CS, *native execution* is the time that it takes to check that an assignment satisfies the constraint system. The plot confirms the quasilinear complexity of the prover; proving times range from fractions of a second to several minutes. Proving time is dominated by the cost of running the underlying IOP prover.

---

[6] The number of variables n also affects performance, but it is usually close to m and so we take $n \approx m$ in our experiments. The number of inputs k in an R1CS instance is at most n, and in typical applications it is much smaller than n, so we do not focus on it.

Argument Size. In Fig. 5 we plot argument size, as absolute cost (top graph) and as relative cost when compared to native witness size (bottom graph). For R1CS, *native witness size* is the number of bytes required to represent an assignment to the constraint system. The plot shows that compression (argument size is smaller than native witness size) occurs for $m \geq 4000$. The plot also shows that argument size ranges from 50 kB to 250 kB, and is dominated by the cryptographic digests to authenticate query answers.

Verifier Running Time. In Fig. 6 we plot the running time of the verifier, as absolute cost (top graph) and as relative cost when compared to native execution (bottom graph). The plot shows that verification times range from milliseconds to seconds, and confirms that our implementation incurs a constant multiplicative overhead over native execution.

4.2 Comparison of Ligero, Stark, and Aurora

In Figs. 7, 8 and 9 we compare costs (proving time, argument size, and verification time) on R1CS instances for three IOP-based SNARGs: Ligero [4], Stark [13], and Aurora (this work). As in Sect. 4.1, we plot costs as the number of constraints m increases (and with $n \approx m$ variables as explained in Footnote 6); we also set security to the standard level of 128 bits and use the binary field $\mathbb{F}_{2^{192}}$.

Comparison of Ligero and Aurora. Ligero natively supports R1CS so a comparison with Aurora is straightforward. Figure 8 shows that argument size in Aurora is much smaller than in Ligero, even for a relatively small number of constraints. The gap between the two grows bigger as the number of constraints increases, as Aurora's argument size is polylogarithmic while Ligero's is only sublinear (an exponential gap).

Comparison of Stark and Aurora. Stark does not natively support the NP-complete relation R1CS but instead natively supports an NEXP-complete relation known as *Algebraic Placement and Routing* (APR). These two relations are quite different, and so to achieve a meaningful comparison, we consider an APR instance that *simulates* a given R1CS instance. We thus plot the costs of Stark on a hand-optimized APR instance that simulates R1CS instances. Relying on the reductions described in [13], we wrote an APR instance that realizes a simple abstract computer that checks that a variable assignment satisfies each one of the rank-1 constraints in a given R1CS instance.

Figure 8 shows that argument size in Aurora is much smaller than in Stark, even if both share the same asymptotic growth. This is due to the fact that R1CS and APR target different computation models (explicit circuits vs. uniform computations), so Stark incurs significant overheads when used for R1CS. Figure 9 shows that verification time in Stark grows linearly with the number of constraints (like Ligero and Aurora); indeed, the verifier must read the description of the statement being proved, which is the entire constraint system.

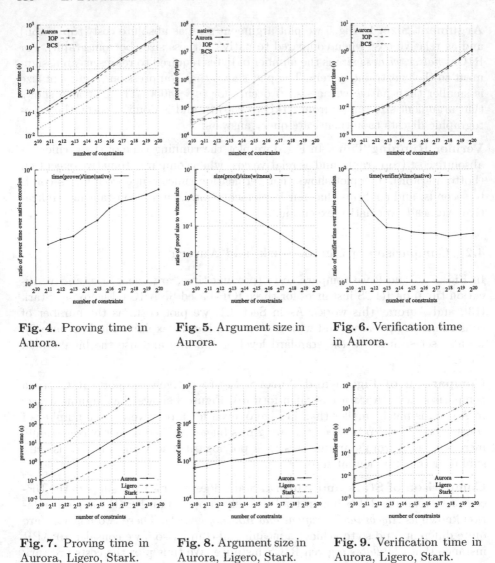

Fig. 4. Proving time in Aurora.

Fig. 5. Argument size in Aurora.

Fig. 6. Verification time in Aurora.

Fig. 7. Proving time in Aurora, Ligero, Stark.

Fig. 8. Argument size in Aurora, Ligero, Stark.

Fig. 9. Verification time in Aurora, Ligero, Stark.

5 libiop: A Library for IOP-Based Non-interactive Arguments

We provide libiop, a codebase that enables the design and implementation of IOP-based non-interactive arguments. The codebase uses the C++ language and has three main components: (1) a library for writing IOP protocols; (2) a realization of the [21] transformation, mapping any IOP written with our library to a corresponding non-interactive argument; (3) a portfolio of IOP protocols, including our new IOP protocol for R1CS and IOP protocols from [4,13]. We discuss each of these components in turn.

5.1   Library for IOP Protocols

We provide a library that enables a programmer to write IOP protocols. Informally, the programmer provides a blueprint of the IOP by specifying, for each round, the number and sizes of oracle messages (and non-oracle messages) sent by the prover, as well as the number of random bytes subsequently sent by the verifier. For the prover, the programmer specifies how each message is to be computed. For the verifier, the programmer specifies how oracle queries are generated and, also, how the verifier's decision is computed based on its random choices and information received from the prover. Notable features of our library include:

- Support for writing new IOPs by using other IOPs as sub-protocols. This includes juxtaposing or interleaving selected rounds of these sub-protocols. This latter feature not only facilitates reducing round complexity in complex IOP constructions but also makes it possible to take advantage of optimizations such as column hashing (discussed in Sect. 5.2) when constructing a non-interactive argument.
- A realization of the transformation described in the full version, which constructs an IOP by combining an 'encoded' IOP and a low-degree test. This is a powerful paradigm (it applies to essentially all published IOP protocols) that reduces the task of writing an IOP to merely providing suitable choices of these two simpler ingredients.

5.2   BCS Transformation

We realize the transformation of [21], by providing code that maps any IOP written in our library into a corresponding non-interactive argument (which consists of a prover algorithm and a verifier algorithm).

We use BLAKE2b [8] to instantiate the random oracle in the [21] transformation (our code allows to conveniently specify alternative instantiations). This hash function is an improvement to BLAKE (a finalist in the SHA-3 competition) [7], and its performance on all recent x86 platforms is competitive with the most performant (and often hardware-accelerated) hash functions [42]. Moreover, BLAKE2b can be configured to output digests of any length between 1 and 64 bytes (between 8 and 512 bits in multiples of 8). When aiming for a security level of λ bits, we only need the hash function to output digests of 2λ bits, and our code automatically sets this length.

Our code incorporates additional optimizations that, while simple, are generic and effective.

One is *column hashing*, which informally works as follows. In many IOP protocols (essentially all published ones, including Ligero [4] and Stark [13]), the prover sends multiple oracles over the same domain in the same round, and the verifier accesses all of them at the same index in the domain. The prover can then build a Merkle tree over *columns* consisting of corresponding entries of the oracles, rather than building separate Merkle trees for each or a single Merkle tree over their concatenation. This reduces a non-interactive proof's length, because the proof only has to contain a *single* authentication path for the desired column, rather than authentication paths corresponding to the indices across all the oracles.

Another optimization is *path pruning*. When providing multiple authentication paths relative to the same root (in the non-interactive argument), some digests become redundant and can thus be omitted. For example, if one considers the authentication paths for all leaves in a particular sub-tree, then one can simple provide the authentication path for the root of the sub-tree. A simple way to view this optimization is to provide the smallest number of digests to authenticate a *set* of leaves.

5.3   Portfolio of IOP Protocols and Sub-Components

We use our library to realize several IOP protocols:

- **Aurora:** our IOP protocol for R1CS (specifically, the one provided in Fig. 12 in Sect. 6).
- **Ligero:** an adaptation of the IOP protocol in [4] to R1CS. While the protocol(s) in [4] are designed for (boolean or arithmetic) circuit satisfiability, the same ideas can be adapted to support R1CS *at no extra cost*. This simplifies comparisons with R1CS-based arguments, and confers additional expressivity.
- **Stark:** the IOP protocol in [13] for *Algebraic Placement and Routing* (APR), a language that is a "succinct" analogue of algebraic satisfaction problems such as R1CS. (See [13] for details.)

Each of the above IOPs is obtained by specifying an encoded IOP and a low-degree test. As explained in Sects. 5.1 and 5.2, our library compiles these into an IOP protocol, and the latter into a non-interactive argument. This toolchain enables specifying protocols with few lines of code (see Fig. 10), and also enhances code auditability.

The IOP protocols above benefit from several algebraic components that our library also provides.

- *Finite field arithmetic.* We support prime and binary fields. Our prime field arithmetic uses Montgomery representation [64]. Our binary field arithmetic uses the carryless multiplication instructions [53]; these are ubiquitous in x86 CPUs and, being used in AES-GCM computations, are highly optimized.
- *FFT algorithms.* The choice of FFT algorithm depends on whether the R1CS instance (and thus the rest of the protocol) is defined over a prime or binary field. In the former case, we use the radix-2 FFT (whose evaluation domain

is a multiplicative coset of order 2^a for some a) [38]. In the latter case, we use an additive FFT (whose evaluation domain is an affine subspace of the binary field) [28,37,44,58,59]. We also provide the respective inverse FFTs, and variants for cosets of the base domains.

Remark 2. Known techniques can be used to reduce given programs or general machine computations to low-level representations such as R1CS and APR (see, e.g., [13,23,78]). Such techniques have been compared in prior work, and our library does not focus on these.

encoded IOP protocol	lines of code	low-degree test	lines of code
Stark	321	FRI	416
Ligero	1281	direct	212
Aurora	1165		

Fig. 10. Lines of code to express various sub-components in our library.

6 Aurora: An IOP for Rank-One Constraint Satisfaction (R1CS)

We describe the IOP for R1CS that comprises the main technical contribution of this paper, and also underlies the SNARG for R1CS that we have designed and built (more about this in Sect. 5).

For the discussions below, we introduce notation about the low-degree test in [14], known as "Fast Reed–Solomon IOPP" (FRI): given a subspace L of a binary field $\mathbb{F}$ and rate $\rho \in (0,1)$, we denote by $\varepsilon_i^{\mathrm{FRI}}(\mathbb{F}, L)$ and $\varepsilon_q^{\mathrm{FRI}}(L, \rho, \delta)$ the soundness error of the interactive and query phases in FRI (respectively) when testing proximity of a δ-far function to RS $[L, \rho]$.

We first provide a "barebones" statement with constant soundness error and no zero knowledge.

Theorem 3. *There is an IOP for the R1CS relation over binary fields $\mathbb{F}$ that, given an R1CS instance having n variables and m constraints, letting $\rho \in (0,1)$ be a constant and L be any subspace of $\mathbb{F}$ such that $2\max(m, n+1) \leq \rho|L|$, has the following parameters:*

$$
\begin{array}{lll}
\textit{alphabet} & \Sigma & = \mathbb{F} \\
\textit{number of rounds } \mathsf{k} & & = O(\log|L|) \\
\textit{proof length} & \mathsf{p} & = (5 + \tfrac{1}{3})|L| \\
\textit{query complexity } \mathsf{q}_\pi & & = O(\log|L|) \\
\textit{randomness} & (\mathsf{r_i}, \mathsf{r_q}) & = (O(\log|L| \cdot \log|\mathbb{F}|), O(\log|L|)) \\
\textit{soundness error} & (\varepsilon_i, \varepsilon_q) & = \left(\frac{m+1}{|\mathbb{F}|} + \frac{|L|}{|\mathbb{F}|} + \varepsilon_i^{\mathrm{FRI}}(\mathbb{F}, L), \varepsilon_q^{\mathrm{FRI}}(L, \rho, \delta) \right) \\
\textit{prover time} & \mathsf{t_p} & = O(|L| \cdot \log(n+m) + \|A\| + \|B\| + \|C\|) + 17 \cdot \mathrm{FFT}(\mathbb{F}, |L|) \\
\textit{verifier time} & \mathsf{t_v} & = O(\|A\| + \|B\| + \|C\| + n + m + \log|L|)
\end{array}
,
$$

where $\delta := \min(\frac{1-2(\rho/2)}{2}, \frac{1-(\rho/2)}{3}, 1 - \rho)$.

Next, we provide a statement that additionally has parameters for controlling the soundness error, is zero knowledge, and includes other (whitebox) optimizations; the proof is analogous except that we use zero knowledge components. The resulting IOP protocol, fully specified in Fig. 12, underlies our SNARG for R1CS (see Sect. 5).

Theorem 4. *There is an IOP for the R1CS relation over binary fields $\mathbb{F}$ that, given an R1CS instance having n variables and m constraints, letting $\rho \in (0,1)$ be a constant and L be any subspace of $\mathbb{F}$ such that $2\max(m, n+1) + 2\mathsf{b} \leq \rho|L|$, is zero knowledge against b queries and has the following parameters:*

$$
\begin{array}{lll}
\textit{alphabet} & \Sigma & = \mathbb{F} \\
\textit{number of rounds} & \mathsf{k} & = O(\log|L|) \\
\textit{proof length} & \mathsf{p} & = (4 + 2\lambda_i + \lambda_i'\lambda_i^{\mathrm{FRI}}/3)|L| \\
\textit{query complexity} & \mathsf{q}_\pi & = O(\lambda_i\lambda_i^{\mathrm{FRI}}\lambda_q^{\mathrm{FRI}}\log|L|) \\
\textit{randomness} & (\mathsf{r}_i, \mathsf{r}_q) & = (O((\lambda_i\lambda_i' + \lambda_i^{\mathrm{FRI}}\log|L|)\log|\mathbb{F}|), O(\lambda_q^{\mathrm{FRI}}\log|L|)) \\
\textit{soundness error} & (\varepsilon_i, \varepsilon_q) & = \left(\left(\frac{m+1}{|\mathbb{F}|}\right)^{\lambda_i} + \left(\frac{|L|}{|\mathbb{F}|}\right)^{\lambda_i'} + \varepsilon_i^{\mathrm{FRI}}(\mathbb{F}, L)^{\lambda_i^{\mathrm{FRI}}}, \varepsilon_q^{\mathrm{FRI}}(L, \rho, \delta)^{\lambda_q^{\mathrm{FRI}}}\right) \\
\textit{prover time} & \mathsf{t}_\mathsf{p} & = \lambda_i \cdot (O(|L| \cdot (\log(n+m) + \|A\| + \|B\| + \|C\|) \\
& & \quad + 18 \cdot \mathrm{FFT}(\mathbb{F}, |L|)) + O(\lambda_i'\lambda_i^{\mathrm{FRI}}|L|) \\
\textit{verifier time} & \mathsf{t}_\mathsf{V} & = \lambda_i \cdot O(\|A\| + \|B\| + \|C\| + n + m + \log|L|) \\
& & \quad + O(\lambda_i'\lambda_i^{\mathrm{FRI}}\lambda_q^{\mathrm{FRI}}\log|L|)
\end{array}
$$

where $\delta := \min(\frac{1-2\rho}{2}, \frac{1-\rho}{3}, 1-\rho)$. Setting $\mathsf{b} \geq \mathsf{q}_\pi$ ensures honest-verifier zero knowledge.

polynomial	degree	values that define the polynomial
p_α	$t-1$	$\hat{p}_\alpha(a) = \begin{cases} \alpha^{\gamma(a)} & \text{for } a \in H_1 \\ 0 & \text{for } a \in (H_1 \cup H_2) \setminus H_1 \end{cases}$
$p_\alpha^{(M)}$	$t-1$	$\hat{p}_\alpha^{(M)}(b) = \begin{cases} \sum_{a \in H_1} M_{a,b} \cdot \alpha^{\gamma(a)} & \text{for } b \in H_2 \\ 0 & \text{for } b \in (H_1 \cup H_2) \setminus H_2 \end{cases}$
codeword	**code**	**polynomial that defines the codeword**
f_w	$\mathrm{RS}\left[L, \frac{n-k+\mathsf{b}}{|L|}\right]$	random polynomial $\bar{f}_w$ of degree less than $n - k + \mathsf{b}$ such that,
		for all $b \in H_2$ with $k < \gamma(b) \leq n$, $\bar{f}_w(b) = \frac{w_{\gamma(b)-k} - \hat{f}_{(1,v)}(b)}{\mathbb{Z}_{H_2^{\leq k}}(b)}$
f_{Mz}	$\mathrm{RS}\left[L, \frac{m+\mathsf{b}}{|L|}\right]$	random polynomial $\bar{f}_{Az}$ of degree less than $m + \mathsf{b}$ such that,
		for all $a \in H_1$, $\bar{f}_{Az}(a) = \sum_{b \in H_2} M_{a,b} \cdot z_{\gamma(b)} = (Mz)_a$

Fig. 11. Polynomials and codewords used in the IOP protocol given in Fig. 12.

Given an R1CS instance $(\mathbb{F}, k, n, m, A, B, C, v)$, we fix subspaces $H_1, H_2 \subseteq \mathbb{F}$ such that $|H_1| = m$ and $|H_2| = n + 1$ (padding to the nearest power of 2 if necessary) with $H_1 \subseteq H_2$ or $H_2 \subseteq H_1$, and a sufficiently large affine subspace $L \subseteq \mathbb{F}$ such that $L \cap (H_1 \cup H_2) = \varnothing$. We let $t := |H_1 \cup H_2| = \max(m, n+1)$. Figure 11 below gives polynomials and codewords used in Fig. 12. We also define $\xi := \sum_{a \in H_1 \cup H_2} a^{t-1}$.

$P((\mathbb{F}, k, n, m, A, B, C, v), w)$ $V(\mathbb{F}, k, n, m, A, B, C, v)$

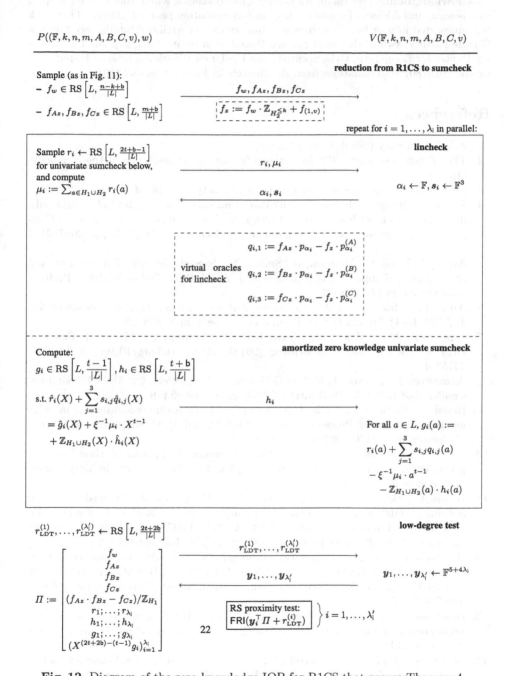

Fig. 12. Diagram of the zero knowledge IOP for R1CS that proves Theorem 4.

Acknowledgments. We thank Alexander Chernyakhovsky and Tom Gur for helpful discussions, and Aleksejs Popovs for help in implementing parts of `libiop`. This work was supported in part by: the Ethics and Governance of Artificial Intelligence Fund; a Google Faculty Award; the Israel Science Foundation (grant 1501/14); the UC Berkeley Center for Long-Term Cybersecurity; the US-Israel Binational Science Foundation (grant 2015780); and donations from the Interchain Foundation and Qtum.

References

1. ZCash Company (2014). https://z.cash/
2. The Zcash Ceremony (2016). https://z.cash/blog/the-design-of-the-ceremony.html
3. Zero knowledge proof standardization (2017). https://zkproof.org/
4. Ames, S., Hazay, C., Ishai, Y., Venkitasubramaniam, M.: Ligero: lightweight sublinear arguments without a trusted setup. In: Proceedings of the 24th ACM Conference on Computer and Communications Security, CCS 2017, pp. 2087–2104 (2017)
5. Arora, S., Lund, C., Motwani, R., Sudan, M., Szegedy, M.: Proof verification and the hardness of approximation problems. J. ACM **45**(3), 501–555 (1998). Preliminary version in FOCS 1992
6. Arora, S., Safra, S.: Probabilistic checking of proofs: a new characterization of NP. J. ACM **45**(1), 70–122 (1998). Preliminary version in FOCS 1992
7. Aumasson, J.-P., Meier, W., Phan, R.C.-W., Henzen, L.: The Hash Function BLAKE. ISC. Springer, Heidelberg (2014). https://doi.org/10.1007/978-3-662-44757-4
8. Aumasson, J.P., Neves, S., Wilcox-O'Hearn, Z., Winnerlein, C.: BLAKE2: simpler, smaller, fast as MD5 (2013). https://blake2.net/blake2.pdf
9. Babai, L., Fortnow, L., Levin, L.A., Szegedy, M.: Checking computations in polylogarithmic time. In: Proceedings of the 23rd Annual ACM Symposium on Theory of Computing, STOC 1991, pp. 21–32 (1991)
10. Babai, L., Fortnow, L., Lund, C.: Non-deterministic exponential time has two-prover interactive protocols. Comput. Complex. **1**, 3–40 (1991). Preliminary version appeared in FOCS 1990
11. Baum, C., Bootle, J., Cerulli, A., del Pino, R., Groth, J., Lyubashevsky, V.: Sub-linear lattice-based zero-knowledge arguments for arithmetic circuits. In: Shacham, H., Boldyreva, A. (eds.) CRYPTO 2018. LNCS, vol. 10992, pp. 669–699. Springer, Cham (2018). https://doi.org/10.1007/978-3-319-96881-0_23
12. Ben-Sasson, E., et al.: Computational integrity with a public random string from quasi-linear PCPs. In: Coron, J.-S., Nielsen, J.B. (eds.) EUROCRYPT 2017. LNCS, vol. 10212, pp. 551–579. Springer, Cham (2017). https://doi.org/10.1007/978-3-319-56617-7_19
13. Ben-Sasson, E., Bentov, I., Horesh, Y., Riabzev, M.: Scalable, transparent, and post-quantum secure computational integrity. Cryptology ePrint Archive, Report 2018/046 (2018)
14. Ben-Sasson, E., Bentov, I., Horesh, Y., Riabzev, M.: Fast Reed-Solomon interactive Oracle proofs of proximity. In: Proceedings of the 45th International Colloquium on Automata, Languages and Programming, ICALP 2018, pp. 14:1–14:17 (2018)

15. Ben-Sasson, E., Chiesa, A., Forbes, M.A., Gabizon, A., Riabzev, M., Spooner, N.: Zero knowledge protocols from succinct constraint detection. In: Kalai, Y., Reyzin, L. (eds.) TCC 2017. LNCS, vol. 10678, pp. 172–206. Springer, Cham (2017). https://doi.org/10.1007/978-3-319-70503-3_6

16. Ben-Sasson, E., Chiesa, A., Gabizon, A., Riabzev, M., Spooner, N.: Interactive Oracle Proofs with constant rate and query complexity. In: Proceedings of the 44th International Colloquium on Automata, Languages and Programming, ICALP 2017, pp. 40:1–40:15 (2017)

17. Ben-Sasson, E., Chiesa, A., Gabizon, A., Virza, M.: Quasi-linear size zero knowledge from linear-algebraic PCPs. In: Kushilevitz, E., Malkin, T. (eds.) TCC 2016-A. LNCS, vol. 9563, pp. 33–64. Springer, Heidelberg (2016). https://doi.org/10.1007/978-3-662-49099-0_2

18. Ben-Sasson, E., et al.: Zerocash: decentralized anonymous payments from Bitcoin. In: Proceedings of the 2014 IEEE Symposium on Security and Privacy, SP 2014, pp. 459–474 (2014)

19. Ben-Sasson, E., Chiesa, A., Genkin, D., Tromer, E., Virza, M.: SNARKs for C: verifying program executions succinctly and in zero knowledge. In: Canetti, R., Garay, J.A. (eds.) CRYPTO 2013. LNCS, vol. 8043, pp. 90–108. Springer, Heidelberg (2013). https://doi.org/10.1007/978-3-642-40084-1_6

20. Ben-Sasson, E., Chiesa, A., Green, M., Tromer, E., Virza, M.: Secure sampling of public parameters for succinct zero knowledge proofs. In: Proceedings of the 36th IEEE Symposium on Security and Privacy, S&P 2015, pp. 287–304 (2015)

21. Ben-Sasson, E., Chiesa, A., Spooner, N.: Interactive Oracle proofs. In: Hirt, M., Smith, A. (eds.) TCC 2016-B. LNCS, vol. 9986, pp. 31–60. Springer, Heidelberg (2016). https://doi.org/10.1007/978-3-662-53644-5_2

22. Ben-Sasson, E., Chiesa, A., Tromer, E., Virza, M.: Scalable zero knowledge via cycles of elliptic curves. In: Garay, J.A., Gennaro, R. (eds.) CRYPTO 2014. LNCS, vol. 8617, pp. 276–294. Springer, Heidelberg (2014). https://doi.org/10.1007/978-3-662-44381-1_16. Extended version at http://eprint.iacr.org/2014/595

23. Ben-Sasson, E., Chiesa, A., Tromer, E., Virza, M.: Succinct non-interactive zero knowledge for a von Neumann architecture. In: Proceedings of the 23rd USENIX Security Symposium, Security 2014, pp. 781–796 (2014). Extended version at http://eprint.iacr.org/2013/879

24. Ben-Sasson, E., Kaplan, Y., Kopparty, S., Meir, O., Stichtenoth, H.: Constant rate PCPs for Circuit-SAT with sublinear query complexity. In: Proceedings of the 54th Annual IEEE Symposium on Foundations of Computer Science, FOCS 2013, pp. 320–329 (2013)

25. Ben-Sasson, E., Kopparty, S., Saraf, S.: Worst-case to average case reductions for the distance to a code. In: Proceedings of the 33rd ACM Conference on Computer and Communications Security, CCS 2018, pp. 24:1–24:23 (2018)

26. Ben-Sasson, E., Sudan, M.: Robust locally testable codes and products of codes. Random Struct. Algorithms **28**(4), 387–402 (2006)

27. Ben-Sasson, E., Sudan, M.: Short PCPs with Polylog query complexity. SIAM J. Comput. **38**(2), 551–607 (2008). Preliminary version appeared in STOC 2005

28. Bernstein, D.J., Chou, T.: Faster binary-field multiplication and faster binary-field MACs. In: Joux, A., Youssef, A. (eds.) SAC 2014. LNCS, vol. 8781, pp. 92–111. Springer, Cham (2014). https://doi.org/10.1007/978-3-319-13051-4_6

29. Bitansky, N., Chiesa, A., Ishai, Y., Paneth, O., Ostrovsky, R.: Succinct non-interactive arguments via linear interactive proofs. In: Sahai, A. (ed.) TCC 2013. LNCS, vol. 7785, pp. 315–333. Springer, Heidelberg (2013). https://doi.org/10.1007/978-3-642-36594-2_18

30. Boneh, D., Ishai, Y., Sahai, A., Wu, D.J.: Lattice-based SNARGs and their application to more efficient obfuscation. In: Coron, J.-S., Nielsen, J.B. (eds.) EUROCRYPT 2017. LNCS, vol. 10212, pp. 247–277. Springer, Cham (2017). https://doi.org/10.1007/978-3-319-56617-7_9

31. Bootle, J., Cerulli, A., Chaidos, P., Groth, J., Petit, C.: Efficient zero-knowledge arguments for arithmetic circuits in the discrete log setting. In: Fischlin, M., Coron, J.-S. (eds.) EUROCRYPT 2016. LNCS, vol. 9666, pp. 327–357. Springer, Heidelberg (2016). https://doi.org/10.1007/978-3-662-49896-5_12

32. Bowe, S., Gabizon, A., Green, M.: A multi-party protocol for constructing the public parameters of the Pinocchio zk-SNARK. Cryptology ePrint Archive, Report 2017/602 (2017)

33. Bowe, S., Gabizon, A., Miers, I.: Scalable multi-party computation for zk-SNARK parameters in the random beacon model. Cryptology ePrint Archive, Report 2017/1050 (2017)

34. Brassard, G., Chaum, D., Crépeau, C.: Minimum disclosure proofs of knowledge. J. Comput. Syst. Sci. $37(2)$, 156–189 (1988)

35. Bünz, B., Bootle, J., Boneh, D., Poelstra, A., Wuille, P., Maxwell, G.: Bulletproofs: short proofs for confidential transactions and more. In: Proceedings of the 39th IEEE Symposium on Security and Privacy, S&P 2018, pp. 315–334 (2018)

36. Byott, N.P., Chapman, R.J.: Power sums over finite subspaces of a field. Finite Fields Appl. $5(3)$, 254–265 (1999)

37. Cantor, D.G.: On arithmetical algorithms over finite fields. J. Comb. Theor. Series A $50(2)$, 285–300 (1989)

38. Cooley, J.W., Tukey, J.W.: An algorithm for the machine calculation of complex Fourier series. Math. Comput. 19, 297–301 (1965)

39. Cormode, G., Mitzenmacher, M., Thaler, J.: Practical verified computation with streaming interactive proofs. In: Proceedings of the 4th Symposium on Innovations in Theoretical Computer Science, ITCS 2012, pp. 90–112 (2012)

40. Costello, C., et al.: Geppetto: versatile verifiable computation. In: Proceedings of the 36th IEEE Symposium on Security and Privacy, S&P 2015, pp. 250–273 (2015)

41. Cramer, R., Damgård, I.: Zero-knowledge proofs for finite field arithmetic, or: can zero-knowledge be for free? In: Krawczyk, H. (ed.) CRYPTO 1998. LNCS, vol. 1462, pp. 424–441. Springer, Heidelberg (1998). https://doi.org/10.1007/BFb0055745

42. eBACS: ECRYPT Benchmarking of Cryptographic Systems: Measurements of hash functions, indexed by machine (2017). https://bench.cr.yp.to/results-hash.html

43. Feige, U., Goldwasser, S., Lovász, L., Safra, S., Szegedy, M.: Interactive proofs and the hardness of approximating cliques. J. ACM $43(2)$, 268–292 (1996). Preliminary version in FOCS 1991

44. Gao, S., Mateer, T.: Additive fast Fourier transforms over finite fields. IEEE Trans. Inf. Theory $56(12)$, 6265–6272 (2010)

45. Gennaro, R., Gentry, C., Parno, B., Raykova, M.: Quadratic span programs and succinct NIZKs without PCPs. In: Johansson, T., Nguyen, P.Q. (eds.) EUROCRYPT 2013. LNCS, vol. 7881, pp. 626–645. Springer, Heidelberg (2013). https://doi.org/10.1007/978-3-642-38348-9_37

46. Gentry, C., Wichs, D.: Separating succinct non-interactive arguments from all falsifiable assumptions. In: Proceedings of the 43rd Annual ACM Symposium on Theory of Computing, STOC 2011, pp. 99–108 (2011)

47. Goldreich, O., Håstad, J.: On the complexity of interactive proofs with bounded communication. Inf. Process. Lett. $67(4)$, 205–214 (1998)

48. Goldwasser, S., Kalai, Y.T., Rothblum, G.N.: Delegating computation: interactive proofs for Muggles. J. ACM **62**(4), 27:1–27:64 (2015)
49. Goldwasser, S., Micali, S., Rackoff, C.: The knowledge complexity of interactive proof systems. SIAM J. Comput. **18**(1), 186–208 (1989). Preliminary version appeared in STOC 1985
50. Groth, J.: Short pairing-based non-interactive zero-knowledge arguments. In: Abe, M. (ed.) ASIACRYPT 2010. LNCS, vol. 6477, pp. 321–340. Springer, Heidelberg (2010). https://doi.org/10.1007/978-3-642-17373-8_19
51. Groth, J.: On the size of pairing-based non-interactive arguments. In: Fischlin, M., Coron, J.-S. (eds.) EUROCRYPT 2016. LNCS, vol. 9666, pp. 305–326. Springer, Heidelberg (2016). https://doi.org/10.1007/978-3-662-49896-5_11
52. Groth, J., Maller, M.: Snarky signatures: minimal signatures of knowledge from simulation-extractable SNARKs. In: Katz, J., Shacham, H. (eds.) CRYPTO 2017. LNCS, vol. 10402, pp. 581–612. Springer, Cham (2017). https://doi.org/10.1007/978-3-319-63715-0_20
53. Gueron, S.: Intel carry-less multiplication instruction and its usage for computing the GCM mode (2011). https://software.intel.com/en-us/articles/intel-carry-less-multiplication-instruction-and-its-usage-for-computing-the-gcm-mode
54. Ishai, Y., Kushilevitz, E., Ostrovsky, R.: Efficient arguments without short PCPs. In: Proceedings of the Twenty-Second Annual IEEE Conference on Computational Complexity, CCC 2007, pp. 278–291 (2007)
55. Ishai, Y., Mahmoody, M., Sahai, A., Xiao, D.: On zero-knowledge PCPs: limitations, simplifications, and applications (2015). http://www.cs.virginia.edu/~mohammad/files/papers/ZKPCPs-Full.pdf
56. Kalai, Y.T., Raz, R.: Interactive PCP. In: Aceto, L., Damgård, I., Goldberg, L.A., Halldórsson, M.M., Ingólfsdóttir, A., Walukiewicz, I. (eds.) ICALP 2008. LNCS, vol. 5126, pp. 536–547. Springer, Heidelberg (2008). https://doi.org/10.1007/978-3-540-70583-3_44
57. Kilian, J.: A note on efficient zero-knowledge proofs and arguments. In: Proceedings of the 24th Annual ACM Symposium on Theory of Computing, STOC 1992, pp. 723–732 (1992)
58. Lin, S., Al-Naffouri, T.Y., Han, Y.S.: FFT algorithm for binary extension finite fields and its application to Reed-Solomon codes. IEEE Trans. Inf. Theory **62**(10), 5343–5358 (2016)
59. Lin, S., Chung, W.H., Han, Y.S.: Novel polynomial basis and its application to Reed-Solomon erasure codes. In: Proceedings of the 55th Annual IEEE Symposium on Foundations of Computer Science, FOCS 2014, pp. 316–325 (2014)
60. Lipmaa, H.: Succinct non-interactive zero knowledge arguments from span programs and linear error-correcting codes. In: Sako, K., Sarkar, P. (eds.) ASIACRYPT 2013. LNCS, vol. 8269, pp. 41–60. Springer, Heidelberg (2013). https://doi.org/10.1007/978-3-642-42033-7_3
61. Lund, C., Fortnow, L., Karloff, H.J., Nisan, N.: Algebraic methods for interactive proof systems. J. ACM **39**(4), 859–868 (1992)
62. Meir, O.: Combinatorial PCPs with short proofs. In: Proceedings of the 26th Annual IEEE Conference on Computational Complexity, CCC 2012 (2012)
63. Micali, S.: Computationally sound proofs. SIAM J. Comput. **30**(4), 1253–1298 (2000). Preliminary version appeared in FOCS 1994
64. Montgomery, P.L.: Modular multiplication without trial division. Math. Comput. **44**(170), 519–521 (1985)
65. Nakamoto, S.: Bitcoin: a peer-to-peer electronic cash system (2009). http://www.bitcoin.org/bitcoin.pdf

66. NIST: Post-quantum cryptography (2016). https://csrc.nist.gov/Projects/Post-Quantum-Cryptography
67. Parno, B., Gentry, C., Howell, J., Raykova, M.: Pinocchio: Nearly practical verifiable computation. In: 2013 Proceedings of the 34th IEEE Symposium on Security and Privacy, Oakland, pp. 238–252 (2013)
68. Polishchuk, A., Spielman, D.A.: Nearly-linear size holographic proofs. In: Proceedings of the 26th Annual ACM Symposium on Theory of Computing, STOC 1994, pp. 194–203 (1994)
69. Reingold, O., Rothblum, R., Rothblum, G.: Constant-round interactive proofs for delegating computation. In: Proceedings of the 48th ACM Symposium on the Theory of Computing, STOC 2016, pp. 49–62 (2016)
70. SCIPR Lab: libsnark: a C++ library for zkSNARK proofs. https://github.com/scipr-lab/libsnark
71. Setty, S., Braun, B., Vu, V., Blumberg, A.J., Parno, B., Walfish, M.: Resolving the conflict between generality and plausibility in verified computation. In: Proceedings of the 8th EuoroSys Conference, EuroSys 2013, pp. 71–84 (2013)
72. Shamir, A.: IP = PSPACE. J. ACM **39**(4), 869–877 (1992)
73. Thaler, J.: Time-optimal interactive proofs for circuit evaluation. In: Canetti, R., Garay, J.A. (eds.) CRYPTO 2013. LNCS, vol. 8043, pp. 71–89. Springer, Heidelberg (2013). https://doi.org/10.1007/978-3-642-40084-1_5
74. Thaler, J.: A note on the GKR protocol (2015). http://people.cs.georgetown.edu/jthaler/GKRNote.pdf
75. Thaler, J., Roberts, M., Mitzenmacher, M., Pfister, H.: Verifiable computation with massively parallel interactive proofs. CoRR abs/1202.1350 (2012)
76. Wahby, R.S., Howald, M., Garg, S.J., Shelat, A., Walfish, M.: Verifiable ASICs. In: Proceedings of the 37th IEEE Symposium on Security and Privacy, S&P '16, pp. 759–778 (2016)
77. Wahby, R.S., et al.: Full accounting for verifiable outsourcing. In: Proceedings of the 24th ACM Conference on Computer and Communications Security, CCS 2017, pp. 2071–2086 (2017)
78. Wahby, R.S., Setty, S., Ren, Z., Blumberg, A.J., Walfish, M.: Efficient RAM and control flow in verifiable outsourced computation. In: Proceedings of the 22nd Annual Network and Distributed System Security Symposium, NDSS 2015 (2015)
79. Wahby, R.S., Tzialla, I., Shelat, A., Thaler, J., Walfish, M.: Doubly-efficient zkSNARKs without trusted setup. Cryptology ePrint Archive, Report 2017/1132 (2017)
80. Walfish, M., Blumberg, A.J.: Verifying computations without reexecuting them. Commun. ACM **58**(2), 74–84 (2015)
81. Wee, H.: On round-efficient argument systems. In: Caires, L., Italiano, G.F., Monteiro, L., Palamidessi, C., Yung, M. (eds.) ICALP 2005. LNCS, vol. 3580, pp. 140–152. Springer, Heidelberg (2005). https://doi.org/10.1007/11523468_12
82. Zhang, Y., Genkin, D., Katz, J., Papadopoulos, D., Papamanthou, C.: vSQL: verifying arbitrary SQL queries over dynamic outsourced databases. In: Proceedings of the 38th IEEE Symposium on Security and Privacy, S&P 2017, pp. 863–880 (2017)
83. Zhang, Y., Genkin, D., Katz, J., Papadopoulos, D., Papamanthou, C.: A zero-knowledge version of VSQL. Cryptology ePrint Archive, Report 2017/1146 (2017)

The Double Ratchet: Security Notions, Proofs, and Modularization for the Signal Protocol

Joël Alwen[2(✉)], Sandro Coretti[1], and Yevgeniy Dodis[1]

[1] New York University, New York, USA
{corettis,dodis}@nyu.edu
[2] Wickr Inc., San Francisco, USA
jalwen@wickr.com

Abstract. Signal is a famous secure messaging protocol used by billions of people, by virtue of many secure text messaging applications including Signal itself, WhatsApp, Facebook Messenger, Skype, and Google Allo. At its core it uses the concept of "double ratcheting," where every message is encrypted and authenticated using a fresh symmetric key; it has many attractive properties, such as forward security, post-compromise security, and "immediate (no-delay) decryption," which had never been achieved in combination by prior messaging protocols.

While the formal analysis of the Signal protocol, and ratcheting in general, has attracted a lot of recent attention, we argue that none of the existing analyses is fully satisfactory. To address this problem, we give a clean and general definition of *secure messaging*, which clearly indicates the types of security we expect, including forward security, post-compromise security, and immediate decryption. We are the first to explicitly formalize and model the immediate decryption property, which implies (among other things) that parties seamlessly recover if a given message is permanently lost—a property not achieved by any of the recent "provable alternatives to Signal."

We build a modular "generalized Signal protocol" from the following components: (a) *continuous key agreement (CKA)*, a clean primitive we introduce and which can be easily and generically built from public-key encryption (not just Diffie-Hellman as is done in the current Signal protocol) and roughly models "public-key ratchets;" (b) *forward-secure authenticated encryption with associated data (FS-AEAD)*, which roughly captures "symmetric-key ratchets;" and (c) a two-input hash function that is a pseudorandom function (resp. generator with input) in its first (resp. second) input, which we term PRF-PRNG. As a result, in addition to instantiating our framework in a way resulting in the existing,

J. Alwen—Partially supported by the European Research Council under ERC Consolidator Grant (682815 - TOCNeT).
S. Coretti—Supported by NSF grants 1314568 and 1319051.
Y. Dodis—Partially supported by gifts from VMware Labs, Facebook and Google, and NSF grants 1314568, 1619158, 1815546.

© International Association for Cryptologic Research 2019
Y. Ishai and V. Rijmen (Eds.): EUROCRYPT 2019, LNCS 11476, pp. 129–158, 2019.
https://doi.org/10.1007/978-3-030-17653-2_5

widely-used Diffie-Hellman based Signal protocol, we can easily get post-quantum security and not rely on random oracles in the analysis.

1 Introduction

Signal [23] is a famous secure messaging protocol, which is—by virtue of many secure text messaging applications including Signal itself, WhatsApp [29], Facebook Messenger [11], Skype [21] and Google Allo [22]—used by billions of people. At its core it uses the concept of *double ratcheting*, where every message is encrypted and authenticated using a fresh symmetric key. Signal has many attractive properties, such as forward security and post-compromise security, and it supports immediate (no-delay) decryption. Prior to Signal's deployment, these properties had never been achieved in combination by messaging protocols.

Signal was designed by practitioners and was implemented and deployed well before any security analysis was obtained. In fact, a clean description of Signal has been posted by its inventors Marlinspike and Perrin [23] only recently. The write-up does an excellent job at describing the double-ratchet protocol, gives examples of how it is run, and provides security intuition for its building blocks. However, it lacks a formal definition of the secure messaging problem that the double ratchet solves, and, as a result, does not have a formal security proof.

IMMEDIATE DECRYPTION AND ITS IMPORTANCE. One of the main issues any messaging scheme must address is the fact that messages might arrive out of order or be lost entirely. Additionally, parties can be offline for extended periods of time and send and receive messages asynchronously. Given these inherent constraints, immediate decryption is a very attractive feature. Informally, it ensures that when a legitimate message is (eventually) delivered to the recipient, the recipient can not only immediately decrypt the message but is also able to place it in the correct spot in relation to the other messages already received. Furthermore, immediate decryption also ensures an even more critical liveness property, termed *message-loss resilience (MLR)* in this work: if a message is permanently lost by the network, parties should still be able to communicate (perhaps realizing at some point that a message has never been delivered). Finally, even in settings where messages are eventually delivered (but could come out of order), giving up on immediate decryption seems cumbersome: should out-of-order messages be discarded or buffered by the recipient? If discarded, how will the sender (or the network) know that it should resend the message later? If buffered, how to prevent denial-of-service attacks and distinguish legitimate out-of-order messages (which cannot be immediately decrypted) from fake messages? While these questions could surely be answered (perhaps by making additional timing assumptions about the network), it appears that the simplest answer would be to design a secure messaging protocol which support immediate decryption. Indeed, to the best of our knowledge, all secure messaging services deployed in practice do have this feature (and, hence, MLR).

ADDITIONAL PROPERTIES. In practice, parties' states might occasionally leak. To address this concern, a secure messaging protocol should have the following two properties:

- *Forward secrecy (FS):* if the state of a party is leaked, none of the previous messages should get compromised (assuming they are erased from the state, of course).
- *Post-compromise security (PCS) (aka channel healing):* once the exposure of the party's state ends, security is restored after a few communication rounds.

In isolation, fulfilling either of these desirable properties is well understood: FS is achieved by using basic steam ciphers (aka pseudorandom generators (PRGs)) [4], while PCS [8] is achieved by some form of key agreement executed after the compromise, such as Diffie-Hellman. Unfortunately, these techniques, both of which involve some form of *key evolution*, are clearly at tension with immediate decryption when the network is fully asynchronous. Indeed, the main elegance of Signal, achieved by its *double-ratchet* algorithm, comes from the fact that FS and PCS are not only achieved together, but also *without sacrificing immediate decryption and MLR.*

GOALS OF THIS WORK. One of the main drawbacks of all formal Signal-related papers [3,10,16,26], following the initial work of [7], is the fact that they all achieve FS and PCS by *explicitly giving up* not only on immediate decryption, but also MLR. (This is not merely a definitional issue as their constructions indeed cease any and all further functionality when, say, a single message is dropped in transit.) While such a drastic weakening of the liveness/correctness property considerably simplifies the algorithmic design for these provably secure alternatives to Signal, it also made them insufficient for settings where message loss is indeed possible. This can occur, in practice, due to a variety of reasons. For example, the protocol may be using an unreliable transport mechanism such as SMS or UDP. Alternatively, traffic may be routed (via more reliable TCP) through a central back-end server so as to facilitate asynchronous communication between end-points (as is very common for secure messaging deployments in practice). Yet, even in this setting, packet losses can still occur as the server itself may end up dropping messages due to a variety of unintended events such as due to outages or being subject to a heavy work/network load (say, because of an ongoing (D)DOS attack, partial outages, or worse yet, an emergency event generating sudden high volumes of traffic). With the goal of providing resilient communication even under these and similar realistic conditions, the main objectives of this work are to:

(a) propose formal definitions of *secure messaging* as a cryptographic primitive that *explicitly mandates immediate decryption and MLR*; and
(b) to provide an analysis of *Signal itself* in a well-defined *general* model for secure messaging.

Our work is the first to address either of these natural goals. Moreover, in order to improve the general understanding of secure messaging and to develop alternative (e.g., post-quantum secure) solutions, this paper aims at

(c) generalizing and abstracting out the reliance on the specific Diffie-Hellman key exchange (making the current protocol insecure in the post-quantum world) as well as clarifying the role of various cryptographic hash functions used inside the current Signal instantiation. That is, the idea is to build a "generalized Signal" protocol of which the current instantiation is a special case, but where other instantiations are possible, including those which are post-quantum secure and/or do not require random oracles.

1.1 Our Results

Addressing the points (a)–(c) above, this paper's main contributions are the following:

- Providing a clean definition of *secure messaging* that clearly indicates the expected types of security, including FS, PCS, and—for the first time— immediate decryption.
- Putting forth a modular *generalized* Signal protocol from several simpler building blocks:
  (1) *forward-secure authenticated encryption with associated data (FS-AEAD)*, which can be easily built from a regular PRG and AEAD and roughly models the so-called symmetric-key ratchet of Signal;
  (2) *continuous key agreement (CKA)*, which is a clean primitive that can easily be built generically from public-key encryption and roughly models the so-called public-key ratchet of Signal;
  (3) a two-input hash function, called *PRF-PRNG*, which is a pseudorandom function (resp. generator) in its first (resp. second) input and helps to "connect" the two ratchet types.
- Instantiating the framework such that we obtain the existing Diffie-Hellman- based protocol and observing that one can easily achieve post-quantum secu- rity (by using post-quantum-secure public-key encryption, such as [1,6,15]) and/or not rely on random oracles in the analysis.
- Extending the design to include other forms of "fine-grained state com- promise" recently studied by Poettering and Rösler [26] and Jaeger and Stepanovs [16] but, once more, without sacrificing the immediate decryption property.

THE SECURE MESSAGING DEFINITION. The proposed secure messaging (SM) definition encompasses, in one clean game, (Fig. 1) all desired properties, includ- ing FS as well as PCS and immediate decryption. The attacker in the definition is very powerful, has full control of the order of sending and receiving messages, can corrupt parties' state multiple times, and even controls the randomness used for encryption.[1] In order to avoid trivial and unpreventable attacks, a few restric- tions need to be placed on an attacker $\mathcal{A}$. In broad strokes, the definition requires the following properties:

---

[1] Namely, good randomness is only needed to achieve PCS, while all other security properties hold even with the adversarially controlled randomness (when parties are not compromised).

– When parties are uncompromised, i.e., when their respective states are unknown to $\mathcal{A}$, the protocol is secure against *active* attacks. In particular, the protocol must detect injected ciphertexts (not legitimately sent by one of the parties) and properly handle legitimate ciphertexts delivered in arbitrary order (capturing correctness and immediate decryption).
– When parties are uncompromised, messages are protected even against future compromise of either the sender or the receiver, modeling *forward security*.
– When one or both parties are compromised and the attacker remains *passive*, security is restored "quickly," i.e., within a few rounds of back-and-forth, which models *PCS*.

While the proposed definition is still rather complex, we believe it to be intuitive and *considerably shorter and easier to understand* compared to the recent works of [16,26], which are discussed in more detail in Sect. 1.2.

It should be stressed that the basic SM security in this paper only requires PCS against a passive attacker. Indeed, when an active attacker compromises the state of, say, party A, it can always send ciphertexts to the partner B in A's name (thereby even potentially hijacking A's communication with B and removing A from the channel altogether) or decrypt ciphertexts sent by B immediately following state compromise. As was observed by [16,26] at CRYPTO'18, one might achieve certain limited forms of fine-grained security against active attacks. For example, it is not a priori clear if the attacker should be able to decrypt ciphertexts sent by A to B (if A uses good randomness) or forge legitimate messages from B to A (when A's state is exposed). We comment on these possible extensions in the full version of this paper [2] but notice that they are still rather limited, given that the simple devastating attacks mentioned above are inherently non-preventable against active attackers immediately following state compromise. Thus, our main SM security notion simply *disallows all active attacks for* Δ_{SM} *epochs immediately following state compromise* where Δ_{SM} is the number of rounds of communication required to refresh a compromised state.

THE BUILDING BLOCKS.   Since the original Signal protocol is quite subtle and somewhat tricky to understand, one of the main contributions of this work is to distill out three basic and intuitive building blocks used inside the double ratchet.

The first block is *forward-secure authenticated encryption with associated data (FS-AEAD)* and models secure messaging security inside a single so-called *epoch*; an epoch should be thought of as a unidirectional stream of messages sent by one of the parties, ending once a message from the other party is received. As indicated by the name, an FS-AEAD protocol must provide forward secrecy, but also immediate decryption. Capturing this makes the definition of FS-AEAD somewhat non-trivial (cf. Fig. 3), but still simpler than that of general SM; in particular, no PCS is required (which allows us to define FS-AEAD as a deterministic primitive and not worry about poor randomness).

Building FS-AEAD turns out to be rather easy: in essence, one uses message counters as associated data for standard AEAD and a PRG to immediately refresh the secret key of AEAD after every message successfully sent or received. This is exactly what is done in Signal.

The second block is a primitive called *continuous key agreement (CKA)* (cf. Fig. 2), which could be viewed as an abstraction of the DH-based public-key ratchet in Signal. CKA is a *synchronous* and *passive* primitive, i.e., parties A and B speak in turns, and no adversarial messages or traffic mauling are allowed. With each message sent or received, a party should output a fresh key such that (with "sending" keys generated by A being equal to "receiving" keys generated by B and vice versa). Moreover, CKA guarantees its own PCS, i.e., after a potential state exposure, security is restored within two rounds. Finally, CKA must be forward-secure, i.e., past keys must remain secure when the state is leaked. Forward security is governed by a parameter $\Delta_{\mathsf{CKA}} \geq 0$, which, informally, guarantees that all keys older than Δ_{CKA} rounds remain secure upon state compromise.

Not surprisingly, minimizing Δ_{CKA} results in faster PCS for secure messaging.[2] Fortunately, optimal CKA protocols achieving optimal $\Delta_{\mathsf{CKA}} = 0$ can be built generically from key-encapsulation mechanisms. Interestingly, the elegant DH-based CKA used by Signal achieves slightly sub-optimal $\Delta_{\mathsf{CKA}} = 1$, which is due to how long parties need to hold on to their secret exponents. However, the Signal CKA saves about a factor of 2 in communication complexity, which makes it a reasonable trade-off in practice.

The third and final component of the generalized Signal protocol is a two-argument hash function P, called a PRF-PRNG, which is used to produce secret keys for FS-AEAD epochs from an entropy pool refreshed by CKA keys. More specifically, with each message exchanged using FS-AEAD, the parties try to run the CKA protocol "on the side," by putting the CKA messages as associated data. Due to asynchrony, the party will repeat a given CKA message until it receives a legitimate response from its partner, after which the CKA moves forward with the next message. Each new CKA key is absorbed into the state of the PRF-PRNG, which is then used to generate a new FS-AEAD key.

Informally, a PRF-PRNG takes as inputs a state σ and a CKA key I and produces a new state σ' and a (FS-AEAD) key k. It satisfies a *PRF* property saying that if σ is random, then $\mathsf{P}(\sigma, \cdot)$ acts like a PRF (keyed by σ) in that outputs (σ', k) on *adversarially chosen* inputs I are random. Moreover, it also acts like a PRNG in that, if the input I is random, then so are the resulting state σ' and key k. Observe that standard hash functions are assumed to satisfy this notion; alternatively, one can also very easily build a PRF-PRNG from any PRG and a pseudorandom permutation (cf. Sect. 4.3).

GENERALIZED SIGNAL. Putting the above blocks together properly yields the generalized Signal protocol (cf. Fig. 6). As a special case, one can easily obtain the existing Signal implementation[3] by using the standard way of building FS-AEAD

---

[2] Specifically, the healing time of the generic Signal protocol presented in this work is $\Delta_{\mathsf{SM}} = 2 + \Delta_{\mathsf{CKA}}$.

[3] For syntactic reasons having to do with our abstractions, our protocol is a minor variant of Signal, but is logically equivalent to Signal in every aspect.

from PRG and AEAD, CKA using the Diffie-Hellman based public-key ratchet mentioned above, and an appropriate cryptographic hash function in place of PRF-PRNG. However, many other variants become possible. For example, by using a generic CKA from DH-KEM, one may trade communication efficiency (worse by a factor of 2) for a shorter healing period Δ_{SM} (from 3 rounds to 2). More interestingly, using any post-quantum KEM, such as [1,6,15] results in a *post-quantum secure variant of Signal.* Finally, we also believe that our generalized double ratcheting scheme is much more intuitive than the existing DH-based variant, as it abstracts precisely the cryptographic primitives needed, including the two types ratchets, and what security is needed from each primitive.

BEYOND DOUBLE RATCHETING TO FULL SIGNAL. Following most of the prior (and concurrent) work [3,10,16,26] (discussed in the next section), this paper primarily concerned with formalizing the double-ratchet aspect of the Signal protocol. This assumes that any set of two parties can correctly and securely agree on the initial secret key. The latter problem is rather non-trivial, especially (a) in the multi-user setting, when a party could be using a global public key to communicate with multiple recipients, some of which might be malicious, (b) when the initial secret key agreement is required to be non-interactive, and (c) when state compromise (including that of the master secret for the PKI) is possible, and even frequent. Some of those subtleties are discussed and analyzed by Cohn-Gordon et al. [7], but, once again, in a manner specific to the existing Signal protocol (rather than a general secure messaging primitive). Signal also suggests using the X3DH protocol [24] as one particular way to generate the initial shared key. Certainly, studying (and even appropriately defining) secure messaging *without idealized setup*, and analyzing "full Signal" in this setting, remains an important area for future research.

1.2 Related Work

The OTR (off-the-record) messaging protocol [5] is an influential predecessor of Signal, which was the first to introduce the idea of the DH-based double ratchet to derive fresh keys for each encrypted message. However, it was mainly suitable for synchronous back-and-fourth conversations, so Signal's double ratchet algorithm had to make a number of non-trivial modifications to extend the beautiful OTR idea into a full-fledged asynchronous messaging protocol.

Following the already discussed rigorous description of DH-based double ratcheting by Marlinspike and Perrin [23], and the protocol-specific analysis by Cohn-Gordon et al. [7], several formal analyses of ratcheting have recently appeared [3,10,16,26]; they design definitions of various types of ratcheting and provide schemes meeting these definitions. As previously mentioned, all these works have the drawback of no longer satisfying immediate decryption.

Bellare et al. [3] looked at the question of *unidirectional ratcheting.* In this simplified variant of double (or bidirectional) ratcheting, the receiver is never corrupted, and never needs to update its state. Coupled with giving up immediate decryption, this allowed the authors to obtain a rather simple solution

Unfortunately, extending their ideas to the case of bidirectional communication appeared non-trivial and was left to future work.

Bidirectionality has been achieved in work by Jaeger and Stepanovs [16] and Poettering and Rösler [26]. The papers differ in syntax (one treats secure messaging while the other considers key exchange) and hence use different definitions. However, in spirit both papers attempt to model a bidirectional channel satisfying FS and PCS (but not immediate decryption). Moreover, both consider "fine-grained" PCS requirements which are not met by Signal's double ratchet protocol (and not required by the SM definition in this work). The extra security appears to come at a steep price: both papers use growing (and potentially unbounded) state as well as heavy techniques from public-key cryptography, including hierarchical identity-based encryption [12] (HIBE). More discussion can be found in the full version of this paper [2], including an (informally stated) extension to Signal which achieves a slightly weaker form of fine-grained compromise than [16,26], yet still using only constant sized states, bandwidth and computation as well as comparatively lightweight primitives.

Finally, the notion of immediate decryption is reminiscent in spirit to the zero round trip time (0-RTT) communication with forward secrecy which was recently studied by [9,14]. However, the latter primitive is stateless on the sender side, making it more difficult to achieve (e.g., the schemes of [9,14] use a heavy tool called *puncturable encryption* [13]).

CONCURRENT AND INDEPENDENT WORK. We have recently become aware of two concurrent and independent works by Durak and Vaudenay [10] and Jost, Maurer and Mularczyk [17]. Like other prior works, these works (1) designed their own protocols and did not analyze Signal; and (2) do not satisfy immediate decryption or even message-loss resilience (in fact, they critically rely on receiving messages from one party in order). Both works also provide formal notions of security, including privacy, authenticity, and a new property called unrecoverability by [10] and post-impersonation authentication by [17]: if an active attacker sends a fake message to the recipient immediately following state compromise of the sender, the sender can, by design, never recover (and, thus, will notice the attack by being unable to continue the conversation).

2   Preliminaries

2.1   Game-Based Security and Notation

All security definitions in this work are game-based, i.e., they consider games executed between a challenger and an adversary. The games have one of the following formats:

- *Unpredictability games:* First, the challenger executes the special init procedure, which sets up the game. Subsequently, the attacker is given access to a set of oracles that allow it to interact with the scheme in question. The goal of the adversary is to provoke a particular, game-specific *winning* condition.

The *advantage* of an adversary $\mathcal{A}$ against construction C in an unpredictability game Γ^C is

$$\mathrm{Adv}_\Gamma^C(\mathcal{A}) := \mathsf{P}[\mathcal{A} \text{ wins } \Gamma^C] .$$

- *Indistinguishability games:* In addition to setting up the game, the init procedure samples a secret bit $b \in \{0,1\}$. The goal of the adversary is to determine the value of b. Once more, upon completion of init, the attacker interacts arbitrarily with all available oracles up to the point where it outputs a guess bit b'. The adversary *wins* the game if $b = b'$. The *advantage* of an adversary $\mathcal{A}$ against construction C in an indistinguishability game Γ is

$$\mathrm{Adv}_\Gamma^C(\mathcal{A}) := 2 \cdot \left| \mathsf{P}[\mathcal{A} \text{ wins } \Gamma^C] - 1/2 \right| .$$

With the above in mind, to describe a any security (or correctness) notion, one need only specify the init oracle and the oracles available to $\mathcal{A}$. The following special keywords are used to simplify the exposition of the security games:

- **req** is followed by a condition; if the condition is not satisfied, the oracle/procedure containing the keyword is exited and all actions by it are undone.
- **win** is used to declare that the attacker has won the game; it can be used for both types of games above.
- **end** disables all oracles and returns all values following it to the attacker.

Moreover, the descriptions of some games/schemes involve *dictionaries*. For ease of notation, these dictionaries are described with the *array-notation* described next, but it is important to note that they are to be implemented by a data structure whose size grows (linearly) with the number of elements *in* the dictionary (unlike arrays):

- *Initialization:* The statement $D[\cdot] \leftarrow \lambda$ initializes an *empty* dictionary D.
- *Adding elements:* The statement $D[i] \leftarrow v$ adds a value v to dictionary D with key i, overriding the value previously stored with key i if necessary.
- *Retrieval:* The expression $D[i]$ returns the value v with key i in the dictionary; if there are no values with key i, the value λ is returned.
- *Deletion:* The statement $D[i] \leftarrow \lambda$ *deletes* the value v corresponding to key i.

Finally, sometimes the random coins of certain probabilistic algorithms are made explicit. For example, $y \leftarrow A(x; r)$ means that A, on input x and with random tape r, produces output y. If r is not explicitly stated, is assumed to be chosen uniformly at random; in this case, the notation $y \leftarrow\!\!\$\ A(x)$ is used.

2.2 Cryptographic Primitives

This paper makes use of the following cryptographic primitives:

AEAD. An *authenticated encryption with associated data (AEAD) scheme* is a pair of algorithms AE = (Enc, Dec) with the following syntax:

- *Encryption:* Enc takes a key K, associated data a, and a message m and produces a ciphertext $e \leftarrow \mathsf{Enc}(K, a, m)$.
- *Decryption:* Dec takes a key K, associated data a, and a ciphertext e and produces a message $m \leftarrow \mathsf{Dec}(K, a, e)$.

All AEAD schemes in this paper are assumed to be deterministic, i.e., all randomness stems from the key K.

KEMs. A *key-encapsulation mechanism* (KEM) is a public-key primitive consisting of three algorithms KEM = (KG, Enc, Dec) with the following syntax:

- *Key generation:* KG takes a (implicit) security parameter and outputs a fresh key pair $(\mathsf{pk}, \mathsf{sk}) \leftarrow_\$ \mathsf{KG}$.
- *Encapsulation:* Enc takes a public key pk and produces a ciphertext and a symmetric key $(c, k) \leftarrow_\$ \mathsf{Enc}(\mathsf{pk})$.
- *Decapsulation:* Dec takes a secret key sk and a ciphertext c and recovers the symmetric key $k \leftarrow \mathsf{Dec}(\mathsf{sk}, c)$.

3 Secure Messaging

A *secure messaging (SM)* scheme allows two parties A and B to communicate securely bidirectionally and is expected to satisfy the following informal requirements:

- *Correctness:* If no attacker interferes with the transmission, B outputs the messages sent by A in the correct order and vice versa.
- *Immediate decryption and message-loss resilience (MLR):* Messages must be decrypted as soon as they arrive and may not be buffered; if a message is lost, the parties do not stall.
- *Authenticity:* While the parties' states are uncompromised (i.e., unknown to the attacker), the attacker cannot change the messages sent by them or inject new ones.
- *Privacy:* While the parties' states are uncompromised, an attacker obtains no information about the messages sent.
- *Forward secrecy (FS):* All messages sent and received prior to a state compromise of either party (or both) remain hidden to an attacker.
- *Post-compromise security (PCS, aka "healing"):* If the attacker remains passive (i.e., does not inject any corrupt messages), the parties recover from a state compromise (assuming each has access to fresh randomness).

– *Randomness leakage/failures:* While the parties' states are uncompromised, all the security properties above except PCS hold even if the attacker completely controls the parties' local randomness. That is, good randomness is only required for PCS.

This section presents the syntax of and a formal security notion for SM schemes.

3.1 Syntax

Formally, an SM scheme consists of two initialization algorithms, which are given an initial shared key k, as well as a sending algorithm and a receiving algorithm, both of which keep (shared) state across invocations. The receiving algorithm also outputs a so-called epoch number and an index, which can be used to determine the order in which the sending party transmitted their messages.

Definition 1. *A secure-messaging (SM) scheme* consists of four probabilistic algorithms SM = (Init-A, Init-B, Send, Rcv), *where*

– Init-A *(and similarly* Init-B*) takes a key k and outputs a state $s_A \leftarrow$* Init-A(k),
– Send *takes a state s and a message m and produces a new state and a ciphertext $(s', c) \leftarrow\$ $* Send$(s, m)$, *and*
– Rcv *takes a state s and a ciphertext c and produces a new state, an epoch number, an index, and a message $(s', t, i, m) \leftarrow$* Rcv(s, c).

3.2 Security

Basics. The security notion for SM schemes considered in this paper is intuitive in principle. However, formalizing it is non-trivial and somewhat cumbersome due to a number of subtleties that naturally arise and cannot be avoided if the criteria put forth at the beginning of Sect. 3 are to be met. Therefore, before presenting the definition itself, this section introduces some basic concepts that will facilitate understanding of the definition.

EPOCHS. SM schemes proceed in so-called *epochs*, which roughly correspond the "back-and-forth" between the two parties A and B. By convention, odd epoch numbers t are associated with A sending and B receiving, and the other way around for even epochs. Note, however, that SM schemes are completely asynchronous, and, hence, epochs overlap to a certain extent. Correspondingly, consider two epoch counters t_A and t_B for A and B, respectively, satisfying the following properties:

– The two counters are never more than one epoch apart, i.e., $|t_A - t_B| \leq 1$ at all times.
– When A receives an epoch-t message from B for $t = t_A + 1$, it sets $t_A \leftarrow t$ (even). The next time A sends a message, t_A is incremented again (to an odd value).
– Similarly, when B receives an epoch-t message from A for $t = t_B + 1$, it sets $t_B \leftarrow t$ (odd). The next time B sends a message, t_B is incremented again (to an even value).

MESSAGE INDICES. Within an epoch, messages are identified by a simple counter. To capture the property of immediate decryption and MLR, the receive algorithm of an SM scheme is required to output the correct epoch number and index *immediately* upon reception of a ciphertext, even when messages arrive out of order.

CORRUPTIONS AND THEIR CONSEQUENCES. Since SM schemes are required to be forward-secure and to recover from state compromise, any SM security game must allow the attacker to learn the state of either party at any given time. Moreover, to capture authenticity and privacy, the attacker should be given the power to inject malicious ciphertexts and to call a (say) left-or-right challenge oracle, respectively. These requirements, however, interfere as follows:

- When either party is in a compromised state, the attacker cannot invoke the challenge oracle since this would allow him to trivially distinguish.
- When either party is in a compromised state, the attacker can trivially forge ciphertexts and must therefore be barred from calling the inject oracle.
- When the receiver of messages in transmission is compromised, these messages lose all security, i.e., the attacker learns their content and can replace them by a valid forgery. Consequently, while any challenge ciphertext is in transmission, the recipient may not be corrupted. Similarly, an SM scheme must be able to deal with forgeries of compromised messages (once the parties have healed).

These issues require that the security definition keep track of ciphertexts *in transmission*, of *challenge* ciphertexts, and of *compromised* ciphertexts; this will involve some (slightly cumbersome) record keeping.

NATURAL SM SCHEMES. For simplicity, SM schemes in this work are assumed to satisfy the natural requirements below.[4]

Definition 2. *An SM scheme* SM = (Init-A, Init-B, Send, Rcv) *is* natural *if the following criteria are satisfied:*

(A) Whenever Rcv *outputs* $m = \bot$, *the state remains unchanged.*
(B) Any given ciphertext corresponds to an epoch t *and an index* i, *i.e., the values* (t, i) *output by* Rcv *are an (efficiently computable) function of* c.
(C) Algorithm Rcv *never accepts two messages corresponding to the same pair* (t, i).
(D) A party always rejects ciphertexts corresponding to an epoch in which the party does not act as receiver
(E) If a party, say A, *accepts a ciphertext corresponding to an epoch* t, *then* $t_A \geq t - 1$.

The Security Game. The security game, which is depicted in Fig. 1, consists of an initialization procedure **init** and of

---

[4] The reader may skip over this definition on first read. The properties are referenced where they are needed.

- two "send" oracles, **transmit-A** (normal transmission) and **chall-A** (challenge transmission);
- two "receive" oracles, **deliver-A** (honest delivery) and **inject-A** (for forged ciphertexts); and
- a corrupt oracle **corr-A**

pertaining to party A, and of the corresponding oracles pertaining to B. Moreover, Fig. 1 also features an epoch-management function **ep-mgmt**, a function **sam-if-nec** explained below, and two record-keeping functions **record** and **delete**; these functions cannot be called by the attacker. The game is parametrized by Δ_{SM}, which relates to how fast parties recover from a state compromise. All components are explained in detail below, following the intuition laid out above.

The advantage of $\mathcal{A}$ against an SM scheme SM is denoted by $\mathrm{Adv}^{\mathsf{SM}}_{\mathrm{sm},\Delta_{\mathsf{SM}}}(\mathcal{A})$. The attacker is parameterized by its running time t, the total number of queries q it makes, and the maximum number of epochs q_{ep} it runs for.

Definition 3. *A secure-messaging scheme* SM *is* $(t, q, q_{\mathsf{ep}}, \Delta_{\mathsf{SM}}, \varepsilon)$-secure *if for all* (t, q, q_{ep})-*attackers* $\mathcal{A}$,

$$\mathrm{Adv}^{\mathsf{SM}}_{\mathrm{sm},\Delta_{\mathsf{SM}}}(\mathcal{A}) \leq \varepsilon .$$

INITIALIZATION AND STATE. The initialization procedure chooses a random key and initializes the states s_A and s_B of A and B, respectively. Moreover, it defines several variables to keep track of the execution: (1) t_A and t_B are the epoch counters for A and B, respectively; (2) variables i_A and i_B count how many messages have been sent by each party in their respective current epochs; (3) t_L records the last time either party's state was leaked to the attacker and is used, together with t_A and t_B, to preclude trivial attacks; (4) the sets trans, chall, and comp will contain records and allow to track ciphertexts in transmission, challenge ciphertexts, and compromised ciphertexts, respectively; (5) the bit b is used to create the challenge.

SAMPLING IF NECESSARY. The send oracles **transmit-A** and **chall-A** allow the attacker to possibly control the random coins r of Send. If $r = \bot$, the function samples $r \leftarrow\!\!\$ \; \mathcal{R}$ (from some appropriate set $\mathcal{R}$), and returns (r, good), where good indicates that fresh randomness is used. If, on the other hand, $r \neq \bot$, the function returns (r, bad), indicating, via bad, that adversarially controlled randomness is used.

EPOCH MANAGEMENT. The epoch management function **ep-mgmt** advances the epoch of the calling party if that party's epoch counter has a "receiving value" (even for A; odd for B) and resets the index counter. The flag argument is to indicate whether fresh or adversarial randomness is used. If a currently corrupted party starts a new epoch with bad randomness, the new epoch is considered corrupted. However, if it does not start a new epoch, bad randomness does not make the ciphertext corrupted. This captures that randomness should only be used for PCS (but for none of the other properties mentioned above).

Security Game for Secure Messaging

init
- $k \leftarrow\!\!\$\ \mathcal{K}$
- $s_A \leftarrow \mathsf{Init\text{-}A}(k)$
- $s_B \leftarrow \mathsf{Init\text{-}B}(k)$
- $(t_A, t_B) \leftarrow (0,0)$
- $i_A, i_B \leftarrow 0$
- $t_L \leftarrow -\infty$
- $\mathsf{trans, chall, comp} \leftarrow \emptyset$
- $b \leftarrow\!\!\$\ \{0,1\}$

corr-A
- req $B \notin$ chall
- comp $\stackrel{+}{\leftarrow}$ trans(B)
- $t_L \leftarrow \max(t_A, t_B)$
- return s_A

transmit-A (m, r)
- $(r, \mathsf{flag}) \leftarrow$
- sam-if-nec(r)
- ep-mgmt(A, flag)
- $i_A{+}{+}$
- $(s_A, c) \leftarrow \mathsf{Send}(s_A, m; r)$
- record(A, norm, m, c)
- return c

chall-A (m_0, m_1, r)
- $(r, \mathsf{flag}) \leftarrow$
- sam-if-nec(r)
- ep-mgmt(A, flag)
- req safe-ch$_A$ and
- $|m_0| = |m_1|$
- $i_A{+}{+}$
- $(s_A, c) \leftarrow \mathsf{Send}(m_b; r)$
- record$(A, \mathsf{chall}, m_b, c)$
- return c

deliver-A (c)
- req $(B, t, i, m, c) \in$ trans
-    for some t, i, m
- $(s_A, t', i', m') \leftarrow \mathsf{Rcv}(s_A, c)$
- if $(t', i', m') \neq (t, i, m)$
-   | win
- if $(t, i, m) \in$ chall
-   | $m' \leftarrow \bot$
- $t_A \leftarrow \max(t_A, t)$
- delete(t, i)
- return (t', i', m')

inject-A (c)
- req $(B, c) \notin$ trans and safe-inj
- $(s_A, t', i', m') \leftarrow \mathsf{Rcv}(s_A, c)$
- if $m' \neq \bot$ and $(B, t', i') \notin$ comp
-   | win
- $t_A \leftarrow \max(t_A, t')$
- delete(t', i')
- return (t', i', m')

ep-mgmt (P, flag)
- if $P = A$ and t_P *even or*
- $P = B$ and t_P *odd*
-   | if flag = bad *and*
-     $\neg$safe-ch$_P$
-     | $t_L \leftarrow t_P + 1$
-   | $t_P{+}{+}$
-   | $i_P \leftarrow 0$

sam-if-nec (r)
- flag $\leftarrow$ bad
- if $r = \bot$
-   | $r \leftarrow\!\!\$\ \mathcal{R}$
-   | flag $\leftarrow$ good
- return (r, flag)

record (P, flag, m, c)
- rec $\leftarrow (P, t_P, i_P, m, c)$
- trans $\stackrel{+}{\leftarrow}$ rec
- if $\neg$safe-ch$_P$
-   | comp $\stackrel{+}{\leftarrow}$ rec
- if flag = chall
-   | chall $\stackrel{+}{\leftarrow}$ rec

delete (t, i)
- rec $\leftarrow (P, t, i, m, c)$
-    for some P, m, c
- trans, chall, comp $\stackrel{-}{\leftarrow}$ rec

safe-ch$_P$ $:\Longleftrightarrow\ t_P \geq t_L + \Delta_{SM}$

safe-inj
$:\Longleftrightarrow\ \min(t_A, t_B) \geq t_L + \Delta_{SM}$

Fig. 1. Oracles corresponding to party A of the SM security game for a scheme SM = (Init-A, Init-B, Send, Rcv); the oracles for B are defined analogously.

RECORD KEEPING. The game keeps track of ciphertexts in transmission, of challenge ciphertexts, and of compromised ciphertexts. Records have the format (P, t_P, i_P, m, c), where P is the sender, t_P the epoch in which the message was sent, i_P the index within the epoch, m the message itself, and c the ciphertext.

Whenever **record** is called, the new record is added to the set trans. If a party is not in a safe state, the record is also added to the set of compromised ciphertexts comp. If the function is called with flag = chall, the record is added to chall. The function **delete** takes an epoch number and an index and removes the corresponding record from all three record keeping sets trans, chall, and comp.

Sometimes, it is convenient to refer to a particular record (or a set thereof) by only specifying parts of it. For example, the expression B $\notin$ chall is equivalent to there not being any record (B, $*, *, *, *$) in the set chall. Similarly, trans(B) is the set of all records of this type in trans.

SEND ORACLES.   Both send oracles, **transmit-A** and **chall-A**, begin with **sam-if-nec**, which samples fresh randomness if necessary, followed by a call to **ep-mgmt**. Observe that the flag argument is set to flag $\leftarrow$ good by **sam-if-nec** if fresh randomness is used, and to flag $\leftarrow$ bad otherwise. Subsequently:

- **transmit-A** increments i_A, executes Send, and creates a record using flag = norm, indicating that this is not a challenge ciphertext. Observe that if A is not currently in a safe state, the record is added to comp.
- **chall-A** works similarly to **transmit-A**, except that one of the two inputs is selected according to b, and the record is saved with flag = chall, which will cause it to be added to the challenges chall. Note that **chall-A** can only be called when A is not in a compromised state, which is captured by the statement **req** safe-ch$_A$.

The oracles for B are defined analogously.

RECEIVE ORACLES.   Two oracles are available by which the attacker can get A to receive a ciphertext: **deliver-A** is intended for honest delivery, i.e., to deliver ciphertexts created by B, whereas **inject-A** is used to inject forgeries. These rules are enforced by checking (via **req**) the set trans.

- **deliver-A**: The ciphertext is first passed through Rcv, which must correctly identify the values t, i, and m recorded when c was created; if it fails to do so, the correctness property is violated and the attacker immediately wins the game. In case c was a challenge, the decrypted message is replaced by $\perp$ in order to avoid trivial attacks. Before returning the output of Rcv, t_A is incremented if t is larger than t_A, and the record corresponding to c is deleted.
- **inject-A**: Again, the ciphertext is first passed through Rcv. Unless the ciphertext corresponds[5] to $(t, i) \in$ comp, algorithm Rcv must reject it; otherwise, authenticity is violated and the attacker wins the game. The final instructions are as in **deliver-A**. Oracle **inject-A** may only be called if neither party is currently recovering from state compromise, which is taken care of by flag safe-inj.

The oracles for B are defined analogously.

By deleting records at the end of **deliver-A** and **inject-A**, the game enforces that no replay attacks take place. For example, if a ciphertext c that at some point is in trans is accepted twice, the second time counts as a forgery. Similarly,

---

[5] cf. Property (B) in Definition 2.

if two forgeries for a compromised pair (t, i) are accepted, the attacker wins as well. Note, however, that natural schemes do not allow replay attacks (cf. Property (C) in Definition 2).

CORRUPTION ORACLES. The corruption oracle for A, **corr-A**, can be called whenever no challenges are in transit from B to A, i.e., when B $\notin$ chall. If corruption is allowed, all ciphertexts in transit sent by B become compromised. Before returning A's state, the oracle updates the time of the most recent corruption. The corruption oracle **chall-B** for B is defined similarly.

4   Building Blocks

The SM scheme presented in this work is a modular construction and uses three components: continuous key-agreement (CKA), forward-secure authenticated encryption with associated data (FS-AEAD) and—for lack of a better name—PRF-PRNGs. These primitives are presented in isolation in this section before combining them into an SM scheme in Sect. 5.

4.1   Continuous Key Agreement

This work distills out the public-ratchet part of the Signal protocol and casts it as a separate primitive called *continuous key agreement (CKA)*. This step is not only useful to improve the intuitive understanding of the various components of the Signal protocol and their interdependence, but it also increases modularity, which, for example, would—once the need arises—allow to replace the current CKA mechanism based on DDH by one that is post-quantum secure.

Defining CKA. At a high level, CKA is a synchronous two-party protocol between A and B. Odd rounds i consist of A sending and B receiving a message T_i, whereas in even rounds, B is the sender and A the receiver. Each round i also produces a key I_i, which is output by the sender upon sending T_i and by the receiver upon receiving T_i.

Definition 4. *A continuous-key-agreement (CKA) scheme is a quadruple of algorithms* CKA = (CKA-Init-A, CKA-Init-B, CKA-S, CKA-R), *where*

- CKA-Init-A *(and similarly* CKA-Init-B*) takes a key k and produces an initial state* $\gamma^A \leftarrow$ CKA-Init-A(k) *(and* γ^B*),*
- CKA-S *takes a state* γ, *and produces a new state, message, and key* $(\gamma', T, I) \leftarrow$ $ CKA-S(γ), *and*
- CKA-R *takes a state* γ *and message* T *and produces new state and a key* $(\gamma', I) \leftarrow$ CKA-R(γ, T).

Denote by $\mathcal{K}$ *the space of initialization keys k and by* $\mathcal{I}$ *the space of CKA keys I.*

Security Game for CKA

init (t^*)
 | $k \leftarrow\!\!\$ \; \mathcal{K}$
 | $\gamma^A \leftarrow$ CKA-Init-A(k)
 | $\gamma^B \leftarrow$ CKA-Init-B(k)
 | $t_A, t_B \leftarrow 0$
 | $b \leftarrow\!\!\$ \; \{0, 1\}$

corr-A
 | req allow-corr or
 | finished$_A$
 | return γ^A

send-A
 | t_A ++
 | $(\gamma, T_{t_A}, I_{t_A}) \leftarrow\!\!\$$ CKA-S(γ)
 | return (T_{t_A}, I_{t_A})

send-A'(r)
 | t_A ++
 | req allow-corr
 | $(\gamma, T_{t_A}, I_{t_A}) \leftarrow$ CKA-S$(\gamma; r)$
 | return (T_{t_A}, I_{t_A})

receive-A
 | t_A ++
 | $(\gamma^A, *) \leftarrow$ CKA-R(γ^A, T_{t_A})

chall-A
 | t_A ++
 | req $t_A = t^*$
 | $(\gamma, T_{t_A}, I_{t_A}) \leftarrow\!\!\$$ CKA-S(γ)
 | if $b = 0$
 | | return (T_{t_A}, I_{t_A})
 | else
 | | $I \leftarrow\!\!\$ \; \mathcal{I}$
 | | return (T_{t_A}, I)

---

allow-corr$_P$:$\iff$ $\max(t_A, t_B) \leq t^* - 2$

finished$_P$:$\iff$ $t_P \geq t^* + \Delta_{CKA}$

Fig. 2. Oracles corresponding to party A of the CKA security game for a scheme CKA = (CKA-Init-A, CKA-Init-B, CKA-S, CKA-R); the oracles for B are defined analogously.

CORRECTNESS. A CKA scheme is correct if in the security game in Fig. 2 (explained below), A and B always, i.e., with probability 1, output the same key in every round.

SECURITY. The basic property a CKA scheme must satisfy is that conditioned on the transcript $T_1, T_2, \ldots$, the keys $I_1, I_2, \ldots$ look uniformly random and independent. An attacker against a CKA scheme is required to be passive, i.e., may not modify the messages T_i. However, it is given the power to possibly (1) control the random coins used by the sender and (2) leak the current state of either party. Correspondingly, the keys I_i produced under such circumstances need not be secure. The parties are required to recover from a state compromise *within* 2 *rounds*.[6]

The formal security game for CKA is provided in Fig. 2. It begins with a call to the **init** oracle, which samples a bit b, initializes the states of both parties, and defines epoch counters t_A and t_B. Procedure **init** takes a value t^*, which determines in which round the challenge oracle may be called.

Upon completion of the initialization procedure, the attacker gets to interact arbitrarily with the remaining oracles, as long as *the calls are in a "ping-pong" order*, i.e., a call to a send oracle for A is followed by a receive call for B, then by a send oracle for B, etc. The attacker only gets to use the challenge oracle for epoch t^*. No corruption or using bad randomness (**send-A'** and **send-B'**) is allowed less than two epochs before the challenge is sent (allow-corr).

---

[6] Of course, one could also parametrize the number of rounds required to recover (all CKA schemes in this work recover within two rounds, however).

The game is parametrized by Δ_{CKA}, which stands for the number of epochs that need to pass after t^* until the states do not contain secret information pertaining to the challenge. Once a party reaches epoch $t^* + \Delta_{\mathsf{CKA}}$, its state may be revealed to the attacker (via the corresponding corruption oracle). The game ends (not made explicit) once both states are revealed after the challenge phase. The attacker wins the game if it eventually outputs a bit $b' = b$.

The advantage of an attacker $\mathcal{A}$ against a CKA scheme CKA with $\Delta_{\mathsf{CKA}} = \Delta$ is denoted by $\mathrm{Adv}^{\mathsf{CKA}}_{\mathrm{ror},\Delta}(\mathcal{A})$. The attacker is parameterized by its running time t.

Definition 5. *A CKA scheme* CKA *is* (t, Δ, ε)-secure *if for all* t-attackers $\mathcal{A}$,

$$\mathrm{Adv}^{\mathsf{CKA}}_{\mathrm{ror},\Delta}(\mathcal{A}) \leq \varepsilon .$$

Instantiating CKA. This paper presents several instantiations of CKA: First, a generic CKA scheme with $\Delta = 0$ based on any key-encapsulation mechanism (KEM). Then, by considering the ElGamal KEM and observing that an encapsulated key can be "reused" as public key, one obtains a CKA scheme based on the decisional Diffie-Hellman (DDH) assumption, where the scheme saves a factor of 2 in communication compared to a straight-forward instantiation of the generic scheme. However, the scheme has $\Delta = 1$.

CKA FROM KEMs. A CKA scheme with $\Delta = 0$ can be built from a KEM in natural way: in every epoch, one party sends a public key pk of a freshly generated key pair and an encapsulated key under the key pk' received from the other party in the previous epoch. Specifically, consider a CKA scheme CKA = (CKA-Init-A, CKA-Init-B, CKA-S, CKA-R) that is obtained from a KEM KEM as follows:

- The initial shared state $k = (\mathsf{pk}, \mathsf{sk})$ consists of a (freshly generated) KEM key pair. The initialization for A outputs $\mathsf{pk} \leftarrow \mathsf{CKA\text{-}Init\text{-}A}(k)$ and that for B outputs $\mathsf{sk} \leftarrow \mathsf{CKA\text{-}Init\text{-}B}(k)$.
- The send algorithm CKA-S takes as input the current state $\gamma = \mathsf{pk}$ and proceeds as follows: It
  1. encapsulates a key $(c, I) \leftarrow_{\$} \mathsf{Enc}(\mathsf{pk})$;
  2. generates a new key pair $(\mathsf{pk}, \mathsf{sk}) \leftarrow_{\$} \mathsf{KG}$;
  3. sets the CKA message to $T \leftarrow (c, \mathsf{pk})$;
  4. sets the new state to $\gamma \leftarrow \mathsf{sk}$; and
  5. returns (γ, T, I).
- The receive algorithm CKA-R takes as input the current state $\gamma = \mathsf{sk}$ as well as a message $T = (c, \mathsf{pk})$ and proceeds as follows: It
  1. decapsulates the key $I \leftarrow \mathsf{Dec}(\mathsf{sk}, c)$;
  2. sets the new state to $\gamma \leftarrow \mathsf{pk}$; and
  3. returns (γ, I).

The full version of this paper [2] shows that the above scheme is a secure CKA protocol by reducing its security to that of the underlying KEM.

CKA FROM DDH. Observe that if one instantiates the above KEM-based CKA scheme with the ElGamal KEM over some group G, both the public key and the encapsulated key are elements of G. Hence, the Signal protocol uses an optimization of the ElGamal KEM where a single group element first serves as an encapsulated key sent by, say, A and then as the public key B uses to encapsulate his next key. Interestingly, this comes at the price of having $\Delta = 1$ (as opposed to $\Delta = 0$) due to the need for parties to hold on to their exponents (which serve both as secret keys and encapsulation randomness) longer.

Concretely, a CKA scheme $\mathsf{CKA} = (\mathsf{CKA\text{-}Init\text{-}A}, \mathsf{CKA\text{-}Init\text{-}B}, \mathsf{CKA\text{-}S}, \mathsf{CKA\text{-}R})$ can be obtained from the DDH assumption[7] in a cyclic group $G = \langle g \rangle$ as follows:

- The initial shared state $k = (h, x_0)$ consists of a (random) group element $h = g^{x_0}$ and its discrete logarithm x_0. The initialization for A outputs $h \leftarrow \mathsf{CKA\text{-}Init\text{-}A}(k)$ and that for B outputs $x_0 \leftarrow \mathsf{CKA\text{-}Init\text{-}B}(k)$.
- The send algorithm $\mathsf{CKA\text{-}S}$ takes as input the current state $\gamma = h$ and proceeds as follows: It
  1. chooses a random exponent x;
  2. computes the corresponding key $I \leftarrow h^x$;
  3. sets the CKA message to $T \leftarrow g^x$;
  4. sets the new state to $\gamma \leftarrow x$; and
  5. returns (γ, T, I).
- The receive algorithm $\mathsf{CKA\text{-}R}$ takes as input the current state $\gamma = x$ as well as a message $T = h$ and proceeds as follows: It
  1. computes the key $I = h^x$;
  2. sets the new state to $\gamma \leftarrow h$; and
  3. returns (γ, I).

The full version of this paper [2] shows that the above scheme is a secure CKA protocol if the DDH assumption holds in group G.

4.2 Forward-Secure AEAD

Defining FS-AEAD. *Forward-secure authenticated encryption with associated data* is a stateful primitive between a sender A and a receiver B and can be considered a single-epoch variant of an SM scheme, a fact that is also evident from its security definition, which resembles that of SM schemes.

Definition 6. Forward-secure authenticated encryption with associated data (FS-AEAD) *is a tuple of algorithms* $\mathsf{FS\text{-}AEAD} = (\mathsf{FS\text{-}Init\text{-}S}, \mathsf{FS\text{-}Init\text{-}R}, \mathsf{FS\text{-}Send}, \mathsf{FS\text{-}Rcv})$, *where*

- $\mathsf{FS\text{-}Init\text{-}S}$ *(and similarly* $\mathsf{FS\text{-}Init\text{-}R}$*) takes a key* k *and outputs a state* $v_A \leftarrow \mathsf{FS\text{-}Init\text{-}S}(k)$,

---

[7] The DDH assumption states that it is hard to distinguish DH triples (g^a, g^b, g^{ab}) from random triples (g^a, g^b, g^c), where a, b, and c are uniformly random and independent exponents.

- FS-Send *takes a state v, associated data a, and a message m and produces a new state and a ciphertext* $(v', e) \leftarrow$ FS-Send(v, a, m), *and*
- FS-Rcv *takes a state v, associated data a, and a ciphertext e and produces a new state, an index, and a message* $(v', i, m) \leftarrow$ FS-Rcv(v, a, e).

Observe that all algorithms of an FS-AEAD scheme are deterministic.

MEMORY MANAGEMENT. In addition to the basic syntax above, it is useful to define the following two functions FS-Stop (called by the sender) and FS-Max (called by the receiver) for memory management:

- FS-Stop, given an FS-AEAD state v, outputs how many messages have been received and then "erases" the FS-AEAD session corresponding to v form memory; and
- FS-Max, given a state v and an integer ℓ, remembers ℓ internally such that the session corresponding to v is erased from memory as soon as ℓ messages have been received.

These features will be useful in the full protocol (cf. Sect. 5) to be able to terminate individual FS-AEAD sessions when they are no longer needed. Providing a formal requirement for these additional functions is omitted. Moreover, since an attacker can infer the value of the message counter from the behavior of the protocol anyway, there is no dedicated oracle included in the security game below.

CORRECTNESS AND SECURITY. Both correctness and security are built into the security game depicted in Fig. 3. The game is the single-epoch analogue of the SM security game (cf. Fig. 1) and therefore has similarly defined oracles and similar record keeping. A crucial difference is that as soon as the receiver B is compromised, the game ends with a full state reveal as no more security can be provided. If only the sender A is compromised, the game continues and uncompromised messages must remain secure.

The advantage of an attacker $\mathcal{A}$ against an FS-AEAD scheme FS-AEAD is denoted by the expression $\mathrm{Adv}_{\text{fs-aead}}^{\text{FS-AEAD}}(\mathcal{A})$. The attacker is parameterized by its running time t and the total number of queries q it makes.

Definition 7. *An FS-AEAD scheme* FS-AEAD *is* (t, q, ε)-*secure if for all* (t, q)-*attackers* $\mathcal{A}$,
$$\mathrm{Adv}_{\text{fs-aead}}^{\text{FS-AEAD}}(\mathcal{A}) \leq \varepsilon .$$

Instantiating FS-AEAD. An FS-AEAD scheme can be easily constructed from two components:

- an AEAD scheme $\mathsf{AE} = (\mathsf{Enc}, \mathsf{Dec})$, and
- a PRG $G : \mathcal{W} \rightarrow \mathcal{W} \times \mathcal{K}$, where $\mathcal{K}$ is the key space of the AEAD scheme.

Security Game for FS-AEAD

init
| $k \leftarrow\!\!\$\ \mathcal{K}$
| $v_\mathsf{A} \leftarrow \mathsf{FS\text{-}Init\text{-}S}(k)$
| $v_\mathsf{B} \leftarrow \mathsf{FS\text{-}Init\text{-}R}(k)$
| $i_\mathsf{A} \leftarrow 0$
| $\mathsf{corr}_\mathsf{A} \leftarrow \mathsf{false}$
| $\mathsf{trans}, \mathsf{chall}, \mathsf{comp} \leftarrow \emptyset$
| $b \leftarrow\!\!\$\ \{0,1\}$

corr-A
| $\mathsf{corr}_\mathsf{A} \leftarrow \mathsf{true}$
| return v_A

corr-B
| req $\mathsf{chall} = \emptyset$
| end $(v_\mathsf{A}, v_\mathsf{B})$

transmit-A (a, m)
| $i_\mathsf{A} \mathrel{+}\!\!\!+$
| $(v_\mathsf{A}, e) \leftarrow \mathsf{FS\text{-}Send}(v_\mathsf{A}, a, m)$
| $\mathsf{record}(\mathsf{good}, a, m, e)$
| return c

chall-A (a, m_0, m_1)
| req $\neg\mathsf{corr}_\mathsf{A}$ and $|m_0| = |m_1|$
| $i_\mathsf{A} \mathrel{+}\!\!\!+$
| $(v_\mathsf{A}, e) \leftarrow \mathsf{FS\text{-}Send}(v_\mathsf{A}, a, m_b)$
| $\mathsf{record}(\mathsf{chall}, a, m_b, e)$
| return e

deliver-A (a, e)
| req $(i, a, m, e) \in \mathsf{trans}$
| for some i, m
| $(v_\mathsf{A}, i', m') \leftarrow \mathsf{FS\text{-}Rcv}(v_\mathsf{A}, a, e)$
| if $(i', m') \neq (i, m)$
| | win
| if $(i, m) \in \mathsf{chall}$
| | $m' \leftarrow \perp$
| $\mathsf{delete}(i)$
| return (i', m')

inject-A (a, e)
| req $(a, e) \notin \mathsf{trans}$
| $(v_\mathsf{A}, i', m') \leftarrow \mathsf{FS\text{-}Rcv}(v_\mathsf{A}, a, e)$
| if $m' \neq \perp$ and $(\mathsf{B}, i') \notin \mathsf{comp}$
| | win
| $\mathsf{delete}(i')$
| return (i', m')

record (flag, a, m, e)
| $\mathsf{rec} \leftarrow (i_\mathsf{A}, a, m, e)$
| $\mathsf{trans} \xleftarrow{+} \mathsf{rec}$
| if $\mathsf{flag} = \mathsf{bad}$ or corr_A
| | $\mathsf{comp} \xleftarrow{+} \mathsf{rec}$
| if $\mathsf{flag} = \mathsf{chall}$
| | $\mathsf{chall} \xleftarrow{+} \mathsf{rec}$

delete (i)
| $\mathsf{rec} \leftarrow (i, a, m, e)$ for m, a, e
| s.t. $(i, a, m, e) \in \mathsf{trans}$
| $\mathsf{trans}, \mathsf{chall}, \mathsf{comp} \xleftarrow{-} \mathsf{rec}$

Fig. 3. Oracles corresponding to party A of the FS-AEAD security game for a scheme FS-AEAD $= (\mathsf{FS\text{-}Init\text{-}S}, \mathsf{FS\text{-}Init\text{-}R}, \mathsf{FS\text{-}Send}, \mathsf{FS\text{-}Rcv})$; the oracles for B are defined analogously.

The scheme is described in Fig. 4. For simplicity the states of sender A and receiver B are is not made explicit; it consists of the variables set during initialization. The main idea of the scheme, is that the A and B share the state w of a PRG G. State w is initialized with a pre-shared key $k \in \mathcal{W}$, which is assumed to be chosen uniformly at random. Both parties keep local counters i_A and i_B, respectively.[8] A, when sending the ith message m with associated data (AD) a, uses G to expand the current state to a new state and an AEAD key $(w, K) \leftarrow G(w)$ and computes an AEAD encryption under K of m with AD $h = (i, a)$.

Since B may receive ciphertext out of order, whenever he receives a ciphertext, he first checks whether the key is already stored in a dictionary $\mathcal{D}$. If the index of the message is higher than expected (i.e., larger than $i_\mathsf{B} + 1$), B skips the PRG ahead and stores the skipped keys in $\mathcal{D}$. In either case, once the key is

---

[8] For ease of description, the FS-AEAD state of the parties is not made explicit as a variable v.

Forward-Secure AEAD

Init-A (k)
| $w \leftarrow k$
| $i_A \leftarrow 0$

Init-B (k)
| $w \leftarrow k$
| $i_B \leftarrow 0$
| $\mathcal{D}[\cdot] \leftarrow \lambda$

try-skipped (i)
| $K \leftarrow \mathcal{D}[i]$
| $\mathcal{D}[i] \leftarrow \perp$
| **return** K

FS-Send (a, m)
| i_A ++
| $(w, K) \leftarrow G(w)$
| $h \leftarrow (i_A, a)$
| $e \leftarrow \mathsf{Enc}(K, h, m)$
| **return** (i_A, e)

skip (u)
| **while** $i_B < u - 1$
| | i_B ++
| | $(w, K) \leftarrow G(w)$
| | $\mathcal{D}[u] \leftarrow K$

FS-Rcv (a, c)
| $(i, e) \leftarrow c$
| $K \leftarrow$ try-skipped(i)
| **if** $K = \perp$
| | skip(i)
| | $(w, K) \leftarrow G(w)$
| | $i_B \leftarrow i$
| $h \leftarrow (i, a)$
| $m \leftarrow \mathsf{Dec}(K, h, e)$
| **if** $m = \perp$
| | error
| **return** (i, m)

Fig. 4. FS-AEAD scheme based on AEAD and a PRG.

obtained, it is used to decrypt. If decryption fails, FS-Rcv throws an exception (**error**), which causes the state to be rolled back to where it was before the call to FS-Rcv.

In the full version of this work [2], it is shown that, based on the security of the AEAD scheme and the PRG, the above yields a secure FS-AEAD scheme.

4.3 PRF-PRNGs

Defining PRF-PRNGs. A *PRF-PRNG* resembles both a pseudo-random function (PRF) and a pseudorandom number generator with input (PRNG)—hence the name. On the one hand, as a PRNG would, a PRF-PRNG (1) repeatedly accepts inputs I and uses them to refresh its state σ and (2) occasionally uses the state, provided it has sufficient entropy, to derive a pseudo-random pair of output R and new state; for the purposes of secure messaging, it suffices to combine properties (1) and (2) into a single procedure. On the other hand, a PRF-PRNG can be used as a PRF in the sense that if the state has high entropy, the answers to various inputs I *on the same state* are indistinguishable from random and independent values.

Definition 8. *A* PRF-PRNG *is a pair of algorithms* P = (P-Init, P-Up), *where*

- P-Init *takes a key* k *and produces a state* $\sigma \leftarrow$ P-Init(k), *and*
- P-Up *takes a state* σ *and an input* I *and produces a new state and an output* $(\sigma', R) \leftarrow$ P-Up(σ, I).

SECURITY. The simple intuitive security requirement for a double-seed PRG is that P-Init(σ, I) produce a pseudorandom value if the state σ is uncorrupted (i.e., has high entropy) or the input I is chosen uniformly from some set $\mathcal{S}$. Moreover, if the state is uncorrupted, it should have the PRF property described above. This is captured by the security definition described by Fig. 5:

Security Game for PRF-PRNG		

init
  $k \leftarrow_\$ \mathcal{K}$
  $\sigma \leftarrow \text{P-Init}(k)$
  corr $\leftarrow$ false
  prng, prf $\leftarrow$ false
  $b \leftarrow_\$ \{0, 1\}$

corr
  req $\neg$prf
  return σ

process (I)
  $I \leftarrow \text{sam-if-nec}(I)$
  $(\sigma, R) \leftarrow \text{P-Up}(\sigma, I)$
  return R

chall-prf (I)
  req $\neg$corr and $\neg$prng
  prf $\leftarrow$ true
  $(\sigma', R) \leftarrow \text{P-Up}(\sigma, I)$
  if $b = 1$
    $R \leftarrow_\$ \mathcal{R}$
  return (σ', R)

chall-prng (I)
  $I \leftarrow \text{sam-if-nec}(I)$
  req $\neg$corr and $\neg$prf
  prng $\leftarrow$ true
  $(\sigma, R) \leftarrow \text{P-Up}(\sigma, I)$
  if $b = 1$
    $R \leftarrow_\$ \mathcal{R}$
  return R

sam-if-nec (I)
  if $I = \bot$
    $I \leftarrow_\$ \mathcal{I}$
    corr $\leftarrow$ false
  return I

Fig. 5. Oracles of the PRF-PRNG security game for a scheme P = (P-Init, P-Up).

- *Initialization:* Procedure **init** chooses a random bit b, initializes the PRF-PRNG with a random key, and sets two flags prng and prf to false: the PRNG and PRF modes are mutually exclusive and only one type of challenge may be called; the flags keep track of which.
- *PRNG mode:* The oracle **process** can be called in two ways: either I is an input specified by the attacker and is simply absorbed into the state, or $I = \bot$, in which case the game chooses it randomly (inside **sam-if-nec**) and absorbs it into the state, which at this point becomes uncorrupted. Oracle **chall-prng** is works in the same fashion but creates a challenge.
- *PRF mode:* Once the state is uncompromised the attacker can decide to obtain PRF challenges by calling **chall-prf**, which simply evaluates the (adversarially chosen) input on the current state without updating it and creates a challenge.
- *Corruption:* At any time, except after asking for PRF challenges, the attacker may obtain the state by calling **corr**.

The advantage of $\mathcal{A}$ in the PRF-PRNG game is denoted by $\text{Adv}_{PP}^P(\mathcal{A})$. The attacker is parameterized by its running time t.

Definition 9. *An PRF-PRG* P *is* (t, ε)-*secure if for all* t-*attackers* $\mathcal{A}$,

$$\text{Adv}_{PP}^P(\mathcal{A}) \leq \varepsilon .$$

Instantiating PRF-PRNGs. Being a PRF-PRNG is a property the HKDF function used by Signal is assumed to have; in particular, Marlinspike and Perrin [23] recommend the primitive be implemented with using HKDF [19] with SHA-256 or SHA-512 [25] where the state σ is used as HKDF salt and I as

HKDF input key material. This paper therefore merely reduces the security of the presented schemes to the PRF-PRNG security of whatever function is used to instantiate it.

Alternatively, a simple standard-model instantiation (whose rather immediate proof is omitted) can be based on a pseudorandom *permutation* (PRP) $\Pi : \{0,1\}^n \times \{0,1\}^n \to \{0,1\}^n$ and a PRG $G : \{0,1\}^n \to \{0,1\}^n \times \mathcal{K}$ by letting the state be the PRP key $s \in \{0,1\}^n$ and

$$(s', R) \leftarrow \mathsf{P\text{-}Up}(s, I) = G(\Pi_s(I)) .$$

5 Secure Messaging Scheme

This section presents a Signal-based secure messaging (SM) scheme and establishes its security under Definition 3. The scheme suitably and modularly combines continuous key-agreement (CKA), forward-secure authenticated encryption with associated data (FS-AEAD), and PRF-PRNGs; these primitives are explained in detail in Sect. 4.

5.1 The Scheme

The scheme is inspired by the Signal protocol, but differs from it in a few points, as explained in Sect. 5.2. The main idea of the scheme is that the parties A and B keep track of the same PRF-PRG (aka the "root RNG"), which they use to generate keys for FS-AEAD instances as needed. The root RNG is continuously refreshed by random values output by a CKA scheme that is run "in parallel."

STATE. Scheme SM keeps an internal state s_A (resp. s_B), which is initialized by Init-A (resp. Init-B) and used as well as updated by Send and Rcv. The state s_A of SM consists of the following values:

- an ID field with id = A,
- the state σ_{root} of the root RNG,
- states $v[0], v[1], v[2], \ldots$ of the various FS-AEAD instances,
- the state γ of the CKA scheme,
- the current CKA message T_{cur}, and
- an epoch counter t_A.

In order to remove expired FS-AEAD sessions from memory, there is also a variable ℓ_{prv} that remembers the number of messages sent in the second most recent epoch . Recall (cf. Sect. 4.2) that once the maximum number of messages has been set via FS-Max, a session "erases" itself from the memory, and similarly for calling FS-Stop on a particular FS-AEAD session. For simplicity, removing the corresponding $v[t]$ from memory is not made explicit in either case. The state s_B is defined analogously.

Signal-Based Secure-Messaging Scheme

Init-A (k)
  id $\leftarrow$ A
  $(k_{root}, k_{CKA}) \leftarrow k$
  $\sigma_{root} \leftarrow$ P-Init(k_{root})
  $v[\cdot] \leftarrow \lambda$
  $(\sigma_{root}, k) \leftarrow$ P-Up(σ_{root}, λ)
  $v[0] \leftarrow$ FS-Init-R(k)
  $\gamma \leftarrow$ CKA-Init-A(k_{CKA})
  $T_{cur} \leftarrow \lambda$
  $\ell_{prv} \leftarrow 0$
  $t_A \leftarrow 0$

Send-A (m)
  if t_A is even
    $\ell_{prv} \leftarrow$ FS-Stop$(v[t_A - 1])$
    t_A ++
    $(\gamma, T_{cur}, I) \leftarrow\$ CKA-S(γ)
    $(\sigma_{root}, k) \leftarrow$ P-Up(σ_{root}, I)
    $v[t_A] \leftarrow$ FS-Init-S(k)
  $h \leftarrow (t_A, T_{cur}, \ell_{prv})$
  $(v[t_A], e) \leftarrow$ FS-Send$(v[t_A], h, m)$
  return (h, e)

Rcv-A (c)
  $(h, e) \leftarrow c$
  $(t, T, \ell) \leftarrow h$
  req t even and $t \leq t_A + 1$
  if $t = t_A + 1$
    t_A ++
    FS-Max$(v[t - 2], \ell)$
    $(\gamma, I) \leftarrow$ CKA-R(γ, T)
    $(\sigma_{root}, k) \leftarrow$ P-Up(σ_{root}, I)
    $v[t] \leftarrow$ FS-Init-R(k)
  $(v[t], i, m) \leftarrow$ FS-Rcv$(v[t], h, e)$
  if $m = \perp$
    error
  return (t, i, m)

Fig. 6. Secure-messaging scheme based on a FS-AEAD, a CKA scheme, and a PRF-PRNG.

THE ALGORITHMS. The algorithms of scheme SM are depicted in Fig. 6 and described in more detail below. For ease of description, the algorithms Send and Rcv are presented as Send-A and Rcv-A, which handle the case where id = A; the case id = B works analogously. Moreover, to improve readability, the state s_A is not made explicit in the description: it consists of the variables set by the initialization algorithm.

– *Initialization:* The initialization procedure Init-A expects a key k shared between A and B; k is assumed to have been created at some point before the execution during a trusted setup phase and to consist of initialization keys k_{root} and k_{CKA} for the root RNG and the CKA scheme, respectively. In a second step, the root RNG is initialized with k. Then, it is used to generate a key for FS-AEAD epoch $v[0]$; A acts as receiver in $v[0]$ and all subsequent even epochs and as sender in all subsequent odd epochs. Furthermore, Init-A also initializes the CKA scheme and sets the initial epoch $t_A \leftarrow 0$ and T_{cur} to a default value.[9]

As pointed out above, scheme SM runs a CKA protocol in parallel to sending its messages. To that end, A's first message includes the first message T_1 output by CKA-S. All subsequent messages sent by A include T_1 until some message received from B includes T_2. At that point A would run CKA-S again and include T_3 with all her messages, and so on (cf. Sect. 4.1).

Upon either sending or receiving T_i for odd or even i, respectively, the CKA protocol also produces a random value I_i, which A absorbs into the root RNG. The resulting output k is used as key for a new FS-AEAD epoch.

---

[9] B also starts in epoch $t_B \leftarrow 0$.

– *Sending messages:* Procedure Send-A allows A to send a message to B. As a first step, Send-A determines whether it is A's turn to send the next CKA message, which is the case if t_A is even. Whenever it is A's turn, Send-A runs CKA-S to produce the her next CKA message T and key I, which is absorbed into the root RNG. The resulting value k is used as a the key for a new FS-AEAD epoch, in which A acts as sender. The now old epoch is terminated by calling FS-Stop and the number of messages in the old epoch is stored in ℓ_{prv}, which will be sent along inside the header for every message of the new epoch.

Irrespective of whether it was necessary to generate a new CKA message and generate a new FS-AEAD epoch, Send-A creates a header $h = (t_A, T_{cur}, \ell_{prv})$, and uses the current epoch $v[t_A]$ to get a ciphertext for (h, m) (where h is treated as associated data).

– *Receiving messages:* When a ciphertext $c = (h, e, \ell)$ with header $h = (t, T, \ell)$ is processed by Rcv-A, there are two possibilities:

  • $t \leq t_A$ (and t even): In this case, ciphertext c pertains to an existing FS-AEAD epoch, in which case FS-Send is simply called on $v[t]$ to process e. If the maximum number of messages has been received for session $v[t]$, the session is removed from memory.

  • $t = t_A + 1$ and t_A odd: Here, the receiver algorithm advances t_A by incrementing it and processes T with CKA-R. This produces a key I, which is absorbed into the PRF-PRG to obtain a key k with which to initialize a new epoch $v[t_A]$ as receiver. Then, e is processed by FS-Rcv on $v[t_A]$. Note that Rcv also uses FS-Max to store ℓ as the maximum number of messages in the previous receive epoch.

Irrespective of whether a new CKA message was received and a new epoch created, if e is rejected by FS-Rcv, the algorithm raises an exception (**error**), which causes the entire state s_A to be rolled back to what it was before Rcv-A was called.

5.2 Differences to Signal

By instantiating the building blocks as shown below, one obtains an SM scheme that is very close to the actual Signal protocol (cf. [23, Section 5.2] for more details):

– *CKA:* the DDH-based CKA scheme from Sect. 4.1 using Curve25519 or Curve448 as specified in [20];
– *FS-AEAD:* FS-AEAD scheme from Sect. 4.2 with HMAC [18] with SHA-256 or SHA-512 [25] for the PRG, and an AEAD encryption scheme based on either SIV or a composition of CBC with HMAC [27,28];
– *PRF-PRNG:* HKDF [19] with SHA-256 or SHA-512 [25], used as explained in Sect. 4.3.

We now detail the main differences:

DEFERRED RANDOMNESS FOR SENDING. Deployed Signal implementations generate a new CKA message and absorb the resulting key into the RNG in Rcv, as opposed to taking care of this inside Send, as done here. The way it is done here is advantageous in the sense that the new key is not needed until the Send operation is actually initiated, so there is no need to risk its exposure unnecessarily (in case the state is compromised in between receiving and sending). Indeed, this security enhancement to Signal was explicitly mentioned by Marlinspike and Perrin [23] (cf. Section 6.5), and we simply follow this suggestion for better security.

EPOCH INDEXING. In our scheme we have an explicit epoch counter t to index a given epoch. In Signal, one uses the uniqueness of latest CKA message (of the form g^x) to index an epoch. This saves an extra counter from each party's state, but we find our treatment of having explicit epoch counters much more intuitive, and not relying on any particular structure of CKA messages. In fact, indexing a dictionary becomes slightly more efficient when using a simple counter than the entire CKA message (which could be long for certain CKA protocols; e.g., post-quantum from lattices).

FS-AEAD ABSTRACTION. Unlike the SM proposed from this section, Signal does not use the FS-AEAD abstraction. Instead, each party maintains a sending and a receiving PRG that are kept in sync with the other party's receiving and sending PRG, respectively. Moreover, when receiving the first message of a new epoch, the current receive PRG is skipped ahead appropriately depending on the value ℓ, and the skipped keys are stored in a *single*, global dictionary. The state of the receive PRG is then overwritten with the new state output by the root RNG. Then, upon the next send operation new randomness for the CKA message is generated, and the sending RNG is also overwritten by the state output from updating the root RNG again. This is logically equivalent to our variant of Signal with the particular FS-AEAD implementation in Fig. 4, except we will maintain multiple dictionaries (one for each epoch t). However, merging these dictionaries into one global dictionary (indexed by epoch counter in addition to the message count within epoch) becomes a simple efficiency optimization of the resulting scheme. Moreover, once this optimization is done, there is no need to store an array of FS-AEAD instances $v[t]$. Instead, we can only remember the latest sending and receiving FS-AEAD instance, overwriting them appropriately with each new epoch. Indeed, storing old message keys from not-yet-delivered messages is the only information one needs to remember from the prior FS-AEAD instances. So once this information is stored in the global dictionary, we can simply overwrite the remaining information when moving to the new epoch. With these simple efficiency optimizations, we arrive to (almost) precisely what is done by Signal (cf. Fig. 7).

To sum up, blindly using the FS-AEAD abstraction results in a slightly less efficient scheme, but (1) we feel our treatment is more modular and intuitive; (2) when using a concrete FS-AEAD scheme from Sect. 4.2, getting actual Signal

Signal Scheme

Init-A (k)
 id $\leftarrow$ A
 $(k_{root}, k_{CKA}) \leftarrow k$
 $\sigma_{root} \leftarrow$ P-Init(k_{root})
 $(\sigma_{root}, \sigma_B) \leftarrow$ P-Up(σ_{root}, λ)
 $\gamma \leftarrow$ CKA-Init-A(k_{CKA})
 $T_{cur} \leftarrow \lambda$
 $\ell_{prv} \leftarrow 0$
 $t_A, i_A, i_B \leftarrow 0$
 $\mathcal{D}[\cdot] \leftarrow \lambda$

skip (t, u)
 while $i_B < u$
  i_B ++
  $(\sigma_B, K) \leftarrow G(\sigma_B)$
  $\mathcal{D}[t, i_B] \leftarrow K$

Send-A (m)
 if t_A is even
  t_A ++
  $\ell_{prv} \leftarrow i_A$
  $i_A \leftarrow 0$
  $(\gamma, T_{cur}, I) \leftarrow\!\!\$\ $CKA-S$(\gamma)$
  $(\sigma_{root}, \sigma_A) \leftarrow$ P-Up(σ_{root}, I)
 i_A ++
 $h \leftarrow (t_A, i_A, T_{cur}, \ell_{prv})$
 $(\sigma_A, K) \leftarrow G(\sigma_A)$
 $e \leftarrow$ Enc(K, h, m)
 return (h, e)

try-skipped (t, i)
 $K \leftarrow \mathcal{D}[t, i]$
 $\mathcal{D}[t, i] \leftarrow \perp$
 return K

Rcv-A (c)
 $(h, e) \leftarrow c$
 $(t, i, T, \ell) \leftarrow h$
 req t even and $t \leq t_A + 1$
 if $t = t_A + 1$
  skip$(t - 2, \ell_{prv})$
  t_A ++, $i_B \leftarrow 0$
  $(\gamma, I) \leftarrow$ CKA-R(γ, T)
  $(\sigma_{root}, \sigma_B) \leftarrow$ P-Up(σ_{root}, I)
 $K \leftarrow$ try-skipped(t, i)
 if $K = \perp$
  skip$(t, i - 1)$
  i_B ++
  $(\sigma_B, K) \leftarrow G(\sigma_B)$
 $m \leftarrow$ Dec(K, h, e)
 if $m = \perp$
  error
 return (t, i, m)

Fig. 7. Signal scheme without the FS-AEAD abstraction, based on a CKA scheme, a PRF-PRNG, authenticated encryption, and a regular PRG. The figure only shows the algorithms for A; B's algorithms are analogous, with the roles of i_A and i_B switched.

becomes a simple efficiency optimization of the resulting scheme. In particular, the security of Signal itself still follows from our framework.

INITIAL KEY AGREEMENT. As mentioned in the introduction, our modeling only addresses the double-ratchet aspect of the Signal protocol, and does not tackle the challenging problem of the generation of the initial shared key k. One thing this also allows us to do is to elegantly side-step the issue that natural CKA protocols are *unkeyed*, and do not generate shared a shared key I_0 from the initial message T_0. Instead, we model CKA as a *secret key* primitive, where the initial key k_{CKA} effectively generates the first message T_0 of "unkeyed CKA" protocol, but now shared keys $I_1, I_2, \ldots$ get generated right away from subsequent messages $T_1, T_2, \ldots$. In other words, rather than having k only store the root key k_{root}, in our protocol we let it store a tuple (k_{root}, k_{CKA}), and then use k_{CKA} to solve the syntactic issue of having a special treatment for the first CKA message T_0.

In most actual Signal implementations, the initial shared key k will only contain the value k_{root}, and it is the receiver B who stores several initial CKA messages T_0 (called "one-time prekeys") on the Signal server for new potential senders A. When such A comes along, A would take one such one-time prekey value T_0 from the Signal server, and (optionally) *use it* to generate the initial shared key k_{root} using the X3DH Key Agreement Protocol [24]. This creates slight circularity, and we leave it to the future work to properly model and analyze such generation of the initial key k_{root}.

5.3 Security of the SM Scheme

The proof of the following main theorem can be found in the full version of this paper [2].

Theorem 1. *Assume that*

- CKA *is a* $(t', \Delta_{\mathsf{CKA}}, \varepsilon_{\mathrm{cka}})$-*secure CKA scheme,*
- FS-AEAD *is a* $(t', q, \varepsilon_{\mathrm{fs\text{-}aead}})$-*secure FS-AEAD scheme, and*
- P *is a* $(t', \varepsilon_{\mathrm{p}})$-*secure PRF-PRNG.*

Then, the SM *construction above is* $(t, q, q_{\mathsf{ep}}, \Delta_{\mathsf{SM}}, \varepsilon)$-*SM-secure for* $t \approx t'$, $\Delta_{\mathsf{SM}} = 2 + \Delta_{\mathsf{CKA}}$, *and*

$$\varepsilon \leq 2q_{\mathsf{ep}}^2 \cdot (\varepsilon_{\mathrm{cka}} + q \cdot \varepsilon_{\mathrm{fs\text{-}aead}} + \varepsilon_{\mathrm{p}}) \,.$$

References

1. Alkim, E., Ducas, L., Pöppelmann, T., Schwabe, P.: Post-quantum key exchange - a new hope. In: Holz, T., Savage, S., (eds.) 25th USENIX Security Symposium, pp. 327–343. USENIX Association (2016)
2. Alwen, J., Coretti, S., Dodis, Y.: The double ratchet: security notions, proofs, and modularization for the signal protocol. Cryptology ePrint Archive, Report 2018/1037 (2018). https://eprint.iacr.org/2018/1037
3. Bellare, M., Singh, A.C., Jaeger, J., Nyayapati, M., Stepanovs, I.: Ratcheted encryption and key exchange: the security of messaging. In: Katz, J., Shacham, H. (eds.) CRYPTO 2017, Part III. LNCS, vol. 10403, pp. 619–650. Springer, Cham (2017). https://doi.org/10.1007/978-3-319-63697-9_21
4. Bellare, M., Yee, B.: Forward-security in private-key cryptography. In: Joye, M. (ed.) CT-RSA 2003. LNCS, vol. 2612, pp. 1–18. Springer, Heidelberg (2003). https://doi.org/10.1007/3-540-36563-X_1
5. Borisov, N., Goldberg, I., Brewer, E.A.: Off-the-record communication, or, why not to use PGP. In: Proceedings of the 2004 ACM Workshop on Privacy in the Electronic Society, WPES 2004, 28 October 2004, pp. 77–84 (2004)
6. Bos, J., et al.: Crystals - kyber: a CCA-secure module-lattice-based KEM. Cryptology ePrint Archive, Report 2017/634 (2017). https://eprint.iacr.org/2017/634
7. Cohn-Gordon, K., Cremers, C.J.F., Dowling, B., Garratt, L., Stebila, D.: A formal security analysis of the signal messaging protocol. In: 2017 IEEE European Symposium on Security and Privacy, EuroS&P 2017, pp. 451–466. IEEE (2017)
8. Cohn-Gordon, K., Cremers, C.J.F., Garratt, L.: On post-compromise security. In: IEEE 29th Computer Security Foundations Symposium, CSF 2016, pp. 164–178. IEEE Computer Society (2016)
9. Derler, D., Jager, T., Slamanig, D., Striecks, C.: Bloom filter encryption and applications to efficient forward-secret 0-RTT key exchange. In: Nielsen, J.B., Rijmen, V. (eds.) EUROCRYPT 2018, Part III. LNCS, vol. 10822, pp. 425–455. Springer, Cham (2018). https://doi.org/10.1007/978-3-319-78372-7_14
10. Durak, F.B., Vaudenay, S.: Bidirectional asynchronous ratcheted key agreement without key-update primitives. Cryptology ePrint Archive, Report 2018/889 (2018). https://eprint.iacr.org/2018/889

11. Messenger secret conversations: Technical whitepaper. https://fbnewsroomus.files.wordpress.com/2016/07/secret_conversations_whitepaper-1.pdf
12. Gentry, C., Silverberg, A.: Hierarchical ID-based cryptography. In: Zheng, Y. (ed.) ASIACRYPT 2002. LNCS, vol. 2501, pp. 548–566. Springer, Heidelberg (2002). https://doi.org/10.1007/3-540-36178-2_34
13. Green, M.D., Miers, I.: Forward secure asynchronous messaging from puncturable encryption. In: 2015 IEEE Symposium on Security and Privacy, SP 2015, pp. 305–320 (2015)
14. Günther, F., Hale, B., Jager, T., Lauer, S.: 0-RTT key exchange with full forward secrecy. In: Coron, J.-S., Nielsen, J.B. (eds.) EUROCRYPT 2017, Part III. LNCS, vol. 10212, pp. 519–548. Springer, Cham (2017). https://doi.org/10.1007/978-3-319-56617-7_18
15. Hülsing, A., Rijneveld, J., Schanck, J., Schwabe, P.: High-speed key encapsulation from NTRU. In: Fischer, W., Homma, N. (eds.) CHES 2017. LNCS, vol. 10529, pp. 232–252. Springer, Cham (2017). https://doi.org/10.1007/978-3-319-66787-4_12
16. Jaeger, J., Stepanovs, I.: Optimal channel security against fine-grained state compromise: the safety of messaging. In: Shacham, H., Boldyreva, A. (eds.) CRYPTO 2018, Part I. LNCS, vol. 10991, pp. 33–62. Springer, Cham (2018). https://doi.org/10.1007/978-3-319-96884-1_2
17. Jost, D., Maurer, U., Mularczyk, M.: Efficient ratcheting: almost-optimal guarantees for secure messaging. In: Ishai, Y., Rijmen, V. (eds.) EUROCRYPT 2019, LNCS, vol. 11476, pp. 159–188 (2019). https://eprint.iacr.org/2018/954
18. Krawczyk, H., Bellare, M., Canetti, R.: HMAC: keyed-Hashing for Message Authentication. RFC 2104, February 1997
19. Krawczyk, H., Eronen, P.: HMAC-based Extract-and-Expand Key Derivation Function (HKDF). RFC 5869, May 2010
20. Langley, A., Hamburg, M., Turner, S.: Elliptic Curves for Security. RFC 7748, January 2016
21. Lund, J.: Signal partners with Microsoft to bring end-to-end encryption to Skype. https://signal.org/blog/skype-partnership/
22. Marlinspike, M.: Open whisper systems partners with Google on end-to-end encryption for Allo. https://signal.org/blog/allo/
23. Marlinspike, M., Perrin, T.: The double Ratchet algorithm, November 2016. https://whispersystems.org/docs/specifications/doubleratchet/doubleratchet.pdf
24. Marlinspike, M., Perrin, T.: The double Ratchet algorithm, November 2016. https://signal.org/docs/specifications/x3dh/x3dh.pdf
25. National Institute of Standards and Technology (NIST). FIPS 180–4. secure hash standard. Technical report, US Department of Commerce, August 2015
26. Poettering, B., Rösler, P.: Asynchronous ratcheted key exchange. Cryptology ePrint Archive, Report 2018/296 (2018). https://eprint.iacr.org/2018/296
27. Rogaway, P.: Authenticated-encryption with associated-data. In: CCS 2002, Washington, DC, 18–22 November 2002, pp. 98–107 (2002)
28. Rogaway, P., Shrimpton, T.: A provable-security treatment of the key-wrap problem. In: Vaudenay, S. (ed.) EUROCRYPT 2006. LNCS, vol. 4004, pp. 373–390. Springer, Heidelberg (2006). https://doi.org/10.1007/11761679_23
29. Whatsapp encryption overview: Technical white paper, December 2017. https://www.whatsapp.com/security/WhatsApp-Security-Whitepaper.pdf

Efficient Ratcheting: Almost-Optimal Guarantees for Secure Messaging

Daniel Jost[✉][iD], Ueli Maurer, and Marta Mularczyk

Department of Computer Science, ETH Zurich, 8092 Zurich, Switzerland
{dajost,maurer,mumarta}@inf.ethz.ch

Abstract. In the era of mass surveillance and information breaches, privacy of Internet communication, and messaging in particular, is a growing concern. As secure messaging protocols are executed on the not-so-secure end-user devices, and because their sessions are long-lived, they aim to guarantee strong security even if secret states and local randomness can be exposed.

The most basic security properties, including forward secrecy, can be achieved using standard techniques such as authenticated encryption. Modern protocols, such as Signal, go one step further and additionally provide the so-called backward secrecy, or healing from state exposures. These additional guarantees come at the price of a moderate efficiency loss (they require public-key primitives).

On the opposite side of the security spectrum are the works by Jaeger and Stepanovs and by Poettering and Rösler, which characterize the optimal security a secure-messaging scheme can achieve. However, their proof-of-concept constructions suffer from an extreme efficiency loss compared to Signal. Moreover, this caveat seems inherent.

This paper explores the area in between: our starting point are the basic, efficient constructions, and then we ask how far we can go towards the optimal security without losing too much efficiency. We present a construction with guarantees much stronger than those achieved by Signal, and slightly weaker than optimal, yet its efficiency is closer to that of Signal (only standard public-key cryptography is used).

On a technical level, achieving optimal guarantees inherently requires key-updating public-key primitives, where the update information is allowed to be public. We consider secret update information instead. Since a state exposure temporally breaks confidentiality, we carefully design such secretly-updatable primitives whose security degrades gracefully if the supposedly secret update information leaks.

M. Mularczyk—Research was supported by the Zurich Information Security and Privacy Center (ZISC).

Y. Ishai and V. Rijmen (Eds.): EUROCRYPT 2019, LNCS 11476, pp. 159–188, 2019.
https://doi.org/10.1007/978-3-030-17653-2_6

1 Introduction and Motivation

1.1 Motivation

The goal of a secure-messaging protocol is to allow two parties, which we from now on call Alice and Bob, to securely exchange messages over asynchronous communication channels in any arbitrary interleaving, without an adversary being able to read, alter, or inject new messages.

Since mobile devices have become a ubiquitous part of our lives, secure-messaging protocols are almost always run on such end-user devices. It is generally known, however, that such devices are often not very powerful and vulnerable to all kinds of attacks, including viruses which compromise memory contents, corrupted randomness generators, and many more [13,14]. What makes it even worse is the fact that the sessions are usually long-lived, which requires storing the session-related secret information for long periods of time. In this situation it becomes essential to design protocols that provide some security guarantees even in the setting where the memory contents and intermediate values of computation (including the randomness) can be exposed.

The security guarantee which is easiest to provide is *forward secrecy*, which, in case of an exposure, protects confidentiality of previously exchanged messages. It can be achieved using symmetric primitives, such as stateful authenticated encryption [2].

Further, one can consider *healing* (also known as post-compromise recovery or backward secrecy). Roughly, this means that if after a compromise the parties manage to exchange a couple of messages, then the security is restored.[1] Providing this property was the design goal for some modern protocols, such as OTR [5] and Signal [15]. The price for additional security is a loss of efficiency: in both of the above protocols the parties regularly perform a Diffie-Hellman key exchange (public-key cryptography is necessary for healing). Moreover, the above technique does not achieve optimal post-compromise recovery (in particular, healing takes at least one full round-trip). The actual security achieved by Signal was recently analyzed by Cohn-Gordon et al. [7].

This raises a more conceptual question: what security guarantees of secure messaging are even possible to achieve? This question was first formulated by Bellare et al. [4], who abstract the concept of *ratcheting* and formalize the notions of ratcheted key exchange and communication. However, they only consider a very limited setting, where the exposures only affect the state of one of the parties. More recently, Jaeger and Stepanovs [11], and Poettering and Rösler [16] both formulated the optimal security guarantees achievable by secure messaging. To this end, they start with a utopian definition, which cannot be satisfied by any correct scheme. Then, one by one, they disable all generic attacks, until they end with a formalization for which they can provide a proof-of-concept construction. (One difference between the two formalizations is that [11] considers exposing intermediate values used in the computation, while [16] does not.) The resulting optimal security

---

[1] Of course, for the healing to take effect, the adversary must remain passive and not immediately use the compromised state to impersonate a party.

implies many additional properties, which were not considered before. For example, it requires *post-impersonation security*, which concerns messages sent after an active attack, where the attacker uses an exposed state to impersonate a party (we will say that the partner of the impersonated party is *hijacked*).

Unfortunately, these strong guarantees come at a high price. Both constructions [11,16] use very inefficient primitives, such as hierarchical identity-based encryption (HIBE) [9,10]. Moreover, it seems that an optimally-secure protocol would in fact imply HIBE.

This leads to a wide area of mostly unexplored trade-offs with respect to security and efficiency, raising the question how much security can be obtained at what efficiency.

1.2   Contributions

In this work we contribute to a number of steps towards characterizing the area of sub-optimal security. We present an efficient secure-messaging protocol with almost-optimal security in the setting where *both* the memory and the intermediate values used in the computation can be exposed.

Unlike the work on optimal security [11,16], we start from the basic techniques, and gradually build towards the strongest possible security. Our final construction is based on standard digital signatures and CCA-secure public-key encryption. The ciphertext size is constant, and the size of the secret state grows linearly with the number of messages sent since the last received message (one can prove that the state size cannot be constant). We formalize the precise security guarantees achieved in terms of game-based security definitions.

Intuitively, the almost-optimal security comes short of optimal in that in two specific situations we do not provide post-impersonation security. The first situation concerns exposing the randomness of one of *two* specific messages,[2] and in the second, the secret states of both parties must be exposed at almost the same time. The latter scenario seems rather contrived: if the parties were exposed at *exactly* the same time, then any security would anyway be impossible. However, one could imagine that the adversary suddenly loses access to one of the states, making it possible to restore it. Almost-optimal guarantees mean that the security need not be restored in this case.

It turns out that dealing with exposures of the computation randomness is particularly difficult. For example, certain subtle issues made us rely on a circularly-secure encryption scheme. Hence, we present our overall proof in the random oracle model. We stress, however, that the random oracle assumption is only necessary to provide additional guarantees when the randomness can leak.

1.3   Further Related Work

Most work on secure messaging [4,8,11,16], including this paper, considers the situation where messages can only be decrypted in order (so out-of-order messages must be either buffered or dropped). In a recent work, Alwen, Coretti and

---

[2] Namely, the messages sent right before or right after an active impersonation attack.

Dodis [1] consider a different setting in which it is required that any honestly-generated message can be immediately decrypted. The authors motivate this property by practical aspects, as for example immediate decryption is necessary to prevent certain denial-of-service attacks. Moreover, immediate decryption is actually achieved by Signal. This setting requires different definitions of both authenticity and correctness. Moreover, requiring the ability to immediately decrypt messages appears to incur a significant hit on the post-impersonation security a protocol can guarantee.

We find it very interesting to analyze the optimal and sub-optimal security guarantees in the setting of [1], and how providing them impacts the efficiency. However, this is not the focus of this work. Note that most practical secure messengers buffer the messages on a central server, so that even if parties are intermittently offline, they receive all their messages once they go online. Hence, not handling out-of-order messages should not significantly affect practicality.

In a recent concurrent and independent work, Durak and Vaudenay [8] also present a very efficient asynchronous communication protocol with sub-optimal security. However, their setting, in contrast to ours, explicitly excludes exposing intermediate values used in computation, in particular, the randomness. Allowing exposure of the randomness seems much closer to reality. Why would we assume that the memory of a device can be insecure, but the sampled randomness is *perfect*? Our construction provides strong security if the randomness fully leaks, while [8] gives no guarantees even if a very small amount of partial information is revealed. In fact, it is not clear how to modify the construction of [8] to work in the setting with randomness exposures. We note that the proof of [8], in contrast to ours, is in the standard model. On the other hand, we only need the random oracle to provide the additional guarantees not considered in [8].

2   Towards Optimal Security Guarantees

In this section we present a high-level overview of the steps that take us from the basic security properties (for example, those provided by Signal) towards the almost-optimal security, which we later implement in our final construction. We stress that all constructions use only standard primitives, such as digital signatures and public-key encryption. The security proofs are in the random oracle model.

2.1   Authentication

We start with the basic idea of using digital signatures and sequence numbers. These simple techniques break down in the presence of state exposures: once a party's signing key is exposed, the adversary can inject messages at any time in the future. To prevent this and guarantee healing in the case where the adversary remains passive, we can use the following idea. Each party samples a fresh signing and verification key with each message, sends along the new (signed) verification key, and stores the fresh signing key to be used for the next message. If either of the parties' state gets exposed, say Alice's, then Eve obtains her current signing key that she can use to impersonate Alice towards Bob at this point in time. If, however, Alice's next message containing a fresh verification key has already been delivered, then the signing key captured by the adversary becomes useless thereby achieving the healing property.

The above technique already allows to achieve quite meaningful guarantees: in fact, it only ignores post-impersonation security. We implement this idea and formalize the security guarantees of the resulting construction in Sect. 3.

2.2 Confidentiality

Assume now that all communication is authentic, and that none of the parties gets impersonated (that is, assume that the adversary does not inject messages when he is allowed to do so). How can we get forward secrecy and healing?

Forward secrecy itself can be achieved using standard forward-secure authenticated encryption in each direction (this corresponds to Signal's symmetric ratcheting layer). However, this technique provides no healing.

Perfectly Interlocked Communication. The first, basic idea to guarantee healing is to use public-key encryption, with separate keys per direction, and constantly exchange fresh keys. The protocol is sketched in Fig. 1. Note that instead of using a PKE scheme, we could also use a KEM scheme and apply the KEM-DEM principle, which is essentially what Signal does for its asymmetric ratcheting layer.

Let us consider the security guarantees offered by this solution. Assume for the moment that Alice and Bob communicate in a completely interlocked manner, i.e., Alice sends one message, Bob replies to that message, and so on. This situation is depicted in Fig. 1. Exposing the state of a party, say Alice, right after sending a message (dk_A^1, ek_B^0 in the figure) clearly allows to decrypt the next message (m_2), which is unavoidable due to the correctness requirement. However, it no longer affects the confidentiality of any other messages. Further, exposing the state right after receiving a message has absolutely no effect (note that a party can delete its secret key immediately after decrypting, since it will no longer be used). Moreover, exposing the sending or receiving randomness is clearly no worse than exposing both the state right before and after this operation. Hence, our scheme obtains optimal confidentiality guarantees (including forward-secrecy and healing) when the parties communicate in such a turn-by-turn manner.

The Unidirectional Case. The problems with the above approach arise when the communication is not perfectly interlocked. Consider the situation when Alice sends many messages without receiving anything from Bob. The straightforward solution to encrypt all these messages with the same key breaks forward secrecy: Bob can no longer delete his secret key immediately after receiving a message, so exposing his state would expose many messages received by him in the past. This immediately suggests using forward-secure public-key encryption [6], or the closely-related HIBE [9,10] (as in the works by Jaeger et al. and Poettering et al.). However, we crucially want to avoid using such expensive techniques.

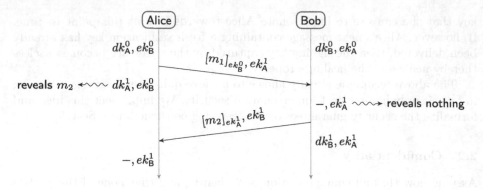

Fig. 1. Constantly exchanging fresh public-keys achieves optimal security when communication is authenticated and in a strict turn-by-turn fashion.

The partial solution offered by Signal is the symmetric ratcheting. In essence, Alice uses the public key once to transmit a fresh shared secret, which can then be used with forward-secure authenticated encryption. However, this solution offers very limited healing guarantees: when Alice's state is exposed, all messages sent by her in the future (or until she receives a new public key from Bob) are exposed. Can we do something better?

The first alternative solution which comes to mind is the following. When encrypting a message, Alice samples a fresh key pair for a public-key encryption scheme, transmits the secret key encrypted along with the message, stores the public key and deletes the secret key. This public key is then used by Alice to send the next message. This approach is depicted in Fig. 2. However, this solution totally breaks if the sending randomness does leak. In essence, exposing Alice's randomness causes a large part of Bob's next state to be exposed, hence, we achieve roughly the same guarantees as Signal's symmetric ratcheting.

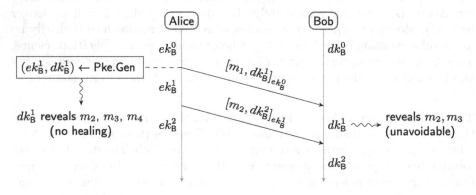

Fig. 2. First attempt to handle asynchronous messages, where one party (here Alice) can send multiple messages in a row. This solution breaks totally when the randomness can leak.

Hence, our approach will make the new decryption key depend on the previous decryption key, and not solely on the update information sent by Alice. We note that, for forward secrecy, we still rely on the update information being transmitted confidentially. This technique achieves optimal security up to impersonation (that is, we get the same guarantees as for simple authentication). The solution is depicted in Fig. 3. At a high level, we use the ElGamal encryption, where a key pair of Bob is (b_0, g^{b_0}) for some generator g of a cyclic group. While sending a message, Alice sends a new secret exponent b_1 encrypted under g^{b_0}, the new encryption key is $g^{b_0}g^{b_1}$, and the new decryption key is $b_0 + b_1$.[3] This idea is formalized in Sect. 4.

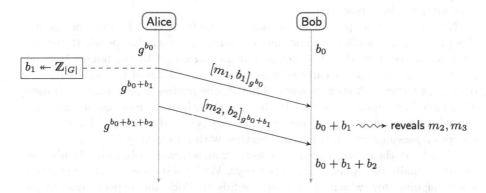

Fig. 3. Second attempt to handle asynchronous messages, where one party (here Alice) can send multiple messages in a row.

2.3   A First Efficient Scheme

Combining the solutions for authentication and confidentiality from the previous subsections already yields a very efficient scheme with meaningful guarantees. Namely, we only give up on the post-impersonation security. That is, we achieve the optimal guarantees up to the event that an adversary uses the exposed state of a party to inject a message to the other party.

One may argue that such a construction is in fact the one that should be used in practice. Indeed, the only guarantees we can hope for after such an active impersonation concern the party that gets impersonated, say Alice, towards the other one, say Bob: Alice should not accept any messages from Bob or the adversary anymore, and the messages she sends should remain confidential. Observe that the former guarantee potentially enables Alice to detect the attack by the lack of replies to her messages. However, providing those guarantees to their full extent seems to inherently require very inefficient tools, such as HIBE, in contrast to the quite efficient scheme outlined above.

---

[3] Looking ahead, it turns out that in order to prove the security of this construction, we need circular-secure encryption. We achieve this in the random oracle model.

In the next subsections we make further steps towards our final construction, which provides some, but not all, after-impersonation guarantees, thereby compromising between efficiency and security.

2.4 Post-Impersonation Authentication

Consider the situation where the adversary exposes the state of Alice and uses it to impersonate her towards Bob (that is, he hijacks Bob). Clearly, due to the correctness requirement, the adversary can now send further messages to Bob. For the optimal security, we would require that an adversary cannot make Alice accept any messages from Bob anymore, even given Bob's state exposed at any time after the impersonation.

Note that our simple authentication scheme from Sect. 2.1 does not achieve this property, as Bob's state contains the signing key at this point. It does not even guarantee that Alice does not accept messages sent by the honest Bob anymore. The latter issue we can easily fix by sending a hash of the communication transcript along with each message. That is, the parties keep a value h (initially 0), which Alice updates as $h \leftarrow \text{Hash}(h \parallel m)$ with every message m she sends, and which Bob updates accordingly with every received message. Moreover, Bob accepts a message only if it is sent together with a matching hash h.

To achieve the stronger guarantee against an adversary obtaining Bob's state, we additionally use ephemeral signing keys. With each message, Alice generates a new signing key, which she securely sends to Bob, and expects Bob to sign his next message with. Intuitively, the adversary's injection "overwrites" this ephemeral key, rendering Bob's state useless. Note that for this to work, we need the last message received by Bob before hijacking to be confidential. This is not the case, for example, if the sending randomness leaks.[4] For this reason, we do not achieve optimal security. In the existing optimal constructions [11,16] the update information can be public, which, unfortunately, seems to require very strong primitives, such as forward-secure signatures.

2.5 Post-Impersonation Confidentiality

In this section we focus on the case where the adversary impersonates Alice towards Bob (since this is only possible if Alice's state exposed, we now consider her state to be a public value).

Consider once more the two approaches to provide confidentiality in the unidirectional case, presented in Sect. 2.2 (Figs. 2 and 3). Observe that if we assume that the randomness cannot be exposed, then the first solution from Fig. 2, where Alice sends (encrypted) a fresh decryption key for Bob, already achieves very good guarantees. In essence, during impersonation the adversary has to choose a new decryption key (consider the adversary sending $[m_3, \bar{dk}_\mathsf{B}^3]_{ek_\mathsf{B}^2}$ in the figure), which overwrites Bob's state. Hence, the information needed to

---

[4] Note that this also makes the choice of abstraction levels particularly difficult, as we need confidentiality, in order to obtain authentication.

decrypt the messages sent by Alice from this point on (namely, dk_B^2) is lost.[5] In contrast, the second solution from Fig. 3 provides no guarantees for post-impersonation messages: after injecting a message and exposing Bob's state, the adversary can easily compute Bob's state from right before the impersonation and use it to decrypt Alice's messages sent after the attack.

While the former idea has been used in [8] to construct an efficient scheme with almost-optimal security for the setting where the randomness generator is perfectly protected, we aim at also providing guarantees in the setting where the randomness can leak. To achieve this, we combine the two approaches, using both updating keys from the latter scheme and ephemeral keys from the former one, in a manner analogous to how we achieved post-impersonation authentication. More concretely, Alice now sends (encrypted), in addition to the exponent, a fresh ephemeral decryption key, and stores the corresponding encryption key, which she uses to additionally encrypt her next message. Now the adversary's injected message causes the ephemeral decryption key of Bob to be overwritten.

As was the case for authentication, this solution does not provide optimal security, since we rely on the fact that the last message, say c, received before impersonation, is confidential. Moreover, in order to achieve confidentiality we also need the message sent by Alice right after c to be confidential.

2.6 The Almost-Optimal Scheme

Using the ideas sketched above, we can construct a scheme with almost-optimal security guarantees. We note that it is still highly non-trivial to properly combine these techniques, so that they work when the messages can be arbitrarily interleaved (so far we only considered certain idealized settings of perfectly interlocked and unidirectional communication).

The difference between our almost-optimal guarantees and the optimal ones [11,16] is in the imperfection of our post-impersonation security. As explained in the previous subsections, for these additional guarantees we need two messages sent by the impersonated party (Alice above) to remain confidential: the one right before and the one right after the attack. Roughly, these messages are *not* confidential either if the encryption randomness is exposed for one of them, or if the state of the impersonated party is exposed right before receiving the last message before the attack. Note that the latter condition basically means that both parties are exposed at almost the same time. If they were exposed at exactly the same time, any security would anyway be impossible.

In summary, our almost-optimal security seems a very reasonable guarantee in practice.

---

[5] We can assume that Alice sends this value confidentially. It makes no sense to consider Bob's state being exposed, as this would mean that both parties are exposed at the same time, in which case, clearly, we cannot guarantee any security.

3   Unidirectional Authentication

In this section we formalize the first solution for achieving authentication, sketched informally in Sect. 2.1. That is, we consider the goal of providing authentication for the communication from a sender (which we call the signer) to a receiver (which we call the verifier) in the presence of an adversary who has full control over the communication channel. Additionally, the adversary has the ability to expose secrets of the communicating parties. In particular, this means that for each party, its internal state and, independently, the randomness it chose during operations may leak.

We first intuitively describe the properties we would like to guarantee:

- As long as the state and sampled randomness of the *signer* are secret, the communication is authenticated (in particular, all sent messages, and only those, can only be received in the correct order). We require that leaking the state or the randomness of the *verifier* has no influence on authentication.
- If the state right before signing the i-th message or the randomness used for this operation is exposed, then the adversary can trivially replace this message by one of her choice. However, we want that if she remains passive (that is, if she delivers sufficiently many messages in order), and if new secrets do not leak, then the security is eventually restored. Concretely, if only the state is exposed, then *only* the i-th message can be replaced, while if the signing randomness is exposed, then only two messages (i and $i+1$) are compromised.

Observe that once the adversary decides to inject a message (while the signer is exposed), security cannot be restored. This is because from this point on, she can send any messages to the verifier by simply executing the protocol. We will say that in such case the adversary *hijacks* the channel, and is now communicating with the verifier.

The above requirements cannot be satisfied by symmetric primitives, because compromising the receiver should have no effect on security. Moreover, in order to protect against deleting and reordering messages, the algorithms need to be stateful. Hence, in the next subsection, we define a new primitive, which we call *key-updating signatures*. At a high level, a key-updating signature scheme is a stateful signature scheme, where the signing key changes with each signing operation, and the verification key changes with each verification. We require that the verification algorithm is deterministic, so that leaking the randomness of the verifier trivially has no effect.

3.1   Key-Updating Signatures

Syntax. A key-updating signature scheme KuSig consists of three polynomial-time algorithms (KuSig.Gen, KuSig.Sign, KuSig.Verify). The probabilistic algorithm KuSig.Gen generates an initial signing key sk and a corresponding verification key vk. Given a message m and sk, the signing algorithm outputs an updated signing key and a signature: $(sk', \sigma) \leftarrow \mathsf{KuSig.Sign}(sk, m)$. Similarly, the verification algorithm outputs an updated verification key and the result v of verification: $(vk', v) \leftarrow \mathsf{KuSig.Verify}(vk, m, \sigma)$.

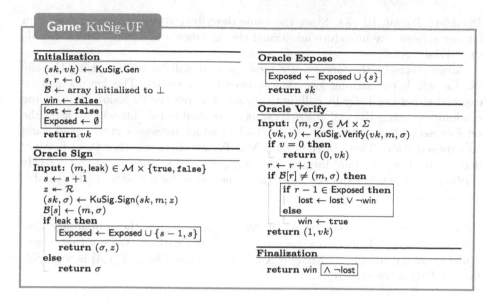

Fig. 4. The strong unforgeability game for key-updating signatures.

Correctness. Let (sk_0, vk_0) be any output of KuSig.Gen, and let $m_1, \ldots, m_k$ be any sequence of messages. Further, let $(sk_i, \sigma_i) \leftarrow$ KuSig.Sign(sk_{i-1}, m_i) and $(vk_i, v_i) \leftarrow$ KuSig.Verify$(vk_{i-1}, m_i, \sigma_i)$ for $i = 1, \ldots, k$. For correctness, we require that $v_i = 1$ for all $i = 1, \ldots, k$.

Security. The security of KuSig is formalized using the game KuSig-UF, described in Fig. 4. For simplicity, we define the security in the single-user setting (security in the multi-user setting can be obtained using the standard hybrid argument).

The Game Interface. The game without the parts of the code marked by boxes defines the interface exposed to the adversary.

At a high level, the adversary wins if he manages to set the internal flag win to **true** by providing a message with a forged signature. To this end, he interacts with three oracles: **Sign**, **Verify** and **Expose**. Using the oracle **Sign**, he can obtain signatures and update the secret signing key, using the oracle **Verify**, he can update the verification key (or submit a forgery), and the oracle **Expose** reveals the secret signing key.

A couple of details about the above oracles require further explanation. First, the verification key does not have to be kept secret. Hence, the updated key is always returned by the verification oracle. Second, we extend the signing oracle to additionally allow "insecure" queries. That is, the adversary learns not only the signature, but also the randomness used to generate it.

Disabling Trivial Attacks. Since the game described above can be trivially won for any scheme, we introduce additional checks (shown in boxes), which disable the trivial "wins".

More precisely, the forgery of a message that will be verified using the key vk, for which the signing key sk was revealed is trivial. The key sk can be exposed either explicitly by calling the oracle **Expose**, or by leaking the signing randomness using the call $\text{Sign}(m, \text{true})$. To disable this attack, we keep the set Exposed, which, intuitively, keeps track of which messages were signed using an exposed state. Then, in the oracle **Verify**, we check whether the adversary decided to input a trivial forgery (this happens if the index $r - 1$ of currently verified message is in Exposed). If so, the game can no longer be won (the variable lost is set to true).[6]

Advantage. For an adversary $\mathcal{A}$, let $\text{Adv}_{\text{KuSig}}^{\text{ku-suf}}(\mathcal{A})$ denote the probability that the game KuSig-UF returns true after interacting with $\mathcal{A}$. We say that a key-updating signature scheme KuSig is KuSig-UF secure if $\text{Adv}_{\text{KuSig}}^{\text{ku-suf}}(\mathcal{A})$ is negligible for any PPT adversary $\mathcal{A}$.

3.2 Construction

We present a very simple construction of a KuSig, given any one-time signature scheme Sig, existentially-unforgeable under chosen-message attack. The construction is depicted in Fig. 5. The high-level idea is to generate a new key pair for Sig with each signed message. The message, together with the new verification key and a counter,[7] is then signed using the old signing key, and the new verification key is appended to the signature. The verification algorithm then replaces the old verification key by the one from the verified signature.

Theorem 1. *Let* Sig *be a signature scheme. The construction of Fig. 5 is* KuSig-UF *secure, if* Sig *is 1-SUF-CMA secure.*

A proof of Theorem 1 is presented in the full version of this work [12].

3.3 Other Definitions of Key-Updating Signatures

Several notions of signatures with evolving keys are considered in the literature. For example, in forward-secure signatures [3] the signing key is periodically updated. However, in such schemes the verification key is fixed. Moreover, the goal of forward secrecy is to protect the past (for signatures, this means that there exists some notion of time and exposing the secret key does not allow to forge signatures for the past time periods). On the other hand, we are interested in protecting the future, that is, the scheme should "heal" after exposure.

---

[6] The adversary knows which states are exposed, and hence can check himself before submitting a forgery attempt, whether this will make him lose the game.

[7] In fact, the counter is not necessary to prove security of the construction, since every message is signed with a different key. However, we find it cleaner to include it.

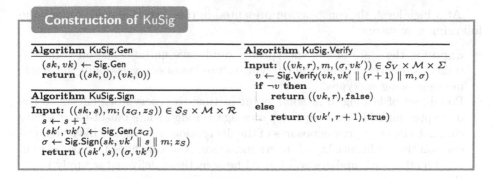

Fig. 5. The construction of key-updating signatures.

The notion closest to our setting is that of key-updateable digital signatures [11]. Here the difference is that their notion provides stronger guarantees (hence, the construction is also less efficient). In particular, in key-updateble digital signatures the signing key can be updated with *any (even adversarially chosen) public* information. In contrast, in our definition the secret key is updated secretly by the signer, and only part of the information used to update it is published as part of the signature.[8]

Relaxing the requirements of key-updateble digital signatures allows us to achieve a very efficient construction ([11] uses rather inefficient forward-secure signatures as a building block). On the other hand, the stronger guarantee seems to be necessary for the optimal security of [11].

4   Unidirectional Confidentiality

In this section we formalize the second solution for achieving confidentiality in the unidirectional setting, where the sender, which we now call the encryptor generates some secret update information and communicates it (encrypted) to the receiver, which we now call the decryptor. In the following, we assume that the secret update information is delivered through an idealized secure channel.

The setting is similar to the one we considered for authentication: the secret states and the randomness of the encryptor and of the decryptor can sometimes be exposed. However, now we also assume that the communication is authenticated. We assume authentication in the sense of Sect. 3, however, we do not consider hijacking the channel. In this section we give no guarantees if the channel is hijacked.

---

[8] For example, in our construction the public part of the update is a fresh verification key, and the secret part is the corresponding signing key. This would not satisfy the requirements of [11], since there is no way to update the signing key using only the fresh verification key.

At a high level, the construction presented in this section should provide the following guarantees:

- Exposing the state of the encryptor should have no influence on confidentiality. Moreover, leaking the encryption randomness reveals only the single message being encrypted.
- Possibility of healing: if at some point in time the encryptor delivers to the decryptor an additional (update) message through some out-of-band secure channel, then any prior exposures of the decryption state should have no influence on the confidentiality of future messages. (Looking ahead, in our overall construction such updates will indeed be sometimes delivered securely.)
- Weak forward secrecy: exposing the decryptor's state should not expose messages sent before the last securely delivered update.

For more intuition about the last two properties, consider Fig. 6. The states 1 to 7 correspond to the number of updates applied to encryption or decryption keys. The first two updates are not delivered securely (on the out-of-band channel), but the third one is. Exposing the decryption key at state 5 (after four updates) causes all messages encrypted under the public keys at states 4, 5 and 6 to be exposed. However, the messages encrypted under keys at states 1 to 3 are not affected.

Fig. 6. Intuition behind the confidentiality guarantees.

To formalize the above requirements, we define a new primitive, which we call *secretly key-updatable public-key encryption* (SkuPke).

4.1   Secretly Key-Updatable Public-Key Encryption

At a high level, a secretly key-updatable public-key encryption scheme is a public-key encryption scheme, where both the encryption and the decryption key can be (independently) updated. The information used to update the encryption key can be public (it will be a part of the encryptor's state, whose exposure comes without consequences), while the corresponding update information for the decryption key should be kept secret (this update will be sent through the out-of-band secure channel).

In fact, for our overall scheme we need something a bit stronger: the update information should be generated independently of the encryption or decryption keys. Moreover, the properties of the scheme should be (in a certain sense) preserved even when the same update is applied to many independent key pairs. The reason for these requirements will become more clear in the next section, when we use the secretly key-updatable encryption to construct a scheme for the sesqui-directional setting.

The security definition presented in this section is slightly simplified and it does not consider the above additional guarantees. However, it is sufficient to understand our security goals. In the proof of the overall construction we use the full definition presented in the full version [12], which is mostly a straightforward extension to the multi-instance setting.

Syntax. Formally, a secretly key-updatable public-key encryption scheme SkuPke consists of six polynomial-time algorithms (SkuPke.Gen, SkuPke.Enc, SkuPke.Dec,   SkuPke.UpdateGen, SkuPke.UpdateEk, SkuPke.UpdateDk).   The probabilistic algorithm SkuPke.Gen generates an initial encryption key ek and a corresponding decryption key dk. Then, the probabilistic encryption algorithm can be used to encrypt a message m as $c \leftarrow$ SkuPke.Enc(ek, m), while the deterministic decryption algorithm decrypts the message: $m \leftarrow$ SkuPke.Dec(dk, c).

Furthermore, the probabilistic algorithm SkuPke.UpdateGen generates public update information u_e and the corresponding secret update information u_d, as $(u_e, u_d) \leftarrow$ SkuPke.UpdateGen. The former can then be used to update an encryption key $ek' \leftarrow$ SkuPke.UpdateEk(u_e, ek), while the latter can be used to update the corresponding decryption key $dk' \leftarrow$ SkuPke.UpdateDk(u_d, dk).

Correctness. Let (ek_0, dk_0) be the output of SkuPke.Gen, and let $(ue_1, ud_1), \ldots, (ue_k, ud_k)$ be any sequence of outputs of SkuPke.UpdateGen. For $i = 1 \ldots k$, let $ek_i \leftarrow$ SkuPke.UpdateEk(ue_i, e_{i-1}) and $dk_i \leftarrow$ SkuPke.UpdateDk(ud_i, d_{i-1}). A SkuPke is called correct, if SkuPke.Dec$(dk_k,$ SkuPke.Enc$(ek_k, m)) = m$ for any message m with probability 1.

Security. Figure 7 presents the single-instance security game for a SkuPke scheme, which we describe in the following paragraphs.

The Game Interface. The interface exposed to the adversary is defined via the part of the code not marked by boxes.

We extend the standard notion of IND-CPA for public-key encryption, where the adversary gets to see the initial encryption key ek and has access to a left-or-right **Challenge** oracle. Furthermore, the adversary can generate new update information by calling the oracle **UpdateGen**, and later apply the generated updates to the encryption and decryption key, by calling, respectively, the oracles **UpdateEk** and **UpdateDk**. In our setting the adversary is allowed to expose the randomness and the state of parties. The encryption state is considered public information, hence, the key ek and the public update $\mathcal{U}_e[\mathsf{ind}]$ are always returned by the corresponding oracles. The decryption key dk can be revealed by calling the **Expose** oracle[9], and the secret decryption updates—by setting the randomness for the oracle **UpdateGen**. Finally, the **Challenge** oracle encrypts the message together with the previously generated secret update information, chosen by the adversary (recall the idea sketched in Sect. 2.2).

---

[9] For technical reasons, we only allow one query to the **Expose** oracle.

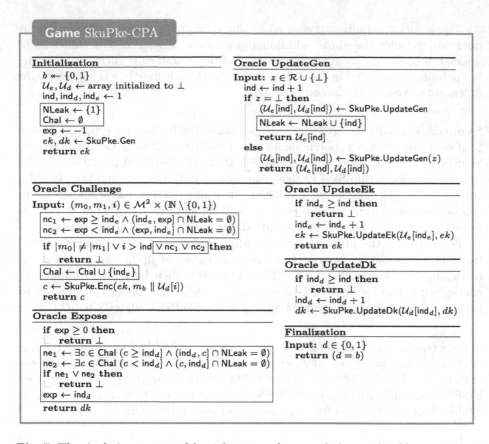

Fig. 7. The single-instance confidentiality game for secretly key-updatable encryption.

Disabling Trivial Attacks. In essence, in the presence of exposures, it is not possible to protect the confidentiality of all messages. As already explained, we allow an exposure of the secret key to compromise secrecy of all messages sent between two consecutive secure updates. Hence, the game keeps track of the following events: generating a secure update (the set NLeak), exposing the secret key (the variable exp), and asking for a challenge ciphertext (the set Chal). Then, the adversary is not allowed to ask for a challenge generated using the encryption key, corresponding to a decryption key, which is in the "exposed" interval (that is, if all updates between the decryption key and the exposed state are insecure). An analogous condition is checked by the **Expose oracle**.

Advantage. Recall that in this section we present the single-instance security game, but in the proofs later we need the multi-instance version SkuPke-MI-CPA defined in the full version [12]. Hence, we define security using the multi-instance game. For an adversary $\mathcal{A}$, let $\mathsf{Adv}^{\mathsf{sku\text{-}cpa}}_{\mathsf{SkuPke}}(\mathcal{A}) := 2\Pr[\mathcal{A}^{\mathsf{SkuPke\text{-}MI\text{-}CPA}} \Rightarrow \mathbf{true}] - 1$, where $\Pr[\mathcal{A}^{\mathsf{SkuPke\text{-}MI\text{-}CPA}} \Rightarrow \mathbf{true}]$ denotes the probability that the game

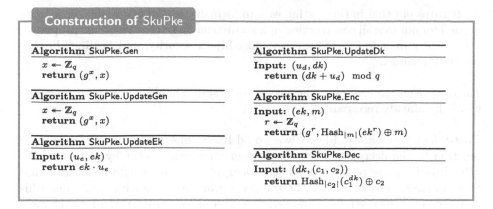

Fig. 8. The construction of secretly key-updatable encryption.

SkuPke-MI-CPA returns **true** after interacting with an adversary $\mathcal{A}$. We say that a secretly key-updatable encryption scheme SkuPke is SkuPke-MI-CPA secure if $\mathsf{Adv}^{\mathsf{sku\text{-}cpa}}_{\mathsf{SkuPke}}(\mathcal{A})$ is negligible for any PPT adversary $\mathcal{A}$.

4.2 Construction

We present an efficient construction of SkuPke, based on the ElGamal cryptosystem. At a high level, the key generation, encryption and decryption algorithms are the same as in the ElGamal encryption scheme. To generate the update information, we generate a new ElGamal key pair, and set the public and private update to, respectively, the new public and private ElGamal keys. To update the encryption key, we multiply the two ElGamal public keys, while to update the decryption key, we add the ElGamal secret keys. Finally, in order to deal with encrypting previously generated update information, we need the hash function $\mathsf{Hash}_l(\cdot)$, where l is the output length.

The construction is defined in Fig. 8. We let G be a group of prime order q, generated by g. These parameters are implicitly passed to all algorithms.

A proof of the following theorem is presented in the full version [12].

Theorem 2. *The construction of Fig. 8 is* SkuPke-MI-CPA *secure in the random oracle model, if CDH is hard.*

5 Sesquidirectional Confidentiality

The goal of this section is to define additional confidentiality guarantees in the setting where also an authenticated back channel from the decryptor to the encryptor exists (but we still focus only the properties of the unidirectional from the encryptor to the decryptor). That is, we assume a perfectly-authenticated back channel and a forward channel, authenticated in the sense of Sect. 3 (in particular, we allow hijacking the decryptor).

It turns out that in this setting we can formalize all confidentiality properties needed for our overall construction of a secure channel. Intuitively, the properties we consider include forward secrecy, post-hijack security, and healing through the back channel.

Forward Secrecy. Exposing the decryptor's state should not expose messages which he already received.

Post-hijack Guarantees. Ideally, we would like to guarantee that if the communication to the decryptor is hijacked, then all messages sent by the encryptor *after* hijacking are secret, even if the decryptor's state is exposed (note that these messages cannot be read by the decryptor, since the adversary caused his state to be "out-of-sync"). However, this guarantee turns out to be extremely strong, and seems to inherently require HIBE. Hence, we relax it by giving up on the secrecy of post-hijack messages in the following case: a message is sent insecurely (for example, because the encryption randomness is exposed), the adversary *immediately* hijacks the communication, and at some later time the decryptor's state is exposed. We stress that the situation seems rather contrived, as explained in the introduction.

Healing Through the Back Channel. Intuitively, the decryptor will update his state and send the corresponding update information on the back channel. Once the encryptor uses this information to update his state, the parties heal from past exposures. At a high level, this means that we require the following additional guarantees:

- Healing: messages sent after the update information is delivered are secret, irrespective of any exposures of the decryptor's state, which happened before the update was generated.
- Correctness: in the situation where the messages on the back channel are delayed, it should still be possible to read the messages from the forward channel. That is, it should be possible to use a decryption key after i updates to decrypt messages encrypted using an "old" encryption key after $j < i$ updates.

Challenges. It turns out that the setting with both the back channel, and the possibility of hijacking, is extremely subtle. For example, one may be tempted to use an encryption scheme which itself updates keys and provides some form of forward secrecy, and then simply send on the back channel a fresh key pair for that scheme. With this solution, in order to provide correctness, every generated secret key would have to be stored until a ciphertext for a newer key arrives. Unfortunately, this simple solution does not work. Consider the following situation: the encryptor sends two messages, one before and one after receiving an update on the back channel, and these messages are delayed. Then, the adversary hijacks the decryptor by injecting an encryption under the *older* of the two keys. However, if now the decryptor's state is exposed, then the adversary will learn the message encrypted with the *new* key (which breaks the post-hijack guarantees we wish to provide). Hence, it is necessary that receiving a message updates all decryption keys, also those for future messages. Intuitively, this is why we require that the same update for SkuPke can be applied to many keys.

5.1 Healable and Key-Updating Public-Key Encryption

To formalize the requirements sketched above, we define healable and key-updating public-key encryption (HkuPke). In a nutshell, a HkuPke scheme is a *stateful* public-key encryption scheme with additional algorithms used to generate and apply updates, sent on the back channel.

Syntax. A healable and key-updating public-key encryption scheme HkuPke consists of five polynomial-time algorithms (HkuPke.Gen, HkuPke.Enc, HkuPke.Dec, HkuPke.BcUpEk, HkuPke.BcUpDk).

The probabilistic algorithm HkuPke.Gen generates an initial encryption key ek and a corresponding decryption key dk. Encryption and decryption algorithms are stateful. Moreover, for reasons which will become clear in the overall construction of a secure channel, they take as input additional data, which need not be kept secret.[10] Formally, we have $(ek', c) \leftarrow$ HkuPke.Enc(ek, m, ad) and $(dk', m) \leftarrow$ HkuPke.Dec(dk, c, m), where ek' and dk' are the updated keys and ad is the additional data. The additional two algorithms are used to handle healing through the back channel: the operation $(dk', upd) \leftarrow$ HkuPke.BcUpDk(dk) outputs the updated decryption key dk' and the information upd, which will be sent on the back channel. Then, the encryption key can be updated by executing $ek' \leftarrow$ HkuPke.BcUpEk(ek, upd).

Correctness. Intuitively, we require that if all ciphertexts are decrypted in the order of encryption, and if the additional data used for decryption matches that used for encryption, then they decrypt to the correct messages. Moreover, decryption must also work if the keys are updated in the meantime, that is, if an arbitrary sequence of HkuPke.BcUpDk calls is performed and the ciphertext is generated at a point where only a prefix of the resulting update information has been applied to the encryption key using HkuPke.BcUpEk. A formal definition of correctness is given in the full version [12].

Security. The security of HkuPke is formalized using the game HkuPke-CPA, described in Fig. 9. Similarly to the unidirectional case, we extend the IND-CPA game.

---

[10] Roughly, the additional data is needed to provide post-hijack security of the final construction: changing the additional data means that the adversary decided to hijack the channel, hence, the decryption key should be updated.

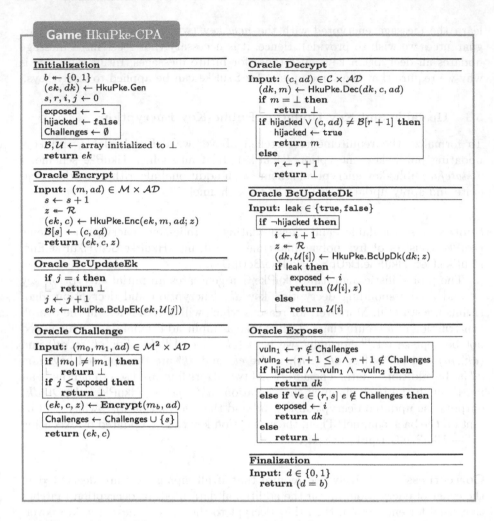

Fig. 9. The confidentiality game for healable and key-updating encryption.

The Interface. Consider the (insecure) variant of our game without the parts of the code marked in boxes. As in the IND-CPA game, the adversary gets to see the encryption key ek and has access to a left-or-right **Challenge** oracle. Since HkuPke schemes are stateful, we additionally allow the adversary to update the decryption key through the calls to the **Decrypt** oracle (which for now only returns $\perp$). The encryption key is updated using the calls to the **Encrypt** oracle (where the encrypted message is known) and to the **Challenge** oracle.

Furthermore, in our setting the adversary is allowed to expose the randomness and the state. To expose the state (that is, the decryption key), he can query the **Expose** oracle. To expose the randomness of any randomized oracle, he can set the input flag leak to true.

Finally, the adversary can access two oracles corresponding to the back channel: the oracle **BcUpdateDk** executes the algorithm HkuPke.BcUpDk and returns the update information to the adversary (this corresponds to sending on the back channel), and the oracle **BcUpdateEk executes** HkuPke.BcUpEk with the next generated update (since the channel is authenticated, the adversary has no influence on which update is applied).

Disabling Trivial Attacks. Observe that certain attacks are disabled by the construction itself. For example, the randomness used to encrypt a challenge ciphertext cannot be exposed.

Furthermore, the game can be trivially won if the adversary asks for a challenge ciphertext and, before calling Decrypt with this ciphertext, exposes the decryption key (by correctness, the exposed key can be used to decrypt the challenge). We disallow this by keeping track of when the adversary queried a challenge in the set Challenges, and adding corresponding checks in the **Expose** oracle. Similarly, in the **Challenge** oracle we return $\perp$ whenever the decryption key corresponding to the current encryption key is known to the adversary. Finally, the decryptor can be hijacked, which the game marks by setting hijacked to **true**. Once this happens, the **Decrypt** oracle "opens up" and returns the decrypted message.

Moreover, ideally, exposing the secret key after hijacking would not reveal anything about the messages (the adversary gets to call Expose "for free", without setting exposed). However, as already mentioned, we relax slightly the security. In particular, exposing is free only when hijacking did not occur immediately after leaking encryption randomness. This is checked using the conditions vuln_1 and vuln_2.

Advantage. In the following, let $\mathsf{Adv}_{\mathsf{HkuPke}}^{\mathsf{hku\text{-}cpa}}(\mathcal{A}) := 2\Pr[\mathcal{A}^{\mathrm{HkuPke\text{-}CPA}} \Rightarrow \mathbf{true}] - 1$, where $\Pr[\mathcal{A}^{\mathrm{HkuPke\text{-}CPA}} \Rightarrow \mathbf{true}]$ denotes the probability that the game HkuPke-CPA returns **true** after interacting with an adversary $\mathcal{A}$. We say that a healable and key-updating encryption scheme HkuPke is HkuPke-CPA secure if $\mathsf{Adv}_{\mathsf{HkuPke}}^{\mathsf{hku\text{-}cpa}}(\mathcal{A})$ is negligible for any PPT adversary $\mathcal{A}$.

5.2 Construction

To construct a HkuPke scheme, we require two primitives: a secretly key-updatable encryption scheme SkuPke from Sect. 4, and an IND-CCA2 secure public-key encryption scheme *with associated data* PkeAd. Intuitively, the latter primitive is a public-key encryption scheme, which additionally takes into account non-secret associated data, such that the decryption succeeds if and only if the associated data has not been modified. A bit more formally, in the corresponding security game the decryption oracle is only blinded if the adversary requests to decrypt the challenge ciphertext together with the associated data provided with the challenge. It will decrypt the challenge for any other associated data. A formal description of this notion, together with a simple construction in the random oracle model, is presented in the full version [12].

Construction of HkuPke

Algorithm HkuPke.Gen

$DK^{upd}, DK^{eph}, U_e \leftarrow$ array initialized to $\perp$
$(ek^{upd}, DK^{upd}[0]) \leftarrow$ SkuPke.Gen
$(ek^{eph}, DK^{eph}[0]) \leftarrow$ PkeAd.Gen
$s, r, i, j, tr_s, tr_r \leftarrow 0$
$i_{ack} \leftarrow -1$
return $((ek^{upd}, ek^{eph}, s, j, U_e, tr_s),$
$\qquad (DK^{upd}, DK^{eph}, r, i, i_{ack}, tr_r))$

Algorithm HkuPke.Enc

Input: $((ek^{upd}, ek^{eph}, s, j, U_e, tr_s), m, ad;$
$\qquad (z_1, \dots, z_4)) \in \mathcal{EK} \times \mathcal{M} \times \mathcal{AD} \times \mathcal{R}$
$s \leftarrow s + 1$
$(U_e[s], u_d) \leftarrow$ SkuPke.UpdateGen(z_1)
$\hat{c} \leftarrow$ SkuPke.Enc$(ek^{upd}, (m, u_d, z_2); z_3)$
$c \leftarrow$ PkeAd.Enc$(ek^{eph}, \hat{c}, ad; z_4)$
$tr_s \leftarrow$ Hash$(tr_s \parallel (c, j, ad))$
$ek^{upd} \leftarrow$ SkuPke.UpdateEk$(U_e[s], ek^{upd})$
$(ek^{eph}, \_) \leftarrow$ PkeAd.Gen$($Hash$(tr_s \parallel z_2))$
return $((ek^{upd}, ek^{eph}, s, j, U_e, tr_s), (c, j))$

Algorithm HkuPke.BcUpDk

Input: $((DK^{upd}, DK^{eph}, r, i, i_{ack}, tr_r);$
$\qquad (z_1, z_2)) \in \mathcal{DK} \times \mathcal{R}$
$i \leftarrow i + 1$
$(ek^{upd}, \hat{dk}^{upd}) \leftarrow$ SkuPke.Gen(z_1)
$(ek^{eph}, \hat{dk}^{eph}) \leftarrow$ PkeAd.Gen(z_2)
$DK^{upd}[i] \leftarrow \hat{dk}^{upd}$
$DK^{eph}[i] \leftarrow \hat{dk}^{eph}$
return $((DK^{upd}, DK^{eph}, r, i, i_{ack}, tr_r),$
$\qquad (\hat{ek}^{upd}, \hat{ek}^{eph}, r))$

Algorithm HkuPke.BcUpEk

Input: $((ek^{upd}, ek^{eph}, s, j, U_e, tr_s),$
$\qquad (\hat{ek}^{upd}, \hat{dk}^{eph}, r^{msg})) \in \mathcal{EK} \times \mathcal{U}$
if $r^{msg} \geq s$ **then**
$\quad ek^{eph} \leftarrow \hat{ek}^{eph}$
$\quad ek^{upd} \leftarrow \hat{ek}^{upd}$
for $\ell \leftarrow (r^{msg} + 1), \dots, s$ **do**
$\quad ek^{upd} \leftarrow$ SkuPke.UpdateEk$(U_e[\ell], ek^{upd})$
return $(ek^{upd}, ek^{eph}, s, j + 1, U_e, tr_s)$

Algorithm HkuPke.Dec

Input: $((DK^{upd}, DK^{eph}, r, i, i_{ack}, tr_r), (c, i_{msg}), ad) \in \mathcal{DK} \times \mathcal{C} \times \mathcal{AD}$
if $i_{msg} \geq i_{ack} \wedge i_{msg} > i$ **then**
$\quad \hat{c} \leftarrow$ PkeAd.Dec$(DK^{eph}[i_{msg}], c, ad)$
$\quad$ **if** $\hat{c} \neq \perp$ **then**
$\quad\quad \hat{m} \leftarrow$ SkuPke.Dec$(DK^{upd}[i_{msg}], \hat{c})$
$\quad\quad$ **if** $\hat{m} \in \mathcal{M} \times$ SkuPke.$\mathcal{U} \times$ PkeAd.$\mathcal{DK}$ **then**
$\quad\quad\quad (m, u_d, z) \leftarrow \hat{m}$
$\quad\quad\quad tr_r \leftarrow$ Hash$(tr_r \parallel (c, i_{msg}, ad))$
$\quad\quad\quad (\_, \hat{dk}^{eph}) \leftarrow$ PkeAd.Gen$($Hash$(tr_r \parallel z_2))$
$\quad\quad\quad$ **for** $\ell \leftarrow 1 \dots i$ **do**
$\quad\quad\quad\quad$ **if** $\ell < i_{msg}$ **then**
$\quad\quad\quad\quad\quad DK^{eph}[\ell] \leftarrow \perp$
$\quad\quad\quad\quad\quad DK^{upd}[\ell] \leftarrow \perp$
$\quad\quad\quad\quad$ **else**
$\quad\quad\quad\quad\quad DK^{eph}[\ell] \leftarrow \hat{dk}^{eph}$
$\quad\quad\quad\quad\quad DK^{upd}[\ell] \leftarrow$ SkuPke.UpdateDk$(u_d, DK^{upd}[\ell])$
$\quad\quad\quad$ **return** $((DK^{upd}, DK^{eph}, r + 1, i, i_{msg}), m)$
return $((DK^{upd}, DK^{eph}, r, i, i_{ack}, tr_r), \perp)$

Fig. 10. The construction of healable and key-updating encryption.

At the core of our construction, in order to encrypt a message m, we generate an update u_e, d_d for an SkuPke scheme and encrypt the secret update information u_d together with m. This update information is then used during decryption to update the secret key.

Unfortunately, this simple solution has a few problems. First, we need the guarantee that after the decryptor is hijacked, his state cannot be used to decrypt messages encrypted afterwards. We achieve this by adding a second layer of encryption, using a PkeAd. We generate a new key pair during every encryption, and send the new decryption key along with m and u_d, and store the corresponding encryption key for the next encryption operation. The decryptor will use his current such key to decrypt the message and then completely overwrite it with the new one he just received. Therefore, we call those keys "ephemeral". The basic idea is of course that during the hijacking, the adversary has to provide a different ciphertext containing a new ephemeral key, which will then be useless for him when exposing the receiver afterwards. In order to make this idea sound, we have to ensure that this key is not only different from the previous one, but unrelated. To achieve this, we actually do not send the new encryption key directly, but send a random value z instead and then generate the key pairs using Hash$(tr \parallel z)$ as randomness. Here tr stands for a hash chain of ciphertexts and associated data sent/received so far, including the current one. Overall, an encryption of m is PkeAd.Enc$(ek^{\mathsf{eph}}, \mathsf{SkuPke.Enc}(ek^{\mathsf{upd}}, (m, u_d, z_2)), ad)$, for some associated data ad.

Second, we need to provide healing guarantees through the back channel. This is achieved by generating fresh key pairs for both, the updating and the ephemeral, encryption schemes. For correctness, the encryptor however might have to ignore the new ephemeral key, if he detects that it will be overwritten by one of his updates in the meantime. He can detect this by the decryptor explicitly acknowledging the number of messages he received so far as part of the information transmitted on the backward-channel.

Third, observe that for correctness, the decryptor needs to store all decryption keys generated during the back-channel healing, until he receives a ciphertext for a newer key (consider the back-channel messages being delayed). In order to still guarantee post-hijack security, we apply the SkuPke update u_d to *all* secret keys he still stores. This also implies that the encryptor has to store the corresponding public update information and apply them the new key he obtains from the backward-channel, if necessary.

Theorem 3. *Let SkuPke be a secretly key-updatable encryption scheme, and let PkeAd be an encryption scheme with associated data. The scheme of Fig. 10 is HkuPke-CPA secure in the random oracle model, if the SkuPke scheme is SkuPke-MI-CPA secure, and the PkeAd is IND-CCA2-AD secure.*

A proof of Theorem 3 is presented in the full version of this work [12].

6   Overall Security

So far, we have constructed two intermediate primitives that will help us build a secure messaging protocol. First, we showed a unidirectional authentication scheme that provides healing after exposure of the signer's state. Second, we introduced a sesqui-directional confidentiality scheme that achieves forward secrecy, healing after the exposure of the receiver's state, and it also provides post-hijack confidentiality.

The missing piece, except showing that the schemes can be securely plugged together, is post-hijack authentication: with the unidirectional authentication scheme we introduced, exposing a hijacked party's secret state allows an attacker to forge signatures that are still accepted by the other party. This is not only undesirable in practice (the parties lose the chance of detecting the hijack), but it actually undermines post-hijack confidentiality as well. More specifically, an attacker might trick the so far uncompromised party into switching over to adversarially chosen "newer" encryption key, hence becoming a man-in-the-middle after the fact.

In contrast to confidentiality, one obtains healing of authentication in the unidirectional setting, but post-hijack security requires some form of bidirectional communication: receiving a message must irreversibly destroy the signing key. Generally, we could now follow the approach we took when dealing with the confidentiality and define a sesqui-directional authentication game. We refrain from doing so, as we believe that this does not simplify the exposition. As the reader will see later, our solution for achieving post-hijack authentication guarantees requires that the update information on the backward-channel is transmitted confidentially. This breaks the separation between authentication and confidentiality. More concretely, in order for a sesqui-directional authentication game to serve as a useful intermediate abstraction on which one could then build upon, it would now have to model the partial confidential channel of HkuPke in sufficient details. Therefore, we avoid such an intermediate step, and build our overall secure messaging scheme directly. First, however, we formalize the precise level of security we actually want to achieve.

6.1   Almost-Optimal Security of Secure Messaging

Syntax. A *secure messaging scheme* SecMsg consists of the following triple of polynomial-time algorithms (SecMsg.Init, SecMsg.Send, SecMsg.Receive). The probabilistic algorithm SecMsg.Init generates an initial pair of states st_A and st_B for Alice and Bob, respectively. Given a message m and a state st_u of a party, the probabilistic sending algorithm outputs an updated state and a ciphertext c: $(st_u, c) \leftarrow$ SecMsg.Send$(st_u, m; z)$. Analogously, given a state and a ciphertext, the receiving algorithms outputs an updated state and a message m: $(st_u, m) \leftarrow$ SecMsg.Send(st_u, c).

Correctness. Correctness of a secure messaging scheme SecMsg requires that if all sent ciphertext are received in order (per direction), then they decrypt to the correct message. More formally, we say the scheme is correct if no adversary can win the correctness game SecMsg-Corr, depicted in Fig. 11, with non-negligible probability. For simplicity, we usually consider perfect correctness, i.e., even an unbounded adversary must have probability zero in winning the game.

Security. The security of SecMsg is formalized using the game SecMsg-Sec, described in Fig. 12.

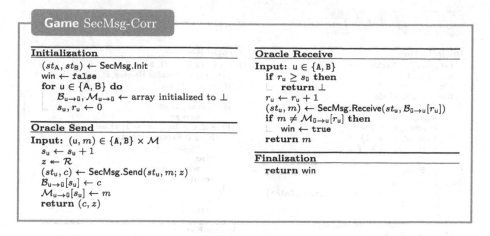

Game SecMsg-Corr

Initialization
$(st_A, st_B) \leftarrow$ SecMsg.Init
win $\leftarrow$ **false**
for $u \in \{A, B\}$ **do**
    $\mathcal{B}_{u \rightarrow \bar{u}}, \mathcal{M}_{u \rightarrow \bar{u}} \leftarrow$ array initialized to $\perp$
    $s_u, r_u \leftarrow 0$

Oracle Send
Input: $(u, m) \in \{A, B\} \times \mathcal{M}$
$s_u \leftarrow s_u + 1$
$z \twoheadleftarrow \mathcal{R}$
$(st_u, c) \leftarrow$ SecMsg.Send$(st_u, m; z)$
$\mathcal{B}_{u \rightarrow \bar{u}}[s_u] \leftarrow c$
$\mathcal{M}_{u \rightarrow \bar{u}}[s_u] \leftarrow m$
return (c, z)

Oracle Receive
Input: $u \in \{A, B\}$
    if $r_u \geq s_{\bar{u}}$ **then**
        return $\perp$
$r_u \leftarrow r_u + 1$
$(st_u, m) \leftarrow$ SecMsg.Receive$(st_u, \mathcal{B}_{\bar{u} \rightarrow u}[r_u])$
if $m \neq \mathcal{M}_{\bar{u} \rightarrow u}[r_u]$ **then**
    win $\leftarrow$ **true**
return m

Finalization
    return win

Fig. 11. The correctness game for a secure messaging scheme.

In general, the game composes the aspects of the security game for key-updating signature scheme KuSig-UF, depicted in Fig. 4, with the sesqui-directional confidentiality game HkuPke-CPA, depicted in Fig. 9. Nevertheless, there are a few noteworthy points:

- The game can be won in two ways: either by guessing the bit b, i.e., breaking confidentiality, or by setting the flag win to **true**, i.e., being able to inject messages when not permitted by an appropriate state exposure. Note that in contrast to the unidirectional authentication game, the game still has to continue after a permitted injection, hence no lost flag exists, as we want to guarantee post-hijack security.
- In contrast to the sesqui-directional confidentiality game, the **Send** oracle takes an additional flag as input modeling whether the randomness used during this operations leaks or not. This allows us to capture that a message might not remain confidential because the receivers decryption key has been exposed, yet it contributes to the healing of the reverse direction (which is not the case if the freshly sampled secret key already leaks again).
- Observe that r_u stops increasing the moment the user u is hijacked. Hence, whenever hijacked$_u$ is **true**, r_u corresponds to the number of messages he received before.
- The two flags vuln$_1$ and vuln$_2$ correspond to the two situations in which we cannot guarantee proper post-hijack security. First, vuln$_1$ corresponds to the situation that the last message from $\bar{u}$ to u before u got hijacked was not transmitted confidentially. This can have two reasons: either the randomness of the encryption of $\bar{u}$ leaked, or u has been exposed just before receiving that message. Observe that in order to hijack u right after that message, the state of $\bar{u}$ needs to be exposed right after sending that message. So in a model where randomness does not leak, vuln$_1$ implies that both parties' state have been compromised almost at the same time. Secondly, vuln$_2$ implies that the next message by $\bar{u}$ was not sent securely either.

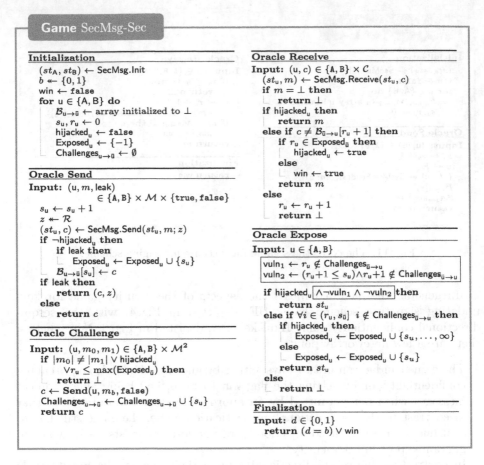

Fig. 12. The game formalizing almost-optimal security of a secure messaging scheme. The indicate the differences in comparison to the game with optimal security.

6.2 Construction

Our Basic Scheme. As the first step, consider a simplified version of our scheme depicted in Fig. 13. This construction works by appropriately combining one instance of our unidirectional key-updating signature scheme SkuSig, and one instance of our healable and key-updating confidentiality scheme HkuPke, per direction.

Adding Post-hijack Authenticity. The scheme depicted in Fig. 13 does not provide any post-hijack authenticity, which we now add.

Observe that in order to achieve such a guarantee, we have to resort to sesqui-directional techniques, i.e., we have to send some update information on the channel from u to ū that affects the signing key for the other direction. Given that this update information must "destroy" the signing key in case of a hijack,

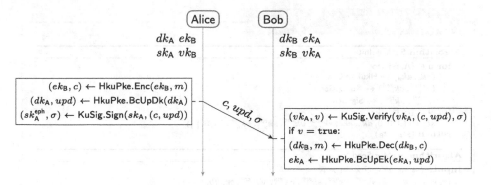

Fig. 13. The scheme obtained by plugging our HkuPke and our SkuSig schemes together. Note how the keys are only used for the corresponding direction, except the update information for the encryption key of our sesqui-directional confidentiality scheme, which is sent along the message.

we will use the following simple trick: the update information is simply a fresh signing key under which the other party has to sign, whenever he acknowledges the receipt of this message. Note that the signer only has to keep the latest such signing key he received, and can securely delete all previous ones. Hence,

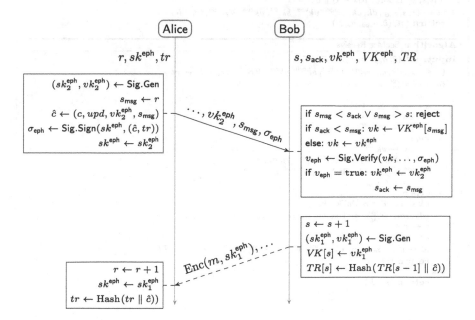

Fig. 14. Handling of the additional signature keys for the communication from Alice to Bob. Each message additionally includes an index s_{msg}, indicating the number of messages Alice received so far, which allows Bob to look up the corresponding verification key. Moreover, they also maintain include a hash of the transcript in each signature.

Construction SecChan **of** SecMsg

Algorithm SecMsg.Init

---
for $u \in \{A, B\}$ **do**
  $(ek_u, dk_u) \leftarrow$ HkuPke.Gen
  $(sk_u^{upd}, vk_u^{upd}) \leftarrow$ KuSig.Gen
  $(sk_u^{eph}, vk_u^{eph}) \leftarrow$ Sig.Gen
for $u \in \{A, B\}$ **do**
  $st_u \leftarrow (0, 0, 0, dk_u, ek_{\bar{u}}, sk_u^{upd}, vk_u^{upd}, sk_u^{eph}, vk_{\bar{u}}^{eph}, [\,], 0, [\,])$
return (st_A, st_B)

Algorithm SecMsg.Send

Input: $(st, m; z) \in \mathcal{S} \times \mathcal{M} \times \mathcal{R}$
$(r, s, s_{ack}, dk, ek, sk^{upd}, vk^{upd}, sk^{eph}, vk^{eph}, VK^{eph}, tr, TR) \leftarrow st$
$(sk_1^{eph}, vk_1^{eph}) \leftarrow$ Sig.Gen(z_1) ▷ The key pair for the backwards channel.
$(sk_2^{eph}, vk_2^{eph}) \leftarrow$ Sig.Gen(z_2) ▷ The key pair for the forwards channel.

▷ Encrypt.
$(dk, upd) \leftarrow$ HkuPke.BcUpDk$(dk; z_3)$
$(ek, c) \leftarrow$ HkuPke.Enc$(ek, (m, sk_1^{eph}), (upd, vk_2^{eph}, r); z_4)$

▷ Sign.
$\hat{c} \leftarrow (c, upd, vk_2^{eph}, r)$
$(sk^{upd}, \sigma_{upd}) \leftarrow$ KuSig.Sign$(sk^{upd}, (\hat{c}, tr); z_5)$
$\sigma_{eph} \leftarrow$ Sig.Sign$(sk^{eph}, (\hat{c}, tr); z_6)$

▷ Update the state.
$s \leftarrow s + 1$
$VK[s] \leftarrow vk_1^{eph}$
$TR[s] \leftarrow$ Hash$(TR[s-1] \parallel \hat{c})$
$st \leftarrow (r, s, s_{ack}, dk, ek, sk^{upd}, vk^{upd}, sk_2^{eph}, vk^{eph}, VK^{eph}, tr, TR)$
return $(st, (\hat{c}, \sigma_{upd}, \sigma_{eph}))$

Algorithm SecMsg.Receive

Input: $(st, (\hat{c}, \sigma_{upd}, \sigma_{eph})) \in \mathcal{S} \times \mathcal{C}$
$(r, s, s_{ack}, dk, ek, sk^{upd}, vk^{upd}, sk^{eph}, vk^{eph}, VK^{eph}, tr, TR) \leftarrow st$
$(c, upd, vk_{msg}^{eph}, s_{msg}) \leftarrow \hat{c}$
$v \leftarrow$ **false**
if $s_{ack} \leq s_{msg} \leq s$ **then**
  if $s_{msg} > s_{ack}$ **then**
    $vk \leftarrow VK^{eph}[s_{msg}]$
  else
    $vk \leftarrow vk^{eph}$
  $v_{eph} \leftarrow$ Sig.Verify$(vk, (\hat{c}, TR[s_{msg}]), \sigma_{eph})$
  $(vk^{upd}, v_{upd}) \leftarrow$ KuSig.Verify$(vk^{upd}, \hat{c}, \sigma_{upd})$
  $v \leftarrow v_{eph} \wedge v_{upd}$
if v **then**
  $ek \leftarrow$ HkuPke.BcUpEk(ek, upd)
  $(dk, (m, sk_{msg}^{eph})) \leftarrow$ HkuPke.Dec$(dk, c, (upd, vk_{msg}^{eph}, s_{msg}))$
  $r \leftarrow r + 1$
  $tr \leftarrow$ Hash$(tr \parallel \hat{c})$
  $st \leftarrow (r, s, s_{msg}, dk, ek, sk^{upd}, vk^{upd}, sk_{msg}^{eph}, vk_{msg}^{eph}, VK^{eph}, tr, TR)$
  return (st, m)
else
  return $(st, \perp)$

Fig. 15. The construction of an almost-optimally secure messaging scheme.

whenever he gets hijacked, the signing key that he previously stored, and that he needs to sign his next message, gets irretrievably overwritten. This, of course, requires that those signing keys are transmitted securely, and hence will be included in the encryption in our overall scheme. However, the technique as described so far does not heal properly. In order to restore the healing property, we will simply ratchet this key as well in the usual manner: whenever we use it, we sample a fresh signing key and send the verification key along. In short, the additional signature will be produced with the following key:

- If we acknowledge a fresh message, i.e., we received a message since last sending one, we use the signing key included in that message (only the last one in case we received multiple messages).
- Otherwise, we use the one we generated during sending the last message.

To further strengthen post-hijack security, the parties also include a hash of the communication transcript in each signature. This ensures that even if the deciding message has not been transferred confidentially, at least the receiver will not accept any messages sent by the honest but hijacked sender. A summary of the additional signatures, the key handling, and the transcript involved in the communication form Alice to Bob is shown in Fig. 14. Of course, the actual scheme is symmetric and these additional signatures will be applied by both parties. See Fig. 15 for the full description of our overall scheme.

Theorem 4. *Let* HkuPke *be a healable and key-updating encryption scheme, let* KuSig *be a key-updating signature scheme, and let* Sig *be a signature scheme. The scheme* SecChan *of Fig. 15 is* SecMsg-Sec *secure, if* HkuPke *scheme is* HkuPke-CPA *secure,* KuSig *is* KuSig-UF *secure, and* Sig *is* 1-SUF-CMA *secure.*

A proof of Theorem 4 can be found in the full version [12].

References

1. Alwen, J., Coretti, S., Dodis, Y.: The double ratchet: security notions, proofs, and modularization for the signal protocol. In: Ishai, Y., Rijmen, V. (eds.) EURO-CRYPT 2019, LNCS, vol. 11476, pp. 129–158. Springer, Heidelberg (2019)
2. Bellare, M., Kohno, T., Namprempre, C.: Breaking and provably repairing the SSH authenticated encryption scheme: a case study of the Encode-then-Encrypt-and-MAC paradigm. ACM Trans. Inf. Syst. Secur. 7(2), 206–241 (2004)
3. Bellare, M., Miner, S.K.: A forward-secure digital signature scheme. In: Wiener, M. (ed.) CRYPTO 1999. LNCS, vol. 1666, pp. 431–448. Springer, Heidelberg (1999). https://doi.org/10.1007/3-540-48405-1_28
4. Bellare, M., Singh, A.C., Jaeger, J., Nyayapati, M., Stepanovs, I.: Ratcheted encryption and key exchange: the security of messaging. In: Katz, J., Shacham, H. (eds.) CRYPTO 2017. LNCS, vol. 10403, pp. 619–650. Springer, Cham (2017). https://doi.org/10.1007/978-3-319-63697-9_21
5. Borisov, N., Goldberg, I., Brewer, E.: Off-the-record communication, or, why not to use PGP. In: Proceedings of the 2004 ACM Workshop on Privacy in the Electronic Society, WPES 2004, pp. 77–84. ACM, New York (2004)

6. Canetti, R., Halevi, S., Katz, J.: A forward-secure public-key encryption scheme. In: Biham, E. (ed.) EUROCRYPT 2003. LNCS, vol. 2656, pp. 255–271. Springer, Heidelberg (2003). https://doi.org/10.1007/3-540-39200-9_16
7. Cohn-Gordon, K., Cremers, C., Dowling, B., Garratt, L., Stebila, D.: A formal security analysis of the signal messaging protocol. In: 2nd IEEE European Symposium on Security and Privacy, EuroS and P 2017, pp. 451–466 (2017)
8. Durak, F.B., Vaudenay, S.: Bidirectional asynchronous ratcheted key agreement without key-update primitives. Cryptology ePrint Archive, Report 2018/889 (2018). https://eprint.iacr.org/2018/889
9. Gentry, C., Silverberg, A.: Hierarchical ID-based cryptography. In: Zheng, Y. (ed.) ASIACRYPT 2002. LNCS, vol. 2501, pp. 548–566. Springer, Heidelberg (2002). https://doi.org/10.1007/3-540-36178-2_34
10. Horwitz, J., Lynn, B.: Toward hierarchical identity-based encryption. In: Knudsen, L.R. (ed.) EUROCRYPT 2002. LNCS, vol. 2332, pp. 466–481. Springer, Heidelberg (2002). https://doi.org/10.1007/3-540-46035-7_31
11. Jaeger, J., Stepanovs, I.: Optimal channel security against fine-grained state compromise: the safety of messaging. In: Shacham, H., Boldyreva, A. (eds.) CRYPTO 2018. LNCS, vol. 10991, pp. 33–62. Springer, Cham (2018). https://doi.org/10.1007/978-3-319-96884-1_2
12. Jost, D., Maurer, U., Mularczyk, M.: Efficient ratcheting: almost-optimal guarantees for secure messaging. Cryptology ePrint Archive, Report 2018/954 (2018). https://eprint.iacr.org/2018/954. (full version of this paper)
13. Kaplan, D., Kedmi, S., Hay, R., Dayan, A.: Attacking the Linux PRNG on android: weaknesses in seeding of entropic pools and low boot-time entropy. In: Proceedings of the 8th USENIX Conference on Offensive Technologies, WOOT 2014, p. 14. USENIX Association, Berkeley (2014)
14. Li, Y., Shen, T., Sun, X., Pan, X., Mao, B.: Detection, classification and characterization of Android malware using API data dependency. In: Thuraisingham, B., Wang, X.F., Yegneswaran, V. (eds.) SecureComm 2015. LNICST, vol. 164, pp. 23–40. Springer, Cham (2015). https://doi.org/10.1007/978-3-319-28865-9_2
15. Open Whisper Systems. Signal protocol library for java/android. GitHub repository (2017). https://github.com/WhisperSystems/libsignal-protocol-java. Accessed 01 Oct 2018
16. Poettering, B., Rösler, P.: Towards bidirectional ratcheted key exchange. In: Shacham, H., Boldyreva, A. (eds.) CRYPTO 2018. LNCS, vol. 10991, pp. 3–32. Springer, Cham (2018). https://doi.org/10.1007/978-3-319-96884-1_1

Obfuscation

Indistinguishability Obfuscation Without Multilinear Maps: New Methods for Bootstrapping and Instantiation

Shweta Agrawal[✉]

IIT Madras, Chennai, India
shweta.a@cse.iitm.ac.in

Abstract. Constructing indistinguishability obfuscation (iO) [17] is a central open question in cryptography. We provide new methods to make progress towards this goal. Our contributions may be summarized as follows:

1. **Bootstrapping.** In a recent work, Lin and Tessaro [71] (LT) show that iO may be constructed using (i) Functional Encryption (FE) for polynomials of degree L, (ii) Pseudorandom Generators (PRG) with *blockwise locality* L and polynomial expansion, and (iii) Learning With Errors (LWE). Since there exist constructions of FE for quadratic polynomials from standard assumptions on bilinear maps [16,68], the ideal scenario would be to set $L = 2$, yielding iO from widely believed assumptions

   Unfortunately, it was shown soon after [18,73] that PRG with block locality 2 and the expansion factor required by the LT construction, concretely $\Omega(n \cdot 2^{b(3+\epsilon)})$, where n is the input length and b is the block length, do not exist. In the worst case, these lower bounds rule out 2-block local PRG with stretch $\Omega(n \cdot 2^{b(2+\epsilon)})$. While [18,73] provided strong negative evidence for constructing iO based on bilinear maps, they could not rule out the possibility completely; a tantalizing gap has remained. Given the current state of lower bounds, the existence of 2 block local PRG with expansion factor $\Omega(n \cdot 2^{b(1+\epsilon)})$ remains open, although this stretch does not suffice for the LT bootstrapping, and is hence unclear to be relevant for iO.

   In this work, we improve the state of affairs as follows.

   (a) *Weakening requirements on Boolean PRGs:* In this work, we show that the narrow window of expansion factors left open by lower bounds *do* suffice for iO. We show a new method to construct FE for NC_1 from (i) FE for degree L polynomials, (ii) PRGs of block locality L and expansion factor $\tilde{\Omega}(n \cdot 2^{b(1+\epsilon)})$, and (iii) LWE (or RLWE).

   (b) *Broadening class of sufficient randomness generators*: Our bootstrapping theorem may be instantiated with a broader class of pseudorandom generators than hitherto considered for iO, and may circumvent lower bounds known for the arithmetic degree of iO-sufficient PRGs [18,73]; in particular, these may admit instantiations with arithmetic degree 2, yielding iO with the additional

© International Association for Cryptologic Research 2019
Y. Ishai and V. Rijmen (Eds.): EUROCRYPT 2019, LNCS 11476, pp. 191–225, 2019.
https://doi.org/10.1007/978-3-030-17653-2_7

assumptions of SXDH on Bilinear maps and LWE. In more detail, we may use the following two classes of PRG:

    i. *Non-Boolean PRGs*: We may use pseudorandom generators whose inputs and outputs need not be Boolean but may be integers restricted to a small (polynomial) range. Additionally, the outputs are not required to be pseudorandom but must only satisfy a milder indistinguishability property (We note that our notion of non Boolean PRGs is qualitatively similar to the notion of Δ RGs defined in the concurrent work of Ananth, Jain and Sahai [9]. We emphasize that the methods of [9] and the present work are very different, but both works independently discover the same notion of weak PRG as sufficient for building iO.).

    ii. *Correlated Noise Generators*: We introduce an even weaker class of pseudorandom generators, which we call correlated noise generators (CNG) which may not only be non-Boolean but are required to satisfy an even milder (seeming) indistinguishability property than Δ RG.

(c) *Assumptions and Efficiency.* Our bootstrapping theorems can be based on the hardness of the Learning With Errors problem or its ring variant (LWE/RLWE) and can compile FE for degree L polynomials directly to FE for NC_1. Previous work compiles FE for degree L polynomials to FE for NC_0 to FE for NC_1 to iO [12,45,68,72].

Our method for bootstrapping to NC_1 does not go via randomized encodings as in previous works, which makes it simpler and more efficient than in previous works.

2. **Instantiating Primitives.** In this work, we provide the first direct candidate of FE for constant degree polynomials from new assumptions on lattices. Our construction is new and does not go via multilinear maps or graded encoding schemes as all previous constructions. Together with the bootstrapping step above, this yields a **completely new candidate for** iO (as well as FE for NC_1), which makes no use of multilinear or even bilinear maps. Our construction is based on the ring learning with errors assumption (RLWE) as well as new untested assumptions on NTRU rings.

We provide a detailed security analysis and discuss why previously known attacks in the context of multilinear maps, especially zeroizing and annihilation attacks, do not appear to apply to our setting. We caution that our construction must yet be subject to rigorous cryptanalysis by the community before confidence can be gained in its security. However, we believe that the significant departure from known multilinear map based constructions opens up a new and potentially fruitful direction to explore in the quest for iO.

Our construction is based entirely on lattices, due to which one may hope for post quantum security. Note that this feature is not enjoyed by instantiations that make any use of bilinear maps even if secure instances of weak PRGs, as identified by the present work, the follow-up by Lin and Matt [69] and the independent work by Ananth, Jain and Sahai [9] are found.

1   Introduction

Indistinguishability Obfuscation. Program obfuscation aims to make a program "unintelligible" while preserving its functionality. Indistinguishability obfuscation [17] (iO) is a flavour of obfuscation, which converts a circuit C to an obfuscated circuit $\mathcal{O}(C)$ such that any two circuits that have the same size and compute the same function are indistinguishable to a computationally bounded adversary.

While it is non-obvious at first glance what this notion is useful for, recent work has demonstrated the tremendous power of iO. iO can be used to construct almost any cryptographic object that one may desire – ranging (non-exhaustively) from classical primitives such as one way functions [63], trapdoor permutations [22], public key encryption [82] to deniable encryption [82], fully homomorphic encryption [31], functional encryption [45], succinct garbling schemes [20, 30, 64, 70] and many more.

The breakthrough work of Garg et al. [45] presented the first candidate construction of iO from the beautiful machinery of graded encoding schemes [43]. This work heralded substantial research effort towards understanding iO: from cryptanalysis to new constructions to understanding and weakening underlying assumptions to applications. On the cryptanalysis front, unfortunately, all known candidate graded encoding schemes [39, 43, 51] as well as several candidates of iO have been broken [13, 33–35, 37, 38, 60, 75]. Given the power of iO, a central question in cryptography is to construct iO from better understood hardness assumptions.

Functional Encryption. Functional encryption (FE) [80, 81] is a generalization of public key encryption in which secret keys correspond to programs rather than users. In more detail, a secret key embeds inside it a circuit, say f, so that given a secret key SK_f and ciphertext $\mathsf{CT_x}$ encrypting a message $\mathbf{x}$, the user may run the decryption procedure to learn the value $f(\mathbf{x})$. Security of the system guarantees that nothing beyond $f(\mathbf{x})$ can be learned from $\mathsf{CT_x}$ and SK_f. Recent years have witnessed significant progress towards constructing functional encryption for advanced functionalities, even from standard assumptions [4, 5, 19, 24, 26, 27, 32, 36, 45, 46, 52, 57–59, 62, 66, 83]. However, most constructions supporting general functionalities severely restrict the attacker in the security game: she must request only a bounded number of keys [8, 55, 56], or may request unbounded number of keys but from a restricted space[1] [2, 58]. Schemes that may be proven secure against a general adversary are restricted to compute linear or quadratic functions [1, 6, 16, 68].

Constructing iO *from* FE. Recent work [10, 11, 23] provided an approach for constructing iO via FE. While we do not have any candidate constructions for FE that satisfy the security and efficiency requirements for constructing iO (except constructions that themselves rely on graded encoding schemes or iO [45, 47]), FE is a primitive that is closer to what cryptographers know to construct and brings iO nearer the realm of reachable cryptography.

---

[1] Referred to in the literature as "predicate encryption".

An elegant sequence of works [12,67,68,71,72] has attempted to shrink the functionality of FE that suffices for iO, and construct FE for this minimal functionality from graded encoding schemes or multilinear maps. Concretely, the question is: what is the smallest L such that FE supporting polynomials of degree L suffices for constructing iO? At a high level, these works follow a two step approach described below:

1. **Bootstrapping FE to iO.** The so called "bootstrapping" theorems have shown that general purpose iO can be built from one of the following: (i) sub-exponentially secure FE for NC_1 [10,11,21,23], or (ii) sub-exponentially secure FE for NC_0 and PRG in NC_0 [72] (iii) PRGs with locality L and FE for computing degree L polynomials [68] or (iv) PRGs with *blockwise* locality L and FE for computing degree L polynomials [71].

   At a high level, all bootstrapping theorems make use of *randomized encodings* [14,61] to reduce computation of a polynomial sized circuit to computation of low degree polynomials.

2. **Instantiating Primitives.** Construct FE supporting degree L polynomials based on graded encodings or multilinear maps [68,72].

1.1 Bootstrapping, The Ideal

Since we have candidates of FE for quadratic polynomials from standard assumptions on bilinear maps [16,68], a dream along this line of work would be to reduce the degree required to be supported by FE all the way down to 2, yielding iO, from bilinear maps and other widely believed assumptions (like LWE and PRG with constant locality). The recent work of Lin and Tessaro [71] (LT) came closest to achieving this, by leveraging a new notion of PRG they termed *blockwise local* PRG. A PRG has blockwise locality L and block-size b, if when viewing the input seed as a matrix of b rows and n columns, every output bit depends on input bits in at most L columns. As mentioned above, they showed that PRGs with blockwise locality L and certain polynomial stretch, along with FE for computing degree L polynomials and LWE suffice for iO.

Unfortunately, it was shown soon after [18,73] that PRG with block locality 2 and the stretch required by the LT construction, concretely $\Omega(n \cdot 2^{b(3+\epsilon)})$, do not exist. In the worst case, these lower bounds rule out 2-block local PRG with stretch $\Omega(n \cdot 2^{b(2+\epsilon)})$. On the other hand, these works suggest that 3 block local PRG are likely to exist, thus shrinking the iO-sufficient degree requirement on FE to 3.

While [18,73] provided strong negative evidence for constructing iO based on 2 block local PRG and hence bilinear maps, they could not rule out the possibility completely; a tantalizing gap has remained. Roughly speaking, the construction of candidate PRG (first suggested by Goldreich [53]) must choose a hyper-graph with variables on vertices, then choose predicates that are placed on each hyper-edge of the graph and output the values of the edge-predicates on the vertex-variables. The lower bounds provided by [18,73] vary depending on how the graph and predicates are chosen in the above construction: in particular whether the graph is chosen randomly or could be constructed in some arbitrary "worst case" way, whether the predicate is chosen randomly or arbitrarily, and whether the same predicate is used for each hyper-edge or different predicates may be used per hyper-edge or output bit. The following table by [73] summarises our current understanding on the existence of 2 block local PRG:

Stretch	Worst case versus Random Predicate	Worst case versus Random Graph	Different versus Same Predicate per output bit	Reference
$\tilde{\Omega}(n \cdot 2^{b(1+\epsilon)})$	Random	Random	Different	[18]
$\tilde{\Omega}(n \cdot 2^{b(2+\epsilon)})$	Worst case	Worst case	Different	[18]
$\tilde{\Omega}(n \cdot 2^{b(1+\epsilon)})$	Worst case	Worst case	Same	[73]
$\tilde{\Omega}(n \cdot 2^{b(1+\epsilon)})$	Worst case	Worst case	Different	Open

As we see in the above table, the existence of 2 block local PRG with carefully chosen graph and predicates with stretch $\tilde{\Omega}(n \cdot 2^{b(1+\epsilon)})$ is open. However, even in the case that these exist, it is not clear whether its even useful, since the Lin-Tessaro compiler requires larger stretch $\Omega(n \cdot 2^{b(3+\epsilon)})$, which is ruled out by row 2 above. In the current version of their paper, [71] remark that "Strictly speaking, our results leave a narrow window of expansion factors open where block-wise PRGs could exist, but we are not aware whether our approach could be modified to use such low-stretch PRGs."

Bootstrapping: Our Results. In this work, we show that the narrow window of expansion factors left open by lower bounds *do* suffice for iO. Moreover, we define a larger class of pseudorandomness generators than those considered so far, which may admit lower degree instantiations. We then show that these generators with the same expansion suffice for iO. We discuss each of these contributions below.

Weakening requirements on PRGs: We show a new method to construct FE for NC_1 from FE for degree L polynomials, sub-exponentially secure PRGs of block locality L and LWE (or RLWE). Since FE for NC_1 implies iO for P/Poly [10, 21, 23], this suffices for bootstrapping all the way to iO for P/Poly. Our transformation requires the PRG to only have expansion $n \cdot 2^{b(1+\epsilon)}$ which is not ruled out as discussed above. This re-opens the possibility of realizing 2 block local PRG with our desired expansion factor (see below for a detailed discussion), which would imply iO from 2 block local PRG, SXDH on Bilinear maps and LWE. A summary of the state of art in PRG based bootstrapping is provided in Fig. 1.

Broadening class of sufficient PRGs : Our bootstrapping theorem may be instantiated with a broader class of pseudorandom generators than hitherto considered for iO, and may circumvent lower bounds known for the arithmetic degree of iO-sufficient PRGs [18, 73]; in particular, these may admit instantiations with arithmetic degree 2, yielding iO along with the additional assumptions of SXDH on Bilinear maps and LWE. In more detail, we may use the following two classes of PRG:

1. *Non-Boolean PRGs*: We may use pseudorandom generators whose inputs and outputs need not be Boolean but may be integers restricted to a small (polynomial) range. Additionally, the outputs are not required to be pseudorandom but must only satisfy a milder indistinguishability property[2]. We tentatively propose initializing these PRGs using the multivariate quadratic assumption MQ which has been widely studied in the literature [41,74,84] and against the general case of which, no efficient attacks are known.
2. *Correlated Noise Generators*: We introduce an even weaker class of pseudorandom generators, which we call correlated noise generators (CNG) which may not only be non-Boolean but are required to satisfy an even milder (seeming) indistinguishability property.

Assumptions and Efficiency. Our bootstrapping theorems can be based on the hardness of LWE or its ring variant RLWE and compiles FE for degree L polynomials directly to FE for NC_1. Our method for bootstrapping to NC_1 does not go via randomized encodings as in previous works. Saving the transformation to randomized encodings makes bootstrapping to NC_1 more efficient than in previous works. For instance, [71] require the encryptor to choose Q PRG seeds,

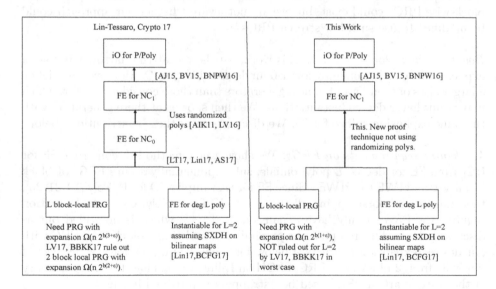

Fig. 1. State of the Art in Bootstrapping FE to iO. In the present work, we may bootstrap directly to FE for NC_1 without going through NC_0.

---

[2] For the knowledgeable reader, we do not require the polynomials computing our PRGs to be sparse and hence the general attack of [18] does not rule out existence of degree 2 instantiations to the best of our knowledge.

where Q is the (polynomial) length of random tapes needed by the randomized encodings. On the other hand we only need 2 PRG seeds, since we avoid using randomized encodings, yielding a ciphertext that is shorter by a factor of Q, as well as (significantly) simpler pre-processing.

1.2   Instantiation: The Ideal

To instantiate iO via FE for constant degree polynomials, [71] rely on the FE for degree L polynomials constructed by Lin [68], which relies on SXDH on noiseless algebraic multilinear maps of degree L, for which no candidates of degree greater than 2 are known to exist. As discussed by [68], instantiating her construction with noisy multilinear maps causes the proof to fail, in addition to the SXDH assumption itself being false on existing noisy multilinear map candidates. We refer the reader to [68, 71] for a detailed discussion.

Evidently, one ideal instantiation for iO would be to construct noiseless multilinear maps of degree at least 3^3, on which the SXDH assumption is believed to hold. At the moment, we have no evidence that such objects exist. Another ideal instantiation would be to provide a direct construction of FE for constant degree polynomials from well understood hardness assumptions, satisfying the requisite compactness properties for implication to iO. Constructing FE from well-understood hardness assumptions has received significant attention in recent years, and for the moment we do not have any constructions that suffice for iO excepting those that themselves rely on multilinear maps or iO.

Thus, at present, all concrete instantiations of the FE to iO compiler must go via noisy multilinear maps on which SXDH fails.

Instantiation: Our Results. In our work, we take a different approach to the question of instantiation. We propose to construct FE *directly*, without going through multilinear maps or graded encoding schemes, and use this FE to instantiate the transformation to iO. We believe this new approach has the following advantages:

1. **May be Simpler:** Construction of iO-sufficient FE might not need the full power of asymmetric multilinear maps, since FE is not known to imply asymmetric multilinear maps equipped with SXDH to the best of our knowledge[4]. Hence, constructing FE directly may be simpler.
2. **Yield New and Possibly More Robust Assumptions:** Attempts to construct FE directly for low degree polynomials yield new hardness assumptions which are likely different from current assumptions on noisy multilinear maps. This direction may yield more resilient candidates than those that go via multilinear maps.

---

[3] Ideally degree 5, so as to remove the reliance on even blockwise local PRG and rely directly on 5 local PRG which are better understood.

[4] A line of work can traverse the route of FE to iO to PiO (probabilistic iO) to *symmetric* multilinear maps (see [42] and references therein) using multiple complex subexponential reductions, still not yielding *asymmetric* multilinear maps with SXDH.

In this work, we provide the first direct candidate of symmetric key FE for constant degree polynomials from new assumptions on lattices. Let $\mathcal{F}$ be the class of circuits with depth d and output length ℓ. Then, for any $f \in \mathcal{F}$, our scheme achieves $\mathsf{Time}(\mathsf{KeyGen}) = O(\mathrm{poly}(\kappa, |f|))$, and $\mathsf{Time}(\mathsf{Enc}) = O(|\mathbf{x}| \cdot 2^d \cdot \mathrm{poly}(\kappa))$ where κ is the security parameter. This suffices to instantiate the bootstrapping step above. Our construction is based on the ring learning with errors assumption (RLWE) as well as new untested assumptions on NTRU rings. We provide a detailed security analysis and discuss why currently known attacks in the multilinear map setting do not appear to apply. We also provide a proof in a restricted security game where the adversary is allowed to request only one ciphertext, based on a new assumption. While such a security game is too limited to be reasonable, we view this as a first step to provable security. We caution that the assumptions underlying our construction must yet be subject to rigorous cryptanalysis by the community. However, our approach is fundamentally different and we hope it inspires other candidates.

1.3 Our Techniques: Bootstrapping

Let us start by restating the goal: we wish to construct FE for the function class NC_1 such that the size of the ciphertext depends only sublinearly on the size of the function. Previous work [10,23] shows that such an FE suffices to construct iO. At a high level, to compute $f(\mathbf{x})$, our FE scheme will make use of a fully homomorphic encryption scheme (FHE) to evaluate the function f on the FHE ciphertext $\mathsf{CT}_\mathbf{x}$ of $\mathbf{x}$ to obtain a "functional" ciphertext $\mathsf{CT}_{f(\mathbf{x})}$ and then perform FHE decryption on $\mathsf{CT}_{f(\mathbf{x})}$ to obtain $f(\mathbf{x})$.

Agrawal and Rosen [8] show how to instantiate the above blueprint from the LWE assumption, but incur large ciphertext size that does not suffice for bootstrapping to iO. Their construction is the starting point of our work. Below, we assume familiarity of the reader with RLWE and Regev's public key encryption scheme [52,79]. Although our bootstrapping can also be based on standard LWE, we describe it using RLWE here since it is simpler.

"FE-*compatible*" *Homomorphic Encryption by* [8]. The main technical contribution of [8] may be seen as developing a special "FE-compatible" FHE scheme that lends itself to the *constrained* decryption required by FE. Note that FHE enables an evaluator to compute arbitrary functions on the ciphertext. In contrast, FE requires that given a ciphertext $\mathsf{CT}_\mathbf{x}$, decryption is constrained to some function f for which the decyptor possess a secret key SK_f. Thus, decryption must reveal $f(\mathbf{x})$ alone, and leak no other function of $\mathbf{x}$.

To address this issue, [8] design new algorithms for FHE encryption and ciphertext evaluation, inspired by an FHE by Brakerski and Vaikuntanathan [29]. The evaluator/decryptor, given the encoding of some input $\mathbf{x}$ and some (arithmetic) circuit $f \in \mathsf{NC}_1$ can execute the ciphertext evaluation algorithm, which we denote by $\mathsf{Eval}_{\mathsf{CT}}$, to compute a "functional" ciphertext $\mathsf{CT}_{f(\mathbf{x})}$ that encodes $f(\mathbf{x})$. The functional ciphertext can then by decrypted by SK_f alone, to reveal $f(\mathbf{x})$ and nothing else.

The encryption algorithm of [8] is "levelled" in that given input $\mathbf{x}$, it outputs a set of encodings $\mathcal{C}^i$ for $i \in [d]$ where d is the depth of the circuit being computed. The functional ciphertext $\mathsf{CT}_{f(\mathbf{x})} = \mathsf{Eval}_{\mathsf{CT}}(\underset{i \in [d]}{\cup} \mathcal{C}^i, f)$ of [8] has the following useful structure:

$$\mathsf{CT}_{f(\mathbf{x})} = \langle \mathsf{Lin}_f, \mathcal{C}^d \rangle + \mathsf{Poly}_f(\mathcal{C}^1, \dots, \mathcal{C}^{d-1})$$

for some f-dependent linear function Lin_f and polynomial Poly_f. Moreover, upon decrypting $\mathsf{CT}_{f(\mathbf{x})}$, we get

$$f(\mathbf{x}) + \mathsf{noise}_{f(\mathbf{x})} = \langle \mathsf{Lin}_f, \mathcal{M}^d \rangle + \mathsf{Poly}_f(\mathcal{C}^1, \dots, \mathcal{C}^{d-1}) \tag{1.1}$$

where $\mathcal{M}^d$ is the message vector encoded in level d encodings $\mathcal{C}^d$. Here, $f(\mathbf{x}) \in R_{p_0}$ for some ring R_{p_0} and $\mathsf{noise}_{f(\mathbf{x})}$ is the noise term that results from FHE evaluation which may be removed using standard techniques to recover $f(\mathbf{x})$ as desired.

Using Linear FE *and Noise Flooding.* Given the above structure, an approach to compute $f(\mathbf{x})$ is to leverage functional encryption for linear functions [1,6], denoted by LinFE to compute the term $\langle \mathsf{Lin}_f, \mathcal{M}^d \rangle$. Recall the functionality of LinFE: the encryptor provides a ciphertext $\mathsf{CT}_{\mathbf{z}}$ for some vector $\mathbf{z} \in R^n$, the key generator provides a key $\mathsf{SK}_{\mathbf{v}}$ for some vector $\mathbf{v} \in R^n$ and the decryptor learns $\langle \mathbf{z}, \mathbf{v} \rangle \in R$. Thus, we may use LinFE to enable the decryptor to compute $\langle \mathsf{Lin}_f, \mathcal{M}^d \rangle$, let the decryptor compute $\mathsf{Poly}_f(\mathcal{C}^1, \dots, \mathcal{C}^{d-1})$ herself to recover $f(\mathbf{x}) + \mathsf{noise}_{f(\mathbf{x})}$. Surprisingly, constrained decryption of a *linear* function on secret values suffices, along with additional public computation, to perform constrained decryption of a function in NC_1.

Unfortunately, this approach is insecure as is, as discussed in [8]. For bounded collusion FE, the authors achieve security by having the encryptor encode a fresh, large noise term $\mathsf{noise}_{\mathsf{fld}}$ for each requested key f which "floods" $\mathsf{noise}_{f(\mathbf{x})}$. This noise is forcibly added to the decryption equation so that the decryptor recovers $f(\mathbf{x}) + \mathsf{noise}_{f(\mathbf{x})} + \mathsf{noise}_{\mathsf{fld}}$, which by design is statistically indistinguishable from $f(\mathbf{x}) + \mathsf{noise}_{\mathsf{fld}}$. [8] show that with this modification the scheme can be shown to achieve strong simulation style security, by relying just on security of LinFE. However, encoding a fresh noise term per key causes the ciphertext size to grow at least linearly with the number of function keys requested, or in the case of single key FE, with the output length of the function. As noted above, this renders their FE insufficient for iO.

Noisy Linear Functional Encryption. In this work, we show that the approach of [8] can be extended to construct a single key FE for NC_1 with ciphertext size sublinear in the output length, by replacing linear functional encryption LinFE with *noisy linear functional encryption*, denoted by NLinFE. Noisy linear functional encryption is like regular linear functional encryption [1,6], except that the function value is recovered only up to some bounded additive error/noise, and indistinguishability holds even if the challenge messages evaluated on any function key are only "approximately" and not exactly equal. The functionality of NLinFE is as follows: given a ciphertext CT_z which encodes vector $z \in R^n$ and a secret key SK_v which encodes vector $v \in R^n$, the decryptor recovers $\langle z, v \rangle + \text{noise}_{z,v}$ where $\text{noise}_{z,v}$ is specific to the message and function being evaluated.

Let $f \in NC_1$ and let the output of f be of size ℓ. Let $f_1, \ldots, f_\ell$ be the functions that output the ith bit of f for $i \in [\ell]$. At a high level, our FE for NC_1 will enable the decryptor to compute $\langle \text{Lin}_{f_i}, \mathcal{M}^d \rangle + \text{noise}_{\text{fld}i}$ as in [8] but instead of having the encryptor encode ℓ noise terms during encryption, we use NLinFE to compute and add noise terms $\text{noise}_{\text{fld}i}$ into the decryption equation. Given NLinFE with sublinear ciphertext size, we can then construct FE for NC_1 with sublinear ciphertext size, which suffices for bootstrapping to iO. In more detail, we show:

Theorem 1.1. *(Informal) There exists an* FE *scheme for the circuit class* NC_1 *with sublinear ciphertext and satisfying indistinguishability based security, assuming:*

- *A noisy linear FE scheme* NLinFE *with sublinear ciphertext satisfying indistinguishability based security.*
- *The Learning with Errors (*LWE*) Assumption.*
- *A pseudorandom generator (*PRG*).*[5]

The formal theorem is provided in Sect. 4. Note that while [8] argue simulation based security of FE for NC_1 using simulation security of LinFE in the bounded key setting, we must argue *indistinguishability* based security of FE for NC_1 assuming indistinguishability based security NLinFE. This is significantly more complex and requires new proof techniques, which we develop in this work. Please see Sect. 4 for details.

The key question that remains is how does NLinFE construct the noise term to be added to the decryption equation? As discussed next, NLinFE is a primitive flexible enough to admit multiple instantiations, which in turn yield FE for NC_1 from diverse assumptions, improving the state of art.

Constructing NLinFE. Next, we discuss multiple methods to construct NLinFE, which imply FE for NC_1 from various assumptions. Together with the bootstrapping of NLinFE to FE for NC_1 described above, this suffices for applying the FE to iO compiler of [10,23]. Before we describe our constructions, we provide a summary via the following theorem:

Theorem 1.2. *(Informal) Noisy linear functional encryption (*NLinFE*) with sublinear ciphertext and satisfying indistinguishability based security may be constructed using:*

1. *(i) An* FE *scheme supporting evaluation of degree L polynomials and satisfying indistinguishability based security, (ii) sub-exponentially secure* PRG *with*

---

[5] We actually only need a randomness generator that is weaker than a standard PRG but do not discuss this here. for the formal statement.

block locality L and expansion $n \cdot 2^{b(1+\epsilon)}$, *where n is the input length and b is the block length, and (iii)* LWE *(or* RLWE*).*

2. *(i) An* FE *scheme supporting evaluation of degree L polynomials and satisfying indistinguishability based security, (ii) sub-exponentially secure weak randomness generators called "Correlated Noise Generators" (CNG) (iii)* LWE *(or* RLWE*).*

3. *(i) An* FE *scheme supporting evaluation of degree L polynomials and satisfying indistinguishability based security, (ii) sub-exponentially secure weak randomness generators called "non-Boolean* PRG*" (iii)* LWE *(or* RLWE*).*

4. *New lattice assumptions on NTRU rings.*

Note that the above instantiations of NLinFE have several desirable features as discussed below:

1. The first instantiation uses a block local PRG with smaller expansion factor than that required by the Lin-Tessaro compiler [71]. More importantly, 2-local PRG with the above expansion factor is not ruled out in the worst case by [18,73]. Thus, if PRG with block locality 2 and the above expansion factor exist, we may instantiate FE for $L = 2$ using SXDH on bilinear maps [68].

2. The notion of correlated noise generators CNG is new to our work, and appears significantly weaker than a standard Boolean PRG, as discussed below. Prior to our work, the only notions of PRG that were used to construct iO are standard Boolean PRG and block local PRG [71].

3. Our notion of non-Boolean PRG interpolates CNG and standard Boolean PRG. This notion is qualitatively the same as the notion of Δ-RG discovered concurrently and independently by [9].

4. Our direct construction of NLinFE makes no use of bi/multi-linear maps, and provides a candidate for post quantum iO. We note that most (if not all) candidates of iO are vulnerable in the post quantum setting [77].

The "Right" Abstraction. Noisy linear functional encryption provides the right abstraction (in our opinion) for the smallest functionality that may be bootstrapped to FE for NC_1 using our methods. NLinFE captures the precise requirements on noise that is required for the security of FE for NC_1 and integrates seamlessly with our new proof technique. Moreover, as discussed above, NLinFE may be constructed in different ways from different assumptions, and properties such as ciphertext size and collusion resistance achieved by NLinFE are inherited by FE for NC_1. We remark that the follow up work by Lin and Matt [69] also uses our notion of NLinFE in essentially the same manner for their overall construction.

Put together, an overview of our transformation is provided in Fig. 2.

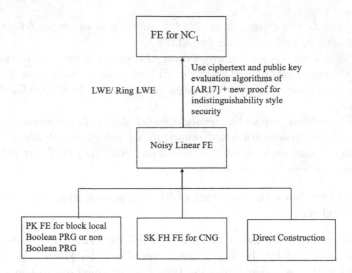

Fig. 2. Overview of our transformation. Above, FH refers to function hiding.

Weaker Requirements on PRG. Next, we discuss the requirements on the PRG used by the first three instantiations of NLinFE discussed above.

PRG *with smaller expansion factor.* Our first method makes use of a compact FE scheme which is powerful enough to compute PRG/blockwise local PRG [71]. Let PrgFE be a functional encryption scheme that supports evaluation of a PRG with polynomial stretch. Then, we may construct NLinFE and hence FE for NC_1 by leveraging PrgFE to compute the noise to be added by NLinFE as the output of a PRG.

In more detail, by the discussion above, we would like the decryptor to compute:

$$f(\mathbf{x}) + \mathsf{noise}_{f(\mathbf{x})} + \mathsf{noise}_{\mathsf{fld}} = \langle \mathsf{Lin}_f, \ \mathcal{M}^d \rangle + \mathsf{noise}_{\mathsf{fld}} + \mathsf{Poly}_f(\mathcal{C}^1, \ldots, \mathcal{C}^{d-1})$$

where $\mathsf{noise}_{f(\mathbf{x})} + \mathsf{noise}_{\mathsf{fld}}$ is indistinguishable from $\mathsf{noise}_{\mathsf{fld}}$. Say that the norm of $\mathsf{noise}_{f(\mathbf{x})}$ may be bounded above by value $\mathsf{B}_{\mathsf{nse}}$. Then, it suffices to sample a uniformly distributed noise term $\mathsf{noise}_{\mathsf{fld}}$ of norm bounded by $\mathsf{B}_{\mathsf{fld}}$, where $\mathsf{B}_{\mathsf{fld}}$ is superpolynomially larger than $\mathsf{B}_{\mathsf{nse}}$ for the above indistinguishability to hold. This follows from security of PRG and a standard statistical flooding argument. We will use PrgFE to compute $\mathsf{noise}_{\mathsf{fld}}$.

In more detail, let G be a PRG with polynomial stretch which outputs ℓ uniform ring elements of norm bounded by $\mathsf{B}_{\mathsf{fld}}$, and let G_i be the function that selects the ith output symbol of G, namely $G_i(\mathsf{seed}) = G(\mathsf{seed})[i]$ where seed is the seed of the PRG. Then,

1. The encryptor may provide PrgFE encryptions of $(\mathcal{M}^d, \mathsf{seed})$ along with encodings $\underset{i \in [d-1]}{\cup} \mathcal{C}^i$,
2. The key generator may provide a PrgFE secret key for polynomial $P_i(\mathbf{z}_1, \mathbf{z}_2) = \langle \mathsf{Lin}_{f_i}, \ \mathbf{z}_1 \rangle + G_i(\mathbf{z}_2)$

3. The decryptor may compute $\langle \mathsf{Lin}_{f_i}, \quad \mathcal{M}^d \rangle + G_i(\mathsf{seed})$ as well as $\mathsf{Poly}_f(\mathcal{C}^1, \ldots, \mathcal{C}^{d-1})$, to recover $f(\mathbf{x}) + \mathsf{noise}_{f(\mathbf{x})} + G_i(\mathsf{seed})$ by Eq. 1.1 as desired.

It is crucial to note that the degree of the polynomial P_i is equal to the degree required to compute G_i, because Lin_{f_i} is a linear function. Moreover, the degree is unchanged even if we make use of a standard PRG with binary range. To see this, take a binary PRG that requires degree L to compute, and apply the standard (linear) powers of two transformation to convert binary output to larger alphabet.

Thus, an FE scheme that supports polynomials of degree L, where L is the degree required to compute a PRG, suffices to construct NLinFE and hence FE for NC_1. Moreover, we may pre-process the seed of the PRG as in [71] to leverage "blockwise locality"; our construction allows the PRG to have smaller expansion factor than that required by the Lin-Tessaro construction, as discussed next.

Analyzing the Expansion Factor of the PRG. Above, the polynomial P_i we are required to construct must compute a function that is degree 1 in the PRG output plus an additional linear function of the encoded messages. By comparison, the underlying degree L FE in [68,71,72] must natively compute a polynomial which has degree 3 in output of a PRG. In more detail, the FE of [68,71] must compute a randomizing polynomial [15], which contains terms of the form $r_i r_j r_k$ where r_i, r_j, r_k are random elements, each computed using a PRG. Thus, if L is the locality of the PRG, the total degree of the polynomial is $3L$, which is reduced to L using a clever precomputing trick developed by Lin [68]. In contrast, our FE must compute a polynomial of the form $\langle \mathsf{Lin}_f, \mathcal{M}^d \rangle + \mathsf{PRG}(\mathsf{seed})$ as discussed above, thus natively yielding a polynomial of degree L.

Further, Lin and Tessaro [71] construct a method to leverage the blockwise locality L of the PRG, which is believed to be smaller than locality as discussed above. Recall that the PRG seed is now a matrix of b rows and n columns, and each output bit depends on input bits in at most L columns. Then, to begin, [71] suggest computing all possible monomials in any block, for all blocks, resulting in $O(n \cdot 2^b)$ total monomials. At this point, the output of a PRG can be computed using a degree L polynomial. However, since the final polynomial has degree 3 in the PRG outputs, LT further suggest precomputing all degree 3 products of the monomials constructed so far, leading to a total of $O(n \cdot 2^{3b})$ terms. To maintain ciphertext compactness then, the expansion of the PRG must be at least $\Omega(n \cdot 2^{b(3+\epsilon)})$. We refer the reader to [18, Appendix B] for an excellent overview of the LT construction, and to [71] for complete details.

Since the polynomial in our FE must compute has degree only 1 rather than 3 in PRG output, we need not compute degree 3 products in $O(n \cdot 2^b)$ monomials as required by [71], thus requiring to encode a total number of $O(n \cdot 2^b)$ monomials. Hence, to maintain ciphertext compactness (which is necessary for implication to iO [10,23]), the expansion of the PRG in our case must be $\Omega(n \cdot 2^{b(1+\epsilon)})$. Thus, our transformation requires a lower expansion factor than that of [71]. Moreover, by sidestepping the need to compute randomized encodings, our bootstrapping becomes simpler and more efficient than that of [68,71,72].

Correlated Noise Generators. As discussed above, FE for implementing PRG suffices to construct NLinFE and hence FE for NC_1. However, examining carefully the requirements on the noise that must be added to decryption by NLinFE, reveals that using a PRG to compute the noise is wasteful; a weaker object appears to suffice. Specifically, we observe that the noise terms $\mathsf{noise}_{f_i(\mathbf{x})}$ for $i \in [\ell]$, which must be flooded are low entropy, correlated random variables, constructed as polynomials in $O(L)$ noise terms where $L = |\bigcup_{k \in [d]} C_k|$. A PRG mimics ℓ i.i.d noise terms where $\ell > L$, i.e. $O(\ell)$ bits of entropy, whereas the random variables that must be flooded have only $O(L)$ bits of entropy.

Indeed, even to flood ℓ functions on L noise terms *statistically*, only L fresh noise terms are needed. For instance, let us say that we are required to flood $(f_i(\boldsymbol{\mu}))_{i \in [\ell]}$ for $\boldsymbol{\mu} \in R^L$. Then, it suffices to choose $\boldsymbol{\beta} \in R^L$ such that $\boldsymbol{\beta}$ is superpolynomially larger than $\boldsymbol{\mu}$ to conclude that

$$\mathsf{SD}\Big(\boldsymbol{\beta}, \; \boldsymbol{\beta} + \boldsymbol{\mu}\Big) = \mathsf{negl}(\kappa)$$

This implies that

$$\mathsf{SD}\Big(\big(f_1(\boldsymbol{\beta}), \dots, f_\ell(\boldsymbol{\beta})\big), \; \big(f_1(\boldsymbol{\beta} + \boldsymbol{\mu}), \dots, f_\ell(\boldsymbol{\beta} + \boldsymbol{\mu})\big)\Big) = \mathsf{negl}(\kappa)$$

Considering that we must only generate $O(L)$ bits of pseudoentropy, can we make do with something weaker than a PRG?

To make this question precise, we define the notion of a correlated noise generator, denoted by CNG. A CNG captures computational flooding of correlated noise, to mimic statistical flooding described above. In more detail, we will use a CNG to generate flooding terms $g_1(\boldsymbol{\beta}), \dots, g_\ell(\boldsymbol{\beta})$ such that to any computationally bounded adversary Adv, it holds that

$$(g_1(\boldsymbol{\beta}), \dots, g_\ell(\boldsymbol{\beta})) \stackrel{c}{\approx} \big((g_1(\boldsymbol{\beta}) + f_1(\boldsymbol{\mu}), \dots g_\ell(\boldsymbol{\beta}) + f_\ell(\boldsymbol{\mu}))$$

Note that if we denote by g_i the function for computing the ith element of the PRG output, then by choosing the range of PRG superpolynomially larger than $|f_i(\boldsymbol{\mu})|$, the above condition is satisfied. Thus, a PRG implies a CNG. However, implication in the other direction does not hold, since CNG only generates strictly fewer bits of pseudoentropy than a PRG.

Moreover, matters are even nicer in the case of CNG, because g_i can be chosen after seeing f_i, and the distribution of $\boldsymbol{\mu}$ is known at the time of choosing g_i. So each g_i can be different depending on what it needs to "swallow". Additionally, we may leverage the fact that a CNG posits that a distribution must be indistinguishable from *itself* plus a fixed function, not indistinguishable from uniform.

Our hope is that since a CNG appears weaker than a PRG, it may sidestep the lower bounds known for the blockwise-locality of polynomial stretch PRGs, thereby providing a new route to iO from bilinear maps. Suggesting candidates for CNG that have lower degree than PRG is outside the scope of this work but we believe it is useful to identify the weakest object that suffices for bootstrapping to iO. For the precise definition of CNG, please see Sect. 3.

Non Boolean PRG. A notion of randomness generators that interpolates CNG and Boolean PRG is that of non-Boolean PRG, which allows the inputs and outputs to lie in a bounded (polynomial) sized interval over the integers and must only satisfy the computational flooding property described above. Taking a step back, we note that in prior work [12, 68, 71], Boolean PRGs were required in order to compute the binary randomness needed for constructing randomizing polynomials. In our case, the PRG output need not be binary since we do not require these as input to randomized encodings. Additionally, they must satisfy a much weaker property than indistinguishability to uniform i.i.d random variables as discussed above. In more detail, say we can bound $|f_i(\boldsymbol{\mu})| \leq \epsilon$ for $i \in [\ell]$. Then we require the PRG output $G_i(\boldsymbol{\beta})$ to computationally flood $f_i(\boldsymbol{\mu})$ for $i \in [\ell]$, i.e. $G_i(\boldsymbol{\beta}) + f_i(\boldsymbol{\mu})$ must be computationally indistinguishable from $G_i(\boldsymbol{\beta})$.

We note that the above notion of non Boolean PRGs is qualitatively the same as the notion of ΔRGs defined in the concurrent work of Ananth et al. [9] except that ΔRG are weaker, in that they allow the adversary to win the game with $1/$poly probability whereas we require that the adversary only wins with standard negligible probability. By relying on the security amplification theorem of [9], our notion can be weakened and made equivalent to ΔRG.

1.4   Related Work: Bootstrapping

In this section, we provide a detailed comparison with works that are most closely related to ours.

Predicate encryption [2, 28, 58] and reusable garbled circuits [2, 54]. A successful approach to constructing functional encryption schemes from standard assumptions is the predicate encryption scheme by Gorbunov et al. [58] and its extensions [2, 28]. Roughly speaking, these schemes make use of an attribute based encryption (ABE) scheme for circuits [25, 57] in conjunction with a fully homomorphic encryption scheme to achieve a system where the input **x** is hidden only as long as the adversary's key requests obey a certain "one sided" restriction w.r.t the challenge messages. In more detail, security holds as long as the adversary does not obtain keys SK_f for any circuit f such that $f(\mathbf{x}) = 0$. Given an adversary who obeys this "one sided" restriction, functionality is general, i.e. the adversary may request for a key corresponding to any polynomial sized circuit. However, in the general "two sided" security game, the schemes are shown to be insecure [2, 58]. The reusable garbled schemes of [2, 54] satisfy general two sided security but do not achieve compact ciphertext required for bootstrapping to iO.

The techniques of the aforementioned line of work and the present work are fundamentally different. While [55, 58] also use FHE in order to hide the attributes in an FE scheme, the building block of ABE necessitates the restriction of one sided security due to the basic structure of ciphertexts and secret keys. As discussed in [2], if the predicate encryption scheme [58] is subject to a general two sided adversary, the adversary may requests keys for related functions which can lead to the recovery of a secret lattice basis, leading to a complete break in security. We emphasize that this attack exploits the structure of the secret keys and ciphertext in the underlying ABE scheme and is in distinct from the attack implied by the leakage of the FHE noise learnt by the attacker upon decryption – indeed, the follow-up work of [28] shows how to construct predicate encryption that does not contain the FHE noise leakage. Thus, despite supporting powerful functionality, current techniques for generalizing ABE to FE get stuck in the quicksand of one sided security.

To overcome this challenge, we insist on a two sided adversary at any cost to functionality. We follow the approach of [8] which starts with the modest functionality of linear functional encryption [1,6] satisfying two sided security, and makes use of special properties of the FHE scheme of Brakerski and Vaikuntanathan [29] to decompose function computation into a "deep" public computation performed using FHE and a "shallow" (linear) private computation, performed using linear FE [1,6]. The public FHE computation is performed by the decryptor outside any FE scheme, namely, without any guarantee of constrained decryption. This is in contrast to [2,28,58] where the entire function evaluation is performed within the confines of the ABE evaluation, which constrains decryption of the final FHE ciphertext and renders futile any attempts to tamper with the functionality. However, in [8] the only constrained decryption is via the modest functionality of linear FE but the authors argue that constraining a linear function suffices to constrain computation in NC_1, at the cost of non-compact ciphertext.

In the present work, to achieve security as well as succinct ciphertext, we look at the mildest possible strengthening of this functionality, namely one that supports computation of linear functions plus a noise term which satisfies a relatively mild statistical property, as formalized via the notion of noisy linear functional encryption (NLinFE). We then show that this notion of NLinFE may be bootstrapped all the way to iO. Our bootstrapping uses NLinFE in a black box way, and when NLinFE is instantiated using bilinear maps, the results in an interesting "hybrid" scheme which uses FHE to perform deep computation in the clear and then performs a careful FHE decryption in the exponent.

Comparison with [8]. Even though the present work uses the ciphertext and public key evaluation algorithms developed by Agrawal and Rosen [8], our construction of FE for NC_1 and particularly our proof technique are quite different. Firstly, [8] is in the bounded collusion setting with non-compact ciphertext, and achieves a simulation based security which is known to be impossible in our setting where compact ciphertext is crucial. Hence, we must give an indistinguishability style proof which is significantly harder, and requires using a new proof technique developed in this work. Moreover, [8] adds statistical large flooding noise which is oblivious of the distribution of noise it needs to drown, whereas we will analyze and leverage the distribution carefully. Most importantly, [8] can make do with linear FE whereas we crucially need noisy linear FE[6]. Finally, we give many instantiations of NLinFE using bilinear maps and weak randomness generators as well as directly using new assumptions.

---

[6] We remark that a weak version of NLinFE in the bounded collusion setting was developed in an earlier version of [8] (see [7]) but was found to be redundant and was subsequently removed. The current, published version of [8] relies on LinFE alone. Our definition of NLinFE is significantly more general.

Independent and concurrent work. In an independent and concurrent work, Ananth, Jain and Sahai [9] also provide an approach to construct iO without multilinear maps. They rely on (subexponentially-secure) bilinear maps, LWE, block-locality 3 PRGs, and a new type of randomness generator, which they call perturbation resilient generator, denoted as ΔRG. Their techniques and overall construction are extremely different from ours. However, we find it very interesting that both works intersect in identifying a very similar new type of PRG as sufficing to fill the gap between assumptions we believe and iO. Their notion of ΔRG is almost exactly the same as the non-Boolean PRG that (is one of the types of PRGs) we identify – both notions require the generation of some noise N such that N is indistinguishable from $N + e$ for some bounded e. However, they only require a weak form of indistinguishability, namely the adversary is allowed to distinguish between N and $N + e$ with $1/\text{poly}$ probability in their case, whereas we require standard negligible distinguishing probability. They also provide a generic security amplification theorem, which transforms FE for NC_1 which satisfies this weak indistinguishability to FE with standard indistinguishability. Their security amplification theorem can be used black box in our construction to also rely on ΔRG (or the weaker notion of CNG) with similar weak indistinguishability.

We may also use their security amplification theorem to weaken the requirement on the underlying quadratic FE scheme so that it can be instantiated using existing constructions [16,68]. In more detail, we use quadratic FE to compute a noise term which must be natively superpolynomial in size to argue security. However, existing constructions of quadratic functional encryption schemes [16,68] perform decryption "brute force", by computing a discrete logarithm in the end, restricting the space of decryptable values to be polynomial in size. To align with known constructions of quadratic FE, we choose our flooding noise to be polynomial in size – this overcomes the above issue but results in $1/\text{poly}$ advantage to the adversary. This advantage can be made negligible by leveraging the security amplification theorem of [9] in a black box manner.

Aside from significantly different techniques, the final results obtained by the two works are also different. First, we define the abstraction of noisy linear FE, and bootstrap this to iO. The instantiation of noisy linear FE using bilinear maps and ΔRG is only one of the ways of achieving iO; we also define an even weaker type of PRG, namely correlated noise generators (denoted by CNG, please see Sect. 3) which suffices for iO. On the other hand, their security requirement from their randomness generators is significantly weaker – they only require $1/\text{poly}$ security as discussed above. Moreover, they provide a general security amplification theorem which we do not. The details of the techniques in the two works are vastly different: [9] define and instantiate the notion of tempered cubic encodings which do not have any analogue in our work. Also, we provide a direct construction of NLinFE from new lattice assumptions, which they do not. In our instantiation that uses bilinear maps, we rely on the SXDH assumption in the standard model, whereas they argue security in the generic bilinear map model.

We remark that in [9], the special PRG, namely ΔRG needs to be computable by a cubic polynomial that degree 1 in a public seed component and degree 2 in the secret seed components. In the present work, as well as [69], the special PRG output must be computed using quadratic polynomials.

Follow-up work. In a follow-up work[7], Lin and Matt [69] leverage our techniques to provide a different construction of iO from bilinear maps, LWE and weak pseudorandom objects, which they term *Pseudo Flawed-smudging Generators* (PFG). The high level structure of their construction is very similar to ours: they also use special properties of the FHE scheme of Brakerski and Vaikuntanathan [29] to split the functional computation into a deep public computation and a shallow private computation, the former being done by the decryptor in the clear, and the latter being performed in the exponent of a bilinear group using quadratic operations. To argue security, they must, similarly to us, perform noise flooding in the exponent. The main difference from our work is that the choice of noise in our setting is natively super-polynomial as discussed above, and we must use security amplification to make do with polynomial noise. On the other hand, [69] can make do with polynomial noise via their notion of Pseudo Flawed-smudging Generators (PFG). We remark that in contrast to the present work, [69] construct FE for NC_0 and then rely on randomized encodings to bootstrap this to FE for NC_1, as in prior work [68,71,72]. On the other hand, we use techniques from [29] in a non blackbox way to bootstrap all the way to NC_1 directly.

1.5   Our Techniques: Direct Construction of NLinFE

Next, we provide a direct construction of NLinFE based on new (untested) assumptions that strengthen ring learning with errors (RLWE) and NTRU. Our construction is quite different from known constructions and does not rely on multilinear maps or graded encoding schemes.

As discussed above, flooding correlated noise terms appears qualitatively easier than generating uniform pseudorandom variables. Recall that ℓ is the output length of the function and L is sublinear in ℓ. In this section, we discuss a method to provide L encodings of a seed vector β in a way that the decryptor can compute ℓ encodings of $\{g_i(\beta)\}_{i\in[\ell]}$ on the fly. The careful reader may suspect that we are going in circles: if we could compute encodings of $g_i(\beta)$ on the fly, could we not just compute encodings of $f_i(\mathbf{x})$ on the fly?

We resolve this circularity by arguing that the demands placed on noise in lattice based constructions are significantly weaker than the demands placed on messages. In particular, while computation on messages must maintain integrity, noise need only create some perturbation, the exact value of this perturbation is not important. Therefore if, in our attempt to compute an encoding $g_i(\beta)$, we instead compute an encoding of $g_i'(\beta')$, this still suffices for functionality. Intuitively $g_i'(\beta')$ will be a polynomial equation of β designed to flood $f_i(\mu)$.

---

[7] We had shared an earlier version with the authors several months ago.

In order to construct FE that supports the computation of noisy linear equations, we begin with an FE that supports computation of linear equations, denoted by LinFE, provided by [1,6]. All our constructions use the blueprint provided in [6] to support linear equations, and develop new techniques to add noise. In order to interface with the LinFE construction of [6], we are required to provide encodings of noise terms β such that:

1. Given encodings of β and g_i the decryptor may herself compute on these to construct an encoding of $g_i(\beta)$.
2. The functional encoding of $g_i(\beta)$ must have the form $h_{g,i} \cdot s + \text{noise}_{g,i} + g_i(\beta)$ where $h_{g,i}$ is computable by the key generator given only the public/secret key and the function value. In particular, $h_{g,i}$ should not depend on the ciphertext.

In order to compute efficiently on encodings of noise, we introduce a strengthening of the RLWE and NTRU assumptions. Let $R = \mathbb{Z}[x]/\langle x^n + 1 \rangle$ and $R_{p_1} = R/(p_1 \cdot R)$, $R_{p_2} = R/(p_2 \cdot R)$ for some primes $p_1 < p_2$. Then, the following assumptions are necessary (but not sufficient) for security of our scheme:

1. *We assume that the* NTRU *assumption holds even if multiple samples have the same denominator.* This assumption has been discussed by Peikert [76, 4.4.4], denoted as the *NTRU learning problem* and is considered a reasonable assumption. Moreover, this assumption is also used in the multilinear map constructions [44] and has never been subject to attack despite extensive cryptanalysis.
   In more detail, for $i \in \{1, \ldots, w\}$, sample f_{1i}, f_{2i} and g_1, g_2 from a discrete Gaussian over ring R. If g_1, g_2 are not invertible over R_{p_2}, resample. Set

   $$h_{1i} = \frac{f_{1i}}{g_1}, \quad h_{2i} = \frac{f_{2i}}{g_2} \in R_{p_2}$$

   We assume that the samples $\{h_{1i}, h_{2j}\}$ for $i, j \in [w]$ are indistinguishable from random. Note that NTRU requires the denominator to be chosen afresh for each sample, i.e. h_{1i} (resp. h_{2i}) should be constructed using denominator g_{1i} (resp. g_{2i}), for $i \in [w]$.
2. *We assume that* RLWE *with small secrets remains secure if the noise terms in* RLWE *samples live in some secret ideal.* In more detail, for $i \in [w]$, let $\widehat{\mathcal{D}}(\Lambda_2), \widehat{\mathcal{D}}(\Lambda_1)$ be discrete Gaussian distributions over lattices Λ_2 and Λ_1 respectively. Then, sample

   $$e_{1i} \leftarrow \widehat{\mathcal{D}}(\Lambda_2), \quad \text{where } \Lambda_2 \triangleq g_2 \cdot R. \text{ Let } e_{1i} = g_2 \cdot \xi_{1i} \in \text{small},$$

   $$e_{2i} \leftarrow \widehat{\mathcal{D}}(\Lambda_1), \quad \text{where } \Lambda_1 \triangleq g_1 \cdot R. \text{ Let } e_{2i} = g_1 \cdot \xi_{2i} \in \text{small},$$

   Above, small is a place-holder term that implies the norm of the relevant element can be bounded well below the modulus size, $p_2/5$, say. We use it for intuition when the precise bound on the norm is not important. Hence, for $i, j \in [w]$, it holds that:

   $$h_{1i} \cdot e_{2j} = f_{1i} \cdot \xi_{2j}, \quad h_{2j} \cdot e_{1i} = f_{2j} \cdot \xi_{1i} \in \text{small}$$

Now, sample small secrets t_1, t_2 and for $i \in [w]$, compute

$$d_{1i} = h_{1i} \cdot t_1 + p_1 \cdot e_{1i} \in R_{p_2}$$
$$d_{2i} = h_{2i} \cdot t_2 + p_1 \cdot e_{2i} \in R_{p_2}$$

We assume that the elements d_{1i}, d_{2j} for $i, j \in [w]$ are pseudorandom. The powerful property that this assumption provides is that the product of the samples $d_{1i} \cdot d_{2j}$ do not suffer from large cross terms for any $i, j \in [w]$ – since the error of one sample is chosen to annihilate with the large element of the other sample, the product yields a well behaved RLWE sample whose label is a product of the original labels. In more detail,

$$d_{1i} \cdot d_{2j} = \left(h_{1i} \cdot h_{2j} \right) \cdot (t_2 \, t_2) + p_1 \cdot \mathsf{noise}$$

where $\mathsf{noise} = p_1 \cdot \left(f_{1i} \cdot \xi_{2j} \cdot t_1 + f_{2j} \cdot \xi_{1i} \cdot t_2 + p_1 \cdot g_1 \cdot g_2 \cdot \xi_{1i} \cdot \xi_{2j} \right) \in \mathsf{small}$

If we treat each d_{1i}, d_{2j} as an RLWE sample, then we may use these samples to encode noise terms so that direct multiplication of samples is well behaved. Note that the noise terms we wish to compute on, are the messages encoded by the "RLWE sample" d_{1i}, hence d_{1i} must contain two kinds of noise: the noise required for RLWE security and the noise that behaves as the encoded message. This requires some care, but can be achieved by nesting these noise terms in different ideals as:

$$d_{1i} = h_{1i} \cdot t_1 + p_1 \cdot \tilde{e}_{1i} + p_0 \cdot e_{1i} \in R_{p_2}$$
$$d_{2i} = h_{2i} \cdot t_2 + p_1 \cdot \tilde{e}_{2i} + p_0 \cdot e_{2i} \in R_{p_2}$$

Here, $(p_1 \cdot \tilde{e}_{1i},\ p_1 \cdot \tilde{e}_{2i})$ behave as RLWE noise and $(p_0 \cdot e_{1i},\ p_0 \cdot e_{2i})$ behave as the encoded messages. Both $\tilde{e}_{1i}, e_{1i}$ as well as $\tilde{e}_{2i}, e_{2i}$ are chosen from special ideals as before. Now, we may compute quadratic polynomials on the encodings "on-the-fly" as $\sum_{i,j} d_{1i} d_{2j}$ to obtain a structured-noise RLWE sample whose label is computable by the key generator. If we treat this dynamically generated encoding as an RLWE encoding of correlated noise, then we can use this to add noise to the NLinFE decryption equation by generalizing techniques from [6]. The decryptor can, using all the machinery developed so far, recover $f_i(\mathbf{x}) + \mathsf{noise}_{f(\mathbf{x})} + \mathsf{noise}_{\mathsf{fld}\,i}$ where $\mathsf{noise}_{\mathsf{fld}\,i}$ is constructed as a quadratic polynomial of noise terms that live in special ideals.

Mixing Ideals. While it suffices for functionality to choose the correlated noise term as a polynomial evaluated on noise living in special ideals, the question of security is more worrisome. By using the new "on-the-fly" encodings of noise, the decryptor recovers noise which lives in special, secret ideals, and learning these ideals would compromise security. In more detail, the noise term we constructed above is a random linear combination of terms $(g_1 \cdot g_2)$, $\{f_{1i}\}_i$, $\{f_{2j}\}_j$, which must be kept secret for semantic security of d_{1i}, d_{2j} to hold. Indeed, if we oversimplify and assume that the attacker can recover noise terms that live in the ideal generated by $g_1 \cdot g_2$, then recovering $g_1 \cdot g_2$ from these terms becomes an instance of the *principal ideal problem* [40,50].

While the principal ideal problem has itself resisted efficient classical algorithms so far, things in our setting can be made significantly better by breaking the ideal structure using additional tricks. We describe these next.

1. *Mixing ideals.* Instead of computing a single set of pairs $\Big((h_{1i}, d_{1i}), (h_{2i}, d_{2i}) \Big)$, we now compute k of them, for some polynomial k fixed in advance. Thus, we sample f_{1i}^j, f_{2i}^j and g_1^j, g_2^j for $i \in \{1, \ldots, w\}$, $j \in \{1, \ldots, k\}$, where w, k are fixed polynomials independent of function output length ℓ, and set

$$h_{1i}^j = \frac{f_{1i}^j}{g_1^j}, \quad h_{2i}^j = \frac{f_{2i}^j}{g_2^j} \in R_{p_2}$$

The encoding of a noise term constructed corresponding to the $(i, j)^{th}$ monomial is $d_{ii'} = \sum_{j \in [k]} d_{1i} d_{2i'}$. Thus, the resultant noise term that gets added to the decryption equation looks like:

$$p_0 \cdot \left[\sum_{j \in [k]} \left(g_2^j \cdot g_1^j \cdot \left(p_0 \cdot (\xi_{1i}^j \cdot \xi_{2i'}^j) \right) + \left(f_{1i}^j \cdot \xi_{2i'}^j \cdot t_1 + f_{2i'}^j \cdot \xi_{1i}^\ell \cdot t_2 \right) \right) \right] \quad (1.2)$$

Thus, by adding together noise terms from multiple ideals, we "spread" it out over the entire ring rather than restricting it to a single secret ideal. Also, we note that it is only the higher degree noise terms that must live in special ideals; if the polynomial contains linear terms, these may be chosen from the whole ring without any restrictions. In more detail, above, we computed a noise term corresponding to a quadratic monomial which required multiplying and summing encodings. If we modify the above quadratic polynomial to include a linear term, we will need to add a degree 1 encoding into the above equation. The degree 1 encoding which does not participate in products, may encode noise that is chosen without any restrictions, further randomizing the resultant noise.

2. *Adding noise generated collectively by ciphertext and key.* Aside from computing polynomials over structured noise terms encoded in the ciphertext, we suggest an additional trick which forces noise terms into the decryption equation. These noise terms are quadratic polynomials where each monomial is constructed jointly by the encryptor and the key generator. This trick relies on the structure of the key and ciphertext in our construction. We describe the relevant aspects of the key and ciphertext here. The key for function f_i is a short vector $\mathbf{k}$ such that:

$$\langle \mathbf{w}, \ \mathbf{k} \rangle = u_{f_i}$$

where $\mathbf{w}$ is part of the master secret, and u_{f_i} is computed by the key generator. The encryptor provides an encoding

$$\mathbf{c} = \mathbf{w} \cdot s + p_1 \cdot \mathsf{noise}_0$$

As part of decryption, the decryptor computes $\langle \mathbf{k}, \mathbf{c} \rangle$ to obtain $u_{f_i} \cdot s + p_1 \cdot \langle \mathbf{k}, \mathsf{noise}_0 \rangle$. Moreover by running $\mathsf{Eval}_{\mathsf{CT}}(\mathcal{C}^1, \ldots, \mathcal{C}^d)$, she also obtains $u_{f_i} \cdot s + f(\mathbf{x}) + p_0 \cdot \mathsf{noise} + p_1 \cdot \mathsf{noise}'$. Subtracting these and reducing modulo p_1 and then modulo p_0 yields $f(\mathbf{x})$ as desired.

Intuitively, the structured noise computed above is part of the noise in the sample computed by $\mathsf{Eval}_{\mathsf{CT}}$, i.e. part of $\left(p_0 \cdot \mathsf{noise} + p_1 \cdot \mathsf{noise}' \right)$ in the notation above. Our next trick shows how to add noise to $\langle \mathbf{k}, \mathbf{c} \rangle$.

We modify KeyGen so that instead of choosing a single $\mathbf{k}$, it now chooses a pair $(\mathbf{k}^1, \mathbf{k}^2)$ such that:

$$\langle \mathbf{w}, \mathbf{k}^1 \rangle = u_{f_i} + p_0 \cdot \Delta^1 + p_1 \cdot \tilde{\Delta}^1$$
$$\langle \mathbf{w}, \mathbf{k}^2 \rangle = u_{f_i} + p_0 \cdot \Delta^2 + p_1 \cdot \tilde{\Delta}^2$$

Here, $\Delta^1, \Delta^2, \tilde{\Delta}^1, \tilde{\Delta}^2$ are discrete Gaussians sampled by the key generator *unique to the key for f_i*. Additionally, the encryptor splits $\mathbf{c}$ as:

$$\mathbf{c}_{01} = \mathbf{w} \cdot s_1 + p_1 \cdot \boldsymbol{\nu}_1$$
$$\mathbf{c}_{02} = \mathbf{w} \cdot s_2 + p_1 \cdot \boldsymbol{\nu}_2$$

where $s_1 + s_2 = s$ and s_1, s_2 are small, then,

$$\langle \mathbf{k}^1, \mathbf{c}_{01} \rangle + \langle \mathbf{k}^2, \mathbf{c}_{02} \rangle = u_{f_i} \cdot s + p_0 \cdot \left(\Delta^1 \cdot s_1 + \Delta^2 \cdot s_2 \right)$$
$$+ p_1 \cdot \left(\tilde{\Delta}^1 \cdot s_1 + \tilde{\Delta}^2 \cdot s_2 \right) + p_1 \cdot \mathsf{noise}$$

Thus, we forced the quadratic polynomial $p_0 \cdot \left(\Delta^1 \cdot s_1 + \Delta^2 \cdot s_2 \right) + p_1 \cdot \left(\tilde{\Delta}^1 \cdot s_1 + \tilde{\Delta}^2 \cdot s_2 \right)$ into the noise, where Δ^1, Δ^2 and $\tilde{\Delta}^1, \tilde{\Delta}^2$ are chosen by the key generator for the particular key request and the terms s_1 and s_2 are chosen by the encryptor unique to that ciphertext. Note that $\mathbf{w}$ can be hidden from the view of the adversary since it is not required for decryption, hence the adversary may not compute $\langle \mathbf{w}, \mathbf{k}^1 \rangle$, $\langle \mathbf{w}, \mathbf{k}^2 \rangle$ in the clear. For more details, please see the full version [3].

1.6   Related Work: Instantiation

To the best of our knowledge, all prior work constructing FE for degree $L \geq 3$ polynomials rely on either iO itself [45] or multilinear maps [48] or bilinear maps and weak pseudorandomness [12,68,71] as discussed above. Since our direct construction also makes use of NTRU lattice assumptions, we discuss here some high level differences between multilinear map based approaches and our approach.

Let us describe the main ideas behind the multilinear map construction of [44]. Our description follows the summary of [65]. Similarly to us, the authors consider the polynomial rings $R = \mathbb{Z}[x]/\langle x^n + 1 \rangle$ and $R_q = R/qR$. They generate a small secret $g \in R$ and set $I = \langle g \rangle$ to be the principal ideal over R generated by g. Next, they sample a uniform $z \in R_q$ which stays secret. The "plaintext" is an element of R/I, and is encoded via a division by z in R_q: to encode a coset of R/I, give element $[c/z]_q$ where c is an arbitrary small coset representative. Since g is hidden, the authors provide another public parameter y, which is an encoding of 1 and the encoding of the coset is chosen as $[e \cdot y]_q$ where e is a small coset representative. At level $i \neq 1$, the encoding has the form $[c/z^i]_q$. The encodings are additively and multiplicatively homomorphic, and for testing whether an element in the last level D (say) encodes 0, the authors provide a "zero test parameter" $p_{zt} = hz^D g^{-1} \bmod q$ where h is an element of norm approximately $\sqrt{q}$. The parameters are set so that if an element encodes 0, its product with this parameter is "small" otherwise it is "large".

Known attacks against multilinear maps and obfuscators operate on the following broad principle: (i) perform algebraic manipulations on some initial encodings, (ii) apply the zero test to each top level encoding, (iii) perform an algebraic computation on the results of the zero testing so as to obtain an element in the ideal $\langle g \rangle$, (iv) use this somehow break the scheme. Once an element in $\langle g \rangle$ is obtained, different attacks work in different ways, but in the "weak multilinear map model" [49], being able to recover an element in $\langle g \rangle$ is considered a successful attack. Thus, the unique element g must crucially be kept secret.

In our work, decryption of the FE scheme also results in a high degree polynomial containing secret elements $f_{1i} g_1, f_{2i} g_2$ for $i \in [\mathrm{poly}]$ along with fresh random elements per ciphertext. However, unlike the multilinear map template where there is a single secret g, there are a polynomial number of secret elements that play (what appears to us) qualitatively the same role as g in our construction. Moreover, these are "spread out" in the recovered polynomial which makes obtaining any term isolating any one secret element via algebraic manipulations seem improbable. Additionally, annihilation attacks [75] crucially make use of the fact that the unstructured elements that are unique to every encoding are *linear*, which assists in the computation of the annihilation polynomial. In contrast, unstructured elements in our recovered polynomial that are unique to the encoding are high degree and seem much harder to annihilate.

Our construction of NLinFE appears much simpler in design than the construction of multilinear map based obfuscators, we refer the reader to [75,78] for a clean description of an abstract obfuscator. Unlike current candidate obfuscators, we do not need to use straddling sets for handling mixed input attacks, eliminating a vulnerability recently exploited by [78]. This is because mixed input attacks seem very hard to launch in our construction, since we do not use branching programs and all parts of a given input are tied together using an LWE secret (albeit with a non-standard LWE assumption). Moreover, the function keys in our FE construction have a different structure than the ciphertext and do not seem amenable to mix and match attacks.

2   Noisy Linear Functional Encryption

We refer the reader to the full version [3] for definitions and preliminaries. In this section, we define the notion of noisy linear functional encryption. At a high level, noisy linear functional encryption is like regular linear functional encryption [1,6], except that the function value is recovered only up to some bounded additive error (which we informally call noise), and indistinguishability holds even if the challenge messages evaluated on all the function keys are only "approximately" equal, i.e. they differ by an additive term of low norm.

On the noise added by NLinFE. Recall from Sect. 1.3, that NLinFE must add a noise term $\mathsf{noise}_{\mathsf{fld}}$ which "floods" the noise term $\mathsf{noise}_{f(\mathbf{x})}$ for security. Also recall that in our setting, $\mathsf{noise}_{f(\mathbf{x})}$ is the noise term that results from evaluating the circuit f on the FHE encodings of $\mathbf{x}$. We denote by $\mathcal{D}$ the distribution from which the noise terms in FHE are sampled and by $\mathcal{F}$ the class of circuits that are used to compute on the FHE noise terms, resulting in $\mathsf{noise}_{f(\mathbf{x})}$. Thus, NLinFE must add a noise term that wipes out the leakage resulting from the adversary learning $\mathsf{noise}_{f(\mathbf{x})}$. In general, $\mathcal{F}$ represents the class of functions that NLinFE can be used to bootstrapped to. In more detail, if $\mathcal{F} = \mathsf{NC}_1$, NLinFE enables bootstrapping to FE for NC_1, whereas when $\mathcal{F} = \mathsf{NC}_0$, it enables bootstrapping to FE for NC_0.

Definition of NLinFE. In our constructions, R is a ring, instantiated either as the ring of integers $\mathbb{Z}$ or the ring of polynomials $\mathbb{Z}[x]/f(x)$ where $f(x) = x^n + 1$ for n a power of 2. We let $R_q = R/qR$ for some prime q. Let $\mathcal{D}$ be a distribution over R, $\mathcal{F}$ be a class of functions $\mathcal{F} : R^\ell \to R$ and $B \in \mathbb{R}^+$ a bounding value on the norm of the decryption error. In general, we require $B << q$. We are ready for the formal definition.

Definition 2.1. *A* $(\mathcal{D}, \mathcal{F}, B)$-*noisy linear functional encryption scheme* FE *is a tuple* FE $=$ (FE.Setup, FE.Keygen, FE.Enc, FE.Dec) *of four probabilistic polynomial-time algorithms with the following specifications:*

- FE.Setup($1^\kappa, R_q^\ell$) *takes as input the security parameter* κ *and the space of message and function vectors* R_q^ℓ *and outputs the public key and the master secret key pair* (PK, MSK).
- FE.Keygen(MSK, $\mathbf{v}$) *takes as input the master secret key* MSK *and a vector* $\mathbf{v} \in R_q^\ell$ *and outputs the secret key* $\mathsf{SK}_{\mathbf{v}}$.
- FE.Enc(PK, $\mathbf{z}$) *takes as input the public key* PK *and a message* $\mathbf{z} \in R_q^\ell$ *and outputs the ciphertext* $\mathsf{CT}_{\mathbf{z}}$.
- FE.Dec($\mathsf{SK}_{\mathbf{v}}, \mathsf{CT}_{\mathbf{z}}$) *takes as input the secret key of a user* $\mathsf{SK}_{\mathbf{v}}$ *and the ciphertext* $\mathsf{CT}_{\mathbf{z}}$, *and outputs* $y \in R_q \cup \{\perp\}$.

Definition 2.2 (Approximate Correctness). *A noisy linear functional encryption scheme* FE *is correct if for all* $\mathbf{v}, \mathbf{z} \in R_q^\ell$,

$$\Pr\left[\begin{array}{l}(\mathsf{PK}, \mathsf{MSK}) \leftarrow \mathsf{FE.Setup}(1^\kappa); \\ \mathsf{FE.Dec}\Big(\mathsf{FE.Keygen}(\mathsf{MSK}, \mathbf{v}), \mathsf{FE.Enc}(\mathsf{PK}, \mathbf{z})\Big) = \langle \mathbf{v}, \mathbf{z} \rangle + \mathsf{noise}_{\mathsf{fld}}\end{array}\right] = 1 - \mathrm{negl}(\kappa)$$

where $\mathsf{noise}_{\mathsf{fld}} \in R$ *with* $\|\mathsf{noise}_{\mathsf{fld}}\| \leq B$ *and the probability is taken over the coins of* FE.Setup, FE.Keygen, *and* FE.Enc.

Security. Next, we define the notion of Noisy-IND security.

Definition 2.3 (Noisy-INDSecurity Game). *We define the security game between the challenger and adversary as follows:*

1. **Public Key:** *Challenger returns* PK *to the adversary.*
2. **Pre-Challenge Queries:** Adv *may adaptively request keys for any functions* $\mathbf{v}_i \in R_q^\ell$ *for* $i \in [k]$ *for some polynomial* k. *Along with* $\mathbf{v}_i$, Adv *also submits a function* $f_i \in \mathcal{F}$ *which must satisfy some constraints discussed later. In response,* Adv *is given the corresponding keys* SK$(\mathbf{v}_i)$.
3. **Challenge Ciphertexts:** Adv *outputs the challenge message pairs* $(\mathbf{z}_0^i, \mathbf{z}_1^i) \in R_q^\ell \times R_q^\ell$ *for* $i \in [Q]$, *where* Q *is some polynomial, to the challenger. Along with the challenger pair* $(\mathbf{z}_0^i, \mathbf{z}_1^i)$, *the adversary also outputs* $\boldsymbol{\mu}^i \leftarrow \mathcal{D}^\ell$, *which must satisfy some constraints discussed later. The challenger chooses a random bit* b, *and returns the ciphertexts* $\{\mathsf{CT}(\mathbf{z}_b^i)\}_{i \in [Q]}$.
4. **Post-Challenge Queries:** Adv *may request additional keys for functions of its choice and is given the corresponding keys.* Adv *may also output additional challenge message pairs which are handled as above.*
5. **Guess.** Adv *outputs a bit* b', *and succeeds if* $b' = b$.

The advantage of Adv *is the absolute value of the difference between the adversary's success probability and* $1/2$.

   In the selective *game, the adversary must announce the challenge in the first step, before receiving the public key. In the* semi-adaptive *game, the adversary must announce the challenge after seeing the public key but before making any key requests.*

   We next define the notion of admissible adversary.

Definition 2.4 (Admissible Adversary). *We say an adversary is* $\mathcal{F}$-*admissible if for any pair of challenge messages* $\mathbf{z}_0, \mathbf{z}_1 \in R_q^\ell$ *and its corresponding vector* $\boldsymbol{\mu}^i \leftarrow \mathcal{D}^\ell$, *any queried key* $\mathbf{v}_i \in R_q^\ell$ *and corresponding function* $f_i \in \mathcal{F}$, *it holds that* $\langle \mathbf{v}_i, \mathbf{z}_0 - \mathbf{z}_1 \rangle = f_i(\boldsymbol{\mu})$.

Definition 2.5 (Noisy-IND security). *A* $(\mathcal{D}, \mathcal{F}, B)$ *noisy linear FE scheme* NLinFE *is Noisy-IND secure if for all* $\mathcal{F}$-*admissible probabilistic polynomial-time adversaries* Adv, *the advantage of* Adv *in the* Noisy-IND *security game is negligible in the security parameter* κ.

Remark 2.6. In most of our constructions of NLinFE, the precise distribution of $f_i(\boldsymbol{\mu})$ will not be important, and it will suffice that $\|f_i(\boldsymbol{\mu})\| < B_{\mathsf{leak}}$ for some bound B_{leak}, to perform the noise flooding. While it may appear strange to restrict the adversary to choosing messages and functions that satisfy a strong constraint such as the above, such a restricted adversary suffices for our main application in Sect. 4.

3 Broader Classes of Randomness Generators

In this section we define broader classes of randomness generators that suffice for our bootstrapping.

3.1 Correlated Noise Generators

In this section we define the notion of a correlated noise generator, which we denote by CNG. We denote by R the ring of integers $\mathbb{Z}$ or the ring of polynomials $\mathbb{Z}[x]/f(x)$ where $f(x) = x^d + 1$. Let $\mathcal{D}_1$ be a distribution over R and $\mathcal{F} : R^w \to R$ be a set of deterministic functions. Let $\mathsf{Dom}_{\mathsf{Cng}}, \mathsf{Rg}_{\mathsf{Cng}}$ be finite subsets of R, let $\mathcal{G} : \mathsf{Dom}_{\mathsf{Cng}}^n \to \mathsf{Rg}_{\mathsf{Cng}}^m$ be a family of deterministic functions and $\mathcal{D}_2$ be a distribution over $\mathsf{Dom}_{\mathsf{Cng}}$. We require that n be linear in w, i.e. $n = O(w, \mathrm{poly}(\kappa))$.

Definition 3.1 ($(\mathcal{D}_1, \mathcal{F})$-Correlated Noise Generator). *We say that $(\mathcal{D}_2, \mathcal{G})$ is a $(\mathcal{D}_1, \mathcal{F})$- Correlated Noise Generator (CNG) if the advantage of any P.P.T adversary $\mathcal{A}$ is negligible in the following game:*

1. *Challenger chooses n i.i.d samples $\boldsymbol{\beta} \leftarrow \mathcal{D}_2^n$.*
2. *The adversary $\mathcal{A}$ does the following:*
   (a) *It chooses m functions $f_1, \ldots, f_m \in \mathcal{F}$.*
   (b) *It samples $\boldsymbol{\mu} \leftarrow \mathcal{D}_1^w$.*
   (c) *It returns $(\{f_i\}_{i \in [m]}, \boldsymbol{\mu})$ to the challenger.*
3. *The challenger chooses m functions $G_1, \ldots, G_m \in \mathcal{G}$. It tosses a coin b. If $b = 0$, it returns $\{f_i(\boldsymbol{\mu}) + G_i(\boldsymbol{\beta})\}_{i \in [m]}$, else it returns $\{G_i(\boldsymbol{\beta})\}_{i \in [m]}$.*
4. *The adversary outputs a guess for the bit b and wins if correct.*

We will refer to $\boldsymbol{\beta}$ as the seed of the CNG. We say that an CNG has polynomial stretch if $m = n^{1+c}$ for some constant $c > 0$.

3.2 Non Boolean Pseudorandom Generators

As discussed in Sect. 1, in prior work [68,71], Boolean PRGs were required in order to compute the binary randomness needed for constructing randomizing polynomials. In our case, the PRG output must satisfy a much weaker property. Say we can bound $\|f_i(\boldsymbol{\mu})\| \leq \epsilon$ for $i \in [m]$. Then we require the PRG output $G_i(\boldsymbol{\beta})$ to computationally flood $f_i(\boldsymbol{\mu})$ for $i \in [m]$, i.e., $G_i(\boldsymbol{\beta}) + f_i(\boldsymbol{\mu}) \stackrel{c}{\approx} G_i(\boldsymbol{\beta}), \ \forall i \in [m]$.

4 Functional Encryption for NC_1

In this section, we construct a functional encryption scheme for NC_1, denoted by FeNC_1, from a correlated noise generator CNG, the RLWE assumption and a noisy linear functional encryption scheme NLinFE.

Background. Let $R = \mathbb{Z}[x]/(\phi)$ where $\phi = x^d + 1$ and d is a power of 2. Let $R_p \triangleq R/pR$ for any large prime p satisfying $p = 1 \mod 2n$.

We consider arithmetic circuits $\mathcal{F} : R_{p_0}^w \to R_{p_0}$ of depth d, consisting of alternate addition and multiplication layers. For circuits with long output, say ℓ, we consider ℓ functions, one computing each output bit. For $k \in [d]$, layer k of the circuit is associated with a modulus p_k. For an addition layer at level k, the modulus p_k will be the same as the previous modulus p_{k-1}; for a multiplication layer at level k, we require $p_k > p_{k-1}$. Thus, we get a tower of moduli $p_1 < p_2 = p_3 < p_4 = \ldots < p_d$. We define encoding functions $\mathcal{E}^k$ for $k \in [d]$ such that $\mathcal{E}^k : R_{p_{k-1}} \to R_{p_k}$. The message space of the scheme FeNC_1 is R_{p_0}.

At level k, the encryptor will provide L^k encodings, denoted by $\mathcal{C}^k$, for some $L^k = O(2^k)$. For $i \in [L^k]$ we define

$$\mathcal{E}^k(y_i) = u_i^k \cdot s + p_{k-1} \cdot \eta_i^k + y_i.$$

Here $u_i^k \in R_{p_k}$ is called the "label" or "public key" of the encoding, η_i^k is noise chosen from some distribution χ_k, $s \leftarrow R_{p_1}$ is the RLWE secret, and $y_i \in R_{p_{k-1}}$ is the message being encoded. We will refer to $\mathcal{E}^k(y_i)$ as the *Regev encoding* of y_i. We denote:

$$\mathsf{PK}\big(\mathcal{E}^k(y_i)\big) \triangleq u_i^k, \quad \mathsf{Nse}(\mathcal{C}^k) \triangleq p_{k-1} \cdot \eta_i^k$$

The messages encoded in level k encodings $\mathcal{C}^k$ are denoted by $\mathcal{M}^k$.

Agrawal and Rosen [8] show that at level k, the decryptor is able to compute a Regev encoding of functional message $f^k(\mathbf{x})$ where f^k is the circuit f restricted to level k. Formally:

Theorem 4.1 [8]. *There exists a set of encodings $\mathcal{C}^i$ for $i \in [d]$, such that:*

1. **Encodings have size sublinear in circuit.** $\forall i \in [d]$ $|\mathcal{C}^i| = O(2^i)$.
2. **Efficient public key and ciphertext evaluation algorithms.** *There exist efficient algorithms $\mathsf{Eval}_{\mathsf{PK}}$ and $\mathsf{Eval}_{\mathsf{CT}}$ so that for any circuit f of depth d, if $\mathsf{PK}_f \leftarrow \mathsf{Eval}_{\mathsf{PK}}(\mathsf{PK}, f)$ and $\mathsf{CT}_{(f(\mathbf{x}))} \leftarrow \mathsf{Eval}_{\mathsf{CT}}(\bigcup_{i \in [d]} \mathcal{C}^i, f)$, then $\mathsf{CT}_{(f(\mathbf{x}))}$ is a "Regev encoding" of $f(\mathbf{x})$ under public key PK_f. Specifically, for some LWE secret s, we have:*

$$\mathsf{CT}_{(f(\mathbf{x}))} = \mathsf{PK}_f \cdot s + p_{d-1} \cdot \eta_f^{d-1} + \mu_{f(\mathbf{x})} + f(\mathbf{x}) \tag{4.1}$$

where $p_{d-1} \cdot \eta_f^{d-1}$ is RLWE noise and $\mu_{f(\mathbf{x})} + f(\mathbf{x})$ is the desired message $f(\mathbf{x})$ plus some noise $\mu_{f(\mathbf{x})}$[8].
3. **Ciphertext and public key structure.** *The structure of the functional ciphertext is as:*

$$\mathsf{CT}_{f(\mathbf{x})} = \mathsf{Poly}_f(\mathcal{C}^1, \ldots, \mathcal{C}^{d-1}) + \langle \mathsf{Lin}_f, \mathcal{C}^d \rangle \tag{4.2}$$

---

[8] Here $\mu_{f(\mathbf{x})}$ is clubbed with the message $f(\mathbf{x})$ rather than the RLWE noise $p_{d-1} \cdot \eta_f^{d-1}$ since $\mu_{f(\mathbf{x})} + f(\mathbf{x})$ is what will be recovered after decryption of $\mathsf{CT}_{f(\mathbf{x})}$.

where $\mathsf{Poly}_f(\mathcal{C}^1, \ldots, \mathcal{C}^{d-1}) \in R_{p_{d-1}}$ *is a degree* d *polynomial and* $\mathsf{Lin}_f \in R_{p_d}^{L_d}$ *computed by* $\mathsf{Eval}_{\mathsf{PK}}(\mathsf{PK}, f)$ *is a linear function. We also have*

$$f(\mathbf{x}) + \mu_{f(\mathbf{x})} = \mathsf{Poly}_f(\mathcal{C}^1, \ldots, \mathcal{C}^{d-1}) + \langle \mathsf{Lin}_f, \mathcal{M}^d \rangle \tag{4.3}$$

where $\mathcal{M}^d$ *are the messages encoded in* $\mathcal{C}^d$.
The public key for the functional ciphertext is structured as:

$$\mathsf{PK}(\mathsf{CT}_{f(\mathbf{x})}) = \left\langle \mathsf{Lin}_f, \left(\mathsf{PK}(\mathcal{C}_1^d), \ldots, \mathsf{PK}(\mathcal{C}_{L_d}^d) \right) \right\rangle \tag{4.4}$$

4. **Noise Structure.** *The term* $\mu_{f(\mathbf{x})}$ *is the noise resulting from FHE evaluation of function* f *on the encodings of* $\mathbf{x}$. *Moreover,* $\mu_{f(\mathbf{x})}$ *can be expressed as a linear combination of noise terms, each noise term being a multiple of* p_k *for* $k \in \{0, \ldots, d-1\}$.

The Encodings. The encodings $\mathcal{C}^k$ for $k \in [d]$ are defined recursively as:

1. $\mathcal{C}^1 \triangleq \{\mathcal{E}^1(x_i), \mathcal{E}^1(s)\}$
2. If k is a multiplication layer, $\mathcal{C}^k = \{\mathcal{E}^k(\mathcal{C}^{k-1}), \mathcal{E}^k(\mathcal{C}^{k-1} \cdot s), \mathcal{E}^k(s^2)\}$[9]. If k is an addition layer, let $\mathcal{C}^k = \mathcal{C}^{k-1}$.

We will use NLinFE to enable the decryptor to compute $\langle \mathsf{Lin}_f, \mathcal{M}^d \rangle + G_f(\boldsymbol{\beta})$ where $G_f(\boldsymbol{\beta})$ is a large noise term that floods functional noise $\mu_{f(\mathbf{x})}$. She may then compute $\mathsf{Poly}_f(\mathcal{C}^1, \ldots, \mathcal{C}^{d-1})$ herself and by Eq. 4.3 recover $f(\mathbf{x}) + \mu_{f(\mathbf{x})} + G_f(\boldsymbol{\beta})$.

4.1   Construction

Next, we proceed to describe the construction. The construction below supports a single function of output length ℓ or equivalently ℓ functions with constant size output (however, in this case ℓ must be fixed in advance and input to all algorithms).

$\mathsf{FeNC}_1.\mathsf{Setup}(1^\kappa, 1^w, 1^d)$: Upon input the security parameter κ, the message dimension w, and the circuit depth d, do:

1. For $k \in [d]$, let $L_k = |\mathcal{C}^k|$ where $\mathcal{C}^k$ is as defined in Theorem 4.1. For $k \in [d-1]$, $i \in [L_k]$, choose uniformly random $u_i^k \in R_{p_k}$. Denote $\mathbf{u}^k = (u_i^k) \in R_{p_k}^{L_k}$.
2. Invoke $\mathsf{NLinFE}.\mathsf{Setup}(1^\kappa, 1^{L_d}, p_d)$ to obtain $\mathsf{PK} = \mathsf{NLinFE}.\mathsf{PK}$ and $\mathsf{MSK} = \mathsf{NLinFE}.\mathsf{MSK}$.
3. Sample a CNG seed $\boldsymbol{\beta} \leftarrow \mathcal{D}_{\mathsf{seed}}^n$. Sample $t_0, \ldots, t_{L_d} \leftarrow R_{p_{d-1}}$ and let $\mathbf{t} = (t_0, \ldots, t_{L_d})$.
4. Output PK $= (\mathbf{u}^1, \ldots, \mathbf{u}^{d-1}, \mathsf{NLinFE}.\mathsf{PK})$ and $\mathsf{MSK} = (\mathsf{NLinFE}.\mathsf{MSK}, \boldsymbol{\beta}, \mathbf{t})$.

---

[9] Here, we use the same secret s for all RLWE samples, but this is for ease of exposition – it is possible to have a different secret at each level so that circular security need not be assumed. We do not describe this extension here.

$FeNC_1.KeyGen(MSK, f)$: Upon input the master secret key $NLinFE.MSK$, CNG seed β and a circuit $f : R_{p_0}^w \rightarrow R_{p_0}$ [10] of depth d, do:

1. Let $Lin_f \leftarrow Eval_{PK}(PK, f) \in R_{p_d}^{L_d}$ as described in Eq. 4.4.
2. Let G_f denote the CNG chosen corresponding to function f as described in the full version [3].
3. Compute $key_f = \langle Lin_f, t\rangle - G_f(\beta)$.
4. Let $SK_f = NLinFE.KeyGen(MSK, Lin_f\|key_f)$.

$FeNC_1.Enc(\mathbf{x}, PK)$: Upon input the public key and the input $\mathbf{x}$, do:

1. Compute the encodings $\mathcal{C}^k$ for $k \in [d-1]$ as defined in Theorem 4.1. Denote by s the RLWE secret used for these encodings.
2. Define $\mathcal{M}^d = (\mathcal{C}^{d-1}, \ \mathcal{C}^{d-1} \cdot s, \ s^2) \in R_{p_d}^{L_d}$. Compute $\mathcal{C}^d = NLinFE.Enc(NLinFE.PK, \mathcal{M}^d)$.
3. Output $CT_\mathbf{x} = (\{\mathcal{C}^k\}_{k\in[d]})$.

$FeNC_1.Dec(PK, CT_\mathbf{x}, SK_f)$: Upon input a ciphertext $CT_\mathbf{x}$ for vector $\mathbf{x}$, and a secret key SK_f for circuit f, do:

1. Compute $CT_{f(\mathbf{x})} = Eval_{CT}(\{\mathcal{C}^k\}_{k\in[d-1]}, f)$. Express $CT_{f(\mathbf{x})} = Poly_f(\mathcal{C}^1, \ldots, \mathcal{C}^{d-1}) + \langle Lin_f, \mathcal{C}^d\rangle$ as described in Eq. 4.2.
2. Compute $NLinFE.Dec(SK_f, \mathcal{C}^d)$ to obtain $\langle Lin_f, \mathcal{M}^d\rangle + \eta_f$ for some noise η_f added by NLinFE.
3. Compute $Poly_f(\mathcal{C}^1, \ldots, \mathcal{C}^{d-1}) + \langle Lin_f, \mathcal{M}^d\rangle + \eta_f \mod p_d \mod p_{d-1}, \ldots,$ $\mod p_0$ and output it.

4.2 Correctness

Correctness follows from correctness of $Eval_{PK}$, $Eval_{CT}$ and NLinFE. We have by correctness of $Eval_{PK}$, $Eval_{CT}$ that:

$$CT_{f(\mathbf{x})} = \langle Lin_f, \mathcal{C}^d\rangle + Poly_f(\mathcal{C}^1, \ldots, \mathcal{C}^{d-1})$$

$$Poly_f(\mathcal{C}^1, \ldots, \mathcal{C}^{d-1}) + NLinFE.Dec(SK_f, \mathcal{C}^d) = Poly_f(\mathcal{C}^1, \ldots, \mathcal{C}^{d-1}) + \langle Lin_f, \mathcal{M}^d\rangle + \eta_f$$

$$= f(\mathbf{x}) + \mu_{f(\mathbf{x})} + \eta_f \text{ by theorem 4.1}$$

$$= f(\mathbf{x}) \mod p_d \mod p_{d-1}, \ldots, \mod p_0$$

where the last step follows since:

1. $\mu_{f(\mathbf{x})}$ and η_f are linear combinations of noise terms, each noise term being a multiple of p_k for $k \in \{0, \ldots, d-1\}$. For details regarding the structure of the noise terms $\mu_{f(\mathbf{x})}$. The noise term η_f is chosen by NLinFE to flood $\mu_{f(\mathbf{x})}$ as discussed in Sect. 2, and hence is also a linear combination of noise terms, each noise term being a multiple of p_k for $k \in \{0, \ldots, d-1\}$.
2. We set the parameters so that p_i is sufficiently larger than p_{i-1} for $i \in [d]$, so that over R_{p_i}, any error term which is a multiple of p_{i-1} may be removed by reducing modulo p_{i-1}. Thus the successive computation of $\mod p_i$, for $i = d, \ldots, 0$, results in $f(\mathbf{x}) \mod p_0$ in the end.

---

[10] We will let the adversary request ℓ functions.

4.3   Efficiency and Security

The size of the ciphertext is $|\bigcup_{k\in[d-1]} \mathcal{C}^k| + |\mathsf{NLinFE.CT}(\mathcal{M}^d)|$. Note that $|\bigcup_{k\in[d-1]} \mathcal{C}^k| = O(2^d)$ and $|\mathcal{M}^d| = O(2^d)$ by Theorem 4.1. All our constructions of NLinFE will have compact ciphertext, hence the ciphertext of the above scheme is also sublinear in circuit size. We refer the reader to the full version [3] for our constructions of NLinFE.

In the full version [3], we prove the following security theorem:

Theorem 4.2. *Assume the noisy linear FE scheme* NLinFE *satisfies semi-adaptive indistinguishability based security as in Definition 2.5 and that G is a secure* CNG *as defined in Definition 3.1. Then, the construction* $\mathsf{FeNC_1}$ *achieves semi-adaptive indistinguishability based security.*

References

1. Abdalla, M., Bourse, F., De Caro, A., Pointcheval, D.: Simple Functional Encryption Schemes for Inner Products. In: Katz, J. (ed.) PKC 2015. LNCS, vol. 9020, pp. 733–751. Springer, Heidelberg (2015). https://doi.org/10.1007/978-3-662-46447-2_33

2. Agrawal, S.: Stronger security for reusable garbled circuits, general definitions and attacks. In: Katz, J., Shacham, H. (eds.) CRYPTO 2017, Part I. LNCS, vol. 10401, pp. 3–35. Springer, Cham (2017). https://doi.org/10.1007/978-3-319-63688-7_1

3. Agrawal, S.: Indistinguishability obfuscation without multilinear maps: new methods for bootstrapping and instantiation. Cryptology ePrint Archive, Report 2018 (2018)

4. Agrawal, S., Boneh, D., Boyen, X.: Efficient lattice (H)IBE in the standard model. In: Gilbert, H. (ed.) EUROCRYPT 2010. LNCS, vol. 6110, pp. 553–572. Springer, Heidelberg (2010). https://doi.org/10.1007/978-3-642-13190-5_28

5. Agrawal, S., Freeman, D.M., Vaikuntanathan, V.: Functional encryption for inner product predicates from learning with errors. In: Lee, D.H., Wang, X. (eds.) ASIACRYPT 2011. LNCS, vol. 7073, pp. 21–40. Springer, Heidelberg (2011). https://doi.org/10.1007/978-3-642-25385-0_2

6. Agrawal, S., Libert, B., Stehle, D.: Fully secure functional encryption for linear functions from standard assumptions, and applications. In: Crypto (2016)

7. Agrawal, S., Rosen, A.: Online offline functional encryption for bounded collusions. Eprint/2016 (2016)

8. Agrawal, S., Rosen, A.: Functional encryption for bounded collusions, revisited. In: Kalai, Y., Reyzin, L. (eds.) TCC 2017, Part I. LNCS, vol. 10677, pp. 173–205. Springer, Cham (2017). https://doi.org/10.1007/978-3-319-70500-2_7

9. Ananth, P., Jain, A., Sahai, A.: Indistinguishability obfuscation without multilinear maps: iO from LWE, bilinear maps, and weak pseudorandomness. Cryptology ePrint Archive, Report 2018/615 (2018)

10. Ananth, P., Jain, A.: Indistinguishability obfuscation from compact functional encryption. In: Gennaro, R., Robshaw, M. (eds.) CRYPTO 2015, Part I. LNCS, vol. 9215, pp. 308–326. Springer, Heidelberg (2015). https://doi.org/10.1007/978-3-662-47989-6_15

11. Ananth, P., Jain, A., Sahai, A.: Achieving compactness generically: indistinguishability obfuscation from non-compact functional encryption. IACR Cryptol. ePrint Arch. **2015**, 730 (2015)

12. Ananth, P., Sahai, A.: Projective arithmetic functional encryption and indistinguishability obfuscation from degree-5 multilinear maps. In: Coron, J.-S., Nielsen, J.B. (eds.) EUROCRYPT 2017, Part I. LNCS, vol. 10210, pp. 152–181. Springer, Cham (2017). https://doi.org/10.1007/978-3-319-56620-7_6

13. Apon, D., Döttling, N., Garg, S., Mukherjee, P.: Cryptanalysis of indistinguishability obfuscations of circuits over ggh13. eprint 2016 (2016)

14. Applebaum, B., Ishai, Y., Kushilevitz, E.: Computationally private randomizing polynomials and their applications. Comput. Complex. **15**(2), 115–162 (2006)

15. Applebaum, B., Ishai, Y., Kushilevitz, E.: How to garble arithmetic circuits. In: IEEE 52nd Annual Symposium on Foundations of Computer Science, FOCS 2011, Palm Springs, 22–25 October 2011, pp. 120–129 (2011)

16. Baltico, C.E.Z., Catalano, D., Fiore, D., Gay, R.: Practical functional encryption for quadratic functions with applications to predicate encryption. In: Katz, J., Shacham, H. (eds.) CRYPTO 2017, Part I. LNCS, vol. 10401, pp. 67–98. Springer, Cham (2017). https://doi.org/10.1007/978-3-319-63688-7_3

17. Barak, B., et al.: On the (Im)possibility of obfuscating programs. In: Kilian, J. (ed.) CRYPTO 2001. LNCS, vol. 2139, pp. 1–18. Springer, Heidelberg (2001). https://doi.org/10.1007/3-540-44647-8_1

18. Barak, B., Brakerski, Z., Komargodski, I., Kothari, P.K.: Limits on low-degree pseudorandom generators (or: sum-of-squares meets program obfuscation). In: Nielsen, J.B., Rijmen, V. (eds.) EUROCRYPT 2018, Part II. LNCS, vol. 10821, pp. 649–679. Springer, Cham (2018). https://doi.org/10.1007/978-3-319-78375-8_21

19. Bethencourt, J., Sahai, A., Waters, B.: Ciphertext-policy attribute-based encryption. In: IEEE Symposium on Security and Privacy, pp. 321–334 (2007)

20. Bitansky, N., Garg, S., Lin, H., Pass, R., Telang, S.: Succinct randomized encodings and their applications. In: Proceedings of the Forty-Seventh Annual ACM on Symposium on Theory of Computing, STOC 2015, Portland, 14–17 June 2015, pp. 439–448 (2015)

21. Bitansky, N., Nishimaki, R., Passelègue, A., Wichs, D.: From cryptomania to obfustopia through secret-key functional encryption. In: Hirt, M., Smith, A. (eds.) TCC 2016, Part II. LNCS, vol. 9986, pp. 391–418. Springer, Heidelberg (2016). https://doi.org/10.1007/978-3-662-53644-5_15

22. Bitansky, N., Paneth, O., Wichs, D.: Perfect structure on the edge of chaos. In: Kushilevitz, E., Malkin, T. (eds.) TCC 2016, Part I. LNCS, vol. 9562, pp. 474–502. Springer, Heidelberg (2016). https://doi.org/10.1007/978-3-662-49096-9_20

23. Bitansky, N., Vaikuntanathan, V.: Indistinguishability obfuscation from functional encryption. FOCS 2015, 163 (2015). http://eprint.iacr.org/2015/163

24. Boneh, D., Franklin, M.: Identity-based encryption from the weil pairing. In: Kilian, J. (ed.) CRYPTO 2001. LNCS, vol. 2139, pp. 213–229. Springer, Heidelberg (2001). https://doi.org/10.1007/3-540-44647-8_13

25. Boneh, D., et al.: Fully key-homomorphic encryption, arithmetic circuit abe and compact garbled circuits. In: Nguyen, P.Q., Oswald, E. (eds.) EUROCRYPT 2014. LNCS, vol. 8441, pp. 533–556. Springer, Heidelberg (2014). https://doi.org/10.1007/978-3-642-55220-5_30

26. Boneh, D., Waters, B.: Conjunctive, subset, and range queries on encrypted data. In: Vadhan, S.P. (ed.) TCC 2007. LNCS, vol. 4392, pp. 535–554. Springer, Heidelberg (2007). https://doi.org/10.1007/978-3-540-70936-7_29

27. Boyen, X., Waters, B.: Anonymous hierarchical identity-based encryption (without random oracles). In: Dwork, C. (ed.) CRYPTO 2006. LNCS, vol. 4117, pp. 290–307. Springer, Heidelberg (2006). https://doi.org/10.1007/11818175_17

28. Brakerski, Z., Tsabary, R., Vaikuntanathan, V., Wee, H.: Private constrained PRFs (and More) from LWE. In: Kalai, Y., Reyzin, L. (eds.) TCC 2017, Part I. LNCS, vol. 10677, pp. 264–302. Springer, Cham (2017). https://doi.org/10.1007/978-3-319-70500-2_10

29. Brakerski, Z., Vaikuntanathan, V.: Fully homomorphic encryption from ring-LWE and security for key dependent messages. In: Rogaway, P. (ed.) CRYPTO 2011. LNCS, vol. 6841, pp. 505–524. Springer, Heidelberg (2011). https://doi.org/10.1007/978-3-642-22792-9_29

30. Canetti, R., Holmgren, J., Jain, A., Vaikuntanathan, V.: Succinct garbling and indistinguishability obfuscation for RAM programs. In: Proceedings of the Forty-Seventh Annual ACM on Symposium on Theory of Computing, STOC 2015, Portland, 14–17 June 2015, pp. 429–437 (2015)

31. Canetti, R., Lin, H., Tessaro, S., Vaikuntanathan, V.: Obfuscation of probabilistic circuits and applications. In: Dodis, Y., Nielsen, J.B. (eds.) TCC 2015, Part II. LNCS, vol. 9015, pp. 468–497. Springer, Heidelberg (2015). https://doi.org/10.1007/978-3-662-46497-7_19

32. Cash, D., Hofheinz, D., Kiltz, E., Peikert, C.: Bonsai trees, or how to delegate a lattice basis. In: Gilbert, H. (ed.) EUROCRYPT 2010. LNCS, vol. 6110, pp. 523–552. Springer, Heidelberg (2010). https://doi.org/10.1007/978-3-642-13190-5_27

33. Cheon, J.H., Han, K., Lee, C., Ryu, H., Stehlé, D.: Cryptanalysis of the multilinear map over the integers. In: Oswald, E., Fischlin, M. (eds.) EUROCRYPT 2015, Part I. LNCS, vol. 9056, pp. 3–12. Springer, Heidelberg (2015). https://doi.org/10.1007/978-3-662-46800-5_1

34. Cheon, J.H., Fouque, P.-A., Lee, C., Minaud, B., Ryu, H.: Cryptanalysis of the new CLT multilinear map over the integers. In: Fischlin, M., Coron, J.-S. (eds.) EUROCRYPT 2016, Part I. LNCS, vol. 9665, pp. 509–536. Springer, Heidelberg (2016). https://doi.org/10.1007/978-3-662-49890-3_20

35. Cheon, J.H., Jeong, J., Lee, C.: An algorithm for NTRU problems and cryptanalysis of the GGH multilinear map without a low level encoding of zero. Eprint 2016/139 (2016)

36. Cocks, C.: An identity based encryption scheme based on quadratic residues. In: Honary, B. (ed.) Cryptography and Coding 2001. LNCS, vol. 2260, pp. 360–363. Springer, Heidelberg (2001). https://doi.org/10.1007/3-540-45325-3_32

37. Coron, J.-S., et al.: Zeroizing without low-level zeroes: new MMAP attacks and their limitations. In: Gennaro, R., Robshaw, M. (eds.) CRYPTO 2015, Part I. LNCS, vol. 9215, pp. 247–266. Springer, Heidelberg (2015). https://doi.org/10.1007/978-3-662-47989-6_12

38. Coron, J.-S., Lee, M.S., Lepoint, T., Tibouchi, M.: Zeroizing attacks on indistinguishability obfuscation over CLT13. In: Fehr, S. (ed.) PKC 2017, Part I. LNCS, vol. 10174, pp. 41–58. Springer, Heidelberg (2017). https://doi.org/10.1007/978-3-662-54365-8_3

39. Coron, J.-S., Lepoint, T., Tibouchi, M.: Practical multilinear maps over the integers. In: Canetti, R., Garay, J.A. (eds.) CRYPTO 2013, Part I. LNCS, vol. 8042, pp. 476–493. Springer, Heidelberg (2013). https://doi.org/10.1007/978-3-642-40041-4_26

40. Cramer, R., Ducas, L., Peikert, C., Regev, O.: Recovering short generators of principal ideals in cyclotomic rings. In: Fischlin, M., Coron, J.-S. (eds.) EUROCRYPT 2016, Part II. LNCS, vol. 9666, pp. 559–585. Springer, Heidelberg (2016). https://doi.org/10.1007/978-3-662-49896-5_20

41. Ding, J., Yang, B.Y.: Multivariate public key cryptography. In: Bernstein D.J., Buchmann J., Dahmen E. (eds) Post-Quantum Cryptography, pp. 193–241. Springer, Heidelberg (2009). https://doi.org/10.1007/978-3-540-88702-7_6

42. Farshim, P., Hesse, J., Hofheinz, D., Larraia, E.: Graded encoding schemes from obfuscation. In: Abdalla, M., Dahab, R. (eds.) PKC 2018, Part II. LNCS, vol. 10770, pp. 371–400. Springer, Cham (2018). https://doi.org/10.1007/978-3-319-76581-5_13

43. Garg, S., Gentry, C., Halevi, S.: Candidate multilinear maps from ideal lattices. In: Johansson, T., Nguyen, P.Q. (eds.) EUROCRYPT 2013. LNCS, vol. 7881, pp. 1–17. Springer, Heidelberg (2013). https://doi.org/10.1007/978-3-642-38348-9_1

44. Garg, S., Gentry, C., Halevi, S.: Candidate multilinear maps from ideal lattices. In: Johansson, T., Nguyen, P.Q. (eds.) EUROCRYPT 2013. LNCS, vol. 7881, pp. 1–17. Springer, Heidelberg (2013). https://doi.org/10.1007/978-3-642-38348-9_1

45. Garg, S., Gentry, C., Halevi, S., Raykova, M., Sahai, A., Waters, B.: Candidate indistinguishability obfuscation and functional encryption for all circuits. In: FOCS, pp. 40–49 (2013). http://eprint.iacr.org/

46. Garg, S., Gentry, C., Halevi, S., Sahai, A., Waters, B.: Attribute-based encryption for circuits from multilinear maps. In: Canetti, R., Garay, J.A. (eds.) CRYPTO 2013, Part II. LNCS, vol. 8043, pp. 479–499. Springer, Heidelberg (2013). https://doi.org/10.1007/978-3-642-40084-1_27

47. Garg, S., Gentry, C., Halevi, S., Zhandry, M.: Fully secure functional encryption without obfuscation. In: IACR Cryptology ePrint Archive. vol. 2014, p. 666 (2014). http://eprint.iacr.org/2014/666

48. Garg, S., Gentry, C., Halevi, S., Zhandry, M.: Functional encryption without obfuscation. In: Kushilevitz, E., Malkin, T. (eds.) Theory of Cryptography (2016)

49. Garg, S., Miles, E., Mukherjee, P., Sahai, A., Srinivasan, A., Zhandry, M.: Secure obfuscation in a weak multilinear map model. In: Hirt, M., Smith, A. (eds.) TCC 2016, Part II. LNCS, vol. 9986, pp. 241–268. Springer, Heidelberg (2016). https://doi.org/10.1007/978-3-662-53644-5_10

50. Gentry, C.: Fully homomorphic encryption using ideal lattices. In: STOC, pp. 169–178 (2009)

51. Gentry, C., Gorbunov, S., Halevi, S.: Graph-induced multilinear maps from lattices. In: Dodis, Y., Nielsen, J.B. (eds.) TCC 2015, Part II. LNCS, vol. 9015, pp. 498–527. Springer, Heidelberg (2015). https://doi.org/10.1007/978-3-662-46497-7_20

52. Gentry, C., Peikert, C., Vaikuntanathan, V.: Trapdoors for hard lattices and new cryptographic constructions. In: STOC, pp. 197–206 (2008)

53. Goldreich, O.: Foundations of Cryptography: Basic Tools. Cambridge University Press, New York (2000)

54. Goldwasser, S., Kalai, Y.T., Popa, R., Vaikuntanathan, V., Zeldovich, N.: Reusable garbled circuits and succinct functional encryption. In: Proceedings of STOC, pp. 555–564. ACM Press, New York (2013)

55. Goldwasser, S., Kalai, Y.T., Popa, R.A., Vaikuntanathan, V., Zeldovich, N.: Reusable garbled circuits and succinct functional encryption. In: STOC, pp. 555–564 (2013)

56. Gorbunov, S., Vaikuntanathan, V., Wee, H.: Functional encryption with bounded collusions via multi-party computation. In: Safavi-Naini, R., Canetti, R. (eds.) CRYPTO 2012. LNCS, vol. 7417, pp. 162–179. Springer, Heidelberg (2012). https://doi.org/10.1007/978-3-642-32009-5_11

57. Gorbunov, S., Vaikuntanathan, V., Wee, H.: Attribute based encryption for circuits. In: STOC (2013)

58. Gorbunov, S., Vaikuntanathan, V., Wee, H.: Predicate encryption for circuits from LWE. In: Gennaro, R., Robshaw, M. (eds.) CRYPTO 2015, Part II. LNCS, vol. 9216, pp. 503–523. Springer, Heidelberg (2015). https://doi.org/10.1007/978-3-662-48000-7_25

59. Goyal, V., Pandey, O., Sahai, A., Waters, B.: Attribute-based encryption for fine-grained access control of encrypted data. In: ACM Conference on Computer and Communications Security, pp. 89–98 (2006)

60. Hu, Y., Jia, H.: Cryptanalysis of GGH map. Cryptology ePrint Archive: Report 2015/301 (2015)

61. Ishai, Y., Kushilevitz, E.: Randomizing polynomials: A new representation with applications to round-efficient secure computation. In: FOCS (2000)

62. Katz, J., Sahai, A., Waters, B.: Predicate encryption supporting disjunctions, polynomial equations, and inner products. In: Smart, N. (ed.) EUROCRYPT 2008. LNCS, vol. 4965, pp. 146–162. Springer, Heidelberg (2008). https://doi.org/10.1007/978-3-540-78967-3_9

63. Komargodski, I., Moran, T., Naor, M., Pass, R., Rosen, A., Yogev, E.: One-way functions and (im)perfect obfuscation. In: 55th IEEE Annual Symposium on Foundations of Computer Science, FOCS, pp. 374–383 (2014)

64. Koppula, V., Lewko, A.B., Waters, B.: Indistinguishability obfuscation for turing machines with unbounded memory. In: STOC, pp. 419–428 (2015)

65. Langlois, A., Stehlé, D., Steinfeld, R.: GGHLite: more efficient multilinear maps from ideal lattices. In: Nguyen, P.Q., Oswald, E. (eds.) EUROCRYPT 2014. LNCS, vol. 8441, pp. 239–256. Springer, Heidelberg (2014). https://doi.org/10.1007/978-3-642-55220-5_14

66. Lewko, A., Okamoto, T., Sahai, A., Takashima, K., Waters, B.: Fully secure functional encryption: attribute-based encryption and (hierarchical) inner product encryption. In: Gilbert, H. (ed.) EUROCRYPT 2010. LNCS, vol. 6110, pp. 62–91. Springer, Heidelberg (2010). https://doi.org/10.1007/978-3-642-13190-5_4

67. Lin, H.: Indistinguishability obfuscation from constant-degree graded encoding schemes. In: Fischlin, M., Coron, J.-S. (eds.) EUROCRYPT 2016, Part I. LNCS, vol. 9665, pp. 28–57. Springer, Heidelberg (2016). https://doi.org/10.1007/978-3-662-49890-3_2

68. Lin, H.: Indistinguishability obfuscation from SXDH on 5-linear maps and locality-5 PRGs. In: Katz, J., Shacham, H. (eds.) CRYPTO 2017, Part I. LNCS, vol. 10401, pp. 599–629. Springer, Cham (2017). https://doi.org/10.1007/978-3-319-63688-7_20

69. Lin, H., Matt, C.: Pseudo flawed-smudging generators and their application to indistinguishability obfuscation. Cryptology ePrint Archive, Report 2018/646 (2018)

70. Lin, H., Pass, R., Seth, K., Telang, S.: Output-compressing randomized encodings and applications. In: Kushilevitz, E., Malkin, T. (eds.) TCC 2016, Part I. LNCS, vol. 9562, pp. 96–124. Springer, Heidelberg (2016). https://doi.org/10.1007/978-3-662-49096-9_5

71. Lin, H., Tessaro, S.: Indistinguishability obfuscation from trilinear maps and block-wise local PRGs. In: Katz, J., Shacham, H. (eds.) CRYPTO 2017, Part I. LNCS, vol. 10401, pp. 630–660. Springer, Cham (2017). https://doi.org/10.1007/978-3-319-63688-7_21

72. Lin, H., Vaikuntanathan, V.: Indistinguishability obfuscation from DDH-like assumptions on constant-degree graded encodings. In: FOCS, pp. 11–20 (2016)

73. Lombardi, A., Vaikuntanathan, V.: On the non-existence of blockwise 2-local prgs with applications to indistinguishability obfuscation. In: TCC (2018)

74. Matsumoto, T., Imai, H.: Public quadratic polynomial-tuples for efficient signature-verification and message-encryption. In: Barstow, D., Brauer, W., Brinch Hansen, P., Gries, D., Luckham, D., Moler, C., Pnueli, A., Seegmüller, G., Stoer, J., Wirth, N., Günther, C.G. (eds.) EUROCRYPT 1988. LNCS, vol. 330, pp. 419–453. Springer, Heidelberg (1988). https://doi.org/10.1007/3-540-45961-8_39

75. Miles, E., Sahai, A., Zhandry, M.: Annihilation attacks for multilinear maps: crypt-analysis of indistinguishability obfuscation over GGH13. In: Robshaw, M., Katz, J. (eds.) CRYPTO 2016, Part II. LNCS, vol. 9815, pp. 629–658. Springer, Heidelberg (2016). https://doi.org/10.1007/978-3-662-53008-5_22

76. Peikert, C.: A Decade of Lattice Cryptography, vol. 10, pp. 283–424, March 2016

77. Pellet-Mary, A.: Quantum attacks against indistinguishablility obfuscators proved secure in the weak multilinear map model. In: Shacham, H., Boldyreva, A. (eds.) CRYPTO 2018, Part III. LNCS, vol. 10993, pp. 153–183. Springer, Cham (2018). https://doi.org/10.1007/978-3-319-96878-0_6

78. Pellet-Mary, A.: Quantum attacks against indistinguishablility obfuscators proved secure in the weak multilinear map model. In: Shacham, H., Boldyreva, A. (eds.) CRYPTO 2018, Part III. LNCS, vol. 10993, pp. 153–183. Springer, Cham (2018). https://doi.org/10.1007/978-3-319-96878-0_6

79. Regev, O.: On lattices, learning with errors, random linear codes, and cryptography. J. ACM 56(6), 34 (2009). (extended abstract in STOC 2005)

80. Sahai, A., Waters, B.: Functional encryption:beyond public key cryptography. Power Point Presentation (2008). http://userweb.cs.utexas.edu/~bwaters/presentations/files/functional.ppt

81. Sahai, A., Waters, B.: Fuzzy identity-based encryption. In: Cramer, R. (ed.) EURO-CRYPT 2005. LNCS, vol. 3494, pp. 457–473. Springer, Heidelberg (2005). https://doi.org/10.1007/11426639_27

82. Sahai, A., Waters, B.: How to use indistinguishability obfuscation: Deniable encryption, and more. In: STOC, pp. 475–484 (2014). http://eprint.iacr.org/2013/454.pdf

83. Waters, B.: Functional encryption for regular languages. In: Safavi-Naini, R., Canetti, R. (eds.) CRYPTO 2012. LNCS, vol. 7417, pp. 218–235. Springer, Heidelberg (2012). https://doi.org/10.1007/978-3-642-32009-5_14

84. Wolf, C.: Multivariate Quadratic Polynomials In Public Key Cryptography. Ph.D. thesis, katholieke universiteit leuven (2005)

Sum-of-Squares Meets Program Obfuscation, Revisited

Boaz Barak[1]([⊠]), Samuel B. Hopkins[2], Aayush Jain[3], Pravesh Kothari[4], and Amit Sahai[3]

[1] Harvard University, Cambridge, USA
b@boazbarak.org
[2] University of California, Berkeley, USA
hopkins@berkeley.edu
[3] University of California, Los Angeles, USA
{aayushjain,sahai}@cs.ucla.edu
[4] Princeton University and the Institute for Advanced Study, Princeton, USA
kothari@cs.princeton.edu

Abstract. We develop attacks on the security of variants of pseudorandom generators computed by quadratic polynomials. In particular we give a general condition for breaking the one-way property of mappings where every output is a quadratic polynomial (over the reals) of the input. As a corollary, we break the degree-2 candidates for security assumptions recently proposed for constructing indistinguishability obfuscation by Ananth, Jain and Sahai (ePrint 2018) and Agrawal (ePrint 2018). We present conjectures that would imply our attacks extend to a wider variety of instances, and in particular offer experimental evidence that they break assumption of Lin-Matt (ePrint 2018).

Our algorithms use semidefinite programming, and in particular, results on low-rank recovery (Recht, Fazel, Parrilo 2007) and matrix completion (Gross 2009).

1 Introduction

In this work, we initiate the algorithmic study of cryptographic hardness that may exist in general expanding families of low-degree polynomials over $\mathbb{R}$. As a result, we obtain strong attacks on certain pseudorandom generators whose output is a "simple" function of the input. Such "simple" pseudorandom generators are interesting in their own right, but have recently become particularly important because of their role in candidate constructions for *Indistinguishabily Obfuscators*.

The question of whether Indistinguishabily Obfuscators (iO) exist is one of the most consequential open questions in cryptography. On one hand, a sequence of works [14,29] has shown that iO, if it exists, would imply a huge variety of cryptographic objects, several of which we know of no other way to achieve. On the other hand, the current candidate constructions for iO's are not based

© International Association for Cryptologic Research 2019
Y. Ishai and V. Rijmen (Eds.): EUROCRYPT 2019, LNCS 11476, pp. 226–250, 2019.
https://doi.org/10.1007/978-3-030-17653-2_8

on well-studied standard assumptions, and there have been several attacks on several iO constructions as well as underlying primitives.

A promising line of works [3,19,20,22,23] has aimed at basing iOs on more standard assumptions, and in particular Lin and Tessaro [22] reduced constructing iO to the combination following three assumptions:

1. The *learning with errors (LWE)* assumption.
2. Existence of *three local* pseudorandom generators with sufficiently large super linear stretch. These are pseudorandom generators $G : \{0,1\}^n \to \{0,1\}^{n^{1+\varepsilon}}$ (for arbitrarily small $\varepsilon > 0$) such that if we think of the input as split into n/k blocks of length k each (for some $k = n^{o(1)}$) then every output of G depends on at most three blocks of the input.
3. Existence of *trilinear maps* satisfying certain strengthening of the Decisional-Diffie-Hellman assumption.

Of the three assumptions, the learning with errors assumption is well studied and widely believed. The existence of local pseudorandom generators has also been recently extensively studied; it also relates to questions on random constraint satisfaction problems that have been looked at by various communities. Based on our current knowledge, it is reasonable to assume that such three-local generators exist with stretch, say, $n^{1.1}$ which would be sufficient for the Lin-Tessaro construction.

The most problematic assumption is the existence of trilinear maps. Since the seminal work of Garg, Gentry and Halevi [13], there have been some candidate constructions for (noisy) k-linear maps for $k > 2$, but these are not based on any standard assumption, and in fact there have been several *attacks* [6–10,17, 18,25,26] showing that these construction fail to satisfy natural analogs of the Decisional Diffie Hellman assumption. This is in contrast to the $k = 2$ or *bilinear* case, where we have had constructions for almost 20 years that are believed to be secure (with respect to classical polynomial-time algorithms) based on elliptic curve groups that admit certain *pairing* operations [5]. Thus a main open question has been whether one can achieve iO based only on cryptographic *bilinear maps* as well as local (or otherwise "simple") pseudorandom generators that can be reasonably conjectured to be secure.

1.1 Basing iO on Bilinear Maps and Our Results

In the first version of their manuscript, Lin and Tessaro [22] gave a construction of iO based on *two local* generators with a certain stretch, and a candidate construction for the latter object based on a random two-local map with a certain nonlinear predicate. Alas, Barak, Brakerski, Komargodski and Kothari, [4], as well as Lombardi and Vaikuntanathan [24] showed that the Lin-Tessaro candidate construction, as well as *any* generator with their required parameters, can be broken using semidefinite programming, and specifically the degree two *sum of squares* program [4].

Very recently, the work of Ananth, Jain, and Sahai [2], followed shortly by the independent works of Agrawal [1] and Lin and Matt [21], proposed a way around that hurdle, obtaining constructions for iO where the role of the trilinear map is replaced with objects that:

1. Satisfy security notions that are weaker than being a full fledged pseudorandom generators.
2. Satisfy structural properties that are weaker than being two-local, and in particular requiring the outputs only to be a *degree two*[1] *polynomial* of the input.

As such, these objects do not automatically fall under the attacks described by [4,24]. However, in this work we show that:

- The specific candidate objects in all these three works (based on random polynomials) can be broken using a distinguisher built on the same sum-of-squares semidefinite program.
- Moreover, this results extends to other families of constructions, including ones that are not based on random polynomials. In fact, we do not know of *any* degree-2 construction that does not fall prey to a variant of the same attack.

The Ananth-Jain-Sahai "Cubic Assumption". The work of [2] also obtained a construction of iO based on a (variant of a) pseudorandom generator where every output is a *cubic* polynomial of the input, but where some information about the input is "leaked" in a way "masked" using instances of LWE[2]. Our attacks in their current form are not applicable to this new construction. The question of whether secure degree-3 ΔRGs exist, or whether an extended form of the sum of squares algorithm can be applied to it, is one that deserves further study. More generally, understanding the structure of hard distributions for expanding families of constant-degree polynomials over the integers, is a fascinating and important area of study, which is strongly motivated by the problem of securely constructing iO. Taking inspiration from SoS lower bounds [15,30], we also suggest a candidate for the same. Our candidate is inspired by the hardness of refuting random satisfiable 3SAT instances. For further details, see Sect. 7.

1.2   Our Results

We consider the following general hypothesis that, if true, would rule out not just the three proposed approaches based on quadratic polynomials for obtaining iO, but also a great many potential generalizations of them. Below we say that an n-variate polynomial q is Λ-*bounded* if all of q's coefficients are integers in the interval $[-\Lambda, +\Lambda]$. We say that a distribution $\mathcal{X}$ over $\mathbb{Z}^n$ is Λ *bounded* if it is supported over $[-\Lambda, +\Lambda]^n$.

---

[1] The work of Ananth, Jain, and Sahai [2] also considered degree-3 polynomials. We do not have attacks on such degree-3 polynomials; we discuss this further below.

[2] This cubic version of their assumption was made explicit in an update to [2].

Hypothesis 1 (No Expanding Weak Quadratic Pseudorandom Generators). *For every $\varepsilon > 0$, polynomial $\Lambda(n)$, sufficiently large $n \in \mathbb{N}$, if:*

- *$q_1, \ldots, q_m : \mathbb{R}^n \to \mathbb{R}$ are quadratic $\Lambda(n)$-bounded polynomials for $m \geqslant n^{1+\varepsilon}$*
- *$\mathcal{X}$ is a $\Lambda(n)$-bounded distribution over $\mathbb{Z}^n$*
- *For every i, Δ_i is a $\Lambda(n)$ bounded distribution over $\mathbb{Z}$ such that $\mathbb{P}[\Delta_i = z] < 0.9$ for every $z \in \mathbb{Z}$.*

then there exists an algorithm $\mathcal{A}$ that can distinguish between the following distributions with $\Omega(1)$ bias:

- *$(q_1, \ldots, q_m, q_1(x), \ldots, q_m(x))$ for $x \sim \mathcal{X}$.*
- *$(q_1, \ldots, q_m, q_1(x) + \delta_1, \ldots, q_m(x) + \delta_m)$ where for every i, δ_i is drawn independently from Δ_i.*

Note that this hypothesis would be violated by the existence of a pseudorandom generator $G : \{0,1\}^n \to \{0,1\}^{n^{1+\varepsilon}}$ whose outputs are degree two polynomials. It would also be violated if the distribution $G(x)$ is indistinguishable from the distribution $(G(x) + \delta_1', \ldots, G(x) + \delta_m')$ where $\delta_1', \ldots, \delta_m'$ are drawn independently from some distribution Δ over integers that satisfies $\mathbb{P}[\Delta = 0] \leqslant 0.9$.

An efficient algorithm to recover x from $q_1(x), \ldots, q_m(x)$ would allow to distinguish between the two distributions. However, generally speaking, it need not even be information theoretically possible to recover x from this information. Even if it is information-theoretically possible this can be computationally intractable, as recovering x from $q_1(x), \ldots, q_m(x)$ is an instance of the NP hard problem of *Quadratic Equations*.

In Hypothesis 1, the polynomials are arbitrary. However, the candidate constructions of pseudorandom generators considered so far used $q_1, \ldots, q_m$ that are sampled *independently* from some distribution $\mathcal{Q}$. This is natural, as intuitively if we want $q_1(x), \ldots, q_m(x)$ to look like a product distribution, then the more randomness in the choice of the q_i's the better.

However, in this work we give general attacks on candidates that have this form. As these are some of the most natural approaches to refute Hypothesis 1, our work can be seen as providing some (partial) evidence to its veracity. To state our result, we need the following definition of "nice" distributions.

Definition 1 (Nice Distributions). *Let $\mathcal{Q}$ be a distribution over n-variate quadratic polynomials with integer coefficients. We say that $\mathcal{Q}$ is nice if it satisfies that:*

- *There is a constant $C = C(\mathcal{Q}) = O(1)$ such that $\mathcal{Q}$ is supported on homogeneous (i.e. having no linear term) degree-2 polynomials q with $\|q\|_2^2 \leqslant C \mathbb{E}\|q\|_2^2$, where $\|q\|$ is the ℓ_2-norm of the vector of coefficients of q.*
- *$Var(Q_{i,j}) = 1$ where $Q_{i,j}$ denotes the coefficient of $x_i x_j$ in a polynomial Q sampled from $\mathcal{Q}$.*
- *If $\{i, j\} \neq \{k, \ell\}$ then the random variables $Q_{i,j}$ and $Q_{k,\ell}$ are independent.*

Roughly speaking, a distribution over quadratic polynomials is nice if it satisfies certain normalization properties as well as *pairwise independence* of the vectors of coefficients. Many natural distributions on polynomials are nice, and in particular random dense as well as random sparse polynomials are nice.

The following theorem shows that it is always possible to recover x from a superlinear number of quadratic observations, if the latter are chosen from a nice distribution.

Theorem 2 (Recover from Random Quadratic Observations). *There is a polynomial-time algorithm $\mathcal{A}$ (based on the sum of squares algorithm) with the following guarantees. For every nice distribution $\mathcal{Q}$ and every $t \leqslant n^{O(1)}$, for large-enough n, with probability at least $1 - n^{-\log(n)}$ over $x \sim \{-t, -t+1, \ldots, 0, \ldots, t-1, t\}^n$ and $q_1, \ldots, q_m \sim \mathcal{Q}$, if $m \geqslant n(\log n)^{O(1)}$ then $\mathcal{A}(q_1, \ldots, q_m, q_1(x), \ldots, q_m(x)) = x$.*

On "niceness". The definition of "nice" distributions above is fairly natural, and captures examples such as when the polynomials are chosen with all coefficients as independent Gaussian or Bernoulli variables. In particular as a corollary of Theorem 2 we break the candidate pseudorandom generator of Ananth et al (and even its Δ-RG property). Moreover, we obtain such results even in the *sparse* case where most of the coefficients of the polynomials $q_1, \ldots, q_m$ are zero.

At the moment however our theoretical analysis does not extend to the "blockwise random" polynomials that were used by Lin and Matt which can be thought of as a sum of random dense polynomial and a random sparse polynomial. While this combination creates theoretical difficulty in the analysis, we believe that it can be overcome and that it is possible to recover in this case as well. In particular, we also have *experimental results* showing that we can break the Lin-Matt generator as well.

Finally, we note that by Markov's inequality for any $\mathcal{Q}$ we have $\mathbb{P}(\|q\|^2 \geqslant C\,\mathbb{E}\|q\|^2) \leqslant 1/C$. Our niceness assumption just has the effect of restricting $\mathcal{Q}$ to this relatively high-probability event. If $\mathcal{Q}$ is not pathological – that is, it is not dominated by events with probability $\ll 1/C$ for a large constant C – then this kind of truncation will result in a nice distribution.[3]

---

[3] Along the same lines, we note that if $\mathcal{Q}$ is nice and $\mathbb{E}_{q \sim \mathcal{Q}} q = 0$ (as we observe later, the latter can be enforced without loss of generality) then $\mathcal{Q}$ is also $\Lambda(n)$-bounded for $\Lambda(n) \leqslant O(n)$. The reason is that if $\mathcal{Q}$ is nice and has $\mathbb{E}\, q = 0$ then

$$\mathbb{E}\|q\|^2 = \sum_{i,j \leqslant n} \mathbb{E}\, Q_{ij}^2 = \sum_{i,j \leqslant n} Var(Q_{ij}) = n^2 \,.$$

For every i, j and every q in the support of $\mathcal{Q}$, we have by niceness that $|q_{ij}| \leqslant \|q\|_2 \leqslant Cn$. Hence $\mathcal{Q}$ is $O(n)$-bounded.

One implication is that $\mathcal{Q}$ cannot be a distribution on where the all-zero polynomial appears with probability, say, $1 - 1/n$, as otherwise its support would also have to contain polynomials with coefficients $\gg n$. Our main theorem could not apply to such a distribution, since clearly at least $\Omega(n^2)$ independent samples would be needed to get enough information to recover x from $\{q_i, q_i(x)\}$, while we assume $m \leqslant n(\log n)^{O(1)} \ll n^2$.

On the distribution of x. For concreteness, we phrase Theorem 2 so that the distribution of x is uniform over $\{-t, \ldots, t\}^n$. However, the proof of the theorem allows x to be a more general $\mathbb{R}^n$-valued random variable. In particular, x may be any n-dimensional real-valued random vector which has $\mathbb{E}\,x = 0$ and is $O(\mathbb{E}\|x\|^2/n)$-sub-Gaussian. The coordinates of x need not even be independent: for instance, x may be drawn from the uniform distribution on the unit sphere.

Experiments. We implement the sum-of-squares attack and verify that indeed it efficiently breaks random dense quadratic polynomials. Furthermore, we implement a variant of the attack that efficiently breaks the Lin-Matt candidate: the Lin-Matt candidate is, roughly speaking, a sum of two independent polynomials, where one is dense and one is sparse. Since the planted solution must be composed of polynomially-bounded integers, we observe that it is possible to efficiently guess the squared L2 norm of the portion of the planted solution that corresponds to the sparse part of the polynomial. Given this guess, we can introduce a new constraint into the semidefinite program that fixes the trace of the portion of the semidefinite matrix that corresponds to the sparse matrix. We show experimentally that this attack breaks the Lin-Matt candidate for moderate values of n. In particular, in Fig. 1 we plot the correlation between the recovered solution with the planted solution, where the x-axis is labeled by the ratio m/n showing the expansion needed for the attack to work, for $n = 60$ total variables. More details can be found in Sect. 6.

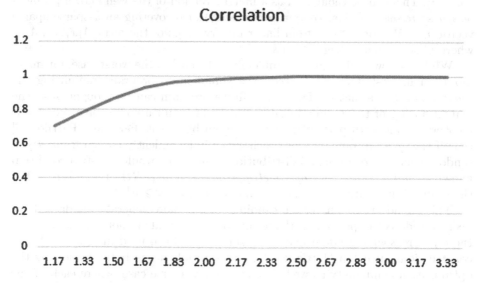

Fig. 1. Experimentally breaking Lin-Matt candidate. Graph shows quality of recovered solution vs. planted solution, for various values of m/n shown in the x-axis. Let $\mathbf{v}$ be the eigen vector with largest eigen value of the optimum matrix returned by the SDP. Let $\mathbf{x}$ be the planted solution. Quality of solution is defined as $\dfrac{\langle \mathbf{v}, \mathbf{x} \rangle}{\langle \mathbf{v}, \mathbf{v} \rangle^{\frac{1}{2}} \langle \mathbf{x}, \mathbf{x} \rangle^{\frac{1}{2}}}$

In particular, we are not aware of any candidate construction of weak pseudo-random generator computed by quadratic polynomials that is not broken experimentally by our algorithms.

2   Our Techniques

Our algorithms use essentially the same semidefinite program constraints that were used in the work of [4], namely the *sum of squares* program. However, we use a different, simpler, objective function, and moreover we crucially use a different *analysis* (which also inspired some tweaks to the algorithm that seem to help in experiments). Specifically, consider the task of recovering an unknown vector $x \in \mathbb{R}^n$ from the values $(q_1(x), \ldots, q_m(x))$ where $q_1, \ldots, q_m$ are quadratic polynomials. We focus on the case that the q_i's are *homogenous* polynomials, which means that (thinking of x as a column vector), $q_i(x) = x^\top Q_i x$ for some $n \times n$ matrix Q_i. Another way to write this is that $q_i(x) = \langle Q_i, X \rangle$ where X is the rank one matrix $xx^\top$.

In the above notation, our problem becomes the task of recovering a rank one matrix X from the observations

$$\langle Q_1, X \rangle, \ldots, \langle Q_m, X \rangle \tag{1}$$

for some known $n \times n$ matrices $Q_1, \ldots, Q_m$ where $m \geqslant n^{1+\varepsilon}$. Luckily, this task has been studied in the literature and is known as the *low rank recovery problem* [28]. This can be thought of as a matrix version of the well known problem of *sparse recovery* (a.k.a. *compressed sensing*) of recovering an k-sparse sparse vector $x \in \mathbb{R}^n$ (for $n \gg k$) from linear observations of the form $A_1 x, \ldots, A_{k'} x$ where k' is not much bigger than k.

While the low rank recovery problem is NP hard in the worst-case, for many inputs of interest it can be solved by a semidefinite program minimizing the *nuclear norm* of a matrix. This semidefinite program can be thought of as the matrix analog of the L_1 minimization linear program used to solve the sparse recovery problem. In particular, it was shown by Recht, Fazel and Parrilo [28] that if the Q_i's are *random* (with each entry independently chosen from, say, a random Gaussian or Bernoulli distribution), then they would satisfy a condition known as *matrix restricted isoperimetry property (matrix RIP)* that ensures that the semidefinite matrix recovers X in our regime of $m \geqslant n^{1+\varepsilon}$.

This already rules out certain candidates, but more general candidates have been considered. In particular, the results of Recht et al are not applicable when the Q_i's are *sparse* random matrices, which have been used in some of the iO constructions such Lin-Matt's. Luckily, this problem has been studied by the optimization community as well. The extremely sparse case, where each of the Q_i's has just a single nonzero coordinate, is particularly well studied. In this case, recovering X from (1) corresponds to *completing* X using m observations of its entries, and is known as the *matrix completion* problem.

Specifically, a beautiful paper of Gross [16] gave quite general bounds that in some sense interpolate between these two extremes. Specifically, Gross showed that it is possible to recover X from (1) as long as these observations $Q_1, \ldots, Q_m$ are sampled independently from a collection $\{Q^1, \ldots, Q^N\}$ that satisfies certain "isotropy" and "incoherence with respect to X" properties. We show that under the "niceness" conditions of Theorem 2, we can "massage" our input so that it is of the form where Gross's theorem applies. Once we do so we can appeal to this theorem to obtain our result. A key property that we use in our proof is that in the cryptographic setting, we do not need to recover $X = xx^\top$ for *every* $x \in \mathbb{Z}^n$ but rather only for *most* x's. This allows us to achieve the incoherence property even in settings where it would not hold for a worst-case choice of a vector.

3   Preliminaries

For a matrix X, we write $\|X\|$ for its operator norm: $\sup_{v:\|v\|_2=1} |\langle v, Xv \rangle|$. We use the standard inner product on the Hilbert space of $n \times n$ matrices: $\langle A, B \rangle = \mathrm{tr}(AB)$. The *nuclear* norm of a matrix X is defined by $\|X\|_* = \sup_{A:\|A\|\leqslant 1} \langle A, X \rangle$. For a positive semidefinite matrix X, $\|X\|_* = \mathrm{tr}(X)$.

For any matrix $Q \in \mathbb{R}^{n \times n}$, $\mathsf{vec}(Q)$ denotes "vectorization" of the matrix Q as a n^2 dimensional vector.

For a matrix $M \in \mathbb{R}^{n \times n}$, we define the operator norm (also called the spectral norm) of M as $\max_{x \in \mathbb{R}^n} \|Mx\|/\|x\|$. The Frobenius norm of M is $\|M\|_F = \sqrt{\sum_{ij\leqslant n} M_{ij}^2}$.

For a matrix M, we write $M \in (1 \pm \varepsilon)\mathrm{Id}$ if $\|M - \mathrm{Id}\| \leqslant \varepsilon$, where $\|\cdot\|$ is the operator norm.

3.1 ΔRGs (Ananth-Jain-Sahai)

Ananth-Jain-Sahai proposed a variant of (integer valued) PRG such that it is hard to distinguish between the output of a PRG and a small perturbation of it. Specifically, the following definition describes the object they proposed.

Definition 2 $((n, \lambda, B, \chi)$-$\Delta\mathbf{RG})$. *Let $f : \chi^n \to \mathbb{Z}^m$ be an integer valued function with the ith output described by $f_i : \chi^n \to \mathbb{Z}$ and at any $x \in \chi^n$, $f_i(x) = q_i(x)$ for quadratic polynomials q_i for $1 \leqslant i \leqslant m$.*

f is said to be a ΔRG, if for distributions D_1, D_2 on $\mathbb{Z}^m$ defined below and for any circuit $\mathcal{A}$ of size 2^λ,

$$| \mathop{\mathbb{P}}_{z \sim D_1} [\mathcal{A}(z) = 1] - \mathop{\mathbb{P}}_{z \sim D_2} [\mathcal{A}(z) = 1]| < 1 - 2/\lambda$$

Distribution D_1
Sample $x \leftarrow \chi$. Output $\{q_i, q_i(x)\}_{i \in [m]}$

Distribution D_2
Sample $x \leftarrow \chi$. Output $\{q_i, q_i(x) + \delta_i\}_{i \in [m]}$
Here $\delta_i \in \mathbb{Z}$ are arbitrary perturbations such that $|\delta_i| < B$ for all $i \in [m]$.

Concurrently and independently, [21] proposed Pseudo-Flawed Smudging Generators which have similar security guarantees.

4   Candidates for Quadratic PRGs

In this section we formally describe the candidate polynomial and input distributions proposed by [1,2,21] to realize corresponding notions of pseudo-random generators of $\mathbb{Z}$.

Note that any algorithm that given the polynomials $q_1, \ldots, q_m$ and measurements $q_1(x), \ldots q_m(x)$ when x, $q_1, .., q_m$ are sampled from required distributions of the pseudorandom generator, successfully recovers x, also breaks the corresponding candidate for the pseudorandom generator.

To be precise, we describe the candidate polynomials and input distributions proposed by:

- Ananth et al. [2] to instantiate ΔRGs.
- Lin-Matt [21] to instantiate Pseudo Flawed-Smudging Generators.
- Agarwal [1] to instantiate Non-boolean PRGs.

Along with assumptions on cryptographic bilinear maps, learning with error assumption and PRGs with constant block locality, either of these three assumptions imply iO.

4.1   Candidate for ΔRG

Ananth-Jain-Sahai proposed the following candidate construction for a ΔRG. Let χ be the uniform distribution in $[-B_1, B_1]$. Choose $m = n^{1+\varepsilon}$ for some small enough constant $\varepsilon > 0$. Let C be some constant positive integer and B_1 be a polynomial in λ, the security parameter.

Distribution Q: Sample each polynomial as follows. Let $q(x_1, ..., x_n) = \Sigma_{i \neq j} c_{i,j} \cdot x_i \cdot x_j$, where each coefficient $c_{i,j}$ is chosen uniformly from $[-C, C]$.

Distribution X: Inputs are sampled as follows. Sample x_i for $i \in [n]$ uniformly from $[-B_1, B_1]$. Output $x = (x_1, ..., x_n)$. Implicitly, [1] also considered these polynomials for their notion of a non-boolean PRG.

4.2   Candidate for Pseudo Flawed-Smudging Generators (Lin-Matt)

Lin and Matt [21] proposed a variant of pseudorandom generators with security properties closely related to the notion of ΔRGs above. Here, we recall their candidate polynomials.

Distribution Q: For each $j \in [m]$,

$$q_j(x_1, ..., x_n, x'_1, ..., x'_{n'}) = S_j(x_1, ..., x_n) + MQ_j(x'_1, ..., x'_{n'})$$

Here we write more about polynomials S_j, MQ_j.

1. MQ_j **Polynomials:** MQ_j are random quadratic polynomials over $(x'_1, ..., x'_{n'})$, where the coefficients of each degree two monomial $x'_i x'_k$ and degree one monomial x'_i are integers chosen independently at random from $[-C, C]$.

2. S_j **Polynomials:** S_j are random quadratic polynomials over $(x_1, ..., x_n)$ of the form:

$$S_j(x_1, ..., x_n) = \Sigma_{i=1}^{n/2} \alpha_i x_{\sigma_j(2 \cdot i)} x_{\sigma_j(2 \cdot i - 1)} + \Sigma_{i=1}^{n} \beta_i x_i + \gamma$$

Here each coefficient α_i, β_i and γ are random integers chosen independently from $[-C, C]$. Here, σ_j is a random permutation from $[n]$ to $[n]$.

Distribution X:

1. Each x_i for $i \in [n]$ is chosen as a random integer sampled independently from the distribution χ_{B_1, B_2}. χ_{B_1, B_2} samples a random integer from $[-B_1, B_1]$ with probability 0.5 and from $[-B_2, B_2]$ with probability 0.5.
2. **Distribution of inputs** $x_1', ..., x_{n'}'$: Each x_i' is chosen as a random integer sampled independently from the distribution $\chi_{B'}$. $\chi_{B'}$ samples a random integer from $[-B', B']$.

   Parameters: Set B_1, B_2, B', n, n' as follows:

   – Set $n = n'$ and $m = n^{1+\varepsilon}$, for some $\varepsilon > 0$.
   – B_1, B' and C are set arbitrarily.
   – Set $B_2 = \Omega(nB^2 + nBB_1)$.

Here B is some polynomial in the security parameter.

All the pseudorandom generators we consider are maps from $\mathbb{Z}^n$ into $\mathbb{Z}^m$ where each of the m output is computed by a degree 2 polynomial with integer coefficients in the input. Since any degree two polynomial in $\mathbb{R}^n$ can be seen as a linear map on $\mathbb{R}^{n \times n^4}$, one can equivalently think of such PRGs as linearly mapping symmetric rank 1 matrices into $\mathbb{R}^m$.

5   Inverting Linear Matrix Maps

In this section, we describe the main technical tool that we rely on in this work - an algorithm based on semidefinite programming for inverting *linear matrix maps*.

Definition 3 (Linear Matrix Maps). *A linear matrix map* $\mathcal{A} : \mathbb{R}^{n \times n} \to \mathbb{R}^m$ *described by a collection of* $n \times n$ *matrices* $Q_1, Q_2, ..., Q_m$ *is a linear map that maps any matrix* $X \in \mathbb{R}^{n \times n}$ *to the vector* $\mathcal{A}(X) \in \mathbb{R}^m$ *such that* $\mathcal{A}(X)_i = \langle Q_i, X \rangle$.

We will use calligraphic letters such as $\mathcal{A}$ *and* $\mathcal{B}$ *to denote such maps.*

We are interested in the algorithmic problem of *inverting* such maps, that is, finding X given $\mathcal{A}(X)$. If Q_is are linearly independent and $m \gg n^2$, then this can be done by linear equation solvers. Our interest is in inverting such maps

---

[4] For any $q(x) = \sum_{i,j} q_{i \leqslant j} x_i x_j$, we define $Q : \mathbb{R}^{n \times n} \to \mathbb{R}$ by $Q_{i,j} = Q_{j,i} = q_{i,j}/2$. Then, $Q(X) = \text{tr}(QX) = \langle Q, X \rangle$ is a linear map on $\mathbb{R}^{n \times n}$.

for low rank matrices X with the "number of measurements" $m \ll n^2$. Indeed, our results will show that for various classes of linear maps $\mathcal{A}$, we can efficiently find a low-rank solution to $\mathcal{A}(X) = z$, whenever it exists, for $m = \tilde{O}(n)$.

Such problems have been well-studied in the literature and rely on a primitive based on semidefinite programming called "nuclear norm minimization". We will use this algorithm and rely on various known results about the success of this algorithm in our analysis.

Algorithm 3 (Trace Norm Minimization).

Given: – $\mathcal{A}$ described by $Q_1, Q_2, \ldots, Q_m \in \mathbb{R}^{n \times n}$.
  – $z \in \mathbb{R}^m$.
Operation: *Output* $X = \arg \min\limits_{\substack{X \succeq 0 \\ \mathcal{A}(X) = z}} \operatorname{tr}(X)$.

In what follows, we will give an analysis of this algorithm for a class of linear matrix maps.

5.1 Incoherent Linear Measurements

In this section we describe a remarkably general result due to Gross on a class of instances $x, Q_1, \ldots, Q_m$ for which trace norm minimization recovers x [16]. These instances are called *incoherent*. Gross's result is the main tool in the proof of our main theorem, which will ultimately show that "nice" distributions $\mathcal{Q}$ produce incoherent instances of trace norm minimization.

We note that many other sufficient conditions for the success of trace norm minimization have been discussed in the literature. One prominent condition is matrix-RIP (Restricted Isometry Property), analyzed in [27]. The restricted isometry property is not known to apply in many natural settings for which we would like to apply our main theorem – for example, if $Q_1, \ldots, Q_m$ have independent entries with on average 1 nonzero entry per row.

Definition 4 (Incoherent Overcomplete Basis). *Let* $\mathcal{B} = \{B_1, B_2, \ldots, B_N\}$ *be a collection of matrices in* $\mathbb{R}^{n \times n}$. *For any rank 1 matrix* $X \in \mathbb{R}^{n \times n}$, $\mathcal{B}$ *is said to be* ν-*incoherent basis for* X *if the following holds:*

1. $(1 - o(1))1/n^2 I_{n^2 \times n^2} \preceq 1/N \sum_{i=1}^{N} \mathsf{vec}(B_i)\mathsf{vec}(B_i)^\top \preceq (1 + o(1))1/n^2 I_{n^2 \times n^2}$.
2. *For each* $i \leqslant N$, $|\langle X, B_i \rangle| \leqslant \nu/n \cdot \|X\|_F$.

We can now define a ν-incoherent measurement.

Definition 5 (Incoherent Measurement). *Let* $\mathcal{B}$ *be a* ν-*incoherent overcomplete basis for an* $n \times n$ *rank 1 matrix* X, *and suppose* $\mathcal{B}$ *has size* $N = \operatorname{poly}(n)$. *Let* $\mathcal{A} : \mathbb{R}^{n \times n} \to \mathbb{R}^m$ *be a map obtained by choosing* Q_i *for each* $i \leqslant m$ *to be a uniformly random and independently chosen element of* $\mathcal{B}$. *Then,* $\mathcal{A}$ *is said to be a* ν-*incoherent measurement of* X.

The following result follows directly from the Proof of Theorem 3 in the work of Gross [16]. While that work focuses on $\mathcal{B}$ being orthonormal - the proof extends to approximately orthonormal basis (i.e., part 1 in the above definition) in a straightforward way.

Theorem 4. *Let $\mathcal{B}$ be a ν-incoherent basis for a rank 1 matrix X of size $N = \mathrm{poly}(n)$. Let $\mathcal{A} : \mathbb{R}^{n \times n} \to \mathbb{R}^m$ be a map obtained by choosing Q_i for each $i \leqslant m$ to be a uniformly random and independently chosen element of $\mathcal{B}$. Then, for large enough $m = \Theta(\nu n \,\mathrm{poly} \log n)$, Algorithm 3, when given input $\mathcal{A}$ and $\mathcal{A}(X)$ recovers X, with probability at least $1 - n^{-10 \log(n)}$ over the choice of $\mathcal{A}$.*

5.2   Invertible Linear Matrix Maps

In this section we prove Theorem 2 on solving random quadratic systems.

Proof (Proof of Theorem 2). Fix $t \leqslant n^{O(1)}$ and a nice distribution $\mathcal{Q}$.

Centering. We may assume that $\mathbb{E}_{\mathcal{Q}} Q = 0$. Otherwise, we can replace $\mathcal{Q}$ with $\mathcal{Q}'$ where $Q' = \frac{1}{\sqrt{2}}(Q_0 - Q_1)$ for independent draws $Q_0, Q_1 \sim \mathcal{Q}$. This is because $\mathcal{Q}'$ remains nice if $\mathcal{Q}$ is, clearly $\mathbb{E} Q' = 0$, and given $q_1, \ldots, q_m, q_1(x), \ldots, q_m(x)$ our algorithm can pair i to $i + 1$ (for even i) and instead consider $m/2$ samples of the form $(1/\sqrt{2})(q_i + q_{i+1}), (1/\sqrt{2})(q_i(x) + q_{i+1}(x))$. Thus for the remainder of the proof we assume $\mathbb{E} Q = 0$.

Our goal is to establish that there is $N \leqslant n^{O(1)}$ such that if $Q_1, \ldots, Q_N$ are i.i.d. draws from $\mathcal{Q}$, then $(1/n)Q_1, \ldots, (1/n)Q_N$ are ν-incoherent with respect to most $x \in [-t, t]^n$.

Incoherence Part One: Orthogonal Basis. First observe that since $\mathbb{E} Q = 0$ and $\mathbb{E} Q_{ij}^2 = 1$ and our pairwise independence assumption, we have

$$\mathbb{E} \,\mathrm{vec}(Q)\mathrm{vec}(Q)^\top = \mathrm{Id}_{n^2 \times n^2} .$$

Also, by niceness, every $|Q_{ij}| \leqslant O(n)$ with probability 1, for every i, j. Fix $i, j, k, \ell \leqslant n$. By the Bernstein inequality, given N independent draws $Q^{(1)}, \ldots, Q^{(N)}$, for any $s \geqslant 0$,

$$\mathbb{P}\left\{ \left| \frac{1}{N} \sum_{a \leqslant N} Q_{ij}^{(a)} Q_{k\ell}^{(a)} - \mathbb{E}\, Q_{ij} Q_{k\ell} \right| > s \right\} \leqslant \exp\left(\frac{-CNs^2}{n^4 + sn^2} \right)$$

for some universal constant C. Take $s = 1/n^4$ and $N = n^{10}$, this probability is at most $\exp(-O(n^2))$. Taking a union bound over $i, j, k, \ell \in [n]$, we find that with probability at least $1 - \exp(-O(n^2))$,

$$(1 - o(1))\mathrm{Id}_{n^2 \times n^2} \preceq \frac{1}{N} \sum_{u \leqslant N} \mathrm{vec}(Q^{(a)})\mathrm{vec}(Q^{(a)})^\top \preceq (1 + o(1))\mathrm{Id}_{n^2 \times n^2} .$$

Incoherence Part Two: Small Inner Products. Next we establish the other part of incoherence: that $\frac{1}{n^2}\langle x, Q^{(a)}x\rangle \leqslant \nu/n$ for all $a \leqslant N$. The coordinates of the vector x are independent, and each is bounded by t. Thus x is sub-Gaussian, with variance proxy $O(t^2)$. Since the coordinates of x have $\mathbb{E}\,x_i^2 \geqslant \Omega(t^2)$, the random vector y with coordinates $x_i/\sqrt{\mathbb{E}\,x_i^2}$ has sub-Gaussian norm $O(1)$.

Consider a fixed matrix $M \in \mathbb{R}^{n\times n}$, where M has Frobenius norm $\|M\|_F$ and spectral norm $\|M\|$. By the Hansen-Wright inequality, for any $s \geqslant 0$,

$$\mathbb{P}_y\left\{|y^\top M y - \mathbb{E}\,y^\top M y| > s\right\} \leqslant \exp\left(-Cs^2/(\|M\|_F^2 + s\|M\|)\right)$$

for some constant C.

If Q is any matrix in the support of $\mathcal{Q}$, then $\|Q/n\|_F \leqslant O(1)$ by niceness, and $\|Q\| \leqslant \|Q\|_F$. So for any such Q,

$$\mathbb{P}_y\left\{|y^\top (Q/n)y - \mathbb{E}\,y^\top (Q/n)y| > s\right\} \leqslant \exp\left(-Cs^2/(1+O(s))\right).$$

Taking $s = (\log n)^4$, this probability is at most $n^{-(\log n)^2}$ for large-enough n. Taking a union bound over $N \leqslant n^{O(1)}$ samples $Q^{(a)}$, with probability at least $1 - n^{-(\log n)^{1.5}}$ over y (for large enough n), every $Q^{(a)}$ has

$$\left|x^\top \cdot \frac{Q^{(a)}}{n} \cdot x\right| \leqslant \frac{(\log n)^{O(1)}}{n} \cdot \|xx^\top\|_F.$$

Putting it together, for $N = n^{O(1)}$, with probability at least $1 - n^{-(\log n)^{1.4}}$ for big-enough n, if $x \sim \{-t,\ldots,t\}^n$ then $Q^{(1)}/n,\ldots,Q^{(N)}/n$ are a $(\log n)^{O(1)}$-incoherent basis for x. Thus with probability at least $1 - n^{-10\log n}$ over $Q^{(1)},\ldots,Q^{(N)}$, we have $\mathbb{P}_x(Q^{(1)}/n,\ldots,Q^{(N)}/n$ is ν-incoherent for $x) \geqslant 1 - n^{-10\log n}$ (again for large enough n).

From Incoherence to Recovery. We can simulate the procedure of sampling $Q_1,\ldots,Q_m$ as in the theorem statement by first sampling $Q_1,\ldots,Q_N$, then randomly subsampling m of the Q's. If $Q_1,\ldots,Q_N$ are $(\log n)^{O(1)}$-incoherent for $xx^\top$, then Theorem 4 shows that with probability $1 - n^{-\log n}$ over the second sampling step, trace norm minimization recovers x, so long as the number of samples m is at least $n(\log n)^{O(1)}$. This finishes the proof.

6   Experiments

In this section, we describe the experiments that we performed on various classes of polynomials and how well do they perform in practice. All the codes were run and analysed on a MacBook Air (2013) laptop with 4 GB 1600 Mhz DDR3 RAM and an intel i5 processor with clock speed of 1.3 Ghz. We used Julia as our programming language and the package "Mosek" for the implementation of an SDP solver.

6.1 Experimental Cryptanalysis of Dense or Sparse Polynomials

First, we describe the setting of multivariate quadratic polynomials over the integers where the coefficients of each monomial is chosen independently at random from some distribution $\mathcal{D}$. Such dense polynomials were considered in [1,2]. We denote such polynomials by MQ.

The function `genmatrixDMQ` takes as input number of variables n and a coefficient bound C, and does the following:

1. For every monomial $x_i x_j$ where $i, j \in [n]$ and $j \geqslant i$, it samples a coefficient as a uniformly random integer in $[-C, C]$.
2. This coefficient is stored as $V[i][j]$ inside the matrix V.
3. The entire coefficient matrix is then made symmetric by just computing sum of itself with its transpose. Note that this quadratic form is the same as the one obtained in step 2.

The code can be found in Sect. A

Having described how to sample a polynomial, now we turn to the procedure to sample the input.

The function `genxMQ` on input number of variables n and a bound B, and does the following:

1. It samples an input vector $(x[1], ..., x[n])$ where each $x[i]$ is a sampled independently as a random integer between $[-B, B]$

The code of this function can be found in Sect. A.

Once we know how to sample polynomials and inputs we generate observations.

The function `genobsMQ` takes as input the number of input variables n, number of random polynomials m, coefficient bound C and bound on the planted input B. The function does the following:

1. It generates m polynomials randomly as per the distribution given by function `genmatrixDMQ` and stores them inside the vector L.
2. Then, it samples a planted input vector $\mathbf{x} = (x_1, ..., x_n)$ given by the distribution `genxMQ`.
3. Finally, it creates m observations of the form $obs[i] = \mathbf{x}^T L[i] \mathbf{x}$ for $i \in [m]$ where $\mathbf{x}^T$ is the transpose of vector $\mathbf{x}$.
4. It outputs polynomials, input and the observations.

This code can also be found in Sect. A

Once we have the observation we compute the function `recoverMQ` which implements the attack.

This function `recoverMQ` takes as m input observations as a vector obs along with the polynomial vector L. Then it finds a semi-definite matrix X constrained to the linear constraints that $Tr(L[i] * X) = obs[i]$ for $i \in [m]$, with the objective to minimize $Tr(X)$. Clearly, such an SDP is feasible as $X = \mathbf{x} \cdot \mathbf{x}^T$ (product of input vector with its transpose) satisfies the constraints.

Our experiments support the theorems given earlier in this paper. Indeed, for $m > 3n$, the SDP successfully recovers x for MQ polynomials. We similarly conducted experiments for sparse polynomials, where again the SDP successfully recovers x for $m > 3n$ in all experiments. We omit details of the sparse case to avoid redundancy.

6.2 Attacking [Lin-Matt18] Candidate Polynomials

In this section, we mount an attack on systems of quadratic polynomials with special structure. In particular, we consider the quadratic polynomials conjectured to provide security by [21]. Recall that the polynomials described in [21] are of the following structure. For each $j \in [m]$,

$$q_j(x_1, ..., x_n, x'_1, ..., x'_{n'}) = S_j(x_1, ..., x_n) + MQ_j(x'_1, ..., x'_{n'})$$

Here we write more about polynomials S_j, MQ_j as well as the input vector $(x_1, ..., x_n, x'_1, ..., x'_{n'})$.

1. MQ_j **Polynomials:** MQ_j are random quadratic polynomials over $(x'_1, ..., x'_{n'})$, where the coefficients of each degree two monomial $x'_i x'_k$ and degree one monomial x'_i are integers chosen independently at random from $[-C, C]$.
2. S_j **Polynomials:** S_j are random quadratic polynomials over $(x_1, ..., x_n)$ of the form:

$$S_j(x_1, ..., x_n) = \Sigma_{i=1}^{n/2} \alpha_i x_{\sigma_j(2*i)} x_{\sigma_j(2*i-1)} + \Sigma_{i=1}^{n} \beta_i x_i + \gamma$$

Here each coefficient α_i, β_i and γ are random integers chosen independently from $[-C, C]$. Here, σ_j is a random permutation from $[n]$ to $[n]$.
3. **Distribution of inputs** $x_1, ..., x_n$: Each x_i is chosen as a random integer sampled independently from the distribution χ_{B_1, B_2}. χ_{B_1, B_2} samples a random integer from $[-B_1, B_1]$ with probability 0.5 and from $[-B_2, B_2]$ with probability 0.5.
4. **Distribution of inputs** $x'_1, ..., x'_{n'}$: Each x'_i is chosen as a random integer sampled independently from the distribution $\chi_{B'}$. $\chi_{B'}$ samples a random integer from $[-B', B']$.
5. Set B_1, B_2, B', n, n' as follows:
   - Set $n = n'$ and $m = n^{1+\varepsilon}$, for some $\varepsilon > 0$.
   - B_1, B' and C are set arbitrarily.
   - Set $B_2 = \Omega(nB^2 + nBB_1)$.
   Here B is some polynomial in the security parameter.

The function `genmatrixsmq` generates polynomials of the form $S + MQ$. Then, we sample inputs using the function `genxdiscsmq`. Note that, this function samples input of length $n + n' + 2$. Two special variables $x[1]$ and $x[n+2]$ are set to 1 to achieve linear terms in the polynomials (as such $\mathbf{x}^T V \mathbf{x}$ is a homogeneous degree two polynomial in $\mathbf{x}$). Now we generate observation using the function `genobssmq`, which is implemented similarly.

Changing the SDP. To attack these special polynomials, we modify the SDP to introduce new constraints that help capture the structure of the polynomial. Specifically, because we know that the values $x_1, \ldots, x_n$ take small polynomially bounded values, we can enumerate over all possible "guesses" for $\Sigma_{i \in [n]} x_i^2$, and be sure that one of these will be correct. Let val1 be this guess. As such, we can add a constraint that $\Sigma_{i \in [n]} X[i, i] = $ val1 to the SDP, where X is the SDP matrix variable of size $n + n'$ by $n + n'$, and then solve. The code is formally described in Sect. A.

The Fig. 2 shows the plot of the ratio m/n versus the correlation of the recovered solution with the planted solution, for $n + n' = 60$ total variables, where $n = n' = 30$. Larger values of n that were still experimentally feasible, such as $n + n' = 120$ yielded similar graphs. We also remark that similar experimental observations can be made if we replace polynomials S with randomly generated sparse polynomials with $O(n)$ monomials. As before, for $m > 3(n + n')$, in our experiments, we always recover the correct solution.

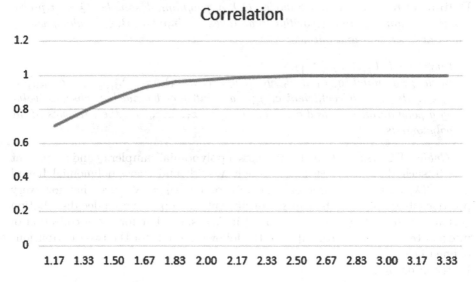

Correlation

1.2	
1	
0.8	
0.6	
0.4	
0.2	
0	

1.17 1.33 1.50 1.67 1.83 2.00 2.17 2.33 2.50 2.67 2.83 3.00 3.17 3.33

Fig. 2. Experimentally breaking Lin-Matt candidate. Graph shows quality of recovered solution vs. planted solution, for various values of m/n shown in the x-axis. Let $\mathbf{v}$ be the eigen vector with largest eigen value of the optimum matrix returned by the SDP. Let $\mathbf{x}$ be the planted solution. Quality of solution is defined as $\dfrac{\langle \mathbf{v}, \mathbf{x} \rangle}{\langle \mathbf{v}, \mathbf{v} \rangle^{\frac{1}{2}} \langle \mathbf{x}, \mathbf{x} \rangle^{\frac{1}{2}}}$

6.3 Attacking Polynomials of the Form $S + S + MQ$

Now we consider attacking a more general form of systems where each polynomial $q_j(\mathbf{x}_1, \mathbf{x}_2, \mathbf{x}_3)$ is of the following form:

- q_j takes as input three input vectors $\mathbf{x}_\ell = (x_{\ell,1}, \ldots, x_{\ell,n})$ for $\ell \in [3]$.
- Then, $q_j = S_{j,1}(\mathbf{x}_1) + S_{j,2}(\mathbf{x}_2) + MQ_j(\mathbf{x}_3)$

Inputs $\mathbf{x}_\ell$ for $\ell \in [3]$ are chosen as in the previous section. We observe that when we constrain the sum $\Sigma_{\ell \in [2], j \in [n]} x_{\ell,n}^2$, then SDP successfully recovers the planted solution using about same number of samples (for the same size of input) as for the previous case. This code of the recovery function recoverspecialssm is given in Sect. A. Note that in the code, the sum val1 + val2 is used to constrain this sum. This seems to generalise. If we consider a family where the polynomials q are of the form $S_1 + \ldots + S_k + MQ_1 + \ldots + MQ_k$, for values of k we could experimentally try (specifically $k \in \{1, 2, 3\}$) constraining the sum of trace corresponding to inputs of S polynomials leads to a break with probability 1.

7   Cubic Assumption

In this section, we discuss the cubic assumption proposed by [2]. Let us first recall the cubic version of the ΔRG assumption considered by [2]. First, we define a notion of a polynomial sampler Q.

Definition 6. *(Polynomial Sampler Q.) A polynomial sampler Q is a probabilistic polynomial time algorithm that takes as input $n, B, \in \mathbb{N}$ along with a constant $1 > \varepsilon > 0$ and outputs:*

- *Polynomials $(q_1, ..., q_{\lfloor n^{1+\varepsilon} \rfloor})$.*
- *Each polynomial $q_j(e_1, ..., e_n, y_1, ..., y_n, z_1, ..., z_n) = \Sigma_{i_1, i_2, i_3 \in [n]} c_{i_1, i_2, i_3} e_{i_1} y_{i_2} z_{i_3}$. Here, each coefficient c_{i_1, i_2, i_3} are integers bounded in absolute value by a polynomial in n and $e_1, ..., e_n, y_1, .., y_n, z_1, ..., z_n$ are the variables of the polynomials.*

Cubic ΔRG Assumption. There exists a polynomial sampler Q and a constant $\varepsilon > 0$, such that for every large enough $\lambda \in \mathbb{N}$, and every polynomial bound $B = B(\lambda)$ there exist large enough polynomial $n_B = \lambda^c$ such that for every positive integer $n > n_B$ there exists an efficiently samplable bounded distribution χ that is bounded by some polynomial in λ, n such that for every collection of integers $\{\delta_i\}_{i \in [\lfloor n^{1+\varepsilon} \rfloor]}$ with $|\delta_i| \leqslant B$, the following holds for the two distributions defined below:
Distribution dist$_1$:

- Fix a prime modulus $p = O(2^\lambda)$.
- **(Sample Polynomials.)** Run $Q(n, B, \varepsilon) = (q_1, ..., q_{\lfloor n^{1+\varepsilon} \rfloor})$.
- **(Sample Secret.)** Sample a secret $\mathbf{s} \leftarrow \mathbb{Z}_p^\lambda$
- Sample $a_i \leftarrow \mathbb{Z}_p^\lambda$ for $i \in [n]$.
- **(Sample LWE Errors.)** For every $i \in [n]$, sample $e_i, y_i, z_i \leftarrow \chi$. χ is a bounded distribution with a bound $poly(n)$ such that LWE assumption holds with error distribution χ, modulus p and dimension λ.
- Output $\{a_i, \langle a_i, s \rangle + e_i \mod p\}_{i \in [n]}, \{q_j, q_j(e_1, .., e_n, y_1, ..., y_n, z_1, ..., z_n)\}_{j \in [\lfloor n^{1+\varepsilon} \rfloor]}$

Distribution dist_2:

- Fix a prime modulus $p = O(2^\lambda)$.
- (**Sample Polynomials.**) Run $Q(n, B, \varepsilon) = (q_1, ..., q_{\lfloor n^{1+\varepsilon} \rfloor})$.
- (**Sample Secret.**) Sample a secret $\mathbf{s} \leftarrow \mathbb{Z}_p^\lambda$
- Sample $a_i \leftarrow \mathbb{Z}_p^\lambda$ for $i \in [n]$
- (**Sample LWE Errors**). For every $i \in [n]$, sample $e_i, y_i, z_i \leftarrow \chi$. χ is a bounded distribution with a bound $poly(\lambda)$ such that LWE assumption holds with error distribution χ, modulus p and dimension λ.
- Output $\{a_i, \langle a_i, s \rangle + e_i \mod p\}_{i \in [n]}, \{q_j, q_j(e_1, ..., e_n, y_1, ..., y_n, z_1, ..., z_n) + \delta_j\}_{j \in [\lfloor n^{1+\varepsilon} \rfloor]}$

The assumption states that there exists a constant $\varepsilon_{adv} > 0$ such that for any adversary $\mathcal{A}$ of size $2^{\lambda^{\varepsilon_{adv}}}$, the following holds:

$$|\mathbb{P}[\mathcal{A}(\text{dist}_1) = 1] - \mathbb{P}[\mathcal{A}(\text{dist}_2) = 1]| < 1 - 1/\lambda$$

Linearization Attack for n^2 stretch. The assumption above is only required to hold for stretch $n^{1+\varepsilon}$ for any small constant ε. However, we observe that the assumption described above suffers from an attack if the stretch is $O(\lambda \cdot n^2)$. The attack is simple and is described below.

Theorem 5. *The cubic ΔRG assumption does not hold with $m = O(\lambda \cdot n^2)$ polynomials $q_1, ... q_m$.*

Proof. Here is the breaking algorithm. For notational convenience, we only consider homogenous degree-3 polynomials. In this case, we can set $m = n^2(\lambda + 1)$. However, the algorithm trivially generalizes to all degree-3 polynomials with $m = n^2(\lambda + 3) + 2n + \lambda$.

1. Consider a polynomial $q_\ell(e_1, ..., e_n, y_1, ..., y_n, z_1, ..., z_n) = \Sigma_{i,j,k} c_{i,j,k,\ell} e_i y_j z_k$.
2. Rewrite $q_\ell(e_1, ..., e_n, y_1, ..., y_n, z_1, ..., z_n) = \Sigma_{i,j,k} c_{i,j,k,\ell}(\langle a_i, s \rangle + e_i - \langle a_i, s \rangle) y_j z_k$. Now note that a_i and $b_i = \langle a_i, s \rangle + e_i$ is given. Set $y_j z_k = w_{j,k}$ and $s_{\text{ind}} y_j z_k = w_{\text{ind},j,k}$ for $\text{ind} \in [\lambda], j \in [n], k \in [n]$.
3. Note that since $q_\ell(e_1, ..., e_n, y_1, ..., y_n, z_1, ..., z_n) = \Sigma_{i,j,k} c_{i,j,k,\ell}(b_i - \langle a_i, s \rangle) y_j z_k$ in $\mathbb{Z}_p$, the entire system of $m = n^{1+\varepsilon}$ samples can be written as a system of linear equations over $\mathbb{Z}_p$ in $(\lambda + 1)n^2$ variables $w_{\text{ind},j,k}$ and $w_{j,k}$. A simple gaussian elimination then recovers the solution.

On the existence of hard degree-3 polynomials. Feige [12] conjectured that its hard to distinguish a satisfiable random 3-SAT instance from a random 3-SAT instance with $C \cdot n$ clauses. Each disjunction $x_1 \vee x_2 \vee x_3$ corresponds to the polynomial $1 - (1 - x_1)(1 - x_2)(1 - x_3)$. This intuition gives rise to a set of candidate polynomials $q_{i,j}$, which depends on three randomly chosen variables and maps $\{0, 1\}^n$ to $\{0, 1\}$. Each $q_{i,j}$ has at most 8 monomials. Intuitively speaking, to choose clauses, instead of choosing clauses at random – something that is known to lead to weak RANDOM 3SAT instances – we first choose a planted *boolean* solution $x^* \in \{0, 1\}^n$, and always choose clauses such that exactly one or all three literals in the clause evaluate to true. This has the property that each clause individually induces a uniform constraint on any pair of variables x_i and x_j. In the boolean case, this distribution of clauses is believed to give rise to hard distributions, which suggests that the expanding polynomial systems corresponding to these clauses should be hard to solve in general.

To construct obfuscation, we need the stretch to be at least $n^{1+\varepsilon}$ for any constant $\varepsilon > 0$. All known algorithms take exponential time as long as $n^{1.5-\varepsilon}$ clauses are given out. This leads to a conjecture, which is also related to the work of [11]. As a result, we conjecture that the following candidate expanding family of degree-3 polynomials is hard to solve.

3SAT Based Candidate. Let $t = B^2\lambda$. Here, $B(\lambda)$ is the magnitude of the perturbations allowed. Sample each polynomial q_i' for $i \in [\eta]$ as follows. $q_i'(\mathbf{x}_1, \ldots, \mathbf{x}_t, \mathbf{y}_1, \ldots, \mathbf{y}_t, \mathbf{z}_1, \ldots, \mathbf{z}_t) = \Sigma_{j \in [t]} q_{i,j}'(\mathbf{x}_j, \mathbf{y}_j, \mathbf{z}_j)$. Here $\mathbf{x}_j \in \chi^{d \times n}$ and $\mathbf{y}_j, \mathbf{z}_j \in \chi^n$ for $j \in [t]$. In other words, q_i' is a sum of t polynomials $q_{i,j}'$ over t disjoint set of variables. Let χ denote a discrete gaussian random variable with mean 0 and standard deviation n. Now we describe how to sample $q_{i,j}'$ for $j \in [\eta]$.

1. Sample randomly inputs $\mathbf{x}^*, \mathbf{y}^*, \mathbf{z}^* \in \{0,1\}^n$.
2. To sample $q_{i,j}'$ do the following. Sample three indices randomly and independently $i_1, i_2, i_3 \leftarrow [n]$. Sample three signs $b_{1,i,j}, b_{2,i,j}, b_{3,i,j} \in \{0,1\}$ uniformly such that $b_{1,i,j} \oplus b_{2,i,j} \oplus b_{3,i,j} \oplus \mathbf{x}^*[i_1] \oplus \mathbf{y}^*[i_2] \oplus \mathbf{z}^*[i_3] = 1$.
3. Set $q_{i,j}'(\mathbf{x}_j, \mathbf{y}_j, \mathbf{z}_j) = 1 - (b_{1,i,j} \cdot \mathbf{x}_j[i_1] + (1 - b_{1,i,j}) \cdot (1 - \mathbf{x}_j[i_1])) \cdot (b_{2,i,j} \cdot \mathbf{y}_j[i_2] + (1 - b_{2,i,j}) \cdot (1 - \mathbf{y}_j[i_2])) \cdot (b_{3,i,j} \cdot \mathbf{z}_j[i_3] + (1 - b_{3,i,j}) \cdot (1 - \mathbf{z}_j[i_3]))$

Acknowledgements. Boaz Barak was supported by NSF awards CCF 1565264 and CNS 1618026 and a Simons Investigator Fellowship. Samuel B. Hopkins was supported by a Miller Postdoctoral Fellowship and NSF award CCF 1408673. Pravesh Kothari was supported in part by Ma fellowship from the Schmidt Foundation and Avi Wigderson's NSF award CCF-1412958. Amit Sahai and Aayush Jain were supported in part from a DARPA/ARL SAFEWARE award, NSF Frontier Award 1413955, and NSF grant 1619348, BSF grant 2012378, a Xerox Faculty Research Award, a Google Faculty Research Award, an equipment grant from Intel, and an Okawa Foundation Research Grant. Aayush Jain was also supported by Google PhD Fellowship 2018, in the area of Privacy and Security. This material is based upon work supported by the Defense Advanced Research Projects Agency through the ARL under Contract W911NF-15-C-0205. The views expressed are those of the authors and do not reflect the official policy or position of the Department of Defense, the National Science Foundation, the U.S. Government or Google.

A   Julia Code

```
function genmatrixDMQ(n, C)
    V = randn(n,n)
    for i in 1:n
        for j in 1:n
            V[i,j] = 0
        end
    end
```

```
    for i in 1:n
        for j in i:n
            V[i,j] = nr.randint(-C, high=C+1)
        end
    end

    (V'+V)/2
end


function genxMQ(n ,B)
    x = randn(n,1)
    for i in 1:n
        x[i] = nr.randint(-B, high=B+1)
    end
    x
end

function genobsMQ(n,m,C,B)
    L = [genmatrixDMQ(n,C) for i in 1:m]
    x = genxMQ(n,B)
    obs = [x'*L[i]*x for i in 1:m]
    L,obs,x
end

function recoverMQ(L,obs)

    n = size(L[1])[1]
    m = length(L)

    model = Model(solver = MosekSolver())
    @variable(model, X[1:n,1:n], SDP)

    # let's maximize the trace
    @objective(model, Min, trace(X))

    # this makes the constraints
    for i in 1:m
        @constraint(model, trace(L[i]*X).==obs[i])
    end

    # this solves the problem
    solve(model)
    getvalue(X)

end


function genmatrixsmq(n, n2, nprime, C)
    V = randn(n+nprime+2,n+nprime+2)
    Z = randn(n+nprime+2,n+nprime+2)
```

```
    for i in 1:n+nprime+2
        for j in 1:n+nprime+2
            V[i,j] = 0
        end
    end

    a=randperm(n)

    #sparse monomials

    for i in 1:n2
        V[ a[2*i-1]+1,a[2*i]+1] = rand(-C:C)
    end



    #MQ monomials
    for i in n+3:n+nprime+2
        for j in n+3:n+nprime+2
            V[i,j] = rand(-C:C)
        end
    end

    #Linear terms in S


    for j in 2:n+1
        V[1,j]=rand(-C:C)

    end

    #Linear terms in MQ

    for j in n+3:n+nprime+2
        V[n+2,j]=rand(-C:C)
    end



    Z=V'+V



    Z

end

function genxdiscsmq(n,n2,nprime,B1,B2,Bprime)
    x = randn(n+nprime+2,1)
    x[1]=1
    x[n+2]=1
    for i in n+3:n+nprime+2
```

```
            x[i] = rand(-Bprime:Bprime)
        end

        for i in 2:n+1
            temp1=rand(-B1:B1)
            temp2=rand(-B2:B2)
            temp3=rand(0:1)
            x[i] = (temp3*temp1+(1-temp3)*temp2 )
        end


        x
    end

    function genobssmq(n,n2,nprime,m,C,B1,B2,Bprime)
        L = [genmatrixsmq(n,n2,nprime,C) for i in 1:m]
        x = genxdiscsmq(n,n2, nprime,B1,B2,Bprime)
        obs = [x'*L[i]*x for i in 1:m]
        L,obs,x
    end


    function recoverspecialsmq(L,obs,n,n2,nprime,m,val1)




        model = Model(solver = MosekSolver())
        @variable(model, X[1:nprime+n+2,1:nprime+n+2], SDP)

        # let's maximize the trace
        @objective(model, Min, trace(X))

        # this makes the constraints
        for i in 1:m
            @constraint(model, trace(L[i]*X).==obs[i])
        end

        @constraint(model, X[1,1]==1)
        @constraint(model, X[n+2,n+2]==1)


        @constraint(model, trace(X[1:n+1,1:n+1])=val1[1])



        # this solves the problem
        solve(model)
        getvalue(X)

    end

    function recoverspecialssm(L,obs,n,n2,nprime,m,val1, val2,val3)
        model = Model(solver = MosekSolver())
        @variable(model, X[1:nprime+2*n+3,1:nprime+2*n+3], SDP)
```

```
# let's maximize the trace
@objective(model, Min, trace(X))

# this makes the constraints
for i in 1:m
    @constraint(model, trace(L[i]*X).==obs[i])
end


@constraint(model, X[1,1]==1)
@constraint(model, X[n+2,n+2]==1)
@constraint(model, X[n*2+3,2*n+3]==1)


@constraint(model, trace(X[1:n+1,1:n+1]) + trace(X[n+2:2*n+2,n+2:2*n+2])
>= val1[1] + val2[1])
 # this solves the problem
solve(model)
getvalue(X)

end
```

References

1. Agrawal, S.: New methods for indistinguishability obfuscation: Bootstrapping and instantiation. IACR Cryptology ePrint Archive 2018, 633 (2018). https://eprint.iacr.org/2018/633
2. Ananth, P., Jain, A., Sahai, A.: Indistinguishability obfuscation without multilinear maps: iO from LWE, bilinear maps, and weak pseudorandomness. IACR Cryptology ePrint Archive 2018, 615 (2018). https://eprint.iacr.org/2018/615
3. Ananth, P., Sahai, A.: Projective arithmetic functional encryption and indistinguishability obfuscation from degree-5 multilinear maps. In: Coron, J.-S., Nielsen, J.B. (eds.) EUROCRYPT 2017, Part I. LNCS, vol. 10210, pp. 152–181. Springer, Cham (2017). https://doi.org/10.1007/978-3-319-56620-7_6
4. Barak, B., Brakerski, Z., Komargodski, I., Kothari, P.K.: Limits on low-degree pseudorandom generators (or: sum-of-squares meets program obfuscation). In: Nielsen, J.B., Rijmen, V. (eds.) EUROCRYPT 2018, Part II. LNCS, vol. 10821, pp. 649–679. Springer, Cham (2018). https://doi.org/10.1007/978-3-319-78375-8_21
5. Boneh, D., Silverberg, A.: Applications of multilinear forms to cryptography. Contemp. Math. **324**, 71–90 (2002)
6. Boneh, D., Wu, D.J., Zimmerman, J.: Immunizing multilinear maps against zeroizing attacks. IACR Cryptology ePrint Archive 2014, 930 (2014). http://eprint.iacr.org/2014/930
7. Brakerski, Z., Gentry, C., Halevi, S., Lepoint, T., Sahai, A., Tibouchi, M.: Cryptanalysis of the quadratic zero-testing of GGH. Cryptology ePrint Archive, Report 2015/845 (2015). http://eprint.iacr.org/

8. Cheon, J.H., Han, K., Lee, C., Ryu, H., Stehlé, D.: Cryptanalysis of the multilinear map over the integers. In: Oswald, E., Fischlin, M. (eds.) EUROCRYPT 2015, Part I. LNCS, vol. 9056, pp. 3–12. Springer, Heidelberg (2015). https://doi.org/10.1007/978-3-662-46800-5_1

9. Cheon, J.H., Lee, C., Ryu, H.: Cryptanalysis of the new clt multilinear maps. Cryptology ePrint Archive, Report 2015/934 (2015). http://eprint.iacr.org/

10. Coron, J.-S., et al.: Zeroizing without low-level zeroes: new MMAP attacks and their limitations. In: Gennaro, R., Robshaw, M. (eds.) CRYPTO 2015, Part I. LNCS, vol. 9215, pp. 247–266. Springer, Heidelberg (2015). https://doi.org/10.1007/978-3-662-47989-6_12

11. Daniely, A., Linial, N., Shalev-Shwartz, S.: From average case complexity to improper learning complexity. In: STOC, pp. 441–448. ACM (2014)

12. Feige, U.: Relations between average case complexity and approximation complexity. In: STOC, pp. 534–543. ACM (2002)

13. Garg, S., Gentry, C., Halevi, S.: Candidate multilinear maps from ideal lattices. In: Johansson, T., Nguyen, P.Q. (eds.) EUROCRYPT 2013. LNCS, vol. 7881, pp. 1–17. Springer, Heidelberg (2013). https://doi.org/10.1007/978-3-642-38348-9_1

14. Garg, S., Gentry, C., Halevi, S., Raykova, M., Sahai, A., Waters, B.: Candidate indistinguishability obfuscation and functional encryption for all circuits. In: 54th Annual IEEE Symposium on Foundations of Computer Science, FOCS 2013, 26–29 October, 2013, Berkeley, pp. 40–49 (2013)

15. Grigoriev, D.: Linear lower bound on degrees of positivstellensatz calculus proofs for the parity. Theor. Comput. Sci. **259**(1–2), 613–622 (2001)

16. Gross, D.: Recovering low-rank matrices from few coefficients in any basis. IEEE Trans. Inform. Theory **57**(3), 1548–1566 (2011). https://doi.org/10.1109/TIT.2011.2104999

17. Halevi, S.: Graded encoding, variations on a scheme. IACR Cryptol. ePrint Archive **2015**, 866 (2015)

18. Hu, Y., Jia, H.: Cryptanalysis of GGH map. IACR Cryptol. ePrint Archive **2015**, 301 (2015)

19. Lin, H.: Indistinguishability obfuscation from constant-degree graded encoding schemes. In: Fischlin, M., Coron, J.-S. (eds.) EUROCRYPT 2016, Part I. LNCS, vol. 9665, pp. 28–57. Springer, Heidelberg (2016). https://doi.org/10.1007/978-3-662-49890-3_2

20. Lin, H.: Indistinguishability obfuscation from SXDH on 5-linear maps and locality-5 PRGs. In: Katz, J., Shacham, H. (eds.) CRYPTO 2017, Part I. LNCS, vol. 10401, pp. 599–629. Springer, Cham (2017). https://doi.org/10.1007/978-3-319-63688-7_20

21. Lin, H., Matt, C.: Pseudo flawed-smudging generators and their application to indistinguishability obfuscation. IACR Cryptology ePrint Archive 2018, 646 (2018). https://eprint.iacr.org/2018/646

22. Lin, H., Tessaro, S.: Indistinguishability obfuscation from bilinear maps and blockwise local PRGs. Cryptology ePrint Archive, Report 2017/250 (2017). http://eprint.iacr.org/2017/250

23. Lin, H., Vaikuntanathan, V.: Indistinguishability obfuscation from DDH-like assumptions on constant-degree graded encodings. In: IEEE 57th Annual Symposium on Foundations of Computer Science, FOCS 2016, 9–11 October 2016, Hyatt Regency, New Brunswick, pp. 11–20 (2016)

24. Lombardi, A., Vaikuntanathan, V.: On the non-existence of blockwise 2-local prgs with applications to indistinguishability obfuscation. IACR Cryptology ePrint Archive 2017, 301 (2017). http://eprint.iacr.org/2017/301

25. Miles, E., Sahai, A., Zhandry, M.: Annihilation attacks for multilinear maps: crypt-analysis of indistinguishability obfuscation over GGH13. In: Robshaw, M., Katz, J. (eds.) CRYPTO 2016, Part II. LNCS, vol. 9815, pp. 629–658. Springer, Heidelberg (2016). https://doi.org/10.1007/978-3-662-53008-5_22
26. Minaud, B., Fouque, P.A.: Cryptanalysis of the new multilinear map over the integers. Cryptology ePrint Archive, Report 2015/941 (2015). http://eprint.iacr.org/
27. Recht, B.: A simpler approach to matrix completion. J. Mach. Learn. Res. **12**, 3413–3430 (2011)
28. Recht, B., Fazel, M., Parrilo, P.A.: Guaranteed minimum-rank solutions oflinear matrix equations via nuclear norm minimization. SIAM Rev. **52**(3), 471–501 (2010). https://doi.org/10.1137/070697835
29. Sahai, A., Waters, B.: How to use indistinguishability obfuscation: deniable encryption, and more. In: Symposium on Theory of Computing, STOC 2014, New York, May 31 - June 03, 2014, pp. 475–484 (2014)
30. Schoenebeck, G.: Linear level lasserre lower bounds for certain k-CSPs. In: 49th Annual IEEE Symposium on Foundations of Computer Science, FOCS 2008, 25–28 October 2008, Philadelphia, pp. 593–602 (2008)

How to Leverage Hardness of Constant-Degree Expanding Polynomials over $\mathbb{R}$ to build $i\mathcal{O}$

Aayush Jain[1]($\boxtimes$), Huijia Lin[2], Christian Matt[3], and Amit Sahai[1]

[1] UCLA, Los Angeles, USA
{aayushjain,sahai}@cs.ucla.edu
[2] University of Washington, Seattle, USA
rachel@cs.washington.edu
[3] Concordium, Zurich, Switzerland
cm@concordium.com

Abstract. In this work, we introduce and construct D-restricted Functional Encryption (FE) for any constant $D \geq 3$, based only on the SXDH assumption over bilinear groups. This generalizes the notion of 3-restricted FE recently introduced and constructed by Ananth et al. (ePrint 2018) in the generic bilinear group model.

A $D = (d + 2)$-restricted FE scheme is a secret key FE scheme that allows an encryptor to efficiently encrypt a message of the form $M = (\boldsymbol{x}, \boldsymbol{y}, \boldsymbol{z})$. Here, $\boldsymbol{x} \in \mathbb{F}_{\mathbf{p}}^{d \times n}$ and $\boldsymbol{y}, \boldsymbol{z} \in \mathbb{F}_{\mathbf{p}}^n$. Function keys can be issued for a function $f = \Sigma_{I=(i_1,..,i_d,j,k)}\, c_I \cdot \boldsymbol{x}[1, i_1] \cdots \boldsymbol{x}[d, i_d] \cdot \boldsymbol{y}[j] \cdot \boldsymbol{z}[k]$ where the coefficients $c_I \in \mathbb{F}_{\mathbf{p}}$. Knowing the function key and the ciphertext, one can learn $f(\boldsymbol{x}, \boldsymbol{y}, \boldsymbol{z})$, if this value is bounded in absolute value by some polynomial in the security parameter and n. The security requirement is that the ciphertext hides $\boldsymbol{y}$ and $\boldsymbol{z}$, although it is not required to hide $\boldsymbol{x}$. Thus $\boldsymbol{x}$ can be seen as a public attribute.

D-restricted FE allows for useful evaluation of constant-degree polynomials, while only requiring the SXDH assumption over bilinear groups. As such, it is a powerful tool for leveraging hardness that exists in constant-degree expanding families of polynomials over $\mathbb{R}$. In particular, we build upon the work of Ananth et al. to show how to build indistinguishability obfuscation ($i\mathcal{O}$) assuming only SXDH over bilinear groups, LWE, and assumptions relating to weak pseudorandom properties of constant-degree expanding polynomials over $\mathbb{R}$.

1 Introduction

Program obfuscation transforms a computer program P into an equivalent program $O(P)$ such that any secrets present within P are "as hard as possible" to extract from $O(P)$. This property can be formalized by the notion of indistinguishability obfuscation ($i\mathcal{O}$) [9,32]. Formally, $i\mathcal{O}$ requires that given any two

---

This paper is a merge of two independent works, one by Jain and Sahai, and the other by Lin and Matt.

© International Association for Cryptologic Research 2019
Y. Ishai and V. Rijmen (Eds.): EUROCRYPT 2019, LNCS 11476, pp. 251–281, 2019.
https://doi.org/10.1007/978-3-030-17653-2_9

equivalent programs P_1 and P_2 of the same size, a computationally bounded adversary cannot distinguish $O(P_1)$ from $O(P_2)$. $i\mathcal{O}$ has far-reaching application [26, 50], significantly expanding the scope of problems to which cryptography can be applied [12, 16, 19, 25, 28, 31, 34, 35, 38, 50].

The work of [26] gave the first mathematical candidate $i\mathcal{O}$ construction, and since then several additional candidates have been proposed and studied [3, 5, 7, 8, 13–15, 17, 18, 20–24, 29, 33, 36, 39, 43, 44, 46, 47, 49].

Constructing $i\mathcal{O}$ without MMaps. Until 2018, all known constructions relied on multilinear maps [21, 22, 24, 29]. Unfortunately, multilinear map constructions are complex and surviving multilinear map security models [11, 27, 45] are themselves complex and difficult to analyze, as they have had to be modified in light of a sequence of attacks on multilinear map candidates [13, 14, 17, 18, 20, 33, 36, 46, 47].

This state of affairs is troubling scientifically, as we would like to be able to reduce the security of $i\mathcal{O}$ to problems that are simple to state, and where the underlying mathematics has a long history of study.

Everything old is new again: low-degree polynomials over the reals. Humanity has studied solving systems of (low-degree) polynomials over the reals for hundreds of years. Is it possible to use *hardness* associated with polynomial systems over the reals cryptographically? Surprisingly, despite hundreds of years of study, remarkably little is known about average-case hardness corresponding to *expanding* polynomial systems, where the number of real variables is n, and the polynomial equations over them is $n^{1+\epsilon}$ for $\epsilon > 0$.

The recent works of [1, 4, 42] introduced a new way constructing $i\mathcal{O}$ without relying on multilinear maps, by looking to hardness that may be present in degree two [1, 4, 42] or degree three [4] expanding polynomial systems over the reals.

The primary goal of our work is to extend the approach proposed by [4] to be able to use hardness associated with suitable expanding polynomial systems of *any constant degree.*

Leveraging low degree pseudorandomness over Z to build $i\mathcal{O}$. The key idea behind the work of [4] is to posit the existence of weak pseudorandom objects that are closely related to polynomials of degree 2 or 3 over the integers. They then introduce the crucial notion of 3-restricted functional encryption, which is a notion of functional encryption that allows for a *restricted* but still useful evaluation of degree-3 polynomials. This notion allows for the *natural* application of expanding families of degree-3 polynomials. (See below for further discussion on restricted-FE and its uses.).

Departing from previous work [5, 40, 43] that required at least trilinear maps to construct any meaningful FE for degree-3 functions, [4] show how to construct 3-restricted FE using only *bilinear maps.* Finally, by combining 3-restricted FE with the weak pseudorandom objects mentioned above, they achieve $i\mathcal{O}$ (also assuming LWE).

The goals of our present work are two-fold:

- To show how to extend the above approach beyond degree 3, to any constant degree D for $D \geq 3$. To do so, the key ingredient we construct is D-restricted FE, again *only* using bilinear maps regardless of the constant D.
- Furthermore, we construct D-restricted FE assuming only the SXDH assumption to hold over the bilinear map groups, instead of the generic bilinear model that was needed in [4].

We now elaborate.

D-restricted FE. A D-restricted FE scheme naturally generalizes the notion of 3-restricted FE scheme from [4]. We will write $D = d+2$ for notational convenience. Such a scheme is a secret key FE scheme that allows an encryptor to encrypt a message of the form $M = (\boldsymbol{x}, \boldsymbol{y}, \boldsymbol{z})$, where $\boldsymbol{x} \in \mathbb{F}^{d \times n}$ and $\boldsymbol{y}, \boldsymbol{z} \in \mathbb{F}_{\mathbf{p}}^n$. Function keys can be issued for a function $f = \Sigma_{\boldsymbol{I}=(i_1,..,i_d,j,k)} \, c_I \cdot \boldsymbol{x}[1, i_1] \cdots \boldsymbol{x}[d, i_d] \cdot \boldsymbol{y}[j] \cdot \boldsymbol{z}[k]$ with coefficients $c_I \in \mathbb{F}_{\mathbf{p}}$. Knowing the key and the ciphertext, one can learn $f(\boldsymbol{x}, \boldsymbol{y}, \boldsymbol{z})$, if this value is bounded in absolute value by some polynomial in the security parameter and n. The security requirement is that the ciphertext hides $\boldsymbol{y}$ and $\boldsymbol{z}$, although it is not required to hide $\boldsymbol{x}$. Thus $\boldsymbol{x}$ can be seen as a public attribute. For implications to $i\mathcal{O}$, we require that encryption complexity should grow only linearly in n (up to a polynomial factor in the security parameter).

Observe that for a given family of degree-D polynomials Q fixed in a function key, the notion of D-restricted FE allows an encryptor to choose the values of all variables $\boldsymbol{x}, \boldsymbol{y}, \boldsymbol{z}$ at the time of encryption, and the decryptor will obtain $Q(\boldsymbol{x}, \boldsymbol{y}, \boldsymbol{z})$. This allows for the most natural use of degree-D polynomials. We stress this point because other, less natural uses, are possible without using D-restricted FE, but these are unsatisfactory: One example would be where along with the polynomial Q the values of all variables $\boldsymbol{x}$ would also be fixed inside the function key. This would reduce the degree-D polynomials Q to quadratic polynomials, and just quadratic FE would then suffice (see, e.g., [1, 42]). However, again, this latter, less natural, approach would not allow $\boldsymbol{x}$ to be chosen freshly with each encryption. With our notion of D-restricted FE, such an unnatural setting – where some variables are fixed but others are freshly chosen with each encryption – can be avoided completely.

Why is it important to go beyond degree 3? At the core of the new works that construct $i\mathcal{O}$ without multilinear maps is the following key question: For some constant D, do there exist "expanding" distributions of polynomials $q_1, \ldots, q_m$ of degree D, where $m = n^{1+\epsilon}$ with polynomially-bounded coefficients, such that if one obtains $\boldsymbol{x} = (x_1, \ldots, x_n) \in \mathbb{Z}^n$ by sampling each x_i from a "nice" distribution with polynomially-bounded support, then is it hard to solve for $\boldsymbol{x}$ given $q_1(\boldsymbol{x}), \ldots, q_m(\boldsymbol{x})$? Remarkably, even though this question has a many-hundred year history within mathematics and nearly every branch of science, surprisingly little is known about *hardness* in this setting! And yet the hardness of such inversion problems is necessary (though not sufficient, see below) for this new line of work on constructing $i\mathcal{O}$.

Recently, [10] gave evidence that such problems may *not* be hard for $D = 2$. The case for $D = 3$ is less studied, and seems related to questions like the hardness of RANDOM 3-SAT. However, it seems that increasing D to larger constants should give us more confidence that hard distributions exist. For example, for $D = 5$ and larger, this becomes related to the hardness of natural generalizations of the Goldreich PRG [30,48]. It is also likely that as D grows, hardness "kicks in" for smaller values of n, similar to how the hardness of RANDOM k-SAT for constant $k > 3$ can be observed experimentally for much smaller values of n, than for RANDOM 3-SAT. Thus, our study could impact efficiency, as well.

Since studying the hardness of solving expanding families of polynomial equations over $\mathbb{R}$ is an exciting new line of cryptanalytic research, it is particularly important to study what values of D are cryptographically interesting. Before our work, only $D = 2$ and $D = 3$ were known to lead to $i\mathcal{O}$; our work shows that hardness for any constant degree D is interesting and cryptographically useful.

We stress that ensuring the hardness of solving for $\boldsymbol{x}$ given $q_1(\boldsymbol{x}), \ldots, q_m(\boldsymbol{x})$ is just the first step. Our work also clarifies the actual hardness assumptions that we need to imply $i\mathcal{O}$ as the following two assumptions. Since $D > 2$, let $D = d + 2$ for the rest of the discussion.

Weak LWE with leakage. This assumption says that there exists distributions χ over the integers and Q over families of multilinear degree-D polynomials such that the following two distributions are weakly indistinguishable, meaning that no efficient adversary can correctly identify the distribution from which a sample arose with probability above $\frac{1}{2} + 1/4\lambda$.

Distribution $\mathcal{D}_1$: Fix a prime modulus $\mathbf{p} = O(2^\lambda)$. Run $Q(n, B, \epsilon)$ to obtain polynomials $(q_1, ..., q_{\lfloor n^{1+\epsilon} \rfloor})$. Sample a secret $\boldsymbol{s} \leftarrow \mathbb{Z}_p^\lambda$ and sample $\boldsymbol{a}_{j,i} \leftarrow \mathbb{Z}_p^\lambda$ for $j \in [d], i \in [n]$. Finally, for every $j \in [d], i \in [n]$, sample $e_{j,i}, y_i, z_i \leftarrow \chi$, and write $\boldsymbol{e}_j = (e_{j,1}, \ldots, e_{j,n})$, $\boldsymbol{y} = (y_1, \ldots, y_n)$, $\boldsymbol{z} = (z_1, \ldots, z_n)$. Output:

$$\{\boldsymbol{a}_{j,i}, \langle \boldsymbol{a}_{j,i}, \boldsymbol{s} \rangle + e_{j,i} \bmod p\}_{j \in [d], i \in [n]}$$

along with

$$\{q_k, q_k(\boldsymbol{e_1}, \ldots, \boldsymbol{e_d}, \boldsymbol{y}, \boldsymbol{z})\}_{k \in [n^{1+\epsilon}]}$$

Distribution $\mathcal{D}_2$ is the same as $\mathcal{D}_1$, except that we additionally sample $e'_{j,i} \leftarrow \chi$ for $j \in [d], i \in [n]$. The output is now

$$\{\boldsymbol{a}_{j,i}, \langle \boldsymbol{a}_{j,i}, \boldsymbol{s} \rangle + e'_{j,i} \bmod p\}_{j \in [d], i \in [n]}$$

along with

$$\{q_k, q_k(\boldsymbol{e_1}, \ldots, \boldsymbol{e_d}, \boldsymbol{y}, \boldsymbol{z})\}_{k \in [n^{1+\epsilon}]}$$

We can think of the polynomials $q_k(\boldsymbol{e_1}, \ldots, \boldsymbol{e_d}, \boldsymbol{y}, \boldsymbol{z})$ as "leaking" some information about the LWE errors $e_{j,i}$. The assumption above states that such leakage provides only a limited advantage to the adversary. Critically, the fact that there are $n^2 > n^{1+\epsilon}$ quadratic monomials involving just $\boldsymbol{y}$ and $\boldsymbol{z}$ above, which are not used in the LWE samples at all, is crucial to avoiding linearization attacks over

$\mathbb{Z}_p$ in the spirit of Arora-Ge [6]. For more discussion of the security of the above assumption in the context of $D = 3$, see [10].

The second assumption deals only with expanding degree-D polynomials over the reals, and requires that these polynomials are weakly perturbation resilient.

Weak Perturbation-Resilience. The second assumption is that there exists polynomials that for the same parameters above the following two distributions are weakly indistinguishable. By weakly indistinguishability we mean that no efficient adversary can correctly identify the distribution from which a sample arose with probability above $1 - 2/\lambda$. Let $\delta_i \in \mathbb{Z}$ be such that $|\delta_i| < B(\lambda, n)$ for some polynomial B and $i \in [n^{1+\epsilon}]$:

Distribution $\mathcal{D}_1$ consists of the evaluated polynomial samples. That is, we output:

$$\{q_k, q_k(e_1, \ldots, e_d, y, z)\}_{k \in [n^{1+\epsilon}]}$$

Distribution $\mathcal{D}_2$ consists of the evaluated polynomial samples with added perturbations δ_i for $i \in [n^{1+\epsilon}]$. That is, we output:

$$\{q_k, q_k(e_1, \ldots, e_d, y, z) + \delta_k\}_{k \in [n^{1+\epsilon}]}$$

These assumptions are sketched here informally; the formal definitions are given in Sect. 5.

Our Results: Our results can be summarized as follows. First, we construct a $(d + 2)$ restricted FE scheme from the SXDH assumption.

Theorem 1. *Assuming SXDH over bilinear maps, there is a construction of a $(d + 2)$ restricted FE scheme for any constant $d \geq 1$.*

Then, we give candidates of perturbation resilient generators that can be implemented using a $(d + 2)$ restricted FE scheme. Finally, using such a perturbation resilient generator and $(d + 2)$ restricted FE, we construct $i\mathcal{O}$ via the approach given by [4]. Here is our final theorem.

Theorem 2. *For any constant integer $d \geq 1$, two distinguishing gaps $\mathsf{adv}_1, \mathsf{adv}_2$, if $\mathsf{adv}_1 + \mathsf{adv}_2 \leq 1 - 2/\lambda$ then assuming,*

- *Subexponentially hard LWE.*
- *Subexponentially hard SXDH.*
- *PRGs with*
  - *Stretch of $k^{1+\epsilon}$ (length of input being k bits) for some constant $\epsilon > 0$.*
  - *Block locality $d + 2$.*
  - *Security with distinguishing gap bounded by adv_1 against adversaries of sub-exponential size.*
- *dΔRG with distinguishing gap bounded by adv_2 against adversaries of size 2^λ. Details about the notion of dΔRG can be found in Sects. 5 and 6.*

there exists a secure $i\mathcal{O}$ scheme for P/poly.

We additionally note that the work of [42] provides a construction of $i\mathcal{O}$ from a different notion of weak randomness generators called pseudo flawed-smudging generators, and a partially hiding FE scheme that can compute them. Their notion of partially hiding FE is implied by our degree $(d + 2)$ restricted FE. Therefore, if using our candidates of perturbation resilient generators as candidates of pseudo flawed-smudging generators, we can obtain $i\mathcal{O}$ via the the approach of [42], as summarized in the theorem below.

Theorem 3. *For any constant integer $d \geq 1$, assuming,*

- *LWE,*
- *SXDH,*
- *PRGs with*
  - *Stretch of $k^{1+\epsilon}$ (length of input being k bits) for some constant $\epsilon > 0$,*
  - *Constant locality and additional mild structural properties (see [42] for details),*
- *Pseudo flawed-smudging generators with degree d public computation and degree 2 private computation. Details about the notion of pseudo flawed-smudging generators can be found in Sect. 5.2 and [42].*

where all primitives are secure against adversaries of polynomial sizes with subexponentially small distinguishing gaps. Then, there exists a subexponentially secure $i\mathcal{O}$ scheme for P/poly.

For simplicity, we focus on working with the notion of ΔRG here and provide more details on how to work with pseudo flawed-smudging generators in [37].

We now proceed with a more detailed, but still informal, technical overview of our techniques.

2 Technical Overview

$(d + 2)$-*restricted FE.* The key technical tool constructed in this work is the notion of $(d + 2)-$restricted FE (dFE for short) for any constant integer $d \geq 1$. We recall that a dFE scheme over $\mathbb{F}_{\mathbf{p}}$ is a secret key functional encryption scheme for the functions f of the following form: $f : \mathbb{F}_{\mathbf{p}}^{n \times (d+2)} \to \mathbb{F}_{\mathbf{p}}$. To be precise, f takes as input (x, y, z) where $x \in \mathbb{F}_{\mathbf{p}}^{n \times (d)}$ and $y, z \in \mathbb{F}_{\mathbf{p}}^{n}$. Then it computes $f(x, y, z) = \Sigma_{I=(i_1,..,i_d,j,k)} c_I \cdot x[1, i_1] \cdots x[d, i_d] \cdot y[j] \cdot z[k]$ where each coefficient $c_I \in \mathbb{F}_{\mathbf{p}}$. We require the decryption to be efficient only if the output is bounded in norm by a polynomial bound $B(\lambda, n)$. Security of a dFE scheme intuitively requires that a ciphertext only reveals the d public components x and the output of the decryption.

Before we describe our construction, we first recall the construction of 3-restricted FE from [4]:

3-restricted FE [4]. Before getting to 3 restricted FE, we first recap how secret key quadratic functional encryption schemes [41] work at a high level. Let's say that the encryptor wants to encrypt $y, z \in \mathbb{F}_{\mathbf{p}}^{n}$. The master secret key consists of two secret random vectors $\boldsymbol{\beta}, \boldsymbol{\gamma} \in \mathbb{F}_p^n$ that are used for enforcement of computations done on $\boldsymbol{y}$ and $\boldsymbol{z}$ respectively. The idea is that the encryptor encodes $\boldsymbol{y}$ and $\boldsymbol{\beta}$ using some randomness r, and similarly encodes $\boldsymbol{z}$ and $\boldsymbol{\gamma}$ together as well. These encodings are created using bilinear maps in one of the two base groups. These encodings are constructed so that the decryptor can compute an encoding of $[g(\boldsymbol{y}, \boldsymbol{z}) - rg(\boldsymbol{\beta}, \boldsymbol{\gamma})]_t$ in the target group for *any* quadratic function g. The function key for the given function f is constructed in such a manner that it allows the decryptor to compute the encoding $[rf(\boldsymbol{\beta}, \boldsymbol{\gamma})]_t$ in the target group. Thus the output $[f(\boldsymbol{y}, \boldsymbol{z})]_t$ can be recovered in the exponent by computing the sum of $[rf(\boldsymbol{\beta}, \boldsymbol{\gamma})]_t$ and $[f(\boldsymbol{y}, \boldsymbol{z}) - rf(\boldsymbol{\beta}, \boldsymbol{\gamma})]_t$ in the exponent. As long as $f(\boldsymbol{y}, \boldsymbol{z})$ is polynomially small, this value can then be recovered efficiently.

Clearly the idea above only works for degree-2 computations, if we use bilinear maps. However, the work of [4] built upon this idea nevertheless to construct a 3-restricted FE scheme. Recall, in a 3-restricted FE one wants to encrypt three vectors $\boldsymbol{x}, \boldsymbol{y}, \boldsymbol{z} \in \mathbb{F}_p^n$. While $\boldsymbol{y}$ and $\boldsymbol{z}$ are required to be hidden, $\boldsymbol{x}$ is not required to be hidden.

In their scheme, in addition to $\boldsymbol{\beta}, \boldsymbol{\gamma} \in \mathbb{F}_p^n$ in case of a quadratic FE, another vector $\boldsymbol{\alpha} \in \mathbb{F}_p^n$ is also sampled that is used to enforce the correctness of the $\boldsymbol{x}$ part of the computation. As before, given the ciphertext one can compute $[\boldsymbol{y}[j]\boldsymbol{z}[k] - r\boldsymbol{\beta}[j]\boldsymbol{\gamma}[k]]_t$ for $j, k \in [n]$. But this is clearly not enough, as these encodings do not involve $\boldsymbol{x}$ in any way. Thus, in addition, an encoding of $r(\boldsymbol{x}[i] - \boldsymbol{\alpha}[i])$ is also given in the ciphertext for $i \in [n]$. Inside the function key, there are corresponding encodings of $\boldsymbol{\beta}[j]\boldsymbol{\gamma}[k]$ for $j, k \in [n]$ which the decryptor can pair with encoding of $r(\boldsymbol{x}[i] - \boldsymbol{\alpha}[i])$ to form the encoding $[r(\boldsymbol{x}[i] - \boldsymbol{\alpha}[i])\boldsymbol{\beta}[j]\boldsymbol{\gamma}[k]]_t$ in the target group.

Now observe that,

$$\boldsymbol{x}[i] \cdot \big(\boldsymbol{y}[j]\boldsymbol{z}[k] - r\boldsymbol{\beta}[j]\boldsymbol{\gamma}[k]\big) + r(\boldsymbol{x}[i] - \boldsymbol{\alpha}[i]) \cdot \boldsymbol{\beta}[j]\boldsymbol{\gamma}[k]$$
$$= \boldsymbol{x}[i]\boldsymbol{y}[j]\boldsymbol{z}[k] - r\boldsymbol{\alpha}[i]\boldsymbol{\beta}[j]\boldsymbol{\gamma}[k]$$

Above, since $\boldsymbol{x}[i]$ is public, the decryptor can herself take $(\boldsymbol{y}[j]\boldsymbol{z}[k] - r\boldsymbol{\beta}[j]\boldsymbol{\gamma}[k])$, which she already has, and multiply it with $\boldsymbol{x}[i]$ in the exponent. This allows her to compute an encoding of $[\boldsymbol{x}[i]\boldsymbol{y}[j]\boldsymbol{z}[k] - r\boldsymbol{\alpha}[i]\boldsymbol{\beta}[j]\boldsymbol{\gamma}[k]]_t$. Combining these encodings appropriately, she can obtain $[g(\boldsymbol{x}, \boldsymbol{y}, \boldsymbol{z}) - rg(\boldsymbol{\alpha}, \boldsymbol{\beta}, \boldsymbol{\gamma})]_t$ for any degree-3 multilinear function g. Given the function key for f and the ciphertext, one can compute $[rf(\boldsymbol{\alpha}, \boldsymbol{\beta}, \boldsymbol{\gamma})]_t$ which can be used to unmask the output. This is because the ciphertext contains an encoding of r in one of the base groups and the function key contains an encoding of $f(\boldsymbol{\alpha}, \boldsymbol{\beta}, \boldsymbol{\gamma})$ in the other group and pairing them results in $[rf(\boldsymbol{\alpha}, \boldsymbol{\beta}, \boldsymbol{\gamma})]_t$.

The work of [4] shows how to analyze the security of the construction above in a generic bilinear group model.

Towards constructing $(d+2)$-restricted FE. Now let's consider how we can extend the approach discussed above for the case of $d = 2$. Suppose now we want to encrypt $\boldsymbol{u}, \boldsymbol{x}, \boldsymbol{y}$ and $\boldsymbol{z}$. Here $\boldsymbol{y}, \boldsymbol{z}$ are supposed to be private while $\boldsymbol{x}$ and $\boldsymbol{u}$ are not required to be hidden. Let's now also have $\boldsymbol{\phi} \in \mathbb{F}_p^n$ to enforce $\boldsymbol{u}$ part of the computation. How can we generalize the idea above to allow for degree-4 computations? One straightforward idea is to release encodings of $r(\boldsymbol{u}[i_1]\boldsymbol{x}[i_2] - \boldsymbol{\phi}[i_1]\boldsymbol{\alpha}[i_2])$ for $i_1, i_2 \in [n]$ in the ciphertext instead of encodings of $r(\boldsymbol{x}[i_2]-\boldsymbol{\alpha}[i_2])$ like before. This would permit the computation of $[f(\boldsymbol{u},\boldsymbol{x},\boldsymbol{y},\boldsymbol{z}) - rf(\boldsymbol{\phi},\boldsymbol{\alpha},\boldsymbol{\beta},\boldsymbol{\gamma})]_t$. However, such an approach would not be efficient enough for our needs: we require the complexity of encryption to be linear in n. However, the approach above would need to provide n^2 encodings corresponding to $r(\boldsymbol{u}[i_1]\boldsymbol{x}[i_2] - \boldsymbol{\phi}[i_1]\boldsymbol{\alpha}[i_2])$ for every $i_1, i_2 \in [n]$.

Our first idea: A "ladder" of enforcement. Let's now take a step back. Notice that our 3-restricted FE scheme already allows one to compute $[\boldsymbol{x}[i_2]\boldsymbol{y}[j]\boldsymbol{z}[k] - r\boldsymbol{\alpha}[i_2]\boldsymbol{\beta}[j]\boldsymbol{\gamma}[k]]_t$ for any $i_2, j, k \in [n]$. We want to leverage this existing capability to bootstrap to degree-4 computations.

Suppose the decryptor is also able to generate the encoding $[r(\boldsymbol{u}[i_1] - \boldsymbol{\phi}[i_1]) \cdot \boldsymbol{\alpha}[i_2]\boldsymbol{\beta}[j]\boldsymbol{\gamma}[k]]_t$ for any $i_1, i_2, j, k \in [n]$. Then, she can generate the encoding $[\boldsymbol{u}[i_1]\boldsymbol{x}[i_2]\boldsymbol{y}[j]\boldsymbol{z}[k] - \boldsymbol{\phi}[i_1]\boldsymbol{\alpha}[i_2]\boldsymbol{\beta}[j]\boldsymbol{\gamma}[k]]_t$ as follows:

$$r(\boldsymbol{u}[i_1] - \boldsymbol{\phi}[i_1])\boldsymbol{\alpha}[i_2]\boldsymbol{\beta}[j]\boldsymbol{\gamma}[k] + \boldsymbol{u}[i_1] \cdot (\boldsymbol{x}[i_2]\boldsymbol{y}[j]\boldsymbol{z}[k] - r\boldsymbol{\alpha}[i_2]\boldsymbol{\beta}[j]\boldsymbol{\gamma}[k])$$
$$=\boldsymbol{u}[i_1]\boldsymbol{x}[i_2]\boldsymbol{y}[j]\boldsymbol{z}[k] - r\boldsymbol{\phi}[i_1]\boldsymbol{\alpha}[i_2]\boldsymbol{\beta}[j]\boldsymbol{\gamma}[k]$$

Notice that $\boldsymbol{u}$ is public so the decryptor can herself take $(\boldsymbol{x}[i_2]\boldsymbol{y}[j]\boldsymbol{z}[k] - r\boldsymbol{\alpha}[i_2]\boldsymbol{\beta}[j]\boldsymbol{\gamma}[k])$, which she already has, and multiply it with $\boldsymbol{u}[i_1]$ in the exponent. To allow the computation of $[r(\boldsymbol{u}[i_1] - \boldsymbol{\phi}[i_1])\boldsymbol{\alpha}[i_2]\boldsymbol{\beta}[j]\boldsymbol{\gamma}[k]]_t$ we can provide additionally encodings of $(\boldsymbol{u}[i_1] - r\boldsymbol{\phi}[i_1])$ in the ciphertexts for $i_1 \in [n]$ and corresponding encodings of $\boldsymbol{\alpha}[i_2]\boldsymbol{\beta}[j]\boldsymbol{\gamma}[k]$ for $i_2, j, k \in [n]$ in the function key that can be paired together.

What next? As before, the decryptor can homomorphically compute on these encodings and learn $[f(\boldsymbol{u},\boldsymbol{x},\boldsymbol{y},\boldsymbol{z}) - rf(\boldsymbol{\phi},\boldsymbol{\alpha},\boldsymbol{\beta},\boldsymbol{\gamma})]_t$. Finally, the decryptor can compute $[rf(\boldsymbol{\phi},\boldsymbol{\alpha},\boldsymbol{\beta},\boldsymbol{\gamma})]_t$ by pairing an encoding of r given in the ciphertext and and encoding of $f(\boldsymbol{\phi},\boldsymbol{\alpha},\boldsymbol{\beta},\boldsymbol{\gamma})$ given in the function key. Thus, the output can be unmasked in the exponent.

Observe that this solution preserves linear efficiency of the ciphertext. As of now we have not told anything about how security is argued. From computation point of view, this solution indeed turns out to be insightful as this process can now be generalized to form a ladder of enforcement for any constant degree-D computations.

Laddered computations for any constant degree $(d + 2)$. First let's set up some notation. Let $\boldsymbol{x} \in \mathbb{F}_{\mathbf{p}}^{d \times n}$ be the public part of the plain-text and $\boldsymbol{y}, \boldsymbol{z} \in \mathbb{F}_{\mathbf{p}}^n$. Let $\boldsymbol{\alpha} \in \mathbb{F}_{\mathbf{p}}^{d \times n}$ be the vector of random field elements corresponding to $\boldsymbol{x}$. Similarly, $\boldsymbol{\beta}$ and $\boldsymbol{\gamma}$ in $\mathbb{F}_{\mathbf{p}}^n$ be the vector of random elements corresponding to $\boldsymbol{y}$ and $\boldsymbol{z}$ respectively.

The next observation is the following. Suppose the decryptor can generate the following terms by pairing encodings present in the ciphertext and encodings present in the functional key, for every $\boldsymbol{I} = (i_1, .., i_d, j, k) \in [n]^D$.

- $[\boldsymbol{y}[j]\boldsymbol{z}[k] - r\beta_j\gamma_k]_t$ for $j, k \in [n]$.
- $[r(\boldsymbol{x}[d, i_d] - \boldsymbol{\alpha}[d, i_d]) \cdot \boldsymbol{\beta}[j]\boldsymbol{\gamma}[k]]_t$
- $[r(\boldsymbol{x}[d-1, i_{d-1}] - \boldsymbol{\alpha}[d-1, i_{d-1}]) \cdot \boldsymbol{\alpha}[d, i_d]\boldsymbol{\beta}[j]\boldsymbol{\gamma}[k]]_t$

$$- \ldots$$
$$- [r(\boldsymbol{x}[1, i_1] - \boldsymbol{\alpha}[1, i_1]) \cdot \boldsymbol{\alpha}[2, i_2] \cdots \boldsymbol{\alpha}[d, i_d]\boldsymbol{\beta}[j]\boldsymbol{\gamma}[k]]_t$$

As before, the decryptor can also obtain an encoding $[rf(\boldsymbol{\alpha}, \boldsymbol{\beta}, \boldsymbol{\gamma})]_t$ corresponding to the degree-D multilinear function f in the function key.

The main observation to generalize the $D = 4$ case discussed above is then the following. Consider the first two terms: $[\boldsymbol{y}[j]\boldsymbol{z}[k] + r\beta_j\gamma_k]_t$ and $[r(\boldsymbol{x}[d, i_d] - \boldsymbol{\alpha}[d, i_d])\boldsymbol{\beta}[j]\boldsymbol{\gamma}[k]]_t$ and note that:

$$\boldsymbol{x}[d, i_d](\boldsymbol{y}[j]\boldsymbol{z}[k] - r\beta_j\gamma_k) + r(\boldsymbol{x}[d, i_d] - \boldsymbol{\alpha}[d, i_d])\boldsymbol{\beta}[j]\boldsymbol{\gamma}[k]$$
$$= \boldsymbol{x}[d, i_d]\boldsymbol{y}[j]\boldsymbol{z}[k] - r\boldsymbol{\alpha}[d, i_d]\boldsymbol{\beta}[j]\boldsymbol{\gamma}[k]$$

This observation allows the decryptor to compute an encoding

$$\mathsf{Int}_d = [\boldsymbol{x}[d, i_d]\boldsymbol{y}[j]\boldsymbol{z}[k] - r\boldsymbol{\alpha}[d, i_d]\boldsymbol{\beta}[j]\boldsymbol{\gamma}[k]]_t$$

using encodings of the first two types in the list above.

Next observe that using the encoding,

$$[r(\boldsymbol{x}[d - 1, i_{d-1}] - \boldsymbol{\alpha}[d - 1, i_{d-1}]) \cdot \boldsymbol{\alpha}[d, i_d]\boldsymbol{\beta}[j]\boldsymbol{\gamma}[k]]_t$$

and encoding Int_d one can compute

$$\mathsf{Int}_{d-1} = [\boldsymbol{x}[d - 1, i_{d-1}]\boldsymbol{x}[d, i_d]\boldsymbol{y}[j]\boldsymbol{z}[k] - r\boldsymbol{\alpha}[d - 1, i_{d-1}]\boldsymbol{\alpha}[d, i_d]\boldsymbol{\beta}[j]\boldsymbol{\gamma}[k]]_t$$

This is because,

$$\boldsymbol{x}[d - 1, i_{d-1}] \cdot (\boldsymbol{x}[d, i_d]\boldsymbol{y}[j]\boldsymbol{z}[k] - r\boldsymbol{\alpha}[d, i_d]\boldsymbol{\beta}[j]\boldsymbol{\gamma}[k])$$
$$+ r(\boldsymbol{x}[d - 1, i_{d-1}] - \boldsymbol{\alpha}[d - 1, i_{d-1}]) \cdot \boldsymbol{\alpha}[d, i_d]\boldsymbol{\beta}[j]\boldsymbol{\gamma}[k]$$
$$= \boldsymbol{x}[d - 1, i_{d-1}]\boldsymbol{x}[d, i_d]\boldsymbol{y}[j]\boldsymbol{z}[k] - r\boldsymbol{\alpha}[d - 1, i_{d-1}]\boldsymbol{\alpha}[d, i_d]\boldsymbol{\beta}[j]\boldsymbol{\gamma}[k]$$

Continuing this way up a "ladder" the decryptor can compute

$$\mathsf{Mon}_I = [\Pi_{\ell \in [d]}\boldsymbol{x}[\ell, i_\ell]\boldsymbol{y}[j]\boldsymbol{z}[k] - r\Pi_{\ell \in [d]}\boldsymbol{\alpha}[\ell, i_\ell]\boldsymbol{\beta}[j]\boldsymbol{\gamma}[k]]_t$$

Observe that the term $\Pi_{\ell \in [d]}\boldsymbol{x}[\ell, i_\ell]\boldsymbol{y}[j]\boldsymbol{z}[k] - r\Pi_{\ell \in [d]}\boldsymbol{\alpha}[\ell, i_\ell]\boldsymbol{\beta}[j]\boldsymbol{\gamma}[k]$ corresponding to Mon_I can be generated as a linear combination of terms from the list above. Once Mon_I is computed then the decryptor can do the following. Since $f = \Sigma_{I=(i_1,..,i_d,j,k)} c_I \boldsymbol{x}[1, i_1] \cdots \boldsymbol{x}[d, i_d]\boldsymbol{y}[j]\boldsymbol{z}[k]$, the decryptor can then compute:

$$\mathsf{Mon}_f = [f(\boldsymbol{x}, \boldsymbol{y}, \boldsymbol{z}) - rf(\boldsymbol{\alpha}, \boldsymbol{\beta}, \boldsymbol{\gamma})]_t$$

Finally using $[rf(\boldsymbol{\alpha}, \boldsymbol{\beta}, \boldsymbol{\gamma})]_t$ the decryptor can recover $[f(\boldsymbol{x}, \boldsymbol{y}, \boldsymbol{z})]_t$.

How to base security on SXDH? So far, we have just described a potential computation pattern that allows the decryptor to obtain the function output given a function key and a ciphertext. Any scheme that allows constructing the terms described above in the ladder is guaranteed to satisfy correctness. But how do we argue security?

We rely on a primitive called Canonical Function Hiding Inner Product Encryption (cIPE for short). A cIPE scheme allows the decryptor to compute the inner product of a vector encoded in the ciphertext, with a vector encoded in the function key. Also, intuitively, cIPE guarantees that the vector embedded in the function key is also hidden given the function key. More precisely, given any vectors v, v', u, u' such that $\langle u, v \rangle = \langle u', v' \rangle$, no efficient adversary can distinguish between a ciphertext encoding u and a function key encoding v, from a ciphertext encoding u' and a function key encoding v'.

Furthermore, syntactically speaking, in a cIPE scheme, we will require the following to be true:

- The encryption algorithm just computes exponentiation and multiplication operations in G_1. The encryption of a vector $(a_1, .., a_4)$ can just be computed knowing $g_1^{a_i}$ for $i \in [4]$ and the master secret key.
- Key generation algorithm just computes exponentiation and multiplication operation in G_2. The function key for a vector $(b_1, .., b_4)$ can just be computed knowing $g_2^{b_i}$ for $i \in [4]$ and the master secret key.
- The decryption process just computes pairing operations and then computes group multiplications over G_t. The output is produced in G_t. The element g_t^a is represented as $[a]_t$ for the rest of the paper.

Such a cIPE scheme was given by [40], where it was instantiated from SXDH over bilinear maps. That work also used cIPE to build quadratic FE from SXDH. We will also make use of cIPE in our construction of D-restricted FE. Note, however, that unlike in the case of quadratic FE, our construction, and crucially our proof of security, will also need to incorporate the "ladder" enforcement mechanism sketched above. We are able to do so still relying only on the SXDH assumption.

We note that the size of the vectors encrypted using a cIPE scheme cannot grow with n, to achieve linear efficiency. In fact, we just use four-dimensional vectors.

Realizing the Ladder: Warm-up Construction for $d + 2 = 4$. Here is a warm-up construction for the case of $d = 2$ (i.e. $D = 4$).
Setup($1^\lambda, 1^n$): On input security parameter 1^λ and length 1^n,

- Run cIPE setup as follows. $\mathsf{sk}_0 \leftarrow \mathsf{cIPE.Setup}(1^\lambda, 1^4)$. Thus these keys are used to encrypt vectors in $\mathbb{F}_\mathbf{p}^4$.
- Then run cIPE setup algorithm $2 \cdot n$ times. That is, for every $\ell \in [2]$ and $i_\ell \in [n]$, compute $\mathsf{sk}^{(\ell, i_\ell)} \leftarrow \mathsf{cIPE.Setup}(1^\lambda, 1^4)$.
- Sample $\boldsymbol{\alpha} \leftarrow \mathbb{F}_\mathbf{p}^{2 \times n}$. Also sample $\boldsymbol{\beta}, \boldsymbol{\gamma} \leftarrow \mathbb{F}_\mathbf{p}^n$.

- For every set $I = (i_1, i_2, j, k)$ in $[n]^4$ do the following. Let $I' = (i_2, j, k)$ and $I'' = (j, k)$. Compute $\mathsf{Key}_{I'}^{(1,i_1)} =$

$$\mathsf{cIPE.KeyGen}(\mathsf{sk}^{(1,i_1)}, (\alpha[2,i_2]\beta[j]\gamma[k], \alpha[1,i_1]\alpha[2,i_2]\beta[j]\gamma[k], 0, 0))$$

Similarly, compute $\mathsf{Key}_{I''}^{(2,i_2)} =$

$$\mathsf{cIPE.KeyGen}(\mathsf{sk}^{(2,i_2)}, (\beta[j]\gamma[k], \alpha[2,i_2]\beta[j]\gamma[k], 0, 0))$$

- Output $\mathsf{MSK} = (\{\mathsf{sk}^{(\ell,i_\ell)}, \mathsf{Key}_I^{(\ell,i_\ell)}\}_{\ell,i_\ell,I}, \alpha, \beta, \gamma, \mathsf{sk}_0)$

$\underline{\mathsf{Enc}(\mathsf{MSK}, x, y, z)}$: The input message $M = (x, y, z)$ consists of a public attribute $x \in \mathbb{F}_p^{2\times n}$ and private vectors $y, z \in \mathbb{F}_p^n$. Perform the following operations:

- Parse $\mathsf{MSK} = (\{\mathsf{sk}^{(\ell,i_\ell)}, \mathsf{Key}_I^{(\ell,i_\ell)}\}_{\ell,i_\ell,I}, \alpha, \beta, \gamma, \mathsf{sk}_0)$.
- Sample $r \leftarrow \mathbb{F}_p$.
- Compute $\mathsf{CT}_0 = \mathsf{cIPE.Enc}(\mathsf{sk}_0, (r, 0, 0, 0))$.
- Sample $\mathsf{sk}' \leftarrow \mathsf{cIPE.Setup}(1^\lambda, 1^4)$.
- Compute $\mathsf{CTC}_j \leftarrow \mathsf{cIPE.Enc}(\mathsf{sk}, (y[j], \beta[j], 0, 0))$ for $j \in [n]$
- Compute $\mathsf{CTK}_k \leftarrow \mathsf{cIPE.KeyGen}(\mathsf{sk}, (z[k], -r\gamma[k], 0, 0))$ for $k \in [n]$.
- For every $\ell \in [2]$, $i_\ell \in [n]$, compute $\mathsf{CT}^{(\ell,i_\ell)} = \mathsf{cIPE.Enc}(\mathsf{sk}^{(\ell,i_\ell)}, (rx[\ell, i_\ell], -r, 0, 0))$.
- Output $\mathsf{CT} = (x, \mathsf{CT}_0, \{\mathsf{CTC}_j, \mathsf{CTK}_k, \mathsf{CT}^{(\ell,i_\ell)}\}_{\ell\in[2],i_\ell\in[n],j\in[n],k\in[n]})$

$\underline{\mathsf{KeyGen}(\mathsf{MSK}, f)}$: On input the master secret key MSK and function f,

- Parse $\mathsf{MSK} = (\{\mathsf{sk}^{(\ell,i_\ell)}, \mathsf{Key}_I^{(\ell,i_\ell)}\}_{\ell,i_\ell,I}, \alpha, \beta, \gamma, \mathsf{sk}_0)$.
- Compute $\theta_f = f(\alpha, \beta, \gamma)$.
- Compute $\mathsf{Key}_{0,f} = \mathsf{cIPE.KeyGen}(\mathsf{sk}_0, (\theta_f, 0, 0, 0))$
- Output $sk_f = (\mathsf{Key}_{0,f}, \{\mathsf{Key}_I^{(\ell,i_\ell)}\}_{\ell,i_\ell,I})$.

Observe how the computation proceeds. This scheme allows to generate all terms in the ladder described above as follows:

Consider all terms associated with the vector $I = (i_1, i_2, j, k) \in [n]^4$.

- $[y[j]z[k] - r\beta_j\gamma_k]_t = \mathsf{cIPE.Dec}(\mathsf{CTK}_k, \mathsf{CTC}_j)$
- $[r(x[2,i_2] - \alpha[2,i_2])\beta[j]\gamma[k]]_t = \mathsf{cIPE.Dec}(\mathsf{Key}_{I''}^{(2,i_2)}, \mathsf{CT}^{(2,i_2)})$ where $I'' = (j, k)$.
- $[r(x[1,i_1] - \alpha[1,i_1])\alpha[2,i_2]\beta[j]\gamma[k]]_t = \mathsf{cIPE.Dec}(\mathsf{Key}_{I'}^{(1,i_1)}, \mathsf{CT}^{(1,i_1)})$ where $I'' = (i_2, j, k)$
- $[rf(\alpha, \beta, \gamma)]_t = \mathsf{cIPE.Dec}(\mathsf{Key}_{0,f}, \mathsf{CT}_0)$.

Thus, we can compute $[f(x, y, z)]_t$. We now briefly describe how security is proven.

Security Proof: Key Points. We use SXDH and function hiding property of the cIPE scheme crucially to argue security. The hybrid strategy is the following.

1. First we switch y to 0 vector in the challenge ciphertext, changing one component at a time.
2. To maintain correctness of output, we simultaneously introduce an offset in the function key to maintain correctness of decryption.
3. Once y is switched, z can be switched to vector 0, due to the function hiding property of the cIPE scheme. This is because the inner products remain the same in both the case as y is always 0 and inner product of any vector with all zero vector is 0. Finally, we are in the hybrid where the challenge ciphertext just depends on x and in particular totally independent of y and z.

Step (1) is most challenging here, and requires careful pebbling and hardwiring arguments made using SXDH and function hiding security property of cIPE. We point the reader to the full version for a detailed proof.

New ΔRG candidates: Our construction of D-restricted FE enables us to meaningfully consider ΔRG candidates that are implementable by D-restricted FE using degree-D polynomials. This enables a much richer class of potential ΔRG candidates than those implementable by 3-restricted FE [4]. In Sect. 6, we describe a few of the new avenues for constructing ΔRG candidates that we open by our construction of D-restricted FE.

Reader's Guide. The rest of the paper is organized as follows. In Sect. 3 we recall the definition of indistinguishability obfuscation and other prerequisites for the paper. In Sect. 4 we define formally the notions of $(d + 2)$ restricted FE. Thereafter, in Sect. 5 perturbation resilient generator (ΔRG for short) is defined. Both primitives are central to this paper. In Sect. 6 we give candidate constructions of ΔRG and show how to implement it using a $(d+2)$ restricted FE scheme. In Sect. 7 we show how to construct $(d + 2)$ restricted FE using SXDH. Finally, in Sect. 8 we stitch all these primitives to show how to build obfuscation.

3 Preliminaries

We denote the security parameter by λ. For a distribution X we denote by $x \leftarrow X$ the process of sampling a value x from the distribution X. Similarly, for a set $\mathcal{X}$ we denote by $x \leftarrow \mathcal{X}$ the process of sampling x from the uniform distribution over $\mathcal{X}$. For an integer $n \in \mathbb{N}$ we denote by $[n]$ the set $\{1, \ldots, n\}$. A function $\mathsf{negl} : \mathbb{N} \to \mathbb{R}$ is negligible if for every constant $c > 0$ there exists an integer N_c such that $\mathsf{negl}(\lambda) < \lambda^{-c}$ for all $\lambda > N_c$.

By $\approx_c$ we denote computational indistinguishability. We say that two ensembles $\mathcal{X} = \{\mathcal{X}_\lambda\}_{\lambda \in \mathbb{N}}$ and $\mathcal{Y} = \{\mathcal{Y}_\lambda\}_{\lambda \in \mathbb{N}}$ are computationally indistinguishable if for every probabilistic polynomial time adversary $\mathcal{A}$ there exists a negligible function negl such that $\left| \Pr_{x \leftarrow \mathcal{X}_\lambda}[\mathcal{A}(1^\lambda, x) = 1] - \Pr_{y \leftarrow \mathcal{Y}_\lambda}[\mathcal{A}(1^\lambda, y) = 1] \right| \leq \mathsf{negl}(\lambda)$ for every sufficiently large $\lambda \in \mathbb{N}$.

For a field element $a \in \mathbb{F}_\mathbf{p}$ represented in $[-p/2, p/2]$, we say that $-B < a < B$ for some positive integer B if its representative in $[-p/2, p/2]$ lies in $[-B, B]$.

Definition 1 (Distinguishing Gap). *For any adversary $\mathcal{A}$ and two distributions $\mathcal{X} = \{\mathcal{X}_\lambda\}_{\lambda \in \mathbb{N}}$ and $\mathcal{Y} = \{\mathcal{Y}_\lambda\}_{\lambda \in \mathbb{N}}$, define $\mathcal{A}$'s distinguishing gap in distinguishing these distributions to be $|\Pr_{x \leftarrow \mathcal{X}_\lambda}[\mathcal{A}(1^\lambda, x) = 1] - \Pr_{y \leftarrow \mathcal{Y}_\lambda}[\mathcal{A}(1^\lambda, y) = 1]|$*

By boldfaced letters such as $\boldsymbol{v}$ we will denote multidimensional matrices. Whenever dimension is unspecified we mean them as vectors.

Throughout, we denote by an adversary an interactive machine that takes part in a protocol with the challenger. Thus, we model such an adversary as a tuple of circuits $(C_1, ..., C_t)$ where t is the number of messages exchanged. Each circuit takes as input the state output by the previous circuit, among other messages. The size of adversary is defined as sum of size of each circuit.

3.1 Indistinguishability Obfuscation ($i\mathcal{O}$)

The notion of indistinguishability obfuscation ($i\mathcal{O}$), first conceived by Barak et al. [9], guarantees that the obfuscation of two circuits are computationally indistinguishable as long as they both are equivalent circuits, i.e., the output of both the circuits are the same on every input. Formally,

Definition 2 (Indistinguishability Obfuscator ($i\mathcal{O}$) for Circuits). *A uniform PPT algorithm $i\mathcal{O}$ is called an indistinguishability obfuscator for a circuit family $\{\mathcal{C}_\lambda\}_{\lambda \in \mathbb{N}}$, where $\mathcal{C}_\lambda$ consists of circuits C of the form $C : \{0,1\}^n \to \{0,1\}$ with $n = n(\lambda)$, if the following holds:*

- **Completeness:** *For every $\lambda \in \mathbb{N}$, every $C \in \mathcal{C}_\lambda$, every input $x \in \{0,1\}^n$, we have that*

$$\Pr[C'(x) = C(x) \ : \ C' \leftarrow i\mathcal{O}(\lambda, C)] = 1$$

- **Indistinguishability:** *For any PPT distinguisher D, there exists a negligible function $\mathsf{negl}(\cdot)$ such that the following holds: for all sufficiently large $\lambda \in \mathbb{N}$, for all pairs of circuits $C_0, C_1 \in \mathcal{C}_\lambda$ such that $C_0(x) = C_1(x)$ for all inputs $x \in \{0,1\}^n$ and $|C_0| = |C_1|$, we have:*

$$\left| \Pr[D(\lambda, i\mathcal{O}(\lambda, C_0)) = 1] - \Pr[D(\lambda, i\mathcal{O}(\lambda, C_1)) = 1] \right| \leq \mathsf{negl}(\lambda)$$

- **Polynomial Slowdown:** *For every $\lambda \in \mathbb{N}$, every $C \in \mathcal{C}_\lambda$, we have that $|i\mathcal{O}(\lambda, C)| = \mathsf{poly}(\lambda, C)$.*

3.2 Bilinear Maps and Assumptions

Let PPGen be a probabilistic polynomial time algorithm that on input 1^λ returns a description $(e, G_1, G_2, G_T, g_1, g_2, \mathbf{p})$ of asymmetric pairing groups where G_1, G_2 and G_T are groups of order $\mathbf{p}$ for a 2λ bit prime $\mathbf{p}$. g_1 and g_2 are generators of G_1 and G_2 respectively. $e : G_1 \times G_2 \to G_T$ is an efficiently computable non-degenerate bilinear map. Define $g_t = e(g_1, g_2)$ as the generator of G_T.

Representation: We use the following representation to describe group elements. For any $b \in \{1, 2, T\}$ define by $[x]_b$ for $x \in \mathbb{F}_p$ as g_b^x. This notation will be used throughout. We now describe SXDH assumption relative to PPGen.

Definition 3. *(SXDH Assumption relative to* PPGen*.) We say that SXDH assumption holds relative to* PPGen*, if* $(e, G_1, G_2, G_T, g_1, g_2, p) \leftarrow$ PPGen*, then for any group* g_ℓ *for* $\ell \in \{1, 2, t\}$*, it holds that, for any polynomial time adversary* $\mathcal{A}$*:*

$$\left| \Pr_{r,s,u \leftarrow \mathbb{F}_p} [\mathcal{A}([r]_\ell, [s]_\ell, [r \cdot s]_\ell) = 1] - \Pr_{r,s,u \leftarrow \mathbb{F}_p} [\mathcal{A}([r]_\ell, [s]_\ell, [u]_\ell) = 1] \right| \leq \mathsf{negl}(\lambda)$$

Further, if $\mathsf{negl}(\lambda)$ *is* $O(2^{-\lambda^c})$ *for some* $c > 0$*, then we say that subexponential SXDH holds relative to* PPGen*.*

3.3   Canonical Function Hiding Inner Product FE

We now describe the notion of a canonical function hiding inner product FE proposed by [40]. A canonical function hiding scheme FE scheme consists of the following algorithms:

- PPSetup$(1^\lambda) \to$ pp. On input the security parameter, PPSetup, outputs parameters pp, which contain description of the groups and the plain text space $\mathbb{Z}_p$.
- Setup$(pp, 1^n) \to$ **sk**. The setup algorithm takes as input the length of vector 1^n and parameters pp and outputs a secret key **sk**. We assume that pp is always implicitly given as input to this algorithm and the algorithms below (sometimes we omit this for ease of notation).
- Enc$(\mathbf{sk}, x) \to$ CT. The encryption algorithm takes as input a vector $x \in \mathbb{Z}_p^n$ and outputs a ciphertext CT.
- KeyGen$(\mathbf{sk}, y) \to \mathbf{sk}_y$. The key generation algorithm on input the master secret key **sk** and a function vector $y \in \mathbb{Z}_p^n$ and outputs a function key $\mathbf{sk}_y$.
- Dec$(1^B, \mathbf{sk}_y, \mathsf{CT}) \to m^*$. The decryption algorithm takes as input a ciphertext CT, a function key $\mathbf{sk}_y$ and a bound B and it outputs a value m^*. Further, it is run in two steps. First step Dec_0, computes $[\langle x, y \rangle]_T$ (if the keys and ciphertexts were issued for x and y) and then the second step, Dec_1, computes its discrete log, if this value lies in $[-B, B]$

A cIPE scheme satisfies linear efficiency, correctness, function hiding security and a canonical structure requirement. All of these are described in the full version.

4   Key Notion 1: $(d+2)-$restricted FE

In this section we describe the notion of a $(d+2)$-restricted functional encryption scheme (denoted by dFE). Let d denote any positive integer constant. Informally, a dFE scheme is a functional encryption scheme that supports homogeneous polynomials of degree $d + 2$ having degree 1 in $d + 2$ input vectors. d out of those $d + 2$ vectors are public. This is a generalization of the notion of a three restricted FE scheme proposed by [4].

Notation: Throughout, we denote by boldfaced letters (multi-dimensional) matrices, where dimensions are either explicitly or implicitly defined.

Function class of interest: Consider a set of functions $\mathcal{F}_{\mathsf{dFE}} = \mathcal{F}_{\mathsf{dFE},\lambda,\mathbf{p},n} = \{f : \mathbb{F}_{\mathbf{p}}^{n(d+2)} \to \mathbb{F}_{\mathbf{p}}\}$ where $\mathbb{F}_{\mathbf{p}}$ is a finite field of order $\mathbf{p}(\lambda)$. Here n is seen as a function of λ. Each $f \in \mathcal{F}_{\lambda,\mathbf{p},n}$ takes as input $d+2$ vectors $(\boldsymbol{x}[1], ..., \boldsymbol{x}[d], \boldsymbol{y}, \boldsymbol{z})$ of length n over $\mathbb{F}_{\mathbf{p}}$ and computes a polynomial of the form $\Sigma c_{i_1,...,i_d,j,k} \cdot \boldsymbol{x}[1, i_1] \cdot ... \cdot \boldsymbol{x}[d, i_d] \cdot \boldsymbol{y}[j] \cdot \boldsymbol{z}[k]$, where $c_{i_1,...,i_d,j,k}$ are coefficients from $\mathbb{F}_{\mathbf{p}}$ for very $i_1, ..., i_d, j, k \in [n]^{d+2}$.

Syntax. Consider the set of functions $\mathcal{F}_{\mathsf{dFE},\lambda,\mathbf{p},n}$ as described above. A $(d+2)-$restricted functional encryption scheme dFE for the class of functions $\mathcal{F}_{\mathsf{dFE}}$ (described above) consists of the following PPT algorithms:

- **Setup, $\mathsf{Setup}(1^\lambda, 1^n)$:** On input security parameter λ (and the number of inputs $n = poly(\lambda)$), it outputs the master secret key MSK.
- **Encryption, $\mathsf{Enc}(\mathsf{MSK}, \boldsymbol{x}[1], ..., \boldsymbol{x}[d], \boldsymbol{y}, \boldsymbol{z})$:** On input the encryption key MSK and input vectors $\boldsymbol{x} \in \mathbb{F}_{\mathbf{p}}^{d \times n}$, $\boldsymbol{y}$ and $\boldsymbol{z}$ (all in $\mathbb{F}_{\mathbf{p}}^n$) it outputs ciphertext CT. Here $\boldsymbol{x}$ is seen as a public attribute and $\boldsymbol{y}$ and $\boldsymbol{z}$ are thought of as private messages.
- **Key Generation, $\mathsf{KeyGen}(\mathsf{MSK}, f)$:** On input the master secret key MSK and a function $f \in \mathcal{F}_{\mathsf{dFE}}$, it outputs a functional key $sk[f]$.
- **Decryption, $\mathsf{Dec}(sk[f], 1^B, \mathsf{CT})$:** On input functional key $sk[f]$, a bound $B = poly(\lambda)$ and a ciphertext CT, it outputs the result *out*.

We define the correctness property below.

    B-Correctness. Consider any function $f \in \mathcal{F}_{\mathsf{dFE}}$ and any plaintext $\boldsymbol{x}, \boldsymbol{y}, \boldsymbol{z} \in \mathbb{F}_{\mathbf{p}}$ (dimensions are defined above). Consider the following process:

- $sk[f] \leftarrow \mathsf{KeyGen}(\mathsf{MSK}, f)$.
- $\mathsf{CT} \leftarrow \mathsf{Enc}(\mathsf{MSK}, \boldsymbol{x}, \boldsymbol{y}, \boldsymbol{z})$
- If $f(\boldsymbol{x}, \boldsymbol{y}, \boldsymbol{z}) \in [-B, B]$, set $\theta = f(\boldsymbol{x}, \boldsymbol{y}, \boldsymbol{z})$, otherwise set $\theta = \bot$.

The following should hold:

$$\Pr\left[\mathsf{Dec}(sk[f], 1^B, \mathsf{CT}) = \theta\right] \geq 1 - \mathsf{negl}(\lambda),$$

for some negligible function negl.

    Linear Efficiency: We require that for any message $(\boldsymbol{x}, \boldsymbol{y}, \boldsymbol{z})$ where $\boldsymbol{x} \in \mathbb{F}_{\mathbf{p}}^{d \times n}$ and $\boldsymbol{y}, \boldsymbol{z} \in \mathbb{F}_{\mathbf{p}}^n$ the following happens:

- Let MSK ← Setup($1^\lambda, 1^n$).
- Compute CT ← Enc(MSK, x, y, z).

The size of encryption circuit computing CT is less than $n \times (d+2) \log_2 \mathbf{p} \cdot poly(\lambda)$. Here *poly* is some polynomial independent of n.

4.1 Semi-functional Security

We define the following auxiliary algorithms.

Semi-functional Key Generation, sfKG(MSK, f, θ): On input the master secret key MSK, function f and a value θ, it computes the semi-functional key $sk[f, \theta]$.

Semi-functional Encryption, sfEnc(MSK, $x, 1^{|y|}, 1^{|z|}$): On input the master encryption key MSK, a public attribute x and length of messages y, z, it computes a semi-functional ciphertext $\mathsf{ct_{sf}}$.

We define two security properties associated with the above two auxiliary algorithms. We will model the security definitions along the same lines as semi-functional FE.

Definition 4 (Indistinguishability of Semi-functional Ciphertexts). *A $(d+2)$-restricted functional encryption scheme* dFE *for a class of functions $\mathcal{F}_{\mathsf{dFE}} = \{\mathcal{F}_{\mathsf{dFE},\lambda,\mathbf{p},n}\}_{\lambda \in \mathbb{N}}$ is said to satisfy the* **indistinguishability of semi-functional ciphertexts property** *if there exists a constant $c > 0$ such that for sufficiently large $\lambda \in \mathbb{N}$ and any adversary $\mathcal{A}$ of size 2^{λ^c}, the probability that $\mathcal{A}$ succeeds in the following experiment is $2^{-\lambda^c}$.*

Expt($1^\lambda, \mathbf{b}$):

1. $\mathcal{A}$ *specifies the following:*
   - *Challenge message $M^* = (x, y, z)$. Here y, z is in $\mathbb{F}_\mathbf{p}^n$ and x is in $\mathbb{F}_\mathbf{p}^{d \times n}$.*
   - *It can also specify additional messages $\{M_k = (x_k, y_k, z_k)\}_{k \in [q]}$ Here y_k, z_k is in $\mathbb{F}_\mathbf{p}^n$ and x_k is in $\mathbb{F}_\mathbf{p}^{d \times n}$. Here q is a polynomial in n, λ.*
   - *It also specifies functions $f_1, \ldots, f_\eta$ and hardwired values $\theta_1, \ldots, \theta_\eta$ where η is a polynomial in n, λ.*
2. *The challenger checks if $\theta_k = f_k(x, y, z)$ for every $k \in [\eta]$. If this check fails, the challenger aborts the experiment.*
3. *The challenger computes the following*
   - *Compute $sk[f_k, \theta_k]$ ← sfKG(MSK, f_k, θ_k), for every $k \in [\eta]$.*
   - *If $\mathbf{b} = 0$, compute CT* ← sfEnc(MSK, $x, 1^{|y|}, 1^{|z|}$). Else, compute CT* ← Enc(MSK, x, y, z).*
   - *CT_i ← Enc(MSK, M_i), for every $i \in [q]$.*
4. *The challenger sends $(\{CT_i\}_{i \in [q]}, CT^*, \{sk[f_k, \theta_k]\}_{k \in [\eta]})$ to $\mathcal{A}$.*
5. *The adversary outputs a bit b'.*

We say that the adversary $\mathcal{A}$ succeeds in $\mathsf{Expt}(1^\lambda, \mathbf{b})$ with probability ε if it outputs $b' = \mathbf{b}$ with probability $\frac{1}{2} + \varepsilon$.

We now define the indistinguishability of semi-functional keys property.

Definition 5 (Indistinguishability of Semi-functional Keys). *A $(d + 2)$-restricted FE scheme dFE for a class of functions $\mathcal{F}_{\mathsf{dFE}} = \{\mathcal{F}_{\mathsf{dFE},\lambda,\mathbf{p},n}\}_{\lambda \in \mathbb{N}}$ is said to satisfy the* **indistinguishability of semi-functional keys property** *if there exists a constant $c > 0$ such that for all sufficiently large λ, any PPT adversary $\mathcal{A}$ of size 2^{λ^c}, the probability that $\mathcal{A}$ succeeds in the following experiment is $2^{-\lambda^c}$.*

$\mathsf{Expt}(1^\lambda, \mathbf{b})$:

1. *$\mathcal{A}$ specifies the following:*
   - *It can specify messages $M_j = \{(\boldsymbol{x}_i, \boldsymbol{y}_i, \boldsymbol{z}_i)\}_{j \in [q]}$ for some polynomial q. Here $\boldsymbol{y}_i, \boldsymbol{z}_i$ is in $\mathbb{F}_{\mathbf{p}}^n$ and $\boldsymbol{x}_i$ is in $\mathbb{F}_{\mathbf{p}}^{d \times n}$.*
   - *It specifies functions $f_1, \ldots, f_\eta \in \mathcal{F}_{\mathsf{dFE}}$ and hardwired values $\theta_1, \ldots, \theta_\eta \in \mathbb{F}_p$. Here η is some polynomial in λ, n.*
2. *Challenger computes the following:*
   - *If $\mathbf{b} = 0$, compute $sk[f_i]^* \leftarrow \mathsf{KeyGen}(\mathsf{MSK}, f_i)$ for all $i \in [\eta]$. Otherwise, compute $sk[f_i]^* \leftarrow \mathsf{sfKG}(\mathsf{MSK}, f_i, \theta_i)$ for all $i \in [\eta]$.*
   - *$\mathsf{CT}_i \leftarrow \mathsf{Enc}(\mathsf{MSK}, M_j)$, for every $j \in [q]$.*
3. *Challenger then sends $\left(\{\mathsf{CT}_i\}_{i \in [q]}, \{sk[f_i]^*\}_{i \in [\eta]}\right)$ to $\mathcal{A}$.*
4. *$\mathcal{A}$ outputs b'.*

The success probability of $\mathcal{A}$ is defined to be ε if $\mathcal{A}$ outputs $b' = \mathbf{b}$ with probability $\frac{1}{2} + \varepsilon$.

If a $(d + 2)$-restricted FE scheme satisfies both the above definitions, then it is said to satisfy semi-functional security.

Definition 6 (Semi-functional Security). *Consider a $(d + 2)$-restricted FE scheme dFE for a class of functions $\mathcal{F}$. We say that dFE satisfies* **semi-functional security** *if it satisfies the indistinguishability of semi-functional ciphertexts property (Definition 4) and the indistinguishability of semi-functional keys property (Definition 5).*

Remark: Two remarks are in order:

1. First, we define sub-exponential security here as that notion is useful for our construction of $i\mathcal{O}$. The definition can be adapted to polynomial security naturally.
2. Semi-functional security implies the indistinguishability based notion naturally. This is pointed out in [5].

5   Key Notion 2: Perturbation Resilient Generator

Now we describe the notion of a Perturbation Resilient Generator (ΔRG for short), proposed by [4]. A ΔRG consists of the following algorithms:

- Setup($1^\lambda, 1^n, B$) $\rightarrow$ (pp, Seed). The setup algorithm takes as input a security parameter λ, the length parameter 1^n and a polynomial $B = B(\lambda)$ and outputs a seed Seed $\in \{0,1\}^*$ and public parameters pp.
- Eval(pp, Seed) $\rightarrow (h_1, ..., h_\ell)$. The evaluation algorithm outputs a vector $(h_1, ..., h_\ell) \in \mathbb{Z}^\ell$. Here ℓ is the stretch of ΔRG.

We have following properties of a ΔRG scheme.

Efficiency: We require for Setup($1^\lambda, 1^n, B$) $\rightarrow$ (pp, Seed) and Eval(pp, Seed) $\rightarrow$ $(h_1, ..., h_\ell)$,

- $|\text{Seed}| = n \cdot poly(\lambda)$ for some polynomial *poly* independent of n. The size of Seed is linear in n.
- For all $i \in [\ell]$, $|h_i| < poly(\lambda, n)$. The norm of each output component h_i in $\mathbb{Z}$ is bounded by some polynomial in λ and n.

Perturbation Resilience: We require that for large enough security parameter λ, for every polynomial B, there exists a large enough polynomial $n_B(\lambda)$ such that for any $n > n_B$, there exists an efficient sampler $\mathcal{H}$ such that for Setup($1^\lambda, 1^n, B$) $\rightarrow$ (pp, Seed) and Eval(pp, Seed) $\rightarrow (h_1, ..., h_\ell)$, we have that for any distinguisher D of size 2^λ and any $a_1, ..., a_\ell \in [-B, B]^\ell$

$$| \Pr[D(x \xleftarrow{\$} \mathcal{D}_1) = 1] - \Pr[D(x \xleftarrow{\$} \mathcal{D}_2) = 1]| < 1 - 2/\lambda$$

Here $\mathcal{D}_1$ and $\mathcal{D}_2$ are defined below:

- Distribution $\mathcal{D}_1$: Compute Setup($1^\lambda, 1^n, B$) $\rightarrow$ (pp, Seed) and Eval(pp, Seed) $\rightarrow (h_1, ..., h_\ell)$. Output $(\text{pp}, h_1, ..., h_\ell)$.
- Distribution $\mathcal{D}_2$: Compute Setup($1^\lambda, 1^n, B$) $\rightarrow$ (pp, Seed) and $\mathcal{H}$(pp, Seed) $\rightarrow (h_1, .., h_\ell)$. Output $(\text{pp}, h_1 + a_1, ..., h_\ell + a_\ell)$.

Remark: Note that one could view ΔRG as a candidate sampler $\mathcal{H}$ itself.

Now we describe the notion of Perturbation Resilient Generator implementable by a $(d + 2)$-restricted FE scheme (dΔRG for short.)

5.1 ΔRG Implementable by $(d + 2)$-Restricted FE

A ΔRG scheme implementable by $(d + 2)$-Restricted FE (dΔRG for short) is a perturbation resilient generator with additional properties. We describe syntax again for a complete specification.

– Setup$(1^\lambda, 1^n, B) \rightarrow$ (pp, Seed). The setup algorithm takes as input a security parameter λ, the length parameter 1^n and a polynomial $B = B(\lambda)$ and outputs a seed Seed and public parameters pp. Here, Seed = (Seed.pub(1), Seed.pub(2), ..., Seed.pub(d), Seed.priv(1), Seed.priv(2)) is a vector on $\mathbb{F}_p$ for a modulus p, which is also the modulus used in $(d+2)$-restricted FE scheme. There are $d + 2$ components of this vector, where d of the $d + 2$ components are public and two components are private, each in $\mathbb{F}_p^{npoly(\lambda)}$. Also each part can be partitioned into subcomponents as follows. Seed.pub(j) = (Seed.pub($j, 1$), ..., Seed.pub(j, n)) for $j \in [d]$, Seed.priv(j) = (Seed.priv($j, 1$), ..., Seed.priv(j, n)) for $j \in [2]$. Here, each sub component is in $\mathbb{F}_p^{poly(\lambda)}$ for some fixed polynomial $poly$ independent of n. Also, pp = (Seed.pub(1), ..., Seed.pub(d), $q_1, .., q_\ell$) where each q_i is a degree $d + 2$ multilinear polynomial described below. We require syntactically there exists two algorithms SetupSeed and SetupPoly such that Setup can be decomposed follows:
   1. SetupSeed$(1^\lambda, 1^n, B) \rightarrow$ Seed. The SetupSeed algorithm outputs the seed.
   2. SetupPoly$(1^\lambda, 1^n, B) \rightarrow q_1, ..., q_\ell$. The SetupPoly algorithm outputs $q_1, .., q_\ell$.
– Eval(pp, Seed) $\rightarrow (h_1, ..., h_\ell)$. The evaluation algorithm outputs a vector $(h_1, ..., h_\ell) \in \mathbb{Z}^\ell$. Here for $i \in [\ell]$, $h_i = q_i$(Seed) and ℓ is the stretch of dΔRG. Here q_i is a homogeneous multilinear degree $d + 2$ polynomial where each monomial has degree 1 in $\{\mathsf{pub}(j)\}_{j \in [d+2]}$ and $\{\mathsf{priv}(j)\}_{j \in [2]}$ components of the seed.

The security and efficiency requirements are the same as before.
Remark: To construct $i\mathcal{O}$ we need the stretch of dΔRG to be equal to $\ell = n^{1+\epsilon}$ for some constant $\epsilon > 0$.

5.2   Pseudo Flawed-Smudging Generators

Related to ΔRGs are pseudo flawed-smudging generators (PFGs) introduced by Lin and Matt [42]. As ΔRGs, PFGs are geared for the purpose of generating a smudging noise $\boldsymbol{Y}$ to hide a small polynomially bounded noise $\boldsymbol{a}$. We first give a high-level description of PFGs and then compare them to ΔRGs. For formal definitions and a further discussion of PFGs, we refer the reader to [42].

Intuitively, the output of a PFG "hides" the noise vector $\boldsymbol{a}$ at *all but a few* coordinates with noticeable probability. More formally, the output distribution of a PFG is indistinguishable to a, so-called, *flawed-smudging distribution* $\mathcal{Y}$. A distribution $\mathcal{Y}$ is flawed-smudging if the following holds with some inverse polynomial probability $\delta = 1/poly(\lambda)$ over the choice of $\boldsymbol{Y} \leftarrow \mathcal{Y}$: For some polynomial $B = poly(\lambda)$, every B-bounded noise vector distribution χ, and $\boldsymbol{Y} \leftarrow \mathcal{Y}$, $\boldsymbol{a} \leftarrow \chi$, there is a random variable I correlated with $\boldsymbol{a}$ and $\boldsymbol{Y}$, representing a small, $|I| = o(\lambda)$, subset of "compromised" coordinates, so that the joint distribution of $(I, \boldsymbol{a}, \boldsymbol{Y} + \boldsymbol{a})$ is statistically close to that of $(I, \boldsymbol{a}', \boldsymbol{Y} + \boldsymbol{a})$, where $\boldsymbol{a}'$ is a fresh sample from χ conditioned on agreeing with $\boldsymbol{a}$ at coordinates in I (i.e., $a'_i = a_i$ for all $i \in I$).

Compared to ΔRGs, there is a "good case" occurring with probability δ, in which most coordinates of $\boldsymbol{a}$ are hidden. On the other hand, the output $\boldsymbol{h}$ of a ΔRG guarantees that $\boldsymbol{h}$ and $\boldsymbol{h} + \boldsymbol{a}$ are *computationally indistinguishable* up to advantage $1 - \delta$. Hence, ΔRGs are weaker in this respect since the guarantee is only computational instead of statistical as for PFGs. However, the output of a PFG may in the good case still reveal $\boldsymbol{a}$ at a few coordinates (i.e., $\boldsymbol{a}$ and $\boldsymbol{a}'$ agree at a few coordinates), whereas the output of a ΔRG hides $\boldsymbol{a}$ completely. In that respect, PFGs are weaker.

Despite the technical differences discussed above, the core guarantees of ΔRGs and PFGs are similar. All candidates discussed in the following are therefore candidates for both notions.

6   dΔRG Candidates

We now describe our candidate for dΔRG implementable by a $(d+2)-$ restricted FE scheme. All these candidates use a large enough prime modulus $\mathbf{p} = O(2^\lambda)$, which is the same as the modulus used by $(d+2)-$restricted FE. Then, let χ be a distribution used to sample input elements over $\mathbb{Z}$. Let Q denote a polynomial sampler. Next we give candidate in terms of χ and Q but give concrete instantiations later.

6.1   dΔRG Candidate

- Setup$(1^\lambda, 1^n, B) \to (\mathsf{pp}, \mathsf{Seed})$. Sample a secret $\boldsymbol{s} \leftarrow \mathbb{F}_{\mathbf{p}}^{1 \times n_{\Delta\mathsf{RG}}}$ for $n_{\Delta\mathsf{RG}} = poly(\lambda)$ such that $\mathsf{LWE}_{n_{\Delta\mathsf{RG}}, n \cdot d, \mathbf{p}, \chi}$ holds. Here χ is a bounded distribution with bound $poly(\lambda)$. Let $\mathcal{Q}$ denote an efficiently samplable distribution of homogeneous degree $(d + 2)$ polynomials (instantiated later). Then proceed with SetupSeed as follows:
  1. Sample $\boldsymbol{a}_{i,j} \leftarrow \mathbb{F}_{\mathbf{p}}^{1 \times n_{\Delta\mathsf{RG}}}$ for $i \in [d], j \in [n]$.
  2. Sample $e_{i,j} \leftarrow \chi$ for $i \in [d], j \in [n]$.
  3. Compute $r_{i,j} = \langle \boldsymbol{a}_{i,j}, \boldsymbol{s} \rangle + e_{i,j} \mod \mathbf{p}$ in $\mathbb{F}_{\mathbf{p}}$ for $i \in [d], j \in [n]$.
  4. Define $\boldsymbol{w}_{i,j} = (\boldsymbol{a}_{i,j}, r_{i,j})$ for $i \in [d], j \in [d]$.
  5. Set $\mathsf{Seed.pub}(j, i) = \boldsymbol{w}_{j,i}$ for $j \in [d], i \in [n]$.
  6. Sample $y_i, z_i \leftarrow \chi$ for $i \in [n]$.
  7. Set $\boldsymbol{t} = (-\boldsymbol{s}, 1)$. Note that $\langle \boldsymbol{w}_{j,i}, \boldsymbol{t} \rangle = e_{j,i}$ for $j \in [d], i \in [n]$.
  8. Set $\boldsymbol{y}_i' = y_i \otimes^d \boldsymbol{t}$. (tensor $\boldsymbol{t}$, d times)
  9. Set $\mathsf{Seed.priv}(1, i) = \boldsymbol{y}_i'$ for $i \in [n]$.
  10. Set $\mathsf{Seed.priv}(2, i) = z_i$ for $i \in [n]$.
  Now we describe SetupPoly. Fix $\eta = n^{1+\epsilon}$.
  1. Write $\boldsymbol{e}_j = (e_{j,1}, \ldots, e_{j,n})$ for $j \in [d]$, $\boldsymbol{y} = (y_1, \ldots, y_n)$ and $\boldsymbol{z} = (z_1, \ldots, z_n)$.
  2. Sample polynomials q_ℓ' for $\ell \in [\eta]$ as follows.
  3. $q_\ell' = \Sigma_{I=(i_1,..,i_d,j,k)} c_I e_{1,i_1} \cdots e_{d,i_d} y_j z_k$ where coefficients c_I are bounded by $poly(\lambda)$. These polynomials are sampled according to $\mathcal{Q}$.

4. Define q_i to be a multilinear homogeneous degree $d + 2$ polynomial that takes as input $\mathsf{Seed} = (\{w_{j,i}\}_{j \in [d], i \in [n]}, y'_1, \ldots, y'_n, z)$. Then it computes each monomial $c_I e_{1,i_1} \cdots e_{d,i_d} y_j z_k$ as follows and then adds all the results (thus computes $q'_i(e_1, \ldots, e_d, y, z)$):

   - Compute $c_I \langle w_{1,i_1}, t \rangle \cdots \langle w_{d,i_d}, t \rangle y_j z_k$. This step requires $y'_i = y_i \otimes^d t$ to perform this computation.

5. Output $q_1, \ldots, q_\eta$.

- $\mathsf{Eval}(\mathsf{pp}, \mathsf{Seed}) \rightarrow (h_1, \ldots, h_\eta)$. The evaluation algorithm outputs a vector $(h_1, \ldots, h_\eta) \in \mathbb{Z}^\eta$. Here for $i \in [\eta]$, $h_i = q_i(\mathsf{Seed})$ and η is the stretch of $d \Delta \mathsf{RG}$. Here q_i is a degree $d + 2$ homogenenous multilinear polynomial (defined above) which is degree 1 in d public and 2 private components of the seed.

Efficiency:

1. Note that Seed contains $n \cdot d$ LWE samples $w_{i,j}$ for $i \in [d], j \in [n]$ of dimension $n_{\Delta \mathsf{RG}}$. Along with the samples, it contains elements $y'_i = y_i \otimes^d t$ for $i \in [n]$ and elements z_i for $i \in [n]$. Note that the size of these elements are bounded by $poly(\lambda)$ and is independent of n.

2. The values $h_i = q_i(\mathsf{Seed}) = \Sigma_{I=(i_1,..,i_d,j,k)} c_I e_{1,i_1} \cdots e_{d,i_d} y_j z_k$. Since χ is a bounded distribution, bounded by $poly(\lambda, n)$, and coefficients c_I are also polynomially bounded, each $|h_i| < poly(\lambda, n)$ for $i \in [m]$.

6.2 Instantiations

We now give various instantiations of Q. Let χ be the discrete gaussian distribution with 0 mean and standard deviation n. The following candidate is proposed by [10] based on the investigation of the hardness of families of expanding polynomials over the reals.

Instantiation 1: 3SAT Based Candidate. Let $t = B^2 \lambda$. Sample each polynomial q'_i for $i \in [\eta]$ as follows. $q'_i(x_1, \ldots, x_t, y_1, \ldots, y_t, z_1, \ldots, z_t) = \Sigma_{j \in [t]} q'_{i,j}(x_j, y_j, z_j)$. Here $x_j \in \chi^{d \times n}$ and $y_j, z_j \in \chi^n$ for $j \in [t]$. In other words, q'_i is a sum of t polynomials $q'_{i,j}$ over t disjoint set of variables. Let $d = 1$ for this candidate.

Now we describe how to sample $q'_{i,j}$ for $j \in [\eta]$.

1. Sample randomly inputs $x^*, y^*, z^* \in \{0,1\}^n$.
2. To sample $q'_{i,j}$ do the following. Sample three indices randomly and independently $i_1, i_2, i_3 \leftarrow [n]$. Sample three signs $b_{1,i,j}, b_{2,i,j}, b_{3,i,j} \in \{0,1\}$ uniformly such that $b_{1,i,j} \oplus b_{2,i,j} \oplus b_{3,i,j} \oplus x^*[i_1] \oplus y^*[i_2] \oplus z^*[i_3] = 1$.
3. Set $q'_{i,j}(x_j, y_j, z_j) = 1 - (b_{1,i,j} \cdot x_j[i_1] + (1 - b_{1,i,j}) \cdot (1 - x_j[i_1]) \cdot (b_{2,i,j} \cdot y_j[i_2] + (1 - b_{2,i,j}) \cdot (1 - y_j[i_2])) \cdot ((b_{3,i,j} \cdot z_j[i_3] + (1 - b_{3,i,j}) \cdot (1 - z_j[i_3]))$.

Remark:

1. Note that any clause of the form $a_1 \vee a_2 \vee a_3$ can be written as $1 - (1 - a_1)(1 - a_2)(1 - a_3)$ over integers where a_1, a_2, a_3 are literals in first case and take values in $\{0,1\}$, and thus any random satisfiable 3SAT formula can be converted to polynomials in this manner.

2. Similarly, the above construction can be generalised to degree $(d+2)$-SAT style construction by considering $(d+2)-$SAT for any constant positive integer d and translating them to polynomials.

Instantiation 2: Goldreich's One-way Function Based Candidate. Goldreich's one-way function [30] consists of a predicate P involving $d+2$ variables and computes a boolean function that can be expressed a degree $d+2$ polynomial over the integers. Our candidate $q'_{i,j}(\boldsymbol{x}_j, \boldsymbol{y}_j, \boldsymbol{z}_j)$ consists of the following step.

1. Sample $d+2$ indices $i_1, ..., i_{d+2} \in [n]$.
2. Output $q'_{i,j} = P(\boldsymbol{x}_j[1, i_1], \ldots, \boldsymbol{x}_j[d, i_d], \boldsymbol{y}_j[i_{d+1}], \boldsymbol{z}_j[i_{d+2}])$.

For $d = 3$, [48] provided with the following candidate.

$P(a_1, .., a_5) = a_1 \oplus a_2 \oplus a_3 \oplus a_4 a_5$ where each $a_i \in \{0, 1\}$. Note that this can be naturally converted to a polynomial as follows. Rewrite $a \oplus b = (1 - a)b + (1 - b)a$ and this immediately gives rise to a polynomial over the integers.

6.3 Simplifying Assumptions

In this section, we remark that the $d\Delta$RG assumption can be simplified from being an exponential family of assumptions to two simpler assumptions as follows. We provide two sub-assumptions, which together imply $d\Delta$RG assumptions.

LWE with degree $d + 2$ leakage. There exists a polynomial sampler Q and a constant $\epsilon > 0$, such that for every large enough $\lambda \in \mathbb{N}$, and every polynomial bound $B = B(\lambda)$ there exist large enough polynomial $n_B = \lambda^c$ such that for every positive integer $n > n_B$, there exists a $poly(n)-$bounded discrete gaussian distribution χ such that the following two distributions are close (we define closeness later). We define the following two distributions:

Distribution $\mathcal{D}_1$:

- Fix a prime modulus $p = O(2^\lambda)$.
- (**Sample Polynomials.**) Run $Q(n, B, \epsilon) = (q_1, ..., q_{\lfloor n^{1+\epsilon} \rfloor})$.
- (**Sample Secret.**) Sample a secret $\boldsymbol{s} \leftarrow \mathbb{Z}_p^\lambda$.
- Sample $\boldsymbol{a}_{j,i} \leftarrow \mathbb{Z}_p^\lambda$ for $j \in [d], i \in [n]$.
- (**Sample LWE Errors.**) For every $j \in [d], i \in [n]$, sample $e_{j,i}, y_i, z_i \leftarrow \chi$. Write $\boldsymbol{e}_j = (e_{j,1}, \ldots, e_{j,n})$ for $j \in [d]$, $\boldsymbol{y} = (y_1, \ldots, y_n)$ and $\boldsymbol{z} = (z_1, \ldots, z_n)$.
- Output $\{\boldsymbol{a}_{j,i}, \langle \boldsymbol{a}_{j,i}, \boldsymbol{s} \rangle + e_{j,i} \mod p\}_{j \in [d], i \in [n]}$ and $\{q_k, q_k(\boldsymbol{e}_1, \ldots, \boldsymbol{e}_d, \boldsymbol{y}, \boldsymbol{z})\}_{k \in [\lfloor n^{1+\epsilon} \rfloor]}$

Distribution $\mathcal{D}_2$:

- Fix a prime modulus $p = O(2^\lambda)$.
- (**Sample Polynomials.**) Run $Q(n, B, \epsilon) = (q_1, ..., q_{\lfloor n^{1+\epsilon} \rfloor})$.
- (**Sample Secret.**) Sample a secret $\boldsymbol{s} \leftarrow \mathbb{Z}_p^\lambda$.
- Sample $\boldsymbol{a}_{j,i} \leftarrow \mathbb{Z}_p^\lambda$ for $j \in [d], i \in [n]$.

- (**Sample independent LWE Errors.**) For every $j \in [d], i \in [n]$, sample $e_{j,i}, e'_{j,i}, y_i, z_i \leftarrow \chi$.[1] Write $e'_j = (e'_{j,1}, \ldots, e'_{j,n})$, $e_j = (e_{j,1}, \ldots, e_{j,n})$ for $j \in [d]$, $\boldsymbol{y} = (y_1, \ldots, y_n)$ and $\boldsymbol{z} = (z_1, \ldots, z_n)$.
- Output $\{a_{j,i}, \langle a_{j,i}, s \rangle + e'_{j,i} \mod p\}_{j \in [d], i \in [n]}$ and $\{q_k, q_k(e_1, \ldots, e_d, \boldsymbol{y}, \boldsymbol{z})\}_{k \in [\lfloor n^{1+\epsilon} \rfloor]}$

The assumption states that there exists a constant $\epsilon_{adv} > 0$ such that for any adversary $\mathcal{A}$ of size $2^{\lambda^{\epsilon_{adv}}}$, the following holds:

$$|\Pr[\mathcal{A}(\mathcal{D}_1) = 1] - \Pr[\mathcal{A}(\mathcal{D}_2) = 1]| < 1/2\lambda$$

Remark. This assumption says that to a bounded adversary, the advantage of distinguishing the tuple consisting of polynomials samples, along with correlated LWE samples with tuple consisting of polynomials samples, along with uncorrelated LWE samples is bounded by $1/2\lambda$. Second assumption says that the tuple of polynomial samples looks close to independent discrete gaussian variables with a large enough variance and 0 mean. Below we define the notion of a (B, δ)−smooth distribution.

Definition 7. (B, δ)−*Smooth distribution* $\mathcal{N}$ *is an efficiently samplable distribution over* $\mathbb{Z}$ *with the property that* $\Delta(\mathcal{N}, \mathcal{N} + b) \leq \delta$ *for any* $b \in [-B, B]$.

Weak Pseudo-Independence Generator Assumption [2,42]. For the parameters defined above, the assumption states that there exists a constant $\epsilon_{adv} > 0$ such that for any adversary $\mathcal{A}$ of size $2^{\lambda^{\epsilon_{adv}}}$, the following holds:

$$|\Pr[\mathcal{A}(\mathcal{D}_1) = 1] - \Pr[\mathcal{A}(\mathcal{D}_2) = 1]| < 1 - 3/\lambda$$

where distributions are defined below.
Distribution $\mathcal{D}_1$:

- Fix a prime modulus $p = O(2^\lambda)$.
- (**Sample Polynomials.**) Run $Q(n, B, \epsilon) = (q_1, \ldots, q_{\lfloor n^{1+\epsilon} \rfloor})$.
- For every $j \in [d], i \in [n]$, sample $e_{j,i}, y_i, z_i \leftarrow \chi$. Write $e_j = (e_{j,1}, \ldots, e_{j,n})$ for $j \in [d]$, $\boldsymbol{y} = (y_1, \ldots, \boldsymbol{y}_n)$ and $\boldsymbol{z} = (z_1, \ldots, z_n)$.
- Output $\{q_k, q_k(e_1, \ldots, e_d, \boldsymbol{y}, \boldsymbol{z})\}_{k \in [\lfloor n^{1+\epsilon} \rfloor]}$

Distribution $\mathcal{D}_2$:

- Fix a prime modulus $p = O(2^\lambda)$.
- (**Sample Polynomials.**) Run $Q(n, B, \epsilon) = (q_1, \ldots, q_{\lfloor n^{1+\epsilon} \rfloor})$.
- Output $\{q_k, h_k \leftarrow \mathcal{N}\}_{k \in [\lfloor n^{1+\epsilon} \rfloor]}$

Here $\mathcal{N}$ is a $(B, \frac{1}{n^2\lambda})$−smooth distribution.
Thus we have,

---

[1] Thus, we can observe that χ should be a distribution such that LWE assumption holds with respect to χ and parameters specified above.

Claim. Assuming,

1. LWE with degree $d + 2$ leakage.
2. Weak Pseudo-Independence Generator Assumption

There exists a dΔRG scheme.

Proof. (Sketch.) This is immediate and the proof goes in three hybrids. First, we use LWE with degree $d + 2$ leakage assumption with $1/2\lambda$ security loss. In the next hybrid, we sample from $\mathcal{N}$ given to us by Weak Pseudo-Independence Generator Assumption. With that, we have another $1 - 3/\lambda$ loss in the security. Finally, we move to a hybrid where all perturbations are 0. This leads to a security loss of $n^{1+\epsilon} \times \frac{1}{n^2\lambda} < \frac{1}{n^{1-\epsilon}\lambda}$ due to statistical distance. Adding these security losses, we prove the claim. Thus $\mathcal{H}$ just uses $\mathcal{N}$ to sample each component independently.

7   Constructing $(d + 2)$ Restricted FE from Bilinear Maps

In this section we describe our construction for a $d + 2-$restricted FE scheme.
We now describe our construction as follows:

7.1   Construction

Ingredients: Our main ingredient is a secret-key canonical function-hiding inner product functional encryption scheme cIPE (see Sect. 3.3).

Notation: We denote by $\mathbb{F}_\mathbf{p}$ the field on which the computation is done in slotted encodings.

1. By boldfaced letters, we denote (multi-dimensional) matrices, where dimensions are specified. Messages are of the form $(\boldsymbol{x}, \boldsymbol{y}, \boldsymbol{z})$. Here, $\boldsymbol{x} \in \mathbb{F}_\mathbf{p}^{d \times n}$. $\boldsymbol{y}, \boldsymbol{z} \in \mathbb{F}_\mathbf{p}^n$.
2. **Function class of interest:** We consider the set of functions $\mathcal{F}_{\mathsf{dFE}} = \mathcal{F}_{\mathsf{dFE}, \lambda, \mathbf{p}, n} = \{f : \mathbb{F}_\mathbf{p}^{n(d+2)} \to \mathbb{F}_\mathbf{p}\}$ where $\mathbb{F}_\mathbf{p}$ is a finite field of order $\mathbf{p}(\lambda)$. Here n is seen as a function of λ. Each $f \in \mathcal{F}_{\lambda, \mathbf{p}, n}$ takes as input $d + 2$ vectors $(\boldsymbol{x}[1], \ldots, \boldsymbol{x}[d], \boldsymbol{y}, \boldsymbol{z})$ over $\mathbb{F}_\mathbf{p}$ and computes a polynomial of the form $\Sigma c_{i_1, \ldots, i_d, j, k} \cdot \boldsymbol{x}[1, i_1] \cdots \boldsymbol{x}[d, i_d] \cdot \boldsymbol{y}[j] \cdot \boldsymbol{z}[k]$, where $c_{i_1, \ldots, i_d, j, k}$ are coefficients from $\mathbb{F}_\mathbf{p}$.

Notation. For a secret key generated for the cIPE encryption algorithm, by using primed variables such as sk' we denote the secret key that is not generated during the setup of the dFE scheme but during its encryption algorithm. We describe the construction below.

Setup($1^\lambda, 1^n$): On input security parameter 1^λ and length 1^n,

- Sample pp $\leftarrow$ cIPE.PPSetup(1^λ). We assume pp $= (e, G_1, G_2, G_T, g_1, g_2, \mathbb{Z}_\mathbf{p})$.
- Run cIPE setup as follows. $\mathbf{sk}_0 \leftarrow$ cIPE.Setup(pp, 1^4). Thus these keys are used to encrypt vectors in $\mathbb{F}_\mathbf{p}^4$.
- Then run cIPE setup algorithm $n \cdot d$ times. That is, for every $\ell \in [d]$ and $i_\ell \in [n]$, compute $\mathbf{sk}^{(\ell, i_\ell)} \leftarrow$ cIPE.Setup(pp, 1^4).
- Sample $\boldsymbol{\alpha} \leftarrow \mathbb{F}_\mathbf{p}^{d \times n}$. Also sample $\boldsymbol{\beta}, \boldsymbol{\gamma} \leftarrow \mathbb{F}_\mathbf{p}^n$.
- For $\ell \in [d]$, $i_\ell \in [n]$ and every set $\boldsymbol{I} = (i_{\ell+1}, \ldots, i_d, j, k) \in [n]^{d-\ell+2}$, compute $\mathsf{Key}_{\boldsymbol{I}}^{(\ell, i_\ell)} =$ cIPE.KeyGen($\mathbf{sk}^{(\ell, i_\ell)}, (\boldsymbol{\alpha}[\ell+1, i_{\ell+1}] \cdots \boldsymbol{\alpha}[d, i_d]\boldsymbol{\beta}[j]\boldsymbol{\gamma}[k], \boldsymbol{\alpha}[\ell, i_\ell] \cdots \boldsymbol{\alpha}[d, i_d]\boldsymbol{\beta}[j]\boldsymbol{\gamma}[k], 0, 0)$).
- Output MSK $= (\{\mathbf{sk}^{(\ell, i_\ell)}, \mathsf{Key}_{\boldsymbol{I}}^{(\ell, i_\ell)}\}_{\ell, i_\ell, \boldsymbol{I}}, \boldsymbol{\alpha}, \boldsymbol{\beta}, \boldsymbol{\gamma}, \mathbf{sk}_0)$

KeyGen(MSK, f): On input the master secret key MSK and function f,

- Parse MSK $= (\{\mathbf{sk}^{(\ell, i_\ell)}, \mathsf{Key}_{\boldsymbol{I}}^{(\ell, i_\ell)}\}_{\ell, i_\ell, \boldsymbol{I}}, \boldsymbol{\alpha}, \boldsymbol{\beta}, \boldsymbol{\gamma}, \mathbf{sk}_0)$.
- Compute $\theta_f = f(\boldsymbol{\alpha}, \boldsymbol{\beta}, \boldsymbol{\gamma})$.
- Compute $\mathsf{Key}_{0, f} =$ cIPE.KeyGen($\mathbf{sk}_0, (\theta_f, 0, 0, 0)$)
- Output $sk_f = (\mathsf{Key}_{0, f}, \{\mathsf{Key}_{\boldsymbol{I}}^{(\ell, i_\ell)}\}_{\ell, i_\ell, \boldsymbol{I}})$

Enc(MSK, $\boldsymbol{x}, \boldsymbol{y}, \boldsymbol{z}$): The input message $M = (\boldsymbol{x}, \boldsymbol{y}, \boldsymbol{z})$ consists of a public attribute $\boldsymbol{x} \in \mathbb{F}_\mathbf{p}^{d \times n}$ and private vectors $\boldsymbol{y}, \boldsymbol{z} \in \mathbb{F}_\mathbf{p}^n$. Perform the following operations:

- Parse MSK $= (\{\mathbf{sk}^{(\ell, i_\ell)}, \mathsf{Key}_{\boldsymbol{I}}^{(\ell, i_\ell)}\}_{\ell, i_\ell, \boldsymbol{I}}, \boldsymbol{\alpha}, \boldsymbol{\beta}, \boldsymbol{\gamma}, \mathbf{sk}_0)$.
- Sample $r \leftarrow \mathbb{F}_\mathbf{p}$.
- Compute $\mathsf{CT}_0 =$ cIPE.Enc($\mathbf{sk}_0, (r, 0, 0, 0)$)
- Sample $\mathbf{sk}' \leftarrow$ cIPE.Setup(pp, 1^4).
- Compute $\mathsf{CTC}_j \leftarrow$ cIPE.Enc($\mathbf{sk}', (\boldsymbol{y}[j], \boldsymbol{\beta}[j], 0, 0)$) for $j \in [n]$
- Compute $\mathsf{CTK}_k \leftarrow$ cIPE.KeyGen($\mathbf{sk}', (\boldsymbol{z}[k], -r\boldsymbol{\gamma}[k], 0, 0)$) for $k \in [n]$.
- For $\ell \in [d]$, $i_\ell \in [n]$, compute $\mathsf{CT}^{(\ell, i_\ell)} =$ cIPE.Enc($\mathbf{sk}^{(\ell, i_\ell)}, (r\boldsymbol{x}[\ell, i_\ell], -r, 0, 0)$).
- Output $\mathsf{CT} = (\boldsymbol{x}, \mathsf{CT}_0, \{\mathsf{CTC}_j, \mathsf{CTK}_k, \mathsf{CT}^{(\ell, i_\ell)}\}_{\ell \in [d], i_\ell \in [n], j \in [n], k \in [n]})$

Dec($1^B, sk_f, \mathsf{CT}$):

- Parse $\mathsf{CT} = (\boldsymbol{x}, \mathsf{CT}_0, \{\mathsf{CTC}_j, \mathsf{CTK}_k, \mathsf{CT}^{(\ell, i_\ell)}\}_{\ell \in [d], i_\ell \in [n], j \in [n], k \in [n]})$.
- Parse $sk_f = \{\mathsf{Key}_{0, f}, \mathsf{Key}_{\boldsymbol{I}}^{(\ell, i_\ell)}\}_{\ell, i_\ell, \boldsymbol{I}}$.
- For every $\ell \in [d]$ and $\boldsymbol{I} = (i_\ell, \ldots, i_d, j, k) \in [n]^{d-\ell+3}$ do the following. Let $\boldsymbol{I}'$ be such that $\boldsymbol{I} = i_\ell || \boldsymbol{I}'$. In other words, $\boldsymbol{I}'$ has all but first element of $\boldsymbol{I}$. Compute $\mathsf{Mon}_{\boldsymbol{I}'}^{(\ell, i_\ell)} =$ cIPE.Dec($\mathsf{Key}_{\boldsymbol{I}'}^{(\ell, i_\ell)}, \mathsf{CT}^{(\ell, i_\ell)}) = [r(\boldsymbol{x}[\ell, i_\ell] - \boldsymbol{\alpha}[\ell, i_\ell])\boldsymbol{\alpha}[\ell - 1, i_{\ell-1}] \cdots \boldsymbol{\alpha}[d, i_d]\boldsymbol{\beta}[j]\boldsymbol{\gamma}[k]]_T$.
- Compute $\mathsf{Mon}_0 =$ cIPE.Dec($\mathsf{Key}_{0, f}, \mathsf{CT}_0) = [rf(\boldsymbol{\alpha}, \boldsymbol{\beta}, \boldsymbol{\gamma})]_T$.
- Compute $\mathsf{Mon}^{(j, k)} =$ cIPE.Dec($\mathsf{CTK}_k, \mathsf{CTC}_j) = [\boldsymbol{y}[j]\boldsymbol{z}[k] - r\boldsymbol{\beta}[j]\boldsymbol{\gamma}[k]]_T$.
- Let $f = \Sigma_{\boldsymbol{I}=(i_1, \ldots, i_d, j, k)} c_{\boldsymbol{I}} \boldsymbol{x}[1, i_1] \cdots \boldsymbol{x}[d, i_d]\boldsymbol{y}[j]\boldsymbol{z}[k]$. Now fix $\boldsymbol{I} = (i_1, \ldots, i_d, j, k)$. For the monomial corresponding to $\boldsymbol{I}$ compute $\mathsf{Int}_{\boldsymbol{I}} = [\boldsymbol{x}[1, i_1] \cdots \boldsymbol{x}[d, i_d]\boldsymbol{y}[j]\boldsymbol{z}[k] - r\boldsymbol{\alpha}[1, i_1] \cdots \boldsymbol{\alpha}[d, i_d]\boldsymbol{\beta}[j]\boldsymbol{\gamma}[k]]_T$ as follows.

1. For $v \in [d]$, denote $\boldsymbol{I}_v = (i_v, \ldots, i_d, j, k)$ and $\boldsymbol{I}_v' = (i_{v+1}, \ldots, i_d, j, k)$.
2. Compute $\mathsf{Int}_I = \Pi_{v \in [d]} \mathsf{Mon}^{(v,i_v)}{}_{\boldsymbol{I}_v'}{}^{\rho_{\boldsymbol{I}_v}}$. We describe $\rho_{\boldsymbol{I}_v}$ shortly.
3. We want these coefficients $\rho_{\boldsymbol{I}_v}$ such that $\mathsf{Int}_I = [\Sigma_{v \in [d]} \rho_v (\boldsymbol{x}[v, i_v] \boldsymbol{\alpha}[v + 1, i_{v+1}] \cdots \boldsymbol{\alpha}[d, i_d] \boldsymbol{\beta}[j] \boldsymbol{\gamma}[k] - r\boldsymbol{\alpha}[v, i_v] \cdots \boldsymbol{\alpha}[d, i_d] \boldsymbol{\beta}[j] \boldsymbol{\gamma}[k])]_T$.
4. This defines $\rho_{\boldsymbol{I}_1} = 1$ and $\rho_{\boldsymbol{I}_v} = \boldsymbol{x}[1, i_1], \ldots, \boldsymbol{x}[v - 1, i_{v-1}]$ for $v \in [d]$.
5. This can be verified for $d = 2$ as follows.

$$\boldsymbol{x}[1, i_1]\boldsymbol{x}[2, i_2](\boldsymbol{y}[j]\boldsymbol{z}[k] - r\boldsymbol{\beta}[j]\boldsymbol{\gamma}[k]) + \boldsymbol{x}[1, i_1]r(\boldsymbol{x}[2, i_2]$$
$$-\boldsymbol{\alpha}[i_2])\boldsymbol{\beta}[j]\boldsymbol{\gamma}[k] + r(\boldsymbol{x}[1, i_1] - \boldsymbol{\alpha}[1, i_1])\boldsymbol{\alpha}[i_2]\boldsymbol{\beta}[j]\boldsymbol{\gamma}[k]$$
$$=\boldsymbol{x}[1, i_1]\boldsymbol{x}[2, i_2]\boldsymbol{y}[j]\boldsymbol{z}[k] - r\boldsymbol{\alpha}[1, i_1]\boldsymbol{\alpha}[2, i_2]\boldsymbol{\beta}[j]\boldsymbol{\beta}[k]$$

In this way, the process holds for any d.

- Finally compute $\mathsf{Int}_1 = \Pi_{I=(i_1,\ldots,i_d)} \mathsf{Int}_I^{cI} = [f(\boldsymbol{x}, \boldsymbol{y}, \boldsymbol{z}) - rf(\boldsymbol{\alpha}, \boldsymbol{\beta}, \boldsymbol{\gamma})]_T$.
- Compute $\mathsf{Int}_1 \cdot \mathsf{Mon}_0 = [f(\boldsymbol{x}, \boldsymbol{y}, \boldsymbol{z})]_T$. Using brute force, check if $|f(\boldsymbol{x}, \boldsymbol{y}, \boldsymbol{z})| < B$. If that is the case, output $f(\boldsymbol{x}, \boldsymbol{y}, \boldsymbol{z})$ otherwise output $\perp$.

We now discuss correctness and linear efficiency:

Correctness: Correctness is implicit from the description of the decryption algorithm.

Linear Efficiency: Note that a ciphertext is of the following form:

$$\mathsf{CT} = (\boldsymbol{x}, \mathsf{CT}_0, \{\mathsf{CTC}_j, \mathsf{CTK}_k, \mathsf{CT}^{(\ell, i_\ell)}\}_{\ell \in [d], i_\ell \in [n], j \in [n], k \in [n]})$$

Thus there are $n \times (d + 1) + 1$ cIPE ciphertexts and n cIPE function keys for vectors of length 4. Hence, the claim holds due to the efficiency of cIPE.

Due to lack of space we defer the security proof to the full version. Here is our theorem statement.

Theorem 4. *Assuming SXDH holds relative to* PPGen, *the construction described in Sect. 7 satisfies semi-functional security.*

8 Construction of $i\mathcal{O}$

Following the template of [4] we prove the following theorems. The details can be found in the full version.

Theorem 5. *For any constant integer $d \geq 1$, Assuming*

- *Subexponentially hard LWE.*
- *Subexponentially hard SXDH*
- *PRGs with*
  - *Stretch of $k^{1+\epsilon}$ (length of input being k bits) for some constant $\epsilon > 0$.*
  - *Block locality $d + 2$.*
  - *Security with* negl *distinguishing gap against adversaries of subexponential size.*

– dΔRG with a stretch of $k^{1+\epsilon'}$ for some constant $\epsilon' > 0$.[2]

there exists an $i\mathcal{O}$ scheme for P/poly.

Here is the version with the tradeoff.

Theorem 6. *For any constant integer $d \geq 1$, two distinguishing gaps* $\mathsf{adv}_1, \mathsf{adv}_2$, *if* $\mathsf{adv}_1 + \mathsf{adv}_2 \leq 1 - 2/\lambda$ *then assuming,*

– *Subexponentially hard LWE.*
– *Subexponentially hard SXDH.*
– *PRGs with*
  - *Stretch of $k^{1+\epsilon}$ (length of input being k bits) for some constant $\epsilon > 0$.*
  - *Block locality $d + 2$.*
  - *Security with distinguishing gap bounded by* adv_1 *against adversaries of sub-exponential size.*
– dΔRG *with distinguishing gap bounded by* adv_2 *against adversaries of size 2^{λ}*[3]

there exists a secure $i\mathcal{O}$ scheme for P/poly.

Alternatively, the construction from Sect. 7 can also be used to instantiate a partially hiding FE scheme as in [42]. Together with a pseudo flawed-smudging generator (see Sect. 5.2) that can be computed by that FE scheme, we can follow the approach from [42] to obtain the following theorem. See the full version for details.

Theorem 7. *For any constant integer $d \geq 1$, assuming,*

– *LWE,*
– *SXDH,*
– *PRGs with*
  - *Stretch of $k^{1+\epsilon}$ (length of input being k bits) for some constant $\epsilon > 0$,*
  - *Constant locality and additional mild structural properties (see [42] for details),*
– *Pseudo flawed-smudging generators with degree d public computation and degree 2 private computation.*

where all primitives are secure against adversaries of polynomial sizes with sub-exponentially small distinguishing gaps. Then, there exists a subexponentially secure $i\mathcal{O}$ scheme for P/poly.

Acknowledgements. We would like to thank Prabhanjan Ananth for preliminary discussions on the concept of a $d + 2$ restricted FE scheme. We would also like to thank Pravesh Kothari, Sam Hopkins and Boaz Barak for many useful discussions about our dΔRG Candidates. This work was done in part when both Huijia Lin and Chrisitan Matt were at University of California, Santa Barbara.

---

[2] Instantiations can be found in Sect. 6.2.
[3] Instantiations can be found in Sect. 6.2.

Aayush Jain and Amit Sahai are supported in part from a DARPA/ARL SAFE-WARE award, NSF Frontier Award 1413955, and NSF grant 1619348, BSF grant 2012378, a Xerox Faculty Research Award, a Google Faculty Research Award, an equipment grant from Intel, and an Okawa Foundation Research Grant. This material is based upon work supported by the Defense Advanced Research Projects Agency through the ARL under Contract W911NF-15-C- 0205. Aayush Jain is also supported by a Google PhD Fellowship in Privacy and Security. Huijia Lin and Christian Matt were supported by NSF grants CNS-1528178, CNS-1514526, CNS-1652849 (CAREER), a Hellman Fellowship, the Defense Advanced Research Projects Agency (DARPA) and Army Research Office (ARO) under Contract No. W911NF-15-C-0236, and a subcontract No. 2017-002 through Galois. The views expressed are those of the authors and do not reflect the official policy or position of the Department of Defense, the National Science Foundation, Google, or the U.S. Government.

References

1. Agrawal, S.: New methods for indistinguishability obfuscation: bootstrapping and instantiation. IACR Cryptol. ePrint Archive **2018**, 633 (2018)
2. Ananth, P., Brakerski, Z., Khuarana, D., Sahai, A.: New approach against the locality barrier in obfuscation: pseudo-independent generators. Unpublished Work (2017)
3. Ananth, P., Gupta, D., Ishai, Y., Sahai, A.: Optimizing obfuscation: avoiding Barrington's theorem. In: ACM CCS, pp. 646–658 (2014)
4. Ananth, P., Jain, A., Sahai, A.: Indistinguishability obfuscation without multilinear maps: iO from LWE, bilinear maps, and weak pseudorandomness. IACR Cryptol. ePrint Archive **2018**, 615 (2018)
5. Ananth, P., Sahai, A.: Projective arithmetic functional encryption and indistinguishability obfuscation from degree-5 multilinear maps. In: Coron, J.-S., Nielsen, J.B. (eds.) EUROCRYPT 2017, Part I. LNCS, vol. 10210, pp. 152–181. Springer, Cham (2017). https://doi.org/10.1007/978-3-319-56620-7_6
6. Arora, S., Ge, R.: New algorithms for learning in presence of errors. In: Aceto, L., Henzinger, M., Sgall, J. (eds.) ICALP 2011, Part I. LNCS, vol. 6755, pp. 403–415. Springer, Heidelberg (2011). https://doi.org/10.1007/978-3-642-22006-7_34
7. Badrinarayanan, S., Miles, E., Sahai, A., Zhandry, M.: Post-zeroizing obfuscation: new mathematical tools, and the case of evasive circuits. In: Fischlin, M., Coron, J.-S. (eds.) EUROCRYPT 2016, Part II. LNCS, vol. 9666, pp. 764–791. Springer, Heidelberg (2016). https://doi.org/10.1007/978-3-662-49896-5_27
8. Barak, B., Garg, S., Kalai, Y.T., Paneth, O., Sahai, A.: Protecting obfuscation against algebraic attacks. In: Nguyen, P.Q., Oswald, E. (eds.) EUROCRYPT 2014. LNCS, vol. 8441, pp. 221–238. Springer, Heidelberg (2014). https://doi.org/10.1007/978-3-642-55220-5_13
9. Barak, B., et al.: On the (Im)possibility of obfuscating programs. In: Kilian, J. (ed.) CRYPTO 2001. LNCS, vol. 2139, pp. 1–18. Springer, Heidelberg (2001). https://doi.org/10.1007/3-540-44647-8_1
10. Barak, B., Hopkins, S., Jain, A., Kothari, P., Sahai, A.: Sum-of-squares meets program obfuscation, revisited. Unpublished Work (2018)
11. Bartusek, J., Guan, J., Ma, F., Zhandry, M.: Preventing zeroizing attacks on GGH15. IACR Cryptol. ePrint Archive **2018**, 511 (2018)
12. Bitansky, N., Paneth, O., Rosen, A.: On the cryptographic hardness of finding a nash equilibrium. In: FOCS, pp. 1480–1498 (2015)

13. Boneh, D., Wu, D.J., Zimmerman, J.: Immunizing multilinear maps against zeroizing attacks. IACR Cryptology ePrint Archive 2014, 930 (2014). http://eprint.iacr.org/2014/930

14. Brakerski, Z., Gentry, C., Halevi, S., Lepoint, T., Sahai, A., Tibouchi, M.: Cryptanalysis of the quadratic zero-testing of GGH. Cryptology ePrint Archive, Report 2015/845 (2015). http://eprint.iacr.org/

15. Brakerski, Z., Rothblum, G.N.: Virtual black-box obfuscation for all circuits via generic graded encoding. In: Lindell, Y. (ed.) TCC 2014. LNCS, vol. 8349, pp. 1–25. Springer, Heidelberg (2014). https://doi.org/10.1007/978-3-642-54242-8_1

16. Brzuska, C., Farshim, P., Mittelbach, A.: Indistinguishability obfuscation and UCEs: the case of computationally unpredictable sources. In: Garay, J.A., Gennaro, R. (eds.) CRYPTO 2014, Part I. LNCS, vol. 8616, pp. 188–205. Springer, Heidelberg (2014). https://doi.org/10.1007/978-3-662-44371-2_11

17. Cheon, J.H., Han, K., Lee, C., Ryu, H., Stehlé, D.: Cryptanalysis of the multilinear map over the integers. In: Oswald, E., Fischlin, M. (eds.) EUROCRYPT 2015, Part I. LNCS, vol. 9056, pp. 3–12. Springer, Heidelberg (2015). https://doi.org/10.1007/978-3-662-46800-5_1

18. Cheon, J.H., Lee, C., Ryu, H.: Cryptanalysis of the new CLT multilinear maps. Cryptology ePrint Archive, Report 2015/934 (2015). http://eprint.iacr.org/

19. Cohen, A., Holmgren, J., Nishimaki, R., Vaikuntanathan, V., Wichs, D.: Watermarking cryptographic capabilities. SIAM J. Comput. 47(6), 2157–2202 (2018)

20. Coron, J.-S., et al.: Zeroizing without low-level zeroes: new MMAP attacks and their limitations. In: Gennaro, R., Robshaw, M. (eds.) CRYPTO 2015, Part I. LNCS, vol. 9215, pp. 247–266. Springer, Heidelberg (2015). https://doi.org/10.1007/978-3-662-47989-6_12

21. Coron, J.-S., Lepoint, T., Tibouchi, M.: Practical multilinear maps over the integers. In: Canetti, R., Garay, J.A. (eds.) CRYPTO 2013, Part I. LNCS, vol. 8042, pp. 476–493. Springer, Heidelberg (2013). https://doi.org/10.1007/978-3-642-40041-4_26

22. Coron, J.-S., Lepoint, T., Tibouchi, M.: New multilinear maps over the integers. In: Gennaro, R., Robshaw, M. (eds.) CRYPTO 2015, Part I. LNCS, vol. 9215, pp. 267–286. Springer, Heidelberg (2015). https://doi.org/10.1007/978-3-662-47989-6_13

23. Döttling, N., Garg, S., Gupta, D., Miao, P., Mukherjee, P.: Obfuscation from low noise multilinear maps. IACR Cryptol. ePrint Archive 2016, 599 (2016)

24. Garg, S., Gentry, C., Halevi, S.: Candidate multilinear maps from ideal lattices. In: Johansson, T., Nguyen, P.Q. (eds.) EUROCRYPT 2013. LNCS, vol. 7881, pp. 1–17. Springer, Heidelberg (2013). https://doi.org/10.1007/978-3-642-38348-9_1

25. Garg, S., Gentry, C., Halevi, S., Raykova, M.: Two-round secure MPC from indistinguishability obfuscation. In: Lindell, Y. (ed.) TCC 2014. LNCS, vol. 8349, pp. 74–94. Springer, Heidelberg (2014). https://doi.org/10.1007/978-3-642-54242-8_4

26. Garg, S., Gentry, C., Halevi, S., Raykova, M., Sahai, A., Waters, B.: Candidate indistinguishability obfuscation and functional encryption for all circuits. In: FOCS (2013)

27. Garg, S., Miles, E., Mukherjee, P., Sahai, A., Srinivasan, A., Zhandry, M.: Secure obfuscation in a weak multilinear map model. In: Hirt, M., Smith, A. (eds.) TCC 2016, Part II. LNCS, vol. 9986, pp. 241–268. Springer, Heidelberg (2016). https://doi.org/10.1007/978-3-662-53644-5_10

28. Garg, S., Pandey, O., Srinivasan, A.: Revisiting the cryptographic hardness of finding a nash equilibrium. In: Robshaw, M., Katz, J. (eds.) CRYPTO 2016, Part II. LNCS, vol. 9815, pp. 579–604. Springer, Heidelberg (2016). https://doi.org/10.1007/978-3-662-53008-5_20

29. Gentry, C., Gorbunov, S., Halevi, S.: Graph-induced multilinear maps from lattices. In: Dodis, Y., Nielsen, J.B. (eds.) TCC 2015, Part II. LNCS, vol. 9015, pp. 498–527. Springer, Heidelberg (2015). https://doi.org/10.1007/978-3-662-46497-7_20

30. Goldreich, O.: Candidate one-way functions based on expander graphs. IACR Cryptology ePrint Archive 2000, 63 (2000). http://eprint.iacr.org/2000/063

31. Goldwasser, S., et al.: Multi-input functional encryption. In: Nguyen, P.Q., Oswald, E. (eds.) EUROCRYPT 2014. LNCS, vol. 8441, pp. 578–602. Springer, Heidelberg (2014). https://doi.org/10.1007/978-3-642-55220-5_32

32. Goldwasser, S., Rothblum, G.N.: On best-possible obfuscation. In: Vadhan, S.P. (ed.) TCC 2007. LNCS, vol. 4392, pp. 194–213. Springer, Heidelberg (2007). https://doi.org/10.1007/978-3-540-70936-7_11

33. Halevi, S.: Graded encoding, variations on a scheme. IACR Cryptol. ePrint Archive **2015**, 866 (2015)

34. Hofheinz, D., Jager, T., Khurana, D., Sahai, A., Waters, B., Zhandry, M.: How to generate and use universal samplers. In: Cheon, J.H., Takagi, T. (eds.) ASIACRYPT 2016, Part II. LNCS, vol. 10032, pp. 715–744. Springer, Heidelberg (2016). https://doi.org/10.1007/978-3-662-53890-6_24

35. Hohenberger, S., Sahai, A., Waters, B.: Replacing a random Oracle: full domain hash from indistinguishability obfuscation. In: Nguyen, P.Q., Oswald, E. (eds.) EUROCRYPT 2014. LNCS, vol. 8441, pp. 201–220. Springer, Heidelberg (2014). https://doi.org/10.1007/978-3-642-55220-5_12

36. Hu, Y., Jia, H.: Cryptanalysis of GGH map. IACR Cryptol. ePrint Archive **2015**, 301 (2015)

37. Jain, A., Lin, H., Matt, C., Sahai, A.: How to leverage hardness of constant-degree expanding polynomials over $\mathbb{R}$ to build $i\mathcal{O}$. arXiv (2019)

38. Koppula, V., Lewko, A.B., Waters, B.: Indistinguishability obfuscation for turing machines with unbounded memory. In: STOC (2015)

39. Lin, H.: Indistinguishability obfuscation from constant-degree graded encoding schemes. In: Fischlin, M., Coron, J.-S. (eds.) EUROCRYPT 2016, Part I. LNCS, vol. 9665, pp. 28–57. Springer, Heidelberg (2016). https://doi.org/10.1007/978-3-662-49890-3_2

40. Lin, H.: Indistinguishability obfuscation from SXDH on 5-linear maps and locality-5 PRGs. In: Katz, J., Shacham, H. (eds.) CRYPTO 2017, Part I. LNCS, vol. 10401, pp. 599–629. Springer, Cham (2017). https://doi.org/10.1007/978-3-319-63688-7_20

41. Lin, H.: Indistinguishability obfuscation from SXDH on 5-linear maps and locality-5 PRGs. In: Katz, J., Shacham, H. (eds.) CRYPTO 2017, Part I. LNCS, vol. 10401, pp. 599–629. Springer, Cham (2017). https://doi.org/10.1007/978-3-319-63688-7_20

42. Lin, H., Matt, C.: Pseudo flawed-smudging generators and their application to indistinguishability obfuscation. IACR Cryptol. ePrint Archive **2018**, 646 (2018)

43. Lin, H., Tessaro, S.: Indistinguishability obfuscation from bilinear maps and block-wise local prgs. Cryptology ePrint Archive, Report 2017/250 (2017). http://eprint.iacr.org/2017/250

44. Lin, H., Vaikuntanathan, V.: Indistinguishability obfuscation from DDH-like assumptions on constant-degree graded encodings. In: FOCS, pp. 11–20. IEEE (2016)

45. Ma, F., Zhandry, M.: New multilinear maps from CLT13 with provable security against zeroizing attacks. IACR Cryptol. ePrint Archive **2017**, 946 (2017)
46. Miles, E., Sahai, A., Zhandry, M.: Annihilation attacks for multilinear maps: cryptanalysis of indistinguishability obfuscation over GGH13. In: Robshaw, M., Katz, J. (eds.) CRYPTO 2016, Part II. LNCS, vol. 9815, pp. 629–658. Springer, Heidelberg (2016). https://doi.org/10.1007/978-3-662-53008-5_22
47. Minaud, B., Fouque, P.A.: Cryptanalysis of the new multilinear map over the integers. Cryptology ePrint Archive, Report 2015/941 (2015). http://eprint.iacr.org/
48. Mossel, E., Shpilka, A., Trevisan, L.: On e-biased generators in NC0. In: FOCS, pp. 136–145 (2003)
49. Pass, R., Seth, K., Telang, S.: Indistinguishability obfuscation from semantically-secure multilinear encodings. In: Garay, J.A., Gennaro, R. (eds.) CRYPTO 2014, Part I. LNCS, vol. 8616, pp. 500–517. Springer, Heidelberg (2014). https://doi.org/10.1007/978-3-662-44371-2_28
50. Sahai, A., Waters, B.: How to use indistinguishability obfuscation: deniable encryption, and more. In: Shmoys, D.B. (ed.) Symposium on Theory of Computing, STOC 2014, New York, 31 May – 03 June 2014, pp. 475–484. ACM (2014). https://doi.org/10.1145/2591796.2591825

Block Ciphers

XOR-Counts and Lightweight Multiplication with Fixed Elements in Binary Finite Fields

Lukas Kölsch[✉]

University of Rostock, Rostock, Germany
lukas.koelsch@uni-rostock.de

Abstract. XOR-metrics measure the efficiency of certain arithmetic operations in binary finite fields. We prove some new results about two different XOR-metrics that have been used in the past. In particular, we disprove a conjecture from [10]. We consider implementations of multiplication with one fixed element in a binary finite field. Here we achieve a complete characterization of all elements whose multiplication matrix can be implemented using exactly 2 XOR-operations, confirming a conjecture from [2]. Further, we provide new results and examples in more general cases, showing that significant improvements in implementations are possible.

Keywords: Lightweight cryptography · Linear layer · XOR-count · Multiplication · Finite fields

1 Introduction

In the past years, with the advent of the so called *Internet of Things*, new challenges for cryptography have emerged. Many new devices usually do not have a lot of computational power and memory, but are still required to offer some security by encrypting sensitive data. Consequentially, *lightweight cryptography* has become a major field of research in the past years, mostly focusing on symmetric-key encryption (e.g. [1,5,8]). In particular, linear layers (e.g. [15,16]) and Sboxes (e.g. [3,17]) have been thoroughly investigated as they constitute key components in classical symmetric-key ciphers like AES. The objective here is to try to minimize the cost of storage and the number of operations needed to apply a cryptographic function. Usually, the security properties of cryptographic schemes using finite fields do not depend on a specific field representation (as bit strings) in the actual implementation [4], so the choice of field implementation makes an impact on the performance of the scheme without influencing its security. It is therefore an interesting question which representation minimizes the number of operations needed.

In practice, linear layers are usually $\mathbb{F}_{2^m}$-linear mappings on $\mathbb{F}_{2^m}^n$. Recall that linear mappings are implemented as matrix multiplications. Note that we can

© International Association for Cryptologic Research 2019
Y. Ishai and V. Rijmen (Eds.): EUROCRYPT 2019, LNCS 11476, pp. 285–312, 2019.
https://doi.org/10.1007/978-3-030-17653-2_10

write every $n \times n$ matrix over $\mathbb{F}_{2^m}$ as an $(mn) \times (mn)$ matrix over $\mathbb{F}_2$. As elements in $\mathbb{F}_{2^m}$ are usually represented as bit strings in computers, it is natural to consider only matrices over $\mathbb{F}_2$. Measurements of implementation costs will then only involve the number of bit-operations (XORs) needed. It is an interesting question to evaluate the efficiency of a given matrix. For that purpose two different metrics have been introduced, the *direct XOR-count* (e.g. in [12,15,19,20]) and the *sequential XOR-count* (e.g. [2,10,22]). Roughly speaking, the direct XOR-count counts the number of non-zeros in the matrix, whereas the sequential XOR-count counts the number of elementary row operations needed to transform the matrix into the identity matrix (see Sect. 2 for more precise definitions). Although the sequential XOR-count of a matrix is harder to compute, it often yields a better estimation of the actual optimal number of XOR-operations needed [10], for a simple example see Example 1 in this work. When implementing a linear layer, a field representation can be chosen such that the respective matrix is optimal according to these metrics. In this way, the performance of a given linear layer can be improved (for example by choosing a field representation that results in a sparse diffusion matrix).

Our Contributions. Our goal in this work is to explore some connections and properties of the direct and sequential XOR-count metrics and then to apply these to get some theoretical results regarding optimal implementations of matrices that represent multiplication with a fixed field element $\alpha \in \mathbb{F}_{2^k}$. Optimal choices of these matrices (called *multiplication matrices*) can then be used for local optimizations of matrices over $\mathbb{F}_{2^k}$ (this approach was taken for example in [2,10,15,16,19]). Recently, the focus has shifted to global optimization, as it has become clear that local optimizations are not necessarily also globally optimal [6,13]. However, global optimization techniques currently rely either on tools that improve the XOR-counts of matrices already known to be efficient [13] or exhaustive searches [6,18]. In particular, theoretical results on globally optimal matrices seem to be very hard to obtain. Numerical data suggest that there is a correlation between good local optimizations and good global optimizations (see [13, Figures 2–6]). Because of this correlation, theoretical insights into local optimization are valuable for the search of globally optimal matrices.

In the second section, we compare the direct XOR-count and sequential XOR-count evaluation metrics. We prove some theoretical properties of the sequential XOR-count that can be used to improve algorithms (e.g. an algorithm presented in [2]). We also find an infinite family of matrices that have a lower direct XOR-count than sequential XOR-count, disproving a conjecture in [10]. We want to emphasize that the results presented in this section apply to all invertible matrices, not just multiplication matrices.

In the third section we provide a complete characterisation of finite field elements α where the mapping $x \mapsto \alpha x$ can be implemented with exactly 2 XOR-operations (Theorem 5), which proves a conjecture in [2]. This case is of special interest, since for many finite fields (including the fields $\mathbb{F}_{2^n}$ with $8|n$ that are particularly interesting for many applications) there are no elements for which the mapping $x \mapsto \alpha x$ can be implemented with only 1 XOR-operation [2].

For these fields, our classification gives a complete list of elements α such that multiplication with α can be implemented in the cheapest way possible.

In the fourth section we present some more general results for multiplication matrices with higher XOR-counts. We prove that the number of XOR-operations needed to implement the mapping $x \mapsto \alpha x$ depends on the number of non-zero coefficients of the minimal polynomial of α. In particular, Theorem 6 shows that the gap between the number of XORs used in an optimal implementation and the number of XORs used in the "naive" implementation of a multiplication matrix using the rational canonical form of the mapping $x \mapsto \alpha x$ grows exponentially with the weight of the minimal polynomial of the element. This result shows that there is a large potential for improvement in the implementation of multiplication matrices. Propositions 2 and 3 imply that the bound found in Theorem 6 is optimal.

We conclude our paper with several open problems.

2   XOR-Counts

An XOR-count metric for diffusion matrices was introduced in [12] and then generalized for arbitrary matrices in [20]. It has then subsequently been studied in several works, e.g. [15, 19].

Definition 1. *The direct XOR-count (d-XOR-count) of an invertible $n \times n$ matrix M over $\mathbb{F}_2$, denoted by $\mathrm{wt}_d(M)$ is*

$$\mathrm{wt}_d(M) = \omega(M) - n,$$

where $\omega(M)$ denotes the number of ones in the matrix M.

Note that the d-XOR-count of an invertible matrix is never negative as every row of an invertible matrix needs to have at least one non-zero entry. Moreover, $\mathrm{wt}_d(M) = 0$ if and only if M has exactly one '1' in every row and column, i.e. M is a permutation matrix. The d-XOR-metric only gives an upper bound to the actual minimal implementation cost as the following example shows.

Example 1.

$$\begin{pmatrix} 1\,0\,0\,0 \\ 1\,1\,0\,0 \\ 1\,1\,1\,0 \\ 1\,1\,1\,1 \end{pmatrix} \cdot \begin{pmatrix} a_1 \\ a_2 \\ a_3 \\ a_4 \end{pmatrix} = \begin{pmatrix} a_1 \\ a_1 + a_2 \\ (a_1 + a_2) + a_3 \\ ((a_1 + a_2) + a_3) + a_4 \end{pmatrix}$$

The d-XOR-count of the matrix is 6 but it is easy to see that multiplication with this matrix can actually be implemented with only 3 XOR operations since the results of previous steps can be reused. A metric that allows this was subsequently introduced in [10] and used in further work (e.g. [2, 6, 22]). Let us introduce some notation at first: We denote by I the identity matrix and by $E_{i,j}$ the matrix that has exactly one '1' in the i-th row and j-th column.

Then $A_{i,j} := I + E_{i,j}$ for $i \neq j$ is called an *addition matrix*. Left-multiplication with $A_{i,j}$ adds the j-th row to the i-th row of a matrix, right-multiplication adds the i-th column to the j-th column. Observe that the matrices $A_{i,j}$ are self-inverse over $\mathbb{F}_2$. Let further $\mathcal{P}(n)$ be the set of $n \times n$ permutation matrices and $\mathcal{A}(n)$ the set of all $n \times n$ addition matrices $A_{i,j}$. We will omit the dimension n unless necessary.

Definition 2. *An invertible matrix M over $\mathbb{F}_2$ has a sequential XOR-count (s-XOR-count) of t if t is the minimal number such that M can be written as*

$$M = P \prod_{k=1}^{t} A_{i_k, j_k}$$

where $P \in \mathcal{P}$ and $A_{i_k, j_k} \in \mathcal{A}$. We write $\mathrm{wt}_s(M) = t$.

Note that every invertible matrix can be decomposed as a product of a permutation matrix and addition matrices in the way Definition 2 describes. Indeed, Gauss-Jordan-elimination gives a simple algorithm to do so.

In [22] a similar definition for the s-XOR-count was given that uses a representation of the form $M = \prod_{k=1}^{t} P_k A_{i_k, j_k}$ with permutation matrices P_k. Since products of permutation matrices remain permutation matrices and

$$P A_{i,j} = A_{\sigma^{-1}(i), \sigma^{-1}(j)} P \tag{1}$$

where $\sigma \in S_n$ is the permutation belonging to the permutation matrix P, this definition is equivalent to our definition.

A representation of a matrix M as a product $M = P \prod_{k=1}^{t} A_{i_k, j_k}$ is called an *s-XOR-representation* of M and an s-XOR-representation with $\mathrm{wt}_s(M)$ addition matrices is called an *optimal s-XOR-representation*. Note that optimal s-XOR-representations are generally not unique. Observe that $M = P A_{i_1, j_1} \ldots A_{i_t, j_t}$ is equivalent to $M A_{i_t, j_t} \ldots A_{i_1, j_1} = P$, so the s-XOR-count measures the number of column addition steps that are needed to transform a matrix into a permutation matrix. Because of Eq. (1) the number of column additions needed is equal to the number of row additions needed, so we may also speak about row additions.

Going back to Example 1, it is easy to find an s-XOR-representation with 3 XORs.

$$M = \begin{pmatrix} 1\,0\,0\,0 \\ 1\,1\,0\,0 \\ 1\,1\,1\,0 \\ 1\,1\,1\,1 \end{pmatrix} = I A_{4,3} A_{3,2} A_{2,1}.$$

It is clear that we need at least 3 addition matrices since all rows but the first one need at least one update. Hence, the s-XOR-representation above is optimal and $\mathrm{wt}_s(M) = 3$.

Determining the s-XOR-count of a given matrix is generally not easy. Graph-based algorithms to find an optimal s-XOR-count have been proposed in [22] and (in a slightly different form) in [10]. The algorithms are based on the following observation. Let $G = (V, E)$ be a graph where $G = GL(n, \mathbb{F}_2)$ and $(M_1, M_2) \in E$

if $AM_1 = M_2$ for an $A \in \mathcal{A}$. Then $\mathrm{wt}_s(M) = \min_{P \in \mathcal{P}} d(M, P)$, where $d(M_1, M_2)$ denotes the distance between M_1 and M_2 in the graph G. Thus, the evaluation of the s-XOR-count can be reduced to a shortest-path-problem. Note that because the elementary matrices in $\mathcal{A}$ are all involutory, G is undirected. As the authors of [22] observe, it is possible to reduce the number of vertices by a factor $1/n!$ because matrices with permuted rows can be considered equivalent. Still $(1/n!)|GL(n, \mathbb{F}_2)| = (1/n!)(2^n - 1)(2^n - 2) \ldots (2^n - 2^{n-1})$ and every vertex has $|\mathcal{A}(n)| = n^2 - n$ neighbors, so both the number of vertices and the number of edges grow exponentially. Hence, this approach is impractical unless n is small.

The problem of determining the s-XOR-count is linked with the problem of optimal pivoting in Gauss-Jordan elimination since the number of additions in an optimal elimination process is clearly an upper bound of the s-XOR-count. Pivoting strategies for Gaussian elimination are a classical problem in numerical linear algebra (among lots of examples, see [14]) and the number of steps needed in a Gauss-Jordan elimination process can be used as a heuristic for the s-XOR-count.

Example 1 gives an example of a matrix with lower s-XOR-count than d-XOR-count. Considering this and the fact that the s-XOR-count of a given matrix is generally much harder to determine than the d-XOR-count, it should be clarified whether the s-XOR-count always gives a better estimation of the actual number of XOR operations needed to implement the matrix. In [10] this has been conjectured, i.e. $\mathrm{wt}_s(M) \leq \mathrm{wt}_d(M)$ for all $M \in GL(n, \mathbb{F}_2)$. However, the following theorem gives a counterexample.

Theorem 1. *Let M be as follows:*

$$M = \begin{pmatrix} 1 & 1 & 0 & 0 & 0 & 0 & 0 \\ 0 & 1 & 1 & 0 & 0 & 0 & 0 \\ 0 & 0 & 1 & 1 & 0 & 0 & 0 \\ 0 & 0 & 0 & 1 & 1 & 0 & 0 \\ 0 & 0 & 0 & 0 & 1 & 1 & 0 \\ 0 & 0 & 0 & 0 & 0 & 1 & 1 \\ 1 & 0 & 0 & 1 & 0 & 0 & 1 \end{pmatrix} \in GL(7, \mathbb{F}_2).$$

Then $\mathrm{wt}_s(M) > \mathrm{wt}_d(M)$.

Proof. M is invertible with $\mathrm{wt}_d(M) = 8$. Let $\mathrm{wt}_s(M) = t$, i.e. there are matrices $A_{i_k, j_k} \in \mathcal{A}$ and $P \in \mathcal{P}$ such that $\prod_{k=1}^{t} A_{i_k, j_k} \cdot M = P$. By construction, no two rows and no three rows of M add up to a row with only one non-zero entry. Every row has to be updated at least once to transform M into a permutation matrix. Since no two row vectors add up to a vector with only one non-zero entry, the first row that gets updated (row i_t) needs to get updated at least once more. But as there is also no combination of three vectors adding up to a vector with only one non-zero entry, the second row that is updated (row i_{t-1}) also needs to be updated a second time. So two rows need to get updated at least twice, and all other 5 rows need to get updated at least once, resulting in $\mathrm{wt}_s(M) \geq 9$. $\square$

Remark 1. Note that the structure of the counterexample can be extended to all dimensions $n \geq 7$, the middle '1' in the last row can be in any j-th column with $4 \leq j \leq n-3$. We conclude that there exists a matrix $M \in \mathrm{GL}(n, \mathbb{F}_2)$ with $\mathrm{wt}_s(M) > \mathrm{wt}_d(M)$ for all $n \geq 7$.

Studying the s-XOR-count is an interesting mathematical problem because it has some properties that can be used to get upper bounds of the actual implementation cost of potentially a lot of matrices. The actual number of XOR-operations needed is clearly invariant under permutation of rows and columns. It is therefore desirable that this property is reflected in our XOR-metrics. Obviously, this is the case for the d-XOR-count, i.e. $\mathrm{wt}_d(M) = \mathrm{wt}_d(PMQ)$ for all matrices M and permutation matrices $P, Q \in \mathcal{P}$. The following lemma shows that this also holds for the s-XOR-count. The lemma is a slight modification of a result in [2]. However the proof in [2] has a small gap, so we provide a complete proof here. We denote permutation-similarity with $\sim$, i.e. $M_1 \sim M_2$ if there exists a $P \in \mathcal{P}$ so that $M_1 = P M_2 P^{-1}$.

Lemma 1. *Let $M \in GL(n, \mathbb{F}_2)$. Then $\mathrm{wt}_s(M) = \mathrm{wt}_s(PMQ)$ for $P, Q \in \mathcal{P}$. In particular, if $M_1 \sim M_2$ then $\mathrm{wt}_s(M_1) = \mathrm{wt}_s(M_2)$.*

Proof. Let $\mathrm{wt}_s(M) = t$ and $\sigma \in S_n$ be the permutation belonging to Q. Then, by shifting Q to the left

$$PMQ = PP_2 \prod_{k=1}^{t} A_{i_k, j_k} Q = PP_2 Q \prod_{k=1}^{t} A_{\sigma(i_k), \sigma(j_k)} = P' \prod_{k=1}^{t} A_{\sigma(i_k), \sigma(j_k)}$$

where $P_2, P' \in \mathcal{P}$, so $\mathrm{wt}_s(PMQ) \leq \mathrm{wt}_s(M)$. Since $M = P^{-1}(PMQ)Q^{-1}$ the same argument yields $\mathrm{wt}_s(M) \leq \mathrm{wt}_s(PMQ)$.   □

Based on this result, the following normal form for permutation matrices is proposed in [2]. We introduce a notation for block diagonal matrices. Let $M_1, \ldots, M_d$ be square matrices, then we denote the block matrix consisting of these matrices by

$$\bigoplus_{k=1}^{d} M_k := \begin{pmatrix} M_1 & & & 0 \\ & M_2 & & \\ & & \ddots & \\ 0 & & & M_d \end{pmatrix}.$$

We denote by C_p the companion matrix of a polynomial $p = x^n + a_{n-1}x^{n-1} + \cdots + a_1 x + a_0 \in \mathbb{F}_2[x]$, i.e.

$$C_p = \begin{pmatrix} 0 & \ldots & 0 & a_0 \\ 1 & 0 & 0 & a_1 \\ 0 & \ddots & \ddots & \vdots \\ 0 & \ldots & 1 & a_{n-1} \end{pmatrix}.$$

Lemma 2 ([2, Lemma 2]). *Let $P \in \mathcal{P}(n)$. Then*

$$P \sim \bigoplus_{k=1}^{d} C_{x^{m_k}+1}$$

for some m_k with $\sum_{k=1}^{d} m_k = n$ and $m_1 \geq \cdots \geq m_d \geq 1$.

A permutation matrix of this structure is said to be the *cycle normal form* of P. We can then (up to permutation-similarity) always assume that the permutation matrix of the s-XOR-decomposition is in cycle normal form.

Corollary 1 ([2, Corollary 2]).

$$P \prod_{k=1}^{t} A_{i_k,j_k} \sim P' \prod_{k=1}^{t} A_{\sigma(i_k),\sigma(j_k)}$$

for some permutation $\sigma \in S_n$, where P' is the cycle normal form of P.

We say an s-XOR-representation is in cycle normal form if its permutation polynomial is in cycle normal form. Corollary 1 states that every s-XOR-representation is pemutation-similar to exactly one s-XOR-representation in cycle normal form.

The following theorem gives a connection between the s-XOR-count and optimal s-XOR-representations of a given matrix and that of its inverse.

Theorem 2. *Let M be an invertible matrix with $\mathrm{wt}_s(M) = t$ and*

$$M = P \prod_{k=1}^{t} A_{i_k,j_k} \text{ with } P = \bigoplus_{k=1}^{d} C_{x^{m_k}+1}.$$

Then $\mathrm{wt}_s(M^{-1}) = t$. Moreover,

$$M^{-1} = P A_{\sigma(i_t),\sigma(j_t)} A_{\sigma(i_{t-1}),\sigma(j_{t-1})} \cdots A_{\sigma(i_1),\sigma(j_1)}$$

for some permutation $\sigma \in S_n$ that depends only on P.

Proof. For the inverse matrix we have

$$M^{-1} = A_{i_t,j_t} \cdots A_{i_1,j_1} P^{-1} \sim P^{-1} A_{i_t,j_t} \cdots A_{i_1,j_1},$$

so $\mathrm{wt}_s(M^{-1}) \leq \mathrm{wt}_s(M)$. By symmetry, we get $\mathrm{wt}_s(M^{-1}) = \mathrm{wt}_s(M)$. Observe that $P^{-1} = P^T = \bigoplus_{k=1}^{d} C_{x^{m_k}+1}^T$ where P^T denotes the transpose of P. Let J_r be the $r \times r$ matrix with ones on the counterdiagonal, i.e. $J_{i,j} = 1$ if and only if $j = n-i+1$. Let $Q = \bigoplus_{k=1}^{d} J_{m_k} \in \mathcal{P}$. A direct calculation yields $QP^{-1}Q^{-1} = P$ and thus

$$M^{-1} \sim QP^{-1} \prod_{k=t}^{1} A_{i_k,j_k} Q^{-1} = P \prod_{k=t}^{1} A_{\sigma(i_k),\sigma(j_k)},$$

where $\sigma \in S_n$ denotes the permutation that belongs to Q. $\square$

In particular, Theorem 2 implies that given an optimal s-XOR-representation for a matrix M, an optimal s-XOR-representation of M^{-1} can be determined with very little effort by calculation the permutation σ in the proof. Note that the statement of Theorem 2 does not exist for the d-XOR-count. Indeed, sparse matrices (i.e. matrices with low d-XOR-count) usually have dense inverse matrices (i.e. high d-XOR-count).

The next result also holds for the s-XOR-count only.

Proposition 1. *Let M, N be invertible matrices with $\mathrm{wt}_s(M) = t_1$ and $\mathrm{wt}_s(N) = t_2$. Then $\mathrm{wt}_s(MN) \leq t_1 + t_2$. In particular, $\mathrm{wt}_s(M^k) \leq |k| t_1$ for all $k \in \mathbb{Z}$.*

Proof. Let $M = P \prod_{k=1}^{t_1} A_{i_k, j_k}$ and $N = Q \prod_{k=1}^{t_2} B_{i_k, j_k}$. Then

$$MN = PQ \prod_{k=1}^{t_1} A_{\sigma(i_k), \sigma(j_k)} \prod_{k=1}^{t_2} B_{i_k, j_k},$$

where $\sigma \in S_n$ is the permutation belonging to Q. This implies $\mathrm{wt}_s(MN) \leq t_1 + t_2$. The statement $\mathrm{wt}_s(M^k) \leq |k| t_1$ for $k < 0$ follows from Theorem 2. $\qquad \square$

3   Efficient Multiplication Matrices in Finite Fields

We can consider $\mathbb{F}_{2^n}$ as the n-dimensional vector space $(\mathbb{F}_2)^n$ over $\mathbb{F}_2$. By distributivity, the function $x \mapsto \alpha x$ for $\alpha \in \mathbb{F}_{2^n}$ is linear, so it can be represented as a (left-)multiplication with a matrix in $\mathrm{GL}(n, \mathbb{F}_2)$. This matrix obviously depends on α, but also on the choice of the basis of $(\mathbb{F}_2)^n$ over $\mathbb{F}_2$. We denote the multiplication matrix that represents the function $x \mapsto \alpha x$ with respect to the basis B by $M_{\alpha, B}$. The XOR-count of $M_{\alpha, B}$ generally differs from the XOR-count of $M_{\alpha, B'}$ for different bases B, B'. Our objective here is to find the optimal basis B for a given α, in the sense that the XOR-count of $M_{\alpha, B}$ is minimized. For this, we define the XOR-count metrics from the previous section also for elements from $\mathbb{F}_{2^n}$.

Definition 3. *Let $\alpha \in \mathbb{F}_{2^n}$. We define the s-XOR-count and d-XOR-count of α as follows:*

$$\mathrm{wt}_s(\alpha) = \min_B \mathrm{wt}_s(M_{\alpha, B}), \quad \mathrm{wt}_d(\alpha) = \min_B \mathrm{wt}_d(M_{\alpha, B}),$$

where the minimum is taken over all bases of $\mathbb{F}_2^n$ over $\mathbb{F}_2$. A basis B and matrix $M_{\alpha, B}$ that satisfy the minimum are called s-XOR-optimal and d-XOR-optimal for α, respectively.

In order to find the matrices that optimize the s-XOR-count-metric, an exhaustive search on all matrices with low s-XOR-count is performed in [2]. In this way the s-XOR-count and an optimal s-XOR-matrix of every element $\alpha \in \mathbb{F}_{2^n}$ for $n \leq 8$ was found. Using the results presented in the previous section, the search was restricted to matrices where the permutation matrix is in cycle normal form. The following result was used to determine whether a given matrix is a multiplication matrix for some $\alpha \in \mathbb{F}_{2^n}$ with respect to some basis B. From here on, we denote by $\chi(M) = \det(xI + M)$ the characteristic polynomial of a matrix M and by m_α the minimal polynomial of the finite field element $\alpha \in \mathbb{F}_{2^n}$. Recall that m_α is always irreducible.

Theorem 3 ([2, Theorem 1]). *Let $M \in GL(n, \mathbb{F}_2)$ and $\alpha \in \mathbb{F}_{2^n}$. Then M is a multiplication matrix for α, i.e. $M = M_{\alpha,B}$ with respect to some basis B, if and only if m_α is the minimal polynomial of M.*

Theorem 3 shows in particular that a matrix M is a multiplication for some $\alpha \in \mathbb{F}_{2^n}$ with respect to some basis B if and only if the minimal polynomial of M is irreducible. Additionally, it is clear that two field elements with the same minimal polynomial necessarily have the same XOR-counts.

Remark 2. A direct calculation of the minimal polynomial of the matrix M in Theorem 1 yields $m_M = x^7 + x^6 + x^5 + x^4 + 1$ which is an irreducible polynomial. According to Theorem 3 the matrix M is a multiplication matrix for an element $\alpha \in \mathbb{F}_{2^7}$ with respect to some basis. Hence, there are elements $\alpha \in \mathbb{F}_{2^n}$ such that $\text{wt}_d(\alpha) < \text{wt}_s(\alpha)$. Note that this case does not have to occur for every value of n because the matrices provided in Theorem 1 might have a reducible minimal polynomial. Indeed, an exhaustive search for the cases $n = 4$ and $n = 8$ was conducted in [10], resulting in $\text{wt}_s(\alpha) \leq \text{wt}_d(\alpha)$ for all α in $\mathbb{F}_{2^4}$ and $\mathbb{F}_{2^8}$, respectively. We tested the examples given in Theorem 1 for $n = 16$ without finding any matrices with irreducible minimal polynomial. Hence, we conjecture that $\text{wt}_s(\alpha) \leq \text{wt}_d(\alpha)$ for all $\alpha \in \mathbb{F}_{2^{16}}$. It is an interesting question for which n elements with lower d-XOR-count than s-XOR-count exist.

Corollary 2. *Let $M = P \prod_{k=1}^t A_{i_k,j_k}$ be in cycle normal form. Then M is a multiplication matrix for $\alpha \in \mathbb{F}_{2^n}$ if and only if M^{-1} is a multiplication matrix for $\alpha^{-1} \in \mathbb{F}_{2^n}$. Moreover, M is an optimal s-XOR-matrix for α if and only if M^{-1} is an optimal s-XOR-matrix for α^{-1}.*

Proof. Let p and q be the minimal polynomial of M and M^{-1}, respectively. It is well known that q is then the reciprocal polynomial of p, that is $q(x) = x^n p(1/x)$. Moreover, p is the minimal polynomial of α if and only if q is the minimal polynomial of α^{-1}. The rest follows from Theorem 2. $\qquad\square$

Corollary 2 allows us to determine an s-XOR-optimal matrix for α^{-1} given an s-XOR-optimal matrix M of α. Recall that the cycle normal form of M^{-1} was directly computed in Theorem 2. This allows us to cut the search space (approximately) in half for all algorithms that determine the s-XOR-count by traversing all matrices in $GL(n, \mathbb{F}_2)$.

It is now an interesting question which elements $\alpha \in \mathbb{F}_{2^n}$ have multiplication matrices with low XOR-count. Obviously, the only element that can be implemented with XOR-count 0 is $\alpha = 1$. A simple upper bound on the s-XOR-count and d-XOR-count for elements can be found by considering the rational canonical form of a matrix. Recall that a matrix $M \in \mathrm{GL}(n, \mathbb{F}_2)$ is similar to its (unique) rational canonical form. If M has an irreducible minimal polynomial m with $\deg m = k$ then there exists a $d \geq 1$ so that $kd = n$ and the rational canonical form is $\bigoplus_{i=1}^{d} C_m$. For a polynomial p we denote by $\mathrm{wt}(p)$ the weight of p, that is the number of non-zero coefficients. Note that if $2|\mathrm{wt}(p)$ then 1 is a root of p so the only irreducible polynomial over $\mathbb{F}_2$ with even weight is $x + 1$.

Example 2. Let α be an element of $\mathbb{F}_{2^n}$ with minimal polynomial m_α and $\deg m_\alpha = k$ with $kd = n$ and $d \geq 1$. Then we can find a basis B so that $M_{\alpha,B}$ is in rational canonical form, i.e. $M_{\alpha,B} = \bigoplus_{i=1}^{d} C_{m_\alpha}$. It is easy to check that $\mathrm{wt}_s(M_{\alpha,B}) = \mathrm{wt}_d(M_{\alpha,B}) = d \cdot (\mathrm{wt}(m_\alpha) - 2)$.

This example shows in particular that all $\alpha \in \mathbb{F}_{2^n}$ with $\deg m_\alpha = n$ and $\mathrm{wt}(m_\alpha) = 3$ can be implemented with only one XOR operation. A possible basis for this case is the polynomial basis $\{1, \alpha, \alpha^2, \ldots, \alpha^{n-1}\}$.

As one row-addition on I only produces one extra '1' in the matrix, $\mathrm{wt}_d(M) = 1$ if and only if $\mathrm{wt}_s(M) = 1$, and equivalently, $\mathrm{wt}_d(\alpha) = 1$ if and only if $\mathrm{wt}_s(\alpha) = 1$. In [2] all elements that can be implemented with exactly one XOR-operation are characterized. It turns out, that these cases are exactly those covered by Example 2.

Theorem 4 ([2, Theorem 2]). *Let $\alpha \in \mathbb{F}_{2^n}$. Then $\mathrm{wt}_s(\alpha) = 1$ or $\mathrm{wt}_d(\alpha) = 1$ if and only if m_α is a trinomial of degree n.*

It is an open problem for which n irreducible trinomials of degree n exist. Among other sporadic examples, it is known that there are no irreducible trinomials of degree n if $n \equiv 0 \pmod 8$ [21], so there are no elements α with d/s-XOR-count 1 in these cases. As the case $8|n$ is especially important in practice, it is natural to consider elements that can be implemented with 2 XOR operations. In this case, s-XOR-count and d-XOR count do differ: By simply expanding the product $P A_{i_1,j_1} A_{i_2,j_2} = P(I + E_{i_1,j_1})(I + E_{i_2,j_2})$, it follows that every matrix with $\mathrm{wt}_s(M) = 2$ is of the following form:

$$
M = \begin{cases} P + E_{\sigma^{-1}(i_1),j_1} + E_{\sigma^{-1}(i_2),j_2}, & i_2 \neq j_1 \\ P + E_{\sigma^{-1}(i_1),j_1} + E_{\sigma^{-1}(i_2),j_2} + E_{\sigma^{-1}(i_1),j_2}, & i_2 = j_1, \end{cases} \tag{2}
$$

where σ is the permutation that belongs to P and $i_1 \neq j_1$, $i_2 \neq j_2$. In particular Eq. (2) shows that $\mathrm{wt}_d(M) = 2$ implies $\mathrm{wt}_s(M) = 2$, but there are some matrices with $\mathrm{wt}_s(M) = 2$ and $\mathrm{wt}_d(M) = 3$. In other words, the s-XOR-metric is a better metric for these matrices. In [2] the authors conjecture that $\mathrm{wt}_s(\alpha) = 2$ implies $\mathrm{wt}(m_\alpha) \leq 5$, i.e. the minimal polynomial is a trinomial or a pentanomial. We confirm this conjecture by giving an exact characterization of all elements with $\mathrm{wt}_s(\alpha) = 2$ and their optimal s-XOR-representation in cycle normal form in Theorem 5.

In the proof the following concept from linear algebra is used. We refer the reader to [9] for proofs and more background. Let V be a vector space over a field $\mathbb{F}$ with dimension n, $u \in V$ a vector and M an $n \times n$-matrix over $\mathbb{F}$. The monic polynomial $g(x) \in \mathbb{F}[x]$ with the smallest degree such that $g(M)u = 0$ is called the M-*annihilator* of u. This polynomial divides any polynomial h annihilating u (i.e. $h(M)u = 0$), in particular the minimal polynomial of M. In the case that the minimal polynomial of M is irreducible the M-annihilator of every vector $u \neq 0$ is the minimal polynomial of M. So if we find a polynomial h that annihilates a vector $u \neq 0$ we know that the minimal polynomial divides h. In particular, if h is monic and the degree of h and the minimal polynomial coincide we can infer that h is the minimal polynomial of M.

Theorem 5. *Let $\alpha \in \mathbb{F}_{2^n}$. Then $\mathrm{wt}_s(\alpha) = 2$ if and only if m_α can be written in the form of a pentanomial or the trinomial appearing in Table 1.*

Table 1. Elements with minimal polynomials listed in the left column have s-XOR-count 2. The second column gives an optimal multiplication matrix and the third column points to the corresponding case in the proof.

m_α	optimal matrix representation	Case
$x^n + x^{k_1+k_2} + x^{k_1} + x^{k_2} + 1,$ $k_1 + k_2 \leq n - 2$	$C_{x^n+1} + E_{i_1,j_1} + E_{i_2,j_2}$	(1.3.)
$x^n + x^{n-k_1} + x^{k_2} + x^{k_2-k_1} + 1,$ $k_2 > k1$	$C_{x^n+1} + E_{i_1,j_1} + E_{i_2,j_2}$	(1.4.)
$x^n + x^{k_1+k_2} + x^{k_1} + x^{k_2} + 1$	$C_{x^n+1} + E_{i_1,j_1} + E_{j_1+1,j_2} + E_{i_1,j_2}$	(2.1.)
$x^n + x^{n_1} + x^{n_2} + x^k + 1,$ $k \leq n - 2$	$(C_{x^{n_1}+1} \oplus C_{x^{n_2}+1}) + E_{i_1,j_1} + E_{i_2,j_2}$	(3.2.)
$x^n + x^{n_1+k} + x^{n_2} + x^{n_1} + 1,$ $0 < k < n_2$	$(C_{x^{n_1}+1} \oplus C_{x^{n_2}+1}) + E_{i_1,j_1}$ $+E_{j_1+1 \pmod{n_1},j_2} + E_{i_1,j_2}$	(4.)
$x^{n/2} + x^k + 1$	$(C_{x^{n/2}+1} \oplus C_{x^{n/2}+1}) + E_{i_1,j_1} + E_{i_2,j_2}$	(3.1.)

Proof. Let $M_{\alpha,B}$ be a multiplication matrix for some $\alpha \in \mathbb{F}_{2^n}$ and some basis $B = \{b_1, \ldots, b_n\}$. We can assume that $M_{\alpha,B}$ is in cycle normal form, $M = PA_{i_1,j_1}A_{i_2,j_2}$ with $P = \bigoplus_{k=1}^l C_{x^{m_k}+1}$. As a first step, we show that $l \leq 2$. Assume $l > 2$. As shown in Eq. (2) at most two rows of M have more than one '1' in them. So, by possibly permuting the blocks, P is a triangular block matrix, consisting of two blocks where one block is of the form C_{x^t+1}. So $\chi(C_{x^t+1}) = x^t + 1$ divides $\chi(M)$. But as minimal polynomial and characteristic polynomial share the same irreducible factors, this implies $(x+1)|m_\alpha$ which contradicts the irreducibility of m_α. So $l \leq 2$. We now deal with all possible matrices on a case by case basis, where we differentiate the cases $l \in \{1, 2\}$ and the two cases in Eq. (2).

Case 1. $M = C_{x^n+1} + E_{i_1,j_1} + E_{i_2,j_2}$, $j_1 \neq i_2 - 1$.
We investigate how the matrix operates on the basis $B = \{b_1, \ldots, b_n\}$:

$$\alpha b_1 = b_2$$
$$\vdots$$
$$\alpha b_{j_1-1} = b_{j_1}$$
$$\alpha b_{j_1} = b_{j_1+1} + b_{i_1}$$
$$\alpha b_{j_1+1} = b_{j_1+2} \tag{3}$$
$$\vdots$$
$$\alpha b_{j_2-1} = b_{j_2}$$
$$\alpha b_{j_2} = b_{j_2+1} + b_{i_2}$$
$$\alpha b_{j_2+1} = b_{j_2+2} \tag{4}$$
$$\vdots$$
$$\alpha b_n = b_1.$$

Define $\gamma_1 := b_{j_1+1}$ and $\gamma_2 := b_{j_2+1}$. Then

$$b_{j_1} = \alpha^{n+j_1-j_2-1}\gamma_2, \quad b_{j_2} = \alpha^{j_2-j_1-1}\gamma_1. \tag{5}$$

At first, we show that the minimal polynomial has degree n. Assume $m_\alpha = x^m + \sum_{i=1}^{m-1} c_i x^i + 1$ with $c_i \in \mathbb{F}_2$ and $md = n$ with $d > 1$. In particular, $m \leq n/2$. At least one of $n + j_1 - j_2$ and $j_2 - j_1$ are greater or equal $n/2$. Assume $j_2 - j_1 \geq n/2$. Then $\alpha^i \gamma_1 = b_{j_1+1+i}$ for $i < n/2$. Furthermore, $\alpha^{n/2}\gamma_1 = b_{j_1+1+n/2}$ if $j_2 - j_1 > n/2$ and $\alpha^{n/2}\gamma_1 = b_{j_1+1+n/2} + b_{i_2}$ if $j_2 - j_1 = n/2$. Consequently, $m_\alpha(\alpha)\gamma_1 = \alpha^m \gamma_1 + \sum_{i=1}^{m-1} c_i \alpha^i \gamma_1 + \gamma_1$ is a linear combination of at least one basis element and thus cannot vanish. If $n + j_1 - j_2 \geq n/2$ the same argument holds with γ_2 instead of γ_1. So $\deg m_\alpha = n$. Observe that with the Eqs. (3), (4) and (5)

$$\alpha^{n+j_1-j_2}\gamma_2 = \gamma_1 + b_{i_1} \tag{6}$$
$$\alpha^{j_2-j_1}\gamma_1 = \gamma_2 + b_{i_2}. \tag{7}$$

By plugging γ_2 into the first equation and γ_1 into the second equation, we obtain

$$\alpha^n \gamma_1 + \alpha^{n+j_1-j_2}b_{i_2} + b_{i_1} + \gamma_1 = 0 \tag{8}$$
$$\alpha^n \gamma_2 + \alpha^{j_2-j_1}b_{i_1} + b_{i_2} + \gamma_2 = 0. \tag{9}$$

Case 1.1. $i_1 \in [j_1 + 1, j_2]$ and $i_2 \in [j_1 + 1, j_2]$.
Then $b_{i_1} = \alpha^{t_1}\gamma_1$ and $b_{i_2} = \alpha^{t_2}\gamma_1$ with $t_1 = i_1 - j_1 - 1$ and $t_2 = i_2 - j_1 - 1$ with $t_1 + t_2 < n - 1$. With Eq. (8), we have

$$\alpha^n \gamma_1 + \alpha^{n+j_1-j_2+t_2}\gamma_1 + \alpha^{t_1}\gamma_1 + \gamma_1 = 0$$

So the polynomial $p = x^n + x^{n+j_1-j_2+t_2} + x^{t_1} + 1$ annihilates γ_1. Hence, p is the minimal polynomial of M. But $2|\mathrm{wt}(p)$, so p is not irreducible. We conclude that no matrix of this type can be a multiplication matrix.

Case 1.2. $i_1 \notin [j_1 + 1, j_2]$ and $i_2 \notin [j_1 + 1, j_2]$.
   Then $b_{i_1} = \alpha^{t_1}\gamma_2$ and $b_{i_2} = \alpha^{t_2}\gamma_2$ with $t_1 = i_1 - j_2 - 1 \pmod{n}$ and $t_2 = i_2 - j_2 - 1 \pmod{n}$ with $t_1 + t_2 < n - 1$. With Eq. (9), we have

$$\alpha^n \gamma_2 + \alpha^{j_2-j_1+t_1}\gamma_2 + \alpha^{t_2}\gamma_2 + \gamma_2 = 0$$

As before, the polynomial $p = x^n + x^{j_2-j_1+t_1} + x^{t_2} + 1$ annihilates γ_2, so there is no multiplication matrix of this type.

Case 1.3. $i_1 \in [j_1 + 1, j_2]$ and $i_2 \notin [j_1 + 1, j_2]$.
   Then $b_{i_1} = \alpha^{t_1}\gamma_1$ and $b_{i_2} = \alpha^{t_2}\gamma_2$ with $t_1 = i_1 - j_1 - 1$ and $t_2 = i_2 - j_2 - 1$ $\pmod{n}$ with $t_1 + t_2 < n - 1$. Then by Eq. (6)

$$\gamma_2 = \alpha^{j_2-j_1-n}\gamma_1 + \alpha^{j_2-j_1-n+t_1}\gamma_1$$

and

$$b_{i_2} = \alpha^{j_2-j_1-n+t_2}\gamma_1 + \alpha^{j_2-j_1-n+t_1+t_2}\gamma_1.$$

Using Eq. (8), we obtain

$$\alpha^n \gamma_1 + \alpha^{t_1+t_2}\gamma_1 + \alpha^{t_1}\gamma_1 + \alpha^{t_2}\gamma_1 + \gamma_1 = 0,$$

so $p = x^n + x^{t_1+t_2} + x^{t_1} + x^{t_2} + 1$ is the minimal polynomial of M. Note that we can choose i_1, i_2, j_1, j_2 in a way that t_1 and t_2 take any value from $\{1, \ldots, n-3\}$ as long as $t_1 + t_2 < n - 1$, so every matrix with a minimal polynomial of the form $x^n + x^{a+b} + x^a + x^b + 1$ with $a + b \leq n - 2$ has a multiplication matrix of this type for suitable values of i_1, j_1, i_2, j_2.

Case 1.4. $i_1 \notin [j_1 + 1, j_2]$ and $i_2 \in [j_1 + 1, j_2]$.
   Then $b_{i_1} = \alpha^{t_1}\gamma_2$ and $b_{i_2} = \alpha^{t_2}\gamma_1$ with $t_1 = i_1 - j_2 - 1 \pmod{n}$ and $t_2 = i_2 - j_1 - 1$ with $t_1 + t_2 < n - 1$. Similarly to Case 1.3, Eq. (6) yields

$$\gamma_1 = \alpha^{n+j_1-j_2}\gamma_2 + \alpha^{t_1}\gamma_2$$

and with Eq. (9)

$$\alpha^n \gamma_2 + \alpha^{j_2-j_1+t_1}\gamma_2 + \alpha^{n+j_1-j_2+t_2}\gamma_2 + \alpha^{t_1+t_2}\gamma_2 + \gamma_2 = 0,$$

so $p = x^n + x^{j_2-j_1+t_1} + x^{n+j_1-j_2+t_2} + x^{t_1+t_2} + 1 = x^n + x^{n-k_1} + x^{k_2} + x^{k_2-k_1} + 1$ with $k_1 = j_2 - j_1 - t_2 = j_2 - i_2 - 1 > 0$ and $k_2 = j_2 - j_1 + t_1$. Note that $k_2 > k_1$ for any choice of i_1, i_2, j_1, j_2. Moreover, k_1 can take on every value in $\{1, \ldots, n-3\}$ and k_2 any value greater than k_1.

Case 2. $M = C_{x^n+1} + E_{i_1,j_1} + E_{j_1+1,j_2} + E_{i_1,j_2}$.
If $j_1 = j_2$ then $\mathrm{wt}_s(M) = 1$, so we can assume $j_1 \neq j_2$. Note that the matrix operates on the basis B just as in Case 1, the only difference being that in Eq. (4)

we have $b_{i_1} + b_{j_1+1} = b_{i_1} + \gamma_1$ instead of b_{i_2} on the right hand side. With the same argument as in Case 1 we conclude that the minimal polynomial of M has degree n.

Case 2.1. $i_1 \in [j_1 + 1, j_2]$.
Then $b_{i_1} = \alpha^t \gamma_1$ with $t = i_1 - j_1 - 1$. Similarly to Eq. (8), we obtain

$$\alpha^n \gamma_1 + \alpha^{n+j_1-j_2} \gamma_1 + \alpha^{n+j_1-j_2} b_{i_1} + b_{i_1} + \gamma_1 = 0$$

and thus

$$\alpha^n \gamma_1 + \alpha^{n+j_1-j_2} \gamma_1 + \alpha^{n+j_1-j_2+i_1-j_1-1} \gamma_1 + \alpha^{i_1-j_1-1} \gamma_1 + \gamma_1 = 0.$$

So the minimal polynomial of M is $p = x^n + x^{n+j_1-j_2} + x^{n-j_2+i_1-1} + x^{i_1-j_1-1} + 1$. Set $k_1 = i_1 - j_1 - 1$ and $k_2 = n + j_1 - j_2$ then $p = x^n + x^{k_1+k_2} + x^{k_1} + x^{k_2} + 1$ with $k_1, k_2 \in \{1, \ldots, n-1\}$ and $k_1 + k_2 < n$.

Case 2.2. $i_1 \notin [j_1 + 1, j_2]$.
Then $b_{i_1} = \alpha^t \gamma_2$ with $t = i_1 - j_2 - 1 \pmod{n}$. Similarly to Eq. (7), we have

$$\alpha^{j_2-j_1} \gamma_1 = \gamma_2 + \gamma_1 + \alpha^t \gamma_2.$$

Using Eq. (6) we obtain

$$\alpha^n \gamma_2 + \alpha^{j_2-j_1+t} \gamma_2 + \alpha^{n+j_1-j_2} \gamma_2 + \gamma_2 = 0,$$

so the minimal polynomial of M, $p = x^n + x^{j_2-j_1+t} + x^{n+j_1-j_2} + 1$, is reducible.

Case 3. $M = (C_{x^{n_1}+1} \oplus C_{x^{n_2}+1}) + E_{i_1,j_1} + E_{i_2,j_2}$, $j_1 \neq i_2 - 1$.
If both $i_1, i_2 \leq n_1$ or $i_1, i_2 > n_1$ then M is a triangular block matrix with one block being just a companion matrix. Then $(x+1)|\chi(M) = m_\alpha$, so this case cannot occur. Similarly one of j_1 and j_2 must be less or equal n_1 and the another one greater than n_1. We again investigate how M operates on the basis B:

$$\alpha b_1 = b_2 \qquad\qquad \alpha b_{n_1+1} = b_{n_1+2}$$

$$\vdots \qquad\qquad\qquad \vdots$$

$$\alpha b_{j_1-1} = b_{j_1} \qquad\qquad \alpha b_{j_2-1} = b_{j_2}$$
$$\alpha b_{j_1} = b_{j_1+1} + b_{i_1} \qquad\qquad \alpha b_{j_2} = b_{j_2+1} + b_{i_2}$$
$$\alpha b_{j_1+1} = b_{j_1+2} \qquad\qquad \alpha b_{j_2+1} = b_{j_2+2}$$

$$\vdots \qquad\qquad\qquad \vdots$$

$$\alpha b_{n_1} = b_1 \qquad\qquad \alpha b_n = b_{n_1+1}.$$

We set again $\gamma_1 = b_{j_1+1}$ and $\gamma_2 = b_{j_2+1}$. Then

$$\alpha^{n_1} \gamma_1 = \gamma_1 + b_{i_1} \text{ and } \alpha^{n_2} \gamma_2 = \gamma_2 + b_{i_2}. \tag{10}$$

Case 3.1. $i_1 \in [1, n_1]$ and $i_2 \in [n_1 + 1, n]$.

Then $b_{i_1} = \alpha^{t_1} \gamma_1$ with $t_1 = i_1 - j_1 - 1 \pmod{n_1}$ and $b_{i_2} = \alpha^{t_2} \gamma_2$ with $t_2 = i_2 - j_2 - 1 \pmod{n_2}$. M is a block diagonal matrix: $M = (C_{x^{n_1}+1} + E_{i_1, j_1}) \oplus (C_{x^{n_2}+1} + E_{i_2, j_2}) = B_1 \oplus B_2$. Let m_M, m_{B_1}, m_{B_2} be the minimal polynomial of M, B_1 and B_2. Then $m_M = \mathrm{lcm}(m_{B_1}, m_{B_2})$ and if m_M is irreducible then $m_M = m_{B_1} = m_{B_2}$. This implies that B_1 and B_2 are multiplication matrices with $\mathrm{wt}_s(B_1) = \mathrm{wt}_s(B_2) = 1$. From Theorem 4 we obtain that m_{B_1} and m_{B_2} are trinomials of degree n_1 and n_2, respectively. So $n_1 = n_2 = n/2$ and $m_M = x^{n/2} + x^t + 1$. Using Eq. (10) we can determine the choice for i_1, i_2, j_1, j_2

$$\alpha^{n/2} \gamma_1 = \gamma_1 + \alpha^{t_1} \gamma_1 \text{ and } \alpha^{n/2} \gamma_2 = \gamma_2 + \alpha^{t_2} \gamma_2.$$

Hence i_1, i_2, j_1, j_2 have to be chosen in a way that $t_1 = t_2 = t$. This is possible for every $t \in \{1, \ldots, n/2 - 1\}$.

Case 3.2. $i_1 \in [n_1 + 1, n]$ and $i_2 \in [1, n_1]$.

Then $b_{i_1} = \alpha^{t_1} \gamma_2$ with $t_1 = i_1 - j_2 - 1 \pmod{n_2}$ and $b_{i_2} = \alpha^{t_2} \gamma_1$ with $t_2 = i_2 - j_1 - 1 \pmod{n_1}$. Similarly to Case 1 we can show that the minimal polynomial of M has degree n. Applying Eq. (10) yields

$$\gamma_1 = \alpha^{n_2 - t_2} \gamma_2 + \alpha^{-t_2} \gamma_2$$

and

$$\alpha^{n - t_2} \gamma_2 + \alpha^{n_1 - t_2} \gamma_2 + \alpha^{n_2 - t_2} \gamma_2 + \alpha^{t_1} \gamma_2 + \alpha^{-t_2} \gamma_2 = 0.$$

Multiplying this equation by α^{t_2} we conclude that $p = x^n + x^{n_1} + x^{n_2} + x^{t_1 + t_2} + 1$ annihilates γ_2, so $m_\alpha = p$. Note that $t_1 \in \{0, \ldots, n_2 - 1\}$ and $t_1 \in \{0, \ldots, n_1 - 1\}$ so $t_1 + t_2 \in \{0, \ldots, n - 2\}$.

Case 4. $M = (C_{x^{n_1}+1} \oplus C_{x^{n_2}+1}) + E_{i_1, j_1} + E_{j_1 + 1 \pmod{n_1}, j_2} + E_{i_1, j_2}$.

Again, we can assume $j_1 \neq j_2$. Note that the matrix operates on the basis B just as in Case 3, the only difference being that b_{i_2} is substituted by $b_{i_1} + b_{j_1 + 1} = b_{i_1} + \gamma_1$. This leads to

$$\alpha^{n_2} \gamma_2 = \gamma_2 + \gamma_1 + \alpha^t \gamma_2. \tag{11}$$

With the same argument as before we conclude that the minimal polynomial of M has degree n. If $i_1 \in [1, n_1]$ then M is again a block triangular matrix with one block being a companion matrix, so this case cannot occur. So $i_1 \in [n_1 + 1, n]$ and $b_{i_1} = \alpha^t \gamma_2$ for $t_1 = i_1 - j_2 - 1 \pmod{n_2}$. Similarly to Case 3.2 we get

$$\gamma_2 = \alpha^{n_1 - t} \gamma_1 + \alpha^{-t} \gamma_1.$$

Combining this equation with Eq. (11) we have

$$\alpha^{n - t} \gamma_1 + \alpha^{n_2 - t} \gamma_1 + \alpha^{n_1} \gamma_1 + \alpha^{n_1 - t} \gamma_1 + \alpha^{-t} \gamma_1 = 0$$

and after multiplying with α^t we conclude that $m_\alpha = x^n + x^{n_1 + t} + x^{n_2} + x^{n_1} + 1$, where $t \in \{1, \ldots, n_2 - 1\}$. □

Cases 1 and 3 of Theorem 5 also provide all elements α with $\mathrm{wt}_d(\alpha) = 2$. Moreover, Theorem 4 in [2] is a slightly weaker version of Case 1.3. in Theorem 5.

Remark 3. A suitable choice for the values i_1, j_1, i_2, j_2 in the second column of Table 1 can be found in the proof of the corresponding case.

The following example shows that the cycle normal forms of optimal s-XOR-representations are generally not unique.

Example 3. Let $\alpha \in \mathbb{F}_{2^4}$ with the irreducible minimal polynomial $m_\alpha = x^4 + x^3 + x^2 + x + 1$. Then, by Theorem 5, $\mathrm{wt}_s(\alpha) = \mathrm{wt}_d(\alpha) = 2$ and $M = C_{x^4+1} + E_{2,2} + E_{3,4}$ and $M' = (C_{x^3+1} \oplus C_{x+1}) + E_{3,4} + E_{4,3}$ belong to two different optimal representations, corresponding to Case 1.4. and Case 3.2 of Theorem 5, respectively.

The following corollary is a direct result from Theorem 5 and Example 2.

Corollary 3. *Let $\alpha \in \mathbb{F}_{2^n}$ with $\mathrm{wt}(m_\alpha) = 5$ and $\deg(m_\alpha) = n$. Then $\mathrm{wt}_s(\alpha) = 2$ if f appears in Table 1 and $\mathrm{wt}_s(\alpha) = 3$ otherwise.*

Corollary 3 shows that an implementation via the rational canonical form (as in Example 2) is generally not the best way to implement multiplication in binary finite fields. However, irreducible pentanomials that do not appear in the table in Theorem 5 exist, the examples with the lowest degree are $f = x^8 + x^6 + x^5 + x^4 + 1$ and its reciprocal polynomial (for a table of all s-XOR-counts of finite field elements in $\mathbb{F}_{2^n}$ for $n \leq 8$ see [2]). It is an interesting question for which field elements the "naive" representation using the rational canonical form is optimal.

4   Quantifying the Gap Between the Optimal Implementation and the Naive Implementation

It is now interesting to investigate the gap between the optimal implementation and the "naive" implementation using the rational canonical form. We will give a partial answer to this question in Theorem 6. First, we need some notation and lemmas.

For a square matrix $M = (m_{r,s})$ over $\mathbb{F}_2$ and two index sequences (ordered sets) $I = (i_1, \ldots, i_{l_1}), J = (j_1, \ldots, j_{l_2})$, $l := \min(l_1, l_2)$ we denote by $M^{I,J} = (a_{r,s})$ the matrix that is constructed as follows: All rows in I and all columns in J are filled with zeroes, except the entries $a_{i_1,j_1}, \ldots, a_{i_l,j_l}$ which are set to 1. More precisely:

$$a_{r,s} = \begin{cases} 0, & r = i_k, s \neq j_k \text{ for a } k \in \{1, \ldots, l_1\} \\ 0, & r \neq i_k, s = j_k \text{ for a } k \in \{1, \ldots, l_2\} \\ 1, & r = i_k, s = j_k \text{ for a } k \in \{1, \ldots, l\} \\ m_{r,s}, & \text{otherwise.} \end{cases}$$

The following example illustrates our notation. Let $I = \{2, 4\}$ and $J = \{1, 3\}$.

$$M = \begin{pmatrix} 1 & 1 & 0 & 1 \\ 0 & 1 & 1 & 1 \\ 1 & 1 & 1 & 0 \\ 0 & 1 & 0 & 1 \end{pmatrix}, \quad M^{I,J} = \begin{pmatrix} 0 & 1 & 0 & 1 \\ 1 & 0 & 0 & 0 \\ 0 & 1 & 0 & 0 \\ 0 & 0 & 1 & 0 \end{pmatrix}.$$

In the case that $l_1 \neq l_2$ the matrix $M^{I,J}$ has a zero row/column and is thus not invertible. If $l_1 = l_2$, it is easy to see that $\det(M^{I,J})$ does not depend on the ordering of the index sets I, J and is the same as the determinant of the matrix that is created by deleting all rows of M in I and all columns of M in J. In the case that we are only concerned with the determinant, we will thus just use (unordered) index sets I, J and also talk about determinants of submatrices. If $I = \{i\}$ and $J = \{j\}$ we will also write $M^{(i,j)}$. Moreover, we denote by A_M the characteristic matrix $A_M := xI + M$ of M.

Lemma 3. *Let $M = C_{x^n+1} \in \mathrm{GL}(n, \mathbb{F}_2)$. Then we have $\mathrm{wt}(\det(A_M^{I,J})) \leq 1$ for all possible proper square submatrices $A_M^{I,J}$.*

Proof. The proof is by induction on the size of the submatrix. Clearly, $\det(A_M^{I,J}) \in \{0, 1, x\}$ if $|I| = |J| = n - 1$. Let now $|I| < n - 1$. We denote by c_{ij} the entry in the i-th row and j-th column of A_M. Then

$$c_{ij} = \begin{cases} x, & i = j, \\ 1, & i = j + 1 \pmod{n}, \\ 0, & \text{else}. \end{cases}$$

Let $i \in I$. If $i \notin J$, then $A_M^{I,J}$ has at most one non-zero entry in the i-th column. Then, by Laplace expansion along the i-th column and use of the induction hypothesis, we get $\mathrm{wt}(\det(A_M^{I,J})) \leq 1$. If $i \in J$ and $i + 1 \pmod{n} \notin I$ then the $i + 1 \pmod{n}$-th row has at most one non-zero entry and Laplace expansion along the $i + 1 \pmod{n}$-th row yields $\mathrm{wt}(\det(A_M^{I,J})) \leq 1$. We conclude that $\mathrm{wt}(\det(A_M^{I,J})) \leq 1$ for all I with $|I| < n$. $\qquad\square$

Lemma 4. *Let $M = C_{x^n+1} + \sum_{k=1}^{t} E_{i_k,j_k}$ where i_k, j_k can be chosen arbitrarily. Then we have $\mathrm{wt}(\det(A_M^{I,J})) \leq 2^t$ for all possible proper square submatrices $A_M^{I,J}$.*

Proof. The proof is by induction on t. The case $t = 0$ is covered by Lemma 3. Let now $t \geq 1$. Let $M' = C_{x^n+1} + \sum_{k=1}^{t-1} E_{i_k,j_k}$, so that $M = M' + E_{i,j}$ with $i = i_k$, $j = j_k$. If $i \in I$ or $j \in J$ we have $A_M^{I,J} = A_{M'}^{I,J}$ and thus $\mathrm{wt}(\det(A_M^{I,J})) = \mathrm{wt}(\det(A_{M'}^{I,J})) \leq 2^{t-1}$. If $i \notin I$ and $j \notin J$ then $A_M^{I,J} = A_{M'}^{I,J} + E_{i,j}$ and thus Laplace expansion along the i-th row yields $\det(A_M^{I,J}) \leq \det(A_{M'}^{I,J}) + \det(A_{M'}^{I\cup\{i\},J\cup\{j\}})$ and thus

$$\mathrm{wt}(\det(A_M^{I,J})) \leq \mathrm{wt}(\det(A_{M'}^{I,J})) + \mathrm{wt}(\det(A_{M'}^{I\cup\{i\},J\cup\{j\}})) \leq 2^{t-1} + 2^{t-1} = 2^t$$

by induction hypothesis. $\qquad\square$

Corollary 4. *Let $M = C_{x^n+1} + \sum_{k=1}^{t} E_{i_k,j_k}$ where i_k, j_k can be chosen arbitrarily. Then $\mathrm{wt}(\chi(M)) \leq 2^t + 1$.*

Proof. The proof is by induction on t. The case $t = 0$ holds because $\chi(C_{x^n+1}) = x^n + 1$ by definition of the companion matrix. Let now $t \geq 1$ and $M' = C_{x^n+1} + \sum_{k=1}^{t-1} E_{i_k,j_k}$. Laplace expansion along the i_t-th row yields $\chi(M) = \det(A_M) = \chi(M') + \det(A_{M'}^{(i_t,j_t)})$. We conclude with Lemma 4 and the induction hypothesis that $\mathrm{wt}(\chi(M)) \leq 2^{t-1} + 1 + 2^{t-1} = 2^t + 1$. □

Theorem 6. *Let $\alpha \in \mathbb{F}_{2^n}$ be not contained in a proper subfield of $\mathbb{F}_{2^n}$ and let $M_{\alpha,B}$ be a multiplication matrix of α with respect to some basis B. Then $\mathrm{wt}_d(M_{\alpha,B}) = t$ implies $\mathrm{wt}(m_\alpha) \leq 2^t + 1$.*

Proof. Let B be an optimal (regarding the d-XOR-count) basis and $M := M_{\alpha,B} = \bigoplus_{k=1}^{l} C_{x^{m_k}+1} + \sum_{r=1}^{t} E_{i_r,j_r}$ be an optimal multiplication matrix. The case $l = 1$ is covered in Corollary 4, so we only consider $l > 1$ for the rest of the proof. Since α is not contained in a proper subfield of $\mathbb{F}_{2^n}$, the minimal polynomial of M coincides with its characteristic polynomial. We call the sets

$$\{1,\ldots,m_1\}, \{m_1+1,\ldots,m_2\},\ldots,\{\sum_{k=1}^{l-1} m_k + 1,\ldots,\sum_{k=1}^{l} m_k\}$$

the l *blocks* of M. We can decompose $M = M_1 + M'$ with $M_1 = \bigoplus_{k=1}^{l} C_{x^{m_k}+1} + \sum_{r=1}^{t_1} E_{i_r,j_r}$ and $M' = \sum_{r=1}^{t_2} E_{i_r,j_r}$ in a way that all pairs (i_r, j_r) in M_1 are in the same block and all pairs (i_r, j_r) in M' are in different blocks. M_1 is a block diagonal matrix and with Corollary 4 we get

$$\mathrm{wt}(\chi(M_1)) \leq \prod_{k=1}^{l} (2^{s_k} + 1) \text{ with } \sum_{k=1}^{l} s_k = t_1 \tag{12}$$

where s_k denotes the number of pairs (i_r, j_r) that are in the k-th block. We call $B_1,\ldots,B_l$ the l blocks of M_1 and $m_1,\ldots,m_l$ the size of these blocks. Note that $\chi(M)$ is irreducible which implies that M is not a block triangular matrix and thus $t_2 \geq l$. So we can write $M' = M_2 + M_3$ in a way that (after a suitable permutation of blocks) $M_1 + M_2$ looks like this:

$$M_1 + M_2 = \begin{pmatrix} B_1 & 0 & \ldots & E_{i_l,j_l} \\ E_{i_1,j_1} & B_2 & \ldots & 0 \\ \vdots & \ddots & \ddots & \vdots \\ 0 & \ldots & E_{i_{l-1},j_{l-1}} & B_l \end{pmatrix}. \tag{13}$$

From this, we infer by Laplace expansion along the i_l-th row

$$\chi(M_1 + M_2) = \chi(M_1) + \det(A_{M_1+M_2}^{(i_l,v)}), \tag{14}$$

where $v = \sum_{k=1}^{l-1} m_k + j_l$. We now determine $\mathrm{wt}(\det(A_{M_1+M_2}^{(i_l,v)}))$. We get

$$\det(A_{M_1+M_2}^{(i_l,v)}) = \det \begin{pmatrix} B_1^{(i_l,\emptyset)} & 0 & \cdots & 0 & E_{i_l,j_l} \\ E_{i_1,j_1} & B_2 & \cdots & 0 & 0 \\ \vdots & \ddots & \ddots & \vdots & \vdots \\ 0 & \ddots & E_{i_{l-2},j_{l-2}} & B_{l-1} & 0 \\ 0 & \cdots & 0 & E_{i_{l-1},j_{l-1}} & B_l^{(\emptyset,j_l)} \end{pmatrix}$$

$$= \det \begin{pmatrix} B_1^{(i_l,\emptyset)} & 0 & \cdots & E_{i_{l-1},j_{l-1}} & * \\ E_{i_1,j_1} & B_2 & \cdots & 0 & 0 \\ \vdots & \ddots & \ddots & \vdots & \vdots \\ 0 & \ddots & E_{i_{l-2},j_{l-2}} & B_{l-1} & 0 \\ 0 & \cdots & 0 & 0 & B_l^{(i_{l-1},j_l)} \end{pmatrix}$$

by swapping the i_l-th row with the $\sum_{k=1}^{l-1} m_k + i_{l-1}$-th row. This operation can now be repeated for the upper-left $l-1$ blocks, the result is the following block diagonal matrix

$$\det(A_{M_1+M_2}^{(i_l,v)}) = \det \begin{pmatrix} B_1^{(i_l,j_1)} & * & \cdots & 0 & 0 \\ 0 & B_2^{(i_1,j_2)} & \cdots & 0 & 0 \\ \vdots & \ddots & \ddots & \vdots & \vdots \\ 0 & \ddots & 0 & B_{l-1}^{(i_{l-2},j_{l-1})} & * \\ 0 & \cdots & 0 & 0 & B_l^{(i_{l-1},j_l)} \end{pmatrix}.$$

Lemma 4 then implies $\mathrm{wt}(\det(A_{M_1+M_2}^{(i_l,v)})) \leq \prod_{k=1}^{l} 2^{s_k} = 2^{t_1}$. Equations (12) and (14) yield

$$\mathrm{wt}(\chi(M_1+M_2)) \leq \prod_{k=1}^{l} (2^{s_k}+1) + 2^{t_1}. \tag{15}$$

We now investigate the determinant of the square submatrices of $M_1 + M_2$. Let I, J be index sets and set $I = \bigcup_r I_r$ and $J = \bigcup_r J_r$ where I_r and J_r contain the indices that belong to the r-th block. Observe that $|I| = |J|$. Let us first look at the case $I = I_r$ and $J = J_r$ for some r. Using Lemma 4

$$\mathrm{wt}(\det(A_{M_1+M_2}^{I,J})) \leq 2^{s_r} \prod_{\substack{k \in \{1,\dots,l\} \\ k \neq r}} (2^{s_k}+1) + 2^{t_1}.$$

Similarly, if $|I_r| = |J_r|$ for all $1 \leq r \leq l$ then

$$\mathrm{wt}(\det(A_{M_1+M_2}^{I,J})) \leq \prod_{r:I_r \neq \emptyset} 2^{s_r} \prod_{r:I_r = \emptyset} (2^{s_k}+1) + 2^{t_1}. \tag{16}$$

Let us now assume that there is a block r with $|I_r| \neq |J_r|$. We can assume w.l.o.g. that $r = 1$ and $p := |I_1| < |J_1|$. If $i_1 + m_1, i_l \in I$ or $v, j_1 \in J$ then Eq. (13) implies $\det(A_{M_1 + M_2}^{I,J}) = 0$. We order $I = (a_1, \dots, a_t)$ and $J = (b_1, \dots, b_t)$ in ascending order. Then

$$\det(A_{M_1+M_2}^{I,J}) = \det \begin{pmatrix} B_1^{I_1, J_1} & A \\ C & D \end{pmatrix},$$

with $A = (a_{r,s}) \in \mathbb{F}_2^{m_1 \times (n-m_1)}$, $C = (c_{r,s}) \in \mathbb{F}_2^{(n-m_1) \times m_1}$ with

$$a_{r,s} = \begin{cases} 1, & \text{for } (r,s) = (i_l, v), \\ 0, & \text{else,} \end{cases} \quad c_{r,s} = \begin{cases} 1, & \text{for } (r,s) = (a_k, b_k), k > p, \\ 1, & \text{for } (r,s) = (i_1, j_1), \\ 0, & \text{else.} \end{cases}$$

Swapping the i_l-th row with the a_{p+1}-th row, we obtain

$$\det(A_{M_1+M_2}^{I,J}) = \det \begin{pmatrix} B_1^{I_1 \cup \{i_l\}, J_1} & 0 \\ * & D' \end{pmatrix}$$

and thus $\det(A_{M_1+M_2}^{I,J}) = \det(B_1^{I_1 \cup \{i_l\}, J_1}) \det(D')$. Observe that $\det(A_{M_1+M_2}^{I,J}) = 0$ if $|I_1| \neq |J_1| + 1$. Moreover, $\det(D') = \det(C_{M_1+M_2}^{I',J'})$ where $\{1, \dots, m_1\}$ is a subset of I' and J'. In particular, the number of indices in I' and J' belonging to the first block is the same. By induction, Eq. (16) and Lemma 4, we get

$$\text{wt}(\det(A_{M_1+M_2}^{I,J})) = \text{wt}(\det(B_1^{I_1 \cup \{i_l\}, J_1}) \det(D')) \leq 2^{s_1} \prod_{k=2}^{l} (2^{s_k} + 1) + 2^{t_1}. \quad (17)$$

Equations (16) and (17) imply that for arbitrary index sets I, J, there exists an $r \in \{1, \dots, l\}$ such that

$$\text{wt}(\det(A_{M_1+M_2}^{I,J})) \leq 2^{s_r} \prod_{\substack{k \in \{1, \dots, l\} \\ k \neq r}} (2^{s_k} + 1) + 2^{t_1}. \quad (18)$$

As in the proof of Lemma 4, for arbitrary index sets I, J and $i, j \in \{1, \dots, n\}$ there is an $r \in \{1, \dots, l\}$ such that

$$\text{wt}(\det(A_{M_1+M_2+E_{i,j}}^{I,J})) \leq \text{wt}(\det(A_{M_1+M_2}^{I,J})) + \text{wt}(\det(A_{M_1+M_2}^{I \cup \{i\}, J \cup \{j\}}))$$

$$\leq 2 \cdot \left(2^{s_r} \prod_{\substack{k \in \{1, \dots, l\} \\ k \neq r}} (2^{s_k} + 1) + 2^{t_1} \right)$$

and, inductively, for an arbitrary matrix $M_3 = \sum_{k=1}^{z} E_{i_k,j_k}$ with z non-zero entries

$$\mathrm{wt}(\det(A_{M_1+M_2+M_3}^{I,J})) \leq 2^z \left(2^{s_r} \prod_{\substack{k \in \{1,\dots,l\} \\ k \neq r}} (2^{s_k} + 1) + 2^{t_1} \right)$$

$$< 2^z \left(\prod_{k=1}^{l} (2^{s_k} + 1) + 2^{t_1} \right). \tag{19}$$

We now show by induction that we have for $z \geq 1$

$$\mathrm{wt}(\chi(M_1 + M_2 + M_3)) < 2^z \left(\prod_{k=1}^{l} (2^{s_k} + 1) + 2^{t_1} \right). \tag{20}$$

The case $z = 1$ is dealt with using Eqs. (15) and (18):

$$\mathrm{wt}(\chi(M_1 + M_2 + M_3)) \leq \mathrm{wt}(\chi(M_1 + M_2)) + \mathrm{wt}(\det(A_{M_1+M_2}^{i_1,j_1}))$$

$$< 2 \left(\prod_{k=1}^{l} (2^{s_k} + 1) + 2^{t_1} \right).$$

Let now $z > 1$ and $M_3' = \sum_{k=1}^{z-1} E_{i_k,j_k}$. With the induction hypothesis and Eq. (19) we conclude

$$\mathrm{wt}(\chi(M_1 + M_2 + M_3)) \leq \mathrm{wt}(\chi(M_1 + M_2 + M_3')) + \mathrm{wt}(\det(A_{M_1+M_2+M_3'}^{i_1,j_1}))$$

$$< 2^z \left(\prod_{k=1}^{l} (2^{s_k} + 1) + 2^{t_1} \right),$$

proving Eq. (20). Note that the bound in Eq. (20) depends only on the parameters l, t_2 and $s_k, k = 1, \dots, l$ where $\sum_{k=1}^{l} s_k = t_1$ and $t_1 + t_2 = t = \mathrm{wt}(M)$. For $t_2 > l$ we have

$$\mathrm{wt}(\chi(M_1 + M_2 + M_3)) < 2^{t_2-l} \left(\prod_{k=1}^{l} (2^{s_k} + 1) + 2^{t_1} \right).$$

Using Eq. (15), a matrix N with values $l_N = t_2$ and $s_k = 0$ for $k > l$ yields

$$\mathrm{wt}(\chi(N)) \leq \prod_{k=1}^{l_N} (2^{s_k} + 1) + 2^{t_1}$$

$$= 2^{t_2-l} \prod_{k=1}^{l} (2^{s_k} + 1) + 2^{t_1}.$$

In particular, the upper bound given in Eq. (20) is always worse than the one given in Eq. (15) and we can focus on the case $M_3 = 0$ (or, equivalently, $t_2 = l$) for the rest of this proof. In other words, we just have to find the parameters that give the maximum weight estimation in Eq. (15). A direct calculation yields

$$\prod_{k=1}^{l} (2^{s_k} + 1) \leq (2^{t_1} + 1) \cdot 2^{l-1},$$

i.e. the choice $s_1 = t_1$, $s_i = 0$ for $i > 1$ is optimal. Plugging these parameters into Eq. (15), we get

$$\text{wt}(\chi(M)) \leq 2^{t_1+l-1} + 2^{l-1} + 2^{t_1} = 2^{t-1} + 2^{l-1} + 2^{t-l}. \tag{21}$$

Obviously, the maximum of $2^{l-1} + 2^{t-l}$ for $2 \leq l \leq t$ is attained at $l = t$. The result follows from Eq. (21). $\qquad\square$

We now show that the bound given in Theorem 6 is optimal by giving two examples where the upper bound is attained. Note that the proof of Theorem 6 implies that this can only occur if the number of blocks of the optimal multiplication matrix is 1 or t. We will give examples for both cases in Propositions 2 and 3.

Theorem 7 ([11, Theorem 3.5], [7, Theorem 4.3.9]). *Let R be a (commutative) Euclidean domain and $A \in R^{n \times n}$. Then A can be transformed into an upper triangular matrix using elementary row operations (i.e. a sequence of left-multiplications with matrices $I + rE_{i,j}$ with $r \in R$ and $i \neq j$).*

Proposition 2. *Let $\alpha \in \mathbb{F}_{2^n}$ with an irreducible minimal polynomial f with $\text{wt}(f) = 2^t + 1$ of the form*

$$f = x^n + \prod_{j=1}^{t} \left(x^{i_j} + 1 \right)$$

for arbitrary values of $i_j \in \mathbb{N}$ with $\sum_{j=1}^{t} i_j \leq n - t$. Then there exists a basis B such that the matrix $M := M_{\alpha,B}$ satisfies $\text{wt}_s(M) = \text{wt}_d(M) = t$.

Proof. We show that the matrix $M = C_{x^n+1} + \sum_{k=1}^{t-1} E_{j_k+i_k+1,j_k} + E_{i_t+n-j_t,j_t}$ where the j_k are chosen arbitrarily under the conditions that $j_{k+1} \geq i_k + j_k + 1$ for all $k = 1, \ldots, t-1$ and $i_t < j_1$ has the desired property. It is clear that

$\mathrm{wt}_s(M) = \mathrm{wt}_d(M) = t$. Let $B = \{b_1, \ldots, b_n\}$ be some basis of $(\mathbb{F}_2)^n$ over $\mathbb{F}_2$. We investigate how M (viewed as a transformation matrix) operates on this basis:

$$Mb_1 = b_2$$

$$\vdots$$

$$Mb_{j_1-1} = b_{j_1}$$
$$Mb_{j_1} = b_{j_1+1} + M^{i_1}b_{j_1+1} \tag{22}$$
$$Mb_{j_1+1} = b_{j_1+2}$$

$$\vdots$$

$$Mb_{j_2-1} = b_{j_2}$$
$$Mb_{j_2} = b_{j_2+1} + M^{i_2}b_{j_2+1} \tag{23}$$
$$Mb_{j_2+1} = b_{j_2+2}$$

$$\vdots$$

$$Mb_n = b_1.$$

Set $n_i = j_i - j_{i-1}$ for $2 \le i \le t$ and $n_1 = n + j_1 - j_t$. Note that $\sum_{i=1}^t n_i = n$ and $Mb_{j_k} = M^{n_i}b_{j_{k-1}+1}$. With this and the equations of type (22) and (23) we obtain the following set of equations:

$$\begin{pmatrix} M^{n_2} & M^{i_1}+1 & 0 & \cdots & 0 \\ 0 & M^{n_3} & M^{i_2}+1 & \cdots & 0 \\ & & \ddots & \ddots & \\ 0 & \cdots & 0 & M^{n_t} & M^{i_{t-1}}+1 \\ M^{i_t}+1 & 0 & \cdots & 0 & M^{n_1} \end{pmatrix} \begin{pmatrix} b_{j_1+1} \\ b_{j_2+1} \\ \vdots \\ \vdots \\ b_{j_t+1} \end{pmatrix} = 0. \tag{24}$$

We denote by A the matrix in Eq. (24). A is a matrix over $\mathbb{F}_2[M]$. It is clear that $\mathbb{F}_2[M]$ is isomorphic to the usual polynomial ring $\mathbb{F}_2[x]$ and thus a Euclidean domain. Using the Leibniz formula for determinants, we obtain $\det(A) = f(M)$. By Theorem 7, we can transform A into an upper triangular matrix A' using only elementary row operations. In particular $\det(A') = \prod_{i=1}^n a'_{i,i} = \det(A) = f(M)$ where the $a'_{i,i}$ denote the entries on the diagonal of A'. Since f is irreducible, we obtain $a_{k,k} = f(M)$ for one $1 \le k \le n$ and $a_{i,i} = 1$ for all $i \neq k$, i.e.

$$\begin{pmatrix} 1 & & & * \\ & \ddots & & \\ & & f(M) & * & * \\ & & & \ddots & \\ 0 & & & & 1 \end{pmatrix} \begin{pmatrix} b_{j_1+1} \\ \vdots \\ b_{j_k+1} \\ \vdots \\ b_{j_t+1} \end{pmatrix} = 0.$$

It is clear that all entries $a'_{k,k+1}, \ldots, a'_{k,n}$ can be eliminated by further row additions. Hence, we obtain $f(M)b_{j_k+1} = 0$, i.e. f is the M-annihilator of b_{j_k+1}.

As f is irreducible this implies that the minimal polynomial of M is f and thus M is a multiplication matrix of α.

$\square$

Proposition 3. Let $\alpha \in \mathbb{F}_{2^n}$ with an irreducible minimal polynomial f with $\mathrm{wt}(f) = 2^t + 1$ of the form

$$f = \prod_{j=1}^{t} (x^{n_j} + 1) + x^k$$

for arbitrary values of n_j and $k \leq n - t$ with $\sum_{j=1}^{t} n_j = n$. Then there exists a basis B such that the matrix $M := M_{\alpha,B}$ satisfies $\mathrm{wt}_s(M) = \mathrm{wt}_d(M) = t$.

Proof. The proof is similar to the proof of the previous lemma. Define $\hat{n}_l = \sum_{u=1}^{l-1} n_u$ for $1 \leq l \leq t$. Let r_l be chosen arbitrarily such that $1 \leq r_l \leq n_l$ for $1 \leq l \leq t$ and $\sum_{l=1}^{t} r_l = k$. Further let $j_l := \hat{n}_l + r_l$ for all $1 \leq l \leq t$ and $s_l := i_l + r_{l+1} + 1 \pmod{n_{l+1}}$ for $l < t$ and $s_t := i_t + r_1 + 1 \pmod{n_1}$.

Define now $M = \bigoplus_{i=1}^{t} C_{x^{n_i}+1} + \sum_{k=1}^{t} E_{\hat{n}_k + s_k, j_k}$. Obviously, $\mathrm{wt}_s(M) = \mathrm{wt}_d(M) = t$. Let $B = \{b_1, \ldots, b_n\}$ be some basis of $(\mathbb{F}_2)^n$ over $\mathbb{F}_2$. We investigate how M (viewed as a transformation matrix) operates on this basis:

$$Mb_1 = b_2 \qquad Mb_{n_1+1} = b_{n_1+2} \qquad \ldots \qquad Mb_{n_{t-1}+1} = b_{n_t+2}$$

$$\vdots \qquad\qquad \vdots \qquad\qquad \vdots$$

$$Mb_{j_1-1} = b_{j_1} \qquad Mb_{j_2-1} = b_{j_2} \qquad Mb_{j_t-1} = b_{j_t}$$

$$Mb_{j_1} = b_{j_1+1} + M^{i_1}b_{j_2+1} \quad Mb_{j_2} = b_{j_2+1} + M^{i_2}b_{j_3+1} \qquad Mb_{j_t} = b_{j_t+1} + M^{i_t}b_{j_1+1}$$

$$Mb_{j_1+1} = b_{j_1+2} \qquad Mb_{j_2+1} = b_{j_2+2} \qquad Mb_{j_t+1} = b_{j_t+2}$$

$$\vdots \qquad\qquad \vdots \qquad\qquad \vdots$$

$$Mb_{n_1} = b_1 \qquad Mb_{n_2} = b_{n_1+1} \qquad \cdots \qquad Mb_n = b_{n_{t-1}+1}.$$

Clearly, $Mb_{j_k} = M^{n_k}b_{j_k+1}$, so we get the following set of equations:

$$\begin{pmatrix} M^{n_1}+1 & M^{i_1} & 0 & \cdots & & 0 \\ 0 & M^{n_2}+1 & M^{i_2} & \cdots & & 0 \\ & & \ddots & \ddots & & \\ 0 & \cdots & 0 & M^{n_{t-1}}+1 & M^{i_{t-1}} \\ M^{i_t} & 0 & \cdots & & 0 & M^{n_t}+1 \end{pmatrix} \begin{pmatrix} b_{j_1+1} \\ b_{j_2+1} \\ \vdots \\ \vdots \\ b_{j_t+1} \end{pmatrix} = 0.$$

The determinant of the matrix is exactly $f(M)$. We can now repeat the arguments from the proof of Proposition 2 and obtain that M is a multiplication matrix for α.

$\square$

Observe that the polynomials in Propositions 2 and 3 are generalizations of Case 1.3. and Case 3.2. in Theorem 5.

Note that irreducible polynomials of the types mentioned in Propositions 2 and 3 do exist, examples up to $t = 8$, corresponding to polynomials of weight $2^t + 1$, are compiled in Table 2. The table lists in the second column values for i_l and n that belong to an irreducible polynomial of the type of Proposition 2 and in the third column the values for n_l and k that belong to an irreducible polynomial of the type of Proposition 3. The values listed were found with a simple randomized algorithm. They generally do not correspond to the irreducible polynomial of that type with the least degree. Propositions 2 and 3 together with Theorem 6 show that the gap between the number of XORs used in the optimal implementation and the number of XORs used in the naive implementation of a multiplication matrix using the rational canonical form grows exponentially with the weight of the minimal polynomial of the element.

Table 2. Irreducible polynomials of the form described in Propositions 2 and 3.

t	Values for $i_1, \ldots, i_t; n$	Values for $n_1, \ldots, n_t; k$
2	1,2;5	2,4;1
3	1,2,4;10	4,5,6;1
4	3,5,6,12;30	2,3,6,10;1
5	1,2,4,9,17;39	12,13,15,19,23;9
6	1,12,16,24,31; 123	13,22,26,27,28,30;23
7	2,30,47,56,60,64,91; 357	25,114,174,231,279,281,331;196
8	23,28,41,59,62,106,141,153; 628	44,148,195,357,363,368,386,480;240

Propositions 2 and 3 show that there are elements $\alpha \in \mathbb{F}_{2^n}$ with $wt(m_\alpha) = 2^t + 1$ and $wt_s(\alpha) = t$. We believe that this upper bound is strict, i.e. the bound is the same for s-XOR-count and d-XOR-count.

Conjecture 1. Let $\alpha \in \mathbb{F}_{2^n}$ be not contained in a proper subfield of $\mathbb{F}_{2^n}$ and $M_{\alpha,B}$ a multiplication matrix of α with respect to some basis B. Then $wt_s(M_{\alpha,B}) = t$ implies $wt(\chi(M)) \leq 2^t + 1$.

5   Open Problems

Our investigations open up many possibilities for future research. While Theorem 1 shows that there is an infinite family of matrices with higher s-XOR-count than d-XOR-count, a more precise classification of these cases as well as finding upper/lower bounds is desirable. Because of the nature of the s-XOR-count, answers to these problems would also give insight into optimal Gauss elimination strategies over $\mathbb{F}_2$.

Problem 1. Classify the matrices $M \in \mathrm{GL}(n, \mathbb{F}_2)$ with $wt_d(M) < wt_s(M)$.

Problem 2. Find bounds c, C so that $c\mathrm{wt}_d(M) \leq \mathrm{wt}_s(M) \leq C\mathrm{wt}_d(M)$ for all matrices $M \in \mathrm{GL}(n, \mathbb{F}_2)$.

Finding out if/how the bounds c, C depend on n and $\mathrm{wt}_s(M)$ would greatly improve the understanding of the two XOR-metrics.

As observed in Sect. 3, there are elements $\alpha \in \mathbb{F}_2$ where the optimal implementation of the mapping $x \mapsto \alpha x$ is the rational canonical form in both of the investigated metrics. These elements are (compared to elements with minimal polynomials of the same weight) the most expensive to implement. A more thorough understanding of these elements would be helpful.

Problem 3. Classify the minimal polynomials $m_\alpha \in \mathbb{F}_2[x]$ for which the optimal multiplication matrix is in rational canonical form.

We also want to repeat a problem about elements in subfields mentioned in [2].

Problem 4. Let $\alpha \in \mathbb{F}_{2^n}$ be contained in a subfield $\mathbb{F}_{2^l}$ with $ld = n$. Let M_l be an optimal multiplication matrix of α regarding d- or s-XOR-count. Is $M = \bigoplus_{k=1}^d M_l$ then an optimal multiplication matrix of $\alpha \in \mathbb{F}_{2^n}$ regarding d- or s-XOR-count?

In Sects. 3 and 4 we limited ourselves to optimal XOR-implementations of matrices that are multiplication matrices for a fixed field element (which are exactly those with irreducible minimal polynomial). Investigating a more general case is also an interesting problem.

Problem 5. Let $f: \mathbb{F}_2^n \to \mathbb{F}_2^n$ be a bijective linear mapping and $M_{f,B} \in \mathrm{GL}(n, \mathbb{F}_2)$ the matrix that belongs to f with respect to the basis B. Find a basis B such that the matrix $M_{f,B}$ is the optimal d/s-XOR-count matrix.

In particular, finding optimal matrices $M_{f,B}$ where f denotes the mapping induced by a linear layer of a cryptographic scheme is a very interesting problem.

Acknowledgments. The author wishes to thank the anonymous referees for their comments that improved especially the introduction considerably and helped to set this work into context with existing literature.

I also thank Gohar Kyureghyan for many discussions and help with structuring this paper.

References

1. Babbage, S., Dodd, M.: The MICKEY stream ciphers. In: Robshaw, M., Billet, O. (eds.) New Stream Cipher Designs. LNCS, vol. 4986, pp. 191–209. Springer, Heidelberg (2008). https://doi.org/10.1007/978-3-540-68351-3_15
2. Beierle, C., Kranz, T., Leander, G.: Lightweight multiplication in $GF(2^n)$ with applications to MDS matrices. In: Robshaw, M., Katz, J. (eds.) CRYPTO 2016, Part I. LNCS, vol. 9814, pp. 625–653. Springer, Heidelberg (2016). https://doi.org/10.1007/978-3-662-53018-4_23

3. Canright, D.: A very compact S-box for AES. In: Rao, J.R., Sunar, B. (eds.) CHES 2005. LNCS, vol. 3659, pp. 441–455. Springer, Heidelberg (2005). https://doi.org/10.1007/11545262_32

4. Daemen, J., Rijmen, V.: Correlation analysis in $GF(2^n)$. In: Junod, P., Canteaut, A. (eds.) Advanced Linear Cryptanalysis of Block and Stream Ciphers. Cryptology and Information Security, pp. 115–131. IOS Press (2011)

5. De Cannière, C., Preneel, B.: Trivium. In: Robshaw, M., Billet, O. (eds.) New Stream Cipher Designs. LNCS, vol. 4986, pp. 244–266. Springer, Heidelberg (2008). https://doi.org/10.1007/978-3-540-68351-3_18

6. Duval, S., Leurent, G.: MDS matrices with lightweight circuits. IACR Trans. Symmetric Cryptol. 2018(2), 48–78 (2018). https://doi.org/10.13154/tosc.v2018.i2.48-78

7. Hahn, A., O'Meara, T.: The Classical Groups and K-Theory. Springer, Heidelberg (1989). https://doi.org/10.1007/978-3-662-13152-7

8. Hell, M., Johansson, T., Meier, W.: Grain; a stream cipher for constrained environments. Int. J. Wire. Mob. Comput. 2(1), 86–93 (2007). https://doi.org/10.1504/IJWMC.2007.013798

9. Hoffman, K., Kunze, R.: Linear Algebra. Prentice-Hall, Englewood Cliffs (1961)

10. Jean, J., Peyrin, T., Sim, S.M., Tourteaux, J.: Optimizing implementations of lightweight building blocks. IACR Trans. Symmetric Cryptol. 2017(4), 130–168 (2017)

11. Kaplansky, I.: Elementary divisors and modules. Trans. Amer. Math. Soc. 66, 464–491 (1949). https://doi.org/10.1090/S0002-9947-1949-0031470-3

12. Khoo, K., Peyrin, T., Poschmann, A.Y., Yap, H.: FOAM: searching for hardware-optimal SPN structures and components with a fair comparison. In: Batina, L., Robshaw, M. (eds.) CHES 2014. LNCS, vol. 8731, pp. 433–450. Springer, Heidelberg (2014). https://doi.org/10.1007/978-3-662-44709-3_24

13. Kranz, T., Leander, G., Stoffelen, K., Wiemer, F.: Shorter linear straight-line programs for MDS matrices. IACR Trans. Symmetric Cryptol. 2017(4), 188–211 (2017). https://doi.org/10.13154/tosc.v2017.i4.188-211. https://tosc.iacr.org/index.php/ToSC/article/view/813

14. LaMacchia, B.A., Odlyzko, A.M.: Solving large sparse linear systems over finite fields. In: Menezes, A.J., Vanstone, S.A. (eds.) CRYPTO 1990. LNCS, vol. 537, pp. 109–133. Springer, Heidelberg (1991). https://doi.org/10.1007/3-540-38424-3_8

15. Li, Y., Wang, M.: On the construction of lightweight circulant involutory MDS matrices. In: Peyrin, T. (ed.) FSE 2016. LNCS, vol. 9783, pp. 121–139. Springer, Heidelberg (2016). https://doi.org/10.1007/978-3-662-52993-5_7

16. Liu, M., Sim, S.M.: Lightweight MDS generalized circulant matrices. In: Peyrin, T. (ed.) FSE 2016. LNCS, vol. 9783, pp. 101–120. Springer, Heidelberg (2016). https://doi.org/10.1007/978-3-662-52993-5_6

17. Saarinen, M.-J.O.: Cryptographic analysis of all 4 × 4-bit S-boxes. In: Miri, A., Vaudenay, S. (eds.) SAC 2011. LNCS, vol. 7118, pp. 118–133. Springer, Heidelberg (2012). https://doi.org/10.1007/978-3-642-28496-0_7

18. Sajadieh, M., Mousavi, M.: Construction of lightweight MDS matrices from generalized feistel structures. IACR Cryptology ePrint Archive 2018, 1072 (2018)

19. Sarkar, S., Sim, S.M.: A deeper understanding of the XOR count distribution in the context of lightweight cryptography. In: Pointcheval, D., Nitaj, A., Rachidi, T. (eds.) AFRICACRYPT 2016. LNCS, vol. 9646, pp. 167–182. Springer, Cham (2016). https://doi.org/10.1007/978-3-319-31517-1_9

20. Sim, S.M., Khoo, K., Oggier, F., Peyrin, T.: Lightweight MDS involution matrices. In: Leander, G. (ed.) FSE 2015. LNCS, vol. 9054, pp. 471–493. Springer, Heidelberg (2015). https://doi.org/10.1007/978-3-662-48116-5_23

21. Swan, R.G.: Factorization of polynomials over finite fields. Pacific J. Math. 12(3), 1099–1106 (1962)

22. Zhao, R., Wu, B., Zhang, R., Zhang, Q.: Designing optimal implementations of linear layers (full version). Cryptology ePrint Archive, Report 2016/1118 (2016)

DLCT: A New Tool
for Differential-Linear Cryptanalysis

Achiya Bar-On[1], Orr Dunkelman[2($\boxtimes$)], Nathan Keller[1], and Ariel Weizman[1]

[1] Department of Mathematics, Bar-Ilan University, Ramat Gan, Israel
[2] Computer Science Department, University of Haifa, Haifa, Israel
orrd@cs.haifa.ac.il

Abstract. Differential cryptanalysis and linear cryptanalysis are the two best-known techniques for cryptanalysis of block ciphers. In 1994, Langford and Hellman introduced the differential-linear (DL) attack based on dividing the attacked cipher E into two subciphers E_0 and E_1 and combining a differential characteristic for E_0 with a linear approximation for E_1 into an attack on the entire cipher E. The DL technique was used to mount the best known attacks against numerous ciphers, including the AES finalist Serpent, ICEPOLE, COCONUT98, Chaskey, CTC2, and 8-round DES.

Several papers aimed at formalizing the DL attack, and formulating assumptions under which its complexity can be estimated accurately. These culminated in a recent work of Blondeau, Leander, and Nyberg (Journal of Cryptology, 2017) which obtained an accurate expression under the sole assumption that the two subciphers E_0 and E_1 are independent.

In this paper we show that in many cases, dependency between the two subcipher s significantly affects the complexity of the DL attack, and in particular, can be exploited by the adversary to make the attack more efficient. We present the Differential-Linear Connectivity Table (DLCT) which allows us to take into account the dependency between the two subciphers, and to choose the differential characteristic in E_0 and the linear approximation in E_1 in a way that takes advantage of this dependency. We then show that the DLCT can be constructed efficiently using the Fast Fourier Transform. Finally, we demonstrate the strength of the DLCT by using it to improve differential-linear attacks on ICEPOLE and on 8-round DES, and to explain published experimental results on Serpent and on the CAESAR finalist Ascon which did not comply with the standard differential-linear framework.

1 Introduction

1.1 Background and Previous Work

Cryptanalysis of Block Ciphers. A block cipher is an encryption scheme which accepts an n-bit plaintext and transforms it into an n-bit ciphertext using a k-bit secret key. Block ciphers are the most widely used class of symmetric key primitives nowadays. Most of the modern block ciphers are *iterative*, i.e.,

© International Association for Cryptologic Research 2019
Y. Ishai and V. Rijmen (Eds.): EUROCRYPT 2019, LNCS 11476, pp. 313–342, 2019.
https://doi.org/10.1007/978-3-030-17653-2_11

consist of a sequence of simple operations called rounds repeated multiple times with small alterations. We denote a plaintext/ciphertext pair of a block cipher by (P, C) and the n-bit state at the beginning of the r'th round of the encryption process by X_r.

While the design of block ciphers is a well-developed field and various block cipher designs (most notably, the AES [36]) are widely accepted to provide strong security, there is no block cipher with a security proof that is fast enough for being used in practice. Instead, our confidence in the security of block ciphers stems from analyzing their resistance with respect to all known cryptanalytic techniques. Thus, development of cryptanalytic techniques is the main means for understanding the practical security of block ciphers.

Differential Cryptanalysis and Linear Cryptanalysis. The two central statistical techniques in cryptanalysis of block ciphers are *differential cryptanalysis*, introduced by Biham and Shamir [8], and *linear cryptanalysis*, introduced by Matsui [31].

Differential cryptanalysis studies the development of differences between two encrypted plaintexts through the encryption process. An r-round *differential* with probability p of a cipher is a property of the form $\Pr[X_{i+r} \oplus X'_{i+r} = \Delta_O | X_i \oplus X'_i = \Delta_I] = p$, denoted in short $\Delta_I \xrightarrow{p} \Delta_O$. Differential attacks exploit *long* (with many rounds) differentials with a non-negligible probability.

Linear cryptanalysis studies the development of parities of subsets of the state bits through the encryption process of a single plaintext. An r-round *linear approximation* with bias q is a property of the form $\Pr[X_{i+r} \cdot \lambda_O = X_i \cdot \lambda_I] = \frac{1}{2} + q$, denoted in short $\lambda_I \xrightarrow{q} \lambda_O$. (Recall that the scalar product of $x, y \in \{0, 1\}^n$ is defined as $(\sum_{i=1}^n x_i y_i) \bmod 2$.) Linear attacks exploit "long" approximations with a non-negligible bias.

Differential and linear cryptanalysis were used to mount the best known attacks on numerous block ciphers, most notably DES [35]. As a result, resistance to these two cryptanalytic techniques, and in particular, non-existence of high-probability differentials or high-bias linear approximations spanning many rounds of the cipher, has become a central criterion in block cipher design.

Differential-Linear Cryptanalysis and Other Combined Attacks on Block Ciphers. While precluding long differentials and linear approximations seems to be sufficient for making the cipher immune to differential and linear attacks, it turned out that in many cases, short characteristics and approximations can also be exploited to break the cipher. The first cryptanalytic technique to demonstrate this was *differential-linear cryptanalysis* (in short: *DL technique*), introduced by Langford and Hellman [27] in 1994. Langford and Hellman showed that if the cipher E can be decomposed as a cascade $E = E_1 \circ E_0$, then a high-probability differential for E_0 and a high-bias linear approximation for E_1 can be combined into an efficient distinguisher for the entire cipher E. The DL technique was used to attack many block ciphers, and in particular, yields the best known attacks on the AES finalist Serpent [20,30], the CAESAR [16] candidate ICEPOLE [22], COCONUT98 [4], Chaskey [28], CTC2 [30], etc.

Differential-linear cryptanalysis was followed by several other combined attacks. In particular, boomerang [38], amplified boomerang [24], and rectangle [3] attacks show that high-probability differentials in E_0 an E_1 can also be combined into an attack on the entire cipher. Other combinations include differential-bilinear, higher-order differential-linear, boomerang-linear attacks, etc. [7]. These combined attacks make non-existence of high-probability short differential and linear approximations a desirable (but harder to fulfill) criterion in block cipher design.

An Informal Description of the Differential-Linear Attack. The DL attack works as follows. Assume that we have a differential $\Delta_I \xrightarrow{p} \Delta_O$ for E_0 and a linear approximation $\lambda_I \xrightarrow{q} \lambda_O$ for E_1. In order to distinguish E from a random permutation, the adversary considers plaintext pairs with input difference Δ_I and checks, for each pair, whether the corresponding ciphertexts agree on the parity of the mask λ_O.

Denote the plaintexts by P, P', the ciphertexts by C, C', and the intermediate values between E_0 and E_1 by X, X', respectively. The attack combines three approximations: The values $C \cdot \lambda_O$ and $C' \cdot \lambda_O$ are correlated to $X \cdot \lambda_I$ and $X' \cdot \lambda_I$, respectively, by the linear approximation for E_1. The values $X \cdot \lambda_I$ and $X' \cdot \lambda_I$ are correlated, as consequence of the differential for E_0. Hence, $C \cdot \lambda_O$ is correlated to $C' \cdot \lambda_O$. Figure 1 depicts the relations.

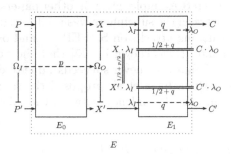

Fig. 1. Differential-linear cryptanalysis

Computation shows that under some randomness assumptions to be discussed below, the equality $C \cdot \lambda_O = C' \cdot \lambda_O$ holds with probability $\frac{1}{2} + 2pq^2$. Hence, if p, q are sufficiently large, then the adversary can distinguish E from a random permutation using $O(p^{-2}q^{-4})$ chosen plaintexts. As usual, the distinguisher can be transformed into a key recovery attack by guessing some key material, performing partial encryption/decryption, and applying the distinguisher.

Randomness Assumptions Behind the DL Attack. The attack analysis described above (initially presented in [4]) crucially depends on two assumptions:

1. Among the cases where the differential is not satisfied, $X' \cdot \lambda_I = X \cdot \lambda_I$ holds in half of the cases, as the cipher behaves randomly.

2. There is independence between the two subciphers E_0 and E_1. In particular, the bias of the linear approximations in E_1 is not affected by the fact that they are applied to two intermediate values which correspond to plaintexts with a fixed difference.

As for the first assumption, already in [4] the authors noted that it may fail in many cases, and suggested to check the overall bias of the approximation experimentally whenever possible. Several subsequent papers aimed at formalizing the assumption and at taking into consideration multiple linear approximations for E_1 instead of a single one. The first of those were by Liu et al. [29] and by Lu [30]. Recently, Blondeau et al. [10] presented a formal treatment of the DL attack, based on a general link between differential and linear attacks introduced by Chabaud and Vaudenay [13] and developed by Blondeau and Nyberg [11]. The formal treatment provides an exact expression for the bias of the approximation under the sole assumption that the two parts of the cipher are independent.

Independence Between the Subciphers in the Boomerang Attack. While the assumption on independence between E_0 and E_1 was not studied in previous works on the DL attack, it was studied for another combined attack – the boomerang attack. In 2011, Murphy [33] showed that in various cases of interest, the dependency between E_0 and E_1 may significantly affect the complexity and even the possibility of success of the boomerang attack. Murphy's claims were supported by several concrete examples given in other papers. In particular, in [9] and [21], dependency between the subciphers was used to significantly reduce the complexity of the boomerang attacks on SAFER++ and on KASUMI, respectively. On the other hand, it was shown in [25] that the boomerang attack on KASUMI presented in [6] fails (i.e., never succeeds), due to dependency between the subciphers. In [21], Dunkelman et al. proposed the sandwich framework in order to take into account the dependency between the subciphers in the attack analysis.

.The Boomerang Connectivity Table (BCT). The inspiration to our work comes from the *boomerang connectivity table* (BCT), proposed by Cid et al. [14] at Eurocrypt'2018 as a new tool for the boomerang attack. The BCT allows computing the complexity of the boomerang attack more accurately, and moreover, enables the adversary to choose the differentials of E_0 and E_1 in a way that exploits the dependency between the subciphers to amplify the overall probability of the boomerang distinguisher. Cid et al. applied the BCT to improve the boomerang attack on the CAESAR finalist Deoxys [23] and to explain an unsolved probability amplification for generating a quartet in the tweakable block cipher SKINNY [2].

1.2 Our Results

In this paper we study the effect of dependency between the subciphers on differential-linear cryptanalysis.

Inaccuracy of Previous Analysis Due to Dependency Between the Subciphers.
We show that in differential-linear attacks on several cryptosystems, due to the
effect of dependency, complexity analysis using the standard DL framework led to
incorrect estimates, which sometimes were very far from the correct value found
experimentally. One concrete example is the attack of Dobraunig et al. [19] on a
5-round variant of the CAESAR finalist Ascon [18]. The authors of [19] state that
while by the theory of the DL attack, the overall bias of their approximation
is expected to be 2^{-20}, experiments show that the bias is significantly larger:
2^{-2}. The discrepancy is attributed in [19] to multiple linear approximations that
affect the overall bias. We show that the huge discrepancy comes mainly from
dependency between the two subciphers, and in fact, when we take dependency
into account using our new tool presented below, the bias estimate is increased
from 2^{-20} all the way to 2^{-5}. (The rest of the difference is indeed explained by
the effect of other approximations, as claimed in [19]).

The Differential-Linear Connectivity Table. In order to (partially) take the
effects of dependency into account, we introduce a new tool: the *differential-
linear connectivity table* (DLCT). For a vectorial Boolean function $S : \{0,1\}^n \rightarrow$
$\{0,1\}^m$ (e.g., an n-to-m bit S-box in a block cipher), the DLCT of S is an $2^n \times 2^m$
table whose rows correspond to input differences to S and whose columns correspond to bit masks of outputs of S. The value in the cell (Δ, λ), where Δ is a
difference and λ is a mask, is

$$DLCT_S(\Delta, \lambda) = |\{x : S(x) \cdot \lambda = S(x \oplus \Delta) \cdot \lambda\}| - 2^{n-1}.$$

We replace the decomposition $E = E_1 \circ E_0$ used in the standard DL attack by
the decomposition $E = E_1' \circ E_m \circ E_0'$, where E_0' is covered by the differential,
E_m is covered by the DLCT, and E_1' is covered by the remainder of the linear
approximation. Usually, E_m covers the first round of E_1 and thus consists of
several DLCTs of individual S-boxes applied in parallel. In this case, when computing the overall bias of the DL distinguisher, we replace the biases computed
in the first round of the linear approximation by the entries of the DLCT in the
corresponding S-boxes. Thus, the DLCT fully addresses the issue of dependency
in the switch between E_0 and E_1. Note however that it does not resolve the
possible effect of other characteristics and approximations, which still has to be
handled using the framework of Blondeau et al. [10] (see Sect. 2).

Relation of the DLCT to the Fourier Transform. We show that each row of the
DLCT is equal (up to normalization) to the Fourier transform of the Boolean
function represented by the corresponding row of the Differential Distribution
Table (DDT) constructed in differential cryptanalysis.

As a result, the DLCT can be computed efficiently using the Fast Fourier
Transform. Specifically, each row of the DLCT can be constructed in time $O(2^n +
m2^m)$ operations (instead of the trivial 2^{m+n}), and thus, the entire DLCT can
be computed in time $O(2^{2n} + m2^{m+n})$ operations. This makes computation of
the DLCT feasible even for larger encryption units (e.g., when one wants to
compute a single row of the 32-bit Super S-box of AES [36]).

Applications of the DLCT. While the basic use of the DLCT is obtaining a more accurate complexity analysis of the DL attack, it can be used to obtain improved DL attacks as well. Indeed, the adversary can use the DLCT to choose the differential for E_0 and the linear approximation for E_1 in a way that exploits the dependency between the subciphers in her favor. We demonstrate this on two concrete ciphers.

Improved DL Attack on ICEPOLE. ICEPOLE [32] is a hardware-oriented authenticated cipher designed by Morawiecki et al. in 2014 and submitted as a candidate to the CAESAR competition. In [22], Huang et al. presented a state-recovery attack in the repeated-nonce settings on 128-bit ICEPOLE with data and time complexity of about 2^{46}, using differential-linear cryptanalysis. This attack is the best known attack on ICEPOLE.

We show that by using better differentials which exploit the dependency between the two underlying subciphers, one can reduce the complexity of the attack to 2^{42}. Furthermore, by exploiting using a better method for choosing the plaintexts, the complexity can be further reduced 2^{36}. We have fully implemented and verified our attack.

Improved DL Attack on 8-round DES. One of the first applications of the DL technique is an attack on 8-round DES [35] presented by Biham et al. [4]. The attack is based on a 7-round differential-linear distinguisher with bias $2^{-5.91}$. By analyzing the DLCT of the DES S-boxes, we show that the differential and the linear approximation used in the attack can be replaced which leads to an improved bias of $2^{-5.6}$, thus reducing the complexity of the attack from about 30,000 plaintexts to about 20,000 plaintexts.

As in the case of ICEPOLE, we have fully implemented and verified the attack. While the improvement of our attack over the result of [4] is rather modest, it is another clear example of the applicability of the DLCT and of its ability to exploit dependency in favor of the adversary.

1.3 Organization of the Paper

The rest of the paper is organized as follows. In Sect. 2 we give an overview of the DL attack, and then we present the DLCT and prove that it can be computed efficiently. We use the newly introduced tool to revisit the cryptanalytic results on Ascon [19] in Sect. 3 and on Serpent [5,20] in Sect. 4, and explain the discrepancy between the theoretical estimate and the experimental results in these two works. In Sects. 5 and 6 we present improved DL attacks on ICE-POLE and reduced-round DES, respectively. We conclude the paper with a few open problems for future research in Sect. 7.

2   The Differential-Linear Connectivity Table

In this section we introduce and discuss the DLCT. We begin with an overview of the DL attack, then we present the DLCT and obtain a new formula for the bias of the DL distinguisher, and finally, we discuss the relation of the DLCT to the Fourier transform and its implications on the DL technique.

2.1   The Differential-Linear Attack

Let E be a cipher that can be decomposed into a cascade $E = E_1 \circ E_0$. Assume that we have a differential $\Delta_I \xrightarrow{p} \Delta_O$ for E_0, i.e., an input difference Δ_I to E_0 leads to an output difference Δ_O from E_0 with probability p, and a linear approximation $\lambda_I \xrightarrow{q} \lambda_O$ for E_1, i.e., for $1/2+q$ of the input/output pairs (I_i, O_i) of E_1 satisfy $\lambda_I \cdot I_i = \lambda_O \cdot O_i$. Denote plaintexts by P, P', ciphertexts by C, C', and intermediate values between E_0 and E_1 by X, X', respectively.

The Procedure of the DL Attack. As mentioned above, the attack procedure is very simple. In order to distinguish E from a random permutation, the adversary considers plaintext pairs (P, P') such that $P \oplus P' = \Delta_I$ and checks whether the corresponding ciphertext pairs (C, C') satisfy $C \cdot \lambda_O = C' \cdot \lambda_O$. Following Blondeau et al. [10], we denote the overall bias of the DL distinguisher by

$$\mathcal{E}_{\Delta_I,\lambda_O} = \Pr[C \cdot \lambda_O = C' \cdot \lambda_O | P \oplus P' = \Delta_I]. \tag{1}$$

Naive Analysis of the Attack Complexity. The attack uses a combination of three approximations:

1. We have $\Pr[C \cdot \lambda_O = X \cdot \lambda_I] = \frac{1}{2} + q$, by the linear approximation for E_1.
2. We have $\Pr[X' \cdot \lambda_I = X \cdot \lambda_I] = \frac{1}{2} \pm \frac{p}{2}$ (where the sign depends on the parity of $\Delta_O \cdot \lambda_I$). This is because by the differential for E_0, for fraction p of the plaintext pairs we have $X \oplus X' = \Delta_O$, and in particular, $X' \cdot \lambda_I = X \cdot \lambda_I \oplus \Delta_O \cdot \lambda_I$, and we assume that among the rest of the plaintext pairs, $X' \cdot \lambda_I = X \cdot \lambda_I$ holds in half of the cases.
3. We have $\Pr[C' \cdot \lambda_O = X' \cdot \lambda_I] = \frac{1}{2} + q$, by the linear approximation for E_1.

Note that the equality $C \cdot \lambda_O = C' \cdot \lambda_O$ holds if among the three equalities (1),(2), (3), either all three hold or exactly one holds. Using Matsui's Piling-up lemma [31] (similar analysis holds also when using correlation matrices [17]), we have

$$\mathcal{E}_{\Delta_I,\lambda_O} = \Pr[C \cdot \lambda_O = C' \cdot \lambda_O] = \frac{1}{2} + 2pq^2. \tag{2}$$

Hence, if p, q are sufficiently large, then the adversary can distinguish E from a random permutation using $O(p^{-2}q^{-4})$ chosen plaintexts (see [10,37] for the exact relation between the data complexity and the success probability of the distinguisher). As usual, the distinguisher can be transformed into a key recovery attack by guessing some key material, performing partial encryption/decryption, and applying the distinguisher.

The Exact Complexity Analysis of Blondeau et al. [10]. As mentioned above, the naive complexity analysis crucially depends on two randomness assumptions. The first is that the equality $X' \cdot \lambda_I = X \cdot \lambda_I$ holds in approximately half of the cases in which the differential in E_0 fails; the second is that E_0 and E_1 are independent. Blondeau et al. [10] provided an exact expression for $\mathcal{E}_{\Delta_I, \lambda_O}$, relying only on the latter assumption. In order to present their result, we need a few more notations (adapted from [10]).

Consider an encryption function E' and denote its inputs by Z, Z' and its outputs by W, W'. We use the notation $\Delta_I \xrightarrow{E'} \Delta_O$ for the differential transition $\Delta_I \to \Delta_O$ through E', and the notation $\lambda_I \xrightarrow{E'} \lambda_O$ for the linear transition $\lambda_I \to \lambda_O$ through E'. For an input difference Δ_I and an output mask λ, we denote

$$\epsilon_{\Delta_I, \lambda}^{E'} = \Pr[W \cdot \lambda = W' \cdot \lambda | Z \oplus Z' = \Delta_I] - \frac{1}{2},$$

and for two masks λ_I, λ_O, we denote

$$c_{\lambda_I, \lambda_O}^{E'} = 2 \left(\Pr[W \cdot \lambda_O = W' \cdot \lambda_O | Z \cdot \lambda = Z' \cdot \lambda] - \frac{1}{2} \right).$$

Note that $c_{\lambda_I, \lambda_O}^{E'}/2$ is the bias of the linear approximation $\lambda_I \xrightarrow{E'} \lambda_O$.

By [10, Theorem 2], assuming only independence between E_0 and E_1, we have:

$$\mathcal{E}_{\Delta_I, \lambda_O} = \sum_{\lambda_I} \epsilon_{\Delta_I, \lambda_I}^{E_0} (c_{\lambda_I, \lambda_O}^{E_1})^2. \tag{3}$$

Of course, the expression (3) is usually hard to evaluate, and thus, in practice one mostly has to rely (at least partially) on randomness assumptions and verify the results experimentally.

2.2   The Differential-Linear Connectivity Table and Its Properties

Definition of the DLCT. Let $S : \{0,1\}^n \to \{0,1\}^m$ be a vectorial Boolean function. The DLCT of S is an $2^n \times 2^m$ table whose rows correspond to input differences to S and whose columns correspond to bit masks of outputs of S. Formally, for $\Delta \in \{0,1\}^n$ and $\lambda \in \{0,1\}^m$, the DLCT entry (Δ, λ) is

$$DLCT_S(\Delta, \lambda) \triangleq \left| \left\{ x \middle| \lambda \cdot S(x) = \lambda \cdot S(x \oplus \Delta) \right\} \right| - 2^{n-1}.$$

Sometimes it will be more convenient for us to use the normalized DLCT entry

$$\overline{DLCT}_S(\Delta, \lambda) \triangleq \frac{DLCT_S(\Delta, \lambda)}{2^n} = \Pr[\lambda \cdot S(x) = \lambda \cdot S(x \oplus \Delta)] - \frac{1}{2}$$

instead of $DLCT_S(\Delta, \lambda)$. The DLCT of Serpent's S-box S0 is given in Table 1.

A natural interpretation of the DLCT is the following. Assume that S is equal to the entire encryption function E. Then $DLCT_S(\Delta, \lambda)$ is equal (up to normalization) to the bias we obtain when we apply to E a DL distinguisher

Table 1. The DLCT of Serpent's S0

$\Delta \backslash \lambda$	0_x	1_x	2_x	3_x	4_x	5_x	6_x	7_x	8_x	9_x	A_x	B_x	C_x	D_x	E_x	F_x
0_x	8	8	8	8	8	8	8	8	8	8	8	8	8	8	8	8
1_x	8	0	-4	0	-4	-4	0	4	0	-4	0	0	0	4	0	0
2_x	8	0	0	0	-4	0	0	-4	-8	0	0	0	4	0	0	4
3_x	8	-4	0	0	4	-4	0	-4	0	0	-4	0	0	4	0	0
4_x	8	0	0	-8	0	0	0	0	-8	0	0	8	0	0	0	0
5_x	8	4	0	0	0	0	-4	0	0	0	4	0	-4	0	-4	-4
6_x	8	-4	-4	0	0	0	0	0	8	-4	-4	0	0	0	0	0
7_x	8	0	4	0	0	0	-4	0	0	4	0	0	-4	0	-4	-4
8_x	8	-4	0	0	-4	0	-4	4	0	0	-4	0	0	0	4	0
9_x	8	0	0	-8	0	0	0	0	0	0	0	0	0	0	0	0
A_x	8	0	-4	0	4	0	-4	-4	0	-4	0	0	0	0	4	0
B_x	8	0	0	0	-4	0	0	-4	0	0	0	-8	4	0	0	4
C_x	8	0	4	0	0	-4	0	0	0	4	0	0	-4	-4	0	-4
D_x	8	-4	-4	8	0	4	4	0	0	-4	-4	0	0	-4	-4	0
E_x	8	4	0	0	0	-4	0	0	0	0	4	0	-4	-4	0	-4
F_x	8	0	0	0	0	4	4	0	0	0	0	-8	0	-4	-4	0

with $\Delta_I = \Delta$ and $\lambda_O = \lambda$ (that is, to the bias $\mathcal{E}_{\Delta,\lambda}$). Thus, if we could construct a DLCT for the entire encryption scheme E, then the DLCT would completely capture the DL attack. As such a construction is mostly infeasible, we construct the DLCT for small components of the cryptosystem (usually, single S-boxes or Super S-boxes) that lie on the boundary between E_0 and E_1, in order to obtain accurate analysis of the transition between the two subciphers.[1]

The DLCT Framework. Like in the sandwich [21] and the BCT [14] frameworks of the boomerang attack, when we use the DLCT, we divide the cipher E into three subciphers: $E = E_1' \circ E_m \circ E_0'$, where E_0' is covered by the differential $\Delta_I \rightarrow \Delta$, E_m is covered by the DLCT (or by several DLCTs applied in parallel), and E_1' is covered by the remainder of the linear approximation $\lambda \rightarrow \lambda_O$. Usually, E_m covers the first round of E_1 and thus it consists of several DLCTs of single S-boxes applied in parallel. However, if it is feasible to cover by the DLCT a larger part of the cipher, this is advantageous, as the DLCT gives the exact result for the part of the cipher it covers. For example, we construct such a (partial) DLCT for three rounds in our improved DL attack on 8-round DES presented in Sect. 6.

---

[1] An important independence assumption on the transition is that the active S-boxes (with non-zero input difference and non-zero output) of the transition are independent of each other.

Complexity Analysis. Assume that we have a differential $\Delta_I \xrightarrow{p} \Delta$ for E_0' and a linear approximation $\lambda \xrightarrow{q'} \lambda_O$ for E_1'. (Note that since E_1' is typically a subcipher of E_1, it is expected that $|q'| > |q|$). Denote the intermediate values after E_0' by X, X' and the intermediate values after E_m by Y, Y'.

Adapting the naive analysis of the DL attack complexity presented above (i.e., Eq. (2)), we obtain

$$\mathcal{E}_{\Delta_I, \lambda_O} = 4p \cdot \frac{DLCT_{E_m}(\Delta, \lambda)}{2^n} \cdot (q')^2 = 4p \cdot \overline{DLCT}_{E_m}(\Delta, \lambda) \cdot (q')^2. \quad (4)$$

Note that in the degenerate case where $E_m = Id$, we have $DLCT_{E_m}(\Delta, \lambda) = 2^{n-1}$ for all (Δ, λ) and $q' = q$, and thus, we obtain $\mathcal{E}_{\Delta_I, \lambda_O} = 2pq^2$ which is equivalent to Eq. (2) above. Interestingly, when $E_1' = Id$, the resulting bias is $\mathcal{E}_{\Delta_I, \lambda_O} = p \cdot \overline{DLCT}_{E_m}(\Delta, \lambda_O)$.

In order to adapt the exact analysis of [10] (i.e., Eq. (3)), a bit more computation is needed. Note that for any λ, we have

$$\epsilon_{\Delta_I, \lambda}^{E_m \circ E_0'} = \sum_{\Delta} (\Pr[\Delta_I \xrightarrow{E_0'} \Delta] \Pr[Y \cdot \lambda = Y' \cdot \lambda | X \oplus X' = \Delta]) - \frac{1}{2}$$

$$= \frac{1}{2^n} \sum_{\Delta} \Pr[\Delta_I \xrightarrow{E_0'} \Delta] \cdot DLCT_{E_m}(\Delta, \lambda) = \sum_{\Delta} \Pr[\Delta_I \xrightarrow{E_0'} \Delta] \cdot \overline{DLCT}_{E_m}(\Delta, \lambda).$$

Plugging this expression into Eq. (3) (where E_1' is used instead of E_1 and $E_m \circ E_0'$ is used instead of E_0), we obtain that the exact bias of the DL distinguisher is

$$\mathcal{E}_{\Delta_I, \lambda_O} = \sum_{\Delta, \lambda} \Pr[\Delta_I \xrightarrow{E_0'} \Delta] \cdot \overline{DLCT}_{E_m}(\Delta, \lambda)(c_{\lambda, \lambda_O}^{E_1'})^2. \quad (5)$$

Note that Eq. (5) is still not free of randomness assumptions; e.g., it relies on round independence within E_0' and E_1' (see [10]). However, it is the most accurate expression for the bias of the DL distinguisher obtained so far. On the other hand, Eq. (5) is usually hard to evaluate, and in the actual applications of the DLCT we do rely on some randomness assumptions and verify our results experimentally.

Properties of the DLCT. A trivial property of the DLCT is that for any S, we have $DLCT_S(0, \lambda) = 2^{n-1}$ for all λ. Indeed, if two inputs to S are equal then the corresponding outputs agree on any bit mask. This means that in the DL attack, if for some S-box in the first round of E_1, the difference Δ_O is zero in the entire S-box, then the linear approximation in that S-box holds for sure, while without the dependency between the intermediate values X, X' it was anticipated to hold only probabilistically. This trivial property corresponds to the *middle round S-box trick* used in [9] to speed up the boomerang attack and covered by the BCT [14]. Surprisingly, this feature was not noted before explicitly in the context of the DL attack. For example, even if we take into account only this trivial type of dependency, the bias of the DL distinguisher of Dobraunig et al. [19] on Ascon increases from 2^{-20} to 2^{-8}. Interestingly, the authors of [19] chose the linear approximation deliberately in such a way that the active S-boxes in its first round correspond to inactive S-boxes in Δ. However, they did not take this dependency into account in the computation of the bias, as it is neglected in the classical DL model.

Another trivial property of the DLCT is that for any S, we have $DLCT_S(\Delta, 0) = 2^{n-1}$ for all Δ's. Indeed, if the the input mask is zero (i.e., no output bits are approximated), then their actual value (and by proxy, their input difference), is of no importance.

Inspection of the DLCT of Serpent's S-box S0 presented in Table 1 shows that it contains the value $\pm 2^{n-1}$ not only in the trivial entries of the form $DLCT_S(0, \lambda)$ or $DLCT_S(\Delta, 0)$, but also in 9 additional places. Moreover, it contains many very high/very low values that can be used by an adversary, if she can adjust the differential and the linear approximation such that these high/low values are used. The existence of such high/low value entries is not surprising, as current design of S-boxes does not take the DLCT into account. Hence, the DLCT can serve as a new design criterion for S-boxes, partially measuring immunity of the cipher with respect to DL attacks.

2.3   Relation of the DLCT to the Fourier Transform

We now show that the DLCT is closely related to the Fourier transform of the DDT and that this relation can be used to efficiently compute the DLCT. We begin with a few preliminaries.

The Fourier-Walsh Transform of Boolean Functions. Let $f : \{0,1\}^m \to \mathbb{R}$ be a Boolean function. (Note that f does not have to be two-valued.) The Fourier-Walsh transform of f is the function $\hat{f} : \{0,1\}^m \to \mathbb{R}$ defined by

$$\hat{f}(y) = \frac{1}{2^m} \sum_{x \in \{0,1\}^m} f(x) \cdot (-1)^{x \cdot y} = \frac{1}{2^m} \left(\sum_{\{x : x \cdot y = 0\}} f(x) - \sum_{\{x : x \cdot y = 1\}} f(x) \right).$$

The DDT and the LAT. The DLCT resembles in its structure the two central tools used in differential and linear cryptanalysis – the Difference Distribution Table (DDT) and the Linear Approximation Table (LAT). For a vectorial Boolean function $S : \{0,1\}^n \to \{0,1\}^m$, the DDT of S is an $2^n \times 2^m$ table whose rows correspond to input differences to S and whose columns correspond to output differences of S. Formally, for $\Delta_I \in \{0,1\}^n$ and $\Delta_O \in \{0,1\}^m$, we have

$$DDT_S(\Delta_I, \Delta_O) = \left| \left\{ x \middle| S(x) \oplus S(x \oplus \Delta_I) = \Delta_O \right\} \right|.$$

Similarly, the LAT of S is an $2^n \times 2^m$ table whose rows correspond to bit masks of inputs to S and whose columns correspond to bit masks of outputs of S. Formally, for $\lambda_I \in \{0,1\}^n$ and $\lambda_O \in \{0,1\}^m$, we have

$$LAT_S(\lambda_I, \lambda_O) = \left| \left\{ x \middle| \lambda_O \cdot S(x) = \lambda_I \cdot x \right\} \right| - 2^{n-1}.$$

Relation of the DLCT to the Fourier-Walsh Transform of the DDT. We assert that each row of the DLCT is equal (up to normalization) to the Fourier-Walsh transform of the corresponding row of the DDT. Formally, for each $\Delta \in \{0,1\}^n$, denote the Boolean function which corresponds to the Δ's row of the DDT by f_Δ. That is, $f_\Delta : \{0,1\}^m \to \mathbb{R}$ is defined by

$$f_\Delta(\Delta') = DDT_S(\Delta, \Delta') = \left| \{x \in \{0,1\}^n \big| S(x) \oplus S(x \oplus \Delta) = \Delta'\} \right|.$$

Proposition 1. *For any $\lambda \in \{0,1\}^m$, we have $DLCT(\Delta, \lambda) = 2^{m-1}\hat{f}_\Delta(\lambda)$.*

Proof. By the definitions of the DLCT and of the Fourier-Walsh transform, we have

$$DLCT_S(\Delta, \lambda) = \left| \left\{ x \big| \lambda \cdot S(x) = \lambda \cdot S(x \oplus \Delta) \right\} \right| - 2^{n-1}$$

$$= \frac{1}{2} \left(\left| \left\{ x \big| \lambda \cdot S(x) = \lambda \cdot S(x \oplus \Delta) \right\} \right| - \left| \left\{ x \big| \lambda \cdot S(x) \neq \lambda \cdot S(x \oplus \Delta) \right\} \right| \right)$$

$$= \frac{1}{2} \left(\left| \left\{ x \big| \lambda \cdot (S(x) \oplus S(x \oplus \Delta)) = 0 \right\} \right| - \left| \left\{ x \big| \lambda \cdot (S(x) \oplus S(x \oplus \Delta)) = 1 \right\} \right| \right)$$

$$= \frac{1}{2} \left(\sum_{\{\Delta':\Delta'\cdot\lambda=0\}} f_\Delta(\Delta') - \sum_{\{\Delta':\Delta'\cdot\lambda=1\}} f_\Delta(\Delta') \right) = 2^{m-1}\hat{f}_\Delta(\lambda),$$

as asserted.   ∎

A Theoretical Implication. The relation of the DLCT to the Fourier-Walsh transform of the DDT yields an interesting theoretical insight on the differential-linear attack.

It is well-known that the DDT and the LAT have the following mathematical interpretations.

- *The DDT:* If we model the evolution of differences through the encryption process of a plaintext pair as a *Markov chain*, then the DDT is simply the *transition matrix* of the chain (see, e.g., [26]). In this regard, a differential attack utilizes a classical probability-theoretic tool for cryptanalysis.
- *The LAT:* For each mask λ, the λ's column of the LAT is equal (up to normalization) to the Fourier-Walsh transform of the Boolean function $x \mapsto \lambda \cdot S(x)$ (see, e.g., [15]). In this regard, linear cryptanalysis studies the function S via its Fourier transform, as is commonly done in Boolean function analysis (see, e.g., [34]).

Proposition 1 shows that each row of the DLCT is equal (up to normalization) to the Fourier-Walsh transform of the corresponding row of the DDT. Since the DLCT of the entire encryption scheme E completely captures DL attacks as shown above, this implies that the differential-linear attack utilizes an interesting combination of probabilistic and Fourier-analytic techniques: it considers the probability-theoretic transition matrix of a Markov chain associated with the function, and studies it via its Fourier-Walsh transform.

A Practical Implication. It is well-known that the Fourier-Walsh transform of a function $f : \{0, 1\}^m \rightarrow \mathbb{R}$ can be computed in time $O(m \cdot 2^m)$ operations. Since each row of the DDT of S can be easily constructed in time $O(2^n)$ operations, Proposition 1 implies that each row of the DLCT can be computed in time $O(2^n + m2^m)$ operations, and that the entire DLCT can be computed in time $O(2^{2n} + m2^{m+n})$ operations. This significantly improves over the trivial algorithm which requires $O(2^{2n+m})$ operations.

This speedup is practically important as it allows us to compute the DLCT for larger parts of the cipher, and thus, obtain a more accurate estimate of the complexity of the DL attack. For example, in the attack on 8-round DES presented in Sect. 6, we compute one DLCT entry for three rounds of DES as a single unit, and so the ability to compute the DLCT efficiently is crucial.

3 Differential-Linear Cryptanalysis of Ascon, Revisited

Ascon [18] is an authenticated encryption algorithm that was recently selected to the final round of the CAESAR [16] competition. In [19], Dobraunig et al. presented a practical differential-linear attack on up to 5 rounds of the Ascon permutation, based on a 4-round DL distinguisher. The authors of [19] stated that while by the theory of the DL attack, the overall bias of the approximation is expected to be 2^{-20}, experiments show that the bias is significantly higher – 2^{-2}. They attributed the difference between practice and the theoretical estimate to multiple linear approximations that affect the overall bias, and used the correct value in their attack. We recompute the bias of the distinguisher using the DLCT and show that a large part of the discrepancy results from dependency between the two subciphers.

In order to recompute the bias, we have to provide some more details on the specific distinguisher used in [19]. We present the distinguisher only schematically.

The Theoretical Analysis of [19]. The DL distinguisher of [19] targets a 4-round reduced variant of Ascon denoted by E and decomposed as $E = E_1 \circ E_0$, where E_0 consists of rounds 1–2 and E_1 consists of rounds 3–4. For E_0, the distinguisher uses a differential characteristic of the form

$$\Delta_0 \xrightarrow[L \circ S]{p_0 = 2^{-2}} \Delta_1 \xrightarrow[L \circ S]{p_1 = 2^{-3}} \Delta_2,$$

where Δ_2 is a truncated difference. For E_1, the distinguisher uses a linear approximation of the form

$$\lambda_0 \xrightarrow[L \circ S]{q_0 = 2^{-7}} \lambda_1 \xrightarrow[L \circ S]{q_1 = 2^{-2}} \lambda_2,$$

where λ_2 consists of a single bit, and all nonzero bits of the mask λ_0 are included in S-boxes in which the all the input bits are known to be zero in Δ_2. Using the naive complexity analysis of the DL attack (i.e., Eq. (2) above), the authors of [19] concluded that the theoretical estimate for the overall bias of the approximation is $2 \cdot 2^{-5} \cdot (2^{-8})^2 = 2^{-20}$. On the other hand, they found experimentally that the bias is as high as 2^{-2}.[2]

---

[2] We emphasize that the results of [19] were not affected by the theoretical estimate, since the authors of [19] used the experimentally verified value instead of the theoretically computed value.

Partial Explanation of the Discrepancy Using Only the Trivial Property of the DLCT. As mentioned above, the linear approximation of E_1 was chosen by Dobraunig et al. in such a way that all nonzero bits of the mask λ_0 are included in S-boxes in which all the input bits are known to be zero in Δ_2. By the trivial property of the DLCT presented in Sect. 2, this implies that the linear approximation in round 3 holds with bias $1/2$, instead of the theoretical bias $q_0 = 2^{-7}$. Therefore, the estimated bias of the approximation is increased to $2 \cdot 2^{-5} \cdot (2^{-2})^2 = 2^{-8}$, which is already much higher than 2^{-20}.

Analysis Using the DLCT. We now obtain a higher bias of 2^{-5} by revisiting the analysis using the DLCT. Let us decompose E into $E = E_2 \circ E_m \circ E_0$, where E_0 consists of rounds 1–3, E_m consists of round 4, and $E_2 = Id$. Note that since in the DL distinguisher of [19], the output mask λ_2 consists of the MSB in the output of S-box no. 9, we are only interested in the entries of the DLCT of that S-box (which we denote by S). For E_0, we use a differential of the form $\Delta_0 \xrightarrow[\text{3 rounds}]{p = 2^{-3}} \Delta_3$, where Δ_0 is the input difference of the DL distinguisher of [19]. In our value of Δ_3, three of the input bits to S-box no. 9 are known to be zero; specifically, the input is of the form ?0?00. (Note that the S-box is from 5 bits to 5 bits). The relevant normalized entries of the DLCT of S satisfy:

$$\overline{DLCT}_S(16, 16) = 0, \overline{DLCT}_S(4, 16) = 0,$$
$$\overline{DLCT}_S(20, 16) = 2^{-1}, \text{ and } \overline{DLCT}_S(0, 16) = 2^{-1}.$$

Hence, assuming that each input difference of S occurs in Δ_3 with the same probability 2^{-2} and using Eq. (4), we obtain the estimate

$$4 \cdot 2^{-3} \cdot 2^{-2}(0 + 0 + 2^{-1} + 2^{-1}) \cdot (2^{-1})^2 = 2^{-5}$$

for the overall bias of the DL distinguisher of [19]. This value is, of course, much lower than the experimentally obtained bias of 2^{-2} (which may be explained by the effect of other differentials and linear approximations). On the other hand, it is significantly higher than the value 2^{-20} which follows from the classical DL framework. This demonstrates the importance of taking the dependency between subciphers into account, which the DLCT facilitates in an easy manner.

4   Differential-Linear Cryptanalysis of Serpent, Revisited

One of the first applications of the DL technique is an attack on the AES finalist Serpent [1] presented in [5]. The attack is based on a 9-round DL distinguisher with bias of 2^{-59} and targets an 11-round variant of the cipher. An improved attack was presented in [20]. The authors of [20] performed experiments with reduced round variants of Serpent, and concluded that the actual bias of the approximation is $2^{-57.75}$ and not 2^{-59}. Using the improved bias, they extended the attack to 12 rounds of Serpent (out of its 32 rounds) yielding the best currently known attack on the cipher.

In [20], the increased bias was attributed to the existence of other approximations that affect the overall bias. In this section we recompute the bias of the distinguisher using the DLCT and obtain the value $2^{-57.68}$, which is very close to the experimental value. Thus, we conclude that the increased bias in the experiment results mostly from the dependency between the two subciphers.

In order to recompute the bias, we have to provide some more details on the specific distinguisher used in [5]. For sake of clarity, we present it only schematically, and refer the reader to [5] for the exact difference and mask values.

The Analysis of [5]. The DL distinguisher of [5] targets a 9-round reduced variant of Serpent that starts with round 2 of the cipher. This variant is denoted by E and decomposed as $E = E_1 \circ E_0$, where E_0 consists of rounds 2–4 and E_1 consists of rounds 5–10. For E_0, the distinguisher uses a differential characteristic of the form

$$\Delta_0 \xrightarrow[\;L \circ S_2\;]{p_0 = 2^{-5}} \Delta_1 \xrightarrow[\;L \circ S_3\;]{p_1 = 2^{-1}} \Delta_2 \xrightarrow[\;L \circ S_4\;]{p_2 = 1} \Delta_3,$$

where Δ_2, Δ_3 are truncated differences. For E_1, the distinguisher uses a linear approximation of the form

$$\lambda_0 \xrightarrow[\;L \circ S_5\;]{q_0 = 2^{-5}} \lambda_1 \xrightarrow[\;5 \text{ rounds}\;]{q_1 = 2^{-23}} \lambda_7,$$

where all nonzero bits of the mask λ_0 are included in the bits that are known to be zero in Δ_3. Using the naive complexity analysis of the DL attack (i.e., Eq. (2) above), the authors of [5] concluded that the overall bias of the approximation is $2 \cdot 2^{-6} \cdot (2^{-27})^2 = 2^{-59}$. (Actually, in their attack they used the lower value of 2^{-60} due to the effect of other differentials.)

The Experimental Results of [20]. The authors of [20] checked experimentally the first 4 rounds of the DL distinguisher of [5] (i.e., a 4-round distinguisher which starts with the difference Δ_0 and ends with the mask λ_1) and found that its bias is $2^{-13.75}$, instead of the theoretical estimate $2 \cdot 2^{-6} \cdot (2^{-5})^2 = 2^{-15}$. They concluded that the overall bias of the 9-round distinguisher is $2^{-57.75}$ instead of 2^{-59}, and used the conclusion to extend the key-recovery attack based on the distinguisher from 11 rounds to 12 rounds.

Analysis Using the DLCT. We considered a 3-round variant of Serpent that starts at round 3, denoted it by E', and computed the normalized DLCT entry $\overline{DLCT}_{E'}(\Delta_1, \lambda_1)$. (Note that computing the entire DLCT for E' is infeasible. However, due to the low diffusion of Serpent, one can compute efficiently part of the entries, including the entry we needed). We obtained $\overline{DLCT}_S(\Delta_1, \lambda_1) = 2^{-8.68}$.

Using Eq. (4) (for the case of $E'_1 = Id$) we conclude that the bias of the 4-round distinguisher examined in [20] is $p_1 \cdot \overline{DLCT}_S(\Delta_1, \lambda_1) = 2^{-5} \cdot 2^{-8.68} = 2^{-13.68}$, which is very close to the experimental result $2^{-13.75}$ of [20].

Note that we obtained an estimate which is very close to the actual value, using only the naive Eq. (4) and not the more accurate Eq. (5) that takes into account the effect of other differentials. This shows that the increased bias found experimentally in [20] follows almost solely from dependency between the subciphers, and demonstrates how the DLCT methodology can be used for obtaining an accurate estimate of the DL attack complexity.

5 Improved Differential-Linear Attack on ICEPOLE

ICEPOLE is an authenticated encryption cipher based on the duplex construction proposed by Morawiecki et al. submitted to the CAESAR competition [32]. The main two versions, ICEPOLE128 and ICEPOLE128a are initialized with a 128-bit key. In addition ICEPOLE128 accepts 128-bit nonce and 128-bit secret message number, in comparison, ICEPOLE128a accepts 96-bit nonce and 0-bit secret message number (to serve as a drop-in replacement for AES-128-GCM).[3] After initialization, the associated data is absorbed into the 1280-bit state. For the processing of the plaintext (encryption and authentication), a block of 1024 bits is extracted (to be XORed to the plaintext) and the plaintext is XORed into 1024 bits of the state.[4] This state is then updated using 6-round Permutation P_6 which iterates a round function P 6 times over 1280 bits. After the entire plaintext has been encrypted, a tag is produced by extracting bits of the internal state. The entire process is depicted in Fig. 2.[5]

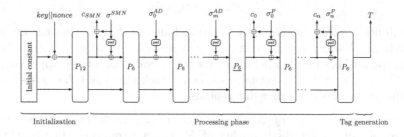

Fig. 2. General structure of ICEPOLE

After ICEPOLE has been introduced, Huang et al. presented a differential-linear attack against ICEPOLE-128 and ICEPOLE-128a [22]. The attack recovers the internal state using a differential-linear attack, where the bias of the differential-linear depends on the value of some bits. Hence, observing the bias in the output allows identifying internal state bits. After full recovery of the internal state, one can extract the secret key (as long as the scheme is not using secret message numbers) or forge new messages (when using secret message numbers).

---

[3] We note that ICEPOLE256a is a variant designed to serve as a drop-in replacement for AES-256-GCM, thus it has the same parameters as AES-256-GCM.

[4] Actually, two additional bits are appended – the frame bit which is set to 0 in all blocks but the last authenticated data block and the last message block, and a padding bit, but their rule and effect on the attack are negligible.

[5] We disregard the exact initialization and the handling of associated data which are of no relevance to this paper. The interested reader is referred to [32] for more information.

5.1   A Short Description of ICEPOLE-128

We first note that there are three variants of ICEPOLE (ICEPOLE-128, ICEPOLE-128a, and ICEPOLE-256), but our attacks and description concern only the 128-bit variants, ICEPOLE-128 and ICEPOLE-128a.

The internal state of ICEPOLE, denoted by S is composed of 20 64-bit words organized in a 4-by-5 matrix. We follow [32] notations: the bit $S[x][y][z]$ is the z'th bit of the word at position (x, y) where $0 \leq x \leq 3$, $0 \leq y \leq 4$, and $0 \leq z \leq 63$. This bit is considered to be bit $64(x + 4y) + z$ of the state. The first n bits of S are denoted by by $S_{\lfloor n \rfloor}$. The z'th *slice* of S is a 4-by-5 binary matrix $(S[x][y][z])_{x,y}$. When z and x are fixed, the 5-bit vector $S([x][y][z])_y$ is called a row.

The round function P is composed of five operations, $P = \kappa \circ \psi \circ \pi \circ \rho \circ \mu$ which are:

- μ operates on each of the 64 slices independently by treating each 20-bit slice as a vector $(Z_0, Z_1, Z_2, Z_3) \in GF(2^5)$ and multiplying this vector by the MDS matrix

$$M = \begin{pmatrix} 2 & 1 & 1 & 1 \\ 1 & 1 & 18 & 2 \\ 1 & 2 & 1 & 18 \\ 1 & 18 & 2 & 1 \end{pmatrix}.$$

The multiplication is done over $GF(2^5)$ (with the irreducible polynomial $x^5 + x^2 + 1$).

- ρ is a cyclic rotation applied to each of the 20 64-bit words of S. Each word (x, y) is rotated by a different constant, i.e., $S[x][y] = S[x][y] \lll \text{offsets}[x][y]$ where the table of offsets$[x][y]$ can be found in [32].
- π reorders the words in S by moving the word $S[x][y]$ into $S[x'][y']$ according to the formula:

$$\begin{cases} x' = (x + y) \bmod 4 \\ y' = (((x + y) \bmod 4) + y + 1) \bmod 5 \end{cases}$$

- ψ applies a 5-bit S-box to each of the 256 rows of the state.
- κ adds a round constant (constant[round]) to $S[0][0]$. The constants are generated by an LFSR and can be found in [32].

For the sake of clarity, we shall denote the linear parts of the round function by $\mathcal{L} = \pi \circ \rho \circ \mu$.

As mentioned before, the internal state S is initialized using a constant which is XORed with $(key\|nonce)$ value. After than P_{12} (12 rounds of P are used to mix the key and nonce into the state). If secret message numbers are used, they are encrypted using the duplex operation. The associated data is chopped into blocks of 1026-bit each (after padding). They are absorbed into the state S, and then the encryption/authentication of the plaintext takes place following the duplex operation using P_6. After the entire plaintext is processed, P_6 is applied to the internal state (or P_{12} in ICEPOLEv2), and the 128-bit tag is computed as $T_0 = S[0][0]$, $T_1 = S[1][0]$.

In the followings, we use e_i to denote a 64-bit word which is 0 in all bits, besides the i'th bit.

5.2 Huang et al.'s Differential-Linear Attack on ICEPOLE-128/ICEPOLE-128a

Huang et al. have presented a differential-linear attack against ICEPOLE-128 and ICEPOLE-128a in the repeated nonce settings [22]. As the attack recovers the internal state, if the scheme is not using a secret message number, then one can obtain the key by inverting the internal state. Otherwise, the recovery of the internal state allows encrypting/authenticating any data.

The attack targets the first application of P_6 after the plaintext is introduced by injecting differences through the plaintext block σ_0 and observing biases in the key stream used to encrypt σ_1. Its general structure is depicted in Fig. 3. For the sake of its description, we denote by I the input to P_6 (after the XOR with the plaintext) and by O the output of P_6. Moreover, we denote by ψ_i, $\mathcal{L}_i$, and κ_i the ψ, $\mathcal{L}$, and κ of the i'th round.

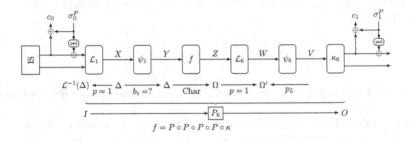

Fig. 3. Differential-linear attack on the "First" P_6 of ICEPOLE

The attack of [22] uses the 1024 bits of I and O which can be easily obtained by knowing the plaintext and ciphertext blocks for a 2-block long message. The attack introduces differences in the first plaintext block, which does not affect the fifth column of I. This difference is transformed into an input difference Δ after the application of the linear layer $\mathcal{L}$ (i.e., the introduced difference into the state is $\mathcal{L}^{-1}(\Delta)$). Due to the MDS property of μ and the zero difference in the fifth column, Δ must have at least two active S-boxes. The behavior of these two active S-boxes, namely, the probability of the differential transition through them highly depends on the actual value of some bit b_i (or two bits). Hence the input difference Δ^* of the differential-linear approximation appears with different probabilities, depending on the value of this bit. Luckily for us, this bit (or pair of bits) is the outcome of XORing an unknown input bit (from the fifth column of I) with known bits (which are controlled by the adversary). This allows the adversary to partition the plaintext pairs into sets according to the possible values of b_i, where for the "correct" set, we expect a significantly higher bias.

The differential-linear approximation $\Delta^* \rightarrow \lambda$ covers the 4 rounds until round 6, and can be extended until the end of the linear layer $\mathcal{L}_6$. We note that in ψ_1, the differential characteristic in use is $\Delta^* \rightarrow \Delta = \Delta^*$. Interestingly,

ψ has a very useful property – given the 4 least significant bits of the output, one can partially recover the input. Table 2 suggests the values, and the probability that partial information can be found given these 4 output bits. Hence, any differential-linear whose output mask can be deduced from the partial information can be used with some probability, which we denote by p_L.

Table 2. Deducing input bits of ψ from the four LSBs of the output

Output	Input				
	MSB	Bit 3	Bit 2	Bit 1	LSB
?0000	1	?	1	?	1
?0001	?	?	?	?	?
?0010	?	0	?	1	0
?0011	?	?	?	?	1
?0100	?	?	?	0	?
?0101	?	0	?	0	?
?0110	?	0	?	1	0
?0111	?	?	?	1	1
?1000	?	1	0	1	0
?1001	?	?	0	?	1
?1010	?	1	0	0	0
?1011	?	?	0	?	1
?1100	?	?	1	0	1
?1101	?	1	1	0	0
?1110	?	1	1	1	0
?1111	0	0	?	?	?
Probability (p_L)	1/8	1/2	1/2	5/8	3/4

? – unknown value

Due to the structure of ICEPOLE, any differential characteristic and any linear approximation can be rotated (by rotating each word, respectively). Hence, the attack is repeated with the 64 rotated versions of the differential-linear approximation. Each time, it recovers the bit b_i (which affects bias of the approximation). This is done by encrypting multiple pairs of plaintexts with input difference Δ with two active S-boxes (and covering all possible values of the recovered bit b_i), and observing the set which has the highest bias.

Actually, instead of taking all ciphertext pairs, only the pairs which can be used to predict the input of ψ_6 from their 1024-bit outputs O are considered (as each ciphertext can be used with probability, this is actually probability p_L^2). The different approach of extending the differential-linear approximation to cover ψ_6 leads to much worse performance (as there are many active S-boxes in the ψ_6 layer).

After studying the differentials and linear approximations that can be used for the attack, Huang et al. [22] identified 5 input difference patterns $\Delta_1, \Delta_2, \ldots, \Delta_5$ that after $\mathcal{L}$ activate only two S-boxes, as well as two good linear approximations $\lambda_1^{mid} \to \lambda_1$ and $\lambda_2^{mid} \to \lambda_2$. Given the word-oriented nature of the permutation P, one can rotate the differences (or masks) by rotating each word of the mask/difference. Hence, for each bit position it is possible to consider all the combinations of Δ_i and λ_j and experimentally calculate the bias of the resulting differential-linear.

The actual attack tries to find the last column of I (as the first four can be trivially deduced). Denote this fifth column by the four words (U_0, U_1, U_2, U_3). The first phase recovers U_0 and U_3, the second phase recovers U_2, and finally the third phase recovers U_1. All phases follow an essentially similar process – a differential-linear approximation is built from a differential characteristic which probability (in the first round) depends on the value of some specific (unknown) bits. Then, by observing enough plaintext/ciphertext pairs and evaluating the bias, one can determine the unknown bits, from which corresponding bits of U_i are computed.

Following the above steps, we give detailed explanation on how to find the first bits of U_0 and U_3 in the attack. For that phase, Huang et al. propose to use the following Δ_2:

$$\Delta_2 = \begin{pmatrix} 0 & 0 & 0 & 0 & 0 \\ 0 & 0 & 0 & 0 & 0 \\ 0 & 0 & 0 & 0 & 0 \\ 0 & e_{10} & e_{41} & 0 & 0 \end{pmatrix}$$

which under $\mathcal{L}^{-1}$ has differences in the LSBs of the words $S[0][2]$, $S[1][0]$, $S[1][1]$, $S[1][2]$, $S[1][3]$, $S[2][1]$, $S[2][3]$, $S[3][0]$, and $S[3][2]$. Another technicality is that the adversary fixes 18 bits (by knowing the first four columns of I she can select the corresponding σ^0), then the probability of the differential transition depends on two unknown bits – for one of the four possibilities it is significantly higher probability than for the rest, as presented in Table 3. The specific fixed bits and their values is given in [22].

The linear mask in use ends with the mask

$$\lambda_1 = \begin{pmatrix} 0 & 0 & 0 & 0 & 0 \\ 0 & e_{33} & 0 & 0 & 0 \\ 0 & 0 & 0 & 0 & 0 \\ 0 & 0 & 0 & 0 & 0 \end{pmatrix} \xrightarrow{\mathcal{L}_6} \lambda_1' = \begin{pmatrix} e_{18} & e_0 & 0 & e_{43} & 0 \\ 0 & e_2 & 0 & 0 & 0 \\ 0 & e_{21} & e_{61} & 0 & 0 \\ 0 & 0 & e_{41} & e_{56} & 0 \end{pmatrix}.$$

Using Table 2, this suggests that with probability $p_L = 2^{-6.45}$ one can compute the output mask from a given output O. As each pair requires the evaluation of two O's, the probability that a pair can be used for the analysis is $p_L^2 = 2^{-11.9}$.

For the specific differential-linear characteristics presented above the two unknown at the entrance of ψ_1 which can be recovered are $b_1 = U_3^{31} \oplus a_0$ and $b_3 = U_0^{43} \oplus U_3^{43} \oplus a_1$, where a_0 and a_1 can be computed from the four known columns of I. Table 3 offers the different biases as a function of b_1 and b_3. These biases were experimentally computed in [22] by taking 2^{30} plaintexts pairs with the required values fixed.

To conclude, given the above differential-linear characteristic, the recovery of two bits b_1 and b_3 is as follows:

1. Collect N plaintext pairs with the required input difference and the 18 bits fixed.
2. For each pair, check whether one can deduce the bits that enter the linear mask of ψ_6 for both ciphertexts.
3. Divide the remaining pairs into four sets according to the value of unknown values of b_1 and b_3.[6]
4. Find the set with the maximal bias (which suggests the correct values of b_1 and b_3). Compute from b_1, b_3 and the known bits the value of the unknown bits.

The analysis shows that when taking $N = 2^{33.9}$ plaintext pairs (with 18 bits fixed) we obtain about $2^{33.9} \cdot p_L^2 = 2^{21}$ pairs which can be analyzed (as we know the input linear mask to ψ_6). These pairs can be divided into four sets of about 2^{19} pairs each, one of which with a bias of $2^{-7.3}$, which is significantly higher than the rest, and thus can be easily detected. The data complexity is thus $2^{33.9}$ pairs of 2-block messages for each pair of bits b_1, b_3 recovered, or a total of $64 \cdot 2 \cdot 2 \cdot 2^{33.9} = 2^{41.9}$ 1024-bit data blocks. We list in Table 3 the different bits recovered in each phase, the relevant p_L, and the data complexity. The full details are available at [22].

5.3   Our New Results on ICEPOLE-128/ICEPOLE-128a

The main reason the attack of [22] used a single-bit mask for the output is to ensure a low hamming weight mask. This was chosen to optimize the two conflicting effects of λ on the complexity of the attack – the more active bits in $\mathcal{L}_6(\lambda)$ there are (which translates to more active bits in λ when λ is of low hamming weight), the lower p_L is. At the same time, λ affects the bias as it sets the output mask of ψ_5, suggesting that constraints on λ may lead to sub-optimal linear approximations.

Moreover, as the actual biases were measured experimentally (rather than analytically) we decided to pick a slightly different approach. Instead of studying single-bit λ we decided to try output masks with a single active S-box. This allowed raising the bias of the transition in ψ_5 from at most 3/16 to 4/16 (which is significant due to the quadratic effect on the bias, which translates to a quadratic effect on the data and time complexities).

The increase in the number of possible output masks carries with it a computational problem – one needs to cover more masks in the process of computing the bias, by a factor of almost 6, for any chosen input difference. Hence, instead

---

[6] We remind the reader that these bits are the XOR of a fixed unknown bits from U_0 and U_3 with already known bits.

Table 3. The different phases of the attack of [22]

Phase	Recovered bits	Value/$\log_2(bias)$	p_L	Data complexity
1	$b_1^i = U_3^{31+i}, b_2^i = U_0^{43+i} \oplus U_3^{43+i}$ $i \in \{0, 1, \ldots 63\}$	$(b_1 = 0, b_3 = 0)\ -13$ $(b_1 = 0, b_3 = 1)\ -7.3$ $(b_1 = 1, b_3 = 0)\ -13.9$ $(b_1 = 1, b_3 = 1)\ -11.9$	$2^{-6.45}$	$64 \cdot 2 \cdot 2 \cdot 2^{33.9} = 2^{41.9}$
2	$b_2^i = U_2^{24+i}$ $i \in \{0, 1, \ldots 63\}$	$b_2 = 0\ -11$ $b_2 = 1\ -15.4$	$2^{-5.86}$	$64 \cdot 2 \cdot 2 \cdot 2^{36.7} = 2^{44.7}$
3	$b_{0,3}^i = U_1^{12+2i}, b_{1,1}^i = U_1^{13+2i}$ $i \in \{0, 1, \ldots 31\}$	$(b_{0,3} = 0, b_{1,1} = 0)\ -11.2$ $(b_{0,3} = 0, b_{1,1} = 1)\ -15.2$ $(b_{0,3} = 1, b_{1,1} = 0)\ -16.4$ $(b_{0,3} = 1, b_{1,1} = 1)\ -14.8$	$2^{-5.86}$	$32 \cdot 2 \cdot 2 \cdot 2^{37.7} = 2^{44.7}$
Total				$2^{45.8}$

of relying on multiple time consuming experiments for each input difference Δ, we use the DLCT of ψ to obtain estimates for the bias of the differential-linear approximation. This is done by taking the input difference Δ and computing for each S-box in ψ_5 the distribution of input differences (i.e., if the input difference is δ, determining p_δ). Then, for all the active S-boxes in the mask leaving ψ_5, for each S-box' mask ω we compute $\sum_\delta p_\delta \cdot DLCT(\delta, \omega)$ to evaluate the probability of the differential-linear transition in ψ_5. As the evaluation of p_d for each S-box is independent of the mask ω, and as the DLCT is computed once, this offers a very efficient procedure.

The result is the discovery of better differential-linear approximation for the second and third phase. We give in Table 4 the new differential-linear approximations used in the second and third phase. Due to the reduced data required in the later phases, we also reduce the data complexity in the first round (to reduce the total data complexity) and change a bit the constraints on the actual values (but they serve the same purpose as in the original attack). One main difference is that the constraints are not on the values, but rather on the parity of some subsets of bits. We list these subsets in Table 5.

Table 4. Our new different-linear approximations for phases 2 and 3 of the attack

Phase	Δ	λ	λ'	p_L	Bias
2	$\begin{pmatrix} e_8 & 0 & 0 & 0 & 0 \\ 0 & 0 & 0 & 0 & 0 \\ 0 & 0 & 0 & 0 & 0 \\ 0 & 0 & e_0 & 0 & 0 \end{pmatrix}$	$\begin{pmatrix} 0 & e_{49} & e_{49} & 0 & 0 \\ 0 & 0 & 0 & 0 & 0 \\ 0 & 0 & 0 & 0 & 0 \\ 0 & 0 & 0 & 0 & 0 \end{pmatrix}$	$\begin{pmatrix} e_3 & e_{49} & 0 & 0 & 0 \\ 0 & e_{51} & 0 & 0 & 0 \\ 0 & 0 & e_{46} & 0 & 0 \\ 0 & 0 & e_{26} & e_{41} & 0 \end{pmatrix}$	$2^{-4.77}$	$2^{-8.88}$
3	$\begin{pmatrix} 0 & 0 & e_0 & 0 & 0 \\ 0 & 0 & 0 & 0 & 0 \\ 0 & 0 & 0 & 0 & 0 \\ 0 & 0 & 0 & e_1 & 0 \end{pmatrix}$	$\begin{pmatrix} 0 & e_{43} & e_{43} & 0 & 0 \\ 0 & 0 & 0 & 0 & 0 \\ 0 & 0 & 0 & 0 & 0 \\ 0 & 0 & 0 & 0 & 0 \end{pmatrix}$	$\begin{pmatrix} e_{61} & e_{43} & 0 & 0 & 0 \\ 0 & e_{45} & 0 & 0 & 0 \\ 0 & 0 & e_{40} & 0 & 0 \\ 0 & 0 & e_{20} & e_{35} & 0 \end{pmatrix}$	$2^{-4.77}$	$2^{-9.49}$

Using the new differential-linear approximations (and bit-fixing) we obtain an attack on the full ICEPOLE in complexity of $2^{41.58}$ data and time. Its phases are listed in Table 6. We note that the first phase has a slightly lower success rate. After presenting a new approach for generating the data that further reduces the data and time (to $2^{35.85}$) we discuss how to mitigate this slightly lower success rate.

Table 5. Bit subsets fixed for our attack

Phase	Subset	Parity (Subset)	Subset	Parity (Subset)
1	$\begin{pmatrix} 0 & e_4 & 0 & 0 & 0 \\ e_4 & 0 & 0 & 0 & 0 \\ 0 & e_4 & 0 & 0 & 0 \\ e_4 & 0 & e_4 & 0 & 0 \end{pmatrix}$	1	$\begin{pmatrix} 0 & e_{35} & 0 & 0 & 0 \\ e_{35} & 0 & 0 & 0 & 0 \\ 0 & e_{35} & 0 & 0 & 0 \\ e_{35} & 0 & e_{35} & 0 & 0 \end{pmatrix}$	1
	$\begin{pmatrix} 0 & 0 & e_{33} & 0 & 0 \\ 0 & 0 & 0 & e_{33} & 0 \\ 0 & 0 & 0 & e_{33} & 0 \\ 0 & 0 & 0 & e_{33} & 0 \end{pmatrix}$	0	$\begin{pmatrix} 0 & 0 & e_0 & 0 & 0 \\ 0 & 0 & 0 & e_0 & 0 \\ 0 & 0 & 0 & e_0 & 0 \\ 0 & 0 & 0 & e_0 & 0 \end{pmatrix}$	0
2	$\begin{pmatrix} 0 & 0 & 0 & e_{27} & 0 \\ 0 & 0 & 0 & e_{27} & 0 \\ 0 & 0 & 0 & 0 & 0 \\ 0 & 0 & e_{27} & 0 & 0 \end{pmatrix}$	1	$\begin{pmatrix} 0 & e_{17} & 0 & 0 & 0 \\ e_{17} & 0 & e_{17} & 0 & 0 \\ e_{17} & 0 & 0 & 0 & 0 \\ 0 & e_{17} & 0 & 0 & 0 \end{pmatrix}$	1
	$\begin{pmatrix} 0 & e_{58} & 0 & 0 & 0 \\ e_{58} & 0 & 0 & 0 & 0 \\ 0 & e_{58} & 0 & 0 & 0 \\ e_{58} & 0 & e_{58} & 0 & 0 \end{pmatrix}$	1	$\begin{pmatrix} 0 & 0 & 0 & 0 & e_8 \\ e_8 & 0 & 0 & 0 & 0 \\ e_8 & 0 & 0 & 0 & 0 \\ e_8 & 0 & 0 & 0 & 0 \end{pmatrix}$	0
	$\begin{pmatrix} 0 & 0 & e_{23} & 0 & 0 \\ 0 & 0 & 0 & e_{23} & 0 \\ 0 & 0 & 0 & e_{23} & 0 \\ 0 & 0 & 0 & e_{23} & 0 \end{pmatrix}$	0		
3	$\begin{pmatrix} 0 & 0 & 0 & 0 & e_0 \\ e_0 & 0 & 0 & 0 & 0 \\ e_0 & 0 & 0 & 0 & 0 \\ e_0 & 0 & 0 & 0 & 0 \end{pmatrix}$	1	$\begin{pmatrix} 0 & 0 & 0 & e_{19} & 0 \\ 0 & 0 & 0 & e_{19} & 0 \\ 0 & 0 & 0 & 0 & e_{19} \\ 0 & 0 & e_{19} & 0 & 0 \end{pmatrix}$	1
	$\begin{pmatrix} 0 & 0 & e_{24} & 0 & 0 \\ 0 & 0 & 0 & e_{24} & 0 \\ 0 & 0 & 0 & e_{24} & 0 \\ 0 & 0 & 0 & e_{24} & 0 \end{pmatrix}$	1	$\begin{pmatrix} 0 & 0 & e_{21} & 0 & 0 \\ 0 & e_{21} & 0 & 0 & 0 \\ 0 & 0 & e_{21} & 0 & 0 \\ 0 & 0 & 0 & e_{21} & 0 \end{pmatrix}$	0
	$\begin{pmatrix} 0 & 0 & e_{55} & 0 & 0 \\ 0 & 0 & e_{55} & 0 & 0 \\ 0 & 0 & 0 & e_{55} & 0 \\ 0 & e_{55} & 0 & 0 & e_{55} \end{pmatrix}$	0		

Efficient Data Generation. We note that each of the three phases of the attack is composed of 64 applications of the same attack up to rotating the differences/masks. Hence, if each such attack requires about N plaintext pairs, a trivial implementation requires $64N$ plaintext pairs. Luckily, there is a more efficient way to do so.

Our key observation is that one can select P_i in advance to satisfy all the conditions of the 64 possible rotations. This can be easily done when there is at least one word which has no conditions/restrictions. For such a P_i we test for each of the 64 rotations whether one can deduce the needed bits at the input of ψ_6. If so, we generate its counterpart P_i^* which satisfies the required difference, and apply the attack as before (with probability of p_L that P_i^* allows recovering the input of ψ_6).

This reduces the data complexity of Phase 1 from $2 \cdot 2 \cdot 64 \cdot 2^{32.8}$ plaintexts to $2 \cdot (2^{32.8} + 64 \cdot 2^{32.8-6.45}) = 2^{34.6}$ plaintexts (as for each of the 64 rotations of the differential-linear approximation there is probability of p_L that P_i is useful). Similar analysis reduces the data complexity of the second phase to $2 \cdot (2^{31.33} + 64 \cdot 2^{31.33-4.77}) = 2^{34.07}$ which is the same also for the third phase. Hence, the total data complexity of the attack is $2^{35.85}$ chosen plaintexts.

We note that the reduced data complexity of the first phase may negatively affect the success rate. Moreover, an error in the first phase is expected to cause errors in the next phases. However, we note that one can easily test the obtained values for correctness. If the recovered internal state is not accurate, the adversary can exhaustively test internal states of hamming distance up to 5 from the extracted one in time of $\binom{256}{5} \approx 2^{33.1}$ recomputations (using simple linear algebra). In other words, as long as the attack has at most 5 wrong bits, the correct internal value can be extracted.

5.4 Experimental Verification of Our Attack

We have experimentally verified our attack. We run the full attack 16 times, using a random key and nonce. The machine was a virtual machine on the Azure infrastructure (instance Standard_F64s_v2). The machine has 64 vCPUs (Intel Xeon 8168 processor) with 128 GB RAM running Ubuntu 18.04.1 TLS.

Table 6. The different phases of the new attack

Phase	Recovered Bits	Value/$\log_2(bias)$	p_L	Data Complexity
1	$b_1^i = U_3^{31+i}, b_2^i = U_0^{43+i} \oplus U_3^{43+i}$ $i \in \{0,1,\ldots 63\}$	$(b_1 = 0, b_3 = 0)$ -13 $(b_1 = 0, b_3 = 1)$ -7.3 $(b_1 = 1, b_3 = 0)$ -13.9 $(b_1 = 1, b_3 = 1)$ -11.9	$2^{-6.45}$	$64 \cdot 2 \cdot 2 \cdot 2^{32.8} = 2^{40.8}$
2	$b_4^i = U_2^{27+i}$ $i \in \{0,1,\ldots 63\}$	$b_4 = 0$ -14.32 $b_4 = 1$ -8.8	$2^{-4.77}$	$64 \cdot 2 \cdot 2 \cdot 2^{31.33} = 2^{39.33}$
3	$b_3^i = U_1^{21+i}$ $i \in \{0,1,\ldots 63\}$	$b_3 = 0$ -9.49 $b_3 = 1$ -13	$2^{-4.77}$	$64 \cdot 2 \cdot 2 \cdot 2^{31.33} = 2^{39.33}$
Total				$2^{41.58}$

We have used the official ICEPOLE code (written in C), while our attack was written in C++. Compiling with gcc-7.3.0 using the optimization flag -O3, each of the attack's instances took about an hour. Its code is available at https:// github.com/cryptobiu.

Out of the 16 experiments, 11 recovered the exact internal state. In 4 of them, a single-bit error took place in phase 3 of the attack (resulting in a single-bit error in the proposed internal state). Finally, in one experiment, a single-bit error took place in the first phase, resulting in three bits error (single-bit error in the second phase and in the third-phase). Of course, once the single-bit error in the first phase is fixed, then the errors in the other phases are resolved as well. Hence, we conclude that all experiments succeeded to recover the internal state (or were sufficiently close to the correct one) using $2^{34.85}$ 2-block plaintexts.

6   Improved Differential-Linear Attack on 8-Round DES

refined analysis of the DLCT, we found out that the attack can be improved by replacing the differential characteristic and the linear approximation with another combination of a characteristic and an approximation, which leads to a higher bias due to dependency between the two underlying subciphers. First we briefly recall the structure of DES and describe the attack of [4], and then we present our improved attack.

In this section we use the DLCT methodology to revisit the DL attack on 8-round DES [35] presented by Biham et al. [4]. We show that the attack can be improved by replacing the differential characteristic and the linear approximation with another combination of a characteristic and an approximation, which leads to a higher bias due to dependency between the two underlying subciphers.

6.1   The DL Attack of [4] on 8-Round DES

The attack of [4] is based on a 7-round DL distinguisher. Denote a 7-round variant of DES by E. The distinguisher uses the decomposition $E = E_1 \circ E_0$, where E_0 consists of rounds 1–4 and E_1 consists of rounds 5–7. For E_0, it uses the truncated differential

$$0x00808200\ 60000000 = \Delta_I \xrightarrow[E_0]{p = \frac{14}{64}} \Delta_O = 0x????M???\ 00W0XY0Z,$$

where $M \in \{0, 1, 2, \ldots, 7\}$, $W, X \in \{0, 8\}$, and $Y, Z \in \{0, 2\}$. The characteristic is composed of a 1-round characteristic with probability $\frac{14}{64}$ and a 3-round truncated characteristic with probability 1.

For E_1 (rounds 5–7), it uses the linear approximation

$$0x21040080\ 00008000 = \lambda_I \xrightarrow[E_1]{q = 2 \cdot (\frac{-20}{64})^2} \lambda_I.$$

Note that all nonzero bits of λ_I are included in the bits that are known to be 0 in Δ_O. Using the naive complexity analysis of the DL attack (i.e., Eq. (2) above), the authors of [4] concluded that the overall bias of the approximation is $2pq^2 = 2^{-5.91}$.

6.2   Our Improved DL Attack on 8-Round DES

At a first glance, it seems unlikely that the distinguisher of [4] can be improved. Indeed, the linear approximation it uses is known to be the best 3-round linear approximation of DES, and the only round in the differential characteristic whose probability is less than 1, is almost the best possible (the highest possible probability being $\frac{16}{64}$). In fact, we verified experimentally that for any other combination of a differential with probability p' and a linear approximation with bias q', we have $2p'(q')^2 < 2^{-5.91}$.

Nevertheless, we obtain a higher bias, using the dependency between the subciphers. We decompose E as $E = E_1' \circ E_m \circ E_0'$, where E_0' consists of rounds 1–2, E_m consists of rounds 3–5, and E_1' consists of rounds 6–7. For E_0', we use the differential

$$0x00200008\ 00000400 = \Delta_I \xrightarrow[E_0']{p' = \frac{16}{64}} \Delta_O = 0x60000000\ 00000000.$$

For E_1', we use the linear approximation

$$0x00808202\ 00000000 = \lambda_I \xrightarrow[E_1']{q' = \frac{-18}{64}} \lambda_O = 00808202\ 80000000.$$

For E_m, we use the DLCT entry[7]

$$\overline{DLCT}_{E_m}(0x60000000\ 00000000, 0x00808202\ 00000000) \approx 0.26.$$

Using Eq. (4) above, we find that the overall bias of our approximation is

$$4p' \cdot \overline{DLCT}_{E_m}(\Delta_O, \lambda_I) \cdot (q')^2 = 4 \cdot \left(\frac{16}{64}\right)^2 \cdot 0.26 \cdot \left(\frac{-18}{64}\right)^2 = 2^{-5.6}. \quad (6)$$

Since the data complexity of the DL attack is quadratic in the bias, the improvement from $2^{-5.91}$ to $2^{-5.6}$ reduces the data complexity of the attack of [4] on 8-round DES by a factor of about 1.5.

Comparison Between our Distinguisher and the Distinguisher of [4]. In order to compare our distinguisher to that of [4], we present the latter within the DLCT framework. It is composed of the differential $0x00200008\ 00000400 \xrightarrow{p' = \frac{14}{64}} 0x00000400\ 00000000$ for E_0, the linear approximation $0x21040080\ 00000000 \xrightarrow{q' = \frac{-18}{64}} 0x21040080\ 00008000$ for E_2, and the DLCT entry $\overline{DLCT}_{E_m}(0x00000400\ 00000000, 0x21040080\ 00000000) \approx 0.24$. Using Eq. (4), its overall bias is $2^{-5.81}$. Note that while the value $p'(q')^2$ in the distinguisher of [4] is larger than the corresponding value in our distinguisher, the overall bias we obtain is higher due to the larger value in the DLCT. This emphasizes that the advantage of our new DL distinguisher stems mainly from dependency between the two subciphers, reflected in the DLCT.

---

[7] This entry was computed by looking at all 3-round differential characteristics starting at input difference $0x60000000\ 00000000$, computing their output difference δ_i (and probability), and evaluating the bias of $\lambda_I \cdot \delta_i$. After summing over all differential characteristics, we have experimentally verified that this DLCT entry is indeed about 0.26.

Experimental Verification. We experimentally verified the bias of our DL distinguisher, using 100 different keys and 500,000 plaintext pairs for each key. The average bias found in the experiments was $2^{-5.58}$, and the standard deviation was $2^{-10.43}$. This shows that the theoretical estimate of the bias using Eq. (4) is tight in our case, and thus, demonstrates the strength of the DLCT as a tool for accurate evaluation of the DL attack complexity.

For sake of completeness, we verified experimentally also the distinguisher of Biham et al. We checked 100 different keys and 500,000 plaintext pairs for each key. The average bias found in the experiments was $2^{-5.72}$, and the standard deviation was $2^{-10.56}$. In addition, we verified that our DL distinguisher has the maximal bias among all 7-round DL distinguishers that start and end with a single active S-box. While we could not check 7-round DL distinguishers with more than one active S-box in the input difference or in the output bias, it seems highly unlikely that such a distinguisher will have a higher bias, even if it exploits the dependency between the subciphers.

Another 7-round DL Distinguisher Used in [4]. The authors of [4] present another 7-round DL distinguisher of DES, which they use in the key recovery attack on 9-round DES. (Its bias is somewhat lower, but it activates less S-boxes in the round before the distinguisher). We checked this distinguisher using the DLCT framework and found that its bias is $2^{-5.95}$, instead of $2^{-6.13}$ computed in [4]. We verified experimentally this result as well, and obtained average bias of $2^{-5.94}$ and standard deviation of $2^{-10.53}$. This slightly improved bias reduces the data complexity of the attack of [4] on 9-round DES by a factor of about 1.3.

7 Summary and Conclusions

In this paper we studied the effect of the dependency between the subciphers on the differential-linear attack. We showed that in various cases of interest, including previously published DL attacks on Ascon and Serpent, the dependency significantly affects the attack's complexity. We presented a new tool – the *differential-linear connectivity table* (DLCT) – which allows to (partially) take the dependency into account and to use it for making DL attacks more efficient. We showed a relation of the DLCT to the Fourier transform and deduced from it a new theoretical insight on the differential-linear attack. Finally, we demonstrated the strength of our new tool, by improving previously published DL attacks against ICEPOLE and 8-round DES.

Our objective in this paper was to introduce the DLCT and to present a few initial applications. Thus, several natural research directions are left for future work. The first is formalizing the relation of the DLCT with consideration of multiple linear approximations, as was done for the basic DL framework by Blondeau, Leander, and Nyberg [10]. The second is finding a way to extend the DLCT methodology so that it will cover more rounds at the boundary between E_0 and E_1. The third direction is studying properties of the DLCT, in a similar way to the way the properties of the BCT were recently studied by Boura and Canteaut [12]. The fourth direction is finding other applications of the DLCT. We believe that the DLCT is a useful generic tool, and so, we expect more applications to be found.

Acknowledgements. The research was partially supported by European Research Council under the ERC starting grant agreement n. 757731 (LightCrypt) and by the BIU Center for Research in Applied Cryptography and Cyber Security in conjunction with the Israel National Cyber Bureau in the Prime Minister's Office. Orr Dunkelman was supported in part by the Israel Ministry of Science and Technology, the Center for Cyber, Law, and Policy in conjunction with the Israel National Cyber Bureau in the Prime Minister's Office and by the Israeli Science Foundation through grant No. 880/18.

References

1. Anderson, R., Biham, E., Knudsen, L.R.: Serpent: a proposal for the advanced encryption standard. In: NIST AES Proposal (1998)
2. Beierle, C., et al.: The SKINNY family of block ciphers and its low-latency variant MANTIS. In: Robshaw, M., Katz, J. (eds.) CRYPTO 2016. LNCS, vol. 9815, pp. 123–153. Springer, Heidelberg (2016). https://doi.org/10.1007/978-3-662-53008-5_5
3. Biham, E., Dunkelman, O., Keller, N.: The rectangle attack — rectangling the Serpent. In: Pfitzmann, B. (ed.) EUROCRYPT 2001. LNCS, vol. 2045, pp. 340–357. Springer, Heidelberg (2001). https://doi.org/10.1007/3-540-44987-6_21
4. Biham, E., Dunkelman, O., Keller, N.: Enhancing differential-linear cryptanalysis. In: Zheng, Y. (ed.) ASIACRYPT 2002. LNCS, vol. 2501, pp. 254–266. Springer, Heidelberg (2002). https://doi.org/10.1007/3-540-36178-2_16
5. Biham, E., Dunkelman, O., Keller, N.: Differential-linear cryptanalysis of Serpent. In: Johansson, T. (ed.) FSE 2003. LNCS, vol. 2887, pp. 9–21. Springer, Heidelberg (2003). https://doi.org/10.1007/978-3-540-39887-5_2
6. Biham, E., Dunkelman, O., Keller, N.: A related-key rectangle attack on the full KASUMI. In: Roy, B. (ed.) ASIACRYPT 2005. LNCS, vol. 3788, pp. 443–461. Springer, Heidelberg (2005). https://doi.org/10.1007/11593447_24
7. Biham, E., Dunkelman, O., Keller, N.: New combined attacks on block ciphers. In: Gilbert, H., Handschuh, H. (eds.) FSE 2005. LNCS, vol. 3557, pp. 126–144. Springer, Heidelberg (2005). https://doi.org/10.1007/11502760_9
8. Biham, E., Shamir, A.: Differential cryptanalysis of DES-like cryptosystems. J. Cryptol. 4(1), 3–72 (1991)
9. Biryukov, A., De Cannière, C., Dellkrantz, G.: Cryptanalysis of SAFER++. In: Boneh, D. (ed.) CRYPTO 2003. LNCS, vol. 2729, pp. 195–211. Springer, Heidelberg (2003). https://doi.org/10.1007/978-3-540-45146-4_12
10. Blondeau, C., Leander, G., Nyberg, K.: Differential-linear cryptanalysis revisited. J. Cryptol. 30(3), 859–888 (2017)
11. Blondeau, C., Nyberg, K.: New links between differential and linear cryptanalysis. In: Johansson, T., Nguyen, P.Q. (eds.) EUROCRYPT 2013. LNCS, vol. 7881, pp. 388–404. Springer, Heidelberg (2013). https://doi.org/10.1007/978-3-642-38348-9_24

12. Boura, C., Canteaut, A.: On the boomerang uniformity of cryptographic S-boxes. IACR Trans. Symmetric Cryptol. **3**, 2018 (2018)
13. Chabaud, F., Vaudenay, S.: Links between differential and linear cryptanalysis. In: De Santis, A. (ed.) EUROCRYPT 1994. LNCS, vol. 950, pp. 356–365. Springer, Heidelberg (1995). https://doi.org/10.1007/BFb0053450
14. Cid, C., Huang, T., Peyrin, T., Sasaki, Y., Song, L.: Boomerang connectivity table: a new cryptanalysis tool. In: Nielsen, J.B., Rijmen, V. (eds.) EUROCRYPT 2018. LNCS, vol. 10821, pp. 683–714. Springer, Cham (2018). https://doi.org/10.1007/978-3-319-78375-8_22
15. Collard, B., Standaert, F.-X., Quisquater, J.-J.: Improving the time complexity of Matsui's linear cryptanalysis. In: Nam, K.-H., Rhee, G. (eds.) ICISC 2007. LNCS, vol. 4817, pp. 77–88. Springer, Heidelberg (2007). https://doi.org/10.1007/978-3-540-76788-6_7
16. The CAESAR committee: CAESAR: competition for authenticated encryption: security, applicability, and robustness (2014). http://competitions.cr.yp.to/caesar.html
17. Daemen, J., Govaerts, R., Vandewalle, J.: Correlation matrices. In: Preneel, B. (ed.) FSE 1994. LNCS, vol. 1008, pp. 275–285. Springer, Heidelberg (1995). https://doi.org/10.1007/3-540-60590-8_21
18. Dobraunig, C., Eichlseder, M., Mendel, F., Schläffer, M.: Ascon. Submission to the CAESAR competition (2014). http://ascon.iaik.tugraz.at
19. Dobraunig, C., Eichlseder, M., Mendel, F., Schläffer, M.: Cryptanalysis of Ascon. In: Nyberg, K. (ed.) CT-RSA 2015. LNCS, vol. 9048, pp. 371–387. Springer, Cham (2015). https://doi.org/10.1007/978-3-319-16715-2_20
20. Dunkelman, O., Indesteege, S., Keller, N.: A differential-linear attack on 12-round Serpent. In: Chowdhury, D.R., Rijmen, V., Das, A. (eds.) INDOCRYPT 2008. LNCS, vol. 5365, pp. 308–321. Springer, Heidelberg (2008). https://doi.org/10.1007/978-3-540-89754-5_24
21. Dunkelman, O., Keller, N., Shamir, A.: A practical-time related-key attack on the KASUMI cryptosystem used in GSM and 3G telephony. J. Cryptol. **27**(4), 824–849 (2014)
22. Huang, T., Tjuawinata, I., Wu, H.: Differential-linear cryptanalysis of ICEPOLE. In: Leander, G. (ed.) FSE 2015. LNCS, vol. 9054, pp. 243–263. Springer, Heidelberg (2015). https://doi.org/10.1007/978-3-662-48116-5_12
23. Jean, J., Nikolić, I., Peyrin, T., Seurin, Y.: Deoxys v1.41. Submission to the CAESAR competition (2016)
24. Kelsey, J., Kohno, T., Schneier, B.: Amplified boomerang attacks against reduced-round MARS and Serpent. In: Goos, G., Hartmanis, J., van Leeuwen, J., Schneier, B. (eds.) FSE 2000. LNCS, vol. 1978, pp. 75–93. Springer, Heidelberg (2001). https://doi.org/10.1007/3-540-44706-7_6
25. Kim, J., Hong, S., Preneel, B., Biham, E., Dunkelman, O., Keller, N.: Related-key boomerang and rectangle attacks: theory and experimental analysis. IEEE Trans. Inf. Theory **58**(7), 4948–4966 (2012)
26. Lai, X., Massey, J.L., Murphy, S.: Markov ciphers and differential cryptanalysis. In: Davies, D.W. (ed.) EUROCRYPT 1991. LNCS, vol. 547, pp. 17–38. Springer, Heidelberg (1991). https://doi.org/10.1007/3-540-46416-6_2
27. Langford, S.K., Hellman, M.E.: Differential-linear cryptanalysis. In: Desmedt, Y.G. (ed.) CRYPTO 1994. LNCS, vol. 839, pp. 17–25. Springer, Heidelberg (1994). https://doi.org/10.1007/3-540-48658-5_3

28. Leurent, G.: Improved differential-linear cryptanalysis of 7-round chaskey with partitioning. In: Fischlin, M., Coron, J.-S. (eds.) EUROCRYPT 2016. LNCS, vol. 9665, pp. 344–371. Springer, Heidelberg (2016). https://doi.org/10.1007/978-3-662-49890-3_14

29. Liu, Z., Gu, D., Zhang, J., Li, W.: Differential-multiple linear cryptanalysis. In: Bao, F., Yung, M., Lin, D., Jing, J. (eds.) Inscrypt 2009. LNCS, vol. 6151, pp. 35–49. Springer, Heidelberg (2010). https://doi.org/10.1007/978-3-642-16342-5_3

30. Jiqiang, L.: A methodology for differential-linear cryptanalysis and its applications. Des. Codes Cryptogr. **77**(1), 11–48 (2015)

31. Matsui, M.: Linear cryptanalysis method for DES cipher. In: Helleseth, T. (ed.) EUROCRYPT 1993. LNCS, vol. 765, pp. 386–397. Springer, Heidelberg (1994). https://doi.org/10.1007/3-540-48285-7_33

32. Morawiecki, P., et al.: ICEPOLE: high-speed, hardware-oriented authenticated encryption. In: Batina, L., Robshaw, M. (eds.) CHES 2014. LNCS, vol. 8731, pp. 392–413. Springer, Heidelberg (2014). https://doi.org/10.1007/978-3-662-44709-3_22

33. Murphy, S.: The return of the cryptographic boomerang. IEEE Trans. Inf. Theory **57**(4), 2517–2521 (2011)

34. O'Donnell, R.: Analysis of Boolean Functions. Cambridge University Press, Cambridge (2014)

35. US National Bureau of Standards. Data Encryption Standard, Federal Information Processing Standards publications no. 46 (1977)

36. US National Institute of Standards and Technology. Advanced Encryption Standard, Federal Information Processing Standards publications no. 197 (2001)

37. Selçuk, A.A.: On probability of success in linear and differential cryptanalysis. J. Cryptol. **21**(1), 131–147 (2008)

38. Wagner, D.: The boomerang attack. In: Knudsen, L. (ed.) FSE 1999. LNCS, vol. 1636, pp. 156–170. Springer, Heidelberg (1999). https://doi.org/10.1007/3-540-48519-8_12

Linear Equivalence of Block Ciphers with Partial Non-Linear Layers: Application to LowMC

Itai Dinur[1]([✉]), Daniel Kales[2], Angela Promitzer[3], Sebastian Ramacher[2], and Christian Rechberger[2]

[1] Department of Computer Science, Ben-Gurion University, Beersheba, Israel
dinuri@cs.bgu.ac.il
[2] Graz University of Technology, Graz, Austria
[3] Graz, Austria

Abstract. LowMC is a block cipher family designed in 2015 by Albrecht et al. It is optimized for practical instantiations of multi-party computation, fully homomorphic encryption, and zero-knowledge proofs. LowMC is used in the PICNIC signature scheme, submitted to NIST's post-quantum standardization project and is a substantial building block in other novel post-quantum cryptosystems. Many LowMC instances use a relatively recent design strategy (initiated by Gérard et al. at CHES 2013) of applying the non-linear layer to only a part of the state in each round, where the shortage of non-linear operations is partially compensated by heavy linear algebra. Since the high linear algebra complexity has been a bottleneck in several applications, one of the open questions raised by the designers was to reduce it, without introducing additional non-linear operations (or compromising security).

In this paper, we consider LowMC instances with block size n, partial non-linear layers of size $s \leq n$ and r encryption rounds. We redesign LowMC's linear components in a way that preserves its specification, yet improves LowMC's performance in essentially every aspect. Most of our optimizations are applicable to all SP-networks with partial non-linear layers and shed new light on this relatively new design methodology.

Our main result shows that when $s < n$, each LowMC instance belongs to a large class of equivalent instances that differ in their linear layers. We then select a *representative instance* from this class for which encryption (and decryption) can be implemented much more efficiently than for an arbitrary instance. This yields a new encryption algorithm that is equivalent to the standard one, but reduces the evaluation time and storage of the linear layers from $r \cdot n^2$ bits to about $r \cdot n^2 - (r-1)(n-s)^2$. Additionally, we reduce the size of LowMC's round keys and constants and optimize its key schedule and instance generation algorithms. All of these optimizations give substantial improvements for small s and a reasonable choice of r. Finally, we formalize the notion of linear equivalence of block ciphers and prove the optimality of some of our results.

Comprehensive benchmarking of our optimizations in various LowMC applications (such as PICNIC) reveals improvements by factors that typically range between 2x and 40x in runtime and memory consumption.

© International Association for Cryptologic Research 2019
Y. Ishai and V. Rijmen (Eds.): EUROCRYPT 2019, LNCS 11476, pp. 343–372, 2019.
https://doi.org/10.1007/978-3-030-17653-2_12

Keywords: Block cipher · LowMC · Picnic signature scheme ·
Linear equivalence

1 Introduction

LowMC is a block cipher family designed by Albrecht et al. [2], and is heavily optimized for practical instantiations of multi-party computation (MPC), fully homomorphic encryption (FHE), and zero-knowledge proofs. In such applications, non-linear operations incur a higher penalty in communication and computational complexity compared to linear ones. Due to its design strategy, LowMC is a popular building block in post-quantum designs that are based on MPC and zero-knowledge protocols (cf. [6,7,9,10,14]). Most notably, it is used in the Picnic signature algorithm [8] which is a candidate in NIST's post-quantum cryptography standardization project.[1]

Instances of LowMC are designed to perform well in two particular metrics that measure the complexity of non-linear operations over GF(2). The first metric is multiplicative complexity (MC), which simply counts the number of multiplications (AND gates in our context) in the circuit. The second metric is the multiplicative (AND) depth of the circuit.

The relevance of each metric depends on the specific application. For example, in the context of MPC protocols, Yao's garbled circuits [20] with the free-XOR technique [16] (and many of their variants) have a constant number of communication rounds. The total amount of communication depends on the MC of the circuit as each AND gate requires communication, whereas XOR operations can be performed locally. In an additional class of MPC protocols (e.g., GMW [13]), the number of communication rounds is linear in the ANDdepth of the evaluated circuit. The performance of these protocols depends on both the MC and ANDdepth of the circuit.

In order to reduce the complexity of non-linear operations for a certain level of security, LowMC combines very dense linear layers over $GF(2)^n$ (where n is the block size) with simple non-linear layers containing 3×3 Sboxes of algebraic degree 2. The LowMC block cipher family includes a huge number of instances, where for each instance, the linear layer of each round is chosen independently and uniformly at random from all invertible $n \times n$ matrices.

The design strategy of LowMC attempts to offer flexibility with respect to both the MC and ANDdepth metrics. In particular, some LowMC instances minimize the MC metric by applying only a *partial non-linear layer* to the state of the cipher at each round, while the linear layers still mix the entire state. In general, this approach requires to increase the total number of rounds in the scheme in order to maintain a certain security level, but this is compensated by the reduction in the size of the non-linear layers and the total AND count is generally reduced. The global parameters of LowMC that are most relevant for

---

[1] https://csrc.nist.gov/Projects/Post-Quantum-Cryptography/Round-1-Submissions.

this paper are (1) the block size of n bits, (2) the number of rounds r (which is determined according to the desired security level), and (3) a parameter s which denotes the domain length of each non-linear layer, namely, the number of bits on which it operates (which may be smaller than n).[2]

While LowMC's design aims to minimize the non-linear complexity of the scheme at the expense of using many linear algebra (XOR) operations, in several practical applications, XORs do not come for free and may become a bottleneck in the implementation. This phenomenon was already noted and demonstrated in the original LowMC paper. Indeed, due to the large computational cost of LowMC's dense linear layers, one of the open problems raised by its designers was to reduce their computational cost, presumably by designing more specific linear layers that offer the same security level with improved efficiency.

More recently, the high cost of LowMC's linear operations influenced the design of the PICNIC signature algorithm, where the most relevant metric is the MC that affects the signature size. In order to minimize the AND count (and the signature size), the LowMC instances used by PICNIC should have a very small partial non-linear layer in each round (perhaps using only a single 3×3 Sbox). However, such an instance has a large number of rounds r and each encryption requires computation of r matrix-vector products that increase the signing and verification times. Consequently, the PICNIC designers settled for non-linear layers of intermediate size in order to balance the signature size on one hand and the signing and verification times on the other.

In fact, in PICNIC there is another source of inefficiency due to the heavy cost of the linear operations in LowMC's key schedule: the computation of LowMC inside PICNIC involves splitting the LowMC instance to 3 related instances which are evaluated with a fresh share of the key in each invocation. Therefore, in contrast to standard applications, the key schedule has to be run before each cipher invocation and it is not possible to hard-code the round keys into the LowMC instance in this specific (and very important) application. In LowMC, each of the $r + 1$ round keys is generated by applying an independent $n \times \kappa$ random linear transformation to the κ-bit master key. Therefore, the total complexity of the key schedule is $(r + 1) \cdot n \cdot \kappa$ in both time and memory, which is a substantial overhead on the signing and verification processes in PICNIC.

Our Contribution. In this paper we revisit the open problem of the LowMC designers to reduce the complexity of its linear operations, focusing on instances with partial non-linear layers (i.e., $s < n$). We consider a generalized LowMC construction in which the r linear layers are selected uniformly at random from the set of all invertible matrices and the non-linear layers are arbitrary and applied to s bits of the n-bit internal state in each of the r rounds. Our results are divided into several parts.

1. The round keys and constants of a generalized LowMC cipher require memory of $(r + 1) \cdot n$ bits. We compress them to $n + r \cdot s$ bits. We then consider

---

[2] The LowMC specification denotes by m the number of 3×3 Sboxes in each non-linear layer and therefore $s = 3m$ in our context.

LowMC's linear key schedule (with a master key of size κ bits) and reduce its complexity from $(r+1) \cdot n \cdot \kappa$ to $n \cdot \kappa + r \cdot (s \cdot \kappa)$. This has a substantial effect on the performance of PICNIC, as described above.

2. The linear algebra of the encryption (and decryption) algorithm requires matrices of size $r \cdot n^2$ bits and performs matrix-vector products with about the same complexity. We describe a new algorithm that uses matrices requiring only $r \cdot n^2 - (r-1)(n-s)^2$ bits of storage and about the same linear algebra time complexity (using standard matrix-vector products[3]).

3. We consider the complexity of generating a generalized LowMC instance, assuming its linear layers are sampled at random. We devise a new sampling algorithm that reduces this complexity[4] from about $r \cdot n^3$ to $n^3 + (r-1) \cdot (s^2 \cdot n)$. Our sampling algorithm further reduces the number of uniform (pseudo) random bits required to sample the linear layers from about $r \cdot n^2$ to $n^2 + (r-1) \cdot (n^2 - (n-s)^2)$. These optimizations are useful in applications that require frequent instance generation, e.g. for the RASTA design strategy [11].

4. We address the question of whether the linear layer description we use during encryption is optimal (i.e., minimal) or can be further compressed. Indeed, it may seem that the formula $n^2 + (r-1)(n^2 - (n-s)^2)$ is suboptimal, and the formula $n^2 + (r-1) \cdot s \cdot n^2$ is more reasonable, as it is linear in s (similarly to the reduction in the size of the round keys). However, we prove (under two assumptions which we argue are natural) that no further optimizations that reduce the linear layer sizes are possible without changing their functionality.

Table 1 summarizes our improvements and the assumptions under which they can be applied to an SP-network with partial non-linear layers. Surprisingly, although the open problem of the LowMC designers presumably involved changing the specification of LowMC's linear layers to reduce its linear algebra complexity, our improvements achieve this without any specification change. All of these improvements are significant for $s \ll n$ and r that is not too small.

We stress that our optimized encryption algorithm is applicable to any SP-network with partial non-linear layers (such as Zorro[5] [12]) since it does not assume any special property of the linear or non-linear layers. Yet, if the linear layers are not selected uniformly at random, the question of whether our algorithm is more efficient compared to the standard one depends on the specific design. On the other hand, when designing new SP-networks with partial non-linear layers, one may use our optimized linear layers as a starting point for additional improvements. We further note that the reduced complexity of the linear layer evaluation during encryption is also useful for adversaries that attempt to break LowMC instances via exhaustive search.

---

[3] Optimizations in matrix-vector multiplications (such as the "method of four Russians" [1]) can be applied to both the standard and to our new encryption algorithm.

[4] Using asymptotically fast matrix multiplication and invertible matrix sampling algorithms will reduce the asymptotic complexity of both the original and our new algorithm. Nevertheless, it is not clear whether they would reduce their concrete complexity for relevant choices of parameters.

[5] Although Zorro is broken [3,18,19], its general design strategy remains valid.

Table 1. Improvements in time/memory/randomness $(T/M/R)$ and assumptions under which they are applicable (RK = round keys, RC = round constants, KS = key schedule, LL = linear layer).

	Metric	Unoptimized	Optimized	Sect.	Assumption
RK and RC	M	$(r+1) \cdot n$	$n + s \cdot r$	3.1	None
KS	T/M	$(r+1) \cdot (n \cdot \kappa)$	$n \cdot \kappa + r \cdot (s \cdot \kappa)$	3.2	Linear KS
LL evaluation	T/M	$r \cdot n^2$	$n^2 + (r-1) \cdot (n^2 - (n-s)^2)$	5	None
LL sampling	T	$r \cdot n^3$	$n^3 + (r-1) \cdot (s^2 \cdot n)$	7	Random LL
	R	$r \cdot n^2$	$n^2 + (r-1) \cdot (n^2 - (n-s)^2)$		sampling

Table 2. Multiplicative gains (previous/new) in memory consumption and in runtimes for LowMC encryption and Picnic signing and verification.

Parameters			Memory	Runtime	
n	s	r		LowMC	Picnic
128	30	20	2.38x	1.41x	1.34x
192	30	30	3.99x	2.48x	1.72x
256	30	38	4.84x	2.82x	2.01x
128	3	182	16.51x	6.57x	4.74x
192	3	284	31.85x	11.50x	7.97x
256	3	363	39.48x	16.18x	10.83x

Table 2 compares[6] the size of LowMC's linear layers in previous implementations to our new encryption algorithm for several instances. The first three instances are the ones used by the Picnic signature algorithm and for them we obtain a multiplicative gain of between 2.38x and 4.84x in memory consumption. Runtime-wise we obtain an improvement of a factor between 1.41x to 2.82x for LowMC encryption and by a factor between 1.34x to 2.01x for Picnic.

Even more importantly, prior to this work, reducing s (in order to optimize the MC metric) while increasing r (in order to maintain the same security level for a LowMC instance) increased the linear algebra complexity proportionally to the increase in the number of rounds, making those instances impractical. One of the main consequences of this work is that such a reduction in s now also reduces the linear algebra complexity per round, such that the larger number of rounds is no longer a limiting factor. In particular, the last three instances in Table 2 correspond to a choice of parameters with a minimal value of s that minimizes signature sizes in Picnic. For those instances, we reduce the size of the linear layers by a factor between 16.51x to 39.48x and improve runtimes by up to a factor of 16x. Moreover, compared to the *original* Picnic instances that use $s = 30$, using our optimizations, instances with $s = 3$ reduce memory consumption and achieve comparable runtime results.

---

[6] For key size and the allowed data complexity, we refer to the full version.

Our Techniques. The first step in reducing the size of the round keys and constants is to exchange the order of the key and constant additions with the application of the linear layer in a round of the cipher. While this is a common technique in symmetric cryptography, we observe that in case $s < n$, after reordering, the constant and key additions of consecutive rounds can be merged through the $n - s$ bits of the state that do not go through the non-linear transformation. Applying this process recursively effectively eliminates all the key and constant additions on $n - s$ bits of the state (except for the initial key and constant additions). We then exploit the linear key schedule of LowMC and compute the reduced round keys more efficiently from the master key.

In order to derive our new encryption algorithm, we show that each (generalized) LowMC instance belongs to a class of equivalent instances which is of a very large size when $s \ll n$. We then select a representative member of the equivalence class that can be implemented efficiently using linear algebra optimizations which apply matrices with a special structure instead of random matrices (yet the full cipher remains equivalent). This requires a careful examination of the interaction between linear operations in consecutive rounds which is somewhat related to (but more complex than) the way that round keys and constants of consecutive rounds interact. After devising the encryption algorithm, we show how to sample a representative member of an equivalence class more efficiently than a random member. Our new sampling algorithm breaks dependencies among different parts of the linear layers in a generalized LowMC cipher, shedding further light on its internal structure.

Finally, we formalize the notion of linear equivalence among generalized LowMC ciphers. This allows us to prove (based on two natural assumptions) that we correctly identified the linear equivalence classes and hence our description of the linear layers is optimal in size and we use the minimal amount of randomness to sample it. The formalization requires some care and the proof of optimality is somewhat non-standard (indeed, the claim that we prove is non-standard).

Related Work. Previous works [4,5] investigated equivalent representations of AES and other block ciphers obtained by utilizing the specific structure of their Sboxes (exploiting a property called self-affine equivalence [5]). On the other hand, our equivalent representation and encryption algorithm is independent of the non-linear layer and can be applied regardless of its specification. Yet we only deal with block ciphers with partial non-linear layers in this paper.

Paper Organization. The rest of the paper is organized as follows. We describe some preliminaries in Sect. 2. Our first optimizations regarding round keys, constants, and the key schedule are described in Sect. 3. In Sect. 4, we prove basic linear algebra properties, which are then used in our optimized encryption algorithm, described in Sect. 5. Our evaluation of LowMC implementations that make use of these optimization are detailed in Sect. 6. Next, our optimized instance generation algorithm for sampling the linear layers is given in Sect. 7.

Finally, we prove the optimality of our description of the linear layers in Sect. 8 and conclude in Sect. 9.

2 Preliminaries

2.1 Notation

Given a string of bits $x \in \{0,1\}^n$, denote by $x[|d]$ its d most significant bits (MSBs) and by $x[d|]$ its d least significant bits (LSBs). Given strings x, y, denote by $x\|y$ their concatenation. Given a matrix A, denote by $A[*,i]$ its i'th column, by $A[*,d|]$ its first d columns and by $A[*,|d]$ its last d columns. Given two matrices $A \in \mathrm{GF}(2)^{d_1 \times d_2}$ and $B \in \mathrm{GF}(2)^{d_1 \times d_3}$ denote by $A\|B \in \mathrm{GF}(2)^{d_1 \times (d_2+d_3)}$ their concatenation. Denote by $I_d \in GF(2)^{d \times d}$ the identity matrix.

Throughout this paper, addition $x + y$ between bit strings $x, y \in \{0,1\}^n$ is performed bit-wise over $\mathrm{GF}(2)^n$ (i.e., by XORing them).

2.2 Generalized LowMC Ciphers

We study generalized LowMC (GLMC) ciphers where the block size is n bits, and each non-linear layer operates on $s \leq n$ bits of the state. Each instance is characterized by a number of rounds r, round keys k_i for $i \in \{0,\dots,r\}$ and round constants C_i, for $i \in \{0,\dots,r\}$. The cipher consists of r (partial) invertible non-linear layers $S_i : \{0,1\}^s \to \{0,1\}^s$ and r invertible linear layers $L_i \in \mathrm{GF}(2)^{n \times n}$ for $i \in \{1,\dots,r\}$.

A GLMC instance is generated by choosing each L_i independently and uniformly at random among all invertible $n \times n$ matrices.[7] However, we note that the main encryption algorithm we devise in Sect. 5 is applicable regardless of the way that the linear layers are chosen. We do not restrict the invertible non-linear layers.

The encryption procedure manipulates n-bit words that represent GLMC states, while breaking them down according to their s LSBs (which we call "part 0 of the state") and $n - s$ MSBs (which we call "part 1 of the state"). To simplify our notation, given any n-bit string x, we denote $x^{(0)} = x[s|]$ and $x^{(1)} = x[|n - s]$.

The basic GLMC encryption procedure is given in Algorithm 1. Decryption is performed by applying the inverse operations to a ciphertext.

Key Schedule. The key schedule optimization of Sect. 3.2 assumes that round keys are generated linearly from the master key (as in LowMC) and we now define appropriate notation. The master key k is of length κ bits. It is used to generate round key k_i for $i \in \{0,1,\dots,r\}$ using the matrix $K_i \in GF(2)^{n \times \kappa}$, namely, $k_i = K_i \cdot k$. During instance generation, each matrix $\{K_i\}_{i=0}^r$ is chosen uniformly at random among all $n \times \kappa$ matrices.

---

[7] Alternatively, they can be selected in a pseudo-random way from a short seed, as in LowMC.

```
Input   : x₀
Output : x_{r+1}
begin
    x₁ ← x₀ + k₀ + C₀
    for i ∈ {1, 2, ..., r} do
        yᵢ ← Sᵢ(xᵢ^{(0)})‖xᵢ^{(1)}
        x_{i+1} ← Lᵢ(yᵢ) + kᵢ + Cᵢ
    end
    return x_{r+1}
end
```

Algorithm 1: Basic encryption.

2.3   Breaking Down the Linear Layers

Given L_i (which is an $n \times n$ matrix), we partition its n-bit input into the first s LSBs (part 0 of the state that is output by S_i) and the remaining $n - s$ bits (part 1 of the state). Similarly, we partition its n-bit output into the first s LSBs (that are inputs of S_{i+1}) and the remaining $n - s$ bits. We define 4 sub-matrices of L_i that map between the 4 possible pairs of state parts:

$$L_i^{00} \in \mathrm{GF}(2)^{s \times s}, L_i^{01} \in \mathrm{GF}(2)^{s \times (n-s)},$$
$$L_i^{10} \in \mathrm{GF}(2)^{(n-s) \times s}, L_i^{11} \in \mathrm{GF}(2)^{(n-s) \times (n-s)}.$$

Thus, in our notation L_i^{ab} for $a, b \in \{0, 1\}$ maps the part of the state denoted by b to the part of the state denoted by a.

$$L_i = \left[\begin{array}{c|c} L_i^{00} & L_i^{01} \\ \hline L_i^{10} & L_i^{11} \end{array} \right] \begin{array}{l} \} s \\ \} n - s \end{array}$$
$$\underbrace{\qquad}_{s} \underbrace{\qquad}_{n-s}$$

We extend our notation L_i^{ab} by allowing $a, b \in \{0, 1, *\}$, where the symbol '*' denotes the full state. Therefore,

$$L_i^{0*} \in \mathrm{GF}(2)^{s \times n}, L_i^{1*} \in \mathrm{GF}(2)^{(n-s) \times n}, L_i^{*0} \in \mathrm{GF}(2)^{n \times s}, L_i^{*1} \in \mathrm{GF}(2)^{n \times (n-s)},$$

are linear transformations which are sub-matrices of L_i, as shown below.

$$L_i = \left[\begin{array}{c} L_i^{0*} \\ \hline L_i^{1*} \end{array} \right], L_i = \left[L_i^{*0} \mid L_i^{*1} \right]$$

2.4   Complexity Evaluation

In this paper, we analyze the complexity of the linear layers of generalized LowMC schemes. We will be interested in the two natural measures of time complexity (measured by the number of bit operations) and memory complexity (measured by the number of stored bits) of a single encryption (or decryption) of an arbitrary plaintext (or ciphertext). The linear layers are naturally represented by matrices, and thus evaluating a linear layer on a state is a simply a matrix-vector product. Since the time and memory complexities of evaluating and storing the linear layers are proportional in this paper, we will typically refer to both as the linear algebra complexity of the linear layers. For algorithms that generate GLMC instances, we will be interested in time complexity and in the number of random bits (or pseudo-random bits) that they use.

3   Optimized Round Key Computation and Constant Addition

In this section we optimize the round key computation and constant addition in a GLMC cipher. First, we show how to compress the round keys and constants and then we optimize the key schedule of the cipher, assuming it is linear. These optimizations are significant in case we need to run the key schedule for every cipher invocation (which is the case in PICNIC).

3.1   Compressing the Round Keys and Constants

We combine the last two linear operations in encryption Algorithm 1 and obtain $x_{i+1} \leftarrow L_i(y_i) + k_i + C_i$. Moreover, $y_i \leftarrow S_i(x_i^{(0)}) \| x_i^{(1)}$, namely S_i only operates on the first s bits of the state and does not change $x_i^{(1)}$. Based on this observation, we perform the following:

- Modify $x_{i+1} \leftarrow L_i(y_i) + k_i + C_i$ to $x_{i+1} \leftarrow L_i(y_i + L_i^{-1} \cdot k_i) + C_i$.
- Split $L_i^{-1} \cdot k_i$ into the lower s bits (the "non-linear part", i.e., $(L_i^{-1} \cdot k_i)^{(0)}$) and the upper $n - s$ bits (the "linear part", i.e., $(L_i^{-1} \cdot k_i)^{(1)}$) and move the addition of the upper $n - s$ bits before the Sbox layer.

Figure 1 demonstrates one round of the cipher with the above modifications (which do not change its output).

Next, we observe that the addition of $(L_i^{-1} \cdot k_i)^{(1)}$ at the beginning of the round can be combined with the addition of k_{i-1} in the previous round. We can now perform similar operations to round $i - 1$ and continue recursively until all additions to the linear part of the state have been moved to the start of the algorithm. In general, starting from the last round and iterating this procedure down to the first, we eliminate all additions of the linear parts of the round keys and move them before the first round. For each round $i \geq 1$, we are left with a reduced round key of size s.

In total, the size of the round keys is reduced from $n \cdot (r + 1)$ to $n + s \cdot r$. We remark that the same optimization can be performed to the constant additions, reducing their size by the same amount. We denote the new reduced round key of round i by k_i' and the new reduced round constant by C_i'. The new encryption procedure is given in Algorithm 2. Observe that all the values $\{k_i' + C_i'\}_{i=0}^{r}$ can be computed and stored at the beginning of the encryption and their total size is $n + s \cdot r$.

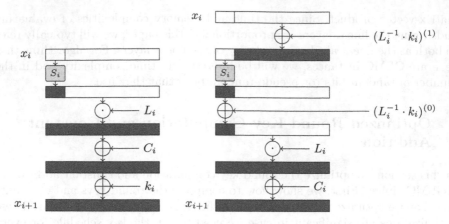

Fig. 1. One round before (left) and after (right) splitting the round key addition.

```
Input  : x₀
Output : x_{r+1}
begin
    x₁ ← x₀ + k'₀ + C'₀
    for i ∈ {1, 2, ..., r} do
        y_i ← (S_i(x_i^{(0)}) + k'_i + C'_i)‖x_i^{(1)}
        x_{i+1} ← L_i(y_i)
    end
    return x_{r+1}
end
```

Algorithm 2: Encryption with reduced round keys and constants.

3.2 Optimizing the Key Schedule

We now deal with optimizing the round key computation of Algorithm 2, assuming a linear key schedule. The original key schedule applies $r + 1$ round key matrices K_i to the κ-bit key k in order to compute the round keys $k_i = K_i \cdot k$. It therefore has a complexity of $(r+1) \cdot (n \cdot \kappa)$ (using a similar amount of memory). We show how to reduce this complexity to $n \cdot \kappa + r \cdot (s \cdot \kappa)$.

The main observation is that all transformations performed in Sect. 3.1 in order to calculate the new round keys from the original ones are linear. These linear transformations can be composed with the linear transformations K_i in order to define linear transformations that compute the new round keys directly from the master key k. Since the total size of the round keys is $n + s \cdot r$ bits, we can define matrices of total size $n \cdot \kappa + r \cdot (s \cdot \kappa)$ that calculate all round keys from the master κ-bit key.

More specifically, we define the matrix $\overline{L_i^{-1}}$ which is the inverse of the linear layer matrix L_i, with the first s rows of this inverse set to 0. Applying the iterative procedure defined in Sect. 3.1 from round r down to round i, we obtain

$$P_{N,i} = \sum_{j=i}^{r} \left(\prod_{\ell=i}^{j} \overline{L_\ell^{-1}} \right) \cdot K_j.$$

For $i \geq 1$, the new round key k_i' (for the non-linear part of the state) is computed by taking the s least significant bits of $P_{N,i} \cdot k$. Using the notation of Sect. 2.3, we have

$$k_i' = (P_{N,i})^{0*} \cdot k.$$

Observe that the total size of all $\{(P_{N,i})^{0*}\}_{i=1}^r$ is $r \cdot (s \cdot \kappa)$ bits. Finally, the new round key k_0' is calculated by summing the contributions from the linear parts of the state, using the matrix

$$P_L = K_0 + \sum_{j=1}^{r} \left(\prod_{\ell=1}^{j} \overline{L_\ell^{-1}} \right) \cdot K_j.$$

Therefore, we have $k_0' = P_L \cdot k$, where P_L is an $n \times \kappa$ matrix. All matrices $\{(P_{N,i})^{0*}\}_{i=1}^r, P_L$ can be precomputed after instance generation and we do not need to store the original round key matrices K_i.

4   Linear Algebra Properties

In this section we describe the linear algebra properties that are relevant for the rest of this paper. We begin by describing additional notational conventions.

4.1   General Matrix Notation

The superscript of L_i^{ab} introduce in Sect. 2.3 has a double interpretation, as specifying both the dimensions of the matrix and its location in L_i. We will use this notation more generally to denote sub-matrices of some $n \times n$ matrix A, or simply to define a matrix with appropriate dimensions (e.g., $A^{01} \in \text{GF}(2)^{s \times (n-s)}$ may be defined without defining A and this should be clear from the context). Therefore, dimensions of the matrices in the rest of the paper will be explicitly specified in superscript as A^{ab}, where $a, b \in \{0, 1, *\}$ (we do not deal with matrices of other dimensions). In case the matrix A^{ab} is a sub-matrix of a larger matrix A, the superscript has a double interpretation as specifying both the dimensions of A^{ab} and its location in A. When no superscript is given, the relevant matrix is of dimensions $n \times n$. There will be two exceptions to this rule which will be specified separately.

4.2   Invertible Binary Matrices

Denote by α_n the probability that an $n \times n$ uniformly chosen binary matrix is invertible. We will use the following well-known fact:

Fact 1 *[[15], page 126, adapted] The probability that an $n \times n$ uniform binary matrix is invertible is $\alpha_n = \prod_{i=1}^{n}(1 - 1/2^i) > 0.2887$. More generally, for positive integers $d \leq n$, the probability that a $d \times n$ binary matrix, chosen uniformly at random, has full row rank of d is $\prod_{i=n-d+1}^{n}(1 - 1/2^i) = \left(\prod_{i=1}^{n}(1 - 1/2^i)\right)/\left(\prod_{i=1}^{n-d}(1 - 1/2^i)\right) = \alpha_n/\alpha_{n-d}$.*

We will be interested in invertibility of matrices of a special form, described in the following fact (which follows from basic linear algebra).

Fact 2 *An $n \times n$ binary matrix of the form*

$$\left[\begin{array}{c|c} A^{00} & A^{01} \\ \hline A^{10} & I_{n-s} \end{array}\right]$$

is invertible if and only if the $s \times s$ matrix $B^{00} = A^{00} + A^{01}A^{10}$ is invertible and its inverse is given by

$$\left[\begin{array}{c|c} (B^{00})^{-1} & -(B^{00})^{-1} \cdot A^{01} \\ \hline -A^{10} \cdot (B^{00})^{-1} & I_{n-s} - A^{10} \cdot (B^{00})^{-1} \cdot A^{01} \end{array}\right].$$

Finally, we prove (in the full version) a simple proposition regarding random matrices.

Proposition 1. *Let $A \in \mathrm{GF}(2)^{n \times n}$ be an invertible matrix chosen uniformly at random and let $B^{11} \in \mathrm{GF}(2)^{(n-s) \times (n-s)}$ be an arbitrary invertible matrix (for $s \leq n$) that is independent from A. Then the matrix*

$$C = \left[\begin{array}{c|c} A^{00} & A^{01} \cdot B^{11} \\ \hline A^{10} & A^{11} \cdot B^{11} \end{array}\right]$$

is a uniform invertible matrix.

4.3   Normalized Matrices

Definition 1. *Let A^{1*} be a Boolean matrix with full row rank of $n - s$ (and therefore it has $n - s$ linearly independent columns). Let $\mathrm{COL}(A)$ denote the first set of $n - s$ linearly independent columns of A^{1*} in a fixed lexicographic ordering of columns sets. Then, these columns form an $(n - s) \times (n - s)$ invertible matrix which is denoted by $\dot{A}$, while the remaining columns form an $(n - s) \times s$ matrix which is denoted by $\ddot{A}$. Moreover, denote $\hat{A} = \dot{A}^{-1} \cdot A^{1*} \in \mathrm{GF}(2)^{(n-s) \times}$ (in this matrix $\hat{A}$, the columns of $\mathrm{COL}(A)$ form the identity matrix).*

Remark 1. The only exception to the rule of Sect. 4.1 has to do with Definition 1 (and later with the related Definition 2). In this paper, the decomposition of Definition 1 is always applied to matrices $A^{1*} \in \mathrm{GF}(2)^{(n-s) \times n}$ (in case A^{1*} is a sub-matrix of A, it contains the bottom $n - s$ rows of A). Hence the resulting matrices $\dot{A} \in \mathrm{GF}(2)^{(n-s) \times (n-s)}$, $\ddot{A} \in \mathrm{GF}(2)^{(n-s) \times s}$ and $\hat{A} \in \mathrm{GF}(2)^{(n-s) \times n}$ have fixed dimensions and do not need any superscript. On the other hand, we will use superscript notation to denote sub-matrices of these. For example $\hat{A}^{10} \in \mathrm{GF}(2)^{(n-s) \times s}$ is a sub-matrix of $\hat{A}$, consisting of its first s columns.

It will be convenient to consider a lexicographic ordering in which the columns indices of A^{1*} are reversed, i.e., the first ordered set of $n - s$ columns is $\{n, n - 1, \ldots, s + 1\}$, the second is $\{n, n - 1, \ldots, s + 2, s\}$, etc. To demonstrate the above definition, assume that $COL(A) = \{n, n - 1, \ldots, s + 1\}$ is a consecutive set of linearly independent columns. Then, the matrix A^{1*} is shown below.

$$A^{1*} = [\; \underbrace{\ddot{A}}_{s} \mid \underbrace{\dot{A}}_{n-s} \;] \; \} n - s$$

We can write $A = (\dot{A} \cdot \dot{A}^{-1}) \cdot A = \dot{A} \cdot (\dot{A}^{-1} \cdot A) = \dot{A} \cdot \hat{A}$, where

$$\hat{A} = \dot{A}^{-1} \cdot A^{1*} = [\; \underbrace{\dot{A}^{-1} \cdot \ddot{A}}_{s} \mid \underbrace{I_{n-s}}_{n-s} \;] \; \} n - s. \tag{1}$$

Normalized Equivalence Classes. Given an invertible matrix $A \in GF(2)^{n \times n}$, define

$$N(A) = \left[\begin{array}{c} A^{0*} \\ \hline \hat{A} \end{array} \right] = \left[\begin{array}{c} A^{0*} \\ \hline \dot{A}^{-1} \cdot A^{1*} \end{array} \right] = \left[\begin{array}{c|c} I_s & 0^{01} \\ \hline 0^{10} & \dot{A}^{-1} \end{array} \right] \cdot A.$$

The transformation $N(\cdot)$ partitions the set of invertible $n \times n$ boolean matrices into *normalized equivalence classes*, where A, B are in the same normalized equivalence class if $N(A) = N(B)$. We denote $A \leftrightarrow_N B$ the relation $N(A) = N(B)$.

Proposition 2. *Two invertible $n \times n$ boolean matrices A, B satisfy $A \leftrightarrow_N B$ if and only if there exists an invertible matrix C^{11} such that*

$$A = \left[\begin{array}{c|c} I_s & 0^{01} \\ \hline 0^{10} & C^{11} \end{array} \right] \cdot B.$$

For the proof of Proposition 2, we refer the reader to the full version.

Let $\Phi = \{N(A) \mid A \in GF(2)^{n \times n} \text{ is invertible}\}$ contain a representative from each normalized equivalence class. Using Fact 1 and Proposition 2, we deduce the following corollary.

Corollary 1. *The following properties hold for normalized equivalence classes:*

1. *Each member of Φ represents a normalized equivalence class whose size is equal to the number of invertible $(n - s) \times (n - s)$ matrices C^{11}, which is $\alpha_{n-s} \cdot 2^{(n-s)^2}$.*
2. *The size of Φ is*

$$|\Phi| = \frac{\alpha_n \cdot 2^{n^2}}{\alpha_{n-s} \cdot 2^{(n-s)^2}} = \alpha_n / \alpha_{n-s} \cdot 2^{n^2 - (n-s)^2}.$$

4.4   Matrix-Vector Product

Definition 2. *Let A^{1*} and B^{*1} be two Boolean matrices such that A^{1*} has full row rank of $n - s$. Define $\check{B}_A = B \cdot \dot{A} \in GF(2)^{n \times (n-s)}$.*

When A is understood from the context, we simply write $\check{B}$ instead of $\check{B}_A$.

Remark 2. The notational conventions that apply to Definition 1 also apply Definition 2 (see Remark 1), as it is always applied to matrices $A^{1*} \in \mathrm{GF}(2)^{(n-s) \times n}$ and $B^{*1} \in \mathrm{GF}(2)^{n \times (n-s)}$, where $\check{B} \in \mathrm{GF}(2)^{n \times (n-s)}$ (and its sub-matrices are denoted using superscript).

Proposition 3. *Let A^{1*} and B^{*1} be two Boolean matrices such that A^{1*} has full row rank of $n - s$. Let $C = B^{*1} \cdot A^{1*} \in \mathrm{GF}(2)^{n \times n}$. Then, after preprocessing A^{1*} and B^{*1}, C can be represented using $b = n^2 - s^2 + n$ bits. Moreover, given $x \in \mathrm{GF}(2)^n$, the matrix-vector product Cx can be computed using $O(b)$ bit operations.*

Note that the above representation of the $n \times n$ matrix C is more efficient than the trivial representation that uses n^2 bits (ignoring the additive lower order term n). It is also more efficient than a representation that uses the decomposition $C = B^{*1} \cdot A^{1*}$ which requires $2n(n - s) = (n^2 - s^2) + (n - s)^2 \geq n^2 - s^2$ bits.

Proof. The optimized representation is obtained by "pushing" linear algebra operations from A^{1*} into B^{*1}, which "consumes" them, as formally described next. Note that since A^{1*} has full row rank of $n - s$, we use Definitions 1 and 2, and write $C = B^{*1} \cdot A^{1*} = B^{*1} \cdot (\dot{A} \cdot \dot{A}^{-1}) \cdot A^{1*} = (B^{*1} \cdot \dot{A}) \cdot (\dot{A}^{-1} \cdot A^{1*}) = \check{B} \cdot \hat{A}$, where $\check{B}$ and $\hat{A}$ can be computed during preprocessing. Let us assume that the last $n - s$ columns of A^{1*} are linearly independent (namely, $\mathrm{COL}(A^{1*}) = \{n, n-1, \ldots, s+1\}$). Then due to (1), $\hat{A}$ can be represented using $s(n - s)$ bits and the matrix-vector product Cx can be computed using $O(s(n - s) + n(n - s)) = O(n^2 - s^2)$ bit operations by computing $\hat{A}x = (\hat{A}^{-1} \cdot \ddot{A}) \cdot x[s|] + x[|n - s]$.

We assumed that the last $n - s$ columns of A^{1*} are linearly independent. If this is not the case, then $\mathrm{COL}(A^{1*})$ can be specified explicitly (to indicate the columns of $\hat{A}$ that form the identity) using at most n additional bits. The product $\hat{A}x$ is computed by decomposing x according to $\mathrm{COL}(A^{1*})$ (rather than according to its s LSBs). ∎

Remark 3. Consider the case that A^{1*} is selected uniformly at random among all matrices of full row rank. Then, using simple analysis based on Fact 1, $n - s$ linearly independent columns of A^{1*} are very likely to be found among its $n-s+3$ last columns. Consequently, the additive low-order term n in the representation size of C can be reduced to an expected size of about $3 \log n$ (specifying the 3 indices among are final $n - s + 3$ that do not belong in $\mathrm{COL}(A^{1*})$). Moreover, computing the product $\hat{A}x$ requires permuting only 3 pairs of bits of x on average (and then decomposing it as in the proof above).

5   Optimized Linear Layer Evaluation

In this section, we describe our encryption algorithm that optimizes the linear algebra of Algorithm 2. We begin by optimizing the implementation of a 2-round GLMC cipher and then consider a general r-round cipher.

It will be convenient to further simplify Algorithm 2 by defining $k_0'' = k_0' + C_0'$. For $i > 0$, we move the addition of $k_i' + C_i'$ into S_i by redefining $S_i''(x_i^{(0)}) = S_i(x_i^{(0)}) + k_i' + C_i'$. This makes the Sbox key-dependent, which is not important

Input   : x_0
Output : x_{r+1}
begin
$\quad\big|\quad x_1 \leftarrow x_0 + k_0$
$\quad\big|\quad$ **for** $i \in \{1, 2, \ldots, r\}$ **do**
$\quad\big|\quad\big|\quad y_i \leftarrow S_i(x_i^{(0)}) \| x_i^{(1)}$
$\quad\big|\quad\big|\quad x_{i+1} \leftarrow L_i(y_i)$
$\quad\big|\quad$ **end**
$\quad\big|\quad$ **return** x_{r+1}
end

Algorithm 3: Simplified encryption.

for the rest of the paper. Finally, we abuse notation for simplicity and rename k_0'' and S_i'' back to k_0 and S_i, respectively. The outcome is given in Algorithm 3.

5.1   Basic 2-Round Encryption Algorithm

We start with a basic algorithm that attempts to combine the linear algebra computation of two rounds. This computation can be written as

$$\begin{pmatrix} x_3^{(0)} \\ x_3^{(1)} \end{pmatrix} = \left[\begin{array}{c|c} L_2^{00} & L_2^{01} \\ \hline L_2^{10} & L_2^{11} \end{array}\right] \begin{pmatrix} y_2^{(0)} \\ y_2^{(1)} \end{pmatrix}, \quad \begin{pmatrix} x_2^{(0)} \\ x_2^{(1)} \end{pmatrix} = \left[\begin{array}{c|c} L_1^{00} & L_1^{01} \\ \hline L_1^{10} & L_1^{11} \end{array}\right] \begin{pmatrix} y_1^{(0)} \\ y_1^{(1)} \end{pmatrix}.$$

Note that $x_2^{(0)}$ and $y_2^{(0)}$ are related non-linearly as $y_2^{(0)} = S_2(x_2^{(0)})$. On the other hand, since $x_2^{(1)} = y_2^{(1)}$ we can compute the contribution of $y_2^{(1)}$ to x_3 at once from y_1 by partially combining the linear operations of the two rounds as

$$\begin{pmatrix} t_3^{(0)} \\ t_3^{(1)} \end{pmatrix} = \left[\begin{array}{c|c} L_2^{01}L_1^{10} & L_2^{01}L_1^{11} \\ \hline L_2^{11}L_1^{10} & L_2^{11}L_1^{11} \end{array}\right] \begin{pmatrix} y_1^{(0)} \\ y_1^{(1)} \end{pmatrix}. \tag{2}$$

The linear transformation of (2) is obtained from the product $L_2 \cdot L_1$ by ignoring the terms involving L_2^{00} and L_2^{10} (that operate on $y_2^{(0)}$). Note that (2) defines an $n \times n$ matrix that can be precomputed.

We are left to compute the contribution of $y_2^{(0)}$ to x_3, which is done directly as in Algorithm 3 by

$$x_2^{(0)} \leftarrow L_1^{0*}(y_1), \; y_2^{(0)} \leftarrow S_2(x_2^{(0)}), \; t_3' \leftarrow L_2^{*0}(y_2^{(0)}). \tag{3}$$

This calculation involves $s \times n$ and $n \times s$ matrices. Finally, combining the contributions of (2) and (3), we obtain

$$x_3 \leftarrow t_3 + t_3'.$$

Overall, the complexity of linear algebra in the two rounds is $n^2 + 2sn$ instead of $2n^2$ of Algorithm 3. This is an improvement provided that $s < n/2$, but is inefficient otherwise.

5.2 Optimized 2-Round Encryption Algorithm

The optimized algorithm requires a closer look at the linear transformation of (2). Note that this matrix can be rewritten as the product

$$\begin{pmatrix} t_3^{(0)} \\ t_3^{(1)} \end{pmatrix} = \begin{bmatrix} L_2^{01} \\ L_2^{11} \end{bmatrix} \begin{bmatrix} L_1^{10} | L_1^{11} \end{bmatrix} \begin{pmatrix} y_1^{(0)} \\ y_1^{(1)} \end{pmatrix}. \tag{4}$$

More compactly, this $n \times n$ linear transformation is decomposed as $L_2^{*1} \cdot L_1^{1*}$, namely, it is a product of matrices with dimensions $(n - s) \times n$ and $n \times (n - s)$. In order to take advantage of this decomposition, we use Proposition 3 which can be applied since L_1^{1*} has full row rank of $n - s$. This reduces linear algebra complexity of $L_2^{*1} \cdot L_1^{1*}$ from n^2 to $n(n - s) + n(n - s) - (n - s)^2 = n^2 - s^2$, ignoring an additive low order term of $3 \log n$, as computed in Remark 3.

Input : x_0
Output: x_3
begin
$\quad x_1 \leftarrow x_0 + k_0$
$\quad y_1 \leftarrow S_1(x_1^{(0)}) \| x_1^{(1)}$
$\quad x_2^{(0)} \leftarrow L_1^{0*}(y_1)$
$\quad y_2^{(0)} \leftarrow S_2(x_2^{(0)})$
$\quad x_3 \leftarrow L_2^{*0}(y_2^{(0)})$
$\quad x_3 \leftarrow x_3 + \check{L}_2(\hat{L}_1(y_1))$
$\quad$ **return** x_3
end

Algorithm 4: Optimized 2-round encryption.

Input : x_0
Output: x_3
begin
$\quad x_1 \leftarrow x_0 + k_0$
$\quad y_1 \leftarrow S_1(x_1^{(0)}) \| x_1^{(1)}$
$\quad x_2^{(0)} \leftarrow L_1^{0*}(y_1)$
$\quad z_2^{(1)} \leftarrow \hat{L}_1(y_1)$
$\quad y_2^{(0)} \leftarrow S_2(x_2^{(0)})$
$\quad x_3 \leftarrow L_2^{*0}(y_2^{(0)})$
$\quad x_3 \leftarrow x_3 + \check{L}_2(z_2^{(1)})$
$\quad$ **return** x_3
end

Algorithm 5: Refactored 2-round encryption.

Algorithm 4 exploits the decomposition $L_2^{*1} \cdot L_1^{1*} = \check{L}_2 \cdot \hat{L}_1$. Altogether, the linear algebra complexity of 2 rounds is reduced to

$$n^2 + 2sn - s^2 = 2n^2 - (n - s)^2$$

(or $2n^2 - (n - s)^2 + 3 \log n$ after taking Remark 3 into account). This is an improvement by an additive factor of about s^2 compared to the basic 2-round algorithm above and is an improvement over the standard complexity of $2n^2$ for essentially all $s < n$.

5.3 Towards an Optimized r-Round Encryption Algorithm

The optimization applied in the 2-round algorithm does not seem to generalize to an arbitrary number of rounds in a straightforward manner. In fact, there is more than one way to generalize this algorithm (and obtain improvements over the standard one in some cases) using variants of the basic algorithm of Sect. 5.1

which directly combines more that two rounds. These variants are sub-optimal since they do not exploit the full potential of Proposition 3.

The optimal algorithm is still not evident since the structure of the rounds of Algorithm 4 does not resemble their structure in Algorithm 3 that we started with. Consequently, we rewrite it in Algorithm 5 such that $z_2^{(1)} = \hat{L}_1(y_1)$ is computed already in round 1 instead of round 2. The linear algebra in round 2 of Algorithm 5 can now be described using the $n \times n$ transformation

$$\begin{pmatrix} x_3^{(0)} \\ x_3^{(1)} \end{pmatrix} = \begin{bmatrix} L_2^{00} & \check{L}_2^{01} \\ L_2^{10} & \check{L}_2^{11} \end{bmatrix} \begin{pmatrix} y_2^{(0)} \\ z_2^{(1)} \end{pmatrix}.$$

Note that $z_2^{(1)}$ is a value that is never computed by the original Algorithm 3.

When we add additional encryption rounds, we can apply Proposition 3 again and "push" some of the linear algebra of round 2 into round 3, then "push" some of the linear algebra of round 3 into round 4, etc. The full algorithm is described in detail next.

5.4 Optimized r-Round Encryption Algorithm

In this section, we describe our optimized algorithm for evaluating r rounds of a GLMC cipher. We begin by defining the following sequence of matrices.

$$\text{For } i = 1: \qquad\qquad R_1^{1*} = L_1^{1*}$$
$$\hat{R}_1 = (\dot{R}_1)^{-1} \cdot R_1^{1*}.$$
$$\text{For } 2 \leq i \leq r-1: \qquad\qquad \check{T}_i = L_i^{*1} \cdot \dot{R}_{i-1}$$
$$R_i^{1*} = L_i^{10} \| \check{T}_i^{11}.$$
$$\hat{R}_i = (\dot{R}_i)^{-1} \cdot R_i^{1*}.$$
$$\text{For } i = r: \qquad\qquad \check{T}_r = L_r^{*1} \cdot \dot{R}_{r-1}.$$

Basically, the matrix $\check{T}_i$ combines the linear algebra of round i with the linear algebra that is pushed from the previous round (represented by $\dot{R}_{i-1}$). The matrix $\hat{R}_i$ is the source of optimization, computed by normalizing the updated round matrix (after computing $\check{T}_i$). The byproduct of this normalization is $\dot{R}_i$, which is pushed into round $i + 1$, and so forth.

Before we continue, we need to prove the following claim (the proof is given in the full version).

Proposition 4. *The matrix R_i^{1*} has full row rank of $n - s$ for all $i \in \{1, \dots, r - 1\}$, hence $(\dot{R}_i)^{-1}$ exists.*

The general optimized encryption algorithm is given in Algorithm 6. At a high level, the first round can be viewed as mapping the "real state" $(y_1^{(0)}, y_1^{(1)})$ into the "shadow state" $(x_2^{(0)}, z_2^{(1)})$ using the linear transformation

$$\begin{pmatrix} x_2^{(0)} \\ z_2^{(1)} \end{pmatrix} = \begin{bmatrix} L_1^{00} & L_1^{01} \\ \hat{R}_1^{10} & \hat{R}_1^{11} \end{bmatrix} \begin{pmatrix} y_1^{(0)} \\ y_1^{(1)} \end{pmatrix}.$$

Input : x_0
Output : x_{r+1}
begin
    $x_1 \leftarrow x_0 + k_0$
    $y_1 \leftarrow S_1(x_1^{(0)}) \| x_1^{(1)}$ ▷ Round 1
    $x_2^{(0)} \leftarrow L_1^{0*}(y_1)$
    $z_2^{(1)} \leftarrow \hat{R}_1(y_1)$
    for $i \in \{2, \ldots, r-1\}$ **do**
        $y_i^{(0)} \leftarrow S_i(x_i^{(0)})$ ▷ Round i
        $x_{i+1}^{(0)} \leftarrow L_i^{00}(y_i^{(0)}) + \check{T}_i^{01}(z_i^{(1)})$
        $z_{i+1}^{(1)} \leftarrow \hat{R}_i(y_i^{(0)} \| z_i^{(1)})$
    end
    $y_r^{(0)} \leftarrow S_r(x_r^{(0)})$ ▷ Round r
    $x_{r+1} \leftarrow L_r^{*0}(y_r^{(0)}) + \check{T}_r(z_r^{(1)})$
    return x_{r+1}
end

Algorithm 6: Optimized r-round encryption.

In rounds $i \in \{2, \ldots, r-1\}$, the shadow state $(y_i^{(0)}, z_i^{(1)})$ (obtained after applying $S_i(x_i^{(0)})$) is mapped to the next shadow state $(x_{i+1}^{(0)}, z_{i+1}^{(1)})$ using the linear transformation

$$\begin{pmatrix} x_{i+1}^{(0)} \\ z_{i+1}^{(1)} \end{pmatrix} = \left[\begin{array}{c|c} L_i^{00} & \check{T}_i^{01} \\ \hline \hat{R}_i^{10} & \hat{R}_i^{11} \end{array} \right] \begin{pmatrix} y_i^{(0)} \\ z_i^{(1)} \end{pmatrix}.$$

Finally, in round r, the shadow state $(y_r^{(0)}, z_r^{(1)})$ is mapped to the final real state $(x_{r+1}^{(0)}, x_{r+1}^{(1)})$ using the linear transformation

$$\begin{pmatrix} x_{r+1}^{(0)} \\ x_{r+1}^{(1)} \end{pmatrix} = \left[\begin{array}{c|c} L_r^{00} & \check{T}_r^{01} \\ \hline L_r^{10} & \check{T}_r^{11} \end{array} \right] \begin{pmatrix} y_r^{(0)} \\ z_r^{(1)} \end{pmatrix}.$$

Complexity Evaluation. As noted above, Algorithm 6 applies r linear transformation, each of dimension $n \times n$. Hence, ignoring the linear algebra optimizations for each $\hat{R}_i$, the linear algebra complexity of each round is n^2, leading to a total complexity of $r \cdot n^2$. Taking the optimizations into account, for each $i \in \{1, \ldots, r-1\}$, the actual linear algebra complexity of $\hat{R}_i$ is reduced by $(n-s)^2$ to $n^2 - (n-s)^2$ (as $\hat{R}_i$ contains the $(n-s) \times (n-s)$ identity matrix). Therefore, the total linear algebra complexity is

$$r \cdot n^2 - (r-1)(n-s)^2.$$

Taking Remark 3 into account, we need to add another factor of $3(r-1)\log n$.

Remark 4. Note that Algorithm 6 is obtained from Algorithm 3 independently of how the instances of the cipher are generated. Hence, Algorithm 6 is applicable in principle to all SP-networks with partial non-linear layers.

Correctness. We now prove correctness of Algorithm 6 by showing that its output value is identical to a standard implementation of the scheme in Algorithm 3. For each $i \in \{0, 1, \ldots, r+1\}$, denote by $\bar{x}_i$ the state value at the beginning of round i in a standard implementation and by $\bar{y}_i$ the state after the application of S_i. The proof of Proposition 5 are given in the full version.

Proposition 5. *For each $i \in \{1, \ldots, r-1\}$ in Algorithm 6, $y_i^{(0)} = \bar{y}_i^{(0)}, x_{i+1}^{(0)} = \bar{x}_{i+1}^{(0)}$ and $z_{i+1}^{(1)} = (\dot{R}_i)^{-1}(\bar{x}_{i+1}^{(1)})$.*

Proposition 6. *Algorithm 6 is correct, namely $x_{r+1} = \bar{x}_{r+1}$.*

Proof. By Algorithm 6 and using Proposition 5,

$$x_{r+1} = L_r^{*0}(y_r^{(0)}) + \check{T}_r(z_r^{(1)}) = L_r^{*0}(\bar{y}_r^{(0)}) + L_r^{*1} \cdot \dot{R}_{r-1}\big((\dot{R}_{r-1})^{-1}(\bar{x}_r^{(1)})\big)$$
$$= L_r^{*0}(\bar{y}_r^{(0)}) + L_r^{*1}(\bar{y}_r^{(1)}) = L_r(\bar{y}_r) = \bar{x}_{r+1}.$$

■

6   Applications to LowMC in Picnic and Garbled Circuits

To verify the expected performance and memory improvements, we evaluate both suggested optimizations in three scenarios: LowMC encryption, the digital signature scheme Picnic, and in the context of Yao's garbled circuits. We discuss the details on the choice of LowMC instances and how LowMC is used in Picnic and garbled circuits and their applications in the full version. Throughout this section, we benchmark LowMC instances with block size n, non-linear layer size s and r rounds and simply refer to them as LowMC-n-s-r. For the evaluation in the context of Picnic, we integrated our optimizations in the SIMD-optimized implementation available on GitHub.[8] For the evaluation in a garbled circuit framework, we implement it from scratch. All benchmarks presented in this section were performed on an Intel Core i7-4790 running Ubuntu 18.04.

6.1   LowMC

We first present benchmarking results for encryption of LowMC instances selected for the Picnic use-case, i.e., with data complexity 1, and $s = 3$, as well as the instances currently used in Picnic with $s = 30$. While the optimized round key computation and constant addition (ORKC, Sect. 3) already reduces the runtime of a single encryption by half, which we would also obtain by pre-computing the round keys (when not used inside Picnic), the optimized linear layer evaluation (OLLE, Sect. 5) significantly reduces the runtime even using a SIMD optimized implementation. For $s = 30$, we achieve improvements by a factor up to 2.82x and for $s = 3$ up to a factor of 16.18x, bringing the performance of the instances with only one Sbox close to ones with more Sboxes.

---

[8] See https://github.com/IAIK/Picnic for the integration in Picnic and https://github.com/IAIK/Picnic-LowMC for the matrix generation.

Table 3. Benchmarks (R) of LowMC-n-s-r instances using SIMD, without optimization, with ORKC, and OLLE (in μs). Sizes (S) of matrices and constants stored in compiled implementation (in KB).

LowMC-n-s-r		w/o opt.	With ORKC	With OLLE	Improv. (old/new)
128-30-20	R	3.29	2.36	2.33	1.41x
	S	84.2	55.0	35.4	2.38x
192-30-30	R	10.03	5.64	4.04	2.48x
	S	369.8	211.2	92.8	3.99x
256-30-38	R	16.41	9.21	5.81	2.82x
	S	620.8	353.5	128.3	4.84x
128-3-182	R	30.93	17.13	4.71	6.57x
	S	749.9	383.9	45.4	16.51x
192-3-284	R	90.99	47.32	7.91	11.50x
	S	3449.5	1743.2	108.3	31.85x
256-3-363	R	167.05	78.64	10.32	16.18x
	S	5861.4	2963.7	148.5	39.48x

Table 4. Benchmarks of PICNIC-n-s-r using SIMD without optimizations, with ORKC, and OLLE (in ms).

Parameters	w/o opt.		With ORKC		With OLLE		Improv. (old/new)	
	Sign	Verify	Sign	Verify	Sign	Verify	Sign	Verify
PICNIC-128-30-20	3.56	2.41	2.71	1.89	2.65	1.87	1.34x	1.29x
PICNIC-192-30-30	10.91	7.76	7.52	5.22	6.33	4.44	1.72x	1.75x
PICNIC-256-30-38	22.80	15.63	15.41	10.82	11.37	7.88	2.01x	1.98x
PICNIC-128-3-182	20.49	14.23	11.78	8.28	4.32	3.11	4.74x	4.57x
PICNIC-192-3-284	80.76	58.23	42.85	29.94	10.13	7.29	7.97x	7.99x
PICNIC-256-3-363	192.65	139.62	91.77	64.45	18.47	12.89	10.43x	10.83x

Memory-wise we observe huge memory reductions for the instances used in PICNIC. While ORKC reduces the required storage for the LowMC matrices and constants to about a half, OLLE further reduces memory requirements substantially. As expected, the instances with a small number of Sboxes benefit most significantly from both optimizations. For example, for LowMC-256-10-38 the matrices and constants shrink from 620.8 KB to 128.3 KB, a reduction by 79%, whereas for LowMC-256-1-363 instead of 5861.4 KB encryption requires only 148.5 KB, i.e., only 2.5% of the original size. The full benchmark results and sizes of the involved matrices and constants are given in Table 3.

6.2 Picnic

We continue with evaluating our optimizations in PICNIC itself. In Table 4 we present the numbers obtained from benchmarking PICNIC with the original

Table 5. Benchmarks of LowMC-n-s-r instances with standard linear layer using method of four Russians (M4RM) and OLLE (in seconds for 2^{10} circuit evaluations).

Parameters	w/o opt.	With M4RM	With OLLE	Improv. (old/new)
LowMC-128-3-287	8.46	8.01	0.69	12.26x
LowMC-192-3-413	25.26	20.59	1.54	16.40x
LowMC-256-3-537	66.50	40.88	2.69	24.72x

LowMC instances, as well as those with $s = 3$.[9] For instances with 10 Sboxes we achieve an improvement of up to a factor of 2.01x. For the extreme case using only 1 Sbox, even better improvements of up to a factor of 10.83x are possible. With OLLE those instances are close to the performance numbers of the instances with 10 Sboxes, reducing the overhead from a factor 8.4x to a factor 1.6x. Thus those instances become practically useful alternatives to obtain the smallest possible signatures.

6.3 Garbled Circuits

Finally, we evaluated LowMC in the context of garbled circuits, where we compare an implementation using the standard linear layer and round-key computation (utilizing the method of four Russians to speed up the matrix-vector products) to an implementation using our optimizations. In Table 5 we present the results of our evaluation. We focus on LowMC instances with 1 Sbox, since in the context of garbled circuits, the number of AND gates directly relates to the communication overhead. Instances with only 1 Sbox thus minimize the size of communicated data. In terms of encryption time, we observe major improvements of up to a factor of 24.72x when compared to an implementation without any optimizations, and a factor of 15.9x when compared to an implementation using the method of four Russians. Since in this type of implementation we have to operate on a bit level instead of a word or 256-bit register as in PICNIC, the large reduction of XORs has a greater effect in this scenario, especially since up to 99% of the runtime of the unoptimized GC protocol is spent evaluating the LowMC encryption circuit.

7 Optimized Sampling of Linear Layers

In this section we optimize the sampling of linear layers of generalized LowMC ciphers, assuming they are chosen uniformly at random from the set of all invertible matrices. Sampling the linear layers required by Algorithm 6 in a straightforward manner involves selecting r invertible matrices and applying additional

---

[9] PICNIC instances may internally use the Fiat-Shamir (FS) or Unruh (UR) transforms. However, as both evaluate LowMC exactly in the same way, only numbers for PICNIC instances using the FS transform are given. Namely, improvements to LowMC encryption apply to PICNIC-FS and PICNIC-UR in the same way.

linear algebra operations that transform them to normalized form. This increases
the complexity compared to merely sampling these r matrices in complexity
$O(r \cdot n^3)$ using a simple rejection sampling algorithm (or asymptotically faster
using the algorithm of [17]) and encrypting with Algorithm 3.

We show how to reduce the complexity from $O(r \cdot n^3)$ to[10]

$$O(n^3 + (r-1)(s^2 \cdot n)).$$

We also reduce the amount of (pseudo) random bits requires to sample the linear
layers from about $r \cdot n^2$ to about $r \cdot n^2 - (r-1)\big((n-s)^2 - 2(n-s)\big)$. We note that
similar (yet simpler) optimizations can be applied to sampling the key schedule
matrices of the cipher (in case it is linear and its matrices are selected at random,
as considered in Sect. 3.2).

The linear layer sampling complexity is reduced in three stages. The first
stage breaks the dependency between matrices of different rounds. The second
stage breaks the dependency in sampling the bottom part of each round matrix
(containing $n-s$ rows) from its top part. Finally, the substantial improvement in
complexity for small s is obtained in the third stage that optimizes the sampling
of the bottom part of the round matrices. Although the first two stages do not
significantly reduce the complexity, they are necessary for applying the third
stage and are interesting in their own right.

7.1   Breaking Dependencies Among Different Round Matrices

Recall that for $i \in \{2, \ldots, r\}$, the linear transformation of round i is generated
from the matrix

$$\begin{bmatrix} L_i^{00} & \check{T}_i^{01} \\ L_i^{10} & \check{T}_i^{11} \end{bmatrix} \tag{5}$$

where

$$\check{T}_i = L_i^{*1} \cdot \dot{R}_{i-1}.$$

For $i = r$, this gives the final linear transformation, while for $i < r$, the final
transformation involves applying the decomposition of Definition 1 to $L_i^{10} \| \check{T}_i^{11}$.
Since $\check{T}_i$ depends on the invertible $(n-s) \times (n-s)$ matrix $\dot{R}_{i-1}$ (computed in the
previous round), a naive linear transformation sampling algorithm would involve
computing the linear transformations in their natural order by computing $\dot{R}_{i-1}$
in round $i-1$ and using it in round i. However, this is not required, as the linear
transformation of each round can be sampled independently. Indeed, by using
Proposition 1 with the invertible matrix $B^{11} = \dot{R}_{i-1}$, we conclude that in round i
we can simply sample the matrix given in (5) as a uniform invertible $n \times n$ matrix
without ever computing $\dot{R}_{i-1}$. Therefore, the linear transformation sampling for
round r simplifies to selecting a uniform invertible $n \times n$ matrix, L_r. For rounds
$i \in \{1, \ldots, r-1\}$, we can select a uniform invertible $n \times n$ matrix, L_i, and then
normalize it and discard $\dot{R}_i$ after the process. This simplifies Algorithm 6, and
it can be rewritten as in Algorithm 7. Note that we have renamed the sequence
$\{z_i^{(1)}\}$ to $\{x_i^{(1)}\}$ for convenience.

---

[10] Further asymptotic improvements are possible using fast matrix multiplication.

Input : x_0
Output : x_{r+1}
begin
  $\quad x_1 \leftarrow x_0 + k_0$
  $\quad$ **for** $i \in \{1, \ldots, r-1\}$ **do**
  $\quad\quad y_i \leftarrow S_i(x_i^{(0)}) \| x_i^{(1)}$ $\qquad\qquad\qquad\qquad\qquad$ ▷ Round i
  $\quad\quad x_{i+1} \leftarrow L_i^{0*}(y_i) \| \hat{L}_i(y_i)$
  $\quad$ **end**
  $\quad y_r \leftarrow S_r(x_r^{(0)}) \| x_r^{(1)}$ $\qquad\qquad\qquad\qquad\qquad\quad$ ▷ Round r
  $\quad x_{r+1} \leftarrow L_r(y_r)$
  $\quad$ **return** x_{r+1}
end

Algorithm 7: Simplified and optimized r-round encryption.

7.2 Reduced Sampling Space

We examine the sample space of the linear layers more carefully.

For each of the first $r-1$ rounds, the sampling procedure for Algorithm 7 involves selecting a uniform invertible matrix and then normalizing it according to Definition 1. However, by Corollary 1, since each normalized equivalence class contains the same number of $\alpha_{n-s} \cdot 2^{(n-s)^2}$ invertible matrices, this is equivalent to directly sampling a uniform member from Φ to represent its normalized equivalence class. If we order all the matrices in Φ, then sampling from it can be done using $\log |\Phi|$ uniform bits. However, encrypting with Algorithm 7 requires an explicit representation of the matrices and using an arbitrary ordering is not efficient in terms of complexity. In the rest of this section, our goal is to optimize the complexity of sampling from Φ, but first we introduce notation for the full sampling space.

Let the set Λ_r contain r-tuples of matrices defined as

$$\Lambda_r = \Phi^{r-1} \times \{A \in \mathrm{GF}(2)^{n \times n} \text{ is invertible}\},$$

where $\Phi^{r-1} = \underbrace{\Phi \times \Phi \ldots \times \Phi}_{r-1 \text{ times}}.$

The following corollary is a direct continuation of Corollary 1.

Corollary 2. *The following properties hold:*

1. *Each r-tuple $(L_1, \ldots, L_{r-1}, L_r) \in \Lambda_r$ represents a set of size $(\alpha_{n-s})^{r-1} \cdot 2^{(r-1)(n-s)^2}$ containing r-tuples of matrices $(L_1', \ldots, L_{r-1}', L_r')$ such that*

$$\big(N(L_1'), \ldots, N(L_{r-1}'), L_r'\big) = (L_1, \ldots, L_{r-1}, L_r).$$

2. *Λ_r contains*

$$|\Lambda_r| = \frac{(\alpha_n)^r \cdot 2^{n^2}}{(\alpha_{n-s})^{r-1} \cdot 2^{(r-1)(n-s)^2}} = (\alpha_n)^r / (\alpha_{n-s})^{r-1} \cdot 2^{r \cdot n^2 - (r-1)(n-s)^2}$$

r-tuples of matrices.

As noted above, sampling from Λ_r reduces to sampling the first $r-1$ matrices uniformly from Φ and using a standard sampling algorithm for the r'th matrix.

7.3 Breaking Dependencies Between Round Sub-Matrices

We describe how to further simplify the algorithm for sampling the linear layers by breaking the dependency between sampling the bottom and top sub-matrices in each round. From this point, we will rename the round matrix L_i to a general matrix $A \in \mathrm{GF}(2)^{n \times n}$ for convenience. In order to sample from Φ, the main idea is to sample the bottom $n - s$ linearly independent rows of A first, apply the decomposition of Definition 1 and then use this decomposition in order to efficiently sample the remaining s linearly independent rows of A. Therefore, we never directly sample the larger $n \times n$ matrix, but obtain the same distribution on output matrices as the original sampling algorithm.

Sampling the Bottom Sub-Matrix. We begin by describing in Algorithm 8 how to sample and compute $\hat{B}$ (which will be placed in the bottom $n - s$ rows of A) and $\mathrm{COL}(B^{1*})$ using simple rejection sampling. It uses the sub-procedure $GenRand(n_1, n_2)$ that samples an $n_1 \times n_2$ binary matrix uniformly at random.

Correctness of the algorithm follows by construction. In terms of complexity, we keep track of the span of $\dot{B}$ using simple Gaussian elimination. Based on Fact 1, the expected complexity of (a naive implementation of) the algorithm until it succeeds is $O((n - s)^3 + s^2(n - s))$ due to Gaussian elimination and matrix multiplication.

The Optimized Round Matrix Sampling Algorithm. Let us first assume that after application of Algorithm 8, we obtain $\hat{B}, \mathrm{COL}(B^{1*})$ such that $\mathrm{COL}(B^{1*})$ includes the $n - s$ last columns (which form the identity matrix in $\hat{B}$). The matrix A is built by placing $\hat{B}$ in its bottom $n - s$ columns, and in this case it will be of the block form considered in Fact 2. There is a simple formula (stated in Fact 2) that determines if such matrices are invertible, and we can use this formula to efficiently sample the top s rows of A, while making sure that the full $n \times n$ matrix is invertible. In case $\mathrm{COL}(B^{1*})$ does not include the $n - s$ last columns, then a similar idea still applies since A would be in the special form after applying a column permutation determined by $\mathrm{COL}(B^{1*})$. Therefore, we assume that A is of the special form, sample the top s rows accordingly and then apply the inverse column permutation to these rows. Algorithm 9 gives the details of this process. It uses a column permutation matrix, denoted by P (computed from $\mathrm{COL}(B^{1*})$), such that $\hat{B} \cdot P = ((\dot{B})^{-1} \cdot \ddot{B}) \| I_{n-s}$ is of the required form. The algorithm also uses two sub-procedures:

1. $GenRand(n_1, n_2)$ samples an $n_1 \times n_2$ binary matrix uniformly at random.
2. $GenInv(n_1)$ samples a uniform invertible $n_1 \times n_1$ matrix.

The complexity of the algorithm is $O((n-s)^3 + s^2(n-s) + s^3 + s^2(n-s) + sn) = O((n - s)^3 + s^2(n - s) + s^3)$ (using naive matrix multiplication and invertible

Output : $\hat{B}, \text{COL}(B^{1*})$
begin

  $B^{1*} \leftarrow \mathbf{0}^{(n-s) \times n}, \dot{B} \leftarrow \mathbf{0}^{(n-s) \times (n-s)}$
  $\text{COL}(B^{1*}) \leftarrow \emptyset, rank \leftarrow 0$
  for $i \in \{n, n-1, \ldots, 1\}$ **do**
   $B^{1*}[*, i] \leftarrow GenRand(n-s, 1)$
   if $rank = n-s$ **or**
   $B^{1*}[*, i] \in \text{span}(\dot{B})$ **then**
    continue
   end
   $rank \leftarrow rank + 1$
   $\text{COL}(B^{1*}) \leftarrow \text{COL}(B^{1*}) \cup \{i\}$
   $\dot{B}[*, rank] \leftarrow B^{1*}[*, i]$
  end
  if $rank = n-s$ **then**
   $\hat{B} \leftarrow (\dot{B})^{-1} \cdot B^{1*}$
   return $\hat{B}, \text{COL}(B^{1*})$
  else
   return FAIL
  end
end

Algorithm 8: *SampleBottom()* iteration

Output : Round matrix for
Algorithm 7
begin

  $\hat{B}, \text{COL}(B^{1*}) \leftarrow$
  SampleBottom()
  $A^{1*} \leftarrow \hat{B}$
  $C^{00} \leftarrow GenInv(s)$
  $A'^{01} \leftarrow GenRand(s, n-s)$
  $D^{10} \leftarrow (\hat{B} \cdot P)^{10}$
  $A'^{00} \leftarrow C^{00} + A'^{01} \cdot D^{10}$
  $A^{0*} \leftarrow (A'^{00} \| A'^{01}) \cdot P^{-1}$
  return A
end

Algorithm 9: Optimized round matrix sampling.

matrix sampling algorithms), where the dominant factor for small s is $(n-s)^3$. The algorithm requires about $sn + n(n-s) = n^2$ random bits.

Proposition 7. *Algorithm 9 selects a uniform matrix in Φ, namely, the distribution of the output A is identical to the distribution generated by sampling a uniform invertible $n \times n$ matrix and applying the transformation of Definition 1 to its bottom $n - s$ rows.*

For the proof of Proposition 7 we refer the reader to the full version.

7.4   Optimized Sampling of the Bottom Sub-Matrix

For small values of s, the complexity of Algorithm 9 is dominated by Algorithm 8 (*SampleBottom()*), whose complexity is $O((n-s)^3 + s^2(n-s))$. We now show how to reduce this complexity to $O(s(n-s))$ on average. Thus, the total expected complexity of Algorithm 9 becomes

$$O(s^2(n-s) + s^3) = O(s^2 \cdot n)$$

(using naive matrix multiplication and invertible matrix sampling algorithms). Moreover, the randomness required by the algorithm is reduced from about $sn + n(n-s) = n^2$ to about

$$sn + (s+2)(n-s) = n^2 - (n-s)^2 + 2(n-s).$$

Below, we give an overview of the algorithm. Its formal description and analysis are given in the full version.

Recall that the output of $SampleBottom()$ consists of $\hat{B}, \mathrm{COL}(B^{1*})$, where $\hat{B}$ contains I_{n-s} and s additional columns of $n - s$ bits. The main idea is to directly sample $\hat{B}$ without ever sampling the full B^{1*} and normalizing it. In order to achieve this, we have to artificially determine the column set $\mathrm{COL}(B^{1*})$ (which contains the identity matrix in $\hat{B}$), and the values of the remaining s columns. The optimized algorithm simulates $SampleBottom()$ (Algorithm 8). This is performed by maintaining and updating the $\mathrm{COL}(B^{1*})$ and $rank$ variables as in $SampleBottom()$ and sampling concrete vectors only when necessary. For example, the columns of $\mathrm{COL}(B^{1*})$ are not sampled at all and will simply consist of the identity matrix in the output of the algorithm. There are 3 important cases to simulate in the optimized algorithm when considering column i:

1. In $SampleBottom()$, full rank is not reached (i.e., $rank < n - s$) and column i is added to $\mathrm{COL}(B^{1*})$. Equivalently, the currently sampled vector in $SampleBottom()$ is not in the subspace spanned by the previously sampled vectors (whose size is 2^{rank}). This occurs with probability $1 - 2^{rank}/2^{n-s} = 1 - 2^{(n-s)-rank}$ and can be simulated exactly by (at most) $(n - s) - rank$ coin tosses in the optimized algorithm (without sampling any vector).

2. In $SampleBottom()$, full rank is not reached (i.e., $rank < n - s$) and column i is not added to $\mathrm{COL}(B^{1*})$. This is the complementary event to the first, which occurs with probability $2^{(n-s)-rank}$. In $SampleBottom()$, such a column i is sampled uniformly from the subspace spanned by the previously sampled vectors whose size is 2^{rank}. The final multiplication with $(\dot{B})^{-1}$ is a change of basis which transforms the basis of the previously sampled columns to the last $rank$ vectors in the standard basis $e_{(n-s)-rank+1}, e_{(n-s)-rank+2}, \ldots, e_{n-s}$. Hence, column i is a uniform vector in the subspace spanned by $e_{(n-s)-rank+1}, e_{(n-s)-rank+2}, \ldots, e_{n-s}$ and the optimized algorithm samples a vector from this space (using $rank$ coin tosses).

3. In $SampleBottom()$, full rank is reached (i.e., $rank = n - s$). The optimized algorithm samples a uniform column using $n - s$ coin tosses. This can be viewed as a special case of the previously considered one, for $rank = n - s$.

Note that no linear algebra operations are performed by the optimized algorithm and it consists mainly of sampling operations.

8   Optimality of Linear Representation

In this section, we prove that the representation of the linear layers used by Algorithm 7 for a GLMC cipher is essentially optimal. Furthermore, we show that the number of uniform (pseudo) random bits used by the sampling algorithm derived in Sect. 7 is close to optimal. More specifically, we formulate two assumptions and prove the following theorem under these assumptions, recalling the value of $|\Lambda_r|$ from Corollary 2.

Theorem 1. *Sampling an instance of a GLMC cipher with uniform linear layers must use at least*

$$b = \log |\Lambda_r| = \log \left((\alpha_n)^r / (\alpha_{n-s})^{r-1} \cdot 2^{r \cdot n^2 - (r-1)(n-s)^2} \right)$$

$$\geq r \cdot n^2 - (r-1)(n-s)^2 - 3.5r.$$

uniform random bits and its encryption (or decryption) algorithm requires at least b bits of storage on average. Moreover, if a secure PRG is used to generate the randomness for sampling, then it must produce at least b pseudo-random bits and the encryption (and decryption) process requires at least b bits of storage on average, assuming that it does not have access to the PRG.

We mention that the theorem does not account for the storage required by the non-linear layers. The theorem implies that the code size of Algorithm 7 is optimal up to an additive factor of about $r \cdot (3.5 + 3 \log n)$, which is negligible (less than $0.01 \cdot b$ for reasonable choices of parameters).

8.1 Basic Assumptions

The proof relies on the following two assumptions regarding a GLMC cipher, which are further discussed in the full version.

1. If a PRG is used for the sampling process, it is not used during encryption.
2. The linear layers are stored in a manner which is independent of the specification of the non-linear layers. Namely, changing the specification of the non-linear layers does not affect the way that the linear layers are stored.

8.2 Model Formalization

We now define our model which formalizes the assumptions above and allows to prove the optimality of our representation.

Definition 3. *Given a triplet of global parameters (n, s, r), a (simplified) standard representation of a GLMC cipher is a triplet $\mathcal{R} = (k_0, \mathcal{S}, \mathcal{L})$ such that $k_0 \in \{0,1\}^n$, $\mathcal{S} = (S_1, S_2, \ldots, S_r)$ is an r-tuple containing the specifications of r non-linear invertible layers $S_i : \{0,1\}^s \to \{0,1\}^s$ and $\mathcal{L} = (L_1, L_2, \ldots, L_r)$ is an r-tuple of invertible matrices $L_i \in \mathrm{GF}(2)^{n \times n}$. The r-tuple $\mathcal{L}$ is called a standard linear representation.*

To simplify notation, given a standard representation $\mathcal{R} = (k_0, \mathcal{S}, \mathcal{L})$, we denote the encryption algorithm defined by Algorithm 3 as $E_{\mathcal{R}} : \{0,1\}^n \to \{0,1\}^n$.

Definition 4. *Two standard cipher representations $\mathcal{R}, \mathcal{R}'$ are equivalent (denoted $\mathcal{R} \equiv \mathcal{R}'$) if for each $x \in \{0,1\}^n$, $E_{\mathcal{R}}(x) = E_{\mathcal{R}'}(x)$.*

Definition 5. *Two standard linear representations $\mathcal{L}, \mathcal{L}'$ are equivalent (denoted $\mathcal{L} \equiv \mathcal{L}'$) if for each tuple of non-linear layers $\mathcal{S}$, and key k_0, $(k_0, \mathcal{S}, \mathcal{L}) \equiv (k_0, \mathcal{S}, \mathcal{L}')$.*

The requirement that $(k_0, \mathcal{S}, \mathcal{L}) \equiv (k_0, \mathcal{S}, \mathcal{L}')$ for *any* $\mathcal{S}, k_0$ captures the second assumption of Sect. 8.1 that a standard representation of the linear layers is independent of the non-linear layers (and the key).

Clearly, the linear equivalence relation partitions the r-tuples of standard linear representations into linear equivalence classes. It is important to mention that Theorem 1 does not assume that the encryption algorithm uses Algorithm 3 or represents the linear layers as an r-tuple of matrices. These definitions are merely used in its proof, as shown next.

8.3 Proof of Theorem 1

We will prove the following lemma regarding linear equivalence classes, from which Theorem 1 is easily derived.

Lemma 1. *For any* $\mathcal{L} \neq \mathcal{L}' \in \Lambda_r$, $\mathcal{L} \not\equiv \mathcal{L}'$.

The lemma states that each r-tuple of Λ_r is a member of a distinct equivalence class, implying that we have precisely identified the equivalence classes.

Proof (of Theorem 1). Lemma 1 asserts that there are at least $|\Lambda_r|$ linear equivalence classes. Corollary 2 asserts that each r-tuple in Λ_r represents a set of linear layers of size $(\alpha_{n-s})^{r-1} \cdot 2^{(r-1)(n-s)^2}$, hence every r-tuple in Λ_r has the same probability weight when sampling the r linear layers uniformly at random. The theorem follows from the well-known information theoretic fact that sampling and representing a uniform string (an r-tuple in Λ_r) chosen out of a set of 2^t strings requires at least t bits on average (regardless of any specific sampling or representation methods).   ∎

The proof of Lemma 1 relies on two propositions which are implications of the definition of equivalence of standard linear representations (Definition 5).

Proposition 8. *Let* $\mathcal{L} \equiv \mathcal{L}'$ *be two equivalent standard linear representations. Given* $k_0, \mathcal{S}$, *let* $\mathcal{R} = (k_0, \mathcal{S}, \mathcal{L})$ *and* $\mathcal{R}' = (k_0, \mathcal{S}, \mathcal{L}')$. *Fix any* $x \in \{0,1\}^n$ *and* $i \in \{0, 1, \ldots, r+1\}$, *and denote by* x_i *(resp.* x'_i) *the value* $E_{\mathcal{R}}(x)$ *(resp.* $E_{\mathcal{R}'}(x)$) *at the beginning of round* i. *Then* $x_i^{(0)} = x_i'^{(0)}$.

Namely, non-linear layer inputs (and outputs) have to match at each round when encrypting the same plaintext with ciphers instantiated with equivalent standard linear representations (and use the same key and non-linear layers).

Proposition 9. *Let* $\mathcal{L} \equiv \mathcal{L}'$ *be two equivalent standard linear representations. Given* $k_0, \mathcal{S}$, *let* $\mathcal{R} = (k_0, \mathcal{S}, \mathcal{L})$ *and* $\mathcal{R}' = (k_0, \mathcal{S}, \mathcal{L}')$. *Fix any* $x \in \{0,1\}^n$ *and* $i \in \{0, 1, \ldots, r+1\}$, *and denote by* x_i *(resp.* x'_i) *the value* $E_{\mathcal{R}}(x)$ *(resp.* $E_{\mathcal{R}'}(x)$) *at the beginning of round* i. *Moreover, fix* $\bar{x} \neq x$ *such that* $\bar{x}_i = \bar{x}_i^{(0)}, \bar{x}_i^{(1)}$, *where* $\bar{x}_i^{(0)} \neq x_i^{(0)}$, *but* $\bar{x}_i^{(1)} = x_i^{(1)}$. *Then,* $\bar{x}_i'^{(1)} = x_i'^{(1)}$.

The proposition considers two plaintexts x and $\bar{x}$ whose encryptions under the first cipher in round i differ only in the 0 part of the state. We then look at the second cipher (formed using equivalent standard linear representations) and claim that the same property must hold for it as well. Namely, the encryptions of x and $\bar{x}$ under the second cipher in round i differ only on the 0 part of the state. For the proofs of Propositions 8 and 9 and Lemma 1, we refer the reader to the full version.

9   Conclusions

SP-networks with partial non-linear layers (i.e., $s < n$) have shown to be beneficial in several applications that require minimizing the AND count of the cipher. Initial cryptanalytic results analyzing ciphers built with this recent design strategy contributed to our understanding of their security. In this paper, we contribute to the efficient implementation of these SP-networks. In particular, we redesign the linear layers of LowMC instances with $s < n$ in a way that does not change their specifications, but significantly improves their performance. We believe that our work will enable designing even more efficient SP-networks with $s < n$ by using our optimizations as a starting point, allowing to use this design strategy in new applications.

Acknowledgements. We thank Tyge Tiessen for interesting ideas and discussions on optimizing LowMC's round key computation. I. Dinur has been supported by the Israeli Science Foundation through grant n°573/16 and by the European Research Council under the ERC starting grant agreement n°757731 (LightCrypt). D. Kales has been supported by IOV42. S. Ramacher, and C. Rechberger have been supported by EU H2020 project PRISMACLOUD, grant agreement n°644962. S. Ramacher has additionally been supported by A-SIT. C. Rechberger has additionally been supported by EU H2020 project PQCRYPTO, grant agreement n°645622.

References

1. Albrecht, M.R., Bard, G.V., Hart, W.: Algorithm 898: efficient multiplication of dense matrices over GF(2). ACM Trans. Math. Softw. **37**(1), 9:1–9:14 (2010)
2. Albrecht, M.R., Rechberger, C., Schneider, T., Tiessen, T., Zohner, M.: Ciphers for MPC and FHE. In: Oswald, E., Fischlin, M. (eds.) EUROCRYPT 2015. LNCS, vol. 9056, pp. 430–454. Springer, Heidelberg (2015). https://doi.org/10.1007/978-3-662-46800-5_17
3. Bar-On, A., Dinur, I., Dunkelman, O., Lallemand, V., Keller, N., Tsaban, B.: Cryptanalysis of SP networks with partial non-linear layers. In: Oswald, E., Fischlin, M. (eds.) EUROCRYPT 2015. LNCS, vol. 9056, pp. 315–342. Springer, Heidelberg (2015). https://doi.org/10.1007/978-3-662-46800-5_13
4. Barkan, E., Biham, E.: In how many ways can you write Rijndael? In: Zheng, Y. (ed.) ASIACRYPT 2002. LNCS, vol. 2501, pp. 160–175. Springer, Heidelberg (2002). https://doi.org/10.1007/3-540-36178-2_10

5. Biryukov, A., De Cannière, C., Braeken, A., Preneel, B.: A toolbox for cryptanalysis: linear and affine equivalence algorithms. In: Biham, E. (ed.) EUROCRYPT 2003. LNCS, vol. 2656, pp. 33–50. Springer, Heidelberg (2003). https://doi.org/10.1007/3-540-39200-9_3

6. Boneh, D., Eskandarian, S., Fisch, B.: Post-quantum group signatures from symmetric primitives. IACR ePrint **2018**, 261 (2018)

7. Chase, M., et al.: Post-quantum zero-knowledge and signatures from symmetric-key primitives. In: CCS, pp. 1825–1842. ACM (2017)

8. Chase, M., et al.: The picnic signature algorithm specification (2017). https://github.com/Microsoft/Picnic/blob/master/spec.pdf

9. Derler, D., Ramacher, S., Slamanig, D.: Generic double-authentication preventing signatures and a post-quantum instantiation. In: Baek, J., Susilo, W., Kim, J. (eds.) ProvSec 2018. LNCS, vol. 11192, pp. 258–276. Springer, Cham (2018). https://doi.org/10.1007/978-3-030-01446-9_15

10. Derler, D., Ramacher, S., Slamanig, D.: Post-quantum zero-knowledge proofs for accumulators with applications to ring signatures from symmetric-key primitives. In: Lange, T., Steinwandt, R. (eds.) PQCrypto 2018. LNCS, vol. 10786, pp. 419–440. Springer, Cham (2018). https://doi.org/10.1007/978-3-319-79063-3_20

11. Dobraunig, C., Eichlseder, M., Grassi, L., Lallemand, V., Leander, G., List, E., Mendel, F., Rechberger, C.: Rasta: a cipher with low ANDdepth and few ANDs per bit. In: Shacham, H., Boldyreva, A. (eds.) CRYPTO 2018. LNCS, vol. 10991, pp. 662–692. Springer, Cham (2018). https://doi.org/10.1007/978-3-319-96884-1_22

12. Gérard, B., Grosso, V., Naya-Plasencia, M., Standaert, F.-X.: Block ciphers that are easier to mask: how far can we go? In: Bertoni, G., Coron, J.-S. (eds.) CHES 2013. LNCS, vol. 8086, pp. 383–399. Springer, Heidelberg (2013). https://doi.org/10.1007/978-3-642-40349-1_22

13. Goldreich, O., Micali, S., Wigderson, A.: How to play any mental game or a completeness theorem for protocols with honest majority. In: STOC, pp. 218–229. ACM (1987)

14. Katz, J., Kolesnikov, V., Wang, X.: Improved non-interactive zero knowledge with applications to post-quantum signatures. In: CCS, pp. 525–537. ACM (2018)

15. Kolchin, V.F.: Random Graphs. Cambridge University Press, Cambridge (1999)

16. Kolesnikov, V., Schneider, T.: Improved garbled circuit: free XOR gates and applications. In: Aceto, L., Damgård, I., Goldberg, L.A., Halldórsson, M.M., Ingólfsdóttir, A., Walukiewicz, I. (eds.) ICALP 2008. LNCS, vol. 5126, pp. 486–498. Springer, Heidelberg (2008). https://doi.org/10.1007/978-3-540-70583-3_40

17. Randall, D.: Efficient generation of random nonsingular matrices. Random Struct. Algorithms **4**(1), 111–118 (1993)

18. Rasoolzadeh, S., Ahmadian, Z., Salmasizadeh, M., Aref, M.R.: Total break of Zorro using linear and differential attacks. ISeCure ISC Int. J. Inf. Secur. **6**(1), 23–34 (2014)

19. Wang, Y., Wu, W., Guo, Z., Yu, X.: Differential cryptanalysis and linear distinguisher of full-round Zorro. In: Boureanu, I., Owesarski, P., Vaudenay, S. (eds.) ACNS 2014. LNCS, vol. 8479, pp. 308–323. Springer, Cham (2014). https://doi.org/10.1007/978-3-319-07536-5_19

20. Yao, A.C.: How to generate and exchange secrets (extended abstract). In: FOCS, pp. 162–167. IEEE Computer Society (1986)

Differential Privacy

Distributed Differential
Privacy via Shuffling

Albert Cheu[1]([⊠]), Adam Smith[2], Jonathan Ullman[1],
David Zeber[3], and Maxim Zhilyaev[4]

[1] Khoury College of Computer Sciences, Northeastern University, Boston, USA
cheu.a@husky.neu.edu, jullman@ccs.neu.edu
[2] Computer Science Department, Boston University, Boston, USA
ads22@bu.edu
[3] Mozilla Foundation, Mountain View, USA
dzeber@mozilla.com
[4] Mountain View, USA

Abstract. We consider the problem of designing scalable, robust protocols for computing statistics about sensitive data. Specifically, we look at how best to design differentially private protocols in a distributed setting, where each user holds a private datum. The literature has mostly considered two models: the "central" model, in which a trusted server collects users' data in the clear, which allows greater accuracy; and the "local" model, in which users individually randomize their data, and need not trust the server, but accuracy is limited. Attempts to achieve the accuracy of the central model without a trusted server have so far focused on variants of cryptographic multiparty computation (MPC), which limits scalability.

In this paper, we initiate the analytic study of a *shuffled model* for distributed differentially private algorithms, which lies between the local and central models. This simple-to-implement model, a special case of the ESA framework of [5], augments the local model with an anonymous channel that randomly permutes a set of user-supplied messages. For sum queries, we show that this model provides the power of the central model while avoiding the need to trust a central server and the complexity of cryptographic secure function evaluation. More generally, we give evidence that the power of the shuffled model lies strictly between those of the central and local models: for a natural restriction of the model, we show that shuffled protocols for a widely studied *selection* problem require exponentially higher sample complexity than do central-model protocols.

1 Introduction

The past few years has seen a wave of commercially deployed systems [17,29] for analysis of users' sensitive data in the *local model of differential privacy (LDP)*. LDP systems have several features that make them attractive in practice, and

The full version of this paper is accessible on arXiv.

© International Association for Cryptologic Research 2019
Y. Ishai and V. Rijmen (Eds.): EUROCRYPT 2019, LNCS 11476, pp. 375–403, 2019.
https://doi.org/10.1007/978-3-030-17653-2_13

limit the barriers to adoption. Each user only sends private data to the data collector, so users do not need to fully trust the collector, and the collector is not saddled with legal or ethical obligations. Moreover, these protocols are relatively simple and scalable, typically requiring each party to asynchronously send just a single short message.

However, the local model imposes strong constraints on the utility of the algorithm. These constraints preclude the most useful differentially private algorithms, which require a *central model* where the users' data is sent in the clear, and the data collector is trusted to perform only differentially private computations. Compared to the central model, the local model requires enormous amounts of data, both in theory and in practice (see e.g. [20] and the discussion in [5]). Unsurprisingly, the local model has so far only been used by large corporations like Apple and Google with billions of users.

In principle, there is no dilemma between the central and local models, as any algorithm can be implemented without a trusted data collector using cryptographic *multiparty computation (MPC)*. However, despite dramatic recent progress in the area of practical MPC, existing techniques still require large costs in terms of computation, communication, and number of rounds of interaction between the users and data collector, and are considerably more difficult for companies to extend and maintain.

In this work, we initiate the analytic study of an intermediate model for distributed differential privacy called the *shuffled model*. This model, a special case of the ESA framework of [5], augments the standard model of local differential privacy with an anonymous channel (also called a shuffler) that collects messages from the users, randomly permutes them, and then forwards them to the data collector for analysis. For certain applications, this model overcomes the limitations on accuracy of local algorithms while preserving many of their desirable features. However, under natural constraints, this model is dramatically weaker than the central model. In more detail, we make two primary contributions:

- We give a simple, non-interactive algorithm in the shuffled model for estimating a single Boolean-valued statistical query (also known as a counting query) that essentially matches the error achievable by centralized algorithms. We also show how to extend this algorithm to estimate a bounded real-valued statistical query, albeit at an additional cost in communication. These protocols are sufficient to implement any algorithm in the *statistical queries model* [22], which includes methods such as gradient descent.
- We consider the ubiquitous *variable-selection problem*—a simple but canonical optimization problem. Given a set of counting queries, the variable-selection problem is to identify the query with nearly largest value (i.e. an "approximate argmax"). We prove that the sample complexity of variable selection in a natural restriction of the shuffled model is exponentially larger than in the central model. The restriction is that each user send only a single message into the shuffle, as opposed to a set of messages, which we call this the *one-message* shuffled model. Our positive results show that the sample complexity in the shuffled model is polynomially smaller than in the local

model. Taken together, our results give evidence that the central, shuffled, and local models are strictly ordered in the accuracy they can achieve for selection. Our lower bounds follow from a structural result showing that any algorithm that is private in the one-message shuffled model is also private in the local model with weak, but non-trivial, parameters.

In concurrent and independent work, Erlingsson et al. [16] give conceptually similar positive results for local protocols aided by a shuffler. We give a more detailed comparison between our work and theirs after giving a thorough description of the model and our results (Sect. 2.3)

1.1 Background and Related Work

Models for Differentially Private Algorithms. Differential privacy [14] is a restriction on the algorithm that processes a dataset to provide statistical summaries or other output. It ensures that, no matter what an attacker learns by interacting with the algorithm, it would have learned nearly the same thing whether or not the dataset contained any particular individual's data [21]. Differential privacy is now widely studied, and algorithms satisfying the criterion are increasingly deployed [1,17,24].

There are two well-studied models for implementing differentially-private algorithms. In the *central model*, raw data are collected at a central server where they are processed by a differentially private algorithm. In the *local model* [14,18,33], each individual applies a differentially private algorithm locally to their data and shares only the output of the algorithm—called a report or response—with a server that aggregates users' reports. The local model allows individuals to retain control of their data since privacy guarantees are enforced directly by their devices. It avoids the need for a single, widely-trusted entity and the resulting single point of security failure. The local model has witnessed an explosion of research in recent years, ranging from theoretical work to deployed implementations. A complete survey is beyond the scope of this paper.

Unfortunately, for most tasks there is a large, unavoidable gap between the accuracy that is achievable in the two models. [4] and [8] show that estimating the sum of bits, one held by each player, requires error $\Omega(\sqrt{n}/\varepsilon)$ in the local model, while an error of just $O(1/\varepsilon)$ is possible the central model. [12] extended this lower bound to a wide range of natural problems, showing that the error must blowup by at least $\Omega(\sqrt{n})$, and often by an additional factor growing with the data dimension. More abstractly, [20] showed that the power of the local model is equivalent to the *statistical query model* [22] from learning theory. They used this to show an exponential separation between the accuracy and sample complexity of local and central algorithms. Subsequently, an even more natural separation arose for the variable-selection problem [12,31], which we also consider in this work.

Implementing Central-Model Algorithms in Distributed Models. In principle, one could also use the powerful, general tools of modern cryptography, such as multiparty computation (MPC), or secure function evaluation, to simulate central model algorithms in a setting without a trusted server [13], but such algorithms currently impose bandwidth and liveness constraints that make them impractical for large deployments. In contrast, Google [17] now uses local differentially private protocols to collect certain usage statistics from hundreds of millions of users' devices.

A number of specific, efficient MPC algorithms have been proposed for differentially private functionalities. They generally either (1) focus on simple summations and require a single "semi-honest"/"honest-but-curious" server that aggregates user answers, as in [6,9,26] ; or (2) allow general computations, but require a network of servers, a majority of whom are assumed to behave honestly, as in [11]. As they currently stand, these approaches have a number of drawbacks: they either require users to trust that a server maintained by a service provided is behaving (semi-)honestly, or they require that a coalition of service providers collaborate to run protocols that reveal to each other who their users are and *what computations they are performing on their users' data.* It is possible to avoid these issues by combining anonymous communication layers and MPC protocols for universal circuits but, with current techniques, such modifications destroy the efficiency gains relative to generic MPC.

Thus, a natural question—relevant no matter how the state of the art in MPC evolves—is to identify simple (and even minimal) primitives that can be implemented via MPC in a distributed model and are expressive enough to allow for sophisticated private data analysis. In this paper, we show that shuffling is a powerful primitive for differentially private algorithms.

Mixnets. One way to realize the shuffling functionality is via a mixnet. A *mix network*, or *mixnet*, is a protocol involving several computers that takes as input a sequence of encrypted messages, and outputs a uniformly random permutation of those messages' plaintexts. Introduced by [10], the basic idea now exists in many variations. In its simplest instantiation, the network consists of a sequence of servers, whose identities and ordering are public information.[1] Messages, each one encrypted with all the servers' keys, are submitted by users to the first server. Once enough messages have been submitted, each server in turn performs a *shuffle* in which the server removes one layer of encryption and sends a permutation of the messages to the next server. In a *verifiable shuffle*, the server also produces a cryptographic proof that the shuffle preserved the multi-set of messages. The final server sends the messages to their final recipients, which might be different for each message. A variety of efficient implementations of mixnets with verifiable shuffles exist (see, e.g., [5,23] and citations therein).

Another line of work [19,30] shows how to use differential privacy *in addition* to mixnets to make communication patterns differentially private for the purposes of anonymous computation. Despite the superficial similarly, this line of work is orthogonal to ours, which is about how to use mixnets themselves to achieve (more accurate) differentially private data analysis.

---

[1] Variations on this idea based on *onion routing* allow the user to specify a secret path through a network of mixes.

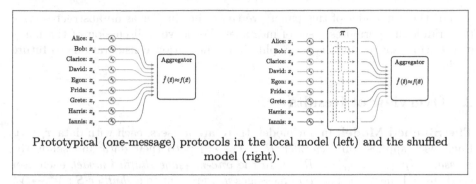

Prototypical (one-message) protocols in the local model (left) and the shuffled model (right).

Shufflers as a Primitive for Private Data Analysis. This paper studies how to use a shuffler (e.g. a mixnet) as a cryptographic primitive to implement differentially-private algorithms. Bittau et al. [5] propose a general framework, dubbed *encode-shuffle-analyze* (or *ESA*), which generalizes the local and central models by allowing a local randomized encoding step E performed on user devices, a permutation step S in which encrypted encodings are shuffled, and a final randomized process A that analyzes the permuted encodings. We ask what privacy guarantee can be provided if we rely only on the local encoding E and the shuffle S—the analyst A is untrusted. In particular, we are interested in protocols that are substantially more accurate than is possible in the local model (in which the privacy guarantee relies entirely on the encoding E). This general question was left open by [5].

One may think of the shuffled model as specifying a highly restricted MPC primitive on which we hope to base privacy. Relative to general MPC, the use of mixnets for shuffling provides several advantages: First, there already exist a number of highly efficient implementations. Second, their trust model is simple and robust—as long as a single one of the servers performs its shuffle honestly, the entire process is a uniformly random permutation, and our protocols' privacy guarantees will hold. The architecture and trust guarantees are also easy to explain to nonexperts (say, with metaphors of shuffled cards or shell games). Finally, mixnets automatically provide a number of additional features that are desirable for data collection: they can maintain secrecy of a company's user base, since each company's users could use that company's server as their first hop; and they can maintain secrecy of the company's computations, since the specific computation is done by the analyst. Note that we think of a mixnet here as operating on large batches of messages, whose size is denoted by n. (In implementation, this requires a fair amount of latency, as the collection point must receive sufficiently many messages before proceeding—see Bittau et al. [5]).

Understanding the possibilities and limitations of shuffled protocols for private data analysis is interesting from both theoretical and practical perspectives. It provides an intermediate abstraction, and we give evidence that it lies strictly between the central and local models. Thus, it sheds light on the minimal cryptographic primitives needed to get the central model's accuracy. It also provides an attractive platform for near-term deployment [5], for the reasons listed above.

For the remainder of this paper, we treat the shuffler as an abstract service that randomly permutes a set of messages. We leave a discussion of the many engineering, social, and cryptographic implementation considerations to future work.

2   Overview of Results

The Shuffled Model. In our model, there are n users, each with data $x_i \in \mathcal{X}$. Each user applies some *encoder* $R : \mathcal{X} \to \mathcal{Y}^m$ to their data and sends the *messages* $(y_{i,1}, \ldots, y_{i,m}) = R(x_i)$. In the *one-message shuffled model*, each user sends $m = 1$ message. The $n \cdot m$ messages $y_{i,j}$ are sent to a *shuffler* $S : \mathcal{Y}^* \to \mathcal{Y}^*$ that takes these messages and outputs them in a uniformly random order. The shuffled set of messages is then passed through some *analyzer* $A : \mathcal{Y}^* \to \mathcal{Z}$ to estimate some function $f(x_1, \ldots, x_n)$. Thus, the protocol P consists of the tuple (R, S, A). We say that the algorithm is (ε, δ)-*differentially private in the shuffled model* if the algorithm $M_R(x_1, \ldots, x_n) = S(\cup_{i=1}^n R(x_i))$ satisfies (ε, δ)-differential privacy. For more detail, see the discussion leading to Definition 8.

In contrast to the local model, differential privacy is now a property of all n users' messages, and the (ε, δ) may be functions of n. However, if an adversary were to inject additional messages, then it would not degrade privacy, provided that those messages are independent of the honest users' data. Thus, we may replace n, in our results, as a *lower bound* on the number of honest users in the system. For example, if we have a protocol that is private for n users, but instead we have $\frac{n}{p}$ users of which we assume at least a p fraction are honest, the protocol will continue to satisfy differential privacy.

2.1   Algorithmic Results

Our main result shows how to estimate any bounded, real-valued linear statistic (a *statistical query*) in the shuffled model with error that nearly matches the best possible utility achievable in the central model.

Theorem 1. *For every $\varepsilon \in (0, 1)$, and every $\delta \gtrsim \varepsilon n 2^{-\varepsilon n}$ and every function $f :$ $\mathcal{X} \to [0, 1]$, there is a protocol P in the shuffled model that is (ε, δ)-differentially private, and for every n and every $X = (x_1, \ldots, x_n) \in \mathcal{X}^n$,*

$$\mathbb{E}\left[\left| P(X) - \sum_{i=1}^n f(x_i) \right| \right] = O\left(\frac{1}{\varepsilon} \log \frac{n}{\delta} \right).$$

Each user sends $m = \Theta(\varepsilon \sqrt{n})$ one-bit messages.

For comparison, in the central model, the Laplace mechanism achieves $(\varepsilon, 0)$-differential privacy and error $O(\frac{1}{\varepsilon})$. In contrast, error $\Omega(\frac{1}{\varepsilon}\sqrt{n})$ is necessary in the local model. Thus, for answering statistical queries, this protocol essentially has the best properties of the local and central models (up to logarithmic factors).

In the special case of estimating a sum of bits (or a Boolean-valued linear statistic), our protocol has a slightly nicer guarantee and form.

Theorem 2. *For every $\varepsilon \in (0,1)$, and every $\delta \gtrsim 2^{-\varepsilon n}$ and every function f : $\mathcal{X} \to \{0,1\}$, there is a protocol P in the shuffled model that is (ε, δ)-differentially private, and for every n and every $X = (x_1, \ldots, x_n) \in \mathcal{X}^n$,*

$$\mathbb{E}\left[\left|P(X) - \sum_{i=1}^{n} f(x_i)\right|\right] = O\left(\frac{1}{\varepsilon}\sqrt{\log \frac{1}{\delta}}\right).$$

Each user sends a single one-bit message.

The protocol corresponding to Theorem 2 is extremely simple:

1. For some appropriate choice of $p \in (0,1)$, each user i with input x_i outputs $y_i = x_i$ with probability $1 - p$ and a uniformly random bit y_i with probability p. When ε is not too small, $p \approx \frac{\log(1/\delta)}{\varepsilon^2 n}$.
2. The analyzer collects the shuffled messages $y_1, \ldots, y_n$ and outputs

$$\frac{1}{1-p}\left(\sum_{i=1}^{n} y_i - \tfrac{p}{2}\right).$$

Intuition. In the local model, an adversary can map the set of observations $\{y_1, \ldots, y_n\}$ to users. Thus, to achieve ε-differential privacy, the parameter p should be set close to $\frac{1}{2}$. In our model, the attacker sees only the anonymized set of observations $\{y_1, \ldots, y_n\}$, whose distribution can be simulated using only $\sum_i y_i$. Hence, to ensure that the protocol is differentially private, it suffices to ensure that $\sum_i y_i$ is private, which we show holds for $p \approx \frac{\log(1/\delta)}{\varepsilon^2 n} \ll \frac{1}{2}$.

Communication Complexity. Our protocol for real-valued queries requires $\Theta(\varepsilon\sqrt{n})$ bits per user. In contrast, the local model requires just a single bit, but incurs error $\Omega(\frac{1}{\varepsilon}\sqrt{n})$. A generalization of Theorem 1 gives error $O(\frac{\sqrt{n}}{r} + \frac{1}{\varepsilon}\log\frac{r}{\delta})$ and sends r bits per user, but we do not know if this tradeoff is necessary. Closing this gap is an interesting open question.

2.2 Negative Results

We also prove negative results for algorithms in the *one-message* shuffled model. These results hinge on a structural characterization of private protocols in the one-message shuffled model.

Theorem 3. *If a protocol $P = (R, S, A)$ satisfies (ε, δ)-differential privacy in the one-message shuffled model, then R satisfies $(\varepsilon + \ln n, \delta)$-differential privacy. Therefore, P is $(\varepsilon + \ln n, \delta)$-differentially private in the local model.*

Using Theorem 3 (and a transformation of [7] from (ε, δ)-DP to $(O(\varepsilon), 0)$-DP in the local model), we can leverage existing lower bounds for algorithms in the local model to obtain lower bounds on algorithms in the shuffled model.

Variable Selection. In particular, consider the following *variable selection problem*: given a dataset $x \in \{0,1\}^{n \times d}$, output $\widehat{J}$ such that

$$\sum_{i=1}^{n} x_{i,\widehat{J}} \geq \left(\max_{j \in [d]} \sum_{i=1}^{n} x_{i,j} \right) - \frac{n}{10}.$$

(The $\frac{n}{10}$ approximation term is somewhat arbitrary—any sufficiently small constant fraction of n will lead to the same lower bounds and separations.)

Any local algorithm (with $\varepsilon = 1$) for selection requires $n = \Omega(d \log d)$, whereas in the central model the exponential mechanism [25] solves this problem for $n = O(\log d)$. The following lower bound shows that for this ubiquitous problem, the one-message shuffled model cannot match the central model.

Theorem 4. *If P is a $(1, \frac{1}{n^{10}})$-differentially private protocol in the one-message shuffled model that solves the selection problem (with high probability) then $n = \Omega(d^{1/17})$. Moreover this lower bound holds even if x is drawn iid from a product distribution over $\{0,1\}^d$.*

In Sect. 6, we also prove lower bounds for the well studied *histogram problem*, showing that any one-message shuffled-model protocol for this problem must have error growing (polylogarithmically) with the size of the data domain. In contrast, in the central model it is possible to release histograms with no dependence on the domain size, even for *infinite* domains.

We remark that our lower bound proofs do not apply if the algorithm sends multiple messages through the shuffler. However, we do not know whether beating the bounds is actually possible. Applying our bit-sum protocol d times (together with differential privacy's composition property) shows that $n = \tilde{O}(\sqrt{d})$ samples suffice in the general shuffled model. We also do not know if this bound can be improved. We leave it as an interesting direction for future work to fully characterize the power of the shuffled model.

2.3   Comparison to [16]

In concurrent and independent work, Erlingsson et al. [16] give conceptually similar positive results for local protocols aided by a shuffler. Specifically, they prove a general amplification result: adding a shuffler to any protocol satisfying local differential privacy improve the privacy parameters, often quite significantly. This amplification result can be seen as a partial converse to our transformation from shuffled protocols to local protocols (Theorem 3).

Their result applies to *any* local protocol, whereas our protocol for bit-sums (Theorem 2) applies specifically to the one-bit randomized response protocol. However, when specialized to randomized response, their result is quantitatively weaker than ours. As stated, their results only apply to local protocols that satisfy ε-differential privacy for $\varepsilon < 1$. In contrast, the proof of Theorem 2 shows that, for randomized response, local differential privacy $\varepsilon \approx \ln(n)$ can be amplified to $\varepsilon' = 1$. Our best attempt at generalizing their proof to the case of $\varepsilon \gg 1$

does not give any amplification for local protocols with $\varepsilon \approx \ln(n)$. Specifically, our best attempt at applying their method to the case of randomized response yields a shuffled protocol that is 1-differentially private and has error $\Theta(n^{5/12})$, which is just slightly better than the error $O(\sqrt{n})$ that can be achieved without a shuffler.

3   Model and Preliminaries

In this section, we define terms and notation used throughout the paper. We use $\text{Ber}(p)$ to denote the Bernoulli distribution over $\{0, 1\}$, which has value 1 with probability p and 0 with probability $1 - p$. We will use $\text{Bin}(n, p)$ to denote the binomial distribution (i.e. the sum of n independent samples from $\text{Ber}(p)$).

3.1   Differential Privacy

Let $X \in \mathcal{X}^n$ be a *dataset* consisting of elements from some universe $\mathcal{X}$. We say two datasets X, X' are *neighboring* if they differ on at most one user's data, and denote this $X \sim X'$.

Definition 5 (Differential Privacy [14]). *An algorithm $M : \mathcal{X}^* \to \mathcal{Z}$ is (ε, δ)-differentially private if for every $X \sim X' \in \mathcal{X}^*$ and every $T \subseteq \mathcal{Z}$*

$$\mathbb{P}\left[M(X) \in T\right] \le e^\varepsilon \mathbb{P}\left[M(X') \in T\right] + \delta.$$

where the probability is taken over the randomness of M.

Differential privacy satisfies two extremely useful properties:

Lemma 6 (Post-Processing [14]). *If M is (ε, δ)-differentially private, then for every A, $A \circ M$ is (ε, δ)-differentially private.*

Lemma 7 (Composition [14,15]). *If $M_1, \ldots, M_T$ are (ε, δ)-differentially private, then the composed algorithm*

$$\widetilde{M}(X) = (M_1(X), \ldots, M_T(X))$$

is $(\varepsilon', \delta' + T\delta)$-differentially private for every $\delta' > 0$ and $\varepsilon' = \varepsilon(e^\varepsilon - 1)T + \varepsilon\sqrt{2T \log(1/\delta')}$.

3.2   Differential Privacy in the Shuffled Model

In our model, there are n *users*, each of whom holds data $x_i \in \mathcal{X}$. We will use $X = (x_1, \ldots, x_n) \in \mathcal{X}^n$ to denote the *dataset* of all n users' data. We say two datasets X, X' are *neighboring* if they differ on at most one user's data, and denote this $X \sim X'$.

The protocols we consider consist of three algorithms:

- $R : \mathcal{X} \to \mathcal{Y}^m$ is a randomized *encoder* that takes as input a single users' data x_i and outputs a set of m *messages* $y_{i,1}, \ldots, y_{i,m} \in \mathcal{Y}$. If $m = 1$, then P is in the *one-message shuffled model*.
- $S : \mathcal{Y}^* \to \mathcal{Y}^*$ is a *shuffler* that takes a set of messages and outputs these messages in a uniformly random order. Specifically, on input $y_1, \ldots, y_N$, S chooses a uniformly random permutation $\pi : [N] \to [N]$ and outputs $y_{\pi(1)}, \ldots, y_{\pi(N)}$.
- $A : \mathcal{Y}^* \to \mathcal{Z}$ is some *analysis function* or *analyzer* that takes a set of messages $y_1, \ldots, y_N$ and attempts to estimate some function $f(x_1, \ldots, x_n)$ from these messages.

We denote the overall protocol $P = (R, S, A)$. The mechanism by which we achieve privacy is

$$\Pi_R(x_1, \ldots, x_n) = S(\cup_{i=1}^n R(x_i)) = S(y_{1,1}, \ldots, y_{n,m}),$$

where both R and S are randomized. We will use $P(X) = A \circ \Pi_R(X)$ to denote the output of the protocol. However, by the post-processing property of differential privacy (Lemma 6), it will suffice to consider the privacy of $\Pi_R(X)$, which will imply the privacy of $P(X)$. We are now ready to define differential privacy for protocols in the shuffled model.

Definition 8 (Differential Privacy in the Shuffled Model). *A protocol* $P = (R, S, A)$ *is* (ε, δ)-*differentially private if the algorithm* $\Pi_R(x_1, \ldots, x_n) = S(R(x_1), \ldots, R(x_n))$ *is* (ε, δ)-*differentially private (Definition 5).*

In this model, privacy is a property of the entire set of users' messages and of the shuffler, and thus ε, δ may depend on the number of users n. When we wish to refer to P or Π with a specific number of users n, we will denote this by P_n or Π_n.

We remark that if an adversary were to inject additional messages, then it would not degrade privacy, provided that those messages are independent of the honest users' data. Thus, we may replace n, in our results, with an assumed *lower bound* on the number of honest users in the system.

In some of our results it will be useful to have a generic notion of accuracy for a protocol P.

Definition 9 (Accuracy of Distributed Protocols). *Protocol* $P = (R, S, A)$ *is* (α, β)-*accurate for the function* $f : \mathcal{X}^* \to \mathcal{Z}$ *if, for every* $X \in \mathcal{X}^*$, *we have* $\mathbb{P}[d(P(X), f(X)) \leq \alpha] \geq 1 - \beta$ *where* $d : \mathcal{Z} \times \mathcal{Z} \to \mathbb{R}$ *is some application-dependent distance measure.*

As with the privacy guarantees, the accuracy of the protocol may depend on the number of users n, and we will use P_n when we want to refer to the protocol with a specific number of users.

Composition of Differential Privacy. We will use the following useful composition property for protocols in the shuffled model, which is an immediate consequence of Lemma 7 and the post-processing Lemma 6. This lemma allows us to directly compose protocols in the shuffled model while only using the shuffler once, rather than using the shuffler independently for each protocol being composed.

Lemma 10 (Composition of Protocols in the Shuffled Model). *If $\Pi_1 = (R_1, S), \ldots, \Pi_T = (R_T, S)$ for $R_t : \mathcal{X} \to \mathcal{Y}^m$ are each (ε, δ)-differentially private in the shuffled model, and $\widetilde{R} : \mathcal{X} \to \mathcal{Y}^{mT}$ is defined as*

$$\widetilde{R}(x_i) = (R_1(x_i), \ldots, R_T(x_i))$$

then, for every $\delta' > 0$, the composed protocol $\widetilde{\Pi} = (\widetilde{R}, S)$ is $(\varepsilon', \delta' + T\delta)$-differentially private in the shuffled model for $\varepsilon' = \varepsilon^2 + 2\varepsilon\sqrt{T \log(1/\delta')}$.

Local Differential Privacy. If the shuffler S were replaced with the identity function (i.e. if it did not randomly permute the messages) then we would be left with exactly the *local model of differential privacy*. That is, a locally differentially private protocol is a pair of algorithms $P = (R, A)$, and the output of the protocol is $P(X) = A(R(x_1), \ldots, R(x_n))$. A protocol P is differentially private in the local model if and only if the algorithm R is differentially private. In Sect. 6 we will see that if $P = (R, S, A)$ is a differentially private protocol in the one-message shuffled model, then R itself must satisfy local differential privacy for non-trivial (ε, δ), and thus $(R, A \circ S)$ is a differentially private local protocol for the same problem.

4   A Protocol for Boolean Sums

In this section we describe and analyze a protocol for computing a sum of $\{0, 1\}$ bits, establishing Theorem 2 in the introduction.

4.1   The Protocol

In our model, the data domain is $\mathcal{X} = \{0, 1\}$ and the function being computed is $f(x_1, \ldots, x_n) = \sum_{i=1}^{n} x_i$. Our protocol, P_λ, is specified by a parameter $\lambda \in [0, n]$ that allows us to trade off the level of privacy and accuracy. Note that λ may be a function of the number of users n. We will discuss in Sect. 4.3 how to set this parameter to achieve a desired level of privacy. For intuition, one may wish to think of the parameter $\lambda \approx \frac{1}{\varepsilon^2}$ when ε is not too small.

The basic outline of P_λ is as follows. Roughly, a random set of λ users will choose y_i randomly, and the remaining $n - \lambda$ will choose y_i to be their input bit x_i. The output of each user is the single message y_i. The outputs are then shuffled and the output of the protocol is the sum $\sum_{i=1}^{n} y_i$, shifted and scaled so that it is an unbiased estimator of $\sum_{i=1}^{n} x_i$.

The protocol is described in Algorithm 1. The full name of this protocol is $P_\lambda^{0/1}$, where the superscript serves to distinguish it with the real sum protocol

$P_{\lambda,r}^{\mathbb{R}}$ (Sect. 5). Because of the clear context of this section, we drop the superscript. Since the analysis of both the accuracy and utility of the algorithm will depend on the number of users n, we will use $P_{n,\lambda}, R_{n,\lambda}, A_{n,\lambda}$ to denote the protocol and its components in the case where the number of users is n.

Algorithm 1. A shuffled protocol $P_{n,\lambda}^{0/1} = (R_{n,\lambda}^{0/1}, S, A_{n,\lambda}^{0/1})$ for computing the sum of bits

```
// Local Randomizer
```
$R_{n,\lambda}^{0/1}(x)$:
>   **Input:** $x \in \{0,1\}$, parameters $n \in \mathbb{N}, \lambda \in (0,n)$.
>   **Output:** $\mathbf{y} \in \{0,1\}$
>
>   Let $\mathbf{b} \leftarrow Ber(\frac{\lambda}{n})$
>   If $\mathbf{b} = 0:$ **Return** $\mathbf{y} \leftarrow x$;
>   ElseIf $\mathbf{b} = 1:$ **Return** $\mathbf{y} \leftarrow Ber\left(\frac{1}{2}\right)$;

```
// Analyzer
```
$A_{n,\lambda}^{0/1}(y_1,\ldots,y_n)$:
>   **Input:** $(y_1,\ldots,y_n) \in \{0,1\}^n$, parameters $n \in \mathbb{N}, \lambda \in (0,n)$.
>   **Output:** $z \in [0,n]$
>
>   **Return** $z \leftarrow \frac{n}{n-\lambda} \cdot \left(\sum_{i=1}^{n} y_i - \frac{\lambda}{2}\right)$

4.2 Privacy Analysis

In this section we will prove that P_λ satisfies (ε, δ)-differential privacy. Note that if $\lambda = n$ then the each user's output is independent of their input, so the protocol trivially satisfies $(0,0)$-differential privacy, and thus our goal is to prove an upper bound on the parameter λ that suffices to achieve a given (ε, δ).

Theorem 11 (Privacy of P_λ). *There are absolute constants $\kappa_1,\ldots,\kappa_5$ such that the following holds for P_λ. For every $n \in \mathbb{N}$, $\delta \in (0,1)$ and $\frac{\kappa_2 \log(1/\delta)}{n} \leq \varepsilon \leq 1$, there exists a $\lambda = \lambda(n,\varepsilon,\delta)$ such that $P_{n,\lambda}$ is (ε, δ) differentially private and,*

$$\lambda \leq \begin{cases} \frac{\kappa_4 \log(1/\delta)}{\varepsilon^2} & if \ \varepsilon \geq \sqrt{\frac{\kappa_3 \log(1/\delta)}{n}} \\ n - \frac{\kappa_5 \varepsilon n^{3/2}}{\sqrt{\log(1/\delta)}} & otherwise \end{cases}$$

In the remainder of this section we will prove Theorem 11.

The first step in the proof is the observation that the output of the shuffler depends only on $\sum_i y_i$. It will be more convenient to analyze the algorithm C_λ (Algorithm 2) that simulates $S(R_\lambda(x_1),\ldots,R_\lambda(x_n))$. Claim 12 shows that the output distribution of C_λ is indeed the same as that of the output $\sum_i y_i$. Therefore, privacy of C_λ carries over to P_λ.

Algorithm 2. $C_\lambda(x_1 \ldots x_n)$

Input: $(x_1 \ldots x_n) \in \{0,1\}^n$, parameter $\lambda \in (0, n)$.
Output: $\mathbf{y} \in \{0,1,2,\ldots,n\}$

Sample $\mathbf{s} \leftarrow Bin\left(n, \frac{\lambda}{n}\right)$
Define $\mathcal{H}_s = \{H \subseteq [n] : |H| = s\}$ and choose $\mathbf{H} \leftarrow \mathcal{H}_s$ uniformly at random
Return $\mathbf{y} \leftarrow \sum_{i \notin \mathbf{H}} x_i + Bin\left(s, \frac{1}{2}\right)$

Claim 12. *For every $n \in \mathbb{N}$, $x \in \{0,1\}^n$, and every $r \in \{0,1,2,\ldots,n\}$,*

$$\mathbb{P}\left[C_\lambda(X) = r\right] = \mathbb{P}\left[\sum_{i=1}^n R_{n,\lambda}(x_i) = r\right]$$

Proof. Fix any $r \in \{0,1,2,\ldots,n\}$.

$$\mathbb{P}\left[C_\lambda(X) = r\right] = \sum_{H \subseteq [n]} \mathbb{P}\left[C_\lambda(X) = r \cap \mathbf{H} = H\right]$$

$$= \sum_{H \subseteq [n]} \mathbb{P}\left[\sum_{i \notin H} x_i + Bin\left(|H|, \frac{1}{2}\right) = r\right] \cdot \left(\frac{\lambda}{n}\right)^{|H|} \left(1 - \frac{\lambda}{n}\right)^{n-|H|}$$

$$= \sum_{H \subseteq [n]} \mathbb{P}\left[\sum_{i \notin H} x_i + \sum_{i \in H} Ber\left(\frac{1}{2}\right) = r\right] \cdot \left(\frac{\lambda}{n}\right)^{|H|} \left(1 - \frac{\lambda}{n}\right)^{n-|H|}$$

$$\tag{1}$$

Let $\mathbf{G}$ denote the (random) set of people for whom $b_i = 1$ in P_λ. Notice that

$$\mathbb{P}\left[\sum_{i=1}^n R_{n,\lambda}(x_i) = r\right] = \sum_{G \subseteq [n]} \mathbb{P}\left[\sum_i R_{n,\lambda}(x_i) = r \cap \mathbf{G} = G\right]$$

$$= \sum_{G \subseteq [n]} \mathbb{P}\left[\sum_{i \notin G} x_i + \sum_{i \in G} Ber\left(\frac{1}{2}\right) = r\right]$$

$$\cdot \left(\frac{\lambda}{n}\right)^{|G|} \left(1 - \frac{\lambda}{n}\right)^{n-|G|}$$

which is the same as (1). This concludes the proof. □

Now we establish that in order to demonstrate privacy of $P_{n,\lambda}$, it suffices to analyze C_λ.

Claim 13. *If C_λ is (ε, δ) differentially private, then $P_{n,\lambda}$ is (ε, δ) differentially private.*

Proof. Fix any number of users n. Consider the randomized algorithm T : $\{0,1,2,\ldots,n\} \rightarrow \{0,1\}^n$ that takes a number r and outputs a uniformly random string z that has r ones. If C_λ is differentially private, then the output of $T \circ C_\lambda$ is (ε, δ) differentially private by the post-processing lemma.

To complete the proof, we show that for any $X \in \mathcal{X}^n$ the output of $(T \circ C_\lambda)(X)$ has the same distribution as $S(R_\lambda(x_1), \dots R_\lambda(x_n))$. Fix some vector $Z \in \{0,1\}^n$ with sum r

$$\Pr_{T, C_\lambda} [T(C_\lambda(X)) = Z] = \mathbb{P}[T(r) = Z] \cdot \mathbb{P}[C_\lambda(X) = r]$$

$$= \binom{n}{r}^{-1} \cdot \mathbb{P}[C_\lambda(X) = r]$$

$$= \binom{n}{r}^{-1} \cdot \mathbb{P}[f(R_{n,\lambda}(X)) = r] \qquad \text{(Claim 12)}$$

$$= \binom{n}{r}^{-1} \cdot \sum_{Y \in \{0,1\}^n : |Y| = r} \mathbb{P}[R_{n,\lambda}(X) = Y]$$

$$= \sum_{Y \in \{0,1\}^n : |Y| = r} \mathbb{P}[R_{n,\lambda}(X) = Y] \cdot \mathbb{P}[S(Y) = Z]$$

$$= \Pr_{R_{n,\lambda}, S} [S(R_{n,\lambda}(X)) = Z]$$

This completes the proof of Claim 13. $\qquad \square$

We will analyze the privacy of C_λ in three steps. First we show that for *any* sufficiently large H, the final step (encapsulated by Algorithm 3) will ensure differential privacy for some parameters. When then show that for *any* sufficiently large value s and H chosen *randomly* with $|H| = s$, the privacy parameters actually improve significantly in the regime where s is close to n; this sampling of H is performed by Algorithm 4. Finally, we show that when s is chosen *randomly* then s is sufficiently large with high probability.

Algorithm 3. $C_H(x_1 \dots x_n)$

Input: $(x_1 \dots x_n) \in \{0,1\}^n$, parameter $H \subseteq [n]$.
Output: $\mathbf{y}_H \in \{0, 1, 2, \dots, n\}$

Let $\mathbf{B} \leftarrow Bin\left(|H|, \frac{1}{2}\right)$
Return $\mathbf{y}_H \leftarrow \sum_{i \notin H} x_i + \mathbf{B}$

Claim 14. *For any $\delta > 0$ and any $H \subseteq [n]$ such that $|H| > 8 \log \frac{4}{\delta}$, C_H is $(\varepsilon, \frac{\delta}{2})$-differentially private for*

$$\varepsilon = \ln\left(1 + \sqrt{\frac{32 \log \frac{4}{\delta}}{|H|}}\right) < \sqrt{\frac{32 \log \frac{4}{\delta}}{|H|}}$$

Proof. Fix neighboring datasets $X \sim X' \in \{0,1\}^n$, any $H \subseteq [n]$ such that $|H| > 8 \log \frac{4}{\delta}$, and any $\delta > 0$. If the point at which X, X' differ lies within

H, the two distributions $C_H(X), C_H(X')$ are identical. Hence, without loss of generality we assume that $x_j = 0$ and $x'_j = 1$ for some $j \notin H$.

Define $u := \sqrt{\frac{1}{2}|H|\log\frac{4}{\delta}}$ and $I_u := \left(\frac{1}{2}|H| - u, \frac{1}{2}|H| + u\right)$ so that by Hoeffding's inequality, $\mathbb{P}\left[\mathbf{B} \in I_u\right] < \frac{1}{2}\delta$. For any $W \subseteq \{0, 1, 2, \ldots, n\}$ we have,

$$\mathbb{P}\left[C_H(X) \in W\right] = \mathbb{P}\left[C_H(X) \in W \cap \mathbf{B} \in I_u\right] + \mathbb{P}\left[C_H(X) \in W \cap \mathbf{B} \notin I_u\right]$$

$$\leq \mathbb{P}\left[C_H(X) \in W \cap \mathbf{B} \in I_u\right] + \frac{1}{2}\delta$$

$$= \sum_{r \in W \cap I_u} \mathbb{P}\left[\mathbf{B} + \sum_{i \notin H} x_i = r\right] + \frac{1}{2}\delta$$

Thus to complete the proof, it suffices to show that for any H and $r \in W \cap I_u$

$$\frac{\mathbb{P}\left[\mathbf{B} + \sum_{i \notin H} x_i = r\right]}{\mathbb{P}\left[\mathbf{B} + \sum_{i \notin H} x'_i = r\right]} \leq 1 + \sqrt{\frac{32\log\frac{4}{\delta}}{|H|}} \tag{2}$$

Because $x_j = 0, x'_j = 1$ and $j \notin H$, we have $\sum_{i \notin H} x_i = \sum_{i \notin H} x'_i - 1$. Thus,

$$\frac{\mathbb{P}\left[\mathbf{B} + \sum_{i \notin H} x_i = r\right]}{\mathbb{P}\left[\mathbf{B} + \sum_{i \notin H} x'_i = r\right]} = \frac{\mathbb{P}\left[\mathbf{B} + \sum_{i \notin H} x'_i - 1 = r\right]}{\mathbb{P}\left[\mathbf{B} + \sum_{i \notin H} x'_i = r\right]}$$

$$= \frac{\mathbb{P}\left[\mathbf{B} = \left(r - \sum_{i \notin H} x'_i\right) + 1\right]}{\mathbb{P}\left[\mathbf{B} = \left(r - \sum_{i \notin H} x'_i\right)\right]}$$

Now we define $k = r - \sum_{i \notin H} x'_i$ so that

$$\frac{\mathbb{P}\left[\mathbf{B} = \left(r - \sum_{i \notin H} x'_i\right) + 1\right]}{\mathbb{P}\left[\mathbf{B} = \left(r - \sum_{i \notin H} x'_i\right)\right]} = \frac{\mathbb{P}\left[\mathbf{B} = k + 1\right]}{\mathbb{P}\left[\mathbf{B} = k\right]}.$$

Then we can calculate

$$\frac{\mathbb{P}\left[\mathbf{B} = k + 1\right]}{\mathbb{P}\left[\mathbf{B} = k\right]} = \frac{|H| - k}{k + 1} \qquad (\mathbf{B} \text{ is binomial})$$

$$\leq \frac{|H| - \left(\frac{1}{2}|H| - u\right)}{\frac{1}{2}|H| - u + 1} \qquad (r \in I_u \text{ so } k \geq \frac{1}{2}|H| - u)$$

$$< \frac{\frac{1}{2}|H| + u}{\frac{1}{2}|H| - u} = \frac{u^2/(\log\frac{4}{\delta}) + u}{u^2/(\log\frac{4}{\delta}) - u} \qquad (u = \sqrt{\frac{1}{2}|H|\log\frac{4}{\delta}})$$

$$= \frac{u + \log\frac{4}{\delta}}{u - \log\frac{4}{\delta}} = 1 + \frac{2\log\frac{4}{\delta}}{u - \log\frac{4}{\delta}} = 1 + \frac{2\log\frac{4}{\delta}}{\sqrt{\frac{1}{2}|H|\log\frac{4}{\delta}} - \log\frac{4}{\delta}}$$

$$\leq 1 + \frac{4\log\frac{4}{\delta}}{\sqrt{\frac{1}{2}|H|\log\frac{4}{\delta}}} = 1 + \sqrt{\frac{32\log\frac{4}{\delta}}{|H|}} \qquad (|H| > 8\log\frac{4}{\delta})$$

which completes the proof. $\qquad\qquad\qquad\qquad\qquad\qquad\qquad\qquad\qquad\qquad\qquad$ $\square$

Next, we consider the case where H is a *random* subset of $[n]$ with a *fixed* size s. In this case we will use an *amplification via sampling argument* [20,27] to argue that the randomness of H improves the privacy parameters by a factor of roughly $(1 - \frac{s}{n})$, which will be crucial when $s \approx n$.

Algorithm 4. $C_s(x_1, \ldots, x_n)$

Input: $(x_1, \ldots, x_n) \in \{0,1\}^n$, parameter $s \in \{0,1,2,\ldots,n\}$.
Output: $\mathbf{y}_s \in \{0,1,2,\ldots,n\}$

Define $\mathcal{H}_s = \{H \subseteq [n] : |H| = s\}$ and choose $\mathbf{H} \leftarrow \mathcal{H}_s$ uniformly at random
Return $\mathbf{y}_s \leftarrow C_\mathbf{H}(x)$

Claim 15. *For any $\delta > 0$ and any $s > 8 \log \frac{4}{\delta}$, C_s is $(\varepsilon, \frac{1}{2}\delta)$ differentially private for*

$$\varepsilon = \sqrt{\frac{32 \log \frac{4}{\delta}}{s}} \cdot \left(1 - \frac{s}{n}\right)$$

Proof. As in the previous section, fix $X \sim X' \in \{0,1\}^n$ where $x_j = 0, x'_j = 1$. $C_s(X)$ selects $\mathbf{H}$ uniformly from $\mathcal{H}_s$ and runs $C_H(X)$; let H denote the realization of $\mathbf{H}$. To enhance readability, we will use the shorthand $\varepsilon_0(s) := \sqrt{\frac{32 \log \frac{4}{\delta}}{s}}$. For any $W \subset \{0,1,2,\ldots,n\}$, we aim to show that

$$\frac{\mathbb{P}_{\mathbf{H},C_\mathbf{H}} [C_\mathbf{H}(X) \in W] - \frac{1}{2}\delta}{\mathbb{P}_{\mathbf{H},C_\mathbf{H}} [C_\mathbf{H}(X') \in W]} \leq \exp\left(\varepsilon_0(s) \cdot \left(1 - \frac{s}{n}\right)\right)$$

First, we have

$$\frac{\mathbb{P}_{\mathbf{H},C_\mathbf{H}} [C_\mathbf{H}(X) \in W] - \frac{1}{2}\delta}{\mathbb{P}_{\mathbf{H},C_\mathbf{H}} [C_\mathbf{H}(X') \in W]}$$

$$= \frac{\mathbb{P}_{\mathbf{H},C_\mathbf{H}} [C_\mathbf{H}(X) \in W \cap j \in \mathbf{H}] + \mathbb{P}_{\mathbf{H},C_\mathbf{H}} [C_\mathbf{H}(X) \in W \cap j \notin \mathbf{H}] - \frac{1}{2}\delta}{\mathbb{P}_{\mathbf{H},C_\mathbf{H}} [C_\mathbf{H}(X') \in W \cap j \in \mathbf{H}] + \mathbb{P}_{\mathbf{H},C_\mathbf{H}} [C_\mathbf{H}(X') \in W \cap j \notin \mathbf{H}]}$$

$$= \frac{(1-p)\gamma(X) + p\zeta(X) - \frac{1}{2}\delta}{(1-p)\gamma(X') + p\zeta(X')} \tag{3}$$

where $p := \mathbb{P}[j \notin \mathbf{H}] = (1 - s/n)$,

$$\gamma(X) := \mathbb{P}_{C_\mathbf{H}} [C_\mathbf{H}(X) \in W \mid j \in \mathbf{H}] \quad \text{and} \quad \zeta(X) := \mathbb{P}_{C_\mathbf{H}} [C_\mathbf{H}(X) \in W \mid j \notin \mathbf{H}].$$

When user j outputs a uniformly random bit, their private value has no impact on the distribution. Hence, $\gamma(X) = \gamma(X')$, and

$$(3) = \frac{(1-p)\gamma(X) + p\zeta(X) - \frac{1}{2}\delta}{(1-p)\gamma(X) + p\zeta(X')} \tag{4}$$

Since $s = |H|$ is sufficiently large, by Claim 14 we have $\zeta(X) \leq (1 + \varepsilon_0(s)) \cdot \min\{\zeta(X'), \gamma(X)\} + \frac{1}{2}\delta$.

$$(4) \leq \frac{(1-p)\gamma(X) + p \cdot (1 + \varepsilon_0(s)) \cdot \min\{\zeta(X'), \gamma(X)\} + \delta) - \frac{1}{2}\delta}{(1-p)\gamma(X) + p\zeta(X')}$$

$$\leq \frac{(1-p)\gamma(X) + p \cdot (1 + \varepsilon_0(s)) \cdot \min\{\zeta(X'), \gamma(X)\}}{(1-p)\gamma(X) + p\zeta(X')}$$

$$= \frac{(1-p)\gamma(X) + p \cdot \min(\zeta(X'), \gamma(X)) + p \cdot \varepsilon_0(s) \cdot \min\{\zeta(X'), \gamma(X)\}}{(1-p)\gamma(X) + p\zeta(X')}$$

$$\leq \frac{(1-p)\gamma(X) + p\zeta(X') + p \cdot \varepsilon_0(s) \cdot \min\{\zeta(X'), \gamma(X)\}}{(1-p)\gamma(X) + p\zeta(X')}$$

$$= 1 + \frac{p \cdot \varepsilon_0(s) \cdot \min\{\zeta(X'), \gamma(X)\}}{(1-p)\gamma(X) + p\zeta(X')} \tag{5}$$

Observe that $\min\{\zeta(X'), \gamma(X)\} \leq (1-p)\gamma(X) + p\zeta(X')$, so

$$(5) \leq 1 + p \cdot \varepsilon_0(s) = 1 + \varepsilon_0(s) \cdot \left(1 - \frac{s}{n}\right) \leq \exp\left(\varepsilon_0(s) \cdot \left(1 - \frac{s}{n}\right)\right)$$

$$= \exp\left(\sqrt{\frac{32\log(4/\delta)}{s}} \cdot \left(1 - \frac{s}{n}\right)\right)$$

which completes the proof.   □

We now come to the actual algorithm C_λ, where s is not fixed but is random. The analysis of C_s yields a bound on the privacy parameter that increases with s, so we will complete the analysis of C_λ by using the fact that, with high probability, s is almost as large as λ.

Claim 16. *For any $\delta > 0$ and $n \geq \lambda \geq 14\log\frac{4}{\delta}$, C_λ is (ε, δ) differentially private where*

$$\varepsilon = \sqrt{\frac{32\log\frac{4}{\delta}}{\lambda - \sqrt{2\lambda\log\frac{2}{\delta}}}} \cdot \left(1 - \frac{\lambda - \sqrt{2\lambda\log\frac{2}{\delta}}}{n}\right)$$

The proof is in the full version of the paper.

From Claim 13, C_λ and $P_{n,\lambda}$ share the same privacy guarantees. Hence, Claim 16 implies the following:

Corollary 17. *For any $\delta \in (0,1)$, $n \in \mathbb{N}$, and $\lambda \in \left[14\log\frac{4}{\delta}, n\right]$, $P_{n,\lambda}$ is (ε, δ) differentially private, where*

$$\varepsilon = \sqrt{\frac{32\log\frac{4}{\delta}}{\lambda - \sqrt{2\lambda\log\frac{2}{\delta}}}} \cdot \left(1 - \frac{\lambda - \sqrt{2\lambda\log\frac{2}{\delta}}}{n}\right)$$

4.3   Setting the Randomization Parameter

Corollary 17 gives a bound on the privacy of $P_{n,\lambda}$ in terms of the number of users n and the randomization parameter λ. While this may be enough on its own, in order to understand the tradeoff between ε and the accuracy of the protocol, we want to identify a suitable choice of λ to achieve a desired privacy guarantee (ε, δ). To complete the proof of Theorem 11, we prove such a bound.

For the remainder of this section, fix some $\delta \in (0,1)$. Corollary 17 states that for any n and $\lambda \in \left[14 \log \frac{4}{\delta}, n\right]$, $P_{n,\lambda}$ satisfies $(\varepsilon^*(\lambda), \delta)$-differential privacy, where

$$\varepsilon^*(\lambda) = \sqrt{\frac{32 \log \frac{4}{\delta}}{\lambda - \sqrt{2\lambda \log \frac{2}{\delta}}}} \cdot \left(1 - \frac{\lambda - \sqrt{2\lambda \log \frac{2}{\delta}}}{n}\right)$$

Let $\lambda^*(\varepsilon)$ be the inverse of ε^*, i.e. the minimum $\lambda \in [0, n]$ such that $\varepsilon^*(\lambda) \leq \varepsilon$. Note that $\varepsilon^*(\lambda)$ is decreasing as $\lambda \to n$ while $\lambda^*(\varepsilon)$ increases as $\varepsilon \to 0$. By definition, $P_{n,\lambda}$ satisfies (ε, δ) privacy if $\lambda \geq \lambda^*(\varepsilon)$; the following Lemma gives such an upper bound:

Lemma 18. *For all $\delta \in (0,1)$, $n \geq 14 \log \frac{4}{\delta}$, $\varepsilon \in \left(\frac{\sqrt{3456}}{n} \log \frac{4}{\delta}, 1\right)$, $P_{n,\lambda}$ is (ε, δ) differentially private if*

$$\lambda = \begin{cases} \frac{64}{\varepsilon^2} \log \frac{4}{\delta} & \text{if } \varepsilon \geq \sqrt{\frac{192}{n} \log \frac{4}{\delta}} \\ n - \frac{\varepsilon n^{3/2}}{\sqrt{432 \log(4/\delta)}} & \text{otherwise} \end{cases} \tag{6}$$

We'll prove the lemma in two claims, each of which corresponds to one of the two cases of our bound on $\lambda^*(\varepsilon)$. The first bound applies when ε is relatively large.

Claim 19. *For all $\delta \in (0,1)$, $n \geq 14 \log \frac{4}{\delta}$, $\varepsilon \in \left(\sqrt{\frac{192}{n} \log \frac{4}{\delta}}, 1\right)$, if $\lambda = \frac{64}{\varepsilon^2} \log \frac{4}{\delta}$ then $P_{n,\lambda}$ is (ε, δ) private.*

Proof. Let $\lambda = \frac{64}{\varepsilon^2} \log \frac{4}{\delta}$ as in the statement. Corollary 17 states that $P_{n,\lambda}$ satisfies $(\varepsilon^*(\lambda), \delta)$ privacy for

$$\begin{aligned} \varepsilon^*(\lambda) &= \sqrt{\frac{32 \log \frac{4}{\delta}}{\lambda - \sqrt{2\lambda \log \frac{2}{\delta}}}} \cdot \left(1 - \frac{\lambda - \sqrt{2\lambda \log \frac{2}{\delta}}}{n}\right) \\ &\leq \sqrt{\frac{32 \log \frac{4}{\delta}}{\lambda - \sqrt{2\lambda \log \frac{2}{\delta}}}} & (\lambda \leq n) \\ &\leq \sqrt{\frac{64 \log \frac{4}{\delta}}{\lambda}} & (\lambda \geq 8 \log \frac{2}{\delta}) \\ &= \varepsilon \end{aligned}$$

This completes the proof of the claim. □

The value of λ in the previous claim can be as large as n when ε approaches $1/\sqrt{n}$. We now give a meaningful bound for smaller values of ε.

Claim 20. *For all $\delta \in (0,1)$, $n \geq 14 \log \frac{4}{\delta}$, $\varepsilon \in \left(\frac{\sqrt{3456}}{n} \log \frac{4}{\delta}, \sqrt{\frac{192}{n} \log \frac{4}{\delta}} \right)$, if*

$$\lambda = n - \frac{\varepsilon n^{3/2}}{\sqrt{432 \log(4/\delta)}}$$

then $P_{n,\lambda}$ is (ε, δ) private.

Proof. Let $\lambda = n - \varepsilon n^{3/2}/\sqrt{432 \log(4/\delta)}$ as in the statement. Note that for this ε regime, we have $n/3 < \lambda < n$. Corollary 17 states that $P_{n,\lambda}$ satisfies $(\varepsilon^*(\lambda), \delta)$ privacy for

$$
\begin{aligned}
\varepsilon^*(\lambda) &= \sqrt{\frac{32 \log \frac{4}{\delta}}{\lambda - \sqrt{2\lambda \log \frac{2}{\delta}}}} \cdot \left(1 - \frac{\lambda - \sqrt{2\lambda \log \frac{2}{\delta}}}{n} \right) \\
&\leq \sqrt{\frac{64 \log \frac{4}{\delta}}{\lambda}} \cdot \left(1 - \frac{\lambda - \sqrt{2\lambda \log \frac{2}{\delta}}}{n} \right) && (\lambda \geq 8 \log \tfrac{2}{\delta}) \\
&= \sqrt{\frac{64 \log \frac{4}{\delta}}{\lambda}} \cdot \left(\frac{\varepsilon \sqrt{n}}{\sqrt{432 \log(4/\delta)}} + \frac{\sqrt{2\lambda \log \frac{2}{\delta}}}{n} \right) \\
&\leq \sqrt{\frac{64 \log \frac{4}{\delta}}{\lambda}} \cdot \left(\frac{\varepsilon \sqrt{n}}{\sqrt{432 \log(4/\delta)}} + \sqrt{\frac{2 \log \frac{2}{\delta}}{n}} \right) && (\lambda \leq n) \\
&\leq \sqrt{\frac{192 \log \frac{4}{\delta}}{n}} \cdot \left(\frac{\varepsilon \sqrt{n}}{\sqrt{432 \log(4/\delta)}} + \sqrt{\frac{2 \log \frac{2}{\delta}}{n}} \right) && (\lambda \geq n/3) \\
&= \frac{2}{3}\varepsilon + \frac{\sqrt{384 \log \frac{4}{\delta} \log \frac{2}{\delta}}}{n} < \frac{2}{3}\varepsilon + \frac{\sqrt{384}}{n} \log \frac{4}{\delta} \\
&< \frac{2}{3}\varepsilon + \frac{1}{3}\varepsilon = \varepsilon && (\varepsilon > \tfrac{\sqrt{3456}}{n} \log \tfrac{4}{\delta})
\end{aligned}
$$

which completes the proof. $\qquad\square$

4.4   Accuracy Analysis

In this section, we will bound the error of $P_\lambda(X)$ with respect to $\sum_i x_i$. Recall that, to clean up notational clutter, we will often write $f(X) = \sum_i x_i$. As with the previous section, our statements will at first be in terms of λ but the section will end with a statement in terms of ε, δ.

Theorem 21. *For every $n \in \mathbb{N}$, $\beta > 0$, $n > \lambda \geq 2\log\frac{2}{\beta}$, and $x \in \{0,1\}^n$,*

$$\mathbb{P}\left[\left|P_{n,\lambda}(x) - \sum_i x_i\right| > \sqrt{2\lambda \log(2/\beta)} \cdot \left(\frac{n}{n-\lambda}\right)\right] \leq \beta$$

Observe that, using the choice of λ specified in Theorem 11, we conclude that for every $\frac{1}{n} \lesssim \varepsilon \lesssim 1$ and every δ the protocol P_λ satisfies

$$\mathbb{P}\left[\left|P_{n,\lambda}(x) - \sum_i x_i\right| > O\left(\frac{\sqrt{\log(1/\delta)\log(1/\beta)}}{\varepsilon}\right)\right] \leq \beta$$

To see how this follows from Theorem 21, consider two parameter regimes:

1. When $\varepsilon \gg 1/\sqrt{n}$ then $\lambda \approx \frac{\sqrt{\log(1/\delta)}}{\varepsilon^2} \ll n$, so the bound in Theorem 21 is $O(\sqrt{\lambda \log(1/\beta)})$, which yields the desired bound.
2. When $\varepsilon \ll 1/\sqrt{n}$ then $n - \lambda \approx \varepsilon n^{3/2}/\sqrt{\log(1/\delta)} \ll n$, so the bound in Theorem 21 is $O\left(\frac{n^{3/2}\sqrt{\log(1/\beta)}}{n-\lambda}\right)$, which yields the desired bound.

Theorem 2 in the introduction follows from this intuition; a formal proof can be found in the full version.

5   A Protocol for Sums of Real Numbers

In this section, we show how to extend our protocol to compute sums of bounded real numbers. In this case the data domain is $\mathcal{X} = [0,1]$, but the function we wish to compute is still $f(x) = \sum_i x_i$. The main idea of the protocol is to randomly round each number x_i to a Boolean value $b_i \in \{0,1\}$ with expected value x_i. However, since the randomized rounding introduces additional error, we may need to round multiple times and estimate several sums. As a consequence, this protocol is not one-message.

5.1   The Protocol

Our algorithm is described in two parts, an encoder E_r that performs the randomized rounding (Algorithm 5) and a shuffled protocol $P_{\lambda,r}^{\mathbb{R}}$ (Algorithm 6) that is the composition of many copies of our protocol for the binary case, $P_\lambda^{0/1}$. The encoder takes a number $x \in [0,1]$ and a parameter $r \in \mathbb{N}$ and outputs a vector $(b_1, \ldots, b_r) \in \{0,1\}^r$ such that $\mathbb{E}\left[\frac{1}{r}\sum_j b_j\right] = x_j$ and $\mathrm{Var}\left[\frac{1}{r}\sum_j b_j\right] = O(1/r^2)$. To clarify, we give two examples of the encoding procedure:

- If $r = 1$ then the encoder simply sets $b = Ber(x)$. The mean and variance of b are x and $x(1-x) \leq \frac{1}{4}$, respectively.
- If $x = .4$ and $r = 4$ then the encoder sets $b = (1, Ber(.6), 0, 0)$. The mean and variance of $\frac{1}{4}(b_1 + b_2 + b_3 + b_4)$ are $.4$ and $.015$, respectively.

After doing the rounding, we then run the bit-sum protocol $P_\lambda^{0/1}$ on the bits $b_{1,j}, \ldots, b_{n,j}$ for each $j \in [r]$ and average the results to obtain an estimate of the quantity

$$\sum_i \frac{1}{r} \sum_j b_{i,j} \approx \sum_i x_i$$

To analyze privacy we use the fact that the protocol is a composition of bit-sum protocols, which are each private, and thus we can analyze privacy via the composition properties of differential privacy.

Much like in the bit-sum protocol, we use $P_{n,\lambda,r}^{\mathbb{R}}, R_{n,\lambda,r}^{\mathbb{R}}, A_{n,\lambda,r}^{\mathbb{R}}$ to denote the real-sum protocol and its components when n users participate.

Algorithm 5. An encoder $E_r(x)$

Input: $x \in [0,1]$, a parameter $r \in \mathbb{N}$.
Output: $(\mathbf{b}_1, \ldots, \mathbf{b}_r) \in \{0,1\}^r$

Let $\mu \leftarrow \lceil x \cdot r \rceil$ and $p \leftarrow x \cdot r - \mu + 1$
For $j = 1, \ldots, r$

$$\mathbf{b}_j = \begin{cases} 1 & j < \mu \\ Ber(p) & j = \mu \\ 0 & j > \mu \end{cases}$$

Return $(\mathbf{b}_1, \ldots, \mathbf{b}_r)$

Algorithm 6. The protocol $P_{\lambda,r}^{\mathbb{R}} = (R_{\lambda,r}^{\mathbb{R}}, S, A_{\lambda,r}^{\mathbb{R}})$

`// Local randomizer`
$R_{n,\lambda,r}^{\mathbb{R}}(x):$
    Input: $x \in [0,1]$, parameters $n, r \in \mathbb{N}, \lambda \in (0,n)$.
    Output: $(\mathbf{y}_1, \ldots \mathbf{y}_r) \in \{0,1\}^r$

    $(\mathbf{b}_1, \ldots \mathbf{b}_r) \leftarrow E_r(x)$
    Return $(\mathbf{y}_1, \ldots \mathbf{y}_r) \leftarrow \left(R_{n,\lambda}^{0/1}(\mathbf{b}_1), \ldots, R_{n,\lambda}^{0/1}(\mathbf{b}_r) \right)$

`// Analyzer`
$A_{n,\lambda,r}^{\mathbb{R}}(y_{1,1}, \ldots, y_{n,r}):$
    Input: $(y_{1,1}, \ldots, y_{n,r}) \in \{0,1\}^{n \cdot r}$, parameters $n, r \in \mathbb{N}, \lambda \in (0,n)$.
    Output: $z \in [0,n]$

    Return $z \leftarrow \frac{1}{r} \cdot \frac{n}{n-\lambda} \left(\left(\sum_j \sum_i y_{i,j} \right) - \frac{\lambda \cdot r}{2} \right)$

Theorem 22. *For every* $\delta = \delta(n)$ *such that* $e^{-\Omega(n^{1/4})} < \delta(n) < \frac{1}{n}$ *and* $\frac{\text{poly}(\log n)}{n} < \varepsilon < 1$ *and every sufficiently large* n, *there exists parameters* $\lambda \in [0, n], r \in \mathbb{N}$ *such that* $P_{n,\lambda,r}^{\mathbb{R}}$ *is both* (ε, δ) *differentially private and for every* $\beta > 0$, *and every* $X = (x_1, \dots, x_n) \in [0, 1]^n$,

$$\mathbb{P}\left[\left|P_{n,\lambda,r}^{\mathbb{R}}(X) - \sum_{i=1}^{n} x_i\right| > O\left(\frac{1}{\varepsilon} \log \frac{1}{\delta} \sqrt{\log \frac{1}{\beta}}\right)\right] \leq \beta$$

5.2 Privacy Analysis

Privacy will follow immediately from the composition properties of shuffled protocols (Lemma 10) and the privacy of the bit-sum protocol $P_{n,\lambda}$. One technical nuisance is that the composition properties are naturally stated in terms of ε, whereas the protocol is described in terms of the parameter λ, and the relationship between ε, λ, and n is somewhat complex. Thus, we will state our guarantees in terms of the level of privacy that each individual bit-sum protocol achieves with parameter λ. To this end, define the function $\lambda^*(n, \varepsilon, \delta)$ to be the minimum value of λ such that the bit-sum protocol with n users satisfies (ε, δ)-differential privacy. We will state the privacy guarantee in terms of this function.

Theorem 23. *For every* $\varepsilon, \delta \in (0, 1), n, r \in \mathbb{N}$, *define*

$$\varepsilon_0 = \frac{\varepsilon}{\sqrt{8r \log(2/\delta)}} \qquad \delta_0 = \frac{\delta}{2r} \qquad \lambda^* = \lambda^*(n, \varepsilon_0, \delta_0)$$

For every $\lambda \geq \lambda^*$, $P_{n,\lambda,r}^{\mathbb{R}}$ *is* (ε, δ)-*differentially private.*

5.3 Accuracy Analysis

In this section, we bound the error of $P_{\lambda,r}^{\mathbb{R}}(X)$ with respect to $\sum_i x_i$. Recall that $f(X) = \sum_i x_i$.

Observe that there are two sources of randomness: the encoding of the input $X = (x_1, \dots x_n)$ as bits and the execution of $R_{n,\lambda}^{0/1}$ on that encoding. We first show that the bit encoding lends itself to an unbiased and concentrated estimator of $f(X)$. Then we show that the output of $P_{n,\lambda,r}$ is concentrated around any value that estimator takes.

Theorem 24. *For every* $\beta > 0$, $n \geq \lambda \geq \frac{16}{9} \log \frac{2}{\beta}$, $r \in \mathbb{N}$, *and* $X \in [0, 1]^n$,

$$\mathbb{P}\left[\left|P_{n,\lambda,r}^{\mathbb{R}}(X) - f(X)\right| \geq \frac{\sqrt{2}}{r}\sqrt{n \log \frac{2}{\beta}} + \frac{n}{n-\lambda} \cdot \sqrt{2\frac{\lambda}{r} \log \frac{2}{\beta}}\right] < 2\beta$$

The analysis can be found in the full version of the paper, which also argues that setting $r \leftarrow \varepsilon \cdot \sqrt{n}$ suffices to achieve the bound in Theorem 22.

6   Lower Bounds for the Shuffled Model

In this section, we prove separations between central model algorithms and shuffled model protocols where each user's local randomizer is identical and sends one indivisible message to the shuffler (the one-message model).

Theorem 25 (Shuffled-to-Local Transformation). *Let P_S be a protocol in the one-message shuffled model that is*

- *$(\varepsilon_S, \delta_S)$-differentially private in the shuffled model for some $\varepsilon_S \leq 1$ and $\delta_S = \delta_S(n) < n^{-8}$, and*
- *(α, β)-accurate with respect to f for some $\beta = \Omega(1)$.*

Then there exists a protocol P_L in the local model that is

- *$(\varepsilon_L, 0)$-differentially private in the local model for $\varepsilon_L = 8(\varepsilon_S + \ln n)$, and*
- *$(\alpha, 4\beta)$-accurate with respect to f (when n is larger than some absolute constant)*

This means that an impossibility result for approximating f in the local model implies a related impossibility result for approximating f in the shuffled model. In Sect. 6.2 we combine this result with existing lower bounds for local differential privacy to obtain several strong separations between the central model and the one-message shuffled model.

The key to Theorem 25 is to show that if $P_S = (R_S, S, A_S)$ is a protocol in the one-message shuffled model satisfying $(\varepsilon_S, \delta_S)$-differential privacy, then the algorithm R_S itself satisfies $(\varepsilon_L, \delta_S)$-differential privacy without use of the shuffler S. Therefore, the local protocol $P_L = (R_S, A_S \circ S)$ is $(\varepsilon_L, \delta_S)$-private in the local model and has the exact same output distribution, and thus the exact same accuracy, as P_S. To complete the proof, we use (a slight generalization of) a transformation of Bun, Nelson, and Stemmer [7] to turn R into a related algorithm R' satisfying $(8(\varepsilon_S + \ln n), 0)$-differential privacy with only a slight loss of accuracy. We prove the latter result in the full version of the paper.

6.1   One-Message Randomizers Satisfy Local Differential Privacy

The following lemma is the key step in the proof of Theorem 25, and states that for any symmetric shuffled protocol, the local randomizer R must satisfy local differential privacy with weak, but still non-trivial, privacy parameters.

Theorem 26. *Let $P = (R, S, A)$ be a protocol in the one-message shuffled model. If $n \in \mathbb{N}$ is such that P_n satisfies $(\varepsilon_S, \delta_S)$-differential privacy, then the algorithm R satisfies $(\varepsilon_L, \delta_L)$-differential privacy for $\varepsilon_L = \varepsilon_S + \ln n$. Therefore, the symmetric local protocol $P_L = (R, A \circ S)$ satisfies $(\varepsilon_L, \delta_L)$-differential privacy.*

Proof. By assumption, P_n is $(\varepsilon_S, \delta_S)$-private. Let ε be the supremum such that $R : \mathcal{X} \to \mathcal{Y}$ is *not* (ε, δ_S)-private. We will attempt to find a bound on ε. If R is not (ε, δ_S)-differentially private, there exist $Y \subset \mathcal{Y}$ and $x, x' \in \mathcal{X}$ such that

$$\mathbb{P}\left[R(x') \in Y\right] > \exp(\varepsilon) \cdot \mathbb{P}\left[R(x) \in Y\right] + \delta_S$$

For brevity, define $p := \mathbb{P}(R(x) \in Y)$ and $p' := \mathbb{P}(R(x') \in Y)$ so that we have

$$p' > \exp(\varepsilon)p + \delta_S \tag{7}$$

We will show that if ε is too large, then (7) will imply that P_n is *not* $(\varepsilon_S, \delta_S)$-differentially private, which contradicts our assumption. To this end, define the set $\mathcal{W} := \{W \in \mathcal{Y}^n \mid \exists i \ w_i \in Y\}$. Define two datasets $X \sim X'$ as

$$X := (\underbrace{x, \ldots, x}_{n \text{ times}}) \quad \text{and} \quad X' := (x', \underbrace{x, \ldots, x}_{n-1 \text{ times}})$$

Because P_n is $(\varepsilon_S, \delta_S)$-differentially private

$$\mathbb{P}\left[P_n(X') \in \mathcal{W}\right] \leq \exp(\varepsilon_S) \cdot \mathbb{P}\left[P_n(X) \in \mathcal{W}\right] + \delta_S \tag{8}$$

Now we have

$$\mathbb{P}\left[P_n(X) \in \mathcal{W}\right]$$

$$= \mathbb{P}\left[S(\underbrace{R(x), \ldots, R(x)}_{n \text{ times}}) \in \mathcal{W}\right]$$

$$= \mathbb{P}\left[(\underbrace{R(x), \ldots, R(x)}_{n \text{ times}}) \in \mathcal{W}\right] \qquad (\mathcal{W} \text{ is symmetric})$$

$$= \mathbb{P}\left[\exists i \ R(x) \in Y\right] \leq n \cdot \mathbb{P}\left[R(x) \in Y\right] \qquad (\text{Union bound})$$

$$= np$$

where the second equality is because the set W is closed under permutation, so we can remove the random permutation S without changing the probability. Similarly, we have

$$\mathbb{P}\left[P_n(X') \in \mathcal{W}\right] = \mathbb{P}\left[(R(x'), \underbrace{R(x) \ldots, R(x)}_{n-1 \text{ times}}) \in \mathcal{W}\right]$$

$$\geq \mathbb{P}\left[R(x') \in Y\right] = p'$$

$$> \exp(\varepsilon)p + \delta_S \qquad (\text{By (7)})$$

Now, plugging the previous two inequalities into (8), we have

$$\exp(\varepsilon)p + \delta_S < \mathbb{P}\left[P_n(X') \in \mathcal{W}\right]$$
$$\leq \exp(\varepsilon_S) \cdot \mathbb{P}\left[P_n(X) \in \mathcal{W}\right]$$
$$\leq \exp(\varepsilon_S)np + \delta_S$$

By rearranging and canceling terms in the above we obtain the conclusion

$$\varepsilon \leq \varepsilon_S + \ln n$$

Therefore R must satisfy $(\varepsilon_S + \ln n, \delta_S)$-differential privacy. $\qquad\qquad\square$

Claim 27. *If the shuffled protocol $P_S = (R, S, A)$ is (α, β)-accurate for some function f, then the local protocol $P_L = (R, A \circ S)$ is (α, β)-accurate for f, where*

$$(A \circ S)(y_1, \ldots, y_N) = A(S(y_1, \ldots, y_N))$$

We do not present a proof of Claim 27, as it is immediate that the distribution of $P_S(x)$ and $P_L(x)$ are identical, since $A \circ S$ incorporates the shuffler.

We conclude this section with a slight extension of a result of Bun, Nelson, and Stemmer [7] showing how to transform any local algorithm satisfying (ε, δ)-differential privacy into one satisfying $(O(\varepsilon), 0)$-differential privacy with only a small decrease in accuracy. Our extension covers the case where $\varepsilon > 2/3$, whereas their result as stated requires $\varepsilon \leq 1/4$.

Theorem 28 (Extension of [7]). *Suppose local protocol $P_L = (R, A)$ is (ε, δ) differentially private and (α, β) accurate with respect to f. If $\varepsilon > 2/3$ and*

$$\delta < \frac{\beta}{8n \ln(n/\beta)} \cdot \frac{1}{\exp(6\varepsilon)}$$

then there exists another local protocol $P'_L = (R', A)$ that is $(8\varepsilon, 0)$ differentially private and $(\alpha, 4\beta)$ accurate with respect to f.

The proof can be found in the full version of the paper. Theorem 25 now follows by combining Theorem 26 and Claim 27 with Theorem 28.

6.2   Applications of Theorem 25

In this section, we define two problems and present known lower bounds in the central and local models. By applying Theorem 25, we derive lower bounds in the one-message shuffled model. These bounds imply large separations between the central and one-message shuffled models.

The Selection Problem. We define the *selection problem* as follows. The data universe is $\mathcal{X} = \{0, 1\}^d$ where d is the *dimension* of the problem and the main parameter of interest. Given a dataset $x = (x_1, \ldots, x_n) \in \mathcal{X}^n$, the goal is to identify a coordinate j such that the sum of the users' j-th bits is approximately as large as possible. That is, a coordinate $j \in [d]$ such that

$$\sum_{i=1}^{n} x_{i,j} \geq \max_{j' \in [d]} \sum_{i=1}^{n} x_{i,j'} - \frac{n}{10} \tag{9}$$

We say that an algorithm *solves the selection problem with probability $1 - \beta$* if for every dataset x, with probability at least $1 - \beta$, it outputs j satisfying (9).

Table 1. Comparisons Between Models. When a parameter is unspecified, the reader may substitute $\varepsilon = 1, \delta = 0, \alpha = \beta = .01$. **All results are presented as the minimum dataset size n for which we can hope to achieve the desired privacy and accuracy as a function of the relevant parameter for the problem.**

Function (Parameters)	Differential privacy model			
	Central	Shuffled (this paper)		Local
		One-Message	General	
Mean, $\mathcal{X} = \{0, 1\}$ (Accuracy α)	$\Theta\left(\frac{1}{\alpha\varepsilon}\right)$	$O\left(\frac{\sqrt{\log(1/\delta)}}{\alpha\varepsilon}\right)$		$\Theta\left(\frac{1}{\alpha^2\varepsilon^2}\right)$
Mean, $\mathcal{X} = [0, 1]$ (Accuracy α)		$O\left(\frac{1}{\alpha^2} + \frac{\sqrt{\log(1/\delta)}}{\alpha\varepsilon}\right)$	$O\left(\frac{\log(1/\delta)}{\alpha\varepsilon}\right)$	
Selection (Dimension d)	$\Theta(\log d)$	$\Omega(d^{\frac{1}{17}})$	$\tilde{O}(\sqrt{d}\log\frac{d}{\delta})$	$\Theta(d\log d)$
Histograms (Domain Size D)	$\Theta\left(\min\left\{\log\frac{1}{\delta}, \log D\right\}\right)$	$\Omega(\log^{\frac{1}{17}} D)$	$O(\sqrt{\log D})$	$\Theta(\log D)$

We would like to understand the minimum n (as a function of d) such that there is a differentially private algorithm that can solve the selection problem with constant probability of failure. We remark that this is a very weak notion of accuracy, but since we are proving a negative result, using a weak notion of accuracy only strengthens our results.

The following lower bound for locally differentially private protocols for selection is from [31], and is implicit in the work of [12].[2]

Theorem 29. *If $P_L = (R_L, A_L)$ is a local protocol that satisfies $(\varepsilon, 0)$-differential privacy and P_L solves the selection problem with probability $\frac{9}{10}$ for datasets $x \in (\{0, 1\}^d)^n$, then $n = \Omega\left(\frac{d\log d}{(e^\varepsilon - 1)^2}\right)$.*

By applying Theorem 25 we immediately obtain the following corollary.

Corollary 30. *If $P_S = (R_S, S, A_S)$ is a $(1, \delta)$-differentially private protocol in the one-message shuffled model, for $\delta = \delta(n) < n^{-8}$, and P_S solves the selection problem with probability $\frac{99}{100}$, then $n = \Omega((d\log d)^{1/17})$.*

Using a multi-message shuffled protocol[3], we can solve selection with $\tilde{O}(\frac{1}{\varepsilon}\sqrt{d})$ samples. By contrast, in the local model $n = \Theta(\frac{1}{\varepsilon^2}d\log d)$ samples are necessary and sufficient. In the central model, this problem is solved by the *exponential mechanism* [25] with a dataset of size just $n = O(\frac{1}{\varepsilon}\log d)$, and this is optimal [2,28]. These results are summarized in Table 1.

---

[2] These works assume that the dataset x consists of independent samples from some distribution $\mathcal{D}$, and define accuracy for selection with respect to mean of that distribution. By standard arguments, a lower bound for the distributional version implies a lower bound for the version we have defined.

[3] The idea is to simulate multiple rounds of our protocol for binary sums, one round per dimension.

Histograms. We define the *histogram problem* as follows. The data universe is $\mathcal{X} = [D]$ where D is the *domain size* of the problem and the main parameter of interest. Given a dataset $x = (x_1, \ldots, x_n) \in \mathcal{X}^n$, the goal is to build a vector of size D such that for all $j \in [D]$ the j-th element is as close to the frequency of j in x. That is, a vector $v \in [0, n]^D$ such that

$$\max_{j \in [D]} \left| v_j - \sum_{i=1}^{n} \mathbb{1}(x_i = j) \right| \leq \frac{n}{10} \tag{10}$$

where $\mathbb{1}(\texttt{conditional})$ is defined to be 1 if $\texttt{conditional}$ evaluates to $\texttt{true}$ and 0 otherwise.

Similar to the selection problem, an algorithm *solves the histogram problem with probability* $1 - \beta$ if for every dataset x, with probability at least $1 - \beta$ it outputs v satisfying (10). We would like to find the minimum n such that a differentially private algorithm can solve the histogram problem; the following lower bound for locally differentially private protocols for histograms is from [3].

Theorem 31. *If $P_L = (R_L, A_L)$ is a local protocol that satisfies $(\varepsilon, 0)$ differential privacy and P_L solves the histogram problem with probability $\frac{9}{10}$ for any $x \in [D]^n$ then $n = \Omega\left(\frac{\log D}{(e^\varepsilon - 1)^2}\right)$*

By applying Theorem 25, we immediately obtain the following corollary.

Corollary 32. *If $P_S = (R_S, S, A_S)$ is a $(1, \delta)$-differentially private protocol in the one-message shuffled model, for $\delta = \delta(n) < n^{-8}$, and P_S solves the histogram problem with probability $\frac{99}{100}$, then $n = \Omega\left(\log^{1/17} D\right)$*

In the shuffled model, we can solve this problem using our protocol for bit-sums by having each user encode their data as a "histogram" of just their value $x_i \in [D]$ and then running the bit-sum protocol D times, once for each value $j \in [D]$, which incurs error $O(\frac{1}{\varepsilon}\sqrt{\log \frac{1}{\delta} \log D})$.[4] But in the central model, this problem can be solved to error $O(\min\{\log \frac{1}{\delta}, \log D\})$, which is optimal (see, e.g. [32]). Thus, the central and one-message shuffled models are qualitatively different with respect to computing histograms: D may be infinite in the former whereas D must be bounded in the latter.

Acknowledgements. AC was supported by NSF award CCF-1718088. AS was supported by NSF awards IIS-1447700 and AF-1763786 and a Sloan Foundation Research Award. JU was supported by NSF awards CCF-1718088, CCF-1750640, CNS-1816028 and a Google Faculty Research Award.

---

[4] Note that changing one user's data can only change two entries of their local histogram, so we only have to scale ε, δ by a factor of 2 rather than a factor that grows with D.

References

1. Abowd, J.M.: The U.S. census bureau adopts differential privacy. In: Proceedings of the 24th ACM SIGKDD International Conference on Knowledge Discovery & Data Mining KDD 2018, pp. 2867–2867. ACM, New York (2018)
2. Bafna, M., Ullman, J.: The price of selection in differential privacy. In: Conference on Learning Theory, pp. 151–168 (2017)
3. Bassily, R., Smith, A.: Local, private, efficient protocols for succinct histograms. In: Proceedings of the Forty-Seventh Annual ACM on Symposium on Theory of Computing, pp. 127–135. ACM (2015)
4. Beimel, A., Nissim, K., Omri, E.: Distributed private data analysis: simultaneously solving how and what. In: Wagner, D. (ed.) CRYPTO 2008. LNCS, vol. 5157, pp. 451–468. Springer, Heidelberg (2008). https://doi.org/10.1007/978-3-540-85174-5_25
5. Bittau, A., et al.: PROCHLO: strong privacy for analytics in the crowd. In: Proceedings of the Symposium on Operating Systems Principles (SOSP) (2017)
6. Bonawitz, K., et al.: Practical secure aggregation for privacy preserving machine learning. IACR Cryptology ePrint Archive (2017)
7. Bun, M., Nelson, J., Stemmer, U.: Heavy hitters and the structure of local privacy. In: ACM SIGMOD/PODS Conference International Conference on Management of Data (PODS 2018) (2018)
8. Chan, T.-H.H., Shi, E., Song, D.: Optimal lower bound for differentially private multi-party aggregation. In: Epstein, L., Ferragina, P. (eds.) ESA 2012. LNCS, vol. 7501, pp. 277–288. Springer, Heidelberg (2012). https://doi.org/10.1007/978-3-642-33090-2_25
9. Chan, T.-H.H., Shi, E., Song, D.: Privacy-preserving stream aggregation with fault tolerance. In: Keromytis, A.D. (ed.) FC 2012. LNCS, vol. 7397, pp. 200–214. Springer, Heidelberg (2012). https://doi.org/10.1007/978-3-642-32946-3_15
10. Chaum, D.L.: Untraceable electronic mail, return addresses, and digital pseudonyms. Commun. ACM **24**(2), 84–90 (1981)
11. Corrigan-Gibbs, H., Boneh, D.: Prio: private, robust, and scalable computation of aggregate statistics. In: Proceedings of the 14th USENIX Conference on Networked Systems Design and Implementation NSDI 2017, pp. 259–282. USENIX Association, Berkeley, CA, USA (2017)
12. Duchi, J.C., Jordan, M.I., Wainwright, M.J.: Local privacy and statistical minimax rates. In: 2013 IEEE 54th Annual Symposium on Foundations of Computer Science (FOCS), pp. 429–438. IEEE (2013)
13. Dwork, C., Kenthapadi, K., McSherry, F., Mironov, I., Naor, M.: Our data, ourselves: privacy via distributed noise generation. In: Vaudenay, S. (ed.) EUROCRYPT 2006. LNCS, vol. 4004, pp. 486–503. Springer, Heidelberg (2006). https://doi.org/10.1007/11761679_29
14. Dwork, C., McSherry, F., Nissim, K., Smith, A.: Calibrating noise to sensitivity in private data analysis. In: Halevi, S., Rabin, T. (eds.) TCC 2006. LNCS, vol. 3876, pp. 265–284. Springer, Heidelberg (2006). https://doi.org/10.1007/11681878_14
15. Dwork, C., Rothblum, G.N., Vadhan, S.P.: Boosting and differential privacy. In: FOCS, pp. 51–60. IEEE (2010)
16. Erlingsson, U., Feldman, V., Mironov, I., Raghunathan, A., Talwar, K., Thakurta, A.: Amplification by shuffling: From local to central differential privacy by anonymity. In: Proceedings of the 30th Annual ACM-SIAM Symposium on Discrete Algorithms. SODA 2019 (2019)

17. Erlingsson, Ú., Pihur, V., Korolova, A.: RAPPOR: randomized aggregatable privacy-preserving ordinal response. In: ACM Conference on Computer and Communications Security (CCS) (2014)

18. Evfimievski, A., Gehrke, J., Srikant, R.: Limiting privacy breaches in privacy preserving data mining. In: PODS, pp. 211–222. ACM (2003)

19. van den Hooff, J., Lazar, D., Zaharia, M., Zeldovich, N.: Vuvuzela: scalable private messaging resistant to traffic analysis. In: Proceedings of the 25th Symposium on Operating Systems Principles SOSP 2015, pp. 137–152. ACM, New York (2015)

20. Kasiviswanathan, S.P., Lee, H.K., Nissim, K., Raskhodnikova, S., Smith, A.: What can we learn privately? In: Foundations of Computer Science (FOCS). IEEE (2008)

21. Kasiviswanathan, S.P., Smith, A.: On the 'semantics' of differential privacy: A bayesian formulation. CoRR arXiv:0803.39461 [cs.CR] (2008)

22. Kearns, M.J.: Efficient noise-tolerant learning from statistical queries. In: STOC, pp. 392–401. ACM, 16–18 May 1993

23. Kwon, A., Lazar, D., Devadas, S., Ford, B.: Riffle: an efficient communication system with strong anonymity. PoPETs **2016**(2), 115–134 (2016)

24. McMillan, R.: Apple tries to peek at user habits without violating privacy. Wall Street J. (2016)

25. McSherry, F., Talwar, K.: Mechanism design via differential privacy. In: IEEE Foundations of Computer Science (FOCS) (2007)

26. Shi, E., Chan, T.H., Rieffel, E.G., Chow, R., Song, D.: Privacy-preserving aggregation of time-series data. In: Proceedings of the Network and Distributed System Security Symposium (NDSS 2011) (2011)

27. Smith, A.: Differential privacy and the secrecy of the sample (2009)

28. Steinke, T., Ullman, J.: Tight lower bounds for differentially private selection. In: 2017 IEEE 58th Annual Symposium on Foundations of Computer Science (FOCS), pp. 552–563. IEEE (2017)

29. Thakurta, A.G., et al.: Learning new words. US Patent 9,645,998, 9 May 2017

30. Tyagi, N., Gilad, Y., Leung, D., Zaharia, M., Zeldovich, N.: Stadium: a distributed metadata-private messaging system. In: Proceedings of the 26th Symposium on Operating Systems Principles SOSP 2017, pp. 423–440. ACM, New York (2017)

31. Ullman, J.: Tight lower bounds for locally differentially private selection. CoRR abs/1802.02638 (2018)

32. Vadhan, S.: The complexity of differential privacy (2016). http://privacytools.seas.harvard.edu/publications/complexity-differential-privacy

33. Warner, S.L.: Randomized response: a survey technique for eliminating evasive answer bias. J. Am. Stat. Assoc. **60**(309), 63–69 (1965)

Lower Bounds for Differentially Private RAMs

Giuseppe Persiano[1,2] and Kevin Yeo[1(✉)]

[1] Google LLC, Mountain View, USA
giuper@gmail.com, kwlyeo@google.com
[2] Università di Salerno, Salerno, Italy

Abstract. In this work, we study privacy-preserving storage primitives that are suitable for use in data analysis on outsourced databases within the differential privacy framework. The goal in differentially private data analysis is to disclose global properties of a group without compromising any individual's privacy. Typically, differentially private adversaries only ever learn global properties. For the case of outsourced databases, the adversary also views the patterns of access to data. Oblivious RAM (ORAM) can be used to hide access patterns but ORAM might be excessive as in some settings it could be sufficient to be compatible with differential privacy and only protect the privacy of individual accesses.

We consider (ϵ, δ)-*Differentially Private RAM*, a weakening of ORAM that only protects individual operations and seems better suited for use in data analysis on outsourced databases. As differentially private RAM has weaker security than ORAM, there is hope that we can bypass the $\Omega(\log(nb/c))$ bandwidth lower bounds for ORAM by Larsen and Nielsen [CRYPTO '18] for storing an array of n b-bit entries and a client with c bits of memory. We answer in the negative and present an $\Omega(\log(nb/c))$ bandwidth lower bound for privacy budgets of $\epsilon = O(1)$ and $\delta \leq 1/3$.

The *information transfer* technique used for ORAM lower bounds does not seem adaptable for use with the weaker security guarantees of differential privacy. Instead, we prove our lower bounds by adapting the *chronogram* technique to our setting. To our knowledge, this is the first work that uses the chronogram technique for lower bounds on privacy-preserving storage primitives.

1 Introduction

In this work, we study *privacy-preserving storage* schemes involving a client and an untrusted server. The goal is to enable the client to outsource the storage of data to the server such that the client may still perform operations on the stored data (e.g. retrieving and updating data). For privacy, the client wishes to keep the stored data hidden from server. One way to ensure the contents of the data remain hidden is for the client to encrypt all data before uploading to the server. However, the server can still view how the encrypted data is accessed as the client performs operations. Previous works such as [4, 20] have shown that the

© International Association for Cryptologic Research 2019
Y. Ishai and V. Rijmen (Eds.): EUROCRYPT 2019, LNCS 11476, pp. 404–434, 2019.
https://doi.org/10.1007/978-3-030-17653-2_14

leakage of patterns of access to encrypted data can be used to compromise the privacy of the encrypted data. Therefore, a very important privacy requirement is also to protect the access patterns.

The traditional way to define the privacy of access pattern is *obliviousness*. An oblivious storage primitive ensures that any adversary that is given two sequences of operations of equal length and observes the patterns of data access performed by one of the two sequences cannot determine which of the two sequences induced the observed access pattern. The most famous oblivious storage primitive is Oblivious RAM (ORAM) that outsources the storage of an array and allows clients to retrieve and update array entries. ORAM was first introduced by Goldreich [16] who presented an ORAM with sublinear amortized bandwidth per operation for clients with constant size memory. Goldreich and Ostrovsky [17] give the first ORAM construction with polylogarithmic amortized bandwidth per operation. In the past decade, ORAM has been the subject of extensive research [18,19,21,27,28,31,32] as well as variants such as statistically secure ORAMs [7,8], parallel ORAMs [2,5,6] and garbled RAMs [14,15,25]. The above references are just a small subset of all the results for ORAM constructions.

Instead, we focus on a different definition of privacy using *differential privacy* [10–12]. The representative scenario for differential privacy is *privacy-preserving data analysis* which considers the problem of disclosing properties about an entire database while maintaining the privacy of individual database records. A mechanism or algorithm is considered differentially private if any fixed disclosure is almost as likely to be outputted for two different input databases that only differ in exactly one record. As a result, an adversary that views the disclosure is unable to determine whether an individual record was part of the input used to compute the disclosure. We consider the scenario of performing privacy-preserving data analysis on data outsourced to an untrusted server. By viewing the patterns of access to the outsourced data, the adversarial server might be able to determine which individual records were used to compute the disclosure compromising differential privacy.

One way to protect the patterns of data access is to outsource the data using an ORAM. However, in many cases, it turns out that the obliviousness guarantees of ORAM may be stronger than required. For example, let's suppose that we wish to disclose a differentially private regression model over a sample of the outsourced data. ORAM guarantees that the identity of all sampled database records will remain hidden from the adversary. On the other hand, the differentially private regression model only provides privacy about whether an individual record was sampled or not. Instead of obliviousness, we want a weaker notion of privacy for access patterns suitable for use with differentially private data analytics. With a weaker notion of privacy, there is hope for a construction with better efficiency than ORAM.

With this in mind, we turn to the notion of *differentially private access* which provides privacy for individual operations but might reveal information about a sequence of many operations. Differentially private access has been previously considered in [33,34]. In particular, this privacy notion ensures that the patterns of data access caused by a fixed sequence of operations is almost as likely to be induced by another sequence of operations of the same length with a single different operation. We define (ϵ, δ)-*differentially private RAM* as a storage primitive that outsources the storage of an array in a manner that allows a client to retrieve and update array entries while providing differentially private access. As this privacy notion is weaker than obliviousness, the $\Omega(\log(n/c))$ lower bounds for ORAMs that store n array entries and clients with c bits of storage by Larsen and Nielsen [23] do not apply. There is hope to achieve a differentially private RAM construction with smaller bandwidth. In this work, we answer in the negative and show that an $\Omega(\log(n/c))$ bandwidth lower bound also exists for differentially private RAM for typical privacy budgets of $\epsilon = O(1)$ and $\delta \leq 1/3$. As differential privacy with budgets of $\epsilon = O(1)$ and $\delta \leq 1/3$ provide weaker security than obliviousness, any ORAM is also a differentially private RAM. Therefore, our lower bounds show that the ORAM constructions by Patel *et al.* [27] and by Asharov *et al.* [1] are, respectively, asymptotically optimal up to an $O(\log \log n)$ factor and asymptotically optimal (ϵ, δ)-differentially private RAM for any constant ϵ and $\delta \leq 1/3$ and any block size b. Our results also prove that Path ORAM [32] is tight, for $b = \Omega(\log^2 n)$.

1.1 Our Results

In this section, we will present our contributions. We first describe the scenarios where our lower bounds apply. Our lower bounds apply to *differentially private* RAMs that process operations in an *online* fashion. The RAM must be both *read-and-write*, that is, the set of permitted operations include both reading and writing array entries. The server that stores the array is assumed to be *passive* in that the server may not perform any computation beyond retrieving and overwriting cells but no assumptions are made on the *storage encoding* of the array. Finally, we assume that the adversary is *computationally bounded*. We now go into detail about each of these requirements.

Differential Privacy. The goal of differential privacy is to ensure that the removal or replacement of an individual in a large population does not significantly affect the view of the adversary. Differential privacy is parameterized by two values $0 \leq \epsilon$ and $\delta \in [0, 1]$. The value ϵ is typically referred to as the *privacy budget*. When $\delta = 0$, the notion is known as *pure* differential privacy while, if $\delta > 0$, the notion is known as *approximate* differential privacy. In our context, an *individual* is a single operation in a sequence (the *population*) of read (also called queries) and write (also called updates) operations over an array of n entries stored on a, potentially adversarial, remote server. For any implementation **DS** and for any sequence Q, we define $\mathbb{V}_{\mathbf{DS}}(Q)$ to be the view of the server when sequence Q is executed by **DS**. A differentially private RAM, **DS**, is defined to ensure that the adversary's view on one sequence of operations should not be significantly different when **DS** executes another sequence of operations which differs for only one operation. We assume that our adversaries are computationally bounded.

Formally, if **DS** is (ϵ, δ)-differentially private, then for any two sequences Q_1 and Q_2 that differ in exactly one operation, it must be that $\Pr[\mathcal{A}(\mathbb{V}_{\mathbf{DS}}(Q_1)) = 1] \leq e^{\epsilon} \Pr[\mathcal{A}(\mathbb{V}_{\mathbf{DS}}(Q_2)) = 1] + \delta$ for any probabilistic polynomial time (PPT) algorithm $\mathcal{A}$. The notion of computational differential privacy was studied by Mironov et al. [26] where various classes of privacy were described. Our lower bounds consider the weakest privacy class and, thus, apply to all privacy classes in [26]. In the majority of scenarios, differential privacy is only considered useful for the cases when $\epsilon = O(1)$ and δ is negligible. This is exactly the scenario where our lower bounds will hold. In fact, our lower bounds hold for any $\delta \leq 1/3$. We note that differential privacy with $\epsilon = O(1)$ and $\delta \leq 1/3$ is a weaker security notion than obliviousness. Obliviousness is equivalent to differential privacy when $\epsilon = 0$ and δ is negligible. Therefore, our lower bounds also hold for ORAM and match the lower bounds of Larsen and Nielsen [23]. We refer the reader to Sect. 2 for a formal definition of differential privacy.

Online RAMs. It is important that we discuss the notion of *online* vs. *offline* processing of operations by RAMs. In the offline scenario, it is assumed that all operations are given before the RAM must start processing updates and answering queries. The first ORAM lower bound by Goldreich and Ostrovsky [17] considered offline ORAMs with "balls-and-bins" encoding and security against an all-powerful adversary. "Balls-and-bins" refers to the encoding where array entries are immutable balls and the only valid operation is to move array entries into various memory locations referred to as bins. Boyle and Naor [3] show that proving an offline ORAM lower bound for non-restricted encodings is equivalent to showing lower bounds in sorting circuits, which is a long-standing problem in complexity. Instead, we consider online RAMs where operations arrive one at a time and must be processed before receiving the next operation. The assumption of online operations is realistic as the majority of RAM constructions consider online operations and almost all applications of RAMs consider online operations. Our lower bounds only apply for online differentially private RAMs.

Read-and-Write RAMs. Traditionally, all ORAM results consider the scenario where the set of valid operations include both reading and writing array entries. A natural relaxation would be to consider *read-only* RAMs where the only valid operation is reading array entries. Any lower bound on read-only RAMs would also apply to read-and-write RAMs. However, in a recent work by Weiss and Wichs [36], it is shown that any lower bounds for read-only ORAMs would imply very strong lower bounds for either sorting circuits and/or locally decodable codes (LDCs). Proving lower bounds for LDCs has, like sorting circuits, been an open problem in the world of complexity theory for more than a decade. As differential privacy is weaker than obliviousness, any lower bounds on read-only, differentially private RAMs also imply lower bounds on read-only ORAMs. To get around these obstacles, our work focuses only on proving lower bounds for read-and-write differentially private RAMs.

Passive Server. In our work, we will assume that the server storing the array is *passive*, which means that the server will not any perform computation beyond retrieving and overwriting the contents of the local memory cell to satisfy the client's requests. This assumption is necessary as there are ORAM constructions that use server computation to achieve constant bandwidth operations [9]. Therefore, our lower bounds on bandwidth only apply to differentially private RAMs with a passive server. For active servers we can reinterpret our results as lower bounds on the amount of server computation required to guarantee differential privacy.

We now informally present our main contribution.

Theorem 1 (informal). *Let* **DS** *be any online, read-and-write RAM that stores n array entries each of size b bits on a passive server without any restrictions on storage encodings. Suppose that the client has c bits of storage. Assume that* **DS** *provides (ϵ, δ)-differential privacy against a computational adversary that views all cell probes performed by the server. If $\epsilon = O(1)$ and $0 \leq \delta \leq 1/3$, then the amortized bandwidth of both reading and writing array entries by* **DS** *is $\Omega(b\log(nb/c))$ bits or $\Omega(\log(nb/c))$ array entries. In the natural scenario where $c \leq b \cdot n^\alpha$ for some $0 \leq \alpha < 1$, then $\Omega(\log n)$ array entries of bandwidth are required.*

1.2   Previous Works

In this section, we present a small survey of previous works on data structure lower bounds. We also describe the first lower bound for data structures that provide privacy guarantees.

The majority of data structure lower bounds are proved using the cell probe model introduced by Yao [37], which only charges for accessing memory and allows unlimited computation. In the case for passive servers that only retrieve and overwrite memory, the costs of the cell probe model directly imply costs in bandwidth. The *chronogram* technique was introduced by Fredman and Saks [13] which can be used to prove $\Omega(\log n/\log\log n)$ lower bounds. Pǎtraşcu and Demaine [30] presented the *information transfer* technique which could be used to prove $\Omega(\log n)$ lower bounds. Larsen [22] presented an $\tilde{\Omega}(\log^2 n)$ lower bound for two-dimensional dynamic range counting, which remains the highest lower bound proven for any $\log n$ output data structures. Recently, Larsen *et al.* [24] presented an $\tilde{\Omega}(\log^{1.5} n)$ lower bound for data structures with single bit outputs which is the highest lower bound for decision query data structures.

For ORAM, Goldreich and Ostrovsky [17] presented an $\Omega(\log_c n)$ lower bound for clients with storage of c array entries. However, Boyle and Naor [3] showed that this lower bound came with the cavaets that the lower bound only for statistical adversaries and constructions in "balls-and-bins" model where array entries could only be moved between memory and not encoded in a more complex manner. Furthermore, Boyle and Naor [3] show that proving lower bounds for offline ORAMs and arbitrary storage encodings imply sorting circuit lower bounds.

In their seminal work, Larsen and Nielsen [23] presented an $\Omega(\log(n/c))$ bandwidth lower bound removing the cavaets such that lower bounds applies to any types of storage encodings and computational adversaries. Recently, Weiss and Wichs [36] show that lower bounds for online, read-only ORAMs would imply lower bounds for either sorting circuits and/or locally decodable codes.

We present a brief overview of the techniques used by Larsen and Nielsen [23], which uses the information transfer technique. We also describe why information transfer does not seem to be of use for differentially private RAM lower bounds. Information transfer first builds a binary tree over $\Theta(n)$ operations where the first operation is assigned to the leftmost leaf, the second operation is assigned to the second leftmost leaf and so forth. Each cell probe is assigned to at most one node of the tree as follows. For a cell probe, we identify the operation that is performing the probe as well as the most recent operation that overwrote the cell that is being probed. The cell probe is assigned to the lowest common ancestor of the leaves associated with the most recent operation to overwrite the cell and the operation performing the probe. Let us fix any node of the tree and consider the subtree rooted at the fixed node. It can be shown that the probes assigned to the root is the entirety of information that can be transferred from the updates of the left subtree to be used to answer queries in the right subtree. Consider the sequence of operations where all leaves in the left subtree write a randomly chosen b-bit string to unique array entries and all leaves in the right subtree read an unique, updated array entry. For any **DS** to return the correct b-bit strings asked by the queries in the right subtree, a large amount of information must be transferred from the left subtree to the right subtree. Thus, many probes should be assigned to the root of this subtree. Suppose that for another sequence of operations, **DS** assigns significantly less probes to the root of this subtree. Then, a computational adversary can count the probes and distinguish between the worst case sequence and any other sequence contradicting obliviousness. As a result, there must be many probes assigned to each node of the information transfer tree. Each cell probe is assigned to at most one node. So, summing up the tree provides a lower bound on the number of cell probes required.

Unfortunately, we are unable to use the information transfer technique to prove lower bounds for differentially private RAMs. The main issue comes from the fact that differentially private RAMs have significantly weaker privacy guarantees compared to ORAMs. When $\epsilon = \Theta(1)$, the probabilistic requirements of the adversary's view when **DS** processes two sequences Q_1 and Q_2 degrade exponentially in the number of operations that Q_1 and Q_2 differ in. On the other hand, the privacy requirements of obliviousness do not degrade when considering two sequences that differ in many operations. Larsen and Nielsen [23] use obliviousness to argue that the adversary's view for the worst case sequence of any subtree cannot differ significantly from any other sequence. However, for any fixed sequence of operations, the worst case sequence for the majority of subtrees differ in many operations (on the order of the number of leaves of the subtree). Applying differential privacy will not yield strong requirements for the number of cell probes assigned to the majority of the nodes in the information transfer binary tree. As a result, we could not adapt the information transfer technique for differentially private RAM lower bounds and resort to other techniques.

1.3   Overview of Our Proofs

In this section, we present an overview of the proof techniques used in Sects. 3 and 4. Our lower bounds use ideas from works by Pătraşcu and Demaine [30] and Pătraşcu [29]. However, we begin by reviewing the original chronogram technique of Fredman and Saks [13].

Consider an ORAM that stores n b-bit array entries in a cell probe model with w-bit cells. We make the reasonable assumption that $w = \Omega(\log n)$ so that a cell can hold the index of an entry. Let t_w and t_r denote the number of cell probes of an update (write) and of a query (read) operation, respectively, and consider a sequence of $\Theta(n)$ update operations followed by a single query. Starting from the query and going backwards in time, updates are partitioned into exponentially increasing epochs at some rate r, so that the i-th epoch will have $\ell_i = r^i$ update operations. Epochs are indexed in reverse time, so the smallest epoch closest to the query is epoch 1. The goal of the chronogram is to prove that there exists a query that requires information from many of the epochs simultaneously. To do this, we first observe that if each update writes a randomly and independently chosen b-bit entry, an update operation preceding epoch i cannot encode any information about epoch i. Therefore, all information about entries updated in epoch i can only be found in cells that have been written as part of the update operations of epoch i or any following epochs, that is epochs $i - 1, \ldots, 1$. Since each update stores b random bits, epoch i encodes $\ell_i \cdot b$ bits in total. On the other hand, the write operations of epochs $i-1, i-2, \ldots, 1$ can probe at most $t_w(r^{i-1} + \ldots + r)$ and by setting $r = (t_w w)^2$, we obtain that $O(\ell_i/(t_w w^2))$ cells can be probed and $O(\ell_i/(t_w w))$ bits can be written. As a result, the majority of the bits encoded by updates in epoch i remain in cells last written in epoch i. Thus, if we construct a random query such that $\Omega(b)$ bits must be transferred from each epoch, then we obtain that $\max\{t_w, t_r\} = \Omega((b/w)\log_r n) = \Omega((b/w)\log n/\log\log n)$.

This lower bound can be improved to $\Omega((b/w)\log n)$ by using an improvement of the chronogram technique by Pătraşcu [29]. In the original chronogram technique, the epochs are fixed since the query's location and the number of updates are fixed. An algorithm may attempt to target an epoch i by having all future update operations encode information only about epoch i. To counteract this, we consider a harder update sequence where epoch locations cannot be predicted by the algorithm. Specifically, we consider a sequence that consists of a random number of update operations followed by a single query operation. For such a sequence, even if an algorithm attempts to target epoch i, it cannot pinpoint the location of epoch i (remember that epochs are indexed starting from the query operation and going back in time) and may only prepare over all possible query locations. We show that any update operation may now only encode $O(t_w w/\log_r n)$ about epoch i where $\log_r n$ is the number of epochs. As a result, future update operations can only encode a $O(1/\log_r n)$ fraction as much information about epoch i as the previous lower bound attempt. This allows us to fix $r = 2$ which increases the number of epochs $\log n$. If we can find a query that requires $\Omega(b)$ bits of information transfer from the majority of epochs, we can prove that $\max\{t_w, t_r\} = \Omega((b/w)\log n)$.

For differentially private RAMs, the update operations enable overwriting a b-bit array entry while the query operations allow retrieving an array entry. We choose our update operations to overwrite unique array entries and each entry is overwritten with a value that is independently and uniformly chosen at random from $\{0,1\}^b$. Focus on an epoch i and consider picking a random query from the ℓ_i array indices updated in epoch i. The majority of these queries must read $\Omega(b)$ bits from cells last written in epoch i as future operations cannot encode all $\ell_i \cdot b$ bits encoded by epoch i. As a result, there exists some query such that $\Omega(b)$ bits must be transferred from epoch i for all sufficiently large epochs. We use differential privacy to show that $\Omega(b)$ bits must be transferred from all sufficiently large epochs. Consider two sequences of operations that only differ in the final query operation and suppose that the first query requires $\Omega(b)$ bits from epoch i. If the latter query transfers $o(b)$ bits from epoch i, the adversary can distinguish between the two sequences with high probability and this contradicts differential privacy as the two sequences only differ in one operations. Therefore, we can prove that $\Omega(b)$ bits have to be transferred from most epochs and thus $\max\{t_w, t_r\} = \Omega((b/w)\log n)$. The proof of this lower bound is found in Sect. 3.

A stronger lower bound is obtained in Sect. 4 using more complex epoch constructions, adapting ideas from [30] and [29]. The lower bound outlined above shows that $\max\{t_w, t_r\} = \Omega((b/w)\log n)$ but it does not preclude the case where $t_w = \Theta((b/w)\log n)$ and $t_r = O(1)$, for example. We show this cannot be the case. In particular, we show that if $\max\{t_w, t_r\} = O((b/w)\log n)$, then it must be the case that $t_w = \Theta((b/w)\log n)$ and $t_r = \Theta((b/w)\log n)$. The idea is to consider different epoch constructions for the cases when t_w and t_r are small, respectively. When $t_w = o((b/w)\log n)$, we know that operations in future epochs cannot encode too much information. We consider an epoch construction where epochs grow by a rate of $\omega(1)$ every r epochs thus increasing the number of epochs to $\omega(\log n)$. In exchange, there are many operations after an epoch i. Since t_w is small, the future operations may not encode too much information about epoch i ensuring most of the information about epoch i remain in cells last written during epoch i. As a result, it can be shown again that $\Omega(b)$ bits must be read from many epochs implying an $t_r = \omega((b/w)\log n)$ lower bound. On the other hand, consider the case when $t_r = o((b/w)\log n)$. We consider epoch constructions that increase exponentially with rate $r = \omega(1)$. As a result, the number of operations after epoch i is a factor of $O(1/r)$ smaller than the ℓ_i operations in epoch i and there are $\Theta(\log_r n)$ epochs. If $t_r = o((b/w)\log_r n)$, then a query operation may not read $\Omega(b)$ from each of the epochs. Instead, update operations must encode a large amount of information about previous epochs to compensate for t_r being so small. As a result, it can be shown that $t_w = \omega((b/w)\log n)$. Combining the above two statements implies that if $\max\{t_w, t_r\} = O((b/w)\log n)$, then $t_w = \Theta((b/w)\log n)$ and $t_r = \Theta((b/w)\log n)$.

2   Differentially Private Cell Probe Model

We start by formalizing the model for which we prove our lower bounds. We rely on the *cell probe model*, first described by Yao [37], and typically used to prove lower bounds for data structures without any requirements for privacy of the stored data and/or the operations performed. In a recent work by Larsen and Nielsen [23], the *oblivious cell probe model* was introduced and used to prove a lower bound for oblivious RAM. The oblivious cell probe model was defined for any data structures where the patterns of access to memory should not reveal any information about the operations performed. We generalize the oblivious cell probe model and present the (ϵ, δ)-*differentially private cell probe model*. In this new model, all data structures are assumed to provide differential privacy for the operations performed with respect to memory accesses viewed by the adversary. The differentially private cell probe model is a generalization of the oblivious cell probe model as obliviousness is equivalent to differential privacy with $\epsilon = 0$ and $\delta = \mathsf{negl}(n)$, that is, any function negligible in the number of items stored in the data structure.

The cell probe model is an abstraction of the interaction between CPUs and word-RAM memory architectures. Memory is defined as an array of *cells* such that each cell contains exactly w bits. Any *operation* of a data structure is allowed to *probe* cells where a probe can consist of either reading the contents of a cell or over-writing the contents of a cell. The running time or cost for any operation of a data structure is measured by the number of cell probes performed. An algorithm is free to do unlimited amounts of computation based on the contents of probed cells.

In this paper we are interested in data structures that provide privacy of the operations performed in a scenario involving two parties denoted the *client* and the *server*. The client outsources the storage of data to the server while maintaining the ability to perform some set of operations over the data efficiently. In addition, the client wishes to hide the operations performed from the adversarial server that views the contents of all cells in memory as well as the sequence of cells probed in memory. The crucial privacy requirement is that the server does not learn about the contents and the sequence of accesses performed by the client's storage. To properly capture the above setting, Larsen and Nielsen [23] defined the *oblivious cell probe model* and proved lower bounds for oblivious RAMs. We introduce the *differentially private cell probe model* that is identical to the oblivious cell probe model of Larsen and Nielsen, except for the simple replacement of obliviousness with differential privacy as the privacy requirement. For a full description of the oblivious cell probe model, we refer the reader to Sect. 2 of [23]. To formally define the differentially private cell probe model, we first describe a *data structure problem* as well as a *differentially private cell probe data structure* for any data structure problem.

Definition 1. *A data structure problem P is defined by a tuple (U, Q, O, f) where*

1. U is the universe of all update operations;
2. Q is the universe of all query operations;

3. O is domain of all possible outputs for all queries;
4. $f : U^* \times Q \to O$ is a function that describes the desired output of any query $q \in Q$ given the history of all updates, $(u_1, u_2, \ldots, u_m) \in U^*$.

A *differentially private cell probe data structure* **DS** for the data structure problem P consists of a randomized algorithm implementing update and query operations for P. **DS** is parameterized by the integers c and w denoting the client storage and cell size in bits respectively. Additionally, **DS** is given a random string $\mathcal{R}$ of finite length r containing all randomness that **DS** will use. Note that $\mathcal{R}$ can be arbitrarily large and, thus, contain all the randomness of a random oracle. Given the random string, our algorithms can be viewed as deterministic. Each algorithm is viewed as a finite decision tree executed by the *client* that probes (read or overwrite) memory cells owned by the *server*. For each $q \in Q$ and $u \in U$, there exists a (possibly) different decision tree. Each node in the decision tree is labelled by an index indicating the location of the server-held memory cell to be probed. For convenience, we will assume that a probe may both read and overwrite cell contents. This only reduces the number of cell probes by a factor of at most 2. Additionally, all leaf nodes are labelled with an element of O indicating the output of **DS** after execution.

Each edge in the tree is labelled by four bit strings. The first bit-string of length w represents the contents of the cell probed. The next w-bit string represents the new cell contents after overwriting. There are two c-bit strings representing the current client storage and the new client storage after performing the probe. Finally, there is a r-bit string representing the random string. The client executes **DS** by traversing the decision tree starting from the root. At each node, the client reads the indicated cell's contents. Using the random string, the current client storage and the cell contents, it finds the edge to the next node and updates the probed cell's contents and client storage accordingly. When reaching a leaf, **DS** outputs the element of O denoted at the leaf.

Note, **DS** is only permitted to use the contents of the previously probed cell, current client storage and the random string as input to generate the next cell probe or produce an output. The running time of **DS** is related to the depth of the decision tree as each edge corresponds to a cell probe. Furthermore, as the servers are passive, the server can only either update or retrieve a cell for the client. As a result, the running time (number of cell probes) multiplied by w (the cell size) gives us the bandwidth of the algorithm in bits. We now define the failure probability of **DS**.

Definition 2. *A* **DS** *for data structure problem* $P = (U, Q, O, f)$ *has failure probability* $0 \le \alpha \le 1$ *if for every sequence of updates* $u_1, \ldots, u_m \in U^*$ *and query* $q \in Q$:

$$\Pr[\mathbf{DS}(u_1, \ldots, u_m, q) \ne f(u_1, \ldots, u_m, q)] \le \alpha$$

where randomness is over the choice of $\mathcal{R}$.

As $\mathcal{R}$ is finite, it might seem that we do not consider algorithms whose failure probability decreases in the running time but may never terminate. Instead, we can consider a variant of the algorithm that may run for an arbitrary long time but must provide an answer once its failure probability is small enough (for example, negligible in the number of item stored). Therefore, by sacrificing failure probability, we can convert such possibly infinitely running algorithms into finite algorithms with slightly larger failure probabilities. As we will prove our lower bounds for **DS** with failure probabilities at most 1/3, we may also consider these kind of algorithms with vanishing failure probabilities and no termination guarantees.

We now move to privacy requirements and define the random variable $\mathbb{V}_{\mathbf{DS}}(Q)$ as the *adversary's view* of **DS** processing a sequence of operations Q where randomness is over the choice of the random string $\mathcal{R}$. The adversary's view contains the sequence of probes performed by **DS** to server-held memory cells. We stress that the view does not include the accesses performed by **DS** to client storage. We now define *differentially private access*.

Definition 3. *DS provides (ϵ, δ)-differentially private access against computational adversaries if for any two sequences $Q = (\mathrm{op}_1, \ldots, \mathrm{op}_m)$ and $Q' = (\mathrm{op}'_1, \ldots, \mathrm{op}'_m)$ such that $|\{i \in \{1, \ldots, m\} \mid \mathrm{op}_i \neq \mathrm{op}'_i\}| = 1$ and any PPT algorithm $\mathcal{A}$, it holds that*

$$\Pr[\mathcal{A}(\mathbb{V}_{\mathbf{DS}}(Q)) = 1] \leq e^\epsilon \cdot \Pr[\mathcal{A}(\mathbb{V}_{\mathbf{DS}}(Q')) = 1] + \delta.$$

Our results focus on *online* data structures where each cell probe may be assigned to a unique operation.

Definition 4. *A DS is online if for any sequence $Q = (\mathrm{op}_1, \ldots, \mathrm{op}_m)$, the adversary's view can be split up as:*

$$\mathbb{V}_{\mathbf{DS}}(Q) = (\mathbb{V}_{\mathbf{DS}}(\mathrm{op}_1), \ldots, \mathbb{V}_{\mathbf{DS}}(\mathrm{op}_m))$$

where each cell probe in $\mathbb{V}_{\mathbf{DS}}(\mathrm{op}_i)$ is performed after receiving op_i and before receiving op_{i+1}.

Finally, we present the definition of an (ϵ, δ)-*differentially private cell probe data structure*. We present a diagram of the model in Fig. 1.

Definition 5. *A DS for problem P is an (ϵ, δ)-differentially private cell probe data structure if DS has failure probability 1/3, provides (ϵ, δ)-differentially private access and is online.*

We comment that the failure probability of 1/3 does not seem to be reasonable for any scenario. However, proving a lower bound for **DS** with failure probability 1/3 results in stronger lower bounds as they also hold for more reasonable situations with zero or negligibly small failure probabilities.

We now present the *array maintenance* problem introduced by Wang *et al.* [35], which crisply defines the online RAM problem.

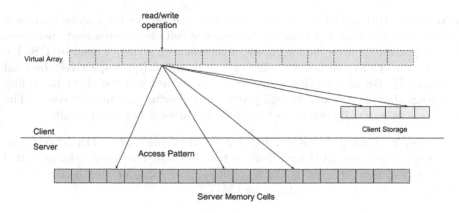

Fig. 1. Diagram of differentially private cell probe model.

Definition 6. *The* array maintenance *problem* AM *is parameterized by two integers* $n, b > 0$ *and defined by the tuple* $(U_{AM}, Q_{AM}, O_{AM}, f_{AM})$ *where*

- $U_{AM} = \{\mathtt{write}(i, B) : i = 0, \dots, n - 1, B \in \{0, 1\}^b\};$
- $Q_{AM} = \{\mathtt{read}(i) : i = 0, \dots, n - 1\};$
- $O_{AM} = \{0, 1\}^b;$

and, for a sequence $Q = (u_1, \dots, u_m)$ *where* $u_1, \dots, u_m \in U^*$, f_{AM} *is:*

$$f_{AM}(Q, \mathtt{read}(i)) = \begin{cases} B, & \text{where } j \text{ is largest index such that } q_j = \mathtt{write}(i, B); \\ 0^b, & \text{if there exists no such } j. \end{cases}$$

In words, the array maintenance problem requires that a data structure to store an array of n entries each of b bits. Each array location is uniquely identified by a number in $[n]$. Typically, it is assumed that a cell is large enough to contain an index. In this case, $w = \Omega(\log n)$. However, in our lower bounds, we will only assume that $w = \Omega(\log \log n)$ to achieve a stronger lower bound. An update operation (also called a *write*) takes as input an integer $i \in [n]$ and a b-bit string B and overwrites the array entry associated with i with the string B. For convenience, we denote a write operation with inputs i and B as $\mathtt{write}(i, B)$. A query operation (also called a *read*) takes as input an integer $i \in [n]$ and returns the current b-bit string of the array entry associated with i. We denote a read operation with input i as $\mathtt{read}(i)$. We will prove lower bounds for (ϵ, δ)-*differentially private RAMs* which are differentially private cell probe data structures for the AM problem.

3   First Lower Bound

Let **DS** be a (ϵ, δ)-differentially private RAM storing n b-bit entries indexed by the integers from 0 to $n - 1$. For any sequence of operations $Q = (\mathtt{op}_1, \dots, \mathtt{op}_m)$, we denote $t_w(Q)$ as the worst case number of cell probes on a write operation and $t_r(Q)$ as the expected amortized number of cell probes on a read operation. Both expectations are over the choice of the random string $\mathcal{R}$ used by **DS**. We write t_w and t_r as the largest value of $t_w(Q)$ and $t_r(Q)$ respectively over all sequences Q. We assume that cells are of size w bits and the client has c bits of storage. In this section, we will prove the following preliminary result. The result will be strengthened in Sect. 4 where we present our main result.

Theorem 2. *Let $\epsilon > 0$ and $0 \leq \delta \leq 1/3$ be constants and let **DS** be an (ϵ, δ)-differentially private RAM for n b-bit entries implemented over w-bit cells that uses c bits of local storage. If **DS** has failure probability at most $1/3$ and $w = \Omega(\log \log n)$, then $t_w + t_r = \Omega\left(\frac{b}{w} \cdot \log\left(\frac{nb}{c}\right)\right)$.*

In terms of block bandwidth, this implies that at least one of read and write has an expected amortized $\Omega(\log(nb/c))$ block bandwidth overhead. The above theorem will be shown when **DS** has to process a sequence Q sampled according to distribution $\mathcal{Q}(0)$. For index $\mathsf{idx} \in \{0, \ldots, n-1\}$, distribution $\mathcal{Q}(\mathsf{idx})$ is defined by the following probabilistic process:

1. Pick m uniformly at random from $\{n/2, n/2 + 1, \ldots, n - 1\}$.
2. Draw $B_1, \ldots, B_m$ independently and uniformly at random from $\{0,1\}^b$.
3. Construct the sequence $U = \mathtt{write}(1, B_1), \ldots, \mathtt{write}(m, B_m)$.
4. Output $Q = (U, \mathtt{read}(\mathsf{idx}))$.

Thus $\mathcal{Q}(\mathsf{idx})$ assigns positive probability to sequences $Q = (U, \mathtt{read}(\mathsf{idx}))$ that consist of a sequence U of m write operation of m b-bit blocks one for each index $1, 2, \ldots, m$, followed by a single read to index idx. It will be useful to define $\mathcal{U}$ to be the distribution of U as determined by the first three steps of the process described above.

In particular, we prove the above theorem using $\mathcal{Q}(0)$, which for convenience, will be denoted by $\mathcal{Q}$ from now on. If privacy is not a concern, sequences in the support of $\mathcal{Q}$ do not seem to require many probes as index 0 is not overwritten. However, the lower bound will, critically, use the fact that the view of any computational adversary cannot differ significantly from the view of sequences whose last operation attempts to read a previously overwritten index $\mathsf{idx} \in \{1, \ldots, m\}$.

We prove the lower bound using the *chronogram* technique first introduced by Fredman and Saks [13] along with the modifications by Pătrașcu [29]. The strategy employed by the chronogram technique when applied to a sequence sampled according to $\mathcal{Q}$ goes as follows. For any choice of m, we consider the $n/2$ $\mathtt{write}$ operations that immediately precede the $\mathtt{read}(0)$ operation and we split them into consecutive and disjoint groups, which we denote as *epochs*. The epochs will grow exponentially in size and are indexed going backwards in time (order of operations performed). That is, the epoch consisting of the $\mathtt{write}$ operations immediately preceding the $\mathtt{read}$ operation will have the smallest index, while the epoch furthest in the past will have the largest index. Note that, when the sequence of operations is sampled according to $\mathcal{Q}$ or, more generally, according to $\mathcal{Q}(\mathsf{idx})$, the set of indices overwritten by the write operations that fall into epoch i is a random variable which we denote by $\mathcal{U}^i$ and that depends on the value of m.

To prove Theorem 2, we consider a simple epoch construction. For $i > 0$, epoch i consists of $\ell_i = 2^i$ write operations and thus there will be $k = \log_2(n/2 + 2) - 1$ epochs. We also define s_i to be the total size of epochs $1, \ldots, i$. In the epoch construction of this section, we have $s_i = 2^{i+1} - 2$. See Fig. 2 for a diagram of the layout of the epochs with regards to a sequence of operations. In Sect. 4, we will derive stronger lower bounds by considering more complex epoch constructions with different parameters.

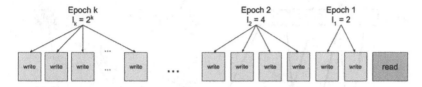

Fig. 2. Diagram of epoch construction of Sect. 3. Operations are performed from left to right.

Defining Random Variables. Since we are considering online data structures, each cell probe performed by **DS** while processing a sequence Q can be uniquely associated to a read or a write operation of Q. Random variable $\mathcal{T}_w(Q)$ is defined as the set of cell probes performed by **DS** while processing the write operations of the sequence Q. Similarly, we define $\mathcal{T}_r(Q)$ as the random variable of the set of cell probes performed by **DS** when processing the read operations of Q. The probability spaces of the two variables are over the choice of $\mathcal{R}$.

The following random variables are specifically defined for sequences $Q = (U, \mathsf{read}(\mathsf{idx}))$ in the support of distribution $\mathcal{Q}(\mathsf{idx})$, for some idx. We remind the reader that these sequences perform a sequence U consisting of m write operations followed by a single $\mathsf{read}(\mathsf{idx})$ operation. The m write operations overwrite entries $1, \ldots, m$ with random b-bit strings. We denote by $\mathcal{T}_w^j(Q)$ the random variable of the cells that are probed during the execution of a write operation of epoch j in Q. We further partition the cell probes in $\mathcal{T}_w^j(Q)$ according to the epoch the cell was last overwritten before being probed in epoch j. Specifically, for $i \geq j$, we define $\mathcal{T}_w^{i,j}(Q)$ as the random variable of the subset of the probes of $\mathcal{T}_w^j(Q)$ performed to a cell that was last overwritten by an operation in epoch i. Note that the sets $\mathcal{T}_w^{i,j}(Q)$ for all pairs (i,j) with $i \geq j$ constitute a partition of $\mathcal{T}_w(Q)$. It will be convenient in the proof to define $\mathcal{T}_w^{<i}(Q) = \mathcal{T}_w^{i,1}(Q) \cup \ldots \cup \mathcal{T}_w^{i,i-1}(Q)$ as the set of probes that are performed by an operation in any of epochs $\{1, \ldots, i-1\}$ to a cell that was last overwritten by an operation in epoch i. Note that if two sequences $Q_1 = (U, \mathsf{idx}_1)$ and $Q_2 = (U, \mathsf{idx}_2)$ share the same initial sequence U of write operations, then, clearly, $\mathcal{T}_w^{<i}(Q_1) = \mathcal{T}_w^{<i}(Q_2)$ if they both use the same random string $\mathcal{R}$.

Finally, we denote the random variable $\mathcal{T}_r^i(Q)$ as the set of probes performed by the read operation of Q to cells that were last overwritten by an operation in epoch i. In Fig. 3, we show a diagram of $\mathcal{T}_w^{<i}(Q)$ and $\mathcal{T}_r^i(Q)$.

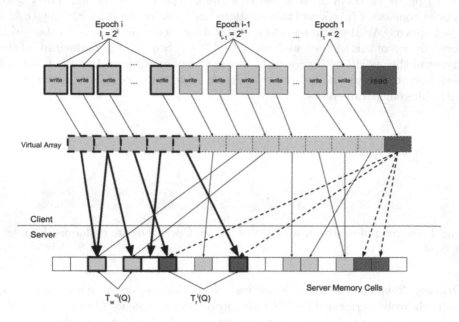

Fig. 3. Diagram of $\mathcal{T}_w^{<i}(Q)$ and $\mathcal{T}_r^i(Q)$.

We extend the definitions above to distributions of sequences in a natural way. For example, $\mathcal{T}_w(\mathcal{Q}(\mathsf{idx}))$ is defined by first picking sequence Q according to $\mathcal{Q}(\mathsf{idx})$ and then sampling a set according to $\mathcal{T}_w(Q)$. Note that, since $\mathcal{T}_w^{<i}(\mathcal{Q}(\mathsf{idx}))$ does not depend on the read operation, we have that for all $\mathsf{idx}_1, \mathsf{idx}_2$ $\mathcal{T}_w^{<i}(\mathcal{Q}(\mathsf{idx}_1)) = \mathcal{T}_w^{<i}(\mathcal{Q}(\mathsf{idx}_2)) = \mathcal{T}_w^{<i}(\mathcal{U})$.

3.1   A Tradeoff Between $\mathcal{T}_w^{<i}(\mathcal{Q})$ and $\mathcal{T}_r^i(\mathcal{Q})$

From a high level, the proof of Theorem 2 is based on the fact that $\mathcal{T}_w^{<i}(\mathcal{Q})$ and $\mathcal{T}_r^i(\mathcal{Q})$ cannot be both small for all epochs i. To see why this must be intuitively true consider distribution $\mathcal{Q}_i$ over query sequences where the last read operation to the 0-th index is replaced with an index chosen uniformly at random from $\mathcal{U}^i$ (remember $\mathcal{U}^i$ are the indices of the array entries that are overwritten by write operations in epoch i). Since each write operation overwrites a distinct entry with a uniformly chosen b-bit string, a sufficiently large number of bits that were encoded by write operations in epoch i must be retrieved by the read operation. There are only three ways that these bits can be retrieved by the read operation. The first way is to probe cells that were last overwritten by any write operation of epoch i which corresponds to $\mathcal{T}_r^i(\mathcal{Q}_i)$. Another way is to probe cells

that were last overwritten by operations that occurred after epoch i; that is, in any epoch $1 \leq j < i$. However, the total number of bits encoded by operations in epochs $1 \leq j < i$ is upper bounded by the number of probes performed in epochs $1 \leq j < i$ to cells that were last overwritten by an operation in epoch i, which corresponds to $\mathcal{T}_w^{<i}(\mathcal{Q}_i)$. The final way to retrieve information from the write operations of epoch i is to encode information in the client's storage of c bits. However, if we consider the case when the number of entries overwritten in epoch i, l_i, is significantly larger than c, then the client's storage is too small to encode any significantly large amount of information compared to the total number of write operations of epoch i. As a result, the total combined size of $\mathcal{T}_w^{<i}(\mathcal{Q}_i)$ and $\mathcal{T}_r^i(\mathcal{Q}_i)$ or, better, a function of the two quantities, can be lower bounded. However, recall that we wish to lower bound the values when processing the random sequence $\mathcal{Q}$ and not $\mathcal{Q}_i$. The only difference between $\mathcal{Q}$ and $\mathcal{Q}_i$ is the index of the read operation performed at the end. By computational differential privacy, any random event that can be verified by a PPT adversary cannot occur with significantly different probabilities when **DS** processes $\mathcal{Q}$ as opposed to $\mathcal{Q}_i$. Since the sets of cell probes can easily be computed in polynomial time, a lower bound on the sum of $|\mathcal{T}_w^{<i}(\mathcal{Q}_i)| + |\mathcal{T}_r^i(\mathcal{Q}_i)|$ also implies a lower bound on the $|\mathcal{T}_w^{<i}(\mathcal{Q})| + |\mathcal{T}_r^i(\mathcal{Q})|$ for a differentially private **DS**.

As explained above, the technical crux of the lower bound on $|\mathcal{T}_w^{<i}(\mathcal{Q}_i)| + |\mathcal{T}_r^i(\mathcal{Q}_i)|$ is an encoding argument that is captured by the following lemma that shows that a certain random variable $\mathcal{Z}_i(\mathcal{Q}(j))$ is "large" with probability at least $1/2$. We say that an epoch i is *large* if $l_i \geq \max\{\sqrt{n}, c^2/b\}$.

Lemma 1. *Assume that* **DS** *has failure probability at most* $1/3$. *Then, for any large epoch* i, *there exists an index* $\mathsf{idx} \in \{1, \ldots, n-1\}$ *such that*

$$\Pr[\mathcal{Z}_i(\mathcal{Q}(\mathsf{idx})) \geq b/8)] \geq 1/2$$

where $\mathcal{Z}_i(\mathcal{Q}(\mathsf{idx}))$ *is*

$$\frac{1}{l_i} \left(|\mathcal{T}_w^{<i}(\mathcal{Q}(\mathsf{idx}))|w + \log \binom{t_w s_{i-1}}{|\mathcal{T}_w^{<i}(\mathcal{Q}(\mathsf{idx}))|} \right) + \left(|\mathcal{T}_r^i(\mathcal{Q}(\mathsf{idx}))|w + \log \binom{t_r}{|\mathcal{T}_r^i(\mathcal{Q}(\mathsf{idx}))|} \right) + \frac{c}{l_i}.$$

The proof of Lemma 1 is found in Sect. 3.4. $\mathcal{Z}_i(\mathcal{Q}(\mathsf{idx}))$ can be viewed as the total average information that the read(idx) operation at the end of $\mathcal{Q}(\mathsf{idx})$ retrieves from the write operations of epoch i. Let us explain the meaning of each term of the value $\mathcal{Z}_i(\mathcal{Q}(\mathsf{idx}))$. The first term of $\mathcal{Z}_i(\mathcal{Q}(\mathsf{idx}))$ measures the average amount of information pertaining to each of the l_i write operations of epoch i that are read by cell probes performed in epochs following epoch i. Each of the cell probes in $\mathcal{T}_w^{<i}(\mathcal{Q}(j))$ reads exactly w bits in a cell. In addition, the choice of which cell probes performed in epochs following epoch i actually belong to $\mathcal{T}_w^{<i}(\mathcal{Q}(j))$ also encodes some information. As there are s_{i-1} write operations epochs following epoch i, there are at most $t_w s_{i-1}$ cell probes and at most $\binom{t_w s_{i-1}}{|\mathcal{T}_w^{<i}(\mathcal{Q}(j))|}$ choices of the cells to probe leading to $\log \binom{t_w s_{i-1}}{|\mathcal{T}_w^{<i}(\mathcal{Q}(j))|}$ bits. Similarly, each probe in $\mathcal{T}_r^i(\mathcal{Q}(j))$ reads w bits in each cell and there are at most $\binom{t_r}{|\mathcal{T}_r^i(\mathcal{Q}(j))|}$ choices of probes when performing read(j) that belong to $\mathcal{T}_r^i(\mathcal{Q}(j))$.

The last term of $\mathcal{Z}_i(\mathcal{Q}(j))$ considers the average amount of information for each of the ℓ_i operations in epoch i that are encoded in the client's storage of c bits. Now, observe that the expected total amount of information that need to be transferred if all ℓ_i possible read operations of $\mathcal{Q}_i$ are performed is $\ell_i \cdot b$ bits. As a result, by taking idx to be the index which requires the most bit transferred of the ℓ_i indices overwritten in epoch i leads us to the above lemma. A formal proof of these ideas is presented in Sect. 3.4.

3.2 Using Differential Privacy

We note that Lemma 1 does not suffice to prove that $t_w + t_r = \omega(1)$. Typically, chronogram lower bounds will find a single sequence that forces a large amount of information transfer from all epochs simultaneously. Instead, Lemma 1 states, that for each epoch, there exists some sequence that forces a large information transfer that the sequences are possibly different for each epoch. In fact, without assuming privacy about a data structure, there can be no single sequence that requires large information from many epochs as there are trivial $\Theta(1)$ data structures that solve the array maintenance problem without any privacy guarantees. As Lemma 1 does not assume privacy for **DS**, we will need to incorporate the fact that **DS** is differentially private to achieve a statement that there exists a single sequence that forces large information transfer from many epochs simultaneously.

Let us now assume that **DS** provides differential privacy against computational adversaries with parameters $\epsilon = O(1)$ and $0 \leq \delta \leq 1/3$. For any fixed sequence Q, we consider any probabilistic event $\mathcal{E}(Q)$ over the randomness of the choice of the random string $\mathcal{R}$ such that there exists a probabilistic polynomial time algorithm that can verify $\mathcal{E}(Q)$ being true or false. Then, computational differential privacy implies that, for any fixed sequence Q_1 and Q_2 that differ in exactly one operation, $\Pr[\mathcal{E}(Q_1) \text{ is true}] \leq e^\epsilon \Pr[\mathcal{E}(Q_2) \text{ is true}] + \delta$. In particular, we can consider the event $\mathcal{E}(Q) = "\mathcal{Z}_i(Q) \geq b/8"$. Note that $\mathcal{Z}_i(Q)$ can be computed by any computational adversary by simply assigning each cell probe performed by **DS** over Q into one of $\{\mathcal{T}_w^{<i}(Q)\}_{i=1,\ldots,k}$ or $\{\mathcal{T}_r^i(Q)\}_{i=1,\ldots,k}$ where assigning a cell probe depends only on the last time the cell was overwritten and the current operation of Q. As a result, we know that for any two fixed sequences Q_1 and Q_2 that differ in exactly one operation, then

$$\Pr[\mathcal{Z}_i(Q_1) \geq b/8] \leq e^\epsilon \Pr[\mathcal{Z}_i(Q_2) \geq b/8] + \delta.$$

Note that $\mathcal{Q}$ and $\mathcal{Q}(\text{idx})$ only differ in the input index to the read operation at the end of the sequence. We use this fact to prove the following lemma that $\mathcal{Z}_i(\mathcal{Q})$ cannot differ significantly from $\mathcal{Z}_i(\mathcal{Q}(\text{idx}))$ for any idx.

Lemma 2. Let **DS** be an (ϵ, δ)-differentially private RAM and let i be a large epoch. Then,
$$\Pr[\mathcal{Z}_i(\mathcal{Q}) \geq b/8] \geq 1/(6e^\epsilon).$$

Proof. By Lemma 1, there exists an index idx such that $\Pr[\mathcal{Z}_i(\mathcal{Q}(\text{idx})) \geq b/8]$ $\geq 1/2$. We define $Q(\text{idx}, B_1, \ldots, B_m) = (\texttt{write}(1, B_1), \ldots, \texttt{write}(m, B_m),$ $\texttt{read}(\text{idx}))$. Then,

$$
\Pr[\mathcal{Z}_i(\mathcal{Q}) \geq b/8] = \sum_{m=n/2}^{n-1} \sum_{B_1, \ldots, B_m \in \{0,1\}^b} \frac{1}{m2^{bm}} \Pr[\mathcal{Z}_i(Q(0, B_1, \ldots, B_m)) \geq b/8]
$$

$$
\geq \sum_{m=n/2}^{n-1} \sum_{B_1, \ldots, B_m \in \{0,1\}^b} \frac{1}{m2^{bm}} \left(\frac{\Pr[\mathcal{Z}_i(Q(\text{idx}, B_1, \ldots, B_m)) \geq b/8] - \delta}{e^\epsilon} \right)
$$

$$
= \frac{\Pr[\mathcal{Z}_i(\mathcal{Q}(\text{idx})) \geq b/8] - \delta}{e^\epsilon} \geq \frac{1/2 - 1/3}{e^\epsilon} = \frac{1}{6e^\epsilon}.
$$

3.3 Completing the Proof of Theorem 2

Lemma 2 resembles the typical desired statement for data structure lower bounds as it guarantees existence of a distribution of query sequences, $\mathcal{Q}$, that forces a large amount of information transfer from all epochs in expectation.

Recall that we consider epoch i consisting of $\ell_i = 2^i$ $\texttt{write}$ operation for a total of $s_i = 2^{i+1} - 2$ $\texttt{write}$ operation in epochs $1, \ldots, i$. We refer the reader to Fig. 2 for a visual reminder of our epoch construction. Using Lemma 2, we will show that $\Omega(b/w)$ bits must be transferred from the *majority* of *large* epochs. In particular, we focus on epochs i for which the number of blocks, ℓ_i, written by the $\texttt{write}$ operations is much larger than the number of blocks that can be stored in client's memory, c/b. For otherwise, the blocks written by the $\texttt{write}$ operations in epoch i may be entirely encoded into client's storage of c bits and thus no information from epoch i is required to be transferred by cell probes of future operations. Concretely, we say that an epoch is *large* if $\ell_i \geq \max\{\sqrt{n}, c^2/b\}$ and note that, by our definition of epochs, we have $\hat{k} := \Theta(\log(nb/c))$ large epochs. We will show that for many large epochs $\Omega(b/w)$ bits must be transferred by cell probes of either $\texttt{write}$ operations of future epochs or the $\texttt{read}$ operation.

To achieve our lower bound, we will analyze the expectation of $\mathcal{Z}_i(\mathcal{Q})$ based on our epoch construction. We will provide a high-level overview of the steps of our analysis in this paragraph before performing a formal analysis. Recall that t_w and t_r are an upper bound on the expected amortized number of cells probed per $\texttt{write}$ and $\texttt{read}$ operation for any sequence. For the majority of epochs i, we cannot expect the $\texttt{read}$ operation of $\mathcal{Q}$ to probe more than $t_r/\hat{k}$ cells containing information about the $\texttt{write}$ operations of epoch i. This provides an upper bound on $|\mathcal{T}_r^i(\mathcal{Q})|$ for the majority of epochs. We want a similar upper bound on the value of $|\mathcal{T}_w^{<i}(\mathcal{Q})|$. Recall that this number corresponds to the number of probes performed by $\texttt{write}$ operations that read cells that encode information about the $\texttt{write}$ operations of epoch i. Our argument will critically use the fact that the sequence $\mathcal{Q}$ is chosen at random. Recall that $\mathcal{Q}$ is chosen to have m $\texttt{write}$ operations where m is chosen uniformly at random from $\{n/2, n/2+1, \ldots, n-1\}$. The data structure **DS** is unable to predict the point in time when the $\texttt{read}$ operation will occur. Instead, the best that **DS** can achieve is to prepare for

all possible epoch configurations. Since there are $\hat{k}$ epochs with size at least $\max\{\sqrt{n}, c^2\}$, each update should be only able to encode $\frac{t_w \cdot w}{\hat{k}}$ about each of these epochs. As a result, we can prove the majority of epochs cannot have very large values of $|\mathcal{T}_w^{<i}(\mathcal{Q})|$ in expectation.

As the two bounds above hold for the majority of epochs, we can show there exists at least one large epoch i such that both the values of $|\mathcal{T}_w^{<i}(\mathcal{Q})|$ and $|\mathcal{T}_r^i(\mathcal{Q})|$ are small. In particular, we show the following:

Lemma 3. *There exists a large epoch i for which* $\mathsf{E}\left[|\mathcal{T}_w^{<i}(\mathcal{Q})|/\ell_i\right] = O\left(t_w/\log(nb/c)\right)$ *and* $\mathsf{E}\left[|\mathcal{T}_r^i(\mathcal{Q})|\right] = O\left(t_r/\log(nb/c)\right)$.

Proof. The lemma is derived from the following two statements:

1. There exists $\hat{k}/2 + 1$ large epochs i such that $\mathsf{E}[|\mathcal{T}_w^{<i}(\mathcal{Q})|/\ell_i] = O(t_w/\log(nb/c))$.
2. There exists $\hat{k}/2 + 1$ large epochs i such that $\mathsf{E}[|\mathcal{T}_r^i(\mathcal{Q})|] = O(t_r/\log(nb/c))$.

Since there are only $\hat{k}$ large epochs, there must exist at least one large epoch where both inequalities hold. We now show the two statements are true.

Let us pick epoch i uniformly at random amongst the $\hat{k}$ large epochs and fix the random string $\mathcal{R}$ as well as the $n-1$ block values $\mathcal{B}_1, \ldots, \mathcal{B}_{n-1}$. We now fix a cell probe **probe** of the execution of **DS** over the **write** operations $\texttt{write}(1, \mathcal{B}_1), \ldots, \texttt{write}(n-1, \mathcal{B}_{n-1})$ and consider the probability that **probe** contributes to $\mathcal{T}_w^{<i}(\mathcal{Q})$ from which we derive a bound on $\mathsf{E}[|\mathcal{T}_w^{<i}(\mathcal{Q})|/\ell_i]$. Note that, having fixed $\mathcal{R}$ and the values $\mathcal{B}_j$'s, the probability space is over the choice of m from $\{n/2, n/2 + 1, \ldots, n-1\}$ and of i. We denote p_r as the index of the **write** operation in $\mathcal{U}$ when **probe** is performed. The value p_w is denoted as the index of the **write** operation in $\mathcal{U}$ when the cell of **probe** was last overwritten. Using p_r and p_w, we can attempt to upper bound the probability that the **probe** belongs to $\mathcal{T}_w^{<i}(\mathcal{Q})$. First, let e be the smallest integer such that $p_r - p_w \leq s_e$. Note that **probe** cannot contribute to $\mathcal{T}_w^{<j}(\mathcal{Q})$ for any epoch $j \leq e - 1$, since there are only s_j operations between the beginning of epoch k and the **read** operation. Since $s_j \leq s_{e-1} < p_r - p_w$, either the read operation has to occur after the **read** operation or the last operation to overwrite the cell probe occurs before the j-th epoch. We remind the reader that the exact locations of epochs is determined by m. The boundary denoting the end of epoch j has to occur after p_w and before p_r meaning there are at most s_e choices from the position of the **read** operation such that this cell probe contributes to $\mathcal{T}_w^{<j}(\mathcal{Q})$. There are $n/2$ choices for m, so the probability is at most $2s_e/n$. We now compute

$$\mathsf{E}[|\mathcal{T}_w^{<i}(\mathcal{Q})|/\ell_i] = \frac{1}{\hat{k}} \sum_{j:\ell_j \geq \max\{\sqrt{n}, c^2\}} \mathsf{E}[|\mathcal{T}_w^{<j}(\mathcal{Q})|/\ell_j].$$

The **probe** only contributes to epochs $j \geq e$. Note, there are at most (in expectation) $t_w \cdot (n-1)$ cell probes performed when processing the **write** operations of $\mathcal{Q}$. By linearity of expectation,

$$\sum_{j:\ell_j \geq \max\{\sqrt{n}, c^2/b\}} \mathsf{E}\left[\frac{|\mathcal{T}_w^{<j}(\mathcal{Q})|}{\ell_j}\right] \leq t_w \cdot n \sum_{j \geq e} \frac{2 \cdot s_e}{n \cdot l_j} \leq 2t_w \cdot \left(\frac{s_e}{l_e} + \frac{s_e}{l_{e+1}} + \ldots\right) \leq 4t_w.$$

As a result, there exists $\hat{k}/2 + 1$ fixed epochs i such that their expectation over the m is at most $12t_w$.

We know that $\sum_i \mathsf{E}[|\mathcal{T}_r^i(\mathcal{Q})|] \leq t_r$. Therefore, there exists $\hat{k}/2+1$ large epochs i such that $\mathsf{E}[|\mathcal{T}_r^i(\mathcal{Q})|] \leq 3t_r$ completing the proof.

We can now achieve our goal of proving Theorem 2 that gives a lower bound on the sum $t_w + t_r$ by plugging the inequalities in Lemma 3 into the expectation of $\mathcal{Z}_i(\mathcal{Q})$ and then using the bound from Lemma 2.

Proof (Theorem 2). First, we analyze the expectation of $\mathcal{Z}_i(\mathcal{Q})$. Note that, for every x, y, $\log \binom{y}{x} = O(x \log(y/x))$. Moreover, for every y, $x \log(y/x)$ is a convex function over x, so we can write the $\mathsf{E}[x \log(y/x)] \leq \mathsf{E}[x] \log(y/\mathsf{E}[x])$ where the expectation is over the choice of x. We now apply this observation to $\mathsf{E}[\mathcal{Z}_i(\mathcal{Q})]$

$$O\left(\frac{\mathsf{E}[|\mathcal{T}_w^{<i}(\mathcal{Q})|]}{\ell_i}\left(w + \log \frac{t_w s_{i-1}}{\mathsf{E}[|\mathcal{T}_w^{<i}(\mathcal{Q})|]}\right) + \mathsf{E}[|\mathcal{T}_r^i(\mathcal{Q})|]\left(w + \log \frac{t_r}{\mathsf{E}[|\mathcal{T}_r^i(\mathcal{Q})|]}\right) + \frac{c}{\ell_i}\right).$$

By Lemma 2, we know that $\mathsf{E}[\mathcal{Z}_i(\mathcal{Q})] = \Omega(b)$. We now pick our epoch i as the one chosen by Lemma 3 and plug in the inequalities to get

$$\frac{t_w}{\log(nb/c)}\left(w + \log \frac{t_w s_{i-1}}{\ell_i t_w / \log(nb/c)}\right) + \frac{t_r}{\log(nb/c)}\left(w + \log \frac{t_r}{t_r / \log(nb/c)}\right) = \Omega(b).$$

Here we have used the fact that epoch i is large and thus $\frac{c}{\ell_i} = O(b)$, since $\ell_i \geq c^2/b$. Also, note that $s_{i-1} = \Theta(\ell_i)$. Therefore, we can simplify and get that $t_w + t_r = \Omega\left((b/(w + \log \log n)) \log(nb/c)\right)$. If we assume that $w = \Omega(\log \log n)$, we can simplify and get the following result $t_w + t_r = \Omega\left((b/w) \log(nb/c)\right)$ which completes the proof.

Therefore, the lower bound of $t_w + t_r$ described in Theorem 2 can be entirely derived from Lemma 1. It remains to prove Lemma 1, which we do next.

3.4 An Encoding Argument Using $\mathcal{T}_w^{<i}(\mathcal{Q})$ and $\mathcal{T}_r^i(\mathcal{Q})$

In this section, we prove Lemma 1. We first give a high level description of the proof. The main idea involves converting any **DS** that solves the array maintenance problem into a one-way communication problem between two parties, for which we have a lower bound on the number of bits that must be sent.

Specifically, we consider the case in which for a fixed epoch $i \in \{1, \ldots, k\}$ and for a sequence drawn according to $\mathcal{Q}$, one party, Alice, receives the m values $\mathcal{B}_1, \ldots, \mathcal{B}_m$ and a random string $\mathcal{R}$ and the other party, Bob, receives the same random string $\mathcal{R}$ as well as $m - \ell_i$ values; that is, all of $\mathcal{B}_1, \ldots, \mathcal{B}_m$ except for the ℓ_i values updated in epoch i of sequence Q. The goal of the protocol is to let Bob obtain the missing ℓ_i values.

As the ℓ_i b-bit values are generated uniformly and independently at random, Alice's input has $\ell_i \cdot b$ bits of entropy conditioned on Bob's input and $\mathcal{R}$. By Shannon's Source Coding Theorem, any protocol for the above problem must have expected communication of at least $\ell_i \cdot b$ bits. We show that if Lemma 1 does not hold, then Shannon's Theorem is contradicted by giving an encoding constructed by simulating **DS** that beats Shannon's bound.

Recall that, for any idx, $\mathcal{Q}(\mathsf{idx})$ is constructed by picking m uniformly at random from $\{n/2, n/2 + 1, \ldots, n - 1\}$ and constructing the sequence of m updates $\mathcal{U} = \mathtt{write}(1, \mathcal{B}_1), \ldots, \mathtt{write}(m, \mathcal{B}_m)$ where each $\mathcal{B}_1, \ldots, \mathcal{B}_m$ is drawn independently and uniformly at random from $\{0, 1\}^b$. We also denote by $\mathcal{U}^i$ the set of $\mathtt{write}$ of epoch i, for $i = 1, \ldots, k$.

Consider the following protocol. Alice and Bob locally execute all $\mathtt{write}$ operations in epochs $k, k - 1, \ldots, i + 1$ using the random string $\mathcal{R}$. Bob keeps a snapshot snap_B of **DS** at this point. Now Bob can learn each of the ℓ_i values $\mathcal{B}_{\mathsf{idx}}$ for $\mathsf{idx} \in \mathcal{U}^i$ written during epoch i, by simulating epoch j, for $j = i, i - 1, \ldots, 1$ followed by the $\mathtt{read}(\mathsf{idx})$ operation. To do this, Bob uses the snapshot snap_B, that gives the state of **DS** before any $\mathtt{write}$ operations of epoch i are executed, and the following information that can be transferred by Alice.

1. The c bits of client storage of **DS** after the $\mathtt{write}$ operations of epoch i have been processed.
2. The location and contents of the cells that are probed by the $\mathtt{write}$ operations of epochs $j = i - 1, \ldots, 1$ and by the $\mathtt{read}(\mathsf{idx})$ operation.

Given this information as well as the random string $\mathcal{R}$, Bob can simulate **DS** by starting from snap_B and executing all the $\mathtt{write}$ operations of $\mathcal{U}$ occurring after epoch i as well as $\mathtt{read}(\mathsf{idx})$ and thus recover $\mathcal{B}_{\mathsf{idx}}$. To encode all ℓ_i block values updated in epoch i, Alice and Bob can repeat the simulation of the $\mathtt{read}$ operation ℓ_i times with idx ranging over the set of the ℓ_i indices that are updated in epoch i. The number of bits that need to be transferred to Bob by Alice depends on the following three values:

1. The number of bits of the client storage, c.
2. The number of probes performed in epochs $j = i - 1, \ldots, 1$ to cells last written in epoch i.
3. The number of probes performed by the ℓ_i $\mathtt{read}$ operations to the ℓ_i indices updated in epoch i.

By Shannon's source coding theorem, we have a lower bound on the number of bits that can be transferred and, consequently, a lower bound on the number of probes performed by **DS**. The rough description above only works for **DS** that never fails but it only requires some small changes to work for failure probability $1/3$. In particular, Alice can indicate the indices idx for which **DS** fails to return $\mathcal{B}_{\mathsf{idx}}$ and explicitly transfer the b bits of $\mathcal{B}_{\mathsf{idx}}$ to Bob in addition to the above protocol. We now present the formal proof of Lemma 1.

Proof (Lemma 1). In our proof, we consider **DS** that have failure probability at most $1/512$. Note that any **DS** with failure probability $1/3$ implies the existence of a **DS** with failure probability $1/512$ as one can execute **DS** a constant number of times with independently chosen randomness and return the most popular result to answer any $\mathtt{read}$ operation. In fact, proving a lower bound for **DS** with failure probability $1/512$ implies any **DS** with failure probability that is a constant greater than $1/2$ using the above method.

Recall that $\mathcal{U}^i$ denotes the set of all ℓ_i indices that are updated by write operations in epoch i. It suffices to prove that $\Pr[\sum_{\mathsf{idx}\in\mathcal{U}^i} \mathcal{Z}_i(\mathcal{Q}(\mathsf{idx})) \geq \ell_i b/8] \geq 1/2$. Since $\mathcal{U}^i$ contains ℓ_i indices, the previous statement implies that there must exist some $\mathsf{idx} \in \mathcal{U}^i$ such that $\Pr[\mathcal{Z}_i(\mathcal{Q}(\mathsf{idx})) \geq b/8] \geq 1/2$ which would complete the proof. Therefore, towards a contradiction, assume that $\Pr[\sum_{\mathsf{idx}\in\mathcal{U}^i} \mathcal{Z}_i(\mathcal{Q}(\mathsf{idx})) \geq \ell_i b/8] < 1/2$ for some data structure **DS** that solves the array maintenance problem with failure probability at most $1/512$. We will present an encoding of $\ell_i \cdot b$ random bits from Alice and Bob using **DS** that uses strictly less than $\ell_i \cdot b$ bits in expectation contradicting Shannon's source coding theorem.

In computing the encoding, Alice receives the m b-bit random values used by the sequence of write operations, $\mathcal{U}$, and a random string $\mathcal{R}$.

Alice's Encoding

1. Alice executes **DS** on the sequence $\mathcal{U}$ using the random string $\mathcal{R}$ up to the final write operation of epoch i. The content of the c bits of client storage after epoch i is completed are added to the encoding.

2. Alice then executes the remaining s_{i-1} write operations of $\mathcal{U}$ of epochs $i - 1, i - 2, \ldots, 1$. While processing these write operations, Alice records the subset $\mathcal{T}_w^{<i}(\mathcal{U})$ of probes to cells that were last written in epoch i as well as their contents. This information is encoded as follows. First the size $|\mathcal{T}_w^{<i}(\mathcal{U})|$ (at most $\log(t_w \cdot s_{i-1})$ bits) is added to the encoding. Then Alice adds an encoding of which $|\mathcal{T}_w^{<i}(\mathcal{U})|$ probes of the at most $t_w(n-1)$ probes over the entire sequence belong to $\mathcal{T}_w^{<i}$ (this costs $\log \binom{t_w \cdot (n-1)}{|\mathcal{T}_w^{<i}(\mathcal{U})|}$ bits). Finally, for each such probe, w bits are added to the encoding to specify the content of the cell probed (for additional $|\mathcal{T}_w^{<i}(\mathcal{U})| \cdot w$ bits).

3. Alice stores a snapshot snap_A of the **DS** after processing all write operations of $\mathcal{U}$. Alice will use this snapshot to simulate the read operations for the ℓ_i entries written in epoch i.

4. For each of the ℓ_i indices $\mathsf{idx} \in \mathcal{U}^i$, Alice executes read($\mathsf{idx}$) on snap_A. Let F be the number of read(idx) operations that return a wrong value (that is, they return a value other than $\mathcal{B}_{\mathsf{idx}}$). Alice adds the value F to the encoding costing $\log n$ bits and an encoding of the subset of the F *failing* indices costing $\log \binom{\ell_i}{F}$ expected bits.

5. For each of the F failing indices $\mathsf{idx} \in \mathcal{U}^i$, Alice adds $\mathcal{B}_{\mathsf{idx}}$ to the encoding costing a total of $F \cdot b$ bits.

6. For each non-failing index $\mathsf{idx} \in \mathcal{U}^i$ (that is, for which read(idx) executed on snap_A with $\mathcal{R}$ successfully returns $\mathcal{B}_{\mathsf{idx}}$), Alice adds the subset of probes performed during read(idx) to the cells in $\mathcal{T}_r^i(\mathcal{Q}(\mathsf{idx}))$ (these are the cells last written in epoch i) as well as their content to the encoding. This costs w bits for each cell in $\mathcal{T}_r^i(\mathcal{Q}(\mathsf{idx}))$ as well as $\log \binom{t_r}{|\mathcal{T}_r^i(\mathcal{Q}(\mathsf{idx}))|}$ bits to encode the subset $\mathcal{T}_r^i(\mathcal{Q}(\mathsf{idx}))$ of the at most t_r probes in read(idx).

7. Alice checks whether either of $\sum_{\mathsf{idx} \in \mathcal{U}^i} \mathcal{Z}_i(\mathcal{Q}(\mathsf{idx})) > \ell_i b/8$ or $F > \ell_i/64$. If either are true, Alice stops and returns an encoding consisting of a 0 bit followed by $\ell_i \cdot b$ bits of the ℓ_i blocks updated by the write operations in $\mathcal{U}^i$.

8. Otherwise, when both $\sum_{\mathsf{idx} \in \mathcal{U}^i} \mathcal{Z}_i(\mathcal{Q}(\mathsf{idx})) \le \ell_i b/8$ and $F \le \ell_i/64$, Alice prepends a 1 bit to the encoding computed in Steps 1–6 and returns it.

In decoding the message sent by Alice, Bob receives the random string $\mathcal{R}$ but does not receive the entirety of $\mathcal{U}$. Instead, Bob receives $\mathcal{U}$ except all block values that are updated in epoch i.

Bob's Decoding

1. Bob checks the first bit of Alice's encoding. If the first bit is a 0, then Bob parses the next $\ell_i \cdot b$ bits as the contents of the ℓ_i block values updated in $\mathcal{U}$ completing the decoding.

2. If the encoding begins with a 1, Bob will execute the write operations in epochs $j = k, k-1, \ldots, i-1$ using random string $\mathcal{R}$. Note that this is straightforward as Bob received all the needed values as input and the indices of the write are fixed.

3. Note that Bob does not have access to the updated array entries of epoch i, and thus will skip it.

4. Next, Bob sets the client storage as specified in the encoding and starts simulating the write operations for epochs $j = i-1, \ldots, 1$. As long as the write operations do not require probing a cell that was last written in epoch i, Bob can simulate **DS** in the exact same way as done by Alice to compute the encoding (note Bob has access to the same $\mathcal{R}$). Whenever **DS** requires probing a cell last written in epoch i, Bob will use the encoding of the cell contents found in the encoding to continue simulation. As a result, Bob can simulate all write operations of $\mathcal{U}$ after epoch i identically to Alice. Bob will now take a (partial) snapshot of **DS** including all cell locations and contents that Bob is aware of.

5. Next, Bob obtains F, the number of failing read, from the encoding along with the indices $\mathsf{idx} \in \mathcal{U}^i$ where read(idx) fails to return $\mathcal{B}_{\mathsf{idx}}$. For each of these F indices, Bob obtains the corresponding value $\mathcal{B}_{\mathsf{idx}}$ from the encoding.

6. For the remaining $\ell_i - F$ indices $\mathsf{idx} \in \mathcal{U}^i$ such that read(idx) returns $\mathcal{B}_{\mathsf{idx}}$, Bob will execute read($\mathsf{idx}$) on the snapshot of **DS**. From the encoding, Bob knows which of the (at most, in expectation) t_r probes performed by read(idx) are to cells last written in epoch i. Bob simulates read(idx) on his snapshot with $\mathcal{R}$ using the cell contents encoded by Alice to retrieve $\mathcal{B}_{\mathsf{idx}}$.

Analysis. It remains to analyze the expected length of Alice's encoding. Recall that we know from Shannon's source coding theorem that Alice's encoding has to be at least $\ell_i \cdot b$ bits long in expectation.

There are two cases to consider. In the first case, when the first bit is a 0, the encoding will be $1 + \ell_i \cdot b$ bits long. Let us now consider the case in which the first bit is 1 and thus $F \le l_i/64$ and $\sum_{\mathsf{idx} \in \mathcal{U}^i} \mathcal{Z}_i(\mathcal{Q}(\mathsf{idx})) \le \ell_i b/8$. The encoding of the failed indices has expected length

$$\log n + \mathsf{E}\left[\log\binom{\ell_i}{F}\right] + b \cdot \mathsf{E}[F] \leq \log n + \mathsf{E}\left[\log\binom{\ell_i}{\frac{\ell_i}{64}}\right] + b \cdot \frac{\ell_i}{64}$$

$$\leq \log n + \frac{\ell_i}{64} \cdot (b + \log(64 \cdot e)) \leq \log n + \frac{9}{64}(\ell_i \cdot b).$$

The second inequality uses Stirling's approximation which states that $\binom{x}{y} \leq (ex/y)^y$. We know Alice's encoding of client storage will always be c bits. We know the expected bits of encoding $\mathcal{T}_w^{<i}(\mathcal{U})$ is

$$\log(t_w \cdot s_{i-1}) + \mathsf{E}\left[\log\binom{t_w \cdot s_{i-1}}{|\mathcal{T}_w^{<i}(\mathcal{U})|} + |\mathcal{T}_w^{<i}(\mathcal{U})| \cdot w\right].$$

Note, $\log(t_w \cdot s_{i-1}) \leq 2\log n$. Similarly, for all $\mathsf{idx} \in \mathcal{U}^i$ that successfully return $\mathcal{B}_{\mathsf{idx}}$, we know that the encoding requires

$$\mathsf{E}\left[|\mathcal{T}_r^i(\mathcal{Q}(\mathsf{idx}))|w + \log\binom{t_r}{|\mathcal{T}_r^i(\mathcal{Q}(\mathsf{idx}))|}\right].$$

Note that

$$\mathsf{E}\left[\log\binom{t_w \cdot s_{i-1}}{|\mathcal{T}_w^{<i}(\mathcal{U})|} + |\mathcal{T}_w^{<i}(\mathcal{U})|w + \sum_{\mathsf{idx}\in\mathcal{U}^i}|\mathcal{T}_r^i(\mathcal{Q}(\mathsf{idx}))|w + \log\binom{t_r}{|\mathcal{T}_r^i(\mathcal{Q}(\mathsf{idx}))|}\right] + c$$

$$\leq \sum_{\mathsf{idx}\in\mathcal{U}^i} \mathcal{Z}_i(\mathcal{Q}(\mathsf{idx})) \leq \frac{1}{8}(\ell_i \cdot b).$$

Summing over all parts of the encoding, we get that

$$3\log n + \frac{9}{64}(\ell_i \cdot b) + \sum_{\mathsf{idx}\in\mathcal{U}^i} \mathcal{Z}_i(\mathcal{Q}(\mathsf{idx})) \leq 3\log n + \frac{17}{64}(\ell_i \cdot b).$$

Finally, we compute the probabilities that Alice places a 0 or a 1 as the first bit of the encoding. By Markov's inequality, $\Pr[F \geq \ell_i/64] \leq 1/8$ and we know that $\Pr[\sum_{\mathsf{idx}\in\mathcal{U}^i} \mathcal{Z}_i(\mathcal{Q}(\mathsf{idx})) \geq \ell_i b/8] < 1/2$ by our initial assumption towards a contradiction. As a result, we know that $\Pr[F \geq \ell_i/64$ or $\sum_{\mathsf{idx}\in\mathcal{U}^i} \mathcal{Z}_i(\mathcal{Q}(\mathsf{idx})) \geq b/8] < 5/8$. So, Alice's expected encoding size is at most

$$1 + 3\log n + \frac{5}{8}(\ell_i \cdot b) + \frac{17}{64}(\ell_i \cdot b) < \ell_i \cdot b$$

contradicting Shannon's source coding theorem when $\ell_i \geq \sqrt{n}$.

4   Main Result

In Sect. 3, we presented a lower bound on the sum of t_w, the worst case bandwidth for write operations, and t_r, the worst case expected amortized bandwidth for read operations that implies that $\max\{t_w, t_r\} = \Omega((b/w)\log(nb/c))$. However, this lower bound does not preclude the existence of a differentially private RAM with $t_w = \Theta((b/w)\log(nb/c))$ and $t_r = o((b/w)\log(nb/c))$ or $t_r = \Theta((b/w)\log(nb/c))$ and $t_w = o((b/w)\log(nb/c))$. In this section, we strenghten our lower bound and prove the following two statements, for (ϵ, δ)-differentially private RAM, for any constant ϵ and $\delta \leq 1/3$,

1. If $t_w = o((b/w)\log(nb/c))$, then $t_r = \omega((b/w)\log(nb/c))$;
2. If $t_r = o((b/w)\log(nb/c))$, then $t_w = \omega((b/w)\log(nb/c))$.

Therefore, since $\max\{t_w, t_r\} = O((b/w)\log(nb/c))$, then it must be the case that both $t_w = \Theta((b/w)\log(nb/c))$ and $t_r = \Theta((b/w)\log(nb/c))$ showing that imbalanced running times for `write` and `read` operations cannot improve the asymptotic efficiency of differentially private RAM constructions.

To achieve these tradeoffs, we revisit our epoch construction of Sect. 3. Let us, first, focus our attention on the first statement where we show that $t_r = \omega((b/w)\log(nb/c))$ when $t_w = o((b/w)\log(nb/c))$. Recall that we constructed epochs that grew exponentially by a factor of 2 for a total of $\Theta(\log n)$ epochs and the number of large epochs (that is with at least $\max\{\sqrt{n}, c^2/b\}$ `write` operations) is $\Theta(\log(nb/c))$. In the techniques used in Sect. 3, we are only able to show that $\Omega(b/w)$ cells must be probed from the majority of the epochs. As there are only $\Theta(\log(nb/c))$ large epochs, there is no hope for us to prove a stronger lower bound t_r with this epoch construction.

Instead, we will use a different epoch construction that is suitable for the scenario where we know that t_w is small. In Lemma 3, we show that, on average, for any large epoch i any `write` operations of future epochs $j \in \{1, \ldots, i-1\}$ can only encode $O(t_w w/\hat{k})$ bits about epoch i where $\hat{k}$ is the number of large epochs. It is important that future `write` operations cannot encode a lot of information about epoch i as it forces the final `read` operation to read sufficient information from epoch i directly. However, as we are assuming that t_w is already small, we may increase the number of future operations after epoch i while simultaneously ensuring that future epochs cannot encode too much information about epoch i. With this observation, we hope that we can increase the number of total epochs which allows us to prove $\omega((b/w)\log(nb/c))$ lower bounds on t_r as desired. We now materialize these ideas in the next section.

4.1   First Epoch Construction

In this section, we consider an epoch construction where epochs grow by the rate r every r epochs, with $r = \omega(1)$ and $r = O(\log n)$. That is, the first r epochs will each have r `write` operations; the next r epochs will each have r^2 `write` operations; the next r epochs will each have r^3 `write` operations and so forth. See Fig. 4 for a diagram of this epoch construction. Once again, we define ℓ_i to be the number of `write` operations of the i-th epoch and s_i to be the total number of `write` operations of epochs $1, \ldots, i$. We note that, by writing $\ell_i = r^f$ for some $f \geq 1$, we have

$$s_{i-1} \leq r \cdot (r^f + r^{f-1} + \ldots + r) \leq 2r \cdot \ell_i.$$

The new epoch construction will potentially give us, for each epoch, r times more future operations in comparison to the epoch construction of Sect. 3. On the other hand, we note that the number of large epochs (that is, with at least $\max\{\sqrt{n}, c^2/b\}$ write operations) is $\hat{k} = \Theta(r\log_r(nb/(rc^2))) = \Theta(r\log_r(nb/c))$, which is larger by a super-constant factor of $\Theta(r/\log r)$ than the number of epochs in the construction of Sect. 3. As a result, this epoch construction matches exactly the requirements that we wanted there to be more epochs which are required to be read by the read operation while only sacrificing that there are more future write operations for any epoch i. We now present a generalization of Lemma 3 which can be applied for the new epoch constructions that are introduced here and in Sect. 4.2.

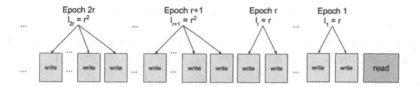

Fig. 4. Diagram of epoch construction of Sect. 4.1.

Lemma 4. *There exists a large epoch i such that* $\mathsf{E}[|\mathcal{T}_w^{<i}(\mathcal{Q})|] = O\left(\frac{t_w}{k} \cdot \max_e \sum_{j \geq e} \frac{s_e}{l_j}\right)$ *and* $\mathsf{E}[|\mathcal{T}_r^i(\mathcal{Q})|] = O\left(\frac{t_r}{k}\right)$.

Proof. Using the same ideas of Lemma 3, we will show that there exists $\hat{k}/2 + 1$ epochs that satisfy the first statement and $\hat{k}/2+1$ epochs that satisfy the second statement. As a result, there exists at least one epoch satisfying both statements.

Pick an epoch i uniformly at random from the $\hat{k}$ large epochs. Fix $\mathcal{B}_1, \ldots, \mathcal{B}_{n-1}$ and $\mathcal{R}$ arbitrarily. We will prove an upper bound on $\mathsf{E}[|\mathcal{T}_w^{<i}(\mathcal{Q})|]$ over the randomness of the location of the read operation and the randomly chosen i. As a result, the expectation's upper bound will hold over any distribution of $\mathcal{B}_1, \ldots, \mathcal{B}_{n-1}$ and $\mathcal{R}$. Fix any cell probe performed by **DS** when processing $\mathsf{write}(1, \mathcal{B}_1), \ldots, \mathsf{write}(n-1, \mathcal{B}_{n-1})$ and suppose that probe occurs when processing the p_r-th write operation to a cell that was last written by the p_w-th write operation. Once again, we pick the smallest e such that $p_r - p_w \leq s_e$. Consider any epoch j where $j \leq e - 1$. Note that there are only s_j operations between the read operation and the beginning of epoch j. But, since $j \leq e-1$, we know that $s_j \leq s_{e-1} < p_r - p_w$ meaning that either the probe occurs after the read operation or the cell was last written before epoch j. When we fix the location of the read operation, we fix the epoch construction. As the boundary of the j-th epoch must occur after the p_w-th operation and before the p_r-th operation, there are at most s_e good locations for the read out of $n/2$ total locations. For any $j \geq e$, this cell probe has probability $2s_e/n$ of contributing to $\mathcal{T}_w^{<j}(\mathcal{Q})$. Therefore, by linearity of expectation over the $(n-1)t_w$ expected cell probes:

$$\mathsf{E}\left[\frac{|\mathcal{T}_w^{<i}(Q)|}{\ell_i}\right] = \frac{1}{\hat{k}} \sum_{\text{epoch } j \text{ is large}} \mathsf{E}\left[\frac{|\mathcal{T}_w^{<j}(Q)|}{\ell_j}\right] \le \frac{t_w}{\hat{k}}\left(\max_e \sum_{j \ge e} \frac{2s_e}{\ell_j}\right).$$

Therefore, there exists $\hat{k}/2 + 1$ fixed epochs i such that $\mathsf{E}[|\mathcal{T}_w^{<i}(Q)|/\ell_i]$ over the choice of the **read** location is at most 3 times the above bound.

As $\sum_i \mathsf{E}[|\mathcal{T}_r^i(Q)|] \le t_r$, there exists $\hat{k}/2 + 1$ epochs i where $\mathsf{E}[|\mathcal{T}_r^i(Q)|] \le 3t_r$ completing the proof.

Theorem 3. *Let* **DS** *be an* (ϵ, δ)-*differentially private RAM for* n *b-bit array entries implemented over* w-*bit cells. Assuming that* $\epsilon = O(1)$ *and* $0 \le \delta \le 1/3$, **DS** *has failure probability at most* $1/3$ *and* $w = \Omega(\log\log n)$, *then*

$$t_w = o((b/w)\log(nb/c)) \implies t_r = \omega((b/w)\log(nb/c)).$$

Proof. Recall we get the following inequality by applying convexity to the inequality of Lemma 2 and noting that $c/\ell_i = O(1)$ for our choices i:

$$\frac{\mathsf{E}[|\mathcal{T}_w^{<i}(Q)|]}{\ell_i}\left(w + \log\frac{t_w s_{i-1}}{\mathsf{E}[|\mathcal{T}_w^{<i}(Q)|]}\right) + \mathsf{E}[|\mathcal{T}_r^i(Q)|]\left(w + \log\frac{t_r}{\mathsf{E}[|\mathcal{T}_r^i(Q)|]}\right) = \Omega(b).$$

By applying Lemma 4, we get that $\mathsf{E}[|\mathcal{T}_r^i(Q)|] = O(t_r \log r/(r\log(nb/c)))$ and $\mathsf{E}[|\mathcal{T}_w^{<i}(Q)|/\ell_i] = O(t_w \log r/\log(nb/c))$ since

$$\left(\frac{s_j}{\ell_j} + \frac{s_j}{\ell_{j+1}} + \dots\right) \le 2r\sum_{j \ge 0}\frac{1}{r^j} = O(r).$$

Plugging into the inequality above and assuming that $w = \Omega(\log\log n)$,

$$t_w + (t_r/r) = \Omega((b/w)\log(nb/c)/\log r) \implies t_r = \Omega((b/w)\log(nb/c)r/\log r)$$

as $t_w = o(b/w)\log(nb/c)$. Since $r/\log r = \omega(1)$, we complete the proof.

4.2 Second Epoch Construction

In this section, we deal with the opposite scenario when we assume that $t_r = o((b/w)\log(nb/c))$ and want to show that $t_w = \omega((b/w)\log(nb/c))$. The same intuition from the previous section can be used for this situation: to show that t_w has to be very large, we will need to require that for any epoch i, the total number of future **write** operations in epochs $j \in \{1,\dots,i-1\}$ is small. If for any epoch i, the number of future **write** operations after epoch i is small and the **read** operation also cannot perform many cell probes into epoch i, then each future **write** operation must encode a large amount of information about epoch i which will be used by the **read** operation. As a result, we can prove a large lower bound on t_w.

Specifically, we consider an epoch construction in which the number of `write` operations in an epoch is larger by a super-constant factor $r = O(\log n)$ compared with the number in the previous epoch. So, the first epoch will have r `write` operations, the second epoch will have r^2 `write` operations, etc. So,

$$s_{i-1} = \ell_{i-1} + \ell_{i-2} + \ldots \leq \frac{1}{r}(\ell_i + \ell_{i-1} + \ldots) \leq 2\ell_i/r.$$

As a result, the number of future operations is $\Theta(1/r)$ times smaller than the epoch construction of Sect. 3. The number of large epochs, that is with at least $\max\{\sqrt{n}, c^2/b\}$ `write`, is $\hat{k} = \Theta(\log_r nb/c)$.

Theorem 4. *Let* **DS** *be an* (ϵ, δ)-*differentially private RAM for* n b-*bit array entries implemented over* w-*bit cells. Assuming that* $\epsilon = O(1)$ *and* $0 \leq \delta \leq 1/3$, **DS** *has failure probability at most* $1/3$ *and* $w = \Omega(\log \log n)$, *then*

$$t_r = o((b/w) \log(nb/c)) \implies t_w = \omega((b/w) \log(nb/c)).$$

Proof. By applying Lemma 4, we get that $\mathsf{E}[|\mathcal{T}_r^i(\mathcal{Q})|] = O(t_r \log r / \log(nb/c))$ and $\mathsf{E}[|\mathcal{T}_w^{<i}(\mathcal{Q})|/\ell_i] = O(t_w \log r / r \log(nb/c))$ since

$$\left(\frac{s_j}{\ell_j} + \frac{s_j}{\ell_{j+1}} + \ldots\right) \leq 2 \sum_{j \geq 0} \frac{1}{r^j} = O(1).$$

Plugging into the inequality of Lemma 2 after applying convexity and noting that $w = \Omega(\log \log n)$ and that, for large epochs, $c/\ell_i = O(1)$, we obtain

$$(t_w/r) + t_r = \Omega((b/w) \log(nb/c)/\log r) \implies t_w = \Omega((b/w) \log(nb/c)r/\log r)$$

since $t_r = o((b/w) \log(nb/c))$. Noting that $r/\log r = \omega(1)$ completes our proof.

5   Discussion

We now discuss three extensions that follow from our lower bound techniques.

Our techniques only enforce the requirements of differential privacy for a single `read` operation. Therefore, our lower bounds would also apply *differentially private-read RAMs* where differential privacy is guaranteed only for sequences of operations that differ in exactly one `read` operation. This might be important in scenarios where the indices of `write` operations are not sensitive (or may be public) but only the indices of `read` operations need to be protected. Once again, this weakening of security does not suffice to get around the $\Omega(\log(nb/c))$ bandwidth overhead lower bounds.

The lower bounds of Sects. 3 and 4 hold for $\delta \leq 1/3$. Most practical scenarios require that δ must be negligible in n, so the above results suffice. For theoretical exploration, we note that our results can be extended to any constant $\delta < 1$. In particular, for any $\rho < 1$, by picking a sufficiently large enough constant C, we can prove that $\Pr[\mathcal{Z}_i(\mathcal{Q}(\mathsf{idx})) \geq b/C] \geq \rho$ which is a variation of Lemma 1.

By using $\rho > \delta$ we can extend Lemma 2 and prove that for all epochs i, $\Pr[\mathcal{Z}_i(\mathcal{Q}) \geq b/C] \geq (\rho - \delta)/e^\epsilon$. This will suffice to extend Theorem 2 to any $\delta < 1$. The main result of Sect. 4 can be similarly extended.

Finally, our lower bound assumes DS has worst time case cost on update operations, but may be extended to worst case amortized update costs. In particular, we only apply Lemma 1 to epochs whose sum of probed cells by update operations is not too much larger than expected. By an averaging argument, it can be shown that a constant fraction of all epochs satisfy this property.

In this work, we show that the $\Omega(\log(nb/c))$ bandwidth overhead lower bound for the array maintenance problem with obliviousness extends to the weaker notion of differential privacy with reasonable privacy budgets of $\epsilon = O(1)$ and $\delta \leq 1/3$. The result is surprising as differentially private RAM provides significantly weaker privacy. This leads to the following natural open question: Does there exist a natural, weaker notion of privacy that enables $o(\log(nb/c))$ bandwidth overhead for the array maintenance problem?

References

1. Asharov, G., Komargodski, I., Lin, W.-K., Nayak, K., Peserico, E., Shi, E.: OptORAMa: Optimal oblivious RAM. ePrint Report 2018/892
2. Boyle, E., Chung, K.-M., Pass, R.: Oblivious parallel RAM and applications. In: Kushilevitz, E., Malkin, T. (eds.) TCC 2016. LNCS, vol. 9563, pp. 175–204. Springer, Heidelberg (2016). https://doi.org/10.1007/978-3-662-49099-0_7
3. Boyle, E., Naor, M.: Is there an oblivious RAM lower bound? In: ITCS 2016, pp. 357–368 (2016)
4. Cash, D., Grubbs, P., Perry, J., Ristenpart, T.: Leakage-abuse attacks against searchable encryption. In: CCS 2015, pp. 668–679 (2015)
5. Chan, T.-H.H., Guo, Y., Lin, W.-K., Shi, E.: Oblivious hashing revisited, and applications to asymptotically efficient ORAM and OPRAM. In: Takagi, T., Peyrin, T. (eds.) ASIACRYPT 2017. LNCS, vol. 10624, pp. 660–690. Springer, Cham (2017). https://doi.org/10.1007/978-3-319-70694-8_23
6. Chen, B., Lin, H., Tessaro, S.: Oblivious parallel RAM: improved efficiency and generic constructions. In: Kushilevitz, E., Malkin, T. (eds.) TCC 2016. LNCS, vol. 9563, pp. 205–234. Springer, Heidelberg (2016). https://doi.org/10.1007/978-3-662-49099-0_8
7. Chung, K.-M., Liu, Z., Pass, R.: Statistically-secure ORAM with $\tilde{O}(\log^2 n)$ overhead. In: Sarkar, P., Iwata, T. (eds.) ASIACRYPT 2014. LNCS, vol. 8874, pp. 62–81. Springer, Heidelberg (2014). https://doi.org/10.1007/978-3-662-45608-8_4
8. Damgård, I., Meldgaard, S., Nielsen, J.B.: Perfectly secure oblivious RAM without random oracles. In: Ishai, Y. (ed.) TCC 2011. LNCS, vol. 6597, pp. 144–163. Springer, Heidelberg (2011). https://doi.org/10.1007/978-3-642-19571-6_10
9. Devadas, S., van Dijk, M., Fletcher, C.W., Ren, L., Shi, E., Wichs, D.: Onion ORAM: a constant bandwidth blowup oblivious RAM. In: Kushilevitz, E., Malkin, T. (eds.) TCC 2016. LNCS, vol. 9563, pp. 145–174. Springer, Heidelberg (2016). https://doi.org/10.1007/978-3-662-49099-0_6
10. Dwork, C.: A firm foundation for private data analysis. Commun. ACM 54(1), 86–95 (2011)

11. Dwork, C., McSherry, F., Nissim, K., Smith, A.: Calibrating noise to sensitivity in private data analysis. In: Halevi, S., Rabin, T. (eds.) TCC 2006. LNCS, vol. 3876, pp. 265–284. Springer, Heidelberg (2006). https://doi.org/10.1007/11681878_14

12. Dwork, C., Roth, A., et al.: The algorithmic foundations of differential privacy. Found. Trends Theor. Comput. Sci. **9**, 211–407 (2014)

13. Fredman, M., Saks, M.: The cell probe complexity of dynamic data structures. In: STOC 1989, pp. 345–354 (1989)

14. Garg, S., Lu, S., Ostrovsky, R., Scafuro, A.: Garbled RAM from one-way functions. In: STOC 2015, pp. 449–458 (2015)

15. Gentry, C., Halevi, S., Lu, S., Ostrovsky, R., Raykova, M., Wichs, D.: Garbled RAM revisited. In: Nguyen, P.Q., Oswald, E. (eds.) EUROCRYPT 2014. LNCS, vol. 8441, pp. 405–422. Springer, Heidelberg (2014). https://doi.org/10.1007/978-3-642-55220-5_23

16. Goldreich, O.: Towards a theory of software protection and simulation by oblivious RAMs. In: STOC 1987, pp. 182–194 (1987)

17. Goldreich, O., Ostrovsky, R.: Software protection and simulation on oblivious RAMs. JACM **43**(3), 431–473 (1996)

18. Goodrich, M.T., Mitzenmacher, M.: Privacy-preserving access of outsourced data via oblivious RAM simulation. In: Aceto, L., Henzinger, M., Sgall, J. (eds.) ICALP 2011. LNCS, vol. 6756, pp. 576–587. Springer, Heidelberg (2011). https://doi.org/10.1007/978-3-642-22012-8_46

19. Goodrich, M.T., Mitzenmacher, M., Ohrimenko, O., Tamassia, R.: Privacy-preserving group data access via stateless oblivious RAM simulation. In: SODA 2012, pp. 157–167 (2012)

20. Islam, M.S., Kuzu, M., Kantarcioglu, M.: Access pattern disclosure on searchable encryption: ramification, attack and mitigation. In: NDSS 2012 (2012)

21. Kushilevitz, E., Lu, S., Ostrovsky, R.: On the (in) security of hash-based oblivious RAM and a new balancing scheme. In: SODA 2012, pp. 143–156 (2012)

22. Larsen, K.G.: The cell probe complexity of dynamic range counting. In: STOC 2012, pp. 85–94 (2012)

23. Larsen, K.G., Nielsen, J.B.: Yes, there is an oblivious RAM lower bound!. In: Shacham, H., Boldyreva, A. (eds.) CRYPTO 2018. LNCS, vol. 10992, pp. 523–542. Springer, Cham (2018). https://doi.org/10.1007/978-3-319-96881-0_18

24. Larsen, K.G., Weinstein, O., Yu, H.: Crossing the logarithmic barrier for dynamic boolean data structure lower bounds. In: STOC 2018, pp. 978–989 (2018)

25. Lu, S., Ostrovsky, R.: Black-Box parallel garbled RAM. In: Katz, J., Shacham, H. (eds.) CRYPTO 2017. LNCS, vol. 10402, pp. 66–92. Springer, Cham (2017). https://doi.org/10.1007/978-3-319-63715-0_3

26. Mironov, I., Pandey, O., Reingold, O., Vadhan, S.: Computational differential privacy. In: Halevi, S. (ed.) CRYPTO 2009. LNCS, vol. 5677, pp. 126–142. Springer, Heidelberg (2009). https://doi.org/10.1007/978-3-642-03356-8_8

27. Patel, S., Persiano, G., Raykova, M., Yeo, K.: PanORAMa: oblivious RAM with logarithmic overhead. In: FOCS 2018, pp. 871–882 (2018)

28. Pinkas, B., Reinman, T.: Oblivious RAM revisited. In: Rabin, T. (ed.) CRYPTO 2010. LNCS, vol. 6223, pp. 502–519. Springer, Heidelberg (2010). https://doi.org/10.1007/978-3-642-14623-7_27

29. Pătraşcu, M.: Lower bound techniques for data structures. Ph.D. thesis. MIT (2008)

30. Pătraşcu, M., Demaine, E.D.: Logarithmic lower bounds in the cell-probe model. SIAM J. Comput. **35**(4), 932–963 (2006)

31. Stefanov, E., Shi, E., Song, D.: Towards practical oblivious RAM. arXiv:1106.3652 (2011)
32. Stefanov, E., et al.: Path ORAM: an extremely simple oblivious RAM protocol. In: CCS 2013, pp. 299–310 (2013)
33. Toledo, R.R., Danezis, G., Goldberg, I.: Lower-cost ϵ-private information retrieval. Proc. Priv. Enhancing Technol. **2016**(4), 184–201 (2016)
34. Wagh, S., Cuff, P., Mittal, P.: Root ORAM: a tunable differentially private oblivious RAM. arXiv:1601.03378 (2016)
35. Wang, X.S., et al.: Oblivious data structures. In: CCS 2014, pp. 215–226 (2014)
36. Weiss, M., Wichs, D.: Is there an Oblivious RAM lower bound for online reads? ePrint report 2018/619
37. Yao, A.C.-C.: Should tables be sorted? JACM **28**(3), 615–628 (1981)

Bounds for Symmetric Cryptography

Beyond Birthday Bound Secure MAC in Faulty Nonce Model

Avijit Dutta$^{(\boxtimes)}$, Mridul Nandi, and Suprita Talnikar

Indian Statistical Institute, Kolkata, India
avirocks.dutta13@gmail.com, mridul.nandi@gmail.com, suprita45@gmail.com

Abstract. Encrypt-then-MAC (EtM) is a popular mode for authenticated encryption (AE). Unfortunately, almost all designs following the EtM paradigm, including the AE suites for TLS, are vulnerable against nonce misuse. A single repetition of the nonce value reveals the hash key, leading to a universal forgery attack. There are only two authenticated encryption schemes following the EtM paradigm which can resist nonce misuse attacks, the GCM-RUP (CRYPTO-17) and the GCM/2$^+$ (INSCRYPT-12). However, they are secure only up to the birthday bound in the nonce respecting setting, resulting in a restriction on the data limit for a single key. In this paper we show that nEHtM, a nonce-based variant of EHtM (FSE-10) constructed using a block cipher, has a beyond birthday bound (BBB) unforgeable security that gracefully degrades under nonce misuse. We combine nEHtM with the CENC (FSE-06) mode of encryption using the EtM paradigm to realize a nonce-based AE, CWC+. CWC+ is very close (requiring only a few more xor operations) to the CWC AE scheme (FSE-04) and it not only provides BBB security but also gracefully degrading security on nonce misuse.

Keywords: Graceful security · Faulty nonce · Mirror theory ·
Extended mirror theory · Expectation method · CWC · GCM

1 Introduction

MESSAGE AUTHENTICATION CODE. It is important to authenticate any digital message or packet transmitted over an insecure communication channel by some cryptographic algorithm. This is achieved by a MAC (Message Authentication Code), a popular primitive in symmetric key cryptography, which enables two legitimate parties (say, Alice and Bob) having access to a shared secret key to authenticate their transmissions. When Alice wants to send a message M, she computes a MAC function that accepts M and the shared secret key K, and possibly an auxiliary variable called IV (initial vector), and obtains an authentication tag T as an output. Then she sends (IV, M, T) to Bob. Upon receiving, Bob verifies the authenticity of (IV, M, T) by computing the MAC using (IV, M) and K to obtain the local tag T', and checks whether T' matches T. If the IV is a nonce (e.g., a counter) this *nonce-based* MAC is said to be stateful.

© International Association for Cryptologic Research 2019
Y. Ishai and V. Rijmen (Eds.): EUROCRYPT 2019, LNCS 11476, pp. 437–466, 2019.
https://doi.org/10.1007/978-3-030-17653-2_15

NONCE MISUSE RESISTANCE SECURITY. The Wegman-Carter (WC) MAC [40] is the first nonce-based MAC that masks the hash value of the message with an encrypted nonce to generate the tag. Although this scheme is optimally secure when the nonce never repeats, the consequences are catastrophic if the nonce repeats even once (as it can leak the hash key). Nonce-based MAC schemes that guarantee security against nonce misuse are therefore desirable, because it becomes challenging in some contexts to maintain the uniqueness of the nonce, e.g., on implementations in a stateless device or in cases where the nonce is chosen randomly from a small set. The nonce may also repeat due to a faulty implementation of the scheme or an occurrence of some other fault (for example, a reset of the nonce). After making an internet-wide scan, Böck et al. [9] found 184 devices that used a duplicate nonce. *Encrypted Wegman-Carter-Shoup* (EWCS) [13] guarantees such security but it only gives a PRF security up to the birthday bound in a nonce-respecting setting, as an adversary making $2^{n/2}$ nonce-respecting queries with the same message will observe no collision in the tag. *Encrypted Wegman-Carter with Davies-Meyer* [13] (or EWCDM) and *Decrypted Wegman-Carter with Davies-Meyer* [16] (or DWCDM) have been proposed with a view to achieve a beyond the birthday bound nonce-respecting security and a reasonable nonce misuse security. However, the security of these constructions falls to the birthday bound with only a single misuse of the nonce. There are other known constructions such as *Dual Encrypted Wegman-Carter with Davies-Meyer* (or EWCDMD) [27,32], *Encrypted Wegman-Carter-Shoup* [13] (or EWCS) and single hash-key variants of CLRW2 [25]. However, these constructions also provide only birthday bound PRF security in nonce-respecting settings.

AE SCHEME AS APPLICATION OF MAC. An authenticated encryption (AE) mode is a cryptographic scheme that guarantees the privacy and authenticity of a message concurrently. Authenticated encryption has received much attention from the cryptographic community mostly due to its application to TLS and many other protocols. The ongoing CAESAR competition [1] which aims to identify a portfolio of authenticated encryption schemes has drafted three use cases, namely *lightweight, high-performance*, and *defense-in-depth*. The competition considers GCM [26] as the baseline algorithm as it is widely adopted (e.g. in TLS 1.2 and in its variant RGCM [6], which shall soon be considered in TLS 1.3 [11]) and standardized. ChaCha20+Poly1305 [7] is a popular alternative for settings where AES-NI is not implemented.

ENCRYPT-THEN-MAC. Both ChaCha20+Poly1305 and GCM follow the Encrypt-then-MAC (EtM) paradigm [5]. Some other popular AE designs following the same paradigm are CWC [24], GCM/2^+ [3], CHM [22], CIP [23], GCM-RUP [4], OGCM1 [41], OGCM2 [41] etc. EtM is a popular design paradigm due to its generic security guarantee. Authors of [12] showed that (stating informally) if $\mathcal{E}$ is a secure symmetric encryption scheme and $\mathcal{I}$ is a secure MAC family then EtM results in secure channels. This has later also been analyzed by [5,31]. However, it turns out that by Joux's "forbidden attack" [2], GCM leaks the hash key whenever an encryption query with a repeated nonce is executed. A similar forgery attack can be applied against all aforementioned AE except GCM-RUP and GCM/2^+, as they use some variants of the WC MAC. GCM-RUP resists this attack as it uses the XEX [38] construction to define the tag. However, in nonce-respecting settings it gives up to birthday bound security. GCM/2^+ resists the birthday bound attack by using the EWCS construction.

1.1 Beyond Birthday Bound Security with Graceful Degradation

Achieving a beyond the birthday bound security would provide a larger data limit for a single key. GCM-RUP can be proven to have at most $\ell q_m^2/2^n$ forging advantage (in the nonce-respecting model), where q_m is the number of encryption queries and ℓ is the maximum number of data blocks a message and an associated data can possess. For example, the GCM-RUP based on AES, which can process a data of size at most $\ell = 2^{32}$ blocks should have a data limit $q_m \leq 2^{32}$ so as to allow an advantage of at most 2^{-32}, a tolerance level much smaller than that provided by beyond birthday security. Therefore, a natural quest is to come up with a nonce-based MAC scheme that provides beyond the birthday bound security that degrades in a graceful manner when the nonce repeats. As a direct application of such a MAC scheme, one can design a nonce-based AE that provides beyond the birthday bound security when repetition of the nonce is limited.

GOAL OF THE PAPER. The main goal of this paper is to find *an efficient MAC which is BBB (beyond birthday bound) secure both as a PRF and a MAC*. Moreover, it should provide *graceful security degradation in a nonce-misuse setting*. It must be mentioned here that there are some deterministic MAC constructions (not requiring any nonce) that provide BBB security. These mainly follow a double-block hash-then-sum approach [14,15] and hence require the computation of two blocks of algebraic hashes (or one pass of block cipher or tweakable block cipher executions). However, a single-block hash (which would be definitely faster than two blocks of hash and require a smaller hash-key size) would be a better option. So, this paper focuses on getting a design based on a single-block algebraic hash (e.g. a single-call of the polynomial hash [30]).

GRACEFUL DEGRADATION OF SECURITY ON NONCE MISUSE. The most popular measure of nonce misuse is the maximum number of multicollisions in nonce values amongst all queries [37]. To the best of our knowledge, none of the existing block cipher-based nonce-based MACs adhere to this notion with BBB security guarantee. We have also explored many other variants of MAC constructions using at most two block cipher calls and a single hash function call. Unfortunately, we found that none of them give beyond birthday bound security in terms of multicollision nonce misuse, even with multicollisions of size 2.

In this paper we instead consider another natural definition of nonce misuse, called the number of faulty nonces. An authentication query is said to be a *faulty query* if there exists a previous MAC query such that their corresponding nonces match. The nonce in a faulty query is called a *faulty nonce*. The notion of a faulty nonce is weaker than multicollision of nonces since although a μ-multicollision also gives $\mu -$ 1 faulty nonces, an occurrence of μ faulty nonces does not mean μ-multicollisions have occurred. When a counter is implemented in an aperiodic manner (e.g. timely nonce [9] used in TLS 1.2), a simple reset does not give a large number of faulty nonces; there are easy countermeasures to prevent a large number of faulty nonce encryptions.

1.2   Our Contribution

Our contribution in this paper is threefold, which we outline as follows:

1. MULTICOLLISION ON UNIVERSAL HASH. We study the probability of occurrence of multicollisions in a universal hash function. In particular, we have shown that the probability of obtaining a $(\xi + 1)$-multicollision tuple amongst q inputs is at most $q^2\epsilon/\xi$ (see Sect. 5). This is clearly an improved bound as compared to a straightforward application of the union bound. We believe that this problem can have independent interest in the cryptographic community and can be used to get improved bounds for other constructions also.

2. BBB SECURE MAC WITH GRACEFUL SECURITY. In [29], a probabilistic MAC EHtM has been analyzed and shown to have roughly $3n/4$-bit MAC security which is also tight [18]. This paper analyzes a construction, which shall be denoted as nEHtM, where (1) the random salt is replaced by the nonce and (2) the two independent pseudorandom functions are replaced by a single-keyed block cipher. Given a data D and a nonce N the tag is computed as follows (see Fig. 1(b)):

$$\mathsf{nEHtM}_{K,K_h}(N, D) \triangleq \mathsf{E}_K(0\|N) \oplus \mathsf{E}_K(1\|\mathsf{H}_{K_h}(D) \oplus N).$$

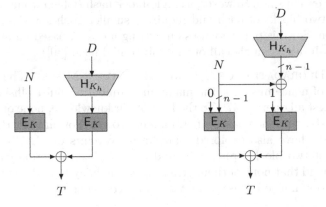

Fig. 1. (a) On the left is the CWC MAC (MAC algorithm used in CWC); (b) on the right is the domain separation variant of *nonce-based Enhanced Hash-then-Mask*.

We have shown that nEHtM is secure roughly up to $2^{2n/3}$ authentication queries and 2^n verification queries in the nonce-respecting setting. Moreover, this security degrades in a graceful manner on introduction of faults in the nonce. The unforgeability of this construction shall be shown through an extended distinguishing game. We apply the expectation method (as it shall later be shown to give a better bound than the coefficients-H technique) to bound the distinguishing advantage of two worlds. In the ideal world, once we realize the random tags T_i, we need to sample the hash key. This would determine all inputs of the underlying block cipher. The equality patterns amongst the nonce values are deterministic and we bound the number of faulty nonces by a parameter μ. However, the equality patterns among other inputs of the form $X \triangleq H_{K_h}(D) \oplus N$ are probabilistic due to randomness of the hash key. As there may not be sufficient entropy in the hash-key (which could be n-bit for polynomial hash), the number of multicollisions amongst the values of X may not be easy to compute. We have tackled this problem using the multicollision result (as stated in the first contribution) of the underlying hash function.

After we limit the multicollisions in the values of both X and N, we shall be in a position to apply mirror theory to show a beyond birthday bound security on the distinguishing advantage of nEHtM. Note that mirror theory cannot give a beyond birthday bound security without restricting the number of multicollisions.

It must be noted here that nEHtM (like all other candidates) is not secure beyond the birthday bound under the notion of multicollision nonce misuse security and the corresponding attack is discussed in the full version of the paper [19].

3. APPLICATION TO A CWC-LIKE AE CONSTRUCTION. We propose CWC+, which is an instance of the EtM composition based on the CENC type encryption with maximum width parameter and the nEHtM MAC. Moreover, we apply an appropriate domain separation to make it a single-keyed construction (even the hash key is generated from the block cipher). The construction is a very close variant of CWC as it requires a few additional xor computations, without requiring any extra calls to the block cipher. Furthermore, CWC+ gives both (1) BBB security and (2) graceful security degradation in the faulty nonce misuse model. In particular, we have the following forging advantage of CWC+:

$$\mathsf{Auth}[\mathsf{CWC+}] = \frac{105\sigma^3\ell}{2^{2n}} + \frac{6\sigma\ell}{2^n} + \frac{2q_d}{2^\rho} + \frac{2q_d\ell}{2^n} + \frac{(2q_e + q_d)2\ell\mu}{2^n} + \left(\frac{5\sigma\ell\mu}{2^n}\right)^2,$$

where q_e and q_d denote the number of encryption and decryption queries, ρ the tag size, ℓ the maximum number of message blocks queried including the associated data blocks, σ the total number of message blocks queried and μ the total number of faulty queries. Moreover, the security of CWC+ gracefully drops to birthday bound when $\ell\mu$ is about $2^{n/2}$. However, when $\ell \leq 2^{n/4}$, then the security bound of CWC+ caps at roughly $2^{7n/12}$, which is strictly greater than the birthday bound. A better bound can be obtained if we assume some restrictions over all the message lengths.

(3) Another notable feature of CWC+ is that the scheme remains secure even with short tag lengths. In GCM, if the tag length is only 32 bits, then an adversary forges the construction with just 1024 verification attempts by querying with a single message consisting of 2^{22} blocks. However, for the same tag size, the authenticity advantage of CWC+ is 2^{-21} when adversary forges the construction with 1024 verification attempts.

2   Preliminaries

BASIC NOTATIONS: For a set $\mathcal{X}$, $X \leftarrow_\$ \mathcal{X}$ denotes that X is sampled uniformly at random from $\mathcal{X}$ and is independent to all other random variables defined so far. $\{0,1\}^n$ denotes the set of all binary strings of length n and $\{0,1\}^*$ denotes the set of all binary strings of finite arbitrary length. We denote 0^n (i.e., n-bit string of all zeroes) by $\mathbf{0}$. For any element $X \in \{0,1\}^*$, $|X|$ denotes the number of bits in x. For any two elements $X, Y \in \{0,1\}^*$, $X\|Y$ denotes the concatenation of X followed by Y. For $X, Y \in \{0,1\}^n$, $X \oplus Y$ denotes the addition modulo 2 of X and Y. For any $X \in \{0,1\}^*$, parse X as $X = X_1\|X_2\|\dots\|X_l$ where for each $i = 1, \dots, l-1$, X_i is an element of $\{0,1\}^n$ and $1 \leq |X_l| \leq n$. We call each X_i a *block*. For a sequence of elements $(X_1, X_2, \dots, X_s) \in \{0,1\}^*$, X_a^i denotes the a-th block of i-th element X_i.

The set of all functions from $\mathcal{X}$ to $\mathcal{Y}$ is denoted as $\mathsf{Func}(\mathcal{X}, \mathcal{Y})$ and the set of all permutations over $\mathcal{X}$ is denoted as $\mathsf{Perm}(\mathcal{X})$. $\mathsf{Func}(\mathcal{X})$ denotes the set of all functions from $\mathcal{X}$ to $\{0,1\}^n$ and Perm denotes the set of all permutations over $\{0,1\}^n$. We often write Func instead of $\mathsf{Func}(\mathcal{X})$ when the domain of the functions is understood from the context. For integers $1 \leq b \leq a$, $(a)_b$ denotes $a(a-1)\dots(a-b+1)$, where $(a)_0 = 1$ by convention. $[q]$ refers to the set $\{1, \dots, q\}$ and $[q_1, q_2]$ to the set $\{q_1, q_1 + 1 \dots, q_2 - 1, q_2\}$.

2.1   Security Definitions

PSEUDO RANDOM FUNCTION (PRF) AND PSUEDO RANDOM PERMUTATION (PRP). A keyed function $\mathsf{F} : \mathcal{K} \times \mathcal{X} \to \mathcal{Y}$ with key space $\mathcal{K}$, domain $\mathcal{X}$ and range $\mathcal{Y}$ is a function for which $\mathsf{F}(K, X)$ shall be denoted by $\mathsf{F}_K(X)$. Given an oracle algorithm A that has oracle access to a function from $\mathcal{X}$ to $\mathcal{Y}$, makes at most q queries in time at most t, and returns a single bit, the prf-advantage of A against the family of keyed functions F is defined as

$$\mathbf{Adv}_\mathsf{F}^{\mathrm{PRF}}(\mathsf{A}) \triangleq \left| \Pr\left[K \leftarrow_\$ \mathcal{K} : \mathsf{A}^{\mathsf{F}_K(\cdot)} = 1 \right] - \Pr\left[\mathsf{RF} \leftarrow_\$ \mathsf{Func}(\mathcal{X}, \mathcal{Y}) : \mathsf{A}^{\mathsf{RF}(\cdot)} = 1 \right] \right|.$$

F is said to be a $(q, \ell, \sigma, t, \epsilon)$-secure PRF if $\mathbf{Adv}_\mathsf{F}^{\mathrm{PRF}}(q, \ell, \sigma, t) \triangleq \max_\mathsf{A} \mathbf{Adv}_\mathsf{F}^{\mathrm{PRF}}(\mathsf{A}) \leq \epsilon$, where the maximum is taken over all adversaries A that make q queries, with a maximum of ℓ data blocks in a single query and the total number of data blocks at most σ, with maximum running time t. Similarly, the prp-advantage of A against a family of keyed permutations E is defined as

$$\mathbf{Adv}_\mathsf{E}^{\mathrm{PRP}}(\mathsf{A}) \triangleq \left| \Pr\left[K \leftarrow_\$ \mathcal{K} : \mathsf{A}^{\mathsf{E}_K(\cdot)} = 1 \right] - \Pr\left[\Pi \leftarrow_\$ \mathsf{Perm}(\mathcal{X}) : \mathsf{A}^{\Pi(\cdot)} = 1 \right] \right|.$$

E is said to be a (q, t, ϵ)-secure PRP if $\mathbf{Adv}_{\mathsf{E}}^{\mathrm{PRP}}(q, t) \stackrel{\Delta}{=} \max_{\mathsf{A}} \mathbf{Adv}_{\mathsf{E}}^{\mathrm{PRP}}(\mathsf{A}) \leq \epsilon$, where maximum is taken over all adversaries A that make q queries and have running time at most t.

MESSAGE AUTHENTICATION CODE (MAC). Let $\mathcal{K}, \mathcal{N}, \mathcal{M}$ and $\mathcal{T}$ be four non-empty finite sets, $\mathsf{F} : \mathcal{K} \times \mathcal{N} \times \mathcal{M} \to \mathcal{T}$ be a nonce-based MAC. For $K \in \mathcal{K}$, let Auth_K be the authentication oracle, which takes as input $(N, M) \in \mathcal{N} \times \mathcal{M}$ and outputs $T = \mathsf{F}(K, N, M)$ and let Ver_K be the verification oracle, which takes as input $(N, M, T) \in \mathcal{N} \times \mathcal{M} \times \mathcal{T}$ and outputs 1 if $\mathsf{F}(K, N, M) = T$ and otherwise outputs 0. An authentication query (N, M) by an adversary A is called a **faulty query** if A has already queried to the first oracle with the same nonce but with a different message.

A (μ, q_m, q_v, t)-adversary against the unforgeability of F is an adversary A with oracle access to Auth_K and Ver_K such that it makes at most μ faulty authentication queries out of at most q_m authentication queries and q_v verification queries, with running time at most t. The adversary is said to be *nonce respecting* if $\mu = 0$ and nonce misusing if $\mu \geq 1$. However, the adversary may repeat nonces in its verification queries. A is said to *forge* F if for any of its verification queries (not obtained through a previous authentication query), the verification oracle returns 1. The advantage of A against the unforgeability of F is defined as

$$\mathbf{Adv}_{\mathsf{F}}^{\mathrm{MAC}}(\mathsf{A}) \stackrel{\Delta}{=} \Pr\left[K \leftarrow_{\$} \mathcal{K} : \mathsf{A}^{\mathsf{Auth}_K(\cdot, \cdot), \mathsf{Ver}_K(\cdot, \cdot, \cdot)} \text{ forges} \right].$$

We write $\mathbf{Adv}_{\mathsf{F}}^{\mathrm{MAC}}(\mu, q_m, q_v, t) \stackrel{\Delta}{=} \max_{\mathsf{A}} \mathbf{Adv}_{\mathsf{F}}^{\mathrm{MAC}}(\mathsf{A})$ where the maximum is taken over all (μ, q_m, q_v, t)-adversaries. In all of these definitions, we skip the parameter t, whenever we maximize over all unbounded adversaries.

ALMOST XOR UNIVERSAL (AXU) HASH FUNCTION. Let $\mathcal{K}$ and $\mathcal{X}$ be two non-empty finite sets and H be a keyed function $\mathsf{H} : \mathcal{K}_h \times \mathcal{X} \to \{0, 1\}^n$. Then, H is said to be an ϵ-almost xor universal hash function if for any distinct $X, X' \in \mathcal{X}$ and for any $Y \in \{0, 1\}^n$,

$$\Pr\left[K_h \leftarrow_{\$} \mathcal{K}_h : \mathsf{H}_{K_h}(X) \oplus \mathsf{H}_{K_h}(X') = Y \right] \leq \epsilon.$$

We say that (X, X') is a colliding pair for a function H_{K_h} if $\mathsf{H}_{K_h}(X) = \mathsf{H}_{K_h}(X')$. H is said to be an ϵ-universal hash function if for any distinct $X, X' \in \mathcal{X}$,

$$\Pr\left[K_h \leftarrow_{\$} \mathcal{K}_h : \mathsf{H}_{K_h}(X) = \mathsf{H}_{K_h}(X') \right] \leq \epsilon.$$

POLYHASH FUNCTION. A general algebraic hash function is a multivariate polynomial. Polyhash [30], one of the most popular examples of an algebraic hash function, is a univariate polynomial over the hash key K_h and its coefficients are the message blocks. For an n-bit hash key K_h, a message $M \in \{0, 1\}^*$ is first padded with 10^* such that the number of bits in the padded message becomes a multiple of n. Let the padded message be $M^* = M_1 \| M_2 \| \ldots \| M_l$, where for each $i = 1, \ldots, l$, $|M_i| = n$. Then the PolyHash function is defined as follows:

$$\mathsf{PH}_{K_h}(M) = M_l K_h \oplus M_{l-1} K_h^2 \oplus \ldots \oplus M_1 K_h^l,$$

where l is the number of n-bit blocks of the padded message M^*. It is a well known result [17] that PolyHash is $\ell/2^n$-universal hash function, where ℓ is the maximum number of message blocks and the hash key is an element of the field $GF(2^n)$.

2.2   A Brief Revisit to the Expectation Method

SYSTEM AND DISTINGUISHER. Consider a computationally unbounded distinguisher A (hence assumed deterministic) that interacts with either of the possibly randomized stateful systems $\mathbf{S}_{re}$ or $\mathbf{S}_{id}$, after which it returns a single bit 0 or 1. For any such system $\mathbf{S}_{re}$ or $\mathbf{S}_{id}$, the interaction between A and the system defines an ordered sequence of queries and responses, $\tau = ((X_1, Y_1), (X_2, Y_2), \ldots, (X_q, Y_q))$ called a *transcript*, where X_i is the i-th query of A and Y_i is the corresponding response from the system. Let X_{re} (resp. X_{id}) be the random variable that takes a transcript resulting from the interaction between A and $\mathbf{S}_{re}$ (resp. A and $\mathbf{S}_{id}$). Then the advantage of A in distinguishing $\mathbf{S}_{re}$ from $\mathbf{S}_{id}$ is bounded from above by the statistical distance between the two random variables X_{re} and X_{id}, which is

$$\Delta(X_{re}, X_{id}) \triangleq \sum_{\tau} \max\{0, \Pr[X_{id} = \tau] - \Pr[X_{re} = \tau]\}.$$

In the following, we briefly state the main result of the *Expectation Method* and show that the *coefficients-H technique* [33] is a special case of the expectation method. Both these techniques are used for bounding the information theoretic distinguishing advantage of two random systems as defined above.

EXPECTATION METHOD. The expectation method was introduced by Hoang and Tessaro to derive a tight multi-user security bound of the key-alternating cipher [20]. Subsequently, this technique has been used for proving the multi-user security of the double encryption method in [21] and recently by Bose et al. to bound the multi-user security of AES-GCM-SIV [10]. This method is a generalization of coefficients-H technique. Let $\phi : \Theta \to [0, \infty)$ be a non-negative function which maps any attainable transcript to a non-negative real value. Suppose there is a set of good transcripts such that for any good transcript τ,

$$\frac{\Pr[X_{re} = \tau]}{\Pr[X_{id} = \tau]} \geq 1 - \phi(\tau). \tag{1}$$

The statistical distance between the two random variables X_{re} and X_{id} can then be bounded as

$$\Delta(X_{re}, X_{id}) \leq \mathbf{E}[\phi(X_{id})] + \Pr[X_{id} \in \Theta_{bad}], \tag{2}$$

where Θ_{bad} is the set of all bad transcripts. In other words, the advantage of A in distinguishing $\mathbf{S}_{re}$ from $\mathbf{S}_{id}$ is bounded by $\mathbf{E}[\phi(X_{id})] + \Pr[X_{id} \in \Theta_{bad}]$. coefficients-H technique can be seen as a simple corollary of the expectation method when ϕ is taken to be a constant function.

3   Design and Security Result of nEHtM and CWC+

In this section we discuss the design and the security result of our proposed nonce-based message authentication code, called nEHtM and a nonce-based authenticated encryption scheme, called CWC+. We begin our discussion with the EtM composition result that combines a standard encryption and a MAC scheme to achieve authenticated encryption.

3.1   Encrypt-then-MAC: Generic Composition Result

Bellare and Namprempre in [5] and Canetti and Krawczyk in [12] explored ways to combine standard encryption schemes with MACs to achieve authenticated encryption schemes. Their results yield three different types of combinations: (a) Encrypt-and-MAC (E&M), (b) MAC-then-Encrypt (MtE) and (c) Encrypt-then-MAC (EtM). In this paper we focus only on EtM.

Let $\mathcal{E} = (\mathcal{E}.\mathsf{KGen}, \mathcal{E}.\mathsf{Enc}, \mathcal{E}.\mathsf{Dec})$ be a nonce-based symmetric key encryption scheme and $\mathcal{I} = (\mathcal{I}.\mathsf{KGen}, \mathcal{I}.\mathsf{Tag}, \mathcal{I}.\mathsf{Ver})$ be a nonce-based message authentication code. The function $\mathcal{E}.\mathsf{Enc} : \mathcal{K}_e \times \mathcal{N} \times \mathcal{M} \to \mathcal{C}$ maps a tuple (K_e, N, M) to a ciphertext C and the decryption function $\mathcal{E}.\mathsf{Dec} : \mathcal{K}_e \times \mathcal{N} \times \mathcal{C} \to \mathcal{M} \cup \{\bot\}$ maps a legitimate tuple (K_e, N, C) to the corresponding message M and otherwise returns the error symbol $\bot$.

For the message authentication code $\mathcal{I}$, $\mathcal{I}.\mathsf{Tag} : \mathcal{K}_m \times \mathcal{N} \times \mathcal{D} \to \mathcal{T}$ maps a tuple (K_m, N, D) to a tag T and the verification function $\mathcal{I}.\mathsf{Ver} : \mathcal{K}_m \times \mathcal{N} \times \mathcal{M} \times \mathcal{T} \to \{\top, \bot\}$ maps a quadruple (K_e, N, D, T) to one of the two symbols $\{\top, \bot\}$ such that if T is the valid tag for the tuple (K_n, N, D) then the verification functions returns $\top$ (i.e., accept the message), otherwise it returns $\bot$ (i.e., reject the message).

Based on these two schemes, we define the EtM authenticated encryption scheme $\mathsf{AE}_{\mathcal{E},\mathcal{I}} = (\mathsf{AE}.\mathsf{KGen}, \mathsf{AE}.\mathsf{Enc}, \mathsf{AE}.\mathsf{Dec})$ where the key-generation algorithm generates a random pair of keys $(K_e, K_m) \in \mathcal{K}_e \times \mathcal{K}_m$. The encryption and decryption algorithms are defined as follows:

$$\mathsf{AE}.\mathsf{Enc}(K_e\|K_m, N, A, M) = \begin{cases} C \leftarrow \mathcal{E}.\mathsf{Enc}(K_e, N, M) \\ T \leftarrow \mathcal{I}.\mathsf{Tag}(K_m, N, A\|C) \end{cases}$$

$$\mathsf{AE}.\mathsf{Dec}(K_e\|K_m, N, A, C, T) = \begin{cases} M \leftarrow \mathcal{E}.\mathsf{Dec}(K_e, N, C), \text{ if } Z = \top \\ \bot, \text{ if } Z = \bot \end{cases}$$

for $Z \leftarrow \mathcal{I}.\mathsf{Ver}(K_m, N, A\|C, T)$. We consider two security notions for the AE scheme: privacy and authenticity. The privacy advantage of AE is defined as follows:

$$\mathbf{Adv}_{\mathsf{AE}}^{\mathrm{priv}}(\mathsf{A}) \triangleq \Pr[(K_e \times K_m) \leftarrow_\$ (\mathcal{K}_e \times \mathcal{K}_m) : \mathsf{A}^{\mathsf{AE}.\mathsf{Enc}(K_e, K_m)} = 1] - \Pr[\mathsf{A}^\$ = 1],$$

where the random oracle $\$$ takes (N, A, M) as input and returns $(C, T) \leftarrow_\$ \{0,1\}^{|M|+\rho}$. We assume that the adversary A is nonce respecting, that is it does not make two queries with the same nonce.

If an adversary A interacts with the encryption and the decryption oracles of the AE, then the authenticity advantage of the AE is defined as follows:

$$\mathbf{Adv}_{\mathsf{AE}}^{\mathrm{auth}}(\mathsf{A}) \triangleq \Pr[(K_e \times K_m) \leftarrow_\$ (\mathcal{K}_e \times \mathcal{K}_m) : \mathsf{A}^{\mathsf{AE.Enc}(K_e, K_m), \mathsf{AE.Dec}(K_e, K_m)} \text{ forges}],$$

where we say that the adversary A forges if the AE.Dec oracle returns a bit string (which is not $\perp$) for a query (N, A, C, T) such that (C, T) was not returned by the AE.Enc oracle as a result of the encryption query (N, A, M). Moreover, we assume that A can repeat nonces in decryption queries and can also use the nonces used in encryption queries.

The security of an AE scheme refers to the sum of its privacy and authenticity advantages. The privacy advantage of a nonce-based encryption scheme $\mathcal{E}$ that forms an AE with a MAC $\mathcal{I}$ is bound by the PRF advantage $\mathcal{E}$ and $\mathcal{I}$, while its authenticity advantage is bound by the forging advantage of the underlying $\mathcal{I}$. The achievement of a beyond birthday bound secure nonce-based AE scheme following the EtM paradigm thus requires a nonce respecting BBB secure nonce-based encryption scheme and a MAC mode that gives beyond birthday bound security for PRF-distinguishability and unforgeability (possibly in the nonce misuse model).

3.2   Encryption Modes Used in Encrypt-then-MAC-based AE

A symmetric encryption scheme is generally defined through a pseudorandom number generator (PRNG) that takes a short master key K and an initial value or nonce N that generates a key stream $(S_1, S_2, \ldots)$. Then the ciphertext is generated from the plaintext and the key stream by applying the one time padding technique.

The counter mode of encryption (CTR) is a popular symmetric key encryption scheme, which gives birthday bound security in terms of the number of blocks, and is used as the underlying encryption scheme in AE constructions such as CWC [24], GCM [26], GCM/2+ [3], GCM-RUP [4]. On the other hand Multi-EDM [41] and Multi-EDMD [41], which give an almost n-bit security, are used as the underlying encryption scheme in OGCM1 [41] and OGMC2 [41] respectively.

CIPHER-BASED ENCRYPTION. Cipher-based encryption [22] (CENC) is parameterized by a fixed non-negative integer w and so can be denoted as CENC_w. The PRNG of CENC_w takes a key K, a nonce ctr and a length l as input and outputs a sequence of fixed length key stream blocks, where the i-th key stream block is defined as

$$S_i \triangleq \mathsf{E}_K(\mathsf{ctr} + j(w+1)) \oplus \mathsf{E}_K(\mathsf{ctr} + j(w+1) + i), \quad j \in [0, l' - 1], i \in [1, w],$$

where $l' = l/w$. The optimal security of CENC_w has been shown in [8] and it is used as the underlying encryption scheme of CHM and CIP AE constructions. An optimally secure nonce-based encryption mode $\mathsf{CENC}_{\max}$ [8], in which w is set to the maximum number of message blocks, is applied as the underlying encryption scheme of mGCM [8].

3.3 MACs Used in Encrypt-then-MAC-based AEs

WEGMAN-CARTER MAC. The Wegman-Carter (WC) MAC [40] is an early and popular nonce-based MAC that authenticates a message by masking its hash value with a random number generated through a pseudorandom function applied on a nonce i.e.

$$\mathsf{WC}[\mathsf{F},\mathsf{H}](N, M) \triangleq \mathsf{F}_K(N) \oplus \mathsf{H}_{k_h}(M).$$

The WC MAC provides $O(\epsilon q_v)$ security when nonces are never reused, where ϵ is the hash differential probability and q_v is the number of verification attempts. However, the construction has no security when the nonce repeats even once. For some constructions, the hash key is revealed and for others, a simple forgery is possible. Different instantiations of the pseudorandom function and hash function gives different instances of the WC MAC. The Wegman-Carter-Shoup (WCS) MAC [39] is a popular instantiation of WC MAC, where the pseudorandom function is replaced by a block cipher. WCS has been used as the underlying MAC in GCM, CHM and CIP. EDM and EDMD are used as instantiations of the PRF in WC MAC and the resultant MACs are used as the underlying MAC algorithms in OGCM1 and OGCM2 respectively. CWC MAC [24] (used as the MAC function in the CWC AE construction) is an another variant of the WC MAC where the pseduorandom function is replaced by a block cipher and the hash function is defined as $E_{K_2}(\mathsf{H}_{K_h}(M))$.

ENCRYPTED WEGMAN-CARTER-SHOUP. The Encrypted Wegman-Carter-Shoup (EWCS) MAC [13] has been proposed as a remedy to the problem of nonce misuse security over the WC MAC. The EWCS MAC encrypts the output of the WCS MAC to generate the tag, and it is then used as the underlying MAC of GCM/2+ construction. EWC gives a security of around $2^{n/2}$ when nonces do not repeat. An attacker can make approximately $2^{n/2}$ queries with distinct nonces but the same message and observe no collisions in the tag.

XOR-ENCRYPT-XOR. Xor-Encrypt-Xor (XEX) was originally proposed as a mode of designing a tweakable block cipher [38]. Luykx et al. [4] used it as the underlying MAC in GCM-RUP. For a nonce N and a message M, XEX works as follows

$$\mathsf{XEX}[\mathsf{E},\mathsf{H}](N, M) \triangleq \mathsf{E}_K(N \oplus \mathsf{H}_{K_h}(M)) \oplus \mathsf{H}_{K_h}(M).$$

XEX is secure upto the birthday bound when nonces do no repeat. It can be easily seen that a collision amongst the values of $N \oplus \mathsf{H}_{K_h}(M)$ leads to a forgery which can be easily detected by finding collision in the values $N \oplus T$.

EWCDM [13] and a single-keyed hash variant of CLRW2 [25] are some possible alternatives of nonce-based MACs that can potentially be applied as the MAC function of any EtM-based AE mode. EWCDM has been proven to be secure upto approximately $2^{2n/3}$ queries when nonces do not repeat [13], and the single-keyed hash variant of CLRW2 can be shown to be birthday bound secure in the nonce respecting setting.

It is to be noted that all these constructions has birthday bound PRF security as an attacker can make $2^{n/2}$ queries with distinct nonces but same message and observes no collision in the tag.

3.4   Security Result of nEHtM: A Nonce-Based Version of EHtM

The previous section demonstrates that the MACs used in the existing AE modes are not secure beyond the birthday bound when nonces repeat just once, making them unsuitable for use in designing an AE that is resilient in the faulty nonce model. This section introduces the *nonce-based Enhanced Hash-then-Mask* nEHtM and gives upto $2n/3$-bit unforgeability in faulty nonce model. The Enhanced Hash-then-Mask (EHtM) proposed by Minematsu [29], is the first BBB secure PRF-based probabilistic MAC that uses only an n-bit random salt and an n-bit PRF. nEHtM is structurally similar to EHtM, except that the random salt is replaced by a nonce and the PRF by a block cipher. Moreover, for the purpose of domain separation, we consider an $(n-1)$-bit nonce and an $(n-1)$-bit keyed hash function. For any message M and nonce N, nEHtM is defined as follows

$$\text{nEHtM}[\text{E}, \text{H}_{K_h}](N, M) \triangleq E_K(0\|N) \oplus E_K(1\|(N \oplus \text{H}_{K_h}(M))).$$

We now state Theorem 1, which bounds the unforgeability of nEHtM in the faulty nonce model. We also demonstrate a birthday bound forging attack on nEHtM when the number of faulty nonces reaches an order of $2^{n/2}$. The underlying idea of the attack is to form an alternating cycle of length 4 in the input of the block cipher; details may be found in [19].

Theorem 1. *Let $\mathcal{M}, \mathcal{K}$ and $\mathcal{K}_h$ be finite and non-empty sets. Let $\text{E} : \mathcal{K} \times \{0,1\}^n \to \{0,1\}^n$ be a block cipher and $\text{H} : \mathcal{K}_h \times \mathcal{M} \to \{0,1\}^{n-1}$ be an ϵ-axu $(n-1)$-bit ϵ-AXU hash function. Let μ be a fixed parameter. Then the forging advantage for any (μ, q_m, q_v, t)-adversary against nEHtM[E, H] that makes authentication queries with at most μ faulty nonces is given by*

$$\mathbf{Adv}_{\text{nEHtM}[\text{E},\text{H}]}^{\text{MAC}}(\mu, q_m, q_v, t) \leq \mathbf{Adv}_{\text{E}}^{\text{PRP}}(\mu, q_m + q_v, t') + \frac{48q_m^3}{2^{2n}} + \frac{12q_m^4\epsilon}{2^{2n}} + \frac{12\mu^2 q_m^2}{2^{2n}}$$
$$+ \frac{q_m + 2q_v}{2^n} + \frac{4q_m^3\epsilon}{2^n} + (2q_m + q_v)\mu\epsilon + q_v\epsilon,$$

where the time parameter t' is of the order of $t + (q_m + q_v)t_\text{H}$ and t_H is the time required for computing the hash function. Assuming $\epsilon \approx 2^{-(n-1)}$ and $q_m \leq \epsilon^{-1}$ simplifies this bound to

$$\mathbf{Adv}_{\text{nEHtM}[\text{Perm},\text{H}]}^{\text{MAC}}(\mu, q_m, q_v, t) \leq \frac{80q_m^3}{2^{2n}} + \left(\frac{12\mu^2 q_m^2}{2^{2n}} + \frac{(4q_m + 2q_v)\mu}{2^n}\right) + \left(\frac{q_m + 4q_v}{2^n}\right).$$

The proof of this theorem is deferred until Sect. 6. The forging advantage of nEHtM for $\mu \leq 2^{n/3}$ and $q_m \leq 2^{2n/3}/9$ is thus given by

$$\mathbf{Adv}_{\text{nEHtM}[\text{Perm},\text{H}]}^{\text{MAC}}(q_m, q_v, t) \leq \frac{18q_m}{2^{2n/3}} + \frac{4q_v}{2^{2n/3}}.$$

Remark 1. EHtM offers $3n/4$-bit security [18], whereas its nonce-based variant offers $2n/3$-bit security. This is because of the need to bound the number of multicollisions in the underlying hash function, for which the only source of randomness present in nEHtM is the hash key whereas EHtM also involves the random salts as an additional source of entropy.

3.5   CWC+: A Beyond Birthday Bound Variant of CWC

We have already seen that $\mathsf{CENC}_{\mathrm{max}}$ is a highly efficient optimally secure nonce respecting encryption scheme and nEHtM is a nonce-based MAC that is secure beyond the birthday bound in the faulty nonce model. Glueing them together using the EtM paradigm, we realize an authenticated encryption scheme, called CWC+, which gives a beyond the birthday bound security in the faulty nonce model. The encryption and decryption functions of CWC+ are shown in Fig. 2. The privacy and the authenticity advantages of CWC+ are stated in the following theorem, the proof of which is deferred until Sect. 7.

Theorem 2 (Privacy and Authenticity Bound of CWC+). *Let* $\mathsf{E} : \mathcal{K} \times \{0,1\}^n \rightarrow \{0,1\}^n$ *be a block cipher and* Poly $: \{0,1\}^n \times \{0,1\}^* \rightarrow \{0,1\}^{n-1}$ *be the* $(n-1)$-*bit truncated PolyHash function which truncates the first bit of the PolyHash output. Let* ρ *and* μ *be two fixed parameters. Then the privacy advantage for any* $(q_e, q_d, \ell, \sigma, t)$-*nonce respecting adversary against* CWC+$[\mathsf{E}, \rho]$ *is given by*

$$\mathbf{Adv}^{\mathrm{priv}}_{\mathsf{CWC+}[\mathsf{E},\rho]}(q_e, q_d, \ell, \sigma, t) \leq \mathbf{Adv}^{\mathrm{PRP}}_{\mathsf{E}}(\sigma + 2q, t') + \frac{105\sigma^3\ell}{2^{2n}} + \frac{6\sigma\ell}{2^n} + \frac{2q_d}{2^\rho} + \frac{2q_d\ell}{2^n}.$$

The authenticity advantage for any $(\mu, q_e, q_d, \ell, \sigma, t)$-*adversary against* CWC+$[\mathsf{E}, \rho]$ *is given by*

$$\mathbf{Adv}^{\mathrm{auth}}_{\mathsf{CWC+}[\mathsf{E},\rho]}(\mu, q_e, q_d, \ell, \sigma, t) \leq \mathbf{Adv}^{\mathrm{PRP}}_{\mathsf{E}}(\sigma + 2q, t') + \frac{105\sigma^3\ell}{2^{2n}} + \frac{6\sigma\ell}{2^n} + \frac{2q_d}{2^\rho} + \frac{2q_d\ell}{2^n}$$
$$+ \frac{(2q_e + q_d)2\ell\mu}{2^n} + \left(\frac{5\sigma\ell\mu}{2^n}\right)^2.$$

We denote $q = q_e + q_d$, *the total number of encryption and decryption queries and* $t' = O(t + qt_{\mathsf{H}} + \sigma + 2q)$, *where* t_{H} *denotes the time for computing the hash function and* μ *denotes the total number faulty encryption queries. The authenticity advantage of* CWC+ *for* $\mu \leq 2^{n/3}, \sigma \leq 2^{2n/3}$ *and* $\sigma \approx q_e\ell$ *is simplified to*

$$\mathbf{Adv}^{\mathrm{auth}}_{\mathsf{CWC+}[\mathsf{E},\rho]}(\mu, q_e, q_d, \ell, \sigma, t) \leq \mathbf{Adv}^{\mathrm{PRP}}_{\mathsf{E}}(\sigma + 2q, t') + \frac{112\sigma\ell}{2^{2n/3}} + \frac{2q_d}{2^\rho} + \frac{4q_d\ell}{2^{2n/3}}.$$

Algorithm CWC+.Enc$_K(N, A, M)$

1. $L \leftarrow \mathsf{E}_K(\mathbf{0})$; $N' \leftarrow N\|0^{n/4-1}$;
2. $l \leftarrow \lceil |M|/n \rceil$;
3. $S \leftarrow \mathsf{CENC}_{\max}(K, 0\|N', l)$;
4. $C \leftarrow M \oplus \mathsf{first}(S, |M|)$;
5. $\tilde{T} \leftarrow \mathsf{nEHtM}[\mathsf{E}, \mathsf{Poly}_{\mathsf{E}_K(\mathbf{0})}](N', C\|A)$;
6. $T \leftarrow \mathsf{chop}_\rho(\tilde{T})$;
7. **return** (C, T)

Algorithm CWC+.Dec$_K(N, A, C, T)$

1. $L = \mathsf{E}_K(\mathbf{0})$; $N' \leftarrow N\|0^{n/4-1}$;
2. $l \leftarrow \lceil |C|/n \rceil$;
3. $\tilde{T}' \leftarrow \mathsf{nEHtM}[\mathsf{E}, \mathsf{Poly}_{\mathsf{E}_K(\mathbf{0})}](N', C\|A)$;
4. **if** $\mathsf{chop}_\rho(\tilde{T}') \neq T$ **then return** $\perp$;
5. $S \leftarrow \mathsf{CENC}_{\max}(K, N', l)$;
6. $M \leftarrow C \oplus \mathsf{first}(S, |C|)$;
7. **return** M

Fig. 2. Encryption and Decryption functions of CWC+. $\mathsf{Poly}_{\mathsf{E}_K(\mathbf{0})}$ denotes the Polyhash function with its n-bit hash key set to the encrypted value of $\mathbf{0}$. $\mathsf{first}(S, |M|)$ denotes the first $|M|$ bits in the sequence S. chop_ρ is a function that truncates the last $n - \rho$ bits of its input.

4   Mirror Theory

Mirror theory, introduced by Patarin in [34], is a technique to provide a lower bound for the number of solutions to a given system of linear (more precisely, affine) bivariate equations and non-equations in a finite field (e.g., GF(2^n)). Solving a system of linear or affine equations is straightforward and a common problem in linear algebra. However, the problem starts complicating when non-equations are included. A special form of problems involving non-equations is to find distinct solutions to all the variables present in the system. If $Y_1, \ldots, Y_s$ are the variables, the system of non-equations $Y_i \oplus Y_j \neq \mathbf{0}$ for all $i \neq j$ essentially restricts the solutions to those in which all variables take distinct values. We call such a solution an *injective solution*. However, Patarin did not consider any other forms of non-equations [34–36]. This has been considered and termed as *extended mirror theory* in a recent work of Datta et al. [16]. In [16], the authors provided a lower bound on the number of injective solutions when the maximum component size $w_{\max}$ (a parameter that shall be defined soon) is three or less. This paper extends their analysis for an arbitrary $w_{\max}$.

INJECTIVE SOLUTION OF EQUATIONS. Let $G = (\mathcal{V} \triangleq \{Y_1, \ldots, Y_\alpha\}, \mathcal{S})$ be a simple acyclic graph with an edge-labelling function $\mathcal{L} : \mathcal{S} \rightarrow \{0,1\}^n$. For an edge $\{Y_i, Y_j\} \in \mathcal{S}$, we write $\mathcal{L}(\{Y_i, Y_j\}) = \lambda_{ij}$ (and so $\lambda_{ij} = \lambda_{ji}$). The system of equations induced by G, denoted $\mathcal{E}_G$, is then defined as:

$$\mathcal{E}_G \triangleq \{Y_i \oplus Y_j = \lambda_{ij}; \ \{Y_i, Y_j\} \in \mathcal{S}\}. \tag{3}$$

That is, each vertex of G denotes a variable in the system of equations and each edge of G denotes an equation in $\mathcal{E}_G$. We denote the set of components in G by $\mathsf{comp}(G) = (\mathcal{C}_1, \ldots, \mathcal{C}_k)$, where k is the number of components in G. w_i denotes the size of (i.e. the number of vertices in) the component C_i, $w_{\max}$ denotes the quantity $\max\{w_1, \ldots, w_k\}$ (also commonly denoted as ξ in Patarin's papers) and σ_i the sum $(w_1 + \cdots + w_i)$ with the convention that $\sigma_0 = 0$.

Definition 1. *With respect to the system of equations $\mathcal{E}_G$ (as defined above), an injective function $\Phi : \mathcal{V} \to \{0, 1\}^n$ is said to be an injective solution if $\Phi(Y_i) + \Phi(Y_j) = \lambda_{ij}$ for all $\{Y_i, Y_j\} \in \mathcal{S}$.*

As the graph G is acyclic, there exists a unique path in the graph between any two vertices Y_s and Y_t in the same connected component, which shall be denoted by P_{st}. Adding all equations induced by the edges of any such path P_{st} gives

$$\mathcal{L}(P_{st}) := \sum_{e \in P_{st}} \mathcal{L}(e) = Y_s \oplus Y_t.$$

So, for an injective solution to exist, the graph G (along with the label function) must satisfy the following property:

NPL (non-zero path label): *For all paths P in graph G, $\mathcal{L}(P) \neq \mathbf{0}$.*

It may be noted here that the NPL condition formalizes the notion of non-degeneracy as mentioned in [28,34]. The restriction on the graph to be acyclic implies that the equations are linearly independent (since otherwise, there is a possibility that the system becomes inconsistent).

Having identified the necessary condition for the existence of an injective solution to $\mathcal{E}_G$ corresponding to any simple edge-labeled undirected acyclic graph G, we now state the following claim due to Patarin [34], which gives a lower bound on the number of injective solutions to $\mathcal{E}_G$. Suppose G has α vertices and q edges. Patarin claimed that the number of injective solutions to $\mathcal{E}_G$ is at least $\frac{(2^n)_\alpha}{2^{nk}}$, provided $\sigma_k(w_{\max} - 1) \leq 2^n/64$. Unfortunately, the proof of this claim is unverifiable. [16] gives a detailed proof for the following lower bound on the number of injective solutions: $\frac{(2^n)_\alpha}{2^{nk}} \cdot (1 - \epsilon)$, with $\epsilon \approx 0$ and $\sigma_k^3 w_{\max}^2 \ll 2^{2n}$.

INJECTIVE SOLUTION TO A SYSTEM OF EQUATIONS AND NON-EQUATIONS. An extended system involving a system of non-equations along with a system of equations shall now be examined. Let $G = (\mathcal{V} \overset{\Delta}{=} \{Y_1, \ldots, Y_\alpha\}, \mathcal{S} \sqcup \mathcal{S}', \mathcal{L})$ be a simple undirected edge-labelled graph ($\mathcal{L}$ is a label function), whose edge set is partitioned into two disjoint sets $\mathcal{S}$ and $\mathcal{S}'$. As before, we simply write $\mathcal{L}(\{Y_i, Y_j\}) = \lambda_{ij}$ for all $\{Y_i, Y_j\} \in \mathcal{S}$ and $\mathcal{L}(\{Y_i, Y_j\}) = \lambda'_{ij}$ for all $\{Y_i, Y_j\} \in \mathcal{S}'$. Let such a graph G induce a system of equations and non-equations $\mathcal{E}_G$ as follows:

$$Y_i \oplus Y_j = \lambda_{ij} \ \forall \ \{Y_i, Y_j\} \in \mathcal{S}, \tag{4}$$

$$Y_i \oplus Y_j \neq \lambda'_{ij} \ \forall \ \{Y_i, Y_j\} \in \mathcal{S}', \tag{5}$$

For a system of equations and non-equations $\mathcal{E}_G$, an injective function $\Phi : \mathcal{V} \to \{0, 1\}^n$ is said to be an *injective solution function* if $\Phi(Y_i) \oplus \Phi(Y_j) = \lambda_{ij}$ for all $\{Y_i, Y_j\} \in \mathcal{S}$ and $\Phi(Y_i) \oplus \Phi(Y_j) \neq \lambda'_{ij}$ for all $\{Y_i, Y_j\} \in \mathcal{S}'$.

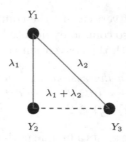

Fig. 3. $\mathcal{E}_G \triangleq \{Y_1 \oplus Y_2 = \lambda_1, Y_1 \oplus Y_3 = \lambda_2, Y_2 \oplus Y_3 \neq \lambda_1 \oplus \lambda_2\}$. The continuous red edges represent equations and the dashed blue edges represent non-equations. Clearly, the system of equations and non-equations is inconsistent. (Color figure online)

GOOD GRAPHS. We shall first investigate the case when $\mathcal{E}_G$ has at least one solution. To ensure this, the subgraph $G^= \triangleq (\mathcal{V}, \mathcal{S}, \mathcal{L}_{|\mathcal{S}})$, where $\mathcal{L}_{|\mathcal{S}}$ is the function $\mathcal{L}$ restricted over the set $\mathcal{S}$, must

 (i) be acyclic (i.e. **No Cycle** or **NC**)
 (ii) satisfy the **NPL** condition and
(iii) satisfy the **NCL (non-zero cycle label)** property which says that *for all cycles C in G such that the edge set of C contains exactly one non-equation edge* $e' \in \mathcal{S}'$, $\mathcal{L}(C) \neq \mathbf{0}$ (see Fig. 3 for an example).

If a graph G satisfies the above three conditions (i)-(iii), it is said to be a **good graph**. In [16], authors have proved the following lower bound for $w_{\max} = 3$. Let $G = (\mathcal{V}, \mathcal{S} \sqcup \mathcal{S}', \mathcal{L})$ be a good graph with $|\mathcal{V}| = \alpha, |\mathcal{S}| = q_m, |\mathcal{S}'| = q_v$. Let $\mathsf{comp}(G^=) = (\mathcal{C}_1, \ldots, \mathcal{C}_k)$ with $|\mathcal{C}_i| = w_i \ (\leq 3)$ and $\sigma_i = (w_1 + \cdots + w_i)$. Let $\mathcal{Z} \subseteq \{0,1\}^n$ such that $|\{0,1\}^n \setminus \mathcal{Z}| = c$. The total number of injective solutions (each solution is chosen from the set $\mathcal{Z}$) for the induced system of equations and non-equations $\mathcal{E}_G$ is at least:

$$\frac{(2^n)_\alpha}{2^{nk}}\left(1 - \frac{5k^3}{2^{2n}} - \frac{q_v + c\alpha}{2^{n-1}}\right). \tag{6}$$

Observe that $q_v + c\alpha$ is the number of non-equations, considering univariate non-equations arising from the constraint of each solution being from the set of size $2^n - c$. Now we state our theorem, which generalizes this result for any $w_{\max}$.

Theorem 3. *Let $G = (\mathcal{V}, \mathcal{S} \sqcup \mathcal{S}', \mathcal{L})$ be a good graph with α vertices such that $|\mathcal{S}| = q_m, |\mathcal{S}'| = q_v$. Let $\mathsf{comp}(G^=) = (\mathcal{C}_1, \ldots, \mathcal{C}_k)$ and $|\mathcal{C}_i| = w_i$, $\sigma_i = (w_1 + \cdots + w_i)$. Then the total number of injective solutions chosen from a set $\mathcal{Z}$ of size $2^n - c$, for some $c \geq 0$, for the induced system of equations and non-equations $\mathcal{E}_G$ is at least:*

$$\frac{(2^n)_\alpha}{2^{nq}}\left(1 - \sum_{i=1}^{k}\frac{6\sigma_{i-1}^2\binom{w_i}{2}}{2^{2n}} - \frac{2(q_v + c\alpha)}{2^n}\right), \tag{7}$$

provided $\sigma_k w_{\max} \leq 2^n/4$.

Proof. We give here a brief sketch of the proof. A detailed proof of the theorem can be found in [19]. The proof proceeds by counting the number of solutions in each of the k components. We denote $\tilde{w}_{ij}$ to be the number of edges from S' connecting vertices between i-th and j-th component of $G^=$ and w'_i to be the number of edges in S' incident on $v_i \in V \setminus G^=(V)$. It is easy to see that the number of solutions for the first component is exactly $(2^n - cw_1)$. We fix a solution and count the number of solutions for the second component which is $(2^n - w_1 w_2 - \tilde{w}_{1,2} - cw_2)$ as it must discard (i) w_1 values $(y_{i_1}, \ldots, y_{i_{w_1}})$ from the first component, (ii) $w_1(w_2 - 1)$ values $(y_{i_1} \oplus \mathcal{L}(P_j), \ldots, y_{i_{w_1}} \oplus \mathcal{L}(P_j))$ for all possible paths P_j from a fixed vertex to any other vertex in the second component and (iii) $cw_2 + \tilde{w}_{12}$ values to compensate for the fact that the set of values is no longer a group. In general, the total number of solutions for the i-th component is at least $\prod_{i=1}^{k} \left(2^n - \sigma_{i-1} w_i - \sum_{j=1}^{i-1} \tilde{w}_{ij} - cw_i \right)$. Suppose there are k' vertices that do not belong to the set of vertices of the subgraph $G^=$. Fix such a vertex $Y_{\sigma_k + i}$ and let us assume that $w'_{\sigma_k + i}$ blue dashed edges are incident on it. If $y_{\sigma_k + i}$ is a valid solution to the variable $Y_{\sigma_k + i}$, then we must have (a) $y_{\sigma_k + i}$ should be distinct from the previous σ_k assigned values, (b) $y_{\sigma_k + i}$ should be distinct from the $(i - 1)$ values assigned to the variables that do not belong to the set of vertices of the subgraph $G^=(V)$ and (c) $y_{\sigma_k + i}$ should not take those $w'_{\sigma_k + i}$ values.

Therefore, the total number of solutions is at least

$$h_\alpha \geq \prod_{i=1}^{k} \left(2^n - \sigma_{i-1} w_i - \sum_{j=1}^{i-1} \tilde{w}_{ij} - cw_i \right) \cdot \prod_{i \in [k']} (2^n - \sigma_k - i + 1 - w'_{\sigma_k + i}). \quad (8)$$

Let us denote $(\tilde{w}_{i1} + \ldots + \tilde{w}_{i,i-1})$ by p_i and $(w'_{\sigma_k + 1} + \ldots + w'_{\sigma_k + k'})$ by q''_v. After a simple algebraic calculation on Eq. (8), we obtain

$$h_\alpha \frac{2^{n q_m}}{(2^n)_\alpha} \geq \underbrace{\prod_{i=1}^{k} \frac{(2^n - \sigma_{i-1} w_i - p_i - cw_i) 2^{n(w_i - 1)}}{(2^n - \sigma_{i-1})_{w_i}}}_{\text{D.1}} \left(1 - \frac{2q''_v}{2^n} \right). \quad (9)$$

Let us denote the expression $\left(\binom{w_i}{2} \sigma_{i-1}^2 + \binom{w_i}{2}(w_i - 1)\sigma_{i-1} + \binom{w_i}{2} \right.$ $\left. \frac{(w_i - 2)(3w_i - 1)}{12} \right)$ by A_i. Expanding $(2^n - \sigma_{i-1})_{w_i}$ along with some simple computations on D.1 gives

$$D.1 \geq \prod_{i=1}^{k} \left(1 - \frac{A_i}{2^{2n} - 2^n(\sigma_{i-1}w_i + \binom{w_i}{2}) + A_i} - \frac{2^n(p_i + cw_i)}{2^{2n} - 2^n(\sigma_{i-1}w_i + \binom{w_i}{2}) + A_i} \right)$$

$$\overset{(4)}{\geq} \prod_{i=1}^{k} \left(1 - \frac{2A_i}{2^{2n}} - \frac{2(p_i + cw_i)}{2^n} \right) \overset{(5)}{\geq} \left(1 - \sum_{i=1}^{k} \frac{6\sigma_{i-1}^2 \binom{w_i}{2}}{2^{2n}} - \sum_{i=1}^{k} \frac{2(p_i + cw_i)}{2^n} \right)$$

$$\overset{(6)}{\geq} \left(1 - \sum_{i=1}^{k} \frac{6\sigma_{i-1}^2 \binom{w_i}{2}}{2^{2n}} - \frac{2q_v'}{2^n} - \frac{2c\alpha}{2^n} \right), \tag{10}$$

where (4) follows from the fact that $2^n(\sigma_{i-1}w_i + \binom{w_i}{2}) - A_i \leq 2^{2n}/2$, which holds true when $\sigma_k w_{\max} \leq 2^n/4$, (5) holds true due to the fact that $A_i \leq 3\sigma_{i-1}^2\binom{w_i}{2}$ and (6) holds true as we denote $(p_1 + \ldots + p_k) = q_v'$, the total number of blue dashed edges across the components of $G^=$ and $w_1 + \ldots + w_k \leq \alpha$. Finally, from Eqs. (9), (10) and $q_v = q_v' + q_v''$, the result follows. $\qquad\square$

5 Mutlicollision in Universal Hash Function

In this section, we study the muticollision advantage of a universal hash function. Suppose H_{K_h} is an ϵ universal hash function where the hash key K_h is chosen uniformly at random from the hash-key space. For any q distinct messages $M_1, \ldots, M_q$, the probability that there exist $i \neq j$, such that M_i and M_j collide under the hash function H_{K_h} is at most $\epsilon\binom{q}{2}$ (by the union bound). Extending this result for multicollisions, we say that $(M_1, \ldots, M_\xi)$ is a ξ-multicollision tuple for H_{K_h} if $\mathsf{H}_{K_h}(M_1) = \mathsf{H}_{K_h}(M_2) = \cdots = \mathsf{H}_{K_h}(M_\xi)$. When H_{K_h} is a ξ-wise independent hash function [40] the probability that a ξ-tuple $(M_1, \ldots, M_\xi)$ is a ξ-multicollision tuple for H_{K_h} is $1/2^{n(\xi-1)}$. Clearly, this cannot be concluded for a universal hash function. In fact, one can easily construct a ξ-tuple of messages such that the multicollision probability under the PolyHash function is $\ell/2^n$.

In the following, we now provide a bound (better than $\epsilon\binom{q}{2}$) on the existence of a multicollision tuple for any given q messages.

Theorem 4 (Multicollision Theorem). *Let $X_1, \ldots, X_q$ be q distinct messages and H_{K_h} be an ϵ-universal hash function. Then for $\xi \in \mathbb{N}$, the probability that a $(\xi + 1)$-multicollision tuple exists in this set of messages is no more than $q^2\epsilon/2\xi$.*

Proof. Let us denote the required probability by P and set $Z_i = \mathsf{H}_{K_h}(X_i)$, $i \in [q]$. Also let $\mathbf{X}$ denote a $(\xi + 1)$-tuple $(X_1, \ldots, X_{\xi+1}) \in \mathcal{V}^{\xi+1}$. Consider the graph $G = (\mathcal{V}, \mathcal{S})$ whose vertex set $\mathcal{V}$ contains each of the q messages. An edge between two nodes exists in $\mathcal{S}$ if and only if the hash values of the corresponding messages collide. Therefore, the event $\mathsf{H}_{K_h}(X_1) = \ldots = \mathsf{H}_{K_h}(X_{\xi+1})$ boils down to the existence of a clique of size $\xi + 1$ in G. Due to Lemma 1, if G has $q^2/2\xi$ edges, then any collection of $\xi + 1$ vertices of the q vertices in $\mathcal{V}$ must contain at least one pair which is in $\mathcal{S}$. i.e. there must exist $\{v_1, \ldots, v_s\} \subseteq [q]$, for $s = q^2/\xi$, such that

$$Z_{i_1} = Z_{i_2} = \ldots = Z_{i_{\xi+1}} \Rightarrow Z_{v_1} = Z_{v_2} \vee Z_{v_3} = Z_{v_4} \vee \ldots \vee Z_{v_{s-1}} = Z_{v_s}, \tag{11}$$

Therefore, the probability P is:

$$\max_{\mathbf{X}} \Pr\left[K_h \xleftarrow{\$} \{0,1\}^n : \exists i_1, \cdots, i_\xi \in [q], \mathsf{H}_{K_h}(X_{i_1}) = \cdots = \mathsf{H}_{K_h}(X_{i_\xi})\right]$$

$$\leq \Pr[Z_{v_1} = Z_{v_2} \vee \ldots \vee Z_{v_{s-1}} = Z_{v_s}] \leq \sum_{i=1}^{s/2} \Pr[Z_{v_{2i-1}} = Z_{v_{2i}}] \leq \frac{s\epsilon}{2} = \frac{q^2\epsilon}{2\xi}.$$

Lemma 1. *Let* $q, \xi \in \mathbb{N}$. *Then for any set* $\mathcal{V}$ *with* $|\mathcal{V}| = q$, *there exists a graph* $G = (\mathcal{V}, \mathcal{S})$ *with* $|\mathcal{S}| = \left\lceil \frac{q^2}{2\xi} \right\rceil$ *such that any collection* C *of* $\xi + 1$ *vertices has at least one edge in* $\mathcal{S}$ *joining two vertices in* C.

Proof. Divide the q vertices into ξ subcollections of size $\left\lceil \frac{q}{\xi} \right\rceil$ each, the last subcollection possibly containing a lesser number of vertices. Construct $\mathcal{S}$ by adding in it, all the edges required to form a clique $C_i, i \in [\xi]$ out of each of the ξ subcollections. Thus, there are at most $\xi \cdot \binom{\lceil \frac{q}{\xi} \rceil - 1}{2}$ edges in all the ξ cliques. Observe that,

$$\xi \cdot \binom{\left\lceil \frac{q}{\xi} \right\rceil - 1}{2} < \xi \cdot \binom{\frac{q}{\xi}}{2} \leq \frac{q^2}{2\xi} \leq \left\lceil \frac{q^2}{2\xi} \right\rceil.$$

Hence, $\mathcal{S}$ must contain more edges, distinct from those involved in the ξ cliques, which must exist between at least one pair of vertices in different cliques C_i and C_j ($i \neq j$). Since there are $\xi + 1$ vertices in C and a total of ξ cliques C_i formed so far in G, it can thus be inferred from the pigeonhole principle that at least one clique C_i contains more than one edge from $\mathcal{S}$, making clear the existence of an edge from $\mathcal{S}$ in C. $\qquad\square$

6 Proof of Theorem 1

In this section, we prove Theorem 1. We shall often refer to the construction nEHtM[E, H] as simply nEHtM when the underlying primitives are assumed to be understood.

The first step of the proof is the standard switch from the computational setting to the information theoretic one by replacing the block cipher E_K with an n-bit uniform random permutation Π at the cost of $\mathbf{Adv}_\mathsf{E}^{\mathrm{PRP}}(q_m + q_v, t')$, where $t' = O(t + (q_m + q_v)t_H)$ and t_H is the time required for computing the hash function. Let us denote this modified construction as nEHtM*[Π, H]. Hence,

$$\mathbf{Adv}_{\mathsf{nEHtM}}^{\mathrm{MAC}}(q_m, q_v, t) \leq \mathbf{Adv}_\mathsf{E}^{\mathrm{PRP}}(q_m + q_v, t') + \underbrace{\mathbf{Adv}_{\mathsf{nEHtM}^*}^{\mathrm{MAC}}(q_m, q_v, t)}_{\delta^*}. \quad (12)$$

To get an upper bound for δ^*, we consider a perfect random oracle Rand, which on input (N, M) returns T, sampled uniformly at random from $\{0,1\}^n$, and an oracle Rej which always returns $\perp$ (i.e., rejects) for all inputs (N, M, T). Now, due to [13, 16, 18] we have

$$\delta^* \leq \max_{\mathsf{D}} \Pr[\mathsf{D}^{\mathsf{TG}[\Pi, \mathsf{H}_{K_h}], \mathsf{VF}[\Pi, \mathsf{H}_{K_h}]} = 1] - \Pr[\mathsf{D}^{\mathsf{Rand}, \mathsf{Rej}} = 1],$$

where the maximum is taken over all non-trivial distinguishers D. This formulation allows us to apply the expectation method [10, 20] to prove that

$$\delta^* \leq \frac{48 q_m^3}{2^{2n}} + \frac{12 q_m^4 \epsilon}{2^{2n}} + \frac{12 \mu^2 q_m^2}{2^{2n}} + \frac{q_m + 2 q_v}{2^n} + \frac{4 q_m^3 \epsilon}{2^n} + (2 q_m + q_v) \mu \epsilon + q_v \epsilon. \quad (13)$$

ATTACK TRANSCRIPT. Henceforth, we fix a deterministic non-trivial (i.e., one that makes no repeated queries) distinguisher D that interacts with either (1) the real oracle $(\mathsf{TG}[\Pi, \mathsf{H}_{K_h}], \mathsf{VF}[\Pi, \mathsf{H}_{K_h}])$ for a uniform random permutation Π and a random hashing key K_h or (2) the ideal oracle $(\mathsf{Rand}, \mathsf{Rej})$ making at most q_m queries to its left (authentication) oracle with at most μ faulty nonces and at most q_v queries to its right (verification) oracle, and returning a single bit. Then

$$\mathbf{Adv}(\mathsf{D}) = \left| \Pr\left[\mathsf{D}^{\mathsf{TG}[\Pi, \mathsf{H}_{K_h}], \mathsf{VF}[\Pi, \mathsf{H}_{K_h}]} = 1 \right] - \Pr\left[\mathsf{D}^{\mathsf{Rand}, \mathsf{Rej}} = 1 \right] \right|.$$

Let $\quad \tau_m \triangleq \{(N_1, M_1, T_1), (N_2, M_2, T_2), \ldots, (N_{q_m}, M_{q_m}, T_{q_m})\}$

be the list of authentication queries made by D and the corresponding responses it receives. Also let

$$\tau_v \triangleq \{(N_1', M_1', T_1', b_1'), (N_2', M_2', T_2', b_2'), \ldots, (N_{q_v}', M_{q_v}', T_{q_v}', b_{q_v}')\}$$

be the list of verification queries made by D and the corresponding responses it receives, where for all j, $b_j' \in \{\top, \perp\}$ denotes the set of accept $(b_j' = \top)$ and reject $(b_j' = \perp)$ responses. The pair $\tau = (\tau_m, \tau_v)$ constitutes the query transcript of the attack. For convenience, we slightly modify the experiment to reveal to the distinguisher (after it made all its queries and obtained the corresponding responses, but before it outputs its decision) the hashing key K_h, if D interacts with the real world, or a uniformly random dummy key K_h if D interacts with the ideal world. Hence, the *extended transcript* of the attack is $\tau' = (\tau, K_h)$ where $\tau = (\tau_m, \tau_v)$, τ_m and τ_v being the tuples of the authentication and verification queries respectively. We shall often simply name a tuple $(N, M, T) \in \tau_m$ an *authentication query*, and a tuple $(N', M', T', b') \in \tau_v$ a *verification query*.

A transcript τ' is said to be an *attainable transcript* (with respect to D) if the probability of realizing this transcript in the ideal world is non-zero. It must be noted that since attainability is with respect to the ideal world, any verification query (N_i', M_i', T_i', b_i') even in an attainable transcript $\tau' = (\tau, K_h)$ is such that $b_i' = \perp$. We denote Θ to be the set of all attainable transcripts and X_{re} and X_{id} to be the random variables that take an extended transcript τ' induced by the real world and the ideal world respectively.

6.1   Definition and Probability of Bad Transcripts

In this section, we define and bound the probability of bad transcripts in the ideal world. For notational simplicity, we denote $N_i \oplus H_{K_h}(M_i)$ as X_i. Note that X_i is an $n - 1$ bit string.

Definition 2 (Bad Transcript). *Given a parameter $\xi \in \mathbb{N}$, where $\xi \geq \mu$, an attainable transcript $\tau' = (\tau_m, \tau_v, K_h)$ is called a **bad** transcript if any one of the following holds:*

- B1 : $\exists\, i \in [q_m]$ *such that $T_i = \mathbf{0}$.*
- B2 : $\exists\, i \neq j, j \neq k$ *such that $N_i = N_j$ and $X_j = X_k$.*
- B3 : $\{i_1, \ldots, i_{\xi+1}\} \subseteq [q_m]$ *such that $X_{i_1} = X_{i_2} = \ldots = X_{i_{\xi+1}}$ (the optimal value of ξ shall be determined later in the proof).*
- B4 $\exists\, a \in [q_v], \exists\, i \in [q_m]$ *such that $N_i = N'_a$, $X_i = X'_a$ and $T_i = T'_a$.*

We denote by Θ_{bad} (resp. Θ_{good}) the set of bad (resp. good) transcripts. We bound the probability of bad transcripts in the ideal world as follows.

Lemma 2. *Let X_{id} and Θ_{bad} be defined as above. Then*

$$\Pr[X_{\text{id}} \in \Theta_{\text{bad}}] \leq \epsilon_{\text{bad}} = \frac{q_m}{2^n} + \frac{q_m^2 \epsilon}{2\xi} + (2q_m + q_v)\mu\epsilon + q_v\epsilon.$$

Proof. By the union bound,

$$\Pr[X_{\text{id}} \in \Theta_{\text{bad}}] \leq \Pr[\text{B1}] + \Pr[\text{B2}] + \Pr[\text{B3}] + \Pr[\text{B4}]. \qquad (14)$$

In the following, we bound the probabilities of all the bad events individually. The lemma will follow by adding the individual bounds. Clearly,

$$\Pr[\text{B1}] \leq \frac{q_m}{2^n}. \qquad (15)$$

Bounding B2. Let $\mathcal{F}$ be the set of all query indices i for which there is a $j \neq i$ such that $N_i = N_j$. It is easy to see that $|\mathcal{F}| \leq 2\mu$. Event B2 occurs if for some $j \in \mathcal{F}$, $H_{K_h}(M_j) = N_k \oplus H_{K_h}(M_k)$ for some $k \neq j$. For any such fixed i, j, k, the probability of the event is at most ϵ. The number of such choices of (j, k) is at most $2\mu q_m$. Hence,

$$\Pr[\text{B2}] \leq 2\mu q_m \epsilon. \qquad (16)$$

Bounding B3. Event B3 occurs if there exist $\xi + 1$ distinct authentication query indices $\{i_1, \ldots, i_{\xi+1}\} \subseteq [q_m]$ such that $X_{i_1} = \ldots = X_{i_{\xi+1}}$. This event is thus a $(\xi+1)$-multicollision on the ϵ universal hash function mapping (N, M) to $H_{K_h}(M) \oplus N$ (as H_{K_h} is an ϵ-almost-xor universal). Therefore, by Theorem 4:

$$\Pr[\text{B3}] \leq q_m^2 \epsilon/2\xi. \qquad (17)$$

Bounding B4. For some $a \in [q_v]$ and $i \in [q_m]$, if $N_i = N'_a$, $X_i = X'_a$ and $T_i = T'_a$, then $M_i \neq M'_a$ (as the adversary does not make any trivial query). Hence the probability that $X_i = X'_a$ holds is at most ϵ. Now, for any a, there can be at most $(\mu + 1)$ indices i such that $N_i = N'_a$. Hence, the required probability is bounded as

$$\Pr[\text{B4}] \leq (\mu + 1)q_v\epsilon. \qquad (18)$$

The proof follows from Eqs. (14)–(18). $\qquad\square$

6.2   Analysis of Good Transcripts

In this section, we show that for a good transcript $\tau' = (\tau, K_h)$, realizing τ' is almost as likely in the real world as in the ideal world.

Consider a good transcript $\tau' = (\tau_m, \tau_v, K_h)$. Since in the ideal world the authentication oracle is perfectly random and the verification oracle always rejects,

$$\Pr[X_{\mathrm{id}} = \tau'] = \frac{1}{|\mathcal{K}_h|} \cdot \frac{1}{2^{nq_m}} \tag{19}$$

We must now lower bound $\Pr[X_{\mathrm{re}} = \tau']$ i.e., the probability of getting τ' in the real world. We say that a permutation Π is *compatible with τ_m* (respectively with τ_v) if (A) (respectively (B)) holds.

(A) $\forall i \in [q_m], \Pi(\widehat{N}_i) \oplus \Pi(\widehat{X}_i) = T_i$,   (B) $\forall a \in [q_v], \Pi(\widehat{N'}_a) \oplus \Pi(\widehat{X'}_a) \neq T'_a$,

where $\widehat{N}_i = 0\|N_i$, $\widehat{X}_i = 1\|X_i$, $\widehat{N'}_a = 0\|N'_a$ and $\widehat{X'}_a = 1\|X'_a$. We simply say that Π is compatible with $\tau = (\tau_m, \tau_v)$ if it is compatible with τ_m and τ_v. We denote by $\mathsf{Comp}(\tau)$ the set of permutations Π that are compatible with τ. Therefore,

$$\Pr[X_{\mathrm{id}} = \tau'] = \frac{1}{|\mathcal{K}_h|} \cdot \Pr[\Pi \xleftarrow{\$} \mathsf{Perm} : \Pi \in \mathsf{Comp}(\tau)]$$

$$= \frac{1}{|\mathcal{K}_h|} \cdot \underbrace{\Pr[\Pi(\widehat{N}_i) \oplus \Pi(\widehat{X}_i) = T_i, \Pi(\widehat{N'}_a) \oplus \Pi(\widehat{X'}_a) \neq T'_a]}_{P_{mv}}. \tag{20}$$

We refer to the system of equations as *"authentication equations"* as they involve only the authentication queries and to the system of non-equations as *"verification non-equations"* as they involve only the verification queries. We denote the system of authentication equations by $\mathcal{E}_m$ and the system of verification non-equations by $\mathcal{E}_v$.

$$(\mathcal{E}_m) = \begin{cases} \Pi(\widehat{N}_1) \oplus \Pi(\widehat{X}_1) = T_1 \\ \Pi(\widehat{N}_2) \oplus \Pi(\widehat{X}_2) = T_2 \\ \vdots \\ \Pi(\widehat{N}_{q_m}) \oplus \Pi(\widehat{X}_{q_m}) = T_{q_m} \end{cases} \qquad (\mathcal{E}_v) = \begin{cases} \Pi(\widehat{N'}_1) \oplus \Pi(\widehat{X'}_1) \neq T'_1 \\ \Pi(\widehat{N'}_2) \oplus \Pi(\widehat{X'}_2) \neq T'_2 \\ \vdots \\ \Pi(\widehat{N'}_{q_v}) \oplus \Pi(\widehat{X'}_{q_v}) \neq T'_{q_v} \end{cases}$$

EQUATION AND NON-EQUATION INDUCING GRAPH. From the above system of bivariate affine equations and non-equations, we induce the edge-labeled undirected graph $G_{\tau'} = (\mathcal{V}, \mathcal{S} \sqcup \mathcal{S}')$, where the set of nodes $\mathcal{V}$ is the set of variables $\{Y_1, \ldots, Y_\alpha\}$, $\mathcal{S}$ is the set of edges corresponding to each authentication equation and $\mathcal{S}'$ is the set of edges corresponding to each verification non-equation. Moreover, if there is an authentication equation $Y_s \oplus Y_t = T_i$, then the corresponding edge $\{Y_s, Y_t\} \in \mathcal{S}$ is labeled T_i. Similarly, if there is a verification non-equation $Y_s \oplus Y_t \neq T'_i$, then the corresponding edge $\{Y_s, Y_t\} \in \mathcal{S}'$ is labeled T'_i. Moreover, $G^{=}_{\tau'} = (\mathcal{V}, \mathcal{S})$ is the subgraph of $G_{\tau'}$.

The proof of the following claim can be found in the full version of the paper [19].

Claim 1. *If the transcript τ' is good, then the induced graph $G_{\tau'}$ is valid.*

Suppose there are k components in the subgraph $G_{\tau'}^{=}$ and the size of the i-th component is W_i. Thus, W_i is a random variable, and so is $W_{\max}$, which denotes the size of the largest component. It is easy to see that $W_{\max} \leq \xi$. As the graph $G_{\tau'}$ is valid (follows from Claim 1), we assume $\xi \leq 2^n/8q_m$ (from the condition of Theorem 3), which allows us to apply Theorem 3 with $c = 0$ to obtain,

$$\mathsf{P}_{mv} \geq \frac{1}{2^{nq_m}} \cdot \left(1 - \sum_{i=1}^{k} \frac{6\sigma_{i-1}^2 \binom{W_i}{2}}{2^{2n}} - \frac{2q_v}{2^n}\right). \tag{21}$$

Therefore, Eqs. (19)–(21) imply that the ratio $\frac{\Pr[X_{re}=\tau']}{\Pr[X_{id}=\tau']}$ is no less than

$$\left(1 - \sum_{i=1}^{k} \frac{6\sigma_{i-1}^2 \binom{W_i}{2}}{2^{2n}} - \frac{2q_v}{2^n}\right) \overset{(1)}{\geq} 1 - \underbrace{\left(\sum_{i=1}^{k} \frac{24q_m^2 \binom{W_i}{2}}{2^{2n}} + \frac{2q_v}{2^n}\right)}_{\phi(\tau')}, \tag{22}$$

where (1) follows due to the inequality $\sigma_{i-1} \leq 2q_m$.

We now compute the expectation of $\phi(X_{id})$ as follows.

$$\mathbf{E}\left[\left(\sum_{i=1}^{k} \frac{24q_m^2 \binom{W_i}{2}}{2^{2n}} + \frac{2q_v}{2^n}\right)\right] = \left(\frac{2q_v}{2^n} + \frac{24q_m^2}{2^{2n}} \mathbf{E}\left[\sum_{i=1}^{k} \binom{W_i}{2}\right]\right). \tag{23}$$

Let $\tilde{W}_i = W_i - 1$ and therefore,

$$\mathbf{E}\left[\sum_{i=1}^{k} \binom{W_i}{2}\right] = \mathbf{E}\left[\sum_{i=1}^{k} \binom{\tilde{W}_i}{2}\right] + \mathbf{E}\left[\sum_{i=1}^{k} \tilde{W}_i\right] \overset{(2)}{\leq} \mathbf{E}\left[\sum_{i=1}^{k} \binom{\tilde{W}_i}{2}\right] + 2q_m \tag{24}$$

where (2) holds as $(\tilde{W}_1 + \ldots \tilde{W}_k) = \sigma_k - k \leq 2q_m$. Let us consider the following two indicator random variables

$$I_{ij} = \begin{cases} 1, & \text{if } X_i = X_j \\ 0, & \text{otherwise} \end{cases} \qquad \tilde{I}_{ij} = \begin{cases} 1, & \text{if } N_i = N_j \\ 0, & \text{otherwise.} \end{cases}$$

Therefore,

$$\mathbf{E}\left[\sum_{i=1}^{k}\binom{\tilde{W}_i}{2}\right] \overset{(3)}{=} \sum_{i\neq j}^{q_m}\mathbf{E}[I_{ij}] + \sum_{i\neq j}^{\mu}\mathbf{E}[\tilde{I}_{ij}]$$

$$\overset{(4)}{=} \sum_{i\neq j}^{q_m}\Pr[\mathsf{H}_{K_h}(M_i) \oplus \mathsf{H}_{K_h}(M_j) = N_i \oplus N_j] + \mu^2/2$$

$$\overset{(5)}{\leq} \binom{q_m}{2}\epsilon + \mu^2/2 \leq q_m^2\epsilon/2 + \mu^2/2, \tag{25}$$

where (3) holds due to the linearity of expectation, (4) holds from the definition of the indicator random variable and (5) holds from the ϵ-almost-xor universal probability of the underlying hash function. Therefore, from Eqs. (23)–(25), we have

$$\mathbf{E}[\phi(X_{\mathrm{id}})] \leq \left(\frac{12q_m^4\epsilon}{2^{2n}} + \frac{12\mu^2 q_m^2}{2^{2n}} + \frac{48q_m^3}{2^{2n}} + \frac{2q_v}{2^n}\right). \tag{26}$$

FINALIZING THE PROOF. We have assumed that $\xi \geq \mu$ and from the condition of Theorem 3, we have $\xi \leq 2^n/8q_m$. By assuming $\mu \leq 2^n/8q_m$ (otherwise the bound becomes vacuously true) we choose $\xi = 2^n/8q_m$. Hence, the result follows by applying Eq. (2), Lemma 2, Eq. (26) and $\xi = 2^n/8q_m$. □

6.3   Security Bound Using the Coefficients-H Technique

We instantiate the underlying hash function of nEHtM by a truncated n-bit PolyHash function that truncates the first bit of the PolyHash output which is $2\ell/2^n$-axu hash function [14], where ℓ is the maximum number of message blocks. Therefore, from Lemma 2, Eq. (22) and the inequality $\sum_{i=1}^{k}\binom{W_i}{2} \leq \xi q_m$, we obtain the following bound using the coefficients-H technique.

$$\delta_{\mathrm{hc}} \leq \frac{q_m + 2q_v}{2^n} + \frac{q_m^2\ell}{2^n\xi} + \frac{(2q_m + q_v)2\ell\mu}{2^n} + \frac{2q_v\ell}{2^n} + \frac{24q_m^3\xi}{2^{2n}}. \tag{27}$$

We choose the optimal value of ξ such that the right hand side of the Eq. (27) gets maximized. To obtain such a value of ξ, we must have $\frac{q_m^2\ell}{2^n\xi} = \frac{24q_m^3\xi}{2^{2n}}$. By solving the equality for ξ, we obtain $\xi_{\mathrm{opt}} = \left(\frac{\ell 2^n}{24q_m}\right)^{\frac{1}{2}}$. Plugging-in this optimal value of ξ_{opt} into Eq. (27) gives

$$\delta_{\mathrm{hc}} \leq \frac{q_m + 2q_v}{2^n} + \frac{(2q_m + q_v)2\ell\mu}{2^n} + \frac{2q_v\ell}{2^n} + 10\left(\frac{q_m^5\ell}{2^{3n}}\right)^{\frac{1}{2}}.$$

The above bound holds true as long as $q \leq 2^{3n/5}/\ell^{1/5} \approx O(2^{3n/5})$, which is weaker than the bound $O(2^{2n/3})$ that we obtained using the expectation method.

7   Proof of Theorem 2

In this section we prove Theorem 2. Instead of separately proving the privacy and the authenticity result of the construction, we bound the distinguishing advantage of the two random systems: (i) the pair of oracles (CWC+.Enc, CWC+.Dec) for a random permutation Π, which is called the real system or the real world and (ii) the pair of oracles (Rand, Rej), which is called the ideal system or the ideal world. The privacy and authenticity bounds of CWC+ then follow as a simple corollary of this result. We prove the following information theoretic bound of CWC+.

$$\delta^* \leq \frac{97\sigma^3\ell}{2^{2n}} + \frac{5\sigma}{2^n} + \frac{\sigma\ell}{2^n} + \frac{8\sigma^3}{2^{2n}} + \frac{2q_d}{2^\rho}\left(1 + \frac{\ell}{2^{n-\rho}}\right) + \frac{(2q_e + q_d)2\ell\mu}{2^n} + \left(\frac{5\sigma\ell\mu}{2^n}\right)^2, \quad (28)$$

where δ^* is the maximum advantage in distinguishing the real world from the ideal world and we assume $q_e\ell \approx \sigma$, $\sigma \leq 2^n/48$. Due to limitations in space, we provide here only a sketch of the proof, and details may be found in [19].

DESCRIPTION OF THE IDEAL WORLD. We begin with the asumption that all the queried messages of an adversary are of length multiple of n and the number of blocks of i-th message is l_i. Now, we consider a deterministic distinguisher A that interacts either with the real world or with the ideal world. Rej simply rejects all the verification attempts of A whereas Rand, on the i-th encryption query (N_i, M_i, A_i) works as shown in Fig. 4.

Algorithm Rand(N_i, A_i, M_i)

    1. if $N_i \in \mathcal{D}$, let $N_i = N$
    2.   if $l_i = l_N$, then $S_i \leftarrow \mathcal{L}(N)$
    3.   if $l_i < l_N$, then $S_i \leftarrow \mathcal{L}(N)[1, nl_i]$
    4.   if $l_i > l_N$, then
    5. $R \leftarrow_\$ (\{0,1\}^n)^{l_i - l_N}, S_i \leftarrow \mathcal{L}(N)\|R$
    6. $l_N = l_i$
    7. else
    8. $S_i \leftarrow_\$ (\{0,1\}^n)^{l_i}, \mathcal{L}(N_i) \leftarrow S_i, l_{N_i} = l_i$
    9. $\mathcal{D} \leftarrow \mathcal{D} \cup \{N_i\}$
    10. $\widetilde{T}_i \leftarrow_\$ \{0,1\}^n; T_i \leftarrow \mathsf{chop}_\rho(\widetilde{T}_i)$
    11. **return** (S_i, T_i)

Fig. 4. Random oracle for the ideal world. l_N denotes the updated number of keystream blocks for nonce N and $\mathcal{L}(N)$ denotes the updated keystream blocks for nonce N of length l_N. $\mathcal{D}$ denotes the domain of the nonce. chop_ρ is a function that truncates the last $n - \rho$ bits of its input.

ATTACK TRANSCRIPT. Let D be a fixed non-trivial computationally unbounded deterministic distinguisher that interacts with either the real world or the ideal world, making at most q_e queries to the left (encryption) oracle with at most μ faulty nonces and at most q_d queries to its right (decryption) oracle, and returning a single bit.

Let $\tau_e \triangleq \{(N_1, M_1, A_1, S_1, T_1), \ldots, (N_{q_e}, M_{q_e}, A_{q_e}, S_{q_e}, T_{q_e})\}$ be the list of encryption queries and $\tau_d \triangleq \{(N_1', A_1', C_1', T_1', Z_1), \ldots, (N_{q_d}', A_{q_d}', C_{q_d}', T_{q_d}', Z_{q_d})\}$ be the list of decryption queries, where $Z_i = M_i \cup \{\perp\}$. Note that the encryption oracle in both the worlds releases the keystream as it determines the ciphertext uniquely. For convenience, we reveal the hash key K_h, which is $\mathsf{E}_K(0)$, if D interacts with the real world or a uniform random element from $\{0,1\}^n$, if D interacts with the ideal world, and also the n-bit tag (without truncating) i.e., $\mathbf{T} \triangleq (\widetilde{T}_1, \ldots, \widetilde{T}_{q_e})$ to the distinguisher after it made all its queries and obtains corresponding responses but before it output its decision and thus the extended query transcript of the attack is $\tau' = (\tau, K_h, \widetilde{\mathbf{T}})$, which is called the *extended transcript*.

BAD TRANSCRIPTS. Recall that N_i is a $3n/4$-bit string. We denote $0\|N_i\|0^{n/4-1}$ as $\widehat{N}_i$ and $1\|X_i$ as $\widehat{X}_i$, where $X_i \triangleq N_i\|0^{n/4-1} \oplus \mathsf{Poly}_{K_h}(M_i)$. Moreover, $S_i[j]$ denotes the j-th keystream block for i-th message. With these notations, we define the bad transcript as follows: a transcript $\tau' = (\tau_e, \tau_d, K_h, \widetilde{\mathbf{T}})$ is called **bad** if any one of the following holds:

- B.1 : $\exists\, i \in [q_e]$ and $j \in [l_i]$ such that $S_i[j] = K_h$.
- B.2 : $\exists\, i \in [q_e]$ and $j \in [l_i]$ such that $S_i[j] = \mathbf{0}$.
- B.3 : $\exists\, i \in [q_e]$ and $j, j' \in [l_i]$ such that $S_i[j] = S_i[j']$.
- B.4 : $\exists\, i \in [q_e]$ such that $\widetilde{T}_i = \mathbf{0}$.
- B.5 : $\exists i \neq j, j \neq k$ such that $\widehat{N}_i = \widehat{N}_j$ and $\widehat{X}_j = \widehat{X}_k$.
- B.6 : $\{i_1, \ldots, i_{\xi+1}\} \subseteq [q_e]$ such that $\widehat{X}_{i_1} = \widehat{X}_{i_2} = \ldots = \widehat{X}_{i_{\xi+1}}$ for some parameter $\xi \geq \mu$.
- B.7 : $\exists\, a \in [q_d], \exists\, i \in [q_e]$ such that $\widehat{N}_i = \widehat{N'}_a$, $\widehat{X}_i = \widehat{X'}_a$ and $\widetilde{T}_i = T_a'$.

Θ_{bad} (resp. Θ_{good}) denotes the set of bad (resp. good) transcripts. Moreover, X_{re} and X_{id} denotes the probability distribution of realizing an extended transcript τ' in the real and the ideal world respectively. We bound the probability of bad transcripts in the ideal world as follows.

Lemma 3. *Let X_{id} and Θ_{bad} be defined as above. Then*

$$\Pr[X_{\mathrm{id}} \in \Theta_{\mathrm{bad}}] \leq \epsilon_{\mathrm{bad}} = \frac{2\sigma}{2^n} + \frac{q_e \ell^2}{2^n} + \frac{q_e}{2^n} + \frac{q_e^2 \ell}{\xi 2^n} + \frac{(2q_e + q_d)2\ell\mu}{2^n} + \frac{2q_d \ell}{2^n}.$$

Proof of the lemma can be found in [19].

GOOD TRANSCRIPTS. Let $\tau' = (\tau_e, \tau_d, K_h, \widetilde{\mathbf{T}})$ be a good transcript. Since in the ideal world the encryption oracle is perfectly random and the decryption oracle always rejects, one simply has

$$\Pr[X_{\mathrm{id}} = \tau'] = \frac{1}{2^n} \cdot \prod_{t=1}^{r} \frac{1}{2^{nl_t}} \cdot \frac{1}{2^{nq_e}} \tag{29}$$

where r is the number of groups of nonces and l_t be the updated number of generated keystream blocks for group t.

REAL INTERPOLATION PROBABILITY. To bound the probability of getting τ' in the real world from below, we model the system of equations and non-equations into the graph theoretic setting to obtain the graph $G_{\tau'}$, where we have $\sigma + q_e$ equations and $2^{n-\rho}q_d$ non-equations. Similar to the analysis of good transcripts in the proof of Theorem 1, one can argue that as τ' is good, $G_{\tau'}$ is valid (i.e., it satisfies NC, NPL and NCL conditions). Thus, we assume $\xi \leq 2^n/8\sigma\ell$ (from the condition of Theorem 3), which allows us to apply Theorem 3 with $c = 1$, $\sigma_{i-1} \leq \sigma_k \leq 2\sigma$ and $\alpha \leq \sigma$ and then dividing by Eq. (29) we have,

$$\frac{\Pr[X_{\mathrm{re}} = \tau']}{\Pr[X_{\mathrm{id}} = \tau']} \geq 1 - \underbrace{\left(\sum_{i=1}^{k} \frac{24\sigma^2 \binom{W_i'}{2}}{2^{2n}} + \frac{2q_d}{2^\rho} + \frac{2\sigma}{2^n} \right)}_{\phi(\tau')}, \tag{30}$$

where k is the number of components of $G_{\tau'}$ and W_i' denotes the size of the i-th component. Note that there are $2^{n-\rho}q_d$ non-equations as the adversary forges with ρ bit tags T_a' and there are $2^{n-\rho}$ tags $\widetilde{T}$s whose first ρ bits match with T_a'. Moreover, we consider $c = 1$ due to the fact that we choose elements from the set $\{0,1\}^n$ excluding the hash key.

FINALIZING THE PROOF. We calculate the expectation of $\phi(\tau')$ as follows:

$$\mathbf{E}[\phi(X_{\mathrm{id}})] = \left(\frac{2q_d}{2^\rho} + \frac{2\sigma}{2^n} + \frac{24\sigma^2}{2^{2n}} \mathbf{E}\left[\sum_{i=1}^{k} \binom{W_i'}{2} \right] \right). \tag{31}$$

It is easy to see that $\binom{W_i'}{2} \leq \binom{W_i}{2}\binom{2\ell}{2}$, where W_i is defined in the proof of Theorem 1. Therefore from Eqs. (24) and (25),

$$\mathbf{E}\left[\sum_{i=1}^{k} \binom{W_i'}{2} \right] \leq \frac{2q_e^2\ell^3}{2^n} + \mu^2\ell^2 + 4q_e\ell^2, \tag{32}$$

where the almost xor universal probability of the truncated PolyHash is at most $2\ell/2^n$. Finally, from Eqs. (31) and (32) we obtain

$$\mathbf{E}[\phi(X_{\mathrm{id}})] \leq \left(\frac{2q_d}{2^\rho} + \frac{2\sigma}{2^n} + \frac{48\sigma^4\ell}{2^{3n}} + \left(\frac{5\sigma\ell\mu}{2^n} \right)^2 + \frac{96\sigma^3\ell}{2^{2n}} \right), \tag{33}$$

where we assume that $\ell q_e \approx \sigma$, the total number of message blocks queried.

FINALIZATION. We have assumed that $\xi \geq \mu$ and from the condition of Theorem 3, we have $\xi \leq 2^n/8\sigma\ell$. By assuming $\mu \leq 2^n/8\sigma\ell$ (otherwise the bound becomes vacuously true) we choose $\xi = 2^n/8\sigma\ell$. Hence, the bound stated in Eq. (28) follows by applying Eq. (2), Lemma 3, Eq. (33), $\xi = 2^n/8\sigma\ell$ and $\sigma \leq 2^n/48$. □

CONCLUDING THE PROOF OF THEOREM 2. The privacy bound of CWC+ is derived from Eq. (28) by setting $\mu = 0$ and the bound stated in Eq. (28) is itself the authenticity bound of CWC+.

Acknowledgements. Authors would like to thank all the reviewers of Eurocrypt, 2019.

References

1. CAESAR: Competition for authenticated encryption: Security, applicability, and robustness
2. Joux, A.: Comments on the draft GCM specification - authentication failures in NIST version of GCM
3. Aoki, K., Yasuda, K.: The security and performance of "GCM" when short multiplications are used instead. In: Kutyłowski, M., Yung, M. (eds.) Inscrypt 2012. LNCS, vol. 7763, pp. 225–245. Springer, Heidelberg (2013). https://doi.org/10.1007/978-3-642-38519-3_15
4. Ashur, T., Dunkelman, O., Luykx, A.: Boosting authenticated encryption robustness with minimal modifications. In: Katz, J., Shacham, H. (eds.) CRYPTO 2017. LNCS, vol. 10403, pp. 3–33. Springer, Cham (2017). https://doi.org/10.1007/978-3-319-63697-9_1
5. Bellare, M., Namprempre, C.: Authenticated encryption: relations among notions and analysis of the generic composition paradigm. In: Okamoto, T. (ed.) ASIACRYPT 2000. LNCS, vol. 1976, pp. 531–545. Springer, Heidelberg (2000). https://doi.org/10.1007/3-540-44448-3_41
6. Bellare, M., Tackmann, B.: The multi-user security of authenticated encryption: AES-GCM in TLS 1.3. In: Robshaw, M., Katz, J. (eds.) CRYPTO 2016. LNCS, vol. 9814, pp. 247–276. Springer, Heidelberg (2016). https://doi.org/10.1007/978-3-662-53018-4_10
7. Bernstein, D.J.: The Poly1305-AES message-authentication code. In: Gilbert, H., Handschuh, H. (eds.) FSE 2005. LNCS, vol. 3557, pp. 32–49. Springer, Heidelberg (2005). https://doi.org/10.1007/11502760_3
8. Bhattacharya, S., Nandi, M.: Revisiting variable output length XOR pseudorandom function. IACR Trans. Symmetric Cryptol. **2018**(1), 314–335 (2018)
9. Böck, H., Zauner, A., Devlin, S., Somorovsky, J., Jovanovic, P.: Nonce-disrespecting adversaries: practical forgery attacks on GCM in TLS. In: 10th USENIX Workshop on Offensive Technologies WOOT 16, Austin, TX, USA, 8–9 August 2016
10. Bose, P., Hoang, V.T., Tessaro, S.: Revisiting AES-GCM-SIV: multi-user security, faster key derivation, and better bounds. In: Nielsen, J.B., Rijmen, V. (eds.) EUROCRYPT 2018. LNCS, vol. 10820, pp. 468–499. Springer, Cham (2018). https://doi.org/10.1007/978-3-319-78381-9_18
11. Smith, B.: Pull request: Removing the AEAD explicit IV. mail to IETF TLS working group (2015)

12. Canetti, R., Krawczyk, H.: Analysis of key-exchange protocols and their use for building secure channels. In: Pfitzmann, B. (ed.) EUROCRYPT 2001. LNCS, vol. 2045, pp. 453–474. Springer, Heidelberg (2001). https://doi.org/10.1007/3-540-44987-6_28

13. Cogliati, B., Seurin, Y.: EWCDM: an efficient, beyond-birthday secure, nonce-misuse resistant MAC. In: Robshaw, M., Katz, J. (eds.) CRYPTO 2016. LNCS, vol. 9814, pp. 121–149. Springer, Heidelberg (2016). https://doi.org/10.1007/978-3-662-53018-4_5

14. Datta, N., Dutta, A., Nandi, M., Paul, G.: Double-block hash-then-sum: a paradigm for constructing BBB secure PRF. IACR Trans. Symmetric Cryptol. **2018**(3), 36–92 (2018)

15. Datta, N., Dutta, A., Nandi, M., Paul, G., Zhang, L.: Single key variant of PMAC_plus. IACR Trans. Symmetric Cryptol. **2017**(4), 268–305 (2017)

16. Datta, N., Dutta, A., Nandi, M., Yasuda, K.: Encrypt or decrypt? to make a single-key beyond birthday secure nonce-based MAC. In: Shacham, H., Boldyreva, A. (eds.) CRYPTO 2018. LNCS, vol. 10991, pp. 631–661. Springer, Cham (2018). https://doi.org/10.1007/978-3-319-96884-1_21

17. Datta, N., Dutta, A., Nandi, M., Yasuda, K.: Encrypt or decrypt? to make a single-key beyond birthday secure nonce-based MAC. Cryptology ePrint Archive, Report 2018/500 (2018)

18. Dutta, A., Jha, A., Nandi, M.: Tight security analysis of EHtM MAC. IACR Trans. Symmetric Cryptol. **2017**(3), 130–150 (2017)

19. Dutta, A., Nandi, M., Talnikar, S.: Beyond birthday bound secure MAC in faulty nonce model. Cryptology ePrint Archive, Report 2019/127 (2019)

20. Hoang, V.T., Tessaro, S.: Key-alternating ciphers and key-length extension: exact bounds and multi-user security. In: Robshaw, M., Katz, J. (eds.) CRYPTO 2016. LNCS, vol. 9814, pp. 3–32. Springer, Heidelberg (2016). https://doi.org/10.1007/978-3-662-53018-4_1

21. Hoang, V.T., Tessaro, S.: The multi-user security of double encryption. In: Coron, J.-S., Nielsen, J.B. (eds.) EUROCRYPT 2017. LNCS, vol. 10211, pp. 381–411. Springer, Cham (2017). https://doi.org/10.1007/978-3-319-56614-6_13

22. Iwata, T.: New blockcipher modes of operation with beyond the birthday bound security. In: Robshaw, M. (ed.) FSE 2006. LNCS, vol. 4047, pp. 310–327. Springer, Heidelberg (2006). https://doi.org/10.1007/11799313_20

23. Iwata, T.: Authenticated encryption mode for beyond the birthday bound security. In: Vaudenay, S. (ed.) AFRICACRYPT 2008. LNCS, vol. 5023, pp. 125–142. Springer, Heidelberg (2008). https://doi.org/10.1007/978-3-540-68164-9_9

24. Kohno, T., Viega, J., Whiting, D.: CWC: a high-performance conventional authenticated encryption mode. In: Roy, B., Meier, W. (eds.) FSE 2004. LNCS, vol. 3017, pp. 408–426. Springer, Heidelberg (2004). https://doi.org/10.1007/978-3-540-25937-4_26

25. Landecker, W., Shrimpton, T., Terashima, R.S.: Tweakable blockciphers with beyond birthday-bound security. In: Safavi-Naini, R., Canetti, R. (eds.) CRYPTO 2012. LNCS, vol. 7417, pp. 14–30. Springer, Heidelberg (2012). https://doi.org/10.1007/978-3-642-32009-5_2

26. McGrew, D.A., Viega, J.: The security and performance of the galois/counter mode (GCM) of operation. In: Canteaut, A., Viswanathan, K. (eds.) INDOCRYPT 2004. LNCS, vol. 3348, pp. 343–355. Springer, Heidelberg (2004). https://doi.org/10.1007/978-3-540-30556-9_27

27. Mennink, B., Neves, S.: Encrypted davies-meyer and its dual: towards optimal security using mirror theory. Cryptology ePrint Archive, Report 2017/473 (2017)

28. Mennink, B., Neves, S.: Encrypted davies-meyer and its dual: towards optimal security using mirror theory. In: Katz, J., Shacham, H. (eds.) CRYPTO 2017. LNCS, vol. 10403, pp. 556–583. Springer, Cham (2017). https://doi.org/10.1007/978-3-319-63697-9_19

29. Minematsu, K.: How to Thwart birthday attacks against MACs via small randomness. In: Hong, S., Iwata, T. (eds.) FSE 2010. LNCS, vol. 6147, pp. 230–249. Springer, Heidelberg (2010). https://doi.org/10.1007/978-3-642-13858-4_13

30. Minematsu, K., Iwata, T.: Building blockcipher from tweakable blockcipher: extending FSE 2009 proposal. In: Chen, L. (ed.) IMACC 2011. LNCS, vol. 7089, pp. 391–412. Springer, Heidelberg (2011). https://doi.org/10.1007/978-3-642-25516-8_24

31. Namprempre, C., Rogaway, P., Shrimpton, T.: Reconsidering generic composition. In: Nguyen, P.Q., Oswald, E. (eds.) EUROCRYPT 2014. LNCS, vol. 8441, pp. 257–274. Springer, Heidelberg (2014). https://doi.org/10.1007/978-3-642-55220-5_15

32. Nandi, M.: Birthday attack on dual EWCDM. Cryptology ePrint Archive, Report 2017/579 (2017). https://eprint.iacr.org/2017/579

33. Patarin, J.: The "Coefficients H" Technique. In: Avanzi, R.M., Keliher, L., Sica, F. (eds.) SAC 2008. LNCS, vol. 5381, pp. 328–345. Springer, Heidelberg (2009). https://doi.org/10.1007/978-3-642-04159-4_21

34. Patarin, J.: Introduction to mirror theory: Analysis of systems of linear equalities and linear non equalities for cryptography. IACR Cryptology ePrint Archive, 2010:287 (2010)

35. Jacques, P.: Security in $o(2^n)$ for the xor of two random permutations – proof with the standard H technique. IACR Cryptology ePrint Archive, 2013:368 (2013)

36. Patarin, J.: Mirror theory and cryptography. IACR Cryptology ePrint Archive, 2016:702 (2016)

37. Peyrin, T., Seurin, Y.: Counter-in-tweak: authenticated encryption modes for tweakable block ciphers. In: Robshaw, M., Katz, J. (eds.) CRYPTO 2016. LNCS, vol. 9814, pp. 33–63. Springer, Heidelberg (2016). https://doi.org/10.1007/978-3-662-53018-4_2

38. Rogaway, P.: Efficient instantiations of tweakable blockciphers and refinements to modes OCB and PMAC. In: Lee, P.J. (ed.) ASIACRYPT 2004. LNCS, vol. 3329, pp. 16–31. Springer, Heidelberg (2004). https://doi.org/10.1007/978-3-540-30539-2_2

39. Shoup, V.: On fast and provably secure message authentication based on universal hashing. In: Koblitz, N. (ed.) CRYPTO 1996. LNCS, vol. 1109, pp. 313–328. Springer, Heidelberg (1996). https://doi.org/10.1007/3-540-68697-5_24

40. Wegman, M.N., Carter, L.: New hash functions and their use in authentication and set equality. J. Comput. Syst. Sci. **22**(3), 265–279 (1981)

41. Zhang, P., Hu, H., Yuan, Q.: Close to optimally secure variants of GCM. Secur. Commun. Netw. **2018**, 9715947:1–9715947:12 (2018)

Tight Time-Memory Trade-Offs
for Symmetric Encryption

Joseph Jaeger[1]($\boxtimes$) and Stefano Tessaro[2]

[1] University of California, San Diego, La Jolla, USA
jsjaeger@eng.ucsd.edu
[2] University of Washington, Seattle, USA
tessaro@cs.washington.edu

Abstract. Concrete security proofs give upper bounds on the attacker's advantage as a function of its time/query complexity. Cryptanalysis suggests however that other resource limitations – most notably, the attacker's memory – could make the achievable advantage smaller, and thus these proven bounds too pessimistic. Yet, handling memory limitations has eluded existing security proofs.

This paper initiates the study of time-memory trade-offs for basic symmetric cryptography. We show that schemes like counter-mode encryption, which are affected by the Birthday Bound, become *more secure* (in terms of time complexity) as the attacker's memory is reduced.

One key step of this work is a generalization of the Switching Lemma: For adversaries with S bits of memory issuing q distinct queries, we prove an n-to-n bit random function indistinguishable from a permutation as long as $S \times q \ll 2^n$. This result assumes a combinatorial conjecture, which we discuss, and implies right away trade-offs for deterministic, stateful versions of CTR and OFB encryption.

We also show an unconditional time-memory trade-off for the security of *randomized* CTR based on a secure PRF. Via the aforementioned conjecture, we extend the result to assuming a PRP instead, assuming only one-block messages are encrypted.

Our results solely rely on standard PRF/PRP security of an underlying block cipher. We frame the core of our proofs within a general framework of indistinguishability for streaming algorithms which may be of independent interest.

Keywords: Provable security · Symmetric cryptography ·
Time-memory trade-offs

1 Introduction

Concrete security proofs upper bound the adversarial advantage ε as a function of the adversary's *resources*. A scheme is deemed secure if the advantage is small for all feasible resource amounts. The classical approach captures such resources in terms of *running time* and/or *description size*.

© International Association for Cryptologic Research 2019
Y. Ishai and V. Rijmen (Eds.): EUROCRYPT 2019, LNCS 11476, pp. 467–497, 2019.
https://doi.org/10.1007/978-3-030-17653-2_16

Time is however not the only resource to determine feasibility of an attack. In particular, the *memory* costs also matter – in the context of provable security, these were first studied by Auerbach et al. [4] and Wang et al. [26], who considered the tightness of reductions with respect to memory usage. Memory-tight reductions lift an assumed time-memory trade-off for the assumption to one for the scheme, and this is particularly important when the underlying assumption does not admit low-memory attacks (e.g., this is true for the LPN problem).

Earlier work on time-memory tradeoffs in symmetric cryptography focused on cryptanalytic attacks [5, 15] or precomputation attacks against primitives like hash functions [6].

SYMMETRIC CRYPTOGRAPHY. Memory tightness is less useful for symmetric cryptography: A typical assumption here is that AES is a PRP for attackers with large time complexity, e.g., $T = 2^{100}$, but the best generic attack is memoryless, so there is generally no trade-off to be assumed.

Still, time-memory trade-offs may affect the actual *modes of operation*. For example, it is well known that (randomized) counter mode (CTR$) allows to encrypt no more than $q = \sqrt{N}$ plaintexts when using an n-bit block cipher (here, $N = 2^n$), yet restricting memory to only store S bits may help. Indeed, let the i-th message m_i be encrypted as $(r_i, c_i = \text{AES}_K(r_i) \oplus m_i)$, where r_i is a random string. The *optimal* distinguishing attack waits for $r_i = r_j$ to occur for $i \neq j$, in which case $c_i \oplus c_j = m_i \oplus m_j$ – which is unlikely to hold if c_i and c_j are random. But this also requires remembering approximately $\sqrt{N}$ r_i's. If we can only store fewer of them, then we need a collision with one of the r_i's we remember, and the attack advantage decrease to $\frac{Sq}{N}$ when q messages are encrypted. However, is this attack the optimal one? – a proof would have to argue *over all possible* adversarial strategies storing S bits of partial information.

Remarkably, despite schemes like CTR$ being decades old, the question of proving bounds that take memory into account has remained open.

OUR RESULTS: OVERVIEW. This paper takes a ground-up approach to *proving* time-memory trade-offs. To this end, we start with exactly those simple symmetric encryption schemes like CTR$ and OFB we ought to understand, and develop proofs and proof techniques – mostly relying on information-theoretic and combinatorial tools – aimed at showing that conjectured trade-offs are optimal.

A common trait of basic encryption schemes is that they are only secure up to the Birthday Bound. For stateless, randomized schemes, this is because inputs to the block cipher are otherwise going to repeat. Also, even when inputs *are* distinct, non-repeating block-cipher outputs become easily distinguishable from random. We will show that this fact is no longer valid if the adversary's memory capacity does *not* exceed $\sqrt{N}$, and more generally, we show a trade-off between the number of encryptions and the attacker's memory.

For example, we revisit the well-known Switching Lemma in the memory-bounded setting: under a combinatorial conjecture (see details below), we show that an adversary making T *distinct* queries to a random function or a random permutation cannot tell them apart with advantage larger than $O(\sqrt{ST/N})$. The special case $S = T$ is the original switching lemma. This gives us bounds

Scheme	Underlying Primitive	Bound
CTR	PRF	ε_{prf}
	PRP	$\varepsilon_{prp} + \mathcal{O}_{sl}(T, S, N)$
OFB	PRF	Insecure when $T \in \Omega(\sqrt{N})$
	PRP	$\varepsilon_{prp} + \mathcal{O}_{sl}(T, S, N) + O(T/N)$
CTR\$	PRF	$\varepsilon_{prf} + O(\sqrt{ST/N})$
1-block CTR\$	weak-PRP	$\varepsilon_{wprp} + 3\mathcal{O}_{sl}(T, S, N)$
Encrypt-then-PRF	INDR and weak-PRF	$\varepsilon_{indr} + \varepsilon_{wprf} + O(\sqrt{ST/N})$

Fig. 1. Encryption schemes we analyze. Schemes with a \$ are randomized, otherwise they are deterministic. If Conjecture 1 holds then $\mathcal{O}_{sl}(T, S, N) \in O(\sqrt{ST/N})$. Bounds are for INDR security. S is the memory bound of the adversary, T is the number of blocks encrypted, and N is the domain size of the family of functions.

for stateful CTR and OFB, assuming the underlying block cipher is a sufficiently secure PRP. We consider the question fundamental enough to justify a partial answer even under a conjecture – moreover, the reduction to this conjecture is highly non-trivial, and a failure of the conjecture is likely to only minimally impact this bound.

We also show a bound of $O(\sqrt{ST\ell/N})$ for randomized CTR\$ based on a pseudorandom *function* (PRF), where ℓ is a bound on the number of blocks per encrypted message. This result does not need any conjecture, beyond PRF security. For the case $\ell = 1$, we show that under the aforementioned conjecture, the result holds when the scheme is based on a PRP, instead of a PRF.

An overview of our results for encryptions schemes is given in Fig. 1. We discuss them in more detail below, but first address an important piece of recent related work.

RELATED WORK. It is worth noting that our work complements a recent paper by Tessaro and Thiruvengadam [25]. Their goal are schemes with security *as high as possible*, well beyond 2^n (where n is the block length of the cipher), provided the cipher is secure enough (e.g., it has a long key), and adversarial memory is bounded. In their work, neither tightness nor practical efficiency is a concern. Here, in contrast, we focus on *tightness* for simple, deployed cryptography. As a result of this, we end up facing different, and somewhat more technically challenging problems.

A FRAMEWORK: STREAMING INDISTINGUISHABILITY. The common denominator of our security proofs is that they reduce to a new, yet natural, setting of memory-bounded streaming algorithms which we refer to as *streaming indistinguishability*. In essence, a memory-bounded algorithm $\mathcal{A}$ is given access, one value at a time, to one of two streams

$$X_1, X_2, \ldots \quad \text{or} \quad Y_1, Y_2, \ldots,$$

with different distributions. The goal is to distinguish them.

To the best of our knowledge, the existing literature on streaming algorithms does not consider this problem explicitly. Rather, the focus is mostly on worst-case complexity (we care about average-case), and search problems. However, one can cast classical problems like building PRGs against space-bounded read-once branching programs (cf. e.g. [21]), as a special case of this setting, where the X_i's are the output bits of the PRG and the Y_i's are random bits.

THE SWITCHING LEMMA. Let us first address our generalized Switching Lemma. It is well known that the advantage of a T-query distinguisher $\mathcal{A}$ trying to tell apart a truly random permutation P from a truly random function F (both from n bits to n bits) is at most T^2/N, which is tight. Also, an optimal distinguisher making $T \approx \sqrt{N}$ can be implemented to only use $S \ll \sqrt{N}$ bits, e.g., with the help of a memory-less collision-finding algorithms (e.g., using Pollard's ρ-method [23,24]). One uses the fact that when accessing P, the algorithm will never succeed in finding a collision.

One observation, however, is that in many useful scenarios, the resulting $\mathcal{A}$ never queries the same input *twice* and it is not hard to see that any memory-less collision-finding strategy *will* query the same input twice.

We show that, assuming the validity of a conjecture we explain next, under non-repeating queries, the Switching Lemma indeed holds with a tradeoff of the form $S \times T = N$. In fact, we prove a more general (and also fundamental) statement about the advantage of distinguishing two streams: The first, $X_1, X_2, \ldots$ samples n-bit values with replacement, the second, $Y_1, Y_2, \ldots$, without.

A CONJECTURE. A proof of a non-trivial bound appears out of reach. Instead, we give a proof that relies on a (plausible) combinatorial conjecture involving *hypergraphs*.

Recall that a k-hypergraph with N vertices is a collection $H = \{e_1, \ldots, e_m\}$, where the e_i's are distinct size-k subsets of $[N] = \{1, 2, \ldots, N\}$. The *degree* $d_H(i)$ of $i \in [N]$ is the number of e_j's such that $i \in e_j$. Then, we look at the maximum $D^2(m)$, over all m-edge hypergraphs H, of the function

$$D^2(H) = \sum_{i=1}^{N} d_H(i)^2.$$

Estimating $D^2(m)$ is challenging: The only known upper bound [9] is loose, and the general question is believed to be out of reach [16]. This is because degree sequences of hypergraphs are poorly understood, even more so when restricted to m edges. Only for the special case of graphs (i.e., $k = 2$) is the question well understood (cf. e.g. [1,10,14,20]), though far from trivial.

Our conjecture will be on the value of $D^2(m)$ when $k > N/2$ for *specific values of m*. We will assume in particular that if $m = \binom{A}{k}$, then the complete hypergraph containing all k-element subsets of $\{1, \ldots, A\}$ achieves $D^2(m)$. We stress that even a slight relaxation of this conjecture would only affect our proof slightly.

RANDOMIZED COUNTER MODE. The above switching lemma for distinct inputs only applies to stateful schemes. Let us look now instead at randomized CTR$

described above and, for simplicity, let us assume that we encrypt single-block plaintexts. Assuming the underlying block cipher is a PRF, the resulting security game can again be cast as a streaming (in)distinguishability setting with

$$X_i = (R_i, Z_i), \qquad Y_i = (R_i, F(R_i)),$$

where F is a random function from n bits to n bits and the R_i, Z_i's are random, independent n-bit strings. We will show a bound of $O(\sqrt{ST/N})$. Interesting, once cast in the right language, the proof is fairly elementary and uses only simple properties of Shannon entropies – what is novel here is the usage of these tools to prove the security of symmetric cryptography, and the fact that they are robust to dealing with memory restrictions.

In practice, of course, F is more likely to be a permutation, as it is built from a block cipher. However, our proof techniques seems not to extend directly to random permutations. We also cannot apply the Switching Lemma *directly*, because R_i's will not be distinct.

We will however do something different – we will apply the streaming indistinguishability result underlying the Switching Lemma to the R_i's first, telling us they can be replaced by random, distinct ones when encrypting single-block plaintexts. This will allow us to ultimately to replace F with a permutation – again by the Switching Lemma – but for a concrete bound, we will need to resort, again to our conjecture. (This can be thought, more generally, as extending the Switching Lemma to the case of random inputs.)

We could of course build a beyond-birthday secure PRF from a block cipher directly, e.g., using the xor construction [7,12,22], but this would require two block-cipher calls per block, or Iwata's CENC [17,18] for better amortized efficiency. We note that we also apply these techniques to analyze the confidentiality of Encrypt-then-PRF.

OUTLINE OF THIS PAPER. Section 2 introduces notation and provides necessary information theoretic and cryptographic background. Section 3.1 introduces our general streaming setting. Sections 3.2 and 4.1 introduce our main streaming theorems which are proven in Sects. 3.3 and 4.2, respectively. In Sects. 3.4 and 4.3 we apply these respective theorems to cryptographic reductions. We emphasize that while the analysis in Sect. 3 requires a conjecture, the results of Sect. 4 are unconditional.

2   Definitions

Let $\mathbb{N} = \{0, 1, 2, \dots\}$. For $N \in \mathbb{N}$ let $[N] = \{1, 2, \dots, N\}$. If S and S' are finite sets, then $\mathsf{Fcs}(S, S')$ denotes the set of all functions $F : S \to S'$ and $\mathsf{Perm}(S)$ denotes the set of all permutations on S. The set of size k subsets of S is $\binom{S}{k}$. Picking an element uniformly at random from S and assigning it to s is denoted by $s \xleftarrow{\$} S$. The set of finite vectors with entries in S is $(S)^*$ or S^*. Thus $\{0,1\}^*$ is the set of finite length strings.

If $M \in \{0,1\}^*$ is a string, then $|M|$ denotes its bitlenth. If $m \in \mathbb{N}$ and $M \in (\{0,1\}^m)^*$, then $|M|_m = |M|/m$ denotes the blocklength of M and M_i denote the i-th m-bit block of M. When using the latter notation, m will be clear from context. The empty string is ε.

Algorithms are randomized when not specified otherwise. If $\mathcal{A}$ is an algorithm, then $y \leftarrow \mathcal{A}^{O_1,\cdots}(x_1,\ldots;r)$ denotes running $\mathcal{A}$ on inputs $x_1,\ldots$ and coins r with access to oracles $O_1,\ldots$ to produce output y. The notation $y \xleftarrow{\$} \mathcal{A}^{O_1,\cdots}(x_1,\ldots)$ denotes picking r at random then running $y \leftarrow \mathcal{A}^{O_1,\cdots}(x_1,\ldots;r)$. The set of all possible outputs of $\mathcal{A}$ when run with inputs $x_1,\ldots$ is $[\mathcal{A}(x_1,\ldots)]$. Adversaries and distinguishers are algorithms. The notation $y \leftarrow O(x_1,\ldots)$ is used for calling oracle O with inputs $x_1,\ldots$ and assigning its output to y (even if the value assigned to y is not deterministically chosen).

Our cryptographic reductions will use pseudocode games (inspired by the code-based framework of [8]). See Fig. 2 for some example games. We let $\Pr[G]$ denote the probability that game G outputs true. The model underlying this pseudocode is the following formalism.

2.1 Model of Computation

COMPUTATIONAL MODEL. Our model is based on those of [2,3,25]. We consider a space-bounded adversary interacting with an oracle O.

The interaction between an adversary and oracle occurs over q stages. In the i-th stage, the adversary deterministically computes, as a function of the state σ_{i-1} and stage number i, a query x_i to O.[1] Then the adversary is give $y_i = O(x_i)$ (with the same inputs as before) based on which it computes the next state σ_i. The state σ_0 is fixed and defined by $\mathcal{A}$. The final output of $\mathcal{A}$ is σ_q. In code, stage i behaves as follows, **Stage** i: $x_i \leftarrow \mathcal{A}(i,\sigma_{i-1})$; $y_i \leftarrow O(x_i)$; $\sigma_i \xleftarrow{\$} \mathcal{A}(i,\sigma_{i-1},y_i)$.

COMPLEXITY MEASURES. An adversary $\mathcal{A}$ is S-bounded if $|\sigma_i| \leqslant S$ holds for all i. The running time of $\mathcal{A}$ is T if it queries at most T bits to its oracle. These complexity measures do not count the local state or time used by $\mathcal{A}$ during a round. This strengthens our main proofs which are information theoretic in nature and only require that the states σ_i and T are bounded in size.

Our applications of these main proofs will involve cryptographic reductions. These complexity measures are not appropriate for this because they could hide a weakness in a reduction that "cheats" by using much more local state or computation time during a round. None of our reductions have such a weakness so we leave reduction efficiency claims informal. See [4] for discussion of what conventions should be used for measuring the memory complexity of a reduction. Our reductions are given via explicit pseudocode so their complexity with respect to particular conventions can easily be extracted.

---

[1] We insist on this computation being deterministic for convenience and because we can think of x_i having been included as part of σ_{i-1}.

2.2   Information-Theoretic Preliminaries

ENTROPIES AND KL-DIVERGENCE. For probability distributions $P, Q : \mathcal{X} \to [0, 1]$ where $Q(x) > 0$ for all $x \in \mathcal{X}$, the Shannon and collision entropies are

$$H(P) = -\sum_{x \in \mathcal{X}} P(x) \log(P(x)) \text{ and } H_2(P) = -\log\left(\sum_{x \in \mathcal{X}} P(x)^2\right).$$

Statistical distance and KL-divergence are defined by

$$\mathsf{SD}(P, Q) = \frac{1}{2} \sum_{x \in \mathcal{X}} |P(x) - Q(x)| \text{ and } \mathsf{KL}(P\|Q) = \sum_{x \in \mathcal{X}} P(x) \log\left(\frac{P(x)}{Q(x)}\right).$$

Pinsker's inequality says that $\mathsf{SD}(P, Q) \leqslant \sqrt{\mathsf{KL}(P\|Q)/2}$.

As usual, for two random variables X and Y with distributions P_X and P_Y, we write $\mathsf{KL}(X\|Y)$ for $\mathsf{KL}(P_X\|P_Y)$ (and the analogous notation for H and H_2).

Lemma 1. *Let X, Y be random variables with range $\mathcal{X}$ with $\Pr[X = x] > 0$ for all $x \in \mathcal{X}$. Let $F : \mathcal{X} \to \{0,1\}^*$ be a (possibly randomized) function. Then,*

$$\mathsf{KL}(F(X)\|F(Y)) \leqslant \mathsf{KL}(X\|Y).$$

Proof. For compactness, denote $\mathsf{P}_Z(x) = \Pr[Z = x]$ for any random variable Z. First, we note that we can consider without loss of generality deterministic F's. Indeed, by convexity (cf. e.g. [11]),

$$\mathsf{KL}(F(X)\|F(Y)) \leqslant \sum_f \Pr[F = f] \cdot \mathsf{KL}(f(X)\|f(Y)).$$

Now fix a function $f : \mathcal{X} \to \{0,1\}^*$. From the log-sum inequality we obtain

$$\mathsf{KL}(F(X)\|F(Y)) = \sum_z \mathsf{P}_{F(X)}(z) \log\left(\frac{\mathsf{P}_{F(X)}(z)}{\mathsf{P}_{F(Y)}(z)}\right)$$

$$= \sum_z \left(\sum_{x \in f^{-1}(z)} \mathsf{P}_X(x)\right) \cdot \log\left(\frac{\sum_{x \in f^{-1}(z)} \mathsf{P}_X(x)}{\sum_{x \in f^{-1}(z)} \mathsf{P}_Y(x)}\right)$$

$$\leqslant \sum_z \sum_{x \in f^{-1}(z)} \mathsf{P}_X(x) \log\left(\frac{\mathsf{P}_X(x)}{\mathsf{P}_Y(x)}\right)$$

$$= \sum_{x \in \mathcal{X}} \mathsf{P}_X(x) \log\left(\frac{\mathsf{P}_X(x)}{\mathsf{P}_Y(x)}\right).$$

The last equality follows because every x is the pre-image of *exactly* one z.   □

Game $G_{F,b}^{prf}(\mathcal{A})$	Game $G_{F,b}^{prp}(\mathcal{A})$	Game $G_{SE,b}^{indr}(\mathcal{A})$
$K \xleftarrow{\$} F.K$	$K \xleftarrow{\$} F.K$	$\sigma \xleftarrow{\$} SE.Sg$
$F \xleftarrow{\$} Fcs(F.Dom, F.Rng)$	$P \xleftarrow{\$} Perm(F.Dom)$	$b' \xleftarrow{\$} \mathcal{A}^{Enc}$
$b' \xleftarrow{\$} \mathcal{A}^{Ror}$	$b' \xleftarrow{\$} \mathcal{A}^{Ror}$	Return $b' = 1$
Return $b' = 1$	Return $b' = 1$	
$Ror(X)$	$Ror(X)$	$Enc(M)$
$Y_1 \leftarrow F.Ev(K, X)$	$Y_1 \leftarrow F.Ev(K, X)$	$(\sigma, C_1) \leftarrow SE.E(\sigma, M)$
$Y_0 \leftarrow F(X)$	$Y_0 \leftarrow P(X)$	$C_0 \xleftarrow{\$} \{0,1\}^{|M|+SE.xl}$
Return Y_b	Return Y_b	Return C_b

Fig. 2. Security games for PRF/PRP security of a family of functions (Left/Middle) and INDR security of an encryption scheme (Right).

2.3 Cryptographic Preliminaries

FAMILY OF FUNCTIONS. A family of functions F specifies algorithms F.K and F.Ev (where the latter of these is deterministic) and sets F.Dom and F.Rng. Key generation algorithm F.K takes no input and outputs a key K. Evaluation algorithm takes as input key K and $X \in F.Dom$ to return $Y \in F.Rng$. We write $K \xleftarrow{\$} F.K$ and $Y \leftarrow F.Ev(K, X)$.

A blockcipher is a family of functions F for which F.Dom = F.Rng and for all $K \in [F.K]$ the function $F.Ev(K, \cdot)$ is a permutation with inverse $F.Inv(K, \cdot)$.

PSEUDORANDOMNESS SECURITY. For security we will consider both pseudorandom function (PRF) and pseudorandom permutation (PRP) security.

Let F be a family of functions. PRF security requires that $F.Ev(K, \cdot)$ looks like a truly random function to somebody who does not know K. Consider the game $G_{F,b}^{prf}(\mathcal{A})$ shown on the left side of Fig. 2. It parameterized by F, a bit $b \in \{0, 1\}$, and an adversary. The adversary is given access to an oracle ROR which on input X either returns F applied to X ($b = 1$) or the output of a random function on X ($b = 0$). The advantage of $\mathcal{A}$ against F is defined by $Adv_F^{prf}(\mathcal{A}) = Pr[G_{F,1}^{prf}(\mathcal{A})] - Pr[G_{F,0}^{prf}(\mathcal{A})]$.

PRP security of a blockcipher F is defined analogously by the game $G_{F,b}^{prp}(\mathcal{A})$ shown in the middle of Fig. 2. This is essentially the same except the random function $F \in Fcs(F.Dom, F.Rng)$ has been replaced by a random permutation $P \in Perm(F.Dom)$. The advantage of $\mathcal{A}$ against F is defined by $Adv_F^{prp}(\mathcal{A}) = Pr[G_{F,1}^{prp}(\mathcal{A})] - Pr[G_{F,0}^{prp}(\mathcal{A})]$.

SYMMETRIC ENCRYPTION. A symmetric encryption scheme SE specifies algorithms SE.Sg, SE.E, and SE.D (where the last of these is deterministic) and set SE.M. State generation algorithm takes no input and outputs state σ which will be used as the initial encryption state σ^e and decryption state σ^d. Encryption algorithm SE.E takes as input σ^e and message $M \in SE.M$. It outputs updated

state σ^e and ciphertext C. We assume there exists a constant expansion length
SE.xl $\in \mathbb{N}$ such that $|C| = |M| + $ SE.xl. Decryption algorithm SE.D takes as input
σ^d and ciphertext C. It outputs updated state σ^d and $M \in$ SE.M $\cup \{\bot\}$. We write
$\sigma \xleftarrow{\$}$ SE.Sg, $(\sigma^e, C) \xleftarrow{\$}$ SE.E (σ^e, M), and $(\sigma^d, M) \leftarrow$ SE.D (σ^d, C).

Correctness requires for all states $\sigma_0^e = \sigma_0^d \in$ [SE.Sg] and all sequences of
messages $\boldsymbol{M} \in$ (SE.M)* that $\Pr[\forall i : \boldsymbol{M}_i = \boldsymbol{M}_i'] = 1$ where the probability is
over the coins of encryption in the operations $(\sigma_i^e, \boldsymbol{C}_i) \xleftarrow{\$}$ SE.E $(\sigma_{i-1}^e, \boldsymbol{M}_i)$ and
$(\sigma_i^d, \boldsymbol{M}_i') \leftarrow$ SE.D $(\sigma_{i-1}^d, \boldsymbol{C}_i)$ for $i = 1, \ldots, |\boldsymbol{M}|$.

This non-standard syntax is used to simultaneously capture *stateful deterministic encryption* and *stateless probabilistic encryption*. For the first of these
SE.E is a deterministic algorithm. For the latter, σ^e and σ^d are equal to some
key K which is never updated.

ENCRYPTION SECURITY. For security we will require that the output of encryption look like a random string. Consider the game $\mathsf{G}^{\mathsf{indr}}_{\mathsf{SE},b}(\mathcal{A})$ shown on the right
side of Fig. 2. It is parameterized by a symmetric encryption scheme SE, adversary $\mathcal{A}$, and bit $b \in \{0, 1\}$. The adversary is given access to an oracle ENC which,
on input a message M, returns either the encryption of that message or a random
string of the appropriate length according to the secret bit b. The advantage of
$\mathcal{A}$ against SE is defined by $\mathsf{Adv}^{\mathsf{indr}}_{\mathsf{SE}}(\mathcal{A}) = \Pr[\mathsf{G}^{\mathsf{indr}}_{\mathsf{SE},1}(\mathcal{A})] - \Pr[\mathsf{G}^{\mathsf{indr}}_{\mathsf{SE},0}(\mathcal{A})]$.

3   The Switching Lemma

How hard is it for a memory-bounded distinguisher to tell apart a random function from a random permutation $[N] \to [N]$? It is easy to do so in a near-memory-less strategy with roughly $\sqrt{N}$ queries, where N is the domain size: The
distinguisher, given access to an oracle $[N] \to [N]$, mounts a classical memory-less collision finding attack – if the attack succeeds, the distinguisher is highly
certain it is interacting with a random function.

However, this attack requires querying the random function at the same
point *twice*. It is not clear if a distinguisher which never repeats a query can still
succeed with low memory and roughly $\sqrt{N}$ queries. We will show that it cannot. This boils down to bounding how well a memory-bounded can distinguish
between a sequence of random values and a sequence of random values without
repetition.

3.1   Streaming Indistinguishability

We consider a streaming setting, where a sequence of random variables

$$X_1, X_2, \ldots, X_q$$

with range $[N]$ is given, one by one, to a (memory-bounded) distinguisher $\mathcal{A}$, which is otherwise computationally unbounded. The distinguisher will need to tell apart this setting from another one, where it is given $(Y_1, Y_2, \ldots, Y_q)$ instead. We are interested in its distinguishing advantage. This is a very natural setting, but we are not aware of this having been considered explicitly.

THE STREAMING MODEL. More formally, in the i-th step (for $i \in [q]$), the distinguisher $\mathcal{A}$ has a state σ_{i-1} and stage number i. Then it asks for the value $V_i \in \{X_i, Y_i\}$ based on which it updates its state to σ_i. We write for notational convenience $\mathcal{A}(i, \sigma_{i-1}, V_i) = \sigma_i$, noting that this mapping can be randomized. We denote in particular $\Sigma_0, \Sigma_1, \ldots, \Sigma_q$ the states during the execution with X^q and $\Gamma_0, \Gamma_1, \ldots, \Gamma_q$ the states during the execution with Y^q. Here $\Sigma_0 = \Gamma_0$ is some a priori fixed value. For the final state (Σ_q or Γ_q) $\mathcal{A}$ outputs a bit, which we denote by $\mathcal{A}(X^q)$ and $\mathcal{A}(Y^q)$, respectively, and we are interested in its advantage

$$\mathsf{Adv}^{\mathsf{dist}}_{X^q, Y^q}(\mathcal{A}) = \Pr[\mathcal{A}(X^q) \Rightarrow 1] - \Pr[\mathcal{A}(Y^q) \Rightarrow 1].$$

It will sometime be convenient to think of this as an interaction between $\mathcal{A}$ and an oracle SAMP which returns V_i's according to one of these distributions (written as $b \xleftarrow{\$} \mathcal{A}^{\mathrm{SAMP}}$).

We will use the following lemma below, for the case where the X_i's are individually uniformly distributed.

Lemma 2. *Let $X^q = X_1, \ldots, X_q$ be independent and uniformly distributed. Then for any $Y^q = Y_1, \ldots, Y_q$,*

$$\mathsf{Adv}^{\mathsf{dist}}_{X^q, Y^q}(\mathcal{A}) \leqslant \frac{1}{\sqrt{2}} \sqrt{q \log N - \sum_{i=1}^{q} \mathsf{H}(Y_i \mid \Gamma_{i-1})}.$$

Proof. Since the final output bit is Σ_q and Γ_q, respectively, we can always upper bound the advantage by the statistical distance of these states, i.e.,

$$\mathsf{Adv}^{\mathsf{dist}}_{X^q, Y^q}(\mathcal{A}) \leqslant \mathsf{SD}(\Sigma_q, \Gamma_q) = \mathsf{SD}(\Gamma_q, \Sigma_q).$$

We will work in the regime of KL-divergence, and thus we also have

$$\mathsf{Adv}^{\mathsf{dist}}_{X^q, Y^q}(\mathcal{A}) \leqslant \frac{1}{\sqrt{2}} \sqrt{\mathsf{KL}(\Gamma_q \| \Sigma_q)}.$$

We note now that for all $i \in [q]$, by Lemma 1,

$$\mathsf{KL}(\Gamma_i \| \Sigma_i) = \mathsf{KL}(\mathcal{A}(i, \Gamma_{i-1}, Y_i) \| \mathcal{A}(i, \Sigma_{i-1}, X_i)) \leqslant \mathsf{KL}((\Gamma_{i-1}, Y_i) \| (\Sigma_{i-1}, X_i)).$$

Write $P(s,x) = \Pr[(\Sigma_{i-1}, X_i) = (s,x)]$, $P(s) = \Pr[\Sigma_{i-1} = s]$ and $P(x|s) = \Pr[X_i = x \mid \Sigma_{i-1} = s]$. Also define analogously $Q(s,x)$, $Q(s)$ and $Q(x|s)$ replacing (Σ_{i-1}, X_i) with (Γ_{i-1}, Y_i). Then,

$$
\begin{aligned}
\mathsf{KL}((\Gamma_{i-1}, Y_i) \| (\Sigma_{i-1}, X_i)) &= \sum_{s,x} Q(s,x) \log\left(\frac{Q(s,x)}{P(s,x)}\right) \\
&= \sum_{s,x} Q(s,x) \log\left(\frac{Q(s)}{P(s)}\right) + \sum_{s,x} Q(s,x) \log\left(\frac{Q(x|s)}{P(x|s)}\right) \\
&= \mathsf{KL}(\Gamma_{i-1} \| \Sigma_{i-1}) + \log N - \sum_s Q(s) \log\left(\frac{1}{Q(x|s)}\right) \\
&= \mathsf{KL}(\Gamma_{i-1} \| \Sigma_{i-1}) + \log N - \mathsf{H}(Y_i \mid \Gamma_{i-1}).
\end{aligned}
$$

Therefore, $\mathsf{KL}(\Gamma_q \mid S_q) \leqslant \mathsf{KL}(\Gamma_0 \| S_0) + q \log N - \sum_{i=1}^{q} \mathsf{H}(Y_i \| \Gamma_{i-1})$, and the lemma follows since $\mathsf{KL}(\Gamma_0 \| S_0) = 0$.   □

3.2 Sampling with and Without Replacement

Consider the streaming indistinguishability of the following natural distributions:

- SAMPLING WITH REPLACEMENT. In the distribution $X^q = (X_1, X_2, \ldots, X_q)$ the X_i's are independent and uniformly distributed over $[N]$.
- SAMPLING WITHOUT REPLACEMENT. In the distribution $Y^q = (Y_1, \ldots, Y_q)$ the Y_i's are sampled uniformly *without* repetition from $[N]$ (thus $q \leqslant N$).

We want to upper bound the advantage in distinguishing these two streams for a memory-bounded distinguisher $\mathcal{A}$ which receives these values one by one. We are going to show a time-memory trade-off for any distinguisher $\mathcal{A}$, assuming a conjecture that we now state. We will discuss the conjecture (and *why* this requires a conjecture) later in Sect. 3.5.

A CONJECTURE ON HYPERGRAPHS. A k-*uniform simple hypergraph* (or henceforth, simply, a k-hypergraph) with N vertices and m edges is a collection $H = \{e_1, e_2, \ldots, e_m\}$ of *distinct* subsets $e_i \subseteq [N]$, each of size k. Conventional graphs correspond to the case $k = 2$. The *degree* $d_H(i)$ of a vertex $i \in [N]$ is

$$
d_H(i) = |\{j \in [m] : i \in e_j\}|,
$$

i.e., the number of edges e_j containing i. By a double-counting argument we have $\sum_{i=1}^{N} d_H(i) = k \cdot m$. We will be interested in the following function of the degrees of a hypergraph,

$$
D^2(H) = \sum_{i=1}^{N} d_H(i)^2.
$$

For example, if H is the complete k-hypergraph, i.e., it contains all $\binom{N}{k}$ possible edges, $d_H(i) = \binom{N-1}{k-1}$ for all $i \in [N]$, and thus $D^2(H) = N \cdot \binom{N-1}{k-1}^2$.

Let $\mathcal{H}_{N,k}(m)$ be the set of all k-hypergraphs with N vertices and m edges. We define in particular,

$$D_{N,k}^2(m) = \max_{H \in \mathcal{H}_{N,k}(m)} D^2(H).$$

The behavior of $D_{N,2}^2(m)$ is fully characterized by a series of papers [1,10,14,20]. However, very little is known about $D_{N,k}^2(m)$ for general k. We will need the following conjecture.

Conjecture 1 (Main conjecture). Let $k > N/2$ and assume further that $m = \binom{A}{k}$ for some $A \geqslant k$. Then, the graph $H = \{e_1, ..., e_m\}$, where $e_1, \ldots, e_m$ are all size k subsets of $\{1, \ldots, A\}$, maximizes $D_{N,k}^2(m)$.

We refer the reader to Sect. 3.5 for an in-depth discussion of why we believe Conjecture 1 to be true, and why it is however hard to provide a full proof. We stress however that even weaker form of the conjecture (e.g., assuming that $D_{N,k}^2(m)$ is at most $(1 + 1/k)$ higher than the value achieved by the complete H) would not invalidate our bound below. Weakening even further would also simply result in a somewhat weaker bound.

INDISTINGUISHABILITY. We are going to now prove the following theorem.

Theorem 1. *Let N be given, $q < N/2$, and $20 \log(e) \leqslant S \leqslant N/4$. Further, let X^q be sampled with replacement and Y^q be sampled without replacement from $[N]$. Then, if Conjecture 1 holds, for every S-bounded distinguisher $\mathcal{A}$, we have*

$$\mathsf{Adv}_{X^q,Y^q}^{\mathsf{dist}}(\mathcal{A}) \leqslant \sqrt{\frac{S \cdot q}{N}}.$$

Let $\mathcal{O}_{\mathsf{sl}}(q, S, N)$ denote the best possible advantage over all S-bounded adversaries. The above result tells us that $\mathcal{O}_{\mathsf{sl}} \in O(\sqrt{S \cdot q/N})$. For the sake of generality our results which use Theorem 1 are stated in terms of $\mathcal{O}_{\mathsf{sl}}$.

3.3   Proof of Theorem 1

We are going to use Lemma 2, and therefore we are going to be concerned solely with showing a lower bound on $\mathsf{H}(Y_i \parallel \Gamma_{i-1})$ for all $i \in [q]$. This involves in particular a random experiment where (1) $Y_1, \ldots, Y_i$ are sampled, and (2) the state Γ_{i-1} is going to be produced, as a function of $Y_1, \ldots, Y_{i-1}$ only (which however, also of course depend on Y_i by being distinct from it).

INTERMEDIATE EXPERIMENT. We note that in the actual random experiment $\mathcal{A}$ has, when outputting Γ_{i-1}, information about $Y_1, \ldots, Y_{i-1}$ which is potentially incomplete, especially if Γ_{i-2} does not allow completely to remember $Y_1, \ldots, Y_{i-2}$, and so on. As a first simplification, we will remove this, and allow an adversary *full* information about $Y_1, \ldots, Y_{i-1}$ when attempting to produce a state Γ_{i-1} with the sole intent of making $\mathsf{H}(Y_i \mid \Gamma_{i-1})$ as small as possible. A second simplification is that, intuitively, the only information $Y_1, \ldots, Y_{i-1}$ give about Y_i is its range, i.e., the set of values Y_i can take.

In particular, for an adversary $\mathcal{B}$, consider the following experiment, producing variables (Y_i, Γ_{i-1}):

$$
\boxed{
\begin{aligned}
&- \text{ Sample } \mathcal{Y} \xleftarrow{\$} \binom{[N]}{N-i+1} \\
&- \text{ Let } \Gamma_{i-1} \xleftarrow{\$} \mathcal{B}(\mathcal{Y}) \\
&- Y_i \xleftarrow{\$} \mathcal{Y} \\
&- \text{ Return } (Y_i, \Gamma_{i-1})
\end{aligned}
}
$$

The additional constraint here is that $|\Gamma_{i-1}| \leqslant S$. Define now $\mathsf{H}^i(\mathcal{B}) = \mathsf{H}(Y_i \mid \Gamma_{i-1})$. We will show the following.

Lemma 3. *For all i, and S-bounded adversary $\mathcal{A}$, there exists a deterministic $\mathcal{B}$ outputting at most S bits such that*

$$
\mathsf{H}(Y_i \mid \Gamma_{i-1}) \geqslant \mathsf{H}^i(\mathcal{B}),
$$

where $\mathsf{H}(Y_i \mid \Gamma_{i-1})$ is with respect to the original experiment.

Proof. We first build a randomized adversary $\mathcal{A}'$ which given $\mathcal{Y}$ first samples a random shuffling $Y_1, \ldots, Y_{i-1}$ of the $i-1$ elements *not* in $\mathcal{Y}$, and then runs $\mathcal{A}$ over $i-1$ rounds feeding $Y_1, \ldots, Y_{i-1}$ to it, to produce Γ_{i-1}, which is then output by $\mathcal{A}'$. Clearly, by construction, $\mathsf{H}(Y_i \mid \Gamma_{i-1}) = \mathsf{H}^i(\mathcal{B})$.

To make $\mathcal{B}$ deterministic, let R be the random coins used by $\mathcal{A}'$, and observe that

$$
\mathsf{H}(Y_i \mid \Gamma_{i-1}) \geqslant \mathsf{H}(Y_i \mid \Gamma_{i-1}, R) = \mathop{\mathbf{E}}_{r \xleftarrow{\$} R} [\mathsf{H}(Y_i \mid \Gamma_{i-1}, R = r)].
$$

Define $\mathcal{B}$ by fixing the coins of $\mathcal{A}'$ to those minimizing $\mathsf{H}(Y_i \mid \Gamma_{i-1}, R = r)$. $\quad\square$

COLLISION ENTROPY AND PROBABILITIES. We take an extra final step to simplify our lower bound, and its connection with Conjecture 1. Namely, we will lower bound

$$
\mathsf{H}_2^i(\mathcal{B}) = \mathop{\mathsf{E}}_{\gamma \xleftarrow{\$} \Gamma_{i-1}} [\mathsf{H}_2(Y_i \mid \Gamma_{i-1} = \gamma)]
$$

since clearly $\mathsf{H}^i(\mathcal{B}) \geqslant \mathsf{H}_2^i(\mathcal{B})$. Also define

$$
\mathsf{Coll}^i(\mathcal{B}) = \mathop{\mathsf{E}}_{\gamma \xleftarrow{\$} \Gamma_{i-1}} \left[\sum_y \Pr[Y_i = y \mid \Gamma_{i-1} = \gamma]^2 \right].
$$

We note here that by Jensen's inequality,

$$
\mathsf{H}_2^i(\mathcal{B}) = \mathop{\mathsf{E}}_{\gamma \xleftarrow{\$} \Gamma_{i-1}} \left[-\log\left(\sum_y \Pr[Y_i = y \mid \Gamma_{i-1} = \gamma]^2 \right) \right] \geqslant -\log \mathsf{Coll}^i(\mathcal{B}),
$$

because $x \mapsto -\log(x)$ is a convex function. Therefore, the rest of the section will be devoted to proving an upper bound for $\mathsf{Coll}^i(\mathcal{B})$. Specifically, we show:

Lemma 4. *For all adversaries $\mathcal{B}$ outputting at most S bits, if Conjecture 1 is true,*

$$\mathsf{Coll}^i(\mathcal{B}) \leqslant \left(1 + \frac{2}{N}\right) \cdot \frac{1}{N - S}.$$

Before we turn to a proof, let us see how this implies the desired result. First off, it immediately implies by the above

$$\mathsf{H}(Y_i \mid \Gamma_{i-1}) \geqslant -\log \mathsf{Coll}^i(\mathcal{B})$$

$$\geqslant -\log\left(1 + \frac{2}{N}\right) + \log(N - S)$$

$$= -\log\left(1 + \frac{2}{N}\right) + \log(N) + \log\left(1 - \frac{S}{N}\right).$$

Now note that $\log(1+x) \leqslant \log(e^x) = x\log(e)$. On the other hand, using the fact that $x = S/N \leqslant 0.25$, we have

$$\log(1 - x) = \frac{1}{\ln 2}\ln(1 - x) \geqslant \frac{1}{\ln 2}\left(-x - x^2/2 - x^3/2\right) \geqslant \frac{-21x}{16\ln 2} \geqslant -1.9x$$

Plugging in gives,

$$\sum_{i=1}^{q} \mathsf{H}(Y_i \mid \Gamma_{i-1}) \geqslant q\left(-\frac{2\log(e)}{N} + \log(N) - \frac{1.9S}{N}\right).$$

Then using Lemma 2 we can complete the proof via

$$\mathsf{Adv}_{X^q,Y^q}^{\mathsf{dist}}(\mathcal{A}) \leqslant \frac{1}{\sqrt{2}}\sqrt{q\log N - \sum_{i=1}^{q}\mathsf{H}(Y_i \mid \Gamma_i)}$$

$$\leqslant \frac{1}{\sqrt{2}}\sqrt{q\left(\frac{2\log(e)}{N} + \frac{1.9S}{N}\right)}$$

$$\leqslant \frac{1}{\sqrt{2}}\sqrt{q\left(\frac{0.1S}{N} + \frac{1.9S}{N}\right)} = \sqrt{\frac{S \cdot q}{N}}.$$

PROOF OF LEMMA 4. We first introduce some more notation. For a k-hypergraph $H = \{e_1, \ldots, e_m\}$ with vertex set $[N]$ where $k := N - i + 1$, consider the distribution p_H which samples a $y \in [N]$ by first picking a random edge e_i, and then letting y be a random element of the set. In particular, $p_H(y) = d_H(y)/m \cdot k$. We also define

$$\mathsf{Coll}(H) = \sum_{y} p_H(y)^2 = \frac{1}{m^2 k^2}D^2(H).$$

Also, let $\mathsf{Coll}_{N,k}(m) = \max_{H \in \mathcal{H}_{N,k}(m)} \mathsf{Coll}(H)$.

Note now that $\mathcal{B}$ assigns sets of size k to every S-bit output γ. For a given γ, we can think of the sets assigned to it as a k-hypergraph, which we denote

$\mathcal{B}^{-1}(\gamma)$. Letting m_γ denote the number of edges in $\mathcal{B}^{-1}(\gamma)$ (and thus $\sum_\gamma m_\gamma = \binom{N}{k}$), we have

$$\mathsf{Coll}(\mathcal{B}) = \frac{1}{\binom{N}{k}} \sum_{\gamma \in \{0,1\}^S} m_\gamma \cdot \mathsf{Coll}(\mathcal{B}^{-1}(\gamma)) \leqslant \frac{1}{\binom{N}{k}} \sum_{\gamma \in \{0,1\}^S} m_\gamma \cdot \mathsf{Coll}_{N,k}(m_\gamma).$$
(1)

We are going to now maximize the right-hand-side of the above inequality over all sets $\{m_\gamma\}_{\gamma \in \{0,1\}^S}$, where $\sum_\gamma m_\gamma = \binom{N}{k}$, using Conjecture 1.[2] We need the following helping lemma, that $\mathsf{Coll}_{N,k}(m_\gamma)$ is a non-increasing function. Its proof is deferred to the full version of this paper [19].

Lemma 5. *For all $m \geqslant 1$, $\mathsf{Coll}_{N,k}(m+1) \leqslant \mathsf{Coll}_{N,k}(m)$.*

Unfortunately, the function $\mathsf{Coll}_{N,k}(m)$ is not "smooth", due to its discrete nature, making our maximization of the RHS of (1) difficult. We will now replace it with a continuous version without too much loss. Concretely, we define

$$A_{N,k}(m) = \frac{1}{\alpha},$$

where $\alpha \in [k, N]$ is the (unique) real number such that

$$\binom{\alpha}{k} = \frac{\alpha(\alpha-1)\cdots(\alpha-k+1)}{k!} = m.$$

We can now use the following lemma.

Lemma 6. *Assume Conjecture 1. For all $m \in \{1, 2, \ldots, \binom{N}{k}\}$, we have*

$$\mathsf{Coll}_{N,k}(m) \leqslant \left(1 + \frac{1}{k}\right) \cdot A_{N,k}(m).$$

Proof. Pick m, and let $m_0 \leqslant m \leqslant m_1$ such that $m_0 = \binom{A}{k}$ and $m_1 = \binom{A+1}{k}$ for a natural number $A \geqslant k$. Then, $A_{N,k}(m) = \frac{1}{\alpha}$ for some $\alpha \in [A, A+1]$, and using Lemma 5 and Conjecture 1,

$$\mathsf{Coll}_{N,k}(m) \leqslant \mathsf{Coll}_{N,k}(m_0) = \frac{1}{A} = \frac{\alpha}{A} A_{N,k}(m) \leqslant \frac{1+A}{A} A_{N,k}(m).$$

The claim follows, because $\frac{1+A}{A} \leqslant 1 + \frac{1}{k}$. □

Therefore, we can now adapt this to (1) as

$$\mathsf{Coll}(\mathcal{B}) \leqslant \left(1 + \frac{1}{k}\right) \frac{1}{\binom{N}{k}} \sum_{\gamma \in \{0,1\}^S} m_\gamma \cdot A_{N,k}(m_\gamma)$$

$$= \left(1 + \frac{1}{k}\right) \frac{1}{\binom{N}{k}} \sum_{\gamma \in \{0,1\}^S} B_{N,k}(m_\gamma),$$
(2)

where $B_{N,k}(m) = m \cdot A_{N,k}(m)$. To conclude the proof, we use the following two lemmas, whose proofs are deferred to the full version of this paper [19].

---

[2] Note that applying this conjecture requires $k > N/2$ which holds because $k = N - i + 1 \geqslant N - q + 1 > N - N/2 + 1$.

Lemma 7. *The function $B_{N,k}(m)$ is concave.*

Lemma 8. *For $N/2 \leqslant k \leq N - S$, we have $\binom{N}{k}/2^S \geqslant \binom{N-S}{k}$.*

Lemma 7 can now be applied to (2) to yield

$$
\begin{aligned}
\mathsf{Coll}(\mathcal{B}) &\leqslant \left(1 + \frac{1}{k}\right) \frac{2^S}{\binom{N}{k}} \frac{1}{2^S} \sum_{\gamma \in \{0,1\}^S} B_{N,k}(m_\gamma) \\
&\leqslant \left(1 + \frac{1}{k}\right) \frac{2^S}{\binom{N}{k}} B_{N,k} \left(\frac{1}{2^S} \sum_{\gamma \in \{0,1\}^S} m_\gamma\right) \\
&= \left(1 + \frac{1}{k}\right) \frac{2^S}{\binom{N}{k}} B_{N,k} \left(\binom{N}{k}/2^S\right) \\
&= \left(1 + \frac{1}{k}\right) \cdot A_{N,k} \left(\binom{N}{k}/2^S\right) \\
&\leqslant \left(1 + \frac{1}{k}\right) \cdot A_{N,k} \left(\binom{N-S}{k}\right) = \left(1 + \frac{1}{k}\right) \frac{1}{N-S},
\end{aligned}
\tag{3}
$$

where for the last inequality we have used Lemma 8 and the fact that $A_{N,k}(\cdot)$ is a non-increasing function.

3.4   Application: The Switching Lemma and Counter-Mode Encryption

THE SWITCHING LEMMA. A classic result in cryptography is the *switching lemma* which says roughly that for any blockcipher F and adversary $\mathcal{A}$ making at most q oracle queries, $\left|\mathsf{Adv}_F^{\mathsf{prf}}(\mathcal{A}) - \mathsf{Adv}_F^{\mathsf{prp}}(\mathcal{A})\right| < q^2/N$ where $N = |\mathsf{F.Dom}|$. The standard proof works by bounding the ability of $\mathcal{A}$ to distinguish a random function from a random permutation by analyzing the probability that the output of a random function repeats. When $\mathcal{A}$ does not repeat its oracle queries we can reduce this to the streaming problem we just analyzed this.

Lemma 9. *Let F be a blockcipher with $\mathsf{F.Dom} = [N]$. Let $\mathcal{A}$ be an S-bounded adversary which makes at most q non-repeating queries to its oracle. Then*

$$
|\mathsf{Adv}_F^{\mathsf{prf}}(\mathcal{A}) - \mathsf{Adv}_F^{\mathsf{prp}}(\mathcal{A})| \leqslant \mathcal{O}_{\mathsf{sl}}(q, S, N).
$$

If Conjecture 1 holds, then we can in turn bound $\mathcal{O}_{\mathsf{sl}}(q, S, N)$ by $\sqrt{S \cdot q/N}$ using Theorem 1. This would make the bound (and others in the section) essentially tight. If an attacker stores S outputs from its oracle, we expect it to see one of these outputs again from a random function after $T \approx N/S$ queries. For a random permutation such a repeat is impossible. In the full version of this paper [19] we provide the (simple) analysis for this attack.

Proof. Without loss of generality, assume that $\mathsf{Adv}^{\mathrm{dist}}_{X^q,Y^q}(\mathcal{A})$ is positive. We claim that $\Pr[\mathsf{G}^{\mathrm{prf}}_{\mathsf{F},0}(\mathcal{A})] = \Pr[\mathcal{A}(X^q) \Rightarrow 1]$ and $\Pr[\mathsf{G}^{\mathrm{prp}}_{\mathsf{F},0}(\mathcal{A})] = \Pr[\mathcal{A}(Y^q) \Rightarrow 1]$. Then the following calculation establishes the result.

$$
\begin{aligned}
|\mathsf{Adv}^{\mathrm{prf}}_{\mathsf{F}}(\mathcal{A}) - \mathsf{Adv}^{\mathrm{prp}}_{\mathsf{F}}(\mathcal{A})| &= |\Pr[\mathsf{G}^{\mathrm{prp}}_{\mathsf{F},0}(\mathcal{A})] - \Pr[\mathsf{G}^{\mathrm{prf}}_{\mathsf{F},0}(\mathcal{A})]| \\
&= |\Pr[\mathcal{A}(Y^q) \Rightarrow 1] - \Pr[\mathcal{A}(X^q) \Rightarrow 1]| \\
&= \mathsf{Adv}^{\mathrm{dist}}_{X^q,Y^q}(\mathcal{A}) \\
&\leqslant \mathcal{O}_{\mathrm{sl}}(q, S, N).
\end{aligned}
$$

The first equality used that games $\mathsf{G}^{\mathrm{prf}}_{\mathsf{F},1}(\mathcal{A})$ and $\mathsf{G}^{\mathrm{prp}}_{\mathsf{F},1}(\mathcal{A})$ are identical. □

COUNTER-MODE ENCRYPTION. Let F be a family of functions with $\mathsf{F}.\mathsf{Dom} = [N]$ for some $N \in \mathbb{N}$ and $\mathsf{F}.\mathsf{Rng} = \{0,1\}^{\mathsf{F}.\mathsf{ol}}$ for some $\mathsf{F}.\mathsf{ol} \in \mathbb{N}$. One classic example of an encryption mode constructed using F is *stateful counter-mode*. Formally this is the encryption scheme $\mathsf{CTR}[\mathsf{F}]$ with $\mathsf{CTR}[\mathsf{F}].\mathsf{M} = (\{0,1\}^{\mathsf{F}.\mathsf{ol}})^*$ and algorithms defined as shown below.

$\mathsf{CTR}[\mathsf{F}].\mathsf{Sg}$	$\mathsf{CTR}[\mathsf{F}].\mathsf{E}(\sigma^e, M)$	$\mathsf{CTR}[\mathsf{F}].\mathsf{D}(\sigma^d, C)$
$K \xleftarrow{\$} \mathsf{F}.\mathsf{K}$	$(i, K) \leftarrow \sigma^e$	$(i, K) \leftarrow \sigma^d$
Return $(0, K)$	For $j = 0, \ldots, \lvert M \rvert_{\mathsf{F}.\mathsf{ol}}$	For $j = 0, \ldots, \lvert C \rvert_{\mathsf{F}.\mathsf{ol}}$
	$\quad C_j \leftarrow M_j \oplus \mathsf{F}.\mathsf{Ev}(K, i + j)$	$\quad M_j \leftarrow C_j \oplus \mathsf{F}.\mathsf{Ev}(K, i + j)$
	$i \leftarrow i + \lvert M \rvert_{\mathsf{F}.\mathsf{ol}}$	$i \leftarrow i + \lvert C \rvert_{\mathsf{F}.\mathsf{ol}}$
	Return $((i, K), C)$	Return $((i, K), M)$

Here addition is mod N. It is trivial to show that if F is a good PRF then, $\mathsf{CTR}[\mathsf{F}]$ is a secure encryption scheme. Consider the following theorem. For simplicity we focus on the case that the attacker queries only 1 block messages.

Theorem 2. *Let F be given with $\mathsf{F}.\mathsf{Dom} = [N]$ and $\mathsf{F}.\mathsf{Rng} = \{0,1\}^{\mathsf{F}.\mathsf{ol}}$. Let $\mathcal{A}$ be an adversary making at most $q < N$ queries to its ENC oracle where each is $\mathsf{F}.\mathsf{ol}$ bits long. Then we can build an adversary $\mathcal{A}_{\mathrm{prf}}$ (Fig. 3) such that*

$$
\mathsf{Adv}^{\mathrm{indr}}_{\mathsf{CTR}[\mathsf{F}]}(\mathcal{A}) = \mathsf{Adv}^{\mathrm{prf}}_{\mathsf{F}}(\mathcal{A}_{\mathrm{prf}}).
$$

Adversary $\mathcal{A}_{\mathrm{prf}}$ is roughly as efficient as $\mathcal{A}$.

Proof. Let $\mathcal{A}_{\mathrm{prf}}$ be the adversary shown in Fig. 3. It uses its ROR oracle to simulate the view of $\mathcal{A}$. We claim that $\Pr[\mathsf{G}^{\mathrm{indr}}_{\mathsf{CTR}[\mathsf{F}],1}(\mathcal{A})] = \Pr[\mathsf{G}^{\mathrm{prf}}_{\mathsf{F},1}(\mathcal{A})]$ and $\Pr[\mathsf{G}^{\mathrm{indr}}_{\mathsf{CTR}[\mathsf{F}],0}(\mathcal{A})] = \Pr[\mathsf{G}^{\mathrm{prf}}_{\mathsf{F},0}(\mathcal{A})]$ from which the stated advantage relationship follows. The former equality holds because in both $\mathcal{A}$ is seeing $\mathsf{CTR}[\mathsf{F}]$ encryptions of M. For the latter equality note that the total block-length of all of $\mathcal{A}$'s queries is less than N so the input to the random function will never repeat. Consequently each value returned by ROR in $\mathsf{G}^{\mathrm{prf}}_{\mathsf{F},0}(\mathcal{A})$ (and thus each $C_j = M_j \oplus \mathrm{ROR}(i + j)$) is a fresh random string. This is identical to the distribution on C returned to $\mathcal{A}$ in $\mathsf{G}^{\mathrm{indr}}_{\mathsf{CTR}[\mathsf{F}],0}(\mathcal{A})$.

The efficiency of $\mathcal{A}_{\mathrm{prf}}$ can be verified by examining its pseudocode. □

Adversary $\mathcal{A}_{\mathsf{prf}}^{\mathrm{RoR}}$	SimEnc(M)
$i \leftarrow 0$	$C \leftarrow M \oplus \mathrm{RoR}(i)$
$b' \xleftarrow{\$} \mathcal{A}^{\mathrm{SimEnc}}$	$i \leftarrow i + 1$
Return b'	Return C

Fig. 3. Adversary for Theorem 2.

Suppose F is a blockcipher (where we identify $[N]$ with $\{0,1\}^{\mathsf{F.ol}}$ in the obvious way). If $q \in \Omega(\sqrt{N})$, then we cannot generically hope that $\mathsf{Adv}_{\mathsf{F}}^{\mathsf{prf}}(\mathcal{A}_{\mathsf{prf}})$ is small because an attacker with unbounded state can remember the outputs of F for every query it made and check if they ever repeated. However, if S is $o(\sqrt{N})$ then we can still meaningfully hope for security because $\mathcal{A}_{\mathsf{prf}}$ cannot remember ever query it made. In particular, by combining Theorem 2 and Lemma 9 we obtain the following corollary.

Corollary 1. *Let* F *be a blockcipher with* $\mathsf{F.Rng} = \{0,1\}^{\mathsf{F.ol}}$. *Let* $\mathcal{A}$ *be an* S-*bounded adversary making at most* $q \leqslant 2^{\mathsf{F.ol}}$ *queries to its* ENC *oracle each of which are* $\mathsf{F.ol}$ *bits long. Then we can build an adversary* $\mathcal{A}_{\mathsf{prf}}$ *(Fig. 3) such that*

$$\mathsf{Adv}_{\mathsf{CTR[F]}}^{\mathsf{indr}}(\mathcal{A}) \leqslant \mathsf{Adv}_{\mathsf{F}}^{\mathsf{prp}}(\mathcal{A}_{\mathsf{prf}}) + \mathcal{O}_{\mathsf{sl}}(q, S, 2^{\mathsf{F.ol}}).$$

Adversary $\mathcal{A}_{\mathsf{prf}}$ *is roughly as efficient as* $\mathcal{A}$.

Proving this requires only observing that $\mathcal{A}_{\mathsf{prf}}$ is S-bounded. Examining the code of $\mathcal{A}_{\mathsf{prf}}$ it may seem like it needs to remember the counter i and M in addition to the state of $\mathcal{A}$. However, as per the computation model in Sect. 2.1, the stage number is given to an adversary during each stage and the i-th message M_i can be deterministically recomputed from $\mathcal{A}$'s state σ_{i-1}.

OUTPUT-FEEDBACK MODE ENCRYPTION. In the full version of this paper [19] we apply our streaming results to analyze the security of stateful output-feedback mode. This mode starts with $Y_0 = 0^{\mathsf{F.ol}}$ and the encrypts each M_i via $Y_i \leftarrow \mathsf{F.Ev}(K, Y_{i-1})$; $C_i \leftarrow M_i \oplus Y_i$ where F is a blockcipher. The analysis of the mode is more involved than the CTR\$ analysis because we cannot a priori assume that the inputs to F will not repeat.

The crux of the proofs lies in considering the streaming problem of distinguishing $1, F(1), F(F(1)), \ldots$ from random where F is a random permutation $[N] \to [N]$. This is exactly what arises from the standard reduction to replace the PRF with a truly random function. In analyzing this streaming problem we first bound the statistical distance between the stated distribution and sampling without replacement. This gives a $O(q/N)$ term corresponding to the probability that 1 is chosen as the output of F for any of first q samples in the distribution. Having done this we can now simply apply a bound on the streaming problem we have been studying in this section. Putting everything together, the reduction from security of the encryption scheme to this new streaming problem is straightforward and gives a bound $\mathsf{Adv}_{\mathsf{OFB[F]}}^{\mathsf{indr}}(\mathcal{A}) = \mathsf{Adv}_{\mathsf{F}}^{\mathsf{prp}}(\mathcal{A}_{\mathsf{prp}}) + \mathcal{O}_{\mathsf{sl}}(q, S, 2^{\mathsf{F.ol}}) + 4q/N$.

Surprisingly, this result *cannot* hold for output-feedback mode with a PRF instead of a PRP. In the full version of this paper [19] we note a low memory attack that with high success probability when the number of encrypted blocks is $\Omega(\sqrt{N})$. The critical difference allowing this attack is that random functions have much shorter cycle lengths than random permutations. The importance of cycle lengths for OFB was first noted by Davies and Parkin [13].

NONCE-BASED ENCRYPTION. A standard way of constructing nonce-based encryption from a randomized encryption scheme is to apply a PRF to the nonce to obtain coins for the underlying encryption scheme. Because nonce repetitions are disallowed in the most basic security definitions for nonce-based encryption we can use Lemma 9 to replace the PRF with a PRP. The proof of this is straightforward and we omit a formalization.

3.5 Validity of Conjecture 1

We now discuss Conjecture 1. First off, we point out that the problem is well understood for the case of graphs, that correspond to $k = 2$.

Additionally, note that the conjecture is not true for all k. For example, take $k = 2$, $m = \binom{4}{2} = 6$ and $N \geqslant 7$. The complete graph over 4 vertices gives $D^2(K_4) = 4 \times 9 = 36$. Yet the star S_6 with edges $\{1,2\}, \{1,3\}, \ldots, \{1,7\}$ has $D^2(S_6) = 6^2 + 6 \times 1 = 42$. In fact, one can show that S_6 is optimal (see below).

THE CASE $k > N/2$. However, this is different for $k > N/2$, and we briefly explain the intuition, by giving an equivalent formulation of our conjecture. The first observation here is that for any k-hypergraph $H = \{e_1, \ldots, e_m\}$, we can define its complement as the $(N - k)$-hypergraph $H' = \{e'_1, \ldots, e'_m\}$, where $e'_i = [N] \setminus e_i$. Now, note that

$$D^2(H) = \sum_{i=1}^{N} d_H(i)^2 = \sum_{i=1}^{N} (m - d_{H'}(i))^2$$

$$= N \cdot m^2 - 2m \cdot \sum_{i=1}^{N} d_{H'}(i) + \sum_{i=1}^{N} d_{H'}(i)^2$$

$$= N \cdot m^2 - 2m^2(N - k) + D^2(H').$$

This in particular implies directly the following: H maximizes $D^2(H)$ over k-hypergraphs with m edges iff H' maximizes $D^2(H')$ over $(N - k)$-hypergraphs with m edges.

In general, if $m = \binom{A}{k}$ for $N/2 < k \leqslant A \leqslant N$, then our conjecture says that the complete k-hypergraph over $[A]$, denoted $K_{A,k}$, maximizes $D^2(H)$. We note that the complement of $K_{A,k}$ is (isomorphic to) $S_{N,N-A,N-k}$, where $S_{N,R,k'}$ for $k' > R$ is the k'-hypergraph with edges

$$\{1, \ldots, R\} \cup e,$$

and e is any subset of size $k' - R$ of $\{R + 1, \ldots, N\}$. Our conjecture is then equivalent to the statement that for any $k' < N/2$ and $m = \binom{A}{N-k'}$, the graph $H = S_{N,R,k'}$ for $R = N - A$ maximizes $D^2(H)$.

Example 1. The conjecture is easily seen to be true for $k = N - 2$, and we are given $m = \binom{N-1}{N-2}$ edges (this is the only non-trivial m). Then, $k' = 2$, and thus $S_{N,N-A,N-k} = S_{N,1,2} = S_N$, the graph which contains exactly all edges $\{i, N\}$ for $i \in [N - 1]$.

Now, we can see that $H = S_N$ maximizes $D^2(H)$. This is because for any k'-hypergraph $H = (e_1, \ldots, e_m)$, let $\mathbf{v}_1, \ldots, \mathbf{v}_m \in \{0, 1\}^N$ be the characteristic vectors of the edges, then

$$D^2(H) = \left(\sum_{i=1}^m \mathbf{v}_i \right)^T \left(\sum_{i=1}^m \mathbf{v}_i \right)$$

$$= \sum_{i=1}^m \mathbf{v}_i^T \mathbf{v}_i + 2 \sum_{i,j} \mathbf{v}_i^T \mathbf{v}_j$$

$$= m \cdot k' + 2 \sum_{i,j} |e_i \cap e_j|.$$

Clearly, for edges of size $k' = 2$, $|e_i \cap e_j|$ is at most 1, and S_N has the property that it is *exactly* one for any $i \neq j$.

The above example, showing the optimality of one simple special case, also shows our intuition. Namely, to maximize $m \cdot k' + 2 \sum_{i,j} |e_i \cap e_j|$, we make every pair of vertices share the highest number of possible vertices, i.e., $N - A$. The number of edges then exactly corresponds to the completion of all edges consisting of all subsets of size A of the remaining vertices.

DUAL GRAPH. We can repeat an analogous analysis of the dual graph of $H = \{e_1, \ldots, e_m\}$. We define this to be the k-hypergraph $\overline{H} = \binom{[N]}{k} \setminus H$. Now, note that

$$D^2(H) = \sum_{i=1}^N d_H(i)^2 = \sum_{i=1}^N \left(\binom{N}{k} - d_{H'}(i) \right)^2$$

$$= N \cdot \binom{N}{k}^2 - 2 \binom{N}{k}^2 (N - k) + D^2(H').$$

This implies that H maximizes $D^2(H)$ over k-hypergraphs with m edges iff H' maximizes $D^2(H')$ over k-hypergraphs with $\binom{N}{k} - m$ edges.

The complement of a k-hypergraph $K_{A,k}$ is isomorphic to $Z_{N,N-A,k}$, where $Z_{N,R,k}$ is the k-hypergraph with all edges $e \in \binom{[N]}{k}$ such that

$$\{1, \ldots, R\} \cap e \neq \emptyset.$$

Our conjecture is then equivalent to the statement that for any $k > N/2$ and $m = \binom{A}{k}$, the graph $H = Z_{N,R,k}$ for $R = N - A$ maximizes $D^2(H)$. Note when $k = 2$, the only S graphs are isomorphic to $S_{N,1,2} = Z_{N,1,2}$. Furthermore, when $k = 2$ for an appropriate generalization of complete graphs and Z graphs (covering when they do not "fits" perfectly for a given m) $D^2(H)$ is *always* maximized by a complete or Z graph.

Complete, S, and Z graphs are very natural ways to try to "pack" a hypergraph. Complete graphs create a uniform packing over a subset of the nodes with no overflow. Both S and Z graphs create very biased packings by making a small subset of the nodes have particularly high degree at the expense of a long tail of nodes that have low, but non-zero degree.

WHY PROVING IT IS HARD? One reason why proving the conjecture is hard is that we are maximizing a function over degree sequences $(d_1, \ldots, d_N)$ of hypergraphs. The structure of this set is however not well understood, even when dropping the restriction that we must have exactly m edges.

4   Randomized Encryption

The general streaming setting introduced in Sect. 3.1 can be used to derive time-memory tradeoff bounds for other encryption schemes by considering other distributions for X^q and Y^q. In this section we study randomized stateless encryption schemes (the only state is an unchanging secret key K). Our main positive result is for randomized counter-mode (CTR\$) with a good PRF. Towards this we start by (in Sect. 4.1) specifying the necessary streaming distribution for analyzing CTR\$. Analyzing this requires different techniques than those used in Sect. 3.3 and is done *unconditionally* (i.e. we do not rely on Conjecture 1).

Note that, unlike in the case of stateful counter-mode, security with a PRF is not trivial because the input to the function may repeat across different encryption queries. We show a $O(\sqrt{Spq/N})$ bound on the adversary's advantage where p is the length of the messages encrypted and q is the number of messages. Note that the running time of an adversary, T, upper bounds $p \cdot q$.

Beyond this we show a generic "switching lemma" between two notions of weak PRF security. In the first an adversary tries to distinguish between $(R, \mathsf{F.Ev}(K, R))$ and $(R, F(R))$ for randomly sampled R and F a random function $[N] \to [N]$. In the other notion, the latter distribution is replaced with (R, Y) where Y is chosen at random. The latter of these is more useful for security, but the former is more plausible achieved with good bounds. We show that there can be at most an $O(\sqrt{ST/N})$ difference between an adversary's advantage in these two games. As an example application of this result we note this can be used to provide a time-memory tradeoff for the INDR security of the Encrypt-then-PRF generic composition.

All of these bounds are essentially tight. If an attacker stores S input-output examples for F, we expect it to see one of these inputs again (allowing it to trivially distinguish from random) after $T \approx N/S$ queries.

4.1   Streaming Distributions for CTR$

Consider the streaming indistinguishability of the following two distributions.

- RAND$[N, M, p, q]$. The distribution $X^q = (X_1, X_2, \ldots, X_q)$ is such that the X_i's are independent and uniformly distributed over $[N] \times [M]^p$.
- CTR$[N, \mathcal{F}, p, q]$. For the distribution $Y^q = (Y_1, \ldots, Y_q)$ first a function F is sampled at random from $\mathcal{F}$. Then $Y_i = (R_i, F(R_i + 1), \ldots, F(R_i + p))$ where R_i's are are independent and uniformly distributed over $[N]$ and addition is modulo N.

To analyze CTR$ with a good PRF we will let $\mathcal{F} = \mathsf{Fcs}(N, M)$. Security with a good PRP could be modeled by letting $N = M$ and $\mathcal{F} = \mathsf{Perm}(N)$.

INDISTINGUISHABILITY. We are going to now prove the following theorem.

Theorem 3. *Let N, M, p, q, and S be given such that $p|N$. Furthermore, let $X^q = \text{RAND}[N, M, p, q]$ and $Y^q = \text{CTR\$}[N, \mathsf{Fcs}(N, M), p, q]$. Then for every S-bounded distinguisher $\mathcal{A}$, we have*

$$\mathsf{Adv}^{\mathsf{dist}}_{X^q, Y^q}(\mathcal{A}) \leqslant \frac{1}{\sqrt{2}} \sqrt{\frac{S \cdot p \cdot q}{N}}.$$

Note that unlike Theorem 1 we prove this result uncategorically, without requiring any conjectures.

For notational convenience we use bold-face to indicate vectors obtained by adding 1 through p to some value. For example, if $R \in [N]$ we will let $\boldsymbol{R} = (R + 1, \ldots, R + p)$. Further, we let $F(\boldsymbol{R}) = (F(R + 1), \ldots, F(R + p))$.

In the proof we will use the chain rule which says $H(X, Y) = H(X|Y) + H(Y)$. We also use that $H(X, Y \mid Z) \leqslant H(X \mid Z) + H(Y \mid Z)$ and $H(X) \leqslant \log \mathcal{X}$ where $\mathcal{X}$ is the support of X with equality when X is uniformly distributed over $\mathcal{X}$. These are standard facts about entropy.

4.2   Proof of Theorem 3

Associating the set $[N] \times [M]^p$ with $[N \cdot M^p]$ we can use Lemma 2 to obtain a bound of,

$$\mathsf{Adv}^{\mathsf{dist}}_{X^q, Y^q}(\mathcal{A}) \leqslant \frac{1}{\sqrt{2}} \sqrt{q \log(N \cdot M^p) - \sum_{i=1}^{q} \mathsf{H}(Y_i \mid \Gamma_i)}.$$

Therefore we are going to be concerned solely with showing a lower bound on $\mathsf{H}(Y_i \mid \Gamma_i)$ for all $i \in [q]$. Recall that Y_i is the tuple $(R_i, F(\boldsymbol{R}_i))$. The chain rule gives that $\mathsf{H}(Y_i \mid \Gamma_i) = H(F(\boldsymbol{R}_i) \mid R_i, \Gamma_i) + H(R_i \mid \Gamma_i)$.

Note that R_i is independent of Γ_i and uniformly sampled from $[N]$ so $H(R_i \mid \Gamma_i) = \log N$. Conditioning over all possible values of R_i gives

$$H(F(\boldsymbol{R}_i) \mid R_i, \Gamma_i) = N^{-1} \cdot \sum_{r \in [N]} H(F(\boldsymbol{r}) \mid \Gamma_{i-1}).$$

Observe that because p divides N the vectors $\boldsymbol{r}$ can be divided into p different partitions of $[N]$. That is for every $j \in [p]$, $\bigsqcup_{k \in [N/p]} \{j + kp + 1, \ldots, j + kp + p\} = [N]$. This observation allows us to continue our calculations as follows,

$$
\begin{aligned}
H(F(\boldsymbol{R}_i) \mid R_i, \Gamma_i) &= N^{-1} \cdot \sum_{j \in [p]} \sum_{k \in [N/p]} H(F(\boldsymbol{j} + \boldsymbol{kp}) \mid \Gamma_{i-1}) \\
&\geqslant N^{-1} \cdot p \cdot H(F \mid \Gamma_{i-1}) \\
&\geqslant N^{-1} \cdot p \cdot (H(F) - H(\Gamma_{i-1})) \\
&\geqslant N^{-1} \cdot p \cdot (N \log M - S)
\end{aligned}
$$

Thence
$$
\begin{aligned}
\sum_{i=1}^{q} \mathsf{H}(Y_i \mid \Gamma_i) &= \sum_{i=1}^{q} H(F(\boldsymbol{R}_i) \mid R_i, \Gamma_i) + H(R_i \mid \Gamma_i) \\
&\geqslant \sum_{i=1}^{q} N^{-1} \cdot p \cdot (N \log M - S) + \log N \\
&= q \log(N \cdot M^p) - Spq/N,
\end{aligned}
$$

from which the result follows. $\square$

4.3 Application: CTR\$ with a PRF and Weak PRFs

RANDOMIZED COUNTER-MODE. We can use Theorem 3 to prove a security result for randomized counter-mode encryption. Let F be a family of functions with $\mathsf{F}.\mathsf{Dom} = [N]$ and $\mathsf{F}.\mathsf{Rng} = \{0,1\}^{\mathsf{F}.\mathsf{ol}}$. Then randomized counter-mode with F is the encryption scheme $\mathsf{CTR\$}[\mathsf{F}]$ with state generation algorithm $\mathsf{CTR\$}[\mathsf{F}].\mathsf{Sg} = \mathsf{F}.\mathsf{K}$, message space $\mathsf{CTR\$}[\mathsf{F}].\mathsf{M} = (\{0,1\}^{\mathsf{F}.\mathsf{ol}})^*$, and encryption/decryption algorithms defined as shown below.

$\mathsf{CTR\$}[\mathsf{F}].\mathsf{E}(K, M)$	$\mathsf{CTR\$}[\mathsf{F}].\mathsf{D}(K, (R, C))$
$R \xleftarrow{\$} [N]$	For $i = 1, \ldots, \lvert C \rvert_{\mathsf{F}.\mathsf{ol}}$
For $i = 1, \ldots, \lvert M \rvert_{\mathsf{F}.\mathsf{ol}}$	$\quad M_i \leftarrow C_i \oplus \mathsf{F}.\mathsf{Ev}(K, R + i)$
$\quad C_i \leftarrow M_i \oplus \mathsf{F}.\mathsf{Ev}(K, R + i)$	Return (K, M)
Return $(K, (R, C))$	

Here $R + i$ is addition mod N. The standard security theorem for $\mathsf{CTR\$}[\mathsf{F}]$ tells us (roughly) that given an adversary $\mathcal{A}$ making q oracle queries we can construct a PRF adversary $\mathcal{A}_{\mathsf{prf}}$ such that $\mathsf{Adv}^{\mathsf{indr}}_{\mathsf{SE}}(\mathcal{A}) \leqslant \mathsf{Adv}^{\mathsf{prf}}_{\mathsf{F}}(\mathcal{A}_{\mathsf{prf}}) + p^2 q^2 / N$. Below is our theorem which takes space into account to provide a better bound when the amount of space used is much less than pq.

Adversary $\mathcal{A}_{\mathsf{prf}}^{\mathrm{RoR}}$	Distinguisher $\mathcal{A}_{\mathsf{dist}}^{\mathrm{SAMP}}$
$b' \xleftarrow{\$} \mathcal{A}^{\mathrm{SimEnc}}$	$b' \xleftarrow{\$} \mathcal{A}^{\mathrm{SimEnc}}$
Return b'	Return b'
$\underline{\mathrm{SimEnc}(M)}$	$\underline{\mathrm{SimEnc}(M)}$
$R \xleftarrow{\$} [N]$	$(R, V_1, \ldots, V_p) \leftarrow \mathrm{SAMP}$
For $i = 1, \ldots, \lvert M \rvert_{\mathsf{F.ol}}$ do	For $i = 1, \ldots, \lvert M \rvert_{\mathsf{F.ol}}$ do
$\quad C_i \leftarrow M_i \oplus \mathrm{RoR}(R + i)$	$\quad C_i \leftarrow M_i \oplus V_i$
Return (R, C)	Return (R, C)

Fig. 4. Adversary for Theorem 4.

Theorem 4. *Let* F *be a family of functions with* $\mathsf{F.Dom} = [N]$ *and* $\mathsf{F.Rng} = \{0,1\}^{\mathsf{F.ol}}$. *Let* $\mathcal{A}$ *be an* S-*bounded adversary making at most* q *queries with lengths at most* $p \cdot \mathsf{F.ol}$ *bits to its oracle. Assume* $p \mid N$. *Then we can build an adversary* $\mathcal{A}_{\mathsf{prf}}$ *(Fig. 4) such that*

$$\mathsf{Adv}_{\mathsf{CTR\$[F]}}^{\mathsf{indr}}(\mathcal{A}) \leqslant \mathsf{Adv}_{\mathsf{F}}^{\mathsf{prf}}(\mathcal{A}_{\mathsf{prf}}) + \frac{1}{\sqrt{2}}\sqrt{\frac{S \cdot p \cdot q}{N}}.$$

Adversary $\mathcal{A}_{\mathsf{prf}}$ *is roughly as efficient as* $\mathcal{A}$.

Proof. (of Theorem 4) Our proof begins with the PRF adversary $\mathcal{A}_{\mathsf{prf}}$ on the left side of Fig. 4. It simulates the view of $\mathcal{A}$ using its own oracle to provide $\mathcal{A}$ with the encryption of messages. Similarly the distinguisher $\mathcal{A}_{\mathsf{dist}}$ shown on the right side of Fig. 4 uses its sample oracle to simulate the view of $\mathcal{A}$.

The claim on the efficiency of $\mathcal{A}_{\mathsf{prf}}$ follow from examination of its code. Note that distinguisher $\mathcal{A}_{\mathsf{dist}}$ is S-bounded because it only needs to store the state of $\mathcal{A}$ during its oracle query (because M can be recomputed from this state).

We claim that the following equalities hold

$$\text{(i)} \ \Pr[\mathsf{G}_{\mathsf{F},1}^{\mathsf{prf}}(\mathcal{A}_{\mathsf{prf}})] = \Pr[\mathsf{G}_{\mathsf{CTR\$[F]},1}^{\mathsf{indr}}(\mathcal{A})],$$
$$\text{(ii)} \ \Pr[\mathsf{G}_{\mathsf{F},0}^{\mathsf{prf}}(\mathcal{A}_{\mathsf{prf}})] = \Pr[\mathcal{A}_{\mathsf{dist}}(Y^q) \Rightarrow 1],$$
$$\text{(iii)} \ \Pr[\mathcal{A}_{\mathsf{dist}}(X^q) \Rightarrow 1] = \Pr[\mathsf{G}_{\mathsf{CTR\$[F]},0}^{\mathsf{indr}}(\mathcal{A})].$$

Here we let $X^q = \mathrm{RAND}[N, 2^{\mathsf{F.ol}}, p, q]$ and $Y^q = \mathrm{CTR\$}[N, \mathsf{Fcs}(N, 2^{\mathsf{F.ol}}), p, q]$.

Claim (i) holds because in both games $\mathcal{A}$ is seeing encryptions of M using $\mathsf{CTR\$[F]}$. Claim (ii) holds because in both games $\mathcal{A}$ is seeing randomized counter-mode encryption of M using a random function F. Claim (iii) holds because in both games $\mathcal{A}$ is seeing random strings.

Game $G_{F,b}^{wprf}(\mathcal{A})$	RoR()
$K \xleftarrow{\$} F.K$	$X \xleftarrow{\$} F.Dom$
$F \xleftarrow{\$} Fcs(F.Dom, F.Rng)$	$Y_1 \leftarrow F.Ev(K, X)$
$b' \xleftarrow{\$} \mathcal{A}^{RoR}$	$Y_0 \leftarrow F(X)$
Return $b' = 1$	$Y_{-1} \xleftarrow{\$} F.Rng$
	Return (X, Y_b)

Fig. 5. Games defining weak pseudorandom function security of a family of functions.

The calculations are then as follows.

$$
\begin{aligned}
\mathsf{Adv}^{indr}_{\mathsf{CTR\$[F]}}(\mathcal{A}) &= \Pr[\mathsf{G}^{indr}_{\mathsf{CTR\$[F]},1}(\mathcal{A})] - \Pr[\mathsf{G}^{indr}_{\mathsf{CTR\$[F]},0}(\mathcal{A})] \\
&= \Pr[\mathsf{G}^{prf}_{F,1}(\mathcal{A}_{prf})] - \Pr[\mathcal{A}_{dist}(X^q) \Rightarrow 1] \\
&= \mathsf{Adv}^{prf}_{F}(\mathcal{A}_{prf}) - \mathsf{Adv}^{dist}_{X^q, Y^q}(\mathcal{A}_{dist}) \\
&\leqslant \mathsf{Adv}^{prf}_{F}(\mathcal{A}_{prf}) + \frac{1}{\sqrt{2}}\sqrt{\frac{S \cdot p \cdot q}{N}}.
\end{aligned}
$$

The final inequality follows by applying Theorem 3 with the distinguisher that outputs the bit $1 \oplus \mathcal{A}^{SAMP}_{dist}$. $\qquad\square$

WEAK PRF. Weak PRF security is a variant of PRF security where the game picks the input to the PRF at random for the adversary. Consider the game $G_{F,b}^{wprf}(\mathcal{A})$ shown in Fig. 5 when $b \in \{0,1\}$. The standard definition of WPRF security is $\mathsf{Adv}_{F}^{wprf}(\mathcal{A}) = \Pr[\mathsf{G}_{F,1}^{wprf}(\mathcal{A})] - \Pr[\mathsf{G}_{F,0}^{wprf}(\mathcal{A})]$. It asks that an adversary cannot distinguish between $F.Ev(K,X)$ and $F(X)$ when X is picked at random and F is a random function.

For proofs a different version of WPRF security is preferable. Consider the game $G_{F,-1}^{prf}(\mathcal{A})$. It differs from $G_{F,0}^{wprf}(\mathcal{A})$ because the RoR oracle returns a fresh random Y even if X's repeat. We define the advantage of $\mathcal{A}$ by $\mathsf{Adv}_{F}^{wprf2}(\mathcal{A}) = \Pr[\mathsf{G}_{F,1}^{prf}(\mathcal{A})] - \Pr[\mathsf{G}_{F,-1}^{prf}(\mathcal{A})]$. We call this WPRF2 security.

A family of functions is deterministic so its output will necessarily repeat on repeated inputs. Thus we can expect better security for the first definition. It is then useful to assume good WPRF security and have a generic proof that WPRF2 security cannot differ from it too much. It is straightforward to show, for example, that $|\mathsf{Adv}_{F}^{wprf}(\mathcal{A}) - \mathsf{Adv}_{F}^{wprf2}(\mathcal{A})| \leqslant q^2/N$. Using our space-bounded techniques we can show the following theorem which improves the bound when the space used by $\mathcal{A}$ is less than the number of queries it makes.

Lemma 10. *Let F be a family of functions with $F.Dom = [N]$. Let $\mathcal{A}$ be an S-bounded adversary making at most q queries to its oracle. Then*

$$
\left|\mathsf{Adv}_{F}^{wprf}(\mathcal{A}) - \mathsf{Adv}_{F}^{wprf2}(\mathcal{A})\right| \leqslant \frac{1}{\sqrt{2}}\sqrt{\frac{S \cdot q}{N}}.
$$

Proof. First note that $|\mathsf{Adv}_\mathsf{F}^{\mathsf{wprf}}(\mathcal{A}) - \mathsf{Adv}_\mathsf{F}^{\mathsf{wprf2}}(\mathcal{A})| = |\Pr[\mathsf{G}_{\mathsf{F},-1}^{\mathsf{wprf}}] - \Pr[\mathsf{G}_{\mathsf{F},0}^{\mathsf{wprf}}(\mathcal{A})]|$ and suppose without loss of generality that this difference in probabilities is positive. Identify F.Rng with $[M]$. In game $\mathsf{G}_{\mathsf{F},-1}^{\mathsf{wprf}}$ the adversary is being given uniformly random samples $(X, Y) \xleftarrow{\$} [N] \times [M]$ and in game $\mathsf{G}_{\mathsf{F},0}^{\mathsf{wprf}}(\mathcal{A})$ it is seeing the same subject to the fact that Y will repeat whenever X does. These views are exactly identical to the view of a distinguisher in the setting of Theorem 3. Applying that result gives the state bound.   □

4.4   CTR$ with a PRP and Weak PRPs

In practice most encryption uses AES - a blockcipher with domain $\{0,1\}^{128}$ which is thus best modeled as a PRP. We do not know how to extend our CTR$ analysis for this case. Our streaming analysis with a random function F used that $H(F) = \log(M^N)$. If F is a random permutation then $H(F) = \log(N!)$ which is not sufficiently large. However, when only one block messages are encrypted, we can using the streaming problem addressed in Sect. 3 to bound the advantage by $O(\mathcal{O}_{\mathsf{sl}})$.

Security of CTR$ for one block messages corresponds closely to the WPRF2 security of the underlying blockcipher. Thus we divide the CTR$ proof into three steps. First we use Theorem 1 to obtain a bound in the streaming setting naturally induced by this problem. Next we use this to prove a generic "switching lemma" between Weak PRP (WPRP) security (defined momentarily) and WPRF2 security analogous to Lemma 10. The security of CTR$ for one block messages follows from this lemma in a straightforward way. The streaming analysis will be presented in full here. The WPRP and CTR$ results are stated, but the (straightforward) proofs are deferred to the full version of this paper [19].

WEAK PRP. WPRP security is defined via the games $\mathsf{G}_{\mathsf{F},b}^{\mathsf{wprp}}$ shown in Fig. 6. The advantage of an adversary $\mathcal{A}$ against blockcipher F is defined by $\mathsf{Adv}_\mathsf{F}^{\mathsf{wprp}}(\mathcal{A}) = \Pr[\mathsf{G}_{\mathsf{F},1}^{\mathsf{wprp}}(\mathcal{A})] - \Pr[\mathsf{G}_{\mathsf{F},0}^{\mathsf{wprp}}(\mathcal{A})]$. The notion is essentially the same as for WPRF security, except the random function has been replaced with a random permutation.

The following lemma bounds the difference between an adversary's WPRP and WPRF2 advantages, allowing one to generically switch between the two. It is an almost immediate implication of the coming streaming analysis.

Lemma 11. *Let F be a family of functions with $\mathsf{F.Dom} = \mathsf{F.Rng} = [N]$. Let $\mathcal{A}$ be an S-bounded adversary making at most q queries to its oracle. Then*

$$\left| \mathsf{Adv}_\mathsf{F}^{\mathsf{wprp}}(\mathcal{A}) - \mathsf{Adv}_\mathsf{F}^{\mathsf{wprf2}}(\mathcal{A}) \right| \leqslant 3\mathcal{O}_{\mathsf{sl}}(q, S, N).$$

RANDOMIZED COUNTER-MODE. The following theorem (proved using Lemma 11) bounds the advantage of an attacker against CTR$ with a blockcipher by the WPRP security of the blockcipher when only one block messages are encrypted.

Game $G_{F,b}^{\mathsf{wprp}}(\mathcal{A})$	ROR()
$K \xleftarrow{\$} \mathsf{F.K}$	$X \xleftarrow{\$} \mathsf{F.Dom}$
$F \xleftarrow{\$} \mathsf{Perm}(\mathsf{F.Dom})$	$Y_1 \leftarrow \mathsf{F.Ev}(K, X)$
$b' \xleftarrow{\$} \mathcal{A}^{\mathrm{ROR}}$	$Y_0 \leftarrow F(X)$
Return $b' = 1$	Return (X, Y_b)

Fig. 6. Games for weak pseudorandom permutation security of a family of functions.

Theorem 5. *Let F be a blockcipher with $\mathsf{F.Dom} = \mathsf{F.Rng} = \{0,1\}^n$. Let $\mathcal{A}$ be an S-bounded adversary making at most q queries of length n to its oracle. Then we can build an adversary $\mathcal{A}_{\mathsf{wprp}}$ such that*

$$\mathsf{Adv}_{\mathsf{CTR\$[F]}}^{\mathsf{indr}}(\mathcal{A}) \leqslant \mathsf{Adv}_F^{\mathsf{wprp}}(\mathcal{A}_{\mathsf{wprp}}) + 3\mathcal{O}_{\mathsf{sl}}(q, S, 2^n).$$

Adversary $\mathcal{A}_{\mathsf{wprp}}$ is roughly as efficient as $\mathcal{A}$.

STEAMING ANALYSIS. In the streaming setting we now analyze $\mathcal{A}$ is given repeated samples (R_i, P_i) where P_i is either random or $F(R_i)$ for a random $F \in \mathsf{Perm}(N)$. We first use $\mathcal{O}_{\mathsf{sl}}$ to switch to R_i being picked without replacement. Now $P_i = F(R_i)$ can be viewed as random samples without replacement; we use $\mathcal{O}_{\mathsf{sl}}$ again to switch P_i to being sampled with replacement. Then we use $\mathcal{O}_{\mathsf{sl}}$ a final time to switch R_i back to being picked with replacement.

Lemma 12. *Let N, q, and S be given. Further, let $W^q = \mathrm{RAND}[N, N, 1, q]$ and $V^q = \mathrm{CTR\$}[N, \mathsf{Perm}(N), 1, q]$. Then for every S-bounded distinguisher $\mathcal{A}$, we have*

$$\mathsf{Adv}_{W^q, V^q}^{\mathsf{dist}}(\mathcal{A}) \leqslant 3\mathcal{O}_{\mathsf{sl}}(q, S, N).$$

Proof. Consider the sequence of game G_0 through G_4 shown in Fig. 7.

In game G_0, each R_i is uniformly and independently sampled and $P_i = F(R_i)$ where F is a random permutation. This is exactly the distribution V^q so $\Pr[G_0] = \Pr[\mathcal{A}(V^q) \Rightarrow 1]$. In game G_4, each R_i and each P_i are uniformly and independently sampled. This is exactly the distribution W^q so $\Pr[G_4] = \Pr[\mathcal{A}(W^q) \Rightarrow 1]$. We can then see that,

$$\mathsf{Adv}_{W^q, V^q}^{\mathsf{dist}}(\mathcal{A}) = \sum_{i=1}^{4} \Pr[G_i] - \Pr[G_{i-1}]$$

Let X^q be sampling with replacement and Y^q be sampling without replacement from $[N]$. We will bound the difference between G_0 and G_4 by using a sequence of distinguishers for (X^q, Y^q), whose advantages we bound with $\mathcal{O}_{\mathsf{sl}}$.

The distinguishers are shown below, where $R_{<i} = \{R_1, \ldots, R_{i-1}\}$. As written, distinguishers $\mathcal{A}_{0,1}$ and $\mathcal{A}_{1,2}$ store large amounts of space. The former stores an entire random permutation $F : [N] \to [N]$. The latter stores a list of q different R_i values. Used naively, this would result in useless advantage bounds.

Games G_0,G_1,G_2,G_3,G_4	$\textsc{Samp}()$
$F \overset{\$}{\leftarrow} \mathrm{Perm}(N)$ $/\!/$ G_0, G_1	$R_i \overset{\$}{\leftarrow} [N]$ $/\!/$ G_0,G_4
$F \overset{\$}{\leftarrow} \mathrm{Fcs}(N,N)$ $/\!/$ G_2	$R_i \overset{\$}{\leftarrow} [N] \smallsetminus \{R_1,\ldots,R_{i-1}\}$ $/\!/$ G_1, G_2, G_3
$i \leftarrow 1$	$P_i \leftarrow F(R_i)$ $/\!/$ G_0, G_1, G_2
$b' \overset{\$}{\leftarrow} \mathcal{A}^{\textsc{Samp}}$	$P_i \overset{\$}{\leftarrow} [N]$ $/\!/$ G_3, G_4
Return $b' = 1$	$i \leftarrow i+1$
	Return (R_i, P_i)

Fig. 7. Games for proof of Lemma 12. Commented lines of code are only included in the indicated games.

However, note that the stored state is sampled *before any oracle queries are made*. Thus we can use a standard coin-fixing argument to upper bound the advantage of these distinguishers by the advantage of distinguishers $\mathcal{A}_{0,1}^*$ and $\mathcal{A}_{1,2}^*$ for which the best choices of F and the R_i values are hardcoded.

The description size of a distinguisher is not included in the bound of their state so we can see that $\mathcal{A}_{0,1}^*$ is S-bounded, $\mathcal{A}_{1,2}^*$ is S-bounded, and $\mathcal{A}_{3,4}$ is S-bounded. Note that $\mathcal{A}_{1,2}^*$ does not need to store the stage counter i for itself because this is provided as input as part of our streaming.

Distinguisher $\mathcal{A}_{0,1}^{\textsc{Samp}}$	Distinguisher $\mathcal{A}_{1,2}^{\textsc{Samp}}$	Distinguisher $\mathcal{A}_{3,4}^{\textsc{Samp}}$
$F \overset{\$}{\leftarrow} \mathrm{Perm}(N)$	For $i = 1,\ldots,q$ do	$b' \overset{\$}{\leftarrow} \mathcal{A}^{\textsc{SimEnc}}$
$b' \overset{\$}{\leftarrow} \mathcal{A}^{\textsc{SimSamp}}$	$\quad R_i \overset{\$}{\leftarrow} [N] \smallsetminus R_{<i}$	Return b'
Return $1 \oplus b'$	$i \leftarrow 1$	
	$b' \overset{\$}{\leftarrow} \mathcal{A}^{\textsc{SimEnc}}$	$\textsc{SimSamp}()$
$\textsc{SimSamp}()$	Return b'	$R \leftarrow \textsc{Samp}$
$R \leftarrow \textsc{Samp}$		$P \overset{\$}{\leftarrow} [N]$
$P \leftarrow F(R)$	$\textsc{SimSamp}()$	Return (R, P)
Return (R, P)	$P \leftarrow \textsc{Samp}$	
	$i \leftarrow i+1$	
	Return (R_i, P)	

Now consider the transition from G_0 to G_1. They differ in whether R_i is sampled with or without replacement. Distinguisher $\mathcal{A}_{0,1}$ tries to use this difference to distinguish between X^q and Y^q using its samples to set R_i and simulating $P = F(R)$ for itself. We have $\Pr[G_1] - \Pr[G_0] = \mathrm{Adv}_{X^q,Y^q}^{\mathrm{dist}}(\mathcal{A}_{0,1})$. Note that $\mathcal{A}_{0,1}$ outputs the bit $1 \oplus b'$ to give the order we want.

Games G_1 and G_2 differ only in whether F is a random permutation or random function. Because they are being fed non-repeating input the values $P_i = F(R_i)$ are distributed according to Y^q in the former case and X^q in the latter. Consequently, we can see that $\Pr[G_2] - \Pr[G_1] = \mathrm{Adv}_{X^q,Y^q}^{\mathrm{dist}}(\mathcal{A}_{1,2})$.

Games G_2 and G_3 are equivalent. They differ in whether each P_i is by $P_i \overset{\$}{\leftarrow} [N]$ or as $F(R_i)$ for a random function F. Because the R_i values are non-repeating these are the same distribution, giving $\Pr[G_3] - \Pr[G_2] = 0$.

Finally, G_3 and G_4 differ in whether R_i is sampled with or without replacement. Via $\mathcal{A}_{3,4}$ we again reduce this to distinguishing between X^q and Y^q. We have $\Pr[G_4] - \Pr[G_3] = \mathsf{Adv}^{\mathsf{dist}}_{X^q, Y^q}(\mathcal{A}_{3,4})$.

Plugging in to Sect. 4.4 and bounding with $\mathcal{A}^*_{0,1}$ and $\mathcal{A}^*_{1,2}$ gives

$$\mathsf{Adv}^{\mathsf{dist}}_{W^q, V^q}(\mathcal{A}) \leqslant \mathsf{Adv}^{\mathsf{dist}}_{X^q, Y^q}(\mathcal{A}^*_{0,1}) + \mathsf{Adv}^{\mathsf{dist}}_{X^q, Y^q}(\mathcal{A}^*_{1,2}) + \mathsf{Adv}^{\mathsf{dist}}_{X^q, Y^q}(\mathcal{A}_{3,4}).$$

The result follows by bounding these advantages with $\mathcal{O}_{\mathsf{sl}}$.   □

4.5 Other Results

ENCRYPT-THEN-PRF. In the full version of this paper [19] we apply the above result to the proving the security of the encrypt-then-PRF construction of an authenticated encryption scheme (for fixed length messages).

NONCE-BASED ENCRYPTION. We note that our CTR$ and encrypt-then-prf theorems composes correctly with the standard way of constructing nonce-based encryption from a randomized encryption scheme by applying a PRF to the nonce to obtain coins for the underlying encryption scheme.

OTHER ENCRYPTION SCHEMES. In the full version of this paper [19] we look at streaming models induced by other randomized encryption schemes (CTR$ with a permutation, OFB$, CBC$, and CFB$). We exhibit straightforward attacks which distinguish length $p \in \Theta(\sqrt{N})$ samples from random with low state, $q = 1$, and good advantage.

Our streaming proof for the model induced by CTR$ with a random function implies such an attack is not possible against it. However, to be clear, these attacks *do not* rule out good time-memory tradeoffs for these other schemes. Instead these very weak attacks indicate that if such bounds are possible, their proofs will require new insights/models. See the full version of this paper [19] for more discussion.

5 Open Questions

Our work leaves open a number of important questions - most directly resolving validity of Conjecture 1 (or a relaxed version thereof which suffices for our final statement). More generally, there is the question of which other encryption schemes admit proofs of tight time-memory trade-offs. Furthermore, we do not know how to prove trade-offs for more complex security games which do not fit within the streaming model, e.g., security in the presence of decryption oracles.

Acknowledgements. We thank Aishwarya Thiruvengadam for insightful discussions in the initial stage of this project. Jaeger was supported in part by NSF grants CNS-1717640 and CNS-1526801, and by NSF grant CNS-1553758 while visiting UC Santa Barbara.

Stefano Tessaro's work was partially supported by NSF grants CNS-1553758 (CAREER), CNS-1719146, CNS-1528178, and IIS-1528041, and by a Sloan Research Fellowship.

References

1. Abrego, B.M., Fernandez-Merchant, S., Neubauer, M.G., Watkins, W.: Sum of squares of degrees in a graph. J. Inequalities Pure Appl. Math. **10**(3) (2009)
2. Alwen, J., Chen, B., Pietrzak, K., Reyzin, L., Tessaro, S.: `Scrypt` is maximally memory-hard. In: Coron, J.-S., Nielsen, J.B. (eds.) EUROCRYPT 2017. LNCS, vol. 10212, pp. 33–62. Springer, Cham (2017). https://doi.org/10.1007/978-3-319-56617-7_2
3. Alwen, J., Serbinenko, V.: High parallel complexity graphs and memory-hard functions. In: Servedio, R.A., Rubinfeld, R. (eds.) 47th ACM STOC, pp. 595–603. ACM Press, June 2015
4. Auerbach, B., Cash, D., Fersch, M., Kiltz, E.: Memory-tight reductions. In: Katz, J., Shacham, H. (eds.) CRYPTO 2017. LNCS, vol. 10401, pp. 101–132. Springer, Cham (2017). https://doi.org/10.1007/978-3-319-63688-7_4
5. Babbage, S.H.: Improved "exhaustive search" attacks on stream ciphers. In: European Convention on Security and Detection, pp. 161–166, May 1995
6. Barkan, E., Biham, E., Shamir, A.: Rigorous bounds on cryptanalytic time/memory tradeoffs. In: Dwork, C. (ed.) CRYPTO 2006. LNCS, vol. 4117, pp. 1–21. Springer, Heidelberg (2006). https://doi.org/10.1007/11818175_1
7. Bellare, M., Krovetz, T., Rogaway, P.: Luby-Rackoff backwards: increasing security by making block ciphers non-invertible. In: Nyberg, K. (ed.) EUROCRYPT 1998. LNCS, vol. 1403, pp. 266–280. Springer, Heidelberg (1998). https://doi.org/10.1007/BFb0054132
8. Bellare, M., Rogaway, P.: The security of triple encryption and a framework for code-based game-playing proofs. In: Vaudenay, S. (ed.) EUROCRYPT 2006. LNCS, vol. 4004, pp. 409–426. Springer, Heidelberg (2006). https://doi.org/10.1007/11761679_25
9. Bey, C.: An upper bound on the sum of squares of degrees in a hypergraph. Discrete Math. **269**(1–3), 259–263 (2003)
10. Cioab, S.M.: Note: sums of powers of the degrees of a graph. Discrete Math. **306**(16), 1959–1964 (2006)
11. Cover, T.M., Thomas, J.A.: Elements of Information Theory. Wiley, New York (2006)
12. Dai, W., Hoang, V.T., Tessaro, S.: Information-theoretic indistinguishability via the chi-squared method. In: Katz, J., Shacham, H. (eds.) CRYPTO 2017. LNCS, vol. 10403, pp. 497–523. Springer, Cham (2017). https://doi.org/10.1007/978-3-319-63697-9_17
13. Davies, D.W., Parkin, G.I.P.: The average cycle size of the key-stream in output feedback encipherment. In: Beth, T. (ed.) EUROCRYPT 1982. LNCS, vol. 149, pp. 263–279. Springer, Heidelberg (1983). https://doi.org/10.1007/3-540-39466-4_19
14. de Caen, D.: An upper bound on the sum of squares of degrees in a graph. Discrete Math. **185**(1–3), 245–248 (1998)
15. Golić, J.D.: Cryptanalysis of alleged A5 stream cipher. In: Fumy, W. (ed.) EUROCRYPT 1997. LNCS, vol. 1233, pp. 239–255. Springer, Heidelberg (1997). https://doi.org/10.1007/3-540-69053-0_17
16. Gruslys, V., Letzter, S., Morrison, N.: Hypergraph Lagrangians: resolving the Frankl-Füredi conjecture. arXiv preprint arXiv:1807.00793 (2018)
17. Iwata, T.: New blockcipher modes of operation with beyond the birthday bound security. In: Robshaw, M. (ed.) FSE 2006. LNCS, vol. 4047, pp. 310–327. Springer, Heidelberg (2006). https://doi.org/10.1007/11799313_20

18. Iwata, T., Mennink, B., Vizár, D.: CENC is optimally secure. Cryptology ePrint Archive, Report 2016/1087 (2016). http://eprint.iacr.org/2016/1087
19. Jaeger, J., Tessaro, S.: Tight time-memory trade-offs for symmetric encryption. Cryptology ePrint Archive, Report 2019/??? (2019). https://eprint.iacr.org/2019/???
20. Nikiforov, V.: Note: the sum of the squares of degrees: sharp asymptotics. Discrete Math. **307**(24), 3187–3193 (2007)
21. Nisan, N.: Pseudorandom generators for space-bounded computation. Combinatorica **12**(4), 449–461 (1992)
22. Patarin, J.: Mirror theory and cryptography. Cryptology ePrint Archive, Report 2016/702 (2016). http://eprint.iacr.org/2016/702
23. Pollard, J.M.: A monte carlo method for factorization. BIT Numer. Math. **15**(3), 331–334 (1975)
24. Quisquater, J.-J., Delescaille, J.-P.: How easy is collision search. New results and applications to DES. In: Brassard, G. (ed.) CRYPTO 1989. LNCS, vol. 435, pp. 408–413. Springer, New York (1990). https://doi.org/10.1007/0-387-34805-0_38
25. Tessaro, S., Thiruvengadam, A.: Provable time-memory trade-offs: symmetric cryptography against memory-bounded adversaries. In: Beimel, A., Dziembowski, S. (eds.) TCC 2018. LNCS, vol. 11239, pp. 3–32. Springer, Cham (2018). https://doi.org/10.1007/978-3-030-03807-6_1
26. Wang, Y., Matsuda, T., Hanaoka, G., Tanaka, K.: Memory lower bounds of reductions revisited. In: Nielsen, J.B., Rijmen, V. (eds.) EUROCRYPT 2018. LNCS, vol. 10820, pp. 61–90. Springer, Cham (2018). https://doi.org/10.1007/978-3-319-78381-9_3

Non-malleability

Non-Malleable Codes Against Bounded Polynomial Time Tampering

Marshall Ball[1]($\boxtimes$), Dana Dachman-Soled[2], Mukul Kulkarni[2], Huijia Lin[3], and Tal Malkin[1]

[1] Columbia University, New York, USA
{marshall,tal}@cs.columbia.edu
[2] University of Maryland, College Park, USA
danadach@ece.umd.edu, mukul@umd.edu
[3] University of Washington, Seattle, USA
rachel@cs.washington.edu

Abstract. We construct efficient non-malleable codes (NMC) that are (computationally) secure against tampering by functions computable in any *fixed* polynomial time. Our construction is in the plain (no-CRS) model and requires the assumptions that (1) **E** is hard for **NP** circuits of some exponential $2^{\beta n}$ ($\beta > 0$) size (widely used in the derandomization literature), (2) sub-exponential trapdoor permutations exist, and (3) **P**-certificates with sub-exponential soundness exist.

While it is impossible to construct NMC secure against *arbitrary* polynomial-time tampering (Dziembowski, Pietrzak, Wichs, ICS '10), the existence of NMC secure against $O(n^c)$-time tampering functions (for any *fixed* c), was shown (Cheraghchi and Guruswami, ITCS '14) via a probabilistic construction. An explicit construction was given (Faust, Mukherjee, Venturi, Wichs, Eurocrypt '14) assuming an *untamperable* CRS with length longer than the runtime of the tampering function. In this work, we show that under computational assumptions, we can bypass these limitations. Specifically, under the assumptions listed above, we obtain non-malleable codes in the plain model against $O(n^c)$-time tampering functions (for any fixed c), with codeword length independent of the tampering time bound.

Our new construction of NMC draws a connection with non-interactive non-malleable commitments. In fact, we show that in the NMC setting, it suffices to have a much weaker notion called *quasi non-malleable commitments*—these are non-interactive, non-malleable commitments in the plain model, in which the adversary runs in $O(n^c)$-time, whereas the honest parties may run in longer (polynomial) time. We then construct a 4-tag quasi non-malleable commitment from any sub-exponential OWF and the assumption that **E** is hard for some exponential size **NP**-circuits, and use tag amplification techniques to support an exponential number of tags.

Y. Ishai and V. Rijmen (Eds.): EUROCRYPT 2019, LNCS 11476, pp. 501–530, 2019.
https://doi.org/10.1007/978-3-030-17653-2_17

1   Introduction

Non-Malleable Codes (NMC) were introduced by Dziembowski, Pietrzak, and Wichs [25] as a modification of error correcting codes, with the goal of achieving security against adversarial tampering functions, that may change every part of a codeword. Informally, a NMC against a class $\mathcal{F}$ guarantees that if a codeword is tampered via the application of a function $f \in \mathcal{F}$, then the decoding of the tampered codeword will either be exactly the original message, or completely unrelated to the original message. As noted in [25], it is impossible to construct NMC against arbitrary tampering functions, since non-malleability can always be broken by a tampering function which first decodes the codeword to learn the underlying message, then re-encodes a related message. In particular, there can be no efficient NMC against arbitrary polynomial time tampering. Thus, to achieve feasibility, we must restrict the class of tampering functions.

A natural way to restrict tampering adversaries is via well-studied computational complexity measures. Several recent works have followed this approach and have developed strong connections between NMC and techniques from computational complexity. For example, Ball et al. [5] constructed NMC against bounded depth circuits with constant fan-in (which includes NC^0), several works [3,6,13] constructed NMC against AC^0 relying on different complexity theoretic techniques, and some works considered (restricted variants of) NMC against space-bounded tampering [6,26]. Specifically, the work of Faust et al. [26] considers space-bounded tampering adversaries in the random oracle model and achieves the security notion of *leaky* continuous non-malleability. The work of Ball et al. [6] is information-theoretic, considers streaming, space-bounded tampering adversaries and achieves standard non-malleability. The current work continues this line of research.

In this paper, we focus on the task of constructing NMC against *bounded* polynomial time tampering, namely tampering functions that are computable in an arbitrary *fixed* polynomial time. This is a very natural class to consider given the impossibility result for (unbounded) polynomial time, and indeed, some of the first works in this line of research have already considered this class. We discuss these next, along with the motivation and goals for our current work.

Cheraghchi and Guruswami [14] gave probabilistic constructions of efficient codes for circuits of size $O(n^c)$ (where an efficient randomized procedure outputs a "good" code with high probability). Faust et al. [27] gave an improved (in terms of the dependence on the error bound) construction against the same class, which is explicit, but relies on a model including an untamperable CRS (common reference string). The presence of CRS is undesirable, as not only must the CRS be generated by a trusted party, the CRS is also a non-tamperable component of the scheme. Moreover, both of these works can be viewed as using limited (t-wise) independence to partially derandomize probabilistic constructions. This approach inherently leads to a CRS whose length is at least as long as the bound on the size of the tampering circuits—meaning the tampering circuits cannot even read the entire CRS. We additionally note that if we allowed other size parameters, in particular the codeword size, to be as large as the runtime of the tampering function, then achieving non-malleability would become trivial. Finally, we note that constructions of NMC against bounded polynomial-time adversaries are trivial in the ideal permutation model, where it is assumed that all parties have access to an ideal, invertible permutation. Since Feistel-based constructions in the random oracle model are indifferentiable from ideal permutations (and indeed ideal cipher) [18–20], the above results hold in the random oracle model as well and can be instantiated in practice based on e.g. SHA-3. However, whereas in the random oracle/ideal permutation/ideal cipher model, non-malleability comes for free, in this work we seek constructions that are based solely on *hardness* assumptions that do not have a non-malleability flavor.

This motivates the following question:

Can we construct efficient NMC against bounded polynomial time adversaries, in the plain *model (i.e. without CRS or random oracles)? Ideally, with codeword length that is independent of the runtime of the adversarial tampering function?*

As we elaborate next, we achieve this by moving to computational security and restricting our attention to uniform adversaries (while [14,27] gave statistical guarantees against non-uniform adversaries). In addition, as explained shortly below, we allow uniform bounded polynomial time tampering adversaries to have an inverse polynomial advantage (as in [14]) as opposed to having only negligible advantage (as in [27]). We emphasize that to the best of our knowledge, there is no transformation that either (a) eliminates the CRS in the NMC of [27] to achieve security against uniform (or non-uniform) adversaries or (b) fully derandomizes the monte carlo construction of [14], *even* under derandomization assumptions. Our techniques highlight interesting new connections to complexity theory.

1.1 Our Results

Our construction requires a complexity theoretic assumption that some language in the complexity class **E** (the class of languages that can be decided by Turing machines running in time $2^{O(n)}$) is hard for **NP** circuits of some exponential $2^{\beta n}$ (for $\beta > 0$) size. As surveyed later, such assumptions are widely used in the derandomization literature, often referred to as *derandomization assumptions*, and have connection with cryptography. Our construction also relies on the following cryptographic assumptions: the existence of subexponential trapdoor permutations and **P**-certificates (**P**-cert) with sub-exponential soundness. **P**-certs (introduced by [15]) are "succinct" non-interactive arguments for languages $\mathcal{L} \in$ P, with proof length which is a fixed polynomial, independent of the time it takes to decide $\mathcal{L}$ (see full version of this paper [4] for a formal definition). We provide more background on these assumptions in Sect. 1.2 below.

Theorem 1 (Informal). *Assuming*

- **E** *is hard for* **NP** *circuits with some exponential size (namely $2^{\beta n}$ for some constant $\beta > 0$)*
- *Existence of sub-exponential trapdoor permutation*
- *Existence of* **P**-*cert with sub-exponential soundness*

for every constant c_A, there is an efficient construction of NMC in the plain (no-CRS) model against uniform, bounded polynomial n^{c_A}-time tampering adversaries, with inverse polynomial indistinguishability (for any polynomial time non-uniform distinguisher). Furthermore, the codeword size is a fixed polynomial independent of n^{c_A}.

A few remarks are in order. First, to formalize that a tampered codeword, if not copied from the orginal codeword, must decode to an independent value, the definition of non-malleability requires that the decoded values, u_1 and u_2, obtained from tampering codewords of different values, v_1 and v_2 respectively, must be indistinguishable (u_b is replaced by same in the case of copying). Our NMC achieves inverse polynomial distinguishing advantage against polynomial-time non-uniform distinguishers

Second, as mentioned before, it is important that the length of the codeword is smaller than the time-bound n^{c_A} of the tampering functions; otherwise, achieving non-malleability becomes trivial. Here, we achieve the ideal case, where the *length* of the codeword is bounded by a fixed polynomial, independent of n^{c_A}. As the adversarial time bound grows, the only parameter that grows is the *run time* of encode/decode. Moreover, this dependence is *necessary* as discussed earlier, since non-malleability is trivially impossible when the class of tampering functions includes the encode/decode functions.

Finally, we note that the assumption of the existence of sub-exponential trapdoor permutation in Theorem 1, can be replaced with the assumption of the existence of ZAPs (public coin, two message witness indistinguishable protocols) [24] with witness indistinguishability against sub-exponential adversaries and the existence of sub-exponential one-way functions (OWF). Note that ZAPs can be constructed from bilinear maps [39], which are not known to imply trapdoor permutations.

Connection between NMC and Non-Malleable Commitments. Our construction of NMC draws a connection with another important notion of non-malleability – non-malleable commitments [21,51]. The only difference between NMC and non-interactive non-malleable commitments is that the former can be decoded efficiently, whereas decommitment of the latter cannot be done efficiently. A few prior works leverage this connection, showing that NMC can be used to obtain improved non-malleable commitments [11,37], and using techniques from the non-malleable commitment literature to obtain NMC [12,58]. However, the latter direction—tapping into the wealth of techniques from the non-malleable commitment literature to construct NMC—has been largely unexplored, perhaps due to the fact that NMCs are typically unconditionally secure.

In our NMC construction, we begin with the framework of Ball et al. [6], which provides a generic way to construct NMC against tampering classes $\mathcal{F}$ for which sufficiently strong average-case hardness results are known, but requires a CRS. We show how to remove the CRS for particular tampering classes, including the class of bounded, poly-time adversaries. One modification is replacing the public key encryption scheme in the framework of [6] (whose pubic keys are contained in the CRS) with a non-interactive, non-malleable commitment scheme NMCom in the plain model.

At a very high(and overly simplified) level, our NMC, like [6], follows the Naor-Yung [56] paradigm that achieves CCA security of encryption, by composing two instances $\mathsf{Encrypt}(\mathrm{PK}, v), \mathsf{Encrypt}(\mathrm{PK}', v)$ of a public key encryption scheme, followed by a NIZK proof of the equality of encrypted values. In the context of NMC, we replace one instance of encryption with an encoding $\mathsf{E}(v)$ that is decodable in some polynomial t time, but has certain complexity theoretic hardness (specified shortly) against the class of circuits of smaller $t' < t$ size. We further replace the other instance of encryption with a non-malleable commitment c to v. Following [6,56], we provide a reduction that can turn any successful tampering adversary A against NMC, into an adversary B able to "maul" an encoding $\mathsf{E}(v)$ of v into a non-malleable commitment $\tilde{c}$ to a related value $\tilde{v}$. The challenge lies in ensuring that the reduction is "simple", namely, can be implemented by a circuit of size t'. Then the complexity theorctic hardness that we rely on is that it is impossible for such a circuit to compress an encoding $\mathsf{E}(v)$ into a much *shorter* string $\tilde{c}$ correlated to v (despite that the correlation may take exponential time to verify). Such an encoding, E, can be based on the incompressible functions of Applebaum et al. [1], which can be constructed based on assumptions that are widely used in the derandomization literature. (For more details on the NMC construction see the technical overview in Sect. 1.3).

Connection between Complexity Theory and Non-Malleable Commitments. Another contribution of this work, is to develop new connections between complexity theory and non-malleable cryptography. We show that derandomization assumptions can be employed to build a new primitive we call Quasi Non-Malleable Commitments, which is weaker than standard non-malleable commitments, but nevertheless suffices for constructing NMC. This allows us to avoid adding the assumptions needed for standard non-interactive NMCom such as time-lock puzzles or hardness amplifiable injective one-way functions [10,50].

Recall that in the non-malleable codes setting, encode/decode can be in a larger complexity class than the adversary, and so standard non-interactive NMCom is an overkill. This motivates our definition of *Quasi Non-Malleable Commitments* in which the adversary runs in $O(n^{c_{com}})$-time, whereas the honest parties may run in longer (polynomial) time. To construct Quasi-NMCom from assumptions widely used in the derandomization literature, observe that these assumptions allow us to construct polynomial-time computable functions ψ for which non-deterministic advice does not help speed up the computation. This stands in stark contrast to the case of inversion of a one-way function ρ, which becomes easy with non-deterministic advice (as the advice can contain a pre-image). Following the framework of [50], we construct two types of commitments that are *harder* than each other in different hardness "axes"—namely "BP-time" (corresponding to probabilistic Turing machines) and "non-deterministic (ND)-size" (corresponding to **NP**-circuits–see Sects. 1.2 and 2.4). Specifically, one type of commitment com_1 are the standard schemes based on one-way functions ρ, and the other com_2 is based on the function ψ given by derandomization assumptions. The com_1 is much harder to break than com_2 in the axis of "BP-time", as inverting one-way function ρ is much harder than computing ψ using probabilistic Turing machines. On other hand, com_2 is much harder to break than com_1 in the axis of "ND-size", where both inverting ρ and computing ψ can be done in poly-size, but computing ψ is significantly harder.

From such mutually harder commitment schemes, we obtain a 4-tag Quasi-NMCom. Then, based on tag-amplification techniques in the literature [10, 46], we increase the number of tags supported to an exponential. It turns out that the *quasi*-setting makes amplification hard, which requires us to introduce a notion of "Double-Agent" adversaries. Informally, double-agent adversaries are probabilistic uniform Turing machines with "large" time complexity, that can also be represented as a distribution over circuits with "small" size complexity (see Sect. 2.1 for additional details). Post-amplification, our final Quasi-NMCom construction employs the same assumptions as Theorem 1. We believe these techniques may be useful for other applications in similar quasi-settings.

1.2 Background on Assumptions

In this section we provide some background on the assumptions that we use.

*On **P**-certificates.* **P**-certificates were introduced by [15] in pursuit of constant-round concurrent zero-knowledge. Loosely, a **P**-certificate is a non-interactive proof system that allows a prover to convince an efficient verifier of the validity of any statement in **P** via a short proof. In particular, both the proof length and the run-time of the verifier are bounded by some fixed polynomial, but the system should work any language in **P** (the prover's efficiency should be comparable to the statement). CS-proofs [54] imply **P**-certificates, but unlike the former, the latter assumption is falsifiable.

*On "**E** requires circuits of exponential size".* A fundamental family of questions in theoretical computer science is when and where randomness helps (vs strictly deterministic procedures). While it is widely believed that **BPP** = **P** (i.e., any efficient, randomized decision procedure can be efficiently simulated by a deterministic procedure), whether the equality indeed holds is still an open problem. This particular question (**BPP** = **P**?) and others in the domain of *derandomization* have deep connections to cryptography.

In the 1980s, Yao [70] showed that one-way permutations suffice to create pseudorandom generators (PRG) for poly-time computation. PRGs expand a small sequence of uniform random bits to a long sequence of pseudorandom bits that "fool" classes of procedures in the sense their behavior is essentially the same as if they were given truly random bits. In this sense, PRGs give a canonical technique for derandomizing decision procedures: running the procedure on multiple outputs of the PRG in parallel and taking majority of the obtained result. Later, it was shown that essentially minimal cryptographic assumptions (one-way functions) suffice for constructing PRGs [42].

However, while most cryptography implies non-trivial derandomization, there seem to be inherent barriers to statements of the converse form. In fact, the so-called "cryptographic" PRG's yield, in two aspects, much more than what is required for derandomization since (a) the output of such PRGs fool *any* polynomial time procedure (including procedures that run in much more time than the PRG itself) and (b) such PRGs guarantee that the behavior of poly-time procedures is only negligibly different from their behavior on true randomness. On the other hand, one-way functions are not known to imply $\mathbf{P} = \mathbf{BPP}$ because known constructions only "stretch" random bits into polynomially many random bits (whereas exponential stretch is required for canonical simulation).

Capitalizing on these observations, Nisan and Wigderson [57] gave a generic means of constructing PRGs which "fool" a certain class of circuits $\mathcal{C}$, from any function that is hard-on-average for a slightly enlarged class of circuits. In particular, this in some sense reduces the problem of explicit derandomization to proving strong circuit lower bounds on explicit functions. To this end, later work showed that, in fact, simply assuming there is a language in $\mathbf{E}$ that does not have circuits size $2^{\beta n}$ for some $\beta > 0$ (for almost all n), is sufficient to derandomize BPP [43,67]. Moreover, because $\mathbf{E}$ has complete problems, this yields explicit pseudorandom generators. However, for reasons alluded to above, this assumption is, to best of our knowledge, incomparable to standard cryptographic assumptions.

This latter (worst-case) conjecture and its generalization has appeared in a variety of contexts pertaining to derandomization and other questions in computational complexity [2,22,29,35,40,41,43,47,53,55,57,64–68]. The conjectures we are concerned with in this work take the following form (following [1]):

Assumption 1 (E is hard for exponential size X-circuit). *There exists a problem $\mathbf{L}$ in $\mathbf{E} = \mathbf{DTIME}(2^{O(n)})$ and a constant $\beta > 0$, such that for every sufficiently large n, X-circuits of size $2^{\beta n}$ fail to compute the characteristic of $\mathbf{L}$ on inputs of length n.*

where X-circuits can be circuits of type, {*non-deterministic, co-non-deterministic*, $\mathbf{NP}$, Σ_i }. See Sect. 2.4, for definitions of these types of circuits. While these types of assumptions are independently interesting, in this work we will utilize some surprising implications outside of derandomization.

Recently, Applebaum et al. [1] presented (explicit) constructions of poly-time computable incompressible functions based on the assumption that $\mathbf{E}$ is hard for exponential size non-deterministic circuits (based on the extractors for samplable distributions of Trevisan and Vadhan [68]). Loosely, a function, ψ is incompressible for a class if no procedure in that class can "shrink" an input to the function, x, such that $\psi(x)$ can later be recovered. Note that, to our knowledge, it is not known how to construct incompressible functions from standard cryptographic assumptions (unlike the case of derandomization).

Barak et al. [8] observed that similar assumptions can be used to construct cryptographic primitives. In particular, they showed that if $\mathbf{E} = \mathbf{DTIME}(2^{O(n)})$ contains a function with *co-non-deterministic* circuit complexity $2^{\Omega(n)}$, then there exists (explicit) non-interactive witness indistinguishable proof systems for $\mathbf{L} \in \mathbf{NP}$ (additionally assuming the existence of trapdoor permutations). They also showed that the same assumption can be used to construct a non-interactive bit commitment scheme from a one-way function.

In this work, we use the above results and demonstrate new connections between these assumptions and non-malleable cryptography. In particular we show that if Assumption 1 holds for $\mathbf{NP}$-circuits and (sub-exponential) one-way functions exist, then we can construct quasi-non-malleable commitment schemes. We combine our construction of such commitment schemes along with NIZK proofs based on the NIWI of [8], as well as the incompressible functions of [1], to obtain our main result: a family of efficient non-malleable codes secure against tampering by uniform algorithms running in time $O(n^c)$.

1.3 Technical Overview

We begin by recalling (a simplified version of) the template for constructing non-malleable codes against complexity class $\mathcal{F}$ (based on the Naor-Yung double encryption paradigm [56]) introduced in the work of Ball et al. [6]:

The CRS contains a public key PK for an encryption scheme $\mathcal{E} = (\mathsf{Gen}, \mathsf{Encrypt}, \mathsf{Decrypt})$, and a CRS crs for a simulation-sound, non-interactive zero knowledge proof (NIZK). For $b \in \{0,1\}$, let $\mathcal{D}_b$ denote disjoint distributions over $x_1 \ldots x_n \in \{0,1\}^n$ such that $\psi(x_1 \ldots x_n) = b$, where ψ is poly-time computable, yet every $f \in \mathcal{F}$ has low correlation with ψ.

To encode a bit b:

1. Randomly choose string $x_1 \ldots x_n$ from $\mathcal{D}_b$
2. Compute $c \leftarrow \mathsf{Encrypt}_{\mathrm{PK}}(b)$.
3. Compute a NIZK proof T of "consistency": $\exists b' \in \{0,1\}$ s.t. $x_1 \ldots x_n$ is in the support of $\mathcal{D}_{b'}$ and b' is the plaintext underlying c.
4. Output $(x_1 \ldots x_n, c, T)$.

To decode $(x_1 \ldots x_n, c, T)$:

1. Verify the NIZK proof T.
2. If it accepts, output $\psi(x_1 \ldots x_n)$.

The proof of [6] proceeds (loosely) as follows: In the first hybrid they switch to simulated proof T'. Then they switch c, in the "challenge" encoding to an encryption of garbage c', and next switch to an alternative decoding algorithm *in* $\mathcal{F}$, which requires the trapdoor SK (corresponding to the public key PK which is contained in the CRS). If, in the final hybrid, decodings of tampered encodings depend on b, a circuit in $\mathcal{F}$ can be constructed, whose output is correlated with the hard function ψ, reaching a contradiction. While [6] do in fact show that the CRS can be removed for constructions against certain classes $\mathcal{F}$ of tampering, naively, their approach requires a CRS in two seemingly inherent ways: First, the CRS allows the use of the secret key trapdoor SK in the alternate decoding algorithm and second, it allows the use of a simulation-sound NIZK, which requires CRS.

In this work, we make two crucial observations that allow us to eliminate the CRS from the above construction. First, we consider a stronger notion of hardness for ψ, known as *incompressibility* (in fact, this hardness notion was already implicitly used in [6] for their multi-bit construction). Briefly, if a function ψ is incompressible by circuit class $\mathcal{C}$, it means that for $t \ll n$, for any (computationally unbounded) Boolean function $F : \{0,1\}^t \to \{0,1\}$ and any $C : \{0,1\}^n \to \{0,1\}^t \in \mathcal{C}$, the output of $F \circ C(x_1, \ldots, x_n)$ is uncorrelated with $\psi(x_1, \ldots, x_n)$ (over uniform choice of $x_1, \ldots, x_n$). Now, since F is allowed to be computationally unbounded, we may consider an F that decrypts the ciphertext $c = \mathsf{Encrypt}_{\mathrm{PK}}(b)$ by brute force search. To elaborate, instead of using SK to efficiently decrypt c in complexity class $\mathcal{C}$, the alternate decoding algorithm D' is split into two parts $\mathsf{D}' = \mathsf{D}'_2 \circ \mathsf{D}'_1$, where D'_1 can be implemented in $\mathcal{F}$, but has short output length, whereas D'_2 is computationally unbounded. Specifically, D'_1 checks the proof T and then outputs the entire ciphertext c (which is fine so long as the length of c is sufficiently smaller than n), and, due to the incompressibility property of ψ, we must still have that the output of $\mathsf{D}' = \mathsf{D}'_2 \circ \mathsf{D}'_1$ is uncorrelated with $\psi(x_1, \ldots, x_n)$. This eliminates the need of providing a trapdoor to the alternate decoding algorithm and so instead of using a public key encryption scheme, we may use a non-interactive statistically binding commitment scheme, which can be constructed from injective one-way function or from derandomization assumptions and any one-way function [8].[1]

Next, we note that it is possible to construct a NIZK proof system in the plain (no-CRS) model (i.e. "One-Message Zero Knowledge"), with soundness against uniform adversaries. To do so, one first constructs a non-interactive witness indistinguishable proof system (NIWI) in the plain model (based on standard cryptographic assumptions and derandomization assumptions [8]) and then converts from witness indistinguishability to full zero knowledge using the well-known FLS paradigm [28]. Specifically, the simulator will be given a trapdoor witness based on a problem that is computationally hard for *uniform* PPT adversaries such as finding a collision in a keyless collision resistant hash function. The problem with this approach is that in the proof sketch outlined above, we actually require *simulation-sound* NIZK, as opposed to regular NIZK. In simulation-sound NIZK, the soundness properties must

---

[1] As we will see, in our setting of non-malleable codes against polynomially-bounded adversaries, our construction requires such derandomization assumptions in any case and so only standard one-way function is required in addition. However, for simplicity we will assume injective one-way function in the remainder of the exposition in this section.

hold, even after the adversary sees a simulated proof of a false statement. Whereas various constructions of (one-time) simulation-sound NIZK rely on embedding a trapdoor within the CRS (cf. [52,63]), our approach to achieve the simulation-soundness property without CRS is to replace the commitment c (which replaced the encryption $\mathsf{Encrypt}_{\mathrm{PK}}(b)$ as described above) with a *non-interactive, non-malleable* commitment scheme. Unfortunately, currently known non-interactive, non-malleable commitment schemes require somewhat non-standard assumptions such as time-lock puzzles or hardness amplifiable injective one-way functions [10,50], whereas our goal is to minimize assumptions. As we will see, the fact that our commitment scheme is only required to be non-malleable against adversaries in a restricted circuit class $\mathcal{F}$, allows us to obtain non-interactive, non-malleable commitments, while reducing assumptions.

Instantiating the Paradigm. In this work we construct non-malleable codes against the class $\mathcal{F}$ of uniform, polynomial-bounded tampering functions. Crucially, we will do so (1) *without* relying on CRS (2) with codeword length that is independent of the polynomial time bound (note that if the codeword length is longer than the polynomial time bound then the adversary does not even get to read the entire input, also it's trivial to construct these) and (3) while reducing computational assumptions, to the extent possible.

Specifically, in addition to standard cryptographic assumptions, we will assume standard derandomization-type assumptions such as those discussed in the previous section. We also require the notion of **P**-certificates, which seem to be necessary to implement the above high-level paradigm, as we discuss next. To see why this is so, note that the statement proved in NIZK proof T, involves proving that $\psi(x_1, \ldots, x_n)$ is equal to some value. Note that ψ is a polynomial-time computable function, but that intrinsic in the approach is choosing ψ that is hard to compute in the specific polynomial time bound $T(n)$ corresponding to tampering class $\mathcal{F}$. Moreover, we require that the size of the proof T be independent of the polynomial time bound $T(n)$, and so in particular the size of the proof T must be independent of the time required to compute ψ. This is now exactly the notion of a **P**-certificate.

We also note that given the above paradigm for encoding of a single bit, it is straightforward to obtain a scheme for the encoding of multiple bits (by individually encoding each bit and then using a single proof T to "wrap" together the individual encodings). The only restriction will be that the number of bits, m, that are encoded, multiplied by the length of a bit commitment, λ, should be sufficiently smaller than n, the input length of the function ψ. See Sect. 3 for additional details.

Instantiation of ψ. Recall that for the above approach to work, we must instantiate ψ with a function that is incompressible against polynomially-bounded adversaries. Fortunately, such a construction was given by [1], based on a derandomization-type assumption. See Sect. 4 for additional details.

Instantiation of NMCom. In fact, as discussed previously, we note that we do not need full-fledged NMCom, but only *Quasi* NMCom, i.e. NMCom with the following two relaxations: (1) The commitment scheme is only secure against bounded-poly (in fact "Double-Agent") adversary and distinguisher (2) The complexity of the honest sender/receiver may be greater than the complexity of the adversary. To construct Quasi-NMCom, we adopt the approach of [50] to initially construct a commitment scheme with small number of tags, and use the fact that the derandomization assumptions that we employ in this work are believed to hold even against *non-deterministic* adversaries. In particular, we employ the well-studied assumption that $\mathbf{E}$ is hard for adversaries represented as exponential size $\mathbf{NP}$-circuits—or circuits with access to a SAT oracle (See Sects. 1.2 and 2.4 for further discussions on these assumptions). To construct our NMCom scheme, we start off with two different types of commitments, Type 0 and Type 1 such that if we get a Type 0 on left, we can extract from Type 1 on the right without breaking the security of Type 0 and vice versa. Each commitment consists of an input x to a Boolean function ψ' (with logarithmic input length) that is hard for $\mathbf{NP}$-circuits of size $2^{\epsilon_3 \cdot \text{input length}}$ to compute as well as the output y of an injective OWF ρ, which is hard for ppt adversaries running in time $2^{\text{input length}^{\epsilon_3}}$.[2] Each of these can be considered as a commitment to a bit (given x, the output of $\psi'(x)$ is the committed value and given y, a hardcore bit of ρ) and the final committed value is the xor of the two bits committed.

Type 0: input length $c_1 \log(n)$ to ψ', input length $n^{\epsilon'_1}$ to ρ.
Type 1: input length $c_2 \log(n)$ to ψ', input length $n^{\epsilon'_2}$ to ρ.

Set $c_2 > c_1 > \epsilon'_1 > \epsilon'_2$ so that (1) $n^{c_1} < n^{\epsilon_3 \cdot c_2}$ and (2) $2^{n^{\epsilon'_2}} < 2^{n^{\epsilon'_3 \cdot \epsilon'_1}}$. We now consider the two possible cases:

Type 0 on left, Type 1 on right. Extract by inverting the injective OWF ρ in deterministic time $2^{n^{\epsilon'_2}}$ and computing ψ' in deterministic time n^{c_2}. Note that this does not allow breaking injective OWF ρ with input length $n^{\epsilon'_1}$, which is secure against time $2^{n^{\epsilon_3 \cdot \epsilon'_1}} > 2^{n^{\epsilon'_2}}$.

Type 1 on left, Type 0 on right. Extract by computing ψ' in deterministic time n^{c_1} and inverting the injective OWF ρ with an $\mathbf{NP}$-circuit of size $n^{\epsilon'_1}$. Note that this does not allow breaking hardness of ψ' with input length $c_2 \log(n)$, which is secure against $\mathbf{NP}$-circuits of size $n^{\epsilon_3 \cdot c_2} > n^{c_1}$.

See Fig. 1 for a summary and [4] for additional details.

The above 2-tag commitment scheme can then be straightforwardly extended to work for 4 tags, at which point amplification techniques from [46] can be applied to obtain NMCom with number of tags exponential in the security parameter. The analysis of the amplified scheme is somewhat different than

---

[2] For this exposition, we assume for simplicity that ψ' can be computed in deterministic time $2^{\text{input length}}$ and that the injective OWF has linear circuit size. Recall that we do not require injective OWF and that any statistically binding, non-interactive commitment scheme is sufficient, but that for simplicity we assuming injective OWF in this exposition.

| | Input length to ψ' | Hardness of ψ' | | Input length to ρ | Hardness of ρ | |
		D	ND		D	ND
Type 0	$c_1 \cdot \log(n)$	$n^{\epsilon_3 \cdot c_1}$	$n^{\epsilon_3 \cdot c_1}$	$n^{\epsilon'_1}$	$2^{n^{\epsilon'_3 \cdot \epsilon'_1}}$	$n^{\epsilon'_1}$
Type 1	$c_2 \cdot \log(n)$	$n^{\epsilon_3 \cdot c_2}$	$n^{\epsilon_3 \cdot c_2}$	$n^{\epsilon'_2}$	$2^{n^{\epsilon'_3 \cdot \epsilon'_2}}$	$n^{\epsilon'_2}$

Fig. 1. ψ' and ρ are the functions described in the paragraph above. D stands for deterministic and ND stands for "non-deterministic" hardness. We set parameters so that $c_2 > c_1 > \epsilon'_1 > \epsilon'_2$.

in prior work, since our analysis must carefully take into account that some assumptions are inherently uniform (One-Message Zero Knowledge) and some assumptions (hardness of ψ') are inherently non-uniform (the adversary in the proof is so limited that it does not have time to *generate* new commitments on its own and therefore must receive them as non-uniform advice when reducing security to the hardness of computing ψ'). To solve this problem, we introduce the notion of "Double Agent" adversaries (as discussed in the introduction) and provide a proof of security of our amplified NMCom scheme against this class of adversaries. See [4] for additional details.

1.4   Related Work

Non-Malleable Codes. Non-malleable codes (NMC) were introduced by Dziembowski, Pietrzak and Wichs in their seminal work [25]. While there has been a long line of important results for various tampering classes, due to space limitations, we discuss here only the results most relevant to this work.

As discussed extensively in the introduction, Faust et al. [27] constructed efficient information-theoretically secure NMC in the CRS model, resilient against tampering function classes $\mathcal{F}$ which can be represented as circuits of size $\mathsf{poly}(n)$. Another important result by Cheraghchi and Guruswami [14] showed the existence of information theoretically secure NMC against tampering families $\mathcal{F}$ of size $|\mathcal{F}| \le 2^{2^{\alpha n}}$ with optimal rate $1 - \alpha$. They achieve error $\varepsilon \in O(1/\mathsf{poly}(n))$ as the run-time of the encoding and decoding algorithms is proportional to $\mathsf{poly}(1/\varepsilon)$ where ε is the error probability.

Ball et al. [5] constructed efficient information theoretic secure NMC against n^δ-local tampering functions, for any constant $\delta > 0$. This class includes tampering functions, which can be represented as constant depth circuits with bounded fan-in i.e NC^0 circuits. Chattopadhyay and Li [13] constructed NMC against AC^0 tampering functions from seedless non-malleable extractors, although the codeword length of this construction is super-polynomial in the message length n. Faust et al. [26] considered non-malleable codes against space bounded tampering adversaries in the random oracle model. The construction achieves a new notion of *leaky* continuous non-malleable codes (with self-destruct property), where the adversary is allowed to learn some bounded $\log(|m|)$ bits of information about the underlying message m.

Recently, Ball et al. [6] presented a generic framework to construct NMC against tampering function classes for which average-case hardness bounds are known. They also instantiated their framework to construct the first efficient, computationally secure multi-bit NMC against tampering functions which can be represented as constant-depth circuits with unbounded fan-in (AC^0 tampering), as well as against tampering functions which can be represented as bounded depth decision tree. Additionally, they showed that the framework can be used to construct information-theoretic NMC against space-bounded streaming tampering. Information-theoretic secure, efficient NMC against AC^0 tampering were subsequently constructed by [3].

Derandomization and Cryptography. The connection between derandomization techniques with cryptography was first explored by Barak et al. [8], who constructed one-message witness indistinguishable proof systems (non-interactive commitment scheme) in the plain model based on trapdoor permutations (one-way functions) in addition to the derandomization assumptions. Recently, Applebaum et al. [1] constructed incompressible functions against the class of bounded polynomial time functions from similar assumptions.

Non-Malleable Commitments. Non-malleable commitments have been studied extensively since their introduction by [21] in their seminal paper. Great progress has been made in reducing the interaction between the sender and the receiver, while minimizing computational assumptions. We list just some of the results in this line of work [7,16,17,36–38,45,48,49,59–61]. Recently, Lin, Pass, and Soni [50] gave a construction of a non-interactive, fully-concurrent, non-malleable commitment scheme secure against uniform adversaries based on sub-exponential non-interactive commitment schemes, non-interactive witness indistingushable proof systems (NIWI), uniform collision resistant hash functions, and time-lock puzzles [62]. When replacing the uniform collision resistant hash functions with a family of collision resistant hash functions, their protocol becomes 2-round. Khurana and Sahai [46] constructed 2-round non-malleable commitments with bounded concurrency from standard sub-exponential assumptions. Bitansky, and Lin [10] gave a construction of a non-interactive, fully-concurrent, non-malleable commitment scheme from multi-collision-resistant keyless hash functions, sub-exponentially-secure time-lock puzzles, and other standard assumptions.

2   Definitions

2.1   Notation

When comparing distribution ensembles $\mathcal{D} = \{\mathcal{D}_n\}_{n\in\mathbb{N}}, \mathcal{D}' = \{\mathcal{D}'_n\}_{n\in\mathbb{N}}$, we use the notation $\mathcal{D} \overset{\mathcal{G},\mathcal{S}}{\approx} \mathcal{D}'$, where $\mathcal{G}, \mathcal{S}$ are sets, to indicate that for sufficiently large n, every distinguisher $D \in \mathcal{G}$ distinguishes $\mathcal{D}_n$ from $\mathcal{D}'_n$ with probability at most $p(n)$, for some $p(\cdot) \in \mathcal{S}$. When comparing functions p, p', we use the notation $p(n) \overset{\mathcal{S}}{\approx} p'(n)$, where $\mathcal{S}$ is a set, to indicate that $|p(n) - p'(n)| \in \mathcal{S}$.

We consider "Double-Agent" adversaries A in computational classes denoted by $\mathbf{BPtime}(T(n)) \cap \mathbf{SIZE}(t(n))$, for some functions $T(\cdot)$, $t(\cdot)$. Intuitively, this computational class contains probabilistic uniform Turing machines A with "large" time complexity $T(n)$, that can also be represented as a distribution over circuits with "small" size complexity $t(n)$. Informally, this is possible since A can be split into subroutines in such a way that subroutines that require "large" time complexity can all be replaced with non-uniform advice. Formally, $A \in \mathbf{BPtime}(T(n)) \cap \mathbf{SIZE}(t(n))$ if the following hold:

- $A = (A_1, A_2)$.
- $A_1 \in \mathbf{BPtime}(T(n))$, $A_2 \in \mathbf{BPtime}(t(n))$.
- A_1 receives only security parameter 1^n as input and produces output of length at most $t(n)$.
- A_2 receives the input of A as its input, along with the output of A_1.

Note that, since A_1 takes only security parameter as input, the output of A_1, can be viewed as non-uniform advice to A_2. Thus, we can convert such a uniform adversary $A = (A_1, A_2)$ into a distribution over non-uniform circuits of size $t(n)$ with identical behavior to A.

2.2 Non-Malleable Codes

Definition 1 (Coding Scheme). *Let $\Sigma, \widehat{\Sigma}$ be sets of strings, and $\kappa, \widehat{\kappa} \in \mathbb{N}$ be some parameters. A coding scheme consists of two algorithms (E, D) with the following syntax:*

- *The encoding algorithm (perhaps randomized) takes input a message in Σ and outputs a codeword in $\widehat{\Sigma}$.*
- *The decoding algorithm takes input a codeword in $\widehat{\Sigma}$ and outputs a message in Σ.*

We require that for any message $\mathsf{msg} \in \Sigma$, $\Pr[\mathsf{D}(\mathsf{E}(\mathsf{msg})) = \mathsf{msg}] = 1$.

Definition 2 ($O(1/p(n))$-Non-malleability [25]). *Let n be the security parameter, $\mathcal{F}$ be some family of functions. For each function $f \in \mathcal{F}$, and $\mathsf{msg} \in \Sigma$, define the tampering experiment:*

$$\mathbf{Tamper}^f_{\mathsf{msg}} \stackrel{\text{def}}{=} \left\{ \begin{array}{c} c \leftarrow \mathsf{E}(\mathsf{msg}), \tilde{c} := f(c), \tilde{\mathsf{msg}} := \mathsf{D}(\tilde{c}). \\ Output: \ \tilde{\mathsf{msg}}. \end{array} \right\},$$

where the randomness of the experiment comes from the encoding algorithm. We say a coding scheme (E, D) is $O(1/p(n))$-non-malleable with respect to $\mathcal{F}$ if for each $f \in \mathcal{F}$, there exists a PPT simulator Sim such that for any message $\mathsf{msg} \in \Sigma$, we have

$$\mathbf{Tamper}^f_{\mathsf{msg}} \stackrel{PPT, O(1/p(n))}{\approx} \mathbf{Ideal}_{\mathsf{Sim}, \mathsf{msg}} \stackrel{\text{def}}{=} \left\{ \begin{array}{c} \tilde{\mathsf{msg}} \cup \{\mathsf{same}^*\} \leftarrow \mathsf{Sim}^{f(\cdot)}. \\ Output : \mathsf{msg} \text{ if output of } \mathsf{Sim} \text{ is } \mathsf{same}^*; \\ \text{otherwise } \tilde{\mathsf{msg}}. \end{array} \right\}$$

Definition 3 ($O(1/p(n))$-Medium Non-malleability). *Let n be the security parameter, $\mathcal{F}$ be some family of functions. Fix* msg $\in \Sigma$. *Let $c \leftarrow$ E(msg) and let $g(\cdot, \cdot)$ be a predicate such that for every $f \in \mathcal{F}$,*

$$\Pr[g(c, f(c)) = 1] \wedge \mathsf{D}(f(c)) \neq \mathsf{msg}] \leq \mathsf{negl}(n).$$

For g as above, each function $f \in \mathcal{F}$, and msg $\in \Sigma$, *define the tampering experiment*

$$\mathbf{MediumNM}^f_{\mathsf{msg},g} \stackrel{\text{def}}{=} \left\{ \begin{array}{c} c \leftarrow \mathsf{E}(\mathsf{msg}), \tilde{c} := f(c), \tilde{\mathsf{msg}} := \mathsf{D}(\tilde{c}) \\ Output: \ same^* \ if \ g(c, \tilde{c}) = 1, \ and \ \tilde{\mathsf{msg}} \ otherwise. \end{array} \right\}$$

The randomness of this experiment comes from the randomness of the encoding algorithm. We say that a coding scheme (E, D) is $O(1/p(n))$-medium non-malleable with respect to $\mathcal{F}$ if there exists a g as above and for any msg, msg' $\in \Sigma$ *and for each $f \in \mathcal{F}$, we have:*

$$\{\mathbf{MediumNM}^f_{\mathsf{msg},g}\}_{n \in \mathbb{N}} \stackrel{PPT, O(1/p(n))}{\approx} \{\mathbf{MediumNM}^f_{\mathsf{msg}',g}\}_{n \in \mathbb{N}}$$

It is straightforward to check that medium non-malleability implies standard non-malleability.

2.3 Non-Interactive Commitment Scheme

Definition 4 (Commitment Scheme). *A (non-interactive) commitment scheme for the message space $\{0, 1\}^m$, is a pair $(\mathsf{Com}, \mathsf{Open})$ such that:*

- *For all* msg $\in \{0, 1\}^m$, *$(c, d) \leftarrow \mathsf{Com}(m)$ is the commitment/opening pair for the message* msg.
- Open(msg, c, d) $\rightarrow \{0, 1\}$, *where 1 indicates that d is a valid opening of c to* msg *and 0 is returned otherwise.*

The commitment scheme must satisfy the standard correctness requirement,

$$\forall m \in \mathbb{N}, \forall \mathsf{msg} \in \{0, 1\}^m, \ \Pr\left[\mathsf{Open}(\mathsf{msg}, \mathsf{Com}(\mathsf{msg})) = 1\right] = 1$$

where the probability is taken over the randomness of Com.

We will consider *statistically* binding commitment schemes. For the formal definitions of the Hiding and Binding properties, see [4].

Well-formed Commitments: Let val$(\cdot)$ be a function which takes an arbitrary commitment c as an input. val outputs msg if $\exists$ unique msg such that Open(msg, $c, \cdot$) = 1, and outputs $\perp$ otherwise.

Definition 5 (Tag-based Commitment Scheme [50]). *A commitment scheme $(\mathsf{Com}, \mathsf{Open})$ is a tag-based commitment scheme with $\tau(m)$ number of tags if, in addition to the the message* msg, *the sender (committer) and receiver also receive a "tag" of length $\mathsf{poly}(\log(\tau(m)))$ as common input.*

If $\tau(m)$ is exponential in security parameter m, we omit the prefix $\tau(m)$ and refer to the commitment scheme as simply a tag-based commitment scheme.

Man In The Middle Execution (MIM): Let (Com, Open) be a tag-based commitment scheme, and A an adversary. For security parameter m, consider the following interactions by $A(1^m)$:

- *Left interaction:* $A(1^m)$ interacts with the sender and receives commitment to a message msg of length m using identity tag as $c \leftarrow \text{Com}(\text{msg}, \text{tag})$.
- *Right interaction:* $A(1^m)$ interacts with the receiver and tries to commit to related message m͂sg using identity t͂ag of its choice. Specifically, for the commitment $\tilde{c}$ sent to the receiver, let m͂sg $= \text{val}(\tilde{c})$. Furthermore, if t͂ag $=$ tag, then we set m͂sg $= \perp$.

Let $\text{mim}_{\mathcal{C}}^A(\text{msg})$ denote the random variable that describes m͂sg that A commits to in the right interaction along with its output in the MIM execution $\text{MIM}_{\mathcal{C}}^A(\text{msg})$ as described above.

Definition 6 ($O(1/p(m))$-Non-Malleability against $\mathcal{G}$ [50]). *A tag-based commitment scheme $\mathcal{C} = (\text{Com}, \text{Open})$ is said to be $O(1/p(m))$-non-malleable against $\mathcal{G}$ if $\forall A \in \mathcal{G}$, the following ensembles are indistinguishable,*

$$\left\{\text{mim}_{\mathcal{C}}^A(\text{msg}_0)\right\}_{m \in \mathbb{N}, \text{msg}_0 \in \{0,1\}^m} \overset{\mathcal{G}, O(1/p(m))}{\approx} \left\{\text{mim}_{\mathcal{C}}^A(\text{msg}_1)\right\}_{m \in \mathbb{N}, \text{msg}_1 \in \{0,1\}^m}.$$

2.4 Incomputable and Incompressible Functions

Definition 7 (Incomputable Function [1]). *A function $\psi : \{0,1\}^n \rightarrow \{0,1\}^m$ is incomputable by a function class $\mathcal{C}$ if ψ is not contained in $\mathcal{C}$. We say that f is ϵ-incomputable by $\mathcal{C}$ if for every function $C : \{0,1\}^n \rightarrow \{0,1\}^m$ in $\mathcal{C}$, $\Pr[C(x) = f(x)] \leq \frac{1}{2^m} + \epsilon$ for uniform random $x \leftarrow \{0,1\}^n$.*

Definition 8 (Incompressible Function [23]). *A function $f : \{0,1\}^n \rightarrow \{0,1\}^m$ is incompressible by a function $C : \{0,1\}^n \rightarrow \{0,1\}^\ell$ if for every function $D : \{0,1\}^\ell \rightarrow \{0,1\}^m$, there exists $x \in \{0,1\}^n$ such that $D(C(x)) \neq f(x)$. We say that f is ϵ-incompressible by C if for every function $D : \{0,1\}^\ell \rightarrow \{0,1\}^m$, $\Pr[D(C(x)) = f(x)] \leq \frac{1}{2^m} + \epsilon$ for uniform random $x \leftarrow \{0,1\}^n$. We say that f is ℓ-incompressible (resp. (ℓ, ϵ)-incompressible) by a class $\mathcal{C}$ if for every $C : \{0,1\}^n \rightarrow \{0,1\}^\ell$ in $\mathcal{C}$, f is incompressible (resp. ϵ-incompressible) by C.*

Definition 9 (Non-deterministic Circuits and NP Circuits [1]). *A non-deterministic circuit C has additional "non-deterministic input wires". We say that the circuit C evaluates to 1 on x if and only if there exists an assignment to the non-deterministic wires that makes C output 1 on x. An oracle circuit $C^{(\cdot)}$ is a circuit which in addition to the standard gates uses an additional gate (potentially with large fan-in). When instantiated with specific boolean function A, C^A is the circuit in which the additional gate is A. Given boolean function $A(x)$, an A-circuit is a circuit that is allowed to use A gates in addition to the standard gates. An **NP**-circuit is a SAT-circuit (where SAT is the satisfiability function).*

The size of all circuits is the total number of wires and gates.

We now present commonly used assumptions in the derandomization literature to *explicitly* construct pseudorandom generators. [2, 8, 22, 29, 35, 40, 47, 55, 57, 64–68]:

Assumption 2 (E is hard for exponential size X-circuits). *There exists a problem* $\mathbf{L}$ *in* $\mathbf{E} = \mathbf{DTIME}(2^{O(n)})$ *and a constant* $\beta > 0$, *such that for every sufficiently large* n, *X-circuits of size* $2^{\beta n}$ *fail to compute the characteristic function of* $\mathbf{L}$ *on inputs of length* n, *where* $X \in \{non\text{-}deterministic, co\text{-}non\text{-}deterministic, \mathbf{NP}\}$.

Theorem 2 (Theorem 1.3, 1.10 [1]). *If* $\mathbf{E}$ *is hard for exponential size X-circuits, where* $X \in \{non\text{-}deterministic, co\text{-}non\text{-}deterministic, \mathbf{NP}\}$ *(Assumption 2), then for every constant* $c > 1$ *there exists a constant* $a > 1$ *such that for every sufficiently large* n, *and every* r *such that* $a \log n \leq r \leq n$ *there is a function* $\psi : \{0,1\}^r \rightarrow \{0,1\}$ *that is* n^{-c}-*incomputable for size* n^c *X-circuits, Furthermore,* ψ *is computable in time* $\mathrm{poly}(n^c)$ *(or* $\mathrm{poly}(n)$).

We define NIZK without CRS against uniform adversaries and NIWI in [4]. In the remainder of this section, we focus on instantiations of the above primitives.

Theorem 3 ([8]). *Assume that* $\mathbf{E}$ *is hard for exponential size co-non-deterministic circuits and that (subexponentially secure) trapdoor permutations (resp. one-way functions) exist. Then every language in* $\mathbf{NP}$ *has a (subexponentially indistinguishable) NIWI proof system (resp. non-interactive commitment scheme).*

Moreover, by correctly setting the output length of the commitment scheme in terms of the security parameter n, we obtain a non-interactive perfectly binding and computationally hiding commitment scheme, such that given a commitment c, the committed message (i.e., $\mathsf{val}(c)$) can be computed by a 2^{n^ϵ}-time algorithm, where n is the security parameter and ϵ is some constant.

To go from NIWI to NIZK, one can apply the well-known FLS technique [28]. The simulator is provided with a trapdoor via non-uniform advice, which is not known to the uniform adversary in the real world. Note that we choose the trapdoor such that it *can* be obtained by a uniform adversary running in super-polynomial (sub-exponential) time. Formally, [9] show how to construct NIZK without CRS against uniform adversaries under the following assumptions:

Assumption A: There exists a NIWI proof system for every language $L \in \mathbf{NP}$ with WI against sub-exponential adversaries.

Assumption B: There exists a non-interactive perfectly binding and computationally hiding commitment scheme, such that given a commitment, the message can be computed by a 2^{n^ϵ}-time algorithm, where n is the security parameter and ϵ is some constant.

Assumption C: There exists a language $\Delta \in P$ and constants $\epsilon_1 < \epsilon_2 < 1$ such that:

  Δ is hard to sample in time $2^{n^{\epsilon_1}}$: For every probabilistic $2^{n^{\epsilon_1}}$-time algorithm A, the probability that $A(1^n) \in \Delta \cap \{0,1\}^n$ is negligible.

  Δ is easy to sample in time $2^{n^{\epsilon_2}}$: There exists a $2^{n^{\epsilon_2}}$ algorithm S_Δ such that for every $n \in N$, $\Pr[S_\Delta(1^n) \in \Delta \cap \{0,1\}^n] = 1$.

Theorem 4 ([9]). *Under Assumptions A, B and C, there exists a NIZK argument system* without *CRS for* **NP** *with soundness against sub-exponential uniform adversaries and zero-knowledge against sub-exponential adversaries.*

Lemma 1. *If* **E** *is hard for exponential size non-deterministic circuits and* **P**-*cert with soundness against sub-exponential adversaries exists, then Assumption C is true.*

The proof of the lemma can be found in [4].

Corollary 1. *Assuming that* **E** *is hard for exponential size (co-)non-deterministic circuits, the existence of sub-exponential trapdoor permutations, and the existence of* **P**-*cert with soundness against sub-exponential adversaries, there exists a NIZK argument system* without *CRS for* **NP** *with soundness against sub-exponential uniform adversaries and zero knowledge against sub-exponential adversaries.*

3   Construction for Multi-Bit Messages

Let $\mathcal{C} = (\mathsf{Com}, \mathsf{Open})$ be a tag-based, non-interactive commitment scheme that is perfectly binding (see Definition 2.3). Let $\Pi^{\mathsf{NI}} = (\mathsf{P}^{\mathsf{NI}}, \mathsf{V}^{\mathsf{NI}}, \mathsf{Sim}^{\mathsf{NI}})$ be a non-interactive simulatable proof system. Let $\mathcal{S} = (\mathsf{Gen}, \mathsf{Sign}, \mathsf{Ver})$ be a one-time signature scheme. Let D_0, D_1 be disjoint distributions over $\{0,1\}^n$. For $b := b^1, \ldots, b^m \in \{0,1\}^m$, D_b denotes a draw from the product distribution $(D_{b^1}, \ldots, D_{b^m})$. We define the following language:

Language $\mathcal{L}$: $s := ([x^i]_{i \in [m]}, c, \mathsf{tag}) \in \mathcal{L}$ iff $\exists b := b^1, \ldots, b^m \in \{0,1\}^m$ such that for $i \in [m]$, $x^i = (x_1^i, \ldots, x_n^i)$ is in the support of D_{b^i} and c is a commitment to b under tag.

  The construction is presented in Fig. 2:

  Let $\Psi(p, x, y, z)$ be defined as a function that takes as input a predicate p, and variables x, y, z. If $p(x, y) = 1$, then Ψ outputs the m-bit string $\mathbf{0}$. Otherwise, Ψ outputs z.

Theorem 5. *Let* (E, D), E_1, E_2, D' *and* g *be as defined in Figs. 2, 3, 4, 5 and 6. Let* $\mathcal{F}$ *be a computational class. If, for every pair of m-bit messages* b_0, b_1 *and for every tampering function* $f \in \mathcal{F}$, *all of the following hold:*

- *Simulation of proofs.*

   1. $\Pr[g(\mathsf{CW}_0, f(\mathsf{CW}_0)) = 1] \stackrel{\mathsf{negl}(n)}{\approx} \Pr[g(\mathsf{CW}_1, f(\mathsf{CW}_1)) = 1],$

$E(\boldsymbol{b} := b^1, \ldots, b^m)$:

1. Choose $(\mathsf{vk}, \mathsf{SK}) \leftarrow \mathsf{Gen}(1^{n'})$, where $n' \ll n$. We assume WLOG $|\mathsf{vk}| = n'$.
2. Sample $\overline{\boldsymbol{x}} := \boldsymbol{x}^1, \ldots, \boldsymbol{x}^m \leftarrow D_b$, where for $i \in [m]$, $\boldsymbol{x}^i = x_1^i, \ldots, x_n^i$.
3. Compute $(\boldsymbol{c}, \boldsymbol{d}) \leftarrow \mathsf{Com}(\boldsymbol{b}, \mathsf{tag} := \mathsf{vk})$.
4. Compute a non-interactive, simulatable proof T proving $([\boldsymbol{x}^i]_{i \in [m]}, \boldsymbol{c}, \mathsf{vk}) \in \mathcal{L}$.
5. Compute $\sigma \leftarrow \mathsf{Sign}(\mathsf{SK}, ([\boldsymbol{x}^i]_{i \in [m]}, \boldsymbol{c}, T))$.
6. Output $\mathsf{CW} := (\mathsf{vk}, [\boldsymbol{x}^i]_{i \in [m]}, \boldsymbol{c}, T, \sigma)$.

$D(\mathsf{CW})$:

1. Parse $\mathsf{CW} := (\mathsf{vk}, [\boldsymbol{x}^i]_{i \in [m]}, \boldsymbol{c}, T, \sigma)$
2. Check that $\mathsf{Ver}(\mathsf{vk}, \sigma, ([\boldsymbol{x}^i]_{i \in [m]}, \boldsymbol{c}, T)) = 1$.
3. Check that V^{NI} outputs 1 on proof T.
4. If yes, output $[b^i]_{i \in [m]}$ such that for all $i \in [m]$, $x_1^i, \ldots, x_n^i$ is in the support of D_{b^i}. If not, output $\boldsymbol{0}$.

Fig. 2. Non-malleable code (E, D), secure against $\mathcal{F}$ tampering.

$E_1(\mathsf{td}, \boldsymbol{b} := b^1, \ldots, b^m)$:

1. Choose $(\mathsf{vk}, \mathsf{SK}) \leftarrow \mathsf{Gen}(1^{n'})$
2. Sample $\overline{\boldsymbol{x}} := \boldsymbol{x}^1, \ldots, \boldsymbol{x}^m \leftarrow D_b$, where for $i \in [m]$, $\boldsymbol{x}^i = x_1^i, \ldots, x_n^i$.
3. Compute $(\boldsymbol{c}, \boldsymbol{d}) \leftarrow \mathsf{Com}(\boldsymbol{b}, \mathsf{tag} := \mathsf{vk})$.
4. Simulate, using td, a non-interactive proof T' proving $s := ([\boldsymbol{x}^i]_{i \in [m]}, \boldsymbol{c}, \mathsf{vk}) \in \mathcal{L}$.
5. Compute $\sigma \leftarrow \mathsf{Sign}(\mathsf{SK}, ([\boldsymbol{x}^i]_{i \in [m]}, \boldsymbol{c}, T'))$.
6. Output $\mathsf{CW} := (\mathsf{vk}, [\boldsymbol{x}^i]_{i \in [m]}, \boldsymbol{c}, T', \sigma)$.

Fig. 3. Encoding algorithm with simulated proof.

2. $\Psi(g, \mathsf{CW}_0, f(\mathsf{CW}_0), D(f(\mathsf{CW}_0))) \overset{PPT, \mathsf{negl}(n)}{\approx}$
   $\Psi(g, \mathsf{CW}_1, f(\mathsf{CW}_1), D(f(\mathsf{CW}_1)))$,
   where $f \in \mathcal{F}$, $\mathsf{CW}_0 \leftarrow E(b_0)$ and $\mathsf{CW}_1 \leftarrow E_1(\mathsf{td}, b_0)$.
- *Simulation of Commitments.*

  1. $\Pr[g(\mathsf{CW}_1, f(\mathsf{CW}_1)) = 1] \overset{\mathsf{negl}(n)}{\approx} \Pr[g(\mathsf{CW}_2, f(\mathsf{CW}_2)) = 1]$,

  2. $\Psi(g, \mathsf{CW}_1, f(\mathsf{CW}_1), D(f(\mathsf{CW}_1))) \overset{PPT, \mathsf{negl}(n)}{\approx}$
     $\Psi(g, \mathsf{CW}_2, f(\mathsf{CW}_2), D(f(\mathsf{CW}_2)))$,
     where $f \in \mathcal{F}$, $\mathsf{CW}_1 \leftarrow E_1(\mathsf{td}, b_0)$ and $\mathsf{CW}_2 \leftarrow E_2(\mathsf{td}, b_0)$.
- *Simulation Soundness.*

  $$\Pr[D(f(\mathsf{CW}_2)) \neq D'(f(\mathsf{CW}_2)) \wedge g(\mathsf{CW}_2, f(\mathsf{CW}_2)) = 0] \in O(1/n^c),$$

$\mathsf{E}_2(\mathsf{td}, \boldsymbol{b} := b^1, \ldots, b^m)$:

1. Choose $(\mathsf{vk}, \mathsf{SK}) \leftarrow \mathsf{Gen}(1^{n'})$
2. Sample $\overline{\boldsymbol{x}} := \boldsymbol{x}^1, \ldots, \boldsymbol{x}^m \leftarrow D_b$, where for $i \in [m]$, $\boldsymbol{x}^i = x_1^i, \ldots, x_n^i$.
3. Compute $(\boldsymbol{c}', \boldsymbol{d}') \leftarrow \mathsf{Com}(0, \mathsf{tag} := \mathsf{vk})$.
4. Simulate, using td, a non-interactive proof T' proving $s := ([\boldsymbol{x}^i]_{i \in [m]}, \boldsymbol{c}', \mathsf{vk}) \in \mathcal{L}$.
5. Compute $\sigma \leftarrow \mathsf{Sign}(\mathsf{SK}, ([\boldsymbol{x}^i]_{i \in [m]}, \boldsymbol{c}', T'))$.
6. Output $\mathsf{CW} := (\mathsf{vk}, [\boldsymbol{x}^i]_{i \in [m]}, \boldsymbol{c}', T', \sigma)$.

Fig. 4. Encoding algorithm with simulated proof and commitments.

$\mathsf{D}'(\mathsf{CW}) := \mathsf{D}_2'(\mathsf{D}_1'(\mathsf{CW}))$:

$\mathsf{D}_1'(\mathsf{CW})$:

1. Parse $\mathsf{CW} := (\mathsf{vk}, [\boldsymbol{x}^i]_{i \in [m]}, \boldsymbol{c}, T, \sigma)$
2. Check that $\mathsf{Ver}(\mathsf{vk}, \sigma, ([\boldsymbol{x}^i]_{i \in [m]}, \boldsymbol{c}, T)) = 1$.
3. Check that V^{NI} outputs 1 on proof T
4. If not, output $\perp$, where $\perp$ is a special symbol.
5. If yes, output $(\boldsymbol{c}, \mathsf{tag} := \mathsf{vk})$.

$\mathsf{D}_2'(\boldsymbol{c}, \mathsf{tag} := \mathsf{vk})$:

1. If $\boldsymbol{c} = \perp$, output $[0]_{i \in [m]}$ and terminate.
2. Otherwise, check if there exists a string $\boldsymbol{d}$ and a string $\widetilde{\boldsymbol{b}}$ such that $\mathsf{Open}(\boldsymbol{d}, \boldsymbol{c}, \mathsf{vk}, \widetilde{\boldsymbol{b}}) = 1$. If yes, output $\widetilde{\boldsymbol{b}}$. Otherwise, output $[0]_{i \in [m]}$.

Fig. 5. Alternate decoding procedure D'.

$g(\mathsf{CW}, \mathsf{CW}^*)$:

1. Parse $\mathsf{CW} = (\mathsf{vk}, [\boldsymbol{x}^i]_{i \in [m]}, \boldsymbol{c}, T, \sigma)$, $\mathsf{CW}^* = (\mathsf{vk}^*, [\boldsymbol{x}^{*i}]_{i \in [m]}, \boldsymbol{c}^*, T^*, \sigma^*)$.
2. If $\mathsf{vk} = \mathsf{vk}^*$ and $\mathsf{Ver}(\mathsf{vk}^*, \sigma^*, ([\boldsymbol{x}^{*i}]_{i \in [m]}, \boldsymbol{c}^*, T^*)) = 1$ then output 1. Otherwise output 0.

Fig. 6. The predicate $g(\mathsf{CW}, \mathsf{CW}^*)$.

where $f \in \mathcal{F}$, $\mathsf{CW}_2 \leftarrow \mathsf{E}_2(\mathsf{td}, \boldsymbol{b}_0)$.

- **Hardness of D_b relative to Alternate Decoding.**

1. $\Pr[g(\mathsf{CW}_2, f(\mathsf{CW}_2)) = 1] \overset{PPT,O(1/n^c)}{\approx} \Pr[g(\mathsf{CW}_3, f(\mathsf{CW}_3)) = 1]$,

2. *For every Boolean function, represented by a circuit F over m variables,*

$$F \circ \mathsf{D}'(f(\mathsf{CW}_2)) \overset{stat,O(1/n^c)}{\approx} F \circ \mathsf{D}'(f(\mathsf{CW}_3)),$$

where $\mathsf{CW}_2 \leftarrow \mathsf{E}_2(\mathsf{td}, \boldsymbol{b}_0)$, and $\mathsf{CW}_3 \leftarrow \mathsf{E}_2(\mathsf{td}, \boldsymbol{b}_1)$.

Then the construction presented in Fig. 2 is a $O(1/n^c)$-non-malleable code for class $\mathcal{F}$.

We present the proof of Theorem 5 in the full version [4].

4   Multi-Bit NMC Against Bounded Poly Adversaries

We describe the underlying components required to instantiate the generic construction. The tampering class $\mathcal{F}$ corresponds to (uniform) tampering functions that run in time $O(n^{c_A})$, where n is security parameter. The length of the encoding is $L := O(n^{c_\ell})$, for some fixed constant c_ℓ. Therefore, the tampering function is allowed to run in time L^{c_A/c_ℓ} with respect to the input length L.

Let n be the input length for the hard distribution described in Sect. 4.1. We fix polynomials $t_\psi(n) = n^{c_\psi}$, $t_{\mathsf{com}}(n) = n^{c_{\mathsf{com}}}$ where c_ψ, c_{com} are constants (both greater than c_A) and superpolynomial time bounds $T_{\mathsf{com}}(n)$, $T'_{NIZK}(n)$, $T_{ZK}(n)$. such that

- $c_\psi \ll c_{\mathsf{com}}$,
- $T'_{NIZK}(n) \ll T_{\mathsf{com}}(n)$,
- $T_{ZK}(n)$ is subexponential.

The distribution described in Sect. 4.1 is hard for $t_\psi(n)$-time adversaries. $m \cdot \lambda \ll n$ is the length of the m-bit commitment using the commitment scheme described in Sect. 4.2, n is set such that $m \cdot \lambda + n' \leq (m+1) \cdot \lambda \in o(n)$ (so n is asymptotically larger than the length of the commitment–$m \cdot \lambda$–plus the length of the tag–n'.). These commitments are hiding for polynomial-time adversaries and quasi-non-malleable for adversaries in $\mathbf{BPtime}(T_{\mathsf{com}}(n)) \bigcap \mathbf{SIZE}(t_{\mathsf{com}}(n))$. The non-interactive simulatable proof system in Sect. 4.3 has soundness against uniform, poly-time adversaries and zero knowledge against $T_{ZK}(n)$ time adversaries.

4.1   The Hard Distribution D_b (instance length n, hard against $t_\psi(n)$-time adversaries)

Theorem 6 ([1]). *If E is hard for exponential size nondeterministic circuits, then for every constant $c_\psi > 1$, there exists a constant $d > 1$ such that for every sufficiently large n, there is a function $\psi : \{0,1\}^n \rightarrow \{0,1\}$ that is (ℓ, n^{-c_ψ})-incompressible for size n^{c_ψ} circuits, where $\ell = n - d \cdot \log n$. Furthermore, ψ is computable in time $\mathrm{poly}(n^{c_\psi}) \in O(n^{c_{\mathsf{com}}})$.*

Setting parameters n, c_ψ, d as above, we let D_b be the uniform distribution over $x \leftarrow \{0,1\}^n$, conditioned on $\psi(x) = b$. The theorem above immediately implies the following:

Claim. Let n, c_ψ, d, ψ be as above, let $\widetilde{F}$ be any Boolean function over $(m+1)\cdot\lambda \leq n - d\cdot\log n < (1-\alpha)n$ variables, and let C be a size n^{c_ψ} circuit with input length n and output length m. Then, over random choice of $x \leftarrow \{0,1\}^n$, $\widetilde{F} \circ C(x)$ has correlation at most $1/n^{-c_\psi}$ with $\psi(x)$.

4.2 Commitment scheme $\mathcal{C} = (\mathsf{Com}, \mathsf{Open})$ (length $\lambda \ll n$, hiding for poly-time adversaries, and quasi non-malleable against adversaries in $\mathbf{BPtime}(T_{\mathsf{com}}(n)) \cap \mathbf{SIZE}(t_{\mathsf{com}}(n))$)

We instantiate the commitment scheme $\mathcal{C} = (\mathsf{Com}, \mathsf{Open})$ with the scheme presented in [4]. Recall that the scheme has the following properties:

- Non-interactive with no-CRS.
- Perfectly binding,
- Quasi-non-malleable against in $\mathbf{BPtime}(T_{\mathsf{com}}(n)) \cap \mathbf{SIZE}(t_{\mathsf{com}}(n))$.

4.3 Non-Interactive Simulatable Proof System (Sound against uniform ppt adversaries, ZK against adversaries running in time $T_{ZK}(n)$)

Let $\Pi = (\mathsf{P}, \mathsf{V}, \mathsf{Sim})$ be a NIZK proof system for NP with no CRS (Construction given in [4]) with soundness against uniform adversaries running in time $T_{NIZK}(n)$. We additionally require that the trapdoor can be extracted by uniform adversaries running in time $T'_{NIZK}(n)$.

Let $\mathcal{C}' = (\mathsf{Com}', \mathsf{Open}')$ be a non-interactive, perfectly binding, commitment scheme with no CRS that can be extracted in time $T_{NIZK}(n)$ and is hiding against adversaries running in time $T_{ZK}(n)$.

We also assume the existence of **P**-certificates with soundness against adversaries running in time $T_{NIZK}(n)$.

We define the proof system $\Pi^{\mathsf{NI}} = (\mathsf{P}^{\mathsf{NI}}, \mathsf{V}^{\mathsf{NI}}, \mathsf{Sim}^{\mathsf{NI}})$ for language $\mathcal{L}$ defined in Sect. 3 as follows:

P^{NI}: Recall that a witness w for statement $s := ([x^i]_{i\in[m]}, c, \mathsf{tag}) \in \mathcal{L}$ consists of a string $b = b^1, \ldots, b^m$ and an opening d such that (1) $\mathsf{Open}(c, b, \mathsf{tag}) = 1$ and (2) for all $i \in [m]$, $\psi(x^i) = b^i$. Given a statement s and witness w, let P be a **P**-certificate that (1) and (2) hold.

Invoke P from proof system Π with the statement $s' = (s, \mathsf{com}) \in \mathcal{L}'$ using proof system Π, where $\mathcal{L}'$ is the language consisting of strings (s, com) such that com is a commitment to (w, P) and P is a **P**-certificate that (1) and (2) hold for (s, w). P outputs a proof π'. P^{NI} outputs proof $\pi = \mathsf{com}\|\pi'$.

V^{NI}: On input statement s, proof π and language $\mathcal{L}$: Parse $\pi := \text{com}||\pi'$. Run the underlying verifier V on π' for statement (s, com) and language $\mathcal{L}'$ and output whatever it does.

Sim^{NI}: On input (td, x), and language $\mathcal{L}$: Set com to a commitment to 0 and invoke the underlying Sim for Π with input $(\text{td}, (s, \text{com}))$ and language $\mathcal{L}'$.

Note that given the **P**-certificate P, computing the NIZK proof using Π^{NI} can be done in fixed polynomial time in the length of the statement (s, com). Moreover, given the trapdoor td, a simulated proof can also be computed in fixed polynomial time. The following claim is straightforward.

Claim. Given the above assumptions, $\Pi^{NI} = (P^{NI}, V^{NI}, \text{Sim}^{NI})$ is a NIZK argument system for language $\mathcal{L}$ with zero knowledge against adversaries running in time $T_{ZK}(n)$ and trapdoor that can be extracted in time $T'_{NIZK}(n)$.

4.4 Main Theorem

Theorem 7. *For any constant $c_A > 1$, $\Pi = (E, D)$ (presented in Fig. 2) is a multi-bit, non-malleable code against (uniform) tampering functions that run in time $O(n^{c_A})$, if parameters $c_\psi, c_{\text{com}}, T_{\text{com}}(n), T'_{NIZK}(n), T_{ZK}(n)$ are chosen as described above and the underlying components are instantiated in the following way:*

- *For $b \in \{0, 1\}$, D_b is the distribution from Sect. 4.1.*
- *$\mathcal{C} := (\text{Com}, \text{Open})$ is the commitment scheme from Sect. 4.2.*
- *$\Pi^{NI} := (P^{NI}, V^{NI}, \text{Sim}^{NI})$ the simulatable proof system from Sect. 4.3.*
- *$\mathcal{S} := (\text{Gen}, \text{Sign}, \text{Ver})$ is any one-time signature scheme secure against PPT adversaries.*

Proof. To prove the theorem, we need to show that the necessary properties from Theorem 5 hold. We next go through these one by one.

Simulation of proofs.

1. $\Pr[g(\text{CW}_0, f(\text{CW}_0)) = 1] \overset{\text{negl}(n)}{\approx} \Pr[g(\text{CW}_1, f(\text{CW}_1)) = 1]$,
2. $\Psi(g, \text{CW}_0, f(\text{CW}_0), D(f(\text{CW}_0))) \overset{PPT, \text{negl}(n)}{\approx} \Psi(g, \text{CW}_1, f(\text{CW}_1), D(f(\text{CW}_1)))$,

where $f \in \mathcal{F}$, $\text{CW}_0 \leftarrow E(b_0)$ and $\text{CW}_1 \leftarrow E_1(\text{td}, b_0)$.
    This follows by ZK property of Π^{NI}.

Simulation of Commitment.

1. $\Pr[g(\text{CW}_1, f(\text{CW}_1)) = 1] \overset{\text{negl}(n)}{\approx} \Pr[g(\text{CW}_2, f(\text{CW}_2)) = 1]$,
2. $\Psi(g, \text{CW}_1, f(\text{CW}_1), D(f(\text{CW}_1))) \overset{PPT, \text{negl}(n)}{\approx} \Psi(g, \text{CW}_2, f(\text{CW}_2), D(f(\text{CW}_2)))$,

where $f \in \mathcal{F}$, $\text{CW}_1 \leftarrow E_1(\text{td}, b_0)$ and $\text{CW}_2 \leftarrow E_2(\text{td}, b_0)$.
    This follows from hiding property of the commitment scheme $\mathcal{C}$.

Simulation Soundness.

$$\Pr_r[\mathsf{D}(f(\mathsf{CW}_2)) \neq \mathsf{D}'(f(\mathsf{CW}_2)) \wedge g(\mathsf{CW}_2, f(\mathsf{CW}_2)) = 0] \in O(1/n^{c_{\text{com}}}),$$

where $f \in \mathcal{F}$, $\mathsf{CW}_2 \leftarrow \mathsf{E}_2(\mathsf{td}, \boldsymbol{b}_0)$.

We begin by defining the following:

$$P_0(n) := \Pr[\mathsf{D}(f(\mathsf{CW}_0)) \neq \mathsf{D}'(f(\mathsf{CW}_0)) \wedge g(\mathsf{CW}_0, f(\mathsf{CW}_0)) = 0],$$

where $f \in \mathcal{F}$, $\mathsf{CW}_0 \leftarrow \mathsf{E}(\boldsymbol{b}_0)$

$$P_1(n) := \Pr_r[\mathsf{D}(f(\mathsf{CW}_1)) \neq \mathsf{D}'(f(\mathsf{CW}_1)) \wedge g(\mathsf{CW}_1, f(\mathsf{CW}_1)) = 0],$$

where $f \in \mathcal{F}$, $\mathsf{CW}_1 \leftarrow \mathsf{E}_1(\mathsf{td}, \boldsymbol{b}_0)$

$$P_2(n) := \Pr_r[\mathsf{D}(f(\mathsf{CW}_2)) \neq \mathsf{D}'(f(\mathsf{CW}_2)) \wedge g(\mathsf{CW}_2, f(\mathsf{CW}_2)) = 0],$$

where $f \in \mathcal{F}$, $\mathsf{CW}_2 \leftarrow \mathsf{E}_2(\mathsf{td}, \boldsymbol{b}_0)$.

We prove the following sequence of claims, which immediately imply the simulation soundness property.

Claim. $P_0(n) \in \mathsf{negl}(n)$.

Since $\mathsf{D}(f(\mathsf{CW}_1)) \neq \mathsf{D}'(f(\mathsf{CW}_1))$ can only occur if the NIZK proof verifies, but the statement being proved is false, this follows from the soundness of the NIZK proof system Π^{NI}.

Claim. $|P_1(n) - P_0(n)| \in \mathsf{negl}(n)$.

This holds due to complexity leveraging–i.e. by appropriately setting parameters, one can check whether the statement being proved is true or false (by deciding whether $\boldsymbol{x}$ is in the support of D_0 or D_1 and by extracting from the commitment scheme) without distinguishing a real from simulated proof since $T_{ZK}(n)$ is subexponential.

Claim. $|P_2(n) - P_1(n)| \in O(1/n^{c_{\text{com}}})$.

Proof. Assume $|P_2(n) - P_1(n)| \notin O(1/n^{c_{\text{com}}})$, we will construct an adversary/distinguisher (A, D) in $\mathbf{BPtime}(T_{\text{com}}(n)) \cap \mathbf{SIZE}(t_{\text{com}}(n))$ that breaks the $O(1/n^{c_{\text{com}}})$-non-malleability of commitment scheme $\mathcal{C}$. Specifically, we must show an adversary A, distinguisher D in $\mathbf{BPtime}(T_{\text{com}}(n)) \cap \mathbf{SIZE}(t_{\text{com}}(n))$ such that D distinguishes the output of $\mathsf{mim}_{\mathcal{C}}^A(\boldsymbol{b}_0)$ from $\mathsf{mim}_{\mathcal{C}}^A(\boldsymbol{0})$ with advantage $a(n) \notin O(1/n^{c_{\text{com}}})$.

$A = (A_1, A_2)$ is specified as follows:
On input security parameter 1^n, A_1 does as follows:

- A_1 generates keys $(\mathsf{vk}, \mathsf{SK}) \leftarrow \mathsf{Gen}(1^n)$
- A_1 runs in uniform time $T'_{NIZK}(n) \leq T_{\text{com}}(n)$ to recover the trapdoor td of the NIZK.

- A_1 outputs $\mathsf{tag} := \mathsf{vk}$ to its challenger as the desired tag and outputs td, SK to A_2.

On input $\mathsf{td}, \mathsf{SK}, \mathsf{vk}, c, A_2$ does as follows:

- For $i \in [m]$, sample $\boldsymbol{x}^i \sim D_{b^i}$ (in time $m \cdot \mathrm{poly}(n^{c_\psi}) \in O(n^{c_\mathrm{com}})$, where poly is a fixed polynomial.
- Use td to generate a simulated proof T in fixed polynomial time and compute $\sigma \leftarrow \mathsf{Sign}(\mathsf{SK}, ([\boldsymbol{x}^i]_{i\in[m]}, c, T))$ in fixed polynomial time.
- Compute $f(\mathsf{vk}, [\boldsymbol{x}^i]_{i\in[m]}, c, T, \sigma) = [\mathsf{vk}', \boldsymbol{x}'^i]_{i\in[m]}, c', T', \sigma')$.
- If the predicate g evaluates to 1, the signature σ' or proof T does not verify, output $\bot$ (this computation takes fixed polynomial time).
- Otherwise, output $(c', \mathsf{out} := [\boldsymbol{x}'^i]_{i\in[m]})$. Note that in this case, $\mathsf{vk}' \neq \mathsf{vk}$ (corresponding to the tag of the commitment) since g evaluates to 0 and σ verifies.

Distinguisher D receives the committed value $\boldsymbol{v}' = v'_1, \ldots, v'_m$ underlying c' (or receives $\bot$) as well as out (the additional output of adversary A). D outputs 0 if for all $i \in [m]$, $v'_i = \psi(\boldsymbol{x}^i)$ (or if its input is $\bot$) and outputs 1 otherwise (computed in time $m \cdot \mathrm{poly}(n^{c_\psi}) \in O(n^{c_\mathrm{com}})$).

Clearly,

$$\Pr_{c \leftarrow \mathsf{Com}(\boldsymbol{b}_0, \mathsf{vk})}[D(\boldsymbol{v}', \mathsf{out}) = 1] = P_2(n), \qquad \text{and}$$

$$\Pr_{c \leftarrow \mathsf{Com}(\boldsymbol{0}, \mathsf{vk})}[D(\boldsymbol{v}', \mathsf{out}) = 1] = P_1(n)$$

Thus, we have that

$$\left| \Pr_{c \leftarrow \mathsf{Com}(\boldsymbol{b}_0, \mathsf{vk})}[D(\boldsymbol{v}', \mathsf{out}) = 1] - \Pr_{c \leftarrow \mathsf{Com}(\boldsymbol{0}, \mathsf{vk})}[D(\boldsymbol{v}', \mathsf{out}) = 1] \right| \notin O(1/n^{c_\mathrm{com}}).$$

Moreover, A, D are in $\mathbf{BPtime}(T_\mathsf{com}(n)) \cap \mathbf{SIZE}(t_\mathsf{com}(n))$. Thus, we obtain a contradiction to the $O(1/n^{c_\mathrm{com}})$ non-malleability of the commitment scheme against adversaries, distinguishers in $\mathbf{BPtime}(T_\mathsf{com}(n)) \cap \mathbf{SIZE}(t_\mathsf{com}(n))$.

Hardness of D_b relative to Alternate Decoding.

1. $\Pr[g(\mathsf{CW}_2, f(\mathsf{CW}_2)) = 1] \overset{O(1/n^{c_\psi})}{\approx} \Pr[g(\mathsf{CW}_3, f(\mathsf{CW}_3)) = 1]$,

2. For every Boolean function, represented by a circuit F over m variables,

$$F \circ \mathsf{D}'(f(\mathsf{CW}_2)) \overset{stat, O(1/n^{c_\psi})}{\approx} F \circ \mathsf{D}'(f(\mathsf{CW}_3)),$$

where $f \in \mathcal{F}$, $\mathsf{CW}_2 \leftarrow \mathsf{E}_2(\mathsf{td}, \boldsymbol{b}_0)$ and $\mathsf{CW}_3 \leftarrow \mathsf{E}_2(\mathsf{td}, \boldsymbol{b}_1)$.

We consider a sequence of distributions where we switch the internal random variables of E_2 from $\boldsymbol{x}^i \leftarrow D_{b_0^i}$, for all $i \in [m]$ to $\boldsymbol{x}^i \leftarrow D_{b_1^i}$, for all $i \in [m]$. Namely, for each $i \in \{0, \ldots, m\}$ we consider a distribution where for $j \leq i$, $\boldsymbol{x}^j \leftarrow D_{b_1^i}$ and for $j > i$, $\boldsymbol{x}^j \leftarrow D_{b_0^i}$.

We must show that (1) and (2) hold for each consecutive pair of distributions. When considering the i-th consecutive pair, fix all random variables except the i-th variable $\boldsymbol{X}^i$ to values $\boldsymbol{x}^1, \ldots, \boldsymbol{x}^{i-1}, \boldsymbol{x}^{i+1}, \ldots, \boldsymbol{x}^m$. Let $\boldsymbol{X}^i$ be a random variable such that with probability $1/2$, $\boldsymbol{X}^i \leftarrow D_{b_0^i}$ and with probability $1/2$, $\boldsymbol{X}^i \leftarrow D_{b_1^i}$. $\boldsymbol{X}^i = \boldsymbol{X}^{i,\gamma}$ where $\gamma \leftarrow \{0,1\}$, and let random variable CW^i denote the output of E_2 when using random variables $\boldsymbol{x}^1, \ldots, \boldsymbol{x}^{i-1}, \boldsymbol{X}^i, \boldsymbol{x}^{i+1}, \ldots, \boldsymbol{x}^m$.

Since proving (1) is similar, but more straightforward than proving (2), we defer the proof of (1) to [4] and proceed to prove (2) next.

To show (2), assume $\mathsf{D}'(f(\mathsf{CW}_2))$ and $\mathsf{D}'(f(\mathsf{CW}_3))$ have greater than $1/n^{c_\psi}$ statistical distance. This implies that there exists a distinguisher F (represented by an m-bit Boolean function) such that $F \circ \mathsf{D}'(f(\mathsf{CW}_2))$ is more than $1/n^{c_\psi}$-far from $F \circ \mathsf{D}'(f(\mathsf{CW}_3))$. This implies that, for some $i \in [m]$, the output of $F \circ \mathsf{D}'(f(\mathsf{CW}^i))$ is $a(n) \notin O(1/n^{c_\psi})$-correlated with $\psi(\boldsymbol{X}^i)$. Note that, by definition, $F \circ \mathsf{D}'(f(\mathsf{CW}^i)) = F \circ \mathsf{D}'_2 \circ \mathsf{D}'_1(f(\mathsf{CW}^i))$, where D'_1 has output length $(m+1) \cdot \lambda$ ($m \cdot \lambda$ for the size of the non-malleable commitment and λ for the length of the tag of the non-malleable commitment). We will show that $\mathsf{D}'_1(f(\mathsf{CW}^i))$ can be computed by a circuit C of size $O(n^{c_\psi})$ (drawn from some distribution $\mathcal{C}$ over circuits) with input $\boldsymbol{X}^i$. We then use Claim 4.1, which says that if C is a size $O(n^{c_\psi})$ circuit taking inputs of length n bits and producing outputs of length $(m+1) \cdot \lambda < (1-\alpha)n$-bits and $\widetilde{F}$ is any $(m+1) \cdot \lambda < (1-\alpha)n$-bit input Boolean function then the output of $\widetilde{F}(C(\boldsymbol{X}^i))$ is at most $O(1/n^{c_\psi})$-correlated with $\psi(\boldsymbol{X}^i)$, instantiating $\widetilde{F} := F \circ \mathsf{D}'_2$. This yields a contradiction. Details follow.

Given non-uniform advice td, f, we construct the distribution of circuits $\mathcal{C}^2_{f,\mathsf{td}}$. A draw $C \sim \mathcal{C}^2_{f,\mathsf{td}}$ as follows:

1. Sample signature keys $(\mathsf{vk}, \mathsf{SK}) \leftarrow \mathsf{Gen}(1^n)$,
2. Sample random commitment to 0^m: $(\boldsymbol{c}', \boldsymbol{d}') \leftarrow \mathsf{Com}(0^m, \mathsf{tag} := \mathsf{vk})$,
3. Sample $\boldsymbol{x}^1, \ldots, \boldsymbol{x}^{i-1}$ from $D_{b_0^i}$, and $\boldsymbol{x}^{i+1}, \ldots, \boldsymbol{x}^m$ from $D_{b_1^i}$.
4. Output the following circuit C that has the following structure:
   - **hardcoded variables:** f, $\boldsymbol{x}^1, \ldots, \boldsymbol{x}^{i-1}$, $\overline{c}'$, $[T_j^{'\beta,i}]_{\beta \in \{0,1\}, i \in [m], j \in [n]}$, $\boldsymbol{x}^1, \ldots, \boldsymbol{x}^{i-1}, \boldsymbol{x}^{i+1}, \ldots, \boldsymbol{x}^m$.
   - **input:** $\boldsymbol{X}^i$.
   - **computes and outputs:** $\mathsf{D}'_1(f(\mathsf{CW}^i))$.

Given all the hardwired variables, computing CW^i can be done in time $O(n^{c_\psi})$ since it only requires computing the simulated proof T and signature σ, which can both be done in fixed polynomial time less than n^{c_ψ}. Additionally, f can be computed in time $n^{c_A} < n^{c_\psi}$, and D'_1 can be computed in fixed polynomial time less than n^{c_ψ}, since it only involves verifying the signature σ and proof T, which both take fixed polynomial time.

Acknowledgments. The first and fifth authors are supported in part by NSF grant #CCF1423306 and the Leona M. & Harry B. Helmsley Charitable Trust. The first author is additionally supported in part by an IBM Research PhD Fellowship. The second and third authors are supported in part by NSF grants #CNS-1840893, #CNS-1453045 (CAREER), by a research partnership award from Cisco and by financial assistance award 70NANB15H328 from the U.S. Department of Commerce, National

Institute of Standards and Technology. The fourth author is supported by NSF grants #CNS-1528178, #CNS-1514526, #CNS-1652849 (CAREER), a Hellman Fellowship, the Defense Advanced Research Projects Agency (DARPA) and Army Research Office (ARO) under Contract No. W911NF-15-C-0236, and a subcontract No. 2017-002 through Galois. The views expressed are those of the authors and do not reflect the official policy or position of the Department of Defense, the National Science Foundation, or the U.S. Government. This work was performed, in part, while the first author was visiting IDC Herzliya's FACT center and supported in part by ISF grant no. 1790/13 and the Check Point Institute for Information Security.

References

1. Applebaum, B., Artemenko, S., Shaltiel, R., Yang, G.: Incompressible functions, relative-error extractors, and the power of nondeterministic reductions. Comput. Complex. **25**(2), 349–418 (2016). https://doi.org/10.1007/s00037-016-0128-9
2. Babai, L., Fortnow, L., Nisan, N., Wigderson, A.: BPP has subexponential time simulations unlessexptime has publishable proofs. Comput. Complex. **3**(4), 307–318 (1993). https://doi.org/10.1007/BF01275486
3. Ball, M., Dachman-Soled, D., Guo, S., Malkin, T., Tan, L.Y.: Non-malleable codes for small-depth circuits. FOCS IEEE Computer Society Press, October 2018 (to appear). https://eprint.iacr.org/2018/207
4. Ball, M., Dachman-Soled, D., Kulkarni, M., Lin, H., Malkin, T.: Non-malleablecodes against bounded polynomial time tampering. Cryptology ePrint Archive, Report 2018/1015 (2018). https://eprint.iacr.org/2018/1015
5. Ball, M., Dachman-Soled, D., Kulkarni, M., Malkin, T.: Non-malleable codesfor bounded depth, bounded fan-in circuits. In: Fischlin and Coron [30], pp. 881–908
6. Ball, M., Dachman-Soled, D., Kulkarni, M., Malkin, T.: Non-malleable codes from average-case hardness: AC^0, decision trees, and streaming space-bounded tampering. In: Nielsen, J.B., Rijmen, V. (eds.) EUROCRYPT 2018, Part III. LNCS, vol. 10822, pp. 618–650. Springer, Cham (2018). https://doi.org/10.1007/978-3-319-78372-7_20
7. Barak, B.: Constant-round coin-tossing with a man in the middle or realizing the shared random string model. In: 43rd FOCS, pp. 345–355. IEEE Computer Society Press, November 2002
8. Barak, B., Ong, S.J., Vadhan, S.: Derandomization in cryptography. SIAM J. Comput. **37**(2), 380–400 (2007). https://doi.org/10.1137/050641958
9. Barak, B., Pass, R.: On the possibility of one-message weak zero-knowledge. In: Naor, M. (ed.) TCC 2004. LNCS, vol. 2951, pp. 121–132. Springer, Heidelberg (2004). https://doi.org/10.1007/978-3-540-24638-1_7
10. Bitansky, N., Lin, H.: One-message zero knowledge and non-malleable commitments. Cryptology ePrint Archive, Report 2018/613 (2018). https://eprint.iacr.org/2018/613
11. Chandran, N., Goyal, V., Mukherjee, P., Pandey, O., Upadhyay, J.: Block-wise non-malleable codes. In: Chatzigiannakis, I., Mitzenmacher, M., Rabani, Y., Sangiorgi, D. (eds.) ICALP 2016. LIPIcs, vol. 55, pp. 31:1–31:14. Schloss Dagstuhl (2016)
12. Chattopadhyay, E., Goyal, V., Li, X.: Non-malleable extractors and codes, withtheir many tampered extensions. In: Wichs and Mansour [69], pp. 285–298

13. Chattopadhyay, E., Li, X.: Non-malleable codes and extractors for small-depth circuits, and affine functions. In: Hatami, H., McKenzie, P., King, V. (eds.) 49th ACM STOC, pp. 1171–1184. ACM Press, June 2017
14. Cheraghchi, M., Guruswami, V.: Capacity of non-malleable codes. In: Naor, M. (ed.) ITCS 2014, pp. 155–168. ACM, January 2014
15. Chung, K.M., Lin, H., Pass, R.: Constant-round concurrent zero knowledge from P-certificates. In: FOCS 2013 [32] , pp. 50–59
16. Ciampi, M., Ostrovsky, R., Siniscalchi, L., Visconti, I.: Concurrent non-malleable commitments (and more) in 3 rounds. In: Robshaw, M., Katz, J. (eds.) CRYPTO 2016, Part III. LNCS, vol. 9816, pp. 270–299. Springer, Heidelberg (2016). https://doi.org/10.1007/978-3-662-53015-3_10
17. Ciampi, M., Ostrovsky, R., Siniscalchi, L., Visconti, I.: Four-round concurrentnonmalleable commitments from one-way functions. In: Katz and Shacham [44], pp. 127–157
18. Coron, J.S., Holenstein, T., Künzler, R., Patarin, J., Seurin, Y., Tessaro, S.: How to build an ideal cipher: the indifferentiability of the Feistel construction. J. Cryptol. **29**(1), 61–114 (2016)
19. Dachman-Soled, D., Katz, J., Thiruvengadam, A.: 10-round Feistel isindifferentiable from an ideal cipher. In: Fischlin and Coron [30], pp. 649–678
20. Dai, Y., Steinberger, J.: Indifferentiability of 8-round feistel networks. In: Robshaw, M., Katz, J. (eds.) CRYPTO 2016, Part I. LNCS, vol. 9814, pp. 95–120. Springer, Heidelberg (2016). https://doi.org/10.1007/978-3-662-53018-4_4
21. Dolev, D., Dwork, C., Naor, M.: Nonmalleable cryptography. SIAM Rev. **45**(4), 727–784 (2003)
22. Drucker, A.: Nondeterministic direct product reductions and the success probability of SAT solvers. In: FOCS 2013 [32], pp. 736–745
23. Dubrov, B., Ishai, Y.: On the randomness complexity of efficient sampling. In: Kleinberg, J.M. (ed.) 38th ACM STOC, pp. 711–720. ACM Press, May 2006
24. Dwork, C., Naor, M.: Zaps and their applications. In: FOCS 2000 [31], pp. 283–293
25. Dziembowski, S., Pietrzak, K., Wichs, D.: Non-malleable codes. In: Yao, A.C.C. (ed.) ICS 2010, pp. 434–452. Tsinghua University Press, January 2010
26. Faust, S., Hostáková, K., Mukherjee, P., Venturi, D.: Non-malleablecodes for space-bounded tampering. In: Katz and Shacham [44], pp. 95–126
27. Faust, S., Mukherjee, P., Venturi, D., Wichs, D.: Efficient non-malleable codes and key-derivation for poly-size tampering circuits. In: Nguyen, P.Q., Oswald, E. (eds.) EUROCRYPT 2014. LNCS, vol. 8441, pp. 111–128. Springer, Heidelberg (2014). https://doi.org/10.1007/978-3-642-55220-5_7
28. Feige, U., Lapidot, D., Shamir, A.: Multiple noninteractive zero knowledge proofs under general assumptions. SIAM J. Comput. **29**(1), 1–28 (1999)
29. Feige, U., Lund, C.: On the hardness of computing the permanent of random matrices. Comput. Complex. **6**(2), 101–132 (1997)
30. Fischlin, M., Coron, J.-S. (eds.): EUROCRYPT 2016, Part II. LNCS, vol. 9666. Springer, Heidelberg (2016). https://doi.org/10.1007/978-3-662-49896-5
31. 41st FOCS. IEEE Computer Society Press, November 2000
32. 54th FOCS. IEEE Computer Society Press, October 2013
33. 58th FOCS. IEEE Computer Society Press (2017)
34. Fortnow, L., Vadhan, S.P. (eds.): 43rd ACM STOC. ACM Press, June 2011
35. Goldreich, O., Wigderson, A.: Derandomization that is rarely wrong from short advice that is typically good. In: Rolim, J.D.P., Vadhan, S. (eds.) RANDOM 2002. LNCS, vol. 2483, pp. 209–223. Springer, Heidelberg (2002). https://doi.org/10.1007/3-540-45726-7_17

36. Goyal, V.: Constant round non-malleable protocols using one way functions. In: Fortnow and Vadhan [34], pp. 695–704
37. Goyal, V., Pandey, O., Richelson, S.: Textbook non-malleable commitments. In: Wichs and Mansour [69], pp. 1128–1141
38. Goyal, V., Richelson, S., Rosen, A., Vald, M.: An algebraic approach to non-malleability. In: 55th FOCS, pp. 41–50. IEEE Computer Society Press, October 2014
39. Groth, J., Ostrovsky, R., Sahai, A.: Non-interactive zaps and new techniques for NIZK. In: Dwork, C. (ed.) CRYPTO 2006. LNCS, vol. 4117, pp. 97–111. Springer, Heidelberg (2006). https://doi.org/10.1007/11818175_6
40. Gutfreund, D., Shaltiel, R., Ta-Shma, A.: Uniform hardness versus randomness tradeoffs for Arthur-Merlin games. Comput. Complex. **12**(3–4), 85–130 (2003)
41. Harnik, D., Naor, M.: On the compressibility of $\mathcal{NP}$ instances and cryptographic applications. SIAM J. Comput. **39**(5), 1667–1713 (2010)
42. Håstad, J., Impagliazzo, R., Levin, L.A., Luby, M.: A pseudorandom generator from any one-way function. SIAM J. Comput. **28**(4), 1364–1396 (1999)
43. Impagliazzo, R., Wigderson, A.: P = BPP if E requires exponential circuits: derandomizing the XOR lemma. In: 29th ACM STOC, pp. 220–229. ACM Press, May 1997
44. Katz, J., Shacham, H. (eds.): CRYPTO 2017, Part II. LNCS, vol. 10402. Springer, Cham (2017). https://doi.org/10.1007/978-3-319-63715-0
45. Khurana, D.: Round optimal concurrent non-malleability from polynomial hardness. In: Kalai, Y., Reyzin, L. (eds.) TCC 2017, Part II. LNCS, vol. 10678, pp. 139–171. Springer, Cham (2017). https://doi.org/10.1007/978-3-319-70503-3_5
46. Khurana, D., Sahai, A.: How to achieve non-malleability in one or two rounds. In: FOCS 2017 [33], pp. 564–575
47. Klivans, A.R., Van Melkebeek, D.: Graph nonisomorphism has subexponential size proofs unless the polynomial-time hierarchy collapses. SIAM J. Comput. **31**(5), 1501–1526 (2002)
48. Lin, H., Pass, R.: Non-malleability amplification. In: Mitzenmacher, M. (ed.) 41st ACM STOC, pp. 189–198. ACM Press, May/June 2009
49. Lin, H., Pass, R.: Constant-round non-malleable commitments from any one-way function. In: Fortnow and Vadhan [34], pp. 705–714
50. Lin, H., Pass, R., Soni, P.: Two-round and non-interactive concurrent non-malleable commitments from time-lock puzzles. In: FOCS 2017 [33], pp. 576–587
51. Lin, H., Pass, R., Venkitasubramaniam, M.: Concurrent non-malleable commitments from any one-way function. In: Canetti, R. (ed.) TCC 2008. LNCS, vol. 4948, pp. 571–588. Springer, Heidelberg (2008). https://doi.org/10.1007/978-3-540-78524-8_31
52. Lindell, Y.: A simpler construction of CCA2-secure public-key encryption under general assumptions. In: Biham, E. (ed.) EUROCRYPT 2003. LNCS, vol. 2656, pp. 241–254. Springer, Heidelberg (2003). https://doi.org/10.1007/3-540-39200-9_15
53. Lipton, R.J.: New directions in testing. In: Feigenbaum, J., Merritt, M. (eds.) Distributed Computing and Cryptography, Proceedings of a DIMACS Workshop, Princeton, New Jersey, USA, 4–6 October 1989, pp. 191–202 (1989)
54. Micali, S.: Computationally sound proofs. SIAM J. Comput. **30**(4), 1253–1298 (2000)
55. Miltersen, P.B., Vinodchandran, N.V.: Derandomizing Arthur-Merlin games using hitting sets. Comput. Complex. **14**(3), 256–279 (2005)
56. Naor, M., Yung, M.: Public-key cryptosystems provably secure against chosen ciphertext attacks. In: 22nd ACM STOC, pp. 427–437. ACM Press, May 1990

57. Nisan, N., Wigderson, A.: Hardness vs randomness. J. Comput. Syst. Sci. **49**(2), 149–167 (1994). https://doi.org/10.1016/S0022-0000(05)80043-1

58. Ostrovsky, R., Persiano, G., Venturi, D., Visconti, I.: Continuously non-malleable codes in the split-state model from minimal assumptions. Cryptology ePrint Archive, Report 2018/542 (2018). https://eprint.iacr.org/2018/542

59. Pass, R., Rosen, A.: Concurrent non-malleable commitments. In: 46th FOCS, pp. 563–572. IEEE Computer Society Press, October 2005

60. Pass, R., Rosen, A.: New and improved constructions of non-malleable cryptographic protocols. In: Gabow, H.N., Fagin, R. (eds.) 37th ACM STOC, pp. 533–542. ACM Press, May 2005

61. Pass, R., Wee, H.: Constant-round non-malleable commitments from sub-exponential one-way functions. In: Gilbert, H. (ed.) EUROCRYPT 2010. LNCS, vol. 6110, pp. 638–655. Springer, Heidelberg (2010). https://doi.org/10.1007/978-3-642-13190-5_32

62. Rivest, R.L., Shamir, A., Wagner, D.A.: Time-lock puzzles and timed-release crypto (1996)

63. Sahai, A.: Non-malleable non-interactive zero knowledge and adaptive chosen-ciphertext security. In: 40th FOCS, pp. 543–553. IEEE Computer Society Press, October 1999

64. Shaltiel, R., Umans, C.: Simple extractors for all min-entropies and a new pseudorandom generator. J. ACM (JACM) **52**(2), 172–216 (2005)

65. Shaltiel, R., Umans, C.: Pseudorandomness for approximate counting and sampling. Comput. Complex. **15**(4), 298–341 (2006)

66. Shaltiel, R., Umans, C.: Low-end uniform hardness versus randomness tradeoffs for AM. SIAM J. Comput. **39**(3), 1006–1037 (2009)

67. Sudan, M., Trevisan, L., Vadhan, S.: Pseudorandom generators without the XOR Lemma. J. Comput. Syst. Sci. **62**(2), 236–266 (2001). http://www.sciencedirect.com/science/article/pii/S0022000000917306

68. Trevisan, L., Vadhan, S.P.: Extracting randomness from samplable distributions. In: FOCS 2000 [31], pp. 32–42

69. Wichs, D., Mansour, Y. (eds.): 48th ACM STOC. ACM Press, June 2016

70. Yao, A.C.: Theory and applications of trapdoor functions (extended abstract). In: 23rd Annual Symposium on Foundations of Computer Science, Chicago, Illinois, USA, 3–5 November 1982, pp. 80–91. IEEE Computer Society (1982). https://doi.org/10.1109/SFCS.1982.45

Continuous Non-Malleable Codes
in the 8-Split-State Model

Divesh Aggarwal[1], Nico Döttling[2], Jesper Buus Nielsen[3(✉)], Maciej Obremski[1], and Erick Purwanto[1]

[1] National University of Singapore, Singapore, Singapore
[2] CISPA Helmholtz Center for Information Security, Saarbrücken, Germany
[3] Aarhus University, Aarhus, Denmark
jbn@cs.au.dk

Abstract. Non-malleable codes (NMCs), introduced by Dziembowski, Pietrzak and Wichs [20], provide a useful message integrity guarantee in situations where traditional error-correction (and even error-detection) is impossible; for example, when the attacker can completely overwrite the encoded message. NMCs have emerged as a fundamental object at the intersection of coding theory and cryptography. In particular, progress in the study of non-malleable codes and the related notion of non-malleable extractors has led to new insights and progress on even more fundamental problems like the construction of multi-source randomness extractors. A large body of the recent work has focused on various constructions of non-malleable codes in the split-state model. Many variants of NMCs have been introduced in the literature, e.g., strong NMCs, super strong NMCs and continuous NMCs. The most general, and hence also the most useful notion among these is that of continuous non-malleable codes, that allows for continuous tampering by the adversary. We present the first efficient information-theoretically secure continuously non-malleable code in the constant split-state model. We believe that our main technical result could be of independent interest and some of the ideas could in future be used to make progress on other related questions.

1 Introduction

Non-malleable codes, introduced by Dziembowski, Pietrzak and Wichs [20], provide a useful message integrity guarantee in situations where traditional error-correction (and even error-detection) is impossible; for example, when the attacker can completely overwrite the encoded message. Non-malleable codes have emerged as a fundamental object at the intersection of coding theory and cryptography.

Informally, given a tampering family $\mathcal{F}$, a non-malleable code $(\mathsf{Enc}, \mathsf{Dec})$ against $\mathcal{F}$ encodes a given message m into a codeword $c \leftarrow \mathsf{Enc}(m)$ in a way

This research was further partially funded by the Singapore Ministry of Education and the National Research Foundation under grant R-710-000-012-135.

Y. Ishai and V. Rijmen (Eds.): EUROCRYPT 2019, LNCS 11476, pp. 531–561, 2019.
https://doi.org/10.1007/978-3-030-17653-2_18

that, if the adversary modifies c to $c' = f(c)$ for some $f \in \mathcal{F}$, then the the message $m' = \mathsf{Dec}(c')$ is either the original message m, or a completely "unrelated value". Formally, we require that if $m' \neq m$, then m' can be simulated using just the tampering function f, but without knowing anything about the tampered codeword c'.

As has been shown by the recent progress [1,3–6,11–14,19–21,23,27] [2,7–9,26], non-malleable codes aim to handle a much larger class of tampering functions $\mathcal{F}$ than traditional error-correcting or error-detecting codes, at the expense of potentially allowing the attacker to replace a given message m by an unrelated message m'. Non-malleable codes are useful in situations where changing m to an unrelated m' is not useful for the attacker (for example, when m is the secret key for a signature scheme.)

Continuous Non-malleable Codes. It is clearly realistically possible that the attacker repeatedly tampers with the device and observes the outputs. The definition in [20] allows the adversary to tamper the codeword *only once*. We call this *one-shot* tampering. Faust et al. [21] consider a stronger model where the adversary can iteratively submit tampering functions f_i and learn $m_i = \mathsf{Dec}(f_i(c))$. We call this the *continuous tampering model.* This stronger security notion is needed in many settings, for instance when using non-malleable codes to make tamper resilient computations on von Neumann architectures [22]. As mentioned in [25], non-malleable codes can provide protection against these kind of attacks if the device is allowed to freshly re-encode its state after each invocation to make sure that the tampering is applied to a fresh codeword at each step. After each execution the entire content of the memory is erased. While such perfect erasures may be feasible in some settings, they are rather problematic in the presence of tampering. Due to this reason, Faust et al. [21] introduced an even stronger notion of non-malleable codes called continuous non-malleable codes where security is achieved against continuous tampering of a single codeword *without* re-encoding. Some additional restrictions are, however, necessary in the continuous tampering model. If the adversary was given an unlimited budget of tampering queries, then, given that the class of tampering functions is sufficiently expressive (e.g. it allows to overwrite single bits of the codeword), the adversary can efficiently learn the entire message just by observing whether tampering queries leave the codeword unmodified or lead to decoding errors, see e.g. [24].

To overcome this general issue, [21] assume a *self-destruct* mechanism which is triggered by decoding errors. In particular, once the decoder outputs a special symbol $\perp$ the device *self-destructs* and the adversary loses access to his tampering oracle. This model still allows an adversary many tamper attempts, as long as his attack remains covert. Jafargholi and Wichs [25] considered the additional aspect of whether tampering is *persistent* in the sense that the tampering is always applied to the current version of the tampered codeword, and all previous versions of the codeword are lost. The alternative definition considers non-persistent tampering where the device resets after each tampering, and the tampering always occurs on the original codeword. In this work, we will exclusively focus on continuous non-malleable codes in the non-persistent self-destruct model. We shorthand such codes by sdCNMC. Note that in the split-state model discussed below, persistent tampering can be simulated by non-persistent tampering by using the tampering function which first reproduces previous tampering and then applies the new tampering function. Hence non-persistent tampering is a strictly stronger model in the split-state model.

Split-State Model. Although any kind of non-malleable codes do not exist if the family of "tampering functions" $\mathcal{F}$ is completely unrestricted,[1] they are known to exist for many large classes of tampering families $\mathcal{F}$.

In [20] the authors considered one such natural family of tampering functions. They gave a construction of an efficient code which is non-malleable with respect to independent, bit-wise tampering. Later works [1,3–6,12–14,19,21,26,27] provided efficient constructions in a stronger model called the t-split state model where the codeword is split into t parts called *states*, which can each be tampered arbitrarily but independently of the other states. If the codeword has length n, then the result of [20] can be seen as a result for the n-state model. The physical motivation for this model is that one might place the different states on physically separated memories and hope this makes it impossible to tamper with one part in a way which depends on the value of the other part. Clearly, one would like t to be as small as possible.

While some of the above-mentioned results achieve security only against computationally bounded adversaries, we focus on security in the information-theoretic setting, i.e., security against unbounded adversaries. The known results in the information-theoretic setting can be summarized as follows. First, [20] showed the existence of (strong) non-malleable codes, and this result was improved by [13] who showed that the optimal rate of these codes is $1/2$. Faust et al. [21] showed the impossibility of continuous non-malleable codes against non-persistent 2-split-state tampering. Later [25] showed that continuous non-malleable codes exist in the split-state model if the tampering is persistent, and [7] gave an efficient construction of such codes.

There have been a series of recent results culminating in constructions of efficient non-malleable codes in the split-state model [4,5,11,12,19,26].

Continuous Non-Malleable Codes in the Split-State Model and Our Result. Faust et al. [21] constructed an sdCNMC in the 2-state model which is secure against computationally bounded adversaries. A recent result [7] gave a construction of non-malleable codes secure against persistent continuous tampering. It was shown in [21] that it is *impossible* to construct an information theoretic sdCNMC for the much more interesting 2-state model with non-persistent tampering. This leaves the following question open.

Question 1. Does there exist a code that is non-malleable in the t-split non-persistent continuous tampering model for some constant $t > 2$?

---

[1] In particular, $\mathcal{F}$ should not include "re-encoding functions" $f(c) = \mathsf{Enc}(f'(\mathsf{Dec}(c)))$ for any non-trivial function f', as $m' = \mathsf{Dec}(f(\mathsf{Enc}(m))) = f'(m)$ is obviously related to m.

In [16] an sdCNMC was constructed in the bit-wise tampering model, which can be seen as an n-state model. However, very little progress has been made towards resolving Question 1. The only result that achieves some sort of non-malleable codes secure against persistent continuous tampering is the result by Chattopadhyay, Goyal, and Li [11]. They achieve this by constructing a so-called many-many non-malleable code in the 2-split state model. Their construction achieves non-malleability as long as the number of rounds of tampering is at most n^γ for some constant $\gamma < 1$, where n is the length of the codeword. Their result has a natural barrier and it is unlikely that their ideas can be used to achieve a construction that allows more than $O(n)$ rounds of tampering. This is both because their construction does not allow self-destruct and is for the 2-split state model, and it is known [7,21] that continuous non-malleable codes with $\omega(n)$ rounds of tampering is impossible both for the two split-state model and for the constant split-state model that does not allow self-destruct.

We construct an information-theoretic sdCNMC for the 8-state model.

Theorem 1 (Informal). *Let k be the security parameter. There exists an efficient, explicit construction of non-persistent self-destruct continuous non-malleable codes which encodes messages of length k bits into 8 states, each of size $O(k \log k)$. The code tolerates $2^{\Omega(k)}$ tampering attempts and is secure except with probability $2^{-\Omega(k)}$.*

Overview of the Construction and Techniques. In this section, we will provide an overview of our construction and the main ideas for its security proof. Our construction combines two Hadamard extractors with a 3-source non-malleable extractor. The construction is given as follows.

Our Construction. Let $\mathbb{K}$ be a finite field of size 2^n, which is an extension field $\mathbb{F}$ of size $2^{n/\ell}$ for an appropriately chosen divisor ℓ of n. Our construction uses the following:

- A three source non-malleable extractor $\mathsf{nmExt} : \mathbb{K}^3 \to \{0,1\}^{3k}$ with $k = \Theta(n/\log n)$, where the min-entropy for each source is required to be at least $(1-\delta)n$, for some constant δ,
- A 2-source Hadamard extractor $\langle \cdot, \cdot \rangle : (\mathbb{K}^3) \times (\mathbb{K}^3) \to \mathbb{K}$, and
- A 2-source Hadamard extractor $\langle \cdot, \cdot \rangle : (\mathbb{F}^{3\ell}) \times (\mathbb{F}^{3\ell}) \to \mathbb{F}$.

Let $\|$ denote concatenation of strings. We define

$$\mathsf{nmExt}' : (\{0,1\}^n)^3 \to \{0,1\}^{3k} \cup \{\bot\}$$

as $\mathsf{nmExt}'(x_1, x_2, x_3) = \mathsf{nmExt}(x_1, x_2, x_3)$ if $\mathsf{nmExt}(x_1, x_2, x_3) = 0^{2k} \| y$ for some $y \in \{0,1\}^k$, and $\bot$, otherwise.

Encoding: Our encoding procedure takes as input a message $m \in \{0,1\}^k$, and does the following.
  - Sample $X = (X_1, X_2, X_3)$ from $(\mathbb{K} \setminus \{0\})^3$ uniformly such that $\mathsf{nmExt}(X) = 0^{2k} \| m$.

- Sample $S = (S_1, S_2, S_3)$ from $(\mathbb{K} \setminus \{0\})^3$ uniformly such that $\mathsf{nmExt}(S) = 0^{2k} \| r$ for some r in $\{0, 1\}^k$.
- $V = \langle X, S \rangle_{\mathbb{K}}$.
- $W = \langle X, S \rangle_{\mathbb{F}}$.
- Output the eight parts $(X_1, X_2, X_3, S_1, S_2, S_3, V, W)$.

Decoding: The decoding procedure is canonical, i.e., on input (x, s, v, w), we first check if x and s pass the two inner product checks and are in the correct domains (i.e. all components non-zero), we try to decode x and s and if neither reports an error we return the decoded value of x.

The adversary, in each round, will choose some functions, $f_1, f_2, f_3, g_1, g_2,$ $g_3, h_1 : \mathbb{K} \to \mathbb{K}, h_2 : \mathbb{F} \to \mathbb{F}$ and will apply these functions to the eight respective parts. Let $f(X)$ denote $(f_1(X_1), f_2(X_2), f_3(X_3))$ and $g(S)$ denote $(g_1(S_1), g_2(S_2), g_3(S_3))$ In order to prove (continuous) non-malleability of the construction, we need to show that even if we collect all the messages obtained after decoding the tampered codewords in multiple rounds excluding any round where all the chosen functions are identity functions (in this case decoding the tampered codeword yields the original message), this should not reveal any useful information about the original message. To formalize this, we define the tampering experiment to output a special symbol same whenever all functions are identity functions. Then, it is required to prove that for any two messages, the output distributions of the corresponding tampering experiments are statistically close to each other. In fact, in this work, we consider a stronger notion of continuous non-malleable codes called super-strong continuous non-malleable codes in which every time the adversary tampers $(c \to c')$, $c' \neq c$, and c' decodes to a valid message, the adversary will learn the whole tampered codeword c'.

Proof Ideas. Before looking at the ideas behind the security of our construction, it is instructive to revisit the reason behind the impossibility of constructions for 2-state information-theoretic continuous non-malleable codes [21]. The main idea behind the attack given in [21] was to find a triple ℓ, r_0, r_1 such that $\mathsf{Dec}(\ell, r_0), \mathsf{Dec}(\ell, r_1) \neq \bot$. Given ℓ, r_0 and r_1, the attack proceeds by overwriting the first state with ℓ, while the second state is overwritten by r_b where b is the first bit of the second state, thereby revealing one bit of information. Repeating this idea for different bits of the codeword, after a linear number of rounds, the adversary will recover the entire codeword.

In our construction, if the adversary decides to preserve a significant amount of entropy of the original codeword when tampering, i.e., the tampering function is close to being bijective, then the non-malleability of nmExt should be sufficient to achieve not just non-malleability but error detection: $\mathsf{nmExt}(f(X))$ is close to being uniform and independent of $\mathsf{nmExt}(X)$ by the non-malleability of nmExt, and hence the tampered codeword decodes to $\bot$ with high probability. However, if the adversary decides to carry only a very small amount of entropy into the tampered codeword, there is nothing preventing him from learning some small amount of information as in the attack by [21] described above. It is not possible to reliably detect such *low entropy tampering*. But we can show that its probability of learning information is always associated with a probability of being detected. Understanding this relation is at the core of the proof.

As mentioned above, the tampering experiment for our code is of the *superstrong* type, i.e., every time the adversary tampers $(C \to C')$, $C' \neq C$, and C' decodes to a valid message, the adversary will learn the whole tampered codeword C'. Notice that given

$$C' = (f_1(X_1), f_2(X_2), f_3(X_3), g_1(S_1), g_2(S_2), g_3(S_3), h_1(V), h_2(W))$$

all the adversary learns is that

- $X_i \in \mathcal{X}_i$ for $i = 1, 2, 3$
- $S_i \in \mathcal{S}_i$ for $i = 1, 2, 3$
- $V \in \mathcal{V}$
- $W \in \mathcal{W}$,

where $\mathcal{X} \times \mathcal{S} \times \mathcal{V} \times \mathcal{W}$ is the preimage of c' for the function $(f_1, f_2, f_3, g_1, g_2, g_3, h_1, h_2)$. In round r of the tampering experiment the adversary will learn that the codeword belongs to some domain $\mathcal{X}^{(r)} \times \mathcal{S}^{(r)} \times \mathcal{V}^{(r)} \times \mathcal{W}^{(r)}$, and will progressively try to make these sets as small as possible. In the [21] attack described above, the domain size is reduced by a factor of two each time, eventually revealing the entire codeword. As long as we can make sure that the domain doesn't become too small, we will be able to argue that if the adversary wants to learn more information (make the set smaller) there is a significant risk of getting detected. We sketch below the idea for showing this for the first round $r = 1$. The argument for the following rounds follows by a slightly tricky inductive argument.

Depending on the functions $f_1, f_2, f_3, g_1, g_2, g_3$, we partition each of $\mathcal{X}_1, \mathcal{X}_2, \mathcal{X}_3, \mathcal{S}_1, \mathcal{S}_2, \mathcal{S}_3$ which induces a partition on the whole domain. For instance $\mathcal{X}_1$ is partitioned into $\ell + 1$ parts for some parameter $\ell = \omega(1)$, as follows.

- $\mathcal{X}_{1,0}$ is the part where the function f_1 is identity, i.e., $\{x \in \mathcal{X}_1 : f_1(x) = x\}$.
- For $i = 1, \ldots, \ell$, $\mathcal{X}_{1,i}$ is defined such that f_1 has between $2^{n(i-1)/\ell}$ and $2^{n \cdot i/\ell}$ preimages.

This implies that for each partition, the entropy of X_1 conditioned on $f_1(X_1)$ is nearly fixed (upto an additive term $n/\ell = o(n)$). The other sets $\mathcal{X}_2, \mathcal{X}_3, \mathcal{S}_1, \mathcal{S}_2, \mathcal{S}_3$ are partitioned similarly. Each partition is of the form

$$\mathcal{X}_{1,i_1}, \mathcal{X}_{2,i_2}, \mathcal{X}_{3,i_3}, \mathcal{S}_{1,j_1}, \mathcal{S}_{2,j_2}, \mathcal{S}_{3,j_3}, V, W.$$

Type−1 corresponds to $i_1 = i_2 = i_3 = j_1 = j_2 = j_3 = 0$.

Type−2 contains all partitions for which the following is true: $(f(X) \neq X$ or $g(S) \neq S)$ and $(f(X), g(S))$ contains almost full information about (X, S), i.e., all tampering functions are close to bijective or identity, but at least one tampering function is not the identity.

Type−3 contains all partitions which do not fall into any of above classes (in particular it means that $(f(X), g(S))$ lost quite a bit information about the original (X, S)), but $(f(X), g(S))$ still carries a substantial/medium amount of information/entropy about (X, S).

Type−4 contains all partitions which do not fall into any of above classes but only at least one of the $(f_i(X_i), g_j(S_j))$ still carries some entropy.

Type−5 contains the partition where $(f(X), g(S))$ is close to constant, i.e., $i_1 = i_2 = i_3 = j_1 = j_2 = j_3 = \ell$.

Analysis of the tampering for each type of partition. In this section, we often implicitly assume that X is independent of S in order to simplify the informal argument, even though there is some limited dependence introduced by the fact that $\langle X, S \rangle_{\mathbb{K}} \in \mathcal{V}$, etc. The full proofs shows how to handle the dependence. We show that when the codeword c falls into either class $2, 3$ or 4, the tampering will be detected with probability $1 - \varepsilon$ for a negligible ε:

In Type−2: On this part of the domain the adversary will attempt to apply close to bijective tampering functions. Either this part of the domain will have negligible size, or the adversary will be detected by the check for nmExt′.

In Type−3: We will argue that the check $\langle f(X), g(S) \rangle_{\mathbb{F}} = h_2(W)$ will fail. To see this, notice that the adversary applied non-bijective tampering, and the vectors $f(X), g(S)$ have a substantial amount of entropy. The argument below follows from the strong extraction properties of the inner-product extractor: The vectors $f(X)$ and $g(S)$ do not carry enough information about X, S, i.e., one of $\tilde{\mathbf{H}}_\infty(X|f(X))$ or $\tilde{\mathbf{H}}_\infty(S|g(S))$ is not too small. Thus $\langle X, S \rangle_{\mathbb{F}}$ and $\langle f(X), g(S) \rangle_{\mathbb{F}}$ are almost independent. However $f(X), g(S)$ have enough entropy to keep $\langle f(X), g(S) \rangle_{\mathbb{F}}$ uniform. The adversary will not be able to guess $\langle f(X), g(S) \rangle_{\mathbb{F}}$ even given $\langle X, S \rangle_{\mathbb{F}}$. Thus he will fail at the check $h_2(\langle X, S \rangle_{\mathbb{F}}) = \langle f(X), g(S) \rangle_{\mathbb{F}}$ and this tampering will be detected.

In Type−4: The reasoning is quite similar to Type−3, but we use the check on $\langle f(X), g(S) \rangle_{\mathbb{K}} = h_1(V)$. The adversary applied far-from-bijective tampering, and the vectors $f(X), g(S)$ still have some small amount of entropy. The argument below follows from the strong extraction properties of the inner product extractor: The vectors $f(X)$ and $g(S)$ only carry a very small amount of information about X, S. Thus $\langle X, S \rangle_{\mathbb{K}}$ and $\langle f(X), g(S) \rangle_{\mathbb{K}}$ are almost independent. However $f(X), g(S)$ still have enough entropy to keep $\langle f(X), g(S) \rangle_{\mathbb{K}}$ unpredictable (not uniform, but with substantial min-entropy). The adversary will not be able to guess $\langle f(X), g(S) \rangle_{\mathbb{K}}$ even given $\langle X, S \rangle_{\mathbb{K}}$, thus he will fail at the check $h_1(\langle X, S \rangle_{\mathbb{K}}) = \langle f(X), g(S) \rangle_{\mathbb{K}}$ and this tampering will be detected.

This leads to the conclusion that the only way that the adversary can learn something and survive (i.e. not get detected) is if the original codeword falls into Type−1 or Type−5. If the codeword was in Type−1, the tampering experiment will output **same** (unless one of the inner product checks fails and the tampered codeword decodes to ⊥). If the codeword was in Type−5, then the output will be some codeword c', and the adversary will learn whether the codeword is Type−1 or Type−5 with respect to the choice of functions f and g. Moreover, on Type−5 there might be close-to-constant but non-constant functions (which, if he does not get detected, potentially provide additional knowledge to the adversary).

Even if the adversary is in a Type-1 or Type-5 partition and succeeds to go to the next round without causing self-destruct, this is not a reason to worry as long as the size of the domain remains large enough. On the other hand, if the adversary can manage to land himself in a small enough domain, this means that the adversary already obtained a lot of information about the codeword, and might be able to recover the message. However, if such small domains are few and scarce, then the probability that the adversary lands in such a domain is quite small. The main cause of concern is if there are many such small domains that cover a significant fraction of the ambient space. In the following, we show that this is not possible.

Type-1 or Type-5: Notice that the adversary is in a Type-1 or a Type-5 partition if either each of $i_1, i_2, i_3, j_1, j_2, j_3$ is 0, or each is equal to ℓ. Since the indices $i_1, i_2, i_3, j_1, j_2, j_3$ are independently distributed random variables, a simple application of the Cauchy-Schwarz inequality shows that $\sqrt{p_1} + \sqrt{p_5} \leq 1$, where p_1 is the probability of being in a Type-1 partition, and p_5 is the probability of being a Type-5 partition.

Type-5: Just like in the case of Type-4 partitions, we have that the vectors $f(X)$ and $g(S)$ only carry a very small amount of information about X, S. Thus $\langle X, S \rangle_{\mathbb{K}}$ and $\langle f(X), g(S) \rangle_{\mathbb{K}}$ are nearly independent. The Type-5 partition corresponds to the domain where each of $f_1, f_2, f_3, g_1, g_2, g_3$ is close to a constant and can be further subdivided such that for each of these subpartitions, each of $f_1, f_2, f_3, g_1, g_2, g_3$ output a fixed value. Intuitively speaking, if say, each of these functions takes two different values then there are potentially 64 different values of $\langle f(X), g(S) \rangle_{\mathbb{K}}$ (although some of these 64 values could be the same), and so the function h_1 cannot guess this value with sufficiently large probability, unless all the inner products magically become equal. Formally, we show in this case that $p_{5,1}^{7/8} + \cdots + p_{5,d}^{7/8} \leq p_5^{7/8}$, where $p_{5,1}, \ldots, p_{5,d}$ are the respective probabilities of being in various subpartitions of Type-5 such that $h_1(\langle X, S \rangle_{\mathbb{K}}) = \langle f(X), g(S) \rangle_{\mathbb{K}}$ holds within these subpartitions.

Together, these results imply that

$$q_1^{7/8} + q_2^{7/8} + \cdots + q_{d+1}^{7/8} \leq 1, \tag{1}$$

where $q_1, \ldots, q_{d+1}$ is a renaming of $p_1, p_{5,1}, \ldots, p_{5,d}$. A simple application of Hölder's inequality implies that for any $\varepsilon \geq 0$,

$$\sum_{q_i \leq \varepsilon} q_i = \sum_{q_i \leq \varepsilon} q_i^{7/8} \cdot q_i^{1/8} \leq \sum_{q_i \leq \varepsilon} q_i^{7/8} \cdot \varepsilon^{1/8} \leq \varepsilon^{1/8}.$$

For an appropriately chosen ε, this implies that it is not possible that there are many small domains on which the decoder does not self-destruct, and their union is large. This concludes the intuitive overview of our proof.

Conclusions and Open Questions. We give a construction of a $2^{-\Omega(k)}$-non-malleable code from k bit messages to $O(k \log k)$ bit codewords in the 8-split state model secure against continuous tampering. The main building block of our construction is a non-malleable 3-source extractor construction from [26], and the Hadamard 2-source extractor.

Prior results achieved continuous non-malleability only for a sublinear number of rounds [11]. The main reason for difficulty in achieving non-malleable codes against continuous tampering is that the adversary can potentially obtain useful information in each round, and even if one bit of information about the codeword is obtained in each round, this is already catastrophic and does not allow non-malleability beyond a linear number of rounds.

Our idea of proving that our construction achieves non-malleability for a large number of rounds is that we ensure that whenever the adversary tampers to gain useful information about the codeword, there is a risk of a decoding error resulting in self-destruct. Central to our proof strategy is what we believe a very novel technique where we obtained and used an inequality of the form (1) to bound the statistical distance between two random experiments. In particular, our main technical result in Theorem 5 where we bound the statistical distance between two random variables by $(\frac{\rho}{q})^c + \varepsilon$, where q is proportional to the size of the domain, c is a constant, and ρ, ε are appropriately chosen parameters, might seem very unusual, but appears naturally in our proof. This, we believe, might be of independent interest.

The following are natural questions left open by our work.

1. Improve the rate of our code.
2. Improve the number of split states to a number smaller than 8.

The first of these questions can be resolved immediately by a non-malleable extractor with parameters (output length) better than the one given in [26]. As for the second question, our construction has a natural barrier and the number of states can likely not be improved by any simple modification. However, we hope that our techniques can lead to new insights that might help resolve this question.

Lastly, in the recent years, progress related to non-malleable codes has led to useful ideas for solving even more fundamental problems like constructing better two-source or multi-source extractors. We hope that our construction and/or techniques can find other similar applications.

2 Preliminaries

All logarithms are to the base 2. For any function $h : \mathcal{X} \to \mathcal{Y}$, we define $h^{-1}(y) := \{x \in \mathcal{X} : h(x) = y\}$. For a set S, we let U_S denote the uniform distribution over S. For an integer $m \in \mathbb{N}$, we let U_m denote the uniform distribution over $\{0, 1\}^m$. We denote two independent bitstrings of length m by U_m, U_m'. For a distribution or random variable X we write $x \leftarrow X$ to denote the operation of sampling a random x according to X. For a set S, we write $s \leftarrow S$ as shorthand for $s \leftarrow U_S$.

For a random variable Z, $f(Z)|_{Z \in \mathcal{C}}$ denotes the distribution $f(Z)$ conditioned on the event that $Z \in \mathcal{C}$.

2.1   Entropy and Statistical Distance

The *min-entropy* of a random variable X is defined as $\mathbf{H}_\infty(X) \stackrel{\text{def}}{=} - \log(\max_x \Pr[X = x])$. We say that X is an (n, k)-*source* if $X \in \{0, 1\}^n$ and $\mathbf{H}_\infty(X) \geq k$. For $X \in \{0, 1\}^n$, we define the *entropy rate* of X to be $\mathbf{H}_\infty(X)/n$. We also define *average (aka conditional) min-entropy* of a random variable X conditioned on another random variable Z as

$$\widetilde{\mathbf{H}}_\infty(X|Z) \stackrel{\text{def}}{=} - \log \left(\mathbb{E}_{z \leftarrow Z} \left[\max_x \Pr[X = x|Z = z] \right] \right)$$
$$= - \log \left(\mathbb{E}_{z \leftarrow Z} \left[2^{-\mathbf{H}_\infty(X|Z=z)} \right] \right)$$

where $\mathbb{E}_{z \leftarrow Z}$ denotes the expected value over $z \leftarrow Z$. We have the following lemma.

Lemma 1 ([18]). *Let (X, W) be some joint distribution. Then,*

- *For any $s > 0$, $\Pr_{w \leftarrow W}[\mathbf{H}_\infty(X|W = w) \geq \widetilde{\mathbf{H}}_\infty(X|W) - s] \geq 1 - 2^{-s}$.*
- *If Z has at most 2^ℓ possible values, then $\widetilde{\mathbf{H}}_\infty(X|(W, Z)) \geq \widetilde{\mathbf{H}}_\infty(X|W) - \ell$.*

Lemma 2. *Let Z be distributed over a set $\mathcal{Z}$ and let h be an arbitrary function. If $|h^{-1}(h(z)) \cap \mathcal{Z}| \leq m$ then $\mathbf{H}_\infty(h(Z)) \geq \log \frac{|\mathcal{Z}|}{m}$.*

Proof. Since for any $h(z)$, for $z \in \mathcal{Z}$, the number of $z' \in \mathcal{Z}$ that maps to $h(z)$ is at most m, we get that the number of distinct $h(z)$ is $\geq \frac{|\mathcal{Z}|}{|h^{-1}(h(z)) \cap \mathcal{Z}|} \geq \frac{|\mathcal{Z}|}{m}$. Thus, $\mathbf{H}_\infty(h(Z)) \geq \log \frac{|\mathcal{Z}|}{m}$. $\square$

The *statistical distance* between two random variables W and Z distributed over some set S is

$$\Delta(W; Z) := \max_{T \subseteq S}(|W(T) - Z(T)|) = \frac{1}{2} \sum_{s \in S} |W(s) - Z(s)|.$$

Note that $\Delta(W; Z) = \max_D(\Pr[D(W) = 1] - \Pr[D(Z) = 1])$, where D is a probabilistic function. We say W is ε-close to Z, denoted $W \approx_\varepsilon Z$, if $\Delta(W; Z) \leq \varepsilon$. We write $\Delta(W; Z|Y)$ as shorthand for $\Delta((W, Y); (Z, Y))$. The following is folklore, and is easy to see.

Lemma 3. *For any two random variables X, Y, and any randomized function f, we have that $\Delta(f(X); f(Y)) \leq \Delta(X; Y)$.*

2.2 Extractors

An extractor [28] can be used to extract uniform randomness out of a weakly-random value which is only assumed to have sufficient min-entropy. Our definition follows that of [18], which is defined in terms of conditional min-entropy.

Definition 1 (Extractors). *An efficient function* $\mathsf{Ext} : \{0,1\}^n \times \{0,1\}^d \to \{0,1\}^m$ *is an (average-case, strong)* (k,ε)*-extractor, if for all* X, Z *such that* X *is distributed over* $\{0,1\}^n$ *and* $\widetilde{\mathbf{H}}_\infty(X|Z) \geq k$, *we get*

$$\Delta(\ (Z, Y, \mathsf{Ext}(X; Y))\ ;\ (Z, Y, U_m)\ ) \leq \varepsilon$$

where $Y \equiv U_d$ *denotes the coins of* Ext *(called the* seed*). The value* $L = k - m$ *is called the* entropy loss *of* Ext, *and the value* d *is called the* seed length *of* Ext.

Definition 2 (Two-Source Extractors). *A function* $\mathsf{Ext} : \mathcal{X}_1 \times \mathcal{X}_2 \to \mathcal{Z}$ *is called a* (k,ε)*-two-source extractor, if it holds for all tuples* $((X_1, Y_1), (X_2, Y_2))$ *for which* (X_1, Y_1) *is independent of* (X_2, Y_2) *and* $\widetilde{\mathbf{H}}_\infty(X_1|Y_1) + \widetilde{\mathbf{H}}_\infty(X_2|Y_2) \geq k$ *that*

$$\Delta(\mathsf{Ext}(X_1, X_2)\ ;\ U_{\mathcal{Z}} \mid Y_1, Y_2) \geq \varepsilon.$$

A well-known flexible two-source extractor is the Hadamard extractor or inner-product extractor.

Lemma 4 ([5,15]). *For any finite field* $\mathbb{F}_q$ *of cardinality* q *and any positive integer* n, *the function* $\mathsf{Ext} : \mathbb{F}_q^n \times \mathbb{F}_q^n \to \mathbb{F}_q$ *given by*

$$\mathsf{Ext}(X_1, X_2) := \langle X_1, X_2 \rangle = X_{1,1} \cdot X_{2,1} + \cdots + X_{1,n} \cdot X_{2,n}$$

is a (k,ε)*-two-source extractor for any* $k \geq (n+1)\log q + 2\log\left(\frac{1}{\varepsilon}\right)$.

We denote the above inner product by $\langle X_1, X_2 \rangle_{\mathbb{F}_q}$. We will drop the subscript if the field is clear from the context.

We will also use non-malleable t-source extractor.

Definition 3 (Non-Malleable t-Source Extractor). *A function* $\mathsf{nmExt} : (\mathcal{X})^t \to \mathcal{Z}$ *is called a* t*-source* (k,ε)*-non-malleable extractor if the following property holds. For all independently distributed tuples* $((X_1, Y_1), (X_2, Y_2), \ldots, (X_t, Y_t))$ *such that* $\widetilde{\mathbf{H}}_\infty(X_i|Y_i) \geq k$, *and for any split-state tampering function* $f = (f_1, \ldots, f_t)$, $f_i : \mathcal{X} \to \mathcal{X}$ *such that there exists* f_i *without fixed points, it holds that*

$$\Delta(\mathsf{nmExt}(X)\ ;\ U_{\mathcal{Z}} \mid \mathsf{nmExt}(f(X)),\ Y_1, \ldots, Y_t) \leq \varepsilon,$$

where $X = (X_1, \ldots, X_t)$, *and* $f(X) = (f_1(X_1), \ldots, f_t(X_t))$.

The following result gives the best known 2-source non-malleable extractor.

Theorem 2 ([26]). *For any finite field* $\mathbb{K}$ *of cardinality* 2^n, *there exists a constant* $\delta^\star \in (0, 1/3)$, *and a function* $\mathsf{nmExt}_2 : \mathbb{K}^2 \to \{0,1\}^{3k}$ *such that the function* nmExt_2 *is a* 2*-source* $((1-\delta^\star)n, 2^{-1000k})$ *non-malleable extractor with* $k = \Theta(n/\log n)$. *Moreover, it is efficiently pre-image sampleable.*

For this paper, we need a 3-source non-malleable extractor. The construction from the above result can be easily modified to obtain a 3-source non-malleable extractor.

Theorem 3. *For any finite field $\mathbb{K}$ of cardinality 2^n, there exists a constant $\delta \in (0, 1/3)$, and a function $\mathsf{nmExt} : \mathbb{K}^3 \to \{0, 1\}^{3k}$ such that the function nmExt is a 3-source $((1 - \delta)n, 2^{-1000k})$ non-malleable extractor with $k = \Theta(n/\log n)$. Moreover, it is efficiently pre-image sampleable.*

Proof. Let $(X_1, Y_1), (X_2, Y_2), (X_3, Y_3)$ be as in Definition 3. Consider the following construction.

$$\mathsf{nmExt}(X_1, X_2, X_3) := \mathsf{nmExt}_2(X_1, X_2) \oplus \mathsf{nmExt}_2(X_2, X_3),$$

where by $\oplus$, we mean the bitwise XOR function. Let the functions applied to the three parts be f_1, f_2, f_3, one of which has no fixed points. Without loss of generality, let f_1 or f_2 be the function with no fixed points. We have that $\widetilde{\mathbf{H}}_\infty(X_1 \mid Y_1) \geq n(1 - \delta^\star)$, and

$$\widetilde{\mathbf{H}}_\infty(X_2 | Y_2, \mathsf{nmExt}_2(X_2, X_3), \mathsf{nmExt}_2(f_2(X_2), f_3(X_3))) \geq n - n \cdot \delta - 6k \geq n(1 - \delta^\star),$$

where we assumed that $\delta = \delta^\star/2$, and $\delta n \geq 12k$. Thus, the statistical distance between $\mathsf{nmExt}_2(X_1, X_2)$ and U_{3k} conditioned on $\mathsf{nmExt}_2(f_1(X_1), f_2(X_2))$, Y_1, Y_2, $\mathsf{nmExt}_2(X_2, X_3)$, and $\mathsf{nmExt}_2(f_2(X_2), f_3(X_3))$ is at most 2^{-1000k}, which implies using Lemma 3 that

$$\Delta\left(\mathsf{nmExt}(X_1, X_2, X_3) \; ; \; U_{3k} \mid \mathsf{nmExt}(f_1(X_1), f_2(X_2), f_3(X_3)) \, Y_1, \, Y_2, \, Y_3\right) \leq 2^{-1000k}.$$

Note that we can sample the pre-image of nmExt efficiently using the sampling procedure of [26]. In order to sample a preimage of $\mu \in \{0, 1\}^{3k}$, we first sample X_1, X_2 uniformly at random from $\mathbb{K}$, and then X_3 is sampled conditioned on the fact that $\mathsf{nmExt}_2(X_2, X_3) = \mathsf{nmExt}_2(X_1, X_2) \oplus \mu$. In particular, by using the randomness of the first sampling procedure in picking X_2 as the first source on the sampling procedure from [26], X_3 is a randomized function of X_2 and the output of the non-malleable extractor. Furthermore, they still satisfy the linear constraints and can be computed and sampled efficiently. $\square$

2.3   Trace Function

We use the following standard fact about trace functions. For a finite field $A = \mathbb{F}_{2^m}$, and for its extension field $B = \mathbb{F}_{2^n}$, and the trace function $\mathrm{tr}_{B \to A} : B \to A$ there is a group isomorphism from $\psi : B^\ell \to A^{n\ell/m}$ such that $\langle \psi(x), \psi(y) \rangle_A = \mathrm{tr}_{B \to A}(\langle x, y \rangle_B)$. We will need this result on many occasions. Using a slight abuse of notation, we will denote $\langle \psi(x), \psi(y) \rangle_A$ by $\langle x, y \rangle_A$. More details appears in the full version.

2.4   Definitions Related to Non-Malleable Codes

Definition 4 (Coding Schemes). *A coding scheme is a pair* (Enc, Dec), *where* Enc : $\mathcal{M} \to \mathcal{C}$ *is a randomized function and* Dec : $\mathcal{C} \to \mathcal{M} \cup \{\perp\}$ *is a deterministic function, such that it holds for all* $M \in \mathcal{M}$ *that* Dec(Enc(M)) = M.

We will now define the continuous super strong tampering experiment. In this experiment the adversary is provided with the tampered codeword C' (instead of the output of the decoder) whenever $C' \neq C$ and the decoder does not output $\perp$.

Definition 5 ((Continuous-) Super Strong Tampering Experiment). *We will define continuous non-persistent self-destruct non-malleable codes analogously to [25]. Fix a coding scheme* (Enc, Dec) *with message space* $\mathcal{M}$ *and codeword space* $\mathcal{C}$. *Also fix a family of functions* $\mathcal{F} : \mathcal{C} \to \mathcal{C}$. *We will first define the tampering oracle* $\mathsf{Tamp}_C^{\mathsf{state}}(f)$, *for which initially* state = alive. *For a tampering function* $f \in \mathcal{F}$ *and a codeword* $c \in \mathcal{C}$ *define the tampering oracle by*

$\mathsf{Tamp}_c^{\mathsf{state}}(f)$:
    If state = dead *output* $\perp$
    $c' \leftarrow f(c)$
    If $c' = c$ *output* same
    $m' \leftarrow \mathsf{Dec}(c')$
    If $m' = \perp$ *set* state $\leftarrow$ dead *and output* $\perp$
    Otherwise output c'

    Fix a codeword $c \in \mathcal{C}$. *We define the continuous tampering experiment* CT_C^r *by*

CT_C^r :
    state $\leftarrow$ alive
    For $i = 1$ *to* r
        Choose functions f
        $v \leftarrow \mathsf{Tamp}_c^{\mathsf{state}}(f)$
        Output v

Definition 6. *Let* (Enc, Dec) *be a coding scheme and* CT *be its corresponding continuous tampering experiment for a class* $\mathcal{F}$ *of tampering functions. We say that* (Enc, Dec) *is an* ε-*secure* r-*round continuously non-malleable code against* $\mathcal{F}$, *if it holds for all tampering adversaries* $\mathcal{A}$ *and all pairs of messages* $m_0, m_1 \in \mathcal{M}$ *that* $\mathsf{CT}_{C_0}^r(\mathcal{A}) \approx_\varepsilon \mathsf{CT}_{C_1}^r(\mathcal{A})$, *where* $C_0 \leftarrow \mathsf{Enc}(m_0)$ *and* $C_1 \leftarrow \mathsf{Enc}(m_1)$.

The only family of tampering functions we are concerned with in this work are split state tampering functions.

Definition 7 (Split State Tampering). *Let* $C = C_1 \times \cdots \times C_s$. *The class of spit state tampering functions* $\mathcal{F}_s$ *consists of all functions* f *of the form* $f = (f_1, \ldots, f_s)$ *where* $f(c_1, \ldots, c_s) = (f_1(c_1), \ldots, f_s(c_s))$ *for all* $(c_1, \ldots, c_s) \in C_1 \times \cdots \times C_s$. *Here the* f_i *are arbitrary functions* $C_i \to C_i$.

2.5   Some Useful Results

Lemma 5 (Deathzone Generation Lemma [10]). *Let $\mathbb{F}$ be a finite field. Let $A_1, \ldots, A_t$, $B_1, \ldots, B_t$ be independent, non-zero random variables. Denote $A = (A_1, \ldots, A_t)$ and $B = (B_1, \ldots, B_t)$. Then*

$$\max_{c \in \mathbb{F}} \sum_{a,b \in \mathbb{F}^t : \langle a,b \rangle_{\mathbb{F}} = c} \left(\Pr\left[(A,B) = (a,b) \right] \right)^{\frac{2t-1}{2t}} \leq 1.$$

Proof. Let us begin with Young's inequality for convolution:

$$\| f_1 * f_2 * \cdots * f_t \|_r \leq \prod_{i=1}^{t} \| f_i \|_{p_i}$$

whenever $\sum_{i=1}^{t} \frac{1}{p_i} = \frac{1}{r} + t - 1$ and $+\infty \geq p_1, \ldots, p_t, r \geq 1$. We will identify random variable A_i with its distribution $A_i(.)$ where $A_i(x) = \Pr[A_i = x]$. We define two convolutions:

$$(A_i *_\times B_i)(z) = \sum_{x,y \,:\, xy = z} A_i(x) \, B_i(y),$$

$$(A_i *_+ B_i)(z) = \sum_{x,y \,:\, x+y = z} A_i(x) \, B_i(y).$$

Notice that for every i, via Young's inequality, we get

$$1 = \| A_i^\alpha(.) \|_{\frac{1}{\alpha}} \cdot \| B_i^\alpha(.) \|_{\frac{1}{\alpha}} \geq \| A_i^\alpha(.) *_\times B_i^\alpha(.) \|_{\frac{1}{2\alpha - 1}}$$

for $1/2 \leq \alpha \leq 1$. Notice again via Young's inequality, we get

$$1 \geq \prod_{i=1}^{t} \| A_i^\alpha(.) *_\times B_i^\alpha(.) \|_{\frac{1}{2\alpha - 1}}$$

$$\geq \| [A_1^\alpha(.) *_\times B_1^\alpha(.)] *_+ \cdots *_+ [A_t^\alpha(.) *_\times B_t^\alpha(.)] \|_{\frac{1}{2t\alpha - (2t-1)}},$$

for $\frac{2t-1}{2t} \leq \alpha \leq 1$. Now we take $\alpha = \frac{2t-1}{2t}$ and we get

$$1 \geq \| [A_i^\alpha(.) *_\times B_i^\alpha(.)] *_+ \cdots *_+ [A_t^\alpha(.) *_\times B_t^\alpha(.)] \|_\infty. \qquad \square$$

Lemma 6. *Suppose $2 \, \Delta(P; Q) = \sum_{i=1}^{m} |p_i - q_i| = \varepsilon$, where $p_i = \Pr[P = x_i]$ and $q_i = \Pr[Q = x_i]$; and $\sum_{i=1}^{m} p_i^r \leq \alpha$, for $r < 1$. Then $\sum_{i=1}^{m} q_i^r \leq \alpha + \varepsilon^r \cdot m^{1-r}$.*

Proof.

$$\sum_{i=1}^{m} q_i^r = \sum_{i=1}^{m} (p_i + |p_i - q_i|)^r \leq \sum_{i=1}^{m} (p_i^r + |p_i - q_i|^r)$$

$$= \sum_{i=1}^{m} p_i^r + \sum_{i=1}^{m} |p_i - q_i|^r \leq \alpha + \sum_{i=1}^{m} |p_i - q_i|^r$$

$$\leq \alpha + \left(\sum_{i=1}^{m} |p_i - q_i| \right)^r \cdot \left(\sum_{i=1}^{m} 1 \right)^{1-r} = \alpha + \varepsilon^r \cdot m^{1-r},$$

where inequality 2 follows from Hölder's inequality. $\qquad \square$

Lemma 7 ([14]). *Let $\mathcal{D}$ and $\mathcal{D}'$ be distributions over the same finite space Ω, and suppose they are ε-close to each other. Let $E \subseteq \Omega$ be any event such that $\mathcal{D}(E) = p$. Then, the conditional distributions $\mathcal{D}|E$ and $\mathcal{D}'|E$ are (ε/p)-close.*

3   The New Construction

Let $\mathbb{K}$ be a finite field of size 2^n. By Theorem 3, we have that there exists a constant c, such that for all n, and $k \leq \frac{c \cdot n}{\log n}$, there is a function

$$\mathsf{nmExt} : \mathbb{K}^3 \to \{0,1\}^{3k}$$

that is a $(1 - \delta, 2^{-1000k})$-non-malleable 3-source extractor. We choose the largest such $k = \Theta(n/\log n)$ such that $\ell = \frac{n}{100k} = O(\log n)$ is an integer. Also, define

$$\mathsf{nmExt}' : \mathbb{K}^3 \to \{0,1\}^{3k} \cup \{\bot\}$$

as $\mathsf{nmExt}'(x_1, x_2, x_3) = \mathsf{nmExt}(x_1, x_2, x_3)$ if $\mathsf{nmExt}(x_1, x_2, x_3) = 0^{2k}\|y$ for some $y \in \{0,1\}^k$, and $\bot$, otherwise.

Let $\mathbb{F}$ be a finite field of size 2^{50k}. Notice that there is a natural bijection between $\mathbb{K}$ and $\mathbb{F}^\ell$. We further assume that $k \leq \min\left(\frac{\delta n}{1000}, \frac{n}{5000}\right)$.

Encoding: Our encoding procedure Enc takes as input a message $m \in \{0,1\}^k$, and does the following.
   - Sample X from $(\mathbb{K} \setminus \{0\})^3$ uniformly such that $\mathsf{nmExt}(X) = 0^{2k}\|m$.
   - Sample S from $(\mathbb{K} \setminus \{0\})^3$ uniformly such that $\mathsf{nmExt}(S) = 0^{2k}\|r$ for some r in $\{0,1\}^k$.
   - $V = \langle X, S \rangle_{\mathbb{K}}$.
   - $W = \langle X, S \rangle_{\mathbb{F}}$.
   - Output (X, S, V, W).

Decoding: Our decoding procedure Dec takes as input some x, s, v, w and does the following.
   - If $(x, s, v, w) \notin (\mathbb{K} \setminus \{0\})^6 \times \mathbb{K} \times \mathbb{F}$, then output $\bot$.
   - If $\mathsf{nmExt}'(x) = \bot$, output $\bot$.
   - If $\mathsf{nmExt}'(s) = \bot$, output $\bot$.
   - If $v \neq \langle x, s \rangle_{\mathbb{K}}$, output $\bot$.
   - If $w \neq \langle x, s \rangle_{\mathbb{F}}$, output $\bot$.
   - Otherwise, output m^*, where $\mathsf{nmExt}(x) = 0^{2k}\|m^*$.

Let $f_1, f_2, f_3, g_1, g_2, g_3, h_1 : \mathbb{K} \to \mathbb{K}$, $h_2 : \mathbb{F} \to \mathbb{F}$ be arbitrary functions, and let $f = (f_1, f_2, f_3)$ and $g = (g_1, g_2, g_3)$.

Definition 8 (Continuous Tampering Experiment). *We will first define the tampering oracle $\mathsf{Tamp}_c^{\mathsf{state}}(f, g, h_1, h_2)$, for $\mathsf{state} \in \{\mathsf{alive}, \mathsf{dead}\}$ and for*

$$c = (x_1, x_2, x_3, s_1, s_2, s_3, \langle x, s \rangle_{\mathbb{K}}, \langle x, s \rangle_{\mathbb{F}}).$$

For a tampering function (f, g, h_1, h_2) define the tampering oracle by

$\mathsf{Tamp}_c^{\mathsf{state}}(f, g, h_1, h_2)$:

    If state $=$ dead *output* $\perp$

    If $(x, s, \langle x, s\rangle_{\mathbb{K}}, \langle x, s\rangle_{\mathbb{F}}) = (f(x), g(s), h_1(\langle x, s\rangle_{\mathbb{K}}), h_2(\langle x, s\rangle_{\mathbb{F}}))$ *output* same

    If $(\mathsf{nmExt}'(f(x)) = \perp)$

        or $(\mathsf{nmExt}'(g(s)) = \perp)$

        or $(\langle f(x), g(s)\rangle_{\mathbb{K}} \neq h_1(\langle x, s\rangle_{\mathbb{K}}))$

        or $(\langle f(x), g(s)\rangle_{\mathbb{F}} \neq h_2(\langle x, s\rangle_{\mathbb{F}}))$

        set state $\leftarrow$ dead *and output* $\perp$

    Otherwise output $(f(x), g(s), h_1(\langle x, s\rangle_{\mathbb{K}}), h_2(\langle x, s\rangle_{\mathbb{F}}))$

Fix some $c = (x, s, v, w)$, *with* $x, s \in \mathbb{K}^3$, $v \in \mathbb{K}$, *and* $w \in \mathbb{F}$. *We define the continuous tampering experiment* CT_c^r *by*

CT_c^r:

    state $\leftarrow$ alive

    For $i = 1$ *to* r

        Choose functions $f_1, f_2, f_3, g_1, g_2, g_3, h_1, h_2$.

        $\psi \leftarrow \mathsf{Tamp}_c^{\mathsf{state}}(f, g, h_1, h_2)$.

        Output ψ

The following result which shows that continuously tampering a codeword for 2^{ck} rounds, for any constant $c < 1$, does not reveal any useful information about the codeword.

Theorem 4. *Let* X, S *be uniform in* $(\mathbb{K} \setminus \{0\})^3$ *conditioned on the event that* $\mathsf{nmExt}'(X) \neq \perp$ *and* $\mathsf{nmExt}'(S) \neq \perp$. *Let* C *be the random variable*

$$(X, S, \langle X, S\rangle_{\mathbb{K}}, \langle X, S\rangle_{\mathbb{F}}).$$

For any integer $r \geq 0$, *we have that*

$$\Delta\big((\mathsf{CT}_C^r, \mathsf{nmExt}(X)); (\mathsf{CT}_C^r, 0^{2k}\|U_k)\big) \leq 2^{-2k} \cdot 10 \cdot r,$$

where U_k *is a uniform* k-*bit string independent from* X, S.

The main result of the paper is obtained as an easy corollary of Theorem 4, as stated below.

Corollary 1. *Let* $m_0, m_1 \in \{0, 1\}^k$, *and let* $C^{(0)} \leftarrow \mathsf{Enc}(m_0)$, *and let* $C^{(1)} \leftarrow \mathsf{Enc}(m_1)$. *For any integer* $r \geq 0$, *we have that*

$$\Delta\left(\mathsf{CT}_{C^{(0)}}^r; \mathsf{CT}_{C^{(1)}}^r\right) \leq 2^{-k} \cdot 20 \cdot r.$$

In particular, for $r = 2^{ck}$, *for any* $c < 1$, *we have that*

$$\Delta\left(\mathsf{CT}_{C^{(0)}}^r; \mathsf{CT}_{C^{(1)}}^r\right) \leq 2^{-\Omega(k)}.$$

Proof. By Theorem 4, for any $r \geq 0$, and the random variable

$$C = (X, S, \langle X, S \rangle_{\mathbb{K}}, \langle X, S \rangle_{\mathbb{F}})$$

we have that

$$\Delta\big((\mathsf{CT}_C^r, \mathsf{nmExt}(X)) ; (\mathsf{CT}_C^r, 0^{2k} \| U_k)\big) \leq 2^{-2k} \cdot 10 \cdot r,$$

where X, S are distributed as in Theorem 4. Thus conditioning on the event that $\mathsf{Dec}(C) = m_i$ for $i = 0, 1$, which is the same as the event that $\mathsf{nmExt}(X) = 0^{2k} \| m_i$ and using Lemma 7, we get that

$$\Delta\big((\mathsf{CT}_C^r, \mathsf{nmExt}(X))|_{\mathsf{nmExt}(X) = 0^{2k} \| m_0} ; (\mathsf{CT}_C^r, 0^{2k} \| U_k)|_{U_k = m_0}\big) = \Delta\big(\mathsf{CT}_{C^{(0)}}^r ; \mathsf{CT}_C^r\big)$$
$$\leq 2^{-k} \cdot 10 \cdot r,$$

and

$$\Delta\big((\mathsf{CT}_C^r, \mathsf{nmExt}(X))|_{\mathsf{nmExt}(X) = 0^{2k} \| m_1} ; (\mathsf{CT}_C^r, 0^{2k} \| U_k)|_{U_k = m_1}\big) = \Delta\big(\mathsf{CT}_{C^{(1)}}^r ; \mathsf{CT}_C^r\big)$$
$$\leq 2^{-k} \cdot 10 \cdot r,$$

The result then follows by the triangle inequality. $\square$

To prove Theorem 4, we will show the more general Theorem 5 which immediately implies Theorem 4. We introduce the following parameters: $\rho = 2^{-40k}$. Also, for any sets $\mathcal{X}, \mathcal{S} \subseteq \mathbb{K}^3$, $\mathcal{V} \subseteq \mathbb{K}$ and $\mathcal{W} \subseteq \mathbb{F}$, we shorthand

$$p[\mathcal{X}, \mathcal{S}, \mathcal{V}, \mathcal{W}] := \Pr[(\widetilde{X}, \widetilde{S}, \langle \widetilde{X}, \widetilde{S} \rangle_{\mathbb{K}}, \langle \widetilde{X}, \widetilde{S} \rangle_{\mathbb{F}}) \in \mathcal{X} \times \mathcal{S} \times \mathcal{V} \times \mathcal{W}]$$

and

$$q[\mathcal{X}, \mathcal{S}, \mathcal{V}, \mathcal{W}] := \Pr[(\widetilde{X}, \widetilde{S}, \langle \widetilde{X}, \widetilde{S} \rangle_{\mathbb{K}}, \langle \widetilde{X}, \widetilde{S} \rangle_{\mathbb{F}}) \in \mathcal{X} \times \mathcal{S} \times \mathcal{V} \times \mathcal{W} \mid$$
$$\mathsf{nmExt}'(\widetilde{X}) \neq \bot, \mathsf{nmExt}'(\widetilde{S}) \neq \bot]$$

where $\widetilde{X}, \widetilde{S}$ are uniform in $(\mathbb{K} \setminus \{0\})^3$.

Remark 1. Our proof will proceed by partitioning the space in a way that the eight parts of our codeword remain independent. We introduced above the definition of the probability of landing in a particular partition. The reason we needed two different definitions depending on whether the codeword is a valid codeword or not is because we want to prove a statement for valid codewords but the proof technique crucially requires us to prove statements assuming that the eight parts of the codeword are independent. The following result shows that as long as $q[\mathcal{X}, \mathcal{S}, \mathcal{V}, \mathcal{W}]$ is not too small, $p[\mathcal{X}, \mathcal{S}, \mathcal{V}, \mathcal{W}]$ and $q[\mathcal{X}, \mathcal{S}, \mathcal{V}, \mathcal{W}]$ are nearly equal. This statement is required only to overcome the above mentioned technical annoyance and the proof appears in the full version.

Lemma 8. *Let* $\mathcal{X}_1, \mathcal{X}_2, \mathcal{X}_3, \mathcal{S}_1, \mathcal{S}_2, \mathcal{S}_3 \subseteq \mathbb{K} \setminus \{0\}$, $\mathcal{V} \subseteq \mathbb{K}$, *and let* $\mathcal{W} \subseteq \mathbb{F}$. *We denote* $\mathcal{X} = (\mathcal{X}_1, \mathcal{X}_2, \mathcal{X}_3)$ *and* $\mathcal{S} = (\mathcal{S}_1, \mathcal{S}_2, \mathcal{S}_3)$. *If* $q[\mathcal{X}, \mathcal{S}, \mathcal{V}, \mathcal{W}] \geq 2^{-800k}$, *then*

$$\frac{p[\mathcal{X}, \mathcal{S}, \mathcal{V}, \mathcal{W}]}{q[\mathcal{X}, \mathcal{S}, \mathcal{V}, \mathcal{W}]} = 1 \pm 2^{-180k},$$

and

$$\frac{\Pr[\widetilde{X} \in \mathcal{X}, \, \widetilde{S} \in \mathcal{S}, \, U_n \in \mathcal{V}, \, \mathrm{tr}_{\mathbb{K} \to \mathbb{F}}(U_n) \in \mathcal{W}]}{q[\mathcal{X}, \mathcal{S}, \mathcal{V}, \mathcal{W}]} = 1 \pm 2^{-180k},$$

where $\widetilde{X}, \widetilde{S}$ *are uniform in* $(\mathbb{K} \setminus \{0\})^3$, *and* U_n *is uniform in* $\mathbb{K}$.

Theorem 5. *Let* $\mathcal{X}_1, \mathcal{X}_2, \mathcal{X}_3, \mathcal{S}_1, \mathcal{S}_2, \mathcal{S}_3 \subseteq \mathbb{K} \setminus \{0\}$, $\mathcal{V} \subseteq \mathbb{K}$, *and let* $\mathcal{W} \subseteq \mathbb{F}$. *We denote* $\mathcal{X} = (\mathcal{X}_1, \mathcal{X}_2, \mathcal{X}_3)$ *and* $\mathcal{S} = (\mathcal{S}_1, \mathcal{S}_2, \mathcal{S}_3)$. *Let* (X, S) *be random variables uniform in* $\mathbb{K}^6$ *conditioned on the event that* $X_i \in \mathcal{X}_i$, $S_i \in \mathcal{S}_i$ *for* $i = 1, 2, 3$, $\mathsf{nmExt}'(X) \neq \bot$, $\mathsf{nmExt}'(S) \neq \bot$, $\langle X, S \rangle_{\mathbb{K}} \in \mathcal{V}$, *and* $\langle X, S \rangle_{\mathbb{F}} \in \mathcal{W}$. *Let* C *be the random variable*

$$(X, S, \langle X, S \rangle_{\mathbb{K}}, \langle X, S \rangle_{\mathbb{F}}).$$

For any integer $r \geq 0$, *we have that*

$$\Delta\left((\mathsf{CT}_C^r, \mathsf{nmExt}(X)) ; (\mathsf{CT}_C^r, 0^{2k} \| U_k)\right) \leq \left(\frac{\rho}{q[\mathcal{X}, \mathcal{S}, \mathcal{V}, \mathcal{W}]}\right)^{\frac{1}{8}} + 9 \cdot r \cdot 2^{-2k}, \quad (2)$$

where U_k *is a uniform k-bit string independent from* X, S.

We will prove Theorem 5 by partitioning the ambient space into appropriate subsets such that Eq. 2 holds for each of these partitions. Theorem 5 can then be shown by the following lemma.

Lemma 9. *Let* $\mathcal{X}_1, \mathcal{X}_2, \mathcal{X}_3, \mathcal{S}_1, \mathcal{S}_2, \mathcal{S}_3 \subseteq \mathbb{K} \setminus \{0\}$, $\mathcal{V} \subseteq \mathbb{K}$, *and let* $\mathcal{W} \subseteq \mathbb{F}$. *We denote* $\mathcal{X} = (\mathcal{X}_1, \mathcal{X}_2, \mathcal{X}_3)$ *and* $\mathcal{S} = \mathcal{S}_1, \mathcal{S}_2, \mathcal{S}_3$. *Let* (X, S) *be random variables uniform in* $\mathbb{K}^6$ *conditioned on the event that* $X_i \in \mathcal{X}_i$, $S_i \in \mathcal{S}_i$ *for* $i = 1, 2, 3$, $\mathsf{nmExt}'(X) \neq \bot$, $\mathsf{nmExt}'(S) \neq \bot$, $\langle X, S \rangle_{\mathbb{K}} \in \mathcal{V}$, *and* $\langle X, S \rangle_{\mathbb{F}} \in \mathcal{W}$. *Let* C *be the random variable*

$$(X, S, \langle X, S \rangle_{\mathbb{K}}, \langle X, S \rangle_{\mathbb{F}}).$$

Let $\mathcal{P}_1, \mathcal{P}_2, \ldots, \mathcal{P}_t$ *be a partitioning of* $\mathcal{X} \times \mathcal{S} \times \mathcal{V} \times \mathcal{W}$. *Then we have that for any integer* $r \geq 0$, *if*

$$\Delta\left((\mathsf{CT}_C^r, \mathsf{nmExt}(X))|_{C \in \mathcal{P}_j} ; (\mathsf{CT}_C^r, 0^{2k} \| U_k)|_{C \in \mathcal{P}_j}\right) \leq \varepsilon_j$$

then

$$\Delta\left((\mathsf{CT}_C^r, \mathsf{nmExt}(X)) ; (\mathsf{CT}_C^r, 0^{2k} \| U_k)\right) \leq \sum_{j=1}^{t} \frac{q[\mathcal{P}_j]}{q[\mathcal{X}, \mathcal{S}, \mathcal{V}, \mathcal{W}]} \cdot \varepsilon_j,$$

where U_k *is a uniform k-bit string independent from* X, S.

Proof. Let $\mathcal{A}$ be the sample space of $(\mathsf{CT}_C^r, \mathsf{nmExt}(X))$. Then, by definition,

$$\Delta = \Delta\left((\mathsf{CT}_C^r, \mathsf{nmExt}(X)); (\mathsf{CT}_C^r, 0^{2k}\|U_k)\right)$$

is given by

$$\Delta = \frac{1}{2} \cdot \sum_{a \in \mathcal{A}} \left| \Pr[(\mathsf{CT}_C^r, \mathsf{nmExt}(X)) = a] - \Pr[(\mathsf{CT}_C^r, 0^{2k}\|U_k) = a] \right|$$

$$= \frac{1}{2} \cdot \sum_{a \in \mathcal{A}} \left| \sum_{j=1}^{t} \Pr[(\mathsf{CT}_C^r, \mathsf{nmExt}(X)) = a, \, C \in \mathcal{P}_j] - \Pr[(\mathsf{CT}_C^r, 0^{2k}\|U_k) = a, \, C \in \mathcal{P}_j] \right|$$

$$\leq \frac{1}{2} \cdot \sum_{a \in \mathcal{A}} \sum_{j=1}^{t} \Pr[C \in \mathcal{P}_j] \cdot \left| \Pr[(\mathsf{CT}_C^r, \mathsf{nmExt}(X)) = a \mid C \in \mathcal{P}_j] - \Pr[(\mathsf{CT}_C^r, 0^{2k}\|U_k) = a \mid C \in \mathcal{P}_j] \right|$$

$$= \frac{1}{2} \cdot \sum_{j=1}^{t} \Pr[C \in \mathcal{P}_j] \cdot \sum_{a \in \mathcal{A}} \left| \Pr[(\mathsf{CT}_C^r, \mathsf{nmExt}(X)) = a \mid C \in \mathcal{P}_j] - \Pr[(\mathsf{CT}_C^r, 0^{2k}\|U_k) = a \mid C \in \mathcal{P}_j] \right|$$

$$= \sum_{j=1}^{t} \frac{q[\mathcal{P}_j]}{q[\mathcal{X}, \mathcal{S}, \mathcal{V}, \mathcal{W}]} \cdot \varepsilon_j.$$

$\square$

We will now partition each of $\mathcal{X}_1, \mathcal{X}_2, \mathcal{X}_3, \mathcal{S}_1, \mathcal{S}_2, \mathcal{S}_3$ which will induce a partitioning of the whole space. The partitions are chosen in a way that if, say, X_i (respectively, S_i) for $i \in \{1, 2, 3\}$ is uniformly distributed over a particular partition of $\mathcal{X}_i$ (respectively, $\mathcal{S}_i$), then this gives a precise estimate of $\widetilde{\mathbf{H}}_\infty(X_i|f_i(X_i))$ (respectively, $\widetilde{\mathbf{H}}_\infty(S_i|g_i(S_i))$).

Definition 9 (Partition). *We partition the set $\mathcal{X}_1 \subseteq \{0,1\}^n$ based on the function f_1 as follows.*

1. $\mathcal{X}_{1,0} = \{x \in \mathcal{X}_1 \, : \, f_1(x) = x\}$.
2. $\mathcal{X}_1 = \mathcal{X}_1 \setminus \mathcal{X}_{1,0}$.
3. *For* $i = 1, \ldots, \ell - 1$, $\mathcal{X}_{1,i} = \{x \in \mathcal{X}_1 \, : \, |f_1^{-1}(f_1(x)) \cap \mathcal{X}_1| \in [2^{100k \cdot (i-1)}, 2^{100k \cdot i})\}$.
4. $\mathcal{X}_{1,\ell} = \{x \in \mathcal{X}_1 \, : \, |f_1^{-1}(f_1(x)) \cap \mathcal{X}_1| \geq 2^{100k \cdot (\ell-1)}\}$

$\mathcal{X}_2, \mathcal{X}_3, \mathcal{S}_1, \mathcal{S}_2, \mathcal{S}_3$ *are partitioned similarly as above.*

We classify the partitions obtained according to the following types.

Definition 10 (Classification of Partitions). *Let* $i_1, i_2, i_3, j_1, j_2, j_3$ *be one of* $\{0, 1, \ldots, \ell\}$. *We then classify the partition*

$$\mathcal{P} := \mathcal{X}_{1,i_1} \times \mathcal{X}_{2,i_2} \times \mathcal{X}_{3,i_3} \times \mathcal{S}_{1,j_1} \times \mathcal{S}_{2,j_2} \times \mathcal{S}_{3,j_3} \times \mathcal{V} \times \mathcal{W}$$

of $\mathcal{X} \times \mathcal{S} \times \mathcal{V} \times \mathcal{W}$ *as follows.*

Type–1: $\mathcal{P}$ *is a Type–1 partition if* $i_1 = i_2 = i_3 = j_1 = j_2 = j_3 = 0$.
Type–2: $\mathcal{P}$ *is a Type–2 partition if*
  1. $\mathcal{P}$ *is not a Type–1 partition, i.e., at least one of* $i_1, i_2, i_3, j_1, j_2, j_3 > 0$.
  2. *Each of* $i_1, i_2, i_3, j_1, j_2, j_3$ *is at most* $\frac{\delta n}{100k} - 1$.
Type–3: $\mathcal{P}$ *is a Type–3 partition if the following hold*
  1. $\mathcal{P}$ *is not a Type–1 or Type–2 partition, i.e., at least one of* $i_1, i_2, i_3, j_1, j_2,$
     $j_3 > \frac{\delta n}{100k} - 1$.
  2. $i_1 + i_2 + i_3 + j_1 + j_2 + j_3 \leq \frac{n}{40k}$.
Type–4: $\mathcal{P}$ *is a Type–4 partition if*
  1. $\mathcal{P}$ *is not a Type–1, 2, or 3 partition, , i.e.,* $i_1 + i_2 + i_3 + j_1 + j_2 + j_3 > \frac{n}{40k}$.
  2. *At least one of* $i_1, i_2, i_3, j_1, j_2, j_3$ *is not* ℓ.
Type–5: $\mathcal{P}$ *is a Type–5 partition if* $i_1 = i_2 = i_3 = j_1 = j_2 = j_3 = \ell$.

In the following we classify partitions of Type–1 and Type–5 further into subpartitions, but before this, we introduce the following definition.

Definition 11. *We define the following subsets of* $\mathcal{V}$.

  - $\mathcal{V}_{\text{same}} = \{v \in \mathcal{V} : h_1(v) = v\}$.
  - $\overline{\mathcal{V}_{\text{same}}} = \mathcal{V} \setminus \mathcal{V}_{\text{same}}$.
  - *For all* $y \in \{0,1\}^n$, $\mathcal{V}_y = \{v \in \mathcal{V} : h_1(v) = y\}$.
  - *For all* $y \in \{0,1\}^n$, $\overline{\mathcal{V}_y} = \mathcal{V} \setminus \mathcal{V}_y$.

We similarly define $\mathcal{W}_{\text{same}}, \overline{\mathcal{W}_{\text{same}}}, \mathcal{W}_z, \overline{\mathcal{W}_z}$ *for all* $z \in \mathbb{F}$ *via the function* h_2.

Using this classification, we now further partition Type–1 and Type–5 partitions.

Definition 12. *Let* $\mathcal{X}_{\text{same}} = \mathcal{X}_{1,0} \times \mathcal{X}_{2,0} \times \mathcal{X}_{3,0}$ *and let* $\mathcal{S}_{\text{same}} = \mathcal{S}_{1,0} \times \mathcal{S}_{2,0} \times \mathcal{S}_{3,0}$

  Type–1a: *We say that* $\mathcal{X}_{\text{same}} \times \mathcal{S}_{\text{same}} \times \mathcal{V}_{\text{same}} \times \mathcal{W}_{\text{same}}$ *is a Type–1a partition.*
  Type–1b: *We say that the following are Type–1b partitions:*
    - $\mathcal{X}_{\text{same}} \times \mathcal{S}_{\text{same}} \times \mathcal{V} \times \overline{\mathcal{W}_{\text{same}}}$.
    - $\mathcal{X}_{\text{same}} \times \mathcal{S}_{\text{same}} \times \overline{\mathcal{V}_{\text{same}}} \times \mathcal{W}_{\text{same}}$.

Definition 13. *For* $\mathbf{a} = (a_1, a_2, a_3) \in \mathbb{K}^3$, *let*

$$\mathcal{X}_{\mathbf{a}} = \{(x_1, x_2, x_3) \in \mathcal{X}_{1,\ell} \times \mathcal{X}_{2,\ell} \times \mathcal{X}_{3,\ell} : f_1(x_1) = a_1, f_2(x_2) = a_2, f_3(x_3) = a_3\}.$$

Similarly, define $\mathcal{S}_{\mathbf{b}}$ *for* $\mathbf{b} = (b_1, b_2, b_3) \in \mathbb{K}^3$.

Type–5a: *We say that* $\mathcal{X}_{\mathbf{a}} \times \mathcal{S}_{\mathbf{b}} \times \mathcal{V}_{\langle \mathbf{a}, \mathbf{b} \rangle_{\mathbb{K}}} \times \mathcal{W}_{\langle \mathbf{a}, \mathbf{b} \rangle_{\mathbb{F}}}$ *is a Type–5a partition.*
Type–5b: *We say that the following are Type–5b partitions:*

$$- \mathcal{X}_\mathbf{a} \times \mathcal{S}_\mathbf{b} \times \mathcal{V} \times \overline{\mathcal{W}_{\langle \mathbf{a}, \mathbf{b} \rangle_\mathbb{F}}}.$$
$$- \mathcal{X}_\mathbf{a} \times \mathcal{S}_\mathbf{b} \times \overline{\mathcal{V}_{\langle \mathbf{a}, \mathbf{b} \rangle_\mathbb{K}}} \times \mathcal{W}_{\langle \mathbf{a}, \mathbf{b} \rangle_\mathbb{F}}.$$

If a partition $\mathcal{P}$ is of Type$-T$, then we denote it as $Type(\mathcal{P}) = T$, where $T \in \{1a, 1b, 2, 3, 4, 5a, 5b\}$.

Before bounding the required statistical distance for each partition, we will prove a few general results.

Lemma 10. *Let* $\mathcal{X}_1, \mathcal{X}_2, \mathcal{X}_3, \mathcal{S}_1, \mathcal{S}_2, \mathcal{S}_3 \subseteq \mathbb{K} \setminus \{0\}$, $\mathcal{V} \subseteq \mathbb{K}$, *and let* $\mathcal{W} \subseteq \mathbb{F}$. *We denote* $\mathcal{X} = (\mathcal{X}_1, \mathcal{X}_2, \mathcal{X}_3)$ *and* $\mathcal{S} = (\mathcal{S}_1, \mathcal{S}_2, \mathcal{S}_3)$. *Let* $|\mathcal{X}_i| \geq 2^{n-100k}$, $|\mathcal{S}_i| \geq 2^{n-100k}$ *for* $i = 1, 2, 3$, *and let* $q[\mathcal{X}, \mathcal{S}, \mathcal{V}, \mathcal{W}] \geq 2^{-800k}$. *Let* (X, S) *be random variables uniform in* $\mathbb{K}^6$ *conditioned on the event that* $X_i \in \mathcal{X}_i$, $S_i \in \mathcal{S}_i$ *for* $i = 1, 2, 3$, $\mathsf{nmExt}'(X) \neq \bot$, $\mathsf{nmExt}'(S) \neq \bot$, $\langle X, S \rangle_\mathbb{K} \in \mathcal{V}$, *and* $\langle X, S \rangle_\mathbb{F} \in \mathcal{W}$. *Then*

$$\Delta \left(\mathsf{nmExt}(X) \, ; \, 0^{2k} \| U_k \right) \leq 2^{-990k},$$

where U_k *is a uniform k-bit string independent from* X, S.

Proof. Notice that if X and S were independent and uniform then this would follow trivially from the fact that nmExt is a 3-source extractor (Notice that we don't need the non-malleability property of nmExt for this part of the proof). Thus, in order to show this, it is sufficient to establish that X and S are nearly independent given partial knowledge about $\langle X, S \rangle_\mathbb{K}$, and $\langle X, S \rangle_\mathbb{F}$. We show this as follows.

Let X', S' be distributed independently and uniform in $\mathcal{X}, \mathcal{S}$, respectively. Notice that $\mathbf{H}_\infty(X') \geq 3n - 300k$, and $\mathbf{H}_\infty(S') \geq 3n - 300k$, and hence $\widetilde{\mathbf{H}}_\infty(X' | \mathsf{nmExt}(X')) \geq 3n - 303k$. By Lemma 4, we get that

$$(\langle X', S' \rangle_\mathbb{K}, \mathsf{nmExt}(X'), \mathsf{nmExt}(S')) \approx_{2^{-2000k}} (U_n, \mathsf{nmExt}(X'), \mathsf{nmExt}(S')),$$

where we assumed that $n \geq 5000k$. Since $\langle X', S' \rangle_\mathbb{F} = \mathrm{tr}_{\mathbb{K} \to \mathbb{F}}(\langle X', S' \rangle_\mathbb{K})$, where $\mathrm{tr}_{\mathbb{K} \to \mathbb{F}}$ is the field trace function, we have that

$$(\langle X', S' \rangle_\mathbb{K}, \langle X', S' \rangle_\mathbb{F}, \mathsf{nmExt}(X'), \mathsf{nmExt}(S')) \approx_{2^{-2000k}} (U_n, \mathrm{tr}_{\mathbb{K} \to \mathbb{F}}(U_n),$$
$$\mathsf{nmExt}(X'), \mathsf{nmExt}(S')).$$

Let $(\widehat{X}, \widehat{S})$ be jointly distributed as (X', S') conditioned on $\langle X', S' \rangle_\mathbb{K} \in \mathcal{V}$, $\langle X', S' \rangle_\mathbb{F} \in \mathcal{W}$. Thus, by Lemma 7, we get that

$$(\mathsf{nmExt}(\widehat{X}), \mathsf{nmExt}(\widehat{S})) \approx_{2^{-1000k}} (\mathsf{nmExt}(X'), \mathsf{nmExt}(S')).$$

Also, since $\mathbf{H}_\infty(X_i') \geq n - 100k \geq n(1 - \delta)$, $\mathbf{H}_\infty(S_i') \geq n - 100k \geq n(1 - \delta)$ for $i = 1, 2, 3$. Thus, by Theorem 3, we have that

$$(\mathsf{nmExt}(X'), \mathsf{nmExt}(S')) \approx_{2 \cdot 2^{-1000k}} (U_{3k}, U'_{3k}).$$

By triangle inequality, we get that

$$(\mathsf{nmExt}(\widehat{X}), \mathsf{nmExt}(\widehat{S})) \approx_{3 \cdot 2^{-1000k}} (U_{3k}, U'_{3k}).$$

Conditioning on $\mathsf{nmExt}'(\widehat{X}) \neq \bot$, and $\mathsf{nmExt}'(\widehat{S}) \neq \bot$, and applying Lemma 7, we obtain the desired result. $\qquad \square$

We now show that for any given partition, if the tampering oracle outputs $\perp$ with high probability, then the desired statistical distance for that particular partition is small.

Lemma 11. *Let* $\mathcal{X}_1, \mathcal{X}_2, \mathcal{X}_3, \mathcal{S}_1, \mathcal{S}_2, \mathcal{S}_3 \subseteq \mathbb{K} \setminus \{0\}$, $\mathcal{V} \subseteq \mathbb{K}$, *and let* $\mathcal{W} \subseteq \mathbb{F}$. *We denote* $\mathcal{X} = (\mathcal{X}_1, \mathcal{X}_2, \mathcal{X}_3)$ *and* $\mathcal{S} = (\mathcal{S}_1, \mathcal{S}_2, \mathcal{S}_3)$. *Let* (X, S) *be random variables uniform in* $\mathbb{K}^6$ *conditioned on the event that* $X_i \in \mathcal{X}_i$, $S_i \in \mathcal{S}_i$ *for* $i = 1, 2, 3$, $\mathsf{nmExt}'(X) \neq \perp$, $\mathsf{nmExt}'(S) \neq \perp$, $\langle X, S \rangle_{\mathbb{K}} \in \mathcal{V}$, *and* $\langle X, S \rangle_{\mathbb{F}} \in \mathcal{W}$. *Let* C *be the random variable*

$$(X, S, \langle X, S \rangle_{\mathbb{K}}, \langle X, S \rangle_{\mathbb{F}}).$$

If

$$\Pr_C[\mathsf{Tamp}_C^{\mathsf{state}}(f, g, h_1, h_2) = \perp] \geq 1 - \varepsilon$$

then for any integer $r \geq 0$

$$\Delta\left((\mathsf{CT}_C^r, \mathsf{nmExt}(X)) ; (\mathsf{CT}_C^r, 0^{2k} \| U_k) \right) \leq \Delta\left(\mathsf{nmExt}(X) ; 0^{2k} \| U_k \right) + 2\varepsilon,$$

where U_k *is a uniform k-bit string independent from* X, S.

Proof. Let T_C denote $\mathsf{Tamp}_C^{\mathsf{state}}(f, g, h_1, h_2)$. Notice that for any $m \in \{0, 1\}^{3k}$, we have that

$$\Pr[T_C = \perp, \mathsf{nmExt}(X) = m] \leq \Pr[\mathsf{nmExt}(X) = m].$$

Since we know that the statistical distance between two random variables A and B is

$$\sum_{a : \Pr[A=a] > \Pr[B=a]} (\Pr[A = a] - \Pr[B = a]),$$

we have that

$$\Delta\left((T_C, \mathsf{nmExt}(X)) ; (\perp, \mathsf{nmExt}(X)) \right) = \Pr[T_C \neq \perp] \leq \varepsilon.$$

This implies that

$$\Delta\left((\mathsf{CT}_C^r, \mathsf{nmExt}(X)) ; (\perp^r, \mathsf{nmExt}(X)) \right) \leq \varepsilon, \tag{3}$$

where by $\perp^r$ we mean the tampering oracle outputs $\perp$ in the first and hence in each of the subsequent rounds. By Eq. 3 and Lemma 3, we have that

$$\Delta\left((\mathsf{CT}_C^r, 0^{2k} \| U_k) ; (\perp^r, 0^{2k} \| U_k) \right) = \Delta\left(\mathsf{CT}_C^r ; \perp^r \right) \leq \varepsilon, \tag{4}$$

By Eqs. 3 and 4, and the triangle inequality, we get the desired result. $\qquad\square$

It is easy to see that when X, S are restricted to belong to a partition of Type$-$1b or 5b, the tampering oracle outputs $\perp$ with probability 1, so for partitions of this type, the corresponding statistical distance can be bounded using Lemmas 11 and 10. We now prove a similar result holds for Type 2, 3, and 4.

Lemma 12. *[Type-2 partition] Let* $\mathcal{X}_{1,i_1}, \mathcal{X}_{2,i_2}, \mathcal{X}_{3,i_3}, \mathcal{S}_{1,j_1}, \mathcal{S}_{2,j_2}, \mathcal{S}_{3,j_3} \subseteq \mathbb{K} \setminus$ $\{0\}$, $\mathcal{V} \subseteq \mathbb{K}$, *and let* $\mathcal{W} \subseteq \mathbb{F}$. *We denote* $\mathcal{X}^{\star} = (\mathcal{X}_{1,i_1}, \mathcal{X}_{2,i_2}, \mathcal{X}_{3,i_3})$ *and* $\mathcal{S}^{\star} = \mathcal{S}_{1,j_1}, \mathcal{S}_{2,j_2}, \mathcal{S}_{3,j_3}$. *Let* $(\mathcal{X}^{\star}, \mathcal{S}^{\star}, \mathcal{V}, \mathcal{W})$ *be a partition of Type–2, and let* $q[\mathcal{X}^{\star}, \mathcal{S}^{\star}, \mathcal{V}, \mathcal{W}] \geq 2^{-45k}$. *Let* (X, S) *be random variables uniform in* $\mathbb{K}^6$ *conditioned on the event that* $X_t \in \mathcal{X}_{t,i_t}$, $S_t \in \mathcal{S}_{t,j_t}$ *for* $t = 1, 2, 3$, $\mathsf{nmExt}'(X) \neq \bot$, $\mathsf{nmExt}'(S) \neq \bot$, $\langle X, S \rangle_{\mathbb{K}} \in \mathcal{V}$, *and* $\langle X, S \rangle_{\mathbb{F}} \in \mathcal{W}$. *Let* C *be the random variable*

$$(X, S, \langle X, S \rangle_{\mathbb{K}}, \langle X, S \rangle_{\mathbb{F}}).$$

Then,
$$\Pr_C[\mathsf{Tamp}_C^{\mathsf{state}}(f, g, h_1, h_2) = \bot] \geq 1 - 2 \cdot 2^{-2k}.$$

Proof. In this lemma, the given partition is of Type−2, which means that at least one of $i_1, i_2, i_3, j_1, j_2, j_3 \neq 0$, and so without loss of generality, let $i_1 > 0$. If X_1, X_2, X_3 were independent random variables then, by the non-malleability property of the non-malleable extractor, and the fact that f, g are nearly bijective functions, $\mathsf{nmExt}(X)$ and $\mathsf{nmExt}(f(X))$ are close to being uniform and independent. However the constraint that $\langle X, S \rangle_{\mathbb{K}} \in \mathcal{V}$ and $\langle X, S \rangle_{\mathbb{F}} \in \mathcal{W}$ might introduce dependence between X_1, X_2, X_3.

To overcome this hurdle, it is sufficient to establish that $X_1, X_2, X_3, S_1, S_2, S_3$ are nearly independent given partial knowledge about $\langle X, S \rangle_{\mathbb{K}}$, and $\langle X, S \rangle_{\mathbb{F}}$. The full proof appears in the full version. $\qquad\qquad\square$

Lemma 13. *[Type-3 partition] Let* $\mathcal{X}_{1,i_1}, \mathcal{X}_{2,i_2}, \mathcal{X}_{3,i_3}, \mathcal{S}_{1,j_1}, \mathcal{S}_{2,j_2}, \mathcal{S}_{3,j_3} \subseteq \mathbb{K} \setminus$ $\{0\}$, $\mathcal{V} \subseteq \mathbb{K}$, *and let* $\mathcal{W} \subseteq \mathbb{F}$. *We denote* $\mathcal{X}^{\star} = (\mathcal{X}_{1,i_1}, \mathcal{X}_{2,i_2}, \mathcal{X}_{3,i_3})$ *and* $\mathcal{S}^{\star} = (\mathcal{S}_{1,j_1}, \mathcal{S}_{2,j_2}, \mathcal{S}_{3,j_3})$. *Let* $(\mathcal{X}^{\star}, \mathcal{S}^{\star}, \mathcal{V}, \mathcal{W})$ *be a partition of Type–3, and let* $q[\mathcal{X}^{\star}, \mathcal{S}^{\star}, \mathcal{V}, \mathcal{W}] \geq 2^{-45k}$. *Let* (X, S) *be random variables uniform in* $\mathbb{K}^6$ *conditioned on the event that* $X_t \in \mathcal{X}_{t,i_t}$, $S_t \in \mathcal{S}_{t,j_t}$ *for* $t = 1, 2, 3$, $\mathsf{nmExt}'(X) \neq \bot$, $\mathsf{nmExt}'(S) \neq \bot$, $\langle X, S \rangle_{\mathbb{K}} \in \mathcal{V}$, *and* $\langle X, S \rangle_{\mathbb{F}} \in \mathcal{W}$. *Let* C *be the random variable*

$$(X, S, \langle X, S \rangle_{\mathbb{K}}, \langle X, S \rangle_{\mathbb{F}}).$$

Then,
$$\Pr_C[\mathsf{Tamp}_C^{\mathsf{state}}(f, g, h_1, h_2) = \bot] \geq 1 - 2^{-4k}.$$

Proof. Since the partition is of Type−3, at least one of $i_1, i_2, i_3, j_1, j_2, j_3 > \frac{\delta n}{100k} - 1$ and
$$i_1 + i_2 + i_3 + j_1 + j_2 + j_3 \leq \frac{n}{40k}.$$

Without loss of generality, let $i_1 > \frac{\delta n}{100k} - 1$.

The intuition behind the proof is that since i_1 is not too small, X has enough entropy given $f(X)$ to ensure that $\langle X, S \rangle_{\mathbb{F}}$ is close to uniform given $f(X), S$ by using the strong extractor property of the inner product. Hence $\langle X, S \rangle_{\mathbb{F}}$ and $\langle f(X), g(S) \rangle_{\mathbb{F}}$ are close to being independent and so the adversary, in order to not decode to $\bot$, should be able to guess $\langle f(X), g(S) \rangle_{\mathbb{F}}$ in the eighth state without having any useful information. Also, since $i_1 + i_2 + i_3 + j_1 + j_2 + j_3$ is not too small,

$f(X), g(S)$ together should have enough entropy to ensure that $\langle f(X), g(S)\rangle_{\mathbb{F}}$ is close to being uniform again because the inner product is a strong two-source extractor. This implies that the probability that the decoder does not decode to $\perp$ after tampering is close to 0. For this argument, we implicitly assumed that X and S are independent and formally we need to take into account the condition that $\langle X, S\rangle_{\mathbb{K}} \in \mathcal{V}$, and $\langle X, S\rangle_{\mathbb{F}} \in \mathcal{W}$ which introduces a limited dependence between X and S. Working out the exact constant is fairly easy. The full proof appears in the full version. □

Lemma 14. *[Type-4 partition] Let* $\mathcal{X}_{1,i_1}, \mathcal{X}_{2,i_2}, \mathcal{X}_{3,i_3}, \mathcal{S}_{1,j_1}, \mathcal{S}_{2,j_2}, \mathcal{S}_{3,j_3} \subseteq \mathbb{K} \setminus \{0\}$, $\mathcal{V} \subseteq \mathbb{K}$ *and let* $\mathcal{W} \subseteq \mathbb{F}$. *We denote* $\mathcal{X}^\star = (\mathcal{X}_{1,i_1}, \mathcal{X}_{2,i_2}, \mathcal{X}_{3,i_3})$ *and* $\mathcal{S}^\star = (\mathcal{S}_{1,j_1}, \mathcal{S}_{2,j_2}, \mathcal{S}_{3,j_3})$. *Let* $(\mathcal{X}^\star, \mathcal{S}^\star, \mathcal{V}, \mathcal{W})$ *be a partition of Type-4, and let* $q[\mathcal{X}^\star, \mathcal{S}^\star, \mathcal{V}, \mathcal{W}] \geq 2^{-45k}$. *Let* (X, S) *be random variables uniform in* $\{0, 1\}^{6n}$ *conditioned on the event that* $X_t \in \mathcal{X}_{t,i_t}$, $S_t \in \mathcal{S}_{t,j_t}$ *for* $t = 1, 2, 3$, $\mathsf{nmExt}'(X) \neq \perp$, $\mathsf{nmExt}'(S) \neq \perp$, $\langle X, S\rangle_{\mathbb{K}} \in \mathcal{V}$, *and* $\langle X, S\rangle_{\mathbb{F}} \in \mathcal{W}$. *Let* C *be the random variable*

$$(X, S, \langle X, S\rangle_{\mathbb{K}}, \langle X, S\rangle_{\mathbb{F}}).$$

Then,

$$\Pr_C[\mathsf{Tamp}_C^{\mathsf{state}}(f, g, h_1, h_2) = \perp] \geq 1 - 2^{-4k}.$$

Proof. Since the partition is of Type-4, at least one of $i_1, i_2, i_3, j_1, j_2, j_3 \neq \ell$ and

$$i_1 + i_2 + i_3 + j_1 + j_2 + j_3 > \frac{n}{40k}.$$

Without loss of generality, let $i_1 \leq \ell - 1$. Also, without loss of generality, let $i_1 + i_2 + i_3 > \frac{n}{80k}$.

The intuition behind the proof is that $i_1 + i_2 + i_3$ is large enough to ensure that X has enough entropy given $f(X)$ to ensure that $\langle X, S\rangle_{\mathbb{K}}$ is close to uniform given $f(X), S$ by using the strong extractor property of the inner product. Hence $\langle X, S\rangle_{\mathbb{K}}$ and $\langle f(X), g(S)\rangle_{\mathbb{K}}$ are close to being independent and so the adversary, in order to decode to a valid message, can only be able to guess $\langle f(X), g(S)\rangle_{\mathbb{K}}$ in the seventh state without having any useful information. Also, since $i_1 \leq \ell - 1$ is not too small, $f_1(X_1)$ has a large amount of entropy which in turn implies that $\langle f(X), g(S)\rangle_{\mathbb{K}}$ has a large amount of entropy since $g_1(S_1) \neq 0$. This implies that the probability that the decoder does not decode to $\perp$ after tampering is close to 0. Of course, for this argument to go through, we implicitly assumed that X and S are independent and formally we need to take into account the condition that $\langle X, S\rangle_{\mathbb{K}} \in \mathcal{V}$, and $\langle X, S\rangle_{\mathbb{F}} \in \mathcal{W}$ which introduces a limited dependence between X and S. Working out the exact constant is fairly easy. The full proof appears in the full version. □

In the above results, we established that the tampering oracle will output $\perp$ with probability very close to 1 for all partitions of Type$-2, 3, 4$ that are not too small. If the size of the partition is extremely small then Lemma 9 guarantees that such a partition does not contribute much to the statistical distance. Also, for a partition of Type$-1b$ and $5b$, the tampering oracle always outputs $\perp$.

The following corollary states the bound on the statistical distance conditioned on X, S in a partition of Type$-1b$, 2, 3, 4, 5b. The proof appears in the full version.

Corollary 2. *Let $\mathcal{X}_1, \mathcal{X}_2, \mathcal{X}_3, \mathcal{S}_1, \mathcal{S}_2, \mathcal{S}_3 \subseteq \mathbb{K} \setminus \{0\}$, $\mathcal{V} \subseteq \mathbb{K}$, and let $\mathcal{W} \subseteq \mathbb{F}$. We denote $\mathcal{X} = (\mathcal{X}_1, \mathcal{X}_2, \mathcal{X}_3)$ and $\mathcal{S} = (\mathcal{S}_1, \mathcal{S}_2, \mathcal{S}_3)$. Let $q[\mathcal{X}, \mathcal{S}, \mathcal{V}, \mathcal{W}] \geq 2^{-40k}$. Let (X, S) be random variables uniform in $\mathbb{K}^6$ conditioned on the event that $X_i \in \mathcal{X}_i$, $S_i \in \mathcal{S}_i$ for $i = 1, 2, 3$, $\mathsf{nmExt}'(X) \neq \bot$, $\mathsf{nmExt}'(S) \neq \bot$, $\langle X, S \rangle_{\mathbb{K}} \in \mathcal{V}$, and $\langle X, S \rangle_{\mathbb{F}} \in \mathcal{W}$. Let C be the random variable*

$$(X, S, \langle X, S \rangle_{\mathbb{K}}, \langle X, S \rangle_{\mathbb{F}}).$$

Then for any integer $r \geq 0$, if

$$\sum_{\mathcal{P}:\, Type(\mathcal{P}) \in \{1b,\, 2,\, 3,\, 4,\, 5b\}} \frac{q[\mathcal{P}]}{q[\mathcal{X}, \mathcal{S}, \mathcal{V}, \mathcal{W}]} \cdot \Delta\big((\mathsf{CT}_C^r, \mathsf{nmExt}(X))|_{C \in \mathcal{P}}\,;$$

$$(\mathsf{CT}_C^r, 0^{2k} \| U_k)|_{C \in \mathcal{P}}\big) \leq 5 \cdot 2^{-2k},$$

where U_k is a uniform k-bit string independent from X, S.

Lemma 15. *[Type-5 partition] Let $\mathcal{X}_{1,\ell}, \mathcal{X}_{2,\ell}, \mathcal{X}_{3,\ell}, \mathcal{S}_{1,\ell}, \mathcal{S}_{2,\ell}, \mathcal{S}_{3,\ell} \subseteq \mathbb{K} \setminus \{0\}$, $\mathcal{V} \subseteq \mathbb{K}$, and let $\mathcal{W} \subseteq \mathbb{F}$. We denote $\mathcal{X}^\star = (\mathcal{X}_{1,\ell}, \mathcal{X}_{2,\ell}, \mathcal{X}_{3,\ell})$ and $\mathcal{S}^\star = (\mathcal{S}_{1,\ell}, \mathcal{S}_{2,\ell}, \mathcal{S}_{3,\ell})$. Let $(\mathcal{X}^\star, \mathcal{S}^\star, \mathcal{V}, \mathcal{W})$ be a partition of Type-5, and let $q[\mathcal{X}^\star, \mathcal{S}^\star, \mathcal{V}, \mathcal{W}] \geq 2^{-45k}$. Let (X, S) be random variables uniform in $\mathbb{K}^6$ conditioned on the event that $X_i \in \mathcal{X}_{1,\ell}$, $S_i \in \mathcal{S}_{i,\ell}$ for $i = 1, 2, 3$, $\mathsf{nmExt}'(X) \neq \bot$, $\mathsf{nmExt}'(S) \neq \bot$, $\langle X, S \rangle_{\mathbb{K}} \in \mathcal{V}$, and $\langle X, S \rangle_{\mathbb{F}} \in \mathcal{W}$. Then,*

$$\sum_{\mathbf{a}, \mathbf{b}} \left(\frac{q[\mathcal{X}_{\mathbf{a}}, \mathcal{S}_{\mathbf{b}}, \mathcal{V}_{\langle \mathbf{a}, \mathbf{b} \rangle_{\mathbb{K}}}, \mathcal{W}_{\langle \mathbf{a}, \mathbf{b} \rangle_{\mathbb{F}}}]}{q[\mathcal{X}_{1,\ell}, \mathcal{S}_{1,\ell}, \mathcal{V}, \mathcal{W}]} \right)^{7/8} \leq \sum_{\mathbf{a}, \mathbf{b}} \Pr[h_1(\langle X, S \rangle_{\mathbb{K}}) = \langle \mathbf{a}, \mathbf{b} \rangle_{\mathbb{K}},\, f(X) = \mathbf{a},$$

$$g(S) = \mathbf{b}]^{\frac{7}{8}}$$

$$\leq 1 + 2^{-50k}.$$

Proof. Since the partition is of Type-5, we have

$$i_1 = i_2 = i_3 = j_1 = j_2 = j_3 = \ell.$$

By Lemma 8, we have that

$$p[\mathcal{X}^\star, \mathcal{S}^\star, \mathcal{V}, \mathcal{W}] \geq 2^{-45k-1},$$

and

$$\Pr[\widetilde{X} \in \mathcal{X}^\star,\, \widetilde{S} \in \mathcal{S}^\star,\, U_n \in \mathcal{V},\, \mathrm{tr}_{\mathbb{K} \to \mathbb{F}}(U_n) \in \mathcal{W}] \geq 2^{-45k-1}.$$

Let X', S' be distributed independently and uniform in $\mathcal{X}^\star, \mathcal{S}^\star$, respectively. We have that

$$\widetilde{\mathbf{H}}_\infty(X'|f(X'), \mathsf{nmExt}(X')) \geq 100k(3\ell - 3) - 3k = 3n - 303k, \quad \text{and}$$

$$\mathbf{H}_\infty(S') \geq 3n - 45k - 1.$$

Thus, by Lemma 4,

$$\Delta\left(\langle X', S'\rangle_\mathbb{K} \; ; \; U_n \mid f(X'), \mathsf{nmExt}(X'), S'\right) \leq 2^{-1000k},$$

where we have used that $n \geq 5000k$. This implies using Lemma 3 that

$$\Delta\left(\langle X', S'\rangle_\mathbb{K} \; ; \; U_n \mid \langle f(X'), g(S')\rangle_\mathbb{K}, \mathsf{nmExt}(X'), \mathsf{nmExt}(S')\right) \leq 2^{-1000k}.$$

Also, $\widetilde{\mathbf{H}}_\infty(X_i'|f_i(X_i')) \geq 100k(\ell-1) \geq n(1-\delta)$, and $\widetilde{\mathbf{H}}_\infty(S_i'|g_i(S_i')) \geq 100k(\ell-1) \geq n(1-\delta)$ for $i = 1, 2, 3$. Thus, by Theorem 3,

$$\Delta\left((\mathsf{nmExt}(X'), \mathsf{nmExt}(S')) \; ; \; (U_{3k}, U_{3k}') \mid \langle f(X'), g(S')\rangle_\mathbb{K}\right) \leq 2 \cdot 2^{-1000k}.$$

Using triangle inequality, we get that

$$\Delta\left((\langle X', S'\rangle_\mathbb{K}, \mathsf{nmExt}(X'), \mathsf{nmExt}(S')); (U_n, U_{3k}, U_{3k}')|\langle f(X'), g(S')\rangle_\mathbb{K}\right) \leq 3 \cdot 2^{-1000k}.$$

Conditioning on $\mathsf{nmExt}'(X') \neq \perp$, $\mathsf{nmExt}'(S') \neq \perp$, $\langle X', S'\rangle_\mathbb{K} \in \mathcal{V}$, and $\mathrm{tr}_{\mathbb{K}\to\mathbb{F}}(\langle X', S'\rangle_\mathbb{K}) \in \mathcal{W}$, by Lemma 7, we get that

$$\Delta\left((\langle f(X), g(S)\rangle_\mathbb{K}, \langle X, S\rangle_\mathbb{K}) \; ; \; (\langle f(X'), g(S')\rangle_\mathbb{K}, V)\right) \leq 2^{-950k}, \tag{5}$$

where V is distributed as U_n conditioned on $U_n \in \mathcal{V}$, and $\mathrm{tr}_{\mathbb{K}\to\mathbb{F}}(U_n) \in \mathcal{W}$.

Now using Lemma 5 on vector pair $(f_1(X_1'), f_2(X_2'), f_3(X_3'), -1)$ and $(g_1(S_1'), g_2(S_2'), g_3(S_3'), h_1(V))$, and $t = 4$, we obtain

$$\sum_{\substack{(a_1,a_2,a_3,b_1,b_2,b_3,c) \; : \\ \langle(a_1,a_2,a_3,-1),(b_1,b_2,b_3,c)\rangle_\mathbb{K}=0}} \Pr[(f(X'), g(S'), h_1(V)) = (\mathbf{a}, \mathbf{b}, c)]^{\frac{7}{8}} \leq 1.$$

Notice that the number of different possible values of the tuple $(a_1, a_2, a_3, b_1, b_2, b_3, c)$ such that $\Pr[(f(X'), g(S'), h_1(V)) = (\mathbf{a}, \mathbf{b}, c)] \neq 0$ is at most 2^{600k}. Thus, using Lemma 6 and the inequality 5, we get that

$$\sum_{\mathbf{a},\mathbf{b}} \Pr[h_1(\langle X, S\rangle_\mathbb{K}) = \langle \mathbf{a}, \mathbf{b}\rangle_\mathbb{K}, \; f(X) = \mathbf{a}, \; g(S) = \mathbf{b}]^{\frac{7}{8}} \leq 1 + 2^{600k \cdot \frac{1}{8}} \cdot 2^{-950k \cdot \frac{7}{8}}$$

$$\leq 1 + 2^{-50k}.$$

Finally,

$$\sum_{\mathbf{a},\mathbf{b}} \left(\frac{q[\mathcal{X}_\mathbf{a}, \mathcal{S}_\mathbf{b}, \mathcal{V}_{\langle \mathbf{a},\mathbf{b}\rangle_\mathbb{K}}, \mathcal{W}_{\langle \mathbf{a},\mathbf{b}\rangle_\mathbb{F}}]}{q[\mathcal{X}_{1,\ell}, \mathcal{S}_{1,\ell}, \mathcal{V}, \mathcal{W}]}\right)^{7/8}$$

$$= \sum_{\mathbf{a},\mathbf{b}} \Pr[h_1(\langle X, S\rangle_\mathbb{K}) = \langle \mathbf{a}, \mathbf{b}\rangle_\mathbb{K}, \; h_2(\langle X, S\rangle_\mathbb{F}) = \langle \mathbf{a}, \mathbf{b}\rangle_\mathbb{F}, \; f(X) = \mathbf{a}, \; g(S) = \mathbf{b}]^{\frac{7}{8}}$$

$$\leq \sum_{\mathbf{a},\mathbf{b}} \Pr[h_1(\langle X, S\rangle_\mathbb{K}) = \langle \mathbf{a}, \mathbf{b}\rangle_\mathbb{K}, \; f(X) = \mathbf{a}, \; g(S) = \mathbf{b}]^{\frac{7}{8}} \leq 1 + 2^{-50k}.$$

$$\square$$

Lemma 16. *[Type-1 or Type-5 partition] Let $\mathcal{X}_1, \mathcal{X}_2, \mathcal{X}_3, \mathcal{S}_1, \mathcal{S}_2, \mathcal{S}_3 \subseteq \mathbb{K} \setminus \{0\}$, $\mathcal{V} \subseteq \mathbb{K}$, and let $\mathcal{W} \subseteq \mathbb{F}$. We denote $\mathcal{X} = (\mathcal{X}_1, \mathcal{X}_2, \mathcal{X}_3)$ and $\mathcal{S} = (\mathcal{S}_1, \mathcal{S}_2, \mathcal{S}_3)$. Let $q[\mathcal{X}, \mathcal{S}, \mathcal{V}, \mathcal{W}] \geq 2^{-40k}$. Let (X, S) be random variables uniform in $\mathbb{K}^6$ conditioned on the event that $X_t \in \mathcal{X}_t$, $S_t \in \mathcal{S}_t$ for $t = 1, 2, 3$, $\mathsf{nmExt}'(X) \neq \perp$, $\mathsf{nmExt}'(S) \neq \perp$, $\langle X, S \rangle_{\mathbb{K}} \in \mathcal{V}$, and $\langle X, S \rangle_{\mathbb{F}} \in \mathcal{W}$. Then,*

$$\Pr[X_t \in \mathcal{X}_{t,0}, S_t \in \mathcal{S}_{t,0} \text{ for } t = 1, 2, 3]^{1/2} + \Pr[X_t \in \mathcal{X}_{t,\ell}, S_t \in \mathcal{S}_{t,\ell} \text{ for } t = 1, 2, 3]^{1/2}$$
$$\leq 1 + 2^{-90k},$$

and hence,

$$\Pr[X_t \in \mathcal{X}_{t,0}, S_t \in \mathcal{S}_{t,0} \text{ for } t = 1, 2, 3]^{7/8} + \Pr[X_t \in \mathcal{X}_{t,\ell}, S_t \in \mathcal{S}_{t,\ell} \text{ for } t = 1, 2, 3]^{7/8}$$
$$\leq 1 + 2^{-90k}.$$

Proof. By Lemma 8, we have that

$$p[\mathcal{X}, \mathcal{S}, \mathcal{V}, \mathcal{W}] \geq 2^{-40k-1},$$

and

$$\Pr[\widetilde{X} \in \mathcal{X}, \widetilde{S} \in \mathcal{S}, U_n \in \mathcal{V}, \mathrm{tr}_{\mathbb{K} \to \mathbb{F}}(U_n) \in \mathcal{W}] \geq 2^{-40k-1}.$$

Let X', S' be distributed independently and uniform in $\mathcal{X}, \mathcal{S}$, respectively. Let $i_1, i_2, i_3, j_1, j_2, j_3 : \mathbb{K} \to \{0, 1, \ldots, \ell\}$ be as defined in the partitioning procedure, i.e., i_1 is a function of X'_1 that indicates the partition in which X'_1 belongs depending on the function f_1, etc.

Since $\widetilde{\mathbf{H}}_\infty(X' | \mathsf{nmExt}(X'), i_1, i_2, i_3) \geq 3n - 40k - 1 - 3\log(\ell+1) \geq 3n - 41k$, using Lemma 4, we have that

$$\Delta\left(\langle X', S' \rangle_{\mathbb{K}} ; U_n \mid \mathsf{nmExt}(X'), \mathsf{nmExt}(S'), i_1, i_2, i_3, j_1, j_2, j_3\right) \leq 2^{-250k}.$$

Additionally, since $\widetilde{\mathbf{H}}_\infty(X'_t | i_t) \geq n - 40k - 1 - \log(\ell+1) \geq n(1 - \delta)$ and $\mathbf{H}_\infty(S'_t | j_t) \geq n - 40k - 1 - \log(\ell+1) \geq n(1 - \delta)$, for $t = 1, 2, 3$, by Theorem 3, we have that

$$\Delta\left((\mathsf{nmExt}(X'), \mathsf{nmExt}(S')) ; (U_{3k}, U'_{3k}) \mid i_1, i_2, i_3, j_1, j_2, j_3\right) \leq 2 \cdot 2^{-1000k}.$$

Thus, the triangle inequality implies that

$$\Delta((\langle X', S' \rangle_{\mathbb{K}}, \mathsf{nmExt}(X'), \mathsf{nmExt}(S')) ; (U_n, U_{3k}, U'_{3k}) \mid i_1, i_2, i_3, j_1, j_2, j_3)$$
$$\leq 3 \cdot 2^{-250k}.$$

Conditioning on $\mathsf{nmExt}'(X') \neq \perp$, $\mathsf{nmExt}'(S') \neq \perp$, $\langle X', S' \rangle_{\mathbb{K}} \in \mathcal{V}$, and $\mathrm{tr}_{\mathbb{K} \to \mathbb{F}}(\langle X', S' \rangle_{\mathbb{K}}) \in \mathcal{W}$ and using Lemma 7, we get that

$$\Delta((i_1(X'_1), i_2(X'_2), i_3(X'_3), j_1(S'_1), j_2(S'_2), j_3(S'_3)) ;$$
$$(i_1(X_1), i_2(X_2), i_3(X_3), j_1(S_1), j_2(S_2), j_3(S_3))) \leq 3 \cdot 2^{-200k}. \quad (6)$$

We introduce the following notation. For $r \in \{0, \ell\}$, let

$$p_r := \Pr[X_t' \in \mathcal{X}_{t,r} \text{ for } t = 1, 2, 3] = \Pr[i_1(X_1') = i_2(X_2') = i_3(X_3') = r],$$

and

$$q_r := \Pr[S_t' \in \mathcal{S}_{t,r} \text{ for } t = 1, 2, 3] = \Pr[i_1(S_1') = i_2(S_2') = i_3(S_3') = r].$$

Then clearly, $p_0 + p_\ell \le 1$, and $q_0 + q_\ell \le 1$. This implies

$$\Pr[X_t' \in \mathcal{X}_{t,0}, S_t' \in \mathcal{S}_{t,0} \text{ for } t = 1, 2, 3]^{1/2} + \Pr[X_t' \in \mathcal{X}_{t,\ell}, S_t' \in \mathcal{S}_{t,\ell} \text{ for } t = 1, 2, 3]^{1/2}$$
$$= \sqrt{p_0 \cdot q_0} + \sqrt{p_\ell \cdot q_\ell}$$
$$\le \sqrt{p_0 \cdot q_0} + \sqrt{(1 - p_0) \cdot (1 - q_0)}$$
$$\le 1,$$

using the Cauchy-Schwarz inequality. Thus, using Lemma 6 and the inequality 6, we get that

$$\Pr[X_t \in \mathcal{X}_{t,0}, S_t \in \mathcal{S}_{t,0} \text{ for } t = 1, 2, 3]^{1/2} + \Pr[X_t \in \mathcal{X}_{t,\ell}, S_t \in \mathcal{S}_{t,\ell} \text{ for } t = 1, 2, 3]^{1/2}$$
$$\le 1 + 2^{\frac{1}{2}} \cdot (3 \cdot 2^{-200k})^{\frac{1}{2}} \le 1 + 2^{-90k}.$$

$\square$

3.1   Proof of Theorem 5

Proof. Now, we prove Theorem 5 by induction on the number of rounds r. For $r = 0$, i.e., when there is no tampering, we need to show that $\mathsf{nmExt}(X)$ is statistically close to $0^{2k} \| U_k$, which follows by Lemma 10. Using Corollary 2, we have that

$$\sum_{\mathcal{P}:\mathsf{Type}(\mathcal{P}) \in \{1b, 2, 3, 4, 5b\}} \frac{q[\mathcal{P}]}{q[\mathcal{X}, \mathcal{S}, \mathcal{V}, \mathcal{W}]} \cdot \Delta \left((\mathsf{CT}_C^r, \mathsf{nmExt}(X))|_{C \in \mathcal{P}}; (\mathsf{CT}_C^r, 0^{2k} \| U_k)|_{C \in \mathcal{P}} \right)$$
$$\le 5 \cdot 2^{-2k}.$$

Let $\mathcal{Q}_1$ be a partition of Type$-$1a (note that there is only one such partition), and let $\mathcal{Q}_2, \ldots, \mathcal{Q}_m$ be partitions of Type$-$5a. Let $\mathcal{X}^\star = (\mathcal{X}_{1,\ell}, \mathcal{X}_{2,\ell}, \mathcal{X}_{3,\ell})$, and $\mathcal{S}^\star = (\mathcal{S}_{1,\ell}, \mathcal{S}_{2,\ell}, \mathcal{S}_{3,\ell})$. We consider two cases.

CASE 1: $q[\mathcal{X}^\star, \mathcal{S}^\star, \mathcal{V}, \mathcal{W}] < 2^{-45k}$. In this case, the total probability of falling in a partition of Type$-$5 is small, and so intuitively the only useful information that can be learnt is by landing in a partition of Type$-$1a. In this case, by Lemma 9 and the induction hypothesis we have that the statistical distance

$\Delta\left((\mathsf{CT}_C^r, \mathsf{nmExt}(X)) ; (\mathsf{CT}_C^r, 0^{2k} \| U_k)\right)$ is upper bounded by

$$\leq 5 \cdot 2^{-2k} + \frac{q[\mathcal{X}^\star, \mathcal{S}^\star, \mathcal{V}, \mathcal{W}]}{q[\mathcal{X}, \mathcal{S}, \mathcal{V}, \mathcal{W}]} 1 + \frac{q[\mathcal{Q}_1]}{q[\mathcal{X}, \mathcal{S}, \mathcal{V}, \mathcal{W}]} \left(\left(\frac{\rho}{q[\mathcal{Q}_1]} \right)^{\frac{1}{8}} + 9 \cdot (r-1) \cdot 2^{-2k} \right)$$

$$\leq 5 \cdot 2^{-2k} + 2^{-5k} + \left(\frac{q[\mathcal{Q}_1]}{q[\mathcal{X}, \mathcal{S}, \mathcal{V}, \mathcal{W}]} \right)^{\frac{7}{8}} \cdot \left(\frac{\rho}{q[\mathcal{X}, \mathcal{S}, \mathcal{V}, \mathcal{W}]} \right)^{\frac{1}{8}} + 9 \cdot (r-1) \cdot 2^{-2k}$$

$$\leq \left(\frac{\rho}{q[\mathcal{X}, \mathcal{S}, \mathcal{V}, \mathcal{W}]} \right)^{\frac{1}{8}} + 9 \cdot r \cdot 2^{-2k}.$$

CASE 2: $q[\mathcal{X}^\star, \mathcal{S}^\star, \mathcal{V}, \mathcal{W}] \geq 2^{-45k}$. In this case, by Lemma 9, and the induction hypothesis we have that the statistical distance $\Delta((\mathsf{CT}_C^r, \mathsf{nmExt}(X)) ; (\mathsf{CT}_C^r, 0^{2k} \| U_k))$ is upper bounded by

$$\leq 5 \cdot 2^{-2k} + \sum_{i=1}^{m} \frac{q[\mathcal{Q}_i]}{q[\mathcal{X}, \mathcal{S}, \mathcal{V}, \mathcal{W}]} \cdot \left(\left(\frac{\rho}{q[\mathcal{Q}_i]} \right)^{\frac{1}{8}} + 9 \cdot (r-1) \cdot 2^{-2k} \right)$$

$$\leq 5 \cdot 2^{-2k} + \sum_{i=1}^{m} \left(\frac{q[\mathcal{Q}_i]}{q[\mathcal{X}, \mathcal{S}, \mathcal{V}, \mathcal{W}]} \right)^{\frac{7}{8}} \cdot \left(\frac{\rho}{q[\mathcal{X}, \mathcal{S}, \mathcal{V}, \mathcal{W}]} \right)^{\frac{1}{8}} + 9 \cdot (r-1) \cdot 2^{-2k}$$

$$\leq 5 \cdot 2^{-2k} + \left(\frac{\rho}{q[\mathcal{X}, \mathcal{S}, \mathcal{V}, \mathcal{W}]} \right)^{\frac{1}{8}} (1 + 2^{-2k}) + 9 \cdot (r-1) \cdot 2^{-2k}$$

$$\leq \left(\frac{\rho}{q[\mathcal{X}, \mathcal{S}, \mathcal{V}, \mathcal{W}]} \right)^{\frac{1}{8}} + 9 \cdot r \cdot 2^{-2k},$$

where the second to last inequality uses Lemmas 15 and 16.

$\square$

References

1. Aggarwal, D.: Affine-evasive sets modulo a prime. Inf. Process. Lett. **115**(2), 382–385 (2015)
2. Aggarwal, D., Agrawal, S., Gupta, D., Maji, H.K., Pandey, O., Prabhakaran, M.: Optimal computational split-state non-malleable codes. In: Kushilevitz, E., Malkin, T. (eds.) TCC 2016. LNCS, vol. 9563, pp. 393–417. Springer, Heidelberg (2016). https://doi.org/10.1007/978-3-662-49099-0_15
3. Aggarwal, D., Briët, J.: Revisiting the Sanders-Bogolyubov-Ruzsa theorem in $\mathbb{f}_p^n$ and its application to non-malleable codes. In: 2016 IEEE International Symposium on Information Theory (ISIT), pp. 1322–1326. IEEE (2016)
4. Aggarwal, D., Dodis, Y., Kazana, T., Obremski, M.: Leakage-resilient nonmalleable codes. In: The 47th ACM Symposium on Theory of Computing (STOC) (2015)
5. Aggarwal, D., Dodis, Y., Lovett, S.: Non-malleable codes from additive combinatorics. In: STOC. ACM (2014)
6. Aggarwal, D., Dziembowski, S., Kazana, T., Obremski, M.: Leakage-resilient non-malleable codes. In: Dodis, Y., Nielsen, J.B. (eds.) TCC 2015. LNCS, vol. 9014, pp. 398–426. Springer, Heidelberg (2015). https://doi.org/10.1007/978-3-662-46494-6_17

7. Aggarwal, D., Kazana, T., Obremski, M.: Inception makes non-malleable codes stronger. In: Kalai, Y., Reyzin, L. (eds.) TCC 2017. LNCS, vol. 10678, pp. 319–343. Springer, Cham (2017). https://doi.org/10.1007/978-3-319-70503-3_10

8. Agrawal, S., Gupta, D., Maji, H.K., Pandey, O., Prabhakaran, M.: A rate-optimizing compiler for non-malleable codes against bit-wise tampering and permutations. In: Dodis, Y., Nielsen, J.B. (eds.) TCC 2015. LNCS, vol. 9014, pp. 375–397. Springer, Heidelberg (2015). https://doi.org/10.1007/978-3-662-46494-6_16

9. Agrawal, S., Gupta, D., Maji, H.K., Pandey, O., Prabhakaran, M.: Explicit non-malleable codes resistant to permutations. In: Advances in Cryptology - CRYPTO (2015)

10. Bogdanov, I.: Deathzone generation lemma (2016). https://mathoverflow.net/questions/252396/inner-product-over-finite-fields

11. Chattopadhyay, E., Goyal, V., Li, X.: Non-malleable extractors and codes, with their many tampered extensions. In: Proceedings of the Forty-Eighth Annual ACM Symposium on Theory of Computing, pp. 285–298. ACM (2016)

12. Chattopadhyay, E., Zuckerman, D.: Non-malleable codes in the constant split-state model. In: FOCS (2014)

13. Cheraghchi, M., Guruswami, V.: Capacity of non-malleable codes. In: ITCS (2014)

14. Cheraghchi, M., Guruswami, V.: Non-malleable coding against bit-wise and split-state tampering. In: Lindell, Y. (ed.) TCC 2014. LNCS, vol. 8349, pp. 440–464. Springer, Heidelberg (2014). https://doi.org/10.1007/978-3-642-54242-8_19

15. Chor, B., Goldreich, O.: Unbiased bits from sources of weak randomness and probabilistic communication complexity. SIAM J. Comput. **17**(2), 230–261 (1988)

16. Coretti, S., Maurer, U., Tackmann, B., Venturi, D.: From single-bit to multi-bit public-key encryption via non-malleable codes. In: Dodis and Nielsen [17], pp. 532–560

17. Dodis, Y., Nielsen, J.B. (eds.): TCC 2015. LNCS, vol. 9014. Springer, Heidelberg (2015). https://doi.org/10.1007/978-3-662-46494-6

18. Dodis, Y., Ostrovsky, R., Reyzin, L., Smith, A.: Fuzzy extractors: how to generate strong keys from biometrics and other noisy data. SIAM J. Comput. **38**(1), 97–139 (2008)

19. Dziembowski, S., Kazana, T., Obremski, M.: Non-malleable codes from two-source extractors. In: Canetti, R., Garay, J.A. (eds.) CRYPTO 2013. LNCS, vol. 8043, pp. 239–257. Springer, Heidelberg (2013). https://doi.org/10.1007/978-3-642-40084-1_14

20. Dziembowski, S., Pietrzak, K., Wichs, D.: Non-malleable codes. In: ICS, pp. 434–452. Tsinghua University Press (2010)

21. Faust, S., Mukherjee, P., Nielsen, J.B., Venturi, D.: Continuous non-malleable codes. In: Lindell, Y. (ed.) TCC 2014. LNCS, vol. 8349, pp. 465–488. Springer, Heidelberg (2014). https://doi.org/10.1007/978-3-642-54242-8_20

22. Faust, S., Mukherjee, P., Nielsen, J.B., Venturi, D.: A tamper and leakage resilient von neumann architecture. In: Katz, J. (ed.) PKC 2015. LNCS, vol. 9020, pp. 579–603. Springer, Heidelberg (2015). https://doi.org/10.1007/978-3-662-46447-2_26

23. Faust, S., Mukherjee, P., Venturi, D., Wichs, D.: Efficient non-malleable codes and key-derivation for poly-size tampering circuits. In: Nguyen, P.Q., Oswald, E. (eds.) EUROCRYPT 2014. LNCS, vol. 8441, pp. 111–128. Springer, Heidelberg (2014). https://doi.org/10.1007/978-3-642-55220-5_7

24. Gennaro, R., Lysyanskaya, A., Malkin, T., Micali, S., Rabin, T.: Algorithmic tamper-proof (ATP) security: theoretical foundations for security against hardware tampering. In: Naor, M. (ed.) TCC 2004. LNCS, vol. 2951, pp. 258–277. Springer, Heidelberg (2004). https://doi.org/10.1007/978-3-540-24638-1_15

25. Jafargholi, Z., Wichs, D.: Tamper detection and continuous non-malleable codes. In: Dodis and Nielsen [17], pp. 451–480

26. Li, X.: Improved non-malleable extractors, non-malleable codes and independent source extractors. In: Proceedings of the 49th Annual ACM SIGACT Symposium on Theory of Computing, pp. 1144–1156. ACM (2017)

27. Liu, F.-H., Lysyanskaya, A.: Tamper and leakage resilience in the split-state model. In: Safavi-Naini, R., Canetti, R. (eds.) CRYPTO 2012. LNCS, vol. 7417, pp. 517–532. Springer, Heidelberg (2012). https://doi.org/10.1007/978-3-642-32009-5_30

28. Nisan, N., Zuckerman, D.: Randomness is linear in space. J. Comput. Syst. Sci. **52**(1), 43–53 (1996)

Correlated-Source Extractors and Cryptography with Correlated-Random Tapes

Vipul Goyal and Yifan Song[✉]

Carnegie Mellon University, Pittsburgh, USA
{vipul,yifans2}@cmu.edu

Abstract. In this paper, we consider the setting where a party uses correlated random tapes across multiple executions of a cryptographic algorithm. We ask if the security properties could still be preserved in such a setting. As examples, we introduce the notion of *correlated-tape zero knowledge*, and, *correlated-tape multi-party computation*, where, the zero-knowledge property, and, the ideal/real model security must still be preserved even if a party uses correlated random tapes in multiple executions.

Our constructions are based on a new type of randomness extractor which we call *correlated-source extractors*. Correlated-source extractors can be seen as a dual of non-malleable extractors, and, allow an adversary to choose several tampering functions which are applied to the randomness source. Correlated-source extractors guarantee that even given the output of the extractor on the tampered sources, the output on the original source is still uniformly random. Given (seeded) correlated-source extractors, and, *resettably-secure* computation protocols, we show how to directly get a positive result for both correlated-tape zero-knowledge and correlated-tape multi-party computation in the CRS model. This is tight considering the known impossibility results on cryptography with imperfect randomness.

Our main technical contribution is an explicit construction of a correlated-source extractor where the length of the seed is independent of the number of tamperings. Additionally, we also provide a (non-explicit) existential result for correlated source extractors with almost optimal parameters.

1 Introduction

Randomness is known to be crucial for cryptography. It is known that several basic tasks in cryptography become impossible in the absence of randomness [GO94, DOPS04]. Given this, a natural and well motivated direction is to develop an understanding of the *extent* to which randomness is necessary. Towards that end, we study the following natural question.

Research supported in part by a grant from Northrop Grumman, a gift from DOS Networks, and, a Cylab seed funding award.

© International Association for Cryptologic Research 2019
Y. Ishai and V. Rijmen (Eds.): EUROCRYPT 2019, LNCS 11476, pp. 562–592, 2019.
https://doi.org/10.1007/978-3-030-17653-2_19

Suppose that a party uses correlated random tapes in multiple executions of a cryptographic algorithm. Can the security still be preserved? As a concrete example, suppose that the prover uses correlated random tapes in multiple executions with an adversarial verifier. Can the zero-knowledge property still be preserved? What about encrypting multiple times (under a randomized encryption scheme) using correlated random tapes? The above question can be motivated by, e.g., a scenario where a party has a defective random number generator which outputs correlated tapes under multiple invocations (even though each individual tape may have high min-entropy or even be close to uniform).

The well-known line of research on resettable security can be seen as a *special case* of our general problem. In resettable zero-knowledge [CGGM00], the prover uses the *same* random tape across multiple executions. By varying the set of parties whose random tape is fixed, one can get various variants such as resettably-sound zero-knowledge [BGGL01], simultanous resettable zero-knowledge [DGS09], and, resettably secure computation [GS09, GM11].

In this work, we initiate a systematic study of the above question. The central object of our study will be a new notion of randomness extractors which we call *correlated-source extractors*. Very informally, a (seeded) correlated source extractor csExt on input a seed s, and a source X produces an output $\text{csExt}(X, s)$ which is guaranteed to be close to uniform even given $\text{csExt}(X_1, s), ..., \text{csExt}(X_t, s)$ where for all i, $X_i \neq X$ and X_i could be arbitrarily correlated with X. One could also view X_i as a result of tampering the original source X. Correlated-source extractors can be seen as a dual of non-malleable extractors [DW09], where the adversary is allowed to tamper the seed instead of the source. Non-malleable extractors have played an important role in cryptography and complexity in problems such as privacy amplification [DW09], designing two-source extractors [CZ16], and in designing non-malleable codes [CG14a, CGL16]. Correlated-source extractors are also closely related to two-source non-malleable extractors [CG14a, CGL16].

1.1   Our Results

We introduce the notion of correlated-tape zero-knowledge. We model correlations among the different random tapes by consider an adversary which may have limited control over the random tape of the honest parties. In correlated-tape zero-knowledge, the adversary is able to specify t tampering functions $f_1, f_2, ..., f_t$ at the beginning of the protocol such that in the i-th execution, the prover uses $f_i(X)$ as its random tape (where X is uniformly random and can be viewed as the original random tape). Other notions like correlated-tape secure multi-party computation (MPC) and correlated-tape secure encryption schemes could also be defined analogously. We also define the main object of our study: correlated-source extractors. Specifically,

Definition 1 (Seeded Correlated-Source Extractor). *A function* csExt : $\{0, 1\}^* \times \{0, 1\}^d \to \{0, 1\}^m$ *is a seeded correlated-source extractor if the following holds: There exists a polynomial* $k(\cdot, \cdot, \cdot)$ *and a negligible function* $\epsilon(\cdot)$*, such that*

for any polynomial $t(\cdot)$, $t = t(d)$ arbitrary functions $\mathcal{A}_1, \mathcal{A}_2, ..., \mathcal{A}_t$, whose output has the same length as the input, with no fixed points, and, a source X with min-entropy $k(t, m, d)$,

$$|\mathsf{csExt}(X, U_d) \circ \{\mathsf{csExt}(\mathcal{A}_i(X), U_d)\}_{i=1}^t \circ U_d - U_m \circ \{\mathsf{csExt}(\mathcal{A}_i(X), U_d)\}_{i=1}^t \circ U_d| < \epsilon(d)$$

where U_m and U_d are uniform strings of length m and d respectively.

Jumping ahead, in our cryptographic applications, the seed will serve as the CRS while the source X will be the local random tape generated by the party. We require the output length, and, the source length (and hence the min-entropy) to be (unbounded) polynomial in the length of the seed. Thus, fixing the CRS (i.e., the seed) doesn't necessarily fix the number of executions (represented by t). One could also define a weaker notion of correlated-source extractors where the seed fixes a bound on the number of executions. Specifically,

Definition 2 (Weak t-Correlated-Source Extractor). *A function* $\mathsf{wcsExt} : \{0,1\}^n \times \{0,1\}^d \to \{0,1\}^m$ *is a weak t-correlated-source extractor for min-entropy k and error ϵ if the following holds: If X is a source in $\{0,1\}^n$ with min-entropy k, and, $\mathcal{A}_1, \mathcal{A}_2, ..., \mathcal{A}_t$ are arbitrary functions whose output has the same length as the input, with no fixed points, then*

$$|\mathsf{wcsExt}(X, U_d) \circ \{\mathsf{wcsExt}(\mathcal{A}_i(X), U_d)\}_{i=1}^t \circ U_d - U_m \circ \{\mathsf{wcsExt}(\mathcal{A}_i(X), U_d)\}_{i=1}^t \circ U_d| < \epsilon$$

where U_m and U_d are uniform strings of length m and d respectively.

Our first main result is a construction of a correlated-source extractor:

Theorem 1. *There exists an explicit correlated-source extractor csExt with*

$$k(t, m, d) = \Theta(t^3 d + t^2 m)$$
$$\epsilon(d) = \Theta(2^{-\sqrt{d}})$$

where m is the length of the output.

Note that it is necessary for the entropy of the source to grow with the number of executions t if the tampered sources may be arbitrarily correlated with the original source. This is because the entropy of the original source may reduce given the output of the extractor on a tampered source. In Sect. 5.4 we generalize the entropy requirements on the sources. In particular, we define what we call *closed-set correlated sources* and show correlated set extractors for such sources. For closed-set correlated source, the entropy of each individual source does not necessarily grow with the number of invocations t. Hence, this would allow us to get constructions where neither the seed length, nor the source length or its entropy grows with the number of invocations.

Going to Correlated-Tape Zero-Knowledge and Secure Computation. We note that correlated-source extractors can only allow us to handle the random tapes where *each random tape differs from every other one.* We relax this constraint by relying on techniques from resettable zero-knowledge [CGGM00,BGGL01], and, resettably secure computation [GS09]. In resettable zero-knowledge, the prover uses the *same* random tape across multiple executions. In our setting, the random tape could either be the same or arbitrarily correlated with another random tape. Very informally, relying on resettable security would allow us to achieve security in case the random tape is the same as another one, and, relying on correlated source extractor would guarantee security in case the random tape is different from every other tape but maybe arbitrarily correlated.

This allows us to obtain positive results for correlated-tape zero-knowledge and multi-party computation in the CRS model where the only (necessary) requirement on the random tape would be sufficient min-entropy; otherwise *each random tape could be arbitrarily correlated to or even the same as other random tapes.* The seed required for the correlated source extractor would be a part of the CRS. Each party in the protocol would first apply correlated-source extractor on its (potentially tampered) random tape, and, use the resulting string as the random tape to execute a resettable secure MPC (or zero-knowledge) protocol. We note that correlated-tape zero-knowledge and similar primitives such as correlated-tape encryption are impossible to obtain in the plain model. This holds even for a single execution and follows from the known impossibility results on cryptography with imperfect and tamperable randomness [DOPS04,ACM+14] (see Sect. 4 for more details). We also give stronger impossibility results in Sect. 4.

Weak Correlated-Source Extractors. Note that basic positive result for *weak* correlated-source extractor follows from the construction of two-source non-malleable extractors in [CGL16]. In fact, two-source non-malleable extractors allow the adversary to also tamper the second source (the random seed) and only requires the second source (the random seed) to have enough min-entropy. However, directly using two-source non-malleable extractor or similar techniques cannot give a positive result for correlated source extractor. This is because two-source non-malleable extractor will require the seed length to be either as long as that of the source or linear in t. Note that this would give a positive result only for *bounded* correlated-tape zero-knowledge and secure computation where the number of executions must be fixed before choosing the CRS. We note that obtaining a construction where the seed length is independent of the number of tamperings has been a challenging problem in this line of research. *In particular, obtaining such an explicit construction for non-malleable extractors still remains an open problem.* An existential result has however been shown very recently [BACD+18].

Correlated-Source Extractors with Almost Optimal Parameters. Next, we turn our attention to the following natural question: what is the optimal entropy and source length that a correlated-source extractor requires? Towards that end, we prove the following existential result:

Theorem 2 (Existence of Correlated-Source Extractor). *There exists a correlated-source extractor* csExt *as long as*

$$k(t, m, d) = \Theta(tm + d) \tag{1}$$
$$\epsilon(d) = \Theta(2^{-d}) \tag{2}$$

where m is the length of the output.

For an overview of our techniques, please refer to Sect. 2.

Related Works. Designing randomness extractors has been a rich line of works. Most relevant to our work are non-malleable extractors [DW09], and, two source non-malleable extractors [CG14a, CGL16]. After the initial constructions, a number of works have focused on improving the entropy requirements and the seed length [Li12a, Li12b, DLWZ14, Li15, CL16, Coh16b, Coh16c, Coh16a, Li16, Li17]. However, all known explicit constructions of non-malleable and two-source non-malleable extractors require the length of the seed to grow with the number of tamperings t. Two-source non-malleable extractors from [CGL16] were used crucially in a recent breakthrough on constructing two-source extractors [CZ16].

A number of works have studied simulating randomized algorithms using weak sources with small min-entropy [VV85, CG88, Zuc96, SSZ95, ACRT97]. Andreev et al. [ACRT97] gave a simulation of any BPP algorithm with an $(n, n^{O(1)})$-source. In contrast, we focus on multiple executions with correlated random tapes and have weaker entropy requirements.

A rich line of works have studied resettable secure protocols [CGGM00], [BGGL01, DGS09, GM11, BP13, CPS16, COPV13, COP+14], where a party may use the same random tape in multiple executions. The class of correlations we handle is more general. Kalai et al. [KLRZ08] introduced network extractor protocols where there are a number of parties each having independent (but imperfect) random tapes. Their result required a strong variant of the Decisional Diffie-Hellman Assumption, and, a polylogarithmic number of parties.

Goldreich and Oren [GO94] showed that constructing zero-knowledge arguments where the prover is deterministic is impossible. Dodis et al. [DOPS04] showed that a number of basic cryptographic primitives like encryption and zero-knowledge are impossible with imperfect randomness. Austrin et al. [ACM+14] similarly showed a number of impossibility results (including for zero-knowledge) in the setting of tampering randomness. These results focus on the plain model and in the setting of a single execution. Moving to the CRS model allows us to bypass these negative results. A line of research also explores cryptography with related keys and related inputs (see [ABP15] and the references therein), typically for a special class of tampering functions (such as affine functions).

2   Technical Overview

Explicit Construction with Fixed Seed Length. In this section, we will give a high level idea of our construction of correlated-source extractors. We use X for the original source and X_i for the tampered source. We use Y for the random seed.

Why Existing Techniques Fail. All the construction ideas related to two-source non-malleable extractors (which imply the existence of *weak* correlated-source extractors) somehow separate the original seed into several independent and uniformly random "slices". A general framework to constructing non-malleable extractors (and two-source non-malleable extractors) followed by several works is based on alternating extraction [DW09] and generating an advice (which is unique w.h.p. across all the tampered executions). A critical step in such constructions is to view the original seed as a second source. In the beginning, a slice of the seed is used to extract from the source. Next, the result is used as a seed to extract from the original seed. Next, the result is again used as a seed to extract from the source, and so on. This technique relies on the length of the original seed to be long enough. In particular, during the analysis, each tampering would "fix" a part of the random seed. This means that, the effective entropy of the seed reduces as the number of tampered executions increase. By using alternating extraction where the seed plays the role of one of the source, it seems that the seed length must be linear in t.

Overview of the Construction. Our idea is to generate two (or multiple) independent sources from the original source itself. A straightforward idea is to generate (X^1, X^2) from X, such that the distribution of $\{X^1, X_1^1, ..., X_t^1\}$ is independent of $\{X^2, X_1^2, ..., X_t^2\}$ (here X_i is the i-th tampering source and (X_i^1, X_i^2) are generated from X_i). Then we may discard the original seed and use a two-source non-malleable extractor on X^1 and X^2. However, we don't know how to prove the joint distributions of two sets are independent. Our starting idea would be to use the given random seed in obtaining such a "decomposition" of the original source. We use one part of the seed Y_1 to generate $X^1 = \mathtt{Ext}(X, Y_1)$ and another part of the seed Y_2 to generate $X^2 = \mathtt{Ext}(X, Y_2)$. By assuming the source X has enough entropy, we can guarantee that, given $\{X^2, X_1^2, ..., X_t^2\}$, X^1 is uniformly random. Note that the joint distribution of $\{X^1, X_1^1, ..., X_t^1\}$ may be dependent on that of $\{X^2, X_1^2, ..., X_t^2\}$, while two-source non-malleable extractors require the adversary tampers both sources separately. Thus, it is not sufficient to use two-source non-malleable extractors.

  We first generate an advice $\mathtt{adv}$ from the source X (and $\mathtt{adv}_i$ from X_i) such that it is unique w.h.p. across all the tampered executions. Let ℓ denote the length of the advice. Then, instead of just generating (X^1, X^2) from X, we generate 2ℓ sources $(X^1, X^2, ..., X^{2\ell})$ from X such that, for every $i \in \{1, ..., 2\ell\}$, X^i is uniformly random given $\{X^j, X_1^j, ..., X_t^j\}_{j \neq i}$.

  Let $\mathtt{adv}^i$ denote the i-th bit of $\mathtt{adv}$. Each bit $\mathtt{adv}^i$ corresponds to a pair of sources (X^{2i-1}, X^{2i}). The extractor first uses one piece of the original seed as the seed and extracts randomness from one source of the first pair (X^1, X^2) decided by the value of $\mathtt{adv}^1$, then uses the result as the seed and extracts randomness from one source of the second pair (X^3, X^4) and so on. Specifically, in the i-th iteration, we choose $X^{2i-1+\mathtt{adv}^i}$. This process can be described by a function $F = F(\mathtt{adv}^i, X^{2i-1}, X^{2i}, Y, Z^{i-1})$, where Z^{i-1} is the result in the last iteration and initially, Z^0 is one piece of the original seed.

Note that, in the case that $\mathtt{adv}$ is different from all tampered $\mathtt{adv}_1, ..., \mathtt{adv}_t$, for all $j \in \{1, ..., t\}$, there exists at least one iteration (denoted by the i-th iteration) such that the i-th bits of $\mathtt{adv}$ and $\mathtt{adv}_j$ are different. We note that $X^{2i-1+\mathtt{adv}^i}$ is in fact independent of $X_j^{2i-1+\mathtt{adv}_j^i}$. Thus, hopefully, we can break the correlation between X and X_j in this iteration, i.e., Z^i is independent of Z_j^i. We also need this independence to be preserved in all later iterations. Therefore, in the end, since $\mathtt{adv}$ is different from all tampered advice, Z^ℓ is independent of $Z_1^\ell, ..., Z_t^\ell$.

Now we are ready to state our construction overview in more detail. It can be divided into two steps.

Step 1: Generating advice adv and limited correlated parts $X^1, ..., X^{2\ell}$

In the beginning, we generate an advice $\mathtt{adv}$ for the source X such that, with high probability, $\mathtt{adv}$ is different from $\mathtt{adv}_1, ..., \mathtt{adv}_t$ (the advice of $X_1, X_2, ..., X_t$). This idea is not new and is widely used in the constructions of non-malleable extractors (e.g. in [Coh15, CGL16]).

Recall that ℓ is the length of $\mathtt{adv}$. We generate $X^1, X^2, ..., X^{2\ell}$ by using a fresh seed Y_1^i for each part X^i. Specifically, $X^i = \mathtt{Ext}(X, Y_1^i)$ (and $X_j^i = \mathtt{Ext}(X_j, Y_1^i)$). Note that the random seeds in all executions are the same.

These sources $X^1, X^2, ..., X^{2\ell}$ directly satisfy our requirement, i.e., for every $i \in \{1, ..., 2\ell\}$, X^i is uniformly random given $\{X^j, X_1^j, ..., X_t^j\}_{j \neq i}$. To see this, let $\mathcal{X}^i = \{X^i, X_1^i, ..., X_t^i\}$. In the case that X has enough min-entropy and $X^1, ..., X^{2\ell}$ are comparatively short, X still has enough min-entropy when fixing $\mathcal{X}^1, ..., \mathcal{X}^{i-1}, \mathcal{X}^{i+1}, ..., \mathcal{X}^{2\ell}$. Also, Y^i is a fresh piece from the seed. Thus, by the property of $\mathtt{Ext}$, X^i is uniformly random and independent of $\mathcal{X}^1, ..., \mathcal{X}^{i-1}, \mathcal{X}^{i+1}, ..., \mathcal{X}^{2\ell}$.

Step 2: Breaking correlation between sources by induction

Let $\mathtt{SAME}^i$ be the set of indices of sources whose advice is different from $\mathtt{adv}$ in at least one bit of the first $(i-1)$ bits but is the same in the i-th bit, and $\mathtt{DIFF}^i$ be the set of indices of sources whose advice is different from $\mathtt{adv}$ in the i-th bit. Then after the $(i-1)$-th iteration, Z^{i-1} should have been uniformly random and independent of $\{Z_j^{i-1}\}_{j \in \mathtt{SAME}^i}$. So for $j \in \mathtt{SAME}^i$, we want this independence to be preserved in the i-th iteration. And we want to further break the correlation with the j-th tampered result where $j \in \mathtt{DIFF}^i$ in the i-th iteration.

In the $(i-1)$-th iteration, we should have already achieved that

$$\left| Z^{i-1} \circ \{Z_j^{i-1}\}_{j \in \mathtt{SAME}^{i-1} \bigcup \mathtt{DIFF}^{i-1}} - U_z \circ \{Z_j^{i-1}\}_{j \in \mathtt{SAME}^{i-1} \bigcup \mathtt{DIFF}^{i-1}} \right| < \epsilon$$

where z is the length of Z^{i-1}. A critical fact is that the above inequality still holds even given $\mathcal{X}^{2i-1}$ and $\mathcal{X}^{2i}$. Because what we need to prove the above property is that X^j is uniformly random when given $\{\mathcal{X}^k\}_{k \neq j}$ for every $j \in \{1, ..., 2i-2\}$.

Fixing $\mathcal{X}^{2i-1}$ and $\mathcal{X}^{2i}$ does not break this condition by the way how we generated $X^1, ..., X^{2i-2}$.

For a series of correlated sources $X, X_1, ..., X_t$, there are two ways to break the correlation. If the random seeds are independent and uniformly random for different sources, then the output is uniformly random and independent of others in the case that X has enough min-entropy. If X given $X_1, X_2, ..., X_t$ still has enough min-entropy, then even if the seeds are not independent, the output of extractor is still uniformly random and independent of others. This idea is also used in the recent construction of non-malleable extractors (e.g. in [CL16]).

We note that, for the executions whose indices $j \in \text{SAME}^i$, they will use $X_j^{2i-1+\text{adv}_j^i}$ where $2i-1+\text{adv}_j^i = 2i-1+\text{adv}^i$. It means that $\{X_j^{2i-1+\text{adv}_j^i}\}_{j \in \text{SAME}^i}$ may be highly correlated with $X^{2i-1+\text{adv}^i}$. However, since $\text{SAME}^i \subseteq \text{SAME}^{i-1} \bigcup \text{DIFF}^{i-1}$, Z^{i-1} is uniformly random and independent of $\{Z_j^{i-1}\}_{j \in \text{SAME}^i}$. If we use Z^{i-1} as the seed to do extraction on $X^{2i-1+\text{adv}^i}$, the result is independent of those of executions whose indices $j \in \text{SAME}^i$. Specifically,

Sub-Step 2.1: Breaking correlation with sources in SAME^i
Let Ext be a strong seeded extractor. Compute $W^i = \text{Ext}(X^{2i-1+\text{adv}^i}, Z^{i-1})$ (and W_j^i for the result in the j-th tampered execution). Then, given $\{W_j^i\}_{j \in \text{SAME}^i}$, W^i is uniformly random.

Now, we want to further break the correlation with the j-th tampered result where $j \in \text{DIFF}^i$. Currently, W^i may be correlated with $\{W_j^i\}_{j \in \text{DIFF}^i}$. However, We note that we can fix $\mathcal{X}^{2i-1+(1-\text{adv}^i)}$ in Sub-Step 2.1 without breaking the property of the result. Since, for $j \in \text{DIFF}^i$, W_j^i only depends on $X_j^{2i-1+\text{adv}_j^i}$ and Z_j^{i-1}, and $X_j^{2i-1+\text{adv}_j^i}$ has already been fixed (because $\text{adv}_j^i = 1 - \text{adv}^i$), W^i will have enough min-entropy even fixing $\{Z_j^{i-1}\}_{j \in \text{DIFF}^i}$ if we choose the length of W^i to be much longer than that of Z^i. It also means that W^i given $\{W_j^i\}_{j \in \text{DIFF}^i}$ still has enough min-entropy. Therefore, we can simply use a fresh piece of the original seed as the seed to do extraction on W^i. The result will be independent of those of executions whose indices are in $\text{SAME}^i \bigcup \text{DIFF}^i$. Specifically,

Sub-Step 2.2: Breaking correlation with sources in DIFF^i
We use a fresh piece Y_2^i from the original random seed Y. Compute $Z^i = \text{Ext}(W^i, Y_2^i)$ as the output of the i-th iteration.

3   Preliminaries

We use capital letters to denote random variables. We use U_r to denote the uniform distribution over $\{0,1\}^r$. For random variable X, we use $x \sim X$ to denote that x is sampled from the distribution of X.

3.1 Statistical Distance, Convex Combination of Distributions and Probability Lemma

Definition 3 (Statistical Distance). *Let D_1 and D_2 be two distributions on a set S. The statistical distance between D_1 and D_2 is defined to be:*

$$|D_1 - D_2| = \max_{T \subseteq S} |D_1(T) - D_2(T)| = \frac{1}{2} \sum_{s \in S} |\Pr[D_1 = s] - \Pr[D_2 = s]|$$

D_1 *is ϵ-close to D_2 if $|D_1 - D_2| \leq \epsilon$.*

Definition 4 (Convex Combination). *A distribution D on a set S is a convex combination of distributions $D_1, D_2, ..., D_\ell$ on S if there exists non-negative constants (called weights) $w_1, w_2, ..., w_\ell$ with $\sum_{i=1}^{l} w_i = 1$ such that $\Pr[D = s] = \sum_{i=1}^{\ell} w_i \Pr[D_i = s]$ for all $s \in S$. We use the notation $D = \sum_{i=1}^{\ell} w_i D_i$ to denote the fact that D is a convex combination of the distributions $D_1, ..., D_\ell$ with weights $w_1, ..., w_\ell$.*

3.2 Min-entropy and Flat Distribution

The min-entropy of a source X is defined as

$$H_\infty(X) = -\log(\max_{x \in \text{support}(X)} (1/\Pr[X = x]))$$

A distribution D is a flat distribution (source) if it is uniformly random over a set S. An (n, k)-source is a distribution over $\{0,1\}^n$ with min-entropy at least k. It is well known that any (n, k)-source is a convex combination of flat sources supported on sets of size 2^k.

3.3 Seeded Extractors, Non-malleable Extractors, Two-Source Non-malleable Extractors and Previous Construction

Definition 5 (Strong seeded Extractor). *A function* $\texttt{Ext} : \{0,1\}^n \times \{0,1\}^d \to \{0,1\}^m$ *is called a strong seeded extractor for min-entropy k and error ϵ if for any (n, k)-source X and an independent uniformly random string U_d, we have*

$$|\texttt{Ext}(X, U_d) \circ U_d - U_m \circ U_d| < \epsilon,$$

where U_m is independent of U_d and m is the output length of $\texttt{Ext}$.

The following definition of t-non-malleable extractors is from [CRS14], which generalizes the definition in [DW09].

Definition 6 (Non-malleable Extractor). *A function* $\texttt{snmExt} : \{0,1\}^n \times \{0,1\}^d \to \{0,1\}^m$ *is a seeded t-non-malleable extractor for min-entropy k and error ϵ if the following holds: If X is a source on $\{0,1\}^n$ with min-entropy k and*

$\mathcal{A}_1, \mathcal{A}_2, ..., \mathcal{A}_t$ are arbitrary (tampering) functions defined on $\{0,1\}^n \to \{0,1\}^n$ with no fixed points, then

$$|\text{snmExt}(X, U_d) \circ \{\text{snmExt}(X, \mathcal{A}_i(U_d))\}_{i=1}^t - U_m \circ \{\text{snmExt}(X, \mathcal{A}_i(U_d))\}_{i=1}^t| < \epsilon,$$

where U_m is independent of U_d and X.

The following definition of two-source non-malleable extractors is from [CGL16], which generalizes the definition in [CG14a].

Definition 7 (Two-source Non-malleable Extractor). *A function* $\text{nmExt} : \{0,1\}^n \times \{0,1\}^n \to \{0,1\}^m$ *is a two-source t-non-malleable extractor for min-entropy* k *and error* ϵ *if the following holds: If* X, Y *are independent sources on* $\{0,1\}^n$ *with min-entropy* k *and* $\mathcal{A}_1 = (f_1, g_1), \mathcal{A}_2 = (f_2, g_2), ..., \mathcal{A}_t = (f_t, g_t)$ *are arbitrary 2-split-state tampering functions where* f_i, g_i *are defined on* $\{0,1\}^n \to \{0,1\}^n$ *such that for any* i, *at least one of* f_i, g_i *has no fixed points, then*

$$|\text{nmExt}(X, Y) \circ \{\text{nmExt}(f_i(X), g_i(X))\}_{i=1}^t - U_m \circ \{\text{nmExt}(f_i(X), g_i(X))\}_{i=1}^t| < \epsilon$$

Theorem 3 ([GUV09]). *For any constant* $\alpha > 0$, *and all integers* $n, k > 0$ *there exists a polynomial time computable strong seeded extractor* $\text{Ext} : \{0,1\}^n \times \{0,1\}^d \to \{0,1\}^m$ *with* $d = O(\log n + \log(1/\epsilon))$ *and* $m = (1-\alpha)k$.

3.4 Conditional Min-entropy

Definition 8. *The average conditional min-entropy is defined as*

$$\tilde{H}_\infty(X|W) = \log\left(\mathbf{E}_{w \sim W}[\max_x \Pr[X = x|W = w]]\right) = -\log\left(\mathbf{E}_{w \sim W}[2^{-H_\infty(X|W=w)}]\right)$$

The following result on conditional min-entropy was proved in [MW97].

Lemma 1. *Let* X, Y *be random variables such that the random variable* Y *takes at most* l *values. Then*

$$\Pr_{y \sim Y}[H_\infty(X|Y = y) \geq H_\infty(X) - \log l - \log(\frac{1}{\epsilon})] > 1 - \epsilon$$

We recall some results on conditional min-entropy from [DRS04].

Lemma 2 ([DRS04]). *If a random variable* B *can take at most* 2^ℓ *values, then* $\tilde{H}_\infty(A|BC) \geq \tilde{H}_\infty(A|C) - \ell$.

Lemma 3 ([DRS04]). *For any* $\delta > 0$, *if* Ext *is a* (k, ϵ)-*extractor then it is also a* $(k + \log(1/\delta), \epsilon + \delta)$ *average case extractor*.

4 Our Model

In this section, we introduce a new model of cryptographic protocol where a party
may be involved in multiple executions with correlated random tapes. We first
focus on zero-knowledge and later generalize to secure multi-party computation.
This captures the setting where an honest party may have a defective random
number generator G which may output highly correlated strings in different
executions. In the worst case, the output of G after the first execution may fully
depend on the output in the first execution. Then an adversary may use the
messages it received in the first execution to get information about the random
tape of the honest parties in the subsequent executions.

We will formalize the above setting by considering an experiment where an
adversary is given limited control of the random tape of the honest party and
can interact with the honest party in multiple sessions.

Correlated-Tape Zero-Knowledge. For every $\{(x_i, w_i)\}_{i=1}^t$ where $w_i \in R_L(x_i)$,
the verifier V^* will sequentially interact with the actual prover P. V^* can
specify t tampering functions $f_1, f_2, ..., f_t$. To overcome known impossibility
results [DOPS04] as discussed later, we also assume the existence of a CRS.
In the beginning, P has a private random tape X distributed uniformly at
random. In the j-th execution, P uses $f_j(X)$ as its random tape. We use
$\tau(P, V^*, \mathsf{CRS}, \{(x_i, w_i)\}_{i=1}^t)$ to denote the transcripts of t consecutive executions
and the total view of V^* where in the j-th execution, P takes $(x_j, w_j), \mathsf{CRS}$ as
input and uses $f_j(X)$ as random tape, and V^* takes x_j, CRS, all previous transcripts and its previous view as input.

Definition 9. *A pair of algorithms (P, V) is a correlated-tape zero-knowledge
proof system for language L, if there exist polynomials $len(\cdot), k(\cdot, \cdot)$ such that for
any polynomial $t(\cdot)$, the following conditions hold:*

- Completeness: *For every security parameter $\kappa, x \in L, w \in R_L(x)$,*

$$\Pr[< P(w, X), V > (x, U_{len(\kappa)}) = 1] = 1$$

  *Here w is the private input for P and X is the private random tape of P.
  By $< P, V > (x, U_{len(\kappa)})$, we denote the output of V when P and V interact
  on the common input x and a common reference string distributed uniformly
  random over $\{0, 1\}^{len(\kappa)}$.*
- Soundness: *For every algorithm A and every $x \notin L$, there exists a negligible
  function $\mu(\cdot)$ such that for every security parameter κ,*

$$\Pr[< A, V > (x, U_{len(\kappa)}) = 1] < \mu(\kappa)$$

- Correlated-Tape Zero Knowledge: *There exists a simulator S, such that for
  any $t = t(\kappa)$ functions $f_1, f_2, ..., f_t$, whose output has the same length as input,
  such that $H_\infty(f_i(X)) \geq k(t(\kappa), \kappa)$, any V^* and (x_i, w_i) where $w_i \in R_L(x_i)$ and
  $i \in [t]$ the following two distributions are computationally indistinguishable:*

$$\{\mathsf{CRS} \sim U_{len(\kappa)} : \tau(P, V^*, \mathsf{CRS}, \{(x_i, w_i)\}_{i=1}^t)\}$$

$$S(\{x_i\}_{i=1}^t, V^*)$$

One could consider a variant of the above definition where there is no CRS. One could also define the complementary setting where the random number generator of the verifier (rather than the prover) is defective (and one would like to ensure soundness in the correlated tape setting).

Correlated-Tape Secure Multi-party Computation. Now let us consider the case of a multi-party computation protocol. We use P_i to denote the i-th party and A to denote the adversary. Suppose there are n parties in total. We use F to represent the desired functionality. Let $T \subseteq [n]$ to be the set of indices of parties that are corrupted.

Our ideal model will be the same as that in a standard definition of MPC. Specifically,

*Ideal Model.*There is a trusted party which computes the desired functionality based on the inputs of all parties. An execution in the ideal model proceeds as follows:

- **Inputs.** All parties (including the corrupted parties) will send their inputs to the trusted party. An honest party always sends its real input. A corrupted party may send modified value depending on the strategy of the adversary. We use x_i to denote the input sent by P_i.
- **Trusted Party Computes the Result.** The trusted party will use the inputs from all parties to compute the desired functionality. Let $(y_1', y_2', ..., y_n') = F(x_1, x_2, ..., x_n)$
- **Trusted Party Sends out the result.** For $i = 1, 2, ..., n$, the trusted party asks A whether it wants to abort. If A does not abort, then the trusted party will send y_i' to P_i. Otherwise, for all $j \geq i$, P_j will receive nothing from the trusted party.
- **Outputs.** An honest party P_i always outputs the response it received from the trusted party (it will output $\perp$ if it receives nothing) together with its input x_i. The adversary A outputs an arbitrary function of its entire view so far (including the views in the previous executions).

We use $\texttt{IDEAL}_{F,A}(\{x_i^j : i \notin T\}_{j=1}^t)$ to represents the outputs of t consecutive sequential executions in the ideal world with functionality F, adversary A and input x_i^j for honest party P_i in the j-th execution.

Real Model. In the real model, the adversary A is allowed to specify t tampering functions for each honest party. We use f_i^j to represent the j-th tampering function for P_i. Then there will be a trusted party generating a uniform string as CRS. In the beginning, each party has a private random tape distributed uniformly random. Denote X_i to be the initial random tape of P_i. In the j-th execution in the real model, an honest party P_i uses $f_i^j(X_i)$ as its random tape. The outputs of a protocol π in the real model include the inputs and outputs of all honest parties together with the full view of the adversary so far (including the view in the previous executions). We use $\texttt{REAL}_{F,A}^{\pi}(\{x_i^j : i \notin T\}_{j=1}^t, \{X_i^j : i \notin T\}_{j=1}^t, \texttt{CRS})$ to represents the outputs of π in t consecutive sequential executions in the real world for functionality F, adversary A, common reference string CRS and input x_i^j for honest party P_i with random tape X_i^j in the j-th execution.

Definition 10. *A protocol π is a secure correlated-tape multi-party computation protocol for functionality F of n parties, if there exist polynomials $len(\cdot), k(\cdot, \cdot)$ such that for any polynomial $t(\cdot)$, adversary A which corrupts $(n - \ell)$ parties with the set of indices $T \subset [n]$ and security parameter κ, there exists an ideal attacker A' such that, for all inputs $\{x_i^j : i \notin T\}_{j=1}^{t(\kappa)}$ and functions $\{f_i^j : i \notin T\}_{j=1}^{t(\kappa)}$, whose output has the same length as input, such that $H_\infty(f_i^j(X_i)) \geq k(t(\kappa), \kappa)$, the following two distributions are computationally indistinguishable:*

$$\{\text{CRS} \sim U_{len(\kappa)} : \text{REAL}_{F,A}(\{x_i^j : i \notin T\}_{j=1}^{t(\kappa)}, \{f_i^j(X_i) : i \notin T\}_{j=1}^{t(\kappa)}, \text{CRS})\}$$

$$\text{IDEAL}_{F,A'}(\{x_i^j : i \notin T\}_{j=1}^{t(\kappa)})$$

Note that in the above definitions, we use the min-entropy condition to constrain the tampering functions that the adversary may choose to avoid known impossibility results on deterministic zero-knowledge [GO94].

Impossibility without CRS. We stress that a common public random string as auxiliary input is necessary for our construction. In the work [DOPS04] of Dodis, Ong, Prabhakaran and Sahai, they studied the model which uses imperfect random tape in a zero knowledge protocol without CRS. The result is negative. Note that in Definition 9 and when we set $t = 1$, one can view $f_1(X)$ as an imperfect random tape and $f_1(X)$ can be all possible flat source with min-entropy $k(t(\kappa), \kappa)$. Thus, if there is no CRS as auxiliary input, it is also impossible to construct a protocol satisfying Definition 9.

Notice that even in the CRS model, the (tampering) functions $f_1, f_2, ..., f_t$ must not *depend on* the CRS to allow for a positive result. We give a proof sketch as following. The idea is very similar to that in [DOPS04].

We focus on the case where $t(\cdot) = 1$. Without loss of generality, we can assume that the length of random tape $N \geq k(1, \kappa) + \kappa$ (It can be achieved by simply padding κ random bits and never use them). Suppose the length of the transcript is bounded by $q(\kappa)$ where q is a polynomial. Consider a distinguisher D_i which just outputs the i-th bit of the transcript. Now for a fixed CRS and a fixed random tape of V^*, we want to show one of the following cases happens:

- There exists two tampering functions f and f' such that $f(U_N)$ and $f'(U_N)$ are both $(N, k(1, \kappa))$–flat sources but two distributions of the transcripts can be distinguished by some D_i with noticeable probability.
- The distribution of the transcript is deterministic except for a negligible probability.

Now consider the distribution of the transcript where the prover just uses the uniform random tape X. If it is deterministic except for a negligible probability, then we are done. Otherwise, there exists some i such that the i-th bit in the transcript is not almost deterministic. Then we may find two sets S, S' of size

$2^{k(1,\kappa)}$ such that for every $X \in S$, the i-th bit of the transcript is always 0, and for every $X \in S'$, the i-th bit of the transcript is always 1. Let f and f' be functions such that $f(U_N)$ is a flat distribution over S and $f'(U_N)$ is a flat distribution over S'. Note that D_i can distinguish these two distributions with probability 1.

Then, we fix the random tape of V^* and consider all possible CRS. We may construct $f(\cdot, \text{CRS})$ and $f'(\cdot, \text{CRS})$ such that one of the above cases happens. We say a CRS is good if the second case happens, i.e. the distribution of the transcript is deterministic except for a negligible probability. We say a CRS is bad otherwise. Note that, for a bad CRS, there exists some D_i such that it will always output 0 when using f and output 1 when using f'. It means that each bad CRS corresponds to one distinguisher. Since there are in total $q(k)$ distinguisher, then there exists D_{i^*} where over $1/q(k)$ bad CRS corresponds to it. If all but a negligible portion of CRS is good, then the prover's behavior almost fully depends on CRS and the random tape of V^*. Otherwise, D_{i^*} may distinguish two distributions with a noticeable probability.

Therefore, except for a negligible probability, the randomness of the prover only comes from CRS and the random tape of V^*, which is impossible for a non-trivial language L.

Correlated-Source Extractors. The construction of correlated tape secure protocols is closely related to the question of designing what we call correlated-source extractors. Informally, correlated-source extractors csExt have power to break correlations between sources with a unique random seed, i.e.,

$$|\mathsf{csExt}(f_i(X), Y) \circ \{\mathsf{csExt}(f_j(X), Y)\}_{j \neq i} \circ Y - U \circ \{\mathsf{csExt}(f_j(X), Y)\}_{j \neq i} \circ Y| < \epsilon,$$

where U is the uniform distribution and we use Y to refer the CRS. With this object and CRS, the prover can obtain a fresh uniformly random tape in each execution. Formally, we define the notion correlated-source extractor as following:

Definition 1 (Seeded Correlated-Source Extractor). *A function* $\mathsf{csExt} : \{0,1\}^* \times \{0,1\}^d \to \{0,1\}^m$ *is a seeded correlated-source extractor if the following holds: There exists a polynomial* $k(\cdot, \cdot, \cdot)$ *and a negligible function* $\epsilon(\cdot)$, *such that for any polynomial* $t(\cdot)$, $t = t(d)$ *arbitrary functions* $\mathcal{A}_1, \mathcal{A}_2, ..., \mathcal{A}_t$, *whose output has the same length as the input, with no fixed points, and, a source* X *with min-entropy* $k(t, m, d)$,

$$|\mathsf{csExt}(X, U_d) \circ \{\mathsf{csExt}(\mathcal{A}_i(X), U_d)\}_{i=1}^{t} \circ U_d - U_m \circ \{\mathsf{csExt}(\mathcal{A}_i(X), U_d)\}_{i=1}^{t} \circ U_d| < \epsilon(d)$$

where U_m *and* U_d *are uniform strings of length* m *and* d *respectively.*

5   Explicit Construction of Correlated-Source Extractor

In this section, we will describe our construction of correlated-source extractors. To this end, we first give an explicit construction of a weak t-correlated source extractor in Sects. 5.1 and 5.2. Then we show that it is indeed a correlated-source extractor in Sect. 5.3. In Sect. 5.4, we introduce a special kind of sources which we can further lower the requirement of min-entropy.

5.1   Explicit Construction of Weak Correlated-Source Extractor

We will frequently use the following lemma in the proof.

Lemma 4. *Suppose* X, X', Y, Y' *are random variables such that* $|X \circ Y - X' \circ Y'| \le \epsilon$. *Then, for any function* $f(x, y)$,

$$|f(X, Y) \circ Y - f(X', Y') \circ Y'| \le \epsilon$$

Especially, when Y is an empty string, we have $|f(X) - f(X')| \le \epsilon$. A formal proof can be found in the full version of this paper [GS19] in Appendix A.

Before we give our construction, we need to point out an important fact about weak correlated-source extractor:

Theorem 4. *If* wcsExt *is a weak* t-correlated-source extractor for min entropy k and output length m, then $(t + 1)m \le k$.

We give a formal proof in the full version [GS19] of this paper in Appendix B.

In Theorem 4, it gives us an upper bound of t, i.e. $t < n$. We will use this fact in our construction.

Theorem 5. *There exists an explicit weak* t-correlated-source extractor wcsExt *for min-entropy* $k \ge O(t^3(\log^2 n + \log^2(1/\epsilon)))$, *seed length* $d = O(\log^2 n + \log^2(1/\epsilon))$ *and output length* $m = O(\log n + \log(1/\epsilon))$.

Proof. Suppose the length of adv is ℓ. We separate Y into several parts. Specifically, let

$$Y = Y_{\text{adv}} \circ Y_1^1 \circ Y_1^2 \circ \dots \circ Y_1^{2\ell} \circ Y_2^1 \circ Y_2^2 \circ \dots \circ Y_2^\ell \circ Y_{\text{start}}$$

The first part Y_{adv} is used to generate adv for X. Then we will use $Y_1^1, Y_1^2, ..., Y_1^{2\ell}$ to generate $X^1, X^2, ..., X^{2\ell}$. $Y_2^1, Y_2^2, ..., Y_2^\ell$ and Y_{start} will be used in the construction of function F. Let $d_{\text{adv}} = |Y_{\text{adv}}|$, $d_1 = |Y_1^i|$, $d_2 = |Y_2^i|$ and $d_{\text{start}} = |Y_{\text{start}}|$.

Step 1: Construction of adv.

We separate X into n/d_{adv} parts such that each part is of length d_{adv}. Suppose $X = X^1 \circ X^2 \circ \dots \circ X^{n/d_{\text{adv}}}$. Construct a polynomial in the field $GF(2^{d_{\text{adv}}})$:

$$F_X(n) = \sum_{i=1}^{n/d_{\text{adv}}} X^i n^{i-1}$$

Let $\mathtt{adv} = F_X(Y_{\mathtt{adv}})$ as the advice of X. Then,

$$|\mathtt{adv}| = \ell = d_{\mathtt{adv}}. \tag{3}$$

For different sources X and X', $F_X(n)$ and $F_{X'}(n)$ are different. Then, $F_X(n) - F_{X'}(n) \neq 0$. It is known that, $F_X(n) - F_{X'}(n) = 0$ has at most $n/d_{\mathtt{adv}}$ roots. Since $Y_{\mathtt{adv}}$ is uniformly random and independent of sources X, X', with probability at most $n/(d_{\mathtt{adv}}2^{d_{\mathtt{adv}}})$, $F_X(Y_{\mathtt{adv}}) = F_{X'}(Y_{\mathtt{adv}})$.

Let $\mathtt{adv}_i = F_{X_i}(Y_{\mathtt{adv}})$ be the advice of the i-th tampering source. By union bound, with probability at least $1 - tn/(d_{\mathtt{adv}}2^{d_{\mathtt{adv}}})$, $\mathtt{adv}$ is different from $\mathtt{adv}_1, ..., \mathtt{adv}_t$. We set

$$d_{\mathtt{adv}} = \log(tn/\epsilon_1) \tag{4}$$

Then $\epsilon_1 = tn/2^{d_{\mathtt{adv}}} > tn/(d_{\mathtt{adv}}2^{d_{\mathtt{adv}}})$. Thus, with probability $1 - \epsilon_1$, we can successfully generate a unique advice for source X.

Let $\mathtt{ADV} = \{\mathtt{adv}, \mathtt{adv}_1, ..., \mathtt{adv}_t, Y_{\mathtt{adv}}\}$. By lemma 1, we have

$$\Pr[H_\infty(X|\mathtt{ADV}) \geq H_\infty(X) - (t+2)d_{\mathtt{adv}} - \log\frac{1}{\epsilon_2}] > 1 - \epsilon_2$$

Thus, by union bound, with probability at least $1 - \epsilon_1 - \epsilon_2$,

$$H_\infty(X|\mathtt{ADV}) \geq H_\infty(X) - (t+2)d_{\mathtt{adv}} - \log\frac{1}{\epsilon_2} \tag{5}$$

and $\mathtt{adv}$ is different from $\mathtt{adv}_1, ..., \mathtt{adv}_t$. We say such $\mathtt{ADV}$ is good.

Now, we fix a good $\mathtt{ADV}$. For simplicity, we omit the condition $\mathtt{ADV}$.

Step 2: Generating $X^1, X^2, ..., X^{2\ell}$.

The idea is very simple, we just apply a strong-seeded extractor with seed Y_1^i to generate X^i. Let q be the length of X^i. According to Theorem 3, there exists a strong-seeded extractor $\mathtt{Ext}_1$ for min-entropy $2q$, $d = c(\log n + \log\frac{1}{\epsilon_3})$ and ϵ_3, where c is some constant. We set

$$d_1 = c(\log n + \log\frac{1}{\epsilon_3}). \tag{6}$$

By Lemma 3, $\mathtt{Ext}_1$ is also a $(2q + \log(1/\epsilon_3), 2\epsilon_3)$ average case extractor. Let

$$X^i = \mathtt{Ext}_1(X, Y_1^i)$$

Recall that $\mathcal{X}^i = \{X^i, X_1^i, ..., X_t^i\}$. For every i, when we fix $\mathcal{X}^1, ..., \mathcal{X}^{i-1}, \mathcal{X}^{i+1}, ..., \mathcal{X}^{2\ell}$ and the seeds $Y_1^1, ..., Y_1^{i-1}, Y_1^{i+1}, ..., Y_1^{2\ell}$, by Lemma 1, (let $\mathtt{Set}_i = \{\mathcal{X}^j, Y_1^j\}_{j\neq i}$)

$$\tilde{H}_\infty(X|\mathtt{Set}_i) \geq H_\infty(X) - (2\ell - 1)(t+1)q - (2\ell - 1)d_1$$

Here we need

$$H_\infty(X) - (2\ell - 1)(t+1)q - (2\ell - 1)d_1 \geq 2q + \log(1/\epsilon_3). \tag{7}$$

Thus,

$$|\text{Ext}_1(X, Y_1^i) \circ Y_1^i \circ \text{Set}_i - U_q \circ Y_1^i \circ \text{Set}_i| \le 2\epsilon_3$$

Step 3: Construction of F.

Now we will construct a suitable function F. Recall that, we use adv^i for the i-th bit of adv. Let $Z^0 = Y_{\text{start}}$ and $Z^i = F(\text{adv}^i, X^{2i-1}, X^{2i}, Y_2^i, Z^{i-1})$.

The function F include two parts. First, we will apply a strong-seeded extractor on the source $X^{2i-1+\text{adv}^i}$ and seed Z_{i-1}. We use W^i to denote the output of the extractor. Then, we apply another strong-seeded extractor on the source W^i and seed Y_2^i. The output will be the final output of $F(\text{adv}^i, X^{2i-1}, X^{2i}, Y_2^i, Z^{i-1})$. Let z be the length of Z^i. Then $z = |Z^0| = |Y_{\text{start}}| = d_{\text{start}}$. Let w be the length of W^i.

According to Theorem 3, there exists a strong-seeded extractor Ext_2 for min-entropy $2w$, $d = c(\log n + \log \frac{1}{\epsilon_4})$ and ϵ_4, where c is some constant. Similarly, there exists a strong-seeded extractor Ext_3 for min-entropy $2z$, $d = c(\log n + \log \frac{1}{\epsilon_5})$ and ϵ_5. (For simplicity, we use the same constant in $\text{Ext}_1, \text{Ext}_2, \text{Ext}_3$. One can choose the largest constant.) By Lemma 3, Ext_2 is also a $(2w + \log(1/\epsilon_4), 2\epsilon_4)$ average case extractor, Ext_3 is also a $(2z + \log(1/\epsilon_5), 2\epsilon_5)$ average case extractor. We set

$$d_{\text{start}} = z = c(\log n + \log \frac{1}{\epsilon_4}) \tag{8}$$

and

$$d_2 = |Y_2^i| = c(\log n + \log \frac{1}{\epsilon_5}). \tag{9}$$

Then

$$F(\text{adv}^i, X^{2i-1}, X^{2i}, Y_2^i, Z^{i-1}) = \text{Ext}_3(\text{Ext}_2(X^{2i-1+\text{adv}^i}, Z^{i-1}), Y_2^i)$$

To show correctness, we will use induction on the length of adv. Note that, we only consider the case that adv is different from $\text{adv}_1, \text{adv}_2, ..., \text{adv}_t$. We have already fixed all advice and Y_{adv}.

We need the following lemma:

Lemma 5. *Suppose we have the following conditions:*

- *For random variables X (of length n) and W, $|X \circ W - U_n \circ W| \le \epsilon_1$*
- *Random variable Z of length z is correlated with X and W*
- *Y is uniformly random and independent of X, W, Z.*

Then, if Ext is a (k, ϵ_2) average case extractor with output length m where $k \le n - z$ and $d \le |Y|$, we have that

$$|\text{Ext}(X, Y) \circ Y \circ W \circ Z - U_m \circ Y \circ W \circ Z| < 2\epsilon_1 + \epsilon_2$$

We give a formal proof in the full version [GS19] of this paper in Appendix C.

Before we state the main lemma, we need to define a class of sets. Let $\text{DIFF}^0 = \text{SAME}^0 = \emptyset$. For $i \ge 1$,

$$\text{DIFF}^i = \{j | \text{adv}_j^i \ne \text{adv}^i\}$$

$$\text{SAME}^i = (\text{DIFF}^{i-1} \bigcup \text{SAME}^{i-1})/\text{DIFF}^i$$

Actually, $\mathtt{DIFF}^i$ is the set of indices of the advice whose i-th bit is different from that of the advice of X. And $\mathtt{SAME}^i$ is the set of indices of the advice whose first $i-1$ bits are different from that of the advice of X, but the i-th bit is the same.

Recall that, $W^i = \mathtt{Ext}_2(X^{2i-1+\mathtt{adv}^i}, Z^{i-1})$ and $F(\mathtt{adv}^i, X^{2i-1}, X^{2i}, Y_2^i, Z^{i-1}) = \mathtt{Ext}_3(W^i, Y_2^i)$. In the i-th step, we want to show that Z^i is uniformly random and independent of $\{Z_j^i\}_{j \in (\mathtt{DIFF}^i \bigcup \mathtt{SAME}^i)}$. By induction hypothesis, we have that Z^{i-1} is uniformly random and independent of $\{Z_j^{i-1}\}_{j \in \mathtt{SAME}^i}$. We hope we can keep this property after computing W^i, i.e., to show that W^i is uniformly random and independent of $\{W_j^i\}_{j \in \mathtt{SAME}^i}$. Then, in the second extraction, we want to break the correlation with Z_i and $\{Z_j^i\}_{j \in \mathtt{DIFF}^i}$.

We have the following main lemma:

Lemma 6. *Suppose* $\mathtt{DIFF}^i$ *and* $\mathtt{SAME}^i$ *are the same as above. Let* $\eta_i = \frac{4^i - 1}{3}(8\epsilon_3 + 4\epsilon_4 + 2\epsilon_5)$. *Then we have*

$$|Z^0 \circ \{\mathcal{X}_s\}_{s=1}^{2\ell} \circ \{Y_1^s\}_{s=1}^{2\ell} - U_z \circ \{\mathcal{X}_s\}_{s=1}^{2\ell} \circ \{Y_1^s\}_{s=1}^{2\ell}| = 0$$

and for every $1 \le i \le \ell$, *we have*

$$|Z^i \circ \{Z_j^i\}_{j \in (\mathtt{DIFF}^i \bigcup \mathtt{SAME}^i)} \circ \{\mathcal{X}_s\}_{s=2i+1}^{2\ell} \circ \{Y_1^s\}_{s=1}^{2\ell} \circ Z^0 \circ \{Y_2^s\}_{s=1}^{i}$$
$$-U_z \circ \{Z_j^i\}_{j \in (\mathtt{DIFF}^i \bigcup \mathtt{SAME}^i)} \circ \{\mathcal{X}_s\}_{s=2i+1}^{2\ell} \circ \{Y_1^s\}_{s=1}^{2\ell} \circ Z^0 \circ \{Y_2^s\}_{s=1}^{i}| \le \eta_i$$

Proof. We prove the lemma by induction. When $i = 0$, $\mathtt{DIFF}^0 \bigcup \mathtt{SAME}^0 = \emptyset$. We want to show that

$$|Z^0 \circ \{\mathcal{X}_s\}_{s=1}^{2\ell} \circ \{Y_1^s\}_{s=1}^{2\ell} - U_z \circ \{\mathcal{X}_s\}_{s=1}^{2\ell} \circ \{Y_1^s\}_{s=1}^{2\ell}| = 0$$

Note that $Z^0 = Y_{\mathtt{start}}$ and $Y_{\mathtt{start}}$ is uniformly random and independent of X and $\{Y_1^s\}_{s=1}^{2\ell}$. Thus, given $\{\mathcal{X}_s\}_{s=1}^{2\ell}$ and $\{Y_1^s\}_{s=1}^{2\ell}$, Z^0 is uniformly random. The statement holds.

For $i = 1$, we want to show that

$$|Z^1 \circ \{Z_j^1\}_{j \in (\mathtt{DIFF}^1 \bigcup \mathtt{SAME}^1)} \circ \{\mathcal{X}_s\}_{s=3}^{2\ell} \circ \{Y_1^s\}_{s=1}^{2\ell} \circ Z^0 \circ Y_2^1$$
$$-U_z \circ \{Z_j^1\}_{j \in (\mathtt{DIFF}^1 \bigcup \mathtt{SAME}^1)} \circ \{\mathcal{X}_s\}_{s=3}^{2\ell} \circ \{Y_1^s\}_{s=1}^{2\ell} \circ Z^0 \circ Y_2^1| \le \eta_1$$

Without loss of generality, assume $\mathtt{adv}^1 = 0$. Then, for $j \in \mathtt{DIFF}^1$, $\mathtt{adv}_j^1 = 1$ and $\mathtt{SAME}^1 = \emptyset$.

In step 2, we have

$$|X^1 \circ \mathcal{X}_2 \circ \{\mathcal{X}_s\}_{s=3}^{2\ell} \circ \{Y_1^s\}_{s=1}^{2\ell} - U_q \circ \mathcal{X}_2 \circ \{\mathcal{X}_s\}_{s=3}^{2\ell} \circ \{Y_1^s\}_{s=1}^{2\ell}| \le 2\epsilon_3$$

Note that Z^0 is uniformly random and independent of $\mathcal{X}_2 \circ \{\mathcal{X}_s\}_{s=3}^{2\ell} \circ \{Y_1^s\}_{s=1}^{2\ell}$. By Lemma 5, (here X is X^1, W is $\mathcal{X}_2 \circ \{\mathcal{X}_s\}_{s=3}^{2\ell} \circ \{Y_1^s\}_{s=1}^{2\ell}$, Z is empty, Y is Z^0 and $\mathtt{Ext}_2$ is a $(2w + \log(1/\epsilon_4), 2\epsilon_4)$ average case extractor),

$$|W^1 \circ Z^0 \circ \mathcal{X}_2 \circ \{\mathcal{X}_s\}_{s=3}^{2\ell} \circ \{Y_1^s\}_{s=1}^{2\ell} - U_w \circ Z^0 \circ \mathcal{X}_2 \circ \{\mathcal{X}_s\}_{s=3}^{2\ell} \circ \{Y_1^s\}_{s=1}^{2\ell}| < 4\epsilon_3 + 2\epsilon_4$$

Here, we require that

$$|X^1| = q \geq 2w + \log(1/\epsilon_4). \tag{10}$$

Since for every $j \in \texttt{DIFF}^1$, $W_j^1 = \texttt{Ext}_2(X_j^2, Z^0)$ is a deterministic function of Z^0 and $\mathcal{X}_2$. Thus

$$|W^1 \circ Z^0 \circ \{W_j^1\}_{j \in \texttt{DIFF}^1} \circ \{\mathcal{X}_s\}_{s=3}^{2\ell} \circ \{Y_1^s\}_{s=1}^{2\ell}$$
$$-U_w \circ Z^0 \circ \{W_j^1\}_{j \in \texttt{DIFF}^1} \circ \{\mathcal{X}_s\}_{s=3}^{2\ell} \circ \{Y_1^s\}_{s=1}^{2\ell}| < 4\epsilon_3 + 2\epsilon_4$$

Note that Y_2^1 is uniformly random and independent of W^1 and $Z^0 \circ \{W_j^1\}_{j \in \texttt{DIFF}^1} \circ \{\mathcal{X}_s\}_{s=3}^{2\ell} \circ \{Y_1^s\}_{s=1}^{2\ell}$. By Lemma 5, (here X is W^1, W is $Z^0 \circ \{W_j^1\}_{j \in \texttt{DIFF}^1} \circ \{\mathcal{X}_s\}_{s=3}^{2\ell} \circ \{Y_1^s\}_{s=1}^{2\ell}$, Z is empty, Y is Y_2^1 and $\texttt{Ext}_3$ is a $(2z + \log(1/\epsilon_5), 2\epsilon_5)$ average case extractor),

$$|Z^1 \circ Y_2^1 \circ Z^0 \circ \{W_j^1\}_{j \in \texttt{DIFF}^1} \circ \{\mathcal{X}_s\}_{s=3}^{2\ell} \circ \{Y_1^s\}_{s=1}^{2\ell}$$
$$-U_z \circ Y_2^1 \circ Z^0 \circ \{W_j^1\}_{j \in \texttt{DIFF}^1} \circ \{\mathcal{X}_s\}_{s=3}^{2\ell} \circ \{Y_1^s\}_{s=1}^{2\ell}| < 8\epsilon_3 + 4\epsilon_4 + 2\epsilon_5$$

Here, we require that

$$|W^1| = w \geq 2z + \log(1/\epsilon_5). \tag{11}$$

Since for every $j \in \texttt{DIFF}^1$, $Z_j^1 = \texttt{Ext}_3(W_j^1, Y_2^1)$ is a deterministic function of Y_2^1 and $\{W_j^1\}_{j \in \texttt{DIFF}^1}$. Thus

$$|Z^1 \circ Y_2^1 \circ Z^0 \circ \{Z_j^1\}_{j \in \texttt{DIFF}^1} \circ \{\mathcal{X}_s\}_{s=3}^{2\ell} \circ \{Y_1^s\}_{s=1}^{2\ell}$$
$$-U_z \circ Y_2^1 \circ Z^0 \circ \{Z_j^1\}_{j \in \texttt{DIFF}^1} \circ \{\mathcal{X}_s\}_{s=3}^{2\ell} \circ \{Y_1^s\}_{s=1}^{2\ell}| < 8\epsilon_3 + 4\epsilon_4 + 2\epsilon_5$$

Note that $\texttt{SAME}^1 = \emptyset$. It is exactly what we want to prove in the case $i = 1$.

Now suppose the lemma is correct for $i-1$, consider the case for i. According to induction hypothesis, we have that

$$|Z^{i-1} \circ \{Z_j^{i-1}\}_{j \in \texttt{SAME}^i} \circ \{\mathcal{X}_s\}_{s=2i-1}^{2\ell} \circ \{Y_1^s\}_{s=1}^{2\ell} \circ Z^0 \circ \{Y_2^s\}_{s=1}^{i-1}$$
$$-U_z \circ \{Z_j^{i-1}\}_{j \in \texttt{SAME}^i} \circ \{\mathcal{X}_s\}_{s=2i-1}^{2\ell} \circ \{Y_1^s\}_{s=1}^{2\ell} \circ Z^0 \circ \{Y_2^s\}_{s=1}^{i-1}| \leq \eta_{i-1}$$

For simplicity, we define

$$\mathfrak{X}_i = \{\mathcal{X}_s\}_{s=2i+1}^{2\ell}, \quad T = \{Y_1^s\}_{s=1}^{2\ell} \circ Z^0, \quad \mathcal{Y}_i = \{Y_2^s\}_{s=1}^{i}$$

Thus, we may rewrite the induction hypothesis for the case $i-1$ by

$$|Z^{i-1} \circ \{Z_j^{i-1}\}_{j \in \texttt{SAME}^i} \circ \mathfrak{X}_{i-1} \circ T \circ \mathcal{Y}_{i-1} - U_z \circ \{Z_j^{i-1}\}_{j \in \texttt{SAME}^i} \circ \mathfrak{X}_{i-1} \circ T \circ \mathcal{Y}_{i-1}| \leq \eta_{i-1}$$

Let Z' be a uniformly random string over $\{0,1\}^z$ and independent of $X, X_1, ..., X_t$ and Y. Then, we may use Z' instead of U_z, i.e.,

$$|Z^{i-1} \circ \{Z_j^{i-1}\}_{j \in \texttt{SAME}^i} \circ \mathfrak{X}_{i-1} \circ T \circ \mathcal{Y}_{i-1} - Z' \circ \{Z_j^{i-1}\}_{j \in \texttt{SAME}^i} \circ \mathfrak{X}_{i-1} \circ T \circ \mathcal{Y}_{i-1}| \leq \eta_{i-1}$$

Without loss of generality, assume $\mathtt{adv}^i = 0$. Then, $\mathtt{adv}^i_j = 1$ for $j \in \mathtt{DIFF}^i$ and $\mathtt{adv}^i_j = 0$ for $j \in \mathtt{SAME}^i$. We have

$$W^i = \mathtt{Ext}_2(X^{2i-1}, Z^{i-1})$$

Note that, in step 2, we have

$$|X^{2i-1} \circ \mathcal{X}_{2i} \circ \mathfrak{X}_i \circ \{Y_1^s\}_{s=1}^{2\ell} - U_q \circ \mathcal{X}_{2i} \circ \mathfrak{X}_i \circ \{Y_1^s\}_{s=1}^{2\ell}| \leq 2\epsilon_3$$

Also note that $\mathcal{Y}_{i-1}$ is independent of $X^{2i-1} \circ \mathcal{X}_{2i} \circ \mathfrak{X}_i \circ \{Y_1^s\}_{s=1}^{2\ell}$. Therefore,

$$|X^{2i-1} \circ \mathcal{X}_{2i} \circ \mathfrak{X}_i \circ \{Y_1^s\}_{s=1}^{2\ell} \circ \mathcal{Y}_{i-1} - U_q \circ \mathcal{X}_{2i} \circ \mathfrak{X}_i \circ \{Y_1^s\}_{s=1}^{2\ell} \circ \mathcal{Y}_{i-1}| \leq 2\epsilon_3$$

Since Z' is independent of X and Y, it is independent of $X^{2i-1} \circ \mathcal{X}_{2i} \circ \mathfrak{X}_i \circ \{Y_1^s\}_{s=1}^{2\ell} \circ \mathcal{Y}_{i-1}$ and $\{W^i_j\}_{j\in\mathtt{SAME}^i} \circ \{Z^{i-1}_j\}_{j\in\mathtt{SAME}^i} \circ Z^0$. By Lemma 5, (here X is X^{2i-1}, W is $\mathcal{X}_{2i} \circ \mathfrak{X}_i \circ \{Y_1^s\}_{s=1}^{2\ell} \circ \mathcal{Y}_{i-1}$, Z is $\{W^i_j\}_{j\in\mathtt{SAME}^i} \circ \{Z^{i-1}_j\}_{j\in\mathtt{SAME}^i} \circ Z^0$, Y is Z' and $\mathtt{Ext}_2$ is a $(2w + \log(1/\epsilon_4), 2\epsilon_4)$ average case extractor),

$$|\mathtt{Ext}_2(X^{2i-1}, Z') \circ Z' \circ \mathcal{X}_{2i} \circ \mathfrak{X}_i \circ T \circ \{W^i_j\}_{j\in\mathtt{SAME}^i} \circ \{Z^{i-1}_j\}_{j\in\mathtt{SAME}^i} \circ \mathcal{Y}_{i-1}$$
$$-U_w \circ Z' \circ \mathcal{X}_{2i} \circ \mathfrak{X}_i \circ T \circ \{W^i_j\}_{j\in\mathtt{SAME}^i} \circ \{Z^{i-1}_j\}_{j\in\mathtt{SAME}^i} \circ \mathcal{Y}_{i-1}| < 4\epsilon_3 + 2\epsilon_4$$

Here, we require that

$$|X^{2i-1}| - |\{W^i_j\}_{j\in\mathtt{SAME}^i} \circ \{Z^{i-1}_j\}_{j\in\mathtt{SAME}^i} \circ Z^0|$$
$$\geq q - (tw + tz + z) \geq 2w + \log(1/\epsilon_4) \tag{12}$$

Recall that,

$$|Z^{i-1} \circ \{Z^{i-1}_j\}_{j\in\mathtt{SAME}^i} \circ \mathfrak{X}_{i-1} \circ T \circ \mathcal{Y}_{i-1} - Z' \circ \{Z^{i-1}_j\}_{j\in\mathtt{SAME}^i} \circ \circ \mathfrak{X}_{i-1} \circ T \circ \mathcal{Y}_{i-1}| \leq \eta_{i-1}$$

Notice that $\{W^i_j\}_{j\in\mathtt{SAME}^i}$ is a deterministic function of $\{Z^{i-1}_j\}_{j\in\mathtt{SAME}^i}$ and $\mathcal{X}_{2i-1}$. Also, $W^i = \mathtt{Ext}_2(X^{2i-1}, Z^{i-1})$.

Thus, by reordering the composition parts, we have

$$|W^i \circ Z^{i-1} \circ \mathcal{X}_{2i} \circ \mathfrak{X}_i \circ T \circ \{W^i_j\}_{j\in\mathtt{SAME}^i} \circ \{Z^{i-1}_j\}_{j\in\mathtt{SAME}^i} \circ \mathcal{Y}_{i-1}$$
$$-\mathtt{Ext}_2(X^{2i-1}, Z') \circ Z' \circ \mathcal{X}_{2i} \circ \mathfrak{X}_i \circ T \circ \{W^i_j\}_{j\in\mathtt{SAME}^i} \circ \{Z^{i-1}_j\}_{j\in\mathtt{SAME}^i} \circ \mathcal{Y}_{i-1}|$$
$$\leq \eta_{i-1}$$

Still, since $\{W^i_j\}_{j\in\mathtt{SAME}^i}$ is a deterministic function of $\{Z^{i-1}_j\}_{j\in\mathtt{SAME}^i}$ and $\mathcal{X}_{2i-1}$, we have

$$|Z' \circ \mathcal{X}_{2i} \circ \mathfrak{X}_i \circ T \circ \{W^i_j\}_{j\in\mathtt{SAME}^i} \circ \{Z^{i-1}_j\}_{j\in\mathtt{SAME}^i} \circ \mathcal{Y}_{i-1}$$
$$-Z^{i-1} \circ \mathcal{X}_{2i} \circ \mathfrak{X}_i \circ T \circ \{W^i_j\}_{j\in\mathtt{SAME}^i} \circ \{Z^{i-1}_j\}_{j\in\mathtt{SAME}^i} \circ \mathcal{Y}_{i-1}|$$
$$\leq |Z' \circ \mathfrak{X}_{i-1} \circ T \circ \{W^i_j\}_{j\in\mathtt{SAME}^i} \circ \{Z^{i-1}_j\}_{j\in\mathtt{SAME}^i} \circ \mathcal{Y}_{i-1}$$
$$-Z^{i-1} \circ \mathfrak{X}_{i-1} \circ T \circ \{W^i_j\}_{j\in\mathtt{SAME}^i} \circ \{Z^{i-1}_j\}_{j\in\mathtt{SAME}^i} \circ \mathcal{Y}_{i-1}|$$
$$= |Z' \circ \mathfrak{X}_{i-1} \circ T \circ \{Z^{i-1}_j\}_{j\in\mathtt{SAME}^i} \circ \mathcal{Y}_{i-1}$$
$$-Z^{i-1} \circ \mathfrak{X}_{i-1} \circ T \circ \{Z^{i-1}_j\}_{j\in\mathtt{SAME}^i} \circ \mathcal{Y}_{i-1}|$$
$$< \eta_{i-1}$$

In total, we have

$$|W^i \circ Z^{i-1} \circ \mathcal{X}_{2i} \circ \mathfrak{X}_i \circ T \circ \{W^i_j\}_{j\in\text{SAME}^i} \circ \{Z^{i-1}_j\}_{j\in\text{SAME}^i} \circ \mathcal{Y}_{i-1}$$
$$-U_w \circ Z^{i-1} \circ \mathcal{X}_{2i} \circ \mathfrak{X}_i \circ T \circ \{W^i_j\}_{j\in\text{SAME}^i} \circ \{Z^{i-1}_j\}_{j\in\text{SAME}^i} \circ \mathcal{Y}_{i-1}|$$
$$\leq |W^i \circ Z^{i-1} \circ \mathcal{X}_{2i} \circ \mathfrak{X}_i \circ T \circ \{W^i_j\}_{j\in\text{SAME}^i} \circ \{Z^{i-1}_j\}_{j\in\text{SAME}^i} \circ \mathcal{Y}_{i-1}$$
$$-\text{Ext}_2(X^{2i-1}, Z') \circ Z' \circ \mathcal{X}_{2i} \circ \mathfrak{X}_i \circ T \circ \{W^i_j\}_{j\in\text{SAME}^i} \circ \{Z^{i-1}_j\}_{j\in\text{SAME}^i} \circ \mathcal{Y}_{i-1}|$$
$$+|\text{Ext}_2(X^{2i-1}, Z') \circ Z' \circ \mathcal{X}_{2i} \circ \mathfrak{X}_i \circ T \circ \{W^i_j\}_{j\in\text{SAME}^i} \circ \{Z^{i-1}_j\}_{j\in\text{SAME}^i} \circ \mathcal{Y}_{i-1}$$
$$-U_w \circ Z' \circ \mathcal{X}_{2i} \circ \mathfrak{X}_i \circ T \circ \{W^i_j\}_{j\in\text{SAME}^i} \circ \{Z^{i-1}_j\}_{j\in\text{SAME}^i} \circ \mathcal{Y}_{i-1}|$$
$$+|U_w \circ Z' \circ \mathcal{X}_{2i} \circ \mathfrak{X}_i \circ T \circ \{W^i_j\}_{j\in\text{SAME}^i} \circ \{Z^{i-1}_j\}_{j\in\text{SAME}^i} \circ \mathcal{Y}_{i-1}$$
$$-U_w \circ Z^{i-1} \circ \mathcal{X}_{2i} \circ \mathfrak{X}_i \circ T \circ \{W^i_j\}_{j\in\text{SAME}^i} \circ \{Z^{i-1}_j\}_{j\in\text{SAME}^i} \circ \mathcal{Y}_{i-1}|$$
$$< \eta_{i-1} + 4\epsilon_3 + 2\epsilon_4$$
$$+|Z' \circ \mathcal{X}_{2i} \circ \mathfrak{X}_i \circ T \circ \{W^i_j\}_{j\in\text{SAME}^i} \circ \{Z^{i-1}_j\}_{j\in\text{SAME}^i} \circ \mathcal{Y}_{i-1}$$
$$-Z^{i-1} \circ \mathcal{X}_{2i} \circ \mathfrak{X}_i \circ T \circ \{W^i_j\}_{j\in\text{SAME}^i} \circ \{Z^{i-1}_j\}_{j\in\text{SAME}^i} \circ \mathcal{Y}_{i-1}|$$
$$\leq 2\eta_{i-1} + 4\epsilon_3 + 2\epsilon_4$$

Note that Y^i_2 is uniformly random and independent of W^i and $Z^{i-1} \circ \mathcal{X}_{2i} \circ \mathfrak{X}_i \circ T \circ \{W^i_j\}_{j\in\text{SAME}^i} \circ \{Z^{i-1}_j\}_{j\in\text{SAME}^i \cup \text{DIFF}^i} \circ \mathcal{Y}_{i-1}$. By Lemma 5, (here X is W^i, W is $Z^{i-1} \circ \mathcal{X}_{2i} \circ \mathfrak{X}_i \circ T \circ \{W^i_j\}_{j\in\text{SAME}^i} \circ \{Z^{i-1}_j\}_{j\in\text{SAME}^i} \circ \mathcal{Y}_{i-1}$, Z is $\{Z^{i-1}_j\}_{j\in\text{DIFF}^i}$, Y is Y^i_2 and Ext_3 is a $(2z + \log(1/\epsilon_5), 2\epsilon_5))$, we have

$$|Z^i \circ Y^i_2 \circ Z^{i-1} \circ \mathcal{X}_{2i} \circ \mathfrak{X}_i \circ T \circ \{W^i_j\}_{j\in\text{SAME}^i} \circ \{Z^{i-1}_j\}_{j\in\text{SAME}^i \cup \text{DIFF}^i} \circ \mathcal{Y}_{i-1}$$
$$-U_z \circ Y^i_2 \circ Z^{i-1} \circ \mathcal{X}_{2i} \circ \mathfrak{X}_i \circ T \circ \{W^i_j\}_{j\in\text{SAME}^i} \circ \{Z^{i-1}_j\}_{j\in\text{SAME}^i \cup \text{DIFF}^i} \circ \mathcal{Y}_{i-1}|$$
$$< 4\eta_{i-1} + 8\epsilon_3 + 4\epsilon_4 + 2\epsilon_5 = \eta_i$$

Here, we need

$$|W^i| - |\{Z^{i-1}_j\}_{j\in\text{DIFF}^i}| \geq w - tz \geq 2z + \log(1/\epsilon_5). \tag{13}$$

Note that $\{Z^i_j\}_{j\in\text{SAME}^i}$ is a deterministic function of Y^i_2 and $\{W^i_j\}_{j\in\text{SAME}^i}$. And $\{Z^i_j\}_{j\in\text{DIFF}^i}$ is a deterministic function of Y^i_2, $\{Z^i_j\}_{j\in\text{DIFF}^i}$ and $\mathcal{X}_{2i}$. Thus, we have (We will discard $\{W^i_j\}_{j\in\text{SAME}^i}$ and Z^{i-1})

$$|Z^i \circ Y^i_2 \circ \{Z^i_j\}_{j\in\text{SAME}^i \cup \text{DIFF}^i} \circ \mathfrak{X}_i \circ T \circ \mathcal{Y}_{i-1}$$
$$-U_z \circ Y^i_2 \circ \{Z^i_j\}_{j\in\text{SAME}^i \cup \text{DIFF}^i} \circ \mathfrak{X}_i \circ T \circ \mathcal{Y}_{i-1}| < \eta_i$$

It is exactly what we want. Thus the statement is true for the case i. $\square$

By assumption, all advice are different. Then $\text{DIFF}^\ell \cup \text{SAME}^\ell$ include all indices of advice. Therefore,

$$|Z^\ell \circ \{Z^\ell_j\}_{j=1}^t \circ \{Y^s_1\}_{s=1}^{2\ell} \circ \{Y^s_2\}_{s=1}^\ell \circ Z^0 - U_z \circ \{Z^\ell_j\}_{j=1}^t \circ \{Y^s_1\}_{s=1}^{2\ell} \circ \{Y^s_2\}_{s=1}^\ell \circ Z^0|$$
$$< \frac{4^\ell - 1}{3}(8\epsilon_3 + 4\epsilon_4 + 2\epsilon_5)$$

So far, we are in the condition that ADV is good. Make everything together,

$$|Z^\ell \circ \{Z_j^\ell\}_{j=1}^t \circ \{Y_1^s\}_{s=1}^{2\ell} \circ \{Y_2^s\}_{s=1}^\ell \circ Z^0 \circ \text{ADV}$$
$$-U_z \circ \{Z_j^\ell\}_{j=1}^t \circ \{Y_1^s\}_{s=1}^{2\ell} \circ \{Y_2^s\}_{s=1}^\ell \circ Z^0 \circ \text{ADV}|$$

$$< \Pr[\text{ADV is bad }] + \Pr[\text{ADV is good }] \cdot \frac{4^\ell - 1}{3}(8\epsilon_3 + 4\epsilon_4 + 2\epsilon_5)$$

$$< \epsilon_1 + \epsilon_2 + \frac{4^\ell - 1}{3}(8\epsilon_3 + 4\epsilon_4 + 2\epsilon_5)$$

The total error is $\epsilon_1 + \epsilon_2 + \frac{4^\ell-1}{3}(8\epsilon_3 + 4\epsilon_4 + 2\epsilon_5)$. Let

$$\epsilon_1 = \epsilon_2 = \frac{\epsilon}{e^2} = O(\epsilon).$$

By (3), (4) and the fact that $t < n$, we have

$$\ell = d_{\text{adv}} = \log(tn/\epsilon_1) = O(\log n + \log(1/\epsilon))$$

Let

$$\epsilon_3 = \epsilon_4 = \epsilon_5 = \frac{\epsilon}{14e^{2\ell}} = O(\frac{\epsilon^3}{t^2 n^2})$$

Then we have

$$\epsilon_1 + \epsilon_2 + \frac{4^\ell - 1}{3}(8\epsilon_3 + 4\epsilon_4 + 2\epsilon_5) < \epsilon$$

By (6), (8), (9) and the fact that $t < n$, we have

$$d_1 = d_2 = d_{\text{start}} = O(\log n + \log(1/\epsilon))$$

Therefore, the length of the seed

$$|Y| = d_{\text{adv}} + 2\ell d_1 + \ell d_2 + d_{\text{start}} = O(\ell(\log n + \log(1/\epsilon))) = O(\log^2 n + \log^2(1/\epsilon))$$

All requirements for the min-entropy of $X|\text{ADV}$ are (7), (10), (11), (12), (13), i.e.

$$H_\infty(X|\text{ADV}) - (2\ell - 1)(t + 1)q - (2\ell - 1)d_1 \geq 2q + \log(1/\epsilon_3)$$
$$q \geq 2w + \log(1/\epsilon_4)$$
$$w \geq 2z + \log(1/\epsilon_5)$$
$$q - (tw + tz + z) \geq 2w + \log(1/\epsilon_4)$$
$$w - tz \geq 2z + \log(1/\epsilon_5)$$

Therefore, we have

$$z = O(\log n + \log(1/\epsilon))$$
$$w = O(t(\log n + \log(1/\epsilon)))$$
$$q = O(t^2(\log n + \log(1/\epsilon)))$$
$$H_\infty(X|\text{ADV}) \geq O(t^3(\log^2 n + \log^2(1/\epsilon)))$$

Note that in (5), for a good ADV, we have

$$H_\infty(X|\text{ADV}) \geq H_\infty(X) - (t+2)d_{\text{adv}} - \log(1/\epsilon_2)$$

Thus, we set

$$H_\infty(X) = O(t^3(\log^2 n + \log^2(1/\epsilon))) + (t+2)d_{\text{adv}} + \log(1/\epsilon_2)$$
$$= O(t^3(\log^2 n + \log^2(1/\epsilon)))$$

Note that the final output is Z^ℓ. Then, the output length $m = z = O(\log n + \log(1/\epsilon))$.

Thus, there exists an explicit construction of wcsExt for

$$H_\infty(X) \geq O(t^3(\log^2 n + \log^2(1/\epsilon)))$$
$$|Y| = O(\log^2 n + \log^2(1/\epsilon))$$
$$m = O(\log n + \log(1/\epsilon))$$

$\square$

Also, we may generalize our result to an average case weak t-correlated-source extractor.

Definition 11 (Average Case weak t-Correlated-Source Extractor). *A function* wcsExt $: \{0,1\}^n \times \{0,1\}^d \to \{0,1\}^m$ *is an average case weak t-correlated-source extractor for average conditional min-entropy k and error ϵ if the following holds: If X is a source in $\{0,1\}^n$, W is some random variable such that $\tilde{H}_\infty(X|W) \geq k$, $\mathcal{A}_1, \mathcal{A}_2, ..., \mathcal{A}_t$ are arbitrary tampering functions defined on $\{0,1\}^n \to \{0,1\}^n$ with no fixed points, then*

$$|\text{wcsExt}(X, U_d) \circ \{\text{wcsExt}(\mathcal{A}_i(X), U_d)\}_{i=1}^t \circ U_d \circ W$$
$$-U_m \circ \{\text{wcsExt}(\mathcal{A}_i(X), U_d)\}_{i=1}^t \circ U_d \circ W| < \epsilon$$

where U_m is independent of U_d and X.

We have the following lemma.

Lemma 7. *For any δ, if* wcsExt *is a weak t-correlated-source extractor for min-entropy k and error ϵ, then it is also an average case t-correlated source extractor for average conditional min-entropy $k + \log 1/\delta$ and error $\epsilon + \delta$.*

The proof can be easily generalized from Lemma 2.3 in [DRS04].

Therefore, combining Theorem 5 and Lemma 7 by setting $\delta = \epsilon$, we have

Theorem 6. *There exists an explicit average case weak t-correlated source extractor* wcsExt *for average conditional min-entropy $k \geq O(t^3(\log^2 n + \log^2(1/\epsilon)))$, seed length $d = O(\log^2 n + \log^2(1/\epsilon))$ and output length $m = O(\log n + \log(1/\epsilon))$.*

5.2   Boosting the Output Length

In the above construction, a major limitation is that the output length is only $O(\log n + \log 1/\epsilon)$. To boost the output length, we separate X into $2\ell + 1$ parts instead of 2ℓ parts in Theorem 5. We may set the length of the last part to be long enough. It can be viewed as the case that we append 0 to all advice. Then the length of the advice becomes $\ell + 1$. Since the last bit is 0, we will never choose $X^{2\ell+2}$. Thus, we only need one more part.

In Lemma 6, we have shown that, for every i, W^i given $\{W^i_j\}_{j\in \mathrm{SAME}^i}$ is uniformly random. When $i = \ell + 1$, $\mathrm{SAME}^{\ell+1} = [t]$. Thus, $W^{\ell+1}$ given $\{W^i_j\}_{j\in[t]}$ is uniformly random. $W^{\ell+1}$ will be the final output of our extractor.

Denote the length of $W^{\ell+1}$ to be m. We need the length of $X^{2\ell+1}$ to be $O(tm)$. Then, the min-entropy requirement for the original source becomes $O(t^3(\log^2 n + \log^2(1/\epsilon)) + t^2 m)$.

We have the following theorem.

Theorem 7. *There exists an explicit weak t-correlated-source extractor* wcsExt *where $k \geq O(t^3(\log^2 n + \log^2(1/\epsilon)) + t^2 m)$ and $d = O(\log^2 n + \log^2(1/\epsilon))$, where m is the output length.*

A formal proof can be found in the full version [GS19] of this paper in Appendix D.

5.3   Explicit Construction of Correlated-Source Extractor

We show that, our explicit construction in Theorem 7 is indeed a correlated-source extractor.

To see this, we set

$$\epsilon(d) = \Theta(2^{-\sqrt{d}})$$

and

$$k(t, m, d) = \Theta(t^3 d + t^2 m)$$

Clearly, $\epsilon(\cdot)$ is a negligible function and $k(\cdot, \cdot, \cdot)$ is a polynomial. Then, we only need to show that $d \geq O(\log^2 n + \log^2(1/\epsilon))$ and $k(t, m) \geq O(t^3(\log^2 n + \log^2(1/\epsilon)) + t^2 m)$. Note that the source length is bounded by a polynomial of d, and $\log^2(1/\epsilon) = \Theta(d)$. Therefore, $d \geq O(\log^2 n + \log^2(1/\epsilon))$. Further, we have $k(t, m, d) = \Theta(t^3 d + t^2 m) \geq O(t^3(\log^2 n + \log^2(1/\epsilon)) + t^2 m)$.

Thus, we have the following theorem.

Theorem 1. *There exists an explicit correlated-source extractor* csExt *with*

$$k(t, m, d) = \Theta(t^3 d + t^2 m)$$
$$\epsilon(d) = \Theta(2^{-\sqrt{d}})$$

where m is the length of the output.

We can also define what we call an average case correlated-source extractor in a similar way.

Definition 12 (Average Case Correlated-Source Extractor). *A function* $\mathtt{csExt} : \{0,1\}^* \times \{0,1\}^d \to \{0,1\}^m$ *is an average case correlated-source extractor if the following holds: There exists a polynomial* $k(\cdot,\cdot,\cdot)$ *and a negligible function* $\epsilon(\cdot)$, *such that for any polynomial* $t(\cdot)$, $t = t(d)$ *arbitrary functions* $\mathcal{A}_1, \mathcal{A}_2, ..., \mathcal{A}_t$, *whose output has the same length as the input, with no fixed points, a source* X *and a random variable* W *such that* $\tilde{H}_\infty(X|W) \geq k(t,m,d)$,

$$|\mathtt{csExt}(X, U_d) \circ \{\mathtt{csExt}(\mathcal{A}_i(X), U_d)\}_{i=1}^t \circ U_d \circ W$$
$$-U_m \circ \{\mathtt{csExt}(\mathcal{A}_i(X), U_d)\}_{i=1}^t \circ U_d \circ W| < \epsilon(d)$$

where U_m *is independent of* U_d *and* X.

If we are using an average case weak t-correlated source extractor, then we will get an average case correlated source extractor.

Theorem 8. *There exists an explicit average case correlated-source extractor* $\mathtt{csExt}$ *with*

$$k(t,m,d) = \Theta(t^3 d + t^2 m)$$
$$\epsilon(d) = \Theta(2^{-\sqrt{d}})$$

where m *is the length of the output.*

5.4 Generalizing the Entropy Requirements

In our construction, the min-entropy requirement on the source (denoted by k) grows with t. This is inherent since the total entropy of all the sources together may only be k (since each source may have zero min-entropy given any other source) which must be at least $t \cdot m$ where m is the size of the output of the extractor. A natural question is: could we place a stronger independence condition on the different sources which allows us to obtain a construction requiring the sources to have lower min-entropy? We outline such an extension in this section.

Definition 13 (Closed-Set Correlated Sources). *We say a sequence of sources* $X_1, X_2, ..., X_\ell$ *is a* (t,k)–*closed-set correlated sources if for every* X_i,

- *There exists a set of sources* S_i *such that* $X_i \in S_i$ *and* $|S_i| \leq t$

- *When given all sources outside* S_i, X_i *still has enough min-entropy, i.e.,*

$$\tilde{H}_\infty(X_i|\{X_j\}_{j=1}^\ell/S_i) \geq k$$

For a (t,k)–closed-set correlated sources, we can use an average case correlated source extractor on the set S_i, viewing X_i as the original source and $X_j \in S_i$ as the tampering source. Thus, we have the following corollary.

Corollary 1. *Let* csExt *be an average case correlated-source extractor constructed in Theorem 6. Let Y be a random seed of length specified in Theorem 8. For a closed-set correlated sources $X_1, X_2, ..., X_\ell$,*

$$|\text{csExt}(X_i, Y) \circ \{\text{csExt}(X_j, Y)\}_{j \neq i} - U_m \circ \{\text{csExt}(X_j, Y)\}_{j \neq i}| < \epsilon$$

6 Constructing Secure Correlated-Tape Multi-party Computation Protocol

We use correlated-source extractor and resettable multi-party computation protocol based on [GS09] as building blocks. Suppose csExt is a correlated-source extractor, π' is a resettably secure multi-party computation protocol for ideal functionality F (in the standard setting). We construct a correlated-tape secure MPC π as follows:

In the protocol π, each party will first run csExt with its secret random tape and CRS. Then use the output of csExt as the new random tape and follow the steps in π'. We have the following theorem.

Theorem 9. *Let π, π' be defined as above. For every security parameter κ, suppose $q(\kappa)$ is the length of the random tape that π' needs. Let csExt be a correlated-source extractor in Theorem 1 with $d = \kappa, m = q(\kappa)$ and polynomials $k'(\cdot, \cdot, \cdot), \epsilon'(\cdot)$. Let $len(\kappa) = \kappa$ and $k(t, \kappa) = k'(t, q(\kappa), \kappa) + t\kappa$. Then π is a correlated-tape multi-party computation protocol.*

Proof. Let T be the set of corrupted parties which controlled by the adversary. We define a pattern $\boldsymbol{S} = (s_1, s_2, ..., s_t)$ where $s_j \in [t]$. If for two patterns $\boldsymbol{S}, \boldsymbol{S}'$, there exists a permutation $p : [t] \rightarrow [t]$ such that $s_j = p(s'_j)$ for every $j \in [t]$, we view them as the same pattern. We say an input X_i is consistent with $\boldsymbol{S}$ respect to $\{f_i^j\}_{j=1}^t$, if for every $j_1, j_2 \in [t]$, $f_i^{j_1}(X_i) = f_i^{j_2}(X_i)$ if and only if $s_{j_1} = s_{j_2}$. Let

$$\text{Pattern}[\boldsymbol{S}, i] = \{X_i| \ X_i \text{ is consistent with } \boldsymbol{S} \text{ respect to } \{f_i^j\}_{j=1}^t\}$$

Note that there are at most $t^t = 2^{t \log t} = 2^{o(t\kappa)}$ patterns in total. Let ratio$[\boldsymbol{S}, i] \in [0, 1]$ be the ratio of X_i which is consistent with $\boldsymbol{S}$ respect to $\{f_i^j\}_{j=1}^t$. Indeed $\{\text{Pattern}[\boldsymbol{S}, i]\}_S$ is a partition of all X_i and thus

$$\sum_{\boldsymbol{S}} \text{ratio}[\boldsymbol{S}, i] = 1$$

After sampling X_i for P_i, we will reveal the pattern information to the adversary. Let

$$\text{BAD}_i = \{\boldsymbol{S} : \text{ratio}[\boldsymbol{S}, i] \leq \frac{1}{2^{t\kappa}}\}$$

Then, we show that, for every $\{S_i : S_i \notin \text{BAD}_i\}_{i \notin T}$, there exists an adversary A' in π' such that the following two distributions are computationally indistinguishable:

$$\{\text{CRS} \sim U_{len(\kappa)} : \text{REAL}_{F,A'}(\{x_i^j : i \notin T\}_{j=1}^t, \{X_i^j : i \notin T\}_{j=1}^t, \text{CRS})\}$$

$$\{\text{CRS} \sim U_{len(\kappa)} : \text{REAL}_{F,A}(\{x_i^j : i \notin T\}_{j=1}^t, \{f_i^j(X_i) : i \notin T\}_{j=1}^t, \text{CRS})\}$$

where $\{X_i^j : i \notin T\}_{j=1}^t$ is sampled based on the strategy of A' we will mention later.

We design A' to follow the strategy: After receiving $\{S_i : S_i \notin \text{BAD}_i\}_{i \notin T}$, for party P_i, in the j-th round, if there exists $j^* \in [j-1]$ such that $(S_i)_j = (S_i)_{j^*}$, then let P_i use the same random tape as the j^*-th round, otherwise let P_i use a fresh random tape.

Since the random tapes of each parties are independent, we only need to show that, for S_i, we have:

$$\{X_i \sim \text{Pattern}[S_i, i], \text{CRS} \sim U_{len(\kappa)} : \{\text{csExt}(f_i^j(X_i), \text{CRS})\}_{j=1}^t\} =_c \{X_i^j\}_{j=1}^t$$

Let $\text{Index}(S) = \{j : \forall j' \in [j-1], s_j \neq s_{j'}\}$. Then it is sufficient to show that

$$|\{\text{csExt}(f_i^j(X_i), \text{CRS})\}_{j \in \text{Index}(S_i)} - U_{q(\kappa)|\text{Index}(S)|}| \leq \mu(\kappa)$$

where $\mu(\cdot)$ is a negligible function. Note that, for every possible output y of f_i^j,

$$\begin{aligned}\Pr[f_i^j(X_i) = y] &\geq \Pr[f_i^j(X_i) = y \text{ and } X_i \in \text{Pattern}[S_i, i]] \\ &= \Pr[f_i^j(X_i) = y| X_i \in \text{Pattern}[S_i, i]]\Pr[X_i \in \text{Pattern}[S_i, i]] \\ &\geq \frac{1}{2^{t\kappa}}\Pr[f_i^j(X_i) = y| X_i \in \text{Pattern}[S_i, i]]\end{aligned}$$

By condition, $H_\infty(f_i^j(X_i)) \geq k(t, \kappa)$. Thus $\Pr[f_i^j(X_i) = y] \leq \frac{1}{2^{k(t,\kappa)}}$. We have

$$\Pr[f_i^j(X_i) = y| X_i \in \text{Pattern}[S_i, i]] \leq \frac{1}{2^{k(t,\kappa)-t\kappa}} = \frac{1}{2^{k'(t,q(\kappa),\kappa)}}$$

Therefore, given $X_i \in \text{Pattern}[S_i, i]$, $f_i^j(X_i)$ still has enough min-entropy to use correlated-source extractor csExt. For every $j \in \text{Index}(S_i)$,

$$\begin{aligned}|\text{csExt}(f_i^j(X_i), \text{CRS}) &\circ \{\text{csExt}(f_i^{j'}(X_i), \text{CRS})\}_{j' \neq j, j' \in \text{Index}(S_i)} \\ -U_{q(\kappa)} &\circ \{\text{csExt}(f_i^{j'}(X_i), \text{CRS})\}_{j' \neq j, j' \in \text{Index}(S_i)}| \leq \epsilon'(\kappa)\end{aligned}$$

By union bound,

$$|\{\text{csExt}(f_i^j(X_i), \text{CRS})\}_{j \in \text{Index}(S_i)} - U_{q(\kappa)|\text{Index}(S)|}| \leq |\text{Index}(S_i)|\epsilon'(\kappa) \leq t\epsilon'(\kappa)$$

Therefore, if $S_i \notin \text{BAD}_i$, the error is bounded by some negligible probability and further, A' satisfies our requirement.

Note that

$$\Pr[X_i \in \mathtt{Pattern}[S, i] \text{ where } S \in \mathtt{BAD}_i] \leq \frac{2^{t \log(t)}}{2^{t\kappa}} = \frac{1}{2^{O(t\kappa)}}$$

Thus, the distinguishable advantage is bounded by the sum of the probability that some $S_i \in \mathtt{BAD}_i$ and the probability that one can distinguish the two distributions generated by A' and A given all $S_i \notin \mathtt{BAD}_i$, which is still a negligible probability over security parameter κ.   □

Correlated-Source Extractors with Almost Optimal Parameters. We give a non-explicit construction of correlated-source extractors with almost optimal parameters. For lack of space, this result can be found in the full version [GS19] of this paper in Appendix E.

References

[ABP15] Abdalla, M., Benhamouda, F., Passelègue, A.: An algebraic framework for pseudorandom functions and applications to related-key security. In: Gennaro, R., Robshaw, M. (eds.) CRYPTO 2015. LNCS, vol. 9215, pp. 388–409. Springer, Heidelberg (2015). https://doi.org/10.1007/978-3-662-47989-6_19

[ACM+14] Austrin, P., Chung, K.-M., Mahmoody, M., Pass, R., Seth, K.: On the impossibility of cryptography with tamperable randomness. In: Garay, J.A., Gennaro, R. (eds.) CRYPTO 2014. LNCS, vol. 8616, pp. 462–479. Springer, Heidelberg (2014). https://doi.org/10.1007/978-3-662-44371-2_26

[ACRT97] Andreev, A.E., Clementi, A.E.F., Rolim, J.D.P., Trevisan, L.: Weak random sources, hitting sets, and BPP simulations. In: Proceedings 38th Annual Symposium on Foundations of Computer Science, pp. 264–272, October 1997

[BACD+18] Ben-Aroya, A., Chattopadhyay, E., Doron, D., Li, X., Ta-Shma, A.: A new approach for constructing low-error, two-source extractors. In: Proceedings of the 33rd Computational Complexity Conference, CCC 2018, Germany, pp. 3:1–3:19. Schloss Dagstuhl-Leibniz-Zentrum fuer Informatik (2018)

[BGGL01] Barak, B., Goldreich, O., Goldwasser, S., Lindell, Y.: Resettably-sound zero-knowledge and its applications. In: Proceedings 2001 IEEE International Conference on Cluster Computing, pp. 116–125, October 2001

[BP13] Bitansky, N., Paneth, O.: On the impossibility of approximate obfuscation and applications to resettable cryptography. In: Proceedings of the Forty-Fifth Annual ACM Symposium on Theory of Computing, STOC 2013, pp. 241–250. ACM, New York (2013)

[CG88] Chor, B., Goldreich, O.: Unbiased bits from sources of weak randomness and probabilistic communication complexity. SIAM J. Comput. **17**(2), 230–261 (1988)

[CG14a] Cheraghchi, M., Guruswami, V.: Non-malleable coding against bit-wise and split-state tampering. In: Lindell, Y. (ed.) TCC 2014. LNCS, vol. 8349, pp. 440–464. Springer, Heidelberg (2014). https://doi.org/10.1007/978-3-642-54242-8_19

[CGGM00] Canetti, R., Goldreich, O., Goldwasser, S., Micali, S.: Resettable zero-knowledge (extended abstract). In: Proceedings of the Thirty-Second Annual ACM Symposium on Theory of Computing, STOC 2000, pp. 235–244. ACM, New York (2000)

[CGL16] Chattopadhyay, E., Goyal, V., Li, X.: Non-malleable extractors and codes, with their many tampered extensions. In: Proceedings of the Forty-Eighth Annual ACM Symposium on Theory of Computing, STOC 2016, pp. 285–298. ACM, New York (2016)

[CL16] Chattopadhyay, E., Li, X.: Explicit non-malleable extractors, multi-source extractors, and almost optimal privacy amplification protocols. In: 2016 IEEE 57th Annual Symposium on Foundations of Computer Science (FOCS), pp. 158–167, October 2016

[Coh15] Cohen, G.: Local correlation breakers and applications to three-source extractors and mergers. In: 2015 IEEE 56th Annual Symposium on Foundations of Computer Science, pp. 845–862, October 2015

[Coh16a] Cohen, G.: Making the most of advice: new correlation breakers and their applications. In: 2016 IEEE 57th Annual Symposium on Foundations of Computer Science (FOCS), pp. 188–196, October 2016

[Coh16b] Cohen, G.: Non-malleable extractors - new tools and improved constructions. In: Raz, R. (ed.) 31st Conference on Computational Complexity (CCC 2016). Leibniz International Proceedings in Informatics (LIPIcs), vol. 50, pp. 8:1–8:29. Schloss Dagstuhl-Leibniz-Zentrum fuer Informatik, Dagstuhl (2016)

[Coh16c] Cohen, G.: Non-malleable extractors with logarithmic seeds. Electron. Colloquium Comput. Complex. (ECCC) 23, 30 (2016)

[COP+14] Chung, K.-M., Ostrovsky, R., Pass, R., Venkitasubramaniam, M., Visconti, I.: 4-round resettably-sound zero knowledge. In: Lindell, Y. (ed.) TCC 2014. LNCS, vol. 8349, pp. 192–216. Springer, Heidelberg (2014). https://doi.org/10.1007/978-3-642-54242-8_9

[COPV13] Chung, K.M., Ostrovsky, R., Pass, R., Visconti, I.: Simultaneous resettability from one-way functions. In: 2013 IEEE 54th Annual Symposium on Foundations of Computer Science, pp. 60–69, October 2013

[CPS16] Chung, K.-M., Pass, R., Seth, K.: Non-black-box simulation from one-way functions and applications to resettable security. SIAM J. Comput. 45(2), 415–458 (2016)

[CRS14] Cohen, G., Raz, R., Segev, G.: Nonmalleable extractors with short seeds and applications to privacy amplification. SIAM J. Comput. 43(2), 450–476 (2014)

[CZ16] Chattopadhyay, E., Zuckerman, D.: Explicit two-source extractors and resilient functions. In: Proceedings of the Forty-Eighth Annual ACM Symposium on Theory of Computing, STOC 2016, pp. 670–683. ACM, New York (2016)

[DGS09] Deng, Y., Goyal, V., Sahai, A.: Resolving the simultaneous resettability conjecture and a new non-black-box simulation strategy. In: 2009 50th Annual IEEE Symposium on Foundations of Computer Science, pp. 251–260, October 2009

[DLWZ14] Dodis, Y., Li, X., Wooley, T.D., Zuckerman, D.: Privacy amplification and nonmalleable extractors via character sums. SIAM J. Comput. 43(2), 800–830 (2014)

[DOPS04] Dodis, Y., Ong, S.J., Prabhakaran, M., Sahai, A.: On the (im)possibility of cryptography with imperfect randomness. In: Annual Symposium on Foundations of Computer Science, pp. 196–205 (2004)

[DRS04] Dodis, Y., Reyzin, L., Smith, A.: Fuzzy extractors: how to generate strong keys from biometrics and other noisy data. In: Cachin, C., Camenisch, J.L. (eds.) EUROCRYPT 2004. LNCS, vol. 3027, pp. 523–540. Springer, Heidelberg (2004). https://doi.org/10.1007/978-3-540-24676-3_31

[DW09] Dodis, Y., Wichs, D.: Non-malleable extractors and symmetric key cryptography from weak secrets. In: Proceedings of the Forty-First Annual ACM Symposium on Theory of Computing, STOC 2009, pp. 601–610. ACM, New York (2009)

[GM11] Goyal, V., Maji, H.K.: Stateless cryptographic protocols. In: 2011 IEEE 52nd Annual Symposium on Foundations of Computer Science, pp. 678–687, October 2011

[GO94] Goldreich, O., Oren, Y.: Definitions and properties of zero-knowledge proof systems. J. Cryptol. **7**(1), 1–32 (1994)

[GS09] Goyal, V., Sahai, A.: Resettably secure computation. In: Joux, A. (ed.) EUROCRYPT 2009. LNCS, vol. 5479, pp. 54–71. Springer, Heidelberg (2009). https://doi.org/10.1007/978-3-642-01001-9_3

[GS19] Goyal, V., Song, Y.: Correlated-source extractors and cryptography with correlated-random tapes. Cryptology ePrint Archive (2019)

[GUV09] Guruswami, V., Umans, C., Vadhan, S.: Unbalanced expanders and randomness extractors from Parvaresh-Vardy codes. J. ACM **56**(4), 20:1–20:34 (2009)

[KLRZ08] Kalai, Y.T., Li, X., Rao, A., Zuckerman, D.: Network extractor protocols. In: 2008 49th Annual IEEE Symposium on Foundations of Computer Science, pp. 654–663, October 2008

[Li12a] Li, X.: Non-malleable extractors, two-source extractors and privacy amplification. In: 2012 IEEE 53rd Annual Symposium on Foundations of Computer Science, pp. 688–697, October 2012

[Li12b] Li, X.: Design extractors, non-malleable condensers and privacy amplification. In: Proceedings of the Forty-Fourth Annual ACM Symposium on Theory of Computing, STOC 2012, pp. 837–854. ACM, New York (2012)

[Li15] Li, X.: Non-malleable condensers for arbitrary min-entropy, and almost optimal protocols for privacy amplification. In: Dodis, Y., Nielsen, J.B. (eds.) TCC 2015. LNCS, vol. 9014, pp. 502–531. Springer, Heidelberg (2015). https://doi.org/10.1007/978-3-662-46494-6_21

[Li16] Li, X.: Improved two-source extractors, and affine extractors for polylogarithmic entropy. In: 2016 IEEE 57th Annual Symposium on Foundations of Computer Science (FOCS), pp. 168–177, October 2016

[Li17] Li, X.: Improved non-malleable extractors, non-malleable codes and independent source extractors. In: Proceedings of the 49th Annual ACM SIGACT Symposium on Theory of Computing, STOC 2017, pp. 1144–1156. ACM, New York (2017)

[MW97] Maurer, U., Wolf, S.: Privacy amplification secure against active adversaries. In: Kaliski, B.S. (ed.) CRYPTO 1997. LNCS, vol. 1294, pp. 307–321. Springer, Heidelberg (1997). https://doi.org/10.1007/BFb0052244

[SSZ95] Saks, M., Srinivasan, A., Zhou, S.: Explicit dispersers with polylog degree. In: Proceedings of the Twenty-Seventh Annual ACM Symposium on Theory of Computing, STOC 1995, pp. 479–488. ACM, New York (1995)

[VV85] Vazirani, U.V., Vazirani, V.V.: Random polynomial time is equal to slightly-random polynomial time. In: 26th Annual Symposium on Foundations of Computer Science (SFCS 1985), pp. 417–428, October 1985

[Zuc96] Zuckerman, D.: Simulating BPP using a general weak random source. Algorithmica 16(4), 367–391 (1996)

Revisiting Non-Malleable Secret Sharing

Saikrishna Badrinarayanan[1][(✉)] and Akshayaram Srinivasan[2]

[1] UCLA, Los Angeles, USA
saikrishna@cs.ucla.edu
[2] UC Berkeley, Berkeley, USA
akshayaram@berkeley.edu

Abstract. A threshold secret sharing scheme (with threshold t) allows a dealer to share a secret among a set of parties such that any group of t or more parties can recover the secret and no group of at most $t - 1$ parties learn any information about the secret. A non-malleable threshold secret sharing scheme, introduced in the recent work of Goyal and Kumar (STOC'18), additionally protects a threshold secret sharing scheme when its shares are subject to tampering attacks. Specifically, it guarantees that the reconstructed secret from the tampered shares is either the original secret or something that is unrelated to the original secret.

In this work, we continue the study of threshold non-malleable secret sharing against the class of tampering functions that tamper each share independently. We focus on achieving greater *efficiency* and guaranteeing a *stronger* security property. We obtain the following results:

- **Rate Improvement.** We give the first construction of a threshold non-malleable secret sharing scheme that has rate > 0. Specifically, for every $n, t \geq 4$, we give a construction of a t-out-of-n non-malleable secret sharing scheme with rate $\Theta(\frac{1}{t \log^2 n})$. In the prior constructions, the rate was $\Theta(\frac{1}{n \log m})$ where m is the length of the secret and thus, the rate tends to 0 as $m \to \infty$. Furthermore, we also optimize the parameters of our construction and give a concretely efficient scheme.
- **Multiple Tampering.** We give the first construction of a threshold non-malleable secret sharing scheme secure in the stronger setting of bounded tampering wherein the shares are tampered by multiple (but bounded in number) possibly different tampering functions. The rate of such a scheme is $\Theta(\frac{1}{k^3 t \log^2 n})$ where k is an apriori bound on the number of tamperings. We complement this positive result by proving that it is impossible to have a threshold non-malleable secret sharing scheme that is secure in the presence of an apriori unbounded number of tamperings.
- **General Access Structures.** We extend our results beyond threshold secret sharing and give constructions of rate-efficient, non-malleable secret sharing schemes for more general monotone access structures that are secure against multiple (bounded) tampering attacks.

© International Association for Cryptologic Research 2019
Y. Ishai and V. Rijmen (Eds.): EUROCRYPT 2019, LNCS 11476, pp. 593–622, 2019.
https://doi.org/10.1007/978-3-030-17653-2_20

1   Introduction

A t-out-of-n threshold secret sharing scheme [Sha79, Bla79] allows a dealer to share a secret among n parties such that any subset of t or more parties can recover the secret but any subset of $t-1$ parties learn no information about the secret. Threshold secret sharing schemes are central tools in cryptography and have several applications such as constructing secure multiparty computation protocols [GMW87, BGW88, CCD88], threshold cryptographic systems [DF90, Fra90, DDFY94] and leakage resilient circuit compilers [ISW03, FRR+10, Rot12] to name a few.

Most of the threshold secret sharing schemes in literature are *linear*. This means that if we multiply each share by a constant c, we get a set of shares that correspond to a new secret that is c times the original secret. This property has in fact, been crucially leveraged in most of the applications including designing secure multiparty computation protocols and constructing threshold cryptosystems. However, this highly desirable feature becomes undesirable if our primary goal is to protect the shares against tampering attacks. More specifically, this linearity property allows an adversary to tamper (or maul) each share independently and output a new set of shares that reconstruct to a related secret (for example, two times the original secret). Indeed, if the shares of the secret are stored on a device such as a smart card, an adversary could potentially tamper with the smart card and change the value of the share that is being stored by overwriting it with a new value or maybe flipping a few bits. Notice that in the above tampering attack, the adversary need not learn the actual secret. However, the adversary is guaranteed to produce a set of shares that reconstruct to a related secret. Such an attack could be devastating when the shares, for example, correspond to a cryptographic secret key (such as a signing key) as it allows an adversary to mount related-key attacks (see [BDL01]). In fact, most of the known constructions of threshold signatures use Shamir's secret sharing to distribute the signing key among the parties and hence they are all susceptible to such attacks.

Non-Malleable Secret Sharing. To protect a secret sharing scheme against such share tampering attacks, Goyal and Kumar [GK18a, GK18b] introduced the notion of Non-Malleable Secret Sharing. Roughly, a secret sharing scheme (Share, Rec) is non-malleable against a tampering function class $\mathcal{F}$ if for every $f \in \mathcal{F}$ and every secret s, Rec(f(shares)) where shares $\leftarrow$ Share(s) is either s or something that is unrelated to s.[1] Of course, we cannot hope to protect against all possible tampering functions as a function can first reconstruct the secret from the shares, multiply it by 2 and then share this value to obtain a valid sharing of a related secret. Thus, the prior works placed restrictions on the set of functions that can tamper the shares. A natural restricted family of tampering functions that we will consider in this work is $\mathcal{F}_{ind}$ which consists of the set of all functions that tamper each share independently.

---

[1] See Sect. 3 for a precise definition.

Connection to Non-Malleable Codes. Non-malleable secret sharing is related to another cryptographic primitive called as Non-Malleable Codes which was introduced in an influential work by Dziembowski, Pietrzak and Wichs [DPW10].[2] A non-malleable code relaxes the usual notion of error correction by requiring that the decoding procedure outputs either the original message or something that is independent of the message when given a tampered codeword as input. A beautiful line of work, starting from [DPW10], has given several constructions of non-malleable codes with security against various tampering function classes [LL12, DKO13, FMNV14, FMVW14, ADL14, AGM+15, FMNV15, JW15, CKR16, CGM+16, AAG+16, CGL16, BDKM16, Li17, KOS17, CL17, KOS18, BDKM18, GMW17, OPVV18, KLT18, BDG+18].

We now elaborate on the connection between non-malleable codes and non-malleable secret sharing. A tampering function family in the literature of non-malleable codes that is somewhat similar to $\mathcal{F}_{ind}$ is the k-split-state function family. A k-split-state function compartmentalizes a codeword into k-parts and applies a tampering function to each part, independent of the other parts. Seeing the similarity between $\mathcal{F}_{ind}$ and k-split-state functions, it might be tempting to conclude that a non-malleable code against a k-split-state function family is in fact a k-out-of-k non-malleable secret sharing. However, as demonstrated in [GK18a], this might not be true in general. In particular, [GK18a] showed that even a 3-split-state non-malleable code need not be a 3-out-of-3 non-malleable secret sharing as non-malleable codes may not always protect the secrecy of the message. In particular, the first few bits of the codeword could reveal some bits of the message and still, this coding scheme could be non-malleable. Nevertheless, for the special case of 2, Aggarwal et al. [ADKO15] showed that any 2-split-state non-malleable code is indeed a 2-out-of-2 non-malleable secret sharing scheme. In the other direction, we note that any k-out-of-k non-malleable secret sharing scheme against $\mathcal{F}_{ind}$ is in fact a k-split-state non-malleable code.

Rate of Non-Malleable Secret Sharing. One of the main efficiency parameters in any secret sharing scheme is its *rate* which is defined as the ratio between the length of the secret and the maximum size of a share. In the prior work, Goyal and Kumar [GK18a] gave an elegant construction of t-out-of-n non-malleable secret sharing from any 2-split-state non-malleable code. However, the rate of this scheme is equal to $O(\frac{1}{n \log m})$ where m is the length of the secret. The rate tends to 0 as the length of the secret m tends to ∞ and hence, a natural question to ask is:

Can we obtain a construction of threshold non-malleable secret sharing with
rate > 0?

The problem of improving the rate was mentioned as an explicit open question in [GK18a].

---

[2] We refer the reader to [GK18a, GK18b] for a thorough discussion on the connection between non-malleable secret sharing and related notions such as verifiable secret sharing [CGMA85] and AMD codes [CDF+08].

Multiple Tamperings. In the real world, a tampering adversary could potentially mount more than one tampering attack. In particular, if each share of a cryptographic secret key is stored on a small device (such as smart cards), the adversary could potentially clone these devices to obtain multiple copies of the shares. The adversary could then apply a different tampering function on each copy and obtain information about related secrets. Thus, a more realistic security definition would be to consider multiple tampering functions $f_1, \ldots, f_k \in \mathcal{F}$, and require that for every secret s, the joint distribution $(\mathsf{Rec}(f_1(\mathsf{shares})), \ldots, \mathsf{Rec}(f_k(\mathsf{shares})))$ where shares $\leftarrow$ Share(s) is independent of s.[3] For the case of non-malleable codes, security against multiple tamperings has already been considered in [FMNV14, JW15, CGL16, OPVV18]. However, for the case of non-malleable secret sharing, the prior work [GK18a] only considered a single tampering function and a natural question would be:

Can we obtain a construction of threshold non-malleable secret sharing against multiple tamperings?

1.1   Our Results

In this work, we obtain the following results.

Rate Improvement. We give the first construction of a threshold non-malleable secret sharing scheme that has rate > 0. Specifically, the rate of our construction is $\Theta(\frac{1}{t \log^2 n})$ where t is the threshold and n is the number of parties. More formally,

Theorem 1. *For any $n, t \geq 4$ and any $\rho > 0$, there exists a construction of t-out-of-n non-malleable secret sharing scheme against $\mathcal{F}_{ind}$ for sharing m-bit secrets for any $m > \log n$ with rate $\Theta(\frac{1}{t \log^2 n})$ and simulation error $2^{-\Omega(\frac{m}{\log^{1+\rho} m})}$. The running times of the sharing and reconstruction algorithms are polynomial in n and m.*

Local Leakage Resilient Secret Sharing. One of the main tools used in proving Theorem 1 (which may be of independent interest) is an efficient construction of *local leakage-resilient* threshold secret sharing scheme [GK18a, BDIR18]. A t-out-of-n secret sharing scheme is said to be local leakage-resilient (parameterized by a leakage bound μ and set size s), if the secrecy holds against any adversary who might obtain at most $t - 1$ shares in the clear and additionally, for any set $S \subseteq [n]$ of size at most s, the adversary obtains μ bits from each share belonging to a party in the set S. Goyal and Kumar [GK18a] gave a construction of a 2-out-of-n local leakage resilient secret sharing scheme. In this work, we give an efficient construction of t-out-of-n local leakage resilient secret sharing

---

[3] As in the case of single tampering, a tampering function could just output the same shares and in which the reconstructed secret will be s. Our definition also captures this property and we refer to Sect. 3 for a precise definition.

scheme when t is a constant. This result must be contrasted with a recent result by Benhamouda et al. [BDIR18] who showed that the Shamir's secret sharing scheme is local leakage resilient when the field size is sufficiently large and the threshold $t = n - o(\log n)$. A more precise statement of our construction of local leakage resilient secret sharing scheme appears below.

Theorem 2. *For any $\epsilon > 0$, $t, n \in \mathbb{N}$, and parameters $\mu \in \mathbb{N}$, $s \le n$, there exists an efficient construction of t-out-of-n secret sharing scheme for sharing m-bit secrets that is (μ, s)-local leakage resilient with privacy error ϵ. The size of each share when t is a constant is $O\left((m + s\mu + \log(\log n/\epsilon))\log n\right)$.*

Concrete Efficiency. A major advantage of our result is its *concrete efficiency*. In the prior work, the constant hidden inside the big-O notation was large and was not explicitly estimated. We have optimized the parameters of our construction and we illustrate the size of shares for various values of (n, t) in Table 1.[4]

Table 1. Share sizes for simulation error of at most 2^{-80}.

(# of Parties, Threshold)	Secret length (in bits)	Share size (in KB)
$(7, 4)$	812	273.73
$(9, 5)$	812	399.85
$(25, 13)$	812	1757.53
$(100, 51)$	812	12.34×10^3
$(7, 4)$	1024	345.19
$(9, 5)$	1024	504.24
$(25, 13)$	1024	2216.40
$(100, 51)$	1024	15.56×10^3

Comparison with [GK18a]. When compared to the result of [GK18a] which could support thresholds $t \ge 2$, our construction can only support threshold $t \ge 4$. However, getting a rate > 0 non-malleable secret sharing scheme for threshold $t = 2$ would imply a 2-split-state non-malleable code with rate > 0 which is a major open problem. For the case of $t = 3$, though we know constructions of 3-split-state non-malleable codes with rate > 0 [KOS18, GMW17], they do not satisfy the privacy property of a 3-out-of-3 secret sharing scheme. In particular, given two states of the codeword, some information about the message is leaked. Thus, getting a 3-out-of-n non-malleable secret sharing scheme with rate > 0 seems out of reach of the current techniques and we leave this as an open problem.

---

[4] 812 bits is the minimal message length that gives 80 bits of security.

Multiple Tampering. We initiate the study of non-malleable secret sharing under multiple tampering. Here, the shares can be subject to multiple (possibly different) tampering functions and we require that the joint distribution of the reconstructed secrets to be independent of s. For this stronger security notion, we first prove a negative result that states that a non-malleable secret sharing cannot exist when the number of tamperings (also called as the tampering degree) is apriori unbounded. This result generalizes a similar result for the case of a split-state non-malleable codes. Formally,

Theorem 3. *For any $n, t \in \mathbb{N}$, there does not exist a t-out-of-n non-malleable secret sharing scheme against $\mathcal{F}_{ind}$ that can support an apriori unbounded tampering degree.*

When the tampering degree is apriori bounded, we get constructions of threshold non-malleable secret sharing scheme. Formally,

Theorem 4. *For any $n, t \geq 4$, and $\mathsf{K} \in \mathbb{N}$, there exists a t-out-of-n non-malleable secret sharing scheme with tampering degree K for sharing m-bit secrets for a large enough[5] m against $\mathcal{F}_{ind}$ with rate $= \Theta(\frac{1}{\mathsf{K}^3 t \log^2 n})$ and simulation error $2^{-m^{\Omega(1)}}$. The running time of the sharing and reconstruction algorithms are polynomial in n and m.*

General Access Structures. We extend our techniques used in the proof of Theorems 1, 4 to give constructions of non-malleable secret sharing scheme for more general monotone access structures rather than just threshold structures. Before we state our result, we give some definitions.

Definition 1. *An access structure $\mathcal{A}$ is said to be monotone if for any set $S \in \mathcal{A}$, any superset of S is also in $\mathcal{A}$. A monotone access structure $\mathcal{A}$ is said to be 4-monotone if for any set $S \in \mathcal{A}$, $|S| \geq 4$.*

We also give the definition of a minimal authorized set.

Definition 2. *For a monotone access structure $\mathcal{A}$, a set $S \in \mathcal{A}$ is a minimal authorized set if any strict subset of S is not in $\mathcal{A}$. We denote t_{max} to be $\max |S|$ where S is a minimal authorized set of $\mathcal{A}$.*

We now state our extension to general access structures.

Theorem 5. *For any $n, \mathsf{K} \in \mathbb{N}$ and 4-monotone access structure $\mathcal{A}$, if there exists a statistically private (with privacy error ϵ) secret sharing scheme for $\mathcal{A}$ that can share m-bit secrets for a large enough m with rate R, there exists a non-malleable secret sharing scheme for sharing m-bit secrets for the same access structure $\mathcal{A}$ with tampering degree K against $\mathcal{F}_{ind}$ with rate $\Theta(\frac{R}{\mathsf{K}^3 t_{max} \log^2 n})$ and simulation error $\epsilon + 2^{-m^{\Omega(1)}}$.*

Thus, starting with a secret sharing scheme for monotone span programs [KW93] or for more general access structures [LV18], we get non-malleable secret sharing schemes for the same access structures with comparable rate.

---

[5] See the main body for the precise statement.

Comparison with [GK18b]. In the prior work [GK18b], the rate of the non-malleable secret sharing for general access structures also depended on the length of the message and thus, even when R is constant, their construction could only achieve a rate of 0. However, unlike our construction, they could support all monotone access structures (and not just 4-monotone) and they could even start with a computational secret sharing scheme for an access structure $\mathcal{A}$ and convert it to a non-malleable secret sharing scheme for $\mathcal{A}$.

Concurrent Work. In a concurrent and independent work, Aggarwal et al. [ADN+18] consider the multiple tampering model and give constructions of non-malleable secret sharing for general access structures in this model. There are three main differences between our work and their work. Firstly, the rate of their construction asymptotically tends to 0 even for the threshold case. However, the rate of our construction is greater than 0 when we instantiate the compiler with a rate > 0 secret sharing scheme. Secondly, their work considers a stronger model wherein each tampering function can choose a different reconstruction set. We prove the security of our construction in a weaker model wherein the reconstruction set is the same for each tampering function. We note that the impossibility result for unbounded tampering holds even if the reconstruction set is the same. Thirdly, their construction can give non-malleable secret sharing scheme for any 3-monotone access structure whereas our construction can only work for 4-monotone access structure. In another concurrent and independent work, Kumar et al. [KMS18] gave a construction of non-malleable secret sharing in a stronger model where the tampering functions might obtain bounded leakage from the other shares.

2 Our Techniques

In this section, we give a high level overview of the techniques used to obtain our results.

2.1 Rate Improvement

Goyal and Kumar [GK18a] *Approach.* We first give a brief overview of the construction of threshold non-malleable secret sharing of Goyal and Kumar [GK18a] and then explain why it could achieve only a rate of 0. At a high level, Goyal and Kumar start with any 2-split-state non-malleable code and convert it into a t-out-of-n non-malleable secret sharing scheme. We only explain their construction for the case when $t \geq 3$, and for the case of $t = 2$, they gave a slightly different construction. For the case when $t \geq 3$, the sharing procedure does the following. The secret is first encoded using a 2-split-state non-malleable code to obtain the two states L and R. L is now shared using any t-out-of-n secret sharing scheme, say Shamir's secret sharing to get the shares $\mathsf{SL}_1, \ldots, \mathsf{SL}_n$ and R is shared using a 2-out-of-n local leakage resilient secret sharing scheme to get the shares $\mathsf{SR}_1, \ldots, \mathsf{SR}_n$. The share corresponding to party i includes $(\mathsf{SL}_i, \mathsf{SR}_i)$.

To recover the secret given at least t shares, the parties first use the recovery procedures of the threshold secret sharing scheme and local leakage resilient secret sharing scheme to recover L and R respectively. Later, the secret is obtained by decoding L and R using the decoding procedure of the non-malleable code. The correctness of the construction is straightforward and to argue secrecy, it can been seen that given any set of $t - 1$ shares, L is perfectly hidden and this follows from the security of Shamir's secret sharing. Now, using the fact that any 2-split-state non-malleable code is a 2-out-of-2 secret sharing scheme, it can be shown that the right state R statistically hides the secret.

To argue the non-malleability of this construction, Goyal and Kumar showed that any tampering attack on the secret sharing scheme can be reduced to a tampering attack on the underlying 2-split-state non-malleable code. The main challenge in designing such a reduction is that the tampering functions against the underlying non-malleable code must be split-state, meaning that the tampering function against L (denoted by f) must be independent of R and the tampering function against R (denoted by g) must be independent of L. To make the tampering function g to be independent of L, [GK18a] made use of the fact that there is an inherent difference in the parameters used for secret sharing L and R. Specifically, since R is shared using a 2-out-of-n secret sharing scheme, the tampered right state $\widetilde{R}$ can be recovered from any two tampered shares, say $\widetilde{SR}_1, \widetilde{SR}_2$. Now, since L is shared using a t-out-of-n secret sharing scheme and $t \geq 3$, the shares SL_1 and SL_2 information theoretically provides no information about L. This, in particular means that we can fix the shares SL_1 and SL_2 independent of L and the tampering function g could use these fixed shares to output the tampered right state $\widetilde{R}$. Now, when f is given the actual L, it can sample $SL_3, \ldots, SL_n$ as a valid secret sharing of L that is consistent with the fixed SL_1, SL_2. This allowed them to argue one-sided independence i.e., g is independent of L. On the other hand, making the tampering function f to be independent of R is a lot trickier. This is because any two shares information theoretically fixes R and in order to recover $\widetilde{L}$, we need at least t (≥ 3) shares. Hence, we may not be able to argue that f is independent of R. To argue this independence, Goyal and Kumar used the fact that R is shared using a *local leakage resilient* secret sharing scheme. In particular, they made the size of SR_i to be much larger than the size of SL_i and showed that even when we leak $|SL_i|$ bits from each share SR_i, R is still statistically hidden. This allowed them to define leakage functions $leak_1, \ldots, leak_n$ where $leak_i$ had SL_i hardwired in its description, it applies the tampering function on (SL_i, SR_i) and outputs the tampered $\widetilde{SL}_i$. Now, from the secrecy of the local leakage resilient secret sharing scheme, the distribution $\widetilde{SL}_1, \ldots, \widetilde{SL}_n$ (which completely determines $\widetilde{L}$) is independent of R and thus $\widetilde{L}$ is independent of R. This allowed them to obtain two-sided independence.

A drawback of this approach is that the rate of this scheme is at least as bad as that of the underlying 2-split-state non-malleable code. As mentioned before, obtaining a 2-split-state non-malleable code with rate > 0 is a major open problem. Thus, this construction could only achieve a rate of 0.

Our Approach. While constructing 2-split-state non-malleable code with rate >0 has been notoriously hard, significant progress has been made for the case of 3-split-state non-malleable codes. Very recently, independent works of Gupta et al. [GMW17] and Kanukurthi et al. [KOS18] gave constructions of 3-split-state non-malleable codes with an explicit constant rate. The main idea behind our rate-improved construction is to use a constant rate, 3-split-state non-malleable code instead of a rate 0, 2-split-state non-malleable code. To be more precise, we first encode the secret using a 3-split-state non-malleable code to get the three states (L, C, R). We then share the first state L using a t-out-of-n secret sharing scheme to get $(SL_1, \ldots, SL_n)$ as before. Then, we share C using a t_1-out-of-n secret sharing scheme to get $(SC_1, \ldots, SC_n)$ and R using a t_2-out-of-n secret sharing scheme to get $(SR_1, \ldots, SR_n)$. Here, t_1, t_2 are some parameters that we will fix later. The share corresponding to party i includes (SL_i, SC_i, SR_i). While the underlying intuition behind this idea is natural, proving that this construction is a non-malleable secret sharing scheme faces several barriers which we elaborate below.

First Challenge. The first barrier that we encounter is, unlike a 2-split-state non-malleable code which is always a 2-out-of-2 secret sharing scheme, a 3-split-state non-malleable code may not be a 3-out-of-3 secret sharing scheme. In particular, we will not be able use the [GK18a] trick of sharing the 3-states using secret sharing schemes with different thresholds to gain one-sided independence. This is because given $t - 1$ shares, complete information about two states will be revealed, and we could use these two states to gain some information about the underlying message. Thus, the privacy of the scheme breaks down. Indeed, as mentioned in the introduction, the constructions of Kanukurthi et al. [KOS18] and Gupta et al. [GMW17] are not 3-out-of-3 secret sharing schemes.

The main trick that we use to solve this challenge is that, while these constructions [KOS18, GMW17] are not 3-out-of-3 secret sharing schemes, we observe that there exist two states (let us call them C and R) such that these two states statistically hide the message. This means that we can potentially share these two states using secret sharing schemes with smaller thresholds and may use it to argue one-sided independence.

Second Challenge. The second main challenge is in ensuring that the tampering functions we design for the underlying 3-split-state non-malleable code are indeed split-state. Let us call the tampering functions that tamper L, C, and R as f, g, and h respectively. To argue that f, g and h are split-state, we must ensure f is independent of C and R and similarly, g is independent of L and R and h is independent of L and C. For the case of 2-split-state used in the prior work, this independence was achieved by using secret sharing with different thresholds and relying on the leakage resilience property. For the case of 3-split-state, we need a more sophisticated approach of *stratifying* the three secret sharing schemes so that we avoid circular dependence in the parameters. We now elaborate more on this solution.

To make g and h to be independent of L, we choose the thresholds t_1 and t_2 to be less than t. This allows us to fix a certain number of shares independent of L and use these shares to extract $\widetilde{C}$ and $\widetilde{R}$. Similarly, to make h to be independent of C, we choose the threshold $t_2 < t_1$. This again allows us to fix certain shares C and use them to extract $\widetilde{R}$. Thus, by choosing $t > t_1 > t_2$, we could achieve something analogous to one-sided independence. Specifically, we achieved independence of g from L and independence of h from (L, C). For complete split-state property, we still need to make sure that f is independent of (C, R) and g is independent of R. To make the tampering function f to be independent of C, we rely on the local leakage resilience property of the t_1-out-of-n secret sharing scheme. That is, we make the size of the shares SC_i to be much larger than SL_i such that, in spite of leaking $|SL_i|$ bits from each share SC_i, the secrecy of C is maintained. We can use this to show that the joint distribution $(\widetilde{SL}_1, \ldots, \widetilde{SL}_n)$ (which completely determines $\widetilde{L}$) is independent of C. Now, to argue that both f and g are independent of R, we rely on the local leakage resilience property of the t_2-out-of-n secret sharing scheme. That is, we make the shares of SR_i to be much larger than (SL_i, SC_i) so that, in spite of leaking $|SL_i| + |SC_i|$ bits from each share SR_i, the secrecy of R is maintained. We then use this property to argue that the joint distribution $(\widetilde{SL}_1, \widetilde{SC}_1), \ldots, (\widetilde{SL}_n, \widetilde{SC}_n)$ is independent of R. Thus, the idea of stratifying the three threshold secret sharing schemes with different parameters as described above allows to argue that f, g and h are split-state. As we will later see, this technique of stratification is very powerful and it allows us to easily extend this construction to more general monotone access structures.

Third Challenge. The third and the more subtle challenge is the following. To reduce the tampering attack on the secret sharing scheme to a tampering attack on the underlying non-malleable code, we must additionally ensure *consistency* i.e., the tampered message output by the split-state functions must be statistically close to the message output by the tampering experiment of the underlying secret sharing scheme. To illustrate this issue in some more detail, let us consider the tampering functions f and g in the construction of Goyal and Kumar [GK18a] for the simple case when $n = t = 3$. Recall that the tampering function g samples SR_1, SR_2 such that it is a valid 2-out-of-n secret sharing of R and uses the fixed SL_1, SL_2 (independent of L) to extract the tampered $\widetilde{R}$ from $(\widetilde{SR}_1, \widetilde{SR}_2)$. However, note that g cannot use any valid secret sharing of SR_1, SR_2 of R. In particular, it must also satisfy the property that the tampering function applied on SL_1, SR_1 gives the exact same $\widetilde{SL}_1$ that f uses in the reconstruction (a similar condition for position 2 must be satisfied). This is crucial, as otherwise there might be a difference in the distributions of the tampered message output by the split-state functions and the message output in the tampering experiment of the secret sharing scheme. In case there is a difference, we cannot hope to use the adversary against the non-malleable secret sharing to break the underlying non-malleable code. This example illustrates this issue for a simple case when $t = n = 3$. To ensure consistency for larger values of n and t, Goyal and Kumar fixed $(\mathsf{SL}_1, \ldots, \mathsf{SL}_{t-1})$ (instead of just fixing $\mathsf{SL}_1, \mathsf{SL}_2$) and the function g ensures consistency of each of the tampered shares $\widetilde{\mathsf{SL}}_1, \ldots, \widetilde{\mathsf{SL}}_{t-1}$. However, this approach completely fails when we move to 3 states. For the case of 3-states, the tampering function, say h, must sample $\mathsf{SR}_1, \ldots, \mathsf{SR}_n$ such that it is consistent with $\widetilde{\mathsf{SL}}_1, \ldots, \widetilde{\mathsf{SL}}_{t-1}$ used by f. However, even to check this consistency, h would need the shares $\mathsf{SC}_1, \ldots, \mathsf{SC}_{t-1}$ which completely determines C. In this case, we cannot argue that h is independent of C.

To tackle this challenge, we deviate from the approach of Goyal and Kumar [GK18a] and have a new proof strategy that ensures consistency and at the same time maintains the split-state property. In this strategy, we only fix the values $(\mathsf{SL}_1, \mathsf{SL}_2, \mathsf{SL}_3)$ for the first secret sharing scheme, $(\mathsf{SC}_1, \mathsf{SC}_2)$ for the second secret sharing scheme and fix SR_3 for the third secret sharing scheme. Note that we consider $t \geq 4$, $t_1 \geq 3$ and $t_2 \geq 2$ and thus, the fixed shares are independent of L, C, and R respectively.[6] We design our split-state functions in such a way that the tampering function f need not do any consistency checks, the tampering function g has to do the consistency check only on $\widetilde{\mathsf{SL}}_3$ (which it can do since SL_3 and SR_3 are fixed) and the function h needs to do a consistency check only on $\{\widetilde{\mathsf{SL}}_i, \widetilde{\mathsf{SC}}_i\}_{i \in [1,2]}$ (which it can do since $\mathsf{SL}_1, \mathsf{SC}_1, \mathsf{SL}_2, \mathsf{SC}_2$ are fixed). This approach of reducing the number of checks to maintain consistency helps us in arguing independence between the tampering functions. However, this approach creates additional problems in extracting $\widetilde{\mathsf{L}}$ as the tampering function f needs to use the shares $(\mathsf{SR}_4, \ldots, \mathsf{SR}_n)$ and $(\mathsf{SC}_4, \ldots, \mathsf{SC}_n)$ (which completely determines C and R respectively). We solve this by letting f extract $\widetilde{\mathsf{L}}$ using shares of some arbitrary values of C and R and we then use the leakage resilience property to ensure that the outputs in the split-state tampering experiment and the secret sharing tampering experiment are statistically close.

Completing the Proof. This proof strategy helps us in getting a rate > 0 construction of a t-out-of-n non-malleable secret sharing scheme for $t \geq 4$. However, there is one crucial block that is still missing. Goyal and Kumar [GK18a] only gave a construction of 2-out-of-n local leakage resilient secret sharing scheme. And, for this strategy to work we also need a construction of t_1-out-of-n local leakage resilient secret sharing scheme for some $t_1 > 2$. As mentioned in the introduction, the recent work by Benhamouda et al. [BDIR18] only gives a construction of local leakage resilient secret sharing when the threshold value is large (in particular, $n - o(\log n)$). To solve this, we give an efficient construction of a t-out-of-n local leakage resilient secret sharing scheme when t is a constant. This is in fact sufficient to get a rate > 0 construction of non-malleable secret sharing scheme. We now give details on the techniques used in this construction.

Local Leakage Resilient Secret Sharing Scheme. The starting point of our construction is the 2-out-of-2 local leakage resilient secret sharing from the work of Goyal and Kumar [GK18a] based on the inner product two-source extractor [CG88]. We first extend it to a k-out-of-k local leakage resilient secret sharing

---

[6] This is the reason why we could only achieve thresholds $t \geq 4$.

scheme for any arbitrary k. Let us now illustrate this for the case when k is even i.e., $k = 2p$. To share a secret s, we first additively secret share s into $s_1, \ldots, s_p$ and we encode each s_i using the 2-out-of-2 leakage resilient secret sharing scheme to obtain the shares $(\text{share}_{2i-1}, \text{share}_{2i})$. We then give share_i to party i for each $i \in [k]$. Note that given $t - 1$ shares, at most $p - 1$ additive secret shares can be revealed. We now rely on the local leakage resilience property of the 2-out-of-2 secret sharing to argue that the final additive share is hidden even when given bounded leakage from the last share. This helps us in arguing the k-out-k local leakage resilience property. The next goal is to extend this to a k-out-of-n secret sharing scheme. Since we are interested in getting good rate, we should not increase the size of the shares substantially. A naïve way of doing this would be to share the secret $\binom{n}{k}$ times (one for each possible set of k-parties) using the k-out-of-k secret sharing scheme and give the respective shares to the parties. The size of each share in this construction would blow up by a factor $\binom{n}{k-1}$ when compared to the k-out-of-k secret sharing scheme. Though, this is polynomial in n when k is a constant, this is clearly sub-optimal when n is large and would result in bad concrete parameters. We note that Goyal and Kumar [GK18a] used a similar approach to obtain a 2-out-of-n local leakage resilient secret sharing.

In this work, we use a very different approach to construct a k-out-of-n local leakage resilient secret sharing from a k-out-of-k local leakage resilient secret sharing. The main advantage of this transformation is that it is substantially more rate efficient than the naïve solution. Our transformation makes use of combinatorial objects called as perfect hash functions [FK84].[7] A family of functions mapping $\{1, \ldots, n\}$ to $\{1, \ldots, k\}$ is said to be a perfect hash function family if for every set $S \subseteq [n]$ of size at most k, there exists at least one function in the family that is injective on S. Let us now illustrate how this primitive is helpful in extending a k-out-of-k secret sharing scheme to a k-out-of-n secret sharing scheme. Given a perfect hash function family $\{h_i\}_{i \in [\ell]}$ of size ℓ, we share the secret s independently ℓ times using the k-out-of-k secret sharing scheme to obtain $(\text{share}_1^i, \ldots, \text{share}_k^i)$ for each $i \in [\ell]$. We now set the shares corresponding to party i as $(\text{share}_{h_1(i)}^1, \ldots, \text{share}_{h_\ell(i)}^\ell)$. To recover the secret from some set of k shares given by $S = \{s_1, \ldots, s_k\}$, we use the following strategy. Given any subset S of size k, perfect hash function family guarantees that there is at least one index $i \in [\ell]$ such that h_i is injective on S. We can now use $\{\text{share}_{h_i(s_1)}^i, \ldots, \text{share}_{h_i(s_k)}^i\} = \{\text{share}_1^i, \ldots, \text{share}_k^i\}$ to recover the secret using the reconstruction procedure of the k-out-of-k secret sharing.

We show that this transformation additionally preserves local leakage resilience. In particular, if we start with a k-out-of-k local leakage resilient secret sharing scheme then we obtain a k-out-of-n local leakage resilient secret sharing. The size of each share in our k-out-of-n leakage resilient secret sharing scheme is ℓ times the share size of k-out-of-k secret sharing scheme. Thus, to minimize

---

[7] We note that using perfect hash function families for constructing threshold secret sharing scheme is not new (see [Bla99, SNW01] for a comprehensive discussion). However, to the best of our knowledge, this is the first application of this technique to construct local leakage resilient secret sharing scheme.

rate we must minimize the size of the perfect hash function family. Constructing perfect hash function family of minimal size for all $k \in \mathbb{N}$ is an interesting and a well-known open problem in combinatorics. In this work, we give an efficient randomized construction (with good concrete parameters) of a perfect hash function family for a constant k with size $O(\log n + \log(1/\epsilon))$ where ϵ is the error probability. Alternatively, we can also use the explicit construction (which is slightly less efficient when compared to the randomized construction) of size $O(\log n)$ (when k is a constant) given by Alon et al. [AYZ95]. Combining either the randomized/explicit construction of perfect hash function family with a construction of k-out-of-k local leakage resilient secret sharing scheme, we get an efficient construction of k-out-of-n local leakage resilient secret sharing scheme when k is a constant.

2.2 Multiple Tampering

We also initiate the study of non-malleable secret sharing under multiple tamperings. As discussed in the introduction, this is a much stronger model when compared to that of a single tampering.

Negative Result. We first show that when the number of tampering functions that can maul the secret sharing scheme is apriori unbounded, there does not exist any threshold non-malleable secret sharing scheme. This generalizes a similar result for the case of split-state non-malleable code (see [GLM+04,FMNV14] for details) and the main idea is inspired by these works. The underlying intuition behind the negative result is simple: we come up with a set of tampering functions such that each tampering experiment leaks one bit of a share. Now, given the outcomes of $t \cdot s$ such tampering experiments where s is the size of the share, the distinguisher can clearly learn every bit of t shares and thus, learn full information about the underlying secret and break non-malleability.

For the tampering experiment to leak one bit of the share of party i, we use the following simple strategy. Let us fix an authorized set of size t say, $\{1, \ldots, t\}$. We choose two sets of shares: $\{\mathsf{share}_1, \ldots, \mathsf{share}_i, \ldots, \mathsf{share}_t\}$ and $\{\mathsf{share}_1, \ldots, \mathsf{share}_i', \ldots, \mathsf{share}_t\}$ such that they reconstruct to two different secrets. Note that the privacy of a secret sharing scheme guarantees that such shares must exist. Whenever the particular bit of the share of party i is 1, the tampering function f_i outputs share_i' whereas the other tampering functions, say f_j will output share_j. On the other hand, if the particular bit is 0 then the tampering function f_i outputs share_i and the other tampering functions still output share_j. Observe that the reconstructed secret in the two cases reveals the particular bit of the share of party i. We can use a similar strategy to leak every bit of all the t shares which completely determine the secret.

Positive Result. We complement the negative result by showing that when the number of tamperings is apriori bounded, we can obtain an efficient construction of a threshold non-malleable secret sharing scheme. A natural approach would be to start with a split-state non-malleable code that is secure against bounded tamperings and convert it into a non-malleable secret sharing scheme. To the best of our knowledge, the only known construction of split-state non-malleable code that is secure in the presence of bounded tampering is that of Chattopadhyay et al. [CGL16]. However, the rate of this code is 0 even when we restrict ourselves to just two tamperings. In order to achieve a better rate, we modify the constructions of Kanukurthi et al. [KOS18] and Gupta et al. [GMW17] such that we obtain a 3-split-state non-malleable code that secure in the setting of bounded tampering. The rate of this construction is $O(\frac{1}{k})$ where k is the apriori bound on the number of tamperings. Fortunately, even in this construction, we still maintain the property that there exists two states that statistically hide the message. We then prove that the same construction described earlier is a secure non-malleable secret sharing under bounded tampering when we instantiate the underlying code with a bounded tampering secure 3-split-state non-malleable codes.

2.3   General Access Structures

To obtain a secret sharing scheme for more general access structures, we start with any statistically secure secret sharing scheme for that access structure, and use it to share L instead of using a threshold secret sharing scheme. We require that the underlying access structure to be 4-monotone so that we can argue the privacy of our scheme. Recall that a 4-monotone access structure is one in which the size of every set in the access structure is at least 4. Even in this more general case, the technique of stratifying the secret sharing schemes allows us to prove non-malleability in almost an identical fashion to the case of threshold secret sharing. We remark that the work of [GK18b] which gave constructions of non-malleable secret sharing scheme for general monotone access structures additionally required their local leakage resilient secret sharing scheme to satisfy a security property called as strong local leakage resilience. Our construction does not require this property and we show that "plain" local leakage resilience is sufficient for extending to more general monotone access structures.

Organization. We give the definitions of non-malleable secret sharing and non-malleable codes in Sect. 3. In Sect. 4, we present the construction of the k-out-of-n leakage resilient secret sharing scheme. In Sect. 5, we describe our rate-efficient threshold non-malleable secret sharing scheme for the single tampering. We give the impossibility result for unbounded many tamperings in the full version. Finally, in Sect. 6, we describe our result on non-malleable secret sharing for general access structures against multiple bounded tampering. Note that the result in Sect. 6 implicitly captures the result for threshold non-malleable secret sharing against bounded tampering. We present this more general result for ease of exposition.

3   Preliminaries

Notation. We use capital letters to denote distributions and their support, and corresponding lowercase letters to denote a sample from the same. Let $[n]$ denote the set $\{1, 2, \ldots, n\}$, and U_r denote the uniform distribution over $\{0,1\}^r$. For any $i \in [n]$, let x_i denote the symbol at the i-th co-ordinate of x, and for any $T \subseteq [n]$, let $x_T \in \{0,1\}^{|T|}$ denote the projection of x to the co-ordinates indexed by T. We write $\circ$ to denote concatenation. We give the standard definitions of min-entropy, statistical distance and seeded extractors in the full version.

3.1   Threshold Non-Malleable Secret Sharing Scheme

We first give the definition of a sharing function, then define a threshold secret sharing scheme and finally give the definition of a threshold non-malleable secret sharing. These three definitions are taken verbatim from [GK18a]. We define non-malleable secret sharing for more general access structures in the full version.

Definition 3 (Sharing Function). *Let* $[n] = \{1, 2, \ldots, n\}$ *be a set of identities of* n *parties. Let* $\mathcal{M}$ *be the domain of secrets. A sharing function* Share *is a randomized mapping from* $\mathcal{M}$ *to* $\mathcal{S}_1 \times \mathcal{S}_2 \times \ldots \times \mathcal{S}_n$*, where* $\mathcal{S}_i$ *is called the domain of shares of party with identity* i*. A dealer distributes a secret* $m \in \mathcal{M}$ *by computing the vector* Share$(m) = (\mathsf{S}_1, \ldots, \mathsf{S}_n)$*, and privately communicating each share* S_i *to the party* i*. For a set* $T \subseteq [n]$*, we denote* Share$(m)_T$ *to be a restriction of* Share(m) *to its* T *entries.*

Definition 4 (($t, n, \epsilon_c, \epsilon_s$)-Secret Sharing Scheme). *Let* $\mathcal{M}$ *be a finite set of secrets, where* $|\mathcal{M}| \geq 2$*. Let* $[n] = \{1, 2, \ldots, n\}$ *be a set of identities (indices) of* n *parties. A sharing function* Share *with domain of secrets* $\mathcal{M}$ *is a* $(t, n, \epsilon_c, \epsilon_s)$*-secret sharing scheme if the following two properties hold:*

- **Correctness:** *The secret can be reconstructed by any* t*-out-of-*n *parties. That is, for any set* $T \subseteq [n]$ *such that* $|T| \geq t$*, there exists a deterministic reconstruction function* Rec $: \otimes_{i \in T} \mathcal{S}_i \to \mathcal{M}$ *such that for every* $m \in \mathcal{M}$*,*

$$\Pr[\mathsf{Rec}(\mathsf{Share}(m)_T) = m] = 1 - \epsilon_c$$

  where the probability is over the randomness of the Share *function. We will slightly abuse the notation and denote* Rec *as the reconstruction procedure that takes in* T *and* Share$(m)_T$ *where* T *is of size at least* t *and outputs the secret.*
- **Statistical Privacy:** *Any collusion of less than* t *parties should have "almost" no information about the underlying secret. More formally, for any unauthorized set* $U \subseteq [n]$ *such that* $|U| < t$*, and for every pair of secrets* $m_0, m_1 \in M$*, for any distinguisher* D *with output in* $\{0,1\}$*, the following holds:*

$$|\Pr[D(\mathsf{Share}(m_0)_U) = 1] - \Pr[D(\mathsf{Share}(m_1)_U) = 1]| \leq \epsilon_s$$

We define the rate of the secret sharing scheme as

$$\lim_{|m|\to\infty} \frac{|m|}{\max_{i\in[n]} |\mathsf{Share}(m)_i|}$$

Definition 5 (Threshold Non-Malleable Secret Sharing [GK18a]). *Let* $(\mathsf{Share}, \mathsf{Rec})$ *be a* $(t, n, \epsilon_c, \epsilon_s)$*-secret sharing scheme for message space* $\mathcal{M}$*. Let* $\mathcal{F}$ *be some family of tampering functions. For each* $f \in \mathcal{F}$*,* $m \in \mathcal{M}$ *and authorized set* $T \subseteq [n]$ *containing* t *indices, define the tampered distribution* $\mathsf{Tamper}_m^{f,T}$ *as* $\mathsf{Rec}(f(\mathsf{Share}(m))_T)$ *where the randomness is over the sharing function* Share*. We say that the* $(t, n, \epsilon_c, \epsilon_s)$*-secret sharing scheme,* $(\mathsf{Share}, \mathsf{Rec})$ *is* ϵ'*-non-malleable w.r.t.* $\mathcal{F}$ *if for each* $f \in \mathcal{F}$ *and any authorized set* T *consisting of* t *indices, there exists a distribution* $D^{f,T}$ *over* $\mathcal{M} \cup \{\mathsf{same}^\star\}$ *such that:*

$$|\mathsf{Tamper}_m^{f,T} - \mathrm{copy}(D^{f,T}, m)| \leq \epsilon'$$

where copy *is defined by* $\mathrm{copy}(x, y) = \begin{cases} x & \textit{if } x \neq \mathsf{same}^\star \\ y & \textit{if } x = \mathsf{same}^\star \end{cases}$.

Many Tampering Extension. We now extend the above definition to capture multiple tampering attacks. Informally, we say that a secret sharing scheme is non-malleable w.r.t. family $\mathcal{F}$ with tampering degree K if for any set of K functions $f_1, \ldots, f_\mathsf{K} \in \mathcal{F}$, the output of the following tampering experiment is independent of the shared message m: (i) we first share a secret m to obtain the corresponding shares, (ii) we tamper the shares using $f_1, \ldots, f_\mathsf{K}$, (iii) we finally, output the K-reconstructed tampered secrets. Note that in the above experiment the message m is secret shared only once but is subjected to K (possibly different) tamperings. We refer to the full version for the formal definition.

3.2 Non-Malleable Codes

Dziembowski, Pietrzak and Wichs [DPW10] introduced the notion of non-malleable codes which generalizes the usual notion of error correction. In particular, it guarantees that when a codeword is subject to tampering attack, the reconstructed message is either the original one or something that is independent of the original message.

Definition 6 (Non-Malleable Codes [DPW10]). *Let* $\mathsf{Enc} : \{0,1\}^m \to \{0,1\}^n$ *and* $\mathsf{Dec} : \{0,1\}^n \to \{0,1\}^m \cup \{\bot\}$ *be (possibly randomized) functions, such that* $\mathsf{Dec}(\mathsf{Enc}(s)) = s$ *with probability 1 for all* $s \in \{0,1\}^m$*. Let* $\mathcal{F}$ *be a family of tampering functions and fix* $\varepsilon > 0$*. We say that* $(\mathsf{Enc}, \mathsf{Dec})$ *is* ε*−non-malleable w.r.t.* $\mathcal{F}$ *if for every* $f \in \mathcal{F}$*, there exists a random variable* D_f *on* $\{0,1\}^m \cup \{\mathsf{same}^\star\}$*, such that for all* $s \in \{0,1\}^m$*,*

$$|\mathsf{Dec}(f(X_s)) - \mathrm{copy}(D_f, s)| \leq \epsilon$$

where $X_s \leftarrow \mathsf{Enc}(s)$ *and copy is defined by* $\mathrm{copy}(x, y) = \begin{cases} x & \textit{if } x \neq \mathsf{same}^\star \\ y & \textit{if } x = \mathsf{same}^\star \end{cases}$. *We call* n *the* length *of the code and* m/n *the* rate.

Chattopadhyay, Goyal and Li [CGL16] defined a stronger notion of non-malleability against multiple tampering. We recall this definition in the full version of the paper.

Split-state Tampering Functions. We focus on the *split-state* tampering model where the encoding scheme splits s into c states: $\mathsf{Enc}(s) = (\mathsf{S}_1, \ldots, \mathsf{S}_c) \in \mathcal{S}_1 \times \mathcal{S}_2 \ldots \times \mathcal{S}_c$ and the tampering family is $\mathcal{F}_{split} = \{(f_1, \ldots, f_c) | f_i : \mathcal{S}_i \to \mathcal{S}_i\}$. We will call such a code as c-split-state non-malleable code.

Augmented Non-Malleable Codes. We recall the definition of augmented, 2-split-state non-malleable codes [AAG+16].

Definition 7 (Augmented Non-Malleable Codes [AAG+16]). *A coding scheme* $(\mathsf{Enc}, \mathsf{Dec})$ *with code length $2n$ and message length m is an augmented 2-split-state non-malleable code with error ϵ if for every function $f, g : \{0, 1\}^n \to \{0, 1\}^n$, there exists a random variable $D_{(f,g)}$ on $\{0, 1\}^n \times (\{0, 1\}^m \cup \{\mathsf{same}^\star\})$ such that for all messages $s \in \{0, 1\}^m$, it holds that*

$$|(\mathsf{L}, \mathsf{Dec}(f(\mathsf{L}), g(\mathsf{R}))) - \mathcal{S}(D_{(f,g)}, s)| \leq \epsilon$$

where $(\mathsf{L}, \mathsf{R}) = \mathsf{Enc}(s)$, $(\mathsf{L}, \widetilde{m}) \leftarrow D_{f,g}$ *and* $\mathcal{S}((\mathsf{L}, \widetilde{m}), s)$ *outputs* (L, s) *if* $\widetilde{m} = \mathsf{same}^\star$ *and otherwise outputs* $(\mathsf{L}, \widetilde{m})$.

Explicit Constructions. We now recall the constructions of split-state non-malleable codes.

Theorem 6 ([Li17]). *For any $n \in \mathbb{N}$, there exists an explicit construction of 2-split-state non-malleable code with efficient encoder/decoder, code length $2n$, rate $O(\frac{1}{\log n})$ and error $2^{-\Omega(\frac{n}{\log n})}$.*

Theorem 7 ([KOS18, GMW17]). *For every $n \in \mathbb{N}$ and $\rho > 0$, there exists an explicit construction of 3-split-state non-malleable code with efficient encoder/decoder, code length $(3 + o(1))n$, rate $\frac{1}{3+o(1)}$ and error $2^{-\Omega(n/\log^{1+\rho}(n))}$.*

Theorem 8 ([CGL16]). *There exists a constant $\gamma > 0$ such that for every $n \in \mathbb{N}$ and $t \leq n^\gamma$, there exists an explicit construction of 2-split-state non-malleable code with an efficient encoder/decoder, tampering degree t, code length $2n$, rate $\frac{1}{n^{\Omega(1)}}$ and error $2^{-n^{\Omega(1)}}$.*

Theorem 9 ([GKP+18]). *There exists a constant $\gamma > 0$ such that for every $n \in \mathbb{N}$ and $t \leq n^\gamma$, there exists an explicit construction of an augmented, split-state non-malleable code with an efficient encoder/decoder, tampering degree t, code length $2n$, rate $\frac{1}{n^{\Omega(1)}}$ and error $2^{-n^{\Omega(1)}}$.*

Theorem 10. *There exists a constant $\gamma > 0$ such that for every $n \in \mathbb{N}$ and $t \leq n^\gamma$, there exists an explicit construction of 3-split-state non-malleable code with an efficient encoder/decoder, tampering degree t, code length $3n$, rate $\Theta(\frac{1}{t})$ and error $2^{-n^{\Omega(1)}}$.*

We give the proof of this theorem in the full version.

Additional Property. We show in the full version that the construction given in [KOS18, GMW17] satisfies the property that given two particular states of the codeword, the message remains statistically hidden.

4 k-out-of-n Leakage Resilient Secret Sharing Scheme

In this section, we give a new, rate-efficient construction of k-out-of-n leakage resilient secret sharing scheme for a constant k. Later, in Sect. 5, we will use this primitive along with a 3-split-state non-malleable code with explicit constant rate (see Theorem 7) from the works of Kanukurthi et al. [KOS18] and Gupta et al. [GMW17] to construct a t-out-of-n non-malleable secret sharing scheme with the above mentioned rate.

We first recall the definition of a leakage resilient secret sharing scheme from [GK18a].

Definition 8 (Leakage Resilient Secret Sharing [GK18a]). *A $(t, n, \epsilon_c, \epsilon_s)$ (for $t \geq 2$) secret sharing scheme (Share, Rec) for message space $\mathcal{M}$ is said to be ϵ-leakage resilient against a leakage family $\mathcal{F}$ if for all functions $f \in \mathcal{F}$ and for any two messages $m_0, m_1 \in \mathcal{M}$:*

$$|f(\mathsf{Share}(m_0)) - f(\mathsf{Share}(m_1))| \leq \epsilon$$

Leakage Function Family. We are interested in constructing leakage resilient secret sharing schemes against the specific function family $\mathcal{F}_{k, \overline{k}, \overrightarrow{\mu}} = \{f_{K, \overline{K}, \overrightarrow{\mu}} : K \subseteq [n], |K| = k, \overline{K} \subseteq K, |\overline{K}| \leq \overline{k}\}$ where $f_{K, \overline{K}, \overrightarrow{\mu}}$ on input $(\mathsf{share}_1, \ldots, \mathsf{share}_n)$ outputs share_i for each $i \in \overline{K}$ in the clear and outputs $f_i(\mathsf{share}_i)$ for every $i \in K \setminus \overline{K}$ such that f_i is an arbitrary function outputting μ_i bits. When we just write μ (without the vector sign), we mean that every function f_i outputs at most μ bits.

Organization. The rest of this section is organized as follows: we first construct a k-out-of-k leakage resilient secret sharing scheme against $\mathcal{F}_{k, k-1, \mu}$ (in other words, $k - 1$ shares are output in the clear and μ bits are leaked from the k-th share) in Sect. 4.1. In Sect. 4.2, we recall the definition of a combinatorial object called as *perfect hash function family* and give a randomized construction of such a family. Next, in Sect. 4.3, we combine the construction of k-out-of-k leakage resilient secret sharing scheme and a perfect hash function family to give a construction of k-out-of-n leakage resilient secret sharing scheme (for a constant k).

4.1 k-out-of-k Leakage Resilient Secret Sharing

In this subsection, we will construct a k-out-k leakage resilient secret sharing scheme against $\mathcal{F}_{k, k-1, \mu}$ for an arbitrary $k \geq 2$ (and not just for a constant k). As a building block, we will use a 2-out-of-2 leakage resilient secret sharing which was constructed in [GK18a]. We first recall the lemma regarding this construction.

Lemma 1 ([GK18a]). *For any $\epsilon > 0$ and $\mu, m \in \mathbb{N}$, there exists a construction of $(2, 2, 0, 0)$ secret sharing scheme for sharing m-bit secrets that is ϵ-leakage resilient against $\mathcal{F}_{2,1,\mu}$ such that the size of each share is $O(m + \mu + \log \frac{1}{\epsilon})$. The running time of the sharing and reconstruction procedures are $\mathrm{poly}(m, \mu, \log(1/\epsilon))$.*

Let us denote the secret sharing scheme guaranteed by Lemma 1 as $(\mathsf{LRShare}_{(2,2)}, \mathsf{LRRec}_{(2,2)})$. We will use this to construct a k-out-of-k leakage resilient secret sharing scheme for $k > 2$.

Lemma 2. *For any $\epsilon > 0$, $k \geq 2$ and $\mu, m \in \mathbb{N}$, there exists a construction of $(k, k, 0, 0)$ secret sharing scheme for sharing m-bit secrets that is ϵ-leakage resilient against $\mathcal{F}_{k,k-1,\mu}$ such that the size of each share is $O(m + \mu + \log \frac{1}{\epsilon})$. The running time of the sharing and the reconstruction procedures are $\mathrm{poly}(m, \mu, k, \log(1/\epsilon))$.*

We give the proof of this Lemma in the full version.

4.2 Perfect Hash Function Family

In this subsection, we recall the definition of the combinatorial objects called as *perfect hash function family* and give an efficient randomized construction for constant k.

Definition 9 (Perfect Hash Function Family [FK84]). *For every $n, k \in \mathbb{N}$, a set of hash functions $\{h_i\}_{i \in [\ell]}$ where $h_i : [n] \to [k]$ is said to be (n, k)-perfect hash function family if for each subset $S \subseteq [n]$ of size k there exists an $i \in [\ell]$ such that h_i is injective on S.*

Before we give the randomized construction, we will state and prove the following useful lemma.

Lemma 3. *For every $\epsilon > 0$, $n, k \in \mathbb{N}$, the set of functions $\{h_i\}_{i \in [\ell]}$ where each h_i is chosen randomly from the set of all functions mapping $[n] \to [k]$ is a perfectly hash function family with probability $1 - \epsilon$ when $\ell = \frac{\log \binom{n}{k} + \log \frac{1}{\epsilon}}{\log \frac{1}{1 - \frac{k!}{k^k}}}$. Specifically, when k is constant, we can set $\ell = O(\log n + \log \frac{1}{\epsilon})$.*

Proof. Let us first fix a subset $S \subseteq [n]$ of size k. Let us choose a function h uniformly at random from the set of all functions mapping $[n] \to [k]$.

$$\Pr[h \text{ is not injective over } S] = 1 - \frac{k!}{k^k}$$

Let us now choose $h_1, \ldots, h_\ell$ uniformly at random from the set of all functions mapping $[n] \to [k]$.

$$\Pr[\forall\ i \in [\ell],\ h_i \text{ is not injective over } S] = (1 - \frac{k!}{k^k})^\ell$$

By union bound,

$$\Pr[\exists\ S\ \text{s.t.},\forall\ i \in [\ell],\ h_i\ \text{is not injective over}\ S] = \binom{n}{k}(1 - \frac{k!}{k^k})^\ell$$

We want $\binom{n}{k}(1 - \frac{k!}{k^k})^\ell = \epsilon$. We get the bound for ℓ by rearranging this equation.

Randomized Construction for Constant k. For any k, n and some error parameter ϵ, set ℓ as in Lemma 3. Choose a function $h_i : [n] \to [k]$ uniformly at random for each $i \in [\ell]$. From Lemma 3, we infer that $\{h_i\}_{i \in [\ell]}$ is a perfect hash function family except with probability ϵ. The construction is efficient since the number of random bits needed for choosing each h_i is $n \log k$ which is polynomial in n when k is a constant.

Explicit Construction. Building on the work of Schmidt and Siegal [SS90], Alon et al. [AYZ95] gave an explicit construction of (n, k)-perfect hash function family of size $2^{O(k)} \log n$. We now recall the lemma from [AYZ95].

Lemma 4 ([AYZ95,SS90]). *For every $n, k \in \mathbb{N}$, there exists an explicit and efficiently computable construction of (n, k)-perfect hash function family $\{h_i\}_{i \in [\ell]}$ where $\ell = 2^{O(k)} \log n$.*

The explicit construction is obtained by brute forcing over a small bias probability space [NN93] and finding such a family is not as efficient as our randomized construction. On the positive side, the explicit construction is error-free unlike our randomized construction.

4.3 Construction of k-out-n Leakage Resilient Secret Sharing

In this subsection, we will use a k-out-of-k leakage resilient secret sharing scheme from Sect. 4.1 and a perfect hash function family from Sect. 4.2 to construct a k-out-of-n leakage resilient secret sharing scheme against $\mathcal{F}_{t,k-1,\overrightarrow{\mu}}$ for an arbitrary $t \leq n$ (recall the definition of $\mathcal{F}_{k,\overline{k},\overrightarrow{\mu}}$ from Definition 8). We give the description in Fig. 1.

Theorem 11. *For every $\epsilon_c, \epsilon_s > 0$, $n, k, m \in \mathbb{N}$ and $\overrightarrow{\mu} \in \mathbb{N}^n$, the construction given in Fig. 1 is a $(k, n, \epsilon_c, 0)$ secret sharing scheme for sharing m-bit secrets that is ϵ_s-leakage resilient against leakage functions $\mathcal{F}_{t,k-1,\overrightarrow{\mu}}$ for any $t \leq n$. The running times of the sharing and reconstruction algorithms are $\mathrm{poly}(n, m, \sum_i \mu_i, \log(1/\epsilon_c\epsilon_s))$ when k is a constant. In particular, when $\epsilon_s = \epsilon_c = 2^{-m}$, the running times are $\mathrm{poly}(n, m, \sum_i \mu_i)$. The size of each share when k is a constant is $O((m + \max_T \sum_{i \in T, T \subseteq [n], |T| = t} \mu_i + \log(\log n/\epsilon_s)) \log n)$.*

We give the proof of this theorem in the full version.

Remark 1. In Fig. 1, we cannot directly set the size $\ell = O(\log n + \log \frac{1}{\epsilon_c})$ and perform a single sampling to find a perfect hash function family. This is because when we want $\epsilon_c = 2^{-m}$, the size of the function family grows with m and this affects the rate significantly. That is why, it is important to set $\epsilon = 1/2$ and do $\log \frac{1}{\epsilon_c}$ independent repetitions in the $\mathsf{LRShare}_{(k,n)}$ function to reduce the error to ϵ_c.

Let $(\mathsf{LRShare}_{(k,k)}, \mathsf{LRRec}_{(k,k)})$ be a k-out-of-k leakage resilient secret sharing scheme.

$\mathsf{LRShare}_{(k,n)}$: To share a secret s:
1. For each trial $\in [1, \log(1/\epsilon_c)]$ do:
   (a) Set $\epsilon = 1/2$ and $\ell = O(\log n)$. Sample a (candidate) (n, k)-perfect hash function family $\{h_i\}_{i \in [\ell]}$ as described in Section 4.2
   (b) Check if $\{h_i\}_{i \in [\ell]}$ is a family of (n, k)-perfect hash functions. That is, for each set $S \subset [n]$ and $|S| = k$, check if there exists an $i \in [\ell]$ such that h_i is injective on S.
   (c) If yes, exit the loop. Otherwise, go to the beginning.
2. If the above loop fails to find a perfect hash function family then abort.
3. For each $i \in [\ell]$, sample $\overline{\mathsf{share}}_{i,1}, \ldots, \overline{\mathsf{share}}_{i,k} \leftarrow \mathsf{LRShare}_{(k,k)}(s)$.
4. For each $j \in [n]$, set $\mathsf{share}_j = (h_1(j), \overline{\mathsf{share}}_{1,h_1(j)}) \circ (h_2(j), \overline{\mathsf{share}}_{2,h_2(j)}) \circ \ldots \circ (h_\ell(j), \overline{\mathsf{share}}_{\ell,h_\ell(j)})$.

$\mathsf{LRRec}_{(k,n)}$: Given the shares $\mathsf{share}_{j_1}, \mathsf{share}_{j_2}, \ldots, \mathsf{share}_{j_k}$ do:
1. Choose an $i \in [\ell]$, such that $\{h_i(j_1), h_i(j_2), \ldots, h_i(j_k)\} = \{1, \ldots, k\}$.
2. Recover s as $\mathsf{LRRec}_{(k,k)}(\overline{\mathsf{share}}_{i,1}, \ldots, \overline{\mathsf{share}}_{i,k})$.

Fig. 1. $(k, n, \epsilon_c, 0)$ Leakage resilient secret sharing scheme

5 Non-Malleable Secret Sharing for Threshold Access Structures

In this section, we give a construction of t-out-of-n (for any $t \geq 4$) Non-Malleable Secret Sharing scheme with rate $\Theta(\frac{1}{t \log^2 n})$ against tampering function family $\mathcal{F}_{\mathsf{ind}}$ that tampers each share independently. We first give the formal description of the tampering function family.

Individual Tampering Family $\mathcal{F}_{\mathsf{ind}}$. Let Share be the sharing function of the secret sharing scheme that outputs n-shares in $\mathcal{S}_1 \times \mathcal{S}_2 \ldots \times \mathcal{S}_n$. The function family $\mathcal{F}_{\mathsf{ind}}$ is composed of functions $(f_1, \ldots, f_n)$ where each $f_i : \mathcal{S}_i \to \mathcal{S}_i$.

5.1 Construction

Building Blocks. The construction uses the following building blocks. We instantiate them with concrete schemes later:

- A 3-split-state non-malleable code $(\mathsf{Enc}, \mathsf{Dec})$ where $\mathsf{Enc} : \mathcal{M} \to \mathcal{L} \times \mathcal{C} \times \mathcal{R}$ and the simulation error of the scheme is ϵ_1. Furthermore, we assume that for any two messages $m, m' \in \mathcal{M}$, $(\mathsf{C}, \mathsf{R}) \approx_{\epsilon_2} (\mathsf{C}', \mathsf{R}')$ where $(\mathsf{L}, \mathsf{C}, \mathsf{R}) \leftarrow \mathsf{Enc}(m)$ and $(\mathsf{L}', \mathsf{C}', \mathsf{R}') \leftarrow \mathsf{Enc}(m')$.
- A $(t, n, 0, 0)$ secret sharing scheme $(\mathsf{SecShare}_{(t,n)}, \mathsf{SecRec}_{(t,n)})$ with perfect privacy for message space $\mathcal{L}$. We will assume that the size of each share is m_1.

- A $(3, n, \epsilon'_3, 0)$ secret sharing scheme $(\mathsf{LRShare}_{(3,n)}, \mathsf{LRRec}_{(3,n)})$ that is ϵ_3-leakage resilient against leakage functions $\mathcal{F}_{t,2,m_1}$[8] for message space $\mathcal{C}$. We assume that the size of each share is m_2.
- A $(2, n, \epsilon'_4, 0)$ secret sharing scheme $(\mathsf{LRShare}_{(2,n)}, \mathsf{LRRec}_{(2,n)})$ for message space $\mathcal{R}$ that is ϵ_4-leakage resilient against leakage functions $\mathcal{F}_{t,1,\vec{\mu}}$ where $\max_T \sum_{i \in T, T \subseteq [n], |T| = t} \mu_i = O(m_2 + tm_1)$. We assume that the size of each share is m_3.

Construction. We give the formal description of the construction in Fig. 2 and give an informal overview below. To share a secret s, we first encode s to $(\mathsf{L}, \mathsf{C}, \mathsf{R})$ using the 3-split-state non-malleable code. We first encode L to $(\mathsf{SL}_1, \ldots, \mathsf{SL}_n)$ using the t-out-of-n threshold secret sharing scheme. We then encode C into $(\mathsf{SC}_1, \ldots, \mathsf{SC}_n)$ using the 3-out-of-n leakage resilience secret sharing scheme $\mathsf{LRShare}_{(3,n)}$. We finally encode R into $(\mathsf{SR}_1, \ldots, \mathsf{SR}_n)$ using the 2-out-of-n leakage resilient secret sharing scheme $\mathsf{LRShare}_{(2,n)}$. We set the i-th share share_i to be the concatenation of $\mathsf{SL}_i, \mathsf{SC}_i$ and SR_i. In order to reconstruct, we using the corresponding reconstruction procedures $\mathsf{SecRec}, \mathsf{LRRec}_{(3,n)}$ and $\mathsf{LRRec}_{(2,n)}$ to compute L, C and R respectively. We finally use the decoding procedure of 3-split-state non-malleable code to reconstruct the secret s from L, C and R.

Theorem 12. *For any arbitrary $n \in \mathbb{N}$ and threshold $t \geq 4$, the construction given in Fig. 2 is a $(t, n, \epsilon'_3 + \epsilon'_4, \epsilon_2)$ secret sharing scheme. Furthermore, it is $(\epsilon_1 + \epsilon_3 + \epsilon_4)$-non-malleable against $\mathcal{F}_{\mathsf{ind}}$.*

We give the proof of this theorem in the full version.

5.2   Rate Analysis

We now instantiate the primitives and provide the rate analysis.

1. We instantiate the three split state non-malleable code from the works of [KOS18,GMW17] (see Theorem 7). Using their construction, the $|\mathsf{L}| = |\mathsf{C}| = |\mathsf{R}| = O(m)$ bits and the error $\epsilon_1 = 2^{-\Omega(m/\log^{1+\rho}(m))}$ for any $\rho > 0$.
2. We use Shamir's secret sharing [Sha79] as the t-out-of-n secret sharing scheme. We get $m_1 = O(m)$ whenever $m > \log n$.
3. We instantiate $(\mathsf{LRShare}_{(3,n)}, \mathsf{LRRec}_{(3,n)})$ and $(\mathsf{LRShare}_{(2,n)}, \mathsf{LRRec}_{(2,n)})$ from Theorem 11. We get $m_2 = O(mt \log n)$ and $m_3 = O(mt \log^2 n)$ by setting ϵ_3 and ϵ_4 to be $2^{-\Omega(m/\log m)}$.

Thus the rate of our construction is $\Theta(\frac{1}{t \log^2 n})$ and the error is $2^{-\Omega(m/\log^{1+\rho}(m))}$.

We defer the concrete optimization of the rate of our construction to the full version of the paper.

---

[8] Recall that this denotes that the function can choose to leak at most m_1 bits from each share in a set of size $t - 2$ apart from the two that are completely leaked.

Share(m) : To share a secret $s \in \mathcal{M}$ do:
1. Encode the secret s as $(L, C, R) \leftarrow \mathsf{Enc}(s)$.
2. Compute the shares

$$(SL_1, \ldots, SL_n) \leftarrow \mathsf{SecShare}_{(t,n)}(L)$$

$$(SC_1, \ldots, SC_n) \leftarrow \mathsf{LRShare}_{(3,n)}(C)$$

$$(SR_1, \ldots, SR_n) \leftarrow \mathsf{LRShare}_{(2,n)}(R)$$

3. For each $i \in [n]$, set share_i as (SL_i, SC_i, SR_i) and output $(\mathsf{share}_1, \ldots, \mathsf{share}_n)$ as the shares.

Rec($\mathsf{Share}(m)_T$) : To reconstruct the secret from the shares in an authorized set T of size t do:
1. Let the shares corresponding to the set T be $(\mathsf{share}_{i_1}, \ldots, \mathsf{share}_{i_t})$.
2. For each $j \in \{i_1, \ldots, i_t\}$, parse share_j as (SL_j, SC_j, SR_j).
3. Reconstruct

$$L := \mathsf{SecRec}_{(t,n)}(SL_{i_1}, \ldots, SL_{i_t})$$

$$C := \mathsf{LRRec}_{(3,n)}(SC_{i_1}, SC_{i_2}, SC_{i_3})$$

$$R := \mathsf{LRRec}_{(2,n)}(SR_{i_1}, SR_{i_2})$$

4. Output the secret s as $\mathsf{Dec}(L, C, R)$.

Fig. 2. Construction of t-out-of-n non-malleable secret sharing scheme

6 NMSS for General Access Structures with Multiple Tampering

We first define non-malleable secret sharing for general access structures in the next subsection and then give the construction in the subsequent subsection.

6.1 Definitions

First, we recall the definition of a secret sharing scheme for a general monotone access structure $\mathcal{A}$ - a generalization of the one defined for threshold access structures in Definition 4.

Definition 10 (($\mathcal{A}, n, \epsilon_c, \epsilon_s$)-Secret Sharing Scheme). *Let $\mathcal{M}$ be a finite set of secrets, where $|\mathcal{M}| \geq 2$. Let $[n] = \{1, 2, \ldots, n\}$ be a set of identities (indices) of n parties. A sharing function* Share *with domain of secrets $\mathcal{M}$ is a $(\mathcal{A}, n, \epsilon_c, \epsilon_s)$-secret sharing scheme with respect to monotone access structure $\mathcal{A}$ if the following two properties hold:*

- **Correctness:** *The secret can be reconstructed by any set of parties that are part of the access structure $\mathcal{A}$. That is, for any set $T \in \mathcal{A}$, there exists a deterministic reconstruction function* Rec $: \otimes_{i \in T} \mathcal{S}_i \rightarrow \mathcal{M}$ *such that for every $m \in \mathcal{M}$,*

$$\Pr[\mathsf{Rec}(\mathsf{Share}(m)_T) = m] = 1 - \epsilon_c$$

where the probability is over the randomness of the Share *function. We will slightly abuse the notation and denote* Rec *as the reconstruction procedure that takes in* $T \in \mathcal{A}$ *and* Share$(m)_T$ *as input and outputs the secret.*

- **Statistical Privacy:** *Any collusion of parties not part of the access structure should have "almost" no information about the underlying secret. More formally, for any unauthorized set* $U \subseteq [n]$ *such that* $U \notin \mathcal{A}$, *and for every pair of secrets* $m_0, m_1 \in M$, *for any distinguisher* D *with output in* $\{0, 1\}$, *the following holds:*

$$| \Pr[D(\text{Share}(m_0)_U) = 1] - \Pr[D(\text{Share}(m_1)_U) = 1]| \le \epsilon_s$$

We define the rate of the secret sharing scheme as $\frac{|m|}{\max_{i \in [n]} |\text{Share}(m)_i|}$

We now define the notion of a non-malleable secret sharing scheme for general access structures which is a generalization of the definition for threshold access structures given in Definition 5.

Definition 11 (Non-Malleable Secret Sharing for General Access Structures [GK18b]). *Let* (Share, Rec) *be a* $(\mathcal{A}, n, \epsilon_c, \epsilon_s)$-*secret sharing scheme for message space* M *and access structure* $\mathcal{A}$. *Let* $\mathcal{F}$ *be a family of tampering functions. For each* $f \in \mathcal{F}$, $m \in M$ *and authorized set* $T \in \mathcal{A}$, *define the tampered distribution* Tamper$_m^{f,T}$ *as* Rec$(f(\text{Share}(m))_T)$ *where the randomness is over the sharing function* Share. *We say that the* $(\mathcal{A}, n, \epsilon_c, \epsilon_s)$-*secret sharing scheme,* (Share, Rec) *is* ϵ'-*non-malleable w.r.t.* $\mathcal{F}$ *if for each* $f \in \mathcal{F}$ *and any authorized set* $T \in \mathcal{A}$, *there exists a distribution* $D^{f,T}$ *over* $M \cup \{\text{same}^\star\}$ *such that:*

$$|\text{Tamper}_m^{f,T} - \text{copy}(D^{f,T}, m)| \le \epsilon'$$

where copy *is defined by* $\text{copy}(x, y) = \begin{cases} x & \text{if } x \ne \text{same}^\star \\ y & \text{if } x = \text{same}^\star \end{cases}$.

Many Tampering Extension. Similar to the threshold case, in the full version, we extend the above definition to capture multiple tampering attacks.

6.2 Construction

In this section, we show how to build a one-many non-malleable secret sharing scheme for general access structures.

First, let (SecShare$_{(\mathcal{A},n)}$, SecRec$_{(\mathcal{A},n)}$) be any statistically private secret sharing scheme with rate R for a 4-monotone access structure $\mathcal{A}$ over n parties. We refer the reader to [KW93, LV18] for explicit constructions.

Let $t_{\max}$ denote the maximum size of a minimal authorized set of $\mathcal{A}$.[9] We give a construction of a Non-Malleable Secret Sharing scheme with tampering degree K for a 4-monotone access structure $\mathcal{A}$ with rate $O(\frac{R}{K^3 t_{\max} \log^2 n})$ with respect to a individual tampering function family $\mathcal{F}_{ind}$.

---

[9] We refer the reader to Definition 1, Definition 2 for definitions of 4-monotone access structures and minimal authorized set.

Building Blocks. The construction uses the following building blocks. We instantiate them with concrete schemes later:

- A one-many 3-split-state non-malleable code (Enc, Dec) where $\mathsf{Enc} : \mathcal{M} \to \mathcal{L} \times \mathcal{C} \times \mathcal{R}$, the simulation error of the scheme is ϵ_1 and the scheme is secure against K tamperings. Furthermore, we assume that for any two messages $m, m' \in \mathcal{M}$, $(\mathsf{C}, \mathsf{R}) \approx_{\epsilon_2} (\mathsf{C}', \mathsf{R}')$ where $(\mathsf{L}, \mathsf{C}, \mathsf{R}) \leftarrow \mathsf{Enc}(m)$ and $(\mathsf{L}', \mathsf{C}', \mathsf{R}') \leftarrow \mathsf{Enc}(m')$.
- A $(\mathcal{A}, n, 0, 0)$ (where $\mathcal{A}$ is 4-monotone) secret sharing scheme $(\mathsf{SecShare}_{(\mathcal{A},n)}, \mathsf{SecRec}_{(\mathcal{A},n)})$ with perfect privacy for message space $\mathcal{L}$.[10] We will assume that the size of each share is m_1.
- A $(3, n, \epsilon_3', 0)$ secret sharing scheme $(\mathsf{LRShare}_{(3,n)}, \mathsf{LRRec}_{(3,n)})$ that is ϵ_3-leakage resilient against leakage functions $\mathcal{F}_{t_{\max}, 2, Km_1}$ for message space $\mathcal{C}$. We assume that the size of each share is m_2.
- A $(2, n, \epsilon_4', 0)$ secret sharing scheme $(\mathsf{LRShare}_{(2,n)}, \mathsf{LRRec}_{(2,n)})$ for message space $\mathcal{R}$ that is ϵ_4-leakage resilient against leakage functions $\mathcal{F}_{t_{\max}, 1, \vec{\mu}}$ where $\max_T \sum_{i \in T, T \in \mathcal{A}, |T| = t_{\max}} \mu_i = O(Km_2 + Kt_{\max}m_1)$. We assume that the size of each share is m_3.

$\mathsf{Share}(m)$: To share a secret $s \in \mathcal{M}$ do:
1. Encode the secret s as $(\mathsf{L}, \mathsf{C}, \mathsf{R}) \leftarrow \mathsf{Enc}(s)$.
2. Compute the shares

$$(\mathsf{SL}_1, \ldots, \mathsf{SL}_n) \leftarrow \mathsf{SecShare}_{(\mathcal{A},n)}(\mathsf{L})$$

$$(\mathsf{SC}_1, \ldots, \mathsf{SC}_n) \leftarrow \mathsf{LRShare}_{(3,n)}(\mathsf{C})$$

$$(\mathsf{SR}_1, \ldots, \mathsf{SR}_n) \leftarrow \mathsf{LRShare}_{(2,n)}(\mathsf{R})$$

3. For each $i \in [n]$, set share_i as $(\mathsf{SL}_i, \mathsf{SC}_i, \mathsf{SR}_i)$ and output $(\mathsf{share}_1, \ldots, \mathsf{share}_n)$ as the set of shares.

$\mathsf{Rec}(\mathsf{Share}(m)_T)$: Given a set of shares in an authorized set $T' \in \mathcal{A}$, let $T \subseteq T'$ denote a minimal authorized set. To reconstruct the secret from the shares in set T, (of size at most $t_{\max}$) do:
1. Let the shares corresponding to the set T be $(\mathsf{share}_{i_1}, \ldots, \mathsf{share}_{i_{t_{\max}}})$.
2. For each $j \in \{i_1, \ldots, i_{t_{\max}}\}$, parse share_j as $(\mathsf{SL}_j, \mathsf{SC}_j, \mathsf{SR}_j)$.
3. Reconstruct

$$\mathsf{L} := \mathsf{SecRec}_{(\mathcal{A},n)}(\mathsf{SL}_{i_1}, \ldots, \mathsf{SL}_{i_{t_{\max}}})$$

$$\mathsf{C} := \mathsf{LRRec}_{(3,n)}(\mathsf{SC}_{i_1}, \mathsf{SC}_{i_2}, \mathsf{SC}_{i_3})$$

$$\mathsf{R} := \mathsf{LRRec}_{(2,n)}(\mathsf{SR}_{i_1}, \mathsf{SR}_{i_2})$$

4. Output the secret s as $\mathsf{Dec}(\mathsf{L}, \mathsf{C}, \mathsf{R})$.

Fig. 3. Construction of non-malleable secret sharing scheme for general access structures against multiple tampering

---

[10] We note that our proof of security goes through even if this secret sharing scheme only has statistical privacy.

Construction. The construction is very similar to the construction of non-malleable secret sharing for threshold access structures given in Sect. 5 with the only difference being that we now use the $(\mathcal{A}, n, 0, 0)$ secret sharing scheme. Note that in the construction we additionally need a procedure to find a minimal authorized set from any authorized set. This procedure is efficient if we can efficiently test the membership in $\mathcal{A}$. We point the reader to [GK18b] for details of this procedure. We give the formal description of the construction in Fig. 3 for completeness.

Theorem 13. *There exists a constant $\gamma > 0$ such that, for any arbitrary $n, \mathsf{K} \in \mathbb{N}$ and 4-monotone access structure $\mathcal{A}$, the construction given in Fig. 3 is a $(\mathcal{A}, n, \epsilon'_3 + \epsilon'_4, \epsilon_2)$ secret sharing scheme for messages of length m where $m \geq \mathsf{K}^\gamma$. Furthermore, it is $(\epsilon_1 + \epsilon_3 + \epsilon_4)$ one-many non-malleable with tampering degree K with respect to tampering function family $\mathcal{F}_{ind}$.*

We give the proof of this theorem and the rate analysis in the full version.

Acknowledgements. The first author's research supported in part by the IBM PhD Fellowship. The first author's research also supported in part from a DARPA /ARL SAFEWARE award, NSF Frontier Award 1413955, and NSF grant 1619348, BSF grant 2012378, a Xerox Faculty Research Award, a Google Faculty Research Award, an equipment grant from Intel, an Okawa Foundation Research Grant, NSF-BSF grant 1619348, DARPA SafeWare subcontract to Galois Inc., DARPA SPAWAR contract N66001-15-1C-4065, US-Israel BSF grant 2012366, OKAWA Foundation Research Award, IBM Faculty Research Award, Xerox Faculty Research Award, B. John Garrick Foundation Award, Teradata Research Award, and Lockheed-Martin Corporation Research Award. This material is based upon work supported by the Defense Advanced Research Projects Agency through the ARL under Contract W911NF-15-C- 0205. The second author's research supported in part from DARPA/ARL SAFEWARE Award W911NF15C0210, AFOSR Award FA9550-15-1-0274, AFOSR YIP Award, DARPA and SPAWAR under contract N66001-15-C-4065, a Hellman Award and research grants by the Okawa Foundation, Visa Inc., and Center for Long-Term Cybersecurity (CLTC, UC Berkeley) of Sanjam Garg. The views expressed are those of the authors and do not reflect the official policy or position of the funding agencies.

The authors thank Pasin Manurangsi for pointing to the work of Alon et al. [AYZ95] for the explicit construction of perfect hash function family. The authors also thank Sanjam Garg, Peihan Miao and Prashant Vasudevan for useful comments on the write-up.

References

[AAG+16] Aggarwal, D., Agrawal, S., Gupta, D., Maji, H.K., Pandey, O., Prabhakaran, M.: Optimal computational split-state non-malleable codes. In: Kushilevitz, E., Malkin, T. (eds.) TCC 2016. LNCS, vol. 9563, pp. 393–417. Springer, Heidelberg (2016). https://doi.org/10.1007/978-3-662-49099-0_15

[ADKO15] Aggarwal, D., Dodis, Y., Kazana, T., Obremski, M.: Non-malleable reductions and applications. In: STOC, pp. 459–468 (2015)

[ADL14] Aggarwal, D., Dodis, Y., Lovett, S.: Non-malleable codes from additive combinatorics. In: STOC, pp. 774–783 (2014)

[ADN+18] Aggarwal, D., et al.: Stronger leakage-resilient and non-malleable secret-sharing schemes for general access structures. Cryptology ePrint Archive, Report 2018/1147 (2018). https://eprint.iacr.org/2018/1147

[AGM+15] Agrawal, S., Gupta, D., Maji, H.K., Pandey, O., Prabhakaran, M.: Explicit non-malleable codes against bit-wise tampering and permutations. In: Gennaro, R., Robshaw, M. (eds.) CRYPTO 2015. LNCS, vol. 9215, pp. 538–557. Springer, Heidelberg (2015). https://doi.org/10.1007/978-3-662-47989-6_26

[AYZ95] Alon, N., Yuster, R., Zwick, U.: Color-coding. J. ACM $42(4)$, 844–856 (1995)

[BDG+18] Ball, M., Dachman-Soled, D., Guo, S., Malkin, T., Tan, L.-Y.: Non-malleable codes for small-depth circuits. In: FOCS (2018, to appear)

[BDIR18] Benhamouda, F., Degwekar, A., Ishai, Y., Rabin, T.: On the local leakage resilience of linear secret sharing schemes. In: Shacham, H., Boldyreva, A. (eds.) CRYPTO 2018, Part I. LNCS, vol. 10991, pp. 531–561. Springer, Cham (2018). https://doi.org/10.1007/978-3-319-96884-1_18

[BDKM16] Ball, M., Dachman-Soled, D., Kulkarni, M., Malkin, T.: Non-malleable codes for bounded depth, bounded fan-in circuits. In: Fischlin, M., Coron, J.-S. (eds.) EUROCRYPT 2016. LNCS, vol. 9666, pp. 881–908. Springer, Heidelberg (2016). https://doi.org/10.1007/978-3-662-49896-5_31

[BDKM18] Ball, M., Dachman-Soled, D., Kulkarni, M., Malkin, T.: Non-malleable codes from average-case hardness: AC^0, decision trees, and streaming space-bounded tampering. In: Nielsen, J.B., Rijmen, V. (eds.) EUROCRYPT 2018. LNCS, vol. 10822, pp. 618–650. Springer, Cham (2018). https://doi.org/10.1007/978-3-319-78372-7_20

[BDL01] Boneh, D., DeMillo, R.A., Lipton, R.J.: On the importance of eliminating errors in cryptographic computations. J. Cryptol. $14(2)$, 101–119 (2001)

[BGW88] Ben-Or, M., Goldwasser, S., Wigderson, A.: Completeness theorems for non-cryptographic fault-tolerant distributed computation (extended abstract). In: STOC, pp. 1–10 (1988)

[Bla79] Blakley, G.R.: Safeguarding cryptographic keys. In: Proceedings of AFIPS 1979 National Computer Conference, vol. 48, pp. 313–317 (1979)

[Bla99] Blackburn, S.R.: Combinatorics and Threshold Cryptography. Chapman and Hall CRC Research Notes in Mathematics, pp. 49–70 (1999)

[CCD88] Chaum, D., Crepeau, C., Damgaard, I.: Multiparty unconditionally secure protocols (extended abstract). In: STOC, pp. 11–19. ACM (1988)

[CDF+08] Cramer, R., Dodis, Y., Fehr, S., Padró, C., Wichs, D.: Detection of algebraic manipulation with applications to robust secret sharing and fuzzy extractors. In: Smart, N. (ed.) EUROCRYPT 2008. LNCS, vol. 4965, pp. 471–488. Springer, Heidelberg (2008). https://doi.org/10.1007/978-3-540-78967-3_27

[CG88] Chor, B., Goldreich, O.: Unbiased bits from sources of weak randomness and probabilistic communication complexity. SIAM J. Comput. $17(2)$, 230–261 (1988)

[CGL16] Chattopadhyay, E., Goyal, V., Li, X.: Non-malleable extractors and codes, with their many tampered extensions. In: Proceedings of the 48th Annual ACM SIGACT Symposium on Theory of Computing, STOC 2016, Cambridge, MA, USA, 18–21 June 2016, pp. 285–298 (2016)

[CGM+16] Chandran, N., Goyal, V., Mukherjee, P., Pandey, O., Upadhyay, J.: Block-wise non-malleable codes. In: ICALP (2016)

[CGMA85] Chor, B., Goldwasser, S., Micali, S., Awerbuch, B.: Verifiable secret sharing and achieving simultaneity in the presence of faults (extended abstract). In: 26th Annual Symposium on Foundations of Computer Science, pp. 383–395. IEEE Computer Society Press, October 1985

[CKR16] Chandran, N., Kanukurthi, B., Raghuraman, S.: Information-theoretic local non-malleable codes and their applications. In: Kushilevitz, E., Malkin, T. (eds.) TCC 2016, Part II. LNCS, vol. 9563, pp. 367–392. Springer, Heidelberg (2016). https://doi.org/10.1007/978-3-662-49099-0_14

[CL17] Chattopadhyay, E., Li, X.: Non-malleable codes and extractors for small-depth circuits, and affine functions. In: Hatami, H., McKenzie, P., King, V. (eds.) 49th Annual ACM Symposium on Theory of Computing, pp. 1171–1184. ACM Press, June 2017

[DDFY94] De Santis, A., Desmedt, Y., Frankel, Y., Yung, M.: How to share a function securely. In: 26th Annual ACM Symposium on Theory of Computing, pp. 522–533. ACM Press, May 1994

[DF90] Desmedt, Y., Frankel, Y.: Threshold cryptosystems. In: Brassard, G. (ed.) CRYPTO 1989. LNCS, vol. 435, pp. 307–315. Springer, New York (1990). https://doi.org/10.1007/0-387-34805-0_28

[DKO13] Dziembowski, S., Kazana, T., Obremski, M.: Non-malleable codes from two-source extractors. In: Canetti, R., Garay, J.A. (eds.) CRYPTO 2013, Part II. LNCS, vol. 8043, pp. 239–257. Springer, Heidelberg (2013). https://doi.org/10.1007/978-3-642-40084-1_14

[DPW10] Dziembowski, S., Pietrzak, K., Wichs, D.: Non-malleable codes. In: Proceedings of Innovations in Computer Science - ICS 2010, Tsinghua University, Beijing, China, 5–7 January 2010, pp. 434–452 (2010)

[FK84] Fredman, M.L., Komlós, J.: On the size of separating systems and families of perfect hash functions. SIAM J. Algebraic Discrete Methods 5(1), 61–68 (1984)

[FMNV14] Faust, S., Mukherjee, P., Nielsen, J.B., Venturi, D.: Continuous non-malleable codes. In: Lindell, Y. (ed.) TCC 2014. LNCS, vol. 8349, pp. 465–488. Springer, Heidelberg (2014). https://doi.org/10.1007/978-3-642-54242-8_20

[FMNV15] Faust, S., Mukherjee, P., Nielsen, J.B., Venturi, D.: A tamper and leakage resilient von neumann architecture. In: Katz, J. (ed.) PKC 2015. LNCS, vol. 9020, pp. 579–603. Springer, Heidelberg (2015). https://doi.org/10.1007/978-3-662-46447-2_26

[FMVW14] Faust, S., Mukherjee, P., Venturi, D., Wichs, D.: Efficient non-malleable codes and key-derivation for poly-size tampering circuits. In: Nguyen, P.Q., Oswald, E. (eds.) EUROCRYPT 2014. LNCS, vol. 8441, pp. 111–128. Springer, Heidelberg (2014). https://doi.org/10.1007/978-3-642-55220-5_7

[Fra90] Frankel, Y.: A practical protocol for large group oriented networks. In: Quisquater, J.-J., Vandewalle, J. (eds.) EUROCRYPT 1989. LNCS, vol. 434, pp. 56–61. Springer, Heidelberg (1990). https://doi.org/10.1007/3-540-46885-4_8

[FRR+10] Faust, S., Rabin, T., Reyzin, L., Tromer, E., Vaikuntanathan, V.: Protecting circuits from leakage: the computationally-bounded and noisy cases. In: Gilbert, H. (ed.) EUROCRYPT 2010. LNCS, vol. 6110, pp. 135–156. Springer, Heidelberg (2010). https://doi.org/10.1007/978-3-642-13190-5_7

[GK18a] Goyal, V., Kumar, A.: Non-malleable secret sharing. In: STOC, pp. 685–698 (2018)

[GK18b] Goyal, V., Kumar, A.: Non-malleable secret sharing for general access structures. In: Shacham, H., Boldyreva, A. (eds.) CRYPTO 2018, Part I. LNCS, vol. 10991, pp. 501–530. Springer, Cham (2018). https://doi.org/10.1007/978-3-319-96884-1_17

[GKP+18] Goyal, V., Kumar, A., Park, S., Richelson, S., Srinivasan, A.: Non-malleable commitments from non-malleable extractors. Manuscript, accessed via personal communication (2018)

[GLM+04] Gennaro, R., Lysyanskaya, A., Malkin, T., Micali, S., Rabin, T.: Algorithmic tamper-proof (ATP) security: theoretical foundations for security against hardware tampering. In: Naor, M. (ed.) TCC 2004. LNCS, vol. 2951, pp. 258–277. Springer, Heidelberg (2004). https://doi.org/10.1007/978-3-540-24638-1_15

[GMW87] Goldreich, O., Micali, S., Wigderson, A.: How to play any mental game or A completeness theorem for protocols with honest majority. In: Aho, A. (ed.) 19th Annual ACM Symposium on Theory of Computing, pp. 218–229. ACM Press, May 1987

[GMW17] Gupta, D., Maji, H.K., Wang, M.: Constant-rate non-malleable codes in the split-state model. Cryptology ePrint Archive, Report 2017/1048 (2017). https://eprint.iacr.org/2017/1048

[ISW03] Ishai, Y., Sahai, A., Wagner, D.: Private circuits: securing hardware against probing attacks. In: Boneh, D. (ed.) CRYPTO 2003. LNCS, vol. 2729, pp. 463–481. Springer, Heidelberg (2003). https://doi.org/10.1007/978-3-540-45146-4_27

[JW15] Jafargholi, Z., Wichs, D.: Tamper detection and continuous non-malleable codes. In: Dodis, Y., Nielsen, J.B. (eds.) TCC 2015, Part I. LNCS, vol. 9014, pp. 451–480. Springer, Heidelberg (2015). https://doi.org/10.1007/978-3-662-46494-6_19

[KLT18] Kiayias, A., Liu, F.-H., Tselekounis, Y.: Non-malleable codes for partial functions with manipulation detection. In: Shacham, H., Boldyreva, A. (eds.) CRYPTO 2018, Part III. LNCS, vol. 10993, pp. 577–607. Springer, Cham (2018). https://doi.org/10.1007/978-3-319-96878-0_20

[KMS18] Kumar, A., Meka, R., Sahai, A.: Leakage-resilient secret sharing. Cryptology ePrint Archive, Report 2018/1138 (2018). https://eprint.iacr.org/2018/1138

[KOS17] Kanukurthi, B., Obbattu, S.L.B., Sekar, S.: Four-state non-malleable codes with explicit constant rate. In: Kalai, Y., Reyzin, L. (eds.) TCC 2017, Part II. LNCS, vol. 10678, pp. 344–375. Springer, Cham (2017). https://doi.org/10.1007/978-3-319-70503-3_11

[KOS18] Kanukurthi, B., Obbattu, S.L.B., Sekar, S.: Non-malleable randomness encoders and their applications. In: Nielsen, J.B., Rijmen, V. (eds.) EUROCRYPT 2018, Part III. LNCS, vol. 10822, pp. 589–617. Springer, Cham (2018). https://doi.org/10.1007/978-3-319-78372-7_19

[KW93] Karchmer, M., Wigderson, A.: On span programs. In: Proceedings of the Eigth Annual Structure in Complexity Theory Conference, San Diego, CA, USA, 18–21 May 1993, pp. 102–111 (1993)

[Li17] Li, X.: Improved non-malleable extractors, non-malleable codes and independent source extractors. In: STOC (2017)

[LL12] Liu, F.-H., Lysyanskaya, A.: Tamper and leakage resilience in the split-state model. In: Safavi-Naini, R., Canetti, R. (eds.) CRYPTO 2012. LNCS, vol. 7417, pp. 517–532. Springer, Heidelberg (2012). https://doi.org/10.1007/978-3-642-32009-5_30

[LV18] Liu, T., Vaikuntanathan, V.: Breaking the circuit-size barrier in secret sharing. In: Diakonikolas, I., Kempe, D., Henzinger, M. (eds.) 50th Annual ACM Symposium on Theory of Computing, pp. 699–708. ACM Press, June 2018

[NN93] Naor, J., Naor, M.: Small-bias probability spaces: efficient constructions and applications. SIAM J. Comput. **22**(4), 838–856 (1993)

[OPVV18] Ostrovsky, R., Persiano, G., Venturi, D., Visconti, I.: Continuously non-malleable codes in the split-state model from minimal assumptions. In: Shacham, H., Boldyreva, A. (eds.) CRYPTO 2018, Part III. LNCS, vol. 10993, pp. 608–639. Springer, Cham (2018). https://doi.org/10.1007/978-3-319-96878-0_21

[Rot12] Rothblum, G.N.: How to compute under $\mathcal{AC}^0$ leakage without secure hardware. In: Safavi-Naini, R., Canetti, R. (eds.) CRYPTO 2012. LNCS, vol. 7417, pp. 552–569. Springer, Heidelberg (2012). https://doi.org/10.1007/978-3-642-32009-5_32

[Sha79] Shamir, A.: How to share a secret. Commun. ACM **22**(11), 612–613 (1979)

[SNW01] Safavi-Naini, R., Wang, H.: Robust additive secret sharing schemes over ZM. In: Lam, K.-Y., Shparlinski, I., Wang, H., Xing, C. (eds.) Cryptography and Computational Number Theory, pp. 357–368. Birkhäuser, Basel (2001)

[SS90] Schmidt, J.P., Siegel, A.: The spatial complexity of oblivious k-probe hash functions. SIAM J. Comput. **19**(5), 775–786 (1990)

Blockchain and Consensus

Multi-party Virtual State Channels

Stefan Dziembowski[1]([✉]), Lisa Eckey[2], Sebastian Faust[2], Julia Hesse[2], and Kristina Hostáková[2]

[1] University of Warsaw, Warsaw, Poland
stefan.dziembowski@crypto.edu.pl
[2] Technische Universität Darmstadt, Darmstadt, Germany
{lisa.eckey,sebastian.faust,julia.hesse,kristina.hostakova}@crisp-da.de

Abstract. Smart contracts are self-executing agreements written in program code and are envisioned to be one of the main applications of blockchain technology. While they are supported by prominent cryptocurrencies such as Ethereum, their further adoption is hindered by fundamental scalability challenges. For instance, in Ethereum contract execution suffers from a latency of more than 15 s, and the total number of contracts that can be executed per second is very limited. *State channel networks* are one of the core primitives aiming to address these challenges. They form a second layer over the slow and expensive blockchain, thereby enabling instantaneous contract processing at negligible costs.

In this work we present the first complete description of a state channel network that exhibits the following key features. First, it supports virtual multi-party state channels, i.e. state channels that can be created and closed *without* blockchain interaction and that allow contracts with any number of parties. Second, the worst case time complexity of our protocol is *constant* for arbitrary complex channels. This is in contrast to the existing virtual state channel construction that has worst case time complexity linear in the number of involved parties. In addition to our new construction, we provide a comprehensive model for the modular design and security analysis of our construction.

1 Introduction

Blockchain technology emerged recently as a promising technique for distributing trust in security protocols. It was introduced by Satoshi Nakamoto in [22] who used it to design *Bitcoin*, a new cryptographic currency which is maintained jointly by its users, and remains secure as long as the majority of computing power in the system is controlled by honest parties. In a nutshell, a blockchain is a system for maintaining a joint database (also called the "ledger") between several users in such a way that there is a *consensus* about its state.

In recent years the original ideas of Nakamoto have been extended in several directions. Particularly relevant to this paper are systems that support so-called *smart contracts* [26], also called *contracts* for short (see Sect. 2.1 for a more detailed introduction to this topic). Smart contracts are self-executing agreements written in a programming language that distribute money according to

© International Association for Cryptologic Research 2019
Y. Ishai and V. Rijmen (Eds.): EUROCRYPT 2019, LNCS 11476, pp. 625–656, 2019.
https://doi.org/10.1007/978-3-030-17653-2_21

the results of their execution. The blockchain provides a platform where such contracts can be written down, and more importantly, be executed according to the rules of the language in which they are encoded. The most prominent blockchain system that offers support for rich smart contracts is *Ethereum*, but many other systems are currently emerging.

Unfortunately, the current approach of using blockchain platforms for executing smart contracts faces inherent scalability limitations. In particular, since all participants of such systems need to reach consensus about the blockchain contents, state changes are costly and time consuming. This is especially true for blockchains working in the so-called *permissionless* setting (like Bitcoin or Ethereum), where the set of users changes dynamically, and the number of participants is typically large. In Ethereum, for example, it can take minutes for a transaction to be confirmed, and the number of maximum state changes per second (the so-called transaction throughput) is currently around 15–20 transactions per second. This is unacceptable for many applications, and in particular, prohibits use-cases such as "microtransactions" or many games that require instantaneous state changes.

Arguably one of the most promising approaches to tackle these problems are *off-chain techniques* (often also called "layer-2 solutions"), with one important example being *payment channels* [2]. We describe this concept in more detail in Sect. 2.1. For a moment, let us just say that the basic idea of a payment channel is to let two parties, say Alice and Bob, "lock" some coins in a smart contract on the blockchain in such a way that the amount of coins that each party owns in the contract can be changed dynamically *without* interacting with the blockchain. As long as the coins are locked in the contract the parties can then update the distribution of these coins "off-chain" by exchanging signatures of the new balance that each party owns in the channel. At some point the parties can decide to close the channel, in which case the latest signed off-chain distribution of coins is realized on the blockchain. Besides for creation and closing, the blockchain is used only in one other case, namely, when there is a *dispute* between the parties about the current off-chain balance of the channel. In this case the parties can send their latest signed balance to the contract, which will then resolve the dispute in a fair way.

This concept can be extended in several directions. *Channel networks* (e.g., the *Lightning network* over Bitcoin [24]) are an important extension which allows to securely "route" transactions over a longer path of channels. This is done in a secure way, which means that intermediaries on the path over which coins are routed cannot steal funds. Another extension is known under the name *state channels* [1]. In a state channel the parties can not only send payments but also execute smart contracts off-chain. This is achieved by letting the channel maintain in addition to the balance of the users a "state" variable that stores the current state of an off-chain contract. Both extensions can be combined resulting into so-called *state channel networks* [5,7,10], where simple state channels can be combined to create longer state channels. We write more about this in Sect. 2.1.

Before we describe our contribution in more detail let us first recall the terminology used in [10] on which our work relies. Dziembowski et al. [10] distinguish
between two variants of two-party state channels – so-called *ledger* and *virtual*
state channels[1]. Ledger state channels are created directly over the ledger, while
virtual state channels are built over multiple existing (ledger/virtual) state channels to construct state channels that span over multiple parties. Technically, this
is done in a recursive way by building a virtual state channel on top of two other
state channels. For instance, given two ledger state channels between Alice and
Ingrid, and Ingrid and Bob respectively, we may create a virtual state channel
between Alice and Bob where Ingrid takes the role of an *intermediary*. Compared
to ledger state channels, the main advantage of virtual state channels is that
they can be opened and closed without interaction with the blockchain.

1.1 Our Contribution

Our main contribution is to propose a new construction for generalized state
channel networks that exhibit several novel key features. In addition, we present
a comprehensive modeling and a security analysis of our construction. We discuss
further details below. The comparison to related work is presented in Sect. 1.2.

Multi-party state channels. Our main contribution is the *first full specification* of multi-party virtual state channels. A multi-party state channel allows
parties to off-chain execute contracts that involve > 2 parties. This greatly
broadens the applicability of state channel networks since many use cases such
as online games or exchanges for digital assets require support for multi-party
contracts. Our multi-party state channels are built "on top" of a network of
ledger channels. Any subset of the parties can form multi-party state channels,
where the only restriction is that the parties involved in the multi-party state
channel are connected via a path in the network of ledger channels. This is an
important distinctive feature of our construction because once a party is connected to the network it can "on-the-fly" form multi-party state channels with
changing subsets of parties. An additional benefit of our construction is that
our multi-party state channels are *virtual*, which allows opening and closing of
the channel without interaction with the blockchain. As a consequence in the
optimistic case (i.e., when there is no dispute between the parties) channels can
be opened and closed instantaneously at nearly zero-costs.

At a more technical level, virtual multi-party state channel are built in a
recursive way using 2-party state channels as a building-block. More concretely,
if individual parties on the connecting path do not wish to participate in the
multi-party state channel, they can be "cut out" via building virtual 2-party
state channels over them.

---

[1] The startup L4 and their project *Counterfactual* [7] use a different terminology:
virtual channels are called "meta channels", but the concepts are the same.

Virtual state channels with direct dispute. The second contribution of our work is to introduce the concept of "direct disputes" to virtual state channels. To better understand the concept of direct disputes let us recall the basic idea of the dispute process from [10]. While in ledger state channels disputes are always directly taken to the ledger, in the 2-party virtual state channels from [10] disputes are first attempted to be resolved by the intermediary Ingrid before moving to the blockchain. There are two advantages of such an "indirect" dispute process. First, it provides "layers of defense" meaning that Alice is forced to go to the blockchain *only* if both Bob and Ingrid are malicious. Second, "indirect" virtual state channels allow for cross-blockchain state channels because the contracts representing the underlying ledger state channels always have to deal with a single blockchain system only.

These features, however, come at the price of an increased worst case time complexity. Assuming a blockchain finality of Δ,[2] the virtual channel construction of [10] has worst case dispute timings of order $O(n\Delta)$ for virtual state channels that span over n parties. We emphasize that these worst case timing may already occur when only a *single* intermediary is corrupt, and hence may frequently happen in state channel networks with long paths.

In this work we build virtual state channels with *direct disputes*. Similar to ledger state channels, virtual state channels with direct dispute allow the members of the channel to resolve conflicts in time $O(\Delta)$, and thus, independent of the number of intermediaries involved. We call our new construction *virtual state channels with direct dispute* to distinguish them from their "indirect" counterpart [10]. To emphasize the importance of this improvement, notice that already for relatively short channels spanning over 13 ledger channels the worst case timings reduce from more than 1 day for the dispute process in [10] to less than 25 min in our construction. A comparison of the two types of two party state channels is presented in the following table.

	Ledger	Direct virtual	Indirect virtual
Creation	on chain	via subchannels	via subchannels
Dispute	on chain	on chain	via subchannels
Closure	on chain	via subchannels	via subchannels

Our final construction generalizes the one of [10] by allowing an arbitrary composition of: (a) 2-party virtual state channels with direct and indirect disputes, and (b) multi-party virtual state channels with direct disputes. We leave the design of multi-party virtual state channels with indirect dispute as an important open problem for future work.

---

[2] In Ethereum typically Δ equal to 6 min is assumed to be safe.

Modeling state channel networks. Our final contribution is a comprehensive security model for designing and analysing complex state channel networks in a modular way. To this end, we use the Universal Composability framework of Canetti [3] (more precisely, its global variant [4]), and a recursive composition approach similar to [10]. One particular nice feature of our modeling approach is that we are able to re-use the ideal state channel functionality presented in [10]. This further underlines the future applicability of our approach to design complex blockchain-based applications in a modular way. Or put differently: our functionalities can be used as subroutines for any protocol that aims at fast and cheap smart contract executions.

1.2   Related Work

One of the first constructions of off-chain channels in the scientific literature was the work of Wattenhofer and Decker [8]. Since then, there has been a vast number of different works constructing protocols for off-chain transactions and channel networks with different properties [12,15–18,25]. These papers differ from our work as they do not consider off-chain execution of arbitrary contract code, but instead focus on payments. Besides academic projects, there are also many industry projects that aim at building state channel networks. Particular relevant to our work is the Counterfactual project of L4 [7], Celer network [5] and Magmo [6]. The whitepapers of these projects typically do not offer full specification of full state channel networks and instead follow a more "engineering-oriented" approach that provides descriptions for developers. Moreover, non of these works includes a formal modeling of state channels nor a security analysis.

To the best of our knowledge, most related to our work is [10], which we significantly extend (as described above), and the recent work of Sprites [21] and its extensions [19,20] on building multi-party *ledger* state channels. At a high-level in [19–21] a set of parties can open a multi-party ledger state channel by locking a certain amount of coins into a contract. Then, the execution of this contract can be taken "off-chain" by letting the parties involved in the channel sign the new states of the contract. In case a dispute occurs among the parties, the dispute is taken on-chain. The main differences to our work are twofold: first [19–21] do not support virtual channels, and hence opening and closing state channels requires interaction with the blockchain. Second, while we support full concurrent execution of multiple contracts in a single channel, [19–21] focuses on the off-chain execution of a single contract. Moreover, our focus is different: while an important goal of our work is formal modeling, [21] aims at improving the worst case timings in payment channel networks, and [19,20] focus on evaluating practical aspects of state channels.

2   Overview of Our Constructions

Before we proceed with the more technical part of this work, we provide some background on the ledger and virtual state channels in Sect. 2.1 (we follow the formalism of [10]). In Sect. 2.2 we give an overview of our construction for handling "direct disputes", while in Sect. 2.3, we describe how we build and maintain multi-party virtual state channels. Below we assume that the parties that interact with the contracts own some coins in their *accounts* on the ledger. We emphasize that the description in this section is very simplified and excludes many technicalities.

2.1   Background on Contracts and State Channels [10]

Contracts. As already mentioned in Sect. 1, contracts are self-executing agreements written in a programming language. More formally, a contract can be viewed as a piece of code that is deployed by one of the parties, can receive coins from the parties, store coins on its account, and send coins to them. A contract never "acts by itself" – in other words: by default it is in an idle state and activates only when it is "woken up" by one of the parties. Technically, this is done by calling a *function* from its code. Every function call can have some coins attached to it (meaning that these coins are deduced from the account of the calling party and added to the contract account).

To be a bit more formal, we use two different terms while referring to a "contract": (i) "contract *code*" C – a static object written in some programming language (and consisting of a number of functions); and (ii) "contract *instance*" ν, which results from deploying the contract code C. Each contract instance ν maintains during its lifetime a (dynamically changing) *storage*, where the current state of the contract is stored. One of the functions in contract code, called a *constructor*, is used to create an instance and its initial storage. These notions are defined formally in Sect. 3. Here, let us just illustrate them by a simple example of a contract C_{sell} for selling a pre-image of some fixed function H. More concretely, suppose that we have two parties: Alice and Bob, and Bob is able to invert H, while Alice is willing to pay 1 coin for a pre-image of H, i.e., for any x such that $H(x) = y$ (where y is chosen by her). Moreover, if Bob fails to deliver x, then he has to pay a "fine" of 2 coins. First, the parties deploy the contract by depositing their coins into it (Alice deposits 1 coins, and Bob deposits 2 coins).[3] Denote the initial storage of the contract instance as G_0. Alice can now challenge Bob by requesting him to provide a pre-image of y. Let G_1 be the storage of the contract after this request has been recorded. If now Bob sends x such that $H(x) = y$ to the contract, $1 + 2 = 3$ coins are paid to Bob, and the contract enters a terminal state of storage G_2. If Bob fails to deliver

---

[3] Technically, this is done by one of the parties, Alice, say, calling a constructor function, and then Bob calling another function to confirm that he agrees to deploy this contract instance. To keep our description simple, we omit these details here.

x in time, i.e. within some time $t > \Delta$, and the contract has still storage G_1, then Alice can request the contract to pay the 3 coins to her, and the contract enters into a terminal state of storage G_3.

The contract code C_{sell} consists of functions used to deploy the contract (see footnote 3), a function that Alice uses to send y to the contract instance, a function used by Bob to send x, and a function that Alice calls to get her coins back if Bob did not send x in time.

Functionality of state channels. State channels allow two parties Alice and Bob to execute instances of some contract code C off-chain, i.e., without interacting with the ledger. These channels offer four sub-protocols that manage their life cycles: (i) *channel create* for opening a new channel; (ii) *channel update* for updating the state of a channel; (iii) *channel execute* for executing contracts off-chain; and finally (iv) *channel close* for closing a channel when it is not needed anymore. In [10] the authors consider two types of state channels: *ledger* state channels and *virtual* state channels. The functionality offered by these two variants is slightly different, which we discuss next.

Ledger state channels. Ledger state channels are constructed directly on the ledger. To this end, Alice and Bob *create* the ledger state channel γ by deploying an instance of a *state channel contract* (denoted SCC) on the ledger. The contract SCC will take the role of a judge, and resolve disputes when Alice and Bob disagree (we will discuss disputes in more detail below). During channel creation, Alice and Bob also lock a certain amount of coins into the contract. These coins can then be used for off-chain contracts. For instance, Alice and Bob may each transfer 10 coins to SCC, and hence in total 20 coins are available in the channel γ. Once the channel γ is established, the parties can *update* the state of γ (without interacting with the state channel contract). These updates serve to create new contract instances "within the channel", e.g., Alice can buy from Bob a pre-image of H and pay for it using her channel funds by deploying an instance of the C_{sell} contract in the channel. At the end the channel is closed, and the coins are transfered back to the accounts of the parties on the ledger. The state channel contract guarantees that even if one of the parties is dishonest she cannot steal the coins of the honest party (i.e.: get more coins than she would get from an honest execution of the protocol). The mechanism behind this is described a bit later (see *"Handling disputes in channels"* on page 8).

Virtual state channels. The main novelty of [10] is the design of *virtual state channels*. A virtual state channel offers the same interface as ledger state channels (i.e.: channel creation, update, execute, and close), but instead of being constructed directly over the ledger, they are built "on top of" other state channels. Consider a setting where Alice and Bob are not directly connected via a ledger state channel, but they both have a ledger channel with an intermediary Ingrid. Call these two ledger state channels α and β, respectively (see Fig. 1, page 8). Suppose now that Alice and Bob want to execute the pre-image selling procedure using the contract C_{sell} according to the same scenario as the one described above. To this end, they can *create* a virtual state channel γ with the help of Ingrid, but without interacting with the ledger. In this process the parties "lock" their coins in channels α and β (so that they cannot be used for any other purpose until γ is closed). The amounts of "locked" coins are as follows: in α Alice locks 1 coin and Ingrid locks 2 coins, and in β Bob locks 2 coins, and Ingrid locks 1 coin. The requirement that Ingrid locks 2 coins in α and 1 coin in β corresponds to the fact that she is "representing" Bob in channel α and "representing" Alice in channel β. Here, by "representing" we mean that she is ready to cover their commitments that result from executing the contract in γ.

Once γ is created, it can be used exactly as a ledger state channel, i.e., Alice and Bob can open a contract instance ν of C_{sell} in γ via the virtual state channel *update* protocol and *execute* it. As in the ledger state channels, when both Alice and Bob are honest, the update and execution of ν can be done without interacting with the ledger or Ingrid. Finally, when γ is not needed anymore, it is *closed*, where closing is first tried *peacefully* via the intermediary Ingrid (in other words: Alice and Bob "register" the latest state of γ at Ingrid).

For example: suppose the execution of C_{sell} ends in the way that Alice receives 0 coins, and Bob receives 3 coins. The effect on the ledger channels is as follows: in channel α Alice receives 0 coins, and Ingrid receives 3 coins, and in channel β Bob receives 3 coins, and Ingrid receives 0 coins. Note that this is financially neutral for Ingrid who always gets backs the coins that she locked (although the distribution of these coins between α and β can be different from the original one). This situation is illustrated on Fig. 1. If the peaceful closing fails, the parties enter into a dispute which we describe next.

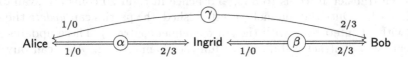

Fig. 1. Virtual channel γ built over ledger channels α and β. The labels "x/y" on the channels denote the fact that a given party locked x coins for the creation of γ, and got y coins as a result of closing γ.

Handling disputes in channels. The description above considered the case when both Alice and Bob are honest. Of course, we also need to take into account conflicts between the parties, e.g., when Alice and Bob disagree on a state update, or refuse to further execute a contract instance. Resolving such conflicts in a fair way is the purpose of the dispute resolution mechanism. The details of this mechanism appear in [10].

In order to better understand the dispute handling, we start by providing some more technical details on the state channel off-chain execution mechanism. Let ν be a contract instance of the pre-image selling contract C_{sell}, say, and denote by G_0 its initial state. To deploy ν in the state channel both parties exchange signatures on $(G_0, 0)$, where the second parameter in the tuple will be called the *version number.* The rest of the execution is done by exchanging signatures on further states with increasing version number. For instance, suppose that in the pre-image selling contract C_{sell} (described earlier in this section) the last state on which both parties agreed on was $(G_1, 1)$ (i.e., both parties have signatures on this state tuple), and Bob wants to provide x such that $H(x) = y$. To this end, he locally evaluates the contract instance to obtain the new state $(G_2, 2)$, and sends it together with his signature to Alice. Alice verifies the correctness of both the computation and the signature, and if both checks pass, she replies with her signature on $(G_2, 2)$.

Let us now move to the dispute resolution for ledger channels and consider a setting where a malicious Alice does not reply with her signature on $(G_2, 2)$ (for example because she wants to avoid "acknowledging" that she received x). In this case, Bob can force the execution of the contract instance ν *on*-chain by *registering* in the state channel contract SCC the latest state on which both parties agreed on. To this end, Bob will send the tuple $(G_2, 2)$ together with the signature of Alice to SCC. Of course, SCC cannot accept this state immediately because it may be the case that Bob cheated by registering an outdated state.[4] Hence, the ledger contract SCC gives Alice time Δ to reply with a more recent signed state (recall that in Sect. 1.1 we defined Δ to be a constant that is sufficiently large so that every party can be sure her transaction is processed by the ledger within this time). When Δ time has passed, SCC finalizes the *state registration* process by storing the version with the highest version number in its storage. Once registration is completed, the parties can continue the execution of the contract instance on-chain.[5]

The dispute process for *virtual* state channels is much more complex than the one for the ledger channels. In particular, in a virtual state channel Alice and Bob first try to resolve their conflicts peacefully via the intermediary Ingrid. That is, both Alice and Bob first send their latest version to Ingrid who takes the role of the judge, and attempts to resolve the conflict. If this does not succeed because a dishonest Ingrid is not cooperating, then the parties resolve their dispute *on*-chain using the underlying ledger state channels α and β (and the *virtual state channel contracts* VSCC).

Longer virtual state channels via recursion. So far, we only considered virtual state channels that can be built on top of 2 ledger state channels. The authors of [10] show how virtual state channels can be used in a recursive way to build virtual state channels that span over n ledger state channels. The key feature that makes this possible is that the protocol presented in [10] is oblivious of whether the channels α or β underlying γ are ledger or virtual state channels. Hence, given a virtual state channel α between P_0 and $P_{n/2}$ and a virtual state channel β between parties $P_{n/2}$ and P_n, we can construct γ, where $P_{n/2}$ takes the role of Ingrid.

---

[4] Notice that SCC is oblivious to what happened inside the ledger state channel γ after it was created.

[5] In the example that we considered, Bob can now force Alice bear the consequences that he revealed x to the contract instance.

As discussed in the introduction, one main shortcoming of the recursive app-roach used by [10] is that even if only one intermediary is malicious[6], the worst-case time needed for dispute resolution is significantly prolonged. Concretely, even a single intermediary that works together with a malicious Alice can delay the execution of a contract instance in γ for up to $\Omega(n\Delta)$ time before it eventually is resolved on the ledger.

2.2   Virtual State Channel with Direct Dispute

The first contribution of this work is to significantly reduce the worst case timings of virtual state channels. To this end, we introduce *virtual state channels with direct dispute*, where in case of disagreement between Alice and Bob the parties do not contact the intermediaries over which the virtual state channel is constructed, but instead directly move to the blockchain. This reduces worst case timings for dispute resolution to $O(\Delta)$, and hence makes it independent of the number of parties over which the virtual channel is spanned. Let us continue with a high-level description of our construction, where we call the virtual state channels constructed in [10] *virtual state channels with indirect dispute* or *indirect virtual state channels* to distinguish them from our new construction.

Overview of virtual state channels with direct dispute. The functionality offered by virtual state channels with direct dispute can be described as a "hybrid" between ledger and indirect virtual state channels. On the one hand – similar to virtual state channels from [10] – *creation* and *closing* involves interaction with the intermediary over which the channel is built. On the other hand – similar to ledger state channels – the *update* and *execution*, in case of dispute between the end parties, is directly moved to the ledger. The latter is the main differ-ence to indirect virtual state channels, where dispute resolution first happens peacefully via an intermediary. The advantage of our new approach is that the result of a dispute is visible to all parties and contracts that are using the same ledger. Hence, the other contracts can use the information about the result of this dispute in order to speed up the execution of their own dispute resolution procedure. This process is similar to the approach used in the Sprites paper [21], but we extend it to the case of virtual (multi-party) channels.

Before we describe in more detail the dispute process, we start by giving a high-level description of the *creation* process. To this end, consider an initial setting with two indirect virtual state channels α and β. Both α and β have length $n/2$, where α is spanned between parties P_0 and $P_{n/2}$, while β is spanned between parties $P_{n/2}$ and P_n (assume that n is a power of 2). Using the channels α and β, parties P_0 and P_n can now create a direct virtual state channel γ of length n. At a technical level this is done in a very similar way to creating an indirect virtual state channel. In a nutshell, with the help of the intermediary

---

[6] While it is sufficient that only one intermediary is malicious, it has to be the inter-mediary that was involved in the last step of the recursion, i.e., in the example from above: party $P_{n/2}$.

$P_{n/2}$ the parties update their subchannels α and β by opening instances of a special so-called *direct virtual state channel contract* dVSCC. The role of dVSCC is similar to the role of the indirect virtual state channel contract presented in [10]. It (i) guarantees balance neutrality for the intermediary (here for $P_{n/2}$), i.e., an honest $P_{n/2}$ will never loose money; and (ii) it ensures that what was agreed on in γ between the end users P_0 and P_n can be realized in the underlying subchannels α and β during closing or dispute.

Once γ is successfully created P_0 and P_n can *update* and *execute* contract instances in γ using a 2-party protocol, which is similar to the protocol used for ledger state channels (i.e., using the version number approach outlined above) as long as P_0 and P_n follow the protocol. The main difference occurs in the dispute process, which we describe next.

Direct dispute via the dispute board. Again, suppose that P_0 and P_n want to execute the pre-image selling procedure. Similarly to the example on page 8 uppose that during the execution of the contract P_0 (taking the role of Alice) refuses to acknowledge that P_n (taking the role of Bob) revealed the pre-image. Unlike in indirect virtual state channels, where P_n would first try to resolve his conflict peacefully via $P_{n/2}$, in our construction P_n registers his latest state directly on the so-called *dispute board* – denoted by $\mathcal{D}$. Since the dispute board $\mathcal{D}$ is a contract running directly on the ledger whose state can be accessed by anyone, we can reduce timings for dispute resolution from $O(n\Delta)$ to $O(\Delta)$. At a technical level, the state registration process on the dispute board is similar to the registration process for ledger channels described above. That is, when P_n registers his latest state regarding channel γ on $\mathcal{D}$, P_0 gets notified and is given time Δ to send her own version to $\mathcal{D}$. While due to the global nature of $\mathcal{D}$ all parties can see the final result of the dispute, only the end parties of γ can dispute the state of γ on $\mathcal{D}$. Our construction for direct virtual state channels uses this novel dispute mechanism also as subroutine during the update. This enables us to reduce the worst case timings of these protocols from $O(n\Delta)$ in indirect virtual state channels to $O(\Delta)$.

The above description omits many technical challenges that we have to address in order to make the protocol design work. In particular, the closing procedure of direct virtual state channels is more complex because sometimes it needs to refer to contents on the public dispute board. Concretely, during closing of channel γ, the end parties P_0 and P_n first try to close γ peacefully via the intermediary. To this end, P_0 and P_n first attempt to update the channels α and β, respectively, in such a way that the updated channels will reflect the last state of γ. If both update requests come with the same version of γ then $P_{n/2}$ confirms the update request, and the closing of γ is completed peacefully. Otherwise $P_{n/2}$ gives the end parties some time to resolve their conflict on the dispute board $\mathcal{D}$, and takes the final result of the state registration from $\mathcal{D}$ to complete the closing of γ. Of course, also this description does not present all the details. For instance, how to handle the case when both P_0 and P_n are malicious and try to steal money from $P_{n/2}$, or a malicious $P_{n/2}$ that does not reply to a closing attempt. Our protocol addresses these issues.

Interleaving direct and indirect virtual state channels. A special feature of our new construction is that users of the system can mix direct and indirect virtual state channels in an arbitrary way. For example, they may construct an indirect virtual γ over two subchannels α and β which are direct (or where α is direct and β is indirect). This allows them to combine the benefits of both direct and indirect virtual channels. If, for instance, γ is indirect and both α and β are direct, then in case of a dispute, P_0 and P_n will first try to resolve it via the intermediary $P_{n/2}$, and only if this fails they use the dispute board. The advantage of this approach is that, as long as $P_{n/2}$ is honest, disputes between P_0 and P_n can be resolved almost instantaneously off-chain (thereby saving fees and time). On the other hand, even if $P_{n/2}$ is malicious, then disputes can be resolved fast, since the next lower level of subchannels α and β are direct, and hence a dispute with a malicious $P_{n/2}$ will be taken directly to the ledger. We believe that the optimal composition of direct and indirect virtual channels highly depends on the use-case and leave a detailed discussion on this topic for future research.

2.3   Multi-party Virtual State Channels

The main novelty of this work is a construction of multi-party virtual state channels. As already mentioned in Sect. 1, multi-party virtual state channels are a natural generalization of 2-party channels presented in the previous sections and have two distinctive features. First, they are *multi-party*, which means that they can execute contracts involving multiple parties. Consider for instance a multi-party extension of C_{sell} – denoted by C_{msell} – where parties $P_1, \ldots P_{t-1}$ each pay 1 coin to P_t for a pre-image of a function H, but if P_t fails to deliver a pre-image, P_t has to pay a "fine" of 2 coins to each of $P_1, \ldots, P_{t-1}$ (and the contract stops). Our construction allows the parties to create an off-chain channel for executing this contract, pretty much in the same way as the standard (bilateral) channels are used for executing C_{sell}. The second main feature of our construction is that our multi-party channels are *virtual*. This means that they are built over 2-party ledger channels, and thus their creation process does not require interaction with the ledger. Our construction has an additional benefit of being highly flexible. Given ledger channels between parties P_i and P_{i+1} for $i \in \{0, \ldots, n-1\}$, we can build multi-party state channels involving any subset of parties. Technically, this is achieved by cutting out individual parties P_j that do not want to participate in the multi-party state channel by building 2-party virtual state channels "over them". Moreover, we show how to generalize this for an arbitrary graph (V, E) of ledger channels, where the vertices V are the parties, and the edges E represent the ledger channels connecting the parties.

An example: a 4-party virtual state channel. To get a better understanding of our construction, we take a look at a concrete example, which is depicted in Fig. 2. We assume that five parties $P_1, \ldots, P_5$, are connected by ledger state channels $(P_1 \Leftrightarrow P_2 \Leftrightarrow P_3 \Leftrightarrow P_4 \Leftrightarrow P_5)$. Suppose P_1, P_3, P_4 and P_5 want to create a 4-party virtual state channel γ. Party P_2 will not be part of the channel

γ but is needed to connect P_1 and P_3. In order to "cut out" P_2, parties P_1 and P_3 first construct a virtual channel denoted by $P_1 \leftrightarrow P_3$.

Now the channel γ can be created on top of the subchannels $P_1 \leftrightarrow P_3$, $P_3 \Leftrightarrow P_4$ and $P_4 \Leftrightarrow P_5$.[7] Assume for simplicity that each party invests one coin into γ. Now in each subchannel, they open an instance of the special "multi-party virtual state channel contract" denoted as mpVSCC, which can be viewed as a "copy" of γ in the underlying subchannels. Note, that some parties have to lock more coins into the subchannel mpVSCC contract instances than others. For example in the channel $P_4 \Leftrightarrow P_5$, party P_4 has to lock three coins while P_5 only locks one coin. This is necessary, since P_4 additionally takes over the role of the parties P_1 and P_3 in this subchannel copy of γ. In other words, we require that in each mpVSCC contract instance, each party has to lock enough coins to match the sum of the investments of all "represented" parties.

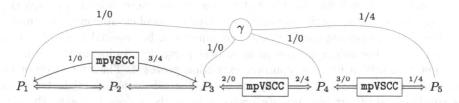

Fig. 2. Example of a multi-party virtual state channel γ between parties P_1, P_3, P_4 and P_5. In each subchannel a contract instance of mpVSCC is opened. Initially every party invests one coin and when the channel is closed, party P_5 owns all coins. The figure depicts the initial/final balance of parties in each of these contract instances.

After γ was successfully created, the parties P_1, P_3, P_4 and P_5 can open and execute multiple contracts ν in γ without talking to P_2. Let us assume that at the end of the channel lifetime party P_5 is the rightful owner of all four coins. Then after γ is successfully closed, the coins locked in the contract instances mpVSCC in the subchannels are unlocked in a way that reflects the final balance of γ. This means, for example, that all coins locked in subchannel $P_4 \Leftrightarrow P_5$ go to P_5. Since party P_4 now lost 3 coins in this subchannel, she needs to be compensated in the subchannel $P_3 \Leftrightarrow P_4$. Hence, the closing protocol guarantees that all four coins locked in $P_3 \Leftrightarrow P_4$ go to P_4. Since P_4 initially locked $2 + 3 = 5$ coins in the subchannels and received $4 + 0 = 4$ coins at the closing of γ, she lost 1 coin which corresponds to the final distribution in γ. As shown in Fig. 2 this process is repeated for the other subchannel $P_1 \leftrightarrow P_3$ as well.

Key ideas of the multi-party state channel update and execution. As for 2-party channels, our multi-party construction consists of 4 sub-protocol and a state registration process that is used by the parties in case of dispute. For registration

---

[7] To keep things simple we do not allow the recursion to build virtual channels on top on n-party channels for $n > 2$. We leave describing this extension as a possible future research direction.

our construction uses the direct dispute process outlined in Sect. 2.2, where all involved parties can register their latest state on the dispute board. One of the main differences between the 2-party and multi-party case is the way in which they handle state channel updates. Recall that in the two party case the initiating party sends an update request to the other party of the state channel, who can then confirm or reject the update request. Hence, in the two-party case it is easy for two honest parties to reach agreement whether an update was successfully completed or not.[8] In the multi-party case the protocol is significantly more complex. When the initiating party, say P_1, requests an update, she sends her update request to all other parties P_3, P_4 and P_5. The challenge is now that a malicious P_1 may for instance send a different update request to P_3 and P_4. At this point honest P_3 and P_4 have a different view on the update request. To resolve this inconsistency we may use standard techniques from the literature on authenticated broadcast protocols [9]. The problem with such an approach, however, is that it is well known [13] that broadcast has communication complexity of $O(n)$ in case most parties are dishonest. Our protocol circumvents this impossibility by a simple approach, where agreement can be reached in $O(1)$ rounds by relying on the ledger as soon as an honest party detects inconsistencies.

Let us now consider the contract execution protocol. The first attempt for constructing a protocol for multi-party state channel execution might be to use a combination of our new update protocol from above together with the contract execution protocol for the 2-party setting. In this case the initiating party P would locally execute the contract instance, and request an update of the multi-party state channel γ according to the new state of the contract instance. Unfortunately, this naive solution does not take into account a concurrent execution from two or more parties. For example, it may happen that P_1 and P_4 simultaneously start different contract instance executions, thereby leading to a protocol deadlock. For 2-party state channels this was resolved by giving each party a different slot when it is allowed to start a contract instance execution. In the multi-party case this approach would significantly decrease the efficiency of our protocol and in particular make its round complexity dependent on the number of involved parties. Our protocol addresses this problem by introducing a carefully designed execution scheduling, which leads to a constant time protocol.

Combining different state channel types. Finally, we emphasize that due to our modular modeling approach, all different state channel constructions that we consider in this paper can smoothly work together in a *fully concurrent* manner. That is, given a network of ledger state channels, parties may at the same time be involved in 2-party virtual state channels with direct or indirect dispute, while also being active in various multi-party state channels. Moreover, our construction guarantees strong fairness and efficiency properties in a fully concurrent setting where all parties except for one are malicious and collude.

---

[8] In case one party behaves maliciously, an agreement is reached via the state registration process.

3   Definitions and Notation

We formally model security of our construction in the Universal Composability framework [3]. Coins are handled by a global ledger $\widehat{\mathcal{L}}(\Delta)$, where Δ is an upper bound for the blockchain delay. We will next present the general notation used in this paper. More details about our model and background on it can be found in the full version of this paper [11].

We assume that the set $\mathcal{P} = \{P_1, \ldots, P_m\}$ of parties that use the system is fixed. In addition, we fix a bijection $\text{Order}_{\mathcal{P}} \colon \mathcal{P} \to [m]$ which on input a party $P_i \in \mathcal{P}$ returns its "order" i in the set $\mathcal{P}$. Following [10,12] we present tuples of values using the following convention. The individual values in a tuple T are identified using keywords called *attributes*, where formally an *attribute tuple* is a function from its set of attributes to $\{0,1\}^*$. The *value of an attribute* identified by the keyword attr in a tuple T (i.e. $T(\text{attr})$) will be referred to as $T.\text{attr}$. This convention will allow us to easily handle tuples that have dynamically changing sets of attributes. We assume the existence of a signature scheme (Gen, Sign, Vrfy) that is existentially unforgeable against a chosen message attack (see, e.g., [14]). The ECDSA scheme used in Ethereum is believed to satisfy this definition.

3.1   Definitions of Multi-party Contracts and Channels

We now present our syntax for describing multi-party contracts and state channels (it has already been introduced informally in Sect. 2.1). We closely follow the notation from [10,12].

Contracts. Let n be the number of parties involved in the contract. A *contract storage* is an attribute tuple σ that contains at least the following attributes: (1) $\sigma.\text{users} \colon [n] \to \mathcal{P}$ that denotes the users involved in the contract (sometimes we slightly abuse the notation and understand $\sigma.\text{users}$ as the set $\{\sigma.\text{users}(1), \ldots, \sigma.\text{users}(n)\}$), (2) $\sigma.\text{locked} \in \mathbb{R}_{\geq 0}$ that denotes the total amount of coins that is locked in the contract, and (3) $\sigma.\text{cash} \colon \sigma.\text{users} \to \mathbb{R}$ that denotes the amount of coins assigned to each user. It must hold that $\sigma.\text{locked} \geq \sum_{P \in \sigma.\text{users}} \sigma.\text{cash}(P)$. Let us explain the above inequality on the following concrete example. Assume that three parties are playing a game where each party initially invests 5 coins. During the game, parties make a bet, where each party puts 1 coin into the "pot". The amount of coins *locked* in the game did not change, it is still equal to 15 coins. However, the amount of coins assigned to each party decreased (each party has only 4 coins now) since it is not clear yet who wins the bet.

We say that a contract storage σ is *terminated* if $\sigma.\text{locked} = 0$. Let us emphasize that a terminated σ does not imply that $\sigma.\text{cash}$ maps to zero for every user. In fact, the concept of a terminated contract storage with non-zero cash values is important for our work since it represents "payments" performed between the users. Consider, for example, a terminated three party contract storage σ with $\sigma.\text{cash}(P_1) = 1$, $\sigma.\text{cash}(P_2) = 1$ and $\sigma.\text{cash}(P_3) = -2$. This means that both P_1 and P_2 paid one coin to P_3.

A *contract code* consists of constructors and functions. They take as input: a contract storage σ, a party $P \in \sigma.$users, round number $\tau \in \mathbb{N}$ and input parameter $z \in \{0,1\}^*$, and output: a new contract storage $\tilde{\sigma}$, information about the amount of unlocked coins add: $\sigma.$users $\rightarrow \mathbb{R}_{\geq 0}$ and some additional output message $m \in \{0,1\}^*$. Importantly, no contract function can ever change the set of users or create new coins. More precisely, it must hold that $\sigma.$users $= \tilde{\sigma}.$users and $\sigma.$locked $- \tilde{\sigma}.$locked $\geq \sum_{P \in \sigma.\text{users}} \text{add}(P)$.

As described already in Sect. 2.1, a *contract instance* represents an instantiation of a contract code. Formally, a contract instance is an attribute tuple ν consisting of the contract storage $\nu.$storage and the contract code $\nu.$code. To allow parties in the protocol to update contract instances off-chain, we also define a *signed contract instance version* of a contract instance which in addition to $\nu.$storage and $\nu.$code contains two additional attributes $\nu.$version and $\nu.$sign. The purpose of $\nu.$version $\in \mathbb{N}$ is to indicate the version of the contract instance. The attribute $\nu.$sign is a function that on input $P \in \nu.$storage.users outputs the signature of P on the tuple $(\nu.$storage$, \nu.$code$, \nu.$version$)$.

Two-party ledger and virtual state channels. Formally, a two-party state channel is an attribute tuple $\gamma = (\gamma.$id$, \gamma.$Alice$, \gamma.$Bob$, \gamma.$cash$, \gamma.$cspace$, \gamma.$length$, \gamma.$Ingrid$, \gamma.$subchan$, \gamma.$validity$, \gamma.$dispute$)$. The attribute $\gamma.$id $\in \{0,1\}^*$ is the identifier of the two-party state channel. The attributes $\gamma.$Alice $\in \mathcal{P}$ and $\gamma.$Bob $\in \mathcal{P}$ identify the two end-parties using γ. For convenience, we also define the set $\gamma.$end$-$users $:= \{\gamma.$Alice$, \gamma.$Bob$\}$ and the function $\gamma.$other$-$party as $\gamma.$other$-$party$(\gamma.$Alice$) := \gamma.$Bob and $\gamma.$other$-$party$(\gamma.$Bob$) := \gamma.$Alice. The attribute $\gamma.$cash is a function mapping the set $\gamma.$end$-$users to $\mathbb{R}_{\geq 0}$ such that $\gamma.$cash(T) is the amount of coins the party $T \in \gamma.$end$-$users has locked in γ. The attribute $\gamma.$cspace is a partial function that is used to describe the set of all contract instances that are currently open in this channel. It takes as input a *contract instance identifier cid* $\in \{0,1\}^*$ and outputs a contract instance ν such that $\nu.$storage.users $= \gamma.$end$-$users. We refer to $\gamma.$cspace(cid) as the *contract instance with identifier cid in* γ. The attribute $\gamma.$length $\in \mathbb{N}$ denotes the length of the two-party state channel.

If $\gamma.$length $= 1$, then we call γ a two-party *ledger state channel*. The attributes $\gamma.$Ingrid and $\gamma.$subchan do not have any meaning in this case and it must hold that $\gamma.$validity $= \infty$ and $\gamma.$dispute $=$ direct. Intuitively, this means that a ledger state channel has no intermediary and no subchannel, there is no a priory fixed round in which the channel must be closed, and potential disputes between the users are resolved directly on the blockchain.

If $\gamma.$length > 1, then we call γ a two-party *virtual state channel* and the remaining attributes have the following meaning. The attribute $\gamma.$Ingrid $\in \mathcal{P}$ denotes the identity of the intermediary of the virtual channel γ. For convenience, we also define the set $\gamma.$users $:= \{\gamma.$Alice$, \gamma.$Bob$, \gamma.$Ingrid$\}$. The attribute $\gamma.$subchan is a function mapping the set $\gamma.$end$-$users to channel identifiers $\{0,1\}^*$. The value of $\gamma.$subchan$(\gamma.$Alice$)$ refers to the identifier of the two-party state channel between $\gamma.$Alice and $\gamma.$Ingrid. Analogously, for the value of $\gamma.$subchan$(\gamma.$Bob$)$. The attribute $\gamma.$validity $\in \mathbb{N}$ denotes the round in which the virtual state channel γ will be closed. Intuitively, the a priory fixed closure round upper bounds the time until when party γ.Ingrid has to play the role of an intermediary of γ.[9] At the same time, the γ.validity lower bounds the time for which the end-users can freely use the channel. Finally, the attribute γ.dispute $\in$ {direct, indirect} distinguishes between virtual state channel with *direct dispute*, whose end-users contact the blockchain immediately in case they disagree with each other, and virtual state channel with *indirect dispute*, whose end-users first try to resolve disagreement via the subchannels of γ.[10]

Multi-party virtual state channel. Formally, an n-party virtual state channel γ is a tuple $\gamma := (\gamma.\text{id}, \gamma.\text{users}, \gamma.\text{E}, \gamma.\text{subchan}, \gamma.\text{cash}, \gamma.\text{cspace}, \gamma.\text{length}, \gamma.\text{validity},$ $\gamma.\text{dispute})$. The pair of attributes $(\gamma.\text{users}, \gamma.\text{E})$ defines an acyclic connected undirected graph, where the set of vertices $\gamma.\text{users} \subseteq \mathcal{P}$ contains the identities of the n parties of γ, and the set of edges $\gamma.\text{E}$ denotes which of the users from $\gamma.\text{users}$ are connected with a two-party state channel. Since $(\gamma.\text{users}, \gamma.\text{E})$ is an undirected graph, elements of $\gamma.\text{E}$ are unordered pairs $\{P, Q\} \in \gamma.\text{E}$. The attribute $\gamma.\text{subchan}$ is a function mapping the set $\gamma.\text{E}$ to channel identifiers $\{0, 1\}^*$ such that $\gamma.\text{subchan}(\{P, Q\})$ is the identifier of the two-party state channel between P and Q. For convenience, we define the function $\gamma.\text{other−party}$ which on input $P \in \gamma.\text{users}$ outputs the set $\gamma.\text{users} \setminus \{P\}$, i.e., all users of γ except for P. In addition, we define a function $\gamma.\text{neighbors}$ which on input $P \in \gamma.\text{users}$ outputs the set consisting of all $Q \in \gamma.\text{users}$ for which $\{P, Q\} \in \gamma.\text{E}$. Finally, we define a function $\gamma.\text{split}$ which, intuitively works as follows. On input the ordered pair (P, Q), where $\{P, Q\} \in \gamma.\text{E}$, it divides the set of users $\gamma.\text{users}$ into two subsets V_P, V_Q. The set V_P contains P and all nodes that are "closer" to P than to Q and the set V_Q contains Q and all nodes that are "closer" to Q than to P. The attribute $\gamma.\text{cash}$ is a function mapping $\gamma.\text{users}$ to $\mathbb{R}_{\geq 0}$ such that $\gamma.\text{cash}(P)$ is the amount of coins the party $P \in \gamma.\text{users}$ possesses in the channel γ. The attributes $\gamma.\text{length}$, $\gamma.\text{cspace}$ and $\gamma.\text{validity}$ are defined as for two-party virtual state channels. The value $\gamma.\text{dispute}$ for multi-party channels will always be equal to direct, since we do not allow indirect multi-party channels. We leave adding this feature to future work. In the following we will for brevity only write multi-party channels instead of virtual multi-party state channels with direct dispute. Additionally, we note that since multi-party channels cannot have intermediaries, the sets $\gamma.\text{users}$ and $\gamma.\text{end−users}$ are equal.

We demonstrate the introduced definitions on two concrete examples depicted in Fig. 3. In the 6-party channel on the left, the neighbors of party P_4 are $\gamma.\text{neighbors}(P_4) = \{P_3, P_5, P_6\}$ and $\gamma.\text{split}(\{P_3, P_4\}) = (\{P_1, P_2, P_3\},$ $\{P_4, P_5, P_6\})$. In the 4-party channel on the right, the neighbors of P_4 are $\gamma.\text{neighbors}(P_4) = \{P_1, P_5, P_6\}$ and $\gamma.\text{split}(\{P_1, P_4\}) = (\{P_1\}, \{P_4, P_5, P_6\})$.

---

[9] In practice, this information would be used to derive fees charged by the intermediary for its service.

[10] Recall from Sect. 2 that disagreements in channels with indirect dispute might require interaction with the blockchain as well. However this happen only in the worst case when all parties are corrupt.

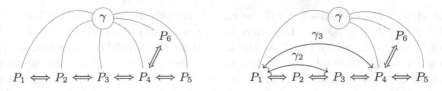

Fig. 3. Examples of multi-party channel setups: A 6-party channel on top of 5 ledger channels (left) and a 4-party channel on 2 ledger and a virtual channel γ_3 (right).

3.2   Security and Efficiency Goals

In the previous section, we formally defined what state channels are. Let us now give several security and efficiency goals that we aim for when designing state channels. The list below can be seen as an extension of the one from [10].

Security goals. We define security goals that guarantee that an adversary cannot steal coins from honest parties, even if he corrupts all parties except for one.

(S1) **Consensus on creation:** A state channel γ can be successfully created only if all users of γ agree with its creation.

(S2) **Consensus on updates:** A contract instance in a state channel γ can be successfully updated (this includes also creation of the contract instances) only if all end-users of γ agree with the update.

(S3) **Guarantee of execution:** An honest end-user of a state channel γ can execute a contract function f of an opened contract instance in any round $\tau_0 < \gamma.\text{validity}$ on an input value z even if all other users of γ are corrupt.

(S4) **Balance security:** If the channel γ has an intermediary, then this intermediary never loses coins even if all end-users of γ are corrupt and collude.

Let us stress that while creation of a state channel has to be confirmed by *all users* of the channel, this includes the intermediary in case of a two-party virtual state channel, the update of a contract instance needs confirmation only from the *end-users* of the state channel. In other words, the intermediary of a two-party virtual state channel has the right to refuse being an intermediary but once he agrees, he can not influence how this channel is being used by the end-users. Let us also emphasize that the last property, (S4), talks only about two-party virtual state channels since, by definition, ledger and multi-party channels do not have any intermediary.

Efficiency goals. We identify four efficiency requirements. Table 1 defines which property is required from what type of channel.

(E1) **Creation in $O(1)$ rounds:** Successful creation of a state channel γ takes a constant number of rounds.

(E2) **Optimistic update/execute in $O(1)$ rounds:** In the optimistic case when all end-users of a state channel γ are honest, they can update/execute a contract instance in γ within a constant number of rounds.

(E3) **Pessimistic update/execute in $O(\Delta)$ rounds:** In the pessimistic case when some end-users of a state channel γ are dishonest, the time complexity of update/execution of a contract instance in γ depends only on the ledger delay Δ but is independent of the channel length.

(E4) **Optimistic closure in $O(1)$ rounds:** In the optimistic case when all users of γ.users are honest, the channel γ is closed in round γ.validity $+ O(1)$.

Table 1. Summary of the efficiency goals for state channels. Above, "Ledger" stands for ledger state channels, "Direct/Indirect" stand for a two party virtual state channels with direct/indirect dispute and "MP" stands for multi-party channels.

	Ledger	Virtual		
		Direct	Indirect	MP
(E1) Creation in $O(1)$		✓	✓	✓
(E2) Opt. update/execute in $O(1)$	✓	✓	✓	✓
(E3) Pess. update/execute in $O(\Delta)$	✓	✓		✓
(E4) Opt. closing in $O(1)$		✓	✓	✓

It is important to note that in the optimistic case when all users of any *virtual state channel* (i.e. multi-party, two-party with direct/indirect dispute) are honest, the time complexity of channel creation, update, execution and closure must be independent of the blockchain delay; hence in this case there cannot be any interaction with the blockchain during the lifetime of the channel.

4   State Channels Ideal Functionalities

Recall that the main goal of this paper is to broaden the class of virtual state channels that can be constructed. Firstly, we want virtual state channels to support direct dispute meaning that end-users of the channel can resolve disputes directly on the blockchain, and secondly, we want to design virtual multi-party state channels that can be built on top of any network of two-party state channels. In order to formalize these goals, we define an ideal functionality $\mathcal{F}_{mpch}^{\widehat{\mathcal{L}}(\Delta)}(i, \mathcal{C})$ which describes what it means to create, maintain and close multi-party as well as two-party state channels of length up to i in which contract instances from the set $\mathcal{C}$ can be opened. The functionality has access to a global ledger functionality $\widehat{\mathcal{L}}(\Delta)$ keeping track of account balances of parties in the system.

The first step towards defining $\mathcal{F}_{mpch}^{\widehat{\mathcal{L}}(\Delta)}(i, \mathcal{C})$ has already been done in [10], where the authors describe an ideal functionality, $\mathcal{F}_{ch}^{\widehat{\mathcal{L}}(\Delta)}(i, \mathcal{C})$, for ledger state channels and two-party virtual state channels with indirect dispute. The second step is to extend the ideal functionality $\mathcal{F}_{ch}^{\widehat{\mathcal{L}}(\Delta)}(i, \mathcal{C})$ such that it additionally describes how virtual state channels with direct dispute are created, maintained and closed. We denote this extended functionality $\mathcal{F}_{dch}^{\hat{\mathcal{L}}(\Delta)}(i, \mathcal{C})$ and describe it in more detail in Sect. 4.1. As a final step, we define how multi-party channels are created, maintained and closed. This is discussed in Sect. 4.2.

Before we proceed with the description of the novel ideal functionalities, let us establish the following simplified notation. In the rest of this paper, we write $\mathcal{F}$ instead of $\mathcal{F}^{\hat{\mathcal{L}}(\Delta)}$, for $\mathcal{F} \in \{\mathcal{F}_{ch}, \mathcal{F}_{dch}, \mathcal{F}_{mpch}\}$.

4.1   Virtual State Channels with Direct Dispute

In this section we introduce our ideal functionality $\mathcal{F}_{dch}(i, \mathcal{C})$ that allows to build any type of two party state channel (ledger state channel, virtual state channel with direct dispute and virtual state channel with indirect dispute) of length up to i in which contract instances with code from the set $\mathcal{C}$ can be opened. The ideal functionality $\mathcal{F}_{dch}(i, \mathcal{C})$ extends the ideal functionality $\mathcal{F}_{ch}(i, \mathcal{C})$ in the following way:

- Messages about ledger state channels and virtual state channels with indirect dispute are handled as in $\mathcal{F}_{ch}(i, \mathcal{C})$.
- Virtual state channels **with direct dispute** are created (resp. closed) using the procedure of $\mathcal{F}_{ch}(i, \mathcal{C})$ for creating (resp. closing) virtual channels with indirect dispute.
- Update (resp. execute) requests of contract instances in channels **with direct dispute** are handled as $\mathcal{F}_{ch}(i, \mathcal{C})$ handles such queries for ledger state channels.

Hence, intuitively, a virtual state channel γ with direct dispute is a "hybrid" between a ledger state channel and a virtual state channel with indirect dispute, meaning that it is created and closed as a virtual state channel with indirect dispute and its contract instances are updated and executed as if γ would be a ledger state channel. In the remainder of this section, we explain how $\mathcal{F}_{dch}(i, \mathcal{C})$ works in more detail and argue that it satisfies all the security and efficiency goals listed in Sect. 3.2. The formal description of the ideal functionality can be found in the full version of this paper [11].

If $\mathcal{F}_{dch}(i, \mathcal{C})$ receives a message about a ledger state channel or a virtual state channel with indirect dispute, then $\mathcal{F}_{dch}(i, \mathcal{C})$ behaves exactly as $\mathcal{F}_{ch}(i, \mathcal{C})$. Since $\mathcal{F}_{ch}(i, \mathcal{C})$ satisfies all the security goals and the efficiency goals (E1)–(E2) (see [10]), $\mathcal{F}_{dch}(i, \mathcal{C})$ satisfies them as well in this case. It is thus left to analyze the properties in the novel case, i.e., for virtual state channels with direct dispute.

Create and close a virtual state channel with direct dispute. The users of the virtual state channel γ, which are the end-users of the channel γ.Alice and γ.Bob and the intermediary γ.Ingrid, express that they want to create γ by sending the message (create, γ) to $\mathcal{F}_{dch}(i, \mathcal{C})$. Once $\mathcal{F}_{dch}(i, \mathcal{C})$ receives such a message, it records it into the memory and locks coins in the corresponding sub-channel. For example, if the sender of the message is γ.Alice, $\mathcal{F}_{dch}(i, \mathcal{C})$ locks

γ.cash(γ.Alice) coins of γ.Alice and γ.cash(γ.Bob) coins of γ.Ingrid in the sub-channel γ.subchan(γ.Alice). If $\mathcal{F}_{dch}(i, \mathcal{C})$ records the message (create, γ) from all three parties within three rounds, then the channel γ is created. The ideal functionality informs both end-users of the channel about the successful creation by sending the message (created, γ) to them. Since all three parties have to agree with the creation of γ, the security goal (S1) is clearly met. The successful creation takes 3 rounds, hence (E1) holds as well.

Once the virtual state channel is successfully created, γ.Alice and γ.Bob can use it (open and execute contract instance) until round γ.validity when the closing of the channel γ begins. In round γ.validity, $\mathcal{F}_{dch}(i, \mathcal{C})$ first waits for τ rounds, where $\tau = 3$ if all users of γ are honest and is set by the adversary otherwise,[11] and then distributes the coins locked in the subchannels according to the final state of the channel γ. It might happen that the final state of γ contains unterminated contract instances, i.e. contract instances that still have locked coins, in which case it is unclear who owns these coins. In order to guarantee the balance security for the intermediary, the property (S4), $\mathcal{F}_{dch}(i, \mathcal{C})$ gives all of these locked coins to γ.Ingrid in *both* subchannels. The goal (E4) is met because γ is closed in round γ.validity $+ 3$ in the optimistic case.

Update a contract instance. A party P that wants to update a contract instance with identifier cid in a virtual state channel γ sends the message (update, γ.id, cid, σ, C) to $\mathcal{F}_{dch}(i, \mathcal{C})$. The parameter σ is the proposed new contract instance storage and the parameter C is the code of the contract instance. $\mathcal{F}_{dch}(i, \mathcal{C})$ informs the party $Q := \gamma$.other$-$party(P) about the update request and completes the update only if Q confirms it. If the party Q is honest, then it has to reply immediately. In case Q is malicious, $\mathcal{F}_{dch}(i, \mathcal{C})$ expects the reply within 3Δ rounds. Let us emphasize that the confirmation time is independent of the channel length. This models the fact that disputes are happening directly on the blockchain and not via the subchannels. In the optimistic case the update procedure takes 2 rounds and in the pessimistic case $2 + 3\Delta$ rounds; hence both update efficiency goals (E2) and (E3) are satisfied. The security property (S2) holds as well since without Q's confirmation the update fails.

Execute a contract instance. When a party P wants to execute a contract instance with identifier cid in a virtual state channel γ on function f and input parameters z, it sends the message (execute, γ.id, cid, f, z) to $\mathcal{F}_{dch}(i, \mathcal{C})$. The ideal functionality waits for τ rounds, where $\tau \leq 5$ in case both parties are honest and $\tau \leq 4\Delta + 5$ in case one of the parties is corrupt. The exact value of τ is determined by the adversary. Again, let us stress that the pessimistic time complexity is independent of channel length which models the fact that registration and force execution takes place directly on the blockchain. After the waiting time is over, $\mathcal{F}_{dch}(i, \mathcal{C})$ performs the function execution and informs both end-users of the channel about the result by outputting the message (execute, γ.id, $cid, \tilde{\sigma}, \mathsf{add}, m$).

---

[11] The value of τ can be set by the adversary as long as it is smaller than some upper bound T which is of order $O(\gamma$.length $\cdot \Delta)$.

Here $\tilde{\sigma}$ is the new contract storage after the execution, add contains information about the amount of coins unlocked from the contract instance and m is some additional output message. Since the adversary can not stop the execution, and only delay it, the guarantee of execution, security property (S3), is satisfied by $\mathcal{F}_{dch}(i, \mathcal{C})$. From the description above it is clear that the two execute efficiency goals (E2) and (E3) are fulfilled as well.

Two-party state channels of length one. Before we proceed to the description of the ideal functionality $\mathcal{F}_{mpch}(i, \mathcal{C})$, let us state one simple but important observation which follows from the fact that the minimal length of a virtual state channel is 2 and the ideal functionality $\mathcal{F}_{dch}(1, \mathcal{C})$ accepts only messages about a state channel of length 1.

Observation 1. *For any set of contract codes $\mathcal{C}$ it holds that $\mathcal{F}_{dch}(1, \mathcal{C})$ is equivalent to $\mathcal{F}_{ch}(1, \mathcal{C})$.*

4.2   Virtual Multi-party State Channels

We now introduce the functionality $\mathcal{F}_{mpch}(i, \mathcal{C})$ which allows to create, maintain and close multi-party as well as two-party state channels of length up to i in which contract instances from the set $\mathcal{C}$ can be opened. Here we provide its high level description and argue that all security and efficiency goals identified in Sect. 3.2 are met.

The ideal functionality $\mathcal{F}_{mpch}(i, \mathcal{C})$ extends the functionality $\mathcal{F}_{dch}(i, \mathcal{C})$, which we described in Sect. 4.1, in the following way. In case $\mathcal{F}_{mpch}(i, \mathcal{C})$ receives a message about a two-party state channel, then it behaves exactly as the functionality $\mathcal{F}_{dch}(i, \mathcal{C})$. Since the functionality $\mathcal{F}_{dch}(i, \mathcal{C})$ satisfies all the security and efficiency goals for two-party state channels, these goals are met by $\mathcal{F}_{mpch}(i, \mathcal{C})$ as well. For the rest of this informal description, we focus on the more interesting case, when $\mathcal{F}_{mpch}(i, \mathcal{C})$ receives a message about a multi-party channel.

Create and close a multi-party channel. Parties express that they want to create the channel γ by sending the message $(\mathsf{create}, \gamma)$ to the ideal functionality $\mathcal{F}_{mpch}(i, \mathcal{C})$. Once the functionality receives such message from a party $P \in \gamma.\mathsf{users}$, it locks coins needed for the channel γ in all subchannels of γ party P is participating in. Let us elaborate on this step in more detail. For every $Q \in \gamma.\mathsf{neighbors}(P)$ the ideal functionality proceeds as follows. Let $(V_P, V_Q) := \gamma.\mathsf{split}(\{P, Q\})$ which intuitively means that V_P contains all the user of γ that are "closer" to P than to Q. Analogously for V_Q. Then $\sum_{T \in V_P} \gamma.\mathsf{cash}(T)$ coins of party P and $\sum_{T \in V_Q} \gamma.\mathsf{cash}(T)$ coins of party Q are locked in the subchannel between P and Q by the ideal functionality. If the functionality receives the message $(\mathsf{create}, \gamma)$ from all parties in $\gamma.\mathsf{users}$ within 4 rounds, then the channel γ is created. The ideal functionality informs all parties about the successful creation by outputting the message $(\mathsf{created}, \gamma)$. Clearly, the security goal (S1) and the efficiency goal (E1) are both met.

Once the multi-party channel is successfully created, parties can use it (open and execute contract instances in it) until the round γ.validity comes. In round γ.validity, the ideal functionality first waits for τ rounds, where $\tau = 3$ if all parties are honest and is set by the adversary otherwise,[12] and then unlocks the coins locked in the subchannels of γ. The coin distribution happens according to the following rules (let $\hat{\gamma}$ denote the final version of γ): If there are no unterminated contract instances in $\hat{\gamma}$.cspace, then the ideal functionality simply distributes the coins back to the subchannels according to the function $\hat{\gamma}$.cash. The situation is more subtle when there are unterminated contract instances in $\hat{\gamma}$.cspace. Intuitively, this means that some coin of the channel are not attributed to any of the users. Our ideal functionality distributes the unattributed coins equally among the users[13] and the attributed coins according to $\hat{\gamma}$.cash. Once the coins are distributed back to the subchannels, the channel γ is closed which is communicated to the parties via the message (closed, γ.id). Since in the optimistic case, γ is closed in round γ.validity $+ 3$, the goal (E4) is clearly met.

Update/Execute a contract instance. The update and execute parts of the ideal functionality $\mathcal{F}_{mpch}(i, \mathcal{C})$ in case of multi-party channels are straightforward generalizations of the update and execute parts of the ideal functionality $\mathcal{F}_{dch}(i, \mathcal{C})$ in case of two-party virtual state *with direct dispute* (see Sect. 4.1).

Towards realizing the ideal functionality. For the rest of the paper, we focus on realization of our novel ideal functionality $\mathcal{F}_{mpch}(i, \mathcal{C})$. Our approach of realizing the ideal functionality $\mathcal{F}_{mpch}(i, \mathcal{C})$ closely follows the modular way we use for *defining* it. On a very high level, we first show how to construct any two party state channel, in other words, how to realize the ideal functionality $\mathcal{F}_{dch}$. This is done in Sect. 5. Thereafter, in Sect. 6, we design a protocol for multi-party channels using two party state channels in a black box way.

5   Modular Approach

In this section, we introduce our approach of realizing $\mathcal{F}_{dch}(i, \mathcal{C})$. We do not want to realize $\mathcal{F}_{dch}(i, \mathcal{C})$ from scratch, but find a modular approach which lets us reuse existing results. We give a protocol $\Pi_{dch}(i, \mathcal{C}, \pi)$ for building two-party state channels supporting direct dispute which uses three ingredients: (1) a protocol π for virtual state channels with indirect dispute up to length i, which was shown in [10] how to build recursively from subchannels, (2) the ideal functionality $\mathcal{F}_{dch}$

---

[12] In case at least one user is corrupt, the value of τ can be set by the adversary as long as it is smaller that some upper bound T which is of order $O(\gamma$.length $\cdot \Delta)$.

[13] Let us emphasize that this design choice does not necessarily lead to a *fair* coin distribution. For example, when users of the multi-party channel play a game and one of the users is "about to win" all the coins when round γ.validity comes. Hence, honest parties should always agree on new contract instances only if they can enforce contract termination before time γ.validity or if they are willing to take this risk.

for virtual channels with direct dispute up to length $i-1$ and (3) an ideal dispute board. $\Pi_{dch}(i, \mathcal{C}, \pi)$ can roughly be described by distinguishing three cases:

Case 1: If a party receives a message about a two-party state channel of length $j < i$, then it forwards the request to $\mathcal{F}_{dch}$.

Case 2: If a party receives a message about a virtual state channel with indirect dispute and of length exactly i, then it behaves as in the protocol π.

Case 3: For the case when a party receives a message about a virtual state channel γ with direct dispute of length exactly i, we describe a new protocol using $\mathcal{F}_{dch}$ and an ideal dispute board $\mathcal{F}_{DB}$ which we will detail shortly. Central element of the new protocol will be a special contract dVSCC used for creating and closing γ.

The protocol is formally described in the full version of this paper [11]. In particular, there we describe the special contract dVSCC whose instances are opened in the subchannels of γ during the creation process and guarantee that the final state of γ will be correctly reflected to the subchannels.

Ideal dispute board. Let us now informally describe our ideal functionality $\mathcal{F}_{DB}(\mathcal{C})$ for directly disputing about contract instances whose code is in some set $\mathcal{C}$. On a high level, the functionality models an ideal judge which allows the users to achieve consensus on the latest valid version of a contract instance. For this, $\mathcal{F}_{DB}(\mathcal{C})$ maintains a public "dispute board", which is a list of contract instances available to all parties. $\mathcal{F}_{DB}(\mathcal{C})$ admits two different procedures: *registration* of a contract instance and *execution* of a contract instance. The registration procedure works as follows: whenever a party determines a dispute regarding a specific instance whose code is in the set $\mathcal{C}$, it can register this contract instance by sending its latest valid version to $\mathcal{F}_{DB}(\mathcal{C})$. The dispute board gives the other party[14] of the contract instance some time to react and send her latest version. $\mathcal{F}_{DB}(\mathcal{C})$ compares both versions and adds the latest valid one to the dispute board. Once a contract instance is registered on the dispute board, a user of the contract instance can execute it via $\mathcal{F}_{DB}(\mathcal{C})$. Upon receiving an execution request, $\mathcal{F}_{DB}(\mathcal{C})$ executes the called function and updates the contract instance on the dispute board according to the outcome. We stress that the other party of the contract instance cannot interfere and merely gets informed about the execution.

Unfortunately, we cannot simply add an ideal dispute board as another hybrid functionality next to one for constructing shorter channels. In a nutshell, the reason is that the balances of virtual channels that are created via subchannels might be influenced by contracts that are in dispute. Upon closing these virtual channel, the dispute board needs to be taken into account. However, in the standard UC model it is not possible that ideal functionalities communicate their state. Thus, we will artificially allow state sharing by merging both ideal functionalities. Technically, this is done by putting a *wrapper* $\mathcal{W}_{dch}$ around both

---

[14] For simplicity, we describe here how $\mathcal{F}_{DB}$ handles a dispute about a two-party contract. $\mathcal{F}_{DB}$ handles disputes about multi-party contracts in a similar fashion.

functionalities, which can be seen just as a piece of code distributing queries to the wrapped functionalities. The formal descriptions of the wrapper as well as the dispute board can be found in the full version of this paper [11].

Now that we described all ingredients, we formally state what our protocol Π_{dch} achieves and what it assumes. On a high level, our protocol gives a method to augment a two-party state channel protocol π with indirect dispute, to also support direct dispute. Our transformation is case-tailored for channel protocols π that are build recursively out of shorter channels. That is, we do not allow an arbitrary protocol π for channels up to length i, but only one that is itself recursively build out of shorter channels.[15]

Theorem 1. *Let $\mathcal{C}_0$ be a set of contract codes, let $i > 1$ and $\Delta \in \mathbb{N}$. Suppose the underlying signature scheme is existentially unforgeable against chosen message attacks. Let π be a protocol that realizes the ideal functionality $\mathcal{F}_{ch}(i, \mathcal{C}_0)$ in the $\mathcal{F}_{ch}(i-1, \mathcal{C}_0')$-hybrid world. Then protocol $\Pi_{dch}(i, \mathcal{C}_0, \pi)$ (cf. [11]) working in the $\mathcal{W}_{dch}(i-1, \mathcal{C}_1, \mathcal{C}_0)$-hybrid model, for $\mathcal{C}_1 := \mathcal{C}_0 \cup \mathcal{C}_0' \cup \mathsf{dVSCC}_i$, emulates the ideal functionality $\mathcal{F}_{dch}(i, \mathcal{C}_1)$.*

Remaining technicalities. Remember that our goal is to add direct dispute to a two-party state channel protocol that is itself recursively build from shorter subchannels. We still need to solve two technicalities. Firstly, note that Theorem 1 yields a protocol realizing $\mathcal{F}_{dch}$ for length up to i, while it requires a *wrapped* $\mathcal{F}_{dch}$ of length up to $i - 1$. Thus, to be able to apply Theorem 1 recursively, we introduce a technical Lemma 2 which shows how to modify the protocol $\Pi_{dch}(i, \mathcal{C}_0, \pi)$ so that it realizes the wrapped $\mathcal{F}_{dch}$. Secondly, we can apply Theorem 1 on any level *except* for ledger channels. In a nutshell, the reason is that Theorem 1 heavily relies on using subchannels, which simply do not exist in case of ledger channels. Fortunately, this can quite easily be resolved by adding our dispute board to a protocol for ledger channels and to its hybrid ideal functionality. In Lemma 1 we show how to do this with a protocol π_1 from [10]. Their ledger channel protocol assumes an ideal functionality $\mathcal{F}_{scc}$ which models state channel contracts on the blockchain.[16] The description of functionality and protocol wrappers as well as the proofs of both lemmas can be found in the full version of this paper [11].

Lemma 1 (The Blue Lemma). *Let $\mathcal{C}$ and $\mathcal{C}_0$ be two arbitrary sets of contract codes and let π_1 be a protocol that UC-realizes the ideal functionality $\mathcal{F}_{ch}(1, \mathcal{C})$ in the $\mathcal{F}_{scc}(\mathcal{C})$-hybrid world. Then the protocol $\mathcal{W}_{prot}(1, \mathcal{C}_0, \Pi_1)$ UC-realizes the ideal functionality $\mathcal{W}_{ch}(1, \mathcal{C}, \mathcal{C}_0)$ in the $\mathcal{W}_{scc}(\mathcal{C}, \mathcal{C}_0)$-hybrid world.*

---

[15] For the sake of correctness, in this section we include details about contract sets that each channel is supposed to handle. In order to understand our modular approach, their relations can be ignored. The reader can just assume that each subchannel can handle all contracts required for building all the longer channels.

[16] Adding the dispute board to any functionality again works by wrapping functionality $\mathcal{F}_x$ and $\mathcal{F}_{DB}$ within a wrapper $\mathcal{W}_x$.

Lemma 2 (The Red Lemma). *Let $i \geq 2$ and let C be a set of contract codes. Let Π_i be a protocol that UC-realizes the ideal functionality $\mathcal{F}_{dch}(i, C)$ in the $\mathcal{W}_{dch}(i - 1, C', C)$-hybrid world for some set of contract codes C'. Then for every $C_0 \subseteq C$ the protocol $\mathcal{W}_{prot}(i, C_0, \Pi_i)$ UC-realizes the ideal functionality $\mathcal{W}_{dch}(i, C, C_0)$ in the $\mathcal{W}_{dch}(i - 1, C', C)$-hybrid world.*

We finish this section with the complete picture of our approach of building any two-party state channel of length up to 3 (Fig. 4). The picture demonstrates how we recursively realize $\mathcal{F}_{dch}$ functionalities of increasing length, as well as their wrapped versions $\mathcal{W}_{dch}$ which additionally comprise the ideal dispute board functionality. While already being required for recursively constructing $\mathcal{F}_{dch}$, $\mathcal{W}_{dch}$ will also serve us as a main building block for our protocol for multi-party channels in the upcoming section.

6   Protocol for Multi-party Channels

In this section we describe a concrete protocol that realizes the ideal functionality $\mathcal{F}_{mpch}(i, C_0)$ for $i \in \mathbb{N}$ and any set of contract codes C_0 in the $\mathcal{W}_{dch}(i, C_1, C_0)$-hybrid world. Recall that $\mathcal{W}_{dch}(i, C_1, C_0)$ is a functionality wrapper (cf. Sect. 5) combining the dispute board $\mathcal{F}_{DB}(C_0)$ and the ideal functionality $\mathcal{F}_{dch}(i, C_1)$ for building two-party state channels of length up to i supporting contract instances whose codes are in C_1. Our strategy of constructing a protocol $\Pi_{mpch}(i, C_0)$ for multi-party channels is to distinguish two cases. These cases also outline the minimal requirements on the set of supported contracts C_1:

Case 1: If a party receives a message about a two-party state channel, it forwards the request to the hybrid ideal functionality. Thus, we require $C_0 \subset C_1$.

Case 2: For the case when a party receives a message about a multi-party channel γ, we design a new protocol that uses (a) the dispute board for fair resolution of disagreements between the users of γ and (b) two-party state channels as a building block that provides monetary guarantees. For (b) we need the subchannels of γ to support contract instances of a special code $\mathtt{mpVSCC}_i$; hence, $\mathtt{mpVSCC}_i \in C_1$.

We now discuss case 2 in more detail, by first describing the special contract code $\mathtt{mpVSCC}_i$ and then the protocol for multi-party channels. Since case 1 is rather straightforward, we refer the reader to the full version of this paper [11] where also the formal description of our protocol can be found.

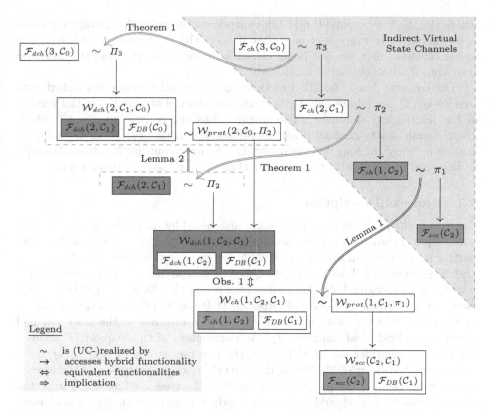

Fig. 4. The complete approach of building virtual state channels with direct dispute of length up to 3 (**top left**), from channels with indirect dispute (**gray background**). Theorem 1 and Lemma 1 allow to add direct dispute to channels. Note that the resulting recursion chain for building longer channels is disconnected due to Theorem 1 requiring $\mathcal{F}_{DB}$. Lemma 2 then reconnects the recursion chain. $\mathcal{C}_0$ is an arbitrary contract set. To build longer channels recursively, we have to allow the necessary channel contracts in each level. Thus, $\mathcal{C}_1 := \mathcal{C}_0 \cup \mathcal{C}'$, where $\mathcal{C}'$ is a special contract used for opening our target channel (i.e., longer channel supporting direct dispute, or multi-party channel). Similarly, $\mathcal{C}_2 := \mathcal{C}_1 \cup \mathcal{C}''$, where again $\mathcal{C}''$ is a special contract that is needed for the target channel. Note that it holds that $\mathcal{C}_0 \subset \mathcal{C}_1 \subset \mathcal{C}_2$, and also that the length of the channels as well as the target contract set have to be known in advance.

6.1 Multi-party Channel Contract

In order to create a multi-party channel γ, parties of the channel need to open a special two-party contract instance in each subchannel of γ (recall the example depicted in Fig. 2 in Sect. 2.3). We denote the code of these instances $\mathtt{mpVSCC}_i$, where $i \in \mathbb{N}$ is the maximal length of the channel in which an instance of $\mathtt{mpVSCC}_i$ can be opened. A contract instance of $\mathtt{mpVSCC}_i$ in a subchannel of γ between two parties P and Q can be understood as a "copy" of γ, where P plays the role of all parties from the set V_P and Q plays the role of parties from the set V_Q, where $(V_P, V_Q) := \gamma.\text{split}(\{P, Q\})$. The purpose of the mpVSCC_i contract instances is to guarantee to every user of γ that he gets the right amount of coins back to his subchannels when γ is being closed in round $\gamma.\text{validity}$. And this must be true even if all other parties collude.[17]

The contract has in addition to the mandatory attributes users, locked, cash (see Sect. 3.1) one additional attribute virtual–channel storing the initial version of the multi-party channel γ. The contract has one constructor $\text{Init}_i^{\text{mp}}$ which given a multi-party channel γ and identities of two parties P and Q as input, creates a "copy" of γ as described above. The only contract function, $\text{Close}_i^{\text{mp}}$, is discussed together with the protocol $\Pi_{dch}(i, \mathcal{C}_0, \pi)$ later in this section.

6.2   Protocol Description

Create a multi-party channel. Parties are instructed by the environment to create a multi-party channel γ via the message (create, γ). As already explained before, parties have to add an instance of mpVSCC_i to every subchannel of γ. This is, on high level, done as follows. Let P and Q be the two parties of a two party channel α which is a subchannel of γ. Let us assume for now that $\text{Order}_\mathcal{P}(P) < \text{Order}_\mathcal{P}(Q)$ (see Sect. 3.1 for the definition of $\text{Order}_\mathcal{P}$). If P receives the message (create, γ) in round τ_0, it requests an update of a contract instance in the state channel α via the hybrid ideal functionality. As parameters of this request, P chooses the channel identifier $cid := P||Q||\gamma.\text{id}$, the contract storage $\text{Init}_i^{\text{mp}}(P, Q, \tau_0, \gamma)$ and contract code mpVSCC_i. Recall that $\text{Init}_i^{\text{mp}}$ is the constructor of the special contract mpVSCC_i. If the party Q also received the message (create, γ) in round τ_0, it knows that it should receive an update request from the hybrid ideal functionality in round $\tau_0 + 1$. If this is indeed the case, Q inspects P's proposal and confirms the update.

Assume that the environment sends (create, γ) to all users of γ in the same round τ_0. If all parties follow the protocol, in round $\tau_0 + 2$ all subchannels of γ should contain a new contract instance with the contract code mpVSCC_i. However, note that a party $P \in \gamma.\text{users}$ only has information about subchannels it is part of, i.e. about subchannels $S_P := \{\alpha \in \gamma.\text{subchan} \mid P \in \alpha.\text{end–users}\}$. To this end, every honest party P sends a message "create–ok" to every other party if all subchannels in S_P contain a new mpVSCC_i instance in round $\tau_0 + 2$. Hence, if all parties are honest, latest in round $\tau_0 + 3$ every party knows that the creation process of γ is completed successfully. However, if there is a malicious party P that sends the "create–ok" to all parties except for one, let us call it Q, then in round $\tau_0 + 3$ only Q thinks that creation failed. In order to reach total consensus on creation among honest parties, Q signals the failure by sending a message "create–not–ok" to all other parties.

To conclude, an honest party outputs $(\text{created}, \gamma)$ to the environment if (1) it received "create–ok" from all parties in round $\tau_0 + 3$ and (2) did not receive any message "create–not–ok" in round $\tau_0 + 4$.

---

[17] This statement assumes that the only contract instances that can be opened in the multi-party channel are the ones whose code allows any user to enforce termination before time $\gamma.\text{validity}$.

Register a contract instance in a multi-party channel. As long as users of the multi-party channel γ behave honestly, they can update/execute contract instances in the channel γ by communicating with each other. However, once the users disagree, they need some third party to fairly resolve their disagreement. The dispute board, modeled by the hybrid ideal functionality $\mathcal{W}_{dch}(i, \mathcal{C}_1, \mathcal{C}_0)$, plays the role of such a judge.

Parties might run into dispute when they update/execute the contract instance or when they are closing the channel γ. In order to avoid code repetition, we define the dispute process as a separate procedure mpRegister(P, id, cid). The input parameter P denotes the initiating party of the dispute process, the parameter id identifies the channel γ and cid is the identifier of the contract instance parties disagree on. The initiating party submits its version of the contract instance, ν^P, to the dispute board which then informs all other parties about P's registration request. If a party Q has a contract instance version with higher version number, i.e. ν^Q.version $> \nu^P$.version, then Q submits this to the dispute board. After a certain time, which is sufficient for other parties to react to P's registration request, any party can complete the process by sending "finalize" to the dispute board which then informs all parties about the result.

Update a contract instance in a multi-party channel. In order to update the storage of a contract instance in a multi-party channel from σ to $\tilde{\sigma}$, the environment sends the message (update, $id, cid, \tilde{\sigma}, \mathsf{C}$) to one of the parties P, which becomes the *initiating party*. Let τ_0 denote the round in which P receives this message. On a high level the update protocol works as follows. P sends the signed new contract storage $\tilde{\sigma}$ to all other parties of γ. Each of these parties $Q \in \gamma.$other$-$party(P) verifies if the update request is valid (i.e., if P's signature is correct) and outputs the update request to the environment. If the environment confirms the request, Q also signs the new contract storage $\tilde{\sigma}$ and sends it as part of the "update$-$ok" message to the other channel parties. In case the environment does not confirm, Q sends a rejection message "update$-$not$-$ok" which contains Q's signature on the original storage σ but with a version number that is increased by two, i.e., if the original version number was w, then Q signs σ with $w + 2$.

If in round $\tau_0 + 2$ a party $P \in \gamma.$users is missing a correctly signed reply from at least one party, it is clear that someone misbehaved. Thus, P initiates the registration procedure to resolve the disagreement via the dispute board.

If P received at least one rejection message, it is unclear to P if there is a malicious party or not. Note that from P's point of view it is impossible to distinguish whether (a) one party sends the "update$-$not$-$ok" message to P and the message "update$-$ok" to all other parties, or (b) one honest party simply does not agree with the update and sends the "update$-$not$-$ok" message to everyone. To resolve this uncertainty, P communicates to all other parties that the update failed by sending the signed message (update$-$not$-$ok, $\sigma, w + 2$) to all other parties. If all honest parties behave as described above, in round $\tau_0 + 3$ party P must have signatures of all parties on the original storage with version number $w + 2$; hence, consensus on rejection is reached. If P does not have all the signatures at this point, it is clear that at least one party is malicious. Thus, P initiates the registration which enforces the consensus via the dispute board.

If P receives a valid "update−ok" from all parties in round $\tau_0 + 2$, she knows that consensus on the updated storage $\tilde{\sigma}$ will eventually be reached. This is because in worst case, P can register $\tilde{\sigma}$ on the dispute board. Still, P has to wait if no other party detects misbehavior and starts the dispute process or sends a reject message in which case P initiates the dispute. If none of this happens, all honest parties output the message "updated" in round $\tau_0 + 3$. Otherwise they output the message after the registration is completed.

Execute a contract instance in a multi-party channel. The environment triggers the execution process by sending the instruction (execute, id, cid, f, z) to a party P in round τ_0. P first tries to perform the execution of the contract instance with identifier cid in a channel γ with identifier id peacefully, i.e. without touching the blockchain. An intuitive design of this process would be to let P compute $f(z)$ locally and send her signature on the new contract storage (together with the environment's instruction) to all other users of γ. Every other user Q would verify this message by recomputing $f(z)$ and confirm the new contract storage by sending her signature on it to the other users of γ.

It is easy to see that this intuitive approach fails when two (or more) parties want to peacefully execute the same contract instance cid in the same round. While in two party channels this can be solved by assigning "time slots" for each party, this idea cannot be generalized to the n-party case, without blowing up the number of rounds needed for peaceful execution from $O(1)$ to $O(n)$. To keep the peaceful execution time constant, we let each contract instance have its own *execution period* which consists of four rounds:

Round 1: If P received (execute, id, cid, f, z) in this or the previous 3 rounds, it sends (peaceful−request, id, cid, f, z, τ_0) to all other parties.

Round 2: P locally sorts[18] all requests it received in this round (potentially including its own from the previous round), locally performs all the executions and sends the signed resulting contract storage to all other parties.

Round 3: If P did not receive valid signatures on the new contract storage from all other parties, it starts the registration process.

Round 4: Unless some party started the registration process, P outputs an execution success message.

If the peaceful execution fails, i.e. one party initiates registration, all execution requests of this period must be performed forcefully via the dispute board.

Close a multi-party channel. The closing procedure of a multi-party channel begins automatically in round γ.validity. Every pair of parties $\{P, Q\} \in \gamma$.E tries to peacefully update the $\mathtt{mpVSCC}_i$ contract instance, let us denote its identifier cid, in their subchannel $\alpha := \gamma$.subchan($\{P, Q\}$). More precisely, both parties locally execute the function $\mathtt{Close}_i^{\mathtt{mp}}$ of contract instance cid with input parameter $z := \gamma$.cspace − the tuple of all contract instances that were ever opened in γ. The function $\mathtt{Close}_i^{\mathtt{mp}}$ adjusts the balances of users in cid according to the provided contract instances in z and unlocks all coins from cid back to α.

---

[18] We assume a fixed ordering on peaceful execution requests.

If the peaceful update fails, then at least one party is malicious and either does not communicate or tries to close the channel γ with a false view on the set γ.cspace. In this case, users have to register all contract instances of γ on the dispute board. This guarantees a fixed global view on γ.cspace. Once the registration process is over, the $\texttt{mpVSCC}_i$ contract instances in the subchannels can be terminated using the execute functionality of $\mathcal{W}_{dch}(i, \mathcal{C}_1, \mathcal{C}_0)$ on function $\texttt{Close}_i^{\texttt{mp}}$. Since the set γ.cspace is now publicly available on the dispute board, the parameter z will be the same in all the $\texttt{mpVSCC}_i$ contract instance executions in the subchannels. Technically, this is taken care of by the wrapper $\mathcal{W}_{ch}(i, \mathcal{C}_1, \mathcal{C}_0)$ which overwrites the parameter z of every execution request with function $\texttt{Close}_i^{\texttt{mp}}$ to the relevant content of the dispute board.

Let us emphasize that the high level description provided in this section excludes some technicalities which are explained in the full version of the paper.

Theorem 2. *Suppose the underlying signature scheme is existentially unforgeable against chosen message attacks. For every set of contract codes $\mathcal{C}_0$, every $i \geq 1$ and every $\Delta \in \mathbb{N}$, the protocol $\Pi_{mpch}(i, \mathcal{C}_0)$ in the $\mathcal{W}_{dch}(i, \mathcal{C}_1, \mathcal{C}_0)$-hybrid model emulates the ideal functionality $\mathcal{F}_{mpch}(i, \mathcal{C}_0)$.*

7   Conclusion

We presented the first full specification and construction of a state channel network that supports multi-party channels. The pessimistic running time of our protocol can be made constant for arbitrary complex channels. While we believe that this is an important contribution by it self, we also think that it is very likely that the techniques developed by us will have applications beyond the area of off-chain channels. In particular, the modeling of multiparty state channels that we have in this paper can be potentially useful in other types of off-chain protocols, e.g., in Plasma [23]. We leave extending our approach to such protocols as an interesting research direction for the future.

Acknowledgments. This work was partly supported by the German Research Foundation (DFG) Emmy Noether Program FA 1320/1-1, the DFG CRC 1119 CROSSING (project S7), the Ethereum Foundation grant *Off-chain labs: formal models, constructions and proofs*, the Foundation for Polish Science (FNP) grant TEAM/2016-1/4, the German Federal Ministry of Education and Research (BMBF) *iBlockchain* project, by the Hessen State Ministry for Higher Education, Research and the Arts (HMWK) and the BMBF within CRISP, and by the Polish National Science Centre (NCN) grant 2014/13/B/ST6/03540, Polish NCBiR Prokrym project.

References

1. Allison, I.: Ethereum's Vitalik Buterin explains how state channels address privacy and scalability (2016)
2. Bitcoin Wiki: Payment Channels (2018). https://en.bitcoin.it/wiki/Payment_channels

3. Canetti, R.: Universally composable security: a new paradigm for cryptographic protocols. In: 42nd FOCS (2001)
4. Canetti, R., Dodis, Y., Pass, R., Walfish, S.: Universally composable security with global setup. In: Vadhan, S.P. (ed.) TCC 2007. LNCS, vol. 4392, pp. 61–85. Springer, Heidelberg (2007). https://doi.org/10.1007/978-3-540-70936-7_4
5. Celer Network (2018). https://www.celer.network
6. Close, T.: Nitro protocol. Cryptology ePrint Archive, Report 2019/219 (2019). https://eprint.iacr.org/2019/219
7. Counterfactual (2018). https://counterfactual.com
8. Decker, C., Wattenhofer, R.: A fast and scalable payment network with bitcoin duplex micropayment channels. In: Pelc, A., Schwarzmann, A.A. (eds.) SSS 2015. LNCS, vol. 9212, pp. 3–18. Springer, Cham (2015). https://doi.org/10.1007/978-3-319-21741-3_1
9. Dolev, D., Strong, H.R.: Authenticated algorithms for Byzantine agreement. SIAM J. Comput. **12**(4), 656–666 (1983)
10. Dziembowski, S., et al.: General state channel networks. In: ACM CCS 2018 (2018)
11. Dziembowski, S., et al.: Multi-party virtual state channels. Cryptology ePrint Archive (2019). https://eprint.iacr.org/2019
12. Dziembowski, S., et al.: Perun: virtual payment hubs over cryptographic currencies. In: Conference Version Accepted to the 40th IEEE Symposium on Security and Privacy (IEEE S&P) 2019 (2017)
13. Garay, J.A., et al.: Round complexity of authenticated broadcast with a dishonest majority. In: 48th FOCS (2007)
14. Katz, J., Lindell, Y.: Introduction to Modern Cryptography (Chapman & Hall/Crc Cryptography and Network Security Series) (2007)
15. Khalil, R., Gervais, A.: NOCUST - a non-custodial 2nd-layer financial intermediary. Cryptology ePrint Archive, Report 2018/642 (2018). https://eprint.iacr.org/2018/642
16. Khalil, R., Gervais, A.: Revive: rebalancing off-blockchain payment networks. In: ACM CCS 2017 (2017)
17. Lind, J., et al.: Teechain: reducing storage costs on the blockchain with offline payment channels. In: Proceedings of the 11th ACM International Systems and Storage Conference, SYSTOR 2018 (2018)
18. Malavolta, G., et al.: Concurrency and privacy with payment-channel networks. In: ACM CCS 2017 (2017)
19. McCorry, P., et al.: Pisa: arbitration outsourcing for state channels. Cryptology ePrint Archive, Report 2018/582 (2018). https://eprint.iacr.org/2018/582
20. McCorry, P., et al.: You sank my battleship! A case study to evaluate state channels as a scaling solution for cryptocurrencies (2018)
21. Miller, A., et al.: Sprites: payment channels that go faster than lightning. CoRR (2017)
22. Nakamoto, S.: Bitcoin: A Peer-to-Peer Electronic Cash System (2009). http://bitcoin.org/bitcoin.pdf
23. Poon, J., Buterin, V.: Plasma: Scalable Autonomous Smart Contracts (2017)
24. Poon, J., Dryja, T.: The bitcoin lightning network: scalable off-chain instant payments. Draft version 0.5.9.2 (2016). https://lightning.network/lightning-network-paper.pdf
25. Roos, S., et al.: Settling payments fast and private: efficient decentralized routing for path-based transactions. In: NDSS (2018)
26. Szabo, N.: Smart contracts: building blocks for digital markets. Extropy Mag. (1996)

Aggregate Cash Systems:
A Cryptographic Investigation
of Mimblewimble

Georg Fuchsbauer[1,2]([☒]), Michele Orrù[1,2], and Yannick Seurin[3]

[1] Inria, Paris, France
[2] École normale supérieure, CNRS, PSL, Paris, France
{georg.fuchsbauer,michele.orru}@ens.fr
[3] ANSSI, Paris, France
yannick.seurin@m4x.org

Abstract. Mimblewimble is an electronic cash system proposed by an anonymous author in 2016. It combines several privacy-enhancing techniques initially envisioned for Bitcoin, such as Confidential Transactions (Maxwell, 2015), non-interactive merging of transactions (Saxena, Misra, Dhar, 2014), and cut-through of transaction inputs and outputs (Maxwell, 2013). As a remarkable consequence, coins can be deleted once they have been spent while maintaining public verifiability of the ledger, which is not possible in Bitcoin. This results in tremendous space savings for the ledger and efficiency gains for new users, who must verify their view of the system.

In this paper, we provide a provable-security analysis for Mimblewimble. We give a precise syntax and formal security definitions for an abstraction of Mimblewimble that we call an *aggregate cash system*. We then formally prove the security of Mimblewimble in this definitional framework. Our results imply in particular that two natural instantiations (with Pedersen commitments and Schnorr or BLS signatures) are provably secure against inflation and coin theft under standard assumptions.

Keywords: Mimblewimble · Bitcoin · Commitments · Aggregate signatures

1 Introduction

Bitcoin and the UTXO model. Proposed in 2008 and launched early 2009, Bitcoin [Nak08] is a decentralized payment system in which transactions are registered in a distributed and publicly verifiable ledger called a blockchain. Bitcoin departs from traditional account-based payment systems where transactions specify an amount moving from one account to another. Instead, each transaction consists of a list of *inputs* and a list of *outputs*.

© International Association for Cryptologic Research 2019
Y. Ishai and V. Rijmen (Eds.): EUROCRYPT 2019, LNCS 11476, pp. 657–689, 2019.
https://doi.org/10.1007/978-3-030-17653-2_22

Each output contains a value (expressed as a multiple of the currency unit, 10^{-8} bitcoin) and a short script specifying how the output can be spent. The most common script is *Pay to Public Key Hash* (P2PKH) and contains the hash of an ECDSA public key, commonly called a Bitcoin address. Each input of a transaction contains a reference to an output of a previous transaction in the blockchain and a script that must match the script of that output. In the case of P2PKH, an input must provide a public key that hashes to the address of the output it spends and a valid signature for this public key.

Each transaction spends one or more previous transaction outputs and creates one or more new outputs, with a total value not larger than the total value of coins being spent. The system is bootstrapped through special transactions called *coinbase* transactions, which have outputs but no inputs and therefore create money (and also serve to incentivize the proof-of-work consensus mechanism, which allows users to agree on the valid state of the blockchain).

To avoid double-spending attacks, each output of a transaction can only be referenced once by an input of a subsequent transaction. Note that this implies that an output must necessarily be spent entirely. As transactions can have multiple outputs, change can be realized by having the sender assign part of the outputs to an address she controls. Since all transactions that ever occurred since the inception of the system are publicly available in the blockchain, whether an output has already been spent can be publicly checked. In particular, every transaction output recorded in the blockchain can be classified either as an *unspent transaction output (UTXO)* if it has not been referenced by a subsequent transaction input so far, or a *spent transaction output (STXO)* otherwise. Hence, the UTXO set "encodes" all bitcoins available to be spent, while the STXO set only contains "consumed" bitcoins and could, in theory, be deleted.

The validation mechanics in Bitcoin requires new users to download and validate the entire blockchain in order to check that their view of the system is not compromised.[1] Consequently, the security of the system and its ability to enroll new users relies on (a significant number of) Bitcoin clients to persistently store the entire blockchain. Once a new node has checked the entire blockchain, it is free to "prune" it[2] and retain only the freshly computed UTXO set, but it will not be able to convince another newcomer that this set is valid.

Consider the following toy example. A coinbase transaction creates an output txo_1 for some amount v associated with a public key pk_1. This output is spent by a transaction T_1 creating a new output txo_2 with amount v associated with a public key pk_2. Transaction T_1 contains a valid signature σ_1 under public key pk_1. Once a node has verified σ_1, it is ensured that txo_2 is valid and the node can therefore delete the coinbase transaction and T_1. By doing this, however, he cannot convince anyone else that output txo_2 is indeed valid.

---

[1] *Simplified Verification Payment (SPV)* clients only download much smaller pieces of the blockchain allowing them to verify specific transactions. However, they are less secure and do not contribute to the general security of the system [GCKG14,SZ16].

[2] This functionality was introduced in Bitcoin Core v0.11, see https://github.com/bitcoin/bitcoin/blob/v0.11.0/doc/release-notes.md#block-file-pruning.

At the time of writing, the size of Bitcoin's blockchain is over 200 GB.[3] Downloading and validating the full blockchain can take up to several days on standard hardware. In contrast, the size of the UTXO set, containing around 60 millions elements, is only a couple of GB.

Bitcoin privacy. Despite some common misconception, Bitcoin offers a very weak level of privacy. Although users can create multiple pseudonymous addresses at will, the public availability of all transaction data often allows to link them and reveals a surprisingly large amount of identifying information, as shown in many works [AKR+13, MPJ+13, RS13, KKM14].

Several protocols have been proposed with the goal of improving on Bitcoin's privacy properties, such as Cryptonote [vS13] (implemented for example by Monero), Zerocoin [MGGR13] and Zerocash [BCG+14]. On the other hand, there are privacy-enhancing techniques compatible with Bitcoin, for example coin mixing [BBSU12, BNM+14, RMK14, HAB+17], to ensure payer anonymity. Below we describe three specific proposals that have paved the way for Mimblewimble.

Confidential Transactions. Confidential Transactions (CT), described by Maxwell [Max15], based on an idea by Back [Bac13] and now implemented by Monero, allow to hide the *values* of transaction outputs. The idea is to replace explicit amounts in transactions by homomorphic commitments: this hides the value contained in each output, but the transaction creator cannot modify this value later on.[4]

More specifically, the amount v in an output is replaced by a Pedersen commitment $C = vH + rG$, where H and G are generators of an (additively denoted) discrete-log-hard group and r is a random value. Using the homomorphic property of the commitment scheme, one can prove that a transaction does not create money out of thin air, i.e., that the sum of the outputs is less than the sum of the inputs. Consider a transaction with input commitments $C_i = v_iH + r_iG$, $1 \leq i \leq n$, and output commitments $\hat{C}_i = \hat{v}_iH + \hat{r}_iG$, $1 \leq i \leq m$. The transaction does not create money *iff* $\sum_{i=1}^{n} v_i \geq \sum_{i=1}^{m} \hat{v}_i$. This can be proved by providing an opening (f, r) with $f \geq 0$ for $\sum_{i=1}^{n} C_i - \sum_{i=1}^{m} \hat{C}_i$, whose validity can be publicly checked. The difference f between inputs and outputs are so-called fees that reward the miner that includes the transaction in a block.

Note that arithmetic on hidden values is done modulo p, the order of the underlying group. Hence, a malicious user could spend an input worth 2 and create two outputs worth 10 and $p-8$, which would look the same as a transaction creating two outputs worth 1 each. To ensure that commitments do not contain large values that cause such mod-p reductions, a non-interactive zero-knowledge (NIZK) proof that the committed value is in $[0, v_{\max}]$ (a so-called *range proof*) is added to each commitment, where $v_{\max}$ is small compared to p.

CoinJoin. When a Bitcoin transaction has multiple inputs and outputs, nothing can be inferred about "which input goes to which output" beyond what is

---

[3] See https://www.blockchain.com/charts/blocks-size.

[4] Commitments are actually never publicly opened; however the opening information is used when spending a coin and remains privy to the participants.

imposed by their values (e.g., if a transaction has two inputs with values 10 BTC and 1 BTC, and two outputs with values 10 BTC and 1 BTC, all that can be said is that at least 9 BTC flowed from the first input to the first output). CoinJoin [Max13a] builds on this technical principle to let different users create a single transaction that combines all of their inputs and outputs. When all inputs and outputs have the same value, this perfectly mixes the coins. Note that unlike CT, CoinJoin does not require any change to the Bitcoin protocol and is already used in practice. However, this protocol is interactive as participants need all input and output addresses to build the transaction. Saxena et al. [SMD14] proposed a modification of the Bitcoin protocol which essentially allows users to perform CoinJoin non-interactively and which relies on so-called *composite* signatures.[5]

Cut-through. A basic property of the UTXO model is that a sequence of two transactions, a first one spending an output txo_1 and creating txo_2, followed by a second one spending txo_2 and creating txo_3, is equivalent to a single *cut-through* transaction spending txo_1 and creating txo_3. While such an optimization is impossible once transactions have been included in the blockchain (as mentioned before, this would violate public verifiability of the blockchain), this has been suggested [Max13b] for *unconfirmed* transactions, i.e., transactions broadcast to the Bitcoin network but not included in a block yet. As we will see, the main added benefit of Mimblewimble is to allow *post-confirmation cut-through*.

Mimblewimble. Mimblewimble was first proposed by an anonymous author in 2016 [Jed16]. The idea was then developed further by Poelstra [Poe16]. At the time of writing, there are at least two independent implementations of Mimblewimble as a cryptocurrency: one is called Grin,[6] the other Beam.[7]

Mimblewimble combines in a clever way CT, a non-interactive version of CoinJoin, and cut-through of transaction inputs and outputs. As with CT, a coin is a commitment $C = vH + rG$ to its value v using randomness r, together with a range proof π. If CT were actually employed in Bitcoin, spending a CT-protected output would require the knowledge of the opening of the commitment *and*, as for a standard output, of the secret key associated with the address controlling the coin. Mimblewimble goes one step further and completely abandons the notion of addresses or more generally scripts: spending a coin *only* requires knowledge of the opening of the commitment. As a result, ownership of a coin $C = vH + rG$ is equivalent to the knowledge of its opening, and the randomness r of the commitment now acts as the *secret key* for the coin.

Exactly as in Bitcoin, a Mimblewimble transaction specifies a list $\mathbf{C} = (C_1, \ldots, C_n)$ of input coins (which must be coins existing in the system) and a list $\hat{\mathbf{C}} = (\hat{C}_1, \ldots, \hat{C}_m)$ of output coins, where $C_i = v_i H + r_i G$ for $1 \le i \le n$

---

[5] An earlier, anonymous version of the paper used the name *one-way aggregate signature* (OWAS), see https://bitcointalk.org/index.php?topic=290971. Composite signatures are very similar to aggregate signatures [BGLS03].

[6] See http://grin-tech.org and https://github.com/mimblewimble/grin/blob/master/doc/intro.md.

[7] See https://www.beam-mw.com.

and $\hat{C}_i = \hat{v}_i H + \hat{r}_i G$ for $1 \le i \le m$. We will detail later how exactly such a transaction is constructed. Leaving fees aside for simplicity, the transaction is balanced (i.e., does not create money) *iff* $\sum \hat{v}_i - \sum v_i = 0$, which, letting $\sum \mathbf{C}$ denote $\sum_{i=1}^{n} C_i$, is equivalent to

$$\sum \hat{\mathbf{C}} - \sum \mathbf{C} = \sum(\hat{v}_i H + \hat{r}_i G) - \sum(v_i H + r_i G) = \left(\sum \hat{r}_i - \sum r_i\right) G.$$

In other words, knowledge of the opening of all coins in the transaction *and* balancedness of the transaction implies knowledge of the discrete logarithm in base G of $E := \sum \hat{\mathbf{C}} - \sum \mathbf{C}$, called the *excess* of the transaction in Mimblewimble jargon. Revealing the opening $(0, r := \sum \hat{r}_i - \sum r_i)$ of the excess E as in CT would leak too much information (e.g., together with the openings of the input coins and of all output coins except one, this would yield the opening of the remaining output coin); however, knowledge of r can be *proved* by providing a valid signature (on the empty message) under public key E using some discrete-log-based signature scheme. Intuitively, as long as the commitment scheme is binding and the signature scheme is unforgeable, it should be infeasible to compute a valid signature for an unbalanced transaction.

Transactions (legitimately) creating money, such as coinbase transactions, can easily be incorporated by letting the *supply* s (i.e., the number of monetary units created by the transaction) be explicitly specified and redefining the excess of the transaction as $E := \sum \hat{\mathbf{C}} - \sum \mathbf{C} - sH$. All in all, a Mimblewimble transaction is a tuple

$$\mathsf{tx} = (s, \mathbf{C}, \hat{\mathbf{C}}, K) \quad \text{with} \quad K := (\boldsymbol{\pi}, \mathbf{E}, \sigma), \tag{1}$$

where s is the supply, $\mathbf{C}$ is the input coin list, $\hat{\mathbf{C}}$ is the output coin list, and K is the so-called *kernel*, which contains the list $\boldsymbol{\pi}$ of range proofs for output coins,[8] the (list of) transaction excesses $\mathbf{E}$ (as there can be several; see below), and a signature σ.[9]

Such transactions can now easily be merged non-interactively *à la* CoinJoin: consider $\mathsf{tx}_0 = (s_0, \mathbf{C}_0, \hat{\mathbf{C}}_0, (\boldsymbol{\pi}_0, E_0, \sigma_0))$ and $\mathsf{tx}_1 = (s_1, \mathbf{C}_1, \hat{\mathbf{C}}_1, (\boldsymbol{\pi}_1, E_1, \sigma_1))$; then the *aggregate transaction* tx resulting from merging tx_0 and tx_1 is simply

$$\mathsf{tx} := \left(s_0 + s_1, \mathbf{C}_0 \parallel \mathbf{C}_1, \hat{\mathbf{C}}_0 \parallel \hat{\mathbf{C}}_1, \left(\boldsymbol{\pi}_0 \parallel \boldsymbol{\pi}_1, (E_0, E_1), (\sigma_0, \sigma_1)\right)\right). \tag{2}$$

Moreover, if the signature scheme supports aggregation, as for example the BLS scheme [BGLS03, BNN07], the pair (σ_0, σ_1) can be replaced by a compact aggregate signature σ for the public keys $\mathbf{E} := (E_0, E_1)$.

An aggregate transaction $(s, \mathbf{C}, \hat{\mathbf{C}}, (\boldsymbol{\pi}, \mathbf{E}, \sigma))$ is valid if all range proofs verify, σ is a valid aggregate signature for $\mathbf{E}$ and if

$$\sum \hat{\mathbf{C}} - \sum \mathbf{C} - sH = \sum \mathbf{E}. \tag{3}$$

---

[8] Since inputs must be coins that already exist in the system, their range proofs are contained in the kernels of the transactions that created them.

[9] A transaction fee can easily be added to the picture by making its amount f explicit and adding fH to the transaction excess. For simplicity, we omit it in this paper.

As transactions can be recursively aggregated, the resulting kernel will contain a list $\mathbf{E}$ of kernel excesses, one for each transaction that has been aggregated.

The main novelty of Mimblewimble, namely cut-through, naturally emerges from the way transactions are aggregated and validated. Assume that some coin C appears as an output in tx_0 and as an input in tx_1; then, one can erase C from the input and output lists of the aggregate transaction tx, and tx will still be valid since (3) will still hold. Hence, each time an output of a transaction tx_0 is spent by a subsequent transaction tx_1, this output can be "forgotten" without losing the ability to validate the resulting aggregate transaction.

In Mimblewimble the ledger is itself a transaction of the form (1), which starts out empty, and to which transactions are recursively aggregated as they are added to the ledger. We assume that for a transaction to be allowed onto the ledger, its input list must be contained in the output list of the ledger (this corresponds to the natural requirement that only coins that exist in the ledger can be spent). Then, it is easy to see that the following holds:

(i) the supply s of the ledger is equal to the sum of the supplies of all transactions added to the ledger so far;
(ii) the input coin list of the ledger is always empty.

Property (i) follows from the definition of aggregation in (2). Property (ii) follows inductively. At the inception of the system the ledger is empty (thus the first transaction added to the ledger must be a transaction with an empty input coin list and non-zero supply, a *minting* transaction). Any transaction tx added to the ledger must have its input coins contained in the output coin list of the ledger; thus cut-through will remove all of them from the joint input list, hence the updated ledger again has no input coins (and the coins spent by tx are deleted from its outputs). The ledger in Mimblewimble is thus a single aggregate transaction whose supply s is equal to the amount of money that was created in the system and whose output coin list $\hat{C}$ is the analogue of the UTXO set in Bitcoin. Its kernel K allows to cryptographically verify its validity. The history of all transactions that have occurred is not retained, and only one kernel excess per transaction (a very short piece of information) is recorded.

Our contribution. We believe it is crucial that protocols undergo a formal security assessment and that the cryptographic guarantees they provide must be well understood before deployment. To this end, we provide a provable-security treatment for Mimblewimble. A first attempt at proving its security was partly undertaken by Poelstra [Poe16]. We follow a different approach: we put forward a general syntax and a framework of game-based security definitions for an abstraction of Mimblewimble that we dub an *aggregate cash system*.

Formalizing security for a cash system requires care. For example, Zerocoin [MGGR13] was recently found to be vulnerable to *denial-of-spending* attacks [RTRS18] that were not captured by the security model in which Zerocoin was proved secure. To avoid such pitfalls, we keep the syntax simple, while allowing to express meaningful security definitions. We formulate two natural properties that define the security of a cash system: *inflation-resistance* ensures that the only way money can be created in a system is explicitly via the supply contained in transactions; *resistance to coin theft* guarantees that no one can spend a user's coins as long as she keeps her keys safe. We moreover define a privacy notion, *transaction indistinguishability*, which states that a transaction does not reveal anything about the values it transfers from its inputs to its outputs.

We then give a black-box construction of an aggregate cash system, which naturally generalizes Mimblewimble, from a homomorphic commitment scheme COM, an (aggregate) signature scheme SIG, and a NIZK range-proof system Π. We believe that such a modular treatment will ease the exploration of post-quantum instantiations of Mimblewimble or related systems.

Note that in our description of Mimblewimble, we have not yet explained how to actually create a transaction that transfers some amount ρ of money from a sender to a receiver. It turns out that this is a delicate question. The initial description of the protocol [Jed16] proposed the following one-round procedure:

- the sender selects input coins $\mathbf{C}$ of total value $v \geq \rho$; it creates *change coins* $\mathbf{C}'$ of total value $v - \rho$ and sends $\mathbf{C}$, $\mathbf{C}'$, range proofs for $\mathbf{C}'$ and the opening $(-\rho, k)$ of $\sum \mathbf{C}' - \sum \mathbf{C}$ to the receiver (over a secure channel);
- the receiver creates additional output coins $\mathbf{C}''$ (and range proofs) of total value ρ with keys (k_i''), computes a signature σ with the secret key $k + \sum k_i''$ and defines $\mathsf{tx} = \left(0, \mathbf{C}, \mathbf{C}' \parallel \mathbf{C}'', \left(\pi, E = \sum \mathbf{C}' + \sum \mathbf{C}'' - \sum \mathbf{C}, \sigma \right) \right)$.

However, a subtle problem arises with this protocol. Once the transaction has been added to the ledger, the change outputs $\mathbf{C}'$ should only be spendable by the sender, who owns them. It turns out that the receiver is also able to spend them by "reverting" the transaction tx. Indeed, he knows the range proofs for coins in $\mathbf{C}$ and the secret key $(-k - \sum k_i'')$ for the transaction with inputs $\mathbf{C}' \parallel \mathbf{C}''$ and outputs $\mathbf{C}$. Arguably, the sender is given back her initial input coins in the process, but (i) she could have deleted the secret keys for these old coins, making them unspendable, and (ii) this violates any meaningful formalization of security against coin theft.

A natural way to prevent such a malicious behavior would be to let the sender and the receiver, each holding a share of the secret key corresponding to public key $E := \sum \mathbf{C}' \parallel \mathbf{C}'' - \sum \mathbf{C}$, engage in a two-party interactive protocol to compute σ. Actually, this seems to be the path Grin is taking, although, to the best of our knowledge, the problem described above with the original protocol has never been documented.

We show that the spirit of the original *non-interactive* protocol can be salvaged, so a sender can make a payment to a receiver without the latter's active involvement. In our solution the sender first constructs a full-fledged transaction tx spending $\mathbf{C}$ and creating change coins $\mathbf{C}'$ as well as a special output coin $C = \rho H + kG$, and sends tx and the opening (ρ, k) of the special coin to the receiver. (Note that, unlike in the previous case, k is now independent from the keys of the coins in $\mathbf{C}$ and $\mathbf{C}'$.) The receiver then creates a second transaction tx' spending the special coin C and creating its own output coins $\mathbf{C}''$ and aggregates tx and tx'. As intended, this results in a transaction with inputs $\mathbf{C}$ and outputs $\mathbf{C}' \parallel \mathbf{C}''$ since C is removed by cut-through. The only drawback of this procedure is that the final transaction, being the aggregate of two transactions, has two kernel excesses instead of one for the interactive protocol mentioned above.

After specifying our protocol $\mathsf{MW}[\mathsf{COM}, \mathsf{SIG}, \Pi]$, we turn to proving its security in our definitional framework. To this end, we first define two security notions, EUF-NZO and EUF-CRO, tying the commitment scheme and the signature scheme together (cf. Page 12). Assuming that proof system Π is simulation-extractable [DDO+01, Gro06], we show that EUF-NZO-security for the pair $(\mathsf{COM}, \mathsf{SIG})$ implies that MW is resistant to inflation, while EUF-CRO-security implies that MW is resistant to coin theft. Transaction indistinguishability follows from zero-knowledge of Π and COM being hiding.

Finally, we consider two natural instantiations of $\mathsf{MW}[\mathsf{COM}, \mathsf{SIG}, \Pi]$. For each, we let COM be the Pedersen commitment scheme [Ped92]. When SIG is instantiated with the Schnorr signature scheme [Sch91], we show that the pair $(\mathsf{COM}, \mathsf{SIG})$ is EUF-NZO- and EUF-CRO-secure under the Discrete Logarithm assumption. When SIG is instantiated with the BLS signature scheme [BLS01], we show that the pair $(\mathsf{COM}, \mathsf{SIG})$ is EUF-NZO- and EUF-CRO-secure under the CDH assumption. Both proofs are in the random-oracle model. BLS signatures have the additional benefit of supporting aggregation [BGLS03, BNN07], so that the ledger kernel always contains a short aggregate signature, independently of the number of transactions that have been added to the ledger. We stress that, unlike Zerocash [BCG+14], none of these two instantiations require a trusted setup.

2   Preliminaries

2.1   General Notation

We denote the (closed) integer interval from a to b by $[a, b]$. We use $[b]$ as shorthand for $[1, b]$. A function $\mu \colon \mathbb{N} \to [0, 1]$ is negligible (denoted $\mu = \mathsf{negl}$) if for all $c \in \mathbb{N}$ there exists $\lambda_c \in \mathbb{N}$ such that $\mu(\lambda) \leq \lambda^{-c}$ for all $\lambda \geq \lambda_c$. A function ν is *overwhelming* if $1 - \nu = \mathsf{negl}$. Given a non-empty finite set S, we let $x \leftarrow_{\$} S$ denote the operation of sampling an element x from S uniformly at random. By $y := M(x_1, \ldots; r)$ we denote the operation of running algorithm M on inputs $x_1, \ldots$ and coins r and letting y denote the output. By $y \leftarrow M(x_1, \ldots)$, we denote letting $y := M(x_1, \ldots; r)$ for random r, and $[M(x_1, \ldots)]$ is the set of values that have positive probability of being output by M on inputs $x_1, \ldots$ If an algorithm calls a subroutine which returns $\perp$, we assume it stops and returns $\perp$ (this does not hold for an adversary calling an *oracle* which returns $\perp$).

A list $\mathbf{L} = (x_1, \ldots, x_n)$, also denoted $(x_i)_{i=1}^{n}$, is a finite sequence. The length of a list $\mathbf{L}$ is denoted $|\mathbf{L}|$. For $i = 1, \ldots, |\mathbf{L}|$, the i-th element of $\mathbf{L}$ is denoted $\mathbf{L}[i]$, or L_i when no confusion is possible. By $\mathbf{L}_0 \parallel \mathbf{L}_1$ we denote the list $\mathbf{L}_0$ followed by $\mathbf{L}_1$. The empty list is denoted $(\,)$. Given a list $\mathbf{L}$ of elements of an additive group, we let $\sum \mathbf{L}$ denote the sum of all elements of $\mathbf{L}$. Let $\mathbf{L}_0$ and $\mathbf{L}_1$ be two lists, each without repetition. We write $\mathbf{L}_0 \subseteq \mathbf{L}_1$ *iff* each element of $\mathbf{L}_0$ also appears in $\mathbf{L}_1$. We define $\mathbf{L}_0 \cap \mathbf{L}_1$ to be the list of all elements that simultaneously appear in both $\mathbf{L}_0$ and $\mathbf{L}_1$, ordered as in $\mathbf{L}_0$. The difference between $\mathbf{L}_0$ and $\mathbf{L}_1$, denoted $\mathbf{L}_0 - \mathbf{L}_1$, is the list of all elements of $\mathbf{L}_0$ that do not appear in $\mathbf{L}_1$, ordered as in $\mathbf{L}_0$. So, for example $(1, 2, 3) - (2, 4) = (1, 3)$. We define the *cut-through* of two lists $\mathbf{L}_0$ and $\mathbf{L}_1$, denoted $\mathsf{cut}(\mathbf{L}_0, \mathbf{L}_1)$, as

$$\mathsf{cut}(\mathbf{L}_0, \mathbf{L}_1) := (\mathbf{L}_0 - \mathbf{L}_1, \mathbf{L}_1 - \mathbf{L}_0).$$

Game $\mathrm{HID}_{\mathsf{COM},\mathcal{A}}(\lambda)$	Oracle $\mathrm{COMMIT}(v_0, v_1)$	Game $\mathrm{BND}_{\mathsf{COM},\mathcal{A}}(\lambda)$
$b \leftarrow_{\$} \{0,1\}$	$r \leftarrow_{\$} \mathcal{R}_{\mathsf{cp}}$	$\mathsf{mp} \leftarrow \mathsf{MainSetup}(1^{\lambda})$
$\mathsf{mp} \leftarrow \mathsf{MainSetup}(1^{\lambda})$	$C := \mathsf{COM}.\mathsf{Cmt}(\mathsf{cp}, v_b, r)$	$\mathsf{cp} \leftarrow \mathsf{COM}.\mathsf{Setup}(\mathsf{mp})$
$\mathsf{cp} \leftarrow \mathsf{COM}.\mathsf{Setup}(\mathsf{mp})$	$\mathbf{return}\ C$	$(v_0, r_0, v_1, r_1) \leftarrow \mathcal{A}(\mathsf{cp})$
$b' \leftarrow \mathcal{A}^{\mathrm{COMMIT}}(\mathsf{cp})$		$C_0 := \mathsf{COM}.\mathsf{Cmt}(\mathsf{cp}, v_0, r_0)$
$\mathbf{return}\ b = b'$		$C_1 := \mathsf{COM}.\mathsf{Cmt}(\mathsf{cp}, v_1, r_1)$
		$\mathbf{return}\ v_0 \neq v_1\ \mathbf{and}\ C_0 = C_1$

Fig. 1. The games for hiding and binding of a commitment scheme COM.

2.2 Cryptographic Primitives

We introduce the three building blocks we will use to construct an aggregate cash system: a commitment scheme COM, an aggregate signature scheme SIG, and a non-interactive zero-knowledge proof system Π. For compatibility reasons, the setup algorithms for each of these schemes are split: a common algorithm $\mathsf{MainSetup}(1^{\lambda})$ first returns *main* parameters mp (specifying e.g. an abelian group), and specific algorithms COM.Setup, SIG.Setup, and Π.Setup take as input mp and return the specific parameters cp, sp, and crs for each primitive. We assume that mp is contained in cp, sp, and crs.

Commitment scheme. A commitment scheme COM consists of the following algorithms:

- $\mathsf{cp} \leftarrow \mathsf{COM}.\mathsf{Setup}(\mathsf{mp})$: the setup algorithm takes as input main parameters mp and outputs commitment parameters cp, which implicitly define a value space $\mathcal{V}_{\mathsf{cp}}$, a randomness space $\mathcal{R}_{\mathsf{cp}}$, and a commitment space $\mathcal{C}_{\mathsf{cp}}$;
- $C := \mathsf{COM}.\mathsf{Cmt}(\mathsf{cp}, v, r)$: the (deterministic) commitment algorithm takes as input commitment parameters cp, a value $v \in \mathcal{V}_{\mathsf{cp}}$ and randomness $r \in \mathcal{R}_{\mathsf{cp}}$, and outputs a commitment $C \in \mathcal{C}_{\mathsf{cp}}$.

In most instantiations, given a value $v \in \mathcal{V}_{\mathsf{cp}}$, the sender picks $r \leftarrow_{\$} \mathcal{R}_{\mathsf{cp}}$ uniformly at random and computes the commitment $C = \mathsf{COM}.\mathsf{Cmt}(\mathsf{cp}, v, r)$. To *open* the commitment, the sender reveals (v, r) so anyone can verify that $\mathsf{COM}.\mathsf{Cmt}(\mathsf{cp}, v, r) = C$.

We require commitment schemes to be *hiding*, meaning that commitment C reveals no information about v, and *binding*, which means that the sender cannot open the commitment in two different ways.

Definition 1 (Hiding). *Let game* HID *be as defined Fig. 1. A commitment scheme* COM *is hiding if for any p.p.t. adversary $\mathcal{A}$:*

$$\mathsf{Adv}^{\mathrm{hid}}_{\mathsf{COM},\mathcal{A}}(\lambda) := 2 \cdot \left| \Pr\left[\mathrm{HID}_{\mathsf{COM},\mathcal{A}}(\lambda) = \mathbf{true} \right] - \tfrac{1}{2} \right| = \mathsf{negl}(\lambda).$$

Definition 2 (Binding). *Let game* BND *be as defined in Fig. 1. A commitment scheme* COM *is binding if for any p.p.t. adversary* $\mathcal{A}$:

$$\mathsf{Adv}_{\mathsf{COM},\mathcal{A}}^{\mathsf{bnd}}(\lambda) := \Pr\left[\mathsf{BND}_{\mathsf{COM},\mathcal{A}}(\lambda) = \mathbf{true}\right] = \mathsf{negl}(\lambda).$$

Lemma 3 (Collision-resistance). *Let* COM *be a (binding and hiding) commitment scheme. Then for any* $(v_0, v_1) \in \mathcal{V}_{\mathsf{cp}}^2$, *the probability that* $\mathsf{Cmt}(\mathsf{cp}, v_0, r_0) = \mathsf{Cmt}(\mathsf{cp}, v_1, r_1)$ *for* $r_0, r_1 \leftarrow_{\$} \mathcal{R}_{\mathsf{cp}}$ *is negligible.*

The proof of the lemma is straightforward: for $v_0 \neq v_1$ this would break binding and for $v_0 = v_1$ it would break hiding.

A commitment scheme is (additively) *homomorphic* if the value, randomness, and commitment spaces are groups (denoted additively) and for any commitment parameters cp, any $v_0, v_1 \in \mathcal{V}_{\mathsf{cp}}$, and any $r_0, r_1 \in \mathcal{R}_{\mathsf{cp}}$, we have:

$$\mathsf{COM.Cmt}(\mathsf{cp}, v_0, r_0) + \mathsf{COM.Cmt}(\mathsf{cp}, v_1, r_1) = \mathsf{COM.Cmt}(\mathsf{cp}, v_0 + v_1, r_0 + r_1).$$

Recursive aggregate signature scheme. An aggregate signature scheme allows to (publicly) combine an arbitrary number n of signatures (from potentially distinct users and on potentially distinct messages) into a single (ideally short) signature [BGLS03, LMRS04, BNN07]. Traditionally, the syntax of an aggregate signature scheme only allows the aggregation algorithm to take as input individual signatures. We consider aggregate signature schemes supporting *recursive* aggregation, where the aggregation algorithm can take as input aggregate signatures (supported for example by the schemes based on BLS signatures [BGLS03, BNN07]). A recursive aggregate signature scheme SIG consists of the following algorithms:

- sp $\leftarrow$ SIG.Setup(mp): the setup algorithm takes as input main parameters mp and outputs signature parameters sp, which implicitly define a secret-key space $\mathcal{S}_{\mathsf{sp}}$ and a public-key space $\mathcal{P}_{\mathsf{sp}}$ (we let the message space be $\{0,1\}^*$);
- (sk, pk) $\leftarrow$ SIG.KeyGen(sp): the key generation algorithm takes signature parameters sp and outputs a secret key sk $\in \mathcal{S}_{\mathsf{sp}}$ and a public key pk $\in \mathcal{P}_{\mathsf{sp}}$;
- $\sigma \leftarrow$ SIG.Sign(sp, sk, m): the signing algorithm takes as input parameters sp, a secret key sk $\in \mathcal{S}_{\mathsf{sp}}$, and a message $m \in \{0,1\}^*$ and outputs a signature σ;
- $\sigma \leftarrow$ SIG.Agg$(\mathsf{sp}, (\mathbf{L}_0, \sigma_0), (\mathbf{L}_1, \sigma_1))$: the aggregation algorithm takes parameters sp and two pairs of public-key/message lists $\mathbf{L}_i = \left((\mathsf{pk}_{i,j}, m_{i,j})\right)_{j=1}^{|\mathbf{L}_i|}$ and (aggregate) signatures σ_i, $i = 0, 1$; it returns an aggregate signature σ;
- *bool* $\leftarrow$ SIG.Ver(sp, $\mathbf{L}$, σ): the (deterministic) verification algorithm takes parameters sp, a list $\mathbf{L} = \left((\mathsf{pk}_i, m_i)\right)_{i=1}^{|\mathbf{L}|}$ of public-key/message pairs, and an aggregate signature σ; it returns **true** or **false**, indicating validity of σ.

Correctness of a recursive aggregate signature scheme is defined recursively. An aggregate signature scheme is *correct* if for every λ, every message $m \in \{0,1\}^*$, every mp $\in [\mathsf{MainSetup}(1^\lambda)]$, sp $\in [\mathsf{SIG.Setup}(\mathsf{mp})]$, (sk, pk) $\in [\mathsf{SIG.KeyGen}(\mathsf{sp})]$ and every $(\mathbf{L}_0, \sigma_0), (\mathbf{L}_1, \sigma_1)$ with $\mathsf{SIG.Ver}(\mathsf{sp}, \mathbf{L}_i, \sigma_i) = \mathbf{true}$ for $i = 0, 1$ we have

Game EUF-CMA$_{\mathsf{SIG},\mathcal{A}}(\lambda)$	Oracle SIGN(m)
$Q := (\,)$; $\mathsf{mp} \leftarrow \mathsf{MainSetup}(1^\lambda)$	$\sigma \leftarrow \mathsf{SIG.Sign}(\mathsf{sk}, m)$
$\mathsf{sp} \leftarrow \mathsf{SIG.Setup(mp)}$; $(\mathsf{sk}, \mathsf{pk}) \leftarrow \mathsf{SIG.KeyGen(sp)}$	$Q := Q \parallel (m)$
$(\mathbf{L}, \sigma) \leftarrow \mathcal{A}^{\mathrm{SIGN}}(\mathsf{pk})$	return σ
return $\big(\exists m : (\mathsf{pk}, m) \in \mathbf{L} \wedge m \notin Q\big)$ and $\mathsf{SIG.Ver}(\mathsf{sp}, \mathbf{L}, \sigma)$	

Fig. 2. The EUF-CMA security game for an aggregate signature scheme SIG.

$$\Pr\big[\mathsf{SIG.Ver}\big(\mathsf{sp}, ((\mathsf{pk}, m)), \mathsf{SIG.Sign}(\mathsf{sp}, \mathsf{sk}, m)\big) = \mathbf{true}\big] = 1 \text{ and}$$
$$\Pr\big[\mathsf{SIG.Ver}\big(\mathsf{sp}, \mathbf{L}_0 \parallel \mathbf{L}_1, \mathsf{SIG.Agg}\big(\mathsf{sp}, (\mathbf{L}_0, \sigma_0), (\mathbf{L}_1, \sigma_1)\big)\big) = \mathbf{true}\big] = 1.$$

Note that for any recursive aggregate signature scheme, one can define an aggregation algorithm $\mathsf{SIG.Agg}'$ that takes as input a list of triples $\big((\mathsf{pk}_i, m_i, \sigma_i)\big)_{i=1}^{n}$ and returns an aggregate signature σ for $\big((\mathsf{pk}_i, m_i)\big)_{i=1}^{n}$, which is the standard syntax for an aggregate signature scheme. Algorithm $\mathsf{SIG.Agg}'$ calls $\mathsf{SIG.Agg}$ recursively $n - 1$ times, aggregating one signature at a time.

The standard security notion for aggregate signature schemes is *existential unforgeability under chosen-message attack* (EUF-CMA) [BGLS03, BNN07].

Definition 4 (EUF-CMA). *Let game* EUF-CMA *be as defined in Fig. 2. An aggregate signature scheme* SIG *is* existentially unforgeable under chosen-message attack *if for any p.p.t. adversary* $\mathcal{A}$,

$$\mathsf{Adv}_{\mathsf{SIG},\mathcal{A}}^{\mathrm{euf\text{-}cma}}(\lambda) := \Pr\big[\mathrm{EUF\text{-}CMA}_{\mathsf{SIG},\mathcal{A}}(\lambda) = \mathbf{true}\big] = \mathsf{negl}(\lambda).$$

Note that any standard signature scheme can be turned into an aggregate signature scheme by letting the aggregation algorithm simply concatenate signatures, i.e., $\mathsf{SIG.Agg}(\mathsf{sp}, (\mathbf{L}_0, \sigma_0), (\mathbf{L}_1, \sigma_1))$ returns (σ_0, σ_1), but this is not compact. Standard EUF-CMA-security of the original scheme implies EUF-CMA-security in the sense of Definition 4 for this construction. This allows us to capture standard and (compact) aggregate signature schemes, such as the ones proposed in [BGLS03, BNN07], in a single framework.

Compatibility. For our aggregate cash system, we require the commitment scheme COM and the aggregate signature scheme SIG to satisfy some "combined" security notions. We say that COM and SIG are *compatible* if they use the same MainSetup and if for any λ, any $\mathsf{mp} \in [\mathsf{MainSetup}(1^\lambda)]$, $\mathsf{cp} \in [\mathsf{COM.Setup(mp)}]$ and $\mathsf{sp} \in [\mathsf{SIG.Setup(mp)}]$, the following holds:

- $\mathcal{S}_{\mathsf{sp}} = \mathcal{R}_{\mathsf{cp}}$, i.e., the secret-key space of SIG is the same as the randomness space of COM;
- $\mathcal{P}_{\mathsf{sp}} = \mathcal{C}_{\mathsf{cp}}$, i.e., the public-key space of SIG is the commitment space of COM;
- $\mathsf{SIG.KeyGen}$ draws $\mathsf{sk} \leftarrow_{\$} \mathcal{R}_{\mathsf{cp}}$ and sets $\mathsf{pk} := \mathsf{COM.Cmt}(\mathsf{cp}, 0, \mathsf{sk})$.

We define two security notions for compatible commitment and aggregate signature schemes. The first one roughly states that only commitments to zero can serve as signature-verification keys; more precisely, a p.p.t. adversary cannot simultaneously produce a signature for a (set of) freely chosen public key(s) *and* a non-zero opening of (the sum of) the public key(s).

Game $\text{EUF-NZO}_{\text{COM},\text{SIG},\mathcal{A}}(\lambda)$

$\text{mp} \leftarrow \text{MainSetup}(1^{\lambda})$; $\text{cp} \leftarrow \text{COM.Setup}(\text{mp})$; $\text{sp} \leftarrow \text{SIG.Setup}(\text{mp})$

$(\mathbf{L}, \sigma, (v, r)) \leftarrow \mathcal{A}(\text{cp}, \text{sp})$

$((X_i, m_i))_{i=1}^{n} := \mathbf{L}$

return $\text{SIG.Ver}(\text{sp}, \mathbf{L}, \sigma)$ **and** $\sum_{i=1}^{n} X_i = \text{COM.Cmt}(\text{cp}, v, r)$ **and** $v \neq 0$

Fig. 3. The EUF-NZO security game for a pair of compatible additively homomorphic commitment and aggregate signature schemes (COM, SIG).

Definition 5 (EUF-NZO). *Let game* EUF-NZO *be as defined in Fig. 3. A pair of compatible homomorphic commitment and aggregate signature schemes* (COM, SIG) *is* existentially unforgeable with non-zero opening *if for any p.p.t. adversary* $\mathcal{A}$,

$$\text{Adv}_{\text{COM},\text{SIG},\mathcal{A}}^{\text{euf-nzo}}(\lambda) := \Pr\left[\text{EUF-NZO}_{\text{COM},\text{SIG},\mathcal{A}}(\lambda) = \textbf{true}\right] = \text{negl}(\lambda).$$

EUF-NZO-security of the pair (COM, SIG) implies that COM is binding, as shown in the full version [FOS18].

The second security definition is more involved. It roughly states that, given a challenge public key C^*, no adversary can produce a signature under $-C^*$. Moreover, we only require the adversary to make a signature under keys $X_1, \ldots, X_n$ of its choice, as long as it knows an opening to the difference between their sum and $-C^*$. This must even hold if the adversary is given a signing oracle for keys related to C^*. Informally, the adversary is faced with the following dilemma: either it picks public keys $X_1, \ldots, X_n$ honestly, so it can produce a signature but it cannot open $\sum X_i + C^*$; or it includes $-C^*$ within the public keys, allowing it to open $\sum X_i + C^*$, but then it cannot produce a signature.

Definition 6 (EUF-CRO). *Let game* EUF-CRO *be as defined in Fig. 4. A pair of compatible homomorphic commitment and aggregate signature schemes* (COM, SIG) *is* existentially unforgeable with challenge-related opening *if for any p.p.t. adversary* $\mathcal{A}$,

$$\text{Adv}_{\text{COM},\text{SIG},\mathcal{A}}^{\text{euf-cro}}(\lambda) := \Pr\left[\text{EUF-CRO}_{\text{COM},\text{SIG},\mathcal{A}}(\lambda) = \textbf{true}\right] = \text{negl}(\lambda).$$

NIZK. Let R be an efficiently computable ternary relation. For triplets $(\text{mp}, u, w) \in \text{R}$ we call u the statement and w the witness. A non-interactive proof system Π for R consists of the following three algorithms:

Game EUF-CRO$_{\mathsf{COM},\mathsf{SIG},\mathcal{A}}(\lambda)$	Oracle $\mathrm{SIGN}'(a, m)$
$\mathsf{mp} \leftarrow \mathsf{MainSetup}(1^\lambda)$; $\mathsf{cp} \leftarrow \mathsf{COM.Setup}(\mathsf{mp})$	$\mathsf{sk}' := a + r^*$
$\mathsf{sp} \leftarrow \mathsf{SIG.Setup}(\mathsf{mp})$; $(r^*, C^*) \leftarrow \mathsf{SIG.KeyGen}(\mathsf{sp})$	**return** $\mathsf{SIG.Sign}(\mathsf{sp}, \mathsf{sk}', m)$
$(\mathbf{L}, \sigma, (v, r)) \leftarrow \mathcal{A}^{\mathrm{SIGN}'}(\mathsf{cp}, \mathsf{sp}, C^*)$; $((X_i, m_i))_{i=1}^n := \mathbf{L}$	
return $\mathsf{SIG.Ver}(\mathsf{sp}, \mathbf{L}, \sigma)$ **and** $\sum_{i=1}^n X_i = -C^* + \mathsf{COM.Cmt}(\mathsf{cp}, v, r)$	

Fig. 4. The EUF-CRO security game for a pair of compatible additively homomorphic commitment and aggregate signature schemes $(\mathsf{COM}, \mathsf{SIG})$.

- $\mathsf{crs} \leftarrow \Pi.\mathsf{Setup}(\mathsf{mp})$: the setup algorithm takes as input main parameters mp and outputs a common reference string (CRS) crs;
- $\pi \leftarrow \Pi.\mathsf{Prv}(\mathsf{crs}, u, w)$: the prover algorithm takes as input a CRS crs and a pair (u, w) and outputs a proof π;
- $bool \leftarrow \Pi.\mathsf{Ver}(\mathsf{crs}, u, \pi)$: the verifier algorithm takes a CRS crs, a statement u, and a proof π and outputs **true** or **false**, indicating acceptance of the proof.

Proof system Π is *complete* if for every λ and every adversary $\mathcal{A}$,

$$\Pr \left[\begin{array}{c} \mathsf{mp} \leftarrow \mathsf{MainSetup}(1^\lambda)\,; \mathsf{crs} \leftarrow \Pi.\mathsf{Setup}(\mathsf{mp}) \\ (u, w) \leftarrow \mathcal{A}(\mathsf{crs})\,; \pi \leftarrow \Pi.\mathsf{Prv}(\mathsf{crs}, u, w) \end{array} : \begin{array}{c} (\mathsf{mp}, u, w) \in \mathsf{R} \Rightarrow \\ \Pi.\mathsf{Ver}(\mathsf{crs}, u, \pi) = \mathbf{true} \end{array} \right] = 1.$$

A proof system Π is *zero-knowledge* if proofs leak no information about the witness. We define a *simulator* $\Pi.\mathsf{Sim}$ for a proof system Π as a pair of algorithms:

- $(\mathsf{crs}, \tau) \leftarrow \Pi.\mathsf{SimSetup}(\mathsf{mp})$: the simulated setup algorithm takes main parameters mp and outputs a CRS together with a trapdoor τ;
- $\pi^* \leftarrow \Pi.\mathsf{SimPrv}(\mathsf{crs}, \tau, u)$: the simulated prover algorithm takes as input a CRS, a trapdoor τ, and a statement u and outputs a simulated proof π^*.

Definition 7 (Zero-knowledge). *Let game* ZK *be as defined in Fig. 5. A proof system Π for relation R is* zero-knowledge *if there exists a simulator $\Pi.\mathsf{Sim}$ such that for any p.p.t. adversary $\mathcal{A}$,*

$$\mathsf{Adv}^{\mathrm{zk}}_{\Pi,\mathsf{R},\mathcal{A}}(\lambda) := 2 \cdot \left| \Pr\left[\mathrm{ZK}_{\Pi,\mathsf{R},\mathcal{A}}(\lambda) = \mathbf{true} \right] - \tfrac{1}{2} \right| = \mathsf{negl}(\lambda).$$

Game ZK$_{\Pi,\mathsf{R},\mathcal{A}}(\lambda)$	Oracle $\mathrm{SIMPROVE}(u, w)$
$b \leftarrow_\$ \{0, 1\}$; $\mathsf{mp} \leftarrow \mathsf{MainSetup}(1^\lambda)$	**if** $\neg\mathsf{R}(\mathsf{mp}, u, w)$ **then return** $\bot$
$\mathsf{crs}_0 \leftarrow \Pi.\mathsf{Setup}(\mathsf{mp})$	$\pi_0 \leftarrow \Pi.\mathsf{Prv}(\mathsf{crs}_0, u, w)$
$(\mathsf{crs}_1, \tau) \leftarrow \Pi.\mathsf{SimSetup}(\mathsf{mp})$	$\pi_1 \leftarrow \Pi.\mathsf{SimPrv}(\mathsf{crs}_1, \tau, u)$
$b' \leftarrow \mathcal{A}^{\mathrm{SIMPROVE}}(\mathsf{crs}_b)$	**return** π_b
return $b = b'$	

Fig. 5. The non-interactive zero-knowledge game for a proof system Π.

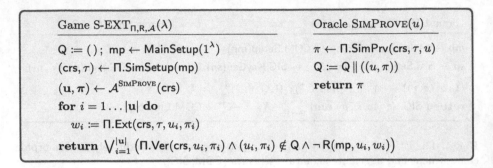

Fig. 6. The (multi-statement) simulation-extractability game for a proof system Π.

Note that the zero-knowledge advantage can equivalently be defined as

$$\mathsf{Adv}^{zk}_{\Pi,R,\mathcal{A}}(\lambda) = \big| \Pr\big[\mathrm{ZK}^1_{\Pi,R,\mathcal{A}}(\lambda) = 1\big] - \Pr\big[\mathrm{ZK}^1_{\Pi,R,\mathcal{A}}(\lambda) = 1\big]\big|,$$

where the game $\mathrm{ZK}^i_{\Pi,R,\mathcal{A}}(\lambda)$ is defined as $\mathrm{ZK}_{\Pi,R,\mathcal{A}}(\lambda)$ except $b \leftarrow_\$ \{0,1\}$ is replaced by $b := i$ and the game returns b'.

The central security property of a proof system is *soundness*, that is, no adversary can produce a proof for a false statement. A stronger notion is *knowledge-soundness*, meaning that an adversary must know a witness in order to make a proof. This is formalized via an extraction algorithm defined as follows:

- $w := \Pi.\mathsf{Ext}(\mathsf{crs}, \tau, u, \pi)$: the (deterministic) extraction algorithm takes a CRS, a trapdoor τ, a statement u, and a proof π and returns a witness w.

Knowledge-soundness states that from a valid proof for a statement u output by an adversary, Π.Ext can extract a witness for u. In security proofs where the reduction simulates certain proofs knowledge-soundness is not sufficient. The stronger notion *simulation-extractability* guarantees that even then, from every proof output by the adversary, Π.Ext can extract a witness. Note that we define a multi-statement variant of simulation extractability: the adversary returns a list of statements and proofs and wins if there is a least one statement such that the corresponding proof is valid and the extractor fails to extract a witness.

Definition 8 (Simulation-Extractability). *Let game S-EXT be as defined in Fig. 6. A non-interactive proof system Π for R with simulator Π.Sim is (multi-statement) simulation-extractable if there exists an extractor Π.Ext such that for any p.p.t. adversary $\mathcal{A}$,*

$$\mathsf{Adv}^{s\text{-}ext}_{\Pi,R,\mathcal{A}}(\lambda) := \Pr\big[\text{S-EXT}_{\Pi,R,\mathcal{A}}(\lambda) = \mathbf{true}\big] = \mathsf{negl}(\lambda).$$

In the instantiation of our cash system, we will deal with *families* of relations, i.e. relations R_δ parametrized by some $\delta \in \mathbb{N}$. For those, we assume that the proof system Π is defined over the family of relations $R = \{R_\delta\}_\delta$ and that the setup algorithm Π.Setup takes an additional parameter δ which specifies the particular relation used during the protocol (and which is included in the returned CRS). For instance, in the case of proofs for a certain range $[0, v_{\max}]$, the proof system will be defined over a relation $\mathsf{R}_{v_{\max}}$, where $v_{\max}$ is the maximum integer allowed.

3 Aggregate Cash System

3.1 Syntax

Coins. The public parameters pp set up by the cash system specify a coin space $\mathcal{C}_{\mathsf{pp}}$ and a key space $\mathcal{K}_{\mathsf{pp}}$. A *coin* is an element $C \in \mathcal{C}_{\mathsf{pp}}$; to each coin is associated a *coin key* $k \in \mathcal{K}_{\mathsf{pp}}$, which allows spending the coin. The *value* v of a coin is an integer in $[0, v_{\max}]$, where $v_{\max}$ is a system parameter. We assume that there exists a function mapping pairs $(v, k) \in [0, v_{\max}] \times \mathcal{K}_{\mathsf{pp}}$ to coins in $\mathcal{C}_{\mathsf{pp}}$; we do not assume this mapping to be invertible or even injective.

Ledger. Similarly to any ledger-based currency such as Bitcoin, an aggregate cash system keeps track of available coins in the system via a ledger. We assume the ledger to be unique and available at any time to all users. How users are kept in consensus on the ledger is outside the scope of this paper. In our abstraction, a ledger Λ simply provides two attributes: a list of all coins available in the system Λ.out, and the total value Λ.sply those coins add up to. We say that a coin C *exists in the ledger* Λ if $C \in \Lambda$.out.

Transactions. Transactions allow to modify the state of the ledger. Formally, a *transaction* tx provides three attributes: a *coin input list* tx.in, a *coin output list* tx.out, and a *supply* tx.sply $\in \mathbb{N}$ specifying the amount of money created by tx. We classify transactions into three types. A transaction tx is said to be:

- a *minting transaction* if tx.sply > 0 and tx.in $= (\ )$; such a transaction creates new coins of total value tx.sply in the ledger;
- a *transfer transaction* if tx.sply $= 0$ and tx.in $\neq (\ )$; such a transaction transfers coins (by spending previous transaction outputs and creating new ones) but does not increase the overall value of coins in the ledger;
- a *mixed transaction* if tx.sply > 0 and tx.in $\neq (\ )$.

Pre-transactions. Pre-transactions allow users to transfer money to each other. Formally, a *pre-transaction* provides three attributes: a *coin input list* ptx.in, a *list of change coins* ptx.chg, and a *remainder* ptx.rmdr. When Alice wants to send money worth ρ to Bob, she selects coins of hers of total value $v \geq \rho$ and specifies the desired values for her change coins when $v > \rho$. The resulting pre-transaction ptx has therefore some input coin list ptx.in with total amount v, a change coin list ptx.chg, and some remainder $\rho = $ ptx.rmdr. Alice sends this pre-transaction (via a secure channel) to Bob, who, in turn, finalizes it into a valid transaction and adds it to the ledger.

Aggregate cash system. An aggregate cash system CASH consists of the following algorithms:

- $(\text{pp}, \Lambda) \leftarrow \text{Setup}(1^\lambda, v_{\max})$: the setup algorithm takes as input the security parameter λ in unary and a maximal coin value $v_{\max}$ and returns public parameters pp and an initial (empty) ledger Λ.
- $(\text{tx}, \mathbf{k}) \leftarrow \text{Mint}(\text{pp}, \mathbf{v})$: the mint algorithm takes as input a list of values $\mathbf{v}$ and returns a minting transaction tx and a list of coin keys $\mathbf{k}$ for the coins in tx.out, such that the supply of tx is the sum of the values $\mathbf{v}$.
- $(\text{ptx}, \mathbf{k}') \leftarrow \text{Send}(\text{pp}, (\mathbf{C}, \mathbf{v}, \mathbf{k}), \mathbf{v}')$: the sending algorithm takes as input a list of coins $\mathbf{C}$ together with the associated lists of values $\mathbf{v}$ and secret keys $\mathbf{k}$ and a list of *change values* $\mathbf{v}'$ whose sum is at most the sum of the input values $\mathbf{v}$; it returns a pre-transaction ptx and a list of keys $\mathbf{k}'$ for the change coins of ptx, such that the remainder of ptx is the sum of the values $\mathbf{v}$ minus the sum of the values $\mathbf{v}'$.
- $(\text{tx}, \mathbf{k}'') \leftarrow \text{Rcv}(\text{pp}, \text{ptx}, \mathbf{v}'')$: the receiving algorithm takes as input a pre-transaction ptx and a list of values $\mathbf{v}''$ whose sum equals the remainder of ptx; it returns a transfer transaction tx and a list of secret keys $\mathbf{k}''$ for the fresh coins in the output of tx, one for each value in $\mathbf{v}''$.
- $\Lambda' \leftarrow \text{Ldgr}(\text{pp}, \Lambda, \text{tx})$: the ledger algorithm takes as input the ledger Λ and a transaction tx to be included in Λ; it returns an updated ledger Λ' or $\bot$.
- $\text{tx} \leftarrow \text{Agg}(\text{pp}, \text{tx}_0, \text{tx}_1)$: the transaction aggregation algorithm takes as input two transactions tx_0 and tx_1 whose input coin lists are disjoint and whose output coin lists are disjoint; it returns a transaction tx whose supply is the sum of the supplies of tx_0 and tx_1 and whose input and output coin list is the cut-through of $\text{tx}_0.\text{in} \,\|\, \text{tx}_1.\text{in}$ and $\text{tx}_0.\text{out} \,\|\, \text{tx}_1.\text{out}$.

We say that an aggregate cash system CASH is correct if its procedures Setup, Mint, Send, Rcv, Ldgr, and Agg behave as expected with overwhelming probability (that is, we allow that with negligible probability things can go wrong, typically, because an algorithm could generate the same coin twice). We give a formal definition that uses two auxiliary procedures: Cons, which checks if a list of coins $\mathbf{C}$ is consistent with respect to values $\mathbf{v}$ and keys $\mathbf{k}$; and Ver, which given as input a ledger or a (pre-)transaction determines if they respect some notion of cryptographic validity.

Definition 9 (Correctness). *An aggregate cash system* CASH *is correct if there exist procedures* $\text{Ver}(\cdot, \cdot)$ *and* $\text{Cons}(\cdot, \cdot, \cdot, \cdot)$ *such that for any* $v_{\max} \in \mathbb{N}$ *and (not necessarily p.p.t.)* $\mathcal{A}_{\text{Mint}}, \mathcal{A}_{\text{Send}}, \mathcal{A}_{\text{Rcv}}, \mathcal{A}_{\text{Agg}}$ *and* $\mathcal{A}_{\text{Ldgr}}$ *the following functions are overwhelming in* λ: $\Pr\left[(\text{pp}, \Lambda) \leftarrow \text{Setup}(1^\lambda, v_{\max}) \; : \; \text{Ver}(\text{pp}, \Lambda) \right]$

$$\Pr\begin{bmatrix} (\text{pp}, \Lambda) \leftarrow \text{Setup}(1^\lambda, v_{\max}) \\ \mathbf{v} \leftarrow \mathcal{A}_{\text{Mint}}(\text{pp}, \Lambda) \\ (\text{tx}, \mathbf{k}) \leftarrow \text{Mint}(\text{pp}, \mathbf{v}) \end{bmatrix} : \; \mathbf{v} \in [0, v_{\max}]^* \Rightarrow \begin{pmatrix} \text{Ver}(\text{pp}, \text{tx}) \wedge \text{tx.in} = (\,) \wedge \\ \text{tx.sply} = \sum \mathbf{v} \wedge \\ \text{Cons}(\text{pp}, \text{tx.out}, \mathbf{v}, \mathbf{k}) \end{pmatrix} \Bigg]$$

$$\Pr\begin{bmatrix} (\text{pp}, \Lambda) \leftarrow \text{Setup}(1^\lambda, v_{\max}) \\ (\mathbf{C}, \mathbf{v}, \mathbf{k}, \mathbf{v}') \leftarrow \mathcal{A}_{\text{Send}}(\text{pp}, \Lambda) \\ (\text{ptx}, \mathbf{k}') \leftarrow \text{Send}(\text{pp}, (\mathbf{C}, \mathbf{v}, \mathbf{k}), \mathbf{v}') \end{bmatrix} : \begin{pmatrix} \text{Cons}(\text{pp}, \mathbf{C}, \mathbf{v}, \mathbf{k}) \wedge \mathbf{v} \,\|\, \mathbf{v}' \in [0, v_{\max}]^* \\ \wedge \sum \mathbf{v} - \sum \mathbf{v}' \in [0, v_{\max}] \end{pmatrix} \\ \Rightarrow \begin{pmatrix} \text{Ver}(\text{pp}, \text{ptx}) \wedge \text{ptx.in} = \mathbf{C} \wedge \\ \text{ptx.rmdr} = \sum \mathbf{v} - \sum \mathbf{v}' \wedge \\ \text{Cons}(\text{pp}, \text{ptx.chg}, \mathbf{v}', \mathbf{k}') \end{pmatrix} \Bigg]$$

Game $\mathrm{INFL}_{\mathsf{CASH},\mathcal{A}}(\lambda, v_{\max})$

$(\mathsf{pp}, \Lambda_0) \leftarrow \mathsf{CASH.Setup}(1^\lambda, v_{\max})$

$(\Lambda, \mathsf{ptx}, \mathbf{v}) \leftarrow \mathcal{A}(\mathsf{pp}, \Lambda_0)$

$(\mathsf{tx}, \mathbf{k}) \leftarrow \mathsf{CASH.Rcv}(\mathsf{pp}, \mathsf{ptx}, \mathbf{v})$

return $\perp \not\leftarrow \mathsf{CASH.Ldgr}(\mathsf{pp}, \Lambda, \mathsf{tx})$ **and** $\Lambda.\mathsf{sply} < \sum \mathbf{v}$

Fig. 7. Game formalizing resistance to inflation of a cash system CASH.

$$\Pr \begin{bmatrix} (\mathsf{pp}, \Lambda) \leftarrow \mathsf{Setup}(1^\lambda, v_{\max}) & (\mathsf{Ver}(\mathsf{pp}, \mathsf{ptx}) \wedge \mathbf{v}'' \in [0, v_{\max}]^* \wedge \mathsf{ptx.rmdr} = \sum \mathbf{v}'') \\ (\mathsf{ptx}, \mathbf{v}'') \leftarrow \mathcal{A}_{\mathsf{Rcv}}(\mathsf{pp}, \Lambda) & \\ (\mathsf{tx}, \mathbf{k}'') \leftarrow \mathsf{Rcv}(\mathsf{pp}, \mathsf{ptx}, \mathbf{v}'') & : \Rightarrow \begin{pmatrix} \mathsf{Ver}(\mathsf{pp}, \mathsf{tx}) \wedge \mathsf{tx.sply} = 0 \wedge \\ \mathsf{tx.in} = \mathsf{ptx.in} \wedge \mathsf{ptx.chg} \subseteq \mathsf{tx.out} \wedge \\ \mathsf{Cons}(\mathsf{pp}, \mathsf{tx.out} - \mathsf{ptx.chg}, \mathbf{v}'', \mathbf{k}'') \end{pmatrix} \end{bmatrix}$$

$$\Pr \begin{bmatrix} (\mathsf{pp}, \Lambda) \leftarrow \mathsf{Setup}(1^\lambda, v_{\max}) & \begin{pmatrix} \mathsf{Ver}(\mathsf{pp}, \mathsf{tx}_0) \wedge \mathsf{tx}_0.\mathsf{in} \cap \mathsf{tx}_1.\mathsf{in} = () \wedge \\ \mathsf{Ver}(\mathsf{pp}, \mathsf{tx}_1) \wedge \mathsf{tx}_0.\mathsf{out} \cap \mathsf{tx}_1.\mathsf{out} = () \end{pmatrix} \\ (\mathsf{tx}_0, \mathsf{tx}_1) \leftarrow \mathcal{A}_{\mathsf{Agg}}(\mathsf{pp}, \Lambda) & \\ \mathsf{tx} \leftarrow \mathsf{Agg}(\mathsf{pp}, \mathsf{tx}_0, \mathsf{tx}_1) & : \Rightarrow \begin{pmatrix} \mathsf{Ver}(\mathsf{pp}, \mathsf{tx}) \wedge \mathsf{tx.sply} = \mathsf{tx}_0.\mathsf{sply} + \mathsf{tx}_1.\mathsf{sply} \wedge \\ \mathsf{tx.in} = (\mathsf{tx}_0.\mathsf{in} \| \mathsf{tx}_1.\mathsf{in}) - (\mathsf{tx}_0.\mathsf{out} \| \mathsf{tx}_1.\mathsf{out}) \wedge \\ \mathsf{tx.out} = (\mathsf{tx}_0.\mathsf{out} \| \mathsf{tx}_1.\mathsf{out}) - (\mathsf{tx}_0.\mathsf{in} \| \mathsf{tx}_1.\mathsf{in}) \end{pmatrix} \end{bmatrix}$$

$$\Pr \begin{bmatrix} (\mathsf{pp}, \Lambda) \leftarrow \mathsf{Setup}(1^\lambda, v_{\max}) & \begin{pmatrix} \mathsf{Ver}(\mathsf{pp}, \Lambda) \wedge \mathsf{Ver}(\mathsf{pp}, \mathsf{tx}) \wedge \\ \mathsf{tx.in} \subseteq \Lambda.\mathsf{out} \wedge \mathsf{tx.out} \cap \Lambda.\mathsf{out} = () \end{pmatrix} \\ (\Lambda, \mathsf{tx}) \leftarrow \mathcal{A}_{\mathsf{Ldgr}}(\mathsf{pp}, \Lambda) & \\ \Lambda' \leftarrow \mathsf{Ldgr}(\mathsf{pp}, \Lambda, \mathsf{tx}) & : \Rightarrow \begin{pmatrix} \Lambda' \neq \perp \wedge \mathsf{Ver}(\mathsf{pp}, \Lambda') \wedge \\ \Lambda'.\mathsf{out} = (\Lambda.\mathsf{out} - \mathsf{tx.in}) \| \mathsf{tx.out} \wedge \\ \Lambda'.\mathsf{sply} = \Lambda.\mathsf{sply} + \mathsf{tx.sply} \end{pmatrix} \end{bmatrix}$$

3.2 Security Definitions

Security against inflation. A sound payment system must ensure that the only way money can be created is via the supply of transactions, typically minting transactions. This means that for any tx the total value of the output coins should be equal to the sum of the total value of the input coins plus the supply tx.sply of the transaction. Since coin values are not deducible from a transaction (this is one of the privacy features of such a system), we define the property at the level of the ledger Λ.

We say that a cash system is resistant to inflation if no adversary can spend coins from $\Lambda.\mathsf{out}$ worth more than $\Lambda.\mathsf{sply}$. The adversary's task is thus to create a pre-transaction whose remainder is strictly greater than $\Lambda.\mathsf{sply}$; validity of the pre-transaction is checked by completing it to a transaction via Rcv and adding it to the ledger via Ldgr. This is captured by the definition below.

Definition 10 (Inflation-resistance). *We say that an aggregate cash system* CASH *is secure against inflation if for any $v_{\max}$ and any p.p.t. adversary $\mathcal{A}$,*

$$\mathsf{Adv}^{\mathsf{infl}}_{\mathsf{CASH},\mathcal{A}}(\lambda, v_{\max}) := \Pr\left[\mathrm{INFL}_{\mathsf{CASH},\mathcal{A}}(\lambda, v_{\max}) = \mathbf{true}\right] = \mathsf{negl}(\lambda),$$

where $\mathrm{INFL}_{\mathsf{CASH},\mathcal{A}}(\lambda, v_{\max})$ *is defined in Fig. 7.*

Game STEAL$_{\text{CASH},\mathcal{A}}(\lambda, v_{\max})$

$(\text{pp}, \Lambda) \leftarrow \text{CASH.Setup}(1^\lambda, v_{\max})$

$\text{Hon}, \text{Val}, \text{Key}, \text{Ptx} := ()$

$\mathcal{A}^{\text{MINT},\text{SEND},\text{RECEIVE},\text{LEDGER}}(\text{pp}, \Lambda)$

return $(\text{Hon} \not\subseteq \Lambda.\text{out})$

Aux function Store$(\mathbf{C}, \mathbf{v}, \mathbf{k})$

$\text{Val}(\mathbf{C}) := \mathbf{v}; \quad \text{Key}(\mathbf{C}) := \mathbf{k}$

Oracle MINT$(\mathbf{v})$

$(\text{tx}, \mathbf{k}) \leftarrow \text{CASH.Mint}(\text{pp}, \mathbf{v})$

$\Lambda \leftarrow \text{CASH.Ldgr}(\text{pp}, \Lambda, \text{tx})$

$\text{Hon} := \text{Hon} \,\|\, \text{tx.out}$

$\text{Store}(\text{tx.out}, \mathbf{v}, \mathbf{k})$

return tx

Oracle SEND$(\mathbf{C}, \mathbf{v}')$

if $\mathbf{C} \not\subseteq \text{Hon}$ **or** $\bigcup_{\text{ptx} \in \text{Ptx}} \text{ptx.in} \cap \mathbf{C} \neq ()$

   return $\perp$ // only honest coins never sent can be queried

$(\text{ptx}, \mathbf{k}') \leftarrow \text{CASH.Send}\big(\text{pp}, (\mathbf{C}, \text{Val}(\mathbf{C}), \text{Key}(\mathbf{C})), \mathbf{v}'\big)$

$\text{Store}(\text{ptx.chg}, \mathbf{v}', \mathbf{k}'); \quad \text{Ptx} := \text{Ptx} \,\|\, (\text{ptx})$

return ptx

Oracle RECEIVE$(\text{ptx}, \mathbf{v})$

$(\text{tx}, \mathbf{k}) \leftarrow \text{CASH.Rcv}(\text{pp}, \text{ptx}, \mathbf{v})$

$\Lambda' \leftarrow \text{LEDGER}(\text{tx})$ // updates Hon

if $\Lambda' = \perp$ **then return** $\perp$

$\text{Hon} := \text{Hon} \,\|\, (\text{tx.out} - \text{ptx.chg})$

$\text{Store}(\text{tx.out} - \text{ptx.chg}, \mathbf{v}, \mathbf{k}); \quad$ **return** tx

---

Oracle LEDGER(tx)

$\Lambda' \leftarrow \text{CASH.Ldgr}(\text{pp}, \Lambda, \text{tx})$

if $\Lambda' = \perp$ **then return** $\perp$ **else** $\Lambda := \Lambda'$

for all $\text{ptx} \in \text{Ptx}$ **do**

   if $\text{ptx.chg} \subseteq \text{tx.out}$

   // if all change of ptx now in ledger

      $\text{Ptx} := \text{Ptx} - (\text{ptx})$

      $\text{Hon} := (\text{Hon} - \text{ptx.in}) \,\|\, \text{ptx.chg}$

   // consider input of ptx consumed

return Λ

Fig. 8. Game formalizing resistance to coin theft of a cash system CASH.

Security against coin theft. Besides inflation, which protects the soundness of the system as a whole, the second security notion protects individual users. It requires that only a user can spend coins belonging to him, where ownership of a coin amounts to knowledge of the coin secret key. This is formalized by the experiment in Fig. 8, which proceeds as follows. The challenger sets up the system and maintains the ledger Λ throughout the game (we assume that the consensus protocol provides this). The adversary can add any valid transaction to the ledger through an oracle LEDGER.

The challenger also simulates an honest user and manages her coins; in particular, it maintains a list Hon, which represents the coins that the honest user expects to own in the ledger. The game also maintains two hash tables Val and Key that map coins produced by the game to their values and keys. We write e.g. $\text{Val}(C) := v$ to mean that the pair (C, v) is added to Val and let $\text{Val}(C)$ denote the value v for which (C, v) is in Val. This naturally generalizes to lists letting $\text{Val}(\mathbf{C})$ be the list $\mathbf{v}$ such that (C_i, v_i) is in Val for all i.

The adversary can interact with the honest user and the ledger using the following oracles:

- MINT is an oracle that mints coins for the honest user. It takes as input a vector of values **v**, creates a minting transaction tx together with the secret keys of the output coins, adds tx to the ledger and appends the newly created coins to Hon.
- RECEIVE lets the adversary send coins to the honest user. The oracle takes as input a pre-transaction ptx and output values **v**; it completes ptx to a transaction tx creating output coins with values **v**, adds tx to the ledger, and appends the newly created coins to Hon.
- SEND lets the adversary make an honest user send coins to it. It takes as input a list **C** of coins contained in Hon and a list of change values **v'**; it also checks that none of the coins in **C** has been queried to SEND before (an honest user does not double-spend). It returns a pre-transaction ptx spending the coins from **C** and creating change output coins with values **v'**. The oracle only produces a pre-transaction and returns it to the adversary, but it does not alter the ledger. This is why the list Hon of honest coins is not altered either; in particular, the sent coins **C** still remain in Hon.
- LEDGER lets the adversary commit a transaction tx to the ledger. If the transaction output contains the (complete) set of change coins of a pre-transaction ptx previously sent to the adversary, then these change coins are added to Hon, while the input coins of ptx are removed from Hon.

Note that the list Hon represents the coins that the honest user should consider hers, given the system changes induced by the oracle calls: coins received directly from the adversary via RECEIVE or as fresh coins via MINT are added to Hon. Coins sent to the adversary in a pre-transaction ptx via SEND are only removed *once all change coins of* ptx *have been added* to the ledger via LEDGER. Note also that, given these oracles, the adversary can simulate transfers between honest users. It can simply call SEND to receive an honest pre-transaction ptx and then call RECEIVE to have the honest user receive ptx.

The winning condition of the game is now simply that Hon does not reflect what the honest user would expect, namely Hon is not fully contained in the ledger (because the adversary managed to spend a coin that is still in Hon, which amounts to stealing it from the honest user).

Definition 11 (Theft-resistance). *We say that an aggregate cash system* CASH *is secure against coin theft if for any* $v_{\max}$ *and any p.p.t. adversary* $\mathcal{A}$,

$$\mathsf{Adv}^{\text{steal}}_{\mathsf{CASH},\mathcal{A}}(\lambda, v_{\max}) := \Pr\left[\text{STEAL}_{\mathsf{CASH},\mathcal{A}}(\lambda, v_{\max}) = \textbf{true}\right] = \mathsf{negl}(\lambda),$$

where $\text{STEAL}_{\mathsf{CASH},\mathcal{A}}(\lambda, v_{\max})$ *is defined in Fig. 8.*

Transaction indistinguishability. An important security feature that Mimblewimble inherits from Confidential Transactions [Max15] is that the amounts involved in a transaction are hidden so that only the sender and the receiver

Game IND-TX$_{\mathsf{CASH},\mathcal{A}}(\lambda, v_{\max})$	Oracle Tx$((\mathbf{v}_0, \mathbf{v}_0', \mathbf{v}_0''), (\mathbf{v}_1, \mathbf{v}_1', \mathbf{v}_1''))$
$b \leftarrow_\$ \{0, 1\}$	**if not** $(\mathbf{v}_0, \mathbf{v}_0', \mathbf{v}_0'', \mathbf{v}_1, \mathbf{v}_1', \mathbf{v}_1'' \in [0, v_{\max}]^*)$
$(\mathsf{pp}, \Lambda) \leftarrow \mathsf{Setup}(1^\lambda, v_{\max})$	$\quad$ **return** $\bot$
$b' \leftarrow \mathcal{A}^{\mathrm{Tx}}(\mathsf{pp}, \Lambda)$	**if** $|\mathbf{v}_0| \neq |\mathbf{v}_1|$ **or** $|\mathbf{v}_0'| + |\mathbf{v}_0''| \neq |\mathbf{v}_1'| + |\mathbf{v}_1''|$
return $b = b'$	$\quad$ **return** $\bot$ // as number of coins is not hidden
	if $\sum \mathbf{v}_0 \neq \sum(\mathbf{v}_0' \,\|\, \mathbf{v}_0'')$ **or** $\sum \mathbf{v}_1 \neq \sum(\mathbf{v}_1' \,\|\, \mathbf{v}_1'')$
	$\quad$ **return** $\bot$ // as transactions must be balanced
	$(\mathsf{tx}, \mathbf{k}) \leftarrow \mathsf{Mint}(\mathsf{pp}, \mathbf{v}_b)$
	$(\mathsf{ptx}, \mathbf{k}') \leftarrow \mathsf{Send}(\mathsf{pp}, (\mathsf{tx}.\mathsf{out}, \mathbf{v}_b, \mathbf{k}), \mathbf{v}_b')$
	$(\mathsf{tx}^*, \mathbf{k}'') \leftarrow \mathsf{Rcv}(\mathsf{pp}, \mathsf{ptx}, \mathbf{v}_b'')$
	return tx^*

Fig. 9. Game formalizing transaction indistinguishability of a cash system CASH.

know how much money is involved. In addition, a transaction completely hides which inputs paid which outputs and which coins were change and which were added by the receiver.

We formalize this via the following game, specified in Fig. 9. The adversary submits two sets of values $(\mathbf{v}_0, \mathbf{v}_0', \mathbf{v}_0'')$ and $(\mathbf{v}_1, \mathbf{v}_1', \mathbf{v}_1'')$ representing possibles values for input coins, change coins and receiver's coins of a transaction. The game creates a transaction with values either from the first or the second set and the adversary must guess which. For the transaction to be valid, we must have $\sum \mathbf{v}_b = \sum \mathbf{v}_b' + \sum \mathbf{v}_b''$ for both $b = 0, 1$. Moreover, transactions do not hide the *number* of input and output coins. We therefore also require that $|\mathbf{v}_0| = |\mathbf{v}_1|$ and $|\mathbf{v}_0'| + |\mathbf{v}_0''| = |\mathbf{v}_1'| + |\mathbf{v}_1''|$ (note that e.g. the number of change coins can differ).

Definition 12 (Transaction indistinguishability). *We say that an aggregate cash system* CASH *is transaction-indistinguishable if for any* $v_{\max}$ *and any p.p.t. adversary* $\mathcal{A}$,

$$\mathsf{Adv}_{\mathsf{CASH},\mathcal{A}}^{\mathrm{tx\text{-}ind}}(\lambda, v_{\max}) := 2 \cdot \left| \Pr\left[\mathrm{TX\text{-}IND}_{\mathsf{CASH},\mathcal{A}}(\lambda, v_{\max}) = \mathbf{true}\right] - \tfrac{1}{2} \right| = \mathsf{negl}(\lambda),$$

where $\mathrm{TX\text{-}IND}_{\mathsf{CASH},\mathcal{A}}(\lambda, v_{\max})$ *is defined in Fig. 9.*

4 Construction of an Aggregate Cash System

4.1 Description

Let COM be an additively homomorphic commitment scheme such that for $\mathsf{cp} \leftarrow \mathsf{COM.Setup}(\mathsf{MainSetup}(1^\lambda))$ we have value space $\mathcal{V}_{\mathsf{cp}} = \mathbb{Z}_p$ with p of length λ (such as the Pedersen scheme). Let SIG be an aggregate signature scheme that is compatible with COM. For $v_{\max} \in \mathbb{N}$, let $R_{v_{\max}}$ be the (efficiently computable) relation on commitments with values at most $v_{\max}$, i.e.,

$$R_{v_{\max}} := \{ (\mathsf{mp}, (\mathsf{cp}, C), (v, r)) \mid \mathsf{mp} = \mathsf{mp}_{\mathsf{cp}} \wedge C = \mathsf{COM.Cmt}(\mathsf{cp}, v, r) \wedge v \in [0, v_{\max}] \}$$

where $\mathsf{mp}_{\mathsf{cp}}$ are the main parameters contained in cp (recall that we assume that for $\mathsf{cp} \in [\mathsf{COM.Setup}(\mathsf{mp})]$, mp is contained in cp). Let Π be a simulation-extractable NIZK proof system for the family of relations $R = \{R_{v_{\max}}\}_{v_{\max}}$:

For notational simplicity, we will use the following vectorial notation for COM, R, and Π: given $\mathbf{C}$, $\mathbf{v}$, and $\mathbf{r}$ with $|\mathbf{C}| = |\mathbf{v}| = |\mathbf{r}|$, we let

$$\mathsf{COM.Cmt}(\mathsf{cp}, \mathbf{v}, \mathbf{r}) := \left(\mathsf{COM.Cmt}(\mathsf{cp}, v_i, r_i) \right)_{i=1}^{|\mathbf{v}|},$$

$$R_{v_{\max}}((\mathsf{cp}, \mathbf{C}), (\mathbf{v}, \mathbf{r})) := \bigwedge_{i=1}^{|\mathbf{C}|} R_{v_{\max}}(\mathsf{mp}_{\mathsf{cp}}, (\mathsf{cp}, C_i), (v_i, r_i)),$$

$$\Pi.\mathsf{Prv}(\mathsf{crs}, (\mathsf{cp}, \mathbf{C}), (\mathbf{v}, \mathbf{r})) := \left(\Pi.\mathsf{Prv}(\mathsf{crs}, (\mathsf{cp}, C_i), (v_i, r_i)) \right)_{i=1}^{|\mathbf{C}|},$$

$$\Pi.\mathsf{Ver}(\mathsf{crs}, (\mathsf{cp}, \mathbf{C}), \boldsymbol{\pi}) := \bigwedge_{i=1}^{|\mathbf{C}|} \Pi.\mathsf{Ver}(\mathsf{crs}, (\mathsf{cp}, C_i), \pi_i),$$

and likewise for $\Pi.\mathsf{SimPrv}$. We also assume that messages are the empty string ε if they are omitted from $\mathsf{SIG.Ver}$ and $\mathsf{SIG.Agg}$; that is, we overload notation and let

$$\mathsf{SIG.Ver}(\mathsf{sp}, (X_i)_{i=1}^n, \sigma) := \mathsf{SIG.Ver}(\mathsf{sp}, ((X_i, \varepsilon))_{i=1}^n, \sigma)$$

and likewise for $\mathsf{SIG.Agg}\left(\mathsf{sp}, ((X_{0,i})_{i=1}^{n_0}, \sigma_0), ((X_{1,i})_{i=1}^{n_1}, \sigma_1)\right)$.

From COM, SIG and Π we construct an aggregate cash system $\mathsf{MW[COM, SIG, \Pi]}$ as follows. The public parameters pp consist of commitment and signature parameters cp, sp, and a CRS for Π. A *coin key* $k \in \mathcal{K}_{\mathsf{pp}}$ is an element of the randomness space $\mathcal{R}_{\mathsf{cp}}$ of the commitment scheme, i.e., $\mathcal{K}_{\mathsf{pp}} = \mathcal{R}_{\mathsf{cp}}$. A *coin* $C = \mathsf{COM.Cmt}(\mathsf{cp}, v, k)$ is a commitment to the value v of the coin using the coin key k as randomness. Hence, $\mathcal{C}_{\mathsf{pp}} = \mathcal{C}_{\mathsf{cp}}$.

A transaction $\mathsf{tx} = (s, \mathbf{C}, \hat{\mathbf{C}}, K)$ consists of a supply $\mathsf{tx.sply} = s$, an input coin list $\mathsf{tx.in} = \mathbf{C}$, an output coin list $\mathsf{tx.out} = \hat{\mathbf{C}}$, and a kernel K. The kernel K is a triple $(\boldsymbol{\pi}, \mathbf{E}, \sigma)$ where $\boldsymbol{\pi}$ is a list of range proofs for the output coins, $\mathbf{E}$ is a non-empty list of signature-verification keys (which are of the same form as commitments) called *kernel excesses*, and σ is an (aggregate) signature. We define the *excess of the transaction* tx, denoted $\mathsf{Exc(tx)}$, as the sum of outputs minus the sum of inputs, with the supply s converted to an input coin with $k = 0$:

$$\mathsf{Exc(tx)} := \sum \hat{\mathbf{C}} - \sum \mathbf{C} - \mathsf{COM.Cmt}(\mathsf{cp}, s, 0). \tag{4}$$

Intuitively, $\mathsf{Exc(tx)}$ should be a commitment to 0, as the committed input and output values of the transaction should cancel out; this is evidenced by giving a signature under key $\mathsf{Exc(tx)}$ (which could be represented as the sum of elements (E_i) due to aggregation; see below).

A transaction $\mathsf{tx} = (s, \mathbf{C}, \hat{\mathbf{C}}, K)$ with $K = (\boldsymbol{\pi}, \mathbf{E}, \sigma)$ is said to be valid if all range proofs are valid, $\mathsf{Exc(tx)} = \sum \mathbf{E}$, and σ is a valid signature for $\mathbf{E}$ (with all messages ε).[10]

---

[10] If $\mathbf{E}$ in a transaction tx consists of a single element, it must be $E = \mathsf{Exc(tx)}$, so E could be omitted from the transaction; we keep it for consistency.

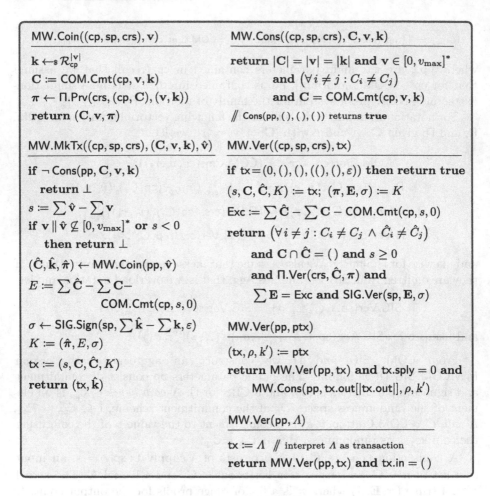

Fig. 10. Auxiliary algorithms for the MW aggregate cash system.

When a user wants to make a payment of an amount ρ, she creates a transaction tx with input coins $\mathbf{C}$ of values $\mathbf{v}$ with $\sum \mathbf{v} \geq \rho$ and with output coins a list of fresh change coins of values $\mathbf{v}'$ so that $\sum \mathbf{v}' = \sum \mathbf{v} - \rho$. She also appends one more special coin of value ρ to the output. The pre-transaction ptx is then defined as this transaction tx, the remainder ptx.rmdr $:= \rho$ and the key for the special coin.

When receiving a pre-transaction ptx $= (\text{tx}, \rho, k)$, the receiver first checks that tx is valid and that k is a key for the special coin $C' := \text{tx.out}[|\text{tx.out}|]$ of value ρ. He then creates a transaction tx$'$ that spends C' (using its key k) and creates coins of combined value ρ. Aggregating tx and tx$'$ yields a transaction tx$''$ with tx$''$.sply $= 0$, tx$''$.in $=$ ptx.in and tx$''$.out containing ptx.chg and the freshly created coins. The receiver then submits tx$''$ to the ledger.

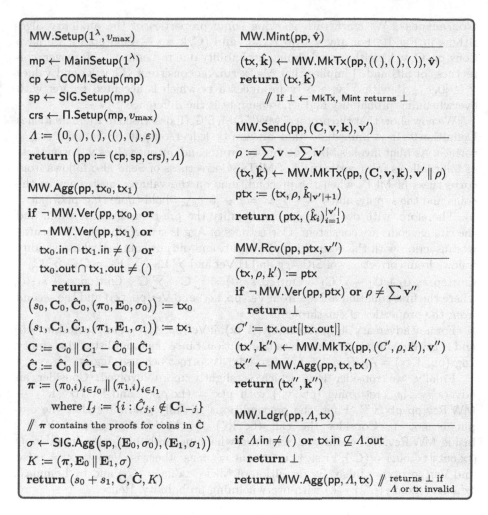

Fig. 11. The MW aggregate cash system. (Recall that algorithms return $\bot$ when one of their subroutines returns $\bot$.)

The ledger accepts a transaction if it is valid (as defined above) and if its input coins are contained in the output coin list of the ledger (which corresponds to the UTXO set in other systems). We do not consider any other conditions related to the consensus mechanism, such as fees being included in a transaction to incentivize its inclusion in the ledger or a proof-of-work being included in a minting transaction.

In Fig. 10 we first define auxiliary algorithms that create coins and transactions and verify their validity by instantiating the procedures Ver and Cons from Definition 9. Using these we then formally define MW[COM, SIG, Π] in Fig. 11.

Correctness. We start with showing some properties of the auxiliary algorithms in Fig. 10. For any $\mathbf{v} \in [0, v_{\max}]^*$ and $(\mathbf{C}, \mathbf{k}, \pi) \leftarrow \mathsf{Coin}(\mathsf{pp}, \mathbf{v})$, we have $\mathsf{Cons}(\mathsf{pp}, \mathbf{C}, \mathbf{v}, \mathbf{k})$ with overwhelming probability due to Lemma 3. Moreover, correctness of SIG and Π implies that MkTx run on consistent $(\mathbf{C}, \mathbf{v}, \mathbf{k})$ and values $\hat{\mathbf{v}} \in [0, v_{\max}]^*$ with $\sum \hat{\mathbf{v}} \geq \sum \mathbf{v}$ produces a tx which is accepted by Ver with overwhelming probability and whose supply is the difference $\sum \mathbf{v} - \sum \hat{\mathbf{v}}$.

We now show that the protocol $\mathsf{MW}[\mathsf{COM}, \mathsf{SIG}, \Pi]$ described in Fig. 11 satisfies Definition 9. It is immediate that an empty ledger output by $\mathsf{Setup}(1^\lambda, v_{\max})$ verifies. As Mint invokes MkTx on empty inputs and output values $\mathbf{v}$, correctness of Mint follows from correctness of MkTx. Correctness of Send also follows from correctness of MkTx when the preconditions on the values, consistency of the coins and the supply, and $\sum \mathbf{v} - \sum \mathbf{v}' = \rho \in [0, v_{\max}]$ hold (note that $\mathsf{ptx}.\mathsf{rmdr} = \rho$). Therefore, with overwhelming probability the pre-transaction is valid, and the change coins are consistent. Correctness of Agg is straightforward: it returns a transaction with the desired supply, input, and output coin list whose validity follows from correctness of $\mathsf{SIG}.\mathsf{Agg}$ and $\Pi.\mathsf{Ver}$ and $\sum \mathbf{E}_0 + \sum \mathbf{E}_1 = \hat{\mathbf{C}}_0 - \sum \mathbf{C}_0 - \mathsf{Cmt}(\mathsf{cp}, s_0, 0) + \hat{\mathbf{C}}_1 - \sum \mathbf{C}_1 - \mathsf{Cmt}(\mathsf{cp}, s_1, 0) = \sum \hat{\mathbf{C}} - \sum \mathbf{C} - \mathsf{Cmt}(\mathsf{cp}, s_0 + s_1, 0)$, where the first equation follows from $\mathsf{Ver}(\mathsf{pp}, \mathsf{tx}_0)$ and $\mathsf{Ver}(\mathsf{pp}, \mathsf{tx}_0)$ and the second from the properties of cut-through.

For any adversary $\mathcal{A}_{\mathsf{Ldgr}}$ returning (Λ, tx), if $\mathsf{Ver}(\mathsf{pp}, \Lambda) = \mathbf{true}$, then $\Lambda.\mathsf{in} = ()$ and Λ is valid when interpreted as a transaction. Since the input list of Λ is empty, $\mathsf{Ldgr}(\mathsf{pp}, \Lambda, \mathsf{tx}) = \mathsf{Agg}(\mathsf{pp}, \Lambda, \mathsf{tx})$ and so Ldgr is correct because Agg is.

Finally, we consider Rcv, which is slightly more involved. Consider an adversary $\mathcal{A}_{\mathsf{Rcv}}$ returning $(\mathsf{ptx}, \mathbf{v}'')$ with $\mathsf{ptx} = (\mathsf{tx}, \rho, k')$ and let $(\mathsf{tx}'', \mathbf{k}'') \leftarrow \mathsf{MW}.\mathsf{Rcv}(\mathsf{pp}, \mathsf{ptx}, \mathbf{v}'')$. First, the preconditions trivially guarantee that the output is not $\bot$. Consider the call $(\mathsf{tx}', \mathbf{k}'') \leftarrow \mathsf{MW}.\mathsf{MkTx}(\mathsf{pp}, (C', \rho, k'), \mathbf{v}'')$ inside $\mathsf{MW}.\mathsf{Rcv}$. We claim that with overwhelming probability, $(\mathsf{tx}.\mathsf{in} \,\|\, \mathsf{tx}'.\mathsf{in}) \cap (\mathsf{tx}.\mathsf{out} \,\|\, \mathsf{tx}'.\mathsf{out}) = (C')$. First, $\mathsf{tx}.\mathsf{in} \cap \mathsf{tx}.\mathsf{out} = ()$, as otherwise $\mathsf{Ver}(\mathsf{pp}, \mathsf{tx}) = \mathbf{false}$ and $\mathsf{Ver}(\mathsf{pp}, \mathsf{ptx}) = \mathbf{false}$. By definition of MkTx, $\mathsf{tx}'.\mathsf{in} = (C')$ and by Lemma 3, $\mathsf{tx}'.\mathsf{out} \cap (\mathsf{tx}.\mathsf{in} \,\|\, (C')) = ()$ with overwhelming probability. Hence,

$$(\mathsf{tx}.\mathsf{in} \,\|\, \mathsf{tx}'.\mathsf{in}) \cap (\mathsf{tx}.\mathsf{out} \,\|\, \mathsf{tx}'.\mathsf{out}) = (C') \cap \mathsf{tx}.\mathsf{out} = (C')$$

and by correctness of Agg, C' is the only coin removed by cut-through during the call $\mathsf{tx}'' \leftarrow \mathsf{MW}.\mathsf{Agg}(\mathsf{pp}, \mathsf{tx}, \mathsf{tx}')$. Thus, the input coin list of tx'' is the same as that of ptx and the change is contained in the output coin list of tx''. The pre-conditions $\mathsf{Ver}(\mathsf{pp}, \mathsf{ptx})$ and $\sum \mathbf{v}'' = \rho$ imply that $\mathsf{tx}.\mathsf{sply} = 0$ and $\mathsf{tx}'.\mathsf{sply} = 0$, respectively. Hence, $\mathsf{tx}''.\mathsf{sply} = 0$ by correctness of Agg. Validity of tx'' and consistency of the new coins follow from correctness of Agg (and validity of the output of MkTx).

4.2   Security

We show that $\mathsf{MW}[\mathsf{COM}, \mathsf{SIG}, \Pi]$ is inflation-resistant, resistant to coin theft and that it satisfies transaction indistinguishability.

Theorem 13 (Inflation-resistance (Definition 10)**).** *Assume that* (COM, SIG) *is EUF-NZO-secure and that* Π *is zero-knowledge and simulation-extractable. Then the aggregate cash system* MW[COM, SIG, Π] *is secure against inflation. More precisely, for any* $v_{\max}$ *and any p.p.t. adversary* $\mathcal{A}$*, there exists a negligible function* $\nu_{\mathcal{A}}$ *and p.p.t. adversaries* $\mathcal{B}$*,* $\mathcal{B}_{\text{zk}}$ *and* $\mathcal{B}_{\text{se}}$ *such that*

$$\text{Adv}^{\text{infl}}_{\text{MW},\mathcal{A}}(\lambda, v_{\max})$$

$$\leq \text{Adv}^{\text{euf-nzo}}_{\text{COM,SIG},\mathcal{B}}(\lambda) + \text{Adv}^{\text{zk}}_{\Pi,\text{R}_{v_{\max}},\mathcal{B}_{\text{zk}}}(\lambda) + \text{Adv}^{\text{s-ext}}_{\Pi,\text{R}_{v_{\max}},\mathcal{B}_{\text{se}}}(\lambda) + \nu_{\mathcal{A}}(\lambda).$$

The full proof can be found in the full version [FOS18]; we give a sketch here. Inflation-resistance follows from EUF-NZO security and *extractability* of Π (we do not actually require *simulation*-extractability, but instead of formally defining extractability we simply relied on Definitions 7 and 8 implying it).

Consider an adversary $\mathcal{A}$ in game INFL$_{\text{MW}}$ in Fig. 7. To win the game, $\mathcal{A}$ must return a valid ledger Λ, a valid ptx and $\mathbf{v}$ with

(i) ptx.in $\subseteq \Lambda$.out and   (ii) $\sum \mathbf{v} = $ ptx.rmdr

(otherwise Rcv and/or Ldgr return $\bot$). All coins in Λ.out, ptx.in and ptx.chg have valid range proofs: the former two in the ledger's kernel $K_\Lambda = (\pi_\Lambda, \mathbf{E}_\Lambda, \sigma_\Lambda)$ (by (i)), and ptx.chg in the kernel of tx$_{\text{ptx}}$ contained in ptx. From these proofs the reduction extracts the values $\mathbf{v}_{\Lambda.\text{out}}, \mathbf{v}_{\text{ptx.in}}, \mathbf{v}_{\text{ptx.chg}} \in [0, v_{\max}]^*$ and keys $\mathbf{k}_{\Lambda.\text{out}}, \mathbf{k}_{\text{ptx.in}}, \mathbf{k}_{\text{ptx.chg}} \in \mathcal{K}^*_{\text{pp}}$ of every coin. We first argue that

(iii) $\sum \mathbf{v}_{\Lambda.\text{out}} - \Lambda$.sply $= 0$ and   (iv) $\sum \mathbf{v}_{\text{ptx.chg}} + $ ptx.rmdr $- \sum \mathbf{v}_{\text{ptx.in}} = 0$.

If (iii) was not the case then $(v^* := \sum \mathbf{v}_{\Lambda.\text{out}} - \Lambda$.sply, $k^* := \sum \mathbf{k}_{\Lambda.\text{out}})$ would be a non-zero opening of the excess Exc of Λ. Since furthermore Exc $= \sum \mathbf{E}_\Lambda$ and σ_Λ is valid for $\mathbf{E}_\Lambda$, the tuple $(\mathbf{E}_\Lambda, \sigma_\Lambda, (v^*, k^*))$ would be an EUF-NZO solution.

Likewise, a non-zero left-hand side of (iv) can be used together with the kernel of tx$_{\text{ptx}}$ to break EUF-NZO. From (i)–(iv) we now get

$$\sum \mathbf{v} \stackrel{\text{(ii)}}{=} \text{ptx.rmdr} \stackrel{\text{(iv)}}{=} \sum \mathbf{v}_{\text{ptx.in}} - \sum \mathbf{v}_{\text{ptx.chg}} \leq \sum \mathbf{v}_{\text{ptx.in}} \stackrel{\text{(i)}}{\leq} \sum \mathbf{v}_{\Lambda.\text{out}} \stackrel{\text{(iii)}}{=} \Lambda.\text{sply},$$

which contradicts the fact that $\mathcal{A}$ won INFL$_{\text{MW}}$, as this requires $\sum \mathbf{v} > \Lambda$.sply. (The function $\nu_{\mathcal{A}}$ accounts for (iii) (or (iv)) only holding over $\mathbb{Z}_p$ but not over $\mathbb{Z}$; this would imply $|\Lambda$.out$| \geq p/v_{\max}$, which can only happen with negligible probability $\nu_{\mathcal{A}}$ for a p.p.t. $\mathcal{A}$.)

Theorem 14 (Theft-resistance (Definition 11)**).** *Assume that the pair* (COM, SIG) *is EUF-CRO-secure and that* Π *is zero-knowledge and simulation-extractable. Then the aggregate cash system* MW[COM, SIG, Π] *is secure against coin theft. More precisely, for any* $v_{\max}$ *and any p.p.t. adversary* $\mathcal{A}$*, which, via its oracle calls, makes the challenger create at most* $h_{\mathcal{A}}$ *coins and whose queries* $(\mathbf{C}, \mathbf{v}')$ *to* SEND *satisfy* $|\mathbf{v}'| \leq n_{\mathcal{A}}$*, there exists a negligible function* ν*, a p.p.t.*

adversary $\mathcal{B}$ making a single signing query, and p.p.t. adversaries $\mathcal{B}_{zk}$ and $\mathcal{B}_{se}$ such that

$$\mathsf{Adv}_{\mathsf{MW},\mathcal{A}}^{\mathsf{steal}}(\lambda, v_{\max})$$
$$\leq h_{\mathcal{A}}(\lambda) \cdot n_{\mathcal{A}}(\lambda) \cdot \left(\mathsf{Adv}_{\mathsf{COM},\mathsf{SIG},\mathcal{B}}^{\mathsf{euf\text{-}cro}}(\lambda) + \mathsf{Adv}_{\Pi,\mathsf{R}_{v_{\max}},\mathcal{B}_{zk}}^{\mathsf{zk}}(\lambda) + \mathsf{Adv}_{\Pi,\mathsf{R}_{v_{\max}},\mathcal{B}_{se}}^{\mathsf{s\text{-}ext}}(\lambda)\right) + \nu(\lambda).$$

The proof can be found in the full version [FOS18]. Here we give some proof intuition. We first assume that all coins created by the challenger are different. By Lemma 3 the probability $\nu(\lambda)$ that two coins collide is negligible.

Since in game STEAL the ledger is maintained by the challenger we have:

(i) the kernel of Λ contains a valid range proof for each coin in $\Lambda.\mathsf{out}$.

In order to win the game, the adversary must at some point *steal* some coin $\widetilde{C}$ from the challenger, by creating a transaction tx^* with $\widetilde{C}$ among its inputs, that is, $\mathsf{tx}^* = (s, \mathbf{C}, \hat{\mathbf{C}}, (\boldsymbol{\pi}, \mathbf{E}, \sigma))$ with $\widetilde{C} \in \mathbf{C}$. For tx^* to be accepted to the ledger, we must have:

(ii) $\mathbf{C} \subseteq \Lambda.\mathsf{out}$;
(iii) tx^* is valid, meaning
  (a) the signature σ verifies under key list $\mathbf{E}$;
  (b) $\sum \mathbf{E} = \sum \hat{\mathbf{C}} - \sum \mathbf{C} - \mathsf{Cmt}(\mathsf{cp}, s, 0)$;
  (c) all proofs $\boldsymbol{\pi}$ for coins $\hat{\mathbf{C}}$ are valid.

From (i), (ii) and (iii)(c) we have that all coins in $\mathbf{C}$ and $\hat{\mathbf{C}}$ have valid proofs, which means we can extract (except for $\widetilde{C}$, as we will see later) their values $\mathbf{v}$ and $\hat{\mathbf{v}}$ and keys $\mathbf{k}$ and $\hat{\mathbf{k}}$. This means, we can write (iii)(b) as:

$$\sum \mathbf{E} = -\widetilde{C} + \mathsf{Cmt}(\mathsf{cp}, \underbrace{\sum \hat{\mathbf{v}} - \sum \mathbf{v} - s}_{=:v^*}, \underbrace{\sum \hat{\mathbf{k}} - \sum \mathbf{k}}_{=:k^*}). \tag{5}$$

Now, if we had set $\widetilde{C} = C^*$ with C^* a challenge for EUF-CRO then (iii)(a) and Eq. (5) together would imply that $(\mathbf{E}, \sigma, (v^*, k^*))$ is a solution for C^* in EUF-CRO. So the basic proof idea is to embed a challenge C^* as one of the honest coins $\widetilde{C}$ created in the system and hope that the adversary will steal $\widetilde{C}$. When $\widetilde{C}$ is first created, it can be during a call to MINT, SEND or RECEIVE, each of which will create a transaction tx using $\mathsf{MW.MkTx}$; we thus set $\mathsf{tx.out}[j] = \widetilde{C}$ for some j. Now tx must contain a range proof for $\widetilde{C}$, which we produce using the zero-knowledge simulator, and a signature under verification key

$$\sum \mathsf{tx.out} - \sum \mathsf{tx.in} = \left(\sum_{i \neq j} \mathsf{tx.out}[i] - \sum \mathsf{tx.in}\right) + \widetilde{C}. \tag{6}$$

The coin keys of $\mathsf{tx.in}$ are input to $\mathsf{MW.MkTx}$ and those of $(\mathsf{tx.out}[i])_{i \neq j}$ are created by it. So we know the secret key a for the expression in parentheses in (6) and can therefore make a query $\mathsf{SIGN}'(a)$ to the related-key signing oracle to obtain the signature.

While this shows that simulating the creation of coin $\widetilde{C}$ is easily dealt with, what complicates the proof is when the adversary queries $\textsc{Send}(\mathbf{C}, \mathbf{v}')$ with $\widetilde{C} \in \mathbf{C}$, which should produce a pre-transaction $\widetilde{\mathsf{ptx}}$. Since $\widetilde{C}$ is a (say the j-th) input of $\widetilde{\mathsf{ptx}}$, this would require a signature related to $-C^*$ for which we cannot use the $\textsc{Sign}'$ oracle. Instead, we pick one random, say the $\tilde{\imath}$-th, change coin $\overline{C}$ and embed the challenge C^* in $\overline{C}$ as well. (If there are no change coins, we abort; we justify this below.) To complete $\widetilde{\mathsf{ptx}}$, we now need a signature for key

$$\sum_{i \neq \tilde{\imath}} \mathsf{tx.out}[i] + \overline{C} - \sum_{i \neq j} \mathsf{tx.in}[i] - \widetilde{C},$$

and since the two occurrences of C^* cancel out, the simulation knows the signing key of the above expression. (The way the reduction actually embeds C^* in a coin $\widetilde{C}$ which in the game is supposed to have value v is by setting $\widetilde{C} := C^* + \mathsf{Cmt}(\mathsf{cp}, v, k)$.)

Let's look again at the transaction tx^* with which the adversary steals $\widetilde{C}$: for tx^* to actually steal $\widetilde{C}$, we must have $\widetilde{\mathsf{ptx}}.\mathsf{chg} \not\subseteq \mathsf{tx}^*.\mathsf{out}$ (where $\widetilde{\mathsf{ptx}}$ was the pre-transaction sending $\widetilde{C}$) as otherwise tx^* could simply be a transaction that completes $\widetilde{\mathsf{ptx}}$. If we were lucky when choosing $\overline{C}$ and $\overline{C}$ is one of the coins that the adversary did not include in $\mathsf{tx}^*.\mathsf{out}$, then tx^* satisfies all the properties in (iii) above, in particular (5), meaning we have a solution to EUF-CRO.

Unfortunately, there is one more complication: the adversary could have included $\overline{C}$ as one of the *inputs* of tx^*, in which case we cannot solve EUF-CRO, since (5) would be of the form

$$\sum \mathbf{E} = -2 \cdot \widetilde{C} + \mathsf{Cmt}(\mathsf{cp}, v^*, k^*). \tag{7}$$

But intuitively, in this case the adversary has also "stolen" $\overline{C}$ and if we had randomly picked $\overline{C}$ when first embedding C^* then we could also solve EUF-CRO.

Unfortunately, "stealing" a change output that has not been added to Hon yet does *not* constitute a win according to game STEAL. To illustrate the issue, consider an adversary making the following queries (where all coins C_1 through C_5 have value 1), which the sketched reduction cannot use to break EUF-CRO:

- $\textsc{Ledger}(\mathsf{tx})$ with $\mathsf{tx} = (2, (\,), (C_1, C_2), K)$ $\quad \to \Lambda.\mathsf{out} = (C_1, C_2), \mathsf{Hon} = (\,)$
- $\textsc{Mint}((1))$, creating coin C_3 $\quad \to \Lambda.\mathsf{out} = (C_1, C_2, C_3), \mathsf{Hon} = (C_3)$
- $\textsc{Send}((C_3), (1,1))$, creating C_4, C_5 $\quad \to \Lambda.\mathsf{out} = (C_1, C_2, C_3), \mathsf{Hon} = (C_3)$
- $\textsc{Ledger}((0, (C_1), (C_4), K'))$ $\quad \to \Lambda.\mathsf{out} = (C_2, C_3, C_4), \mathsf{Hon} = (C_3)$
- $\textsc{Ledger}((0, (C_2), (C_5), K''))$ $\quad \to \Lambda.\mathsf{out} = (C_3, C_4, C_5), \mathsf{Hon} = (C_3)$
- $\textsc{Ledger}((0, (C_3, C_4, C_5), (C_6), K^*) =: \mathsf{tx}^*)$ $\quad \to \Lambda.\mathsf{out} = (C_6), \mathsf{Hon} = (C_3)$

Note that all calls $\textsc{Ledger}(\mathsf{tx}_i)$ leave Hon unchanged, since for ptx created during the $\textsc{Send}$ call we have $(C_4, C_5) = \mathsf{ptx.chg} \not\subseteq \mathsf{tx}_i.\mathsf{out}$. The adversary wins the game since it stole C_3, so the reduction must have set $\widetilde{C} = C_3$; moreover, in order to simulate the $\textsc{Send}$ query, it must set $\overline{C}$ to C_4 or C_5. But now tx^* is of the form as in (7), which the reduction cannot use to break EUF-CRO.

The solution to making the reduction always work is to actually prove a *stronger* security notion, where the adversary not only wins when it spends a coin from Hon (in a way that is not simply a completion of a pre-transaction obtained from SEND), but also if the adversary spends a change output which has not been included in Hon yet. Let us denote the set of all such coins by Chg and stress that if the adversary steals a coin from Chg, which the reduction guessed correctly, then there exists only one coin with the challenge embedded in it and so the situation as in (7) cannot arise.

In the proof of this strengthened notion the reduction now guesses the first coin that was stolen from Chg or Hon and if both happen in the same transaction it only accepts a coin from Chg as the right guess. (In the example above, the guesses $\widetilde{C} = C_4$ or $\widetilde{C} = C_5$ would be correct.)

It remains to argue that the reduction can abort when the adversary makes a query SEND($\mathbf{C}, ()$) with $\widetilde{C} \in \mathbf{C}$: in this case its guess $\widetilde{C}$ must have been wrong: for ptx returned by this oracle call we have ptx.chg $\subseteq$ tx.out for any tx, so ptx.in and thus $\widetilde{C}$ is removed from Hon whenever $\mathcal{A}$ makes a LEDGER call (which it must make in order to steal a coin), assuming w.l.o.g. that the adversary stops as soon as it has made its stealing transaction.

Finally, what happens if the adversary makes a query SEND($\mathbf{C}, \mathbf{v}'$) with $\overline{C} \in \mathbf{C}$? We could embed the challenge a *third* time, in one of the change coins of the pre-transaction we need to simulate. Instead of complicating the analysis, the reduction can actually safely abort if such a query is made, since its guess must have been wrong: SEND must be queried on honest coins, so we must have $\overline{C} \in$ Hon. As only the LEDGER oracle can add existing coins to Hon, it must have been queried with some tx such that $\widetilde{\text{ptx}}$.chg $\subseteq$ tx.out, as then $\widetilde{\text{ptx}}$.chg $\ni \overline{C}$ would be added to Hon; however at the same time this removes $\widetilde{\text{ptx}}$.in $\ni \widetilde{C}$ from Hon, which means that $\widetilde{C}$ cannot be the coin the adversary steals, because $\widetilde{C}$ cannot be included in Hon a second time. (As just analyzed for $\overline{C}$ above, the only way to add an existing coin $\widetilde{C}$ to Hon is if $\widetilde{C}$ was created as change during a query ptx $\leftarrow$ SEND($\mathbf{C}, \mathbf{v}$). But since $\widetilde{C}$ had already been in Hon, there must have been a call LEDGER(tx) with tx completing ptx, after which ptx is discarded from the list Ptx of pre-transactions awaiting inclusion in the ledger; see Fig. 8).

Theorem 15 (Transaction indistinguishability (Definition 12)). *Assume that* COM *is a homomorphic hiding commitment scheme,* SIG *a compatible signature scheme, and* Π *is a zero-knowledge proof system. Then the aggregate cash system* MW[COM, SIG, Π] *is transaction-indistinguishable. More precisely, for any* $v_{\max}$ *and any p.p.t. adversary* $\mathcal{A}$ *which makes at most* $q_{\mathcal{A}}$ *queries to its oracle* TX, *there exist p.p.t. adversaries* $\mathcal{B}_{\mathrm{zk}}$ *and* $\mathcal{B}_{\mathrm{hid}}$ *such that*

$$\mathsf{Adv}^{\mathrm{tx\text{-}ind}}_{\mathrm{MW},\mathcal{A}}(\lambda, v_{\max}) \leq \mathsf{Adv}^{\mathrm{zk}}_{\Pi, R_{v_{\max}}, \mathcal{B}_{\mathrm{zk}}}(\lambda) + q_{\mathcal{A}} \cdot \mathsf{Adv}^{\mathrm{hid}}_{\mathrm{COM}, \mathcal{B}_{\mathrm{hid}}}(\lambda).$$

The proof can be found in the full version [FOS18] and intuitively follows from commitments being hiding and proofs zero-knowledge, and that the coin $C^* = \mathsf{Cmt}(\mathsf{cp}, \rho, k^*)$ that is contained in a pre-transaction together with its key k^* (C^* is then spent by Rcv and eliminated from the final transaction by cut-through) acts as a randomizer between E' and E''. We moreover use the fact that because COM is homomorphic, for any values with $\sum v'_0 + \sum v''_0 - \sum v_0 = \sum v'_1 + \sum v''_1 - \sum v_1$, the tuple

$$\left(\mathbf{C} := \mathsf{Cmt}(\mathsf{cp}, v_b, k), \mathbf{C}' := \mathsf{Cmt}(\mathsf{cp}, v'_b, k') \| \mathbf{C}'' := \mathsf{Cmt}(\mathsf{cp}, v''_b, k''), \overline{k} \right) \quad (8)$$

hides the bit b even though $\overline{k} := \sum k' + \sum k'' - \sum k$ is revealed.

We prove Theorem 15 by showing that transactions returned by oracle Tx when $b = 0$ are indistinguishable from transactions returned when $b = 1$. These are of the form

$$\mathsf{tx}^* = \left(0, \mathbf{C}, \mathbf{C}' \| \mathbf{C}'', (\boldsymbol{\pi}' \| \boldsymbol{\pi}'', (E', E''), \sigma^*) \right), \quad (9)$$

where $E' = \sum \mathbf{C}' + C^* - \sum \mathbf{C}$ and $E'' = \sum \mathbf{C}'' - C^*$, and σ^* is an aggregation of signatures σ' and σ'' under keys $r' := \sum k' + k^* - \sum k$ and $r'' := \sum k'' - k^*$, respectively. We thus have $E' = \mathsf{Cmt}(\mathsf{cp}, 0, r')$ and $E'' = \mathsf{Cmt}(\mathsf{cp}, 0, r'')$.

Together with the fact that k^* is uniform and never revealed, indistinguishability of (8) implies indistinguishability of tx^*, as we can create a tuple as in (9) from a tuple as in (8): simulate the proofs $\boldsymbol{\pi}' \| \boldsymbol{\pi}''$, choose a random r^* and set $E' = \mathsf{Cmt}(\mathsf{cp}, 0, r^*)$, $E'' = \mathsf{Cmt}(\mathsf{cp}, 0, \overline{k} - r^*)$, $\sigma' \leftarrow \mathsf{Sign}(\mathsf{sp}, r^*, \varepsilon)$ and $\sigma'' \leftarrow \mathsf{Sign}(\mathsf{sp}, \overline{k} - r^*, \varepsilon)$ and aggregate σ' and σ''.

5 Instantiations

We consider two instantiations of our system MW. In both of them the commitment scheme is instantiated by the Pedersen scheme PDS. The signature scheme is instantiated either by the Schnorr signature scheme SCH or by the BLS signature scheme BLS. We recall these three schemes, as well as the Discrete Logarithm and the CDH assumptions, on which they rely, in the full version [FOS18]. In contrast to COM and SIG, there are no compatibility or joint security requirements for the proof system. In practice, the Bulletproofs scheme [BBB+18] could be used, although under which assumptions it satisfies Definition 8 remains to be studied.

Security of Pedersen-Schnorr. Our security proofs for the combination Pedersen-Schnorr are in the random oracle model and make use of the standard rewinding technique of Pointcheval and Stern [PS96] for extracting discrete logarithms from a successful adversary. This requires some particular care since in both the EUF-NZO and the EUF-CRO games, the adversary can output multiple signatures for distinct public keys for which the reduction must extract discrete logarithms. Fortunately, a generalized forking lemma by Bagherzandi, Cheon, and Jarecki [BCJ08] shows that for Schnorr signatures, one can perform multiple extractions efficiently. From this, we can prove the following two lemmas, whose proofs can be found in the full version [FOS18].

Lemma 16. *The pair* (PDS, SCH) *is EUF-NZO-secure in the random oracle model under the DL assumption. More precisely, for any p.p.t. adversary $\mathcal{A}$ making at most q_h random oracle queries and returning a forgery for a list* of size at most N, there exists a p.p.t. adversary $\mathcal{B}$ running in time at most $8N^2 q_h/\delta_{\mathcal{A}} \cdot \ln(8N/\delta_{\mathcal{A}}) \cdot t_{\mathcal{A}}$, where $\delta_{\mathcal{A}} = \mathsf{Adv}^{\mathrm{euf\text{-}nzo}}_{\mathsf{PDS},\mathsf{SCH},\mathcal{A}}(\lambda)$ and $t_{\mathcal{A}}$ is the running time of $\mathcal{A}$, such that

$$\mathsf{Adv}^{\mathrm{euf\text{-}nzo}}_{\mathsf{PDS},\mathsf{SCH},\mathcal{A}}(\lambda) \le 8 \cdot \mathsf{Adv}^{\mathrm{dl}}_{\mathsf{GrGen},\mathcal{B}}(\lambda).$$

Lemma 17. *The pair* $(\mathsf{PDS},\mathsf{SCH})$ *is EUF-CRO-secure in the random oracle model under the DL assumption. More precisely, for any p.p.t. adversary* $\mathcal{A}$ *making at most* q_h *random oracle queries and* q_s *signature queries, returning a forgery for a list of size at most* N, *and such that* $\delta_{\mathcal{A}} = \mathsf{Adv}^{\mathrm{euf\text{-}cro}}_{\mathsf{PDS},\mathsf{SCH},\mathcal{A}}(\lambda) \ge 2q_s/p,$ *there exists a p.p.t. adversary* $\mathcal{B}$ *running in time at most* $16N^2(q_h + q_s)/\delta_{\mathcal{A}} \cdot \ln(16N/\delta_{\mathcal{A}}) \cdot t_{\mathcal{A}}$, *where* $t_{\mathcal{A}}$ *is the running time of* $\mathcal{A}$, *such that*

$$\mathsf{Adv}^{\mathrm{euf\text{-}cro}}_{\mathsf{PDS},\mathsf{SCH},\mathcal{A}}(\lambda) \le 8 \cdot \mathsf{Adv}^{\mathrm{dl}}_{\mathsf{GrGen},\mathcal{B}}(\lambda) + \frac{q_s + 8}{p}.$$

Corollary 18. $\mathsf{MW}[\mathsf{PDS},\mathsf{SCH},\Pi]$ *with* Π *zero-knowledge and simulation-extractable is inflation-resistant and theft-resistant in the random oracle model under the DL assumption.*

Security of Pedersen-BLS. The security proofs for the Pedersen-BLS pair are also in the random oracle model but do not use rewinding. They are reminiscent of the proof of [BGLS03, Theorem 3.2] and can be found in the full version [FOS18]. Note that EUF-CRO-security is only proved for adversaries making a constant number of signing queries. Fortunately, adversary $\mathcal{B}$ constructed in Theorem 14 makes a single signing query.

Lemma 19. *The pair* $(\mathsf{PDS},\mathsf{BLS})$ *is EUF-NZO-secure in the random oracle model under the CDH assumption. More precisely, for any p.p.t. adversary* $\mathcal{A}$ *making at most* q_h *random oracle queries and returning a forgery for a list of size at most* N, *there exists a p.p.t. adversary* $\mathcal{B}$ *running in time at most* $t_{\mathcal{A}} + (q_h + N + 2)t_M$, *where* $t_{\mathcal{A}}$ *is the running time of* $\mathcal{A}$ *and* t_M *is the time of a scalar multiplication in* $\mathbb{G}$, *such that*

$$\mathsf{Adv}^{\mathrm{cdh}}_{\mathsf{GrGen},\mathcal{B}}(\lambda) = \mathsf{Adv}^{\mathrm{euf\text{-}nzo}}_{\mathsf{PDS},\mathsf{BLS},\mathcal{A}}(\lambda).$$

Lemma 20. *The pair* $(\mathsf{PDS},\mathsf{BLS})$ *is EUF-CRO-secure in the random oracle model under the CDH assumption. More precisely, for any p.p.t. adversary* $\mathcal{A}$ *making at most* q_h *random oracle queries and* $q_s = O(1)$ *signature queries and returning a forgery for a list of size at most* N, *there exists a p.p.t. adversary* $\mathcal{B}$ *running in time at most* $t_{\mathcal{A}} + (2q_h + 3q_s + N + 2)t_M$, *where* $t_{\mathcal{A}}$ *is the running time of* $\mathcal{A}$ *and* t_M *is the time of a scalar multiplication in* $\mathbb{G}$, *such that*

$$\mathsf{Adv}^{\mathrm{cdh}}_{\mathsf{GrGen},\mathcal{B}}(\lambda) \ge \frac{1}{4 \cdot (2N)^{q_s}} \cdot \mathsf{Adv}^{\mathrm{euf\text{-}cro}}_{\mathsf{PDS},\mathsf{BLS},\mathcal{A}}(\lambda).$$

Corollary 21. $\mathsf{MW}[\mathsf{PDS},\mathsf{BLS},\Pi]$ *with* Π *zero-knowledge and simulation-extractable is inflation-resistant and theft-resistant in the random oracle model under the CDH assumption.*

Acknowledgements. The first author is supported by the French ANR *EfTrEC* project (ANR-16-CE39-0002) and the *MSR-Inria Joint Centre*. The second author is supported by ERC grant 639554 (project *aSCEND*).

References

[AKR+13] Androulaki, E., Karame, G.O., Roeschlin, M., Scherer, T., Capkun, S.: Evaluating user privacy in Bitcoin. In: Sadeghi, A.-R. (ed.) FC 2013. LNCS, vol. 7859, pp. 34–51. Springer, Heidelberg (2013). https://doi.org/10.1007/978-3-642-39884-1_4

[Bac13] Back, A.: Bitcoins with homomorphic value (validatable but encrypted), October 2013. BitcoinTalk post. https://bitcointalk.org/index.php?topic=305791.0

[BBB+18] Bünz, B., Bootle, J., Boneh, D., Poelstra, A., Wuille, P., Maxwell, G.: Bulletproofs: short proofs for confidential transactions and more. In: S&P 2018, pp. 315–334 (2018)

[BBSU12] Barber, S., Boyen, X., Shi, E., Uzun, E.: Bitter to better—how to make Bitcoin a better currency. In: Keromytis, A.D. (ed.) FC 2012. LNCS, vol. 7397, pp. 399–414. Springer, Heidelberg (2012). https://doi.org/10.1007/978-3-642-32946-3_29

[BCG+14] Ben-Sasson, E., et al.: Zerocash: decentralized anonymous payments from Bitcoin. In: S&P 2014, pp. 459–474 (2014)

[BCJ08] Bagherzandi, A., Cheon, J.H., Jarecki, S.: Multisignatures secure under the discrete logarithm assumption and a generalized forking lemma. In: ACM CCS 2008, pp. 449–458 (2008)

[BGLS03] Boneh, D., Gentry, C., Lynn, B., Shacham, H.: Aggregate and verifiably encrypted signatures from bilinear maps. In: Biham, E. (ed.) EUROCRYPT 2003. LNCS, vol. 2656, pp. 416–432. Springer, Heidelberg (2003). https://doi.org/10.1007/3-540-39200-9_26

[BLS01] Boneh, D., Lynn, B., Shacham, H.: Short signatures from the Weil pairing. In: Boyd, C. (ed.) ASIACRYPT 2001. LNCS, vol. 2248, pp. 514–532. Springer, Heidelberg (2001). https://doi.org/10.1007/3-540-45682-1_30

[BNM+14] Bonneau, J., Narayanan, A., Miller, A., Clark, J., Kroll, J.A., Felten, E.W.: Mixcoin: anonymity for Bitcoin with accountable mixes. In: Christin, N., Safavi-Naini, R. (eds.) FC 2014. LNCS, vol. 8437, pp. 486–504. Springer, Heidelberg (2014). https://doi.org/10.1007/978-3-662-45472-5_31

[BNN07] Bellare, M., Namprempre, C., Neven, G.: Unrestricted aggregate signatures. In: Arge, L., Cachin, C., Jurdziński, T., Tarlecki, A. (eds.) ICALP 2007. LNCS, vol. 4596, pp. 411–422. Springer, Heidelberg (2007). https://doi.org/10.1007/978-3-540-73420-8_37

[DDO+01] De Santis, A., Di Crescenzo, G., Ostrovsky, R., Persiano, G., Sahai, A.: Robust non-interactive zero knowledge. In: Kilian, J. (ed.) CRYPTO 2001. LNCS, vol. 2139, pp. 566–598. Springer, Heidelberg (2001). https://doi.org/10.1007/3-540-44647-8_33

[FOS18] Fuchsbauer, G., Orrù, M., Seurin, Y.: Aggregate cash systems: a cryptographic investigation of Mimblewimble. Cryptology ePrint Archive, Report 2018/1039 (2018). https://eprint.iacr.org/2018/1039

[GCKG14] Gervais, A., Capkun, S., Karame, G.O., Gruber, D.: On the privacy provisions of bloom filters in lightweight Bitcoin clients. In: ACSAC 2014, pp. 326–335 (2014)

[Gro06] Groth, J.: Simulation-sound NIZK proofs for a practical language and constant size group signatures. In: Lai, X., Chen, K. (eds.) ASIACRYPT 2006. LNCS, vol. 4284, pp. 444–459. Springer, Heidelberg (2006). https://doi.org/10.1007/11935230_29

[HAB+17] Heilman, E., Alshenibr, L., Baldimtsi, F., Scafuro, A., Goldberg, S.: TumbleBit: an untrusted Bitcoin-compatible anonymous payment hub. In: NDSS (2017)

[Jed16] Jedusor, T.E.: Mimblewimble (2016). https://download.wpsoftware.net/bitcoin/wizardry/mimblewimble.txt

[KKM14] Koshy, P., Koshy, D., McDaniel, P.: An analysis of anonymity in Bitcoin using P2P network traffic. In: Christin, N., Safavi-Naini, R. (eds.) FC 2014. LNCS, vol. 8437, pp. 469–485. Springer, Heidelberg (2014). https://doi.org/10.1007/978-3-662-45472-5_30

[LMRS04] Lysyanskaya, A., Micali, S., Reyzin, L., Shacham, H.: Sequential aggregate signatures from trapdoor permutations. In: Cachin, C., Camenisch, J.L. (eds.) EUROCRYPT 2004. LNCS, vol. 3027, pp. 74–90. Springer, Heidelberg (2004). https://doi.org/10.1007/978-3-540-24676-3_5

[Max13a] Maxwell, G.: CoinJoin: Bitcoin privacy for the real world, August 2013. BitcoinTalk post. https://bitcointalk.org/index.php?topic=279249.0

[Max13b] Maxwell, G.: Transaction cut-through, August 2013. BitcoinTalk post. https://bitcointalk.org/index.php?topic=281848.0

[Max15] Maxwell, G.: Confidential Transactions (2015). https://people.xiph.org/~greg/confidential_values.txt

[MGGR13] Miers, I., Garman, C., Green, M., Rubin, A.D.: Zerocoin: anonymous distributed E-cash from Bitcoin. In: S&P 2013, pp. 397–411 (2013)

[MPJ+13] Meiklejohn, S., et al.: A fistful of Bitcoins: characterizing payments among men with no names. In: Internet Measurement Conference, IMC 2013, pp. 127–140 (2013)

[Nak08] Nakamoto, S.: Bitcoin: A Peer-to-Peer Electronic Cash System (2008). http://bitcoin.org/bitcoin.pdf

[Ped92] Pedersen, T.P.: Non-interactive and information-theoretic secure verifiable secret sharing. In: Feigenbaum, J. (ed.) CRYPTO 1991. LNCS, vol. 576, pp. 129–140. Springer, Heidelberg (1992). https://doi.org/10.1007/3-540-46766-1_9

[Poe16] Poelstra, A.: Mimblewimble (2016). https://download.wpsoftware.net/bitcoin/wizardry/mimblewimble.pdf

[PS96] Pointcheval, D., Stern, J.: Security proofs for signature schemes. In: Maurer, U. (ed.) EUROCRYPT 1996. LNCS, vol. 1070, pp. 387–398. Springer, Heidelberg (1996). https://doi.org/10.1007/3-540-68339-9_33

[RMK14] Ruffing, T., Moreno-Sanchez, P., Kate, A.: CoinShuffle: practical decentralized coin mixing for Bitcoin. In: Kutyłowski, M., Vaidya, J. (eds.) ESORICS 2014. LNCS, vol. 8713, pp. 345–364. Springer, Cham (2014). https://doi.org/10.1007/978-3-319-11212-1_20

[RS13] Ron, D., Shamir, A.: Quantitative analysis of the full Bitcoin transaction graph. In: Sadeghi, A.-R. (ed.) FC 2013. LNCS, vol. 7859, pp. 6–24. Springer, Heidelberg (2013). https://doi.org/10.1007/978-3-642-39884-1_2

[RTRS18] Ruffing, T., Thyagarajan, S.A., Ronge, V., Schröder, D.: Burning zerocoins for fun and for profit: a cryptographic denial-of-spending attack on the zerocoin protocol. IACR Cryptology ePrint Archive, Report 2018/612 (2018)

[Sch91] Schnorr, C.-P.: Efficient signature generation by smart cards. J. Cryptol. 4(3), 161–174 (1991)

[SMD14] Saxena, A., Misra, J., Dhar, A.: Increasing anonymity in Bitcoin. In: Böhme, R., Brenner, M., Moore, T., Smith, M. (eds.) FC 2014. LNCS, vol. 8438, pp. 122–139. Springer, Heidelberg (2014). https://doi.org/10.1007/978-3-662-44774-1_9

[SZ16] Sompolinsky, Y., Zohar, A.: Bitcoin's security model revisited (2016). Manuscript http://arxiv.org/abs/1605.09193

[vS13] van Saberhagen, N.: CryptoNote v 2.0 (2013). Manuscript https://cryptonote.org/whitepaper.pdf

Proof-of-Stake Protocols
for Privacy-Aware Blockchains

Chaya Ganesh[1(✉)], Claudio Orlandi[1], and Daniel Tschudi[1,2]

[1] Department of Computer Science, DIGIT, Aarhus University, Aarhus, Denmark
{ganesh,orlandi,tschudi}@cs.au.dk
[2] Concordium, Aarhus, Denmark

Abstract. *Proof-of-stake (PoS)* protocols are emerging as one of the most promising alternative to the wasteful *proof-of-work (PoW)* protocols for consensus in Blockchains (or distributed ledgers). However, current PoS protocols inherently disclose both the *identity* and the *wealth* of the stakeholders, and thus seem incompatible with privacy-preserving cryptocurrencies (such as ZCash, Monero, etc.). In this paper we initiate the formal study for PoS protocols with privacy properties. Our results include:

1. A (theoretical) feasibility result showing that it is possible to construct a general class of *private PoS (PPoS)* protocols; and to add privacy to a wide class of PoS protocols,
2. A privacy-preserving version of a popular PoS protocol, Ouroboros Praos.

Towards our result, we define the notion of *anonymous verifiable random function*, which we believe is of independent interest.

1 Introduction

Popular decentralized cryptocurrencies like Bitcoin [Nak08] crucially rely on the existence of a distributed ledger, known as the *Blockchain*. The original protocols used to build and maintain the Blockchain were based on *proof-of-work consensus protocols (PoW)*. While blockchain protocols mark a significant breakthrough in distributed consensus, reliance on expensive PoW components result in enormous waste of energy [OM14,CDE+16], therefore it is an important open problem to find alternative consensus mechanisms which are less wasteful than PoW but at the same time maintain the positive features offered by PoW. *Proof-of-stake consensus protocols (PoS)* are one of the most promising technology to replace PoW and still preserve similar robustness properties: while PoW provides robustness assuming that a (qualified) majority of the *computing power* is honest,

This work was supported by the Danish Independent Research Council under Grant-ID DFF-6108-00169 (FoCC), the European Research Council (ERC) under the European Unions's Horizon 2020 research and innovation program under grant agreement No 669255 (MPCPRO) and No 803096 (SPEC), and the Concordium Blockchain Research Center, Aarhus University, Denmark.

Y. Ishai and V. Rijmen (Eds.): EUROCRYPT 2019, LNCS 11476, pp. 690–719, 2019.
https://doi.org/10.1007/978-3-030-17653-2_23

PoS instead relies on the assumption that a majority of the *wealth* in the system is controlled by honest participants. The rationale behind PoS is that users who have significant *stakes* in the system have an economic incentive in keeping the system running according to the protocol specification, as they risk that their stakes will become worthless if trust in the cryptocurrency vanishes. As is usual in this research space, the initial idea of proof-of-stake appeared informally in an online Bitcoin forum [bit11], and since then, there have been a series of candidates for such protocols [KN12, BLMR14, BGM16]. Recently, there have been works on formal models for proof-of-stake and protocols with provable security guarantees [BPS16, KRDO17, GHM+17, DGKR18, BGK+18].

Consensus based on lottery. Very informally, a lottery-based consensus protocol works in the following way: some publicly verifiable "lottery" mechanism is implemented to elect the next committee or a block "leader" who is then allowed to add the next block to the blockchain. The probability of a user being elected is proportional to the amount of some "scarce resource" owned by the user. In PoW, the probability is proportional to the computing power of the user, while in PoS it is proportional to the amount of coins the user owns. Since the resource is scarce and cannot be replicated, *Sybil attacks* are prevented (e.g., a user cannot inflate its probability of become the block leader). This, combined with the assumption that some majority of the resource is controlled by honest parties guarantees that the honest participants are in charge of the blockchain, thus guaranteeing integrity of the stored information.

Proof-of-work. In PoW, such as Bitcoin, for every block, all users try to solve some computationally challenging puzzle. The first user to have solved the puzzle publishes the solution together with the new block, and often together with a new address that allows to collect the transaction fees and the block reward, which act as economic incentives for users to participate in the consensus protocol. All other participants can verify that the received block is valid (e.g., contains a valid solution to the puzzle) and, if so, append it to their local view of the Blockchain. Note that, if we assume that users are connected to each other using anonymous communication channels (e.g., Tor), then PoW provides full anonymity i.e., given two blocks it is hard to tell whether they came from the same user or not.

Proof-of-stake. On the other hand, PoS systems follow a different approach: here we must assume that the Blockchain contains information about the wealth owned by the users in the system. Then, for every block, each user has a way to locally compute (using a pseudorandom process) whether they won the lottery. The lottery has the property that the higher the stakes in the system, the higher the probability of becoming the block leader or committee member. If a user wins, then, in the case of block leadership, they publish a block together with a proof of winning the lottery, and in the case of committee membership publish a message along with proof of lottery win. As in PoW, it is important that the other users in the system can efficiently verify the correctness of this claim, that is, the user claiming to be the block leader has in fact won the lottery.

Unfortunately, in existing PoS protocols, this requires the users to be able to link the newly generated block with some account in the system. Thus, everyone in the system will learn the identity of the block producer (and their wealth).

Privacy in PoS? As privacy-preserving cryptocurrencies (such as ZCash, Monero, Dash, etc.) increase in popularity, it is natural to ask the following question:

Is it possible to design consensus protocols which
are as energy-efficient as PoS, but as private as PoW?

In this paper we address this problem and provide the first positive results in this direction. In particular we offer two contributions: (1) we provide a feasibility result showing that it is possible to construct a *Private PoS (PPoS)* protocol, that is one where the identity of the lottery winner (and their wealth) is kept secret by the protocol; and (2) we show how to adapt a popular PoS protocol, Ouroboros Praos [DGKR18], to satisfy our anonymity requirement. In doing so, we introduce a novel cryptographic primitive – *anonymous verifiable random function (AVRF)* which might be of independent interest.

Related Work. In a recent independent and concurrent work [KKKZ18], the authors present a privacy-preserving proof-of-stake protocol. While our work is more modular and treats privacy of proof-of-stake consensus independently of the cryptocurrency layer, the work of [KKKZ18] builds an overall private transaction ledger system. Additionally, our construction guarantees full privacy (at the cost of assuming anonymous channels), while [KKKZ18] allows for the *leakage* of a function of the stake.

1.1 Technical Overview

All current proof-of-stake proposals rely on stake distribution being available in the clear. As a first step, let us consider how to hide the wealth of the lottery winner: a simple idea that might come to mind is to encrypt the stakes on the blockchain, and to replace the proof of winning the lottery (e.g. in form of a correct block) with a proof that uses the encrypted stakes instead. However, this falls short of our goal in at least two ways. For instance, in case of block leaders: (1) it is still possible to distinguish (for instance) whether two blocks were generated by the same block leader or not and, crucially (2) since the probability of being elected as block leader is proportional to one's wealth, the frequency with which a user wins the lottery indirectly leaks information about their stakes (i.e., a user who is observed to win t out of n blocks has relative stakes in the system close to t/n). Thus, we conclude that even if block leaders are only interested in hiding their wealth, a PPoS must necessarily hide their identity as well. Similar concerns are true for committee based protocols as well, where the number of times a user becomes a committee member reveals information about their wealth.

In Sect. 4 we provide a framework for VRF-based *private stake lottery*. Our framework is parametrized by a lottery mechanism, and allows therefore to construct PPoS protocols for some of the most popular lottery mechanisms used in current (non-private) PoS protocols (e.g., Algorand [GHM+17], Ouroboros Praos [DGKR18], etc.).

After having established the first feasibility result in this area, in Sect. 5 we investigate how to efficiently implement PPoS. Our starting point is one of the main PoS candidates which comes with a rigorous proof of security, namely Ouroboros Praos.

In a nutshell, Ouroboros Praos works as follows: Every user in the system registers a verification key for a *verifiable random function* e.g., a PRF for which it is possible to prove that a given output is in the image of the function relative to the verification key. Then, at every round (or slots), users can apply the VRF to the slot number and thus receive a (pseudorandom) value. If the value is less than (a function of) their wealth, then that user has won the election process and can generate a new block. Thanks to the VRF property, all other users can verify that the VRF has been correctly computed, and since the wealth of the user is public as well, every other user can compare the output of the VRF with the (function of) the user's wealth.

Using the private stake lottery of Sect. 4, it would be possible for the elected leader to prove correctness of all steps above (without revealing any further information) using the necessary zero-knowledge proofs. However, this would result in a very inefficient solution. To see why, we need to say a few more words about the winning condition of Ouroboros Praos: one of the goals of Ouroboros Praos is to ensure that a user cannot artificially increase their probability of winning the lottery, therefore Ouroboros Praos compares the output of the VRF with a function of the wealth that satisfies the "independent aggregation" property i.e., a function such that the probability that two users win the lottery is the same as the probability of winning for a single user who owns the same wealth as the two users combined. In particular, the function used by Ouroboros Praos has the form

$$\phi_f(x) = 1 - (1 - f)^{\mathsf{stk}/\mathsf{Stake}}$$

where stk is wealth of the user, Stake is the total amount of stakes in the system and f is a difficulty parameter. Implementing such a function using the circuit representation required by zero-knowledge proofs would be very cumbersome, due to the non-integer division and the exponentiation necessary to evaluate ϕ. Finally, the variable difficulty level would require to update the circuit in the ZK-proof as the difficulty changes.

One of our insights is to exploit the "independent aggregation" property of the function for efficiency purpose, and in fact our solution uses the function ϕ in a completely *black-box* way, and thus allows to replace the specific function above with any other function that satisfies the "independent aggregation" property. Thanks to the independent aggregation property, we can let the users commit to their wealth in a bit-by-bit fashion, thus effectively splitting their account into a number of "virtual parties" such that party i has wealth 0 or 2^i. Then, the values $V_i = \phi(2^i)$ can be publicly computed (outside of the ZK-proof) and what is left to do for the user is to prove that the output of the VRF (for at least one of the virtual parties i), is less than the corresponding public value V_i (without revealing which one).

The solution as described so far allows to prove that one has won the election for a "committed" stake but, as described above, the frequency with which an account wins the election reveals information about the user's wealth as well. Therefore we need to replace the VRF with an "anonymous VRF" or AVRF. In a nutshell, an AVRF is a VRF in which there exist multiple verification keys for the same secret key, and where it is hard, given two valid proofs for different inputs under different verification keys, to tell whether they were generated by the same secret key or not. We show that it is possible to turn existing efficient VRF constructions into anonymous VRF with an very small efficiency loss, and we believe that AVRF is a natural cryptographic primitive which might have further applications.

2   Preliminaries

Notation. We use $[1, n]$ to represent the set of numbers $\{1, 2, \ldots, n\}$. If A is a randomized algorithm, we use $y \leftarrow \mathsf{A}(x)$ to denote that y is the output of A on x. We write $x \xleftarrow{R} \mathcal{X}$ to mean sampling a value x uniformly from the set $\mathcal{X}$. We write PPT to denote a probabilistic polynomial-time algorithm. Throughout the paper, we use κ to denote the security parameter or level. A function is negligible if for all large enough values of the input, it is smaller than the inverse of any polynomial. We use negl to denote a negligible function. We denote by H a cryptographic hash function.

2.1   Zero-Knowledge Proofs

Let R be an efficiently computable binary relation which consists of pairs of the form (x, w) where x is a statement and w is a witness. Let L be the language associated with the relation R, i.e., $L = \{x \mid \exists w \text{ s.t. } R(x, w) = 1\}$. L is an NP language if there is a polynomial p such that every w in $R(x)$ has length at most $p(x)$ for all x.

A zero-knowledge proof for L lets a prover P convince a verifier V that $x \in L$ for a common input x without revealing w. A proof of knowledge captures not only the truth of a statement $x \in L$, but also that the prover "possesses" a witness w to this fact. This is captured by requiring that if P can convince V with reasonably high probability, then a w can be efficiently extracted from P given x.

Non-interactive Zero-knowledge Proofs. A model that assumes a trusted setup phase, where a string of a certain structure, also called the public parameters of the system is generated, is called the common reference string (CRS) model. Non-interactive zero-knowledge proofs (NIZKs) in the CRS model were introduced in [BFM88]. We give a formal definition of NIZKs in Appendix B.1. In this paper, we will be concerned with non-interactive proofs.

2.2 Commitment Schemes

A commitment scheme for a message space is a triple of algorithms (Setup, Com, Open) such that Setup(1^κ) outputs a public commitment key; Com given the public key and a message outputs a commitment along with opening information. Open given a commitment and opening information outputs a message or $\perp$ if the commitment is not valid. We require a commitment scheme to satisfy correctness, hiding and binding properties. Informally, the hiding property guarantees that no PPT adversary can generate two messages such that it can distinguish between their commitments. The binding property guarantees that, informally, no PPT adversary can open a commitment to two different messages.

2.3 Sigma Protocols

A sigma protocol for a language L is a three round public-coin protocol between a prover P and a verifier V. P's first message in a sigma protocol is denoted by $a \leftarrow P(x; R)$. V's message is a random string $r \in \{0,1\}^\kappa$. P's second message is $e = P(w, a, r, R)$. (a, r, e) is called a transcript, and if the verifier accepts, that is $V(x, a, r, e) = 1$, then the transcript is accepting for s.

Definition 1 (Sigma protocol). *An interactive protocol $\langle P, V \rangle$ between prover P and verifier V is a Σ protocol for a relation R if the following properties are satisfied:*

1. *It is a three move public coin protocol.*
2. *Completeness: If P and V follow the protocol then $\Pr[\langle P(w), V \rangle (x) = 1] = 1$ whenever $(x, w) \in R$.*
3. *Special soundness: There exists a polynomial time algorithm called the extractor which when given s and two transcripts (a, r, e) and (a, r', e') that are accepting for s, with $r \neq r'$, outputs w' such that $(x, w') \in R$.*
4. *Special honest verifier zero knowledge: There exists a polynomial time simulator which on input s and a random r outputs a transcript (a, r, e) with the same probability distribution as that generated by an honest interaction between P and V on (common) input s.*

Sigma protocols and NIZK. The Fiat-Shamir transform [FS87] may be used to compile a Σ protocol into a non-interactive zero-knowledge proof of knowledge in the random oracle model. In this paper, we will be concerned with transformations in the CRS model [Dam00, Lin15]. The transformation of [Dam00] gives a 3-round concurrent zero-knowledge protocol, while [Lin15] is non-interactive.

OR composition of Σ-protocols. In [CDS94], the authors devise a composition technique for using sigma protocols to prove compound OR statements. Essentially, a prover can efficiently show $((x_0 \in \mathcal{L}) \vee (x_1 \in \mathcal{L}))$ without revealing which x_i is in the language. More generally, the OR transform can handle two different relations R_0 and R_1. If Π_0 is a Σ-protocol for R_0 and Π_1 a Σ-protocol for R_1, then there is a Σ-protocol Π_{OR} for the relation R_{OR} given by $\{((x_0, x_1), w) : ((x_0, w) \in R_0) \vee ((x_1, w) \in R_1)\}$.

Pedersen Commitment. Throughout the paper, we use an algebraic commitment scheme that allows proving polynomial relationships among committed values. The Pedersen commitment scheme [Ped92] is one such example that provides computational binding and unconditional hiding properties based on the discrete logarithm problem. It works in a group of prime order q. Given two random generators g and h such that $\log_g h$ is unknown, a value $x \in \mathbb{Z}_q$ is committed to by choosing r randomly from $\mathbb{Z}_q$, and computing $C_x = g^x h^r$. We write $\mathsf{Com}_q(x)$ to denote a Pedersen commitment to x in a group of order q. There are Sigma protocols in literature to prove knowledge of a committed value, equality of two committed values, and so on, and these protocols can be combined in natural ways. In particular, Pedersen commitments allow proving polynomial relationships among committed values: Given $\mathsf{Com}(x)$ and $\mathsf{Com}(y)$, prove that $y = ax + b$ for some public values a and b. In our constructions, we make use of existing sigma protocols for proving statements about discrete logarithms, and polynomial relationships among committed values [Sch91, FO97, CS97, CM99]. Throughout, we use the following notation:

$$\mathsf{PK}\{(x, y, \ldots) : \; statements \text{ about } x, y, \ldots\}$$

In the above, $x, y, \ldots$ are secrets (discrete logarithms), the prover claims knowledge of $x, y, \ldots$ such that they satisfy *statements*. The other values in the protocol are public.

2.4   Merkle Tree

A Merkle tree is a hash based data structure that is used both to generically extend the domain of a hash function and as a succinct commitment to a large string. To construct a Merkle tree from a string $m \in \{0, 1\}^n$, we split the string into blocks $b_i \in \{0, 1\}^k$. Each block is then a leaf of the tree, and we use a hash function H to compress two leaves into an internal node. Again, at the next level, each pair of siblings is compressed into a node using the hash function, and this is continued until a single node is obtained which is the root of the Merkle tree. In order to verify the membership of a block in a string represented by a Merkle tree root, it is sufficient to provide a path from the leaf node corresponding to the claimed block all the way up to the public root node. This is easily verified given the hash values along the path together with the hash values of sibling nodes.

2.5   Decisional Diffie-Hellman Assumption

Let G be the description of cyclic group of prime order q for $q = \Theta(2^\kappa)$ output by a PPT group generator algorithm $\mathcal{G}$ on input 1^κ. Let g be a generator of G. The *decisional Diffie-Hellman* (DDH) problem for G is the following: given group elements (α, β, γ), distinguish whether they are independent and uniformly random in G or whether $\alpha = g^a$ and $\beta = g^b$ are independent and uniformly random and $\gamma = g^{ab}$.

The DDH assumption is said to hold in G if there exists a function negl such that no PPT algorithm $\mathcal{A}$ can win in the above distinguishing game with probability more that $1/2 + \mathsf{negl}(\kappa)$.

3   Model

In this section, we define certain functionalities that are used in our later protocols.

Stake Distribution. We assume that parties have access to a list of (static) stakeholder accounts. Each such account consists of the committed stake or voting-power and a signature verification key. Each stakeholder additionally has access to his own stake value, the signing key, and the randomness used for the stake commitment. In our protocols the functionality $\mathcal{F}_{\mathsf{Init}}^{\mathsf{Com},\mathsf{SIG}}$ is used to provide the static stake information. In practice, this information could for instance be stored in the genesis block of a blockchain.

More generally, the stake distribution is dynamic and can be read from the blockchain. We discuss the extension to dynamic stake in Sect. 4.4.

Functionality $\mathcal{F}_{\mathsf{Init}}^{\mathsf{Com},\mathsf{SIG}}$

The functionality is parametrized by a commitment scheme Com and a signature scheme $\mathsf{SIG} = (\mathsf{KeyGen}, \mathsf{Sign}, \mathsf{Ver})$.

Initialization

The functionality initially contains a list of stakeholder id's pid and their relative stake α_{pid}. For each stakeholder pid, the functionality does:

1. Compute commitment $\mathsf{Com}(\alpha_{\mathsf{pid}}; r_{\mathsf{pid}})$ with fresh randomness r_{pid};
2. Pick a random secret key $\mathsf{sk}_{\mathsf{pid}}$ and compute $\mathsf{vk}_{\mathsf{pid}} = \mathsf{KeyGen}(\mathsf{sk}_{\mathsf{pid}})$.

Information

- Upon receiving input $(\mathrm{GETPRIVATEDATA}, \mathsf{sid})$ from a stakeholder pid (or the adversary in the name of corrupted stakeholder) output $(\mathrm{GETPRIVATEDATA}, \mathsf{sid}, \alpha_{\mathsf{pid}}, r_{\mathsf{pid}}, \mathsf{sk}_{\mathsf{pid}})$.

- Upon receiving $(\mathrm{GETLIST}, \mathsf{sid})$ from a party (or the adversary in the name of corrupted party) output the list $\mathcal{L} = \{(\mathsf{Com}(\alpha_{\mathsf{pid}}), \mathsf{vk}_{\mathsf{pid}})_{\mathsf{pid}}\}$.

Common reference string. In our protocols stakeholders use zero-knowledge proofs to show that they won the stake lottery. The functionality $\mathcal{F}_{\mathsf{crs}}^{\mathcal{D}}$ provides the common reference string required for those zero-knowledge proofs.

Functionality $\mathcal{F}_{crs}^{\mathcal{D}}$

The functionality is parametrized by a distribution $\mathcal{D}$.

- Sample a CRS, $crs \leftarrow \mathcal{D}$
- Upon receiving (Setup, sid) from a party, output (Setup, sid, crs).

Verifiable pseudorandom function. In our protocols, stakeholders use the VRF functionality $\mathcal{F}_{VRF}^{Com}$ to get the randomness in the stake lottery. The functionality allows a stakeholder to generate a key and then evaluate the VRF under that key. The evaluation returns a value and a commitment of that value. The commitment can then used by parties to verify the claimed $\mathcal{F}_{VRF}^{Com}$ evaluation. The functionality also offers VERIFY queries, where anyone can check if a given output of the VRF was computed correctly. Note that the VERIFY queries do not disclose the identity of the party who have generated the output. In other words, VERIFY checks if a given output is in the combined image of all the registered VRF keys.

The VRF functionality $\mathcal{F}_{VRF}^{Com}$ is defined as follows.

Functionality $\mathcal{F}_{VRF}^{Com}$

The functionality maintains a table $T(\cdot, \cdot)$ which is initially empty.

Key Generation

Upon input (KEYGEN, sid) from a stakeholder pid generate a unique "ideal" key vid, record (pid, vid). Return (KEYGEN, sid, vid) to pid.

VRF Evaluation

Upon receiving a message (EVAL, sid, vid, m) from stakeholder pid, verify that pair (pid, vid) has been recorded. If not, ignore the request.

1. If $T(vid, m)$ is undefined, pick random values y, r from $\{0, 1\}^{\ell_{VRF}}$ and set $T(vid, m) = (y, \text{Com}(y; r), r)$.
2. Return (EVALUATED, sid, $T(vid, m)$) to pid.

VRF Verification

Upon receiving a message (VERIFY, sid, m, c) from some party, do:

1. If there exists a vid such that $T(vid, m) = (y, c, r)$ for some y, r then set $f = 1$.
2. Else, set $f = 0$.

Output (VERIFIED, sid, m, c, f) to the party.

Anonymous Broadcast. Stakeholders cannot publish their messages over a regular network as this would reveal their identity. We therefore assume that stakeholders use an anonymous broadcast channel. The functionality $\mathcal{F}_{ABC}$ allows a party to send messages anonymously to all parties. The adversary is allowed to send anonymous messages to specific parties.

Functionality $\mathcal{F}_{ABC}$

Any party can register (or deregister). For each registered party the functionality maintains a message buffer.

Send Message

Upon receiving (SEND, sid, m) from registered party P add m to the message buffers of all registered parties. Output (SENT, sid, m) to the adversary.

Receive Message

Upon receiving (RECEIVE, sid) from registered party P remove all message from P's message buffer and output them to P.

Adversarial Influence

Upon receiving (SEND, sid, m, P') from $\mathcal{A}$ on behalf some *corrupted* registered party add m to the message buffer of registered party P'. Output (SENT, sid, m, P') to the adversary.

---

4   Feasibility of Private Proof-of-Stake

In order to make a proof-of-stake protocol private, a first solution that comes to mind is to have the parties prove in zero-knowledge that they indeed won the lottery (either for a slot or committee membership). This does hide the identity, but it reveals the stake of the winning account. It might seem like one can hide the stake too by having the parties commit to their stakes and give a zero-knowledge proof of winning on committed stake. While this indeed hides the stake in a single proof, it leaks how often a given account wins. One can infer information about the stake in a given account from the frequency with which an account participates in a committee or wins a slot. Therefore, the actual statement that one needs to prove in a private lottery needs to take the list of all accounts as input. Now, a party proves knowledge of corresponding secret key of some public key in a list, and the stake in that account won the lottery. We employ this idea to give a general framework for constructing a private proof-of-stake protocol. The framework applies to proof-of-stake protocols that work with lottery functions which are locally verifiable, that is, a party can locally determine whether it wins or not. The lottery is a function of the party's stake and may depend on other parameters like slot, role etc that we call *entry* parameters. The set $\mathcal{E}$ of entry parameters for the lottery depends on the type of proof of stake. In a slot-based proof-of-stake, for instance, the lottery elects a leader for a particular slot that allows the leader to publish a block for that slot. Ouroboros Praos [KRDO17] is an example of such a slot-based proof-of-stake. In protocols such as Algorand [GHM+17], where the protocol is committee-based, the lottery is for determining a certain role in a committee, and our framework applies to both type of protocols.

4.1   Private Lottery Functionality

The private lottery functionality is an abstraction that we introduce to capture the privacy requirements discussed above. The functionality $\mathcal{F}_{\mathsf{Lottery}}^{\mathsf{LE},\mathcal{E}}$ is parametrized by the set $\mathcal{E}$ of allowed entry parameters, and a predicate function LE. The predicate LE takes as input the relative stake and randomness. It allows stakeholder pid to locally check whether they won the lottery for entry e. If yes, they can publish pairs of the form (e, m) where m is an allowed message as determined and verified by the proof-of-stake protocol that uses the lottery; for instance, when slot-based, m is a block, when committee-based, m is a committee message.

Functionality $\mathcal{F}_{\mathsf{Lottery}}^{\mathsf{LE},\mathcal{E}}$

The functionality is parametrized by a set $\mathcal{E}$ of entries, and a predicate function LE which takes as input the relative stake (represented by a bit string in $\{0,1\}^{\ell_\alpha}$) and randomness (in $\{0,1\}^{\ell_{\mathsf{VRF}}}$).

Registered Parties. The functionality maintains a list $\mathcal{P}$ of registered parties. For each registered party the functionality maintains a message buffer.

Stakeholders. The functionality maintains a list of (the initial) stakeholders $\mathsf{pid}_1, \ldots, \mathsf{pid}_k$ and their relative stakes $\mathsf{pid}_1.\alpha, \ldots, \mathsf{pid}_k.\alpha$. Finally, the functionality also manages a table $T(,)$ where the entry $T(\mathsf{pid}, \mathsf{e})$ defines whether pid is allowed to publish a message relative to entry e. Initially, the table T is empty.

- Upon receiving $(\text{LOTTERY}, \mathsf{sid}, \mathsf{e})$ from a stakeholder pid (or the adversary in the name of corrupted participant pid) do the following:
  1. If $T(\mathsf{pid}, \mathsf{e})$ is undefined sample a random value $r \in \{0,1\}^\ell$ and set $T(\mathsf{pid}, \mathsf{e}) = \mathsf{LE}(\mathsf{pid}.\alpha; r)$.
  2. Output $(\text{LOTTERY}, \mathsf{sid}, \mathsf{e}, T(\mathsf{pid}, \mathsf{e}))$ to pid (or the adversary).

- Upon receiving $(\text{SEND}, \mathsf{sid}, \mathsf{e}, m)$ from a stakeholder pid (or the adversary in the name of corrupted stakeholder pid) do the following:
  1. If $T(\mathsf{pid}, \mathsf{e}) = 1$ add (e, m) to the message buffers of all registered parties and output $(\text{SEND}, \mathsf{sid}, \mathsf{e}, m)$ to the adversary.

- Upon receiving (SEND, sid, e, m, P') from the adversary in the name of corrupted stakeholder pid) do the following:
  1. If $T(\text{pid}, e) = 1$ add (e, m) to the message buffer of P' and output (SEND, sid, e, m, P') to the adversary.
- Upon receiving (FETCH-NEW, sid) from a party P (or the adversary on behalf of P) do the following:
  1. Output the content of P's message buffer and empty it.

Get Information

- Upon receiving (GET-STAKE, sid) from a stakeholder pid (or the adversary on behalf of pid) return (GET-STAKE, sid, pid.α).

4.2   Private Lottery Protocol

The high level idea to implement $\mathcal{F}_{\text{Lottery}}^{\text{LE}, \mathcal{E}}$ is as follows. Parties collect information available on the blockchain about the public keys and the corresponding stake of stakeholders. A list $\mathcal{L} = \{(C_{\text{stk}_1}, \text{vk}_1), \cdots, (C_{\text{stk}_n}, \text{vk}_n)\}$ is compiled with tuples of the form $(C_{\text{stk}}, \text{vk})$ where vk is a verification key for a signature scheme (KeyGen, Sign, Ver), and C_{stk} is a commitment to the stake.

The lottery is defined relative to a lottery predicate LE. A stakeholder pid wins the lottery for entry e, if $\text{LE}(\text{stk}, r(e, \text{pid})) = 1$, where r is randomness that depends on the entry e and stakeholder identity, ensuring that the lottery for different stakeholders is independent. The randomness for the lottery is generated by the VRF functionality $\mathcal{F}_{\text{VRF}}^{\text{Com}}$. Winning the lottery for e allows a stakeholder to publish messages for e.

To ensure privacy, the stakeholder proves in zero-knowledge that he indeed won the lottery, and as part of this proof it is necessary to prove ownership of his stake. We can do this by proving that the tuple containing the same committed stake and a signing key is in the public list $\mathcal{L}$ (without revealing which one it is), and ownership of the key by proving knowledge of the corresponding secret key. The statement to prove is of the form "I know sk, vk, stk such that (vk, sk) is a valid signature key pair, $(C_{\text{stk}}, \text{vk}) \in \mathcal{L}$, and I won the lottery with stake stk for entry e". In addition, there needs to be a signature σ on (e, m) to ensure that no other message can be published with this proof, and this signature is also verified inside the zero-knowledge proof with respect to the verification key in the same tuple. Note that since the proof is used to verify the correctness of the signature, the proof itself (and public values for the statement) are not included in the information that is signed. More formally, the proof is of the following form.

$$PK\{(C_{stk}, stk, \sigma, vk, sk, r) : LE(stk; r) = 1 \wedge C_{stk} = Com(stk)$$
$$\wedge\ Ver_{vk}((e, m), \sigma) = 1 \wedge vk = KeyGen(sk) \wedge (C_{stk}, vk) \in \mathcal{L}\}$$

The published information now consists of entry e, the message m, zero-knowledge proof for the above statement, and certain public values that form the statement. We assume that the zero-knowledge proof requires a CRS which is given by the functionality $\mathcal{F}_{crs}^{\mathcal{D}}$. The actual publication of the message is done via anonymous broadcast $\mathcal{F}_{ABC}$ to protect the identity of the stakeholder.

The detailed construction of the private lottery Lottery Protocol$^{\mathcal{E}, LE}$ is given below. The protocol Lottery Protocol$^{\mathcal{E}, LE}$ is run by parties interacting with ideal functionalities $\mathcal{F}_{ABC}, \mathcal{F}_{Init}^{Com,SIG}, \mathcal{F}_{crs}^{\mathcal{D}}, \mathcal{F}_{VRF}^{Com}$ and among themselves. Let the algorithms (Setup, Prove, Verify) be a non-interactive zero-knowledge argument system. Lottery Protocol$^{\mathcal{E}, LE}$ proceeds as follows.

Protocol Lottery Protocol$^{\mathcal{E}, LE}$

This describes the protocol from the viewpoint of a party P. If the party is a stakeholder, it additionally has stakeholder-id pid.

Initialization

- Send (GETLIST, sid) to $\mathcal{F}_{Init}^{Com,SIG}$ to get the list $\mathcal{L}$ of stakeholders with committed stake and verification key.

- Send (Setup, sid) to $\mathcal{F}_{crs}^{\mathcal{D}}$ and get the crs.

- If you are a stakeholder, send (GETPRIVATEDATA, sid) to functionality $\mathcal{F}_{Init}^{Com,SIG}$ and get $\alpha_{pid}, r_{\alpha,pid}, sk_{pid}$ send (KEYGEN, sid) to functionality $\mathcal{F}_{VRF}^{Com}$ and get vid; and initialize an empty table $V(\cdot)$.

Lottery and Publishing

- As a stakeholder upon receiving (LOTTERY, sid, e) from the environment do the following.
  1. If e is not in $\mathcal{E}$ ignore the request.
  2. If $V(e)$ is undefined:
     (a) Send (EVAL, sid, vid, e) to functionality $\mathcal{F}_{VRF}^{Com}$ and receive response (EVALUATED, sid, (y, c, r)).
     (b) Compute $b = LE(\alpha_{pid}, y)$, and set $V(e) = (b, y, c, r)$.
  3. Return (LOTTERY, sid, e, b) where $V(e) = (b, y, c, r)$.

- As a stakeholder upon receiving (SEND, sid, e, m) from the environment do the following:
  1. Ignore the request if $V(e) = (0, \cdots)$ or is undefined.

2. Let $V(e) = (1, y, c, r)$. Create the tuple (e, m, π_{zk}) in the following way:

   (a) Compute a signature σ on (e, m) under sk_{pid}.

   (b) π_{zk} is a non-interactive zero-knowledge proof of knowledge obtained by running Prove using crs for the following statement.

$$PK\{(\alpha_{pid}, r_{\alpha,pid}, vk_{pid}, sk_{pid}, c_{\alpha,pid}, \sigma, y, r) : Ver_{vk_{pid}}((e, m), \sigma) = 1$$
$$\wedge\, LE(\alpha_{pid}, y) = 1 \wedge vk_{pid} = KeyGen(sk_{pid}) \wedge c = Com(y; r)$$
$$\wedge\, c_{\alpha,pid} = Com(\alpha_{pid}; r_{\alpha,pid}) \wedge (c_{\alpha,pid}, vk_{pid}) \in \mathcal{L}\}$$

3. Send $(\text{SEND}, sid, (e, m, c, \pi_{zk}))$ to $\mathcal{F}_{ABC}$.

- Upon receiving $(\text{FETCH-NEW}, sid)$ from the environment do the following:
  1. Send $(\text{RECEIVE}, sid)$ to $\mathcal{F}_{ABC}$ and receive as message vector $\vec{m}$.
  2. For each $(e, m, c, \pi_{zk}) \in \vec{m}$ do:
     (a) Check that $e \in \mathcal{E}$.
     (b) Send $(\text{VERIFY}, sid, e, c)$ to functionality $\mathcal{F}_{VRF}^{Com}$. For response $(\text{VERIFIED}, sid, e, c, b)$ from $\mathcal{F}_{VRF}^{Com}$, verify that $b = 1$.
     (c) Verify the correctness of zero-knowledge proof π_{zk}, i.e., check that $\text{Verify}(crs, \pi_{zk}) = 1$.
     (d) If all check pass add (e, m, c, π_{zk}) to $\vec{o}$.
  3. Output $(\text{FETCH-NEW}, sid, \vec{o})$.

Get Information

- As a stakeholder upon receiving $(\text{GET-STAKE}, sid)$ from the environment output $(\text{GET-STAKE}, sid, \alpha_{pid})$ where α is your lottery power.

Theorem 1. *The protocol Lottery Protocol$^{\mathcal{E}, LE}$ realizes the $\mathcal{F}_{Lottery}^{LE, \mathcal{E}}$ functionality in the $(\mathcal{F}_{ABC}, \mathcal{F}_{Init}^{Com, SIG}, \mathcal{F}_{crs}^{\mathcal{D}}, \mathcal{F}_{VRF}^{Com})$-hybrid world in the presence of a PPT adversary.*

Proof. Let $\mathcal{S}_{zk} = (\mathcal{S}_1, \mathcal{S}_2)$ be the simulator of the zero-knowledge proof system used in the protocol Lottery-Protocol$^{\mathcal{E}, LE}$.

We construct a simulator $\mathcal{S}_{lottery}$ and argue that the views of the adversary in the simulated execution and real protocol execution are computationally close. Here, we provide a high-level idea of simulation. The full description of the simulator $\mathcal{S}_{lottery}$ can be found in Appendix A. At the beginning, the simulator gets the stake of dishonest stakeholders from $\mathcal{F}_{Lottery}^{LE, \mathcal{E}}$ and internally emulates the $\mathcal{F}_{Init}^{Com, SIG}$. The simulator generates a CRS (with trapdoor) and emulates $\mathcal{F}_{crs}^{\mathcal{D}}$. Similarly, the VRF functionality $\mathcal{F}_{VRF}^{Com}$ is emulated by the simulator. If a dishonest stakeholder ask the VRF for an entry $e \in \mathcal{E}$, the simulator first asks $\mathcal{F}_{Lottery}^{LE, \mathcal{E}}$

if the stakeholder wins for this entry and then samples the output accordingly. The simulation of $\mathcal{F}_{\mathsf{ABC}}$ consists of two parts. First, if the adversary wants to send a message the simulator checks if the message is a valid tuple of the form $m' = (\mathsf{e}, m, c, \pi_{\mathsf{zk}})$. If this is the case, the simulator submits (e, m) to $\mathcal{F}_{\mathsf{Lottery}}^{\mathsf{LE}, \mathcal{E}}$ for publication. Second, if an honest stakeholder publishes a tuple (e, m) the simulator creates a tuple $(\mathsf{e}, m, c, \pi_{\mathsf{zk}})$ which contains a simulated proof π_{zk} and adds it to the message buffers of dishonest parties. $\qquad\square$

4.3   Flavors of Proof-of-Stake

As seen above any proof-of-stake lottery can be made private. In the following we discuss how this process applies to two widely-used types of proof-of-stake lotteries.

Slot-based PoS. In slot-based PoS protocols (e.g., the Ouroboros Praos protocol [KRDO17]), time is divided into slots and blocks are created relative to a slot. Parties with stake can participate in a slot lottery, and winning the lottery allows a stakeholder to create a block in a particular slot. Here, the set of lottery entries are slots, i.e. $\mathcal{E} = \mathbb{N}^+$. An (honest) lottery winner will publish one message in the form of a new block via $\mathcal{F}_{\mathsf{Lottery}}^{\mathsf{LE}, \mathcal{E}}$.

Committee-based PoS. In committee-based PoS protocols, such as Algorand [GHM+17], a stakeholder wins the right to take part in a committee which for example determines the next block. In such a protocol, the set of lottery entries could be of the form (cid, role) where cid is the id of the committee and role is the designated role of the winner. An (honest) lottery winner will then publish his messages for the committee protocol via $\mathcal{F}_{\mathsf{Lottery}}^{\mathsf{LE}, \mathcal{E}}$.

4.4   Dynamic Stake

Our protocol in Sect. 4.2 assumes that the stake distribution is fixed at the onset of the computation (in the form of $\mathcal{F}_{\mathsf{Init}}^{\mathsf{Com}, \mathsf{SIG}}$), which is the *static stake* setting. In the following we give an intuition on how the protocol can be made to support the *dynamic stake* setting where the set of stakeholders and the distribution evolve over time.

Protocol idea. The idea is to collect information about the public keys and the corresponding stake of stakeholders on the blockchain instead of using $\mathcal{F}_{\mathsf{Init}}^{\mathsf{Com}, \mathsf{SIG}}$. We assume that for each entry e the (honest) parties agree on the corresponding stake distribution $\mathcal{L}_{\mathsf{e}}$. This stake distribution might not be known from the beginning of the protocol[1]. We assume that (if defined) $\mathcal{L}_{\mathsf{e}}$ can be computed efficiently from the blockchain. The parties then use $\mathcal{L}_{\mathsf{e}}$ in the lottery protocol when dealing with e.

Observe that computation of $\mathcal{L}_{\mathsf{e}}$ is completely separated from the actual lottery protocol. The protocol therefore remains secure even in the dynamic stake setting.

---

[1] If, for example, the entries are slots (cf. Sect. 4.3) the stake distribution for a particular entry is only defined once the blockchain has grown far enough.

4.5   Rewards

In many proof-of-stake based cryptocurrencies a stakeholder will include some sort of identification (e.g. his verification key) in his messages (e.g. in a new block) so that the rewards such as transaction fees are appropriately paid out. This, of course leaks the identity of the lottery winner and thus also information about his stake. This leakage can be prevented if the cryptocurrency allows for anonymous transactions and anonymous account creation. For instance, one could think of ZCash [BCG+14], which though not based on proof-of-stake allows for such mechanisms. Each stakeholder maintains a list of fresh accounts. Whenever the stakeholder needs to provide information for rewards, the stakeholder uses one of the accounts as the reward destination. Since the account was created anonymously it cannot be linked to the stakeholder. Later on, the stakeholder can anonymously transfer the money from that account to any of its other accounts.

5   Making Ouroboros Praos Private

In this section, we look at the Ouroboros Praos proof-of-stake protocol from [DGKR18], and apply the technique from our private lottery framework. In particular, we describe how the zero-knowledge proofs necessary for π_{zk} are instantiated for the Ouroboros Praos lottery.

5.1   Ouroboros Praos Leader Election

Recall that the VRF leader election scheme in Ouroboros Praos works as follows. The probability p that a stakeholder pid is elected as leader in a slot sl is independent of other stakeholders. It depends only on pid's relative stake $\alpha = $ stk/Stake where Stake is the total stake in the system. More precisely, the probability p is given by,

$$p = \phi_f(\alpha) \triangleq 1 - (1 - f)^\alpha$$

where f is the difficulty parameter. A stakeholder pid can evaluate the VRF using private key k along with a proof of evaluation that can be verified using a public key. To check if they are a leader in slot sl, the stakeholder computes their threshold $T = 2^{\ell_\alpha} p$ where ℓ_α is the output length of the VRF. The stakeholder wins if $y < T$ where $(y, \pi) = \mathsf{VRF}(\mathsf{k}, \mathsf{sl})$. The proof π allows any party to verify pid's claim given pid's verification key. In other words, the LE predicate function for Ouroboros Praos is given by:

$$\mathsf{LE}(\mathsf{stk}; y) = \begin{cases} 1, & \text{if } y < 2^{\ell_\alpha} \cdot \left(1 - (1-f)^{\frac{\mathsf{stk}}{\mathsf{Stake}}}\right) \\ 0, & \text{otherwise} \end{cases} .$$

Ouroboros Praos VRF. Ouroboros Praos uses the 2-Hash VRF of [JKK14] based on the hardness of the computational Diffie-Hellman problem. Let $\mathbb{G} = \langle g \rangle$

be a group of order q. VRF uses two hash functions H_1 and H_2 modeled as random oracles. H_1 has range $\{0,1\}^{\ell_\alpha}$ and H_2 has range $\mathbb{G}$. Given a key $\mathsf{k} \in \mathbb{Z}_q$, the public key is $v = g^\mathsf{k}$, and $(y, \pi) = \mathsf{VRF}(m)$ is given by $y = H_1(m, u)$ where $u = H_2(m)^\mathsf{k}$, and $\pi : \mathsf{PK}\{(\mathsf{k}) : \log_{H_2(m)}(u) = \log_g(v)\}$.

5.2 Anonymous Verifiable Random Function

We define a primitive that we call an *anonymous VRF* that captures a requirement necessary in the proof π_{zk}; which is roughly that verification should not reveal the public key. The high level idea is that there are many public keys associated with a secret key, and two different evaluations (on different messages) under the same secret key cannot be linked to a public key. The verifiability property is still preserved, that is, there is a public key, which allows to verify the correctness of output with respect to a proof. We now give a formal definition.

Definition 2. *A function family* $F_{(\cdot)}(\cdot) : \{0,1\}^k \to \{0,1\}^{\ell(k)}$ *is a family of anonymous VRFs, if there is a tuple of algorithms* (Gen, Update, VRFprove, VRFverify) *such that:* $\mathsf{Gen}(1^k)$ *generates a key pair* (pk, k); Update *takes the public key* pk *and outputs an updated public key* pk'; $\mathsf{VRFprove}_\mathsf{k}(\mathsf{pk}', x)$ *outputs a tuple* $(F_\mathsf{k}(x), \pi_\mathsf{k}(x))$ *where* $\pi_\mathsf{k}(x)$ *is the proof of correct evaluation;* $\mathsf{VRFverify}_\mathsf{pk'}(x, y, \pi)$ *verifies that* $y = F_\mathsf{k}(x)$ *using the proof* π. *We require that the following properties are satisfied.*

- *Pseudorandomness. For any pair of PPT* (A_1, A_2), *the following probability is* $1/2 + \mathsf{negl}(k)$.

$$\Pr\left(\begin{array}{c} b = b' \\ \wedge x \notin Q_1 \cup Q_2 \end{array} \middle| \begin{array}{c} (\mathsf{pk}, \mathsf{k}) \leftarrow \mathsf{Gen}(1^k); (Q_1, x, \mathsf{state}) \leftarrow A_1^{\mathsf{VRFprove}(\cdot)}(\mathsf{pk}); \\ y_0 = F_\mathsf{k}(x); y_1 \leftarrow \{0,1\}^\ell; \\ b \leftarrow \{0,1\}; (Q_2, b') \leftarrow A_2^{\mathsf{VRFprove}(\cdot)}(y_b, \mathsf{state}) \end{array}\right).$$

  The sets Q_1, Q_2 *contain all the queries made to the Prove oracle. The random variable* state *stores information that* A_1 *can save and pass on to* A_2.
- *Uniqueness. There do not exist values* $(\mathsf{pk}, x, y_1, y_2, \pi_1, \pi_2)$ *such that* $y_1 \neq y_2$ *and*
$$\mathsf{VRFverify}_\mathsf{pk}(x, y_1, \pi_1) = \mathsf{VRFverify}_\mathsf{pk}(x, y_2, \pi_2) = 1$$

- *Provability.* $\mathsf{VRFverify}_\mathsf{pk'}(x, y, \pi) = 1$ *for* $(y, \pi) = \mathsf{VRFprove}_\mathsf{k}(\mathsf{pk}', x), \mathsf{pk}' \leftarrow$ Update(pk)
- *Anonymity. For any PPT algorithm* A, *the following probability is* $1/2 + \mathsf{negl}(k)$.

$$\Pr\left(b = b' \middle| \begin{array}{c} (\mathsf{pk}_0, \mathsf{k}_0) \leftarrow \mathsf{Gen}(1^k); (\mathsf{pk}_1, \mathsf{k}_1) \leftarrow \mathsf{Gen}(1^k); \\ x \leftarrow A(\mathsf{pk}_0, \mathsf{pk}_1); \mathsf{pk}_0' \leftarrow \mathsf{Update}(\mathsf{pk}_0); \\ (y_0, \pi_0) = \mathsf{VRFprove}_{\mathsf{k}_0}(\mathsf{pk}_0', x); \mathsf{pk}_1' \leftarrow \mathsf{Update}(\mathsf{pk}_1); \\ (y_1, \pi_1) = \mathsf{VRFprove}_{\mathsf{k}_1}(\mathsf{pk}_1', x); b \leftarrow \{0,1\}; b' \leftarrow A(\mathsf{pk}_b', y_b, \pi_b) \end{array}\right).$$

Intuitively, the above definition says that no adversary can tell which key an output came from, given two public keys.

Anonymous VRF construction. We show how to instantiate the AVRF primitive by adapting the 2-Hash VRF. Let AVRF be the tuple of algorithms (Gen, VRFprove, VRFverify) which are defined as follows.

- Gen(1^k): Choose a generator g of a group of order q such that $q = \Theta(2^k)$, and sample a random $k \in \mathbb{Z}_q$ and output (pk, k), where pk = (g, g^k).
- Update(pk): Let pk be (g, v). Choose a random $r \in \mathbb{Z}_q$, let $g' = g^r, v' = v^r$, set pk$' = (g', v')$, output pk$'$.
- VRFprove$_k$(pk$'$, x): Let pk$'$ be (g, v). Compute $u = H_2(x)^k$, $y = H_1(x, u)$, and π' : PK$\{(k) : \log_{H_2(x)}(u) = \log_g(v)\}$. Output (pk$'$, y, $\pi = (u, \pi')$)
- VRFverify$_{\text{pk}'}$(x, y, π): Output 1 if $y = H_1(x, u)$ and π verifies, and 0 otherwise.

It is clear that the above construction satisfies the standard properties of a VRF. For anonymity, we reduce to DDH; we show that any adversary who breaks anonymity can be used to break DDH. Let $\mathcal{A}$ be the adversary who wins the anonymity game in Definition 2. We now show how to use $\mathcal{A}$ to break DDH. Let $\mathcal{B}$ be an adversary who receives a challenge (g, g^a, g^b, g^c) and has to determine whether it is a DDH tuple or not. $\mathcal{B}$ works as follows: it chooses random $k_0, k_1 \in \mathbb{Z}_q$, sets $k_0 = k_0$, pk$_0 = (g^a, g^{k_0 a})$, $k_1 = k_1$, pk$_1 = (g^b, g^{k_1 b})$ and return pk$_0$, pk$_1$ to $\mathcal{A}$. On receiving a x, $\mathcal{B}$ chooses $\beta \in \{0, 1\}$ at random and returns (pk$'_\beta$, y_β, π_β) to $\mathcal{A}$, where pk$'_\beta = (g^c, g^{k_\beta c})$ and y_β, π_β are evaluated with key k_β. Let β' be the output of $\mathcal{A}$. If $\beta = \beta'$, $\mathcal{B}$ returns DDH tuple, otherwise $\mathcal{B}$ decides not a DDH tuple.

Note that while AVRF gives the anonymous verifiability property, it does not guarantee that the key used to evaluate comes from one of the two keys that the adversary sees at the onset of the game. In applications, it is desirable to satisfy this "key membership" property. Indeed, the $\mathcal{F}_{\text{VRF}}^{\text{Com}}$ functionality that was defined in Sect. 3 has the property that verification does not leak a public key and also guarantees that it is one of the registered keys. The $\mathcal{F}_{\text{VRF}}^{\text{Com}}$ functionality also allows verifiability of y while keeping y secret. We use other techniques on top of the AVRF primitive to realize the $\mathcal{F}_{\text{VRF}}^{\text{Com}}$ functionality; in general, proving membership of the corresponding AVRF secret key in a list of committed secret keys will suffice for membership and we preserve privacy by committing to the output and proving correct evaluation in zero-knowledge. We elaborate on this in the next section.

"Approval Voting" via AVRF. To demonstrate the usefulness of AVRF outside of the context of PPoS, here is a simple example application, namely approval voting. In approval voting, a group of users can vote (e.g., approve) any number of candidates, and the winner of the election is the candidate who is approved by the highest number of voters. To implement such voting with cryptographic techniques, one needs to ensure anonymity of the voters and, at the same time, that each voter can approve each candidate at most once. This can be easily done using our AVRF abstraction: Each user registers an AVRF public key pk$_1, \ldots,$ pk$_n$. To vote on option x, user i publishes pk$' = $ Update(pk$_i$)

and gives a ZK-proof that $\exists i : (\mathsf{pk}', \mathsf{pk}_i) \in L$ (in our AVRF the language L is simply the language of DDH tuple). Then the user computes and publishes $(y, \pi) = \mathsf{VRFprove}_\mathsf{k}(\mathsf{pk}', x)$. If the proof π does not verify or if the value y has already appeared in this poll, then the other users discard this vote. Otherwise, they register a new vote for option x. Now, due to the anonymity and indistinguishability properties of the VRF, it is unfeasible to link any two casted votes, except if the same user tries to approve the same candidate more than once, since the value y is only a function of k and x.

5.3 Private Ouroboros Praos

Recall that our private lottery protocol now needs to prove that $\mathsf{LE}(\mathsf{stk}; y) = 1$ in zero-knowledge. For this, we need to prove $y < T$ in zero knowledge, that is, without revealing y or T^2. Note that, in addition, we need to prove the correct computation of T which involves evaluating ϕ on a secret α involving floating-point arithmetic. Using generic zero-knowledge proofs for a statement like above would be expensive. We show how to avoid this and exploit the specific properties of the statement. In particular, we take advantage of the "independent aggregation" property that is satisfied by the above function ϕ to construct a zero-knowledge proof for leader election i.e., that the function ϕ satisfies the following property:

$$1 - \phi\left(\sum_i \alpha_i\right) = \prod_i (1 - \phi(\alpha_i))$$

The above implies that if a party were to split its stake among virtual parties, the probability that the party is elected for a particular slot is identical to the probability that one of the virtual parties is elected for that slot.

Remark 1. Due to rounding performed when evaluating the predicate LE, the probability of winning is not identical under redistribution of stakes. However, by setting the precision ℓ_α appropriately we can always ensure that the difference between the winning probabilities above is at most negligible.

Proof of correct evaluation of LE predicate π_{LE}. The idea behind our proof is to split the stake among virtual parties and prove that one of the virtual parties wins without revealing which one of them won. We also use the 2-hash AVRF instantiation in the LE since we want to achieve verifiability of correct evaluation without disclosing a public key. More precisely, each stakeholder has a key pair $(\mathsf{vk}, \mathsf{sk})$ of a signature scheme $(\mathsf{KeyGen}, \mathsf{Sign}, \mathsf{Ver})$, and a key pair $(\mathsf{pk}, \mathsf{k})$ for an AVRF family F. To realize the key membership property for the AVRF, we now include the public key for the AVRF in a stakeholder's tuple. Thus, the list $\mathcal{L}$ now consists of tuples $(C_{\mathsf{stk}}, \mathsf{vk}, \mathsf{pk})$.

---

[2] Since T is a direct function of stk, it should be clear why T should stay private. At the same time, revealing the value y and the fact that LE output 1 allows to rule out that $\mathsf{stk} = s$ for any value s such that $\mathsf{LE}(s; y) = 0$.

Let pid be a stakeholder with (absolute) stake stk, and wants to prove that it won the election, that is $\mathsf{LE}(\mathsf{stk}; F_k(\mathsf{sl})) = 1$, where F is the AVRF function. Let $b_i, \forall i \in [0, s-1]$ be the bits of the stake of stakeholder pid, where $\mathsf{stk} = \sum_{i=0}^{s-1} 2^i b_i$ and the maximum stake in the system is represented in s-bits. Now, the stake is split among s virtual parties "in the head" where the stake of the ith virtual party is $2^i b_i$. We now have by the aggregation property, that the probability of winning with stake stk is equal to the probability of winning with one of the above s stakes. Let the probability of winning with stake $2^i b_i$ be p_i, let $(y_i, \pi_i) = \mathsf{VRFprove}_k(\mathsf{pk}_i, i||\mathsf{sl})$, for $\mathsf{pk}_i \leftarrow \mathsf{Update}(\mathsf{pk})$. Let T_i be the threshold corresponding to the ith divided stake, $T_i = 2^{\ell_\alpha} p_i$. We use the AVRF key of the stakeholder to evaluate y_i corresponding to the ith stake by including the index i along with the slot number in the evaluation, and prove that $y_i < T_i$ for at least one i. Note that now, the thresholds T_i in the statement are public values, in contrast to private threshold prior to the stake being split among virtual parties. In addition, the statement only uses the function ϕ in a blackbox way and is independent of the difficulty parameter f. The zero-knowledge proofs, therefore do not have to change with tuning of the difficulty parameter of the leader election function. The proof is for the statement that there exists at least one bit such that the bit is one, the corresponding virtual party won the lottery, bits combine to yield the committed stake and correct evaluation of the AVRF. The following is a proof that LE was evaluated correctly on stk.

$$\mathsf{PK}\{(y_1, \cdots y_s, b_1, \cdots, b_s, i^*, k, \mathsf{stk}) :$$

$$\left(\bigwedge_{i=1}^{s} (b_i \in \{0, 1\}) \right) \wedge (b_{i^*} = 1 \wedge y_{i^*} < T_{i^*} \wedge y_{i^*} = F_k(i^*||\mathsf{sl})) \wedge \mathsf{stk} = \sum 2^j b_j \}$$

Proof π_{LE} is about the correct evaluation of the predicate LE on private stake and randomness, and correct evaluation of the y_i's. The above proof convinces that a committed stake wins the lottery. It is still necessary to prove ownership of this stake. We can do this by proving that the tuple containing the same committed stake, signature verification key, and an APRF key is in the list $\mathcal{L}$, and ownership of the signature key by proving knowledge of the corresponding signing key.

Proof of ownership π_{own}. We represent the list $\mathcal{L}$ as a Merkle tree, where the leaf are the tuples $(C_{\mathsf{stk}_{\mathsf{pid}}}, \mathsf{vk}_{\mathsf{pid}}, \mathsf{pk}_{\mathsf{pid}})_{\mathsf{pid}} \in \mathcal{L}$. We can now prove membership by proving a valid path to the public root given a commitment to a leaf. Let $\mathcal{L}(\mathsf{root})$ denote the Merkle tree representation of the list $\mathcal{L}$. Given the root of a Merkle tree, an AVRF public key, and commitments to the signature verification key, and stake, we want a party proposing a new block to prove that the stake used in the proof of winning lottery corresponds to the signature key and AVRF key it "owns". That is, prove knowledge of $(\mathsf{vk}, \mathsf{stk}, \mathsf{pk})$ such that $C_{\mathsf{stk}}||\mathsf{vk}||\mathsf{pk}$ is a leaf of the Merkle tree with root root. To prove membership, one can reveal the path along with the values of the sibling nodes up to the root. We want to prove membership without disclosing the leaf node and therefore use a zero knowledge proof π_{path} to prove a valid path from a committed leaf to a public root. Let l_i be $C_{\mathsf{stk}}||\mathsf{vk}||\mathsf{pk}$, H be the hash function function used to construct the Merkle tree, and let $\mathsf{sib}_1, \ldots, \mathsf{sib}_t$ be the sibling nodes of the nodes on the path from l_i to the root of a tree with depth t. π_{path} proves $l_i \in \mathcal{L}(\mathsf{root})$.

$$\mathsf{PK}\{(l_i, \mathsf{sib}_1, \ldots, \mathsf{sib}_t) : H(\cdots H(H(l_i||\mathsf{sib}_1)||\mathsf{sib}_2)\cdots) = \mathsf{root}\}$$

Using the above proof π_{path}, we can prove ownership. Given root, we denote by π_{own} the following proof.

$$\mathsf{PK}\{(\mathsf{vk}, \mathsf{stk}, k, \mathsf{pk}, C_{\mathsf{stk}}) : (C_{\mathsf{stk}}||\mathsf{vk}||\mathsf{pk}) \in \mathcal{L}(\mathsf{root}) \wedge \mathsf{pk} = g^k\}$$

Proof of signature on a block under the winning key π_{sig}. π_{zk} also consists of a proof that a block signature verifies under the winning key. π_{sig} denotes the following proof, where M is the public block information that is signed.

$$\mathsf{PK}\{(\mathsf{vk}, \mathsf{sk}, \sigma) : \mathsf{vk} = \mathsf{KeyGen}(\mathsf{sk}) \wedge \mathsf{Ver}_{\mathsf{vk}}(\sigma, M) = 1\}$$

Overall proof. The detailed construction of proof π_{zk} is given below. If the commitment to stake C_{stk} is an extended Pedersen commitment (e.g., $h^r \cdot \Pi_{i=1}^s (g_i)^{b_i}$) where the stakes are already committed bit by bit, the proof π_{LE} is a standard sigma protocol. If instead, it is a Pedersen commitment to the entire stake, one can publish fresh commitments to bits and prove correct recombination. The range proofs that are used in π_{LE} allow one to prove that $x \in [0, R]$ for a public R and committed x. Range proofs may be instantiated using several known techniques [CCs08, Bou00]. More recently, the technique of bulletproof [BBB+18] results in very efficient range proofs when the interval is $[0, 2^n - 1]$ for some n. Since we use SNARKs for other statements, we also implement the range check inside a SNARK resulting in short proofs. The proof π_{LE} also relies on the OR composition of sigma protocols. π_{own} may be realized efficiently using SNARKs when the Merkle tree hash function H is non-algebraic. While it might seems like such a statement would result in inefficient proofs, this can in fact be done efficiently in practice, and is implemented by ZCash's private-pool transactions [BCG+14]. The predicate Eq that tests if two public keys comes from the same key is the following predicate for the concrete 2-Hash AVRF: it outputs 1 if pk and pk' form a DDH tuple. For a public pk' and private pk as in our case, this can be implemented using double discrete logarithm sigma protocol proofs [CS97, MGGR13]. The proof for part of the statement represented as a circuit (the hash functions) in the 2-Hash AVRF can be implemented using SNARKs, and we can use the construction of [AGM18] for SNARK on algebraically committed input and output so we can work with Pedersen commitments and sigma protocols for other parts of the proof. The rest of the proof components may be implemented using standard sigma protocol techniques.

Protocol Constructing π_{zk}

- Given a list $\mathcal{L} = \{(C_{stk_{pid}}, vk_{pid}, pk_{pid})_{pid}\}$, construct a Merkle tree representation. Let root be the root of the tree.
- For stakeholder pid, let $b_1, \ldots, b_s$ represent the bits of the stake stk. Let the private information be $(stk, C_{stk}, C_k, vk, sk, k)$. Let M be the part of the block that is signed. Compute signature $\sigma = \mathsf{Sign}(sk, M)$. To construct a proof π_{zk} for submitting a new block:
  - Compute $pk' \leftarrow \mathsf{Update}(pk)$ and $(y_i, \pi_i) = \mathsf{VRFprove}_k(pk', i||sl)$. Then publish pk'. Compute and publish $C_\sigma = \mathsf{Com}(\sigma), C_{vk} = \mathsf{Com}(vk), C_{sk} = \mathsf{Com}(sk)$. There is a predicate $\mathsf{Eq}(pk_i, pk, k)$ which outputs 1 if pk_i, pk have the same secret key k. Compute proof of correct evaluation of LE predicate π_{LE} :

$$\mathsf{PK}\{(y_1, \cdots y_s, b_1, \cdots, b_s, \pi_j, stk, C_{stk}) :$$
$$(\forall i\, (b_i \in \{0,1\}))$$
$$\wedge (\exists j\, (b_j = 1 \wedge y_j < T_j \wedge \mathsf{VRFverify}_{pk'}(j||sl, y_j, \pi_j) = 1))$$
$$\wedge stk = \sum 2^j b_j \wedge C_{stk} = \mathsf{Com}(stk)\}$$

  - Compute proof of signature on a block under the winning key π_{sig}:

$$\mathsf{PK}\{(vk, sk, \sigma) : vk = \mathsf{KeyGen}(sk) \wedge C_{vk} = \mathsf{Com}(vk) \wedge C_{sk} = \mathsf{Com}(sk)$$
$$\wedge C_\sigma = \mathsf{Com}(\sigma) \wedge \mathsf{Ver}_{vk}(\sigma, M) = 1\}$$

  - Compute proof of ownership of signature and AVRF key π_{own}:

$$\mathsf{PK}\{(vk, stk, k, pk, C_{stk}) : (C_{stk}||vk||pk) \in \mathcal{L}(root) \wedge C_{vk} = \mathsf{Com}(vk)$$
$$\wedge \mathsf{Eq}(pk', pk, k) = 1 \wedge C_{stk} = \mathsf{Com}(stk)\}$$

Set π_{zk} to be $(\pi_{LE}, \pi_{sig}, \pi_{own})$.

Usage of π_{zk} in Ouroboros Praos. If a stakeholder has won the lottery for slot sl, they will create a new block of the form $(pt, sl, st, c, \pi_{zk})$ where pt is a reference to a previous block, st the block payload, c is a commitment to y, the output of the AVRF, and π_{zk} is the proof as described above. The stakeholder then publishes the block using an anonymous broadcast.

Corollary 1. *Ouroboros Praos used with the private lottery protocol results in a private proof-of-stake protocol.*

Proof. The proof easily follows from the properties of the underlying building blocks. Note that overall protocol remains the same as in the original Ouroboros Praos, with only small differences: Instead of using a VRF, a stakeholder uses an AVRF to determine whether they win the slot-lottery. Then, a slot leader will publish a block with a zero-knowledge proof of the above form (instead of adding his verification key and a VRF-proof). Due to the soundness of the zero-knowledge protocol and the uniqueness property of the AVRF, the modified protocol still has the same security properties as Ouroboros Praos i.e., the protocol still reaches consensus under the same security guarantees as the original protocol.

The proof that the resulting protocol is a *private* proof of stake follows directly from the proof of Theorem 1.

We give an estimate of the proof size that determines the overhead that is incurred by privacy preserving Ouroboros Praos compared to the non-private version. The size of π_{LE} is dominated by $O(s)$ group/field elements due to the sigma protocol OR composition, with the rest of the components resulting in succinct SNARK proofs. π_{sig} for the key-evolving signature scheme may be implemented by using SNARK on committed input together with sigma protocols with only a slight overhead in size over the SNARK proof. The size of π_{own} is dominated by the proof size for the predicate Eq which is $O(\kappa)$ elements for a statistical security parameter κ. The size of π_{zk} is therefore roughly (ignoring the size of proofs for statements that use SNARKs and standard sigma protocols), $O(s) + O(\kappa)$ group/field elements where s is the number of bits to represent the stake in the system, κ is the statistical security parameter. We remark that the actual complexity depends on the implementation of the signature scheme, and potentially the hash functions of the VRF.

Appendix

A   Private Proof of Stake Lottery

Theorem 1. *The protocol* Lottery Protocol$^{\mathcal{E},\mathsf{LE}}$ *realizes the* $\mathcal{F}_{\mathsf{Lottery}}^{\mathsf{LE},\mathcal{E}}$ *functionality in the* $(\mathcal{F}_{\mathsf{ABC}}, \mathcal{F}_{\mathsf{Init}}^{\mathsf{Com},\mathsf{SIG}}, \mathcal{F}_{\mathsf{crs}}^{\mathcal{D}}, \mathcal{F}_{\mathsf{VRF}}^{\mathsf{Com}})$-*hybrid world in the presence of a PPT adversary.*

Proof. Let $\mathcal{S}_{\mathsf{zk}} = (\mathcal{S}_1, \mathcal{S}_2)$ be the simulator of the zero-knowledge proof system used in Lottery-Protocol$^{\mathcal{E},\mathsf{LE}}$. We construct a simulator $\mathcal{S}_{\mathsf{lottery}}$ and argue that the views of the adversary in the simulated execution and real protocol execution are computationally close. Consider the simulator $\mathcal{S}_{\mathsf{lottery}}$.

Simulator $\mathcal{S}_{\mathsf{lottery}}$

Initialization

Upon first activation the simulator does the following.

1. For each dishonest stakeholder pid the simulator queries functionality $\mathcal{F}_{\text{Lottery}}^{\text{LE},\mathcal{E}}$ (using (GET-STAKE, sid)) to get the lottery powers α. Then, the simulator creates $(c_{\text{pid}} = \text{Com}(\alpha_{\text{pid}}; r_{\text{pid}}), r_{\text{pid}}, \text{sk}_{\text{pid}}, \text{vk}_{\text{pid}})$ the same way as $\mathcal{F}_{\text{Init}}^{\text{Com,SIG}}$ would do.

2. For each honest stakeholder pid the simulator creates the tuple $(c_{\text{pid}} = \text{Com}(0), r_{\text{pid}}, \text{sk}_{\text{pid}}, \text{vk}_{\text{pid}})$ similar to $\mathcal{F}_{\text{Init}}^{\text{Com,SIG}}$ except that the relative stake is set to 0.

Simulation of $\mathcal{F}_{\text{crs}}^{\mathcal{D}}$

- Call $\mathcal{S}_1$ to generate a simulated CRS scrs and a trapdoor τ.

- Upon receiving (Setup, sid), output (Setup, sid, scrs).

Simulation of $\mathcal{F}_{\text{Init}}^{\text{Com,SIG}}$

- Upon receiving (GETPRIVATEDATA, sid) from the adversary in the name of the dishonest stakeholder pid output (GETPRIVATEDATA, sid, α_{pid}, $r_{\text{pid}}, \text{sk}_{\text{pid}}$).

- Upon receiving (GETLIST, sid) from the adversary in the name of a dishonest party output the list $\mathcal{L} = \{(c_{\text{pid}}, \text{vk}_{\text{pid}})_{\text{pid}}\}$.

Simulation of $\mathcal{F}_{\text{VRF}}^{\text{Com}}$

The simulator maintains a table $T(\cdot, \cdot)$ and a list of vids.

- Upon input (KEYGEN, sid) from the adversary in the name of dishonest stakeholder pid generate a unique key vid, record (pid, vid), and initialise the table $T(\text{vid}, \cdot)$ to be empty. Return (KEYGEN, sid, vid) to the adversary.

- Upon input (EVAL, sid, vid, x) from the adversary in the name of dishonest stakeholder pid do the following:

  Abort and ignore the request if (pid, vid) is undefined.
  if $T(\text{vid}, x)$ is undefined **then**
    if If $x \in \mathcal{E}$ **then**
      Send (LOTTERY, sid, x) in the name of pid to $\mathcal{F}_{\text{Lottery}}^{\text{LE},\mathcal{E}}$.
      Denote by (LOTTERY, sid, x, b) the answer from $\mathcal{F}_{\text{Lottery}}^{\text{LE},\mathcal{E}}$.
      Pick random value y from $\{0,1\}^{\ell_{\text{VRF}}}$ such that $b = \text{LE}(\alpha_{\text{pid}}, y)$.[a]
    else
      Pick random value y from $\{0,1\}^{\ell_{\text{VRF}}}$.
    end if
    Pick random value r from $\{0,1\}^{\ell_{\text{VRF}}}$ and set table $T(\text{vid}, x) = (y, \text{Com}(y; r), r)$.
  end if

Output (EVALUATED, sid, $T(\text{vid}, x)$).

- Upon receiving a message (VERIFY, sid, x, c) from the adversary in the name of a dishonest party do the following:
  1. If there exists a vid such that $T(\text{vid}, x) = (y, c, r)$ for some y, r then set $f = 1$.

  2. Else, set $f = 0$.

  3. Output (VERIFIED, sid, x, c, f) to the adversary.

Simulation of $\mathcal{F}_{\mathsf{ABC}}$

The simulator maintains for each dishonest party a message buffer.

- Upon receiving (SEND, sid, e, m) from $\mathcal{F}_{\mathsf{Lottery}}^{\mathsf{LE},\mathcal{E}}$ do the following:
  1. Create an entry $(\perp, \text{vid})$ with unique vid for the internal $\mathcal{F}_{\mathsf{VRF}}^{\mathsf{Com}}$.

  2. Pick random values y, r from $\{0, 1\}^{\ell_{\mathsf{VRF}}}$ and set table $T(\text{vid}, e) = (y, \mathsf{Com}(y; r), r)$.

  3. Create simulated proof π_{zk} by calling the simulator $\mathcal{S}_2$ on (scrs, τ).

  4. Add $(e, m, \mathsf{Com}(y; r), \pi_{\mathsf{zk}})$ to the message buffers of all dishonest parties.

  5. Output (SEND, sid, $(e, m, \mathsf{Com}(y; r), \pi_{\mathsf{zk}})$) to the adversary.

- Upon receiving (SEND, sid, m') from the adversary do the following:
  1. Add m' to all message buffers of dishonest parties.

  2. If $m' = (e, m, c, \pi_{\mathsf{zk}})$ do:
     (a) Check that $e \in \mathcal{E}$.

     (b) Check that there is vid such that $T(\text{vid}, e) = (y, c, r)$ for some y, r.

     (c) Check that $\mathsf{Verify}(\text{scrs}, \pi_{\mathsf{zk}}) = 1$.

     (d) If all checks pass send (SEND, sid, e, m) to $\mathcal{F}_{\mathsf{Lottery}}^{\mathsf{LE},\mathcal{E}}$.

  3. Output (SENT, sid, m') to the adversary.

- Upon receiving (SEND, sid, m', P') from the adversary do the following:
  1. Add m' to message buffers of dishonest party P'.

  2. If $m' = (e, m, c, \pi_{\mathsf{zk}})$ do:
     (a) Check that $e \in \mathcal{E}$

     (b) Check that there is vid such that $T(\text{vid}, e) = (y, c, r)$ for some y, r.

     (c) Check that $\mathsf{Verify}(\text{scrs}, \pi_{\mathsf{zk}}) = 1$.

> (d) If all checks pass and P' is honest send $(\text{SEND}, \text{sid}, \text{e}, m, P')$ to $\mathcal{F}_{\text{Lottery}}^{\text{LE},\mathcal{E}}$.
>
> 3. Output $(\text{SENT}, \text{sid}, m', P')$ to the adversary.
>
> - Upon receiving $(\text{RECEIVE}, \text{sid})$ from the adversary in the name of corrupted party P. Remove all message from P's message buffer and output them to P.
>
> ---
> [a] This requires that it is possible to efficiently sample randomness r satisfying $\text{LE}(\text{stk}, r) = b$ for given stake stk.

Let HYB_0 be the (distribution) of the protocol execution (in the hybrid world where the auxiliary functionalities are available). We consider the world HYB_1 which is the same as the protocol execution except for the following: calls to $\mathcal{F}_{\text{VRF}}^{\text{Com}}$ are answered as is done by the simulator $\mathcal{S}_{\text{lottery}}$ consistent with the outcome returned by $\mathcal{F}_{\text{Lottery}}^{\text{LE},\mathcal{E}}$. It follows that distributions of HYB_0 and HYB_1 are indistinguishable. We now argue that the world HYB_1 is computationally indistinguishable from the ideal world simulation.

Simulation of $\mathcal{F}_{\text{Init}}^{\text{Com},\text{SIG}}$. The only difference between HYB_1 and the simulation is that the list $\mathcal{L}$ consists of commitments to honest stakes in the protocol, whereas the commitments are to 0 in the interaction with the simulator. By the hiding property of the commitment scheme Com, the two distributions are identical.

Simulation of $\mathcal{F}_{\text{crs}}^{\mathcal{D}}$. The CRS in HYB_1 is distributed the same as in the simulation.

Simulation of $\mathcal{F}_{\text{VRF}}^{\text{Com}}$. The key-generation and evaluation queries by the adversary are distributed the same. The same holds for verification queries where the adversary verifies a commitment which was created by an evaluation query by the adversary. In HYB_1, any other commitment message pair will be verified as true only if the commitment was part of an honest tuple $(\text{e}, m, c, \pi_{\text{zk}})$ which was sent to the adversary via $\mathcal{F}_{\text{ABC}}$. Similarly, in the simulation any other commitment message pair will only be evaluated as true if the commitment was part of a simulated honest tuple.

Simulation of $\mathcal{F}_{\text{ABC}}$. If the adversary sends a tuple $(\text{e}, m, c, \pi_{\text{zk}})$ in HYB_1, parties will accept it only if it is valid with respect to the information of $\mathcal{F}_{\text{Init}}^{\text{Com},\text{SIG}}, \mathcal{F}_{\text{crs}}^{\mathcal{D}}$, and $\mathcal{F}_{\text{VRF}}^{\text{Com}}$. In the ideal world, the simulator does the same checks with respect to the simulated functionalities. The simulator will then submit (e, m) to $\mathcal{F}_{\text{Lottery}}^{\text{LE},\mathcal{E}}$ which will send it to honest parties. The soundness of the zero-knowledge proof system and the binding property of the commitment scheme guarantee that the adversary can only submit tuples $(\text{e}, m, c, \pi_{\text{zk}})$ where the dishonest stakeholder won the lottery for e. Thus the distribution of HYB_1 and the ideal world is indistinguishable.

If in HYB_1 an honest stakeholder wins the lottery for entry e and publishes a message m via $\mathcal{F}_{\mathsf{ABC}}$, the adversary will receive a tuple of the form $(\mathsf{e}, m, c, \pi_{\mathsf{zk}})$. In the ideal world, the simulator gets (e, m) and creates a simulated tuple. By the zero-knowledge property of the proof system the distribution of HYB_1 and the ideal-world is indistinguishable. □

B   Extended Preliminaries

B.1   Non-interactive Zero-Knowledge

Definition 3 (Non-interactive Zero-knowledge Argument). *A non-interactive zero-knowledge argument for an NP relation R consists of a triple of polynomial time algorithms* (Setup, Prove, Verify) *defined as follows.*

- Setup(1^κ) *takes a security parameter κ and outputs a common reference string σ.*
- Prove(σ, x, w) *takes as input the CRS σ, a statement x, and a witness w, and outputs an argument π.*
- Verify(σ, x, π) *takes as input the CRS σ, a statement x, and a proof π, and outputs either 1 accepting the argument or 0 rejecting it.*

The algorithms above should satisfy the following properties.

1. *Completeness. For all $\kappa \in \mathbb{N}$, $(x, w) \in R$,*

$$\Pr\left(\mathsf{Verify}(\sigma, x, \pi) = 1 : \begin{array}{l} \sigma \leftarrow \mathsf{Setup}(1^\kappa) \\ \pi \leftarrow \mathsf{Prove}(\sigma, x, w) \end{array}\right) = 1.$$

2. *Computational soundness. For all PPT adversaries $\mathcal{A}$, the following probability is negligible in κ:*

$$\Pr\left(\begin{array}{l} \mathsf{Verify}(\sigma, \tilde{x}, \tilde{\pi}) = 1 \\ \wedge\ \tilde{x} \notin L \end{array} : \begin{array}{l} \sigma \leftarrow \mathsf{Setup}(1^\kappa) \\ (\tilde{x}, \tilde{\pi}) \leftarrow \mathcal{A}(1^\kappa, \sigma) \end{array}\right).$$

3. *Zero-knowledge. There exists a PPT simulator $(\mathcal{S}_1, \mathcal{S}_2)$ such that $\mathcal{S}_1$ outputs a simulated CRS Σ and trapdoor τ; $\mathcal{S}_2$ takes as input σ, a statement s and τ, and outputs a simulated proof π; and, for all PPT adversaries $(\mathcal{A}_1, \mathcal{A}_2)$, the following probability is negligible in κ:*

$$\left| \Pr\left(\begin{array}{l} (x, w) \in R\ \wedge \\ \mathcal{A}_2(\pi, \mathsf{st}) = 1 \end{array} : \begin{array}{l} \sigma \leftarrow \mathsf{Setup}(1^\kappa) \\ (x, w, \mathsf{st}) \leftarrow \mathcal{A}_1(1^\kappa, \sigma) \\ \pi \leftarrow \mathsf{Prove}(\sigma, x, w) \end{array}\right) - \right.$$

$$\left. \Pr\left(\begin{array}{l} (x, w) \in R\ \wedge \\ \mathcal{A}_2(\pi, \mathsf{st}) = 1 \end{array} : \begin{array}{l} (\sigma, \tau) \leftarrow \mathcal{S}_1(1^\kappa) \\ (x, w, \mathsf{st}) \leftarrow \mathcal{A}_1(1^\kappa, \sigma) \\ \pi \leftarrow \mathcal{S}_2(\sigma, \tau, x) \end{array}\right) \right|.$$

References

[AGM18] Agrawal, S., Ganesh, C., Mohassel, P.: Non-interactive zero-knowledge proofs for composite statements. In: Shacham, H., Boldyreva, A. (eds.) CRYPTO 2018. Part III. LNCS, vol. 10993, pp. 643–673. Springer, Cham (2018). https://doi.org/10.1007/978-3-319-96878-0_22

[BBB+18] Bünz, B., Bootle, J., Boneh, D., Poelstra, A., Wuille, P., Maxwell, G.: Bulletproofs: short proofs for confidential transactions and more. In: 2018 IEEE Symposium on Security and Privacy, pp. 315–334. IEEE Computer Society Press, May 2018. https://doi.org/10.1109/SP.2018.00020

[BCG+14] Ben-Sasson, E., et al.: Zerocash: decentralized anonymous payments from bitcoin. In: 2014 IEEE Symposium on Security and Privacy, pp. 459–474. IEEE Computer Society Press, May 2014. https://doi.org/10.1109/SP.2014.36

[BFM88] Blum, M., Feldman, P., Micali, S.: Non-interactive zero-knowledge and its applications (extended abstract). In: 20th ACM STOC, pp. 103–112. ACM Press, May 1988. https://doi.org/10.1145/62212.62222

[BGK+18] Badertscher, C., Gaži, P., Kiayias, A., Russell, A., Zikas, V.: Ouroboros genesis: composable proof-of-stake blockchains with dynamic availability. Cryptology ePrint Archive, Report 2018/378 (2018). https://eprint.iacr.org/2018/378

[BGM16] Bentov, I., Gabizon, A., Mizrahi, A.: Cryptocurrencies without proof of work. In: Clark, J., Meiklejohn, S., Ryan, P.Y.A., Wallach, D., Brenner, M., Rohloff, K. (eds.) FC 2016. LNCS, vol. 9604, pp. 142–157. Springer, Heidelberg (2016). https://doi.org/10.1007/978-3-662-53357-4_10

[bit11] Proof of stake instead of proof of work, July 2011. https://bitcointalk.org/index.php?topic=27787.0

[BLMR14] Bentov, I., Lee, C., Mizrahi, A., Rosenfeld, M.: Proof of activity: extending bitcoin's proof of work via proof of stake [extended abstract] y. ACM SIGMETRICS Perform. Eval. Rev. **42**(3), 34–37 (2014)

[Bou00] Boudot, F.: Efficient proofs that a committed number lies in an interval. In: Preneel, B. (ed.) EUROCRYPT 2000. LNCS, vol. 1807, pp. 431–444. Springer, Heidelberg (2000). https://doi.org/10.1007/3-540-45539-6_31

[BPS16] Bentov, I., Pass, R., Shi, E.: Snow white: provably secure proofs of stake. Cryptology ePrint Archive, Report 2016/919 (2016). http://eprint.iacr.org/2016/919

[CCs08] Camenisch, J., Chaabouni, R., Shelat, A.: Efficient protocols for set membership and range proofs. In: Pieprzyk, J. (ed.) ASIACRYPT 2008. LNCS, vol. 5350, pp. 234–252. Springer, Heidelberg (2008). https://doi.org/10.1007/978-3-540-89255-7_15

[CDE+16] Croman, K., et al.: On scaling decentralized blockchains. In: Clark, J., Meiklejohn, S., Ryan, P.Y.A., Wallach, D., Brenner, M., Rohloff, K. (eds.) FC 2016. LNCS, vol. 9604, pp. 106–125. Springer, Heidelberg (2016). https://doi.org/10.1007/978-3-662-53357-4_8

[CDS94] Cramer, R., Damgård, I., Schoenmakers, B.: Proofs of partial knowledge and simplified design of witness hiding protocols. In: Desmedt, Y.G. (ed.) CRYPTO 1994. LNCS, vol. 839, pp. 174–187. Springer, Heidelberg (1994). https://doi.org/10.1007/3-540-48658-5_19

718   C. Ganesh et al.

[CM99]   Camenisch, J., Michels, M.: Proving in zero-knowledge that a number is the product of two safe primes. In: Stern, J. (ed.) EUROCRYPT 1999. LNCS, vol. 1592, pp. 107–122. Springer, Heidelberg (1999). https://doi.org/10.1007/3-540-48910-X_8

[CS97]   Camenisch, J., Stadler, M.: Efficient group signature schemes for large groups. In: Kaliski, B.S. (ed.) CRYPTO 1997. LNCS, vol. 1294, pp. 410–424. Springer, Heidelberg (1997). https://doi.org/10.1007/BFb0052252

[Dam00]   Damgård, I.: Efficient concurrent zero-knowledge in the auxiliary string model. In: Preneel, B. (ed.) EUROCRYPT 2000. LNCS, vol. 1807, pp. 418–430. Springer, Heidelberg (2000). https://doi.org/10.1007/3-540-45539-6_30

[DGKR18] David, B., Gaži, P., Kiayias, A., Russell, A.: Ouroboros Praos: an adaptively-secure, semi-synchronous proof-of-stake blockchain. In: Nielsen, J.B., Rijmen, V. (eds.) EUROCRYPT 2018. Part II. LNCS, vol. 10821, pp. 66–98. Springer, Cham (2018). https://doi.org/10.1007/978-3-319-78375-8_3

[FO97]   Fujisaki, E., Okamoto, T.: Statistical zero knowledge protocols to prove modular polynomial relations. In: Kaliski Jr., B.S. (ed.) CRYPTO 1997. LNCS, vol. 1294, pp. 16–30. Springer, Heidelberg (1997). https://doi.org/10.1007/BFb0052225

[FS87]   Fiat, A., Shamir, A.: How to prove yourself: practical solutions to identification and signature problems. In: Odlyzko, A.M. (ed.) CRYPTO 1986. LNCS, vol. 263, pp. 186–194. Springer, Heidelberg (1987). https://doi.org/10.1007/3-540-47721-7_12

[GHM+17] Gilad, Y., Hemo, R., Micali, S., Vlachos, G., Zeldovich, N.: Algorand: scaling byzantine agreements for cryptocurrencies. Cryptology ePrint Archive, Report 2017/454 (2017). http://eprint.iacr.org/2017/454

[JKK14]   Jarecki, S., Kiayias, A., Krawczyk, H.: Round-optimal password-protected secret sharing and T-PAKE in the password-only model. In: Sarkar, P., Iwata, T. (eds.) ASIACRYPT 2014. Part II. LNCS, vol. 8874, pp. 233–253. Springer, Heidelberg (2014). https://doi.org/10.1007/978-3-662-45608-8_13

[KKKZ18] Kerber, T., Kohlweiss, M., Kiayias, A., Zikas, V.: Ouroboros crypsinous: privacy-preserving proof-of-stake. Cryptology ePrint Archive, Report 2018/1132 (2018). To appear at IEEE Symposium on Security and Privacy - S&P 2019. https://eprint.iacr.org/2018/1132

[KN12]   King, S., Nadal, S.: PPcoin: peer-to-peer crypto-currency with proof-of-stake (2012)

[KRDO17] Kiayias, A., Russell, A., David, B., Oliynykov, R.: Ouroboros: a provably secure proof-of-stake blockchain protocol. In: Katz, J., Shacham, H. (eds.) CRYPTO 2017. Part I. LNCS, vol. 10401, pp. 357–388. Springer, Cham (2017). https://doi.org/10.1007/978-3-319-63688-7_12

[Lin15]   Lindell, Y.: An efficient transform from sigma protocols to NIZK with a CRS and non-programmable random Oracle. In: Dodis, Y., Nielsen, J.B. (eds.) TCC 2015. Part I. LNCS, vol. 9014, pp. 93–109. Springer, Heidelberg (2015). https://doi.org/10.1007/978-3-662-46494-6_5

[MGGR13] Miers, I., Garman, C., Green, M., Rubin, A.D.: Zerocoin: anonymous distributed E-cash from Bitcoin. In: 2013 IEEE Symposium on Security and Privacy, pp. 397–411. IEEE Computer Society Press, May 2013. https://doi.org/10.1109/SP.2013.34

[Nak08] Nakamoto, S.: Bitcoin: A Peer-to-Peer Electronic Cash System (2008)

[OM14] O'Dwyer, K.J., Malone, D.: Bitcoin mining and its energy footprint. In: ISSC 2014/CIICT 2014, pp. 280–285 (2014). https://doi.org/10.1049/cp. 2014

[Ped92] Pedersen, T.P.: Non-interactive and information-theoretic secure verifiable secret sharing. In: Feigenbaum, J. (ed.) CRYPTO 1991. LNCS, vol. 576, pp. 129–140. Springer, Heidelberg (1992). https://doi.org/10.1007/3-540-46766-1_9

[Sch91] Schnorr, C.-P.: Efficient signature generation by smart cards. J. Cryptol. 4(3), 161–174 (1991)

Consensus Through Herding

T.-H. Hubert Chan[1(✉)], Rafael Pass[2], and Elaine Shi[2]

[1] The University of Hong Kong, Lung Fu Shan, Hong Kong
hubert@cs.hku.hk
[2] Cornell and Thunder Research, New York, USA
{rafael,elaine}@cs.cornell.edu

Abstract. State Machine Replication (SMR) is an important abstraction for a set of nodes to agree on an ever-growing, linearly-ordered log of transactions. In decentralized cryptocurrency applications, we would like to design SMR protocols that (1) resist adaptive corruptions; and (2) achieve small bandwidth and small confirmation time. All past approaches towards constructing SMR fail to achieve either small confirmation time or small bandwidth under adaptive corruptions (without resorting to strong assumptions such as the erasure model or proof-of-work).

We propose a novel paradigm for reaching consensus that departs significantly from classical approaches. Our protocol is inspired by a social phenomenon called herding, where people tend to make choices considered as the social norm. In our consensus protocol, leader election and voting are coalesced into a single (randomized) process: in every round, every node tries to cast a vote for what it views as the *most popular* item so far: such a voting attempt is not always successful, but rather, successful with a certain probability. Importantly, the probability that the node is elected to vote for v is independent from the probability it is elected to vote for $v' \neq v$. We will show how to realize such a distributed, randomized election process using appropriate, adaptively secure cryptographic building blocks.

We show that amazingly, not only can this new paradigm achieve consensus (e.g., on a batch of unconfirmed transactions in a cryptocurrency system), but it also allows us to derive the first SMR protocol which, even under adaptive corruptions, requires only polylogarithmically many rounds and polylogarithmically many honest messages to be multicast to confirm each batch of transactions; and importantly, we attain these guarantees under standard cryptographic assumptions.

T.-H. Hubert Chan—This research was partially done in a consultancy agreement with Thunder Research.

Electronic supplementary material The online version of this chapter (https://doi.org/10.1007/978-3-030-17653-2_24) contains supplementary material, which is available to authorized users.

© International Association for Cryptologic Research 2019
Y. Ishai and V. Rijmen (Eds.): EUROCRYPT 2019, LNCS 11476, pp. 720–749, 2019.
https://doi.org/10.1007/978-3-030-17653-2_24

1 Introduction

State Machine Replication (SMR), also called consensus, is a core abstraction in distributed systems [5,21,25]: a set of nodes would like to agree on a linearly ordered log of transactions (e.g., in a public ledger or decentralized smart contract application), such that two important security properties, *consistency* and *liveness*, are satisfied. Loosely speaking, consistency requires that all honest nodes' logs are prefixes of one another and that no node's log will ever shrink; and liveness requires that if a client submits a transaction, the transaction will appear in every honest node's log in a bounded amount of time.

The classical literature on distributed systems typically considers deployment of consensus in a single organization (e.g., Google or Facebook), and on a small scale (e.g., a dozen nodes). Typically these nodes are connected through fast, local-area network where bandwidth is abundant. Thus the classical consensus literature typically focuses on optimizating the protocol's *round complexity* which is directly related to the confirmation time—it is well-known that we can design consensus protocols where confirmation happens in expected constant rounds (i.e., independent of the number of players) even in the presence of adaptive corruptions [10,14].

In the past decade, due to new blockchain systems such as Bitcoin and Ethereum, SMR protocols have been deployed in a decentralized setting in an open network. In such a blockchain setting, we typically have a *large* number n of nodes who communicate over a diffusion network (where nodes *multicast* messages to the whole network); and it is simply not practical to have protocols where the number of messages to be multicast grows linearly with the number of nodes. In this paper, we care about achieving SMR in a *communication-efficient* way: we want both the confirmation time and the number of bits multicast (to confirm each batch of transactions) to be polylogarithmic (or even just sublinear) in the number of nodes. More precisely, we refer to an n-party protocol as being communication efficient if the total number the bits multicast is $o(n) \cdot |\mathsf{TXs}| \cdot \kappa$, and the confirmation time is $o(n) \cdot \kappa$ where κ is a security parameter such that the protocol's security must be respected except with negligible in κ probability. Achieving communication efficiency under *static* security is easy: one could randomly elect a small committee of $\mathsf{poly}\log \kappa$ size and next run any SMR protocol that may have polynomial bandwidth overhead to confirm a batch of transactions. If the committee election is random and independent of the choice of corrupt nodes, then except with negligible in κ probability, the committee's corrupt fraction approximates the overall corrupt fraction due to the Chernoff bound. Moreover, under honest majority assumptions, non-committee members can always be convinced of a decision as long as it is vouched for by the majority of the committee.

However, in typical blockchain applications (such as cryptocurrencies where the participating nodes are on an open network), the static corruption model is insufficient for security. Rather, we need to protect the protocol against *adaptive* corruptions, where an attacker may, based on the protocol execution so far, select which parties to attack. The above-mentioned "naïve" committee election approach miserably fails in the presence of adaptive corruptions: the adversary can always corrupt all committee members after having observed who they are and completely violate the security of the protocol. Indeed, obtaining a communication-efficient SMR protocol which withstands adaptive corruptions has been a long-standing open problem:

Does there exists a communication-efficient SMR protocol that withstands adaptive corruptions?

Nakamoto's beautiful blockchain protocol (through its analysis in [11,19,20]) was the first protocol to achieve communication-efficient SMR with adaptive security. This protocol, however, requires using proofs of work [9], and in particular requires honest players to "waste" as much computation as the total computational power of adversarial players. Consequently, in recent years, the research community has focused on removing the use of proofs-of-work and instead rely on standard bounds on the fraction of adversarial players (e.g., honest majority). In particular, the recent work by Chen and Micali [6] (see also David et al. [8], and Pass and Shi [23]) demonstrates communication-efficient SMR protocols with adaptive security without the use of proof-of-work in the so-called *erasures model*: in the erasure model, we assume that honest players have the ability to *completely* erase/dispose of some parts of their local state (such that if some player later on gets corrupted, the erased state cannot be recovered by the attacker). However, as discussed in Canetti et al. [4] (and the references therein) such erasures are hard to perform in software, without resorting to physical intervention, and the security of the heuristics employed in practice are not well understood. As such, solutions in the erasure model may not provide an adequate level of security and thus ideally, we would like to avoid the use of strong erasure assumptions.

In this work, we focus on the design of communication-efficient SMR protocols without proof-of-work and without assuming the possibility of erasures. As far as we know, the design of such protocols is open even in the PKI model and even if assuming, say, 99% of the nodes are honest. We remark that very recently, a communication-efficient "single-shot" version of consensus, referred to as "multi-value agreement" (MVA), was achieved by Abraham et al. [2]; as we discuss in detail in Sect. 1.2, the validity conditions for MVA is much weaker and thus it is not clear how to extend these protocols to SMR.

1.1   Our Results

We propose the first communication-efficient SMR protocol with adaptive security (without assuming erasures or proof-of-work), solving the above-mentioned open problem. Our protocol works in a public-key-infrastructure (PKI) model, assuming a synchronous network, and standard cryptographic assumptions. The protocol tolerates $\frac{1}{3} - \epsilon$ fraction of adaptive corruptions where ϵ is an arbitrarily small constant, moreover to achieve a failure probability that is negligible in the security parameter κ, every transaction tx gets confirmed in poly log $\kappa \cdot \Delta$ time where Δ is the maximum network delay, and requiring at most $|\text{tx}| \cdot \text{poly} \log \kappa$ bits of honest messages to be multicast (assuming that $|\text{tx}|$ is at least as large as a suitable computational security parameter).

Theorem 1.1 (Adaptively secure, communication efficient synchronous state machine replication). *Under standard cryptographic hardness assumptions (more precisely, assuming standard bilinear group assumptions), there exists a synchronous state machine replication protocol which, except with negligible in κ probability, satisfies consistency and confirms transactions in* $\text{poly} \log \kappa \cdot \Delta$ *time where Δ is the maximum network delay—as long as the adversary corrupts no more than $\frac{1}{3} - \epsilon$ fraction of nodes. Moreover, (except with negligible in κ, χ probability) honest nodes only need to multicast* $\text{poly} \log \kappa \cdot (\chi + |\text{TXs}|)$ *bits of messages to confirm every batch of transactions denoted* TXs *where χ is a computational security parameter related to the strength of the cryptographic building blocks involved.*[1]

We remark that our communication complexity bound is asymtotically the same as that achieved by earlier protocols using either proofs of work or in the erasure model.

1.2   Technical Highlights

Why the problem is more challenging in SMR than in single-shot consensus. We stress that achieving communication efficiency under adaptive corruptions is more difficult in SMR than in single-shot consensus. In the latter, a designated sender aims to "broadcast", for once only, a (possibly multi-bit) value to everyone, retaining consistency even when the sender is corrupt, and achieving validity should the sender be honest. Henceforth we refer to the single-shot version as Multi-Valued Agreement (MVA). In MVA, if the sender is adaptively corrupt, typically no validity is necessary (or even attainable depending on the concrete definition). Thus it is not clear how to compose adaptively secure MVA to achieve adaptively secure SMR while preserving communication efficiency: if the "leader" who is supposed to broadcast the next block (of transactions) becomes corrupt, the adversary can cause a "bad block" to be confirmed (e.g., a block that censors some to all outstanding transactions). If only a small number of such "leaders" speak at any point of time, the adversary can continuously corrupt all leaders that speak until it has exhausted its corruption budget—such an attack will cause confirmation time to be large.

For this reason, we stress that *adaptively secure, communication-efficient MVA does NOT lead to adaptively secure, communication-efficient SMR in any* straightforward manner[2]. Also note that even for single-shot consensus, the only

---

[1] If assuming subexponential security of the underlying cryptographic building blocks, χ can be set to $\text{poly} \log \kappa$.

[2] If communication efficiency is not a concern, we could have n broadcast instances (composed either sequentially or in parallel) where everyone is given the chance to act as the leader and suggest the next batch of transactions to confirm; we can then concatenate the outputs of these n broadcasts and treat it as the next block.

known solution that is communication efficient and does not assume erasures is the recent work by Abraham et al. [2].

Defining "batch agreement" with a quality metric. As mentioned, the validity definition in the classical notion of MVA is too weak if we wish to construct communication-efficient state machine replication. One contribution of our paper is to propose a new abstraction which we call "batch agreement"— on the surface it looks very much like MVA since nodes seek to agree on the next batch of transactions. However, batch agreement is defined with a *quality* metric that is lacking in the standard definition of MVA. We say that a block has good quality iff it contains all transactions that have been outstanding for a while; and our batch agreement notion requires that *a batch with good quality be chosen* even when the "leader" is adaptively corrupt (and upon corruption it can inject many blocks).

Constructing batch agreement. To understand the novelty of our approach, we first briefly review existing work. In classical approaches, if only a few number of leaders are elected to speak, all of them can be adaptively corrupt and made to propose multiple blocks in the same round. Even if the adversary is not fast enough to erase the good block that leader already proposed while still honest (i.e., just before it was adaptively corrupted), it can succeed in diverging honest nodes' voting efforts. For example, newly corrupt node may propose many good blocks and delivering them in different order to different honest nodes. At this moment, using classical techniques, it does not seem easy for the honest nodes to coordinate and vote on the same block. As a result, some classical approaches adopt an approach [18] where nodes jointly discover that no block has gained popular votes (e.g., by computing a grade), and then they initiate a binary agreement process to jointly decide to fall back to outputting a default value. Obviously the default value is pre-determined and cannot contain the set of outstanding transactions and thus does not have good quality.

Our approach departs from all known classical approaches: at the core we describe a new randomized process through which the network can jointly make a selection and converge to a good choice, when presented with polynomially many choices. During a batch agreement, a small set of nodes get elected to propose a block. All of these nodes may be adaptively corrupt and then made to propose more good or bad blocks. Thus honest nodes are faced with these polynomially many blocks to choose from, and moreover at any snapshot of time, the set of blocks observed by different honest nodes may differ, since the blocks do not necessarily arrive at the honest nodes at the same time.

To converge to a good choice, we start with an *initial* scoring function that is used to evaluate the quality of each block itself—basically a block that contains all sufficiently long outstanding transactions scores high, and a block that censors some of them will score very low. Since nodes may receive outstanding transactions at slightly different times, honest nodes may end up calculating different initial scores even for the same block—we carefully craft a initial score function to make sure that this difference is not too large as long as transactions are propagated to honest nodes around the same time (indeed, we show that this can be accomplished with small communication too).

Nodes then are randomly elected to vote on the blocks over time; the votes can serve to strengthen a block's score in an additive fashion. At any point of time, an honest node will always try to vote on the *most popular* block in its view (i.e., the one with the highest score), but the voting attempt only succeeds with somewhat small probability such that not too many people need to send votes. If a node is randomly elected to vote for some block B in some time step, it does not mean that it is eligible to vote for other blocks; thus adaptively corrupting a node that is elected to vote does not help the adversary. Since all honest nodes always choose the most popular item so far, honest nodes' voting efforts must be somewhat concentrated. After roughly polylogarithmically many steps, polylogarithmically many honest votes will have been cast. Although at any snapshot of time, nodes may never agree on the precise score or set of votes for each block, (except with negligible probability) it must be the case that everyone sees that the same highest-scoring block at the end, because its final score is significantly larger than the second-best choice. Finally, a block with an initial score that is too low will also not be selected (except with negligible probability) because it is too unlikely for it to ever collect enough votes (even when counting corrupt nodes' votes) to compete with the blocks with good quality.

With our approach, the ability to corrupt a leader on the fly and making it propose many additional blocks (on top of the good block it already proposed) does not help the adversary; nor does adaptively corrupting a voter help as mentioned.

2   Technical Roadmap

In this section, we begin by explaining our construction and proofs (of the primary building block) informally. We then give a more detailed comparison with related work.

2.1   Informal Description of Our Protocol

At the core of our SMR construction is a new abstraction called "batch agreement". Every time a batch agreement instance is invoked, the nodes reach agreement on a set of transactions such that transactions that are sufficiently long-pending are guaranteed to be included (except with negligible probability). The entire SMR protocol simply runs multiple sequential instances of the batch agreement protocol.

As mentioned earlier, although on the surface it seems similar to the classical notion of Multi-Valued Agreement (MVA), our notion has a much stronger validity property that is lacking in classical MVA—specifically, we require that the confirmed block have good quality even when everyone who is randomly elected to speak is adaptively corrupt.

A herding-inspired protocol. Our protocol is inspired a social phenomenon called *herding* where people follow the popular social choice. We show how herding can be leveraged for reaching consensus. Recall that the adversary controls $\frac{1}{3} - \epsilon$ fraction. At a high level, nodes cast votes for batches of transactions over time. Imagine that in any round t, a node has a certain probability $p = \frac{1}{\lambda \Delta n}$ of being elected to vote for a particular batch TXs where λ is sufficiently large such that $\lambda = \omega(\frac{\log \kappa}{\epsilon^2})$—henceforth if a node is elected to vote for TXs in round t, we say that the node "mines" a vote for TXs in round t. Importantly, the probability that a node mines a vote for TXs and round r is independent of its success probability for TXs$'$ and round r' as long as $(\text{TXs}, r) \neq (\text{TXs}', r')$—this is important for achieving adaptive security: if the adversary adaptively corrupts a node that has just cast a vote for TXs, corrupting this node does not make it more or less likely for the adversary to mine a vote for TXs$' \neq$ TXs in the same round than corrupting any other node.

In every round, an honest node would always pick the most popular batch (where popularity will be defined later) in its view, and it will only attempt to vote for this most popular batch—it will not cast a vote for any other batch even if it might be eligible. If a voting attempt for TXs is successful in some round, the node multicasts the new vote as well as all existing votes it has seen observed for TXs to all other nodes. After some time, every node outputs a batch that has collected "ample" number of votes (where "ample" will be defined later in Sect. 2.2); if no such batch is found, output nothing.

Realizing "mining" with cryptography. So far in the above protocol, we did not fully specify how to realize the random eligibility election. As we explain later in the paper, this can be instantiated assuming a Verifiable Random Function (VRF) with appropriate *adaptive security* properties.

Assume that every player i has a VRF public key denoted pk_i that is common knowledge, and the corresponding VRF secret key sk_i is known only to player i. For i to determine its eligibility to vote for TXs in round r, it evaluates $(\mu, \pi) := \text{VRF}(\text{sk}_i, \text{TXs}, r)$ where μ is the VRF evaluation outcome and π is a proof attesting to the evaluation outcome. If $\mu < D_p$ where D_p is an appropriate difficulty parameter, node i is deemed eligible to vote for TXs in round r. While only the secret-key owner can evaluate the VRF, anyone can verify the evaluation outcome. More specifically, any node that receives the tuple $(\text{TXs}, r, \mu, \pi)$ can verify with pk_i that indeed μ is the correct VRF evaluation outcome and verify i's eligibility to vote for TXs in round r. Importantly, a vote received is only considered valid if its purported round is no greater than the current round number (this prevents corrupt nodes from mining into the future).

Later in Sect. 7, we will describe how to instantiate such an adaptively-secure VRF that satisfies our needs, using techniques from Abhraham et al. [2].

Popularity and initial score. It remains to specify how nodes determine the popularity of a batch TXs of transactions. The popularity is the sum of an *initial score* and the *number of valid votes collected* so far for TXs. To make sure that the protocol will preferentially select an all-inclusive batch TXs that omits no long-pending transaction, we design an initial score function that relies on a *discounting* mechanism to punish batches that omit long-pending transactions.

Specifically, we say that node i perceives the *age* of a transaction tx to be α, if at the start of the batch agreement protocol, exactly α rounds have elapsed since node i first observed tx. We assume that the underlying network medium satisfies the following "transaction diffusion" assumption: if any forever honest node observes a transaction tx in round r, then by round $r + \Delta$, all so-far honest nodes must have observed tx too[3]. In this way, we are guaranteed that the perceived age of a transaction tx must be somewhat consistent among all honest nodes. Now, imagine that the maximum initial score a batch can gain is $S_{\max}$ (to be parametrized later). We will discount the initial score of a batch TXs exponentially fast w.r.t. the oldest transaction that TXs omits. Specifically, imagine that node i computes the score of TXs as follows:

$$\text{score}_i(\text{TXs}) := S_{\max} \cdot \left(1 - \frac{1}{\lambda S_{\max}}\right)^{\frac{\alpha^*}{3\Delta}} \tag{1}$$

where α^* is the age (as perceived by node i) of the oldest transaction that is omitted from TXs. Given the transaction diffusion assumption, it is not difficult to see that every two so-far honest nodes' initial score difference for any batch TXs must be less than $\frac{1}{\lambda}$, i.e., honest nodes score every batch somewhat consistently.

2.2   Intuitive Analysis

We can now intuitively argue why such a herding-based protocol satisfies consistency and liveness under appropriate parameters. Imagine that the protocol is parametrized in the following way where λ is chosen such that $\epsilon^2 \lambda = \omega(\log \kappa)$ — for example if ϵ is a(n arbitrarily small) constant, then λ may be any super-logarithmic function:

- in an all-honest execution, in expectation, every $\lambda \Delta$ rounds, some node mines a new vote. This means that each individual mining attempt is successful with probability $\frac{1}{\lambda \Delta n}$;
- the protocol is executed for $T_{\text{end}} := \lambda^2 \Delta$ rounds, i.e., in an all-honest execution, in expectation a total of λ votes are mined; and
- at the end of the protocol, a node would only output a batch that has gained $\frac{2\lambda}{3}$ or more valid votes, i.e., a threshold of $\frac{2\lambda}{3}$ is considered ample.

Consistency. To argue consistency, it suffices to argue that any two different batches TXs and TXs$'$ cannot both gain ample votes by T_{end}. This follows from the following observation: forever honest nodes make only a single mining attempt per round; while eventually corrupt nodes can make a mining attempt

---

[3] As discussed in the Supplemental Materials this assumption can be removed in a synchronous network while preserving communication efficiency.

for TXs and one for TXs′ corresponding for each round r (note that once corrupt, a node can retroactively make mining attempts for past rounds). Thus the total number of mining attempts made for either TXs or TXs′ must be upper bounded by $\frac{4}{3} \cdot n \cdot T_{\text{end}}$. As mentioned earlier, adaptively corrupting a node that has just mined a vote for TXs does not increase the adversary's chance of mining a vote for TXs′ ≠ TXs (for any round). Thus by Chernoff bound, we have that except with $\exp(-\Omega(\epsilon^2\lambda))$ probability (which is negligible in κ), the total number of successfully mined votes (including honest and adversarial) for TXs or TXs′ must be strictly less than $\frac{4\lambda}{3}$—this means that the two different batches cannot both have ample votes. To complete the argument, we need to take a union bound over all pairs of batches. If the adversary and all nodes are polynomially bounded, then the only batches we care about are those that appear in some honest node's view at some point in the execution. Since there are at most polynomially many such batches, the union bound has only polynomial loss.

Liveness. Liveness crucially relies on the fact that the mining difficulty is large enough, such that the average time till some node finds the next vote (set to be $\lambda\Delta$) is much larger than the maximum network delay Δ. Intuitively, this condition is necessary for honest nodes to "concentrate" their voting efforts on the same batch. Recall that honest nodes would score each batch somewhat consistently. This means that if a so-far honest node mines a vote for what he thinks is the most popular batch TXs—if the network delay is small, very soon all so-far honest nodes would find TXs the most popular batch too, and would mine votes only for TXs. As long as all forever honest nodes concentrate their mining efforts, by Chernoff bound some batch would attract ample votes and thus liveness ensues. On the other hand, if the network delay is large w.r.t. to the time it takes to mine a vote, honest nodes will be mining on different batches and likely no batch will gain enough votes at the end. We defer a formal argument to the later technical sections.

Validity. For validity, we would like to argue that any batch that honest nodes agree on cannot omit "long-pending" transactions. To see this, note that except with negligible in κ probability, the total number of valid votes any batch TXs can gain is at most 1.1λ. Now, if we let $S_{\max} := 3\lambda$, then any batch TXs that omits transactions of age $c\lambda^2\Delta$ or higher for an appropriate constant c must have an initial score less than 1.5λ as perceived by any honest node (recall that honest nodes would always assign somewhat consistent scores to every batch). This means that no honest node should ever attempt to mine a vote for such a batch TXs; and thus TXs cannot gain ample votes.

2.3   Additional Related Work

In the past, the only known protocol that achieves both small bandwidth and small confirmation time under adaptive corruptions is the celebrated Nakamoto consensus protocol [11,19,20,22], however, at the price of making very strong, idealized, proof-of-work assumptions. Constrained to making standard cryptographic assumptions, it is known how to construct adaptively-secure SMR that achieves either small confirmation time or small bandwidth, but not both.

First, if we allow many nodes to speak at any point of time (i.e., if we did not care about bandwidth consumption), we can easily construct protocols that achieve small round complexity. Specifically, it is easy to compose multiple instances of small-round MVA protocols [2,10,14] to attain SMR with small confirmation time (while retaining adaptive security). Basically, in every round, we can fork n instances of MVA where each node i acts as the designated sender in the i-th instance, and the log of the SMR is derived by concatenating all instances of all rounds, ordered first by the round and then by the instance within the round. However, even if the underlying MVA achieved small bandwidth [2], the derived SMR protocol would be expensive in bandwidth.

In a second class of approaches, we would like to have only a small number of players speak at any given point of time [1,6–8,15,16,23,26]—this is in fact necessary to achieve our notion of communication efficiency. Past work has suggested multiple ways to construct such protocols:

– One possible approach [1,7,8,15,16,23,26], is inspired by Nakamoto's longest-chain protocol but removing the proof-of-work in a permissioned setting assuming a public-key infrastructure (PKI). Specifically, in such protocols, in every time slot, a node has a chance of being elected leader. When it is elected leader, it signs the next block extending the current longest chain. For such protocols to retain consistency and liveness [1,7,8,15,16,23,26], some additional constraints have to be imposed on the validity of timestamps contained in a blockchain. Among these works, some use a randomized leader election strategy [7,16,16,23]; and some use a deterministic leader election process [1,15,26].
– Another approach, represented by Algorand [6] and improved in subsequent works [2,18], is rely on a classical-style consensus protocol, but in every round, randomly subsample a small, polylogarithmically size committee to cast votes (e.g., by employing a verifiable random function).

No matter which approach is taken, an adaptive adversary can continuously corrupt the small number of players selected to speak until it exhausts its corruption budget. Once corrupt, these players can cast ambiguous votes or propose equivocating blocks (e.g., those that censor certain transactions). In all of the above approaches (without assuming erasure), when such an adaptive-corruption attack is taking place, all blocks confirmed may have bad quality (e.g., censoring certain transactions), causing confirmation time to be at least $\Theta(n/s)$ where s denotes an upper bound on the number of players who speak in every round.

Communication-efficient single-shot consensus. The recent work by Abraham et al. [2] achieves adaptive security and communication efficiency without erasures or PoW, but their approach works only for MVA and does not extend, in any non-trivial fashion, to SMR. As mentioned sequential or parallel repetition of MVA fails to work for this purpose due to the much weaker validity requirement of MVA (see the Supplemental Materials for additional explanations).

As will be obvious soon, although our paper adopts the vote-specific committee election technique from Abraham et al. [2], we require vastly new techniques to simultaneously achieve both adaptive security and communication efficiency for SMR.

3   Protocol Execution Model

A protocol refers to an algorithm for a set of interactive Turing Machines (also called nodes) to interact with each other. The execution of a protocol Π that is directed by an environment $\mathcal{Z}(1^\kappa)$ (where κ is a security parameter), which activates a number of nodes as either *honest* or *corrupt* nodes. Honest nodes faithfully follow the protocol's prescription, whereas corrupt nodes are controlled by an adversary $\mathcal{A}(1^\kappa)$ which reads all their inputs/message and sets their outputs/messages to be sent.

A protocol's execution proceeds in *rounds* that model atomic time steps. At the beginning of every round, honest nodes receive inputs from an environment $\mathcal{Z}$; at the end of every round, honest nodes may send outputs to the environment $\mathcal{Z}$.

Corruption model. $\mathcal{Z}$ spawns n number of nodes upfront, a subset of which may be corrupt upfront, and the remaining are honest upfront. During the execution, $\mathcal{Z}$ may *adaptively* corrupt any honest node. When a node becomes corrupt, $\mathcal{A}$ gets access to its local state, and subsequently, $\mathcal{A}$ controls the corrupt node. Henceforth, at any time in the protocol, nodes that remain honest so far are referred to as *so-far honest* nodes; and nodes that remain honest till the end of the protocol are referred to as *forever honest* nodes[4].

Communication model. We assume that there is a function $\Delta(\kappa, n)$ that is polynomial in κ and n, such that every message sent by a so-far honest node in round r is guaranteed to be received by a so-far honest recipient at the beginning of round $r+\Delta$ (if not earlier). The adversary can delay honest message arbitrarily but up to Δ rounds at the maximum.

All of our protocols will work in the *multicast* model: honest nodes always send the same message M to everyone. We assume that when a so-far honest node i multicasts a message M in some round r, it can immediately become corrupt in the same round and made to send one or more messages in the same round. However, the message M that was already multicast before i became corrupt cannot be retracted, and all nodes that are still honest in round $r + \Delta$ will have received the message M. In our paper we will also account for a protocol's communication efficiency by upper bounding how many bits of honest messages must be multicast during the protocol. Any message that is sent by a so-far honest node is an honest message—but if the node becomes corrupt in the same round and sends another message in the same round, the latter message is treated as a corrupt message. Since corrupt nodes can send any polynomially many messages, we do not seek to bound corrupt messages.

---

[4] Note that "forever honest" is in fact defined w.r.t. the protocol we are concerned with.

In this paper we consider *synchronous* protocols where the protocol is parametrized with Δ, i.e., Δ is hard-wired in the protocol's description.

Notational convention. Protocol execution is assumed to be probabilistic in nature. We would like to ensure that certain security properties such as consistency and liveness hold for almost all execution traces, assuming that both $\mathcal{A}$ and $\mathcal{Z}$ are polynomially bounded.

Henceforth in the paper, we use the notation $\mathsf{EXEC}^{\Pi}(\mathcal{A}, \mathcal{Z}, \kappa)$ to denote a sample of the randomized execution of the protocol Π with $\mathcal{A}$ and $\mathcal{Z}$, and security parameter $\kappa \in \mathbb{N}$. The randomness in the experiment comes from honest nodes' randomness, $\mathcal{A}$, and $\mathcal{Z}$, each sampling of $\mathsf{EXEC}^{\Pi}(\mathcal{A}, \mathcal{Z}, \kappa)$ produces an *execution trace*. We would like that the fraction of execution traces that fail to satisfy relevant security properties be negligibly small in the security parameter κ. A function $\mathsf{negl}(\cdot)$ is said to be negligible if for every polynomial $p(\cdot)$, there exists some κ_0 such that $\mathsf{negl}(\kappa) \leq 1/p(\kappa)$ for every $\kappa \geq \kappa_0$.

Throughout the paper, we assume that n is a polynomial function in κ and Δ is a polynomial function in κ and n—we note in the most general setting, Δ may be dependent on n if, for example, the network layer builds some kind of diffusion tree or graph to propagate messages.

Definition 3.1 $((\rho, \Delta)$-respecting). *We say that $(\mathcal{A}, \mathcal{Z})$ is (ρ, Δ)-respecting w.r.t. protocol Π iff for every $\kappa \in \mathbb{N}$, with probability 1 in $\mathsf{EXEC}^{\Pi}(\mathcal{A}, \mathcal{Z}, \kappa)$, every honest message is delivered within Δ rounds and moreover $(\mathcal{A}, \mathcal{Z})$ adaptively corrupts at most ρ fraction of nodes.*

When the context is clear, we often say that $(\mathcal{A}, \mathcal{Z})$ is (ρ, Δ)-respecting omitting saying which protocol Π is of interest.

4   Scoring Agreement

We define an abstraction called scoring agreement—this is at the of our batch agreement construction. We rely on a herding-based protocol to achieve it. In a scoring agreement protocol, each node starts with an element from some known universe $\mathcal{U}$. Each node can evaluate an initial score for each element from $\mathcal{U}$. The scoring agreement protocol seeks to reach agreement on some element with from $\mathcal{U}$ that is scored relatively highly by (almost) all forever honest nodes.

4.1   Definition of Scoring Agreement

Syntax. A scoring agreement protocol, henceforth denoted Π_{score} is parametrized with a universe $\mathcal{U}$ that defines valid values. Moreover, suppose that there is a publicly known, polynomial-time computable function (also denoted $\mathcal{U}$ for convenience) for verifying whether a value v belongs to $\mathcal{U}$.

The environment $\mathcal{Z}$ instructs all nodes to start the protocol at the same time (treated as round 0 for the current protocol instance). When a node is instructed to start by $\mathcal{Z}$, it additionally receives the following as input from $\mathcal{Z}$:

1. a value $v_i \in \mathcal{U}$;
2. an efficiently computable function $\mathsf{score}_i : \mathcal{U} \to \mathbb{R}$ that can assign an initial, real-valued score for any value $v \in \mathcal{U}$; note that different nodes can receive different scoring functions.

Later when employed in our batch agreement, the value will be blocks of transactions.

Constraints on $\mathcal{Z}$. We require that the following conditions hold with probability 1:

- ϑ-*somewhat-consistent initial scoring*: for every $v \in \mathcal{U}$, for any initially honest i and j, it holds that $|\mathsf{score}_i(v) - \mathsf{score}_j(v)| < \vartheta$.
- *High initial scores*: for every forever honest i, let v_i be i's input—it must be that there is no $v' \in \mathcal{U}$ such that $\mathsf{score}_i(v') > \mathsf{score}_i(v_i)$.

The first condition above requires that initially honest nodes receive relatively consistent scoring functions from $\mathcal{Z}$, i.e., they assign somewhat consistent initial scores for every element in the universe. The second condition requires that every forever honest node's input must be the highest scoring element in the universe (as perceived by the node itself).

Security properties. We want a protocol where nodes reach agreement on a value in $\mathcal{U}$. We say that a protocol Π_{score} (parametrized with $\mathcal{U}$ and Δ) satisfies a certain *property* w.r.t. $(\mathcal{A}, \mathcal{Z})$, iff there exists some negligible function $\mathsf{negl}(\cdot)$ such that for all $\kappa \in \mathbb{N}$, for all but $\mathsf{negl}(\kappa)$ fraction of the execution traces sampled from $\mathbf{EXEC}^{\Pi_{\mathrm{score}}}(\mathcal{A}, \mathcal{Z}, \kappa)$, that property holds. In particular, we care about the following properties.

- *Consistency.* If a so-far honest node i outputs $v \in \mathcal{U}$ and a so-far honest node j outputs $v' \in \mathcal{U}$, it must hold that $v = v'$.
- *d-Validity.* Suppose that for some $B \in \mathbb{R}$, there exists subset S of forever honest nodes of size at least $(\frac{1}{3} + 0.5\epsilon)n$, such that for every $i \in S$, i received an input value v_i satisfying $\mathsf{score}_i(v_i) \geq B$. Then, if any so-far honest node outputs $v^* \in \mathcal{U}$, then there must exist an initially honest node i^* such that $\mathsf{score}_{i^*}(v^*) \geq B - d$.
  In other words, if sufficiently many forever honest nodes receive a high-scoring input value and some honest node outputs v, then it cannot be that all honest nodes assign v a relatively low initial score.
- *T_{end}-Liveness.* Every forever honest node terminates and outputs a value in round T_{end}.

4.2   Message-Specific Random Eligibility Election

To achieve small communication bandwidth, we use a technique proposed in Abraham et al. [2] for vote-specific, random eligibility election. A node with the identifier i should only send a message m if it is determined to be eligible for sending m—otherwise the message m will be discarded by so-far honest nodes.

Random eligibility election with cryptography. Imprecisely speaking such random eligibility election is performed with the help of a Verifiable Random Function (VRF) [17]: assume that every player i has a VRF public key denoted pk_i that is common knowledge, and the corresponding VRF secret key sk_i is known only to player i. For i to determine its eligibility for sending m, it evaluates $(\mu, \pi) := \mathsf{VRF}(\mathsf{sk}_i, \mathsf{m})$ where μ is the VRF evaluation outcome and π is a proof attesting to the evaluation outcome. If $\mu < D_p$ where D_p is an appropriate difficulty parameter, node i is deemed eligible for sending the message m. While only the secret-key owner can evaluate the VRF, anyone can verify the evaluation outcome. More specifically, suppose that node i additionally attaches the pair (μ, π) when sending the message m; then, any node that receives the tuple (m, μ, π) can verify with pk_i that indeed μ is the correct VRF evaluation outcome and verify i's eligibility for m.

Now for technical reasons we will, for the time being, assume that such a VRF exists and moreover can resist adaptive attacks: specifically, even when the adversary can selectively open the secret keys of a subset of the honest nodes, the remaining honest nodes' VRFs will still give pseudo-random evaluation outcomes. Later in Sect. 7, we will describe how to instantiate such an adaptively-secure VRF that satisfies our needs, using techniques from Abraham et al. [2].

Remark 4.1 (Subtleties regarding the use of VRF). Although earlier works such as Algorand [6] and others [8,13,23] also rely on a VRF; they do not use the *vote-specific* election technique; and this is why these earlier works must rely on erasures to achieve adaptive security. Abraham et al. [2] relies on vote-specificity to remove the erasure assumption, but their technique works only for agreement on *a single bit* as explained in Sect. 2.3 and the Supplemental Materials. Finally, although not explicitly noted, Algorand [6] and other prior works [8] also require that the VRF be adaptively secure (i.e., honest VRF evaluations must remain pseudorandom even when the adversary can selectively open honest nodes' keys)—these earlier works rely on a random oracle to achieve such adaptive security. In our work, we instantiate such an adaptively secure VRF without relying on random oracles.

Random eligibility election in an idealized model. Henceforth for simplicity, in our protocol description we will abstract away the cryptographic details and instead assume that an idealized oracle $\mathcal{F}_{\mathrm{mine}}$ exists that takes care of eligibility election—but later in Sect. 7, we will explain how to instantiate $\mathcal{F}_{\mathrm{mine}}$ with adaptively secure cryptographic primitives. Specifically, $\mathcal{F}_{\mathrm{mine}}$ is a trusted (i.e., incorruptible) party that performs the following—we assume that $\mathcal{F}_{\mathrm{mine}}$ has been parametrized with an appropriate probability p:

1. Upon receiving mine(m) from node i, if the coin $\mathsf{Coin}[m, i]$ has not been recorded, flip a random coin b that is 1 with probability p and is 0 with probability $1 - p$. Record $\mathsf{Coin}[m, i] := b$ and return $\mathsf{Coin}[m, i]$
2. Upon receiving verify(m, i) from any node, if the coin $\mathsf{Coin}[m, i]$ has been recorded, return its value; else return 0.

Basically, for node i to check its eligibility for the message m, it calls $\mathcal{F}_{\mathrm{mine}}$.mine(m)—henceforth for simplicity we also call this act "*mining a vote for* m".

4.3 Herding-Based Scoring Agreement Protocol

The protocol Π_{score} is parametrized with some universe $\mathcal{U}$. We describe the protocol in the $\mathcal{F}_{\mathrm{mine}}$-hybrid world, and later in Sect. 7 we show how to remove the $\mathcal{F}_{\mathrm{mine}}$ idealized assumption.

1. *Parameters.* Recall that the adversary controls $\frac{1}{3} - \epsilon$ fraction. Let λ be large enough such that $\epsilon^2 \lambda = \omega(\log \kappa)$; e.g., if ϵ is a(n arbitrarily small) constant then λ can be any super-logarithmic function.
   The mining difficulty parameter is set such that if all nodes were honest, on average exactly 1 vote (among all nodes) would be successfully mined every $\lambda \Delta$ number of rounds. In other words, each mining attempt is successful with probability $\frac{1}{\lambda \Delta n}$ where n is the total number of nodes.
2. *Mining.* In each round t, for every value $v \in \mathcal{U}$ node i has observed so-far, node i computes its *popularity* by adding v's initial score and the number of valid votes seen so far for v. Next, node i picks the most popular value $v \in \mathcal{U}$ that has been observed (breaking ties arbitrarily). The node i then contacts $\mathcal{F}_{\mathrm{mine}}$.mine($v, t$) to mine a vote for the message (v, t)—if successful, node i multicasts (v, t, i) as well as all valid votes that it has observed so far for the value v.
3. *Vote validity.* A node can verify the validity of a received vote (v, t, i), by calling $\mathcal{F}_{\mathrm{mine}}$.verify($v, t, i$). If a vote (v, t, i) is received where t is greater than the node's round number, discard the vote.
4. *Terminate.* Every node runs the protocol for $T_{\mathrm{end}} = \lambda^2 \Delta$ number of rounds, at the end of which the node attempts to output a value based on the following rule: the node has observed at least $\frac{2\lambda}{3}$ valid votes for any value $v \in \mathcal{U}$, then output v; else output nothing.

4.4 Theorem Statements for Scoring Agreement

We summarize this section with the following theorem statements, the proofs of which are deferred to Sect. 8.

Theorem 4.2 (Security of scoring agreement). *Assume that* $\epsilon^2 \lambda = \omega(\log \kappa)$. *The above* $\mathcal{F}_{\mathrm{mine}}$-*hybrid scoring agreement protocol satisfies consistency,* $\frac{2\lambda}{3}$-*validity, and* $\lambda^2 \Delta$-*liveness against any* $(\frac{1}{3} - \epsilon, \Delta)$-*respecting, nonuniform p.p.t.* $(\mathcal{A}, \mathcal{Z})$ *that satisfies the constraints[5] specified in Sect. 4.1.*

---

[5] See the "Syntax" and "Constraints on $\mathcal{Z}$" paragraphs.

When the choice of λ is polylogarithmic in κ, the protocol achieves poly-logarithmic multicast communication complexity, i.e., only polylogarithmically many honest messages are multicast regardless of how $(\mathcal{A}, \mathcal{Z})$ behaves.

Theorem 4.3 (Communication efficiency of scoring agreement). *Suppose that* $\log^{1.1} \kappa \leq \lambda \leq \log^2 \kappa$ *and that* n *is polynomial in* κ. *Then, for any* $(\mathcal{A}, \mathcal{Z})$, *there is a negligible function* $\mathsf{negl}(\cdot)$ *such that except with* $\mathsf{negl}(\kappa)$ *probability over the choice of* $\mathsf{EXEC}^{\Pi_{score}}(\mathcal{A}, \mathcal{Z}, \kappa)$ *where* Π_{score} *denotes the above* $\mathcal{F}_{\mathsf{mine}}$-*hybrid scoring agreement protocol, honest nodes multicast no more than* $\log^3 \kappa \cdot \Theta(\ell + \log \kappa)$ *bits of messages where* ℓ *is the number of bits for encoding each element in* $\mathcal{U}$.

5 Batch Agreement

In this section, we first define a new abstraction called *batch agreement*, a primitive that allows nodes to agree on a batch of transactions, such that long-pending transactions (for some notion of long-pending) must be included in the output batch. Our state machine replication protocol will simply sequentially compose multiple instances of batch agreement to agree on batches of transactions over time (see Sect. 6).

We show that one can construct batch agreement from scoring agreement by choosing an appropriate scoring function that severely discounts batches that omit sufficiently old transactions.

5.1 Formal Definition of Batch Agreement

Syntax. Suppose that nodes receive transactions as input from the environment $\mathcal{Z}$ over time. We assume that $\mathcal{Z}$ respects the following transaction diffusion assumption with probability 1—later in the Supplemental Materials we shall describe how to remove this assumption while preserving communication efficiency:

> *Transaction diffusion assumption:* If some forever honest node receives a transaction tx as input in some round t, then all so-far honest nodes must have received tx as input by the end of round $t + \Delta$.

Remark 5.1 (About the transaction diffusion assumption). One way to remove this assumption is to have every node echo the tx upon first seeing it—in real-world peer-to-peer networks such as those adopted by Bitcoin or Ethereum, everyone echoing the same message should charge only once to the communication cost. Later in the Supplemental Materials we discuss how to remove this assumption for synchronous networks requiring only a small number of so-far honest nodes to echo each tx.

The environment $\mathcal{Z}$ starts all nodes in the same round denoted r_{start}. When starting an initially honest node i, $\mathcal{Z}$ informs node i a set of transactions that are already confirmed—the same set of confirmed transactions, henceforth denoted $TXs_{confirmed}$ must be provided to all honest nodes.

At the end of the batch agreement protocol, every forever honest node must have output a batch of transactions.

Security properties. We require the following security properties. Specifically, there is a negligible function $\mathsf{negl}(\cdot)$ such that for all but $\mathsf{negl}(\kappa)$ fraction of execution traces, the following properties must hold:

- *Consistency.* If a so-far honest node outputs TXs and another so-far honest node outputs TXs' in the batch agreement protocol, it must be that TXs = TXs'.
- T_{end}-*Liveness.* Let $T_{end} = \mathsf{poly}(\kappa, n, \Delta)$ be a polynomial function in κ, n, and Δ. Every node that remains honest in round $r_{start} + T_{end}$ must have output a batch by round $r_{start} + T_{end}$.
- D-*Validity.* If $D \le r_{start}$ and some forever honest node has observed a transaction $\mathsf{tx} \notin TXs_{confirmed}$ by round $r_{start} - D$ where r_{start} is the start of the batch agreement protocol, then tx must appear in any forever honest node's output batch.

5.2   Batch Agreement from Scoring Agreement

Intuition. It is easy to construct a batch agreement protocol from scoring agreement in the synchronous setting. The idea is to rely on a scoring function such that a batch would receive a significant penalty if long-pending transactions were excluded. To obtain liveness, we also need that initially honest nodes assign somewhat consistent initial scores to every batch. This is guaranteed by leveraging transaction diffusion assumption: a fresh transaction is propagated to all nodes at most Δ apart. This means that so-far honest nodes have a somewhat consistent view of any transaction's age. We design our scoring function to make sure that if the transaction diffusion assumption holds, then all honest nodes would assign somewhat consistent initial scores to every batch. Finally, we also need that initially honest nodes receive high-scoring inputs — this is also guaranteed because an honest node always tries to include all pending transactions observed so far in its input batch.

Detailed protocol. We now describe the batch agreement protocol which is build from a scoring agreement instance denoted Π_{score}.

- *Input.* Start a scoring agreement instance denoted Π_{score}, and choose the input to Π_{score} as follows: let TXs be the set of outstanding transactions in the node's view so far. Input $TXs \backslash TXs_{confirmed}$ to Π_{score}.
- *Initial scoring function.* Given a set of transactions TXs, its initial score is computed as the following by node i. Let $\mathsf{tx} \notin TXs \cup TXs_{confirmed}$ be the earliest transaction (not in $TXs \cup TXs_{confirmed}$) which node i has observed so

far, and suppose that node i observed tx in round t—if there is no such tx, we simply let $t := r_{\text{start}}$. Then, the initial score of TXs is computed as:

$$\text{score}_i(\text{TXs}) := 3\lambda \cdot \left(1 - \frac{1}{3\lambda^2}\right)^{\left\lfloor \frac{r_{\text{start}} - t}{3\Delta} \right\rfloor}$$

– *Output.* Now execute the scoring agreement protocol Π_{score} for T_{end} number of rounds, and output whatever it outputs.

Theorem 5.2 (Synchronous batch agreement). *Suppose that $\lambda(\kappa) > 0.5$ for sufficiently large κ. For any $0 < \rho < 1$, any Δ, suppose that $(\mathcal{A}, \mathcal{Z})$ is non-uniform p.p.t. and (ρ, Δ)-respecting and also respects the assumptions stated in Sect. 5.1. Assume that the scoring agreement protocol employed in the above batch agreement construction satisfies consistency, λ-validity, and T_{end}-liveness against $(\mathcal{A}, \mathcal{Z})$. Then the above batch agreement protocol in the $\mathcal{F}_{\text{mine}}$-hybrid world achieves consistency, T_{end}-liveness, and $\Theta(\lambda\Delta)$-validity against $(\mathcal{A}, \mathcal{Z})$.*

Proof. Consistency follows directly from the consistency of the scoring agreement. T_{end}-liveness of the batch agreement would follow if the scoring agreement also satisfies T_{end}-liveness—to show the latter, observe that

1. Due to the transaction diffusion assumption, for any valid batch TXs any two initially honest nodes' scores are at most $\frac{1}{\lambda}$ apart; and
2. All initially honest nodes score their own input 3λ.

It remains to prove $\Theta(\lambda\Delta)$-validity. If some forever honest node has observed a transaction tx $\in$ TXs$_{\text{confirmed}}$ by round $r_{\text{start}} - c\lambda\Delta \geq 0$ for some appropriate constant c, then for any initially honest node i, it must have observed tx by round $t = r_{\text{start}} - c\lambda\Delta + \Delta$, by the transaction diffusion assumption; for any batch TXs that does not contain tx, we have that

$$\text{score}_i(\text{TXs}) = 3\lambda \cdot \left(1 - \frac{1}{3\lambda}\right)^{\left\lfloor \frac{r_{\text{start}} - t}{3\Delta} \right\rfloor} \leq 3\lambda \cdot \left(1 - \frac{1}{3\lambda}\right)^{\frac{c\lambda\Delta - \Delta}{3\Delta}}$$

For an appropriate constant $c = 20$ we have that $\left(1 - \frac{1}{3\lambda}\right)^{(c\lambda - 1)/3} \leq 0.5$ for any $\lambda > 0.5$; therefore $\text{score}_i(\text{TXs}) \leq 1.5\lambda$. Recall that every initially honest node will score its own input value 3λ. Thus $c\lambda\Delta$-validity of the batch agreement follows from the λ-validity of the scoring agreement instance.

Communication efficiency. Suppose that the above synchronous batch agreement adopts the scoring agreement protocol devised in Sect. 4; and further, assume that $\log^{1.1} \kappa \leq \lambda \leq \log^2 \kappa$. Then, due to Theorem 4.3, it is not difficult to see that regardless of $(\mathcal{A}, \mathcal{Z})$'s behavior, except with negligible in κ probability, forever honest nodes multicast no more than $\log^3 \kappa \cdot \Theta(|\text{TX}_{\text{active}}| + \log \kappa)$ bits of messages in the synchronous batch agreement protocol where TX$_{\text{active}} :=$ TXs$_{\text{all}} \setminus$ TXs$_{\text{confirmed}}$ denotes the set of all transactions each of which observed by at least one so-far honest node by the end of the batch agreement protocol (denoted $\mathsf{TXs_{all}}$), subtracting those that were already confirmed prior to the start of the batch agreement instance (denoted $\mathsf{TXs_{confirmed}}$). Recall that the environment $\mathcal{Z}$ informs nodes of the $\mathsf{TXs_{confirmed}}$ set prior to starting a batch agreement instance.

6   SMR from Batch Agreement

6.1   Definition of State Machine Replication

State machine replication has been a central abstraction in the 30 years of distributed systems literature. In a state machine replication protocol, a set of nodes seek to agree on an ever-growing log over time. We require two critical security properties: (1) *consistency*, i.e., all forever honest nodes' logs agree with each other although some nodes may progress faster than others; (2) *liveness*, i.e., transactions received by initially honest nodes as input get confirmed in all forever honest nodes' logs quickly. We now define what it formally means for a protocol to realize a "state machine replication" abstraction.

Syntax. In a state machine replication protocol, in every round, a node receives as input a set of transactions txs from $\mathcal{Z}$ at the beginning of the round, and outputs a LOG collected thus far to $\mathcal{Z}$ at the end of the round. As before, we assume that $\mathcal{Z}$ respects the *transaction diffusion assumption* with probability 1. In other words, two so-far honest nodes observe any transaction tx within Δ rounds apart.

Security. Let $T_{\mathrm{confirm}}(\kappa, n, \Delta)$ be a polynomial function in the security parameter κ, the number of nodes n, and the maximum network delay Δ.

Definition 6.1. *We say that a state machine replication protocol Π satisfies consistency (or T_{confirm}-liveness resp.) w.r.t. some $(\mathcal{A}, \mathcal{Z})$, iff there exists a negligible function $\mathsf{negl}(\cdot)$, such that for any $\kappa \in \mathbb{N}$, except with $\mathsf{negl}(\kappa)$ probability over the choice of view $\leftarrow$ EXEC$^{\Pi}(\mathcal{A}, \mathcal{Z}, \kappa)$, consistency (or T_{confirm}-liveness resp.) is satisfied:*

- *Consistency: A view satisfies consistency iff the following holds:*
  - *Common prefix. Suppose that in view, a so-far honest node i outputs LOG to $\mathcal{Z}$ in round t, and a so-far honest node j outputs LOG$'$ to $\mathcal{Z}$ in round t' (i and j may be the same or different), it holds that either LOG $\preceq$ LOG$'$ or LOG$'$ $\preceq$ LOG. Here the relation $\preceq$ means "is a prefix of". By convention we assume that $\emptyset \preceq x$ and $x \preceq x$ for any x.*
  - *Self-consistency. Suppose that in view, a node i is honest during rounds $[t, t']$, and outputs LOG and LOG$'$ in rounds t and t' respectively, it holds that LOG $\preceq$ LOG$'$.*
- *Liveness: A view satisfies $T_{confirm}$-liveness iff the following holds: if in some round $t \le |\mathsf{view}| - T_{\mathrm{confirm}}$, some forever honest node either received from $\mathcal{Z}$ an input set txs that contains some transaction tx or has tx in its output log to $\mathcal{Z}$ in round t, then, for any node i honest in any round $t' \ge t + T_{confirm}$, let LOG be the output of node i in round t', it holds that tx $\in$ LOG.*

Intuitively, liveness says that transactions input to an initially honest node get included in forever honest nodes' LOGs within $T_{confirm}$ time; and further, if a transaction appears in some forever honest node's LOG, it will appear in every forever honest node's LOG within $T_{confirm}$ time.

6.2   Constructing State Machine Replication from Batch Agreement

It is relatively straightforward how to construct state machine replication from batch agreement: basically, all nodes start a batch agreement instance in round 0 henceforth denoted $\Pi_{batch}[0]$; as soon as the i-th batch agreement instance $\Pi_{batch}[i]$ outputs a batch, a node immediately starts a next batch agreement instance denoted $\Pi_{batch}[i+1]$. At any time, a node outputs a sequential concatenation of all batches that have been output so far by batch agreement instances[6]. For every instance of batch agreement, the confirmed set $TXs_{confirmed}$ provided as input consists of all transactions that have been output by previous instances (i.e., these transactions need not be confirmed again).

Theorem 6.2 (State machine replication). *Let Δ be any polynomial function in κ and n and let $0 < \epsilon < 1$ be any positive constant. Suppose that $(\mathcal{A}, \mathcal{Z})$ is $(\frac{1}{3} - \epsilon, \Delta)$-respecting and moreover respects the assumptions stated in Sect. 6.1. Assume that the batch agreement protocol adopted satisfies consistency, T-liveness, and D-validity w.r.t. $(\mathcal{A}, \mathcal{Z})$, then the above state machine replication protocol satisfies consistency and $(2T + D)$-liveness w.r.t. $(\mathcal{A}, \mathcal{Z})$.*

Proof. Consistency follows directly from the consistency of batch agreement. Moreover, $(2T + D)$-liveness follows from the fact that if $\mathcal{Z}$ inputs a tx to some forever honest node in round r, then consider the first batch agreement instance that is started (by some honest node) in round $r + D$ or after: it takes up to T time till the this batch agreement instance ends by T-liveness of the batch agreement, and the immediate next batch agreement instance will surely output tx if tx is not output earlier by D-validity of the batch agreement.

Communication efficiency. For analyzing communication efficiency, let us assume that we adopt the batch agreement protocol described in Sect. 5; further, assume that ϵ is an arbitrarily small constant and that $\lambda = \log^{1.1} \kappa$. For some transaction tx, suppose that round r is the first round in which some forever honest node observes tx. Then, starting from round r, the transaction tx will be confirmed after poly $\log \kappa$ number of batch agreement instances, and thus it will contribute to the TX_{active} set of poly $\log \kappa$ number of such instances. Recall that

---

[6] The state machine replication protocol above invokes many instances of batch agreement which may then invoke one or more instances of scoring agreement. Recall that each scoring agreement instance calls $\mathcal{F}_{mine}$. For composition, calls to $\mathcal{F}_{mine}$ are tagged with an instance identifier. Here the instance identifier contains a pair: first the identifier of the batch agreement instance and then the identifier of the scoring agreement.

for each batch agreement instance, except with negligible in κ probability, only $\log^3 \kappa \cdot \Theta(|\mathsf{TX}_{\mathrm{active}}| + \log \kappa)$ bits of honest messages are multicast. Thus, the bits of honest messages multicast, amortized to each tx, is bounded by $|\mathsf{tx}| \cdot \mathsf{poly} \log \kappa$ for some suitable polynomial $\mathsf{poly}(\cdot)$ except with negligible in κ probability.

7   Removing the Idealized Functionality $\mathcal{F}_{\mathrm{mine}}$

So far, all our protocols have assumed the existence of an $\mathcal{F}_{\mathrm{mine}}$ ideal functionality. In this section, we describe how to instantiate the protocols in the real world. Our techniques follow the approach described by Abraham et al. [2]. Although this part is not a contribution of our paper, for completeness, we describe all the building blocks and the approach in a self-contained manner, borrowing some text from Abraham et al. [2].

7.1   Preliminary: Adaptively Secure Non-Interactive Zero-Knowledge Proofs

We use $f(k) \approx g(k)$ to mean that there exists a negligible function $\nu(\kappa)$ such that $|f(\kappa) - g(\kappa)| < \nu(\kappa)$.

A non-interactive proof system henceforth denoted nizk for an NP language $\mathcal{L}$ consists of the following algorithms.

- $\mathsf{crs} \leftarrow \mathsf{Gen}(1^\kappa, \mathcal{L})$: Takes in a security parameter κ, a description of the language $\mathcal{L}$, and generates a common reference string crs.
- $\pi \leftarrow \mathsf{P}(\mathsf{crs}, \mathsf{stmt}, w)$: Takes in crs, a statement stmt, a witness w such that $(\mathsf{stmt}, w) \in \mathcal{L}$, and produces a proof π.
- $b \leftarrow \mathsf{V}(\mathsf{crs}, \mathsf{stmt}, \pi)$: Takes in a crs, a statement stmt, and a proof π, and outputs 0 (reject) or 1 (accept).

Perfect completeness. A non-interactive proof system is said to be perfectly complete, if an honest prover with a valid witness can always convince an honest verifier. More formally, for any $(\mathsf{stmt}, w) \in \mathcal{L}$, we have that

$$\Pr\left[\mathsf{crs} \leftarrow \mathsf{Gen}(1^\kappa, \mathcal{L}), \ \pi \leftarrow \mathsf{P}(\mathsf{crs}, \mathsf{stmt}, w) : \mathsf{V}(\mathsf{crs}, \mathsf{stmt}, \pi) = 1\right] = 1$$

Non-erasure computational zero-knowledge. Non-erasure zero-knowledge requires that under a simulated CRS, there is a simulated prover that can produce proofs without needing the witness. Further, upon obtaining a valid witness to a statement a-posteriori, the simulated prover can explain the simulated NIZK with the correct witness.

We say that a proof system $(\mathsf{gen}, \mathsf{P}, \mathsf{V})$ satisfies non-erasure computational zero-knowledge iff there exists a probabilistic polynomial time algorithms $(\mathsf{gen}_0, \mathsf{P}_0, \mathsf{Explain})$ such that

$$\Pr\left[\mathsf{crs} \leftarrow \mathsf{gen}(1^\kappa), \mathcal{A}^{\mathsf{Real}(\mathsf{crs}, \cdot, \cdot)}(\mathsf{crs}) = 1\right] \approx \Pr\left[(\mathsf{crs}_0, \tau_0) \leftarrow \mathsf{gen}_0(1^\kappa), \mathcal{A}^{\mathsf{Ideal}(\mathsf{crs}_0, \tau_0, \cdot, \cdot)}(\mathsf{crs}_0) = 1\right],$$

where $\mathsf{Real}(\mathsf{crs}, \mathsf{stmt}, w)$ runs the honest prover $\mathsf{P}(\mathsf{crs}, \mathsf{stmt}, w)$ with randomness r and obtains the proof π, it then outputs (π, r); $\mathsf{Ideal}(\mathsf{crs}_0, \tau_0, \mathsf{stmt}, w)$ runs the simulated prover $\pi \leftarrow \mathsf{P}_0(\mathsf{crs}_0, \tau_0, \mathsf{stmt}, \varrho)$ with randomness ϱ and without a witness, and then runs $r \leftarrow \mathsf{Explain}(\mathsf{crs}_0, \tau_0, \mathsf{stmt}, w, \varrho)$ and outputs (π, r).

Perfect knowledge extraction. We say that a proof system $(\mathsf{gen}, \mathsf{P}, \mathsf{V})$ satisfies perfect knowledge extraction, if there exists probabilistic polynomial-time algorithms $(\mathsf{gen}_1, \mathsf{Extr})$, such that for all (even unbounded) adversary $\mathcal{A}$,

$$\Pr\left[\mathsf{crs} \leftarrow \mathsf{gen}(1^\kappa) : \mathcal{A}(\mathsf{crs}) = 1\right] = \Pr\left[(\mathsf{crs}_1, \tau_1) \leftarrow \mathsf{gen}_1(1^\kappa) : \mathcal{A}(\mathsf{crs}_1) = 1\right],$$

and moreover,

$$\Pr\left[\begin{array}{l} (\mathsf{crs}_1, \tau_1) \leftarrow \mathsf{gen}_1(1^\kappa); (\mathsf{stmt}, \pi) \leftarrow \mathcal{A}(\mathsf{crs}_1); \\ w \leftarrow \mathsf{Extr}(\mathsf{crs}_1, \tau_1, \mathsf{stmt}, \pi) \end{array} : \begin{array}{l} \mathsf{V}(\mathsf{crs}_1, \mathsf{stmt}, \pi) = 1 \\ \text{but } (\mathsf{stmt}, w) \notin \mathcal{L} \end{array}\right] = 0$$

7.2   Adaptively Secure Non-Interactive Commitment Scheme

An adaptively secure non-interactive commitment scheme consists of the following algorithms:

- $\mathsf{crs} \leftarrow \mathsf{Gen}(1^\kappa)$: Takes in a security parameter κ, and generates a common reference string crs.
- $C \leftarrow \mathsf{com}(\mathsf{crs}, v, \varrho)$: Takes in crs, a value v, and a random string ϱ, and outputs a committed value C.
- $b \leftarrow \mathsf{ver}(\mathsf{crs}, C, v, \varrho)$: Takes in a crs, a commitment C, a purported opening (v, ϱ), and outputs 0 (reject) or 1 (accept).

Computationally hiding under selective opening. We say that a commitment scheme $(\mathsf{gen}, \mathsf{com}, \mathsf{ver})$ is computationally hiding under selective opening, iff there exists a probabilistic polynomial time algorithms $(\mathsf{gen}_0, \mathsf{com}_0, \mathsf{Explain})$ such that

$$\Pr\left[\mathsf{crs} \leftarrow \mathsf{gen}(1^\kappa), \mathcal{A}^{\mathsf{Real}(\mathsf{crs}, \cdot)}(\mathsf{crs}) = 1\right] \approx \Pr\left[(\mathsf{crs}_0, \tau_0) \leftarrow \mathsf{gen}_0(1^\kappa), \mathcal{A}^{\mathsf{Ideal}(\mathsf{crs}_0, \tau_0, \cdot)}(\mathsf{crs}_0) = 1\right]$$

where $\mathsf{Real}(\mathsf{crs}, v)$ runs the honest algorithm $\mathsf{com}(\mathsf{crs}, v, r)$ with randomness r and obtains the commitment C, it then outputs (C, r); $\mathsf{Ideal}(\mathsf{crs}_0, \tau_0, v)$ runs the simulated algorithm $C \leftarrow \mathsf{com}_0(\mathsf{crs}_0, \tau_0, \varrho)$ with randomness ϱ and without v, and then runs $r \leftarrow \mathsf{Explain}(\mathsf{crs}_0, \tau_0, v, \varrho)$ and outputs (C, r).

Perfectly binding. A commitment scheme is said to be perfectly binding iff for every crs in the support of the honest CRS generation algorithm, there does not exist $(v, \varrho) \neq (v', \varrho')$ such that $\mathsf{com}(\mathsf{crs}, v, \varrho) = \mathsf{com}(\mathsf{crs}, v', \varrho')$.

Theorem 7.1 (Instantiation of our NIZK and commitment schemes [12]). *Assume standard bilinear group assumptions. Then, there exists a proof system that satisfies perfect completeness, non-erasure computational zero-knowledge, and perfect knowledge extraction. Further, there exist a commitment scheme that is perfectly binding and computationally hiding under selective opening.*

Proof. The existence of such a NIZK scheme was shown by Groth et al. [12] via a building block that they called *homomorphic proof commitment scheme*. This building block can also be used to achieve a commitment scheme with the desired properties.

7.3   NP Language Used in Our Construction

In our construction, we will use the following NP language $\mathcal{L}$. A pair $(\mathsf{stmt}, w) \in \mathcal{L}$ iff

- parse $\mathsf{stmt} := (\mu, c, \mathsf{crs}_{\mathrm{comm}}, \mathsf{m})$, parse $w := (\mathsf{sk}, s)$;
- it must hold that $c = \mathsf{comm}(\mathsf{crs}_{\mathrm{comm}}, \mathsf{sk}, s)$, and $\mathsf{PRF}_{\mathsf{sk}}(\mathsf{m}) = \mu$.

7.4   Compilation to Real-World Protocols

We can remove the $\mathcal{F}_{\mathrm{mine}}$ oracle by leveraging cryptographic building blocks including a pseudorandom function family, a non-interactive zero-knowledge proof system that satisfies computational zero-knowledge and computational soundness, a perfectly correct and semantically secure public-key encryption scheme, and a perfectly binding and computationally hiding commitment scheme.

Earlier in Sect. 4, we informally have described the intuition behind our approach. In this section we provide a formal description of how to compile our $\mathcal{F}_{\mathrm{mine}}$-hybrid protocols into real-world protocols using cryptography. Using this compilation technique, we can compile our $\mathcal{F}_{\mathrm{mine}}$-hybrid state machine replication protocol to the real world. Our techniques are essentially the same as Abraham et al. [2], but we describe it in full for completeness.

- **Trusted PKI setup.** Upfront, a trusted party runs the CRS generation algorithms of the commitment and the NIZK scheme to obtain $\mathsf{crs}_{\mathrm{comm}}$ and $\mathsf{crs}_{\mathrm{nizk}}$. It then chooses a secret PRF key for every node, where the i-th node has key sk_i. It publishes $(\mathsf{crs}_{\mathrm{comm}}, \mathsf{crs}_{\mathrm{nizk}})$ as the public parameters, and each node i's public key denoted pk_i is computed as a commitment of sk_i using a random string s_i. The collection of all users' public keys is published to form the PKI, i.e., the mapping from each node i to its public key pk_i is public information. Further, each node i is given the secret key (sk_i, s_i). Remark 7.2 later mentions how multiple protocol instances can share the same PKI.
- **Instantiating $\mathcal{F}_{\mathrm{mine}}$.mine.** Recall that in the ideal-world protocol a node i calls $\mathcal{F}_{\mathrm{mine}}$.$\mathsf{mine}(\mathsf{m})$ to mine a vote for a message m. Now, instead, the node i calls $\mu := \mathsf{PRF}_{\mathsf{sk}_i}(\mathsf{m})$, and computes the NIZK proof

$$\pi := \mathsf{nizkP}((\mu, \mathsf{pk}_i, \mathsf{crs}_{\mathrm{comm}}, \mathsf{m}), (\mathsf{sk}_i, s_i))$$

where s_i the randomness used in committing sk_i during the trusted setup. Intuitively, this zero-knowledge proof proves that the evaluation outcome μ is correct w.r.t. the node's public key (which is a commitment of its secret key).

The mining attempt for m is considered successful if $\mu < D_p$ where D_p is an appropriate difficulty parameter such that any random string of appropriate length is less than D_p with probability p—the probability p is selected in the same way as the earlier $\mathcal{F}_{\text{mine}}$-hybrid world protocols.

Recall that earlier in our $\mathcal{F}_{\text{mine}}$-hybrid protocols, every message multicast by a so-far honest node i is a mined message of the form $(\mathsf{m} : i)$ where node i has successfully called $\mathcal{F}_{\text{mine}}.\mathtt{mine}(\mathsf{m})$. Each such *mined* message (m, i) that node i wants to multicast is translated to the real-world protocol as follows: we rewrite $(\mathsf{m} : i)$ as $(\mathsf{m}, i, \mu, \pi)$ where the terms μ and π are those generated by i in place of calling $\mathcal{F}_{\text{mine}}.\mathtt{mine}(\mathsf{m})$ in the real world (as explained above). Note that in our $\mathcal{F}_{\text{mine}}$-hybrid protocols a node $j \neq i$ may also relay a message $(\mathsf{m} : i)$ mined by i—in the real world, node j would be relaying $(\mathsf{m}, i, \mu, \pi)$ instead.

- **Instantiating** $\mathcal{F}_{\text{mine}}.\mathtt{verify}$. In the ideal world, a node would call $\mathcal{F}_{\text{mine}}.\mathtt{verify}$ to check the validity of mined messages upon receiving them, In the real-world protocol, we perform the following instead: upon receiving the mined message $(\mathsf{m}, i, \mu, \pi)$, a node can verify the message's validity by checking:
  1. $\mu < D_p$ where p is an appropriate difficulty parameter that depends on the type of the mined message; and
  2. π is indeed a valid NIZK for the statement formed by the tuple $(\mu, \mathsf{pk}_i, \mathsf{crs}_{\text{comm}}, \mathsf{m})$. The tuple is discarded unless both checks pass.

Remark 7.2 (Protocol composition in the real world). The real-world protocol may invoke multiple instances of scoring agreement each with a unique instance identifier. In the real world, all instances share the same PKI. Recall that in the $\mathcal{F}_{\text{mine}}$-hybrid world, every call to $\mathcal{F}_{\text{mine}}$ is prefixed with the instance identifier. Specifically, in the calls $\mathcal{F}_{\text{mine}}.\mathtt{mine}(\mathsf{m})$ and $\mathcal{F}_{\text{mine}}.\mathtt{verify}(\mathsf{m}, i)$, one can imagine that the m part is tagged with the instance identifier. In the real world, this means that the message m passed to the PRF and the NIZK's prover and verifier is prefixed with the instance identifier too.

Now using the same proofs as Abraham et al. [2], we can prove that the compiled real-world protocols enjoy the same security properties as the $\mathcal{F}_{\text{mine}}$-hybrid protocols. Since the proofs follow identically, we omit the details and simply refer the reader to Abraham et al. [2]. We thus obtain the following theorem, by observing that in the real world protocol, each vote is of the form $(\mathsf{TXs}, r, i, \mu, \pi)$ where μ and π has length χ where χ is a cryptographic security parameter.

Theorem 7.3 (Real-world protocol: synchronous state machine replication). *Under standard cryptographic hardness assumptions (more precisely, the existence of universally composable, adaptively secure non-interactive zero-knowledge and commitments [12]), there exists a synchronous state machine replication protocol that satisfies consistency and $\mathsf{poly}\log \kappa \cdot \Delta$-liveness against any non-uniform p.p.t., $(\frac{1}{3} - \epsilon, \Delta)$-respecting $(\mathcal{A}, \mathcal{Z})$ where $\mathsf{poly}(\cdot)$ is a suitable polynomial function and ϵ is an arbitrarily small positive constant. Moreover,* *honest nodes only need to multicast* $\mathsf{poly}\log\kappa \cdot (\chi + |\mathsf{TXs}|)$ *bits of messages to confirm every batch of transactions denoted* TXs *where* χ *is a computational security parameter related to the strength of the cryptographic building blocks involved.*

Proof. Note our techniques for instantiating $\mathcal{F}_{\mathrm{mine}}$ with actual cryptography is borrowed from Abraham et al. [2]. Their proof for showing that the real-world protocol preserves the security properties proved in the ideal world is immediately applicable to our case.

8 Deferred Proofs for Scoring Agreement

In all of our proofs, we will by default assume that $(\mathcal{A}, \mathcal{Z})$ is non-uniform p.p.t., $(\frac{1}{3} - \epsilon, \Delta)$-respecting, and moreover, respects the assumptions stated in Sect. 4.1 (see the "Syntax" and "Constraints on $\mathcal{Z}$" paragraphs).

8.1 Consistency

We first prove that the protocol satisfies consistency.

Lemma 8.1 (Consistency). *Except with at most* $\mathsf{poly}(\kappa) \cdot \exp(-\Theta(\epsilon^2\lambda))$ *probability, no two so-far honest nodes can output different values in* $\mathcal{U}$.

Proof. We fix two values $v \neq v'$, and give an upper bound on the probability that both values are output by so-far honest nodes. Observe that this happens only if there are at least $\frac{2\lambda}{3}$ votes for each of the values, which means in total there are at least $\frac{4\lambda}{3}$ votes for either value.

We next consider how many mining attempts there can be for value v or v' among the $T := \lambda^2\Delta$ rounds. Observe that each forever honest node attempts to mine for at most one vote labeled with each round, where a (possibly adaptively) corrupted node can mine for both v and v' labeled with each round. Since the fraction of forever honest node is at least $\frac{2}{3} + \epsilon$, the total number of vote mining attempts for either v or v' is at most $(\frac{2}{3} + \epsilon) \cdot n \cdot T + (\frac{1}{3} - \epsilon) \cdot 2n \cdot T = (\frac{4}{3} - \epsilon) \cdot nT$.

Observing that the mining difficulty probability is $\frac{1}{\lambda\Delta n}$ and the mining events over different values, rounds and nodes are independent, by Chernoff Bound, the probability that there are at least $\frac{4\lambda}{3}$ votes for either v or v' is at most $\exp(-\Theta(\epsilon^2\lambda))$.

Independence Remark. Note that the outcome of a vote mining can affect what values an adaptively corrupted node will mine for. However, the important point is that the outcomes of the mining are independent, and there is a sure upper bound on the number of mining attempts, such that the Chernoff Bound can be applied above.

Since each node can perform only polynomial-time computation, and the entire transcript is polynomally bounded, it suffices to take union bound over $\mathsf{poly}(\kappa)$ number of unordered pairs of values to give the desired probability.

8.2 Validity

We next prove that the protocol satisfies d-validity for $d = \frac{2\lambda}{3}$. When an execution is fixed, we can define B to be the largest value such that there exists a subset S of forever-honest nodes of size at least $(\frac{1}{3} + 0.5\epsilon)n$, such that for every $i \in S$, i received an input value v_i satisfying $\mathsf{score}_i(v_i) \geq B$.

Lemma 8.2 (d-**Validity**). *Fix* $d = \frac{2\lambda}{3}$, *the following event happens with at most* $\mathsf{poly}(\kappa) \cdot \exp(-\Theta(\epsilon^2\lambda))$ *probability: there is some so-far honest node that outputs some value* $v^* \in \mathcal{U}$, *but for every initially honest node* j, *it holds that* $\mathsf{score}_j(v^*) < B - d$.

Proof. Fix some value v^* such that some so-far honest node outputs v^*. Suppose S is the set of $(\frac{1}{3} + 0.5\epsilon)n$ forever honest nodes in the hypothesis such that each $i \in S$ has some v_i such that $\mathsf{score}_i(v_i) \geq B$, but $\mathsf{score}_i(v^*) < B - d$. Then, we first show that the event in the lemma implies that there must be at least $d = \frac{2\lambda}{3}$ votes for v^* from nodes outside S.

Observe that for the first node i in S to vote for v^*, there must be at least d votes nodes outside S to compensate for the difference between $\mathsf{score}_i(v_i)$ and $\mathsf{score}_i(v^*)$. On the other hand, if there are no votes from S for v^*, then any so-far honest node that outputs v^* must see at least $\frac{2\lambda}{3} = d$ votes, which must all come from nodes outside S.

Recall that there are at most $(\frac{2}{3} - 0.5\epsilon)n$ nodes outside S. Within the $T = \lambda^2\Delta$ rounds, the mining difficulty for each node for v^* is $\frac{1}{\lambda\Delta n}$. Hence, the expected number of votes for v^* from nodes outside S is at most $(\frac{2}{3} - 0.5\epsilon)\lambda$. By Chernoff Bound, the probability that these nodes can produce at least $d = \frac{2\lambda}{3}$ votes for v^* is at most $\exp(-\Theta(\epsilon^2)\lambda)$.

Taking union bound over $\mathsf{poly}(\kappa)$ possible values contained in the polynomially sized transcript gives the result.

8.3 Liveness

For proving liveness, we assume that the "somewhat consistent initial score" condition is satisfied with the parameter $\vartheta = \frac{1}{\lambda}$ (see also Sect. 4.1).

Convergence Opportunity. We say that a round $t \geq \lambda\Delta$ is a *convergence opportunity*, if there is exactly one so-far honest node that successfully mines a vote in round t; moreover, no so-far honest nodes successfully mine votes within Δ rounds after t. No condition is placed on corrupted nodes.

The following lemma relates the number of convergence opportunities to the number of votes observed by a node for its most popular value.

Henceforth in Lemma 8.3 Corollary 8.4, when the execution we refer to is clear from the context, we use $\mathfrak{I}$ to denote the set of initially honest nodes, and let $M := \min_{j \in \mathfrak{I}} \mathsf{score}_i(v_j)$.

Lemma 8.3 (Convergence Opportunities and Votes). *Consider any execution: for each $r \geq 1$, after Δ rounds following the r-th convergence opportunity, any so-far honest node i has observed at least $M - \mathsf{score}_i(v) + r(1 - \vartheta)$ votes for its most popular value v.*

Proof. We show by induction on r. For $r = 0$, in order for a value v to be the most popular for node i, node i must have observed enough votes to compensate for the difference $\mathsf{score}_i(v_i) - \mathsf{score}_i(v) \geq M - \mathsf{score}_i(v)$.

Assume that claim is true for some $r \geq 0$. Suppose the $(r+1)$-st convergence opportunity due to a vote for value v by the so-far honest node i.

For node i, right before the $(r + 1)$-st convergence opportunity, at least Δ rounds must have passed since the r-th convergence opportunity. The induction hypothesis says that node i has observed at least $M - \mathsf{score}_i(v) + r(1 - \vartheta)$ votes for its most popular value v. Together with the new vote that node i has mined for the $(r + 1)$-st convergence opportunity, node i has observed at least $M - \mathsf{score}_i(v) + 1 + r(1 - \vartheta)$ votes for v. Observe that for any value v' to overtake v to be node i's most popular value, it will need at least $M - \mathsf{score}_i(v') + 1 + r(1 - \vartheta)$ votes; hence, the result holds for node i.

For any other so-far honest node j, after Δ rounds following the $(r + 1)$-st convergence opportunity, all the votes for v associated with the r-th convergence opportunity due to node i will have reached node j. Hence, if $\hat{v}$ is a most popular value for node j at this moment, it must be the case that its popularity implies that the number of votes for $\hat{v}$ observed by node j is at least:

$$(M - \mathsf{score}_i(v) + 1 + r(1 - \vartheta)) + \mathsf{score}_j(v) - \mathsf{score}_j(\hat{v}) \geq M - \mathsf{score}_j(\hat{v}) + (r + 1)(1 - \vartheta)$$

where the last inequality follows because $\mathsf{score}_j(v) - \mathsf{score}_i(v) \geq -\vartheta$. Similarly, for any other value v' to overtake $\hat{v}$ and become node j's most popular value, it must need at least $M - \mathsf{score}_j(v') + (r + 1)(1 - \vartheta)$ votes to be seen by node j. This completes the inductive step and the proof of the lemma.

Corollary 8.4 (Liveness). *Suppose* $\vartheta \leq \frac{1}{\lambda} \leq \Theta(\epsilon)$. *Except with at most* $\exp(-\Theta(\epsilon^2\lambda))$ *probability, every forever-honest honest will have output some value by the end of round* T_{end}.

Proof. For any forever-honest node i, from our "high initial scores" assumption that it will not see any value v' such that $\mathsf{score}_i(v') > \mathsf{score}_i(v_i)$, it follows that for any $u \in \mathcal{U}$, $M - \mathsf{score}_i(u) \geq \mathsf{score}_i(v_i) - \mathsf{score}_i(u) - \vartheta \geq -\vartheta$, where the inequality $M \geq \mathsf{score}_i(v_i) - \vartheta$ follows from the following:

$$M = \min_{j \in \mathcal{J}} \mathsf{score}_j(v_j) \geq \min_{j \in \mathcal{J}}(\mathsf{score}_i(v_j) - \vartheta) \geq \mathsf{score}_i(v_i) - \vartheta$$

where the first inequality follows from the "ϑ-somewhat-consistent initial scoring" assumption.

In view of Lemma 8.3, it suffices to show that except with at most $\exp(-\Theta(\epsilon^2\lambda))$ probability, the number of convergence opportunities is at least $\frac{2\lambda}{3}(1 + \Theta(\epsilon))$, which implies that for each node, its most popular value has received at least $\frac{2\lambda}{3}$ votes.

Observe that if h is the number of so-far honest nodes not in S, the probability that a round is a convergence opportunity is around $hp(1-p)^{h \cdot \Delta + h - 1} \geq \frac{2}{3\lambda\Delta}(1 + \Theta(\epsilon))(1 - \Theta(\frac{1}{\lambda}))$, where the inequality holds because $h \geq (\frac{2}{3} + \frac{1}{3}\epsilon)n$, and $p = \frac{1}{n\lambda\Delta}$

Hence, the expected number of convergence opportunities over the $T_{\text{end}} - \lambda\Delta$ rounds is at least $\frac{2\lambda}{3}(1 + \Theta(\epsilon))$.

Even though that the events of different rounds being convergence opportunities are not independent, Lemma 8.5 shows that a measure concentration result still holds. Setting the parameters $h = (\frac{2}{3} + \epsilon)n$, $H = n$, $p = \frac{1}{\lambda\Delta n}$, $T = \lambda^2\Delta$ in Lemma 8.5, we have that except with $\exp(-\Theta(\epsilon^2\lambda))$ probability, the number of convergence opportunities is at least $\frac{2\lambda}{3}(1 + \Theta(\epsilon))$.

Finally, during the last convergence opportunity by a so-far honest node, all the votes will be multicast to all nodes. Hence, all so-far honest nodes will see some value with at least $\frac{2\lambda}{3}$ votes after Δ more rounds.

Lemma 8.5 (Measure Concentration for Convergence Opportunities).
Suppose in each of $T + \Delta$ rounds, each of at least h but at most H so-far honest nodes mines a vote successfully with probability p.

Then, except with probability at most $\exp(-\Theta(\epsilon^2 T p_0))$, the number of rounds in which there is exactly one successful vote and followed by Δ rounds of no votes is at least $(1 + \Theta(\epsilon) - 2Hp\Delta)Tp_0$, where $p_0 := hp(1-p)^H$.

Proof. The proof is adapted from [24, Lemma 1].

Let Y be the number of rounds within $[1..T]$ in which there is exactly one successful vote. Since the probability that a round has exactly one vote is at least $p_0 := hp(1-p)^H$ and the events for different rounds are independent, by Chernoff Bound, except with probability $\exp(-\Theta(\epsilon^2 T p_0))$, the random variable Y is at least $(1 - \frac{\epsilon}{100})Tp_0$.

For $1 \leq i \leq R := \lceil(1 + \epsilon)Tp_0\rceil$, define the indicator random variable $Z_i \in \{0, 1\}$ that equals 1 *iff* after the i-th round that has exactly one successful vote, there is at least one successful vote within the next Δ rounds. By the union bound, $\Pr[Z_i = 1] \leq Hp\Delta$. Define $Z := \sum_{i=1}^{R} Z_i$.

Next, observe that the random variables Z_i's are independent. The reason is that right after the ith round in which there is exactly one successful vote, when the next successful vote happens will determine the value of Z_i, but the $i + 1$st round with exactly one successful vote will happen afterwards.

By Hoeffding's Inequality, we have that

$$\Pr[Z \geq Hp\Delta R + \frac{\epsilon}{100} \cdot R] \leq \exp(-\Theta(\epsilon^2 R))$$

By the union bound, except with probability at most $\exp(-\Theta(\epsilon^2 T p_0))$, $Y - Z \geq (1 - \frac{\epsilon}{10} - 2Hp\Delta)Tp_0$, which is also a lower bound on the number of convergence opportunities.

8.4   Communication Efficiency

We now prove the communication efficiency of our scoring agreement protocol described in Sect. 4.3, that is, Theorem 4.3.

Recall that $\log^{1.1}\kappa \leq \lambda \leq \log^2\kappa$. By Chernoff bound, for every value $v \in \mathcal{U}$, except with negligible in κ probability. at most 1.1λ votes (honest or adversarial) are successfully mined for the value v during the course of execution.

So-far honest nodes only multicast a message whenever it successfully mines a vote. When it multicasts, it not only multicasts the newly mined vote, but also all votes it has already observed for the relevant value in $\mathcal{U}$—recall that except with negligible in κ probability, there are at most 1.1λ such votes. Obviously, in the $\mathcal{F}_{\mathrm{mine}}$-hybrid world, every vote can be encoded with $\Theta(\log\kappa + \ell)$ bits since both n and the number of rounds are polynomial in κ.

Summarizing the above, except with negligible probability, the total number of honest votes multicast is upper bounded by $\log^3\kappa$ for sufficiently large κ, and thus the total number of bits multicast by so-far honest nodes is upper bounded by $\log^3\kappa \cdot \Theta(\log\kappa + \ell)$.

9 Conclusion and Open Questions

In this paper, we proposed a novel paradigm for reaching consensus that is inspired by a social phenomenon called herding. Through this novel paradigm, we construct a state machine replication protocol that simultaneously achieves communication efficiency and adaptive security—to the best of our knowledge this was previously not possible with classical-style approaches without making strong assumptions such as erasures or the existence of proof-of-work oracles.

Our work naturally leaves open several questions:

1. Can we achieve a similar result for partially synchronous or asynchronous networks?
2. The best known small-round, communication-inefficient synchronous state machine replication protocol can tolerate minority corruptions [3,14,18]. Therefore, another natural question is: can we have a similar result for synchronous networks, but tolerating up to minority corruptions (thus matching the resilience of the best known small-round but communication inefficient protocol)?

References

1. Aura - authority round. https://wiki.parity.io/Aura
2. Abraham, I., et al.: Communication complexity of byzantine agreement, revisited. CoRR, abs/1805.03391 (2018)
3. Abraham, I., Devadas, S., Dolev, D., Nayak, K., Ren, L.: Efficient synchronous byzantine consensus. In: Financial Cryptography (2019)
4. Canetti, R., Eiger, D., Goldwasser, S., Lim, D.-Y.: How to protect yourself without perfect shredding. Cryptology ePrint Archive, Report 2008/291 (2008). https://eprint.iacr.org/2008/291
5. Castro, M., Liskov, B.: Practical byzantine fault tolerance. In: OSDI (1999)
6. Chen, J., Micali, S.: Algorand: the efficient and democratic ledger (2016). https://arxiv.org/abs/1607.01341
7. Daian, P., Pass, R., Shi, E.: Snow white: robustly reconfigurable consensus and applications to provably secure proofs of stake. In: Financial Cryptography (2019). First appeared on Cryptology ePrint Archive, Report 2016/919

8. David, B., Gaži, P., Kiayias, A., Russell, A.: Ouroboros praos: an adaptively-secure, semi-synchronous proof-of-stake blockchain. In: Nielsen, J.B., Rijmen, V. (eds.) EUROCRYPT 2018. LNCS, vol. 10821, pp. 66–98. Springer, Cham (2018). https://doi.org/10.1007/978-3-319-78375-8_3

9. Dwork, C., Naor, M.: Pricing via processing or combatting junk mail. In: Brickell, E.F. (ed.) CRYPTO 1992. LNCS, vol. 740, pp. 139–147. Springer, Heidelberg (1993). https://doi.org/10.1007/3-540-48071-4_10

10. Feldman, P., Micali, S.: An optimal probabilistic protocol for synchronous byzantine agreement. SIAM J. Comput. **26**, 873–933 (1997)

11. Garay, J., Kiayias, A., Leonardos, N.: The bitcoin backbone protocol: analysis and applications. In: Oswald, E., Fischlin, M. (eds.) EUROCRYPT 2015. LNCS, vol. 9057, pp. 281–310. Springer, Heidelberg (2015). https://doi.org/10.1007/978-3-662-46803-6_10

12. Groth, J., Ostrovsky, R., Sahai, A.: New techniques for noninteractive zero-knowledge. J. ACM **59**(3), 11:1–11:35 (2012)

13. Hanke, T., Movahedi, M., Williams, D.: Dfinity technology overview series consensus system. https://dfinity.org/tech

14. Katz, J., Koo, C.-Y.: On expected constant-round protocols for byzantine agreement. J. Comput. Syst. Sci. **75**(2), 91–112 (2009)

15. Kiayias, A., Russell, A.: Ouroboros-BFT: a simple byzantine fault tolerant consensus protocol. Cryptology ePrint Archive, Report 2018/1049 (2018). https://eprint.iacr.org/2018/1049

16. Kiayias, A., Russell, A., David, B., Oliynykov, R.: Ouroboros: a provably secure proof-of-stake blockchain protocol. In: Katz, J., Shacham, H. (eds.) CRYPTO 2017. LNCS, vol. 10401, pp. 357–388. Springer, Cham (2017). https://doi.org/10.1007/978-3-319-63688-7_12

17. Micali, S., Rabin, M., Vadhan, S.: Verifiable random functions. In: FOCS (1999)

18. Micali, S., Vaikuntanathan, V.: Optimal and player-replaceable consensus with an honest majority. MIT CSAIL Technical Report, 2017–004 (2017)

19. Nakamoto, S.: Bitcoin: a peer-to-peer electronic cash system (2008)

20. Pass, R., Seeman, L., Shelat, A.: Analysis of the blockchain protocol in asynchronous networks. In: Coron, J.-S., Nielsen, J.B. (eds.) EUROCRYPT 2017. LNCS, vol. 10211, pp. 643–673. Springer, Cham (2017). https://doi.org/10.1007/978-3-319-56614-6_22

21. Pass, R., Shi, E.: Hybrid consensus: efficient consensus in the permissionless model. In: DISC (2017)

22. Pass, R., Shi, E.: Rethinking large-scale consensus (invited paper). In: CSF (2017)

23. Pass, R., Shi, E.: The sleepy model of consensus. In: Takagi, T., Peyrin, T. (eds.) ASIACRYPT 2017. LNCS, vol. 10625, pp. 380–409. Springer, Cham (2017). https://doi.org/10.1007/978-3-319-70697-9_14

24. Pass, R., Shi, E.: Rethinking large-scale consensus. IACR Cryptology ePrint Archive 2018:302 (2018)

25. Schneider, F.B.: Implementing fault-tolerant services using the state machine approach: a tutorial. ACM Comput. Surv. **22**(4), 299–319 (1990)

26. Shi, E.: Analysis of deterministic longest-chain protocols. https://eprint.iacr.org/2018/1079.pdf

Author Index

Printed in the United States
By Bookmasters